C. C. Coney

STRUCTURE AND FUNCTION
IN THE NERVOUS SYSTEMS
OF INVERTEBRATES

A SERIES OF BOOKS IN BIOLOGY

Editors: Douglas M. Whitaker, Ralph Emerson, Donald Kennedy, George W. Beadle (1946–1961)

Principles of Human Genetics (Second Edition) *Curt Stern*

Experiments in General Biology *Graham DuShane* and *David Regnery*

Principles of Plant Physiology *James Bonner* and *Arthur W. Galston*

General Genetics *Adrian M. Srb* and *Ray D. Owen*

An Introduction to Bacterial Physiology (Second Edition) *Evelyn L. Oginsky* and *Wayne W. Umbreit*

Laboratory Studies in Biology: Observations and their Implications *Chester A. Lawson, Ralph W. Lewis, Mary Alice Burmester,* and *Garrett Hardin*

Plants in Action: A Laboratory Manual of Plant Physiology *Leonard Machlis* and *John G. Torrey*

Comparative Morphology of Vascular Plants *Adriance S. Foster* and *Ernest M. Gifford, Jr.*

Taxonomy of Flowering Plants *C. L. Porter*

Growth, Development, and Pattern *N. J. Berrill*

Biology: Its Principles and Implications *Garrett Hardin*

Animal Tissue Techniques *Gretchen L. Humason*

Microbes in Action: A Laboratory Manual of Microbiology *Harry W. Seeley, Jr.,* and *Paul J. VanDemark*

Botanical Histochemistry: Principles and Practice *William A. Jensen*

Modern Microbiology *Wayne W. Umbreit*

Laboratory Outlines in Biology *Peter Abramoff* and *Robert G. Thomson*

Molecular Biology of Bacterial Viruses *Gunther S. Stent*

Principles of Numerical Taxonomy *Robert R. Sokal* and *Peter H. A. Sneath*

Structure and Function in the Nervous Systems of Invertebrates *Theodore Holmes Bullock* and *G. Adrian Horridge*

Population, Evolution, and Birth Control: A Collage of Controversial Readings *Garrett Hardin,* Editor

Plants in Perspective: A Laboratory Manual of Modern Biology *Eldon H. Newcomb, Gerald C. Gerloff,* and *William F. Whittingham*

Thermophilic Fungi: An Account of Their Biology, Activities, and Classification *Donald G. Cooney* and *Ralph Emerson*

Frontispiece The nervous system of *Aplysia californica* (Gastropoda, Opisthobranchia). *Top:* habit sketch in life. *Upper center:* dorsal exposure, showing the circumesophageal ring of ganglia anteriorly (cerebrals, pleurals, and pedals) and the visceral ganglion posteriorly (see Fig. 23.32 for further details). *Left center:* view of visceral ganglion as seen in the dissecting microscope during an electrophysiological experiment, with two glass microelectrodes inserted in the largest cell; each orange globule is a large nerve cell; the anterior end and the pleurovisceral connectives are to the right. *Right center:* low-power light micrograph of a section of the visceral ganglion, showing large cells (300–400 μ in diameter) above, sending their axons down into the neuropile. *Lower left:* electron micrograph of an axon of about 5 μ in diameter in a pleurovisceral connective, showing shingling of glial processes and deep infoldings of the axonal surface penetrated by glial processes. [Courtesy E. J. Batham.] *Lower right:* example of activity recorded by two intracellular electrodes in two neighboring cells of the visceral ganglion. The upper trace of each of the four pairs is a continuous record from a *pcr* cell, the lower trace from a *gen* cell; both are initially autoactive. Electric shocks at 5/sec, barely above threshold, to afferent fibers in the pleurovisceral connective elicit additional spikes in the *pcr* cell, i.p.s.p.'s and silence in the *gen* cell. Note the repeated cycles following cessation of stimulation, with postexcitatory inhibition in *pcr* and rebound excitation in *gen*; then in the middle of the second row the reverse relation occurs, followed by abrupt silence with hyperpolarization of *pcr* and reciprocal high frequency in *gen*; this cycle is repeated several times. Many distinct types of neuronal potentials are illustrated as well as indications of the complexity of this small ganglion. [Courtesy N. Chalazonitis and A. Arvanitaki.]

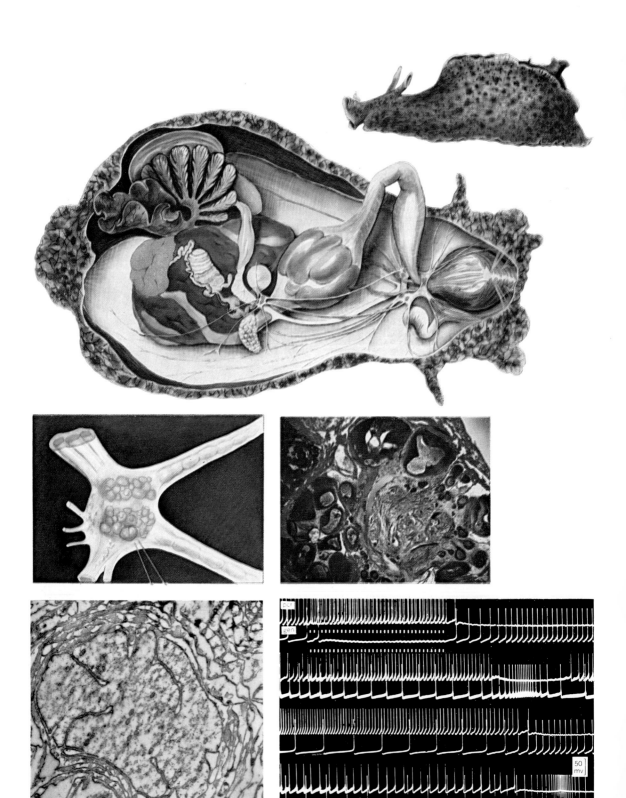

Structure and Function in the Nervous Systems of Invertebrates

VOL. I

THEODORE HOLMES BULLOCK
University of California, Los Angeles
AND
G. ADRIAN HORRIDGE
University of St. Andrews, Scotland

WITH CHAPTERS BY

HOWARD A. BERN, *University of California, Berkeley*
IRVINE R. HAGADORN, *University of North Carolina*
J. E. SMITH, *Queen Mary College, University of London*

W. H. FREEMAN AND COMPANY
SAN FRANCISCO AND LONDON

Copyright © 1965 by
Theodore H. Bullock and G. Adrian Horridge
All rights reserved
Printed in the Netherlands

PREFACE

The time is nearly past when one or even a few authors can embrace the literature and extract its essentials faster than the references multiply in the area covered by this work. But the increasing interest, not only in exploiting invertebrate materials but also in understanding them for themselves, demands that something be done to summarize the present state of knowledge and to bring together the scattered and growing literature. Our nearest forerunner, Hanström's classical work, *Vergleichende Anatomie des Nervensystems der wirbellosen Tiere*, which has for decades served as the only thorough account of invertebrate nervous anatomy, was published in 1928. With respect to the physiology, apart from the short selection of experimental investigations in volume 2 of von Buddenbrock's *Vergleichende Physiologie* (1953) and the incomplete survey of Koshtoyants, in Russian, in volume 2 of his *Fundamentals of Comparative Physiology* (1957), we lack a general summary of what is known.

The present work is an attempt to fill this need; it necessarily represents a compromise in length, fullness of treatment, and areas covered. A reasonable limitation of scope excludes, for the most part, neurochemistry, neuropharmacology, physicochemical or biophysical mechanisms, behavior, learning abilities, and perceptive capacities, except where some of these topics are referred to in connection with known nervous elements. Included, however, are gross microscopic and electron microscopic structure, cellular and organ physiology of central and peripheral nervous systems, the sense organs (but not, for example, physiological optics or visual pigments), and neuroeffector junctions (but not, for example, the viscous properties of muscle). We have had to omit any significant coverage of the ontogenetic development of the nervous system and of its changing pattern and function in developing forms. Most phylogenetic comparisons and judgments are eschewed, with some reluctance, for a proper phylogenetic evaluation would require simultaneous consideration of nonnervous characters and would displace material that is more factual.

The animals considered comprise all the nonchordate groups plus the Tunicata, including for completeness the Protozoa and Porifera even though we conclude that these both lack a true nervous system. The Cephalochordata (*Amphioxus*) are not treated because we feel that these animals can be properly appreciated only when their features are placed in perspective among the Vertebrata. (A comprehensive and critical new survey of the comparative physiology as well as anatomy of the nervous systems in vertebrates is much needed, and would stimulate new research.)

Almost every topic treated in this book—from the cytology of the nerve cells of nemertineans to the pedal nerve supply in snails—has been previously reviewed by a succession of original authors, who give the backgrounds of their own contributions, then by the monographers of such series as the *Fauna und Flora des Golfes von Neapel* and the *Tierwelt der Nord- und Ostsee,* Bronn's and Kükenthal's handbooks, Droogleever Fortuyn (1920), Bütschli (1922), Plate (1922), Hanström (1928), Ariens Kappers (1933), von Buddenbrock (1953), Koshtoyants (1957), and others. It is extremely difficult to escape from the accumulated weight of derived opinion (not so much in the big things as in the little). But this we have tried to do, and the attempt has driven us back to the old original accounts (published before the days of summaries). The harvest was usually meager, except in personal pleasure at the wealth and beauty of the lithographs and of the variety of animal structure. The harvest was nevertheless real and gives us confidence that this is a fresh, not a derivative, review. This book, however, like its predecessors in their day, cannot adequately represent all the old work; it offers only a contemporary view of points that now seem interesting.

As a consequence of our effort to synthesize overlapping accounts and to avoid excessive citations, the reader in search of the source for a particular detail may find himself frustrated. We hope, however, that he will not be too discouraged to examine the original publications, for the present work is extremely condensed and is certainly not infallible. The topographic anatomy given and the figures selected for inclusion are intended to be introductions to the animals concerned, to suggest favorable objects for further study and to illustrate the extent and type of detail available in the original papers. A reader planning an investigation will not rely on our meager selection as an account of the structures and relations. The reservoir of invaluable descriptions and of detailed illustrations in the literature is barely sampled here, and should not be overlooked just because we do not provide an unambiguous answer to a contemporary question.

The work is primarily for reference but it is not an encyclopedia or compendium. Rather, it attempts a synthetic, personal evaluation of the state of our information. A specialist will consider this only as a guide and will make his own appraisal of the papers.

We have tried to steer between two easier paths. On the one hand is the temptation to treat the accumulated literature like a catalogue or annotated bibliography. On the other is the temptation to present a smoothly flowing text that glosses over details of discordant reports and still meaningless fragments. The effort has been to write in each chapter a synthetic, connected account of the state of knowledge as the author of that chapter sees it.

The book is organized on the following plan. The first six chapters survey topics and principles that are relevant to all animals, including vertebrates: the general organization of nervous systems, the microanatomy of the cellular elements, conduction, excitation, transmission, integrative mechanisms, and neurosecretion. The remaining twenty-one chapters are systematic accounts for each of the invertebrate groups, combining anatomical and physiological findings. The treatment of principles in the first six chapters is necessarily more selective and less exhaustive than in the systematic chapters. The aim is to provide substantial introductions to these broad topics. Though useful as entries into the subject and as comparisons of a wide variety of animals with respect to the main features, these chapters will be the first to suffer obsolescence; even as they stand, they are the most vulnerable to criticism because of the larger personal element involved in the choice of what is relevant. The systematic chapters are arranged on a fairly uniform plan, modified according to the volume and direction or drift of the literature. Since this is not an encyclopedic but a synthetic treatment, the reader will need time to become familiar with it; and even so, he will no doubt encounter arbitrary choices where the organization is not obviously dictated by nature or usage.

For the present purpose, redundancy under control is of the essence. For example, the same data must be seen as part of the comparative neurology of transmission (Chapter 4), of comparative cytology (Chapter 2), and of arthropod achievement (Chapters 16–21). However, rather than having too much repetition, we have often been forced by limitations of space to present many facts in only one of the chapters to which they pertain. So far as we were able, however, we have at least referred to the original articles in each context—systematic and topical, general and special—to which they are relevant, though frequently only in the classified lists of references.

Because the problem of finding the information one wants is unusually severe in a work of this nature, we have adopted various devices to assist the reader—including a few that are unconventional. One such device is the organization itself, especially the numerous headings to make it self-explanatory. Chapter summaries are provided and are meant to be read first; they are informative rather than merely indicative. In addition to the index and the outline of contents, runningheads and phrases set in boldface are designed to assist the user in search of a particular subject.

The bibliographies are a major, and perhaps the most enduring, part of the effort to assist the student of the nervous system. Those selected for the first six chapters attempt to assure the serious student an entry into the literature of these broad topics, but space prohibits extensive citations on vertebrates in particular. Those in the systematic chapters approach the goal of being exhaustive; they omit only some preliminary notes, some speculative essays of little historical or neurological interest, and a few other items. (Hanström's bibliographies total about 1,800 citations; that for his annelid chapter, 232; for the arthropods, 724. The present book has about 9,900; 680; and 2,740 in these categories. There is some duplication in both cases. Not all of the difference is in recent literature; for example, the present work cites about 415 references on annelids prior to 1928.) The effort to approach complete coverage and the verification of citations have been major expenses to which the funding agencies mentioned in the Acknowledgments have contributed. To facilitate the use of the bibliographies, lists of authors, each classified by subject, are provided where this seems helpful.

Illustrations carry a special importance in a work of this kind and no pains have been spared to provide them, both in number and in quality. Since this book is a statement of the status of the literature, we have wherever possible selected from the figures of the original authors. Most of the drawings have been redrawn to improve the rendering. In special cases new figures have been created. Defrayment of the cost of preparing figures has been another major contribution of the funding agencies.

Some of the figures lack an indication of magnification or scale. This generally signifies that no indication is found in the original reference. For this reason we have included in the text information taken from other sources which give significant dimensions of structures of interest. Many dimensions of invertebrates—such as the width of a cord or the distance between ganglia—are not meaningful, of course, without additional information about the size of the specimen, whether this is usual or extraordinary for the species or family, and whether the figures were obtained from an extended or a contracted specimen.

The use of terms can be the easiest ground on which to criticize this work. An arbitrary personal preference inevitably intrudes here. We offer the following argument in defense of usages that may annoy authors who have adopted other usages and may feel strongly about them. No general attempt has been made heretofore to define and use terms consistently over the whole range of general neurology. Some definitions that have satisfied their users require alteration when the terms are applied to a more representative sample of the nervous systems presented to us by nature. We have given much thought to the definitions, some of which are found in the Glossary and others, via the Index, in the text. A strong effort has been made to use terms consistently in the text, but the influence of divergent specialized literatures with differing usages and the complexities of multiple authorship have prevented us from quite reaching this goal. We feel, nevertheless, that the use of terms may turn out to be a useful and clarifying feature of this work.

Some critics may feel that we have not provided an over-all simplifying system, that generalizations and theories are avoided, that pronouncements are lacking on many large problems, and even that discussion of most historical speculations on many long-standing issues is absent. In part this is deliberate, because such material would displace material of less evanescent nature. However, it is not true that we lack unifying and simplifying generalizations. Pervading the treatment is the explicit adoption of the main tenets of modern neuroanatomy and neurophysiology, as we now see them. However obsolescent these may become, we assume that they apply as a first approximation to all animal groups and parts; and by the effort to make them explicit we hope to accelerate their amplification, correction, or demise.

At present there is drawn a clear line between neurons and neuroglia in the receipt and transmission of signals, and a rather general line between sensory, internuncial, and motor neurons, for example. Other unifying simplifications are inherent in the positions taken concerning neurofibrils, synaptic potentials, local, pace-maker, and spike potentials, nerve fibers, sheaths, sensory modalities, homology of parts, and so on.

The curious reader is sure to experience an initial disappointment in not finding in immediately available form the answers to many of his questions. How, for instance, does a slug control its prodigious release of slime? How long a memory has a flatworm? Where is the list of the animals and parts of animals with nerve nets? What examples of nervous "wiring diagrams" can be produced? These and many other good questions are introduced, but their treatment is unsatisfactory simply because our knowledge is still too primitive. If recognizing some of the gaps and pointing to the nearest relevant reports is a step forward, this work can pretend to be a contribution.

The humbling discoveries of mistakes in each successive revision of the manuscript leave no doubts that errors remain despite all our efforts. (But some omissions are intentional.) Some errors are probably due to gullibility on the part of the author, as anyone will appreciate who has tried to read and sift old histological monographs without passing over potentially important findings. Although opinions and speculation are identified as carefully as possible, such qualifications as "it seems likely" are perforce scattered among the firmer statements. Inevitably, the boundaries between the softer and the firmer statement may sometimes be difficult to identify.

The authorship of this work is a hybrid between the solo and the orchestral performance, being a studied attempt to use what is of value in each. To be sure, it carries vestiges of the evolution of its final form; starting as a solo project, it acquired expert collaboration as the wisdom and necessity thereof grew more obvious. The virtues of each chapter are due to its author; the distributed shortcomings that arise from the general plan, and the conventions adopted throughout, are the responsibility of the senior author.

Integration of the work of separate authors and uniformity of treatment are virtues, though quite secondary to those of adequacy, accuracy, appropriateness, and integration within each section. Aiming first for virtues, we have also constantly striven for integration, but have not felt it desirable to eliminate all personal differences in approach. What may be taken at first as lack of integration has residual values, we feel. Occasional annoyance at the lack of correspondence in treatment of different groups is the price we willingly pay for such values.

A final personal plea for the sake of our successors and hence for posterity is justified by our experience. Too commonly authors slight their obligation to the compiler and other readers who are pressed for time. In today's flood of literature the ordinary reader cannot read completely through an article unless it is a requirement in his special field. Nor can the reviewer pretend to read everything; he just does the best he can. Titles and introductions deserve extended comment but we speak here only of summaries of papers. Many authors have learned the lesson from Leonardo that a scientist's task is not complete until he has published; they may know the importance of the controlled experiment and of full reporting of techniques. But many still fail to help the reader learn what the contribution is. However important and laboriously won results may be, in successive later and essential reviews a work must be distilled to a sentence or two. The author is not only best situated but is under moral obligation to provide an informative and adequate summary. A phrase such as "The significance of the findings is discussed" means either that the author neglects to perform the needed distillation or that he does not recognize enough significance in the findings to justify being explicit. Even worse are the summaries that omit some of the important facts and relations discovered. If all authors had done themselves justice in this, the crowning section of their papers, the present book, as a mirror of its sources, would not only have been ready years sooner but would be a better book!

THEODORE HOLMES BULLOCK

January 1963   G. ADRIAN HORRIDGE

ACKNOWLEDGMENTS

Patient and meticulous labor in compiling and verifying references, in standardizing their form, and in innumerable similar tasks has been the invaluable contribution of the following much appreciated friends and assistants: J. D. Battenberg, Miss M. A. Biederman, Mrs. Virginia S. Josephson, Miss Lydia Lynn, Mrs. Susan Marcus, Mrs. Eva Sherwood, Mrs. Suzanne Stensaas, and Mrs. Roberta Thomas. F. W. Bloodworth has been a faithful general assistant.

Such a work depends heavily on a critical reading of draft manuscript by experts. We owe a particular debt of gratitude for such service to J. S. Alexandrowicz, B. B. Boycott, H. Grundfest, D. H. Hubel, D. Kennedy, M. S. Laverack, D. Schneider, F. Strumwasser, C. A. G. Wiersma, and J. Z. Young. The scientific names of many insect groups have been checked by W. E. China.

Among many artists, several deserve special mention for their large contribution and patient helpfulness: Charles Bridgman, William B. Schwartz, Miss Frances Thompson, Mrs. Dorothy Miles, Mrs. Emily Reid, Mrs. Maria Jedrzykiewicz, Mrs. Hermine Kavanau, and William F. Stone. The drawings speak for themselves; we can speak of the pleasure of close rapport with these talented friends.

This work would not have been possible, with the best of will, if it had not been for the financial support of five institutions. The normal services of the University of California have, in the aggregate, been very large. The Center for Advanced Study in the Behavioral Sciences, Palo Alto, California, contributed through fellowships and generous services to T. H. B., G. A. H., and H. A. B. The National Institute of Neurological Diseases and Blindness of the National Institutes of Health made available grant funds for bibliographic assistance, art work, and the like. The Office of Scientific Research of the U.S. Air Force aided H. A. B. The National Science Foundation supported the preparation of manuscript, bibliography, and figures with grant funds and T. H. B. and I. R. H. with fellowships.

Thanks for the use of illustrations are due the following sources. Acknowledgment to the authors of the works from which illustrations are taken is made in the figure legends.

Academia Brasileira de Sciencias, Rio de Janeiro
  Annaes
Academic Press, Inc., New York
  Experimental Cell Research; Experimental Neurology; International Review of Cytology; Journal of Theoretical Biology; Journal of Ultrastructure Research; Modern Trends in Physiology and Biochemistry, E. S. G. Barron (ed.)
Académie Royale des Sciences, des Lettres et des Beaux-Arts de Belgique, Brussels
  Bulletin de l'Académie Royale de Belgique. Classe des Sciences
Akademische Verlagsgesellschaft Geest und Portig K.-G., Leipzig
  Bronn's Klassen und Ordnungen des Tierreichs; Zeitschrift für Wissenschaftliche Zoologie; Zeitschrift für Mikroskopisch-anatomische Forschung; Zoologischer Anzeiger
American Institute of Biological Sciences, Washington
  Brain and Behavior, M. A. B. Brazier (ed.); *Molecular Structure and Functional Activity of Nerve Cells*, R. G. Grenell and L. J. Mullins (eds.)
American Medical Association, Chicago
  A. M. A. Archives of Neurology and Psychiatry
The American Physiological Society, Washington
  Journal of Neurophysiology; Physiological Reviews; Handbook of Physiology. Section 1. Neurophysiology, J. Field and H. W. Magoun (eds.)
Athenäum-Verlag Junker und Dunnhaupt K. G., Bonn
  The Love Life of Animals (*Das Liebesleben der Tiere*), W. von Buddenbrock
Birkhäuser Verlag, Basel
  Vergleichende Physiologie, W. von Buddenbrock

E. J. Brill, Leiden
Behaviour; Archives Néerlandaises de Zoologie

Cambridge Philosophical Society, London
Biological Reviews

Cambridge University Press, London
Journal of the Marine Biological Association of the United Kingdom; Parasitology; The Comparative Physiology of the Nervous Control of Muscular Contraction, G. Hoyle; *Endocrine Control in Crustaceans*, D. B. Carlisle and F. Knowles

Centre National de la Recherche Scientifique, Paris
Archives des Sciences Physiologiques

The Company of Biologists Limited, Cambridge
Journal of Experimental Biology; Quarterly Journal of Microscopical Science; Symposia of the Society of Experimental Biology

The Connecticut Academy of Arts and Sciences
Transactions

Cornell University Press, Ithaca
The Control of Growth and Form, V. B. Wigglesworth

Council of the Marine Biological Association, Plymouth
Journal of the Marine Biological Association of the United Kingdom

Édition des Archives de Zoologie Expérimentale, Paris
Archives de Zoologie Expérimentale et Générale

Elsevier Publishing Company, Amsterdam
Biochimica et Biophysica Acta; Electroencephalography and Clinical Neurophysiology; Bioelectrogenesis, C. Chagas and A. Paes de Carvalho (eds.)

Faculty of Science, Okayama University, Okayama
Biological Journal

Faculty of Science, University of Tokyo, Tokyo
Journal

VEB Georg Thieme, Leipzig
Biologisches Zentralblatt

VEB Gustav Fischer Verlag, Jena
Fortschritte der Zoologie; Zoologische Jahrbücher; Allgemeine Zoologie, L. Plate

Igaku Shoin, Ltd., Tokyo
Seitai-no-Kaga-ku (Science of the Living Body)

Johns Hopkins Press, Baltimore
The Physiology of Nerve Cells, J. C. Eccles

Journal of Anatomy, Middlesex

The Journal of Physiology, Cambridge

W. Junk, The Hague
Physiologia Comparata et Oecologia

S. Karger A. G., Basel/New York
Acta Anatomica

Koninklijke Nederlandse Akademie van Wetenschappen, Amsterdam
Proceedings

Librairie Maloine, Paris
Histologie du Système Nerveux de l'Homme et des Vertébrés, S. Ramón y Cajal

J. B. Lippincott Company, Philadelphia
The Elementary Nervous System, G. H. Parker

Little, Brown & Company, Boston
Reticular Formation of the Brain (Henry Ford Hospital International Symposium, Detroit), H. H. Jasper et al. (eds.)

Long Island Biological Association, Inc., Cold Spring Harbor
Cold Spring Harbor Symposia on Quantitative Biology

Macmillan and Company, Ltd., London
Nature

The Marine Biological Laboratory, Woods Hole
Biological Bulletin

The Massachusetts Institute of Technology, Cambridge
Processing Neuroelectric Data, Communications Biophysics Group of Research Laboratory of Electronics and W. M. Siebert; *Sensory Communication*, W. A. Rosenblith (ed.)

Masson et Cie Éditeurs, Paris
Annales des Sciences Naturelles Zoologie; Grassé's Traité de Zoologie

McGraw-Hill Book Company, Inc., New York
The Invertebrates, L. H. Hyman; *Physiological Psychology*, C. Morgan; *Radiation Biology*. Vol. 3, A. Hollaender (ed.)

National Institute of Oceanography, Surrey
'Discovery' Reports

The New York Academy of Sciences, New York
Annals

Ophthalmic Publishing Company, Chicago
American Journal of Ophthalmology

Oslo University Press, Oslo
Nytt Magasin for Zoologi

Pergamon Press, Inc., New York
Inhibition in the Nervous System and Gamma-aminobutyric Acid, E. Roberts et al. (eds.); *Nervous Inhibition*, E. Florey (ed.)

Pergamon Press, Ltd., Oxford
Comparative Biochemistry and Physiology

The Rockefeller Institute Press, New York
Biophysical Journal; Journal of Biophysical and Biochemical Cytology; Journal of General Physiology

The Royal Microscopical Society, London
Journal

The Royal Society of Canada, Ottawa
Transactions

The Royal Society of London, London
Philosophical Transactions; Proceedings

Acknowledgments

The Royal Swedish Academy of Science, Stockholm
Arkiv för Zoologi

The Smithsonian Institute, Washington
Miscellaneous Collection

Société Zoologique de France, Paris
Bulletin

Société Zoologique de la Suisse et du Muséum de l'Histoire Naturelle, Geneva
Revue Suisse de Zoologie

Society for Experimental Biology and Medicine, New York
Proceedings

Springer-Verlag, Berlin
Ergebnisse der Biologie; Naturwissenschaften; Zeitschrift für Morphologie und Ökologie der Tiere; Zeitschrift für Vergleichende Physiologie; Zeitschrift für Zellforschung und Mikroskopische Anatomie; Neue Ergebnisse der Nervenphysiologie, A. von Muralt; *Handbuch der Mikroskopischen Anatomie des Menschen.* Vol. 4, Part 4. Vol. 4, Part 8, W. von Möllendorff and W. Bargmann (eds.)

Stazione Zoologica, Naples
Pubblicazioni della Stazione Zoologica di Napoli

Umschau Verlag, Frankfurt
Die Umschau in Wissenschaft und Technik

Université de Paris, Faculté des Sciences, Paris
Bulletin Biologique de la France et de la Belgique

The University of Chicago Press, Chicago
Physiological Zoölogy

The University of Wisconsin Press, Regents of the University of Wisconsin, Madison
Biological and Biochemical Bases of Behavior, H. F. Harlow and C. N. Woolsey (eds.)

VEB Verlag Volk und Gesundheit, Berlin
Nierenfunktion und Nervensystem, H. Dutz (ed.)

Walter de Gruyter and Company, Berlin
Kükenthal's Handbuch der Zoologie

John Wiley & Sons, Inc., New York
Comparative Endocrinology, A. Gorbman (ed.); *Sensory Communication*, W. A. Rosenblith (ed.); *A Textbook of Comparative Endocrinology*, A. Gorbman and H. A. Bern

The Williams & Wilkins Company, Baltimore
Harvey Lectures; Quarterly Review of Biology

The Wistar Institute of Anatomy and Biology, Philadelphia
The American Journal of Anatomy; The Anatomical Record; Journal of Cellular and Comparative Physiology; The Journal of Comparative Neurology; The Journal of Experimental Zoology; Journal of Morphology

The Yale Journal of Biology and Medicine, Inc., New Haven

Yale University Press, New Haven
The Integrative Action of the Nervous System, C. S. Sherrington

The Zoological Society of London
Proceedings

CONTENTS

PART I
MECHANISMS AND PRINCIPLES

Chapter 1 T. H. BULLOCK

Introduction

Basic Roles of a Nervous System	4
Defining Features of a Nervous System	5
Common Features of Nervous Systems	7
Major Trends and Contrasts	10
Methods of Studying the Nervous System	24

Chapter 2 G. A. HORRIDGE AND T. H. BULLOCK

Comparative Microanatomy of Nervous Elements

General and Comparative Considerations	38
Forms of nerve cells and processes—Distribution of types of neurons among the phyla—Origins and relationships of neuron types	
Special Relations and Components of Nervous Tissue	49
Organization of nervous tissue—Synapses—Nerve cell inclusions—Neuroglia and sheaths—Degeneration and regeneration	

Chapter 3 T. H. BULLOCK

Comparative Neurology of Excitability and Conduction

Introduction	125
Differentiation of Nervous Signals	125
Graded and spike activity—Dendrite and axon	
Excitability and its Primary Parameters	133
Local response excitability—Spike excitability—Rhythmicity and pacemaker excitability	

Chapter 4 T. H. BULLOCK

Comparative Neurology of Transmission

The Concept of the Synapse	181
Nonsynaptic Interactions between Neurons	191
Differences among Synapses	197

Examples of Better-known Junctions	211
Evolution of Nervous Transmission	237

Chapter 5   T. H. BULLOCK

Mechanisms of Integration

Introduction: the Levels of Integration	257
The Unit Level: Integrative Properties of Neurons	258
Integration at the Level of Organized Groups of Neurons	272
Input and its interpretation—Output and its control—Associative levels of moderate complexity	
Spontaneity—Its Sources and Consequences	314
Neurological Deductions from the Study of Behavior	323

Chapter 6   H. A. BERN AND I. R. HAGADORN

Neurosecretion

Phenomenon and Concept of Neurosecretion	356
Nature of Neurosecretion	360
The neurosecretory material—The neurosecretory process—The neuronal properties of neurosecretory cells—The functions of neurosecretion	
Neurosecretory Systems	369

PART II
THE LOWER PHYLA

Chapter 7   T. H. BULLOCK

Protozoa, Mesozoa, and Porifera   433

Chapter 8   T. H. BULLOCK

Coelenterata and Ctenophora   459

Chapter 9   T. H. BULLOCK

Platyhelminthes   535

Chapter 10   T. H. BULLOCK

Nemertinea   579

Chapter 11   T. H. BULLOCK

Pseudocoelomate Phyla   597

Chapter 12   T. H. BULLOCK

Lophophorate Phyla   631

Contents

Chapter 13   T. H. BULLOCK

Sipunculoidea, Echiuroidea, and Priapuloidea   649

Chapter 14   T. H. BULLOCK

Annelida   661

Chapter 15   G. A. HORRIDGE

Onychophora   791

PART III
THE ARTHROPODA

Chapter 16   G. A. HORRIDGE

Arthropoda: General Anatomy   801

Chapter 17   G. A. HORRIDGE

Arthropoda: Nervous Control of Effector Organs   965

Chapter 18   G. A. HORRIDGE

Arthropoda: Receptors Other Than Eyes   1005

Chapter 19   G. A. HORRIDGE

Arthropoda: Receptors for Light, and Optic Lobe   1063

Chapter 20   G. A. HORRIDGE

Arthropoda: Physiology of Neurons and Ganglia   1115

Chapter 21   G. A. HORRIDGE

Arthropoda: Details of the Groups   1165

PART IV
THE MOLLUSCA

Chapter 22   T. H. BULLOCK

Mollusca: Amphineura and Monoplacophora   1273

Chapter 23   T. H. BULLOCK

Mollusca: Gastropoda   1283

Chapter 24   T. H. BULLOCK

Mollusca: Pelecypoda and Scaphopoda   1387

Chapter 25   T. H. BULLOCK

Mollusca: Cephalopoda   1433

PART V
THE DEUTEROSTOMES

Chapter 26   J. E. SMITH
    Echinodermata   1519

Chapter 27   T. H. BULLOCK
    Chaetognatha, Pogonophora, Hemichordata, and Tunicata   1559

Glossary   1593

Indexes   1611

PART I Mechanisms and Principles

CHAPTER I

Introduction

I. Basic Roles of a Nervous System — 4	B. Differentiation of Cells and Tissues — 14
II. Defining Features of a Nervous System — 5	C. Receptors and Sense Organs — 18
III. Common Features of Nervous Systems — 7	D. Effectors and Effector Control — 20
IV. Major Trends and Contrasts — 10	E. Problems of Central Nervous Physiology — 22
A. Morphology and Phylogeny — 10	V. Methods of Studying the Nervous System — 24
1. Origin of the nervous system — 10	Bibliography — 32
2. Centralization — 12	
3. Cephalization — 13	

The animal world, more particularly that of the higher animals, is distinctively signalized by the presence and activities of a nervous system. Most features of animals are shared with other forms of life—for example, metabolism, growth, and inheritance. But the behavior of animals endowed with a nervous system sets them apart by its complexity. The organization and function of the machinery of behavior are surely the highest achievements in the natural world. Before we consider what defines this system, let us recognize its roles or, for those who prefer a formulation from which purpose is omitted, let us enumerate the consequences of its activities.

I. BASIC ROLES OF A NERVOUS SYSTEM

Viewed in one way, there are two distinct spheres of activity of nervous systems: to counteract and to act. They wax and wane in importance from moment to moment, not quite reciprocally and not quite independently. The first is the **domain of regulation.** The nervous system operates as one of the mediators, probably the main one, of reactions that preserve the animal's status quo. Certain of the events or states (stimuli) that tend to displace some functional feature of the organism call forth a compensatory response resulting in the reduction of the displacement and restoration of the norm. These regulatory reactions are mostly simple, prompt, short-lasting, involving only a part of the body or only visceral effectors, "automatic" in the meaning that "volition" or the highest centers are not required—in short, reflexes. By no means are all reflex, however, for even some very complex sequences are both compensatory and stimulated by a need.

The other sphere of functions of nervous systems embraces actions which do not preserve but alter the status quo; it may be called the **domain of initiation.** The system does not simply wait for stimuli and react to cancel their effect; it also initiates or deviates from that status. Especially in higher animals one phase or mood replaces another; exploratory or appetitive behavior is prominent. These can be regarded in a long view as regulatory, but the difference between this domain and the merely compensatory is not only one of time scale. Many of the activities of animals have been found, upon experiment, not to be responses to need, in the sense that stimuli representing needs trigger the action, but spontaneous in the proper meaning of the word (see Glossary). For example, Wells (1950, 1955) found the initiation of movements of various kinds of polychaete worms was not explicable by accumulation of wastes, lack of food, or oxygen —though their movements normally reduced such conditions. Nest building, migration, and even eating and drinking have been shown in given cases not to be performed "in order to" achieve a desirable result, but simply to satisfy an internally arising state (drive) that demands the motions (sometimes even less) which would in nature bring about an adaptive end. The activities in this sphere are usually more complex and prolonged, often rhythmically recurrent, and involve somatic musculature—that which moves the body or its major appendages, often skeletal, voluntary—in short, instincts and higher behavior. By no means are all these activities so complex; some, like the heart beat, are quite elementary. Nor are all instincts adequately described in this way. But the fact that some are precipitated by external events does not remove them from relevance to this category. These events are triggers, often unnecessary if deprivation permits the probability of action to rise high enough (vacuum activity), and they are generally not direct signs of a need or displacement of a bodily state requiring correction. Dichotomy of roles of the nervous system is not final but of fundamental heuristic value. To some, the nervous system is typified by the circuits controlling body temperature, to others it is the organ of behavior; both views are basically inadequate.

Learned behavior can be superimposed on either the homeostatic or the initiating category but it mainly involves the second. It is useful to think of learned behavior as essentially achieving a more adaptive, directing, combining, and timing of species-characteristic acts which tend to occur anyway.

Animals can only learn a certain range of actions within a repertoire, and it is notable how this repertoire increases rapidly in size, complexity, and human interest as we turn to the most complex brains in the three groups of animals, molluscs, arthropods, and vertebrates. Learning is something more than an adaptive change in behavior in response to repeated stimuli; habituation (or lessening of a response), sensitization of a reflex, or long-term effects of the repetition of the reward or punishment can all be adaptive in the sense of being economical. The essential feature of learning, at its most elementary level in the lower animals, is the association of two stimuli. For this at least two sets of receptors are required and a modifiable mechanism which coordinates their common connections in the nervous system. The mechanism and its effects need not be entirely immediate, with an outcome in milliseconds, but could depend in addition on any combination of activities such as growth, hormone secretion, or change of frequency of a regular discharge of impulses.

These generalizations may appear doctrinaire or forced but should not be discounted for that reason alone. They do at least serve to frame some of the questions that exemplify the state of the subject and provide points of departure for new work.

Viewing its roles on a different spectrum, the nervous system performs in such a way as to **extend the speed** (and with it the intricacy) of behavior. This hardly requires elaboration, for the prodigious feats of the pianist, the conversationalist, the hunting wolf, and the worker bee are familiar and are appropriately ascribed to this system. Supporting, maintaining, and cooperating with it in slower and less intricate roles are the other mechanisms of response and coordination, ranging from the specialized endocrine organs to cells which respond directly to certain agents.

Closely related to speed, with the consequent possibility of many steps in a given response, is the role of the nervous system in providing small and hence **numerous units of information handling.** These are conveniently developed from the cells, "invented" phylogenetically on other grounds. But relative to other organ systems, also composed of many cells, the degree to which adjacent cells and even parts of cells can act discretely and can influence each other specifically, beggars comparison. We shall see, as a major conclusion of our survey, that the units and their properties are essentially alike from the lowest to the highest animals with nervous systems. But the number of units, especially of those in between receptor and motor neurons, increases greatly in higher forms; the number and profusion of their branching processes, together with the differentiation of shapes and connections does likewise. This complexity is the structural background which provides for complex manipulation of the signals representing internal and external events; they are coded in a special way assuring high temporal resolution, with many steps per second and with faithful propagation for long distances in the body.

II. DEFINING FEATURES OF A NERVOUS SYSTEM

If these functional generalities are kept in mind, it may be easier to define our object of study. A nervous system may be obvious and easily identified, as in higher groups, at least in its central parts; or it may be extremely difficult either to recognize or to deny, as in the multicellular sponges and the highly coordinated ciliate protozoans (heterotrichs). A **nervous system** may be defined as an organized constellation of cells (neurons) specialized for the repeated conduction of an excited state from receptor sites or from other neurons to effectors or to other neurons. This formulation automatically excludes any coordinating system that may exist in unicellular or acellular forms. We nevertheless devote a chapter to Protozoa because of the intrinsic interest in any specialized conducting organelles that there may be, even though they may have no genetic relation to the systems of cells in metazoans.

The relevance of the definition for sponges, and any other metazoans in which a nervous system is to be demonstrated, is that it provides **anatomical and physiological criteria** to be satisfied by any true nervous system. Thus presumptive nerve cells must be shown to connect receptor sites with effectors or to connect with each other. Cells that send prolongations to end in open tissue or spaces or on spicules are unlikely to be nervous, unless other arguments clearly exist. Cells which reach from a superficial position, presumptively receptive, to a deep position in the neighborhood of shape-changing effectors, can be regarded as indifferent, supporting, or epithelial as well as nervous, unless physiological evidence of specialization for repeated conduction of excitation is adduced. Since the property of excitability is probably general for living material and since any collection of like cells can be called a system, it is the combination of connectedness and specialization for propagating an excited state that we must look for in a nervous system.

Beyond the problem of recognizing the whole system, there are two practical problems involved in treating almost every animal. One is how to prove **whether the nervous system is in fact involved** in a given response. When a jellyfish—or, even worse, a sea anemone—exhibits a coordinated wave of movement, are we witnessing a nervous event? Can properties of the nerve fibers and synapses then be inferred? Or can some non-nervous form of spread be responsible? It is in practice extremely difficult to rule out muscle-to-muscle and other unspecified forms of conduction (such as via mechanical tension) and to prove nervous involvement, in lower forms or in those parts of higher forms where nervous elements are diffusely mixed with muscle, as in the viscera. Even if a nerve net is found by histological methods, further evidence is required before it can be accepted as one of the pathways of conduction (Horridge, p. 463). Where fibers are gathered into nerves, and responses are reflexly mediated through a central system, the matter can more readily be settled by cutting a nerve and noting the consequent deficit. In diffuse systems we must depend on convergence of several lines of circumstantial evidence, building up a case of reasonable presumption of nervous mediation. Anatomical, physiological, sometimes pharmacological tools often permit strongly suggestive evidence; direct electrophysiological micromethods with visual control can be conclusive.

In few mixed systems do we have proof, as is detailed in Chapter 8. Sometimes a muscle-free zone can be found or created or a sheet of tissue continuing the nerve net can be cut across without damaging the muscle. Sometimes distinct refractory periods can be measured by special means, one associated with movement, the other not. Drugs, fatigue, paired shocks applied at separate points, or other means may succeed in divorcing the fact of conduction from the event of contraction. A strong presumption for or against the hypothesis of nervous mediation can sometimes be adduced from the distribution of differentiated parts of the nerve net, the directions and sizes of the nerve fibers, and other anatomical relations. Examples will be developed in Chapter 8.

The other problem requiring treatment is simply **how to recognize a nerve cell**! Granted that it has been shown that a nervous system exists in a given animal, satisfying the criteria above, it may be difficult sometimes to decide whether a certain cell or cell type should be regarded as a nerve cell. For example, how can we decide whether a neurosecretory cell is or is not a nerve cell? The ideal criteria are like those for a nervous system: specialization for conduction and connection with other nerve cells or with receptors or effectors. These criteria may be difficult to use, even apart from the technical obstacles to adequate histological and physiological demonstrations. First, neuroglia, ependymal cells, gland cells, vascular or other elements may make apparently intimate contact with nerve cells and thus reduce the ease of application of the criterion of connection. Second, neuroglia and possibly other types of cells have been reported to exhibit a transient change of membrane potential, like a slow action potential (Tasaki, 1958). Cytological criteria such as the presence of Nissl granules or neurotubules may occasionally be helpful but often are not; similarly, the possession of long processes like dendrites and axons may not be decisive; a unipolar nerve cell, for example, may have a much reduced stem process associated with a neurosecretory function. Some connective tissue and neuroglial cells have long, branched processes. It must be admitted, finally, that no specific criterion is certain. Like the question of nervous mediation, we can only build up a case for reasonable identification by accumulating suggestive evidence. Two signs are perhaps the most basic, even where the axon is greatly reduced: (a) electron microscopic or oscillographic evidence of real functional contact with other cells, themselves clearly neurons, and (b) a reasonably brief action potential and refractory period, permitting repeated signals at moderate frequency. But other indications should also be sought, such as an evolutionary series in related species, suggesting homology with typical nerve cells, ontogenetic stages, and possibly histochemical diagnostic features.

Although a nerve cell can be so hard to recognize, it is remarkable that so few problem cases have arisen. Recent suggestions that glial cells play an active role in nervous events open the possibility of another category of problem cases. At present we may say that glial cells should be distinguished as nonnervous, even though they play a necessary and an active role in brain physiology, so long as it is reasonable to believe that they are not specialized for repeated conduction of signals of excitation and for functional transmission of such signals to other cells. There may well be a spectrum extending from clearly nonnervous cells that provide the essential milieu for neurons to live in, to less clearly nonnervous cells that surround neurons and perhaps alter their responsiveness in a phasic, relatively short term way. Once more the discussion serves to bring out the types of questions that represent today's knowledge of the subject.

III. COMMON FEATURES OF NERVOUS SYSTEMS

To set off the next section on major trends and contrasts in the organization of nervous systems, it will be useful to add to the defining features now before us some notice of other features that are either universal or nearly so.

The neuron and its main properties are universal, within the limitations just discussed. The **neuron doctrine,** which we owe chiefly to Cajal but also to many other workers, some of whom contributed by vigorous opposition (see Cajal, 1937, 1954; Rasmussen, 1947), may be stated thus: all nervous systems consist in essence (whatever other, nonnervous elements may be present) of distinct cells called neurons, which are specialized for nervous functions and which produce prolongations and branches. Each nerve fiber is a cytoplasm-filled tube which extends from the cell body of the neuron of which it is a part. This means that cellular units, and hence embryological and trophic units, comprise the nervous tissue. Like other cells, neurons are generally discrete and separated by cell membranes, though special cases of fusion are perfectly possible and compatible, probably always secondarily derived. Each nerve fiber grows out of a nerve cell and remains organically connected to it. Functional union among neurons is ordinarily by contact or contiguity, rarely and secondarily by continuity of cytoplasm although, as Cajal emphasized, occasional finding of con-

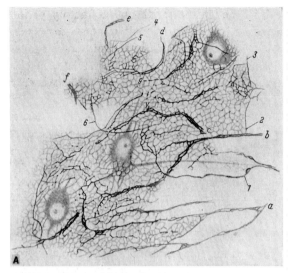

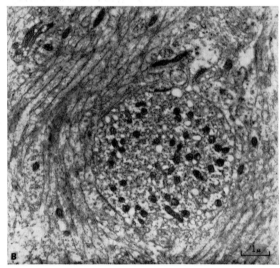

Figure 1.1. Opposing views of the composition of nervous tissue. **A.** The conception of Held (1929) may be regarded as representing the reticularist view. This is the so-called ground net of the cerebellar cortex of man. The lettered elements are considered to be dendrites, the numbered ones neurites whose neurofibrillae penetrate the ground net. Three Purkinje cells are shown. Reduced silver method. **B.** A modern electron micrograph confirms the classical neuron doctrine that nerve cells are in general discontinuous by showing clear membranes surrounding each element. Tangential section of cerebellar cortex, molecular layer; the large central area is a Purkinje cell dendrite cut across and surrounded by granule cell axons called parallel fibers. [Courtesy D. Pease and W. Tallie.]

tinuity is not fatal to the general doctrine. The opposing view is reticularism (Fig. 1.1), which holds that all nerve processes are continuous protoplasmically, so neuropile and central gray matter are without discrete neurons and excitation has complete freedom of spread from any point to any other. Interaction by more remote influence between neurons that do not make contact is not excluded. The neuron doctrine makes no statement as to the direction of normal transmission between neurons; that is the subject of a separate generalization—also due to Cajal—and is considered in the next section. There is no necessary implication that the neuron always acts as a whole, though such a supposition has frequently been made. In a sense the neuron must be a functional unit, being a single cell, but the term is difficult to define further; we know that the whole neuron does not always act when part of it acts. An important contemporary concept of the neuron is that it consists of several or many regions (loci) of different functional capacity, facultatively interacting in complex ways which will be outlined in later sections (see Chapters 3, 4, and 5). Some of these functionally diverse regions correspond to the anatomically distinct parts of the cell.

A corollary of the neuron doctrine is that any process severed from the nucleated portion of the neuron (cell body, soma, perikaryon) can be expected to die. This is the essence of Wallerian degeneration but, besides the fragmentation of the axon, that phenomenon includes also the deterioration of the Schwann cell membrane now known to form the myelin sheath. The reaction of the uninjured Schwann cells when the axon is injured is a specially sensitive example of transcellular or intercellular dependence. Retrograde reaction of the soma or proximal axon and transynaptic degeneration are known to occur in varying degrees but are not predicted by the doctrine.

The properties common to neurons, besides their connections to other neurons, can only be stated on the basis of a small sample studied physiologically. It is not necessary that these properties be universal, and some reasons exist to suspect otherwise. All neurons so far examined are capable of an all-or-none, brief (0.3–10 msec) membrane change called the **nerve impulse** (Fig. 1.2), of repeating this event from dozens to

III. COMMON FEATURES OF NERVOUS SYSTEMS

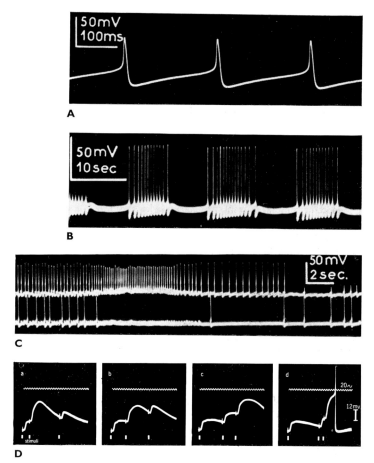

Figure 1.2.

Nerve impulses and prepotentials. **A.** An electrode inside the soma of a visceral ganglion cell of *Aplysia* records in this instance a rhythmic spontaneous firing. The three peaks are spikes or nerve impulses, all-or-none and overshooting the resting potential. Each is followed by a rapid repolarization to the highest membrane potential (the polarity of the connections places high membrane potential, inside negative to outside, downward on the record). There begins immediately a slow steady depolarization called a pacemaker potential, which other observations show to be a local, graded prepotential leading to a critical level that triggers the next spike. **B.** The same, with compressed time scale, shows bouts or bursts of impulses spontaneously ceasing and recurring after a quiescent period during which the membrane potential swings toward high polarization and then gradually depolarizes again. This exemplifies the pulse coded messages that are delivered into the axons. **C.** A similar record taken simultaneously from two separate cells in this ganglion, isolated from the body. Due to spontaneous events in some other neurons in the ganglion, the cell recorded in the upper trace receives influx of excitatory synaptic sort that accelerates its firing rate for a time while the cell recorded in the lower trace is simultaneously silenced. **D.** On a faster time scale and with higher amplification, similar cells show the prepotential resulting from presynaptic activity impinging on the penetrated cell—in this case of an excitatory character, causing therefore graded local depolarizations called excitatory postsynaptic potentials (e.p.s.p.'s); these can summate and, if they reach a critical level, trigger a spike. Here the most effective pattern of three stimuli is found to be that at the right. Visceral ganglion cell of *Aplysia* responding to stimulation of left pleurovisceral connective in trios of shocks. Time calibration signal shows level of zero membrane potential. [**A, B, C,** courtesy L. Tauc; **D,** courtesy J. Segundo.]

hundreds of times per second, and of propagating it without decrement along processes called axons. This property implies also the possession of a sharp threshold for impulse initiation and a refractory period during the impulse. These quantal pulses in axons seem to be the means of conducting excitation over long distances. Since many neurons, which have never been properly examined physiologically, have only short processes, it is not clear whether they all have impulses. All neurons properly examined—a still smaller sample, but a sample from four major phyla (annelids, arthropods, molluscs, and chordates)—exhibit a different form of response called a **postsynaptic potential** (Fig. 1.2), when adequate impulses in other neurons (hence called presynaptic) impinge.

If we except the coelenterates and ctenophores and consider the higher groups with a well-delineated brain and axial central nervous system, beginning with platyhelminths, there are a series of **common, system features**. A central nervous system can be distinguished from a peripheral system (Fig. 1.3). The former contains most of the motor and internuncial cell bodies; the peripheral nervous system contains all the sensory cell bodies (rare exceptions occur), plus local plexuses in the body wall or viscera, local

Figure 1.3. Cajal's scheme, showing the utility of the multiplication of neurons and grouping into central ganglia. **A.** Hypothetical invertebrate possessing only neurons at once sensory and motor. **B.** Invertebrate equivalent to a coelenterate possessing two sorts of neurons, sensory and motor, both dispersed. **C.** Stage represented by all higher animals, in which the motoneurons are concentrated into ganglia to which also come the sensory axons; in actuality all such stages have in addition a third sort of neuron intrinsic to the ganglion, the interneurons. [Based on Cajal, 1909.]

ganglia of either sensory or motor-and-internuncial composition, plus the peripheral axons making up the nerves. No isolated peripheral plexus or ganglion exists, we believe, without connection to the rest of the nervous system. Except in a few special sense organs of vertebrates, all receptor cells are nerve cells. The vertebrate taste buds, acousticolateralis system, and possibly a few others, treated in the next section, employ a novelty in nonnervous receptor cells. By and large, the first-order sensory axons go all the way to the central nervous system, though it is thought that in some cases they relay in peripheral plexuses and fewer second-order axons continue centrally. By and large, effectors are innervated by axons coming all the way from the central nervous system, though here and there it is known or thought that central motoneurons relay by synapses with peripheral motoneurons.

IV. MAJOR TRENDS AND CONTRASTS

Most of this book is devoted to detailing the particulars of nervous structure and function in the great diversity of invertebrate animal types; we are confronted with differences just as much as we are with basically general features and mechanisms. Indeed, the universal mechanisms appear to be few and they seem to get fewer with increased knowledge of individual cases. Even if we possessed a full understanding of those features which are quite general, we would be far from understanding the nervous system of any particular animal, marked as each is by its specific behavioral characters. Pantin (1952) has voiced an impression, based upon comparison of animal groups, that in respect to any given function various groups employ as a rule not one and the same mechanism for fulfilling this function nor do they employ a large number of different mechanisms; we commonly find that several distinct methods have been evolved. The diversity catalogued in this book is sometimes of this either-or type and sometimes in the nature of small or quantitative variation. We call the latter a trend when it proceeds more or less regularly in the same direction as our independently conceived notions of phylogenetic sequence. As we shall see, the evolution of the nervous system, both within major phyla and between them, is notable for the prominence of such trends. It is here, in the machinery of behavior, that most of the basis exists for saying that evolution usually leads to greater complexity and greater control over the environment. "Higher" animals are so called because they have more distinguishable parts and hence events, activities, processes; it would take more information to specify the difference between such an animal and a random assortment of elements than it would for a lower animal.

A. Morphology and Phylogeny

1. Origin of the nervous system

Former texts gave extended consideration to the problem of the origin of the nervous system from prenervous conditions; these discussions have a great intrinsic interest and the reader is referred to Bethe (1903), Parker (1919), Child (1921), Herrick (1924, 1931), Hanström (1928a), Kappers (1929, 1936). Recent years have added virtually nothing, except perhaps a pinch of scepticism and a dash of disinterest. As a starting point it can be asserted that differentiated cells specialized for nervous functions must have been **preceded by**

IV. TRENDS AND CONTRASTS A. Morphology

undifferentiated structures carrying out the same functions in a simpler way. Selection of adequate stimuli, excitability, propagation of change from one part of the organism to another, correlation, and response in adaptive form are all properties of prenervous organisms and nonnervous cells.

As a second assertion, it may be proposed that **effectors** probably differentiated earlier than receptors or nerve cells. This is sometimes challenged and certainly much of the evidence comes from animals already possessing all three. Views depend in part on one's subjective recognition of separation of functions out of undifferentiated cells capable of detection, conduction, and response. The main argument in favor of the proposition is an extrapolation from so-called independent effectors (Fig. 1.4), cells specialized as effectors which do not require receptor or nerve cells to initiate their actions. Parker (1919) and others pointed to many examples, especially in protozoans and sponges but also in higher groups, including oscular myocytes in sponges, nematocysts in coelenterates, cilia, smooth muscles in the iris of some lower vertebrates, and many glands. Certain examples have not been proven unequivocally to be independent and others can under some conditions be influenced by superimposed nervous control. But clearly it is possible for effectors to function without antecedent receptor or nerve cells, whereas the latter can hardly have developed before effectors. Embryological stages, if they are relevant, agree in sequence with the primacy of effectors.

Sensory cells are reasonably derived from epithelial cells. But beyond this, the questions become more controversial. Are epitheliomuscular cells in the direct line of evolution of higher types? Were the ancestral cells sensory and at the same time motor, sending a prolongation to subjacent muscle? Did **internuncial cells** come from sensory cells? The evidence is now hardly greater than that known to the classical authors and space precludes a full exposition.

The **general plan** of a centralized system can be regarded as basically common, so far as it comprises a brain and one or more pairs of longitudinal cords, except in echinoderms and a few

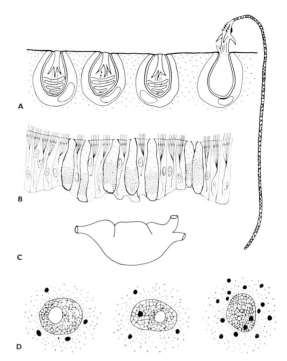

Figure 1.4. Independent effectors. **A.** Nematocysts of the stenotele type in *Hydra*, undischarged and discharged. **B.** A diagrammatic, ciliated epithelium with unicellular glands. Like the nematocysts, both cilia and glands may in some situations receive superimposed nervous control, but they are capable of autonomous action in which each cell determines when it shall act, not by receipt of specific nervous messages but by detecting nonnervous conditions in its environment. **C.** Heart of *Salpa*, representing myogenic hearts, which do not require innervation to maintain a beat. **D.** Dermal pores in the sponge *Stylotella*, showing steps in their closure by the pore membrane. [Partly based on Parker, 1919, *The Elementary Nervous System*, J. B. Lippincott Co., Philadelphia.]

groups with a reduced long axis that consequently lack cords (adult tunicates, entoprocts, ectoprocts, brachiopods). There is in fact sufficient uniformity to have suggested a theory that the cords of the central nervous system are homologous throughout the bilateral phyla. This is the **orthogon theory** (Fig. 1.5), which proposes that primitively there are approximately six or eight longitudinal concentrations of a general superficial plexus, as in ctenophores and some acoel and rhabdocoel turbellarians (Figs. 8.52, 9.1, 9.19), distributed symmetrically around the long axis and connected by transverse commisures, forming an orthogonal pattern (Hanström, 1928a).

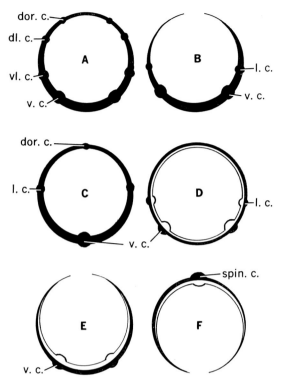

Figure 1.5. The orthogon theory. Diagrams of cross sections of the central nervous systems of: **A.** turbellarians with four pairs of longitudinal connectives; **B.** amphineurans with two pairs; **C.** nematodes; **D.** annelids; **E.** arthropods; and **F.** vertebrates. Of a series of longitudinal cords one or more may be lost, emphasized, or fused. The transverse connections or commissures between the cords form a series of repeated nervous rings, which remain continuous in some of the higher groups but are broken in others. In the three highest groups it is possible to say that at least to some extent the inner (blackened) parts of the cords and of the nerves emerging from them are mainly sensory, the outer mainly motor. dl. c., dorsolateral cord; dor. c., dorsal cord; l.c., lateral cord; spin. c., spinal cord; vl. c., ventrolateral cord; v. c., ventral cord. [Hanström, 1928.]

All other arrangements are derivable from this by emphasis upon some cords, loss of others and fusion of medial pairs in the midline. The theory is tenable and stimulating; whether it is right is impossible to say. The homology of any one organ system between phyla cannot be fully discussed without considering all the evidence, including other systems and ontogeny, as to the common ancestry of the phyla concerned or their polyphyletic origin.

2. Centralization

A conspicuous trend is the development of central nervous systems. Coelenterates have only a diffuse system, though local concentrations and true ganglia (scyphomedusae) or ganglionic rings (hydromedusae) occur (Figs. 8.3, 8.13–8.15). Ctenophores, hemichordates, and pogonophores have slightly more obvious local thickenings but hardly a real central nervous system (Figs. 8.52, 27.5, 27.8, 27.12). Echinoderms have circular and radial cords of just sufficient specialization and functional importance to be called a central nervous system but they have no brain. Platyhelminths and all other groups possess unquestionable central nervous systems and, moreover, a hierarchy within it headed by a cerebral ganglion or brain. Among these groups the central nervous system is **less distinct from the peripheral nervous system** in the lower phyla. The central cords and even the brain lie basiepithelially in the epidermis in lowest forms and are intimately connected with a general peripheral plexus by numerous meshes of the latter; distinct nerves (defined bundles of axons) are not a feature or cannot be found (Fig. 9.4). In forms only a little higher the brain and cords are internal and distinctly delimited, connected to peripheral plexuses or ganglia by long nerves.

These differences are **not distributed in a single series.** They can be found repeated within several phyla and even classes, some of which are quite heterogeneous in these respects (for example turbellarians, annelids, lower molluscs) whereas others are relatively homogeneous (for example nemertineans, nematodes, rotifers). The signs of lower nervous development (Fig. 1.6) go with less complex bodily structure, smaller size, less activity, doubtless a more restricted habit of life and a smaller behavioral repertoire (for example acoel turbellarians, cestodes, enteropneusts, pogonophores, phoronids, archiannelids). Sedentary and sessile forms usually have a notably simple nervous system and few sense organs for their supposed position on the evolutionary tree (for example coelenterate polyps as compared with medusae, ectoprocts, endoprocts, brachiopods, scale insects, pelecypods, amphineurans,

IV. TRENDS AND CONTRASTS   A. Morphology

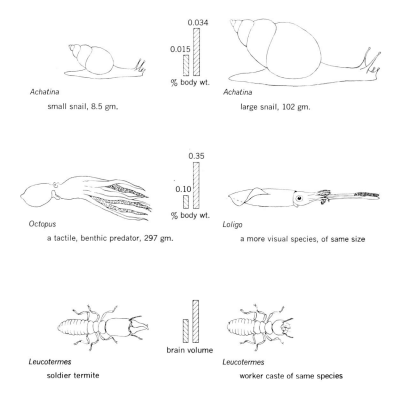

Figure 1.6. Size of the brain in animals of different size or habit of life. Brain weight of equally trimmed dissections of fixed material (formol-acetic-alcohol for snails; formalin for cephalopods) is given as percentage of body weight (living weight, without shell for snails; preserved weight for cephalopods). Brain volume is given in arbitrary units for termites, calculated from pl. 2, figures 11 and 12 of Thompson, 1916, J. comp. Neurol., 26:597.

cirratulid and terebellid polychaetes, but not serpulids or eunicids, cirripedes). Parasites often have a simple nervous system (cestodes, entoconchs, gastropods, sacculinids, mites, pentastomids); on the contrary many cannot readily be said to be reduced (for example trematodes, parasitic nematodes, acanthocephalans, parasitic copepods, fleas, cimicid bugs). Active forms, particularly hunting predators, tend to show signs of advanced nervous development (for example polyclads, nemertineans).

Centralization has clearly been a **function of simultaneous factors**: phylogenetic position (that is to say, the heritage) and habit of life, with its requirements in behavioral performance (but in no simple way) are the broad categories. The phylogenetic factor is not properly represented by the position in a list of phyla, for a group may be placed relatively far along the list because of its affinities, but it may have diverged early and remained little advanced (for example hemichordates, phoronids) or it may have suffered secondary simplification correlated with the habit (cestodes, ectoprocts, pogonophores, tunicates).

The other side of this trend toward greater separation of central from peripheral nervous system is the **reduction of autonomy of the peripheral system**. As cell-strewn plexuses are replaced by nerves carrying sensory impulses into the distant central organ and motor impulses out, the number of synapses in the periphery is reduced. In higher groups they persist only in a few places, notably in the heart and visceral supply. In other words, local autonomy gives way to central reflex control.

3. Cephalization

A more advanced trend is prominent within each of the higher, well-centralized nervous systems, namely, the displacement of both functional responsibilities and masses of nervous tissue more and more anteriorly or cephalically. Independently within the crustaceans, insects, arachnids, gastropods, and cephalopods there has been migration of ganglia forward (Figs. 16.26, 23.3, 25.1). The vertebrates exhibit cephalization in a different way; rather than movement of cell masses, new cell masses or superimposed increments on old ones appear in higher classes, and functions are shifted cephalically while at the same time evolving more complexity. Even within a class like the mammals there is cephalization, as in the functions of the visual cortex of carnivores compared to primates. (Note that there is no linear phylogenetic relation between these groups, but it is assumed that the living carnivores represent a stage in nervous development approximating that which the higher primates went through long ago.)

B. Differentiation of Cells and Tissues

The nerve cell type familiar from textbooks—a mammalian pyramidal cell of the cerebral cortex or a ventral horn cell of the spinal cord—may be taken as a starting point for an introduction to the problems of evolution of nerve cells and tissues (see Chapter 2). It is actually a very special and unusual nerve cell even among advanced nervous systems and is almost peculiar to vertebrates. It is multipolar and heteropolar, the latter term referring to the difference between axon and dendrites.

The **axon** (formerly neurite) is a process specialized to distribute or conduct nerve impulses, generally over considerable distance; it is smooth and only sends off branches (called collaterals) at long intervals, if at all. (See p. 38 for further refinements of definition.) It commonly, but not always, is surrounded by an intimate investment of nonnervous cells, called "neuroglia" inside the central nervous system and "Schwann cells" outside. (The term "Schwann cell" was long applied only to vertebrate nerves but is now used for invertebrates as well, without implying a common nature and origin to all cells given the same name.) Vertebrate neurons usually have only one axon, although the unipolar dorsal root ganglion cells have a single stem process dividing in T-fashion (Fig. 1.7); the axon extending to the peripheral receptor and the axon extending into the spinal cord might be regarded as one or as two. There is some evidence that impulses can normally be initiated centrally and conduct outward ("dorsal root efferents"), presumably in the same neuron that usually conducts sensory impulses centrally. Conduction in either direction is commonplace in invertebrates; there may be two or more sites of impulse initiation along the long axon of unipolar cells (Figs. 1.7, 3.4). Bipolar and multipolar cells in invertebrates often have two or more axons (Figs. 1.7, 2.9).

Cajal enunciated the "law of the dynamic polarity" of the neuron on the basis of observed anatomical relations of axons and dendrites. This principle recognizes that transmission of neural excitation normally occurs from the dendrites toward the axon of the same cell and from axonal terminals to the dendrites of the next neuron.

Dendrites are processes specialized for receiving excitation, usually from other (presynaptic) neurons, but the receptive ends of sensory neurons may also be called dendrites (see p. 40). They differ widely from one cell type to another and rather generally between vertebrates and invertebrates. Vertebrate dendrites in the central nervous system are commonly highly branched, irregular in thickness, thorny, and filled with cytoplasm more like that in the cell soma than that in the axon (Figs. 1.7, 2.4); they are almost always borne on the soma, so there is no clear line between soma and dendrites. The soma is also in the path of the spread of excitation (though not necessarily of impulses) and generally has synaptic endings on it. Dendrites in invertebrates and in vertebrate spinal ganglion neurons are commonly difficult to distinguish from axonal terminations. Invertebrate dendrites are probably much less profusely branched than those in the central nervous system of vertebrates as a rule; they are not characteristically thorny or plasmatic and are borne on the axon or its collaterals (Figs. 1.7, 2.1, 2.2, 2.3). These differences are perhaps of quite profound significance, but if so it remains to be shown (see further, Chapter 2).

From these comparisons it is clear that quite a variety of neurons is possible. The simplest systems consist entirely of bipolar and multipolar neurons, all isopolar except the sensory cells (Figs. 1.7, 8.3, 8.7); aside from distal processes of sense cells, all processes of the cell are axons, and dendrites are presumably at the distal ends of axons, if they are differentiated at all. The role of receiving excitation is perhaps accomplished by unspecialized endings. Already in the

IV. TRENDS AND CONTRASTS B. Differentiation of Cells

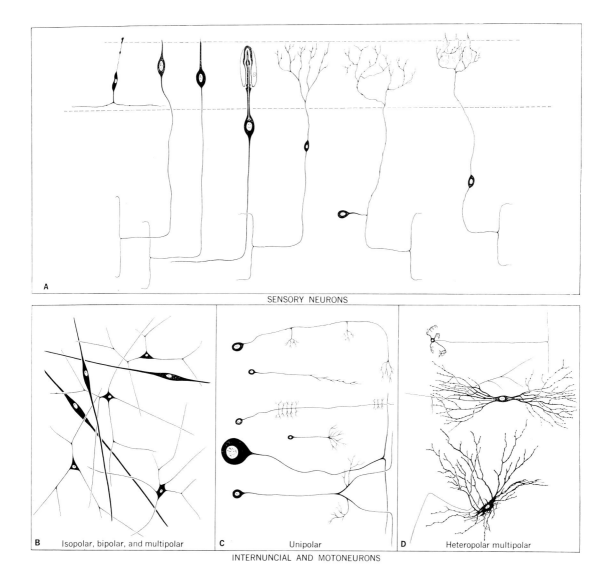

Figure 1.7. Types of neurons based on the number and differentiation of processes. **A.** Sensory neurons. The most primitive (left) send axons into a superficial plexus. In animals with a central nervous system the commonest type is a similar bipolar cell in the epithelium with a short, simple or slightly elaborated (arthropod scolopale) distal process and an axon entering the central nervous system and generally bifurcating into ascending and descending branches. A presumably more derived form is that with a deep-lying cell body and long branching distal process with free nerve endings. In vertebrates such cells secondarily become unipolar and grouped into the dorsal root ganglia. The figure on the right represents a vertebrate vestibular or acoustic sensory neuron that has retained the primitive bipolar form but has adopted (presumably secondarily) a specialized nonnervous epithelial cell as the actual receptor element. **B.** Isopolar, bipolar, and multipolar neurons in the nerve net of medusa. These may be either or both interneurons and motoneurons; differentiated dendrites cannot be recognized. **C.** Unipolar neurons representative of the dominant type in all higher invertebrates. Both interneurons and motoneurons have this form. The upper four are examples of interneurons and the lower two of motoneurons. Dendrites may be elaborate but are not readily distinguished from branching axonal terminals. The number and exact disposition of these two forms of endings and of major branches and collaterals are highly variable. **D.** Heteropolar, multipolar neurons. These are the dominant type in the central nervous system of vertebrates. The upper two represent interneurons and the lower a motoneuron.

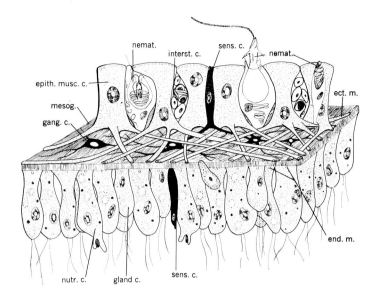

Figure 1.8.

A basiepithelial system. Diagram of the organization of the epithelia and nervous system of *Hydra*. The ectoderm is above; the endoderm below. [Schulze, after Kappers.] *ect. m.*, ectodermal muscle fiber layer; *end. m.*, endodermal muscle fiber layer; *epith. musc. c.*, epithelial muscular cell; *gang. c.*, ganglion cell; *gland c.*, gland cell; *interst. c.*, interstitial cell; *mesog.*, mesoglea; *nemat.*, nematocyst; *nutr. c.*, nutritive cell; *sens. c.*, primary sensory neuron.

platyhelminths, with their distinct brain and central nervous system, a new type of cell is found, destined to be the dominant and nearly exclusive type in the central nervous system of the higher invertebrates. This is the unipolar neuron (Figs. 1.7, 2.1, 2.2). Isopolar nerve cells, bipolar and multipolar, appear here and there among higher invertebrates, particularly in visceral plexuses and ganglia, such as the cardiac ganglion of crustaceans (Figs. 1.7, 2.9, 3.4, 25.20). Cephalopods and some arthropods possess in the brain a few multipolar, heteropolar cells like those in vertebrates (Figs. 2.9, 25.37). The meaning of these differences and patterns of distribution remains to be clarified.

Correlated with cell types is the **histological architecture of nervous tissue.** Whenever it occupies the superficial, basiepithelial position characteristic of the simplest systems (coelenterate, echinoderm, enteropneust, the epidermal plexus of nemerteans, even the brain and cord of some simple acoel flatworms, terebellids, and limicolous oligochaetes), the diagnostic tissue, a stratum of nearly cell-free nerve fibers, lies deep in the epithelium, just external to the basement membrane, if present, and to muscle fibers (Figs. 1.8, 2.11). It is visible even in sections stained by common unselective dyes, as a **layer of punctate texture** but little structure, unless locally there are too few fibers to form a distinct layer.

Whenever nervous tissue forms a ganglion or cord of the central nervous system in the invertebrates, and is therefore dominated by unipolar cells, it presents a constant relation: cell somata are concentrated at the surface, nerve fibers are concentrated in the center of the ganglion. The cells form a layer that may be loosely or tightly packed, one or several cells deep, but which is usually clearly demarcated and can be called the **cell rind.** The single-stem process of each cell is directed inward and, together with their collaterals, dendrites, and many axonal terminals of local and distant cells, forms the fibrous core of the ganglion (Figs. 2.14, 2.15).

A number of degrees and **signs of advancement** can be distinguished, comparing rind and core in various animals. In its lowest degree, the rind is loose and its outer and inner margins are not sharp or even; the core is also loosely textured, without differentiation of regions. In more advanced forms the rind is divided into small clumps of cells, each with its inwardly directed bundle of stems. Very early the core begins to separate into neuropile and tracts. The latter are bundles of parallel fibers of passage. **Neuropile** is a region of nearly pure fibers, with few or no cells but with a concentration of axonal terminations, dendrites, and therefore synapses. We will encounter more about neuropile in several chapters of this book, for here is the seat and secret of many of the most characteristically nervous achievements, especially integrative events.

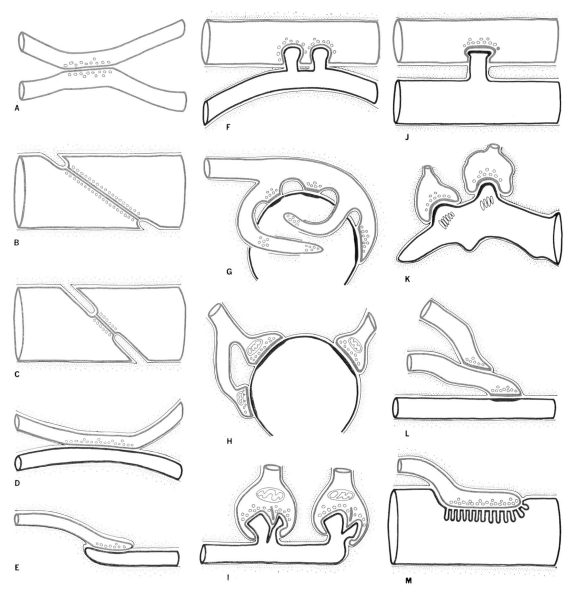

Figure 1.9. Types of synapse revealed by electron microscopy. *Blue,* presynaptic; *red,* postsynaptic; *black,* the innermost extent of the sheath. **A.** Coelenterate axon-axon example from jellyfish ganglion *(Cyanea),* with vesicles on both sides and without adhering sheath cells. **B.** Earthworm septal synapse with close apposition of the membranes of the two cells and sparse vesicles on both sides. **C.** Crustacean septal synapse, as in **B** except that the synaptic area is restricted. Examples **B** and **C** have electrical transmission in either direction. **D.** Axon-axon synapse en passant typical of neuropile in many invertebrate ganglia, often with and often without sheaths. **E.** Axon terminal arborization ending on a fine dendrite in invertebrate neuropile. **F.** Crustacean giant-fiber-to-motoneuron synapse, with postsynaptic motor fiber invaginated into the giant fiber. This example also has electrical transmission, but only in one direction. **G.** Axon arborization to soma synapse typical of vertebrate brain cells but in invertebrates so far only clearly known as inhibitory endings on crustacean peripheral sensory cell of muscle receptor organ. **H.** Boutons terminaux of axon arborizations typical of central cell bodies in vertebrates. **I.** Ribbon synapses between rod cell endings and dendrites of ganglion cells of vertebrate retina, with presynaptic specialization. **J.** Synapse between giant fibers of squid stellate ganglion, postsynaptic invaginated into presynaptic. **K.** Spine synapse (axon-dendrite) from cerebral cortical dendrite of vertebrates with postsynaptic specialization. **L.** Serial synapse. Found so far in spinal cord, cerebral cortex, and plexiform layer of retina in vertebrates, but offering many potentialities for presynaptic inhibition and other complex interaction in neuropile. **M.** Specialized neuromuscular endings found in vertebrate skeletal muscle, with postsynaptic grooves.

One of the differentiated forms of neuropile is composed of **glomeruli,** small knots or balls of denser neuropile, usually with certain definite input pathways and output destinations (Fig. 2.16). The associated rind and somata may be unspecialized. But a higher type of neuropile occurs as a fairly extensive mass of extremely fine and dense texture; therefore it is very smooth in ordinary preparations and always associated with a group of special **globuli cells.** These are small, chromatin-rich and cytoplasm-poor, packed tightly and several to many cells deep (Figs. 1.7, 2.15, 2.16). Glomeruli and globuli cells together form characteristic bodies in the brains of many annelids and arthropods, where they are called mushroom bodies (corpora pedunculata); histologically similar regions occur in the highest centers of several other groups (polyclads, nemertineans, gastropods, cephalopods (Figs. 2.16, 9.10, 10.3, 23.23, 25.6).

Another specialized form of neuropile is **stratified;** up to five or more layers can be distinguished, based on branches mainly of dendrites that confine themselves to a narrow zone at a specific depth, as in the cortex of the optic lobe of *Octopus* (Figs. 2.15, 25.18) and of insects (Figs. 2.2, 19.12). Why optic lobes should have this structure is unknown.

A form of differentiation that has evolved independently many times is the **giant fiber.** Often of colossal proportions relative to other nerve fibers in the same animal, the significance is still generally thought to be simply increased velocity of conduction, together with through-conduction over long distances. Some giant fibers represent the axon (exceeding 400 μ in diameter) of a single cell of only moderately large size (for example that in serpulids, Fig. 14.42); some are but moderately large axons (35–50 μ) from a huge cell (800 μ) (the gastropod *Aplysia*, Fig. 2.6); some are large compound axons resulting from the fusion of many cell processes of unipolar cells of ordinary dimensions (earthworm, squid, Figs. 14.26, 25.24). Giant fibers are known or reported in some flatworms, nemertineans, phoronids, enteropneusts, annelids, arthropods, molluscs, and vertebrates. Repeatedly in each group, we find that related species differ in respect to possession of giant fibers, which must therefore evolve rather readily.

The **varieties of synaptic contact** regions are great. There are many special forms of axonal termination, of dendritic ramification, and of geometric relations between them (Fig. 1.9). In addition, axons end on the cell soma in certain cases among invertebrates and generally among vertebrates. There are cases of axo-axonal synapses though the most classical (squid giant, crayfish giant-to-motor) are better treated as axo-dendritic with very short dendrites (see Chapter 2). Possible somato-dendritic junctions are shown in Fig. 2.20. Chapter 2 gives some details of the types of contacts and others are found in the systematic treatments of the groups.

Scattered widely among invertebrates and vertebrates are nerve cells presenting histological signs of secretion of palpable granules. Physiological experiments have shown this **neurosecretion** to be the mediator of certain important functions. Here is a direction of specialization, perhaps an elaboration of a common property of all nerve cells, of such interest as to require a special chapter surveying its occurrence and varieties in all groups (Chapter 6).

Investigation is active today in almost every one of the areas mentioned. There are abundant opportunities for new work and it seems highly probable that discoveries which will alter general concepts significantly will be made on the wealth of forms available among lower animals.

C. Receptors and Sense Organs

The nervous system connects receptors and effectors; some introduction to each of these is therefore appropriate. In this book the nervous aspects of receptor structure and function are treated in some detail for each group of animals, whereas little attention is given to accessory structures or to the range of sensibilities inferred from behavior (concentrations of sugars, wavelengths of light, and the like).

The specialized structure that transduces the stimulus into an active physiological event capable of initiating nerve impulses in an afferent axon may be either a distal process of the afferent neuron itself or a special, **nonnervous sensory epithelial cell** (Figs. 1.10, 2.25). The former is the rule and the latter an exceptional development known only in a few receptors of vertebrates, namely those of the acousticolateralis system and the taste buds. (Some authors treat the rods and

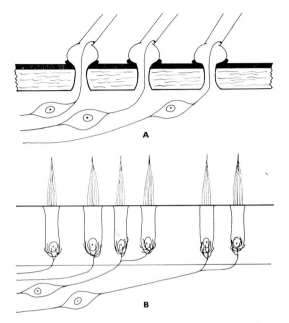

Figure 1.10. Direct and indirect sensory nerve endings. **A.** Primary sense cells are true neurons that are themselves receptors. In this diagrammatic representation of a type of crustacean mechanoreceptor, the distal dendritic process terminates in relation to a long cuticular hair. **B.** Nonnervous sense cells (formerly secondary sense cells) are believed to be the actual receptor elements in some cases and to excite secondarily the endings of the sensory neuron. The diagram represents the inner hair cells and the cochlear neurons of the spiral ganglion of the organ of Corti.

cones of the vertebrate retina as nonnervous sense cells but they may just as reasonably be regarded as nerve cells.) Equivalent receptors in invertebrates are not so far known to employ nonnervous sense cells. No functional significance can yet be attributed to this difference.

Sensory nerve cells differ in the **position of the cell body.** The most primitive form is superficial —that is, in the epithelium; bipolar cells normal to the surface, they generally have a short distal process believed to be the transducer structure (frequently with a modified cilium). Cell bodies lying below the epithelium and having a branched distal process, so-called "free nerve endings," are found less often but tend to occur in some of the smaller, better developed invertebrate groups and in annelids, arthropods, molluscs and vertebrates. In vertebrates the cell bodies are still deeper, usually unipolar with a T-shaped stem process (except those of cranial nerves I, II, and VIII), and gathered into dorsal root ganglia lying close to the cord. Infrequently, sensory nerve cells appear to have moved into the central nervous system (for example the nuchal region of some polychaete brains, coxal mechanoreceptors of decapod crustaceans, and the mesencephalic nucleus of the trigeminal nerve in mammals).

Receptors occur most commonly as **unicellular,** dispersed elements but these may be locally extremely numerous. Clusters of small numbers of superficial sense cells forming simple **multicellular** bodies occur at least in annelids, arthropods, and molluscs. **True sense organs** are ordered groups of sense cells and accessory cells forming a complex of characteristic structure, usually directed toward a limited range of stimuli (Fig. 1.11). Sense organs are already present in

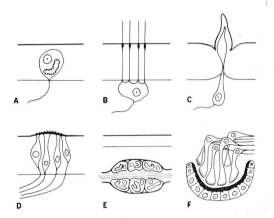

Figure 1.11. Types of receptors vary from single sense cells dispersed in or under epithelium, to groups of sense cells that do not involve other tissues, to true sense organs involving several specialized types of cells in definite relation to each other. **A.** Photoreceptor cell in the skin of leech. **B.** Unicellular receptor in the epidermis of a turbellarian believed to be mechanoreceptive, perhaps detecting water currents. **C.** Specialized, presumed mechanoreceptor in the skin of a polychaete. **D.** Multicellular receptor or sensory bud in the epidermis of an earthworm. **E.** Cluster of photoreceptors along the course of a nerve close to the epidermis of the earthworm. **F.** Simple eye in a flatworm.

some coelenterate medusae—eyes and statocysts—and are said to be the first organs of any kind.

A considerable number of structural and **functional species of receptors** have been distinguished, though probably fewer than actually exist. Older efforts to classify receptors as exteroceptors, interoceptors, and proprioceptors have been largely replaced by a classification according to mode of stimulus—hence mechano-, thermo-, electro-, and osmoreceptors. Strictly, these terms cannot be applied until a given receptor is shown not only to be sensitive to that form of stimulus but to be employed normally by the organism in detecting it. Most receptors are known only anatomically in the less well studied groups. Modalities, a term difficult to define, are more limited; the olfactory and the gustatory modalities both use chemoreceptors. Helmholtz proposed the term as preferable to "qualities" and in usage it approximates "distinguishable classes of sensation" in man. Physiology has shown that a given modality may be a composite of several submodalities or definable receptor types, specific to certain colors, pitch ranges, tastes, quick or slow stretch, and the like. In the skin there is behavioral evidence in man and elsewhere that a number of distinct functional receptor types exist among the apparently undifferentiated free nerve endings (qualitatively distinguishable sensibilities; cold spots, warm spots and the like).

In general there is of course no homology between receptors of the same name in different phyla. Eyes, statocysts, and other sense organs appear again and again, with contrasting structures and in quite different parts of the body, even within a given phylum.

Questions of particularly active interest in recent literature are the physiological mechanisms of transducing and communicating useful messages in afferent pathways. What aspects of the stimulus are measured and coded? And is the code primarily frequency of impulses in every situation? What is the biological meaning of background firing in many afferent lines of varying degrees of rhythmicity, of central control of sensitivity, and how are true signals distinguished from "noise"? These as well as more central problems—the extraction of desired information from raw receptor input, and the analysis, weighting, and mixing of these signals—are developed in Chapter 5 on Integration.

D. Effectors and Effector Control

Muscle, in its various forms, is far and away the most important effector, but it is not the only one. Glands of many kinds, cilia, chromatophores, luminescent organs, and electric organs are also of interest, for themselves and for the principles of control which they may illuminate (Fig. 1.12). Above we spoke of **independent effectors** and these may represent the first of a series in respect to control. Next we may recognize effectors such as certain cilia, which exhibit spontaneity and are thus capable of independent action but which also possess a superimposed nervous control, especially an inhibitory one. Some muscles (Burnstock and Prosser, 1960) and ciliated epithelia manifest a prenervous or neuroid property in the **ability to spread excitation** from cell to cell directly, without nerves; the mechanism is not understood but electrical interaction has been proposed for muscle and mechanical restimulation for cilia.

An apparently early stage in the nervous control of effectors is one in which the train of all-or-none nerve impulses in an efferent fiber calls forth a **response of graded amplitude,** depending on the frequency and number of impulses. "Fast" or twitch-evoking fibers are usually specialized. By building on the simple pattern we reach a condition of **multiple innervation;** two or more (up to five) nerve fibers make synaptic contact with the same muscle cell, generally with quantitatively different effects or even opposite effects. One may inhibit response, the others excite it; one may cause quick contraction and another slow contraction. These modes of control are associated with the presence of many nerve endings scattered at intervals of a few scores of microns along the muscle fiber (Fig. 1.13).

The mode of control in which one nerve impulse causes an **all-or-none muscle contraction,** associated with innervation by a single nerve

IV. TRENDS AND CONTRASTS D. Effectors

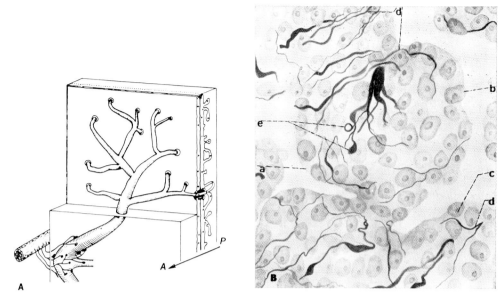

Figure 1.12. Neuroeffector endings on effectors other than muscle. **A.** Semidiagrammatic reconstruction of an electroplaque of the mormyrid fish *Gnathonemus*. The nerve at the left sends several axons to ramify and terminate on a specialized process of the electric organ cell represented by an irregularly outlined flat disc seen in the anteroposterior section. The rectangular blocks represent the connective tissue blocks, each of which contains one transversely flattened electric cell. The thick trunk is the result of anastomosis of small processes of this cell. [Szabo, 1961.] **B.** Innervation of the convoluted tubule of the kidney in the frog, *Rana*. Bielschowsky-Ábrahám silver method. [Ábrahám, 1959.] *a*, wall of the tubule; *b*, epithelial cell; *c*, epithelial cell nucleus; *d*, nerve fiber; *e*, nerve ending.

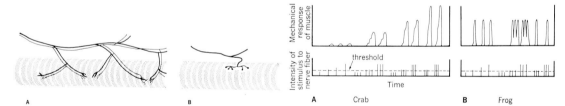

Figure 1.13. Two types of muscle innervation; schematic representation of a few anatomical and physiological contrasts. **A.** Polyneural and multiterminal innervation common in Crustacea, represented by crab skeletal muscle. Typically, single shocks to the nerve fiber—above threshold (indicated by the dotted line)—cause at the most, small excitatory postsynaptic potentials and minute contractions. If the interval between stimuli is shortened, these show facilitation and summation building up to a total response that is a function of frequency of arriving nerve impulses. Since we are stimulating a single nerve fiber, there is of course no effect of stimulus intensity above threshold. **B.** The unineural and uniterminal muscle, represented by the twitch or fast fibers of skeletal muscles of the frog, show an all-or-none response above threshold with no important effect of frequency upon amplitude; the single fiber cannot follow high-frequency stimulation.

fiber via a single junctional region, is widespread in higher vertebrates but must be regarded as a special development; it is not the sole mode even in the striated, skeletal muscles of vertebrates (Hoyle, 1957). It requires new means of grading movements smoothly; these include numerous motor units and recruitment of varying numbers of them by central processes.

A peculiar system is found in **high-frequency fibrillar muscle** of certain insects (flies, bees, cicadas, some beetles, and others) where a few nerve impulses per second are sufficient to maintain a state of excitation in the muscle during which it can, with inertial load, contract at several hundred per second (Chapter 17).

We are far from understanding the control of contraction. Some instances indicate a very indirect relation at the best between muscle membrane depolarization and mechanical events. Muscle has evolved a great deal of diversity, as witness the examples of "catch" muscles in pelecypods, leeches, and elsewhere, that maintain active contraction for long periods (see p. 1417). Strikingly similar to nervously controlled behavior is the abrupt shortening of the stalk of *Vorticella* and related protozoans, which seemingly involves no innervation by a specialized conducting system.

E. Problems of Central Nervous Physiology

Few generalizations can be made in this sphere, particularly of a comparative nature, except that the properties of neurons seem qualitatively to be essentially alike from lowest to highest, so novel achievements in higher forms must be ascribed to quantitative permutations of these properties and to arrangements within constellations of neurons. It may be useful to make some introductory reference to the kinds of problems that represent the state of the subject today and form the substance of much that follows in this book.

It is interesting to see how **integration seems to be more central** in the highest animals, whereas a considerable amount of integration relevant to behavior is peripheral in many lower forms. Muscle control by multiple innervation and graded neuromotor response represents peripheral integration to a greater extent than the all-or-none mode. Through-conduction systems offer a special case in which integrative events tend to be pushed upstream to the earliest stages of analysis of input, and occur before the messages get into the through-conduction pathway.

The physiological **problems at the lower levels** of central nervous events include the properties and potentialities of reflexes as adaptive fragments of normal behavior. Some kind of central switching chooses among competing alternative reflexes and prevents maladaptive intermediate movements. Inhibition and excitation seem to be of several kinds. Autonomy can be seen to have undergone great evolution. Afferent input due to gravity, light, and other stimuli is somehow given a physiologic weight, as shown, for example, by the position of a fish when light comes from the side (Fig. 1.14). The integration of the weighted converging signals is not always simple algebraic summing, but may be more complex.

Discovering the effects of stimulating or removing the brain or more limited higher centers is only one stage in analysis; more difficult is to recognize the relevant features and formulate the **roles of higher centers** in significant terms. This is both important and baffling because nature has piled

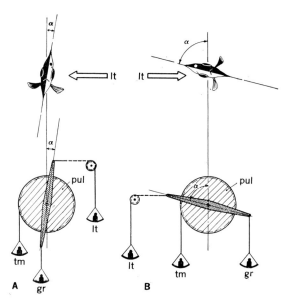

Figure 1.14. Sensory inputs that converge to determine the output of a final common path are evaluated and summed algebraically or otherwise. The diagram represents a fish from which the labyrinth of one side has been extirpated and to which light is delivered from the side. In this experiment of von Holst (1950) the summing is shown to be largely ipsilateral, so light on the intact side has a small but measurable effect as shown by the diagrammatic balance below, whereas light coming on the operated side exerts a large effect, still slightly countered by the contralateral labyrinth. The normal balance in the intact animal between the effect of light *(lt)* and gravity *(gr)* would cause an intermediate position; but the effect of one-sided extirpation is equivalent to adding a constant turning moment, indicated by weight *(tm)* working through the pulley *(pul)*.

level upon level of superimposed nervous structure and at the same time has developed parallel and interlocking systems of qualitatively different meaning, like the so-called specific and nonspecific systems in the mammalian brain (p. 306).

Running through all levels is the question of **how much is achieved by circuitry**, using neurons as elements in complex, specifically connected loops with feedback, parallel paths, self-correcting features, and the like. How much is entrusted to straight communication lines with unmonitored calibrations of the transfer functions between elements? What novel or **emergent properties** inhere in organized populations of neurons? What is the origin and causal significance, for example, of brain waves? This form of central activity presents a curious comparative feature; while virtually uniform in its slow smooth character in vertebrates from fish to man, it is in a wide assortment of invertebrates dominated by large, spiky, fast waves, though in insect optic lobes certain conditions cause transient, smooth, slow waves.

The principles operating in **analysis of input**—recognition of predetermined criteria, decision-making, complex threshold phenomena, and the influence of specific patterns of impulses in time as well as space—are among the motivating problems which will occupy imaginative workers in the future. Similarly, the ways in which the nervous system generates and **commands patterned movements**, from simple to complex, are of fundamental interest and are already known to be of several kinds, sometimes with mixtures in different proportions.

The **nervous organization of normal behavior** is the final objective and it is well to recognize that the gap between our present understanding of the nervous system and an explanation of behavior is probably wider than any other gap between levels of integration in natural science! However, much has been done and can be done at this level. **Localization** of the regions that do and that do not influence chosen behavioral signs, such as the performance of a previously learned task, is one direction. Delineation of activities of units during normal behavior is another. Attempts to characterize in a natural and significant manner alterations in behavior with local brain stimulation (Fig. 1.15) or ablation is another—that is, the recognition of the scaling, division of function, or **parameterization** in natural rather than logical or formal ways. Less studied than learning but in some ways more

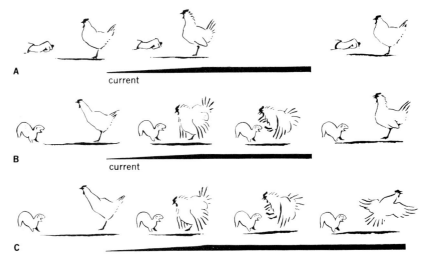

Figure 1.15. Local brain stimulation with electrodes implanted in the diencephalon in a critical locus can release characteristic complex behavior. In this instance from von Holst and St. Paul (1960) the stimulus alone caused only a slight unrest in the absence of a suitable object of reference; in the presence of a fist it caused a slight threatening movement. A motionless stuffed ground predator is threatened and in the presence of an electric current attacked; if neither current nor predator is removed the hen flies away screaming.

hopeful are the relatively stereotyped and species-characteristic activities, including those subsumed under instinct.

It will be clear from the foregoing that the problems of **system operation** chosen for treatment in this book start at the neuronal level and perforce neglect the lower level mechanisms. That is, we do not inquire here about the basis of excitability, how the signals are produced, what permeability changes are relevant in various cases, or what transmitter agents are responsible for junctional transmission. This is a large and active field, as is the whole field of neurochemistry, but limiting lines have had to be drawn; this is a logical level to do so, although the boundary is anything but clear.

V. METHODS OF STUDYING THE NERVOUS SYSTEM

The familiar historical observation that progress in a scientific field goes with the development of new methods often stops short of noting that only some of these depend on technological innovations. An important class of new approaches in the field of the nervous system consists of those that were conceptually novel when they were introduced but could have been used years before, technically. The proof of the nervous nature of the giant fibers of earthworms—Bovard (1918), by section and regeneration, and Yolton (1923), micropuncture—could have been accomplished in the last century while controversy raged over their nature. Many of the experiments showing that neurosecretory structures have functional significance are in the same category. Even when confronted with a momentary technological impasse, imaginative contribution is still possible, often by novel permutations of existing approaches.

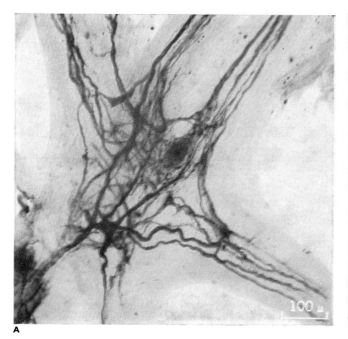
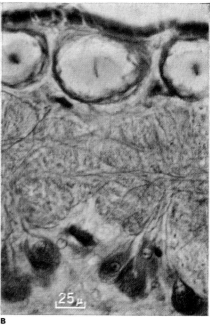

A B

Figure 1.16. Special nerve stains. **A.** Methylene blue intra vitam stain for nerve cells and fibers. The example is a portion of the cardiac ganglion of the lobster *Panulirus*, with one nerve cell body, large axons, and fine endings; the connective tissue matrix is unstained. [Preparation, courtesy of D. M. Maynard.] **B.** Reduced silver impregnation of the ventral nerve cord of the earthworm *Lumbricus*; Bodian protargol procedure. Most tissues are stained to some extent but nerve cells and their processes are brought out; in particular small nerve fibers and neurofibrils are darkly impregnated. [Original.]

V. METHODS

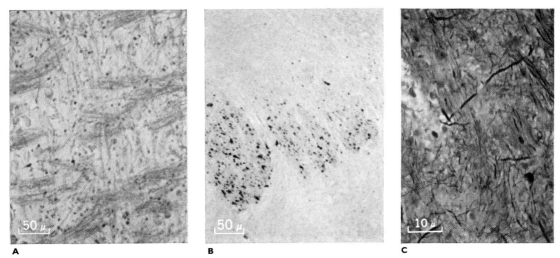

Figure 1.17. Selective stains for special features of nervous tissue. **A.** The Marchi osmium tetroxide stain for degenerating myelin. The example is a degenerating pyramidal tract in the brain stem of the rat after a lesion in the cerebral cortex two weeks before. [Original.] **B.** The Klüver stain for myelin (blue), counterstained with a basic anilin dye for the Nissl bodies in the nerve cell somata. The example is a portion of the reticular formation of the midbrain of a marmoset. (Preparation, courtesy L. Kruger.) **C.** The Nauta silver stain showing degenerating axons and endings, including unmyelinated fibers. The example is from the cat midbrain after a lesion in reticular substance some distance away. [Preparation, courtesy C. Clemente.]

The following paragraphs list briefly a wide selection of the classical and newer methods by which information has been gained on the structure and function of the nervous system. References for every method cannot be detailed here but can be found in the general works at the end of this chapter and in the works cited in subsequent chapters, where the methods were employed.

Anatomical methods begin with the fresh **living specimen;** direct microscopic examination aided by trans- or epi-illumination, have often revealed some features otherwise visualized with difficulty or only after drastic treatment (Figs. 2.7, 8.2). Schultze, Bozler, and others have seen neurofibrils in living axons, thus disposing of suggestions of artifact (see Chapter 2). Patience and repeated observation are required to decipher ghostly images but reward the effort by controlling the artificial images of fixed and prepared tissue. Older workers extensively and profitably used techniques of **dissociation** based on maceration; their methods would bear re-examination and possibly new exploitation (Lee, 1950).

Methylene blue, employed on the still living tissue, in sheets, strips, or fragments rather than on thin sections, continues to hold its unique place as a method for revealing long axonal processes of nerve cells over extended distances, especially in the periphery. Its beauty and utility are shown in a number of classical figures reproduced in this work (Figs. 1.16, 8.7, 10.4, 14.35, 16.30, 16.43, 17.16, 18.14). Capriciously selective, it has seldom or never succeeded on many animal groups and in those cases nothing takes its place.

Reduced silver occasionally offers real advantages in tissue sheets or small fragments but is chiefly employed in conjunction with thin sections. Here it brings out fine detail in cells and processes, only somewhat selectively; it is most used in higher central masses (Figs. 1.16, 2.11, 25.6), but is valuable also in the periphery when interpreted cautiously, because of nonnervous argentophilic cells and fibers. Cajal (1909) wrote a useful general chapter on this and other methods in his classic textbook.

Chrome silver, usually called the "Golgi method," after its inventor, is the only other selective method for visualizing nerve cells and

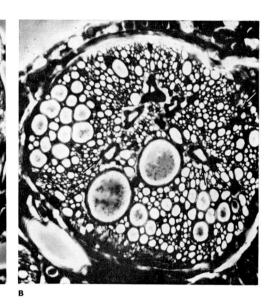

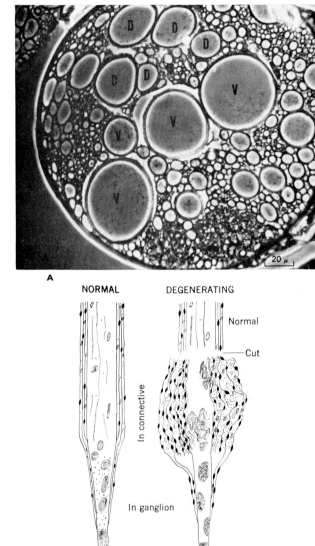

Figure 1.18. Normal and degenerating axons in the central nervous system of an invertebrate. **A.** The normal abdominal connective of the cockroach nerve cord, fixed in Dalton's fluid and imbedded in plastic; phase-contrast micrograph. Dorsal *(D)* and ventral *(V)* fiber groups are marked. **B.** The same anterior to the point of severance of the cord four days previously. The dorsal group and two fibers of the ventral group *(arrows)*, a larger and a smaller, are degenerating. The sheaths of the degenerating nerve fibers are seen to be thickened. Some apparently small fibers are also degenerating as seen in a bundle in the dorsal medial aspect of the connective *(arrow)*. **C.** Diagram showing a normal nerve fiber sheath and termination and the effects of degeneration induced by severance on the peripheral stump of the nerve fiber. [Hess, 1958, 1960.]

their prolongations in the light microscope. Capriciously silhouetting an element here and there against a clear ground of unstained tissue, it provides in unsurpassed clarity the images from which, by patient comparison of thick sections of many preparations, classical analyses of central neuron arrangement have been erected. Besides isolated sense cells in the epidermis and free branched nerve endings, it has been chiefly central structures which have been elucidated, including those which Cajal (1937), the widely experienced neurohistologist and master of silver methods, called the most intricate structures known in the living world—the optic ganglia of insects. As with the last two methods, but in far greater measure, Golgi impregnation is undependable and has yet to be applied for the first time to many major groups of animals.

Special methods—such as the Weigert stain for myelin, the Marchi procedure for degenerating myelin, Cajal's and Hortega's methods for neuroglia, the Nauta method for synaptic endings—are highly developed for mammalian tissues (Lee, 1950; Windle, 1957; Davenport, 1960; Fig. 1.17), but have had little or no application to invertebrate material as yet.

V. METHODS

Histochemical methods are becoming increasingly useful, notably that for cholinesterase (Gomori, 1952; Graumann and Neumann, 1958; Gurr, 1958).

The **electron microscope,** usually regarded as a means of revealing ultrastructure, has added the first major new tool in more than sixty years for recognizing nervous elements and working out histological relations. Used in conjunction with serial sections and three-dimensional reconstruction, a promising new application, relatively low-power electron microscopy makes a distinctive contribution to the study of nervous organization (Figs. 2.4, 2.25, 2.27, 2.28, 2.30, 2.47, 18.6). High-power studies are contributing to the investigation of synapses and inclusions.

Serial sections stained by **ordinary histological and cytological methods** for light microscopy have formed the backbone of most studies on the anatomy and histology of the nervous system in lower forms. These methods do not show individual nerve fibers; masses of nerve fibers give an appearance classically called "punctate" (Figs. 2.15, 9.27, 10.3, 14.24, 23.23). Specialized, dense, fine-textured neuropile can be distinguished from common, loose, or coarse neuropile, and in many cases tracts can be recognized if enough fibers travel closely in parallel. Nerve cell bodies are often well shown except that their processes can seldom be made out and the cell outline is often indiscernible. Where other criteria assure the identification as nerve cells, many inclusions of the soma (or perikaryon) are clearly brought out by general cytological methods. Collected in the rind of ganglia, neuron somata are reasonably distinct and several types can be defined with common methods. Isolated and equivocal cells, however, cannot be identified with security without visualizing the number, types, and connections of the processes. The importance of laborious tracing of small nerves and the like through formidable sets of serial sections, and of slowly becoming really familiar with the particularities of the animal, cannot be overemphasized. Many of the landmark studies in comparative neurology were done in this way.

Physiological methods of a few basic types permit permutation with each other and with

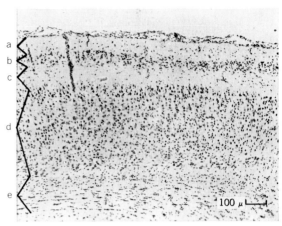

Figure 1.19. Laminar lesion caused by high-energy ionizing radiation (40 Mev alpha particles) focused at a level below the surface of the brain. The photomicrograph of the visual cortex of a rat shows the narrowly restricted area of damage, sparing cortex above the lesion as well as below, illustrating a modern technique for administering controlled limited lesions. [Courtesy L. Kruger.] a. Molecular layer appears normal. b. Part of layer II appears normal. c. The pale stratum is the zone of the lesion, destroying parts of layers II and III. d. Layers IV–VI appear normal. e. White matter appears normal.

terminal anatomical examination. Local stimulation, local depression or abolition of function, local recording of signs of activity, each can be carried out in a variety of ways suitable for the given case (Donaldson, 1958; Bureš et al., 1960). Thus electrical stimuli of relatively long-maintained current (DC) or of brief pulses, more or less frequent (pulsed, "faradic"), or of regularly reversing current (AC) or of slowly rising or other form can be used. Excitatory drugs are sometimes applied topically as by bits of filter paper soaked in a solution of strychnine. Reversible depression can be caused by anodal polarization, by drugs or by a cold thermode (a simple bar of metal chilled at the free end and applied to the tissue at the other end, a fine tube conducting cold water, or a small expansion chamber for compressed CO_2). More commonly employed, however, is surgical ablation or electrolytic lesion (Fig. 1.18), in spite of the disadvantage of irreversibility, because it is possible to assert that such and such area of tissue was certainly destroyed. A promising technique for producing small lesions at depth,

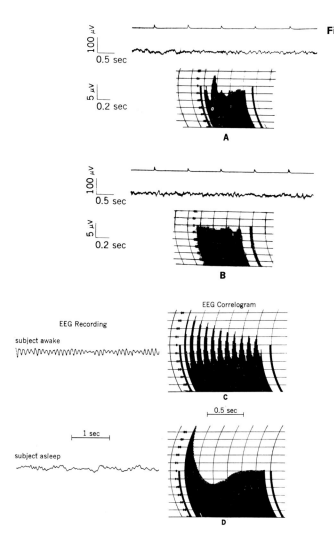

Figure 1.20. Use of computers to extract information from brain wave records. Electrical deflections resulting from periodic clicks delivered to the ear and recorded from the scalp of a human subject are small and buried in background brain wave activity. **A.** Records in the presence of clicks. **B.** Records in the absence of clicks. The upper trace indicates the times of click presentations in **A** and serves merely as a comparable time reference in **B**. The electroencephalograph traces in the two cases are short samples, including much background activity not related to the click response. The shadowgraphs are the output of a computer which has averaged 250 consecutive responses at 1 per second. The characteristic troughs and peaks in **A** build up as the averaging proceeds and the small but consistent deflections due to the clicks summate; the background activity, occurring at random phase relative to the clicks, is smoothed out by failing to summate. **C.** EEG ink recording and autocorrelogram for a subject awake, and **D.** the same subject asleep. The ink traces shown represent $3\frac{1}{2}$ seconds of data; the autocorrelogram was computed from a sample length of 100 seconds of data. The computer compares the voltage at 10 msec. increments of time with the preceding sample moments back for one second and plots the correlation, bringing out features that would not always be obvious to naked eye inspection of the raw EEG. The conspicuous differences between **C** and **D** are due to the presence of relatively fast activity of a periodic sort in the waking subject and of slow irregular activity in the sleeping subject. Electrodes were located in the parietal-occipital area; the subject was in a dark anechoic chamber. [Communications Biophysics Group, 1959.]

without surgical intervention and associated hemorrhage or stasis (octopus), is that of focused high-energy radiation (Fig. 1.19).

Local **recording** is sometimes done with large (approximately 0.2–1.0 mm) metal or wick electrodes, especially on exposed surfaces, when some kind of average potential of a large mass of tissue is desired. Semimicroelectrodes (commonly 10–50 μ, metallic) are often used by insertion, insulated to the tip, and frequently reveal the activity of single neurons, though "extracellular"; this expression means outside the cell in question but not necessarily in some pre-existent intercellular space. Microelectrodes of 2–10 μ, often glass sheathed or otherwise insulated metal or glass capillaries filled with Ringer's fluid or sea water, are used in the same way, particularly to observe unit spike activity; under special circumstances, local prepotentials can be seen. Some giant units can be impaled or cannulated with electrodes of this size. Most intracellular recording at present employs glass capillaries of approximately 0.1–1.0 μ outside diameter, filled with 3M KCl; larger cell somata and axons can be penetrated, but smaller cells and axons—to say nothing of smaller dendrites—are still inaccessible as a rule; in certain cases spikes in very small axons have been recorded (Maturana et al., 1960) but the necessary conditions are not really understood.

Tracing the connections of neurons electrically has been called physiological neuronography (Dusser de Barenne et al., 1941a, 1941b, 1941c; Kerkut and Walker, 1962, see Gastropoda). The

V. METHODS

Figure 1.21.

Analysis of intervals between spikes. The example shows activity of a single unit in the cochlear nucleus during presentation of clicks at various rates as shown. The computer outputs give interval histograms of the same data. The ordinate on all displays is the number of spikes per unit time interval. N is the total number of spikes represented in each histogram. The analysis brings out the degree of success of the unit in firing at a fixed phase relation to the clicks even when it cannot follow every click. [Gerstein and Kiang, 1960.]

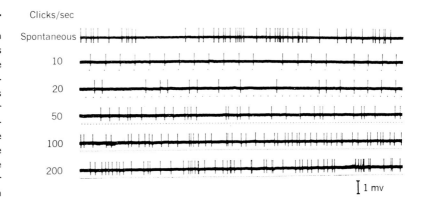

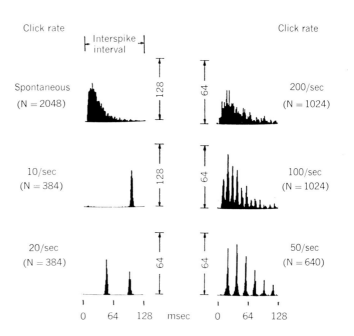

evoked potential technique which is required for this usually involves nearly simultaneous activity in many units in a small area, one or several synaptic relays away from a sudden, synchronizing stimulus. The distribution of the evoked potential may be plotted by separately graphing the potential against time for different spots or may be displayed toposcopically by rising and falling mechanical or luminous analogues of the potential at each spot. If the background activity is large, the proportion of responding units small, or the synchrony too imperfect, evoked potentials are increasingly difficult to detect. Here various forms of average-response-computer (Fig. 1.20) can be used when the same stimulus can be repeated many times and the response is sufficiently faithfully repeated; by means ranging from photographic to digital computer, the successive records are summed and thus very small signals brought to light, while events not associated with the stimulus are averaged out (Brazier et al., 1956, 1960).

Methods cannot be considered here for membrane physiology, such as impedance and specific ion conductance measurements. Similarly pharmacological, neurochemical, histochemical, and related procedures are beyond the present scope.

A wide variety of approaches and some of the most derived and complex techniques are used in **analyzing integrative processes** involving two or more neurons. Discriminators can divide multi-unit spike signals, like those seen by macro-electrodes on an insect ganglion, into several or many amplitude classes for separate analysis. Records of thousands of discrete events like spikes lend themselves to computer analysis of the successive intervals (Fig. 1.21), to examine, for instance, histograms of the distribution of intervals during epochs of some interest such as sleep, attention, patterned stimulation, or drug

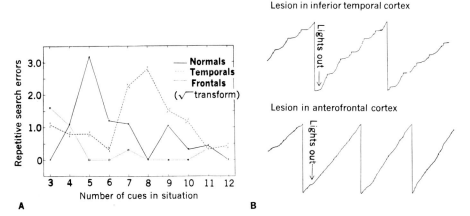

Figure 1.22. Methods of analysis of deficits in performance in learning tasks following brain lesions. **A.** Multiple choice techniques useful in estimating quantitatively the processes altered by brain lesions that determine observed changes in behavior. Monkeys are trained to choose one of several objects placed on a board. The "correct" object is placed over a well that contains a peanut. The objects are placed in different positions on each trial; the positions are chosen according to a random number table. The graph represents results from an experiment in which the number of objects (abscissa) was increased from an initial two to a final twelve. Plotted are the *repetitious* choices of unrewarded objects. Twelve monkeys were used, four with anterofrontal resections, four with inferotemporal ablations, and four were unoperated controls. Note the differences between the curves. The occurrence of peaks (confusion between novel and familiar cues) was accounted for by a sampling model based on statistical learning theory as developed by Estes and Burke and by Bush and Mostellar. The model was tested further by quantitative determinations of the proportion of objects sampled by each group of monkeys. The experimental results were in accord with the predictions made from the model; the paradoxically low error scores made by the operated groups during the first half of the experiment reflected their relatively restricted sampling of objects during this period. [Pribram, 1961.] **B.** Operant conditioning, with performance shown by cumulative recording. The rate of bar pressing by monkeys is recorded on a moving tape by a pen that moves upward a notch each time the bar is depressed; therefore, when the slope is steep, the monkey is bar pressing rapidly, when it plateaus the monkey is not pressing. The momentary downward deflections signal a reward given. In this experiment animals are trained to expect an alternation between one peanut after 40 bar presses ("fixed ratio") and one peanut after 4 minutes ("fixed interval"). When this is learned, the record will appear as a series of steps—steep slopes during the ratio periods, plateaus during the 4 minute intervals. The full-scale vertical deflections merely reset the pen when it has reached the top of its traverse. Note the indiscriminate rate shown by the monkey with an anterofrontal lesion, especially after certain signal lights are turned off. The inferior temporal operate is performing as well as normals. A group of normal monkeys took a mean of $22\frac{1}{2}$ days to reach an 85% criterion of discrimination between the fixed ratio and the fixed interval schedule in the absence of the light signal, two inferior temporal operated animals required 12 days, and two anterofrontal operates did not reach this criterion even in 38 days. The greater speed of the inferior temporal operates over normal animals suggests that these operates rely more than do normals on nonvisual cues, even when the signal was present. This illustrates a method of automatic recording of the performance of an animal that can bring out differences by rapid accumulation of data and simultaneous observation of several animals, especially useful in analysis of environmental determinants of behavior. [Pribram, 1958. Reproduced with the permission of the Regents of the University of Wisconsin.]

action, or to look for autocorrelation, cross correlation, differences in rhythmicity, and other statistical measures (Communications Biophysics Group, 1959, for refs.; Gerstein and Kiang, 1960). Records of continuously changing electrical waves, as from macroelectrodes on the mammalian brain, can be coded for processing by the large general-purpose digital computers, and many tests applied, including laborious and sophisticated ones capable of showing subtle differences in brain waves at stages in learning a simple task (Adey et al., 1960).

Combinations of multichannel recording, stimulation through several sensory modalities, local surgery, and drugs are used in some studies of the localization and distinction of functions in higher brain centers. At the neuronal level much can be learned by monitoring the input and the output in single cells, while delivering stimuli of different temporal patterns through one or more presynaptic fibers. The analysis can be aided by presetting the membrane at different resting potentials with injected current or by the use of drugs. The evoked potential technique on whole brains has been extended by stimulating with a flickering light that sets up easily recognized potentials at the same frequency in many parts of the brain; the distribution of these parts changes greatly during habituation, orientation to a punishment, early and late learning (John and Killam, see p. 347). The use of natural modes of stimulation, such as small moving objects in the visual field, has brought out integrative events not revealed by the most quantitative use of light on and off or even moving stripe stimuli (Maturana et al., 1960).

Electrophysiological methods lend themselves to combination with techniques of studying behavior of the freely moving animal. Controlled ablation or stimulation of local areas of the brain and even recording from single units can usefully accompany tests of discrimination learning (Fig. 1.22), Skinner box sessions, or simply ethological observation of tendency to manifest species-characteristic movements or moods (Pribram, 1960; Holst and St. Paul, 1960).

Control system analysis techniques are particularly useful in observation of whole animals under conditions permitting some task to be evaluated quantitatively before and after interfering with the information flow, as by "opening the loop," frustrating some output or input, reversing it, or imposing false signals. This approach has great power at all levels of biological research, from the molecular to that of the whole animal, where consistent relations between well-defined input and output can be measured and where convenient access exists for external control of input and influential parameters. Especially with regard to gross neural functions and connections immediately underlying stereotyped behavior, the findings are impressive (see p. 326). The results of such analysis (if the system is at all complex) rarely dictate what components must be present, but rather demonstrate, in great detail, precisely what components, if they were present, would suffice to explain the input-output data. Immediately, then, experiments suggest themselves for identification of the inferred components with the biological counterparts, which must be at least mathematically equivalent at the level of input and output.

Bibliography

The accompanying selection of citations includes general works relevant to this book, and some works on sense organs and on techniques. Very few textbooks or ordinary reviews are listed. The former should be sought through the usual library catalog and the latter in recent issues of the review journals. The bibliographies of the general works cited here should be consulted particularly for guides to the older phylogenetic literature.

ABDERHALDEN, E. 1922-1930. Handbuch der biologischen Arbeitsmethoden. Abt. IX. Methoden der Erforschung der Leistungen des tierischen Organismus. Teil 4. Methoden der vergleichenden Physiologie. Urban & Schwarzenberg, Berlin.

ÁBRAHÁM, A. 1938. Der heutige Stand der Neuronlehre. *Állattani Közlemények*, 35:111-130.

ÁBRAHÁM, A. 1959. Morphologische Grundlagen der nervalen Nierenregulation. In: *Nierenfunktion und Nervensystem.* H. Dutz (ed.). Volk und Gesundheit, Berlin.

ADEY, W. R. 1959. The sense of smell. In: *Handbook of Physiology*, Sect. 1. *Neurophysiology*, 1:535-548.

ADEY, W. R., DUNLOP, C. W., and HENDRIX, C. E. 1960. Hippocampal slow waves; distribution and phase relations in the course of approach learning. *Arch. Neurol.*, 3:74-90.

ADRIAN, E. D. 1959. Sensory mechanisms—introduction. In: *Handbook of Physiology*, Sect. 1. *Neurophysiology*, 1:365-367.

AUTRUM, H. 1959. Nonphotic receptors in lower forms. In: *Handbook of Physiology*, Sect. 1. *Neurophysiology*, 1:369-385.

BAGLIONI, S. 1913a. Die Grundlagen der vergleichenden Physiologie des Nervensystems und der Sinnesorgane. *Winterstein's Handb. vergl. Physiol.*, 4:1-22.

BAGLIONI, S. 1913b. Physiologie des Nervensystems. *Winterstein's Handb. vergl. Physiol.*, 4:23-450.

BEKLEMISHEV, V. 1952. *Foundations of Comparative Anatomy of Invertebrates.* Moscow. (In Russian.)

BERITOV, I. 1957. The development of the physiology of the central nervous system in the Soviet Union in the last 40 years. *Sechenov J. Physiol.*, 43:941-955. (*Fiziol. Zh. SSRR*, 43:1021-1036; in Russian.)

BETHE, A. 1903. *Allgemeine Anatomie und Physiologie des Nervensystems.* Thieme, Leipzig.

BICKFORD, R. G., POOLE, E. W., and MCCARTHY, C. E. 1962. Some applications in EEG analysis. *Fed. Proc.*, 21:103-108.

BIEDERMANN, W. 1896-1898. *Electrophysiology.* Macmillan, London. 2 v.

BLEST, A. D. 1961. Some modifications of Holmes's silver method for insect central nervous systems. *Quart. J. micr. Sci.*, 102:413-417.

BOEKE, J. 1940. *Problems of Nervous Anatomy.* Oxford Univ. Press, London.

BOVARD, J. F. 1918. The function of the giant fibers in earthworms. *Univ. Calif. Publ. Zool.*, 18:135-144.

BRADY, J. V. 1960. Temporal and emotional effects related to intracranial electrical self-stimulation. In: *Electrical Studies on the Unanesthetised Brain.* E. R. Ramey and D. S. O'Doherty, (eds.). Hoeber, New York.

BRAZIER, M. A. B. 1959. The historical development of neurophysiology. In: *Handbook of Physiology*, Sect. 1. *Neurophysiology*, 1:1-58.

BRAZIER, M. A. B. 1960. *The Electrical Activity of the Nervous System.* 2nd ed. Macmillan, New York.

BRAZIER, M. A. B. (ed.). 1961. *Brain and Behavior.* Amer. Inst. Biol. Sci., Washington, D.C.

BRAZIER, M. A. B. and BARLOW, J. S. 1956. Some applications of correlation analysis to clinical problems in electroencephalography. *Electroenceph. clin. Neurophysiol.*, 8:325-331.

BRAZIER, M. A. B., KJELLBERG, R. N., SWEET, W. H., and BARLOW, J. S. 1960. Electrographic recording and correlation analysis from deep structures within the human brain. In: *Electrical Studies on the Unanesthetised Brain.* E. R. Ramey and D. S. O'Doherty, (eds.). Hoeber, New York.

BUDDENBROCK, W. VON. 1952. Vergleichende Physiologie, I. Sinnesphysiologie. Birkhäuser, Basel.

BUDDENBROCK, W. VON. 1953. Vergleichende Physiologie, II. Nervenphysiologie. Birkhäuser, Basel.

BULLOCK, T. H. 1954. Comparative aspects of some biological transducers. *Fed. Proc.*, 12:666-672.

BUREŠ, J., PETRÁŇ, M., and ZACHAR, J. 1960. *Electrophysiological Methods in Biological Research.* (Transl. by P. Hahn.) Academic Press, New York.

BURKHARDT, D. 1957. Die Übertragereigenschaften elektrophysiologischer Versuchsanordnungen. *Z. Biol.* 109:297-324.

BURKHARDT, D. 1960. Die Eigenschaften und Funktionstypen der Sinnesorgane. *Ergebn. Biol.*, 22:226-267.

BURNSTOCK, G. and PROSSER, C. L. 1960. Conduction in smooth muscles: comparative electrical properties. *Amer. J. Physiol.*, 199:553-559.

BÜTSCHLI, O. 1921. *Vorlesungen über Vergleichende Anatomie.* Springer, Berlin.

CAJAL, S. R. Y. 1909-1911. *Histologie du Système Nerveux de l'Homme et des Vertébrés.* Paris. (Republished 1952 by Inst. Ramón y Cajal, Madrid.)

CAJAL, S. R. Y. 1937. Recollections of my life. (Transl. by E. H. Craigie.) *Mem. Amer. philos. Soc.*, 8:1-638.

CAJAL, S. R. Y. 1954. *Neuron Theory or Reticular Theory?* (Transl. by M. Ubeda Purkiss and C. A. Fox.) Inst. Ramón y Cajal, C.S.I.S., Madrid.

CARTHY, N. D. 1958. *An Introduction to the Behaviour of Invertebrates.* Macmillan, London.

CATE, J. TEN. 1931. Physiologie der Ganglionsysteme der Wirbellosen. *Ergebn. Physiol.*, 33:137-336.

CHANG, H.-T. 1959. The evoked potentials. In: *Handbook of Physiology*, Sect. 1. *Neurophysiology*, 1:299-313.

CHILD, C. M. 1921. *The Origin and Development of the Nervous System From a Physiological Point of View.* Univ. Chicago Press, Chicago.

COGHILL, G. E. 1929. *Anatomy and the Problem of Behavior.* Macmillan, New York.

Communications Biophysics Group of Research Laboratory of Electronics and SIEBERT, W. M. 1959. Processing neuroelectric data. Techn. Rep. 351. Technol. Press, Mass. Inst. Technol., Cambridge, Mass.

DAVENPORT, H. A. 1960. *Histological and Histochemical Technics.* Saunders, Philadelphia.

DAVIS, H. 1959. Excitation of auditory receptors. In: *Handbook of Physiology*. Sect. 1. *Neurophysiology*, 1:565-584.

DONALDSON, P. E. K. 1958. *Electronic Apparatus for Biological Research.* Butterworths, London.

DUSSER DE BARENNE, J. G., GAROL, H. W., and MCCULLOCH, W. S. 1941a. Functional organization of sensory and adjacent cortex of the monkey. *J. Neurophysiol.*, 4:324-330.

DUSSER DE BARENNE, J. G., GAROL, H. W., and MCCULLOCH, W. S. 1941b. Physiological neuronography of the cortico-striatal connections. *Res. Publ. Ass. nerv. ment. Dis.*, 21:246-266.

DUSSER DE BARENNE, J. G., MARSHALL, C., NIMS, L. F., and STONE, W. E. 1941. The response of the cerebral cortex to local application of strychnine nitrate. *Amer. J. Physiol.*, 132:776-780.

ECCLES, J. C. 1953. *The Neurophysiological Basis of Mind.* Clarendon Press, Oxford.

FARLEY, B. S., FRISHKOPF, L. S., CLARK, W. A., JR., and SILMORE, J. T., JR. 1957. Computer techniques for study of patterns in the electroencephalogram. Mass. Inst. Technol., Res. Lab. Elect. Techn. Rep. No. 337; also Lincoln Lab. Techn. Rep. No. 165.

FIELD, J., MAGOUN H. W., and HALL, V. E. 1959-1960. *Handbook of*

Bibliography

Physiology. Sect. 1. *Neurophysiology.* Amer. Physiol. Soc., Washington, D.C.

FLOREY, E. 1962. Comparative neurochemistry: inorganic ions, amino acids and possible transmitter substances of invertebrates. In: *Neurochemistry.* 2nd. ed. K. A. C. Elliott, I. H. Page, and J. H. Quastel (eds.). Thomas, Springfield, Ill.

FORTUYN Æ. B. D. 1920. *Vergleichende Anatomie des Nervensystems. 1. Die Leitungsbahnen im Nervensystem der wirbellosen Tiere.* Bohn, Haarlem.

FRANK, K. 1959. Identification and analysis of single unit activity in the central nervous system. In: *Handbook of Physiology.* Sect. 1. *Neurophysiology,* 1:261-277.

GALAMBOS, R. 1960. Some neural correlates of conditioning and learning. In: *Electrical Studies on the Unanesthetized Brian.* E. R. Ramey and D. S. O'Doherty (eds.). Hoeber, New York.

GERNANDT, B. E. 1959. Vestibular mechanisms. In: *Handbook of Physiology.* Sect. 1. *Neurophysiology,* 1:549-564.

GERSTEIN, G. L. and KIANG, N. Y.-S. 1960. An approach to the quantitative analysis of electrophysiological data from single neurons. *Biophys. J.,* 1:15-28.

GLASSER, O. 1950. *New Medical Physics.* Year Book, Chicago.

GLEES, P. 1957. *Morphologie und Physiologie des Nervensystems.* Thieme, Stuttgart.

GOMORI, G. 1952. *Microscopic Histochemistry, Principles and Practice.* Univ. Chicago Press, Chicago.

GRANIT, R. 1959. Neural activity in the retina. In: *Handbook of Physiology,* Sect. 1. *Neurophysiology,* 1:693-712.

GRAUMANN, W. and NEUMANN, K. 1958. *Handbuch der Histochemie.* Fischer, Stuttgart.

GRAY, E. G. 1962. A morphological basis for pre-synaptic inhibition? *Nature, Lond.,* 193:82-83.

GURR, E. 1958. *Methods of Analytical Histology and Histochemistry.* Hill, London.

HAGBARTH, K.-E. 1960. Centrifugal mechanisms of sensory control. *Ergebn. Biol.,* 22:47-66.

HANSON, J. and LOWY, J. 1960. Structure and function of the contractile apparatus in the muscles of invertebrate animals. In: *The Structure and Function of Muscle.* Vol. 1. *Structure.* G. H. Bourne (ed.). Academic Press, New York.

HANSTRÖM, B. 1926. Einige Experimente und Reflexionen über Geruch, Geschmack und den allgemeinen chemischen Sinn. *Z. vergl. Physiol.,* 4:528-544.

HANSTRÖM, B. 1928a. *Vergleichende Anatomie des Nervensystems der wirbellosen Tiere.* Springer, Berlin.

HANSTRÖM, B. 1928b. Some points on the phylogeny of nerve cells and of the central nervous system of invertebrates. *J. comp. Neurol.,* 46:475-491.

HARTLINE, H. K. 1959. Vision—Introduction. In: *Handbook of Physiology,* Sect. 1. *Neurophysiology,* 1:615-619.

HEIDER, K. 1914. *Phylogenie der Wirbellosen.* Leipzig, Berlin.

HELD, H. 1929. Die Lehre von dem Neuronen und vom Neurencytium und ihr heutiger Stand. *Fortschr. naturw. Forsch.,* N. F. 8:1-44.

HEMPELMANN, F. 1926. *Tierpsychologie vom Standpunkte des Biologen.* Akademische Verlagsanstalt, Leipzig.

HERRICK, C. J. 1924. *Neurological Foundations of Animal Behavior.* Holt, New York.

HERRICK, C. J. 1931. *An Introduction to Neurology.* Saunders, Philadelphia.

HESS, A. 1958. Experimental anatomical studies of pathways in the severed central nerve cord of the cockroach. *J. Morph.,* 103:479-502.

HESS, A. 1960. The fine structure of degenerating nerve fibers, their sheaths, and their terminations in the central nerve cord of the cockroach (*Periplaneta americana*). *J. biophys. biochem. Cytol.,* 7:339-344.

HOFFMANN, C. 1961. Vergleichende Physiologie des Temperatursinnes und der chemischen Sinne. *Fortschr. Zool.,* 13:190-256.

HOLST, E. VON. 1950. Quantitative Messung von Stimmungen im Verhalten der Fische. *Symp. Soc. exp. Biol.,* 4:143-172.

HOLST, E. VON and ST. PAUL, U. VON. 1960. Vom Wirkungsgefüge der Triebe. *Naturwissenschaften,* 47:409-422.

HOYLE, G. 1957. *Comparative Physiology of the Nervous Control of Muscular Contraction.* Cambridge Univ. Press, London.

HYMAN, L. H. 1940-1959. *The Invertebrates.* McGraw-Hill, New York. 5 v.

JONES, W. C. 1962. Is there a nervous system in sponges? *Biol. Rev.,* 37:1-50.

JORDAN, H. 1919. Die Phylogenese der Leistungen des zentralen Nervensystems. *Biol. Zbl.,* 39:462-474.

JOURDAN, E. 1891. *Die Sinne und Sinnesorgane der niederen Tiere.* (Transl. from French by W. Marshall.) Leipzig.

KAPPERS, C. U. ARIËNS. 1929. *The Evolution of the Nervous System.* de Erven F. Bohn, Haarlem.

KAPPERS, C. U. ARIËNS. 1934. Feinerer Bau und Bahnverbindungen des Zentralnervensystems. Bolk, L. et al., *Handb. vergl. Anat. Wirbeltiere,* 2(2):319-486.

KAPPERS, C. U ARIËNS., HUBER, G. C., and CROSBY, E. C. 1936. *The Comparative Anatomy of the Nervous System of Vertebrates, Including Man.* Macmillan, New York. 2 v.

KAPPERS, J. ARIËNS. 1956. *Progress in Neurobiology.* Elsevier, Amsterdam.

KARAMIAN, A. I. 1956. Evolutionary physiology of the nervous system. (Conference in Leningrad.) *Vest. Akad. Nauk SSSR,* 26(8): 119-122. (In Russian.)

KORSCHELT, E. 1927-1931. *Regeneration und Transplantation.* Borntraeger, Berlin. 2 v.

KOSHTOYANTS, KH. S. 1957. *Fundamentals of Comparative Physiology* Vol. II. *Comparative Physiology of the Nervous System.* Acad. Sci., Moscow. (In Russian.)

KRAUSE, R. 1921-1923. *Mikroskopische Anatomie der Wirbeltiere in Einzeldarstellungen.* de Gruyter, Berlin.

KRUMBACH, T. 1937. Oligomera. In: *Kükenthal's Handb. Zool.,* 3(5):7-66.

LANG, A. 1891-1896. *Textbook of Comparative Anatomy.* (Trans. by H. M. Bernard and M. Bernard.) Macmillan, London. 2v.

LANG, A. 1912-1921. *Handbuch der Morphologie der wirbellosen Tiere.* Vol. 2: Metazoen. Fischer, Jena.

LARSSON, B., LEKSELL, L., REXED, B., SOURANDER, P., MAIR, W., and ANDERSSON, B. 1958. The high-energy proton beam as a neurosurgical tool. *Nature, Lond.,* 182:1222-1223.

LEE, A. B. 1950. *The Microtomists' Vade-Mecum.* 11th. ed. J. B. Gatenby and H. W. Beans, (eds.). Blakiston, Philadelphia.

LIVINGSTON, R. B. 1959. Central control of receptors and sensory transmission systems. In: *Handbook of Physiology.* Sect. 1. *Neurophysiology,* 1:741-760.

LOCATELLI, P. 1926. Rôle du système nerveux dans les phénomènes de régénération. *C. R. Soc. Biol., Paris,* 95:3-28.

LOEB, J. 1900. *Comparative Physiology of the Brain and Comparative Psychology.* Putnam, New York.

MALIS, L. I., LOEVINGER, R., KRUGER, L., and ROSE, J. E. 1957. Production of laminar lesions in the cerebral cortex by heavy ionizing particles. *Science,* 126:302-303.

MATURANA, H. R., LETTVIN, J. Y., MCCULLOCH, W. S., and PITTS, W. H. 1960. Anatomy and physiology of vision in the frog (*Rana pipiens*). *J. gen. Physiol.,* 43:129-175.

MCILWAIN, H. 1955. *Biochemistry and the Central Nervous System.* Little and Brown, Boston.

MEYER, G. F. 1955. Vergleichende Untersuchungen mit der supravitalen Methylenblaufärbung am Nervensystem wirbelloser Tiere. *Zool. Jb. (Anat.),* 74:339-400.

MILLOTT, N. 1957. Animal photosensitivity with special reference to eyeless forms. *Endeavour,* 16:19-28.

MILNE, L. J. and MILNE, M. J. 1956. Invertebrate photoreceptors. In: *Radiation Biology.* Vol. III. *Visible and*

Near-visible Light. A. Hollaender (ed.), McGraw-Hill, New York.
MILNE, L. J. and MILNE, M. 1959. Photosensitivity in invertebrates. In: *Handbook of Physiology*, Sect. 1. *Neurophysiology*, 1:621-645.
MOORE, A. R. 1917. Chemical differentiation of the central nervous system of invertebrates. *Proc. nat. Acad. Sci., Wash.*, 3:598-602.
MURALT, A. VON. 1958. *Neue Ergebnisse der Nervenphysiologie.* Springer, Berlin.
NICOL, J. A. C. 1955. Physiological control of luminescence in animals. In: *The Luminescence of Biological Systems.* H. Johnson (ed.), Amer. Ass. Adv. Sci., Washington, D.C.
OLDS, J. 1960. Differentiation of reward systems in the brain by self-stimulation technics. In: *Electrical Studies on the Unanesthetized Brain.* E. R. Ramey and D. S. S. O'Doherty (eds.). Hoeber, New York.
ONCLEY, J. L. (ed.). 1959. *Biophysical Science—a Study Program.* Wiley, New York.
PANTIN, C. F. A. 1952. The elementary nervous system. *Proc. roy. Soc.*, (B) 140:147-168.
PANTIN, C. F. A. 1956. The origin of the nervous system. *Pubbl. Staz. zool. Napoli*, 28:171-181.
PARKER, G. H. 1910. The phylogenetic origin of the nervous system. *Anat. Rec.*, 4:51-58.
PARKER, G. H. 1911. The origin and significance of the primitive nervous system. *Proc. Amer. phil. Soc.*, 50:217-225.
PARKER, G. H. 1914. The origin and evolution of the nervous system. *Pop. Sci. Mon.*, 84:118-127.
PARKER, G. H. 1916. The sources of nervous activity. *Bull. Scripps Inst.*, 1:11-18; also *Science*, 1917, 45:619-626.
PARKER, G. H. 1918. Some underlying principles in the structure of the nervous system. *Science*, 47:151-162.
PARKER, G. H. 1919. *The Elementary Nervous System.* Lippincott, Philadelphia.
PARKER, G. H. and STABLER, E. M. 1913. On certain distinctions between taste and smell. *Amer. J. Physiol.*, 32:230-240.
PFAFFMANN, C. 1959. The sense of taste. In: *Handbook of Physiology*, Sect. 1 *Neurophysiology*, 1:507-533.
PLATE, L. 1922-1924. *Allgemeine Zoologie und Abstammungslehre.* 1. Nervensystem, etc. II. Die Sinnesorgane. Fischer, Jena.
PRIBRAM, K. H. 1958. Neocortical function in behavior. In: *Biological and Biochemical Bases of Behavior.* H. F. Harlow and C. N. Woolsey (eds.). Univ. of Wisconsin Press, Madison.
PRIBRAM, K. H. 1960. The intrinsic systems of the forebrain. In: *Handbook of Physiology*, Sect. 1. *Neurophysiology*, 2:1323-1344.
PRIBRAM, K. H. 1961. A further experimental analysis of the behavioral deficit that follows injury to the primate frontal cortex. *Exp. Neurol.*, 3:432-466.
PROSSER, C. L. 1946. The physiology of nervous systems of invertebrate animals. *Physiol. Rev.*, 26:337-382.
PROSSER, C. L. 1954. Comparative physiology of nervous systems and sense organs. *Ann. Rev. Physiol.*, 16:103-124.
PROSSER, C. L. 1960. Comparative physiology of activation of muscles, with particular attention to smooth muscles. In: *The Structure and Function of Muscle.* Vol. 2. Biochemistry and Physiology. G. H. Bourne (ed.). Academic Press, New York.
RAMEY, E. R. and O'DOHERTY, D. S. (eds.). 1960. *Electrical Studies on the Unanesthetized Brain.* Hoeber, New York.
RASMUSSEN, A. T. 1947. *Some Trends in Neuroanatomy.* Brown, Dubuque, Iowa.
RENSCH, B. 1954. The relation between the evolution of central nervous function and the body size of animals. In: *Evolution as a Process.* J. S. Huxley, A. C. Hardy, and E. B. Ford, (eds.). Allen and Unwin, London.
ROEDER, K. D. 1955. Spontaneous activity and behavior. *Sci. Mon., N.Y.*, 80:362-370.
ROMANES, G. J. 1885. *Jelly-fish, Starfish and Sea-urchins, Being a Research on Primitive Nervous Systems.* D. Appleton, New York.
ROSE, J. E. and MOUNTCASTLE, V. B. 1959. Touch and kinesthesis. In: *Handbook of Physiology*, Sect. 1. *Neurophysiology*, 1:387-429.
ROSE, J. E., MALIS, L. I., and BAKER, C. P. 1961. Neural growth in the cerebral cortex after lesions produced by monoenergetic deuterons. In: *Sensory Communication.* W. A. Rosenblith (ed.). MIT Press and Wiley, New York.
ROZHANSKII, N. A. 1957. *Studies on the Physiology of the Nervous System.* Medgiz, Leningrad. (In Russian.)
SCHEIBEL, M. E. and SCHEIBEL, A. B. 1958. Structural substrates for integrative patterns in the brain stem reticular core. In: *Reticular Formation of the Brain.* H. H. Jasper et al. (eds.). Little, Brown, Boston.
SCHNEIDER, K. C. 1902. *Lehrbuch der vergleichenden Histologie der Tiere.* Fischer, Jena.
SECHENOV, I. M. 1956. *Collected Works.* Vol. 2. *Physiology of the Nervous System.* Acad. Sci., Moscow. (In Russian.)
STEINER, I. 1890. Die Functionen des Centralnervensystems der wirbellosen Tiere. *S. B. Akad. Wiss. Berlin* 1890(1):39-49.
STEINER, J. 1898. *Die Functionen des Centralnervensystems und ihre Phylogenese.* 3. Abt. Die wirbellosen Tiere. Vieweg, Braunschweig.
SWEET, W. H. 1959. Pain. In: *Handbook of Physiology*, Sect. 1. *Neurophysiology*, 1:459-506.
SZABO, TH. 1961. Les organes électriques des Mormyrides. In: *Bioelectrogenesis.* C. Chagas and A. Paes de Carvalho (eds.). Elsevier, Amsterdam.
TASAKI, I. 1958. Electrical responses of astrocytic glia from the mammalian central nervous system cultivated in vitro. *Experientia*, 14:220-226.
TILNEY, F. and RILEY, H. A. 1938. *The Form and Functions of the Central Nervous System.* 3rd ed. Lewis, London.
UEXKÜLL, J. J. VON. 1909. *Umwelt und Innenwelt der Tiere.* Springer, Berlin.
VERWORN, M. 1913. *Irritability. A Physiological Analysis of the General Effect of Stimuli in Living Substance.* Yale Univ. Press, New Haven.
VOSKRESENSKAYA, A. K. 1958. Materials on the evolution of the functions of the neuro-muscular apparatus. Problems of the evolution of physiological functions. In: "Articles dedicated to the 75th birthday of Academician, L. A. Orbelli." (In Russian.)
VRIES, H. DE. 1956. Physical aspects of the sense organs. *Progr. Biophys.*, 6:207-264.
WARDEN, C. J., JENKINS, T. N., and WARNER, L. H. 1940. *Comparative Psychology.* Vol. 2. *Plants and Invertebrates.* Ronald, New York.
WELLS, G. P. 1950. Spontaneous activity cycles in polychaete worms. *Symp. Soc. exp. Biol.*, 4:127-142.
WELLS, G. P. 1955. *The Sources of Animal Behaviour.* An inaugural lecture delivered at University College, London, 5 May 1955. H. K. Lewis, London.
WINDLE, W. F. 1957. *New Research Techniques of Neuroanatomy.* Thomas, Springfield.
YOLTON, L. W. 1923. The effects of cutting the giant fibers in the earthworm, *Eisenia foetida* (Sav.). *Proc. nat. Acad. Sci., Wash.*, 9:383-385.
YOUNG, J. Z. 1938. The evolution of the nervous system and of the relationship of organism and environment. In: *Evolution; Essays presented to E.S. Goodrich.* G. R. de Beer (ed.). Oxford.
YOUNG, J. Z. 1945. Structure, degeneration and repair of nerve fibres. *Nature, Lond.*, 156:132-136.
ZAWARZIN, A. 1925. Der Parallelismus der Strukturen als ein Grundprinzip der Morphologie. *Z. wiss. Zool.*, 124:118-212.
ZOTTERMAN, Y. 1959. Thermal sensations. In: *Handbook of Physiology*, Sect. 1. *Neurophysiology*, 1:431-458.

CHAPTER 2

Comparative Microanatomy of Nervous Elements

Summary	36
I. General and Comparative Considerations	**38**
A. Forms of Nerve Cells and Processes	38
1. Axons	38
2. Dendrites	40
3. The soma	42
4. Classifications of nerve cells by structure and connections	42
5. Giant cells and axons	43
6. The physical properties of axons and changes with activity	45
B. Distribution of Types of Neurons among the Phyla	45
C. Origins and Relationships of Neuron Types	48
II. Special Relations and Components of Nervous Tissue	**49**
A. Organization of Nervous Tissue	49
1. Solitary nervous elements	49
2. Epithelial nervous systems	50
3. Peripheral plexuses	51
4. Peripheral ganglia	51
5. Nerves	52
6. Connectives and commissures	52
7. Central ganglia and cords	52
B. Synapses	58
1. Varieties of contact at the light microscope level	58
2. Fine structure revealed by the electron microscope	69
C. Nerve Cell Inclusions	81
1. Membrane complexes and lipoidal droplets	84
2. Modified ciliary structures	90
3. Fibrillar structures	91
D. Neuroglia and Sheaths	95
1. The situations in which glia occurs	96
2. The varieties of glial cells	99
3. Axon sheaths	101
E. Degeneration and Regeneration	108
Bibliography	111

Summary

A survey of the forms of nerve cells and their prolongations in all nervously endowed animals leads to a somewhat different perspective, with respect to the common and the derived features of these elements as well as of glia, than is usual in texts based solely upon vertebrates. The following discussion assumes a reading of the Introduction, Chapter 1.

The term **axon** is most useful at present when it is confined to processes specialized anatomically and physiologically for conduction of impulses. There are certain difficulties with this criterion, chief of which is that impulses may arise either before or after the main histological landmarks; in a good many neurons histological criteria are unavailable or ambiguous. **Dendrites** are most usefully defined as receptive processes (but not long afferent fibers such as vertebrate sensory fibers, which are cytologically and physiologically axons). Dendrites are more diverse than axons, especially between vertebrates and invertebrates; in the latter they generally arise from the axon and receive virtually all the synaptic connections. The possibility is emphasized that dendrites, particularly in vertebrates, have functions other than integrating synaptic input. The soma is regarded as a swelling that accommodates the nucleus and associated cytoplasm, which may occur inside either the dendritic or the axonal membrane. In all higher invertebrates, by far the majority of central neurons have the soma at the end of a single stem process, as a quasi-detached appendage, lying in a rind on the surface of the ganglia and removed therefore from the synaptic region.

Nerve cells can be classified as multipolar, bipolar, or unipolar; isopolar or heteropolar; sensory, internuncial, or motor; projection, commissural, or intrinsic; large and plasma-rich, chromatin-poor, small and chromatin-rich, plasma-poor, or intermediate. The small cells tend to form groups (globuli cell masses) or layers of granule cells in higher centers.

Giant nerve cells or fibers are found in many groups of invertebrates and aquatic vertebrates. Giant cells are not the usual source of giant fibers nor are their axons among the largest of giant axons. The ratio of volume of axon to that of soma varies from less than one to more than 10,000. Giant axons commonly are the product of the fusion of axons from several or many nerve cells of moderate size. The biological meaning of giant cells is quite unknown; that of giant axons probably encompasses high velocity of conduction and perhaps also some consequence of the large action current.

Several physical properties and staining affinities of axons have been found to alter measurably under stimulation.

Types of neurons are **distributed unevenly among phyla.** The simplest systems consist mainly of (a) superficial sensory neurons, (b) basiepithelial bipolars without differentiated dendrites, and (c) isopolar multipolars. From the first centralized nervous systems to the highest invertebrates, a new type—unipolar with ramifying side branches and terminals forming a neuropile—becomes the commonest form of central cell; superficial sensory neurons continue throughout all the phyla, but other neurons, with deeper-lying cell bodies and distal, free-branching endings, become common in higher groups and tend to cluster into ganglia. Heteropolar multipolars are found here and there in invertebrates, especially in visceral plexuses, but are the dominant type in the vertebrate central nervous system. Differentiation of nerve cells extends into the dozens of subtypes, each characteristic of its locale. Most neurons are dynamically polarized, such that the excited state spreads from dendritic to axonal terminals; some normally conduct either way in a certain stretch of axon while still obeying this principle.

Nervous tissue is organized in many degrees of structural complexity. Solitary neurons are relatively rare except for sensory elements. Basiepithelial systems of nerve fibers are common in the simplest forms; peripheral plexuses in the interior are common on the gut and elsewhere. Ganglia form a central nervous system in all groups at or above the level of complexity of platyhelminths; in addition, peripheral ganglia become frequent in higher phyla. Nerves, connectives, and commissures develop in higher groups. The ganglia in invertebrates are in general made up of a fibrous core and a cellular rind; the rind contains virtually all the nerve cell bodies, the core virtually all the endings and synapses as well as fibers of passage. These last tracts become distinct from neuropile. **Neuropile** is a concentration of terminals and junctions and displays wide variety of density, texture, organization, and local differentiation, believed to represent levels of complexity. Vertebrate central nervous tissue is made up differently; a primitively central mass of gray matter contains endings and synapses but also cell bodies, while the fibrous white matter is essentially purely tracts. There may be correlated differences in the histologic character and functional properties of dendrites as well as somata of vertebrates and invertebrates.

Summary

The microanatomy of **synapses** is treated as revealed by light and electron microscopy. An elaborate assortment of geometrical arrangements is known, not capriciously distributed but characteristic of each site. Classification based on a few criteria leads to the recognition of dozens of types in light microscopy—baskets, bushes, vines, tufts, interdigitations, and the like can be distinguished; these involve extensive opportunities for contact between pre- and postsynaptic surfaces. Here we may mention only that the following occur, among others: simple contacts between fibers, in passing; parallel, extensive contacts between one axon and one dendrite; right-angled, minimum contacts between one axon and many dendrites; axo-dendritic, axo-somatic, axo-axonal, somato-dendritic, and possibly other combinations; end-to-end macrosynapses or terminal knobs; embracing calyces and others. Neuroeffector junctions and sensory endings add to the array of specialized arrangements, described in part here and in part in the systematic chapters. The electron microscope suggests that most of these types are made up of spots or areas of intricate contact between pre- and postsynaptic membranes, each intact and inviolate. A cleft of about 100–300 Å width, uniform for any given synapse, separates the neurons and may be occupied by material; some presumably electrically transmitting synapses seem to have typically the narrower clefts, and some synapses presumed to be chemically transmitting have the wider clefts, locally dense cell membranes, and sometimes some structure in the cleft material. Synaptic vesicles generally form a cluster on the presynaptic side. They are sometimes present in electrical junctions, on either or both sides, but they tend to be sparse. Specialized inclusions peculiar to certain synapses are known; nevertheless, there is a remarkable degree of similarity between many structures identified as synapses, from coelenterates to man.

Several of the more distinctive forms of nerve cell **inclusions** are treated, with emphasis on recent electron microscope findings concerning the classical objects: Nissl material, Golgi bodies, and neurofibrils. Neurosecretory granules are treated in Chapter 6; a number of minor inclusions are mentioned.

Granular reticulum is seen with the electron microscope and is probably equivalent to the Nissl bodies seen with the light microscope; it is an organized and interconnected series of sheetlike membranes, stacked and clumped, found in the soma and large dendrites of vertebrates and in the soma of some invertebrates, but not in the axoplasm. The granules associated with the membranes are thought to be ribose nucleic acid (RNA).

Agranular reticulum or γ cytomembranes, occurring as groups of closely folded parallel membranes, probably represent the Golgi bodies of at least some classical cytologists. Dictyosomes of invertebrate neurons are closely related. Biochemically, there are probably several kinds of bodies in this category. They are thought to be centers of synthesis of materials, some of which then move down the axon.

Several kinds of lipoidal globules are distinguished in favorable nerve cells. Some inclusions are derivatives of mitochondria, others of cilia; some are not understood at all. Pigment granules are of common occurrence.

Fibrillar structures of several kinds have been found with the electron microscope, especially in axoplasm. In many axons neural filaments are characteristic; in other axons and in vertebrate dendrites neurotubules are common; in a few annelid nerve cells, much coarser and denser material is known. Neurofibrils, defined as in classical light microscopy, may sometimes be explained by long mitochondria or by one of the just mentioned entities, perhaps fused into larger fibrils. But it seems probable that some are not yet satisfactorily explained in terms of electron microscopy.

Neuroglia and sheaths of axons, nerves, and ganglia appear to become more abundant and important in higher animals. Neuroglia cells and their processes make up a variously strong outer sheath of ganglia, connectives, and nerves and fill spaces between bundles of axons, regions of neuropile, and masses of cells. Most axons and many cell bodies have an individual investment by one or several layers of sheath cells. Some large cell bodies receive trabeculae of neuroglia in deep invaginations of the cell surface, forming what is called the "trophospongium." Some large axons in molluscs and arthropods also exhibit invaginations, here in the form of deep, longitudinal folds occupied by glia. No functional meaning has been assigned to these as yet.

In lower forms, distinct categories of glia cells are not recognized but in some arthropods, cephalopods, and especially higher vertebrates, there are several, attaining an advanced level of specialization.

Axon sheaths have been found to vary much more than was acknowledged in the former dichotomy into myelinated and unmyelinated. We define the myelinated rather narrowly, to include only fibers having a private sheath of more than a few layers of tightly wrapped membranes without glial cytoplasm between them. These are common in vertebrates and have been reported once in an invertebrate, complete with Ranvier-type nodes. Simpler sheaths of four or five different configurations are known among invertebrates, loosely wrapped or with cytoplasm or extracellular material between layers or shingled with overlapping glial cells. The simplest sheath is that of axons lying in an invagination of a glial cell, with no duplication or layering of membranes; generally several or many axons share the same glial cell. Such fibers are common in vertebrates and higher invertebrates. Naked axons are known, especially in lower invertebrates, but they are rare in higher forms. The proposition that a naked nerve fiber is incapable of an active spike seems unlikely to be universal, if recent reports on coelenterate axons are correct. Other functions of glia are under investigation, involving active rather than purely passive roles.

Degeneration of injured nerve cells and fibers is known in detail for vertebrates but is extremely poorly known for invertebrates. There are reports of cases of rapid degeneration and other cases of no functional deficit in distal axons weeks after cutting off the cell bodies. Regeneration is apparently unlimited in many animals and severely limited in the central nervous system in others. Limited regeneration is true of higher groups in general but the correlation with phyletic position is poor; frogs can regenerate the retina and optic tract, certain nemertineans cannot regenerate a cerebral ganglion.

This chapter provides a short survey of the cytology and histology of nerve cells and fibers and neuroglial and sheath elements, separately and as they appear in organized tissues. The objective is to bring out common features and contrasts, by introducing the perspective of all animal groups displaying nervous elements. The treatment here can not, of course, take the place of the numerous excellent texts, reviews, and monographs, especially on vertebrate materials (see Cajal, 1909–1911, 1954; Cowdry, 1932; Bumke and Foerster, 1935–37; Sjöstrand, 1956; Maximow and Bloom, 1957; Fernández-Morán and Brown, 1958; Policard and Baud, 1958; Hild, Reiser and Lehmann, 1959; Hager, 1961).

I. GENERAL AND COMPARATIVE CONSIDERATIONS

A. Forms of Nerve Cells and Processes

In Chapter 1 we discussed the application of the terms "nervous system" and "neuron" and introduced the concepts of "axon," "dendrite," and "soma." We should now examine more carefully the meaning and limitations of these last three terms in order to use them without confusion in classifying the forms of neurons.

1. Axons

The axon or neurite may be defined as a process of a neuron (Fig. 2.1), specialized anatomically and physiologically as though for conduction of nerve impulses over considerable distances. Histologically an axon can often be recognized as a long, smooth fiber, branching sparsely if at all; it is usually embedded in an invaginated tube of a sheath cell; usually there are no synaptic endings upon it; there is only one axon in most neurons. Each of these features is subject to qualification and sometimes it is not possible to say whether a given process is an axon or even whether a given cell has an axon. Nerve cells without a demonstrable axon are called amacrine cells (Fig. 2.2). Some axons are relatively short; some give off relatively numerous collaterals (Fig. 2.2). In invertebrates (but rarely in vertebrates) naked axons are found, without any investment of sheath cells. Synaptic endings apparently occur on certain axons; some cells have more than one axon—bipolars, some multipolars (especially isopolars, as in nerve nets), and some unipolars (see pp. 66, 75). The distinction between axon and dendrite can often be made on the basis of structural specializations for conduction or reception, respectively, or on functional grounds, but often neither is unequivocal. The differentiation of the two kinds of processes shows all degrees among animal groups. The main functional characteristic of the axon is the ability to conduct excitation nondecrementally over some distance; that is, to support the all-or-none nerve impulse. But the impulse may arise in either sensory or other neurons a millimeter or more from the beginning

I. GENERAL AND COMPARATIVE A. Forms of Nerve Cells

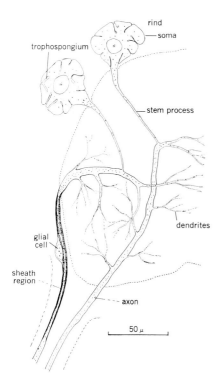

Figure 2.1. Typical form of the neuron as found in most invertebrate ganglia. This is drawn from the central nervous system of *Rhodnius* (Hemiptera). The dashed lines mark the boundary of the nerve trunk, the dotted line that of the dark neuropile. [Wigglesworth, 1959.]

of the anatomically defined axon—for example at the first node of Ranvier. It may be conducted antidromically into the soma and bases of dendrites; rarely, it is thought to arise in these structures. The direction of conduction of impulses can be either toward or away from the soma but is normally in the direction from dendrites to axonal terminals. (In certain invertebrate interneurons it is normally both toward and away from the soma, even at the same instant, resulting in cancellation where impulses meet.) It is not certain whether impulses arise and are conducted at all in neurons with very short neurites; however, electrotonic spread can be effective only over such short distances.

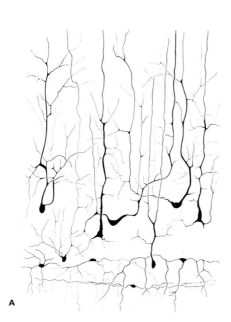
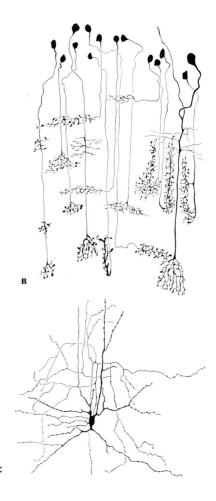

Figure 2.2. Neurons without obvious axon or with very short axon. **A.** Cells of the optic lobe of *Sepia* in which no clear axon can be distinguished. [Cajal, 1917.] **B.** Amacrine cells from the optic medulla of the fly *Calliphora*. [Cajal and Sánchez, 1915.] **C.** Short axon cell of the mammalian cerebral cortex; note the smooth, much-branched axon emerging from the top of the soma. [Cajal, 1909–1911.]

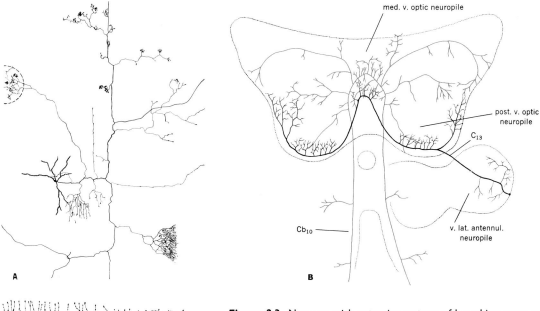

Figure 2.3. Neurons with extensive systems of branching axons or dendrites. **A.** Neuron in the reticular system of the mouse brain, showing the number of terminal patterns and collaterals an axon can have; the area included is a large part of the whole brain stem; Golgi impregnation. [Scheibel and Scheibel, 1958.] **B.** Curiously distributed processes of two internuncial cells in the brain of a crab, *Carcinus*. The cell body of C_{13} is shown. It sends branches into the middle, hind, and lateral part of the inferior optic neuropile of both sides and into the lateral first antennal *(v. lat. antennul.)* neuropile of one side; two other branches run to the medial optic neuropile. The cell body of Cb_{10} is not shown but is medial and superior and sends branches into the posterior optic and second antennal neuropile, into the lateral oculomotor and medial optic neuropile masses, another medially and deep, and perhaps one to the inferior optic neuropile. On the other side it branches in the anterior superior optic, superior tegumentary, and lateral and posterior antennal neuropile areas. Methylene blue. [Bethe, 1897; see Arthropod Bibliography.] **C.** Sensory neuron in the leg of a centipede, *Lithobius*, showing much-branched free nerve endings. Methylene blue. [Meyer, 1955a.]

It should be clearly understood that by histological criteria alone, even in excellent specific stains, it is often difficult or impossible to identify a single process of a cell as an axon (for an example, see the cephalopod optic lobe, p. 1464), or to decide where the boundary is between axon and dendrite (Fig. 2.3). This difficulty is more acute in invertebrates than in vertebrates and reflects the lower degree of differentiation of dendrites.

Axon terminals often arborize but, in spite of their form, are not called dendrites. It is probable that in some cases they do not conduct true impulses but revert to graded, local potentials before finally inducing the transmitter.

2. Dendrites

Dendrites have been variously defined; the most useful current usage is that dendrites are those processes of neurons specialized as though to act as receptive regions for the neuron. They are short (less than a few millimeters and therefore only a few space constants), terminal or subterminal, with endings of presynaptic fibers upon them and generally without sheaths. The distinction between axon and dendrite is more basic than that between either of them and the soma. Dendrites are distinguished with difficulty from the axon in unipolar neurons and bipolar neurons, where they arise from the axon, and

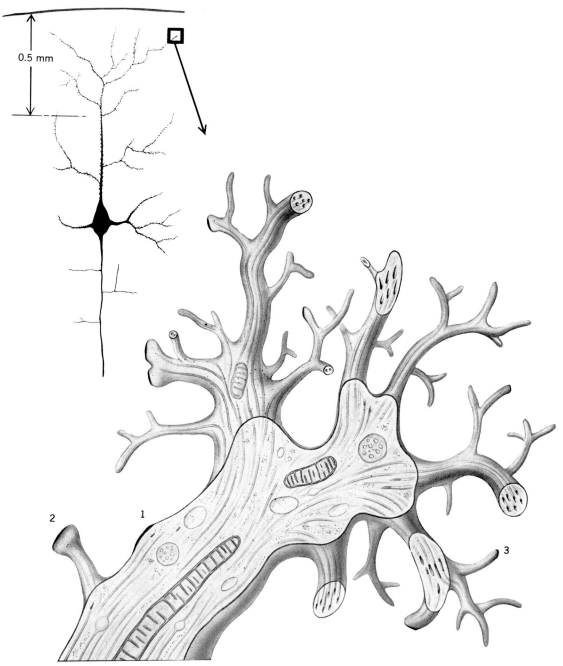

Figure 2.4. Representation of the general morphological characteristics of a small dendrite and its terminal processes in the superficial neuropile of the cerebral cortex. The insert (upper left) shows a typical pyramidal neuron with its apical dendrite in the superficial neocortex (upper 0.5 mm). The central drawing depicts the small segment of the dendrite, within the square shown in the insert, as it might appear in a composite of electron micrographs. Postsynaptic membrane thickenings are shown on the dendritic trunk (1), on a spine (2), and on a terminal process (3). Intracytoplasmic profiles include mitochondria, elements of the endoplasmic reticulum, vacuoles, multivesicular bodies, and dendritic tubules with occasional areas of local dilation. The dendritic tubules extend characteristically into the finest terminal processes. Profiles of these tubules in small processes are shown as they appear in sections of the neuropile. Much of the total dendrite surface is believed not to be involved in synapses. [Pappas and Purpura, 1961.]

from the soma in multipolars, where they arise from the soma (and sometimes from the axon as well). There is no good anatomical criterion for these boundaries (Fig. 2.3), but usage with respect to the multipolar, heteropolar neuron common in vertebrate central nervous systems places it at the base of the dendriform processes. In the vertebrates, dendrites so delimited are usually distinctive in form and inclusions (Fig. 2.4), but these are not general or defining characters. The functional boundary probably varies with respect to the soma. Dendrites are believed to be commonly incapable of propagation of impulses (in some cases there may be propagation in the proximal, thicker portions of the dendriform processes). Synapses and reception are not confined to dendrites but occur on the soma in multipolar heteropolar neurons and even on the axon hillock in some cells; the soma may be regarded in such cases as a swelling in a dendrite to accommodate the nucleus and associated cytoplasm. Reception may not be the only role of dendrites; they integrate converging input and may generate spontaneous slow changes of state, perhaps important in brain waves.

Long, afferent, impulse-propagating fibers, such as those in sensory nerves of vertebrates, are axons, not dendrites, even when they conduct centripetally. But their receptive terminals can be called dendrites, as can sensory endings in general. If these are anatomically differentiated, then, like central dendrites, they may be recognized or delimited to that extent; but they are commonly not distinctive structurally, especially in so-called free nerve endings. For the sake of clarity and consistency it should be said that we explicitly reject the earlier usage which permitted one to say that unipolar neurons are without dendrites or are characterized by axo–axonal synapses.

3. The soma

The soma or nerve cell body or perikaryon is that part of a neuron containing the nucleus and surrounding cytoplasm. The soma may be a gentle swelling along the course of an axon, as in bipolar neurons; or a quasi-detached appendage of the axon, as in unipolars with long stem processes; or terminal or subterminal, as in superficial sensory neurons; or approximately at the junction of dendrites and axon, as in vertebrate multipolars. Therefore the position of the soma is not a fundamental feature, especially in the distinction between axon and dendrite; it is permissible to think of the nucleus and surrounding cytoplasm as a trophic apparatus forming a swelling that may lie either inside that sort of cell membrane functionally of axonal type, or within membrane of dendritic type or where they meet. These and other types of membrane (for example that which is capable only of local potentials) may occur in mosaics or patches (see p. 264) distributed on soma or on processes; the boundaries are functionally quite sharp but may shift from moment to moment. It is not to be expected, then, that there will be clear anatomical boundaries corresponding to the functional ones.

4. Classifications of nerve cells by structure and connections

From the foregoing considerations it is clear that the following classification based on position of the soma would be only one of convenience, except that there is a consistent distribution of types among the phyla. Nerve cells can be **multipolar, bipolar, or unipolar,** according to the number of processes emerging from the soma. If the processes of bipolars or multipolars are indistinguishable in the microscope, the neuron is isopolar; if there is a distinction, as between axonal and dendritic processes, the neuron is heteropolar. The common cell type of the vertebrate spinal cord is a multipolar heteropolar neuron; the common type in the invertebrate central nervous system is unipolar with a branching stem process.

Another classification is based on function but is usually demonstrated anatomically: nerve cells are either **sensory, internuncial, or motor.** The unqualified term "sensory neuron" is now used as equivalent to the earlier and more exact term "primary sensory neuron," meaning a nerve cell that is itself a receptor. Afferent neurons excited by nonnervous receptor cells can be called secondary sensory neurons; one of the few examples is the neuron of the vertebrate acousticolateralis system (vestibular nerve, cochlear nerve, lateral line nerve), believed to be excited by the hair cells. Internuncials are commonly called interneurons; motor nerve cells, motoneurons. The latter may be ultimate or penultimate in the efferent pathway, relaying via a peripheral nerve cell to the effector, as in the vertebrate autonomic system. The expression "ganglion cell" is used loosely for any neuron in a ganglion or in the periphery except superficial and solitary bipolar neurons. A basis for classifying interneurons lies in the kind of connections they make: **projection** neurons send an axon a considerable distance rostrally or caudally in the central nervous system; **commissural** neurons send an axon to corresponding structures on the opposite side; **intrinsic** neurons confine their axons to one side and level. Projection fibers, as well as entering sensory and exiting motor axons, may **decussate** to the contralateral side or remain ipsilateral with respect to their cell bodies.

Another classification of nerve cells can be made on the basis of the **size and nuclear-cytoplasmic ratio** of the cell body or perikaryon (Fig. 2.5). The extreme cases are (a) the **globuli cells** of invertebrates and granule cells of vertebrates, characterized by small size, paucity of cytoplasm in the cell body, and abundance of chromatin in the small (but proportionately large) nucleus, and (b) the **giant cells,** characterized by prodigious size, abundance of cytoplasm, and paucity of chromatin in the large (but proportionately small) nucleus. Globuli cells in small insects may be less than 3 μ in total diameter with a nuclear-cytoplasmic ratio very close to one; giant cells in moderately large gastropods (*Aplysia* of 1 kg) may exceed 800 μ in diameter, with a nuclear-cytoplasmic ratio (of diameters) as low as 0.5. Intermediate cases of all degrees occur; it is common to see discontinuities in the local distribution or in the total numbers of the size classes. Globuli cells often occur in masses without other classes of cells intermingled and the same is only less true of each class; the more advanced the animal or part of the brain, the more pronounced this tendency.

5. Giant cells and axons

Most giant nerve cell bodies are the extremes of a continuous spectrum in the same animal, unlike giant axons, which more commonly occur as discontinuous parts of the size-frequency distribution of nerve fibers in the species concerned (as in earthworm, squid, crayfish, goldfish, but with exceptions like cockroach, *Aplysia*, lamprey, shark). The largest giant cell bodies do not give rise to the largest giant axons; *Aplysia* cells of 800 μ have axons of about 50 μ measured one cell diameter away (Fig. 2.6). The largest axons have cell bodies of no extraordinary size (see the relevant systematic chapters for squid, polychaetes, earthworms, crayfish); this means that the cell bodies have a much smaller volume than the axons. The 300 μ giant fibers of the polychaete *Protula* come from single cells of about 50 μ average diameter; the 800 μ axons of squid come from many cells of about 40 μ. These examples are chosen to show that some giant axons are the product of fusion of the stem processes of several (earthworm) to many (squid) unipolar cells, but that others (*Protula*, crayfish median, teleost Mauthner) are the axons of single cells. The **ratio of axoplasm to somatoplasm** is therefore widely divergent, from a good deal less than one in short axon cells (mostly small cells) to

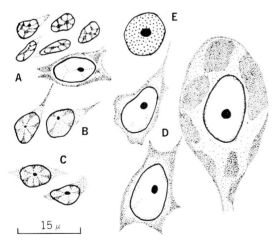

Figure 2.5. Types of nerve cell by size. These are several forms of somata in the brain of *Octopus*, as shown by the Nissl stain. **A.** Granule or globuli cells of the chromophile zone. **B.** Small ganglion cells of the frontal lobe. **C.** Optic lobe granule cells. **D.** Large ganglion cells. **E.** Nucleus of a large cell impregnated with silver. [Bogoraze and Cazal, 1944.]

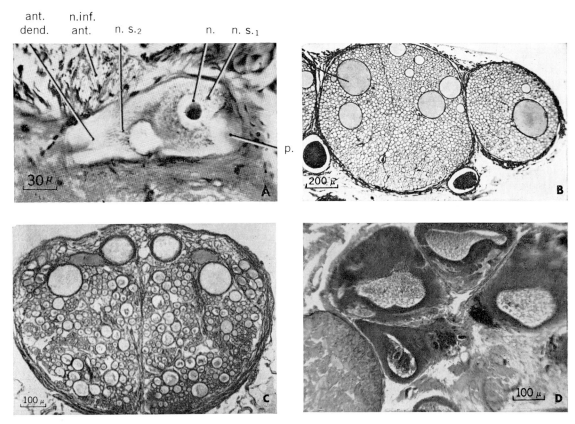

Figure 2.6. Giant cells and fibers. **A.** First-order giant cell in the brain of *Loligo*. Picroformol, hematoxylin and eosin. [Young, 1939.] *ant. dend.*, anterior dendrite; *n.*, nucleus; *n. inf. ant.*, anterior infundibular nerve; *n. s.₁*, central irregular granules of Nissl substance; *n. s.₂*, longitudinally arranged peripheral granules of Nissl substance; *p.*, pathway from axon to dendrites. **B.** Transverse section of nerves of the cuttlefish *Sepia*, showing giant and small fibers. [Courtesy J. Z. Young.] **C.** Transverse section of the ventral cord of a crayfish, fixed by the method of vom Rath. The four giant fibers are seen dorsally and a motor axon is en route out of the cord, on each side, just about to pass dorsal to the lateral giants and to make synaptic contacts with them. [Robertson, 1961.] **D.** The somata of four large cells of the visceral ganglion of the gastropod *Aplysia*. The neuropile is below, right; a connective, which looks like a nerve, is forming on the left. Note the pigment in the large cell cytoplasm and especially the islands and trabeculae of glial tissue penetrating the cytoplasm (trophospongium). [Bullock, 1961.]

something greater than 10,000 (*Protula*). This ratio has been given as 250 for a ventral horn motoneuron of the rhesus monkey and 5000 for the multicellular giant neuron of the polychaete *Myxicola*.

There is no modern theory of the **biological meaning of this great range** and therefore of the significance of giant cell bodies. Giant axons are thought to be of value because of their higher conduction velocity. It has also been suggested that their extremely large spike voltage, as seen by neighboring neurons, may be an important part of their functional meaning (Bullock, 1952); electrotonic spread through specific low resistance connections or through the tissue as a field effect is believed to be adequate to influence the firing of some neurons.

One feature of structure that is well illustrated by some giants, though not peculiar to them, is the sharp **distinction between somatoplasm and axoplasm.** This is conspicuous in large gastropod and crustacean cells in the region of origin of the axon (Fig. 23.13). Staining reactions, inclusions, and density of ground substance all change abruptly at a line in this region.

6. The physical properties of axons and changes with activity

These have been the subjects of several studies. The viscosity of axoplasm varies widely among animals (Schmitt et al., 1936; Chambers, 1947; Flaig, 1947; Chambers and Kao, 1951, 1952; Tobias and Nelson, 1959). In the giant axons of squid it is fluid enough to be expressed from a cut end by squeezing gently, and this opportunity to obtain pure cytoplasm in quantity has been exploited in a good many studies. Viscosity changes measurably with electrical stimulation (Flaig). Turgor has been examined in squid and cuttlefish axons (Young, 1944; Cragg, 1951) and tensile properties in crab nerve (Easton, 1956). The longitudinal orientation of more fluid channels is indicated by movements and shapes of droplets and vesicles (Kao, 1956). Various changes in staining characters, dimensions of nuclei, volume, tension, and light scattering have been reported with stimulation, sometimes of moderate degree, sometimes only after prolonged periods (Dolley, 1913; Golovina, 1949, 1955a, 1955b; Hill, 1950; Ushakov, 1950; Tobias, 1950, 1951, 1958, 1960; Kornakova and Frank, 1952; Chance et al., 1956; Nasonov and Suzdal'skaia, 1957; Tobias and Nelson, 1959; Solomon and Tobias, 1960). Some diffuse suggestions from earlier literature are given by Kappers (1934).

We may now observe examples of the forms of neurons and note their occurrence among animals.

B. Distribution of Types of Neurons among the Phyla

Dendrites seem to appear later than axons, as histologically distinct components. The neurons of the simplest systems, exemplified by coelenterates, are isopolar, except for sensory cells.

The **sensory cells** are remarkably consistently bipolar with the soma in or near an epithelium and therefore with unequal processes—a short distal, sensory one and a long, central axonal one. This type persists and is represented by the olfactory receptors of vertebrates and in our view by rods and cones in the retina (Fig. 2.7). In higher invertebrates, in addition to these superficial, short dendrite sensory cells, there are increasing numbers of cells with the soma lying deeper, beneath the body surface but still singly or in scattered peripheral ganglia; the distal process may be long and may finally branch more or less profusely, and it is then called a free nerve ending. Here the functionally dendritic and axonal parts are difficult to distinguish on morphologic grounds and it seems likely that impulses arise now farther distally, now more centrally, even in the same cell. The T-shaped vertebrate dorsal root ganglion cell (Fig. 2.8) is a slight modification of this, of bipolar ontogenetic origin. It has an exaggeratedly long distal axon, and that it conducts directly into the central axon, without any necessary intervention of the soma, shows the unimportance for neural events of the position of this swelling. The probable ancestry of these unipolars is seen in the bipolar cells of the ganglion of the cochlear nerve as well as in their own embryology.

Internuncial and motor cells are not distinguished in the nerve nets of coelenterates, which are made of isopolars, either bipolar or multipolar elements—in any given net, chiefly one or the other. Internuncials are apparently distinct in the marginal rings and ganglia of medusae. It will not be surprising if some heteropolar elements are found here, or even unipolars, which so far are virtually unknown in this phylum. In Platyhelminthes, the lowest forms with a central nervous system, we already observe a drastic change to an abundance of central unipolar neurons (Figs. 2.7, 9.8). These show a minimal differentiation of branches of the stem process except that one or two are generally quite long and the others shorter and sometimes further branched. There are still many central bipolars and multipolars, both isopolar, in flatworms.

In all the higher invertebrates an overwhelming majority of interneurons and motoneurons are **unipolars**; this cell type is far and away the most widespread form of central neuron in the animal kingdom. An endless variety of subtypes is known. The simplest distinction is between (a) large, up to giant, somata with relatively abundant cytoplasm, large nucleus, and little chromatin, and (b) small cells with a densely chro-

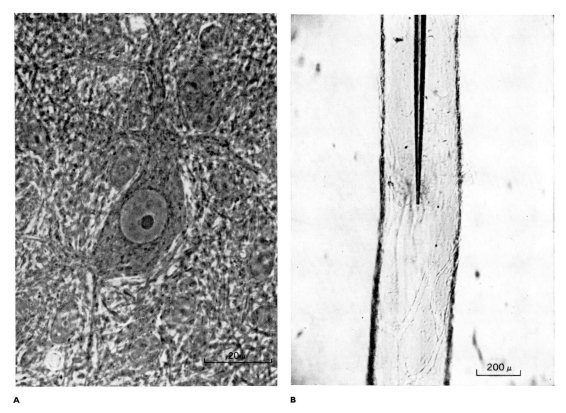

Figure 2.7. Living nerve cell and fiber. **A.** Phase-contrast photomicrograph of a Purkinje cell in a 21-day old tissue culture of the cat cerebellum. Rodlike and filamentous mitochondria can be seen as dark structures in the perikaryon and in the dendrites; Nissl substance is in cloudy flakes near the cell periphery, manifesting a degree of persisting chromatolysis in this young culture. [Hild, 1959.] **B.** A giant axon from *Sepia* with a semimicroelectrode inserted. [Courtesy N. Chalazonitis.]

matic nucleus and a very thin layer of cytoplasm around it. The latter sort are called globuli cells in their advanced form, especially when collected in masses, usually packed closely together. The criteria so far allow a number of permutations such as plasmatic, chromatin-poor, internuncial unipolars of intermediate size, with long decussating, descending axon. But beyond this the differentiation of collaterals, branches, and terminal arborizations permits specialization, reaching a peak in the optic ganglia of arthropods (Figs. 19.12, 19.13, 19.16). Here some 18 types of cells are distinguished by being bushy, bottle-brushed, ropy, or stratified, or having other ramifications in various combinations at receptive and efferent endings. An important feature to note is the common finding, in the ventral cord of articulates, of dendritic branches from a long projection axon, at segmental intervals; physiological evidence shows that the same axon can at one moment conduct ascending impulses from a caudal input and at another a descending impulse from rostral afferents so that two-way conduction is a normal reality in such axons. Unipolar cells are rarely found outside the central nervous system; exceptions are vertebrate spinal ganglion afferents, some invertebrate stomodeal ganglion cells, the optic ganglion of *Helix*, and a number of other special cases. Peripheral motor or internuncial ganglion cells, as in the plexuses on the gut in higher invertebrates, are generally multipolar but occasionally are bipolar and even unipolar.

Multipolar, heteropolar neurons, like the nearly universal type in vertebrate interneurons and motoneurons, are found in invertebrates but in

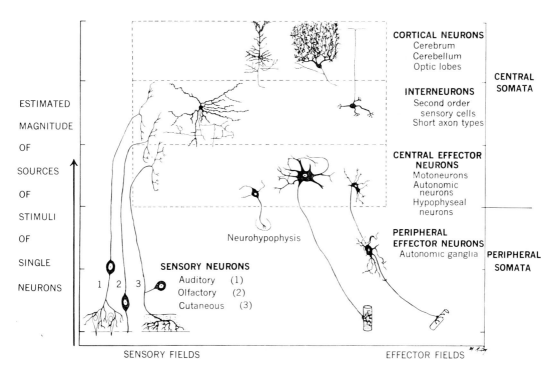

Figure 2.8. Neuron types in the mammalian nervous system, arranged according to general functions and according to probable magnitude of sources of synaptic connections. The cerebellar granule cell *(top right)* is apparently too high on the ordinate scale. [Bodian, 1952.]

small numbers and special places (Fig. 2.9). Some of the cardiac ganglion cells and stretch receptor cells of crustaceans, the first-order giant cells of squids, and certain elements in the stomodeal system of molluscs are examples. With these minor exceptions, this form is essentially a vertebrate development. Unipolars are for all practical purposes absent from the vertebrate central nervous system; bipolars are employed in a few places such as the receptors and the second-

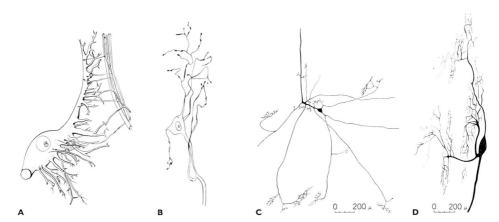

Figure 2.9. Invertebrate multipolar neurons. **A.** *Loligo*, first-order giant cell. [Young, 1939.] **B.** *Astacus*, anteromedian cell of fifth ventral ganglion. [Retzius, 1890.] **C.** *Cancer*, small posterior cell in cardiac ganglion. [Alexandrowicz, 1932, see Arthropod Bibliography.] **D.** *Maia*, large anterior cell in cardiac ganglion. [Alexandrowicz, 1932.].

order neurons of the retina, olfactory receptor cells, and sensory ganglia of the eighth cranial nerve, as well as others in fish.

Differentiation of vertebrate multipolars is well known from the elegant and exhaustive classical studies of Deiters, Gerlach, Kölliker, Golgi, Ranvier, and others, but above all Cajal (see his 1909–1911 summary). Here the anatomical development of extensive dendritic arbors reaches a peak unknown in invertebrates. Only in vertebrates can central dendrites be so regularly distinguished from axons as plasmatic, short, ramified, dichotomous, granular, uneven in thickness, thorny, as well as naked or not privately sheathed. Axons in vertebrates are by contrast finer and lack rough outlines and notches, but they have right-angled collaterals and often a private glial sheath; they are or can be of very great length (Figs. 1.7, 2.20). Not all vertebrate multipolars present a clear distinction between axon and dendrites; one of the classes of neurons is set apart by just this isopolarity (granule cells of the olfactory bulb, certain sympathetic cells of Auerbach's and Meissner's plexuses, and amacrine cells of the retina). Of the others, a simple division can be made into cells with long axons that leave the center or level or side where they arise and cells with short axons that ramify and terminate within the center or cell mass of their origin, often within the range of spread of the dendrites, perhaps less than 1 mm. It suffices to suggest the endless variety of cells characteristic of their respective locations by pointing to a few. The large Purkinje cells of the cerebellar cortex have a profuse and extensive dendritic arbor espaliered in one plane (Fig. 5.30). The small granule cells of the same cortex have a sparse set of dendrites which form peculiar claws grasping the somata of several neighbors (Fig. 2.20). The fusiform, horizontal cells of the cerebral molecular layer have long and much-branched axons contributing to a dense plexus. Large pyramidal cells of the cerebral cortex have an apical dendrite of such length that the influence of distal synapses upon impulse initiation in the axon must be very weak. Stellate cells, mitral cells, bizarre autonomic ganglion cells—the variety suggests physiological specialization not yet adequately studied. Further evidence of this is cited below under forms of synapses.

The principle of the **dynamic polarization of the neuron,** enunciated by Cajal, can be stated in this way: the excited state normally spreads from the dendritic to the axonal terminals of a neuron. In typical vertebrate multipolars this means that the normal or dromic or **orthodromic** direction of propagation in the axon is away from the cell body, cellulifugal. This is simply a consequence of the peculiar anatomical fact that the dendrites in this class of neurons are typically borne on the cell soma. It might be otherwise and certainly is in unipolars of both invertebrates and vertebrates.

C. Origins and Relationships of Neuron Types

Older literature records numerous efforts to deduce an ancestry for the types of neurons (see Chap. 1, pp. 10, 15). These speculations cannot be dismissed today but neither can substantial new light be shed; however, interest has drifted away from such questions.

In Chapter 1 it was proposed that prenervous cells or organelles carry out the functions of the future nervous system in animals or in those parts of animals where nerve cells have not differentiated. That is, the properties of unspecialized cells include a certain level of ability in selection of adequate stimuli, excitability, propagation of change, correlation with the existing state or other stimuli, and adaptive response. Chapter 7 cites examples in Protozoa and Porifera where visible structures differentiated for such functions do not exist but where the activities are nevertheless carried out. Also in Chapter 1 we quoted the classical proposal that the first differentiation in mechanisms of this sphere was that of **independent effectors.** Cilia and oscular myocytes apparently preceded the evolution of nerve cells, and other examples of so-called independent effectors are known in higher animals. Nematocysts, for example, are perfectly capable of normal function without nervous mediation, although the opinion is growing that they may be subject to a superimposed coordination. Cilia here and there are well known to reveal superimposed influence on top of their intrinsic independence, and furthermore this control is nervous; this situation is similar for some glands and smooth muscle. But this does not suggest the origins of nerve cells.

It is generally believed that the **first nerve cells were sensory,** derived from an epithelium and connected to effector cells and therefore in fact not purely sensory but sensorimotor. This hypothesis (Parker, 1919) is attractive, mainly for lack of alternatives. It has not been substantiated by any actual findings of such a connection in any of the many forms studied over many years (Batham, Pantin, and Robson, 1960). The coelenterates give evidence of being too far along in evolution to aid directly in this question. Where a definitive nervous system first appears unequivocally—namely, in the coelenterates—there is already an obvious diversity of neurons, including a readily distinguishable category of bipolar heteropolar sensory cells which are often diverse and specialized. There is always a net of bipolar and multipolar neurons, and in addition there is sometimes a distinct net of specialized bipolar cells with long processes. Judging from its condition throughout the coelenterates, with emphasis on the simplest, the primitive net (as in hydrozoan polyps) includes sensory neurons together with bipolar or multipolar neurons which have the dual function of passing excitation to other nerve cells and also to muscle.

Presumably, from primitive sensorimotor nerve cells there differentiated **first internuncial-and-motor nerve cells** and only later pure interneurons and pure motoneurons. In coelenterates there are no neurons which are known to be restricted only to motor function; even the specialized bipolar neurons of the through-conducting nets in jellyfish or in mesenteries of anemones serve a dual function as both motor and internuncial neurons. Unipolar cells are essentially missing in coelenterates; they are described, but in all cases they are probably superficial sensory cells, many of which bear a minute process connecting them with the surface. Neurons concentrated into ganglia, where they appear for the first time in the animal kingdom—in the marginal bodies of scyphomedusae and in the elongated ganglia of the ring nerve of hydromedusae—form a definitive and distinct class of cells. In jellyfish ganglia their appearance, position, and connections are compatible with the suggestion that they are derived from sensory cells because many (and perhaps all) retain a connection with the surface of the epithelium and some have modified (possibly sensory) cilia. In hydromedusae the varied shapes of the neurons, some of which suggest a trend toward the unipolar type of interneuron (Fig. 8.15), are suggestive of the situation found in all higher phyla, where most of the ganglion cells are unipolar with a T-shaped fiber consisting of axon and dendrite with arborizations.

II. SPECIAL RELATIONS AND COMPONENTS OF NERVOUS TISSUE

The following sections do not give a complete treatment of the special cytology and histology of nervous tissue but do treat those aspects that are peculiar to nervous tissue and consider each in the light of information from lower and higher animals. The references cited above (p. 38) should be consulted for details, especially on vertebrates.

A. Organization of Nervous Tissue

The nerve cells and processes discussed in the previous section occur as isolated elements or as organized tissues in the form of epithelial layers, plexuses, peripheral ganglia, nerves, central ganglia, cords, connectives, and commissures. The general features of histological organization of each of these and their main constituents must now be considered.

1. Solitary nervous elements

These are relatively rare except at sensory and motor endings and in very lowly nervous systems. Most of the nervous system of coelenterates and other groups dominated by nerve nets consists of isolated neurons, making only synaptic contact

CHAPTER 2. MICROANATOMY

and not grouped into nerves or ganglia. Recent electron microscope observations indicate that many coelenterate nerve cells and fibers are even devoid of sheath cells that creep along or cling to them. In addition to the nerve nets, there are concentrations in the marginal rings of hydromedusae and marginal ganglia of scyphomedusae. Isolated nerve cells and axons in higher animals occur at thin places in peripheral plexuses, particularly on the viscera (Fig. 2.10).

2. Epithelial nervous systems

The lowest nervous systems are always mainly or entirely intraepithelial, except for motor axons to underlying effectors. Doubtless this is a primitive position. We see it in coelenterates, enteropneusts (Fig. 2.11), phoronids, simpler turbellarians, oligochaetes and polychaetes and other similar groups, and in most of the nervous system of echinoderms. The organization is always similar in these respects: axons mainly form a nerve fiber layer at the base of the epithelium, just superficial to the basement membrane, if

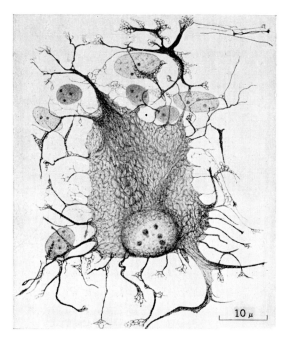

Figure 2.10. Solitary nerve cell in the periphery. Ganglion cell of Dogiel's type I with fibrillar extensions of the short processes; Auerbach's plexus in the rabbit stomach. [Reiser, 1959.]

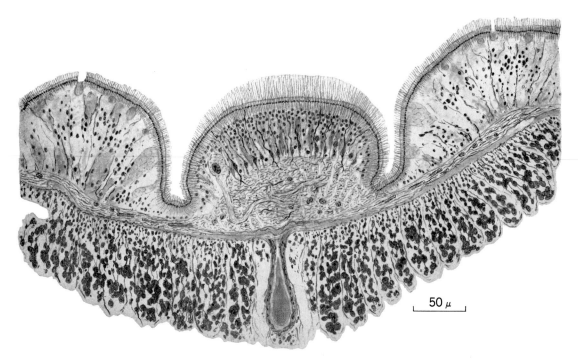

Figure 2.11. Example of a basiepithelial nervous system. The general plexus and dorsal cord of the trunk in an enteropneust, *Balanoglossus* sp. In the epithelium over the cord can be seen bipolar sense cells and one giant cell; in the general plexus very fine fibers contrast with giant (5 μ) axons. Coelomic muscles and dorsal blood vessel, below. Bodian protargol impregnation. [Original.]

3. Peripheral plexuses

The term plexus is more general than nerve net and noncommittal as to the relation between its elements. It therefore includes nerve nets as well as simple tangles of purely sensory axons and their branches, as in crustacean integument. It is the general form of most of the peripheral nervous system in lower groups, including ctenophores, platyhelminths, nemertineans, and the like, where nerves are not yet formed; of important local regions of the peripheral system in higher groups, as in the gastropod foot; and of the gut plexuses in the walls of the alimentary canal of annelids (Fig. 2.12), arthropods, molluscs, and vertebrates. In some of these situations the plexus is functionally semiautonomous and mediates responses; therefore it contains sensory and motor neurons and possibly interneurons. Plexuses are commonly intraepithelial and intergrade completely with interneurons. They may also be subepithelial or internal. There is a minimum of organization; cell bodies are scattered along the strands not segregated from the nerve fibers, except depthwise in epithelia. There is generally no sheath or sharp boundary. These are notoriously difficult objects for physiological analysis (see, for example, gastropod foot, p. 1332).

4. Peripheral ganglia

Higher animals have developed a great variety of small peripheral collections of nerve cells. Some

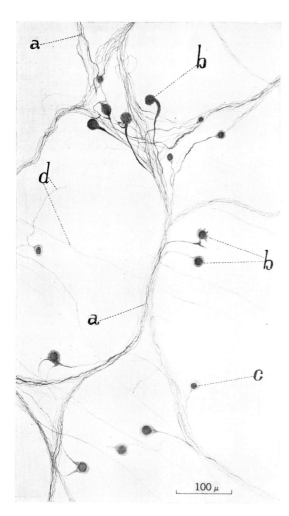

Figure 2.12. A peripheral plexus in the wall of the posterior stomach of the snail *Helix*. Bielschowsky-Gros silver. [Ábrahám, 1940; see Gastropod Bibliography.] *a*, strand of many nerve fibers forming the meshes of the plexus; *b*, unipolar ganglion cells which might be interneurons or motoneurons; *c*, nucleus; *d*, nerve fibers.

there is one, whereas ganglion cell bodies lie among the nuclei of the epithelial cells, just superficial to the nerve fiber layer. Commonly sensory nerve cells are abundant; locally they may be extremely abundant (see echinoderms, p. 1524, and enteropneusts, p. 1575) and their nuclei are then distinct and on the average still more distal. Thickening of the nerve fiber layer is the most noticeable sign of a nerve cord but ganglion cells and sensory neurons are also increased while gland cells are locally decreased in numbers.

Figure 2.13. A small peripheral ganglion. The subacetabular ganglion that lies under each sucker on the arms of *Octopus*. The large nerve cells and the ganglion appear surrounded by a meshwork of fibers identified as axons. Reduced silver. [Rossi and Graziadei, 1956; see Cephalopod Bibliography.]

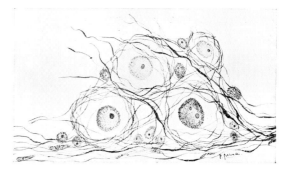

are purely sensory and contain no synapses, and hence have only morphological, embryological, or trophic significance but not physiological meaning for nervous function (dorsal root ganglia of vertebrates, accessory sensory ganglia of gastropod tentacles, annelid nuchal organ, many small clusters in arthropod antennae, and elsewhere). Purely motor cell collections are exemplified by the autonomic ganglia of vertebrates. Mixed ganglia capable of mediating reflexes are known, as in the podial ganglia of polychaetes, the stellate ganglia and the nerve cords in the arms of cephalopods (Fig. 2.13). Understandably, there is not much of a common denominator in the organization of this variety of structures. The smallest have little order, and the larger resemble central ganglia; details are given in the relevant systematic chapters.

5. Nerves

A nerve is a discrete bundle of nerve fibers connecting some central nervous structure with some region of the periphery, and therefore with receptors or effectors or both. Not until groups of the level of differentiation of sipunculids, echiuroids, annelids, and higher are there well-formed bundles of axons distinct from strands of a plexus. There is no clear line; many fascicles of fibers, such as those coming from tentacles of lophophorates or special sense organs of nematodes, are called nerves. In higher forms nerves are free of nerve cell bodies, but in some nerves of molluscs—especially pelecypods and similar lower forms, and in some annelids but rarely in arthropods—there are occasional nerve cell bodies along the course of nerves. These may be deep-lying sensory cells or outlying motor ganglion cells, probably seldom interneurons. It is believed by some (Smith, 1957) that such cells in annelids perform an important function by funneling large numbers of sensory axons into a few afferent fibers that enter the central nervous system and by distributing motor impulses from a few efferent fibers that leave the central ganglia to many thousands of muscle cells (pp. 678, 750). (Horridge cannot confirm this in *Harmothoë*.) Nerves are often composed of fibers of very different sizes and plots of the number of fibers in each size class have functional meaning as well as being signatures of the nerves concerned. The sheath components of nerves are treated below (p. 95).

6. Connectives and commissures

The connectives and commissures are very similar to nerves, though they contain mostly internuncial axons. Connectives join ganglia of the central nervous system, anteroposteriorly separated on the same side; commissures join equivalent structures on the two sides. They may often be not quite cell-free and therefore can sometimes include synaptic relays, but for the most part they are purely axonal.

7. Central ganglia and cords

In all animals with a central nervous system there is a division of its tissue into two zones, fibrous and nucleated. But this primary differentiation takes place in two quite different ways, one found in all invertebrate groups and the other in all vertebrates and only in them! The meaning of this dichotomy would seem to be that the ancestor of vertebrates diverged from other invertebrates very early. The two zones in all living invertebrates are conveniently called the **rind and core,** from their positions (Fig. 2.14); the nerve cell bodies are gathered in a layer on the outside of the ganglion or cord, leaving a central mass of fibers that is almost or quite free of nerve cells. The main feature, however, is that there are, broadly speaking, no fibers ending in the cell rind; the synaptic fields as well as the long pathways are in the fiber core. In contrast, vertebrates present two zones—called **gray matter and white matter;** the gray is primitively inside but in higher brain regions it is also on the outside of the white. The gray contains the nerve cell bodies and also the nerve endings, both dendritic and axonal; the white consists simply of axons en route. Neuroglia but rarely nerve cell nuclei pervade all zones, so that only glial nuclei are seen in the core and in white matter.

This difference between vertebrates and invertebrates is certainly morphologically fundamental and it may perhaps be physiologically important as well. It goes with the prevalence of

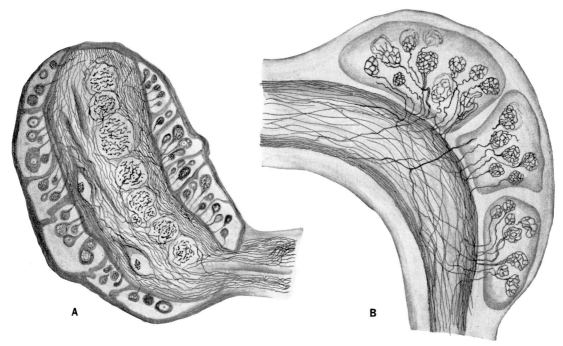

Figure 2.14. Ganglia of the central nervous system in reduced silver impregnation. **A.** Midsagittal section of the subesophageal ganglion of the leech *Haemopis*. The compartments or packets of cells that make up this compound ganglion can be seen, defined by connective tissue walls continuous with the sheath of the ganglion and with an internal capsule that separates the rind from the core. The fibrous core has a row of differentiations containing fibers running transversely. From the lower right emerges the connective to the next ventral ganglion. **B.** One side of a transverse section through the brain of *Hirudo;* the commissure to the other side is at the upper left, the circumesophageal connective continues from the lower right. Besides the packets—of which the brain, like the other ganglia, is composed—one can see clearly the coarse meshed neurofibrillar network in the somata. [Redrawn from Sánchez, 1912.]

synapses on the soma in vertebrates, a feature extremely exceptional in invertebrates. (But it is well known in certain cases: the receptor cell in the muscle receptor organ of crayfish, the cardiac ganglion cell in lobster, the first-order giant cell in squid.) It may also go with a basic difference in the **nature and properties of dendrites;** those of vertebrates are more fleshy, bushy, and thorny, and have characteristic inclusions visible with the electron microscope (Gray, 1959a; Pappas and Purpura, 1961), which are granular, membranous, tubular, and vesicular. Possibly dendrites are responsible for the slow smooth brain waves that are general and conspicuous in vertebrates. The development of full dendrite characters occurs suddenly late in embryonic life or after birth in mammals and correlates with the appearance of characteristic electrical responses (Purpura,

1961). Invertebrate brains manifest great spike activity but no (or very inconspicuous) slow electrical waves. Dendrites in invertebrates almost all arise from the axon and ramify more or less widely as very fine fibers usually without spines; whether the extent or character of the branching or the properties of the membrane are different is not known.

The fiber core of invertebrate ganglia is differentiated except in the lowest forms into **tracts and neuropile.** The tracts, equivalent to vertebrate white matter, are fibers en route without endings or synapses, when they are pure. Neuropile is a plexus of fibers with dendrites, axonal terminations, and synapses. In higher invertebrate forms it is usually easily recognized and distinct from tracts but in annelids and below the separation is incomplete. Vertebrates have local regions of neuropile relatively free of nerve cells, but these are quite limited. Neuropile is therefore the principal region of integrative nervous events in most animal groups. It is still largely a terra incognita but deserves concentrated attention both anatomically and physiologically.

Degrees of differentiation of the neuropile can be found and are more sensitive indicators of level of advancement than the cell rind. The lowest neuropiles are loose and coarse in texture and lack local signs of different texture (Fig. 9.12); this is the general aspect of brains in planarians, chitons, pelecypods, terebellid and cirratulid polychaetes, ectoprocts, brachiopods, nematomorphs, and other groups of low level, as judged by general zoological criteria, including sensory development, behavioral repertoire,

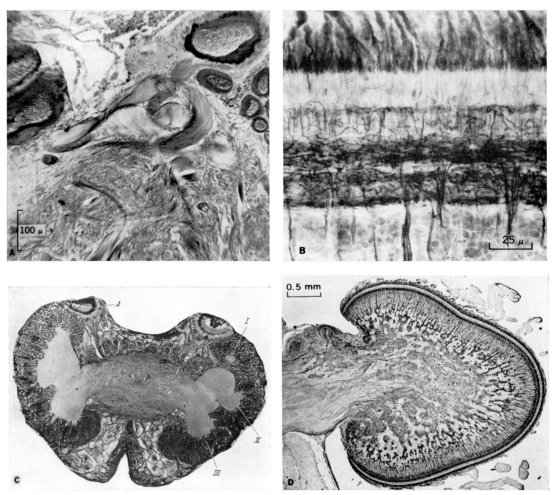

Figure 2.15.   Forms of neuropile. **A.** *Aplysia*, visceral ganglion with unspecialized fiber core, here traversed by the stem processes of giant cells which can be seen at left. Reduced silver. [Bullock, 1961.] **B.** *Octopus*, plexiform layer of the cortex of optic lobe; above and below are the granular layers. The pure neuropile is highly differentiated into strata. [Courtesy J. Z. Young.] **C.** *Harmothoë* (Polychaeta, Polynoidae), transverse section of the brain showing common neuropile in center, three specialized stalks of the corpora pedunculata surmounted by "globuli" cell masses, clumped neuropile or glomeruli, in lateral lobes and tracts entering the circumesophageal connectives below. [Courtesy B. Hanström.] **D.** *Octopus*, optic lobe, low-power view showing the great differences in the neuropiles of the cortex (same as **B**), the outer and inner portions of the lobe and the main tracts connecting with the brain (left). [Courtesy J. Z. Young.]

number of brain cells, and the like. Higher nervous systems have increasing signs of differentiation of texture (Fig. 2.15) and separation of tracts from neuropile. There are certain specialized forms of neuropile that have received names and probably others yet to be recognized. **Glomeruli** are knots or small spherical masses of tighter weave than the surrounding neuropile; good examples are found in the deutocerebral neuropile serving the antennae of crustaceans and in the olfactory bulb of mammals. These should attract more attention as relatively more discrete and simple than other forms of specialization. Another form is stratified neuropile as seen in the plexiform layer of the cephalopod optic lobe and in insect optic ganglia; layers of different density and grain are clearly segregated, representing the planes in which many fine nerve fibers ramify and end (Fig. 2.15). The calyces and stalks of the corpora pedunculata are further types, distinguished especially by the very dense, smooth texture of the stalks, due to their fine, short, packed axons and dendrites (Fig. 2.16).

Differentiation in the cell rind is mainly based on size of nerve cell somata and on the proportions of cytoplasm and chromatin. Signs of specialization are marked uniformity of type of nerve cell, presence of some very large to giant cells (up to 800 μ in *Aplysia;* see Table 23.1), cell masses set off from others, abrupt transition from one cell type to another and individually recognizable cells. The highest centers in invertebrate brains are believed to be those composed of small, cytoplasm-poor, chromatin-rich cells, tightly packed together, called **globuli** masses. These are found already in polyclad flatworms, in nemertineans, more advanced polychaetes, arthropods, and molluscs such as neogastropods, pulmonates, and cephalopods. In cephalopods there is one more sign of high development—islands of nerve cell bodies scattered throughout a large neuropile mass (Figs. 25.15, 25.20).

Differentiation of gray matter in vertebrates is even more pronounced in respect to details of texture, in large part because of the presence of nerve cell somata which themselves show an enormous range of form and branching and orientation. It is noteworthy that, as in invertebrates, many higher levels are characterized in part by relatively smooth, dense histological structure (for example the molecular layer of the cerebral and cerebellar cortices) and by small compact cells. So-called granule cells are much like globuli cells, although multipolar. A relative concentration of nerve cells or a discrete mass of gray matter may be called a nucleus or a ganglion within the brain. In contrast, gray matter of a certain relatively unspecialized texture crossed by many anastomosing strands of white matter is called reticular formation and occurs in places at the margin between gray and white in the spinal cord and brain stem. Other types of gray matter are seen in the substantia gelatinosa, the inferior olive, the red nucleus, the tectum, and the hippocampal cortex, to mention a few of the innumerable sites with distinctive features of architectonics. The details are beyond our present scope but may be found in Cajal (1909–1911). Bumke and Foerster (1935–1937) and Kappers, Huber, and Crosby (1936) are useful for certain aspects.

The **functional meaning of specialized organization** is a challenging problem about which something can be said, even though only a beginning. Gerard among others has pointed out that on a theory of purely digital units or of units specifically connected, as in a telephone system, there is no special value to ordered arrangement, orientation, interdigitation, planes, and strata of elements. The fact of such organization argues for spatial representation—sometimes in essentially two dimensions, as a kind of toposcopic display—of the information of concern to that brain region. This means that handling the information in order to analyze, recognize, and extract relevant aspects and to formulate appropriate outputs is probably to a significant degree an analogue kind of data processing at a multiunit level. It means also that we will be much closer to understanding if we can discover what is displayed, what aspect of information is spatially represented. On the crudest level this is known for area 18 of the occipital cortex and a few similar regions where the visual field or the body surface is topographically represented. Closer examination, such as that of Mountcastle

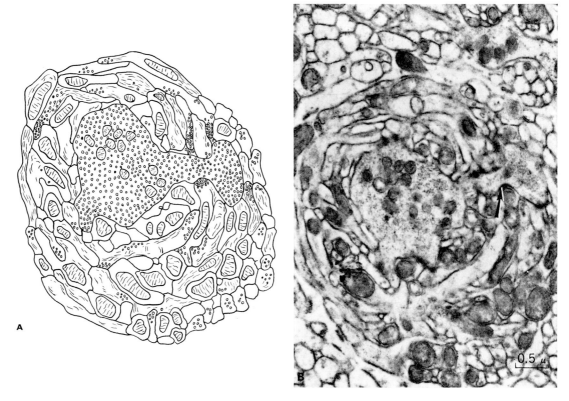

Figure 2.16. Specialized neuropiles; glomeruli. **A.** Center of a glomerulus drawn from an electron micrograph like the following. **B.** Electron micrograph of a typical glomerulus of the calyces of the corpora pedunculata of an insect *(Laplatacris dispar)*. The center is a large fiber filled with vesicles; surrounding it are small concentrically oriented fibers with occasional vesicles and very little evidence of glia. Ordinary neuropile can be seen in the upper and the left lower corners. **C.** Schematic drawing of a micrograph like the following. **D.** Electron micrograph of a portion of a glomerulus in the antennal region of the deutocerebrum of *Laplatacris*. At this level the fibers of the antennal nerve are establishing their synaptic contacts, but presynaptic and postsynaptic fibers cannot be dif-

et al. (p. 348) and Hubel and Wiesel (p. 346), shows even these regions to be more complex; Maturana et al. (1960) describe strata of the frog tectum, representing different kinds of extraction of optic stimulation (p. 280). The point to emphasize here is that beyond the study of properties of units is a domain of great significance in interpreting nervous organization: the uncovering of the kind of information sorted out, segregated, or represented in ordered arrays of central elements.

The **extent of potential interaction** of neurons has one measure in the sheer numbers of countable nerve cell bodies within the area of ramification of the dendrites of a single neuron. Omitting as uncountable the number of other neurons represented in this area only by axonal terminations, Scheibel and Scheibel (1958a) give 4,125 cell bodies within the area of the dendrites of one cell of the magnocellular nucleus of the kitten medulla, 27,500 cells within the area of potential direct interaction of a single long axon of a reticular neuron and its collaterals in the brain stem, and 39,375 cells within the area of terminal branching of one afferent fiber, probably a so-called nonspecific fiber, to the cortex.

Tracts or fascicles of axons differ in several ways. Some are loose, with few axons per unit of cross section, others are compact, with less neuroglia. Some have fibers of widely varying

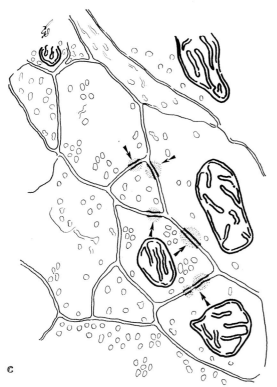
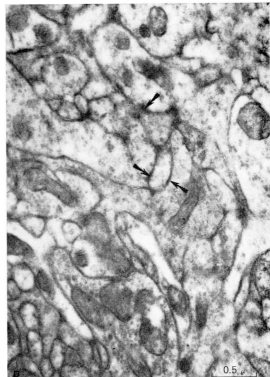
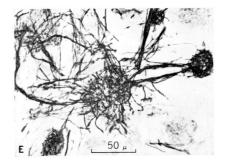

ferentiated. Vesicles common in most fibers. Thickenings or patches of high electron density (*arrows*) may be synaptic regions. [**A–D**, courtesy O. Trujillo-Cenóz and Melamed, 1962.] **E.** Glomeruli as seen in the cardiac ganglion of *Sepia* by Cajal's silver stain and low magnification. The knots of tangled fibers are called glomeruli. [Alexandrowicz, 1960; see Cephalopod Bibliography.]

diameter, others are quite homogeneous in fiber size and hence in velocity. Some tracts show marked parallelism of elements, others a good deal of braiding. Some bundles of fibers are of mixed ascending and descending axons and mixed functional groups, others are quite pure in functional composition. Some have particular, identifiable fibers in particular areas of the cross section; others appear to be haphazard in this respect.

A feature common in all bilateral animals is **decussation** of axons, or the crossing of a median plane by projection fibers. Usually the plane is in the long axis of the body and decussation is from right to left or vice versa. But in optic systems of arthropods and cephalopods there are decussations behind each eye across a plane in the axis of the system. Some decussations are complete, others partial. It seems probable that no one explanation accounts for decussations in general. Some, like the unilateral optic crossings of arthropods and cephalopods, may be the consequence of mechanical factors in development; fibers growing toward a target in the embryo, such as retinal axons growing centrally toward the optic lobe, may be aimed at or attracted to a limited gap in the connective tissue mass and, entering that from different directions, merely grow straight on, thus forming the chiasma. This does not explain the plane of the crossing but supposes that the relevant factor is developmental mechanics. Other decussations have the physiological result that one side of the brain and its sense organs exert a more direct effect on the contralateral body muscles. It is thought that there was adaptive value in such direct control in ancestral forms that gave a rapid withdrawal response with the musculature so activated. This is pretty hypothetical and it must be admitted that the phenomenon still awaits an adequate evolutionary explanation.

B. Synapses

A synapse is an anatomically determinate site at which one nerve cell (the presynaptic) can be demonstrated or inferred to influence another separate nerve cell (the postsynaptic); the site involves a limited part of the neurons and contact or near contact (see p. 182 for a discussion of the implications of this concept). As coined by Sherrington (Foster and Sherrington; 1897; see Fulton, 1949, p. 55), the term referred to hypothetical regions of interactions between nerve cells; it still means this for the physiologist, but beginning with Cajal and increasingly today it has come to refer to actual contacts visible under the light or electron microscope. In vertebrates many synapses are between terminations of axon arborizations and the cell body of the postsynaptic neuron; in invertebrates most are between axon arborizations and dendrite arborizations in a tangle of fine fibers called neuropile. Most, but not all, synapses transmit excitation in only one direction, and the terms pre- and postsynaptic may be so defined physiologically, or, with less certainty, on morphological grounds. The idea that the neuron is polarized and normally receives excitation at its dendritic side and conveys it to its axonal terminals was originally an inference from histology, due to Cajal ("doctrine of dynamic polarization"). With the aid of selective stains light microscopy has built up a great body of information on diverse forms of endings and relations between nerve cells, which we shall briefly survey before going on to more recent electron microscope findings.

1. Varieties of contact at the light microscope level

Although the full significance is far from realized and physiologists are apt to oversimplify the categories, the fabulous variety of characteristic forms of nerve fiber endings and synaptic contacts is one of the most impressive results of anatomical science. It can hardly be functionally meaningless that, within the simple framework of the neuron doctrine, an elaborate assortment of geometric arrangements is found, not capriciously distributed, but so diagnostic that an experienced histologist can recognize at a glance under high power whether he is looking at the plexiform layer of the optic lobe of an octopus, the molecular layer of the vertebrate cerebellum, the nucleus of the trapezoid body, or an insect optic ganglion. Many of the details are given in the systematic chapters, but here a brief overview is desirable—all the more because most of our knowledge is based on vertebrate materials, not elsewhere treated in this book.

When an attempt is made to classify the variety of synapses, the problem of suitable criteria becomes acute. Having concluded above that the position of the soma is not fundamental, the relation of the synapse to this structure does not seem a good criterion, but this is in fact a basis for a first-order classification, applicable with few exceptions only to vertebrates (axo-somatic synapses). A better ground is the relation of processes (axo-dendritic, axo-axonal, dendrodendritic). Another is the relation to other junctions; presynaptic inhibition is a physiological event ascribed to the effect of an inhibitory axon on an excitatory axon just before the latter ends upon the cell it excites (see further, Chapter 4). A distinction between synapses with chemical and those with electrical transmission is possible; only a beginning is at hand in correlating anatomy with this functional distinction. Eccles (1961) points out that with chemical transmission the postsynaptic membrane acts as a generator, whereas with electrical transmission it acts as a channel for current flow. This reasoning implies that there is an optimum width for the synaptic cleft when transmission is chemical, whereas it can be vanishingly small when transmission is electrical. In fact the three best-known examples of electrical transmission do have a synaptic cleft less than 200 Å wide and examples of chemical transmission have a cleft wider than 200 Å where examined (Table 2.1). Another possible basis for classification is the presence of vesicles on one or both sides of the synapse (Fig. 2.17), but this seems to be a variable feature in both chemical and electrical synapses; usually the former have only or predominantly presynaptic vesicles, the latter have fewer and in more variable position; different preparations of the same junction show them on both sides, either side, or neither. Coelenterate synapses do not fall readily into any classification; they have vesicles on both sides and therefore presumably conduct in either direction; the synaptic cleft is about 200 Å, so they might be electrical or chemical.

The greatest number of distinguishable types of synapses is based on the anatomical form of the arriving and the receiving surfaces; at present we cannot do better than to list a series of examples, modified and extended from Cajal's list (1954).

(a) Simple contact en passant between axon terminals and dendrite branches is perhaps the dominant variety in invertebrate neuropile and hence the commonest form of synapse. It may well prove to comprise distinct subtypes, for there is little known of the geometric relations between pre- and postelements. It is probably this type that has been seen in some electron micrographs in earthworm and insect, marked by clusters of vesicles and locally dense membrane (Fig. 16.57), though it would be difficult to prove that in fact these loci are points of functional transmission. This category overlaps with the following several and is only set apart by the negative character of lacking a special configuration of a system of branches on the pre- or postsynaptic side. It is probably important that some simple contacts are single whereas in other cases the two fibers touch, separate, and touch again several times.

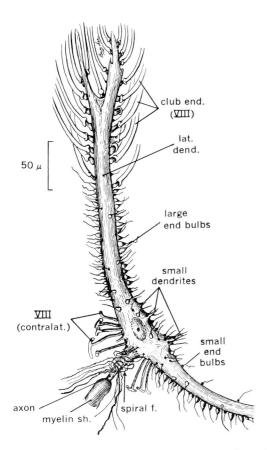

Figure 2.17. Variety of synapses on one neuron. Part of the Mauthner's neuron of the goldfish, drawn from protargol stained sections. [Bodian, 1952.] *club end. (VIII)*, club endings of ipsilateral vestibular nerve fibers; *lat. dend.*, main lateral dendrite; *myelin sh.*, the first internode of the myelin sheath on the axon; *spiral f.*, spiral fibers in the region of the axon cap; *VIII (contralat.)*, contralateral vestibular nerve fibers giving rise to collaterals that terminate as small club endings.

(b) Axo-dendritic connections by climbing fibers (Fig. 2.18). Like ivy entwining a grapevine, this system offers extensive contact and in a serially ordered way, specifically between those two fibers, contrasting in all these features with one or more of the following. Examples are best known in the cerebellum but occur also in the sympathetic ganglia, the lobster cardiac ganglion, and elsewhere. This form illustrates well the far from haphazard relations between nervous elements and also the independence and distinctness of neurons, even when the maximum opportunity for their blending is offered.

Table 2.1. THE MAIN FEATURES OF SYNAPSES IN A VARIETY OF TISSUES, AS SEEN BY THE ELECTRON MICROSCOPE.

Preparation and Reference	Class of Synapse	Structure of Synapse	Associated Specializations
Coelenterate nerve net (1)	Axon-to-axon (en passant) or axon-soma	Lateral contact of two axons, with plain synaptic membranes. Vesicles on both sides	Local osmiophilia of synaptic membrane on both sides
Septal segmental end-unions in giant fibers of earthworm (2)	Axon-to-axon (abutting)	Two unit-membranes apposed over a large area. Sparse vesicles on both sides	Devoid of intervening sheath over a large area
Septal segmental end-unions in giant fibers of crayfish (2)	Axon-to-axon (abutting)	Two unit-membranes apposed only over a limited part of the total possible area. Sparse vesicles on both sides	Sheath cells and membranes intervene between the two cells over much of the possible area; small synaptic windows
Giant fiber-to-motor axon of crayfish (2)	Axon-to-minute dendrite (lateral, en passant); macroscopically axon-to-axon	Postsynaptic axon has dendritic processes which project through the sheath into the presynaptic. Sparse vesicles on both sides or either or neither	None, apart from the invagination
2nd- to 3rd-order giant fiber in the stellate ganglion of squid (2)	Axon-to-minute dendrite; macroscopically axon-to-axon and terminal	As above, but with clustered vesicles on presynaptic side	Osmiophilic membranes and postsynaptic invaginations to accomodate presynaptic processes
Neuropile synapses in insect central nervous system (3)	Axon arborization en passant to dendrite arborization	Apposition of the two unit-membranes, identified by vesicles on one side and increased local density of membrane	Sometimes increased osmiophilia in the synaptic region
Calyciform endings in ciliary ganglion of chick (4)	Enlarged axon terminating around major part of cell body	Two unit-membranes apposed over a relatively enormous area. Vesicles presynaptic	Increased osmiophilia in regions of apposition
Synapses in cerebral cortex of cat (5)	Axon to soma and axon to dendrite; at least 2 types present	Dense region along postsynaptic (type 1) or both (type 2) sides of synaptic membranes, with dark material lying within the synaptic cleft. Vesicles presynaptic	Cleft material a palisade of short filaments. Postsynaptic density partly due to a tenuous subsynaptic web or reticulum
Rod and cone synapses to bipolar retinal ganglion cells of rabbit (6)	Axon to dendrite	Postsynaptic membrane interdigitates deeply into presynaptic cell. Vesicles presynaptic	Specialized finger of postsynaptic membrane projects into a presynaptic cell invagination
As above, guinea pig (7)		Two distinct types of synapses from different rods	Specialized presynaptic ribbon and vacuoles

(1) Horridge, Chapman and MacKay, 1962.
(2) De Lorenzo, 1959, 1960a; Hama (personal communication); Robertson, 1961.
(3) Hess, 1958.
(4) De Lorenzo, 1960b, Martin and Pilar (unpublished).
(5) Gray, 1959; Robertis, 1962.
(6) Robertis, 1958, 1959; Robertis and Franchi, 1956.
(7) Sjöstrand, 1958.

II. SPECIAL ASPECTS B. Synapses

Synaptic Vesicles			Synaptic Cleft (Å)	Transmission	Physiological Features (See further, table 4.1)
Diameter (Å)	Vesicle Wall (Å)	Contents			
500–1000	70–80	Type A, faintly osmiophilic; type B, with a dense central spot	180–220	Short delay; possibly electrical	Probably two-way transmission
150–200 regular in size	Not given	Faint	About 100	Electrical	Two-way transmission. Septa have low electrical resistance
150–250	Not given	Faint	About 100	Electrical	Two-way transmission. Septum is a low-resistance pathway
400–600	Not given	Faint	100 (Hama) 75 (De Lorenzo)	Electrical	One-way transmission. Junction has rectifying properties
500–700	Not given	Faint and variable	About 200	Chemical	No summation; one-to-one transmission. Minimum delay 0.5–1.0 msec
Three types: 300–500	Not given	Small granules	About 200–250	Little information; believed chemical	Some known to be inhibitory; some thought to be cholinergic
1000–1500	Not given	Vesicles proper; dense border, light interior			
1200–2500	Not given	Dense, large, homogeneous granules			
350–450	Not given	Evidently filled	Uniform 300–400	Chemical, cholinergic and electrical, capacitative	One-to-one transmission, two parallel mechanisms
200–600	Thinner than axon membrane	Two-types: pale cholinergic and dark, central spot, adrenergic	200–300; dark material, apparently parallel filaments, occurs in the cleft	Little information; probably diverse	Structure suggests much spatial summation
200–650, mainly 350–400	40–50	Some with contents; diverse within a single presynaptic cell	About 300	Little information; not cholinergic	Some inhibitory, others excitatory in fish retina (Wagner et al., 1960)
300–400	About 70	Evidently filled	250–300		

(Table continues on following page)

(*Continuation of Table 2.1.*) The main features of synapses in a variety of tissues.

Preparation and Reference	Class of Synapse	Structure of Synapse	Associated Specializations
Inhibitory axon to muscle receptor organ sensory cell of crayfish (8)	Axon to soma and axon-dendrite	Apposition of two plain membranes which are especially osmiophilic. Vesicles presynaptic	Lattice of tubules in adjacent glial cytoplasm
Bouton-terminaux of mammalian spinal neuron (9)	Axon arborization to soma and to dendrites	1 μ diameter swellings of axon endings abut on soma membrane. Vesicles presynaptic	No subsynaptic modifications visible
Mammalian motor end plate of intercostal muscle of mouse (10)	Axon arborization terminating on muscle fiber	Axon arborization terminal crammed with vesicles and almost surrounded by folded muscle membrane	Subsynaptic foldings (subneural apparatus of Couteaux, 1955) 800 Å thick and 0.5–0.8 μ long in ordered array
Neuromuscular endings of insect leg muscle (11)	Axon in a groove of the muscle. Axon makes repeated lateral contact with the muscle membrane, en passant	Apposition of plain unit membranes, with increased osmiophilia, all capped by sheath cells. Vesicles presynaptic	No folds of muscle membrane. Postsynaptic aggregates of aposynaptic granules and endoplasmic reticulum of muscle
Neuromuscular endings in indirect flight muscle of insect and tymbal muscle of cicada (11)	Axon in a groove of the muscle	Apposition of plain unit membranes, capped by sheath cells. Vesicles presynaptic	Ramifying layered membranes and aposynaptic granules 50–150 Å in diameter on postsynaptic side. No folds of muscle membrane
Neuromuscular inhibitory synapses of muscle receptor organ of crayfish (8)	Axon to muscle	Apposition of two intensely osmiophilic plain membranes	Many irregular subsynaptic folds continuous with the endoplasmic reticulum of the muscle fibers
Neuroelectroplaque of a variety of electric fish (12)	Axon to modified muscle (in some species modified gland)	Axon has a great variety of relations to the electroplaque cell in different fish	Membrane of the uninnervated surface is continuous with a net of tubules, possibly to increase the surface area

(8) Peterson and Pepe, 1961.
(9) Estable, Reissig and Robertis, 1953; Robertis, 1955b; Palay, 1958.
(10) Couteaux, 1958, 1960; Reger, 1957; Robertson, 1956.
(11) Edwards, Ruska and Harven, 1958a, 1958b.
(12) Luft, 1958; Mathewson, Wachtel and Grundfest, 1961.

(c) Axo-dendritic connections by interdigitation or gears (Fig. 2.19). The rosettes formed by mossy fiber terminal fingers and granule cell dendritic fingers lie in pale islands of the cerebellar granular layer and compel the conclusion that this relation is special and not accidental. The same can be said of the relations in Ammon's horn between granule cell axons and the dendrites of large pyramids. But the most beautiful and diversified examples of gearing or meshing of fine teeth are found in the optic ganglia of insects. Cajal, who himself analyzed this system in great detail, said, "It seems that nature has attempted to show us in the insect nervous system . . . how in minimum space it is possible to organize a maximum of fine and subtle structures . . ." (1954, p. 81). Actually, in consideration of the multifarious forms of endings, both in arthropods and in cephalopods, not to speak of other invertebrates, it is not really justified to force

Synaptic Vesicles			Synaptic Cleft (Å)	Transmission	Physiological Features (See further, table 4.1)
Diameter (Å)	Vesicle Wall (Å)	Contents			
450–480	Not given	Most with uniform osmiophilic contents	About 200	Little information; not cholinergic	Inhibitory; blocked by picrotoxin
200–300	40–50	Faintly osmiophilic	120 (internal measurement)	Chemical	Probably excitatory. Much spatial summation
200–500	40–50	Faintly osmiophilic	About 500, over-all width	Chemical, cholinergic	One-to-one transmission of impulses
About 250 (4000 per μ^2 of synaptic membrane)	Less dense than axon membrane	Lightly osmiophilic	About 120	Presumed chemical	Graded postsynaptic responses; much temporal summation
250–350	Less dense than axon membrane	Evidently filled	140–170	Presumed chemical	Specialized and peculiar; see Chapter 17
About 500	Not given	Most with uniform osmiophilic contents	Not given but evidently very narrow	Presumed chemical	No structural correlate of the variety of inhibitory mechanisms is available
150–250 in *Electrophorus*	Not given	Mostly partially filled	450–500 in *Electrophorus*	Chemical, cholinergic	One surface passive, one surface electrogenic as a general rule. Great variety of responses available

into one category the full range of brushes, tassels, tufts, tap roots, shrubs, clubs, panicles, and other excrescences with which these neurons bristle. Recognition should be given, for example, to the feature of specific, localized terminal arborizations arising at definite sites along the single stem process of these unipolar cells, the form of branching characteristic for each site. Details of many of these cell types are given in the systematic chapters.

(d) Axo-dendritic connections by right-angled arrays, with axons of great length. This curious and provocative synapse is best known in the cerebellum, where smooth, free endings of granule cell axons run for enormous distances, as unbranched, unmyelinated terminal filaments in the molecular layer, parallel to the surface, while Purkinje cell dendrites richly ramify in a more or less perfectly flattened espalier at right angles to them. Thus each granule cell axon makes passing

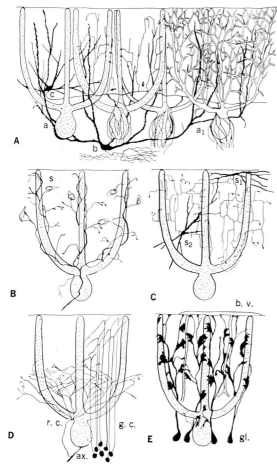

Figure 2.18. Variety of synapses on one neuron. Diagrams of the milieu of Purkinje cell dendrites in the cerebellar cortex; the dendrites greatly simplified, even in the upper right. **A.** Basket cell (c) axon sends descending collaterals to make baskets (a_1) around Purkinje cell somata (a) and ascending collaterals (arrows) to make synapses in the molecular layer. Large Golgi type II ganglion cell (b) of the granular layer sends dendrites widely in the molecular layer. **B.** Climbing fiber is presynaptic not only to the Purkinje dendrites but to many adjacent elements, especially stellate cells (s). **C.** Axonal plexuses of stellate cells (s_1, s_2), which are short-axoned and therefore Golgi type II cells, lie within the dendrite arbor of one Purkinje cell. **D.** Recurrent collaterals (r. c.) from Purkinje axon (ax.) form a plexus in the lower third of the molecular layer. Axons of granule cells (g. c.) ascend and bifurcate to run as parallel fibers at right angles to the plane of the flattened Purkinje dendrite arbor. **E.** Stalks of neuroglia cells (gl.) make extensive contacts with the dendrites and with cortical blood vessels (b. v.). All elements are simultaneously present about each Purkinje dendrite system and these systems extensively overlap. [Scheibel and Scheibel, 1958.]

contact of a minimum sort with thousands of Purkinje cells, and each of the latter is touched by innumerable parallel granule cell axons.

(e) Axo-dendritic connections by **plane, parallel plexuses.** Especially in the retina of vertebrates, cephalopods, and insects there are encountered terminal axon arbors disposed in a narrow plane or stratum interlacing with dendrites likewise confined. Laminar ramifications may be repeated at up to 8 levels and extend to definitely characteristic distances in each plane (Fig. 2.15).

(f) Complex connections embracing **circumscribed cell clusters.** Cajal pointed out the common denominator of several constant forms of junction in the thalamus and cerebral cortex in which the axon ends by exploding into a thicket of fine twigs that tangle up a discrete constellation of postsynaptic cells and their dendrites (Fig. 2.20).

(g) Axo-somatic connections by thick nests. The best example of this class is the cerebellar basket containing a Purkinje cell (Fig. 2.18). The stellate cells of the molecular layer send descending axon collaterals to form more or less densely woven nests or baskets around each of many Purkinje cell somata, including the axon hillock, varying in complexity according to the species of vertebrate (Estable, 1923).

(h) Axo-somatic connections by **sparse nests.** Cajal refers for an example of this distinct variety to the internal granular layer of the avian retina where centrifugal axons make nests poor in branches around cells called associational amacrines. Not at all like the preceding class, these might better be likened to early entwinement by growing vines.

(i) Axo-somatic connections by **calyces of Held.** In the nucleus of the trapezoid body of the mammalian medulla, numerous cells receive axon endings of a peculiar kind, virtually engulfing the soma like a flower with broad petals (Fig. 2.20). Although the functional significance is quite unknown, these objects have received a

II. SPECIAL ASPECTS   B. Synapses

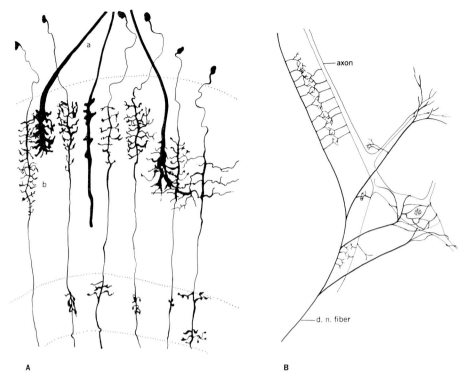

Figure 2.19.   Meshing synapses. **A.** Terminations by interdigitation (meshing or gearing) in the deep retina of the bee: *a*, axonal termination of visual receptor cells; *b*, collateral arborizations of a second-order neuron, presumably dendritic, that is receptive for that neuron. [Cajal, 1954.] **B.** Diagram of the relation of a fiber of the dorsal nerve to a large cell of the cardiac ganglion of a crab: *axon*, the main process of the large cardiac ganglion cell; *d. n. fiber*, dorsal nerve fiber ending in the cardiac ganglion. [Alexandrowicz, 1932; see Arthropod Bibliography.]

good deal of attention and discussion in connection with the eloquent evidence they contribute of the discreteness and independence of neurons.

(j) Axo-somatic connections by **thickened terminal tubercles.** This term covers a heterogeneous assortment of specialized endings with larger or smaller expansions in contact with the soma and larger dendrites. Space permits only brief mention of a few examples. One is the bipolar cell to the ganglion cell junction in the carp retina, especially the case of the giant bipolar for the rods. The axon widens and terminates in a warty tubercle on the soma of a single ganglion cell. Incidentally, this is one of the numerous neurons that lack neurofibrils, as demonstrated by the classical reduced silver methods—a fact overlooked in many recent electron microscope studies. Cells of the tangential nucleus of the medulla in lower vertebrates receive bizarre collaterals of the vestibular nerve that swell and apply themselves as a single suction-cup-like body to the soma. The most familiar and most discussed of all forms of synaptic terminations are doubtless the end feet of Held or "boutons terminaux," abundant on many cells, especially the motor cells of the ventral horn of the spinal cord and medulla (Fig. 2.17). It has been computed that there are some 50,000 of these knobs on the soma and dendrites of one neuron, nearly covering the entire surface. End feet are formed late in development and chiefly upon large neurons. They may take the form of terminal rings with a pale (argentophobe) center; the ring is confirmed by electron microscopy and is a swirl of filaments (Boycott et al., 1960). Repeated claims of neurofibrillar continuity between pre- and postsynaptic elements have as often and more convincingly been denied (Cajal, 1954).

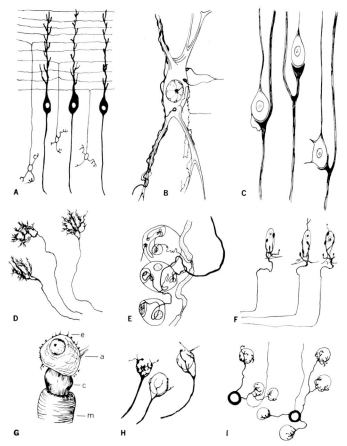

Figure 2.20.
Types of synaptic ending based on light microscopy and special stains; vertebrate central and peripheral nervous systems. **A.** Parallel fibers of cerebellar cortex making right-angle synapses with Purkinje cell dendrites, which lie in the plane vertical to the page. Golgi method. [Cajal, 1954.] **B.** Diagrammatic cell in Clarke's column of the spinal cord with its three types of synapses. (i) The "giant synapses" are seen on the large dendrites at the left, above and below, as entwining fibers from muscle spindle afferents. (ii) End feet on the right come from excitatory interneurons under the influence of skin afferents. (iii) The meshwork of extremely fine fibers on the soma comes from inhibitory interneurons in the pathway from antagonistic muscle afferents. [Szentágothai, 1961.] **C.** Cells of the tangential nucleus of a bird (young kite) receiving synapses from fibers of the vestibular nerve. Cajal method. [Cajal, 1954.] **D.** Terminal ramifications of afferent fibers in the lateral nucleus of the thalamus of a mouse. Golgi method. [Cajal, 1954.] **E.** Endings on intraparietal neurons of the auricle of the heart of a fish. Note the thick unmyelinated fiber terminating in special formations, and the myelinated fiber giving off an unmyelinated collateral from the node of Ranvier. Method of Gros. [Laurent, 1957.] **F.** Nests formed by centrifugal fibers that reach the retina and surround cells, called associational amacrines. Methylene blue. [Cajal, 1954.] **G.** A cell in the reticular formation of the goldfish which—besides small end bulbs *(e)*—receives a large club ending *(c)* from a myelinated fiber *(m)*; *a*, axon of postsynaptic neuron. [Bodian, 1942.] **H.** Chalices of Held in the nucleus of the trapezoid body of a kitten. Golgi method. [Cajal, 1954.] **I.** Possible somatodendritic synapse between granule cells of the cerebellar cortex of a macaque. Dendrite claws make intimate contact with one or more nearby somata and each soma receives claws from several other granule cells, thus affording both convergence and divergence. All the cells shown are granule cells. Golgi method. [Courtesy A. B. Scheibel.]

(k) Axo-axonal connections by simple contact en passant. This type is not common in higher animals. Cajal affirmed that he never saw axon-to-axon synapses. He was familiar with the work of Retzius on crustaceans and annelids which seem to offer such contacts but some scepticism is justified in these groups where dendrites so freely sprout from the side of axons. Bodian (1952) presents as an example of an axo-axonic synapse an intimate contact between a collateral of Mauthner's axon in the goldfish with a nearby myelinated fiber; the contact occurs within the sheath of the latter. Apparently good cases are the bipolar nerve nets in coelenterates, where dendrites have not been found but only simple symmetrical synaptic contacts between axons (Fig. 8.9). A similar relation is described between decussating giant axons in the polychaete *Protula* (Fig. 14.42). The most primitive synapses known are of this class.

(l) Axo-axonal connection by simple end-to-end contact. The giant fibers of earthworms and the lateral giants of crayfish are actually chains

II. SPECIAL ASPECTS B. Synapses

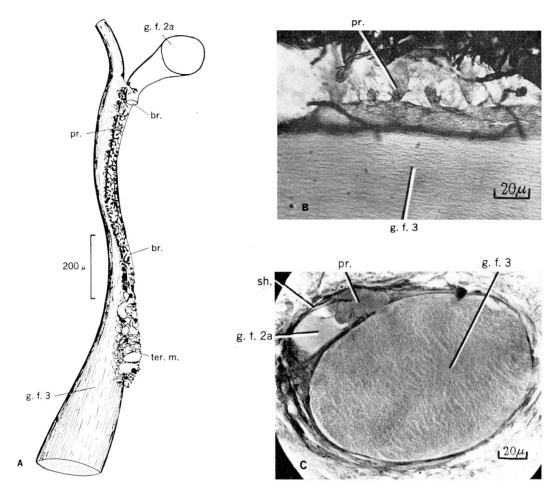

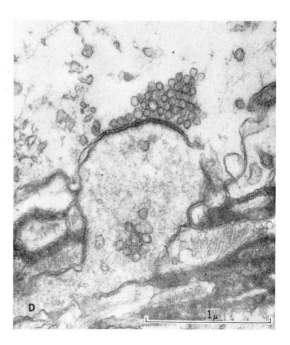

Figure 2.21.

The squid giant synapse between second- and third-order giant fibers. **A.** Drawing made by projection from several sections, showing processes (pr.). The third-order fiber (left, g. f. 3) in the stellate ganglion of Loligo protruding toward and into and wrapping around the end of the second-order fiber (g. f. 2a) to form the distal giant synapse; br., branch of second-order giant fiber; ter. m., terminal mass of processes of third-order giant beyond the end of the second-order fiber. **B.** Synaptic processes (pr.) of the third-order giant seen in a longitudinal section of the fiber. Cajal silver method. **C.** Synaptic process (pr.) in transverse section of g. f. 3. A shrinkage space has developed between the surfaces of synaptic contact; sh. sheath of g. f. 2a. Picric acid, azan stain. [**A**, **B**, **C** from Young, 1939.] **D.** Electron micrograph of a synaptic process, projecting up from below, into the second-order axoplasm above. The synaptic cleft is about 300 Å wide and the membranes are denser in this region; a dense material fills the cleft. Vesicles accumulate on the presynaptic side close to the junction. The glial sheath is seen on either side and below. [Hama, 1961b.]

of segmental units, each bounded by a complete cell membrane. The units join at a simple, symmetrical apposition of membranes, enormous in area, since it extends the entire width of the fiber, and, moreover, is oblique. Further details are given on pp. 701 and 906 (Fig. 14.26).

(m) Pseudo-axo-axonal connections with short dendrites. The synapse between second- and third-order giant fibers of squid (Fig. 2.21) is often called axo-axonal, as are those between central giants and motor axons in the abdominal cord of crayfish. There is no reason for this, since even in the light microscope it was clearly shown that the postsynaptic fiber bristles with specialized projections just in the area of synaptic relation, making intimate contact with the pre-fiber (details below). These are tantamount to dendrites. The synapses upon large unipolar neurons in the visceral ganglion of the gastropod *Aplysia*, now much studied by physiologists, are sometimes referred to as axo-axonal but without reason. The anatomy is not well known in *Aplysia*, but every reason exists to imagine that its neuropile is essentially like other higher invertebrate ganglia, a profuse tangle of dendritic branches and axonal ramifications.

(n) Somato-dendritic connections. The granule cells of the cerebellum send out clawlike dendrites that are said to hold nearby granule cell somata in a tight grip. Whether this relation is dendro-somatic in functional direction or somato-dendritic cannot be decided on the microsopic evidence (Fig. 2.20) and the relation is denied by other authors.

(o) Neuroeffector junctions by free-branched endings. Most neuromuscular junctions, such as those of crustaceans, insects, molluscs, and annelids, as well as most neuroglandular junctions and, so far as known, others on ciliated cells, luminescent organs, and chromatophores, appear to be simple terminals of moderately to extensively branched axons (Figs. 2.22, 2.23). There are probably differentiated kinds of endings on the muscles of orthopterans and *Dytiscus* (Hámori, 1961b).

(p) Neuromuscular junctions with simple end plates. In contrast to the next class, there are end plates in muscle with multiterminal and polyneural innervation and without subsynaptic folds. This is best known in coleopterans (Hámori, 1961b) where—except for *Dytiscus*, which resembles orthopterans in belonging to the preceding class—there are well-circumscribed end plates, often receiving more than one nerve fiber. A cholinesterase type of enzyme is localized in the presynaptic terminals (Hámori, 1961a).

(q) Neuromuscular junctions with **postsynaptic folds.** The junctions on vertebrate skeletal muscle of the twitch type are unique in displaying an expanded terminal apparatus of the axon, not ramifying widely, but spreading more or less like a hand and involving a considerable area of the muscle surface; this is correlated with the usual occurrence of only one such ending or end plate on a given muscle fiber from a given axon, in contrast to many in the just preceding category. More remarkable is the specialization of the postsynaptic membrane; the muscle membrane is thrown into deep folds into which the presynaptic terminals do not follow, so that some space and intercellular material is left (Couteaux, 1958; Policard and Baud, 1958; Katz, 1961). Further details are given in the next section on electron microscopy.

This list is certainly not exhaustive. Some additional special cases are given in the next section on electron microscope evidence. Physiological evidence suggests some kind of relation between the ends of two axons just before they impinge on a third cell in cases of so-called presynaptic inhibition and facilitation (see p. 268). Soma-to-soma connections are not known microscopically but are functionally indicated (see p. 193). Dendro-dendritic contacts are probably commonplace (Loos, 1960) but await recognition and more direct physiological argument implicating them; to the writers it seems probable that they are real and important. Sensory dendrites of various configurations are treated in the sections on receptors in the systematic chapters.

II. SPECIAL ASPECTS B. Synapses

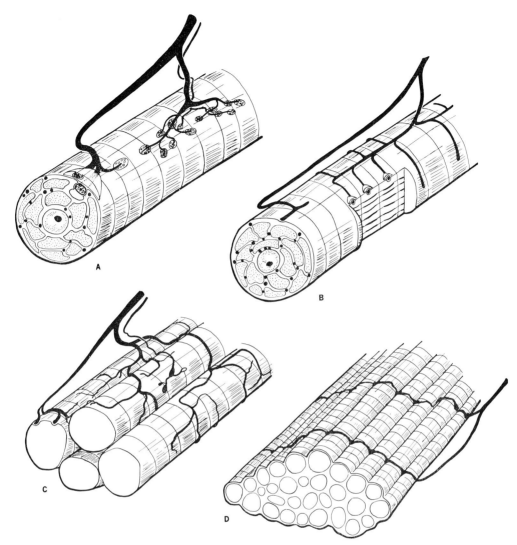

Figure 2.22. Specialized insect neuromuscular endings. [Courtesy J. Szentágothai and J. Hámori.] **A.** The endplate type; general in leg muscles of coleopterans and hymenopterans (also in the proboscis of dipterans, according to Meyer, 1955a); note fibril columns, mitochondria, and glial cell covering nerve ending with vesicles. **B.** The metameric type with circular end branches approximately between every two Z-bands; found in *Dytiscus* (Coleoptera) leg muscle. **C.** The diffuse type; general in Orthoptera; note fast and slow nerve fibers. **D.** The simple transverse type with rarely ramifying circularly disposed end branches; general in flight muscles of coleopterans and hymenopterans. Based on metallic impregnations and electron micrographs.

2. Fine structure revealed by the electron microscope

Generalizations are tentative concerning ultrastructure and its relation to electrophysiological properties, since these are known in only a few types of synapse. Certain salient points stand out when the details discovered by electron microscopy are reviewed. Sheath material or neuroglial processes that separate pre- and postsynaptic elements up to the region of the synapse always finally disappear at the actual contacts; the junctional area in the best-studied cases is a **window in the glial envelope.** However, lessstudied invertebrate neuropile seems to have considerable regions where synapses are not windows because of a local scarcity of glial processes between the fine nervous terminals. It is an out-

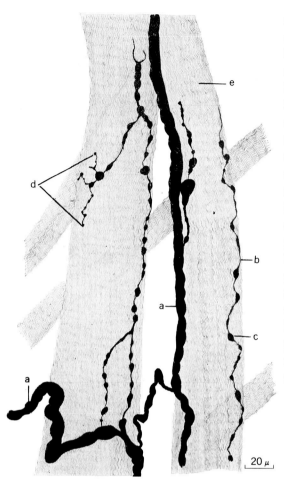

Figure 2.23. Free nerve endings on muscle. Midgut of the leech *Hirudo*, showing innervation of the muscle fibers. Bielschowsky-Ábrahám silver method. [Ábrahám, 1958; see Annelid Bibliography.] *a*, nerve trunk; *b*, nerve fiber; *c*, varicosity on a nerve fiber; *d*, nerve ending; *e*, muscle fiber.

assumption that it contains a solid structure of oriented molecules and that it is not entirely a fluid-filled space. De Robertis (1962) resolves the electron-dense cleft material in certain mammalian brain synapses into a palisade of lines, 300 Å long, normal to the cell surface, as though many short closely spaced filaments connect the pre- and postsynaptic membranes. In the same synapses he also sees the cytoplasm of the postsynaptic cell just under the synapse as a subsynaptic web or reticulum (Fig. 2.24). Robertson (1961) suggests that some synapses, especially the electrically transmitting crayfish junctions, may possibly have no cleft; instead, the unit membranes of the pre- and postsynaptic cells may possibly fuse in some degree, at least here and there. Since each unit membrane is believed to consist of two outer dark lines with a light line between them, probably representing oriented protein and lipid monolayers respectively, fusion means that instead of six distinct lines, five or four or three might be expected, depending on

standing question in the understanding of neuropile whether every contact of naked nerve fibers is a functional contact or whether visible structures always accompany functional contact. The area of contact in window synapses is usually in the range 0.1–10 μ^2 but giant synapses of much larger size are known (for example earthworm giant septa). There is always a definite discontinuity in the form of a membrane; usually this consists of the two unit membranes of the two nerve cells separated by a space of uniform width.

The contents of this intervening **synaptic cleft** are unknown. It is probable that the constant width of this cleft can only be explained by the

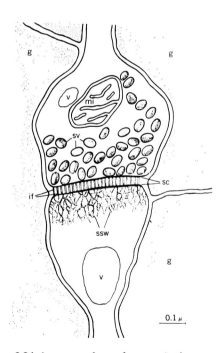

Figure 2.24. A common form of synapse in the mammalian brain, according to de Robertis (1962). The axonal (presynaptic) side above; the dendritic (postsynaptic) side below. *g*, glia; *if*, intersynaptic filaments; *mi*, mitochondria; *sc*, synaptic cleft; *ssw*, subsynaptic web; *sv*, synaptic vesicles; *v*, vesiculate body.

whether the outer protein layers are shared or locally interrupted and, in the latter event, whether the two central lipid layers are still separate or fuse into one. The electron microscope evidence does not yet clearly establish any of these degrees of fusion but certain images and local appearances suggested them to Robertson. If verified, these would be extremely basic classes of junctions.

In contrast to the indications of fusion are the better-studied objects believed to be synapses in mammalian gray matter and in some invertebrate neuropile. At those regions of the nerve cell membrane believed to be synaptic because of a glial window, intimate contact, and synaptic vesicles, there is a characteristic **increase in the density** to an electron beam of the unit membranes on both pre- and postsynaptic sides or only on the postsynaptic side (de Robertis and Bennett, 1955a; de Robertis, 1959; Gray, 1959a). The membrane at the synapse is not usually folded. On one or both sides of the synaptic membranes the cytoplasm commonly has a concentration of synaptic vesicles. Invagination of dendritic processes of the postsynaptic cell into the cytoplasm of the presynaptic cell is known at several specialized synapses, for example in crayfish giant fiber-to-motoneuron synapses, in squid giant fiber synapses, in "spine" synapses of vertebrate dendrites, and, to a remarkable degree, where the visual cells of the vertebrate retina impinge on the second-order cells. Specialized regions of both pre- and postsynaptic membrane are peculiar to some of the examples below. Mitochondria or other cytoplasmic inclusions may consistently appear but are not usually especially abundant at the synapse.

Most of the physiologically known synapses which have been examined under the electron microscope have turned out to have a large number of rounded **synaptic vesicles,** 200–1000 Å in diameter and packed in the terminal cytoplasm of at least the presynaptic axon (Fig. 2.24). So frequent has this feature proved to be that the presence of vesicles is commonly used as an indication of the presence of a synapse. The vesicles are small, rounded bodies, bounded by a membrane, usually 200–500 Å in diameter in vertebrates and up to 1000 Å in some invertebrates (Table 2.1). Some vesicles of some cells have a spot of dense contents but most show no internal structure. The vesicles are thought to contain transmitter substance stored before release. Central synaptic vesicles of vertebrates are characteristically 100–200 Å in diameter and up to 600 Å in lower vertebrate neuromuscular junctions (Katz, 1961). Those of many invertebrates are similar, for example 160 Å in earthworm giant fibers (Hama, 1959), 500 Å in crustacean neuromuscular junction (Peterson and Pepe, 1961), and up to 1000 Å in coelenterates (Horridge, Chapman, and MacKay, 1962).

Evidence that synaptic vesicles are characteristic of synapses is of four kinds. (a) They are, when present, consistently concentrated at synapses and scattered or rare elsewhere within the neuron; no other correlation with their presence or distribution has been offered. (b) Their number in the synaptic region is said to change when the nerve is stimulated. (c) Their presence suggests a basis for the quantal release of transmitter substance in packets as inferred from miniature potentials observed at synapses in both vertebrates and invertebrates (pp. 231, 974). (d) The vesicles can be observed to open into the synaptic cleft as if releasing their contents; this has been claimed in rod synapses on stimulation of the receptors with light (de Robertis, 1959).

But there are difficulties in accepting the notion that vesicles are always and no more than means of transmission. (a) Synapses which transmit electrically nevertheless have vesicles. (b) Vesicles of the same type can be found in glial cells (Schultz et al., 1957). (c) The junctions between specialized nonnervous receptor cells in vertebrates and their innervating sensory terminations are diverse; some seem to have vesicles on the postsynaptic side only. For example, in mammalian taste buds extremely fine postsynaptic nerve fibers, down to 500 Å in diameter, are crammed with vesicles from the point where they leave their sheath and synapse with the apical regions of gustatory cells (De Lorenzo, 1958). In synapses between hair cells and the vestibular nerve and between hair cells and the acoustic nerve, there are in many but not all nerve endings

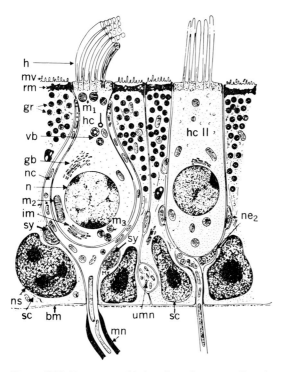

Figure 2.25. Two types of hair cells and nerve endings in the vestibular sensory epithelium; diagram from electron micrographs. [Engström and Wersäll.] *bm*, basilar membrane; *gb*, Golgi complex; *gr*, granules in a supporting cell; *h*, hairs; *hc I*, hair cell of type I; *hc II*, hair cell of type II; *im*, intracellular granulated membranes; m_1, m_2, m_3, m_4, mitochondria; *mn*, myelinated nerve; *mv*, microvilli; *n*, nucleus; *nc*, nerve chalice; ne_2, granulated nerve ending at the base of a hair cell of type II; *ns*, nucleus of a supporting cell; *rm*, reticular membrane; *sc*, supporting cell; *sy*, granulated synapse; *umn*, unmyelinated nerve; *vb*, vesiculated bodies.

similar collections of vesicles (Fig. 2.25); these are only in the nerve terminals (references in De Lorenzo), which are usually thought to be postsynaptic afferent endings. They could, however, be efferent fiber terminals modulating the receptor (Desmedt and Monaco, 1961), although this would imply that efferents far outnumber afferents in the cochlea (Engström and Wersäll, 1958). Vesicles are lined up on the presynaptic side of the presumptive specialized receptor cell in Meissner's corpuscles; in Merkel's corpuscles there are presynaptic granules, not vesicles. Neither vesicles nor granules have been found in receptors with free nerve endings (Cauna, 1962).

Synaptic vesicles in the terminal region of axons can multiply in the region of the synapse, according to observations of de Robertis (1959). Prolonged stimulation of the splanchnic nerve to the adrenal medulla at a high rate (400 per second), known to cause fatigue of the ending and diminished postsynaptic response (output of catechol), causes considerable depletion of vesicles, whereas at 100 per second, known to induce near-maximal postsynaptic response, an increase in the vesicle count was found. Confirmatory observations are required. Synapses fixed during excitation show vesicles fusing with the synaptic membrane and confluent with the synaptic cleft. It is not known whether this is a process of formation of vesicles from the membrane or of release of vesicle contents as transmitter substance. In comment, it should be mentioned that the numbers of nerve impulses used to cause these effects have been very large, and one effect, the increased density of synaptic vesicles, could possibly be interpreted in other ways, as by a loss of fluid from the terminations. Also, other microscopists find variation within control sections equivalent to these reported effects of stimulation; the sampling problem involved calls for further quantitative work. A technique for mass isolation of synaptic endings from brain homogenates gives material which suggests that the vesicles indeed contain transmitter substance (de Robertis et al., 1961; de Robertis, 1962). It is a familiarity that hormones and many other active substances occur in the appropriate cells in an inactivated form and can be activated or "liberated," as if from numerous small vesicles (Birks and McIntosh, 1957, on acetylcholine; Welsh, 1961, on crustacean hormones).

The **origin of synaptic vesicles** has not been established with certainty. Vesicles of 200–500 Å found in the cell body of frog spinal ganglion neurons are similar to those at the corresponding neuromuscular junction. Axonal swellings on the proximal side of ligatures accumulate similar vesicles. Treatment with malononitrile increases the activity of the Golgi complex, which then produces enlarged vesicles in the cell body, found later in the axon. From these observations, Van Breemen et al. (1958) concluded that the Golgi complex originates vesicles, which then pass down the axon. But studies by de Robertis and his co-workers suggest that vesicles might increase in the terminals during prolonged stimulation at a high rate, and that they do originate there. Similar discrepancies as to site of origin occur in consideration of neurosecretory granules (p. 364).

Specialized synaptic structures are common but not universal. Their small size has made them inaccessible to physiological analysis; their varied and peculiar structure helps little in suggesting a significance for their presence. Presynaptically, single large or structurally unique mitochondria are common. The terminations of rod fibers in the eye of the rat have a single mitochondrion, 2 μ long; in the cerebral cortex in cats, mitochondria up to 9 μ long have been seen near junctional regions (Pappas and Purpura, 1961). In the same terminal of the rods there is a synaptic ribbon or lamella, which is a dense curved bar straddling the invaginating bifid dendrite of the bipolar cell and therefore arching over the synaptic union within the swollen rod termination (Ladman, 1958; Sjöstrand, 1958, 1961). Postsynaptic specialization is found on cortical dendrites. The long-known dendritic spines have now been shown to contain dense ribs and bands surrounded by vesicles (Gray, 1959). The spines themselves, 2 μ long and terminating in a synapse, are a specialization of unknown function. Pappas and Purpura (1961) find granular and multivesicular bodies in the dendrite close to synapses. The largest postsynaptic structures are those of the neuromuscular junction of vertebrate striated muscle, where the muscle cell membrane is thrown into subterminal folds, and at a deeper level in the muscle there are complex folds of the endoplasmic reticulum between the synapse and the first rows of muscle filaments (see further below).

It is not to be expected that the several types of junctions recognized above on light microscope evidence would have a one-to-one correspondence with any **classification of synapses** on electron microscope evidence, since the former are based on topography over tens or hundreds of micra and perhaps quite diverse configurations may have the same ultrastructure. The following varieties are certainly incomplete, based as they are on a small sample and a new field of study.

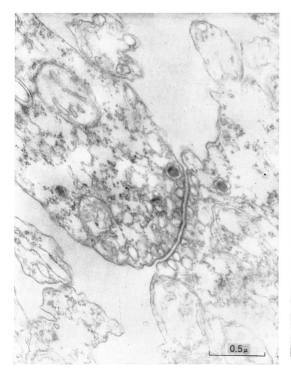
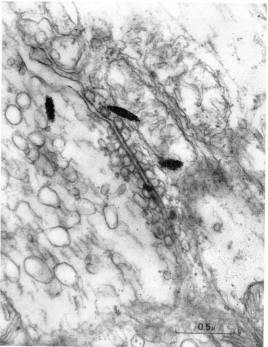

Figure 2.26. Electron micrographs of synapses in a coelenterate. Marginal ganglion of the medusa *Cyanea*. [Horridge, original.]

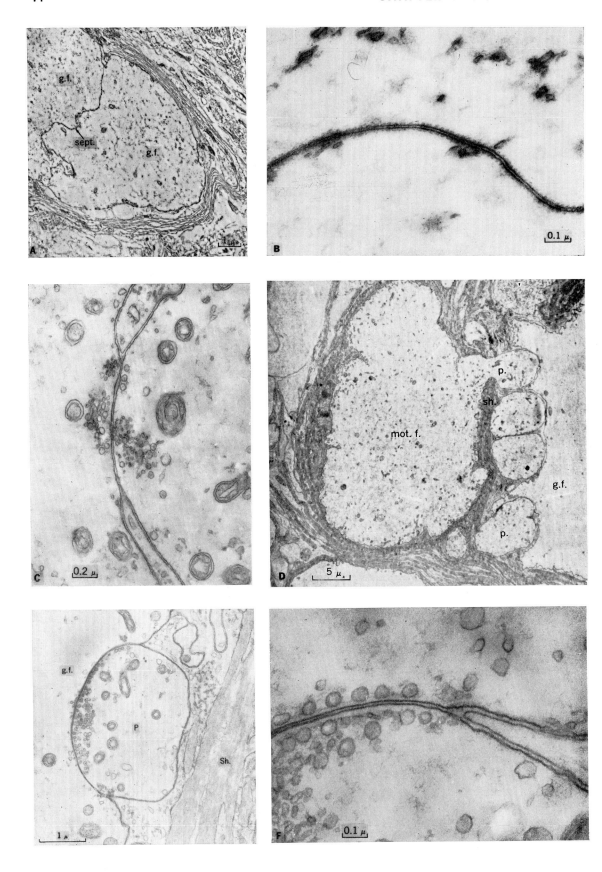

II. SPECIAL ASPECTS B. Synapses

(a) Simple apposition without special structures. A variety of axon-axon synapses which have been shown to have unpolarized electrical transmission are known in detail. The lateral giant fibers in the earthworm and crayfish are divided into segmental units which abut or dovetail into each other along the length of the animal. These segmental septa consist of the axon membranes of the two large axons in close apposition, forming a symmetrical synapse with or without synaptic vesicles on both sides. In the earthworm the two membranes run parallel for many micra, with symmetrically placed vesicles and a cleft with a regular width of about 100 Å. In the crayfish the synaptic areas are only small windows (about 5 μ wide) in a thicker septum (100 μ wide), which includes a sheath of several lamellae. Coelenterate axons have been seen in electron micrographs to make intimate contact by simple apposition of two membranes (Fig. 2.26), with vesicles on both sides and a synaptic cleft of 200 Å, suggesting that here again transmission may be electrical (Horridge et al., 1962).

(b) Pseudo-axo-axonal synapses with short dendrites, few vesicles, and narrow cleft (Fig. 2.27). This category was indistinguishable from the next in the light microscope (Fig. 2.6 C). It is of special interest because, like the preceding category, transmission has been shown to be electrical; in this case, however, transmission is polarized. The example that has been studied both physiologically and ultrastructurally is the crayfish synapse between central giant and segmental motor axon in the abdominal cord. The large postsynaptic axon, supplying flexor muscles, sends processes equivalent to dendrites into indentations of the presynaptic fiber; the processes have the form of ridges and penetrate the windows in the glial sheath otherwise separating the fibers. The membranes are in close contact, with a cleft of a constant 100 Å width, the narrowest known. Robertson (1961) even suggests the possibility that here and there the unit membranes fuse to some degree, as discussed above (p. 70). Vesicles of large size, 400–600 Å in diameter, occur in some views but not in others; they may be on either or both sides and in any event do not form a considerable cluster as they do in the next type.

(c) Pseudo-axo-axonal synapses with short dendrites, clustered vesicles, and wider cleft. This variety is represented by the squid giant synapse, between the second- and third-order giant fibers in the stellate ganglion. This synapse is well known physiologically and is presumably chemically transmitting because electrical transmission has been ruled out. From the postsynaptic axon, processes deserving the name dendrites extend through the membranes of the sheath of the presynaptic fiber. The synaptic membranes are separated by a 200 Å cleft containing a slightly electron-dense material. Synaptic vesicles of 500–700 Å are concentrated on the presynaptic side (Fig. 2.21).

(d) Axo-dendritic connections between fine fibers, with clusters of vesicles. In the neuropile of the jellyfish ganglion, earthworm cord, and

Figure 2.27. Electron microscopy of (electrically transmitting) synapses in earthworm and crayfish giant fiber systems. **A.** Earthworm septal synapse *(sept.)* between successive segmental giant fibers *(g. f.)*. Vesicles do not show at this low magnification. **B.** The same at higher magnification. Note the close apposition of membranes, the single row of uniform (150–200 Å) vesicles, lined up at regular intervals (a feature not seen in every picture), and the symmetrical structure of the synapse. **C.** Crayfish septal synapse. Note glial processes separating axonal membranes except at the synaptic window in the center, where the cleft between them is about 100 Å wide. Vesicles and tubular structures are clumped on both sides. **D.** Low-power view of the junction between lateral giant axon *(g. f.)*, which is presynaptic, and motor axon *(mot. f.)*, which is postsynaptic. Processes of the motor fiber *(p)*, of which four large ones are sectioned, protrude through gaps in the sheath *(sh.)* and make intimate contact with the presynaptic axon. **E.** The same at higher magnification. Note the close apposition of membranes; vesicles are seen in some pictures on both sides, in others more on one side or the other. **F.** The same at still higher magnification to show the membranes, cleft, and vesicles. [Hama, 1959, 1961a.]

insect ganglia numerous localized areas are presumed to be synapses solely on the grounds of contact without intervening glia and of clusters of vesicles on one side, thought to be presynaptic. These are the commonest synapses so far found in neuropile of the invertebrates. Such structures in the earthworm cord were one of the first places where synaptic vesicles were seen, but the argument that the locus concerned is a junction is extremely weak, since many other points of apposition of membranes occur in neuropile, where glial processes are often sparse. Sometimes one other sign of specialization is noted—an increase in the electron density of the membrane of one or both sides, under the vesicle cluster. Trujillo-Cenóz (1959) distinguished three main types of contact in the neuropile of an insect: (i) cross contacts, which may be of minimum area or of large area; (ii) longitudinal contacts; and (iii) end knob contacts, which may touch one or several postsynaptic fibers. Only the end knob contacts are notable for containing vesicles.

(c) Axo-somatic synapses. As represented by spinal motoneurons (Palay and Palade, 1955) and cells of the cerebral cortex (Gray, 1959a), these are relatively simpler structures than the next category. There is less thickening or density increase of the unit membranes and no material in the cleft. The classical boutons terminaux belong here and prove to be swollen by reason of one or more terminal mitochondria as well as the synaptic vesicles (see Fig. 2.29). The problem of deciding when in fact one is looking at a real synapse is very prominent here, as it was in coelenterates and in class **(d)** above. Today there are a number of cases where good reason exists to argue that the structure concerned is a synapse, and that the pre- and postsynaptic sides are correctly identified (Fig. 2.28); this required in some cases patient tracing back to the parent axon and down to an unquestionable dendrite.

The relations are quite similar in the nerve cell bodies of Auerbach's plexus of the mammalian autonomic system (Taxi, 1958) and also in the crayfish stretch receptor nerve cell, where fibers identified with the physiologically known in-

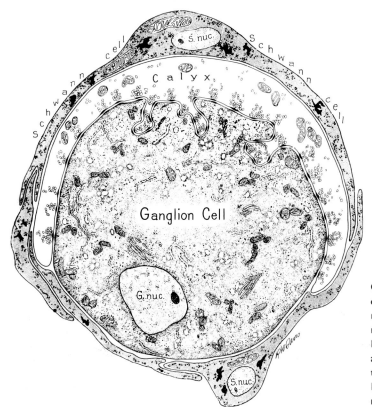

Figure 2.28.

Calyciform synapse in the ciliary ganglion of the chick, as revealed by the electron microscope. Note the locally dense regions of the opposed synaptic membranes, the clusters of synaptic vesicles at these sites on the presynaptic side, the uniform cleft width (300–400 Å). [De Lorenzo, 1960b.] *G. nuc.*, ganglion cell nucleus; *S. nuc.*, Schwann cell nucleus.

II. SPECIAL ASPECTS B. Synapses

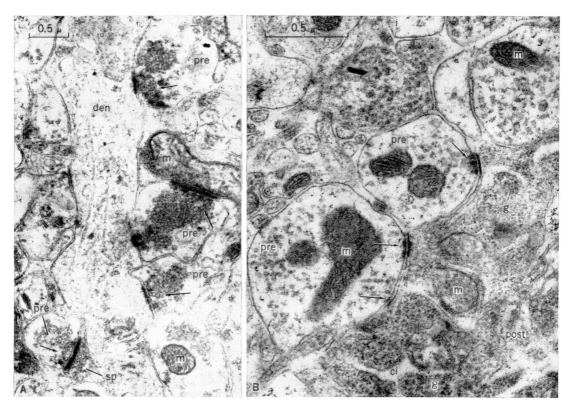

Figure 2.29. Two types of synapse in the cerebral cortex, as seen by electron microscopy. Visual cortex of the rat. **A.** A group of synaptic contacts of type 1 *(arrows)* on a small dendrite *(den.)*. This type exhibits increased thickness and density of the apposed membranes, especially of the postsynaptic side, over a large percentage of the length of close contact; the thickened regions lie farther apart than unthickened apposed surfaces; an intermediate band of material can be seen at high magnification in the cleft. **B.** Type 2 synapses *(arrow)* on a nerve cell body. The cytoplasm of the postsynaptic cell body *(post)* contains characteristic granules *(g)* and cisternae *(ci)* of the endoplasmic reticulum. This type of synapse has a small percentage of the length of apposed membranes thickened, both surfaces are equally thickened, the spacing is little different from adjacent regions, and an intermediate band is not clearly visible. *m*, mitochondria; *pre*, presynaptic ending with cluster of vesicles; *sp*, spine of the small dendrite with a spine synapse. [Gray, 1959.]

hibitor arborize over the sensory cell body and dendrites (Peterson and Pepe, 1961).

(f) Axo-dendritic synapses of the mammalian cortex. Appreciably more complex are certain structures frequently seen in electron micrographs of the cortex and other higher gray masses (Gray, 1959a; de Robertis, 1959, 1962). The common features are the window in the glial envelope, apposition of membranes with a cleft of 200–300 Å, locally increased electron density of membrane, cleft material of appreciable electron density, resolvable according to de Robertis into closely spaced parallel lines, a cluster of synaptic vesicles in the presynaptic cytoplasm, and a tenuous subsynaptic web of osmiophilic material in the postsynaptic cytoplasm. This miniature complex not only goes together in sections of gray matter in many parts of the mammalian brain but can be isolated by centrifugation. The cell membranes break just before and after the junction and this complex stays together, permitting the collection and concentration of quantities of junctions. Slightly different treatment permits isolation and concentration of a component that seems to be at least mainly synaptic vesicles. Chemical characterization of these isolates is already being accomplished and reveals two distinguishable types of vesicles which may be called cholinergic and adrenergic (de Robertis et al., 1961; de Robertis, 1962), perhaps corresponding in their material to light- and dense-center varieties. This and other evidence points to the possibility of microscopic identification of cholinergic and adrenergic endings.

(g) Axo-dendritic synapses on dendrite spines. Similar to the preceding are little complexes upon the spines of dendrites (Fig. 2.29). The spines have long been known, appearing in silver preparations as thorns or spurs decorating the entire surface of many dendrites at close intervals. In electron microscopic preparations they project into depressions on the axonal terminal and can display small ribs or bands as well as a cluster of vesicles (Gray, 1959a). Pappas and Purpura (1961) also find special bodies in the dendritic cytoplasm, often associated intimately with the synaptic complex, the bodies being multivesicular, membrane-bound inclusions, 0.2–0.6 μ in diameter.

(h) Invaginated, ribbon synapses in the vertebrate retina (Fig. 2.17). The efferent end of vertebrate retinal rods is invaginated to receive the dendrite of the bipolar cell (Sjöstrand, 1958, 1961). They also occur in the inner plexiform layer in the vertebrate retina (Kidd, 1961), but they are so far not known from invertebrates. A presynaptic ribbon, 0.1–0.2 μ long and 200–400 Å wide, surrounded by vesicles, lies in the presynaptic ending. Detailed accounts of these extraordinarily specialized synapses are available (Sjöstrand, 1958, 1961). Similar but smaller presynaptic ribbons occur in hair cells of the guinea pig cochlea. Each external hair cell makes synapses with several dendritic processes of two types, but only one type has presynaptic ribbons. Evidently one cell can synthesize different presynaptic structures according to the nature of the postsynaptic fiber. Desmedt and Monaco (1961) reopen the suggestion (Engström and Wersäll, 1958) that cochlear nerve endings with vesicles may be efferents modulating the receptors.

(i) Serial synapses. This name refers to certain tandem relations in which fiber A is presynaptic to fiber B, which in turn is presynaptic to fiber C; they have been found rather sparsely in the cerebrum, in the cord, and in the retina (Kidd, 1961). It is probable that the intermediate fibers are of amacrine cells, which have long been considered exceptional in having no apparent axon. However, it is likely that relationships of this type are common in neuropile of invertebrates. Amacrine cells occur in insect and cephalopod optic lobes. Presynaptic inhibition may employ serial synapses.

(j) Neuromuscular junctions with postsynaptic grooves. The endings on striated muscle in vertebrates are unique in the development of deep grooves in the muscle cell membrane under the axon. The nerve terminals lose their Schwann cell sheath and dip down into a close-fitting indentation or axonal channel of the muscle cell (Fig. 2.30). From the walls and floor of this channel there are invaginated a series of evenly spaced (approximately 0.3 μ), deep (0.6 μ) grooves, narrow at the mouth (500 Å) and wider at the inner, blind end (1000 Å). The wide cleft (500 Å) has a dark band in the middle and two clear bands, one on each side. The dark band follows the muscle cell membrane into the grooves but the axonal membrane does not. Vesicles are abundant and confined to the presynaptic side. A special configuration of endoplasmic reticulum lies in the muscle cytoplasm under the endplate.

(k) Neuromuscular junctions without grooves (Fig. 2.31). Here the true endings of nerve fibers are difficult to find; they have been described in insect leg muscle by Edwards, Ruska, and de Harven (1958a). The nerve fiber rests in an axonal channel roofed by the sheath cell and tracheoblast; the sheath cell is missing where the axon faces the muscle cell. Apposition of the membranes of axon and muscle cells is featured by local increase in osmiophilia of both sides and a narrow cleft of 120 Å. The axon here contains clusters of synaptic vesicles and numerous mitochondria but no neurofilaments. The vesicles are estimated to number 4000 per square micron of synaptic surface. On the postsynaptic side are

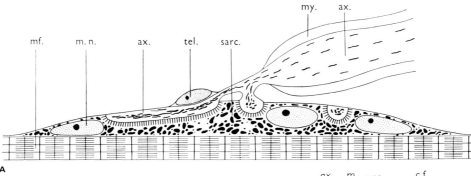

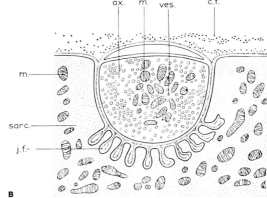

Figure 2.30. The vertebrate, grooved neuromuscular junction as revealed by the electron microscope. **A.** Schematic drawing of a motor end plate at low magnification. **B.** Diagram of a cross section of one synaptic gutter and axon branch. *ax.*, axoplasm; *c. f.*, collagen fibrils; *j. f.*, junctional fold; *m.* mitochondria; *mf.*, myofibrils; *m. n.*, muscle nuclei; *my.*, myelin sheath; *sarc.*, sarcoplasm; *tel.*, teloglia (terminal Schwann cell) nucleus; *ves.*, vesicles. [Couteaux, 1958, modified from Robertson.]

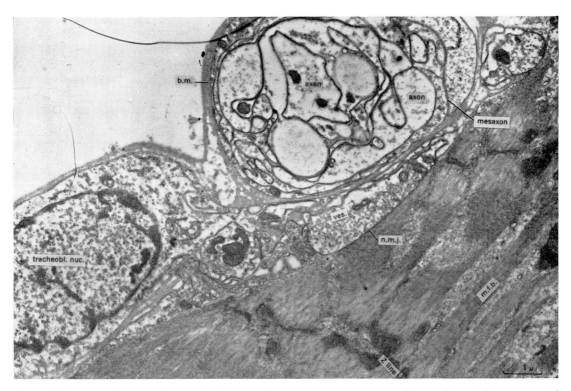

Figure 2.31. Electron micrograph of one sort of insect nerve muscle junction. Ventral abdominal intersegmental muscle of *Blatta*. [Courtesy G. A. Edwards.] *b. m.*, basement membrane; *m. fib.*, myofibril; *n. m. j.*, nerve muscle junction with synaptic vesicles *(ves.)* on the presynaptic side; *tracheobl. nuc.*, nucleus of tracheoblast cell.

abundant large mitochondria, extremely ramified, widened tubules of endoplasmic reticulum and local accumulations of "aposynaptic granules," osmiophilic, 50–150 Å in diameter. By these numerous signs one can tell when a section showing axon lying on muscle is actually through the synaptic region. Nevertheless the structure is simpler than the vertebrate striated muscle junction in the absence of grooves and of cleft material.

For the neurophysiologist it is important to note that one dendrite may bear morphologically diverse synapses; one presynaptic cell may produce morphologically different synapses to other fibers; one process can be both pre- and postsynaptic. In Chapter 4 on Transmission, it is noted that one presynaptic neuron can exert direct inhibitory effects and direct excitatory effects on different postsynaptic cells. We are far from understanding the functional requirements and morphogenetic processes which underlie these differences.

The old controversy of **continuity versus contiguity** is still alive, though largely ignored. It raged about 1890 and appeared to settle down about the turn of the century, but it flared up again about 1909 when Cajal and Golgi, on opposite sides of the uneven fight, shared the Nobel Prize; the former was given the award for his evidence of the independence of the neuron, the latter for his method—which was, ironically, Cajal's main tool. Reticularism, the doctrine that the nerve fibers of the central nervous system form a protoplasmically continuous net, lost adherents, and the neuron doctrine (see p. 1604) became generally accepted. However, followers of Gerlach, Apáthy, and Bethe—for example Held, Boeke, and Stöhr—have continued to publish reports of neurofibrillar continuity between nerve cells in the brain and visceral nervous system of vertebrates and invertebrates (see Hild, Reiser, and Lehmann, 1959). These reports are anything but unified in point of view, each antineuronist adhering to a personal doctrine, and they are contradicted by a mass of convincing observations, beginning with those in the last century of His, Forel, Kölliker, Retzius, van Gehuchten and, above all, the incredibly productive Cajal.

This last author, whose work has stood the test of time, provided a history of the controversy and summary of the evidence for the independent unity of the neuron in his last book, *Neuron Theory or Reticular Theory*, written in the early thirties and translated into English in 1954, which should be consulted for references (see also Cajal, 1909–1911; Rasmussen, 1947). A host of recent investigators, using the light microscope, has richly confirmed this view, among them Bozler (1927); de Castro (1930, 1932, 1942, 1950); Lorente de Nó (1934); Young (1939); Olszewski and Baxter (1954); Zawarzin (1924); Bucy (1944); Rose (1949); Bodian (1952); Chang (1952); Couteaux (1955, 1958); Nauta and Kuypers (1958); Scheibel and Scheibel (1958a). Today it is necessary to admit that the neuron doctrine is stronger than ever, all the more so because of the powerful new evidence from electrophysiology and electron microscopy—Palay and Palade (1955); Palay (1956a, 1956b, 1958a, 1958b); Policard and Baud (1958); de Robertis (1959); Robertson (1961).

Electron microscopy has amply shown the intactness of the nerve cell membrane at the most intimate points of contact, while electrophysiological studies show by activity, time delays, and passive electrical resistance the discontinuity between the most intimate pre- and postsynaptic elements, with special exceptions (see Chapter 4, Transmission). Nevertheless there are still some experienced observers, as there have been since the time of Bethe, Held, Apáthy, Dogiel, and Golgi, who continue to describe reticula, nets of neurofibrils, a "ground-plexus" and the like (Tiegs, 1927; Boeke, 1938, 1949; Stöhr, 1957; Hild, Reiser, and Lehmann, 1959). Since the neuron theory cannot deny that exceptionally anastomosis occurs and indeed some are well accepted (earthworm and squid giant fibers, pp. 702, 1483; interstitial neurons in mammalian viscera, Larentjew, 1934). Each of these reports must be examined on its merits; no one is fatal to the neuron theory or ruled impossible by acceptance of that theory. There is in practice no conclusive way of proving that all the described structures which others claim to be artifacts have no basis in reality, since the electron microscope can prove only the presence of inhomogeneity of electron density, not the absence of structure. The present trend is to trust that the electron microscope is revealing all that is significant, to discount all the neurofibrils of the light microscopist (see further, next section), and to ignore the controversy in order to concentrate effort on aspects where structure and function can be correlated.

A few examples are well documented of **electrotonic connections between neurons,** suggesting either anastomosis or low-resistance synaptic membranes. Lobster cardiac ganglion cells, crayfish and earthworm lateral giant fiber commissural connections, supramedullary ganglion cells of puffer fish, and large medial motoneurons in leech ventral ganglia, all manifest a specific connection favoring slow potentials and more or less attenuating spike potentials. Nothing is known as yet of the anatomical basis of these properties. Anatomical anastomosis is accepted as normal (but without confirmation by electron microscopy) in some coelenterates—for example *Hydra* (Spangenberg and Ham, 1960) and *Porpita* (Mackie, 1959)—and in Nematoda (see Chapter 11). Nerve cells with two nuclei have been described, mainly from vertebrate sensory cells (summarized on p. 214 of Scharf, 1958). Well-defined protoplasmic channels between smooth muscle fibers (Thaemert, 1959) can be invoked as possible routes mediating coordination of electrical activity of the membranes, but it must be admitted that membrane regions of low resistance would serve equally well. The great controversy as to whether neurons are always single cells has been replaced by a welter of complex new information and has been abandoned as a product of the limitations of technique of the light microscope.

C. Nerve Cell Inclusions

We do not as yet know how physiological activity in the membrane around the cell and physiological activity in cytoplasm and inclusions are coupled in any instance. But some workers believe that the contents of nerve cells will be found to be important in controlling behavior of the animal in other ways than as neurosecretions or as merely vegetative support for the cell membrane.

The inclusions of nerve cells, particularly in the invertebrates, have been rich material for discovery. In addition, some reported inclusions have initiated a good deal of denial and subsequent controversy. A great many objects have been described from light microscope studies by a variety of techniques in a great range of material (Fig. 2.32), and there have been many efforts to apply a small number of names to objects which are difficult to define in detail. Golgi bodies, Nissl substance, neurofibrils, dictyosomes, lipochondria, archoplasm, are all names which have been applied over and over again to a variety of objects which, until the past decade, were examined by histological and optical techniques inadequate for the task. The subject had almost reached a point of diminishing returns when the advent of the electron microscope made possible a detailed description of the submicroscopic morphology. There appeared a new crop of names, such as endoplasmic reticulum, γ cytomembranes, agranular reticulum, neurofilaments, and neurotubules, some of which can be referred to the older names.

Any discussion of cytoplasmic inclusions must be preceded by notes of caution. First, the striking resemblances between neurons in different phyla (Fig. 2.33) have led to the supposition that inclusions found in one group of animals will appear in a similar form by use of the same technique elsewhere. This is not necessarily true. Furthermore it does not necessarily follow that one terminology will serve ubiquitously. Second, some of the common techniques of fixation and embedding, for classical as well as electron microscopy, remove much of the organic matter from the neurons. Even by the best techniques the nerve cell cytoplasm can lose 50% of its organic material and the nucleus 80%; most of the lipid is removed by the common methods (Hydén, 1960). Third, many of the special classical techniques for demonstration of nerve cells and their inclusions involve shrinkage and treatment with strong reagents which are likely to introduce artifacts. Fixation in formalin or Carnoy's solution may cause shrinkage of up to 60–80%;

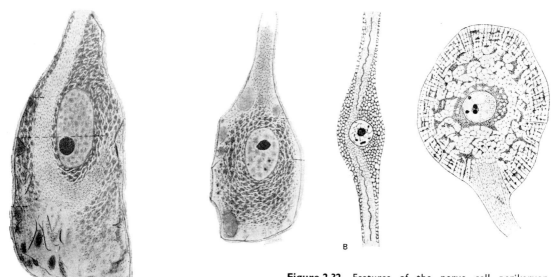

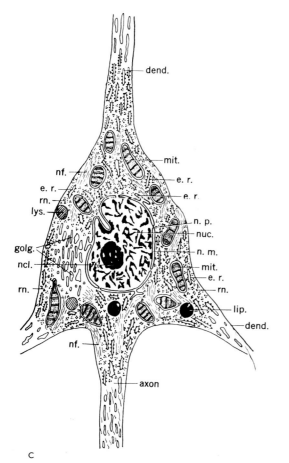

Figure 2.32. Features of the nerve cell perikaryon. **A.** *Astacus*, ganglion cells showing the intrasomatic origin of the axoplasm, portions of trophospongium, Nissl bodies, mitochondria, presumptive neurofibrillae, and part of the glial sheath. The marked difference in composition and orientation of inclusions in the cytoplasm of the axon and of the perikaryon is clear. [Ross, 1922.] **B.** *Ascaris*, ganglion cells showing the same contrast and some of the characteristic features that permit individual identification of every nerve cell in the animal. The unipolar cell with peripheral, concentrically oriented inclusions is cell 23; the bipolar with central fibril in each axon is cell 90, according to the designation of Goldschmidt [1910; see Nematode Bibliography.] **C.** Schematic representation of the main types of inclusions in the perikaryon of a pyramidal neuron of the mammalian cerebral cortex. [Hager, 1961.] *axon*, axon hillock; *dend.*, dendrite; *e. r.*, endoplasmic reticulum; *golg.*, Golgi zone; *lip.*, liposome; *lys.*, lysosome; *mit.*, mitochondria; *nf.*, neurofilaments; *ncl.*, nucleolus; *n. m.*, nuclear membrane; *n. p.*, nuclear pore; *nuc.*, nucleus; *rn.*, ribonucleoprotein granules.

some of the classical methods for demonstration of neurofibrils required treatment with alkali, which can now be inferred to have broken down the endoplasmic reticulum and the associated ribonucleic acid. This may or may not have caused a false view of the structures stained and called neurofibrils. Cytology as a field has not overlooked these dangers and pitfalls in its hundred years and more of intense activity and self criticism; individual workers frequently have, but the best, experienced workers, with an enormous battery of methods giving controls for

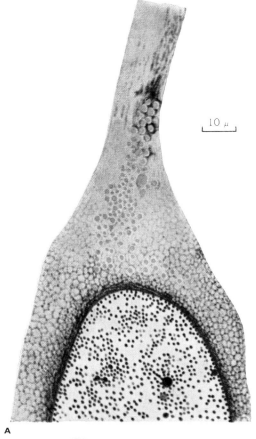
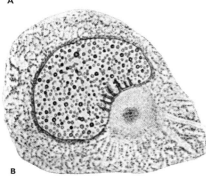
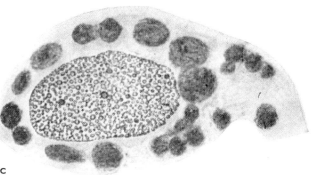

Figure 2.33. Light microscopy of the perikaryon of gastropods. **A.** *Helix pomatia*, large cell of the pedal ganglion. Note the arrangement of the Nissl substance *(blue)* and pigment *(yellow)* at the axonal pole and in the base of the axon. Sublimate fixation, toluidin blue. [Kunze, 1921; see Gastropod Bibliography.] **B.** *Limnaea stagnalis*, nerve cell in unstated central ganglion. Note the basophilic and acidophilic rings around the centrioles and the radial disposition of elements and canals of the trophospongium. Acid alcohol fixation, methylene blue-eosin. [Kolatchev, 1916; see Gastropod Bibliography.] **C.** *Limnaea stagnalis*, nerve cell of a central ganglion. Note the accumulations of glycogen stained by iodophenolxylol after Driessen. Absolute alcohol fixation, borate carmine. [Kolatchev, 1916.]

each other, including observation in vivo, have accumulated a reservoir of useful information on a great range of objects. The adequacy and validity of light microscope descriptions in general decline with the size of the inclusion but it should be remembered that the improvement in our image of some structure by electron microscopy or otherwise does not justify discarding earlier studies as artifacts; every image is a representation and an artifact—including the black and white image on a film, or indeed the flat colored photograph. The distorted and altered structures of earlier studies may still tell us of differences between types of cells or regions of the cell or notify us that something is there, as does a badly reproduced voice on the radio.

A resolution of all the difficulties is not possible but certain propositions seem generally acceptable. First, the future lies in the combination of electron microscope studies with a chemical approach such as the use of enzymes to remove identifiable components, and a functional approach such as observation before and after stimulation, osmotic stress, acclimation, and the like. Part of the success of such a combined approach depends on judicious correlation of electron microscope and light microscope findings; the former is often embarrassingly high-powered and alone cannot find or assure the identification of desired structures; it can overlook configurational changes of profound importance. The light microscope can tell us that there is an object of interest in the stretch receptor or in the difference between hibernating and active snail nerve cells; it can tell us also how to recognize the relevant region. Many of the inclusions of special interest in nerve cells are not beyond the limits of the light microscope, which can therefore assist in validating the electron microscope report by different methods or in vivo, at least as to presence, number, and location. Hagadorn and Nishioka (1961) can distinguish three types of neurons in a leech by light microscopy and tell which is properly neurosecretory, but they cannot see these distinctions in the electron microscope. However, the electron microscope must be given priority at least for positive findings because of its ability to resolve composites and to detect small objects. Confirmation of objects seen after osmic fixation is important—by different fixatives, freeze-drying or otherwise.

Since a general and comprehensive cytology of nervous tissue is inappropriate for the present work, there is treated below only a selection of the inclusions of special interest in nerve cells. A primary object is to call attention to material for new comparative studies. Emphasis is given to newer findings with the electron microscope; the classical light microscope cytology is not comprehensively reviewed.

The important field of **cytochemistry** is likewise neglected here. As part of the large field of neurochemistry, to which references are given in Chapter 1, it is too much to review even briefly in the present work. But it should be noted that techniques and findings are rapidly expanding, permitting the assay and the localization of enzymes, ribonucleic acid (RNA), and other chemically defined entities within the cell. Some of these have special interest in nerve cells, for example cholinesterase and the extraordinarily high content of RNA and rate of synthesis of protein. Besides the general texts and references already cited, the following may be consulted for examples in this area: Maxfield et al. (1951 and later); Koelle and collaborators (1957 and later); Nachmansohn (1959); Schmitt (1959); Hydén (1952, 1960); Kety and Elkes (1961). Invertebrate materials have been studied cytochemically by Libet (1947, 1948); Shafiq and Casselman (1954); Richards (1955); Smallman and Wolfe (1956); Chou (1957); Baker (1959); Malhotra (1960); Maynard and Maynard (1960a, 1960b); Ashhurst (1961); Pipa (1961a, 1961b).

Inclusions of neurons made visible and defined by the use of the electron microscope over the past decade are composed of membranes, particles, vesicles, special synaptic vesicles, lipoidal droplets surrounded by membranes, fibrils, or solid amorphous secreted products. They correspond only in part with objects defined over the years by use of the light microscope. The synaptic vesicles (p. 71) and neurosecretory products (p. 364) are treated elsewhere.

Pigment granules are characteristic of ganglion cells in certain places, animals, and ages, but information is still meager. Many large cells in gastropods display pigment, which apparently accumulates with age; the same is true in parts of the brain stem of mammals.

1. Membrane complexes and lipoidal droplets

These inclusions, occurring alone or associated with membranes, have caused much controversy. The cytomembranes or **endoplasmic reticulum** form systems of double unit membranes, 60 Å thick, which can take varied forms in different neurons and in different animals (fixed with osmic acid). Highly ordered regions sometimes occur, as in vertebrate motor cells, or there may be disordered clumps or scattered membranes. The neurotubules (p. 91) may belong to the same series of membranes. Palay and Palade (1955) distinguish two types of membrane, as follows.

The **granular reticulum** lies in the position where **Nissl bodies** can be seen in life in vertebrate cells and is organized into clumps of interconnected sheetlike configurations compatible with the Nissl bodies of light microscopy, characteristic of the neuron soma and large dendrites but absent in axoplasm (Fig. 2.34). These structures are a special development in nerve cells, especially in vertebrates, but it is not a unique material, for the basophilic components are found commonly elsewhere. The configurations are of unknown significance but are useful because they alter in characteristic, visible ways with pathologic conditions and simplify the recognition of nerve cell cytoplasm in the light microscope because of simple, distinctive staining reactions (so-called Nissl stains). The cytomembranes are rough on one side and lie in pairs, with granules on the sides which point away from each other. The granules are thought to represent RNA, which is known on several lines of evidence to lie in this region of the cell and which can be isolated in the form of granules known as microsomes. However, the visible granules may well be artifacts of fixation and the microsomes may be a product of the technique of breaking down the cytoplasm. A full review of the constitution of the Nissl bodies in vertebrates is given by Hydén (1960). Structures of similar form have been only occasionally described in invertebrate neurons by light microscopists; the RNA is not clumped in such a way that it can be observed as a mass or a network (Boyle, 1937, on *Helix;* Shafiq and Casselman, 1954, on *Locusta;* Wigglesworth, 1960, on *Periplaneta*).

The second type of membrane distinguished by Palay and Palade is associated with vesicles instead of granules. This type forms what is called **agranular reticulum** or γ cytomembranes of Sjöstrand and is distinguished by its appearance as groups of parallel membranes between which there are no granules. The membranes are closely folded, frequently looped at their ends, and sometimes with dilated segments or inflated folds as if vesicles were in process of formation. The appearance in different cells suggests that vesicles indeed grow at these membranes. The whole structure forms a bean-shaped, lamellated mass of a total size well within the resolution of the light microscope and probably corresponding to the **Golgi apparatus** of at least some classical cytologists. Structures of this form are known in many types of animal cells and in plant cells. They occur in many neurons of all animals studied from coelenterates to man. They are evidently secretory in nature, though perhaps only intracellularly, and have been implicated in the formation of synaptic vesicles (p. 72). Although probably having no relation at all with the structures originally described by Golgi (1898), they have come to be called Golgi apparatus or complex by many recent workers on all types of cells.

Whatever the structure in other cells, the electron microscope image of the Golgi region does not always seem to be a good representation of its form in life in neurons. In a critical study on crustacean neurons, Malhotra and Meek (1960) show that structures of this form appearing in electron micrographs can be identified with structures seen as concentric lamellae normally lying around certain spherical bodies which are seen in living cells and which color with neutral red. The spherical "blue" globules of *Helix* neurons also belong to this category of inclusions; they lose their spherical shape after osmic fixation and appear in electron micrographs as crescentic or bean-shaped laminated structures (Chou and Meek, 1958). In invertebrate neurons these structures are commonly called dictyosomes (see below). In biochemical terms, however, there will no doubt prove to be a large variety of bodies with parallel membranes and vesicles.

The material and its configurations treated in the last two and the following six paragraphs do not have any known special relevance or uniqueness in nerve cells. Although details of the chemistry or form are characteristic of certain nerve cells, we are dealing with widespread cellular constituents. The present chapter does not deal with the general cytology of nerve cells, but an exception seems in order here because of the historical prominence of these inclusions in neurocytology and their comparative interest.

The **Golgi complex** of vertebrates, as originally described from the Purkinje neurons of the owl, consists of anastomosing lumpy strands, extending even into the axon (Golgi, 1898). A careful repetition of Golgi's study on the same material

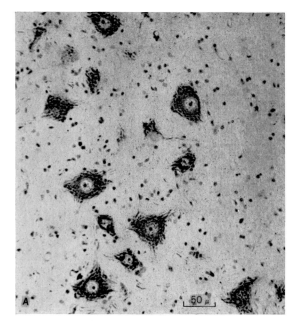
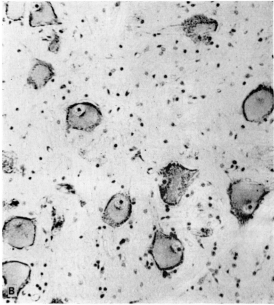

(Malhotra, 1960), confirmed by electron microscopy (Malhotra and Meek, 1960), shows that he described a structure which cannot be justifiably distinguished from folds in the granular endoplasmic reticulum (treated just above with Nissl bodies). The name no longer has its original connotation. Following a tentative (and as it turned out, faulty) identification of agranular reticulum with Golgi's original bodies, the term Golgi region has been used in electron microscope studies of nerve cells in a consistent manner accepted by many authors, referring to a region consisting of a closely folded agranular membrane system associated with vesicles, lying near the nucleus (Hild et al., 1959; Baker, 1959; Picken, 1960; Rosenbluth and Palay, 1961). This is a part of the agranular reticulum of Palay and Palade (1955) and the γ cytomembranes of Sjöstrand (1956); in neurons it is distinguished from Nissl substance by the apparent absence of RNA granules. All these terms refer to parts of the endoplasmic reticulum. The term Golgi region is also commonly used for similar specialized parts of the endoplasmic reticulum that are thought to be the focus of activity in many secretory cells, tissue culture cells, and plant cells and has become so entrenched that it is not likely to be dropped. The region, organelle, or group of organelles characteristically appear as in Fig. 2.35. Broad flattened channels are bounded by membranes 100 Å thick and 200–1000 Å apart, folded repeatedly, often in a characteristic curved or bean-shaped mass. Exhaustive histochemical tests have failed to reveal any special features of this organelle.

The situation with reference to the **Golgi regions and dictyosomes** in invertebrate neurons has been confused and their nature has been in dispute for fifty years. Certain characteristics have been accepted by different authors: (a) discrete filaments or rods around the nucleus; (b) rods or filaments, thought by some workers to originate from fixation artifacts involving mitochondria in particular; (c) short curved rods (batonets) with or without a more weakly staining substance (archoplasm); (d) vacuoles which stain with neutral red but no Golgi networks; (e) only spheroidal bodies and mitochondria; (f) an outer sheath and inner medulla of spheroids, which correspond to the dictyosomes (rods, batonets) and archoplasm of other workers. References will be found in the reviews by Baker (1959) and Lacy (1957). Much of the controversy has centered around interpretation of the inclusions of the large neurons of gastropod molluscs, using the terminology invented for vertebrates.

Under the electron microscope γ cytomembrane or agranular reticulum in invertebrates is

II. SPECIAL ASPECTS C. Nerve Cell Inclusions

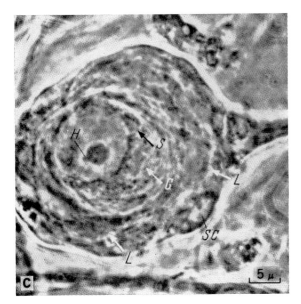
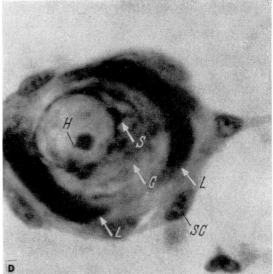
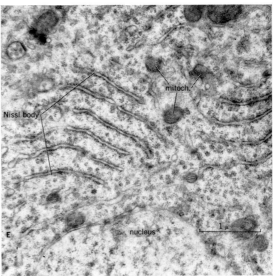

Figure 2.34. Nissl substance in light and electron microscopy. **A.** Normal ventral horn motoneurons of the rhesus monkey. Toluidin blue stain. **B.** Similar cells taken 6 days after section of the relevant ventral root, showing an advanced stage of the retrograde chromatolytic reaction of the perikaryon. Note the nearly complete dissolution of the Nissl bodies, the accumulation of basophilic material near the cell membrane and to a smaller degree near the nuclear membrane, and the eccentric position of the nucleus. [**A** and **B**, courtesy D. Bodian.] **C.** Phase contrast photomicrograph of a living spinal ganglion cell of a chick embryo, cultured in vitro 13 days. In the cytoplasm can be seen large, relatively homogeneous, phase-dense masses (L) and smaller, more granular masses (S and G). Satellite cells (SC) surround the perikaryon. **D.** The same cell after osmic acid fixation and cresyl violet staining (without sectioning). The L masses of figure **A** correspond to large Nissl bodies (L) of the stained preparation; some of the smaller masses (S) form a nuclear cap and the more granular masses (G) show less basophilia. [Deitch and Murray, 1956.] **E.** Electron micrograph of a portion of the perikaryon of a Purkinje neuron from the cerebellar cortex of the rat. [Courtesy S. L. Palay.]

recognizable in neurons and secretory cells in general in many groups, including coelenterates, annelids, crustaceans, and molluscs; it has turned out to be a compact system of membranes remarkably similar to that in the vertebrates. In neurons of *Patella* the chromophilic component of the Golgi apparatus consists of a system of paired membranes which usually enclose an inner, dense substance. The chromophobic component corresponds to a substance lying within dilated regions of this paired membrane. The archoplasm (or faintly staining associated clumps) is found to consist of numerous small vesicles in the surrounding cytoplasm (Lacy, 1957). Similarly, from the work of the Oxford school we can conclude that in neurons of the gastropod *Helix* and the prawn *Leander* the "Golgi apparatus" of light microscopy consists of small groups of multilamellar membranes exactly equivalent to agranular reticulum or γ cytomembranes; they are frequently called **dictyosomes**. However, careful comparison with living

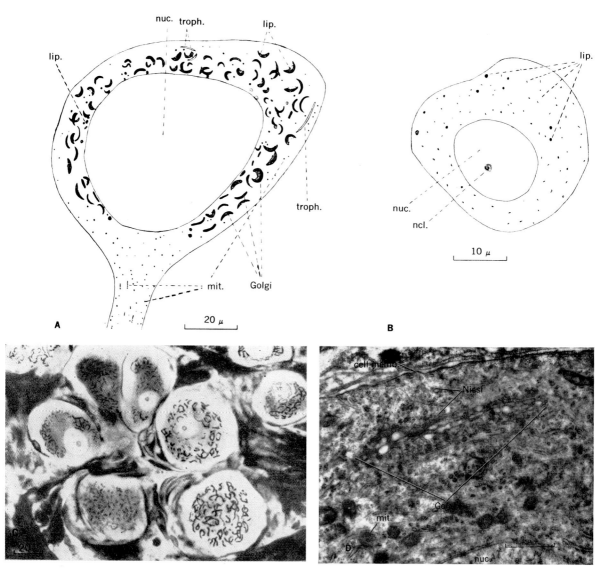

Figure 2.35. Golgi apparatus in light and electron microscopy. **A.** Large cell in *Helix*; Ludford preparation. Golgi elements do not extend into the axon; there mitochondria can be seen arranged in lines; portions of trophospongium are shown. **B.** Medium-sized cell in *Helix*; formalin fixation, Sudan IV stain. In addition to lipoid spheres, rod- and crescent-shaped lipoid bodies can be seen representing the "chromophobe" parts of the Golgi elements. [Boyle, 1937.] *Golgi*, Golgi elements; *lip.*, lipoid material; *mit.*, mitochondria; *ncl.*, nucleolus; *nuc.*, nucleus; *troph.*, trophospongium canals. **C.** Spinal ganglion cell of the hedgehog at a certain stage of chloroform narcosis. Kopsch-Kolatschev method. [Watzka, 1939.] **D.** Electron micrograph of a Purkinje cell of the cerebellum of the rat, showing the agranular reticulum identified as the basis of the Golgi apparatus. [Courtesy S. L. Palay.]

cells shows that in life these are not crescent-shaped "batonets" or rods but are a part of the spherical lipoidal bodies which stain with neutral red. Chou and Meek (for *Helix* neurons, 1958) and Malhotra and Meek (for *Leander*, 1960) show that osmic fixation and subsequent treatment converts the laminated shells of the spheres into structures which then pass as "Golgi apparatus" or dictyosomes. Ashhurst (1961) concludes similarly for dictyosomes of cockroach neurons. This may be true of other cells, but if they have been fractured the crescentic products should have broken ends, whereas they frequently have folded ends, suggesting that they are not broken shells.

Dictyosomes are likely to be highly varied in chemical properties and morphological detail because lipoidal inclusions of well-known neurons, synaptic vesicles, fibrillar structures, and primary neurosecretory products are all associated with them. Dictyosomes of invertebrate neurons are chemically different from the structures (probably Nissl bodies) which form the classical RNA-rich structures of vertebrates described by Golgi (Malhotra, 1960). In other types of cells, especially spermatocytes, there can be found similar small clear spheres surrounded by chromophilic crescents. Following the discovery that the dark crescent consists of membranes similar to the agranular reticulum of mammalian secretory cells and neurons, the conclusion has been commonly expressed that all such structures are equivalent. From this morphological similarity there has followed the conclusion that small bodies up to 3 μ in diameter, which fit the definition of dictyosomes by being composed of crescentic lamellae, are the morphological **equivalent of the Golgi region.** Chou (1957) takes this view with reference to one type of lipoidal inclusion in *Helix* neurons, as do Malhotra and Meek (1960) on crustacean neurons; Wigglesworth (1960b) takes the same view for similar inclusions in cockroach neurons and moreover suggests that they are the ends (or beginnings) of large osmiophobe filaments which fill the axon. In his review, Picken (1960) implies that dictyosomes as originally defined are identical with some of the lipochondria of the Oxford school and treats these and the γ cytomembranes or agranular reticulum as components of the Golgi region, though they are scattered separate units in many invertebrate neurons.

Synthesis in the neuron cell body and subsequent transport of products down the axon seem to be generally accepted for a number of structural components. In most cases the Golgi complex (γ cytomembranes, dictyosomes) is thought to be the initiating source. Inclusions traced back to this origin are (a) synaptic vesicles (Van Breemen et al., 1958), (b) wide osmiophobic fibrillar structures (Wigglesworth, 1960a), and (c) primary neurosecretory particles of several types in the leech (Bern et al., 1961). Cholinesterases also originate in the cell body and are transported down the axon (Lewis and Shute, 1961). On the other hand, there are suggestions that mitochondria are transformed into secretory products. Curious whirled membranes in pericardial organ axons of *Squilla* (Crustacea) can be formed into a series in which mitochondria transform into strings of secretory particles (Knowles, 1958, and unpublished); similar suggestions have been made for transformation of mitochondria in hypophyseal neurons. For the part played by RNA in synthesis in neurons, see the review by Hydén (1960).

Lipoidal globules or lipochondria of living neurons have been carefully compared with those seen in the electron microscope and again with those which can be colored by a variety of histochemical techniques, chiefly by workers of the Oxford school—Chou and Meek (1958), Baker (1959), Malhotra and Meek (1960). In the most thoroughly studied material, the neurons of the snail *Helix*, three kinds of lipid globules are visible in life. One kind, "yellow" because of contained carotenoid, is very electron-dense and is bounded by lamellae. (The "colorless" kind is not so bounded.) The "blue" globules, called so because they stain in life with methylene blue, are spheroids having in their outer layers concentric lamellae which have some resemblance to the γ cytomembranes or Golgi apparatus as described in *Patella* neurons by Lacy (1957), but the membranes are not usually more than 200 Å in thickness. In crustacean neurons, bodies which stain with neutral red are spherical in life; when fixed with osmic acid they appear in electron micrographs as "aggregates of nongranular membranes" which are no longer spherical and they now have the structure of dictyosomes or Golgi complex. Therefore study of rounded lipoid inclusions by the electron microscope alone is not a sufficient guarantee against description of artifacts (Malhotra and Meek, 1960). The third type of globules are colorless in life and arranged in rows like beads. They consist of dense solid masses, about 1 μ in diameter, with a simple bounding

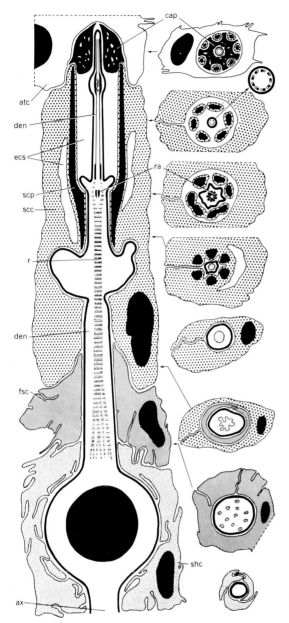

Figure 2.36. Mechanoreceptor neuron with modified cilium (inset, upper right). The hearing organ of the locust as revealed by electron microscopy in longitudinal and transverse sections at the levels indicated. [Gray, 1960.] *atc,* attachment cell; *ax,* axon; *cap,* probably an extracellular structure; *den,* dendrite; *ecs,* extracellular space; *fsc,* fibrous sheath cell; *r,* root; *ra,* root apparatus; *scc,* scolopale cell; *scp,* scolopale pillar; *shc,* sheath cell of the cell body.

membrane. There is no reason why all the great **variety of secreted particles** should be derivatives of γ cytomembranes. It is also apparent that there is no sharp distinction between these structures and those commonly classed as neurosecretory products.

A number of unclassified inclusions so far seen in the electron microscope are **derivatives of mitochondria.** Rosenbluth and Palay (1961) describe conspicuous homogeneous inclusions up to 1 μ in diameter. Curious membranes found in many invertebrate neurons are described by Knowles (1958, and unpublished) as mitochondria in process of transformation into neurosecretory vesicles, even in the terminations of neurosecretory axons of the crustacean pericardial organs. Green and Maxwell (1959) have made similar suggestions for transformation of mitochondria in hypophysial neurons. However, neurosecretory particles originate in γ cytomembranes or the Golgi region in a wide range of animals (Bern et al., 1961) and in plant cells similar membrane systems have been beautifully shown to produce secreted droplets (Mollenhauer et al., 1961).

2. Modified ciliary structures

A surprising finding of several recent studies of sense organs is that the dendrite of the sensory neuron frequently contains modified ciliary structures. These are now known in all kinds of receptors, particularly from arthropods. The mechanoreceptor neurons of the auditory organ of the locust have a complicated structure around and within the sensory dendrite (Gray, 1960); see also p. 1038 and Fig. 2.36. The nine pairs of ciliary fibrils lie within the terminal dendrite membrane; they join proximally at a ciliary base which is distinct from the basal body of motile cilia; proximal again to this is a cross-banded root structure running toward the nucleus. Mechanoreceptor sensilla of the multicellular proprioceptive organs of the leg joints in Crustacea also have a modified cilium in the terminal dendrite (see Fig. 18.6), and the sensilla placodea of the bee's antenna (*Apis*) contain about twenty sensory dendrites, each with a ciliary structure (Slifer, 1961). From this structure Slifer infers these organs are mechanoreceptors because many arthropod mechanoreceptors but none of the known chemoreceptors have some trace of a ciliary structure. Some receptors of light also show traces of being modified cilia. In the vertebrate eye a cilium pushes outward at the growing point of the embryonic rod cell and develops a row of vesicles along its side. Deep infoldings which form in the plasma membrane meet these vesicles, and by a progressive lateral widening they form the primitive rod sacs (Tokuyasu and Yamada, 1959). The cilium persists along one side of the rod. In the distal retina of the eye of *Pecten*, fibrous structures with the typical appearance of cilia in cross section grow from whirl-shaped basal bodies and form a brush border of elongated processes along the margin of the sensory cells (Miller, 1957). Eakin and Westfall (1959) conclude that the reptilian third (parietal) eye is evolved from a ciliary structure. Types of cilia are reviewed by Barnes (1961), who concludes that many sensory cilia (in particular, light receptors) lack the two central fibrils common to motile cilia. She also concludes that central neurosecretory cells of the vertebrate hypophysis have cilia of this peculiar type. No doubt other examples will appear.

Spiral whirls of dense lamellae have been reported intermittently from neurons examined under the electron microscope, particularly in visual cells. The outer retina of the eye of *Pecten* has just been mentioned. In grasshopper retinal sensory neurons it is assumed that they represent intermediate stages in the formation of the microvilli or tubules of the rhabdomere (Fernández-Morán, 1958). Similar whirls occur in neurons of the jellyfish ganglia (Horridge et al., 1962) and in the brain stem of the rat and sensory neurons of the goldfish (Rosenbluth and Palay, 1961). They are somewhat suggestive of the lamellated structure in the rhabdomere, and in many crayfish ganglion cells, but the significance of the whirled form is unknown.

3. Fibrillar structures

Fibrils in the nerve cell cytoplasm and in the axoplasm are of several clearly distinct types (Fig. 2.37).

(a) Neurofilaments or axoplasmic filaments are fine threads 60–100 Å thick, which are commonly found uniformly distributed throughout the cross-section and oriented roughly lengthwise in the axoplasm. The same fibrils can, but less commonly, occur in the cell body, as in the bipolar cells of the goldfish eighth cranial nerve (Rosenbluth and Palay, 1961). Neurofilaments are not universally present in all phyla, nor even in all axons of those in which they occur regularly, such as vertebrates and insects.

(b) Larger and distinctly two-layered structures, the "thick filaments" or **neurotubules,** or canaliculi with an outside diameter of about 200–300 Å, occur in the axoplasm of many animals, including coelenterates, annelids, arthropods, and vertebrates. They are denser structures than the neurofilaments and easily distinguished from them where the two occur together, as in cells of the goldfish eighth cranial nerve or the cat splenic nerve. In his review of electron microscope studies of vertebrate material, Elfvin (1961a) concludes that neurofilaments and neurotubules, as here defined, are found in many nerve cells if suitable staining techniques are employed. There is no doubt that the images represent real structures present in life. Pappas and Purpura (1961) report tubules as characteristic of dendrites in the superficial cerebral cortex of cats. Trujillo-Cenóz (1959) saw tubules in larger fibers of the neuropile of an insect (possibly dendrites) and sometimes they were not uniformly distributed in the fiber but were central. In degenerating axons in the rat, the neurotubules disintegrate within twenty-four hours and after them the neurofilaments disappear within forty-eight hours (Vial, 1958). The neurotubules are composed of membrane material similar to but not identical with that of endoplasmic reticulum. The functions are unknown. Fukuda and Koelle (1959) plot the first reappearance of acetylcholinesterase after it has been removed from the neuron by inhibitors, and interpret their results according to a working theory that this enzyme is synthesized within the endoplasmic reticulum (Nissl substance), then transported via canaliculi of the endoplasmic reticulum (tentatively identified here with neurotubules) to the surface of the cell and the ends of the fibers.

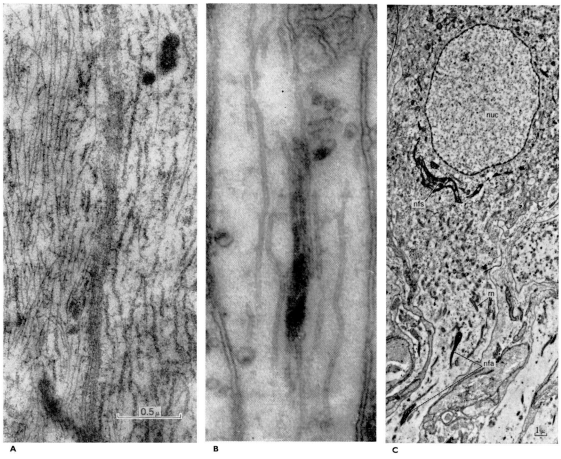

Figure 2.37. Filamentous structures in nerve fibers as seen in electron micrographs. **A.** Thick and thin neurofilaments in axons of the cat splenic nerve. Thin filaments preponderate in the left axon, the thick type in the right axon. PTA staining. [Elfvin, 1961a.] **B.** Neurotubules in axons of *Hirudo*. Some are seen as though agglutinating into coarser fibers. [Couteaux, 1956.] **C.** Heavy filamentous structures in a nerve cell from the brain of the leech *Theromyzon*, seen sparsely in both soma *(nfs)* and axon *(nfa)*; *m,* mitochondria; *nuc,* nucleus. [Courtesy Bern, Nishioka, and Hagadorn.]

Recognizably different fibers can occur in glial cells (Gray, 1959b; Hama, 1959). There have been efforts to isolate proteins from squid giant axoplasm (Schmitt and Geren, 1950; Maxfield, 1953; Maxfield and Hartley, 1957; Schmitt, 1957, 1959); it is a very watery material but at least one filamentous protein is found.

(c) The **classical neurofibril** (Fig. 2.38) is longer and thicker than either the neurofilaments or the tubules, which are seen only in the electron microscope. The common explanation of this classical element is that it represents products of fusion of the smaller elements, which have congealed together. This accounts for many of the instances; for example, it is now clear (Couteaux, 1956) that the large axons of the ventral cord of the leech (*Hirudo*) contain many long structures (200 Å wide) which resemble the tubules in axoplasm of some vertebrate and crustacean neurons and that these tubules can be seen in stages of agglomeration, presumably to form thick elements appearing as neurofibrils when stained by suitable reduced silver impregnation methods (see also Trujillo-Cenóz, 1959). Axons of other annelid cells contain similar neurotubules, which are now as ubiquitous in electron microscope studies as classical neurofibrils have been on a gross scale. However, the exact correspondence between neurofibrils and neurotubules does not hold; for example, examination of the thinnest vertebrate axons with the electron microscope shows that several axons of only 1 μ in diameter can run within a single sheath, and it is probable that some neurofibrils identified in silver impregnated material are in fact thin axons. Other examples of large neurofibrils—for example in the giant axons of the nerve cord in annelids (Smallwood and Holmes, 1927; Ogawa, 1939)—are too definite and consistent to be dismissed as congealed tubules, and yet they have not been described from electron microscope preparations of the same material (Issidorides, 1956; Hama, 1959).

It is important to note that classical neurofibrils are not ubiquitous. They are not found, for example, in squid giant axons or large non-giant axons, though the relevant methods succeed very well in these animals. Nor are they found in many giant fibers in polychaetes or crayfish. Many cell bodies, unlike earthworm, leech, and typical vertebrate neurons, do not reveal a silver impregnating network, for example in roach, crayfish, *Aplysia*, squid. Electron microscopists searching for the substratum of neurofibrils should choose material known to exhibit good neurofibrils.

(d) Different again from the structures previously mentioned are the **osmiophobe strands** described recently by Wigglesworth (1960b) in neurons of insects. The large axons of insects and Crustacea (Ross, 1922) appear to contain many classical neurofibrils which fan out into the cell body, as seen in silver preparations. With improved fixation they are seen to be osmiophobe strands about 0.5 μ in thickness which fill the axon. Possibly these are the spaces between the neurotubules, and it is the latter upon which silver and other heavy metals will aggregate.

(e) **Fibrils in living axons** were observed long ago in fish (Schultze, 1878), jellyfish (Bozler, 1927), lobster (De Rényi, 1929) and chick (Weiss

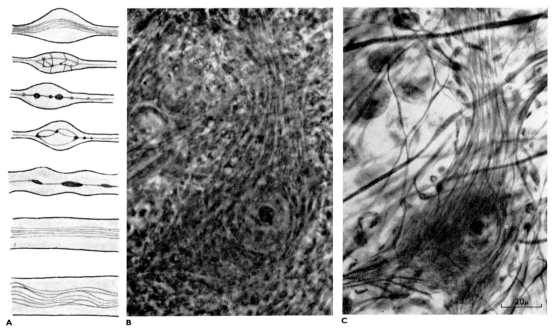

Figure 2.38. Neurofibrils as seen in the light microscope. **A.** Living axons of the crustacean *Squilla*, stained intra vitam with methylene blue. Various appearances encountered are sketched, which perhaps represent stages in alteration of the normal state by the method. At least some of them are probably within the range of the natural state. [Meyer, 1955a.] **B.** Living nerve cell in a 74-day culture from the dentate nucleus of a young cat. Fibrils can be seen especially at the poles of the cell. Phase contrast. **C.** The same cell after fixation and impregnation by Bodian's method. [Hild, 1959.]

and Wang, 1936) and have recently been again clearly demonstrated in vertebrate spinal motor dendrites (David et al., 1961). Their general appearance is compatible with the suggestion that they are composed of clumped neurotubules and neurofilaments. Fibrils in vitally stained neurons, seemingly not to be explained as precipitation of precursors, are observable in annelids and arthropods (Meyer, 1955a). They are relatively few in number per axon and the best explanation of them is that the vital dye is selectively absorbed by the neurotubules and strands of the endoplasmic reticulum, not by the neurofilaments. Something visible in the light microscope, therefore—much larger than single tubules or filaments—is preformed in life (see also Hoerr, 1936), but it is not necessarily always the same entity.

(f) The absolute size of the **smallest axons,** as determined by electron microscopy, is sufficiently small to suggest that many fibers previously described as neurofibrils are in fact axons within a common sheath. Axon arborizations in a variety of vertebrate smooth muscle run down to 2000 Å in diameter and average 4000 Å (Caesar et al., 1957). Remak or C fibers are commonly of a similar or smaller size (Gasser, 1958). Fibers of this size range, when impregnated with silver, could be easily misinterpreted under the light microscope as neurofibrils, and in fact material from the visceral nervous system has provided a large section of the descriptive material on neurofibrils supposed to run between nerve cells (Stöhr, 1957). This explanation is useful only for descriptions of axons; cell somata are classically the best place to visualize neurofibrils.

(g) Large fibrils visible under the electron microscope in the best fixed material, corresponding in size to classical large neurofibrils seen under the light microscope in silver preparations, are known from cells of the ganglia of the leech *Hirudo*. This is one of the sites where workers long ago described clear examples of neurofibrils (Bielschowsky, 1903). Large neurofibrils have been confirmed in these leech cells by a variety of techniques (Meyer, 1955a) and by electron microscope studies (Bern, unpublished; Fig. 2.37), and in polychaetes (Horridge).

Some details of the appearance of light microscope neurofibrils in giant and other large fibers in earthworms and leeches are worth mentioning. They are few, coarse, and central, not fine and evenly distributed through the axoplasm. It is hard to believe in fusion of filaments or tubules which are sparse and distributed through the axoplasm; moreover, the neurofibrils can be seen in life with methylene blue. In longitudinal silver preparations they are sinuous even when the outlines of the axon are not; they act as though they were inelastic elements of constant length, for material fixed under stretch shows more gentle undulations of longer period.

It is impossible to assign at present any **physiological meaning** for neurofibrils. Older theories that they are involved in conduction can be abandoned (although still espoused, for example, by Delov in Bykov, 1958, p. 482), partly because of the weight of evidence that the surface membrane is responsible and partly because of the evidence given in the last paragraph. Taken together with the fact that stretched axons in the same animal conduct at a constant velocity, with less length of neurofibril per millimeter of axon, the evidence does not support suggestions of their conducting role.

In conclusion, classical neurofibrils, prepared by techniques which require strong chemicals, should always be regarded with suspicion until confirmed by modern techniques. Where there has been a recent critical examination of classical neurofibrils there has usually been found a fine filamentous structure or elongated mitochondria (up to 9 μ long according to Pappas and Purpura, 1961) on which an impregnation with heavy metal could have accumulated to form deceptive artifacts. There are extensive accounts by careful workers (for example Stöhr, 1957) who take the opposite view, especially with reference to the autonomic system of vertebrates, that neurofibrils are not only real and extensive structures but are also important for conduction of excitation. These views have not been tested by techniques which could provide definite evidence for or against them. The present authors feel that effort is best spent in the biochemical and functional analysis of well-authenticated fibrillar structures in favorable large axons, rather than in protracted arguments concerning the validity of fibrillar structures observed only when doubtful techniques are employed. (For a partial summary, see Table 2.2.)

D. Neuroglia and Sheaths

For many years, axons in peripheral nerve of invertebrates were thought to be surrounded by a syncytium or a polynucleated fluid (review in Bruno, 1931), and in invertebrate ganglia the numerous nuclei of nonnervous elements were

Table 2.2. COMPARISON OF FIBRILLAR STRUCTURES IN NERVE CELLS BY LIGHT AND ELECTRON MICROSCOPES.

Light Microscope Objects	Electron Microscope Objects
Mitochondria	Mitochondria, sometimes elongated, 1000–5000 Å
Axons	Axons, as small as 750 Å
Classical thin neurofibrils in leech *axons*; seen by many silver methods(1)	Explained as neurotubules or neurofilaments or elongated mitochondria (2)
Classical thick neurofibrils of some *cell bodies* in leech	Not explained away (1); now found with EM (3)
Classical neurofibrils in vertebrates (4)	Most are not investigated by EM. Some are thin axons (C fibers) grouped within sheath cell. Many are probably deposits of silver on neurotubules
Neurofibrils visible in living axons of coelenterates, dendrites of vertebrate motor cells and arthropod axons	Neurotubules and spaces between neurotubules occur in the same place; perhaps some real fibrous structure which does not show in EM
Osmiophobe strands in crustaceans and insects fill the axoplasm	EM studies available but no EM correlate. Perhaps they lie in the spaces between the neurotubules
Classical neurofibrils absent in squid giant axon; present and lying central in earthworm giant axon	EM shows no difference to date; neurofilaments present and diffuse in both

EM = electron microscope.
(1) Meyer (1955); (2) Couteaux (1956); (3) Bern et al. (1961); (4) Stöhr (1957).

A word on terminology is necessary because of uncritical use of the word "neurofibril" by some electron microscopists. The word has a well-established referent in light microscopy, even if we later find that that referent is not a single class of inclusion but is heterogeneous. It should not be used for any structure seen with the electron microscope until a reasonable argument has been supported (that is, by more than compatibility) that the structure corresponds to or accounts for light microscope neurofibrils in the same neurons. For instance, the idea that agglomerated neurofilaments or neurotubules account for neurofibrils is a tenable theory but is rarely supported by any evidence; neurofilaments and neurotubules should not be called neurofibrils.

thought to belong to glial cells similar to those of vertebrates (review in Schneider, 1902). Recent work with the electron microscope has now revealed that glial cells and their relationships with neurons are similar throughout the invertebrates which have been examined except that glial cells are scarce in coelenterates and become more abundant, specialized, and diversified in higher invertebrates and vertebrates.

The **structure of glia** in invertebrates, extensively studied in annelids and gastropods, is now best known in the Arthropoda and yet, until the recent works of Wigglesworth and of Hess, was not well understood. There was a general impression, derived from early accounts by Tiraboschi (1899) and Haller (1905), that a nucleated cytoplasmic syncytium surrounded all central nerve processes. Terminology had long been confused; nonnervous cells within the ganglia have been variously called "Zwischengewebe" (Redikorzew, 1900), "Hüllgewebe" (Schneider, 1902; Bauer, 1904), "Stutzzellen" (Schrader, 1938), and "Stutzgewebezellen" (Kühnle, 1913; Hertweck, 1931). The sheath structure was called "Neurogliahülle" (Haller, 1905) and "Perilemma" (Schrader, 1938; Scharrer, 1939). But a solid body of descriptive information was accumulated, many of the papers being meticulous and well illustrated; new workers should not overlook the details. (Besides those just cited, see Rohde, 1893, 1895, 1896; Joseph, 1902; Cajal, 1904, 1917; Krawany, 1905; Merton, 1907; Boulé, 1908; Sánchez, 1909; Weigl, 1910; Keyl, 1913; Jakubski, 1915; Havet, 1916; Baecker, 1932; Clayton, 1932; Bogoraze and Cazal, 1944.)

The term **neuroglia** is very broad and includes any nonnervous cells of the brain, cords, and ganglia and, in recent usage, even in the peripheral nerves, except for cells comprising blood vessels, trachea, muscle fibers, glands, and epithelia. Roughly, it means connective tissue associated with nervous tissue, though connective tissue in vertebrates has a fairly precise meaning that excludes at least some neuroglia (astrocytes), and the meninges of vertebrates are not considered to be composed of glia cells. There are various forms of neuroglia; but let us first consider the situations in which it is found.

1. The situations in which glia occurs

The ganglia of higher invertebrates, the cords, and larger nerves are covered with a strong **outer sheath** (Fig. 2.39) which consists of an inner layer of cells, properly called the perilemma, and an outer noncellular layer, the neurilemma or neural lamella (Hoyle, 1953; Hess, 1958b). The **neurilemma** of arthropod nerve bundles consists

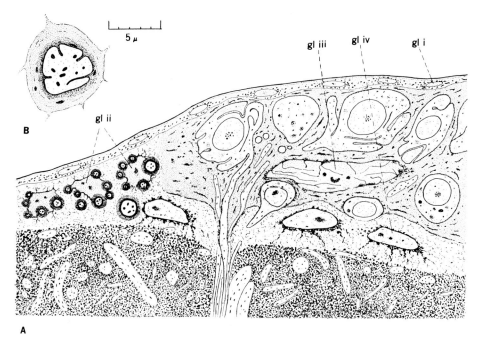

Figure 2.39. Glial cells in an insect ganglion. **A.** Section through the margin of the prothoracic region of the central ganglion of *Rhodnius* (Hemiptera) to show the glial cells; *gl i*, perilemma cells (perineurium) within a thin neural lamella; *gl ii*, glial cell nuclei producing the thick sheath for the lateral motor axons; *gl iii*, giant glial nucleus with invaginated membranes, and cytoplasm extending everywhere between the neurons; *gl iv*, glial nuclei in the zone of fine nerve fibers, sending cytoplasm inward to form the dark substance of the neuropile. **B.** Section through one of the lateral motor axons, showing thick laminated sheath with mitochondria, invaginations of the sheath into the axon, and mitochondria in the axoplasm. [Wigglesworth, 1959.]

of connective tissue fibers presumably secreted by underlying perilemma cells, which are glial cells that lie superficially and may be modified in staining quality and synthetic properties. The outer plasma membrane of the perilemma cell is distinct from the fibrous neurilemma; the nature of the latter has been controversial. In arthropods it consists of concentric lamellae with a crossed fibrous reinforcement (Baccetti, 1955) of some polysaccharide which does not correspond exactly with collagen, reticulin, or elastin. Diffraction patterns show a 2.86 Å line characteristic of collagen, but the intensity of this line implies a collagen concentration of only 10–20%. Imbibition curves of birefringence and the elasticity also indicate that collagen makes up part of the neurilemma sheath (Richards and Schneider, 1958; Ashhurst, 1959). In Crustacea the fibrous neurilemma sheath of nerve bundles is derived from glia cells which are not as distinct from ordinary glia as are ganglionic perilemma cells. In Crustacea, as in insects, the structureless outer sheath stains like collagen. The **perilemma** cells are distinguished from deeper glial cells by staining differences; in insects there is a sharp distinction (Hess, 1958b; Wigglesworth, 1959). Perilemmal cells are cuboid with prominent mitochondria; in insects, these cells are less obviously differentiated over the connectives and nerves than over the ganglia. Other invertebrate groups are known in less detail but it appears from present data that a general consistency of pattern will be found, although divergent cases are known.

Axons in **nerve bundles** have relatively simple relations with glial cells. In peripheral nerves and in central connectives all the spaces between the nervous elements are filled by the cytoplasm of glial cells and their membranes. The glial cells are extensive and of intricate shape, fitting the interstices between the other structures. The larger axons are usually clad individually in unineuronal sheath cells which produce varying amounts of sheath membranes (described below). Other glial cells—often the same polyneuronal glial cells—spread between the axons, filling all the available space and exhibiting contorted and extensive boundaries. Others—or again in some instances the same cells—form an outer sheath around the bundle of nerves. In arthropods the peripheral bundles are not surrounded by a layer of special modified glial or perilemma cells. Again, the above details have been worked out primarily for arthropods but probably apply to other phyla.

The neuropile or the **fiber core of ganglia** in general is sometimes bounded by—and in addition penetrated to varying extents—by neuroglia (Fig. 2.40). In many locations, as for example in the arthropod brain and in the radial nerve of echinoderms, the nuclei of the glial cells are arranged in a particular layer (usually a little below the superficial layer of nerve cell bodies) from which their cytoplasm ramifies deeply and widely among the axons and neuropile below. A peculiarly high development of the layer of glia between rind and core is characteristic of Hirudinea (Fig. 14.9). Typically glial cells do not form a sharp boundary but tend to define clumps and regions of neuropile. The long, branched processes of these cells extend widely and intertwine but, as seen in electron micrographs, the cell membranes are clear and the syncytial appearance under the light microscope is not borne out.

This kind of relationship, in which axon arborizations forming the neuropile are embedded in widely ramifying glial cells, is found in Oligochaeta (de Robertis and Bennett, 1955a; Hama, 1959), Hirudinea (Bern et al., 1961), Gastropoda (Schlote, 1957; Batham, 1961), Cephalopoda and Echinodermata (observations from unpublished figures), Crustacea (papers by McAlear, Kuno, and Hama), Insecta (Hess, 1958; Wigglesworth, 1959), and probably even more generally. Coelenterate ganglia in the marginal bodies of scyphomedusae appear to possess very few formed elements of glial nature (Horridge, Chapman, and MacKay, 1962). There is a strong similarity between the ganglia, with respect to well-developed glia, in Platyhelminthes and Nemertinea and those of higher groups—Annelida, Arthropoda, and Mollusca. The resemblance is markedly less in the groups with simpler nervous histology, for example Hemichordata, Phoronida, Ectoprocta, Rotifera,

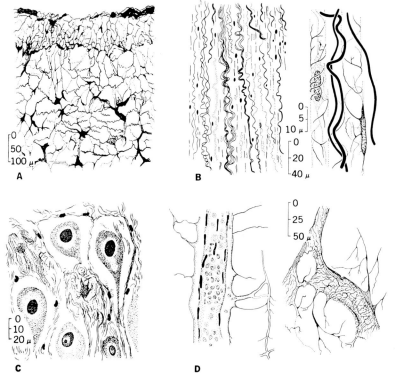

Figure 2.40.

Neuroglia fibers in the octopus. **A.** The meshwork of glia and blood vessels in the subesophageal ganglion as seen in a Bielschowsky silver impregnation. Note the greater density in the upper part, the rind, where the meshes contain large ganglion cells. **B.** Supporting tissue in nerves. At the left a general view; at the right a higher magnification, showing the glial processes wrapped around axons. Rio-Hortega method. **C.** Meshwork of satellite glia in a layer of ganglion cells, showing the capsule for each neuron. Bielschowsky method. **D.** A blood vessel, on the left, showing the glial envelope outside the perivascular space (without muscle fibers); the glia project laterally as pseudovessels. On the right a vessel is seen, solely indicated by its rich glial envelope. Rio Hortega method. [Bogoraze and Cazal, 1944.]

Nematoda, Gastrotricha, and others; here glia is less developed and often impossible to find. Perhaps in the epidermal nerve cords the long, attenuated cell feet of epithelial cells that pass between the nerve fibers to insert on a basement membrane represent a form of glia, at least in separating some of the axons from each other.

The **cell rind** is also permeated to a varying extent by glial cells and their processes. Sometimes these are sparse, as in lower forms and perhaps in globuli cell masses. In other places, such as the ganglia of *Aplysia* and especially where the nerve cells are large, glia is well developed. To a minor extent (except in Hirudinea, where it is major) glia defines cell masses. Mainly, glial cells form investments of individual nerve cell somata by fitting closely as flattened satellite cells, seen long ago as a coat of small, flat nuclei (Retzius, 1890); they tend to build up lamellae on larger neurons.

The **trophospongium** is a term that has been recently revived after half a century of disrepute; it was originally applied by Holmgren (1914) to a protoplasmic network and system of thin channels thought to arise from the neuron sheath cells as an invagination of the membrane. The structure was later confused with the Golgi net and the original descriptions, based on vertebrate spinal ganglion cells, have not been substantiated. However, the term has recently been appropriately applied to a system of blind canals which are invaginations of the cell membrane in polychaete (Fig. 2.41), earthworm (Malhotra, 1957), insect (Hess, 1958b; Wigglesworth, 1959), and gastropod neurons (Bullock, 1961). The hollows of the canals contain glial cell cytoplasm with inclusions, which suggest that they pass nutrients to the neurons—hence justifying the reference to their trophic nature. They have also been called chondriome and intracellular trabeculae, with reference to their supposed function as a mechanical support.

In crustaceans the granular mass of filaments of the trophospongium invade the cell even to the nucleus (Monti, 1915; Ross, 1922; Beams and King, 1932; Lacroix, 1935). In insects (Wigglesworth, 1960a) they contain cytoplasm of surrounding glia cells with many particles, supporting the idea of nutritive function for the neurons. In insects tracheae spread throughout the ganglion (Hilton, 1911) and even penetrate neurons (Ross and Tassell, 1931), but a glial cell membrane still intervenes between neuron and tracheole (Hess, 1958). Ridgelike elongated invaginations of the axon membrane are a feature of many large axons in molluscs (Schlote, 1957) and probably are functionally similar. The inwardly folded axon membranes of large *Aplysia* axons carry with them membrane of the glial cell (Batham, 1961, Frontispiece). This is implied but not reliably demonstrated in earlier studies. It may be related to the conspicuous trophospongium of the cell soma of those axons where the invasion of trabeculae of neuroglia is so extensive on the side of the cell bearing the axon that the surface of the neuron is spongy and enormously increased in area (Bullock, 1961).

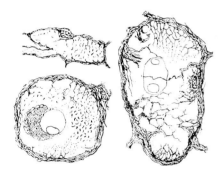

Figure 2.41. Trophospongium, as seen in giant nerve cells of the polychaete *Lanice*. [Dehorne, 1935b.]

Individual axons are very generally (but not universally) surrounded by sheath cells. Around most axons loose folds or overlapping shingles or spiral turns of glial cell membrane form a private sheath. Larger axons have chains of single celled sheaths (vertebrate myelin) or overlapping rows of lamellated sheaths involving several glial cells in any one cross section. Several up to many small axons often travel inside one glial cell. The bundles thus defined have been shown in vertebrates to last for short distances, exchanging fibers at frequent intervals.

2. The varieties of glial cells

The cells in all these situations are not identical, but it has been difficult for authors to agree on categories of neuroglia cells. In many lower invertebrates and even in some Crustacea—for example *Argulus* (Grobben, 1911), *Chirocephalus* (Debaisieux, 1952), and probably other phyla, older studies suggested that the glial cell nuclei are apparently of one kind throughout the central nervous system, whether inside or outside axon sheaths or forming the outer neural lamella sheath. In insects, vertebrates, and perhaps in cephalopods, there are recognizably distinct types of glial cell nuclei but invertebrates differ from vertebrates in having no equivalent of the astrocyte. Perilemma cells are clearly distinct in staining affinities from deeper-lying glial cells in arthropods. The distinction seems to be general: some glial cells happen to lie between nerve fibers and have thin walls, and others lie around the outside of the nerve bundle and have thickened walls.

As an example of more recently discovered **glial cell diversity,** several distinct types occur in the insect *Rhodnius* (Wigglesworth, 1959). (a) A layer of distinctive perilemmal cells (perineurium) around ganglia, but not peripheral nerves, is filled with filamentous mitochondria. (b) The sheath cells of the peripheral nerve bundles (lemnoblast of Edwards et al., 1958a) and of many central fibers is characterized by their large homogeneous nuclei. They form sheaths around the peripheral axon bundles and individually around the fibers themselves. (c) Large glial cells inside the margin of the ganglia have giant nuclei 50–60 μ long (gl iii, Fig. 2.39). These are not found in *Periplaneta* (Pipa, 1961a). The cell body has indefinite limits, meandering around the motor neuron cell bodies. The cytoplasm stains darkly with osmic acid and is rich in lipoid and enzymes, including nonspecific esterase and cholinesterase. (d) Certain motor axons lie within glial cells which have characteristic nuclei and form a sheath in certain definite regions of the axons (gl ii, Fig. 2.39). (e) The centrally lying typical glial cells of the cord and brain have nuclei between the outer rind of ganglion cells and the neuropile. Their cytoplasm spreads inward and fills all the space between the nervous elements, as if their membranes "wet" the nerve cells. Therefore they have a ramifying form which is the shape of the interstices of the neuropile (gl iv, Fig. 2.39). Glial cells supply nutriment to the ganglion cells, apparently by the complex invaginations of the chondriome (which see below) into the cell bodies of the neurons (Wigglesworth, 1960a). In insect brains there are a few giant glial cell nuclei (de Lerma, 1949), but most neuropile areas are surrounded by many small glial nuclei (4–7 μ) which have the typical ramifying cytoplasm. (f) During their researches, Cajal and Sánchez (1915) made some remarkable preparations of two types of glial cells in the optic medulla of Diptera and Hymenoptera: first, multipolar cells with a complex bushy arborization of fine endings, and second, cells with pointed processes. Both types were inextricably intermingled with the neuropile. Specialized forms of glia have also been described in cephalopods (Fig. 2.42) by Bogoraze and Cazal (1944) and others.

The embryological information on the origin of glial cells is poor. In annelids and arthropods they develop from epidermal cells. In many lower annelids the nerve cord and nerves are closely associated with the ectoderm. In arthropods the modified epidermal cells of receptors, the pigment cells of the eye, the tracheoblasts, and the glial cells have many features in common, and each can act as a sheath to small nerve terminals.

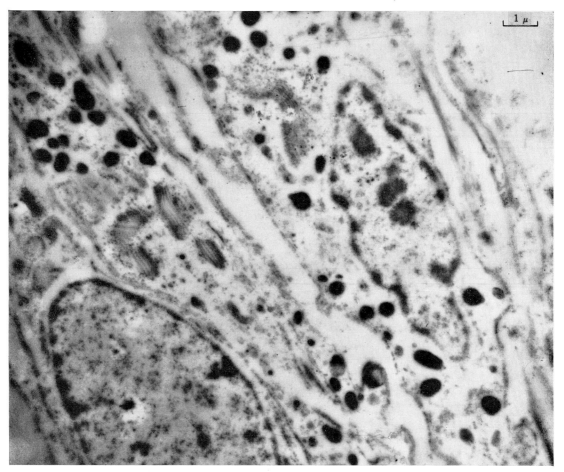

Figure 2.42. Neuroglia cells as seen in the electron microscope. Earthworm ventral cord, showing glial nuclei and processes with large secretory spheres and compact canalicular and vesicular material, which make these processes easy to recognize even when extremely attenuated between axons. [Courtesy E. de Robertis.]

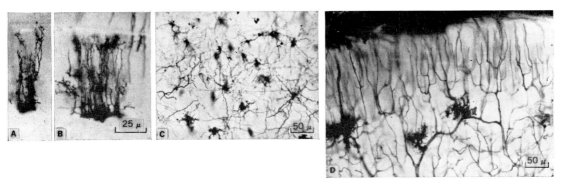

Figure 2.43. Astrocytelike glia in *Octopus*. **A** and **B.** Glial cells in the optic lobe, having radial processes that run through the depth of the plexiform zone. **C.** Section tangential to the surface of the optic lobe, showing glia cells with long processes radiating in the inner tangential layers of the plexiform zone. **D.** Transverse section of the optic lobe to show arterial trees at the surface of the lobe, as well as glial cells. Golgi method. [Courtesy J. Z. Young.]

Sheath cell cytoplasm as visualized with the electron microscope is not everywhere the same. Often, as in earthworm and insects, it is marked by an abundance of granules of high electron density (Fig. 2.43). In other places, as in *Aplysia*, polychaetes, and vertebrates, it is noteworthy for numerous, tight clumps of finely fibrillar material (Frontispiece, Fig. 23.16) or, in insect glia (Trujillo-Cenóz, 1959), tubular material.

Diversity is best known in the **neuroglia cells of vertebrates;** three distinct types in the central nervous system have been achieved in the higher classes but not in the lower. In mammals the ectodermal glia is divided into the astrocytes and the oligodendroglia. A third element, of mesodermal origin and behaving to injury like some of the plastic connective tissue elements, is called microglia. There is a large literature on these remarkable and highly differentiated, not primitive or simple, cells; even to demonstrate them with the light microscope—that is, to see the cell outlines—requires special and extremely specific metallic impregnation methods. The texts and monographs on vertebrate histology and cytology should be consulted for details of these cells (Cowdry, 1932; Penfield, 1932; Maximow and Bloom, 1957). Müller fibers in the vertebrate retina are regarded as glial cells and are packed with 200 Å granules, 70 Å filaments, and irregular 200–1000 Å vesicles and tubules (Kidd, 1961).

3. Axon sheaths

Diversity of axon sheaths is greater than has been heretofore realized (Fig. 2.44). The familiar categories are the unmyelinated and the myelinated axons. Unmyelinated axons are actually of many forms. In addition, we must recognize a normal category of naked fibers, surrounded by intercellular space or other axons, without an intimate glial coat.

Naked axons not surrounded by a glial cell occur but they are not the general rule except in coelenterates, where nerve fibers are without glial cells or sheaths of any kind (Fig. 2.45). Doubtless many other lower groups are similar, especially where the nervous system is intraepithelial; close study with the light microscope shows no sheath cells around enteropneust giant fibers (Bullock, p. 1570), nor with the electron microscope on triclad flatworm nerve fibers (Skaer, 1961). Bundles of axons lying packed together without glial cytoplasm nearby are seen in electron micrographs of *Aplysia* and in earthworm and seem to be naked. There are many examples of sensory dendrites of peripheral receptor cells where no sheath exists round the dendritic termination. However, in some of these cases there are other cells which take the place of sheath cells and may be functionally equivalent to them. Thus the tormogen and trichogen cells of insect hair receptors and the pigment-bearing cells of the compound eye surround otherwise naked nerve

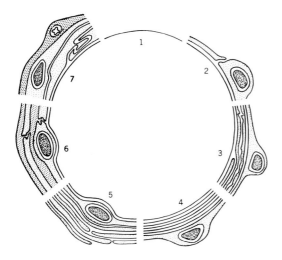

Figure 2.44. Types of nerve sheaths, some of which are commonly found in many phyla, and all of which occur in arthropods. *1.* Naked, without a sheath, as in most axons less than $\frac{1}{2}\mu$ in peripheral nerves, in many central axons, and in coelenterate axons. *2.* Single-sheathed, common in peripheral and central axons of most groups. *3.* Loosely wound myelin with nucleus outside, found in optic nerve and ganglia of crabs, and large motor axons of polychaetes. *4.* Compact, densely wound myelin, as in vertebrates; found in optic nerve of *Cancer* (only this type is called "myelinated" without qualification). *5.* Several participating sheath cells with many shingled processes, as in some crustacean central fibers. *6.* Layered glia processes alternating with thickened amorphous layers ("basement membrane"), found in some peripheral and central arthropod fibers. *7.* Tunicated, with a thick outer amorphous coat, as in some arthropod and mollusc giant fibers. [Courtesy J. H. McAlear.]

processes (Slifer, 1961). In mammalian olfactory mucosa the naked terminations lie between or within nonnervous supporting cells (De Lorenzo, 1957). Müller cells of the vertebrate retina act similarly as sheath cells (Sjöstrand, 1958). Sheath cell membrane adjacent to the axon membrane is not obligatory for conduction, as is shown by the situation in coelenterates as well as by the numerous fine fibers which can run in bundles within a sheath cell; the inner members of the bundle are out of contact with glial cell membrane (Gasser, 1958; Taxi, 1958). In the crustacean

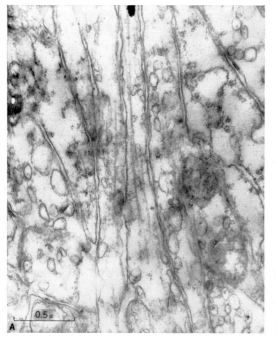
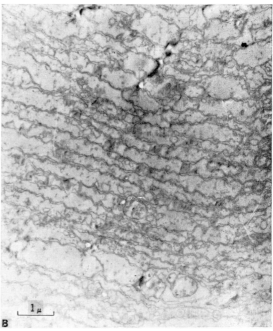

Figure 2.45. Naked axons in coelenterates, as seen in electron micrographs. **A.** *Cyanea*, nerve fibers in the marginal ganglion. Note the tubules running for long distances parallel to the axis of the axon. [Courtesy B. Mackay.] **B.** *Aurelia*, a tract of axons from sensory neurons in a sensory pit, running into the marginal ganglion. Note the absence of glial cells. [Courtesy D. M. Chapman.]

central nervous system and elsewhere there are evidently bundles of axons closely adjacent to each other, as seen in transverse section, lying within quite discrete sheath cells, whereas larger axons have sheath cells restricted entirely to one axon (Uchizono, 1960). Between the sheath cells is a space which appears to be the cytoplasmic area of a large ramifying sheath cell but may be extracellular space.

Axons surrounded by a single glial cell in any cross section, actually a single row of elongated sheath cells, are the rule for unmyelinated fibers of vertebrates and many invertebrates; such axons usually share the sheath cell (or Schwann cell as it is often called). There may be hundreds of axons in bundles in a single Schwann cell, each bundle occupying an invagination of the glial cell membrane and marked by its own mesaxon.

Loosely sheathed axons of several types have been seen (Fig. 2.44). Glial cells may form an unorganized series of lamellae by overlapping like shingles or intertwining their processes. A single cell or several cells may loosely spiral around the axon, leaving visible cytoplasm in the wrappings. A fibrous layer like a basement membrane may be between the cytoplasmic wrappings —for example in central axons of roach, crayfish, lobster, but much less in *Uca* (Fig. 2.46), *Cancer*, and *Callinectes* (Edwards and McAlear, unpublished). The term "tunicated" has been introduced for some of the common insect loosely sheathed axons (Edwards, 1958). As a matter of clarifying terminology, such fibers should not be called "myelinated" without qualification, though "loosely myelinated" should not cause confusion. "Loosely sheathed" is safer until, as is doubtless in store, a new set of terms is established by additional studies revealing valid categories within this assemblage.

Constrictions of the sheath—sometimes, but inappropriately, called nodes—have been described in a few loosely sheathed axons in crustaceans. These are treated in Chapter 16 (p. 902). They are not to be confused with nodes

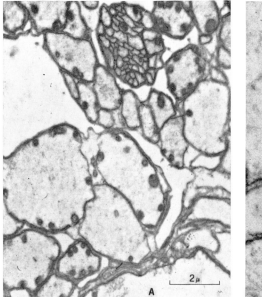
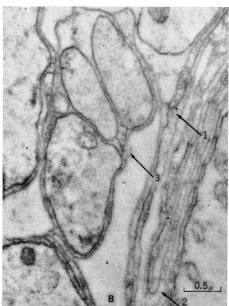

Figure 2.46. Electron micrographs of axon sheaths. Transverse sections of the esophageal connectives of *Uca pugilator*. **A.** Fibers of various sizes. At the top of the picture is a bundle of very fine fibers within a single glial cell; at the bottom is a large axon with three loosely deposited membranes; most of the other fibers have a single giant investing layer. **B.** On the right is a multilayered sheath of a large axon, and the terminations of the spiral structure are shown by the arrows *1* and *2*. Arrow *3* shows the mesaxon of an axon that is thereby suspended from the wall of the glial cell; this glial cell contains two other similar axons. [Courtesy J. H. McAlear.]

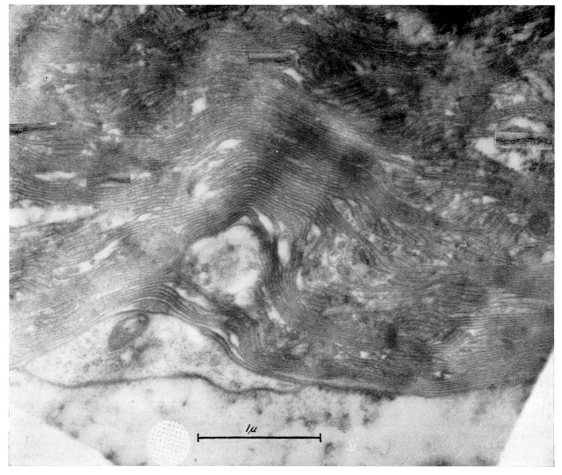

Figure 2.47. The earthworm giant fiber sheath, transverse section with the axon at the bottom. The inner part of the sheath, shown here, is formed of many layers, more or less regularly spaced, with a period of 280–300 Å. The outer region is continuous with the cytoplasm of glial cells; it shows a pattern of double main lines alternating with an intermediate line, with an average period of 1100–1200 Å. The dark lines of the inner part show a fine structure of two segmented lines 20 Å thick, separated 20–30 Å from each other. [Courtesy H. Fernández-Morán and G. Svaetichin.]

of Ranvier of true myelinated fibers and are not known to have a comparable function by permitting saltatory conduction.

Giant fiber sheaths illustrate a wide range of structure. Earthworm giant fibers have a loose sheath of 15–30 layers in the median and 2–15 layers in the lateral fibers (Fig. 2.47). There is not the high degree of order which is so conspicuous in vertebrate myelin sheaths, and the layers are intermingled with cytoplasm and inclusions such as mitochondria (Hama, 1959). Nevertheless, the layers are more regular in thickness than anywhere else in the invertebrates, the sheath is said to be a spiral wrapping, there are very few sheath cells in any cross section, and it is closer to a fully myelinated one than any other known sheath.

The 30 to 50 successive lamellae of the myelin sheath of earthworm giant fibers develop thickened dense regions which are interpreted as providing a mechanical attachment between the lamellae. In the same way membranes of epithelial and other cells may form "attachment zones" (nodes of Bizzozero or desmosomes). Fibrils run through the glial cytoplasm between the thickened regions of the sheath lamellae (Hama, 1959). Comparable desmosomes which appear to bind together the glial membranes are known from vertebrate myelinated fibers (Rosenbluth and Palay, 1961).

Squid giant fibers have a narrow (0.1–0.9 μ thick) but continuous coat of small glial cells, each with a large flattened nucleus, at intervals of about 20 μ. Thus, quite in contrast with myelinated fibers, where also the sheath is one glial cell thick but a single cell completely invests the axons for a length of a millimeter or more, we find in the squid a mosaic one cell thick, with as many as three nuclei in one cross section. As usual where suitably examined, the glial cells are not syncytial but distinct. In the squid giant the glial cell cytoplasm, less than 1 μ thick, contains numerous folded membranes through which occasional contorted channels run, possibly providing an ion pathway from the narrow (72 Å) space between axon membrane and Schwann cell to the basement membrane and intercellular space. On the outer side of the glial cells there is a noncellular fibrous sheath consisting of an amorphous basement membrane, and external to this is connective tissue (Schmitt, 1959; Villegas and Villegas, 1960). Crustacean giant fibers and some quite small central fibers, have many layers of very regular glial membranes (McAlear et al., 1958). Large peripheral crustacean fibers have sheaths of many ordered layers. Large axons of *Aplysia* (Mollusca) have a few loosely investing layers of glial cell membrane (Batham, 1961). Examples from the main invertebrate phyla which are the best known in fine structure and of greatest physiological interest show a general pattern: very roughly speaking, the larger the fiber, the thicker and more regular will be the glial layers. Most axons over 10 μ have something more than the simple single axon-glial cell contact which is typical of fibers thinner than 10 μ. Fibers of 2 μ or less commonly share a glial cell with other fine fibers. However, relatively few examples have been examined and there is no knowing what peculiar features will turn up among the smaller phyla. Certainly the largest giant fibers, those of polychaetes such as sabellids and serpulids, are quite without thick or regular sheaths. Squid giant fibers have very thin and irregular sheaths for their size, compared to earthworms and to some nongiant properly myelinated axons.

Lastly, the **true myelinated fiber** is an axon with many layers of tightly, spirally wrapped glial membrane, having all cytoplasm squeezed out, and in which a single sheath cell provides this wrapping for a length of up to 2 mm and ends abruptly where it meets the next cell to form a node of Ranvier (Fig. 2.48). Nodes occur at fairly regular intervals of the order of 0.5 to 2 mm along all myelinated fibers, central and peripheral. With respect to the central nervous system, this statement is largely an extrapolation; nodes occur there but have not been studied extensively. The axon proceeds through a node without interruption but the wrappings of Schwann cell membrane end, one by one in a regular manner, from the inside out, approaching the node. Interdigitating fingers of the two Schwann cells meet and define a narrow, tortuous path from the space adjacent to the locally naked axon to the interstitial space outside the sheath. The dimensions of the space and the path are quite variable; Fig. 2.48 and Robertson (1957a) should be consulted for details. The functional significance in relation to saltatory conduction, current flow, ion exchange, and the questions of adequacy of the structure known to account for the physiological properties are of great interest (see Tasaki, 1959a; p. 176). Well known and analyzed in vertebrates, nodes of Ranvier have been one of the hallmarks of that group and unknown in invertebrates until recently. It is most remarkable that fibers with all these characters have been found in electron micrographs of crab brain (McAlear et al., 1958). The common ancestors with vertebrates are thought to be so very far down that it must be supposed the myelinated structure and the nodes of Ranvier have evolved independently in arthropod and vertebrate. Even the typical node structure, with layers of sheath cell membrane stopping in an orderly system, is repeated in the crab. Myelinated fibers are not ubiquitous in vertebrates, for the cyclostome *Entosphenus* lacks any sign of them in cross sections of the spinal cord studied with the electron microscope (Schultz, Berkowitz, and Pease, 1956). Elasmobranchs have good myelin and nodes, as do teleosts, where fibers of the same diameter as the

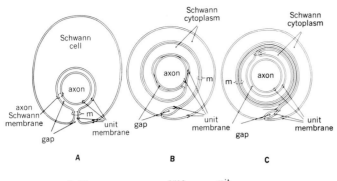

Figure 2.48.
The development and plan of the vertebrate myelin sheath. **A.** Nerve fiber invaginating the surface of a sheath cell. This is the condition of the usual unmyelinated fiber in vertebrates. *m*, mesaxon. **B.** An intermediate stage in development; spiral winding is under way. **C.** Simplified myelinated fiber; normally there are more layers of spiral windings. [Robertson, 1961.] *m*, myelin. **D.** The node of Ranvier seen in a diagrammatic longitudinal section of a myelinated fiber. [Robertson, 1957a.] *coll.*, collagen fibrils of the endoneurium; *d. gr*, dense granules; *mit.*, mitochondria; *proc.*, processes of the Schwann cells; *sch. cyt.*, Schwann cytoplasm; *s. gr.*, small dense bodies.

largest in the cyclostome conduct ten times faster.

Variations in the form of the myelin ultimately depend on differences between the neurons, as suggested by the fact that in vertebrate sensory ganglia a single Schwann cell can produce both loose and compact myelin on different axons. Rosenbluth and Palay (1961) find that the concentration of neurofilaments and tubules is much higher in compactly myelinated neurons than in either loosely myelinated or unmyelinated ones. The glial cells build different sheaths for individually distinct neurons.

There has been controversy as to the thickness and precise **constitution of the glial membranes** which compose the myelin sheath. The most satisfactory view, stemming from measurements on frog peripheral fibers (Robertson, 1957b, 1961), is that the unit is a symmetrical lipid layer bounded on each side, with an over-all thickness of 75 Å. The major dense lines of osmium fixed myelin (probably the protein) are separated by two such unit membranes; the mesaxon also consists of two, and the cell membrane of the neuron (including axon) and the cell membrane of the glial cell consist of one. The structure of the myelin sheaths of vertebrates has been analyzed in detail but is outside the scope of the present work (see the reviews by Schmitt, 1959; Robertson, 1960).

The functional **importance of loose, as against compact, myelin** remains a mystery. Where compact and loose myelin appear adjacent to each other on the same neuron, as in the eighth cranial nerve of the goldfish (Rosenbluth and Palay, 1961), the lamellae of the one run smoothly into the other. Since the complete spectrum of sheaths with few and many, loose or compact, layers occurs in old as well as young specimens in the crab central connective as well as in the goldfish eighth nerve, there is every reason to suppose that these are permanent states and not growth stages. Compact myelin sheaths are apparently always provided with nodes of Ranvier and these have been shown to mediate saltatory conduction, which may be much faster than is otherwise possible with the same axon.

Serious problems are repeatedly encountered in the effort to correlate structural and conducting properties of the axon membrane and its sheath. The first is concerned with the shortage of space outside the axon, which one may consider as the external medium of the axon. The question of an **extracellular space** in invertebrate nervous systems has not been settled, and the problem of where current external to the axon can flow (without disruption of conduction in adjoining axons) is commonly discussed but has not been solved. In electron micrographs of nerves and ganglia of annelids, arthropods, molluscs, and in vertebrates (in fact, in all nervous systems examined except coelenterates), only a minute fraction of the volume is extracellular space and there is no obvious path for the extracellular current which is evidently essential for nervous conduction and transmission at synapses. The crowding is intense in dense synaptic or plexiform areas of neuropile, as in vertebrate cortex (Gray, 1959a, 1959b, 1961) and in enclosed synapses such as that between rods and bipolar cells in the vertebrate retina (Sjöstrand, 1958). One must assume that the ionic medium external to the axons is the cytoplasm of the glial cells; these cells and their processes spread at least very widely if not everywhere between the nerve fibers in annelids, molluscs, and arthropods. Maturana (1960) points out that the ionic content of the minute extracellular space between fine fibers will be dominated by the activity of the nerves and reports that fine fibers in frog optic nerve are reversibly inactivated by stimulation at 40 per second for several seconds. Similarly, fine fibers in *Mya* (Pelecypoda) nerves, conducting at 15–30 cm/sec, are inactivated by stimuli faster than even 1 per second (Horridge, 1958b). Therefore the tortuous path of the fine fibers, never in contact with each other for more than about 1μ, is a feature of the utmost importance, preventing an active fiber from depolarizing the others. Closely related is the problem of finding a path for current flow through the sheaths of axons which have no nodes, if it is decided that in fact current flows out into interstitial spaces and not only through the glial cytoplasm. In the squid giant axon sheath, channels in the glial cell cytoplasm connect the space around the axonal membrane with the extracellular (interstitial) space outside the sheath cell. These have been proposed as channels of current flow through the cytoplasm of the glial cells (Villegas and Villegas, 1960). However, the channels are long, with a lumen of only 60 Å, and it would be a difficult matter to demonstrate such a function for them.

The presence of myelin increases the rate of propagation of the impulse by saltatory conduction in axons which have nodes, but the corresponding mechanism in loosely myelinated axons without nodes remains an enigma. Myelinated fibers sometimes have fissures running through their myelin, visible in the light microscope, known as **Schmidt-Lanterman clefts.** They are faults in the myelin structure which appear as if formed by a stress; they are analogous to the patterns of spaces formed between the pages of a thin paperback book when it is bent into an S-shape and twisted. As pointed out by Robertson (1958), the clefts are extremely long in comparison with pathways available at nodes and are doubtful candidates for ionic current channels. The direct line of the macroscopic cleft is crossed by the same number of myelin layers as is the full thickness of the sheath but appears as a cleft because there are relatively large gaps between them.

Functions of neuroglia are a matter of active speculation today. None is clear. Neuroglia are thought to provide for metabolic requirements of nerve cells and fibers and to dispose of transmitter substances. Glees (1958) adduced some evidence for a nutritive function in vertebrates. The provision of a conducting medium for the external circuit of the nerve impulse has been supposed to be so important that naked axons are thought to be incapable of a conducted event, but this is contradicted by the evidence of naked axons in coelenterates and elsewhere in simpler nervous systems. There is some evidence that neuroglia cytoplasm is unusual in having low potassium and high sodium, as required for an external medium by existing axonology. Speculations of a more active role—even to handling signals—are in print (Galambos, 1961). Trophic roles are discussed in Windle (1958). Hydén (1952, 1960) applied a battery of new ultramicrochemical methods capable of quantitatively assaying the contents of single nerve cells, freed of glia by manual dissection, and of small clumps of glia cells. He found marked differences in both conten and activity between these cell classes and in each class, as the result of physiological activity. The glia in general change reciprocally with nerve cells; for example, the glia lose RNA while nerve cells increase in RNA during sensory stimulation, and vice versa with sensory deprivation. Since in mammals, at least, the glia cells are not only differentiated into specialized types but outnumber nerve cells perhaps tenfold and comprise roughly half the total volume of the brain, it can be expected that they play important and specialized roles. All signs point to the probability in this problem of novel discoveries ahead.

E. Degeneration and Regeneration

Various types of degeneration are distinguished. The simplest and most studied, partly as an anatomical tool, is Wallerian degeneration, which occurs distal to a transection of the nerve fiber (Fig. 2.49). Retrograde reaction in a cell body and its proximal stump after section of a fiber may be reversible or may be serious enough to result in death and degeneration of the whole neuron. Usually only the neuron injured suffers, but in certain cases transsynaptic degeneration has been found to be charactcristic, notably in the optic thalamus after section of the optic nerve. Demyelination occurs in several disease states without frank loss of neurons. Glial reactions, direct morbidity of the nerve cell body, and other special conditions are known. Details of all these types can be found in pathology texts and in the classical monograph of Cajal (1928; see also Penfield, 1932; Bumke and Foerster, 1935; Gray and Hamlyn, 1962). The subject is beyond our present scope except for a short statement of the information available, particularly in lower animals.

Degeneration of myelinated fibers in vertebrates following the separation from the cell body

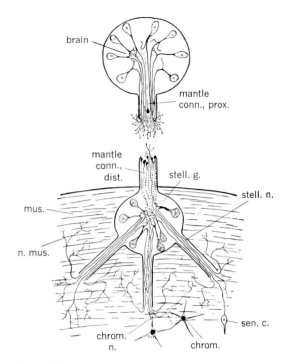

Figure 2.49. Primary degeneration of axons distal to a transection. Diagram shows the changes that take place after section of the mantle connective of an octopus. Degenerating axons are indicated by broken lines; solid lines are normal axons. Note that descending chromatophore fibers degenerate all the way to the skin; preganglionic fibers to the stellate ganglion degenerate to their ends there, but the fibers arising in that ganglion are not injured. Sensory axons are normal below the cut but degenerate in the brain. Sprouts indicate beginning regeneration. [Sereni and Young, 1932.] *chrom.*, chromatophore; *chrom. n.*, chromatophore nerve; *mantle conn., prox.* and *dist.*, proximal and distal stump of mantle connective; *sen. c.*, sensory cell; *stell. g.*, stellate ganglion; *stell. n.*, stellar nerve.

proceeds by a series of well-defined stages that are technically easy to identify (Cajal, 1928). Such fibers are readily traced anatomically by the Marchi method, which is selective for degenerating myelin; many of the tract pathways of the vertebrate central nervous system have been worked out in this way. It should be noted that while the axon itself is degenerating, the myelin sheath is also degenerating, but even more conspicuously. This is now understood in a way unclear to the classical microscopists, since the electron microscope has answered the old question of whether the myelin sheath is a part or a product of the axon, the Schwann cell, or both. We may now say that myelin sheath degeneration represents a transcellular interaction, the Schwann cell membrane suffering when a distant nerve cell body dies or is cut off from its axon; some trophic effect of the axon is necessary to the well-being of the sheath cell. This dependence has not been found in any of the classes of loosely sheathed, invertebrate nerve fibers. In virtually the only detailed study of invertebrate degeneration, Hess (1958c, 1960) found a sheath reaction of thickening and granulation but not loss in the connectives of *Periplaneta*.

Unmyelinated fibers likewise degenerate distal to a cut. There is no obvious sheath cell involvement and methods are lacking for exploiting this degeneration as a neuroanatomical tool except for the Nauta method, which brings out the axonal terminals of killed axons in favorable vertebrate material. In vertebrate nonmyelinated fibers, axon fragmentation begins 6 hours after section; the axons break into small rounded vesicles containing mitochondria and other cytoplasmic inclusions, and they are rejected by the glial cells which formerly enclosed them (Taxi, 1959). This suggests that the normal close contact between glial cell and axon is maintained by an active process for which the morphological integrity of the neuron is essential.

Among the **invertebrates** myelinated fibers are rare, and unmyelinated fibers sometimes do not show signs of degeneration for long periods. Studies are sparse in the invertebrates. Earthworm giant fibers and those of the polychaete *Myxicola* may not degenerate even several weeks after a cut (Nicol, 1948a, 1948b); this is regarded as owing to the presence of cell bodies on both sides of the cut, that is to the syncytial nature of these fibers. But surprisingly, peripheral motor axons of crayfish (Wiersma, 1961) and octopus (Wilson, 1960), presumably without benefit of distal cell bodies, may show no loss of ability to conduct and transmit excitation to muscles for long periods after transection, at least up to a month at normal living temperature! This appears to conflict with the reports that squid mantle muscle loses its nerve fibers in 4 days after stellar nerve section (Young, 1936a) and that *Octopus* in cold weather shows visible degeneration in 7–10 days (Sereni, 1929). These authors used the classical procedure of tracing degeneration in showing the direction and course of fibers in the mantle connective of cephalopods; Wells and Wells (1957) and Young (1961) also used this method in the brain. Small fibers in insects degenerate rapidly (Vowles, 1955; Hess, 1960). Holmes (p. 932) made some observations on crustacean central fiber degeneration, but much remains to be done before we have an adequate picture.

Regeneration is still less understood. It has repeatedly been shown to be extremely rapid in coelenterate nerve nets; at least, restoration of conduction across cuts and establishment of conduction between grafted pieces requires less than a day. Regeneration of nerve cords and even brain is normal in the many lower metazoans that regenerate missing body regions (platyhelminths, nemerteans, annelids, and others); some species can and others closely related cannot regenerate from small pieces or restore an anterior end. Earthworm giant fibers, which do not degenerate (at least for any number of segments), reunite functionally across a cut in a few weeks (Bovard, 1918; Yolton, 1923; Stough, 1930). May (1933) described the formation of new nerve terminals in regenerating cephalopod arms. In arthropods the regeneration of peripheral axons is essentially like the new formation in development, since members regenerate at molts. The powers of central regeneration have been little studied but are apparently poor in that lost ganglion cells are not replaced but cut stumps of axons may grow back and establish functional connections (see Chapter 20).

Regeneration in vertebrates has been extensively studied (Fig. 2.50). The classical account of Cajal (1928) may be consulted, as well as a large experimental literature; a sample is the series on peripheral nerve by Sanders and Young (1944), Simpson and Young (1945), Aitken et al. (1947), Vizoso and Young (1948), Young (1950), and Windle (1955). In the central nervous system the prevailing dogma has been that regeneration does not occur in warm bloods, although spottily spectacular in lower vertebrates—as in complete regeneration of a retina and central optic con-

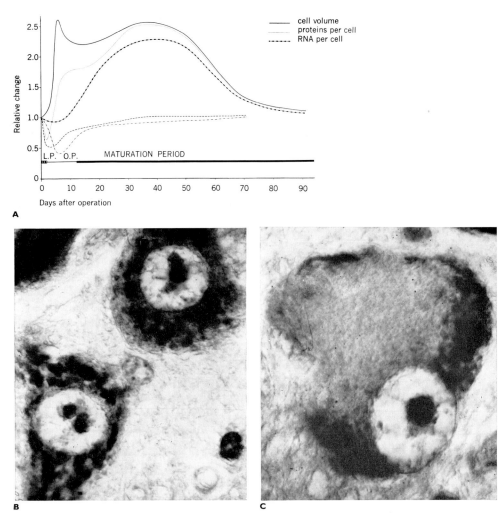

Figure 2.50. Cytochemical reactions in the perikaryon following injury of an axon, through initial reactive and later regenerative phases. **A.** Summary graph of the quantitative changes occurring during the first 90 days after the operation of crushing the axon. *L. P.*, latent period; *O. P.*, outgrowth period. **B.** A normal cell of the hypoglossal nucleus photographed with monochromatic light at 2570 Å. Note the Nissl clumps and their distribution. **C.** A similar cell taken one week after the operation. [Brattgård, Edström, and Hydén, 1958.]

nections in frogs. Recent work in many laboratories, however, has shown that central regeneration of axons in mammals is possible to a limited degree if the glial barrier is prevented. But recovery of any considerable function attributable to regeneration or reunited connections after severe central lesions in mammals has not yet been achieved in adults.

Bibliography

The references assembled here are representative, especially of the literature in lower forms, but are by no means exhaustive. Many older papers in the bibliographies of the systematic chapters include important cytological and histological sections. Neurosecretion is included in the bibliography of Chapter 6. Only a small sampling of papers in cytochemistry is included.

AITKEN, J. T., SHARMAN, M., and YOUNG, J. Z. 1947. Maturation of regenerating nerve fibres with various peripheral connexions. *J. Anat., Lond.*, 81:1-22.

ALEXANDROWICZ, J. S. 1932. The innervation of the heart of the crustacea. I. Decapoda. *Quart. J. micr. Sci.*, 75:181-249.

ALLEN, E. J. 1896. Studies on the nervous system of Crustacea. IV. Further observations on the nerve elements of the embryonic lobster. *Quart. J. micr. Sci.*, 39:33-50.

ANDERSSON-CEDERGREN, E. 1959. Ultrastructure of motor end plate and sarcoplasmic components of mouse skeletal muscle fiber as revealed by three-dimensional reconstructions from serial sections. *J. Ultrastruct. Res. (Suppl.)*, 1:1-191.

ANDRES, K. H. 1961. Untersuchungen über den Feinbau von Spinalganglien. *Z. Zellforsch.*, 55:1-48.

APÁTHY, S. 1897. Das leitende Element des Nervensystems und seine topographischen Beziehungen zu den Zellen. *Mitt. zool. Sta. Neapel*, 12:495-748.

APÁTHY, S. 1898. Ueber Neurofibrillen und über ihre nervös leitende Natur. *Int. Congr. Zool., Proc. IV*, 1898:125-141.

ARVANITAKI, A. and CARDOT, H. 1941a. Observations sur la constitution des ganglions et conducteurs nerveux et sur l'isolement du soma neuronique vivant chez les mollusques gastéropodes. *Bull. Histol. Tech. micr.*, 18:133-144.

ARVANITAKI, A. and CARDOT, H. 1941b. Contribution à la morphologie du système nerveux des gastéropodes. Isolement, à l'état vivant du corps neuroniques. *C.R. Soc. Biol., Paris*, 135:965.

ARVANITAKI, A. and CHALAZONITIS, N. 1961. Excitatory and inhibitory processes initiated by light and infrared radiations in single identifiable nerve cells (giant ganglion cells of *Aplysia*). In: *Nervous Inhibition*. E. Florey (ed.). Pergamon Press, Oxford.

ARVANITAKI, A. and TCHOU, S. H. 1942. Les lois de la croissance relative individuelle des cellules nerveuses chez l'Aplysie. *Bull. Histol. Tech. micr.*, 19:244-256.

ASHHURST, D. E. 1959. The connective tissue sheath of the locust nervous system: a histochemical study. *Quart. J. micr. Sci.*, 100:401-412.

ASHHURST, D. E. 1961. The cytology and histochemistry of the neurones of *Periplaneta americana*. *Quart. J. micr. Sci.*, 102:399-406.

BACCETTI, B. 1955. Ricerche sulla fine struttura del perilemma nel sistema nervoso degli insetti. *Redia*, 40:197-212.

BAECKER, R. 1932. Die Mikromorphologie von *Helix pomatia* und einigen anderen Stylommatophoren. *Ergebn. Anat. Entw. Gesch.*, 29:449-585.

BAKER, J. R. 1944. The structure and chemical composition of the Golgi element. *Quart. J. micr. Sci.*, 85:1-71.

BAKER, J. R. 1949. Further remarks on the Golgi element. *Quart. J. micr. Sci.*, 90:293-307.

BAKER, J. R. 1959. Towards a solution of the Golgi problem: recent developments in cytochemistry and electron microscopy. *J. R. micr. Soc.*, (3) 77:116-129.

BARBER, S. B. 1960. Structure and properties of *Limulus* articular proprioceptors. *J. exp. Zool.*, 143:283-305.

BARNES, B. G. 1961. Ciliated secretory cells in the pars distalis of the mouse hypophysis. *J. Ultrastruct. Res.*, 5:453-467.

BATHAM, E. J. 1961. Infoldings of the nerve fibre membranes in the opisthobranch mollusc *Aplysia californica*. *J. biophys. biochem. Cytol.*, 9:490-491.

BATHAM, E. J., PANTIN, C. F. A., and ROBSON, E. A. 1960. The nerve-net of the sea-anemone *Metridium senile*: the mesenteries and the column. *Quart. J. micr. Sci.*, 101:487-510.

BAUER, V. 1904. Zur innern Metamorphose des Centralnervensystems der Insecten. *Zool. Jb. (Anat.)*, 20:123-152.

BEAMS, H. W. and KING, R. L. 1932. Cytoplasmic structures in the ganglion cells of certain Orthoptera, with special reference to the Golgi bodies, mitochondria, "vacuome", intracellular trabeculae (trophospongium), and neurofibrillae. *J. Morph.*, 53:59-96.

BEAMS, H. W., SEDAR, A. W., and EVANS, T. C. 1953. Studies on the neurons of the grasshopper with special reference to the Golgi bodies, mitochondria and neurofibrillae. *Cellule*, 55:293-304.

BEAR, R. S. and SCHMITT, F. O. 1937. Optical properties of the axon sheaths of crustacean nerves. *J. cell. comp. Physiol.*, 9:275-287.

BEAR, R. S. and SCHMITT, F. O. 1939. Electrolytes in the axoplasm of the giant nerve fibers of the squid. *J. cell. comp. Physiol.*, 14:205-215.

BEAR, R. S., SCHMITT, F. O., and YOUNG, J. Z. 1937a. The sheath components of the giant nerve fibres of the squid. *Proc. roy. Soc. (B)*, 123:496-504.

BEAR, R. S., SCHMITT, F. O., and YOUNG, J. Z. 1937b. The ultrastructure of nerve axoplasm. *Proc. roy. Soc. (B)*, 123:505-519.

BEAR, R. S., SCHMITT, F. O., and YOUNG, J. Z. 1937c. Investigations on the protein constituents of nerve axoplasm. *Proc. roy. Soc. (B)*, 123:520-529.

BERN, H. A., NISHIOKA, R. S., and HAGADORN, I. R. 1961. Association of elementary neurosecretory granules with the Golgi complex. *J. Ultrastruct. Res.*, 5:311-320.

BETHE, A. 1897. Das Nervensystem von *Carcinus maenas*. Ein anatomisch-physiologischer Versuch. I. Theil. I. Mittheilung. *Arch. mikr. Anat.*, 50:460 546.

BETHE, A. 1903. *Allgemeine Anatomie und Physiologie des Nervensystems*. Thieme, Leipzig.

BETHE, A. 1908. Ein neuer Beweis für die leitende Funktion der Neurofibrillen nebst Bemerkungen über die Reflexzeit, Hemmungszeit und Latenzzeit des Muskels beim Blutegel. *Pflüg. Arch. ges. Physiol.*, 122:1-36.

BIAKOWSKA, W. and KULIKOWSKA, Z. 1911. Über den Golgi-Kopschschen Apparat der Nervenzellen bei den *Hirudineen* und *Lumbricus*. *Anat. Anz.*, 38:193-207.

BIEDERMANN, W. 1887. Zur Kenntniss der Nerven und Nervenendungen in den quergestreiften Muskeln der Wirbellosen. *S. B. Akad. Wiss. Wien*, (3) 96:8-39.

BIELSCHOWSKY, M. 1903. Die Silberimprägnation der Neurofibrillen. *Neurol. Centralbl.*, 22:997-1006.

BIRKS, R. I. and MACINTOSH, F. C. 1957. Acetylcholine metabolism at nerve-endings. *Brit. med. Bull.*, 13:157-161.

BLEST, A. D. 1961. Some modifications of Holmes's silver method for insect central nervous systems. *Quart. J. micr. Sci.*, 102:413-417.

BODIAN, D. 1942. Cytological aspects of synaptic function. *Physiol. Rev.*, 22:146-169.

BODIAN, D. 1952. Introductory survey of neurons. *Cold Spr. Harb. Symp. quant. Biol.*, 17:1-13.

BODIAN, D. and MELLORS, R. C. 1945. The regenerative cycle of motoneurons, with special reference to phosphatase activity. *J. exp. Med.*, 81:469-488.

BOEKE, J. 1938a. Über die Verbindungen der Nervenzellen untereinander und mit den Erfolgsorganen. *Anat. Anz.*, 85 (Suppl. H): 111-141.

BOEKE, J. 1938b. Sympathetic groundplexus and reticuline fibres. *Anat. Anz.*, 86:150-162.

BOEKE, J. 1949. The sympathetic endformation, its synaptology, the interstitial cells, the periterminal network and its bearing on the neurone theory. *Acta anat.*, 8:18-61.

BOELL, E. J. and NACHMANSOHN, D. 1940. Choline esterase in nerve fibers. *Biol. Bull., Woods Hole*, 79:357.

BOGORAZE, D. and CAZAL, P. 1944. Recherches histologiques sur le système nerveux du poulpe. Les neurones, le tissu interstitiel et les éléments neuricrines. *Arch. Zool. exp. gén.*, 83:412-444.

BOULÉ, L. 1908. Recherches sur le système nerveux central normal du lombric. *Névraxe*, 10:15-59.

BOVARD, J. F. 1918. The function of the giant fibers in earthworms. *Univ. Calif. Publ. Zool.*, 18:135-144.

BOYCOTT, B. B., GRAY, E. G., and GUILLERY, R. W. 1960. A theory to account for the absence of boutons in silver preparations of the cerebral cortex, based on a study of axon terminals by light and electron microscopy. *J. Physiol.*, 152:3P-5P.

BOYLE, W. 1937. The cytology of the neurones of *Helix aspersa*. *J. R. micr. Soc.*, (3) 57:243-254.

BOZLER, E. 1927. Untersuchungen über das Nervensystem der Coelenteraten. I. Kontinuität oder Kontakt zwischen den Nervenzellen? *Z. Zellforsch.*, 5:244-262.

BRACHET, J. and MIRSKY, A. E. (eds.). 1959-1961. *The Cell*. Academic Press, New York. 5 v.

BRATTGÅRD, S.-O., EDSTRÖM, J. E., and HYDÉN, H. 1958. The productive capacity of the neuron in retrograde reaction. *Exp. Cell Res.*, Suppl. 5: 185-200.

BREITSCHNEIDER, L. H. 1952. The fine structure of protoplasm. In: *Surv. biol. Progr.*, 2:223-257.

BRUNO, G. 1931. La cellula nervosa dei gangli cerebrali degli ortotteri. *Boll. Soc. ital. Biol. sper.*, 6:1001-1004.

BRYANT, S. H. and TOBIAS, J. M. 1955. Optical and mechanical concomitants of activity in *Carcinus* nerve. I. Effect of sodium on the optical response. II. Shortening of the nerve with activity. *J. cell. comp. Physiol.*, 46:71-95.

BUCY, P. C. (ed.). 1944. *The Precentral Motor Cortex*. Univ. Illinois Press, Urbana, Illinois.

BULLOCK, T. H. 1952. The invertebrate neuron junction. *Cold Spr. Harb. Symp. quant. Biol.*, 17:267-273.

BULLOCK, T. H. 1961. On the anatomy of the giant neurons of the visceral ganglion of *Aplysia*. In: *Nervous Inhibition*. E. Florey (ed.). Pergamon Press, New York.

BUMKE, O. and FOERSTER, O. (eds.). 1935-1937. *Handbuch der Neurologie*. Springer, Berlin. 17 v.

BYKOV, K. M. (ed.). 1958. *Textbook of Physiology*. (Transl. by S. Belsky and D. Myshne.) Foreign Lang. Publ. House, Moscow.

CAESAR, R., EDWARDS, G. A., and RUSKA, H. 1957. Architecture and nerve supply of mammalian smooth muscle tissue. *J. biophys. biochem. Cytol.*, 3:867-877.

CAIN, A. J. 1948. The accumulation of carotenoids in the Golgi apparatus of neurones of *Helix*, *Planorbis* and *Limnaea*. *Quart. J. micr. Sci.*, 89:421-428.

CAJAL, S. R. Y. 1903. Sobre la existencia de un aparato tubuliforme en el protoplasma de las celulas nerviosas y epiteliales de la lombriz de tierra. *Bol. Soc. esp. Hist. nat.*, 3:395-398.

CAJAL, S. R. Y. 1904. Variaciones morfológicas del retículo nervioso de invertebrados y vertebrados sometidos á la acción de condiciones naturales (nota preventiva). *Trab. Lab. Invest. biol. Univ. Madr.*, 3:287-297.

CAJAL, S. R. Y. 1909. Nota sobre la estructura de la retina de la mosca. *Trab. Lab. Invest. biol. Univ. Madr.*, 7:217-257.

CAJAL, S. R. Y. 1909-1911. *Histologie du Système Nerveux de l'Homme et des Vertébrés*. Edition française revue et mise à jour par l'auteur. Traduite de l'espagnol par L. Azoulay. Maloine, Paris. 2 v. (Republished 1952 by Consejo Superior de Investigaciones Científicas, Madrid.)

CAJAL, S. R. Y. 1917. Contribución al conocimiento de la retina y centros ópticos de los cefalópodos. *Trab. Lab. Invest. biol. Univ. Madr.*, 15:1-82.

CAJAL, S. R. Y. 1918. Observaciones sobre la estructura de los ocelos y vias nerviosas ocelares de algunos insectos. *Trab. Lab. Invest. biol. Univ. Madr.*, 16:109-139.

CAJAL, S. R. Y. 1928. *Degeneration and Regeneration in the Nervous System*. Oxford Univ. Press, Oxford. 2 v.

CAJAL, S. R. Y. 1934. Les preuves objectives de l'unité anatomique des cellules nerveuses. *Trav. Lab. Rech. biol. Univ. Madr.*, 29:1-137.

CAJAL, S. R. Y. 1954. *Neuron Theory or Reticular Theory?* (Transl. by M. Ubeda Purkiss and C. A. Fox, from the Spanish version, 1934.) Inst. Ramón y Cajal, C.S.I.S., Madrid.

CAJAL, S. R. Y and SÁNCHEZ, D. 1915. Contribución al conocimiento de los centros nerviosos de los insectos. *Trab. Lab. Invest. biol. Univ. Madr.*, 13:1-164.

CARLISLE, D., DUPONT-RAABE, M., and KNOWLES, F. 1955. Recherches préliminaires relatives à la séparation et à la comparaison des substances chromactives des crustacés et des insectes. *C. R. Acad. Sci., Paris*, 240:665-667.

CASTRO, F. DE. 1930. Recherches sur la dégénération du système nerveux sympathique. *Trab. Lab. Invest. biol. Univ. Madr.*, 26:357-456.

CASTRO, F. DE. 1932. Sympathetic ganglia, normal and pathological. In: *Cytology and Cellular Pathology of the Nervous System*. W. Penfield (ed.). Vol. 1. Hoeber, New York.

CASTRO, F. DE. 1942. Nuevas ideas sobre la sinapsis. *Trab. Lab. Invest. biol. Univ. Madr.*, 34:217-301.

CASTRO, F. DE. 1950. Die normale Histologie des peripheren vegetativen Nervensystems. Das Synapsen-Problem: anat. exp. Untersuchungen. *Verh. dtsch. Ges. Path.* (34 Tagg), 1950:1-52.

CAUNA, N. 1962. The submicroscopical relationship between nervous and non-nervous elements of the cutaneous receptor organs and its significance. *J. Anat., Lond., (Proc.)*, 96:17-19.

CAUSEY, G. and BARTON, A. A. 1958. Synapses in the superior cervical ganglion and their changes under experimental conditions. In: *The Submicroscopic Organization and Function of Nerve Cells*. H. Fernández-Morán and R. Brown (eds.). Academic Press, New York. (Exp. Cell Res., Suppl. 5)

CHALAZONITIS, N. 1961. Chemopotentials in giant nerve cells (*Aplysia fasciata*). In: *Nervous Inhibition*. E. Florey (ed.). Pergamon Press, Oxford.

CHALAZONITIS, N. and LENOIR, J. 1959. Ultrastructure et organisation du neurone d'*Aplysia*. Étude au microscope électronique. *Bull. Inst. océanogr. Monaco*, 56(1144):1-11.

CHAMBERS, R. 1947. The shape of oil drops injected into the axoplasm of the giant nerve of the squid. *Biol. Bull., Woods Hole*, 93:191.

CHAMBERS, R. and KAO, C.-Y. 1951. The physical state of the axoplasm in situ in the nerve of the squid mantle. *Biol. Bull., Woods Hole*, 101:206.

CHAMBERS, R. and KAO, C.-Y. 1952. The effect of electrolytes on the physical state of the nerve axon of the squid and of *Stentor*, a protozoan. *Exp. Cell Res.*, 3:564-573.

CHANCE, M. R. A., LUCAS, A. J., and WATERHOUSE, J. A. H. 1956. Changes in the dimensions of the nuclei of neurones with activity. *Nature, Lond.*, 177:1081-1082.

CHANG, H.-T. 1952. Cortical neurons with particular reference to the apical dendrites. *Cold Spr. Harb. Symp. quant. Biol.*, 17:189-201.

Bibliography

CHANG, P.-I. 1951. The action of DDT on the Golgi bodies in insects nervous tissue. *Ann. ent. Soc. Amer.*, 44:311-326.

CHOU, J. T. Y. 1957. The cytoplasmic inclusions of the neurones of *Helix aspersa* and *Limnaea stagnalis*. *Quart. J. micr. Sci.*, 98:47-58.

CHOU, J. T. Y. and MEEK, G. A. 1958. The ultra-fine structure of lipid globules in the neurones of *Helix aspersa*. *Quart. J. micr. Sci.*, 99:279-284.

CLAYTON, D. E. 1932. A comparative study of the non-nervous elements in the nervous system of invertebrates. *J. Ent. Zool.*, 24:3-22.

CLEMENTE, C. D. 1955. Structural regeneration in the mammalian central nervous system and the role of neuroglia and connective tissue. In: *Regeneration in the Central Nervous System*. W. F. Windle (ed.). Thomas, Springfield, Ill.

Cold Spring Harbor Symposia on Quantitative Biology 1952. Vol. 17. *The Neuron*. Long Island Biol. Ass., New York.

CONEL, J. L. 1939-1959. *The Postnatal Development of the Human Cerebral Cortex*. Harvard Univ. Press, Cambridge. 6 v.

COUTEAUX, R. 1955. Les jonctions intertissulaires. In: *Problèmes de Structures, d'Ultrastructures et de Fonctions Cellulaires*. J. A. Thomas (ed.). Masson, Paris.

COUTEAUX, R. 1956. Neurofilaments et neurofibrilles dans les fibres nerveuses de la sangsue. In: *Electron Microscopy*, Proc. Stockholm Conf., Sept. 1956. F. S. Sjöstrand and J. Rhodin (eds.). Almqvist & Wiksell, Stockholm.

COUTEAUX, R. 1958. Morphological and cytochemical observations on the post-synaptic membrane at motor end-plates and ganglionic synapses. *Exp. Cell Res. (Suppl.)*, 5:294-322.

COUTEAUX, R. 1960. Motor end-plate structure. In: *The Structure and Function of Muscle*. Vol. 1. Structure. G. H. Bourne (ed.). Academic Press, New York.

COWDRY, E. V. 1932. *Special Cytology*. 2nd ed. Vol. 3. Hoeber, New York.

CRAGG, B. G. 1951. The turgidity of giants axons. *J. Physiol.*, 114:234-239.

DANN, L. 1938. La grandezza delle cellule nervose in vari individui della stessa specie. Ricerche su feti umani a termine. *Riv. Biol. Firenze*, 26:413-424.

DAVID, G. B., BROWN, A. W., and MALLION, K. B. 1961. On the density of the "neurofibrils", "Nissl complex", "Golgi apparatus", and "trophospongium" in the neurones of vertebrates. *Quart. J. micr. Sci.*, 102:481-494.

DEBAISIEUX, P. 1952. Histologie et histogenèse chez *Chirocephalus diaphanus* Prev. *Cellule*, 54:253-294.

DEHORNE, A. 1935a. Sur le neuroplasme des fibres géantes des polychètes. *C. R. Soc. Biol., Paris*, 119:1253-1256.

DEHORNE, A. 1935b. Sur le trophosponge des cellules nerveuses géantes de *Lanice conchylega*, Pallas. *C. R. Soc. Biol., Paris*, 120:1188-1190.

DEHORNE, A. 1936. Analyse de quelques aspects des cellules nerveuses de *Nephthys* et de *Nereis*. *C. R. Soc. Biol., Paris*, 121:757-760.

DEITCH, A. D. and MURRAY, M. R. 1956. The Nissl substance of living and fixed spinal ganglion cells. *J. biophys. biochem. Cytol.*, 2:433-444.

DE LORENZO A. J. 1957. Electron microscopic observations of the olfactory mucosa and olfactory nerve. *J. biophys. biochem. Cytol.*, 3:839-850.

DE LORENZO, A. J. 1958. Electron microscopic observations on the taste buds of the rabbit. *J. biophys. biochem. Cytol.*, 4:143-150.

DE LORENZO, A. J. 1959. The fine structure of synapses. *Biol. Bull., Woods Hole*, 117:390.

DE LORENZO, A. J. 1960a. Electron microscopy of electrical synapses in the crayfish. *Biol. Bull., Woods Hole*, 119:325.

DE LORENZO, A. J. 1960b. The fine structure of synapses in the ciliary ganglion of the chick. *J. biophys. biochem. Cytol.*, 7:31-36.

DE RÉNYI, G. S. 1929 Observations of neurofibrils in the living nervous tissue of the lobster (*Homarus americanus*). *J. comp. Neurol.*, 48:441-457.

DESMEDT, J. E. and MONACO, P. 1961. Mode of action of the efferent olivocochlear bundle on the inner ear. *Nature, Lond.*, 192:1263-1265.

DOLLEY, D. H. 1913. The morphology of functional activity in the ganglion cells of the crayfish, *Cambarus virilis*. *Arch. Zellforsch.*, 9:485-551.

DUDEL, J. and KUFFLER, S. W. 1961. The quantal nature of transmission and spontaneous miniature potentials at the crayfish neuromuscular junction. *J. Physiol.*, 155:514-529.

EAKIN, R. M. and WESTFALL, J. A. 1959. Fine structure in the retina of the reptilian third eye. *J. biophys. biochem. Cytol.*, 6:133-134.

EASTON, D. M. 1950. Synthesis of acetylcholine in crustacean nerve and nerve extract. *J. biol. Chem.*, 185(2):813-816.

EASTON, D. M. 1956. Some tensile properties of nerve. *J. cell. comp. Physiol.*, 48:87-94.

ECCLES, J. C. 1961. The mechanism of synaptic transmission. *Ergebn. Physiol.*, 51:299-430.

ECCLES, J. C. and JAEGER, J. C. 1958. The relationship between the mode of operation and the dimensions of the junctional regions at synapses and motor end-organs. *Proc. roy. Soc. (B)*, 148:38-56.

EDWARDS, G. A. 1957. Electron microscope observations of annelid muscle and nerve. *Anat. Rec.*, 128:542-543.

EDWARDS, G. A. 1958. Comparative studies on the fine structure of motor units. *Int. Congr. Electr. Micr.*, 4:301-308.

EDWARDS, G. A. 1959. The fine structure of a multiterminal innervation of an insect muscle. *J. biophys. biochem. Cytol.*, 5:241-244.

EDWARDS, G. A., RUSKA, H., and HARVEN, E. DE. 1958a. Electron microscopy of peripheral nerves and neuromuscular junctions in the wasp leg. *J. biophys. biochem. Cytol.*, 4:107-114.

EDWARDS, G.A., RUSKA, H., and HARVEN, E. DE. 1958b. Neuromuscular junctions in flight and tymbal muscles of the cicada. *J. biophys. biochem. Cytol.*, 4:251-256.

EISNER, T. 1953. The histology of a sense organ in the labial palps of Neuroptera. *J. Morph.*, 93:109-121.

ELFVIN, L.-G. 1961a. Electron-microscopic investigation of filament structures in unmyelinated fibers of cat splenic nerve. *J. Ultrastruct. Res.*, 5:51-64.

ELFVIN, L.-G. 1961b. The ultrastructure of the nodes of Ranvier in cat sympathetic nerve fibers. *J. Ultrastruct. Res.*, 5:374-387.

ENGSTRÖM, A. and FINEAN, J. B. 1958. *Biological Ultrastructure*. Academic Press, New York.

ENGSTRÖM, H. and WERSÄLL, J. 1958. The ultrastructural organization of the organ of Corti and of the vestibular sensory epithelia. *Exp. Cell Res. (Suppl.)*, 5:460-492.

ERHARD, H. 1912. Studien über Nervenzellen. I. Allgemeine Grössenverhältnisse, Kern, Plasma und Glia. Nebst einem Anhang: Das Glykogen im Nervensystem. *Arch. Zellforsch.*, 8:442-547.

ESTABLE, C. 1923. Notes sur la structure comparative de l'écorce cérébelleuse, et dérivées physiologiques possibles. *Trab. Lab. Invest. biol. Univ. Madr.*, 21:169-256.

ESTABLE, C., REISSIG, M., and ROBERTIS, E. DE. 1953. The microscopic and submicroscopic structure of the synapses in the ventral ganglion of the acoustic nerve. *J. appl. Physics*, 24:1421-1422.

FEDOROW, B. 1927. Über den Bau der Riesenganglienzellen der Lumbriconereinen. *Z. mikr-anat. Forsch.*, 12:347-370.

FERNÁNDEZ-MORÁN, V. H. 1953. La organización submicroscópica del segmento interanular de las fibras nerviosas meduladas en los vertebrados. Estudio electrono-microscópico de fibras nerviosas del adulto humano,

de la rata, el gato y la rana. Litografia del Comercio, Caracas.
FERNÁNDEZ-MORÁN, H. 1954. The submicroscopic structure of nerve fibres. *Progr. Biophys.*, 4:112-146.
FERNÁNDEZ-MORÁN, H. 1956. Fine structure of the insect retinula as revealed by electron microscopy. *Nature, Lond.*, 177:742-743.
FERNÁNDEZ-MORÁN, H. 1957. *Electron Microscopy of Nerve Tissue.* Instituto Venezolano de Neurologia e Investigaciones Cerebrales, Caracas.
FERNÁNDEZ-MORÁN, H. 1958. Fine structure of the light receptors in the compound eyes of insects. *Exp. Cell Res.*, 5:586-644.
FERNÁNDEZ-MORÁN, H. and BROWN, R. (eds.). 1958. *The Submicroscopic Organization and Function of Nerve Cells.* Exp. Cell Res., Suppl. 5. Academic Press, New York.
FERNÁNDEZ-MORÁN, H. and FINEAN, J. B. 1957. Electron microscope and low-angle x-ray diffraction studies of the nerve myelin sheath. *J. biophys. biochem. Cytol.*, 3:725-748.
FESSARD, A. 1955. Ultra-structures et fonctions du neurone. In: *Problèmes de Structures, d'Ultrastructures et de Fonctions Cellulaires.* J. A. Thomas (ed.). Masson, Paris.
FLAIG, J. V. 1947. Viscosity changes in axoplasm under stimulation. *J. Neurophysiol.*, 10:211-221.
FOSTER, M. and SHERRINGTON, C. S. 1897. The central nervous system. In: *A Text Book of Physiology.* 7th ed. Pt. III. Macmillan, London.
FRANKENHAEUSER, B. and HODGKIN, A. L. 1956. The after-effects of impulses in the giant nerve fibres of *Loligo. J. Physiol.*, 131:341-376.
FREIDENFELT, T. 1905. Über den feineren Bau des Visceralganglions von *Anodonta. Acta Univ. lund.*, Ad. 2, 40(5):1-28.
FUKUDA, T. and KOELLE, G. B. 1959. The cytological localization of intracellular neuronal acetylcholinesterase. *J. biophys. biochem. Cytol.*, 5:433-440.
FULTON, J. F. 1949. *Physiology of the Nervous System.* 3rd ed. Oxford University Press, New York.
GALAMBOS, R. 1961. The glial-neural theory of brain function. *Proc. nat. Acad. Sci., Wash.*, 47:129-136.
GASSER, H. S. 1952. Discussion. The hypothesis of saltatory conduction. *Cold Spr. Harb. Symp. quant. Biol.*, 17:32-36.
GASSER, H. S. 1958. Comparison of the structure, as revealed with the electron microscope, and the physiology of the unmedullated fibers in the skin nerves and in the olfactory nerves. In: *The Submicroscopic Organization and Function of Nerve Cells.* H. Fernández-Morán and R. Brown (eds.). Academic Press, New York. (Exp. Cell Res., Suppl. 5.)

GATENBY, J. B. 1929. Study of Golgi apparatus and vacuolar system of *Cavia, Helix* and *Abraxas*, by intravital methods. *Proc. roy. Soc. (B)*, 104:302-321.
GATENBY, J. B., MOUSSA, T. A., ELBANHAWY, M., and GORNALL, J. I. K. 1953. Ciaccio bodies and the life of the neurone. *Cellule*, 55:139-164.
GEMELLI, A. 1905a. Sopra le neurofibrille delle cellule nervose dei vermi secondo un nuovo metodo di dimostrazione *Anat. Anz.*, 27:449-462.
GEMELLI, A. 1905b. Su di una fine particolarità di struttura delle cellule nervose dei vermi. *Riv. Fis. Mat. Sci. nat.*, 6:518-532.
GEREN, B. B. and SCHMITT, F. O. 1954. The structure of the Schwann cell and its relation to the axon in certain invertebrate nerve fibers. *Proc. nat. Acad. Sci.*, 40:863-870.
GLEES, P. 1958. The biology of the neuroglia: a summary. In: *Biology of Neuroglia.* W. F. Windle (ed.). Thomas, Springfield, Ill.
GOLGI, C. 1898. Sur la structure des cellules nerveuses. *Arch. ital. Biol.*, 30:60-71.
GOLOVINA, N. V. 1949. Substantial changes of the nerve fiber during excitation. Diss., Inst. Eksp. Med. SSSR, Akad. Nauk, Leningrad. (In Russian).
GOLOVINA, N. V. 1955a. Change of absorptive properties of cerebrovisceral connective in *Anodonta* under excitation. *Sechenov J. Physiol. (Fiziol. Zh. SSSR)*, 41:822-829. (In Russian.)
GOLOVINA, N. V. 1955b. Changes in nerve fibers on excitation. *C. R. Acad. Sci. URSS (Dokl. Akad. Nauk)*, 105:1378-1381. (In Russian.)
GRAY, E. G. 1959a. Axo-somatic and axo-dendritic synapses of the cerebral cortex: an electron microscope study *J. Anat., Lond.*, 93:420-433.
GRAY, E. G. 1959b. Electron microscopy of neuroglial fibrils of the cerebral cortex. *J. biophys. biochem. Cytol.*, 6:121-122.
GRAY, E. G. 1960. The fine structure of the insect ear. *Philos. Trans. (B)*, 243:75-94.
GRAY, E. G. 1961. The granule cells, mossy synapses and Purkinje spine synapses of cerebellum, light and electron microscope observations. *J. Anat., Lond.*, 95:345-356.
GRAY, E. G. 1962. A morphological basis for pre-synaptic inhibition? *Nature, Lond.*, 193:82-83.
GRAY, E. G. and HAMLYN, L. H. 1962. Electron microscopy of experimental degeneration in the optic tectum of the chicken. *J. Physiol.*, 162: 39P-41P.
GRAZIADEI, P. 1960a. Osservazioni sulla struttura delle cellule nervose nel ganglio stellato di *Sepia officinalis. R. C. Accad. Lincei*, (8) 28:686-690.

GRAZIADEI, P. 1960b. Sulla struttura delle cellule nervose e dell'intreccio neuropilare nel ganglio stellato di *Sepia officinalis. Arch. ital. Anat. Embriol.*, 65:269-283.
GREEN, J. D. and MAXWELL, D. S. 1959. Comparative anatomy of the hypophysis and observations on the mechanism of neurosecretion. In: *Comparative Endocrinology.* A. Gorbman (ed.). Wiley, New York.
GRESSON, R. A. R., THREADGOLD, L. T., and STINSON, N. E. 1956. The Golgi elements of the neurones of *Helix, Locusta* and *Lumbricus. Cellule.* 58:7-16.
GRIMSTONE, A. V., HORNE, R. W., PANTIN, C. F. A., and ROBSON, E. A. 1958. The fine structure of the mesenteries of the sea anemone *Metridium senile. Quart. J. micr. Sci.*, 99:523-540.
GROBBEN, K. 1911. Die Bindesubstanzen von *Argulus.* Ein Beitrag zur Kenntnis der Bindesubstanz der Arthropoden. *Arb. zool. Inst. Univ. Wien*, 19:74-98.
HAGADORN, I. R. and NISHIOKA, R. S. 1961. Neurosecretion and granules in neurones of the brain of the leech. *Nature. Lond.*, 191:1013-1014.
HAGER, H. 1961. Ergebnisse der Elektronenmikroskopie am zentralen, peripheren und vegetativen Nervensystem. *Ergebn. Biol.*, 24:106-154.
HALLER, B. 1905. Ueber den allgemeinen Bauplan des Tracheatensyncerebrums. *Arch. mikr. Anat.*, 65:181-279.
HAMA, K. 1953. A cytological study of the ganglion cells of the snail. *Kyûshû Mem. Med. Sci.*, 4:17-23.
HAMA, K. 1959. Some observations on the fine structure of the giant nerve fibers of the earthworm, *Eisenia foetida. J. biophys. biochem. Cytol.*, 6:61-66.
HAMA, K. 1961a. Some observations on the fine structure of the giant fibers of the crayfishes (*Cambarus virilis* and *Cambarus clarkii*) with special reference to the submicroscopic organization of the synapses. *Anat. Rec.*, 141:275-294.
HAMA, K. 1961b. The fine structure of some electrical synapses. *Sci. Living Body*, 12:72-84. (In Japanese.)
HAMA, K. 1962. Some observations on the fine structure of the giant synapse in the stellate ganglion of the squid. *Z. Zellforsch.*, 56:437-444.
HAMMARSTEN, O. D. and RUNNSTRÖM, J. 1926. Ein Beitrag zur Diskussion über die Verwandtschaftsbeziehungen der Mollusken. *Acta zool., Stockh.*, 7:1-67.
HÁMORI, J. 1961a. Cholinesterases in insect muscle innervation with special reference to insecticide effects of DDT and DFP. *Bibl. anat.*, 2:194-206.
HÁMORI, J. 1961b. Innervation of insect

Bibliography

leg muscle. *Acta biol. hung.*, 12:219-230.

HANSTRÖM, B. 1928. Some points on the phylogeny of nerve cells, and of the central nervous system of invertebrates. *J. comp. Neurol.*, 46:475-493.

HARVEN, E. DE and COËRS, C. 1959. Electron microscope study or the human neuromuscular junction. *J. biophys. biochem. Cytol.*, 6:7-10.

HAVET, J. 1916. Contribution à l'étude de la névroglie des invertébrés. *Trab. Lab. Invest. biol. Univ. Madr.*, 14:35-85.

HERTWECK, H. 1931. Anatomie und Variabilität des Nervensystems und der Sinnesorgane von *Drosophila melanogaster* (Meigen). *Z. wiss. Zool.*, 139:559-663.

HESS, A. 1958a. The fine structure and morphological organization of the peripheral nerve-fibres and trunks of the cockroach (*Periplanata americana*). *Quart. J. micr. Sci.*, 99:333-340.

HESS, A. 1958b. The fine structure of nerve cells and fibers, neuroglia, and sheaths of ganglion chain in the cockroach (*Periplaneta americana*). *J. biophys. biochem. Cytol.*, 4:731-742.

HESS, A. 1958c. Experimental anatomical studies of pathways in the severed central nerve cord of the cockroach. *J. Morph.*, 103:479-502.

HESS, A. 1960. The fine structure of degenerating nerve fibers, their sheaths, and their terminations in the central nerve cord of the cockroach (*Periplaneta americana*). *J. biophys. biochem. Cytol.*, 7:339-344.

HILD, W., REISER, K. A., and LEHMANN, H. J. 1959. *Handbuch der Mikroskopischen Anatomie des Menschen. IV. Nervensystem* Part 4. *Das Neuron. Die Nervenzelle. Die Nervenfaser.* Springer, Berlin.

HILL, D. K. 1950. The volume change resulting from stimulation of a giant nerve fibre. *J. Physiol.* 111:304-327.

HILLARP, N.-Å. 1946-1947. Structure of the synapse and the peripheral innervation apparatus of the autonomic nervous system. *Acta anat.*, 2 (Suppl. 4):1-153.

HILLARP, N.-Å. 1959. The construction and functional organization of the autonomic apparatus. *Acta physiol. scand.*, 46 (Suppl. 157):1-38.

HILTON, W. A. 1911. The structure of the nerve cells of an insect. *J. comp. Neurol.*, 21:373-382.

HODGE, M. H. and CHAPMAN, G. B. 1958. Some observations on the fine structure of the sinus gland of a land crab *Gecarcinus lateralis*. *J. biophys. biochem. Cytol.*, 4:571-574.

HOERR, N. L. 1936. The preexistence of neurofibrillae and their disposition in the nerve fiber. *Anat. Rec.*, 66:81-90.

HOLMES, W., PUMPHREY, R. J., and YOUNG, J. Z. 1941. The structure and conduction velocity of the medullated nerve fibres of prawns. *J. exp. Biol.*, 18:50-54.

HOLMGREN, E. 1914. Trophospongium und Apparato reticulare der spinal Ganglienzellen. *Anat. Anz.*, 46:127-138.

HÖPKER, W. 1953. Über den Nucleolus der Nervenzelle. *Z. Zellforsch.*, 38:218-229.

HORRIDGE, A. 1954. Observations on the nerve fibres of *Aurellia aurita*. *Quart. J. micr. Sci.*, 95:85-92.

HORRIDGE, A. 1956. The nervous system of the ephyra larva of *Aurellia aurita*. *Quart. J. micr. Sci.*, 97:59-74.

HORRIDGE, G. A. 1958a. The co-ordination of the responses of *Cerianthus* (Coelenterata). *J. exp. Biol.*, 35:369-382.

HORRIDGE, G. A. 1958b. Transmission of excitation through the ganglia of *Mya* (Lamellibranchiata). *J. Physiol.*, 143:553-572.

HORRIDGE, G. A. 1961. The organization of the primitive central nervous system as suggested by examples of inhibition and the structure of neuropile. In: *Nervous Inhibition.* E. Florey (ed.). Pergamon Press, Oxford.

HORRIDGE, G. A., CHAPMAN, D. M., and MACKAY, B. 1962. Naked axons and symmetrical synapses in an elementary nervous system. *Nature, Lond.*, 193:899-900.

HOSSELET, C. 1929a. Chondriome à formes d'éléments golgiens dans la cellule nerveuse des insectes. *C. R. Soc. Biol., Paris*, 100:1075-1077.

HOSSELET, C. 1929b. Les éléments du chondriome dans les espaces nerveux intercellulaires et dans le nerf, chez les insectes. *C. R. Soc. Biol., Paris*, 101:85-87.

HOYLE, G. 1953. Potassium ions and insect nerve muscle. *J. exp. Biol.*, 30:121-135.

HYDÉN, H. 1952. *Chemische Komponenten der Nervenzelle und ihre Veränderungen im Alter und während der Funktion.* Mosbacher Gespräche. Springer, Berlin.

HYDÉN, H. 1960. The neuron. In: *The Cell.* Vol. 4. J. Brachet and A. E. Mirsky (eds.). Academic Press, New York.

HYDÉN, H. and LANGE, P. 1961. Differences in the metabolism of oligodendroglia and nerve cells in the vestibular area. In: *Regional Neurochemistry.* S. S. Kety and J. Elkes (eds.). Pergamon Press, New York.

ISSIDORIDES, M. 1956. Ultrastructure of the synapse in the giant axons of the earthworm. *Exp. Cell Res.*, 11:423-436.

JAKUBSKI, A. W. 1912. Zur Kenntnis des Gliagewebes im Nervensystem der Mollusken. *Int. Congr. Zool. 8th. Graz*, 936-939.

JAKUBSKI, A. W. 1913. Studien über das Gliagewebe der Mollusken. I. Lamellibranchiata und Gastropoda. *Z. wiss. Zool.*, 104:81-118.

JAKUBSKI, A. W. 1915. Studien über das Gliagewebe der Mollusken. II. Teil. Cephalopoda. *Z. wiss. Zool.*, 112:48-69.

JONES, W. C. 1962. Is there a nervous system in sponges? *Biol. Rev.*, 37:1-50.

JOSEPH, H. 1902. Untersuchungen über die Stützsubstanzen des Nervensystems, nebst Erörterungen über deren histogenetische und phylogenetische Deutung. *Arb. zool. Inst. Univ. Wien*, 13:335-400.

KAO, C. Y. 1956. Structure of squid axoplasm as revealed by spontaneously formed vesicles. *J. comp. Neurol.*, 104:373-383.

KAPPERS, C. U. ARIËNS. 1934. Differences in the effect of various impulses on the structure of the central nervous system. *Irish J. med. Sci.*, (6)105:495-519.

KAPPERS, C. U. ARIËNS, HUBER, G. C., and CROSBY, E. C. 1936. *The Comparative Anatomy of the Nervous System of Vertebrates, Including Man.* Macmillan, New York. 2 v.

KATZ, B. 1961. The terminations of the afferent nerve fibre in the muscle spindle of the frog. *Philos. Trans. (B)*, 243:221-240.

KETY, S. S. and ELKES, J. 1961. *Regional Neurochemistry.* Pergamon, Oxford.

KEYL, I. 1913. Beiträge zur Kenntnis von *Branchiura sowerbyi* Beddard. *Z. wiss. Zool.*, 107:199-308.

KEYNES, R. D. and LEWIS, P. R. 1951. The sodium and potassium content of cephalopod nerve fibres. *J. Physiol.*, 114:151-182.

KEYNES, R. D. and LEWIS, P. R. 1956. The intracellular calcium contents of some invertebrate nerves. *J. Physiol.*, 134:399-407.

KIDD, M. 1962. Electron microscopy of the inner plexiform layer of the retina in the cat and the pigeon. *J. Anat., Lond.*, 96:179-187.

KING, R. L. and BEAMS, H. W. 1932. The Golgi bodies in the male germ cells and nerve cells of mantids. *Anat. Rec.*, 54(suppl.):94.

KLUSS, B. C. 1958. Light and electron microscope observations on the photogenic organ of the firefly, *Photuris pennsylvanica*, with special reference to the innervation. *J. Morph.*, 103:159-185.

KNOWLES, F. G. W. 1958. Electron microscopy of a crustacean neurosecretory organ. In: *II Internat. Symp. Neurosekretion.* W. Bargmann et al. (eds.). Springer, Berlin.

KOELLE, G. B. 1957. Histochemical demonstration of reversible anticholinesterase action at selective cellular sites in vivo. *J. Pharmacol.*, 120:488-503.

KOELLE, G. B. 1959. Neurohumoral agents as a mechanism of nervous integration. In: *Evolution of Nervous Control from Primitive Organisms to Man.* A. D. Bass (ed.). Amer. Ass. Adv. Sci., Washington, D.C.

KOELLE, G. B. 1961. Evidence for differences in primary functions of acetylcholinesterase at different synapses and neuroeffector junctions. In: *Regional Neurochemistry.* S. S. Kety and Elkes (eds.). Pergamon Press, New York.

KOELLE, W. A. and KOELLE, G. B. 1959. The localization of external or functional acetylcholinesterase at the synapses of autonomic ganglia. *J. Pharmacol.*, 126:1-8.

KORNAKOVA, E. V. and FRANK, G. M. 1952. Change in the mechanical properties of nerve during its stimulation. *C.R. Acad. Sci. URSS (Dokl. Akad. Nauk)*, 87:555-558. (In Russian.)

KOWALSKI, J. 1909. Contribution à l'étude des neurofibrilles chez le lombric. *Cellule*, 25:290-346.

KRAUSE, R. 1921-23. *Mikroskopische Anatomie der Wirbeltiere in Einzeldarstellungen.* De Gruyter, Berlin.

KRAWANY, J. 1905. Untersuchungen über das Zentralnervensystem des Regenwurms. *Arb. zool. Inst. Univ. Wien*, 15:281-316.

KÜHNLE, K. F. 1913. Vergleichende Untersuchungen über das Gehirn, die Kopfnerven und die Kopfdrüsen des gemeinen Ohrwurms (*Forficula auricularia* L.) usf. *Jena. Z. Naturw.*, 50:147-276.

KUMAMOTO, T. and SHIMIZU, N. 1955. Histological studies on alkaline phosphatase and argyrophil fibers of the capsule of spinal ganglion cells. *Zool. Mag., Tokyo*, 64:141.

KUNTZ, A. 1934. *The Autonomic Nervous System.* Lea and Febiger, Philadelphia.

LACROIX, P. 1932. Axone et cellule nerveuse chez les invertébrés. *Bull. Acad. Belg. Cl. Sci.*, (5) 18:282-294.

LACROIX, P. 1935. Recherches cytologiques sur les centres nerveux chez les invertébrés. II. *Astacus fluviatilis. Cellule*, 44:251-270.

LACY, D. 1957. The Golgi apparatus in neurons and epithelial cells of the common limpet *Patella vulgata. J. biophys. biochem. Cytol.*, 3:779-796.

LACY, D. and HORNE, R. 1956. A cytological study of the neurones of *Patella vulgata* by light and electron microscopy. *Nature, Lond.*, 178:976-978.

LADMAN, A. J. 1958. The fine structure of the rod-bipolar cell synapse in the retina of the albino rat. *J. biophys. biochem. Cytol.*, 4:459-465.

LARENTJEW, B. I. 1934. Experimentell-morphologische Studien über den feineren Bau des autonomen Nervensystems. IV. *Z. mikr.-anat. Forsch.*, 35:71-118.

LAURENT, P. 1957. L'Innervation auriculaire du coeur des téléostéens. *Arch. Anat. micr. Morph. exp.*, 46:503-520.

LEGENDRE, R. 1909. Contribution à la connaissance de la cellule nerveuse. La cellule nerveuse d'*Helix pomatia. Arch. Anat. micr.*, 10:287-554.

LEGHISSA, S. 1942. Peculiari conessioni nervose nelle catena ganglionare di *Carausius morosus. Arch. zool. (ital.), Napoli*, 30:289-310.

LENHOSSÉK, M. 1895. *Der feinere Bau des Nervensystems im Lichte neuester Forschungen.* 2nd ed. Fischer, Berlin.

LERMA, B. DE. 1949. Sulla presenza di cellule giganti nel cerebron e nella massa gangliare sottoesofagea dei Coleotteri. *Boll. Zool.*, 16:169-177.

LEVER, J. 1958. On the occurrence of a paired follicle gland in the lateral lobes of the cerebral ganglia of some Ancylidae. *Proc. Acad. Sci. Amst. (C)*, 61:235-242.

LEWIS, M. 1896. Centrosome and sphere in certain of the nerve cells of an invertebrate. *Anat. Anz.*, 12:291-299.

LEWIS, P. R. and SHUTE, C. C. D. 1961. Intra-axonal localisation of acetylcholinesterase in normal brains and distributional changes following tractotomy. In: *Cytology of Nervous Tissue.* Taylor and Francis, London.

LIBET, B. 1947. Localization of adenosinetriphosphatase (ATP-ase) in the giant nerve fiber of the squid. *Biol. Bull., Woods Hole*, 93:219-220.

LIBET, B. 1948. Enzyme localization in the giant nerve fiber of the squid. *Biol. Bull., Woods Hole*, 95:277-278.

LOOS, H. VAN DER. 1960. On dendro-dendritic junctions in the cerebral cortex. In: *Structure and Function of the Cerebral Cortex.* D. B. Tower and J. P. Schadé (eds.). Elsevier, Amsterdam.

LORENTE DE NÓ, R. 1934. Studies on the structure of the cerebral cortex. *J. Psychol. Neurol.*, 45:381-438.

LÜERS, T. 1955. Zur Frage eines Geschlechtsunterschiedes in den Nervenzellkernen von *Drosophila. Z. Naturf.*, 10B:166-168.

LÜERS, T., KÖPF, H., and LÜERS, H. 1954. Über Nervenzellveränderungen bei *Drosophila* nach DDT-Vergiftung. *Biol. Zbl.*, 73:203-212.

LÜERS, T., KÖPF, H., BOCHNIG, V., and LÜERS, H. 1954. Vergleichend histologische Untersuchungen am Zentralnervensystem DDT-resistenter und sensibler Stämme von *Drosophila melanogaster* nach DDT-Einwirkung. *Verh. dtsch. zool. Ges.*, 48:408-412.

LUFT, J. H. 1958. The fine structure of electric tissue. In: *The Submicroscopic Organization and Function of Nerve Cells.* H. Fernández-Morán and R. Brown (eds.). Academic Press, New York. (*Exp. Cell Res.*, Suppl. 5:168-182.)

MACKIE, G. O. 1959. The evolution of the *Chondrophora* (Siphonophora-Disconanthe): new evidence from behavioural studies. *Trans. roy. Soc. Canada (V)*, (3)53:7-20.

MACKIE, G. O. 1960. The structure of the nervous system in *Velella. Quart. J. micr. Sci.*, 101:119-131.

MALHOTRA, S. K. 1955. Golgi bodies in nerve cells of insects. *Nature, Lond.*, 176:886-887.

MALHOTRA, S. K. 1957. The cytoplasmic inclusions of the neurones of the earthworm. *Res. Bull. Punjab Univ.*, 118:367-381.

MALHOTRA, S. K. 1960. The cytoplasmic inclusions of the neurones of crustacea. *Quart. J. micr. Sci.*, 101:75-93.

MALHOTRA, S. K. and MEEK, G. A. 1960. An electron microscope study of some cytoplasmic inclusions of the neurones of the prawn, *Leander serratus. J. R. micr. Soc.*, 80:1-8.

MATHEWSON, R., WACHTEL, A., and GRUNDFEST, H. 1961. Fine structure of electroplaques. In: *Bioelectrogenesis.* C. Chagas and A. Paes de Carvalho (eds.). Elsevier, Amsterdam.

MATURANA, H. R. 1961. The fine anatomy of the optic nerve of anurans—an electron microscope study. *J. biophys. biochem. Cytol.*, 7:107-120.

MATURANA, H. R., LETTVIN, J. Y., MCCULLOCH, W. S., and PITTS, W. H. 1960. Anatomy and physiology of vision in the frog (*Rana pipiens*). *J. gen. Physiol.*, 43:129-175.

MAXFIELD, M. 1951. Studies of nerve proteins. Isolation and physicochemical characterization of a protein from lobster nerve. *J. gen. Physiol.*, 34:853-863.

MAXFIELD, M. 1953. Axoplasmic proteins of the squid giant nerve fiber with particular reference to the fibrous protein. *J. gen. Physiol.*, 37:201-216.

MAXFIELD, M. and HARTLEY, R. W., Jr. 1955. Two proteins purified from lobster nerve extract. Isolation and physiochemical characterization. *J. biophys. biochem. Cytol.*, 1:279-286.

MAXFIELD, M. and HARTLEY, R. W., Jr. 1957. Dissociation of the fibrous protein of nerve. *Biochem. biophys. Acta*, 24:83-87.

MAXIMOW, A. A. and BLOOM, W. 1957. *A Textbook of Histology.* 7th ed. Saunders, Philadelphia.

MAY, R. M. 1933. La formation des terminaisons nerveuses dans les ventouses du bras régénéré du céphalopode *Octopus vulgaris* Lam. *Ann. Sta. oceanogr. Salammbô*, 7:1-15.

MAYNARD, E. A. and MAYNARD, D. M. 1960a. Cholinesterases in the nervous system of the lobster, *Homarus americanus. Anat. Rec.*, 137:380.

MAYNARD, E. A. and MAYNARD, D. M. 1960b. Cholinesterase in the crustacean muscle receptor organ. *J. Histochem. Cytochem.*, 8:376-379.

McAlear, J. H. and Edwards, G. A. 1959. Continuity of plasma membrane and nuclear membrane. *Exp. Cell. Res.*, 16:689-692.

McAlear, J. H., Milburn, N. S., and Chapman, G. B. 1958. The fine structure of Schwann cells, nodes of Ranvier and Schmidt-Lanterman incisures in the central nervous system of the crab, *Cancer irroratus*. *J. Ultrastruct. Res.*, 2:171-176.

Meisenheimer, J., 1898. Entwicklungsgeschichte von *Limax maximus*. II. Die Larvenperiode. *Z. wiss. Zool.*, 63:573-664.

Merton, H. 1907. Über den feineren Bau der Ganglienzellen aus dem Centralnervensystem von *Tethys leporina* Cuv. *Z. wiss. Zool.*, 88:327-357.

Meyer, G. F. 1951. Versuch einer Darstellung von Neurofibrillen im zentralen Nervensystem verschiedener Insekten. *Zool. Jb. (Anat)*, 71:413-426.

Meyer, G. F. 1952. Der feinere Bau der Nervenzellen der Insekten. *Mikrokosmos*, 41:270-273.

Meyer, G. F. 1955a. Vergleichende Untersuchungen mit der supravitalen Methylenblaufärbung am Nervensystem wirbelloser Tiere. (Mit besonderer Berücksichtigung Neurofibrillen bei Evertebraten und Vertebraten, sowie des Bindegewebes der Insekten.) *Zool. Jb. (Anat)*, 74:339-400.

Meyer, G. F. 1955b. Altersveränderungen an Nervenzellen sozialer Insekten. *Mikrokosmos*, 44:209-211.

Meyer, G. F. 1956. Feinhistologische Untersuchungen an Insekten-Neuronen. (Unter Bezugnahme auf *Hirudo medicinalis* L.) *Zool. Jb. (Anat.)*, 75:389-400.

Meyer, G. F. 1957. Elektronenmikroskopische Untersuchungen an den Apáthyschen Neurofibrillen von *Hirudo medicinalis*. *Z. Zellforsch.*, 45:538-542.

Miller, W. H. 1957. Derivatives of cilia in the distal sense cells of retina of *Pecten*. *J. biophys. biochem. Cytol.*, 4:227-228.

Mollenhauer, H. H., Whaley, W. G., and Leech, J. H. 1961. A function of the Golgi apparatus in outer root cap cells. *J. Ultrastruct. Res.*, 5:193-200.

Monné, L. 1930. Vergleichende Untersuchungen über den Golgi-Apparat and das Vacuome in Soma- und Geschlechts-Zellen einiger Gastropoden (*Helix, Paludina, Cerithium*). *Bull. Int. Polska Akad. Umiej.*, *Wyd. Mat.-Prz. (B) 1930*:179-238.

Monti, R. 1915. I condriosomi e gli apparati di Golgi nelle cellule nervosi. *Arch. ital. Anat. Embriol.*, 14:1-45.

Nachmansohn, D. 1959. *Chemical and Molecular Basis of Nerve activity*. Academic Press, New York.

Nansen, F. 1887. *The Structure and Combination of the Histological Elements of the Central Nervous System*. Bergen.

Nasonov, D. N. and Suzdal'skaia, I. P. 1957. Changes in the cytoplasm of myelinated nerve fibres during excitation. *Sechenov J. Physiol.*, 43:617-624.

Nauta, W. J. H. and Kuypers, G. J. M. 1958. Some ascending pathways in the brain stem reticular formation. In: *Reticular Formation of the Brain*. H. H. Jasper et al. (eds.). Little, Brown, Boston.

Nicol, J. A. C. 1948a. The giant nerve-fibres in the central nervous system of *Myxicola* (Polychaeta, Sabellidae). *Quart. J. micr. Sci.*, 89:1-45.

Nicol, J. A. C. 1948b. Giant axons of *Eudistylia vancouveri* (Kinberg). *Trans. roy. Soc. Canada*, 42:107-124.

Ogawa, F. 1939. The nervous system of earthworm (*Pheretima communissima*) in different ages. *Sci. Rep. Tôhoku Univ.*, (4) 13:395-488.

Olszewski, J. and Baxter, D. 1954. *Cytoarchitecture of the Human Brain Stem*. Karger, New York.

Oncley, J. L. (ed.). 1959. *Biophysical Science—a Study Program*. Wiley, New York.

Palay, S. L. 1956a. Structure and function in the neuron. In: *Neurochemistry*, S. R. Korey and J. I. Nurnberger (eds.). Hoeber-Harper, New York.

Palay, S. L. 1956b. Synapses in the central nervous system. *J. biophys. biochem. Cytol.*, Suppl. 2:193-202.

Palay, S. L. 1958a. *Frontiers in Cytology*. Yale Univ. Press, New Haven.

Palay, S. L. 1958b. The morphology of synapses in the central nervous system. In: *The Submicroscopic Organization and Function of Nerve Cells*. H. Fernández-Morán and R. Brown (eds.). Academic Press, New York. (*Exp. Cell Res.*, Suppl. 5:275-293.)

Palay, S. L. and Palade, G. E. 1955. The fine structure of neurons. *J. biophys. biochem. Cytol.*, 1:69-88.

Pappas, G. D. and Purpura, D. P. 1961. Fine structure of dendrites in the superficial neocortical neuropil. *Exp. Neurol.*, 4:507-530.

Parker, G. H. 1909. The origin of the nervous system and its appropriation of effectors. *Pop. Sci. Monthly*, 75:56-64, 137-146, 253-263, 338-345.

Parker, G. H. 1914. The origin and evolution of the nervous system. *Pop. Sci. Mon.*, 84:118-127. (See others in bibliography for Introduction.)

Parker, G. H. 1919. *The Elementary Nervous System*. Lippincott, Philadelphia.

Parker, G. H. 1929a. What are neurofibrils? *Amer. Nat.*, 63:97-117.

Parker, G. H. 1929b. The neurofibril hypothesis. *Quart. Rev. Biol.*, 4:155-148.

Pelseneer, P. 1938. L'indivisibilité et la croissance continue des neurones. *Ann. Soc. zool. Belg.*, 69:187-193.

Penfield, W. (ed.). 1932. *Cytology and Cellular Pathology of the Nervous System*. Vol. 1. Hoeber, New York.

Peters, A. 1960. The structure of myelin sheaths in the central nervous system of *Xenopus laevis* (Daudin). *J. biophys. biochem. Cytol.*, 7:121.

Peterson, R. P. and Pepe, F. A. 1961. The fine structure of inhibitory synapses in the crayfish. *J. biophys. biochem. Cytol.*, 11:157-169.

Pflücke, M. 1895. Zur Kenntnis des feineren Baues der Nervenzellen bei Wirbellosen. *Z. wiss. Zool.*, 60:500-542.

Picken, L. E. R. 1960. *The Organization of Cells and Other Organisms*. Oxford Univ. Press, Oxford.

Pipa, R. L. 1961a. Studies on the hexapod nervous system. III. Histology and histochemistry of cockroach neuroglia. *J. comp. Neurol.*, 116:15-26.

Pipa, R. L. 1961b. Studies on the hexapod nervous system. IV. A cytological and cytochemical study of neurons and their inclusions in the brain of a cockroach, *Periplaneta americana*. *Biol. Bull.*, 121:521-534.

Policard, A. and Baud, C. A. 1958. *Les Structures Inframicroscopiques Normales et Pathologiques des Cellules et des Tissus*. Masson, Paris.

Popoff, M. 1906. Zur Frage der Homologisierung des Binnennetzes der Ganglienzellen mit den Chromidien der Geschlechtszellen. *Anat. Anz.*, 29:249-258.

Prentiss, C. W. 1904. The nervous structures in the palate of the frog: the peripheral networks and the nature of their cells and fibers. *J. comp. Neurol.*, 14:93-117.

Purpura, D. P. 1961. Analysis of axodendritic synaptic organizations in immature cerebral cortex. *Ann. N.Y. Acad. Sci.*, 94:604-654.

Ranvier, L. 1880. Leçons d'anatomie générale. Appareils nerveux terminaux des muscles de la vie organique. Paris.

Rasmussen, A. T. 1947. *Some Trends in Neuroanatomy*. Brown, Dubuque, Iowa.

Redikorzew, W. 1900. Untersuchungen über den Bau der Ocellen der Insekten. *Z. wiss. Zool.*, 68:581-624.

Reger, J. F. 1957. The ultrastructure of normal and denervated neuromuscular synapses in mouse gastrocnemius muscle. *Exp. Cell Res.*, 12:662-665.

Reiser, K. A. 1959. Die Nervenzelle. In: *Handbuch der Mikroskopischen*

Anatomie des Menschen. Bd. IV. *Nervensystem.* Teil 4. W. v. Möllendorff and W. Bargmann (eds.). Springer, Berlin.

Rényi, G. S. de. 1929. The structure of cells in tissues as revealed by microdissection. IV. Observations on neurofibrils in the living nervous tissue of the lobster (*Homarus americanus*). *J. comp. Neurol.*, 48: 441-457.

Retzius, G. 1890. Zur Kenntnis des Nervensystems der Crustaceen. *Biol. Untersuch.*, N.F. 1:1-50.

Retzius, G. 1902. Weiteres zur Kenntnis der Sinneszellen der Evertebraten. *Biol. Untersuch.*, N.F. 10: 24-33.

Retzlaff, E. and Fontaine, J. 1960. Reciprocal inhibition as indicated by a differential staining reaction. *Science*, 131:104-105.

Richards, A. G. 1944. The structure of living insect nerves and nerve sheaths as deduced from the optical properties. *J. N. Y. ent. Soc.*, 52:286-310.

Richards, A. G. 1955. Structure, chemistry and pathology of the central nervous system of arthropods. *Neurochem.*, 31:818-843.

Richards, A. G. and Schneider, D. 1958. Über den komplexen Bau der Membranen des Bindegewebes von Insekten. *Z. Naturf.*, 13B:680-687.

Richards, A. G., Steinbach, H. B., and Anderson, T. F. 1943. Electron microscope studies of squid giant nerve axoplasm. *J. cell. comp. Physiol.*, 21:129-143.

Rieser, P. 1949. The protoplasmic viscosity of muscle and nerve. *Biol. Bull., Woods Hole*, 97:245-246.

Robertis, E. D. P. de. 1954. The nucleo-cytoplasmic relationship and the basophilic substance (ergastoplasm) of nerve cells (electron microscope observations). *J. Histochem. Cytochem.*, 2:341-345.

Robertis, E. D. P. de. 1955a. La relation nucléo-plasmatique et la substance basophile de la cellule nerveuse C. R. Soc. Biol., Paris, 149:1700-1710.

Robertis, E. de. 1955b. Submicroscopic organization of some synaptic regions. *Acta neurol. Latinoamer.*, 1:3-15.

Robertis, E. de. 1958. Submicroscopic morphology and function of the synapse. In: *The Submicroscopic Organization and Function of Nerve Cells.* H. Fernández-Morán and R. Brown (eds.). Academic Press, New York. (*Exp. Cell Res.*, Suppl. 5: 347-369.)

Robertis, E. D. P. de. 1959. Submicroscopic morphology of the synapse. *Int. Rev. Cytol.*, 8:61-96.

Robertis, E. de. 1962. Ultrastructure and chemical organization of synapses in the central nervous system. In: *Brain and Behavior.* M. A. B. Brazier (ed.). American Institute of Biological Sciences, Washington, D.C.

Robertis, E. D. P. de and Bennett, H. S. 1954. Submicroscopic vesicular components in the synapse. *Fed. Proc.*, 13:35.

Robertis, E. D. P. de and Bennett, H. S. 1955a. Some features of the submicroscopic morphology of synapses in frog and earthworm. *J. biophys. biochem. Cytol.*, 1:47-58.

Robertis, E. D. P. de and Bennett, H. S. 1955b. Some features of fine structure of cytoplasm of cells in the earthworm nerve cord. In: *Fine Structure of Cells.* Noordhoff, Groningen.

Robertis, E. de and Ferreira, A. Vaz. 1957. Submicroscopic changes of the nerve endings in the adrenal medulla after stimulation of the splanchnic nerve. *J. biophys. biochem. Cytol.*, 3:611-614.

Robertis, E. de and Franchi, C. M. 1956. Electron microscope observations on synaptic vesicles in synapses of the retinal rods and cones. *J. biophys. biochem. Cytol.*, 2:307-318.

Robertis, E. D. P. de and Schmitt, F. O. 1948. An electron microscope analysis of certain nerve axon constituents. *J. cell. comp. Physiol.*, 31:1-23.

Robertis, E. de, Pellegrino de Iraldi, A., Rodriguez de Lores Arnaiz, G., and Salganicoff, L. 1961. On the isolation of nerve endings and synaptic vesicles. *J. biophys. biochem. Cytol.*, 9:229-235.

Robertson, J. D. 1953. Ultrastructure of two invertebrate synapses. *Proc. Soc. exp. Biol., N. Y.*, 82:219-223.

Robertson, J. D. 1955a. Recent electron microscope observations on the ultrastructure of the crayfish median-to-motor giant synapse. *Exp. Cell. Res.*, 8:226-229.

Robertson, J. D. 1955b. The ultrastructure of adult vertebrate peripheral myelinated fibers in relation to myelogenesis. *J. biophys. biochem. Cytol.* 1:271-278.

Robertson, J. D. 1956. The ultrastructure of a reptilian myoneural junction. *J. biophys. biochem. Cytol.*, 2:381-394.

Robertson, J. D. 1957a. The ultrastructure of nodes of Ranvier in frog nerve fibres. *J. Physiol.*, 137: 8P-9P.

Robertson, J. D. 1957b. New observations on the ultrastructure of the membranes of frog peripheral nerve fibers. *J. biophys. biochem. Cytol.*, 3:1043-1047.

Robertson, J. D. 1958. The ultrastructure of Schmidt-Lanterman clefts and related shearing defects of the myelin sheath. *J. biophys. biochem. Cytol.*, 4:39-44.

Robertson, J. D. 1960. The molecular structure and contact relationships of cell membranes. *Progr. Biophys.*, 10:343-418.

Robertson, J. D. 1961. Ultrastructure of excitable membranes of the crayfish median-giant synapse. *Ann. N. Y. Acad. Sci.*, 94:339-389.

Rohde, E. 1893. Ganglienzelle und Neuroglia. *Arch. mikr. Anat.*, 42: 423-441.

Rohde, E. 1895. Ganglienzelle, Axencylinder, Punktsubstanz und Neuroglia. *Arch. mikr. Anat.*, 45: 384-411.

Rohde, E. 1896. Ganglienzellkern und Neuroglia. *Arch. mikr. Anat.*, 47: 121-135.

Rohde, E. 1898. Die Ganglienzelle. *Z. wiss. Zool.*, 64:697-727.

Rose, J. E. 1949. The cellular structure of the auditory region of the cat. *J. comp. Neurol.*, 91:409-439.

Rosenbluth, J. and Palay, S. L. 1961. The fine structure of the nerve cell bodies and their myelin sheaths in the eighth nerve ganglion of the goldfish. *J. biophys. biochem. Cytol.*, 9:853-877.

Ross, L. S. 1922. Cytology of the large nerve cells of the crayfish (*Cambarus*). *J. comp. Neurol.*, 34:37-71.

Ross, L. S. and Tassell, R. R. 1931. Tracheation of grasshopper nerve ganglia. *J. comp. Neurol.*, 52:347-352.

Ruska, H., Edwards, G. A., and Caesar, R. 1958. A concept of intracellular transmission of excitation by means of the endoplasmic reticulum. *Experentia*, 14:117-120.

Sánchez, D. 1909. El sistema nerviosa de los hirudineos. *Trab. Lab. Invest. biol. Univ. Madr.*, 7:31-199.

Sánchez, D. 1912. El sistema nerviosa de los hirudineos. *Trab. Lab. Invest. biol. Univ. Madr.*, 10:1-143.

Sánchez y Sánchez, D. 1935. Contribution à l'étude de l'origine et de l'évolution de certains types de neuroglie chez les insectes. *Trab. Lab. Invest. biol. Univ. Madr.*, 30:299-353.

Sanders, F. K. and Young, J. Z. 1944. The role of the peripheral stump in the control of fibre diameter in regenerating nerves. *J. Physiol.*, 103:119-136.

Schäfer, E. A. 1879. Observations on the nervous system of *Aurelia aurita*. *Philos. Trans.*, 169:563-575.

Scharf, J.-H. 1953. Die Beziehungen der Lipoide zu den perizellulären Strukturen der Ganglienzellen bei einigen Wirbellosen im Vergleich zu Wirbeltieren. *Z. Zellforsch.*, 38: 526-570.

Scharf, J. H. 1958. Sensible Ganglien. In: *Handbuch der Mikroskopischen Anatomie des Menschen.* Bd. IV. *Nervensystem.* Teil 3. W. Möllendorff and W. Bargmann (eds.). Springer, Berlin.

Scharrer, B. C. J. 1939. The differ-

Bibliography

entiation between neuroglia and connective tissue sheath in the cockroach (*Periplaneta americana*). *J. comp. Neurol.*, 70:77-88.

SCHEIBEL, M. E. and SCHEIBEL, A. B. 1958a. Structural substrates for integrative patterns in the brain stem reticular core. In: *Reticular Formation of the Brain.* H. H. Jasper et al. (eds.). Little, Brown, Boston.

SCHEIBEL, M. E. and SCHEIBEL, A. B. 1958b. A symposium on dendrites. Formal discussion. *Electroenceph. clin. Neurophysiol.*, Suppl. 10:43-50.

SCHLOTE, F.-W. 1957. Submikroskopische Morphologie von Gastropodennerven. *Z. Zellforsch.*, 45:543-568.

SCHMITT, F. O. 1950. The structure of the axon filaments of the giant nerve fibers of *Loligo* and *Myxicola*. *J. exp. Zool.*, 113:499-512.

SCHMITT, F. O. 1957. The fibrous protein of the nerve axon. *J. cell. comp. Physiol.*, 49(Suppl. 1):165-174.

SCHMITT, F. O. 1958. Axon-satellite cell relationships in peripheral nerve fibers. In: *The Submicroscopic Organization and Function of Nerve Cells.* H. Fernández-Morán and R. Brown (eds.). Academic Press, New York. (*Exp. Cell Res.*, Suppl. 5.)

SCHMITT, F. O. 1959. Molecular organization of the nerve fiber. In: *Biophysical Science—A Study Program.* J. L. Oncley (ed.). Wiley, New York.

SCHMITT, F. O. and BEAR, R. S. 1939. The ultrastructure of the nerve axon sheath. *Biol. Rev.*, 14:27-50.

SCHMITT, F. O. and GEREN, B. B. 1950. The fibrous structure of the nerve axon in relation to the localization of "neurotubules." *J. exp. Med.*, 91:499-504.

SCHMITT, F. O., BEAR, R. S., and SILBER, R. H. 1939. Organic and inorganic electrolytes in lobster nerves. *J. cell. comp. Physiol.*, 14:351-356.

SCHMITT, F. O., BEAR, R. S., and YOUNG, J. Z. 1936. Some physical and chemical properties of the axis cylinder of the giant axons of the squid, *Loligo pealii*. *Biol. Bull., Woods Hole*, 71:402-403. (Abs.)

SCHNEIDER, K. C. 1902. *Lehrbuch der vergleichenden Histologie der Tiere.* Fisher, Jena.

SCHRADER, K. 1938. Untersuchungen über die Normalentwicklung des Gehirns und Gehirntransplantationen bei der Mehlmotte *Ephestia kühniella* Zeller nebst einigen Bemerkungen über das Corpus allatum. *Biol. Zbl.*, 58:52-90.

SCHUCHARDT, E. 1958. Zur funktionellen Organisation des Nervengewebes. (Eine Arbeitshypothese.) *Z. mikr. anat. Forsch.*, 64:258-266.

SCHULTZ, R., BERKOWITZ, E. C., and PEASE, D. C. 1956. The electron microscopy of the lamprey spinal cord. *J. Morph.*, 98:251-274.

SCHULTZ, R. L., MAYNARD, E. A., and PEASE, D. C. 1957. Electron microscopy of neurons and neuroglia of cerebral cortex and corpus callosum. *Amer. J. Anat.*, 100:369-388.

SCHULTZE, H. 1878. Axencylinder und Ganglienzelle. Mikroskopische Studien über die Structur der Nervenfaser und Nervenzelle bei Wirbelthieren. *Arch. Anat. Physiol., Lpz.*, 1878:259-287.

SCHWARZACHER, H. G. (ed.). 1961. Histochemistry of cholinesterase. *Bibl. anat.*, 2:1-255. (Symposium, Basel, 1960. Karger, Basel.)

SEITE, R. 1961. Données récentes de cytophysiologie nerveuse. La transmission synaptique de l'excitation et le problème de l'acétylcholine liée (essai sur certaines données biochimiques, physiologiques et morphologiques), *Année biol.*, 37:217-253.

SERENI, E. 1929. Fenomeni fisiologici consecutivi alla sezione dei nervi nei cefalopodi. *Boll. Soc. ital. Biol. sper.*, 4:736-740.

SERENI, E. and YOUNG, J. Z. 1932. Nervous degeneration and regeneration in cephalopods. *Pubbl. Staz. zool. Napoli*, 12:173-208.

SHAFIQ, S. A. 1953. Cytological studies of the neurones of *Locusta migratoria*. Part I. Cytoplasmic inclusions of the motor neurones of the adult. *Quart. J. micr. Sci.*, 94:319-328.

SHAFIQ, S. A. and CASSELMAN, W. G. B. 1954. Cytological studies of the neurones of *Locusta migratoria*. III. Histochemical investigations with special reference to the lipochondria. *Quart. J. micr. Sci.*, 95:315-320.

SIMPSON, S. A. and YOUNG, J. Z. 1945. Regeneration of fibre diameter after cross-unions of visceral and somatic nerves. *J. Anat., Lond.*, 79:48-65.

SJÖSTRAND, F. S. 1956. The ultrastructure of cells as revealed by the electron microscope. *Int. Rev. Cytol.*, 5:455-533.

SJÖSTRAND, F. S. 1958. Ultrastructure of retinal rod synapses of the guinea pig eye as revealed by three-dimensional reconstructions from serial sections. *J. Ultrastruct. Res.*, 2:122-170.

SJÖSTRAND, F. S. 1961a. Electron microscopy of the retina. In: *The Structure of the Eye.* G. K. Smelser (ed.). Academic Press, New York.

SJÖSTRAND, F. S. 1961b. Topographic relationship between neurons, synapses and glia cells. In: *The Visual System: Neurophysiology and Psychophysics.* Springer, Heidelberg.

SKAER, R. J. 1961. Some aspects of the cytology of *Polycelis nigra*. *Quart. J. micr. Sci.*, 102:295-317.

SLIFER, E. H. 1961. The fine structure of insect sense organs. *Int. Rev. Cytol.*, 11:125-159.

SMALLMAN, B. N. and WOLFE, L. S. 1956. Soluble and particulate cholinesterase in insects. *J. cell. comp. Physiol.*, 48:197-213.

SMALLWOOD, W. M. and HOLMES, M. T. 1927. The neurofibrillar structure of the giant fibers in *Lumbricus terrestris* and *Eisenia foetida*. *J. comp. Neurol.*, 43:327-345.

SMALLWOOD, W. M. and ROGERS, C. G. 1909. Studies on nerve cells. II. The comparative cytology and physiology of some of the metabolic bodies in the cytoplasm of invertebrate nerve cells. *Folia neuro-biol., Lpz.*, 3:11-20.

SMALLWOOD, W. M. and ROGERS, C. G. 1910. Studies on nerve cells. III. Some metabolic bodies in the cytoplasm of nerve cells of gasteropods, a cephalopod, and an annelid. *Anat. Anz.*, 36:226-232.

SMITH, C. A. and SJÖSTRAND, F. S. 1961. A synaptic structure in the hair cells of the guinea pig cochlea. *J. Ultrastruct. Res.*, 5:184-192.

SMITH, J. E. 1957. The nervous anatomy of the body segments of nereid polychaetes. *Philos Trans.*, (B) 240:135-196.

SMITH, S. W. 1959. "Reticular" and "areticular" Nissl bodies in sympathetic neurons of a lizard. *J. biophys. biochem. Cytol.*, 6:77-83.

SOLOMON, S. and TOBIAS, J. M. 1960. Thixotropy of axoplasm and effect of activity on light emerging from an internally lighted giant axon. *J. cell. comp. Physiol.*, 55:159-166.

SOSA, J. M. and MENEGAZZI, J. A. 1939. L'appareil de Golgi des neurones de *Helix aspersa*. *Arch. Soc. Biol. Montevideo*, 9:157-164.

SPANGENBERG, D. B. and HAM, R. G. 1960. The epidermal nerve net of *Hydra*. *J. exp. Zool.*, 143:195-202.

SPEIDEL, C. C. 1935. Studies of living nerves. III. Phenomena of nerve irritation and recovery, degeneration and repair. *J. comp. Neurol.*, 61:1-80.

STAMMER, A., MINKER, E., HORVATH, I., and ERDÉLYI, L. 1958. The structure of the peripheral transmission apparatuses and the forms of their connection. *Acta. biol. hung. (Suppl.).* 2:32.

STÖHR, P., Jr. 1957. Mikroskopische Anatomie des vegetativen Nervensystems. In: *Handbuch der Mikroskopischen Anatomie des Menschen.* Bd. V. *Verdauungsapparat.* Teil 5. W. von Möllendorff (ed.). Springer, Berlin.

STOUGH, H. B. 1930. Polarization of the giant nerve fibers of the earthworm. *J. comp. Neurol.*, 50:217-229.

SZENTÁGOTHAI, J. 1961. Anatomical aspects of inhibitory pathways and synapses. In: *Nervous Inhibition.* E. Florey (ed.). Pergamon Press, New York.

SZENTÁGOTHAI, J. and SZÉKELY, GY. 1956. Zum Problem der Kreuzung der Nervenbahnen. *Acta biol. hung.,* 6:215-229.

TASAKI, I. 1959. Conduction of the nerve impulse. In: *Handbook of Physiology.* Sect. 1. *Neurophysiology,* 1:75-121.

TAXI, J. 1957. Étude au microscope électronique de ganglions sympathiques de mammifères. *C. R. Acad. Sci., Paris,* 245:564-567.

TAXI, J. 1958. Sur la structure du plexus d'Auerbach de la souris étudié au microscope électronique. *C. R. Acad. Sci., Paris,* 246:1922-1925.

TAXI, J. 1959. Étude au microscope électronique de la dégénérescence Wallérienne des fibres nerveuses amyéliniques. *C. R. Acad. Sci., Paris,* 248:2796-2798.

TAYLOR, G. W. 1940. The optical properties of the earthworm nerve fiber sheath as related to fiber size. *J. cell. comp. Physiol.,* 15:363-371.

THAEMERT, J. C. 1959. Intercellular bridges as protoplasmic anastomoses between smooth muscle cells. *J. biophys. biochem. Cytol.,* 6:67-70.

THOMAS, O. L. 1947a. Some observations with the phase-contrast microscope on the neurones of *Helix aspersa. Quart. J. micr. Sci.,* 88: 269-273.

THOMAS, O. L. 1947b. The cytology of the neurones of *Helix aspersa. Quart. J. micr. Sci.,* 88:445-462.

THOMAS, O. L. 1954. The cytoplasmic inclusions of worm ganglion cells. *Cellule,* 56:229-240.

TIEGS, O. W. 1927. A critical review of the evidence on which is based the theory of discontinous synapses in the spinal cord. *Aust. J. exp. Biol. med. Sci.,* 4:193-212.

TIEGS, O. W. 1953. Innervation of voluntary muscle. *Physiol. Rev.,* 33: 90-144.

TIRABOSCHI, C. 1899. Contributo allo studio della cellula nervosa in alcuni invertebrati e specialmente negli insetti. *Boll. Soc. Stud. Zool., Roma,* 8:53-65, 143-151.

TOBIAS, J. M. 1950. Electrically induced optical and dimensional changes in single axons including the squid. Preliminary observations on ion effects. *Biol. Bull., Woods Hole,* 99:345.

TOBIAS, J. M. 1951. Qualitative observations on visible changes in single frog, squid and other axones subjected to electrical polarization. Implications for excitation and conduction. *J. cell. comp. Physiol.,* 37:91-105.

TOBIAS, J. M. 1958. Experimentally altered structure related to function in the lobster axon with an extrapolation to molecular mechanisms in excitation. *J. cell. comp. Physiol.,* 52(1):89-125.

TOBIAS, J. M. 1960. Further studies on the nature of the excitable system in nerve. I. Voltage-induced axoplasm movement in squid axons. II. Penetration of surviving, excitable axons by proteases. III. Effects of proteases and of phospholipases on lobster giant axon resistance and capacity. *J. gen. Physiol.,* 43(Suppl.):57-71.

TOBIAS, J. M. and NELSON, P. G. 1959. Structure and function in nerve. In: *A Symposium on Molecular Biology,* R. E. Zirkle (ed.). Univ. of Chicago Press, Chicago.

TOKUYASU, K. and YAMADA, E. 1959. The fine structure of the retina studied with the electron microscope. IV. Morphogenesis of outer segments of retinal rods. *J. biophys. biochem. Cytol.,* 6:225-230.

TREHERNE, J. E. 1961. The kinetics of sodium transfer in the central nervous system of the cockroach, *Periplaneta americana. J. exp. Biol.,* 38:737-746.

TRUJILLO-CENÓZ, O. 1957. Electron microscope study of the rabbit gustatory bud. *Z. Zellforsch.,* 46: 272-280.

TRUJILLO-CENÓZ, O. 1959. Study on the fine structure of the central nervous system of *Pholus labruscoe* L. (Lepidoptera). *Z. Zellforsch.,* 49: 432-446.

TRUJILLO-CENÓZ, O. 1960. The fine structure of a special type of nerve fiber found in the ganglia of *Armadillidium vulgare* (Crustacea-Isopoda). *J. biophys. biochem. Cytol.,* 7:185-186.

TRUJILLO-CENÓZ, O. 1961. Electron microscope observations on chemo- and mechano-receptor cells of fishes. *Z. Zellforsch.,* 54:654-676.

TRUJILLO-CENÓZ, O. 1962. Some aspects of the structural organization of the arthropod ganglia. *Z. Zellforsch.,* 56:649-682.

TRUJILLO-CENÓZ, O. and MELAMED, J. 1962. Electron microscope observations on the calyces of the insect brain. *J. Ultrastruct. Res.,* 7:389-398.

UCHIZONO, K. 1960. Comparative studies of myelinated and non-myelinated nerve fibers. In: *Electrical Activity of Single Cells.* Y. Katsuki (ed.). Igaku Shoin, Tokyo.

USHAKOV, B. P. 1950. Change in the staining properties of nerve of the crab *Hyas araneus* with excitation. *C. R. Acad. Sci. URSS (Dokl. Akad. Nauk SSSR),* 71:205-208.

VAN BREEMEN, V. L., ANDERSON, E., and REGER, J. F. 1958. An attempt to determine the origin of synaptic vesicles. *Exp. Cell. Res.,* Suppl. 5:153-167.

VIAL, J. D. 1958. The early changes in the axoplasm during Wallerian degeneration. *J. biophys. biochem. Cytol.,* 4:551-556.

VILLEGAS, C. M. and VILLEGAS, R. 1960. The ultrastructure of the giant nerve fibre of the squid: axon-Schwann cell relationship. *J. Ultrastruct. Res.,* 3:362-373.

VIZOSO, A. D. and YOUNG, J. Z. 1948. Internode length and fibre diameter in developing and regenerating nerves. *J. Anat., Lond.,* 82:110-134.

VOÏNOV, D. 1934. Structures ergastoblastiques, dictyosomes (ergastoblastes), parasomes, corps vitellins de Balbiani, interprétation de l'appareil réticulaire interne de Golgi. *Arch. Zool. exp. gén.,* 76:399-491.

VOWLES, D. M. 1955. The structure and connexions of the corpora pedunculata in bees and ants. *Quart. J. micr. Sci.,* 96:239-255.

WALTER, G. 1863. *Mikroskopische Studien über das Centralnervensystem wirbelloser Thiere.* Bonn.

WATZKA, M. 1939. Veränderungen des Golgi-Netzapparates nach Chloroformnarkose. *Z. Mikr.-anat. Forsch.,* 46:622-626.

WEIGL, R. 1910. Über den Golgi-Kopschschen Apparat in den Ganglienzellen der Cephalopoden. *Bull. int. Acad. Cracovie,* 1910B:691-710.

WEIGL, R. 1913. Vergleichend-zytologische Untersuchungen über den Golgi-Kopsch'schen Apparat und dessen Verhältnis zu anderen Strukturen in den somatischen Zellen und Geschlechtszellen verschiedener Tiere. *Anz. Akad. Wiss. Krakau, Math.-Nat. Kl. (Bull. int. Acad. Cracovie) (B),* 1912:417-447.

WEISS, P. 1961. The concept of perpetual neuronal growth and proximo-distal substance convection. In: *Regional Neurochemistry.* S. S. Kety and J. Elkes (eds.). Pergamon Press, New York.

WEISS, P. and HISCOE, H. B. 1948. Experiments on the mechanism of nerve growth. *J. exp. Biol.,* 107: 315-395.

WEISS, P. and WANG, H. 1936. Neurofibrils in living ganglion cells of the chick, cultivated in vitro. *Anat. Rec.,* 67:105-117.

WELLS, M. J. and WELLS, J. 1957. The effect of lesions to the vertical and optic lobes on tactile discrimination in *Octopus. J. exp. Biol.,* 34:378-393.

WELSH, J. H. 1961. Neurohumors and neurosecretion. In: *The Physiology of Crustacea.* Vol. II. T. H. Waterman (ed.). Academic Press, New York.

WIERSMA, C. A. G. 1961. Reflexes and the central nervous system. In: *The Physiology of Crustacea.* Vol II. T. H. Waterman (ed.). Academic Press, New York.

WIGGLESWORTH, V. B. 1958. The distribution of esterase in the nervous system and other tissues of the insect *Rhodnius prolixus. Quart. J. micr. Sci.,* 99:441-450.

WIGGLESWORTH, V. B. 1959. The histology of the nervous system of an insect *Rhodnius prolixus* (Hemiptera). II. The central ganglia. *Quart. J. micr. Sci.*, 100:299-314.

WIGGLESWORTH, V. B. 1960a. The nutrition of the central nervous system in the cockroach *Periplaneta americana* L. *J. exp. Biol.*, 37:500-512.

WIGGLESWORTH, V. B. 1960b. Axon structure and the dictyosomes (Golgi bodies) in the neurones of the cockroach, *Periplaneta americana*. *Quart. J. micr. Sci.*, 101:381-388.

WILSON, D. M. 1960. Nervous control of movement in cephalopods. *J. exp. Biol.*, 37:57-72.

WINDLE, W. F. (ed.). 1955. *Regeneration in the Central Nervous System*. Thomas, Springfield, Ill.

WINDLE, W. F. (ed.). 1958. *Biology of Neuroglia*. Thomas, Springfield, Ill.

WOLKEN, J. J. 1956. Photoreceptor structures. I. Pigment monolayers and molecular weight. *J. cell. comp. Physiol.*, 48:349-370.

WOLKEN, J. J. 1957. A comparative study of photoreceptors. *Trans. N.Y. Acad. Sci.*, 19:315-327.

WOLKEN, J. J. 1958. Retinal structure, Mollusc cephalopods: *Octopus*, *Sepia*. *J. biophys. biochem. Cytol.*, 4:835-838.

WOLKEN, J. J., CAPENOS, J., and TURANO, A. 1957. Photoreceptor structures. III. *Drosophila melanogaster*. *J. biophys. biochem. Cytol.*, 3:441-448.

WOLKEN, J. J., MELLON, A. D., and CONTIS, G. 1957. Photoreceptor structures. II. *Drosophila melanogaster*. *J. exp. Zool.*, 134:383-410.

YOLTON, L. W. 1923. The effects of cutting the giant fibers in the earthworm, *Eisenia foetida* (Sav.). *Proc. nat. Acad. Sci., Wash.*, 9:383-385.

YOUNG, J. Z. 1932. On the cytology of the neurons of cephalopods. *Quart. J. micr. Sci.*, 75:1-47.

YOUNG, J. Z. 1936a. The structure of nerve fibres and synapses in some invertebrates. *Cold Spr. Harb. Symp. quant. Biol.*, 4:1-6.

YOUNG, J. Z. 1936b. The structure of nerve fibres in cephalopods and Crustacea. *Proc. roy. Soc. (B)*, 121:319-337.

YOUNG, J. Z. 1939. Fused neurons and synaptic contacts in the giant nerve fibres of cephalopods. *Philos. Trans.*, (B)229:465-503.

YOUNG, J. Z. 1944. Contraction, turgor and the cytoskeleton of nerve fibres. *Nature, Lond.*, 153:333-335.

YOUNG, J. Z. 1950. The determination of the characteristics of nerve fibers. In: *Genetic Neurology*. P. Weiss (ed.). Univ. of Chicago Press, Chicago.

YOUNG, J. Z. 1961. Learning and discrimination in the octopus. *Biol. Rev.*, 36:32-96.

ZAWARZIN, A. 1924. Zur Morphologie der Nervenzentren. Das Bauchmark der Insekten. Ein Beitrag zu vergleichenden Histologie. (Histologische Studien über Insekten VI.) *Z. wiss. Zool.*, 122:323-424.

ZEIGER, K. 1950. Zur Problematik des Golgi-Apparates. *Zool. Anz.*, 145 (Ergänzungsb.):1140-1154.

ZEIGER, K. and HARDERS, H. 1951. Über vitale Fluorochromfärbung des Nervengewebes. *Z. Zellforsch.*, 36:62-78.

CHAPTER 3

Comparative Neurology of Excitability and Conduction

Summary	124	B. Spike Excitability	135
		1. Sodium theory	135
I. Introduction	**125**	2. Strength-duration relation	135
		3. Strength-slope relation	140
II. Differentiation of Nervous Signals	**125**	4. Recovery cycle	142
		5. Form of spike	144
A. Graded and Spike Activity	125	6. Afterpotentials	145
		7. Velocity and its correlates	148
B. Dendrite and Axon	131	C. Rhythmicity and Pacemaker Excitability	157
III. Excitability and its Primary Parameters	**133**	Classified References	164
A. Local Response Excitability	133	Bibliography	166

Summary

This chapter reviews the parameters of excitability and of propagation of signals as distinguished from transmission across junctions, emphasizing the comparison of different kinds of neurons and animals. The subject of basic mechanisms accounting for the magnitude and form of the observed potentials in terms of membrane processes is not undertaken in this book beyond a short enumeration of the principal variables presently identified.

Already in jellyfish and hydroids, **all-or-none spikes** testify to the presence of nerve impulses. Except for a few special cases, such as retinula cells in *Limulus* and rods and cones in vertebrates, neurons are not known which lack impulses, although it is believed that this was the primitive condition and that many short-axon cells may normally act without impulses today. Spikes are regarded as a specialized form of neuronal activity characteristic of axons and useful in long distance propagation. The form of excitability measured by spike threshold is normally important only at the locus of initiation of spikes, near the dendrite-axon boundary.

Several types of **graded, local activity** are known. Autogenic or pacemaker potentials are forms of spontaneity. A rhythmic sinusoidal type, a more or less rhythmic relaxation oscillation type, and a random, miniature, quantal type at certain endings are known. Exogenic or transducer potentials are receptor and synaptic responses and will be found in Chapter 4. Endogenic or internal response potentials are secondary reactions to one of the first two classes in the same neuron; they include local, spike, and afterpotentials and are almost confined to axonal, electrically excitable membrane. Dendrites and some axonal terminals typically employ graded, not spike activity. These possess several different excitabilities measured, not by a threshold, but by nonlinear input-output curves.

The **strength-duration relation** of spike threshold is known for many cases but not usually in a fully satisfactory way; still, it seems to vary significantly among them. The **strength-slope relation** gives the spike threshold at different rates of rise of stimulus; an inverse relation is often seen but should be called "apparent accommodation" until confirmed by monitoring depolarization threshold. Carefully studied axons show little or no true accommodation, but some nerve cell bodies do, and some show an accommodative fall in threshold stimulus during maintained hyperpolarization. There is sometimes a minimal slope for excitation.

Refractoriness is absolute for about the duration of the spike, which varies approximately from 0.4 or less to 10 msec; relatively refractory periods are longer and more variable. Later supernormal and subnormal periods are known in a few cases only.

Spike height does not vary systematically (50–120 mv) measured across the membrane, but as seen by external electrodes is a function of diameter.

Afterpotentials of several kinds are known and are characteristic of fiber types but are widely different between them and between axon and soma.

Velocity of propagation is a first-power function of diameter in many comparisons but probably not in others. It correlates also with conductivity of external medium, sheath thickness, the development of Ranvier nodes, and other intrinsic factors. Highly stretch-tolerant axons maintain velocity though decreasing in diameter under elongation.

Tables are given comparing values of time constant, resting, action, and overshoot potentials, rheobase, chronaxie, recovery cycle, accommodation, velocity, diameter, and pacemaker sensitivity.

Pacemaker sensitivity is probably the most important form of excitability in typical central neurons; it is measured by the change in frequency of autochthonous, on-going firing (not driven one-to-one) with applied steady current. The tendency to fire repeatedly varies over a wide range among neurons, and this form of excitability is relevant in proportion to that tendency. Absolute values are available in a few cases and show meaningful signaling of stimuli below rheobase in the form of change in frequency.

Departure from perfect **rhythmicity,** measured as standard deviation of successive intervals, varies widely and, in a given neuron, increases in percentage with average interval—that is, at low frequency.

Slow alterations in excitability are significant in certain cases and may be caused by the milieu, temperature, light, osmotic pressure, and hormones.

I. INTRODUCTION

Conduction, as distinct from transmission, is the spread of a nervously significant change of state within the confines of a neuron. (Transmission, the communication between units, is treated in the next chapter.) One of the prime questions of comparative neurology is whether the changes of state which carry nervous messages ("signals") are similar for the great range of animals or show meaningful differences. Accordingly, the present survey seeks to discern the phenomena and the variables presented by animals. It cannot treat the fundamental nature or the mechanisms underlying the operations performed; such a treatment belongs to a lower level of analysis.

II. DIFFERENTIATION OF NERVOUS SIGNALS

As yet, little can be said definitely about the actual evolutionary history of the signals employed in nervous tissue except that the most characteristic of these, the nerve impulse, is already present in the simplest nervous systems yet studied—those of jellyfish. Nevertheless some reasonable inferences may be put forth and they will at the same time serve as a framework for introducing the types of signals and their relations to each other.

A. Graded and Spike Activity

The main distinction to be made among the modes of neuronal activity is between the **nerve impulse** and all other forms. The impulse is a brief, all-or-none event which propagates itself along competent parts of the neuron, notably the axon. Time after time, for thousands of repetitions, it exhibits exactly the same amplitude and wave form with respect to voltage change, though these can vary locally under some conditions while retaining the all-or-none character. The impulse is like a fuse which continues to burn, without dying out in a somewhat thinned or slightly damp spot. The nerve impulse in a single nerve fiber is hence a quantal event. It is spoken of as regenerative because the mechanism involves a self-re-excitation by electrical current flowing from the near-by unexcited portions of the nerve fiber into the portion occupied by the impulse at the moment; the more current that flows, the more current is made to flow—by an "active" drop in the resistance of the cell membrane—up to a limit which determines the height of the all-or-none voltage change. This height is many times the voltage change necessary to start the chain reaction leading to regenerative discharge ("firing") of the next portion of the axon; the excess is said to be the safety factor. The nerve impulse is doubtless a complex of constituent and sequential processes. Certainly it brings about a chain of physical and chemical changes, though it is difficult to say how many of these are essential parts of the impulse and how many are consequences or recovery processes. The details may be found in recent texts of general physiology. Clearly the electrical changes occupy a unique position for they are the earliest known and are regarded in current theory as causal; that is, the flow of current itself apparently brings about the change in the state of the membrane which leads to more flow of current. Although the totality of the concept of the nerve impulse cannot be identified with the electrical sign, it is the only sign we are sure belongs to the essence of the impulse; and, since the electrical change is sudden and brief, it is commonly called the **spike** because of its shape in visual records, which plot potential or current against time. Henceforth we may speak interchangeably of the nerve impulse or the spike. The term "action potential" has often been used for the same phenomenon but we avoid it because of the possible confusion with other, graded forms of active response (Fig. 3.1).

Since the impulse is quantal and is followed by a **refractory period** during which another impulse cannot be elicited, the propagated signals in a nerve fiber become a series of pulses, all essentially alike. The information is carried by the

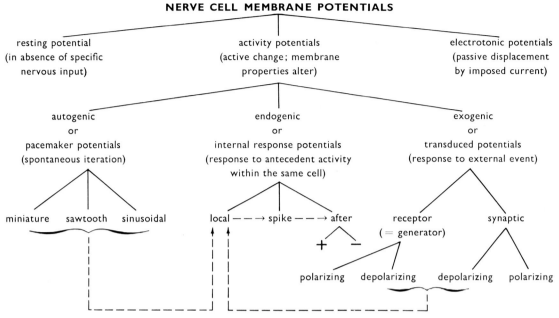

Figure 3.1. Diagram of the types of nerve cell membrane potentials, with special reference to the potentials of activity. *Arrows* indicate that a sufficient level of one potential may cause the initiation of another; *plus* and *minus* signify that there are those which increase the membrane potential and those which decrease the membrane potential, equivalent to polarizing (or hyperpolarizing) and depolarizing.

presence, the number, and the intervals between these pulses. Sometimes the technical term digital as used for computing machines is applied to this principle but this is not strictly correct: the intervals between pulses are significant and are, so far as is known, continuously graded rather than made up of periods of fixed duration during which an impulse is either present or not present. The principle could be called a **pulse-coded analog** principle.

In contrast, there are other forms of response in the neuron which are graded in amplitude and can be thought of as **uncoded analog responses,** in which the intensity of the response is a continuous function of the intensity of the cause or stimulus. Since they are nonregenerative and not all-or-none, these graded responses are typically small compared to spikes; indeed, if a graded response rises to the critical height which represents the threshold of the latter, a spike is triggered—the membrane is said to fire. In most conductile membranes the critical height or spike threshold is about one-tenth or one-fifth of the height of the spike itself. To give typical values, the spike amounts to 90–120 mv when the resting potential difference between inside and outside of the cell membrane is 60–70 mv; the spike completely depolarizes and then goes on to reverse-polarize the membrane briefly—this is called the overshoot. The graded forms of response in such typical examples could be from just detectable up to 10 or 20 mv, at which level they trigger a spike. But if one or another agent raises the threshold, they can rise higher, even to overshooting. A consequence of the nonregenerative character is that such forms of graded response cannot propagate or, more accurately, can only propagate decrementally and therefore for a limited distance, dying out gradually as they go. This feature confines them to the region of the stimulus or antecedent causal event, and they are therefore local responses. In typical cases the distance at which local response has declined to half amplitude is 0.3–1.5 mm.

The **principal types of graded activity** are the pacemaker (autogenic) potentials, treated in Chapter 5; the transducer or generator (exogenic) potentials, comprising synaptic and receptor potentials (Chapter 4); and the internal response (endogenic) local potentials (Figs. 3.1, 3.2). The

II. DIFFERENTIATION OF SIGNALS A. Graded and Spike

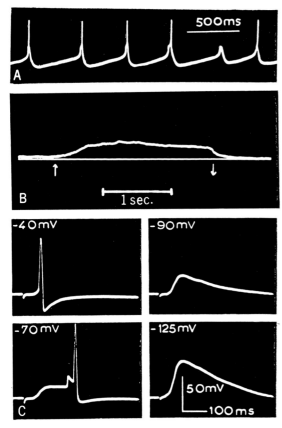

Figure 3.2. Examples of types of potentials: pacemaker, generator, local, synaptic, and spike potentials. **A.** Spontaneous firing of a cardiac ganglion cell in the lobster heart, with sloping pacemaker potentials leading at a critical level to steeper local potentials and these in turn (except in one case) leading at a critical level to spike potentials. The spike potentials are in this case very small in amplitude, indicating that the all-or-none event is arising some distance away and does not invade the region of the electrode. [Bullock and Terzuolo, 1957.] **B.** Generator potential of 6 to 7 mv in a crayfish stretch receptor cell giving subthreshold stretch, between arrows. [Eyzaguirre and Kuffler, 1955a.] **C.** Spike and synaptic potential in visceral ganglion cell of *Aplysia*. A presynaptic nerve is stimulated at the moment of the first small deflection, which is an artifact of the stimulus, causing transmission of excitation across a junction leading to a spike in the penetrated cell, at normal membrane potential levels. But at higher membrane potential levels, which are preset by a separate polarizing electrode inserted into the same cell, the transmitted event reaches spike threshold after a long delay and hesitation (which may involve an intermediate event not yet understood, called a pseudospike), or fails altogether to elicit a spike. In the absence of a spike a slow and graded potential called the excitatory postsynaptic potential (e.p.s.p.) is seen in isolation. [Tauc, 1958.]

last are essentially coextensive with spikes in respect to the portions of the neuron occupied or available, but occur perhaps in some finer processes where spikes do not. As is explained more fully in the next chapter, the spike (and perhaps the endogenic local potential) seemingly cannot occur in the same part of the neuronal surface membrane as the synaptic potentials, at least in the usual case. In the more familiar examples the surface membrane is electrically inexcitable and is only aroused to response by specific external events. Such transducer responses play their role in transmission rather than conduction. Similarly, the local potentials which arise in electrically excitable membrane in response to antecedent potential changes in the same cell (hence endogenic)—for example, following synaptic or pacemaker potentials or spikes—are heterogeneous. There is as yet, however, no clear separation into categories, in contrast to the synaptic potentials, which are subdivided.

The local potential, like the others just named, is an **active as opposed to a passive** potential change. This means that the recorded event involves some change in the properties of the membrane—actually the conductance or permeability to ions—and is not simply the potential change caused by the antecedent causal event or stimulus, distorted and spread by the electrical properties of the resting membrane. A potential change of the passive type, easily recognized because it is perfectly proportional to the stimulus and of constant form with smaller and smaller measuring current, is known as an **electrotonic potential** or passive potential. Its form in response to a square-fronted stimulus proves to be the simple rise and exponential fall of a voltage applied to an electrical condenser with a resistance in parallel to it. The form permits the application of the equations used for leaky cables, and satisfactory predictions of the distortion of the form with distance result: the peak is rounded and progressively delayed as well as reduced in height with distance from the focus of the change. Figure 3.3 shows the appearance of electrotonic potentials. They have the same form whether the applied current is inward or outward (increasing or decreasing the

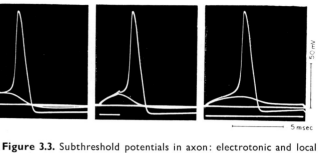
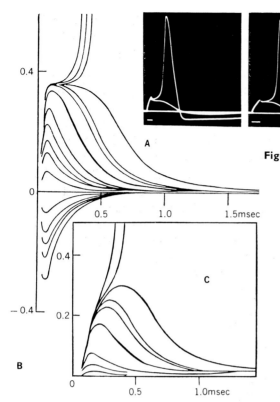

Figure 3.3. Subthreshold potentials in axon: electrotonic and local potentials. **A.** Intracellular records from the squid giant axon during and after stimulation by a rectangular current pulse applied through a second intracellular electrode. The duration of the stimulus is shown by the bar in each record. Two stimuli and responses are given in each frame, one just supraliminal and the other just subliminal and leaving a convex, slowly subsiding local potential. Note the hesitation and then slow development of the spike. [Hagiwara and Oomura, 1959.] **B.** Electrical changes at a stimulating electrode produced by shocks with relative strengths successively from above 1.00 (upper six curves), 0.96, 0.85, 0.71, 0.57, 0.43, 0.21, —021, —0.43, —0.57, —0.71, —1.00. The ordinate scale gives the potential as a fraction of the propagated spike, which was about 40 mv in amplitude. The 0.96 curve is thicker than the others because the local response had begun to fluctuate very slightly at this strength; the width of the line indicates the extent of fluctuation. Extracellular recording from isolated axons of crab leg nerve. [Hodgkin, 1938.] **C.** Responses produced by shocks with strength, successively from above, 1.00 (upper five curves), 0.96, 0.85, 0.71, 0.57; obtained from curves in **B** by subtracting anodal changes from corresponding cathodal curves. Two of the anodal curves necessary for this analysis were recorded but are not shown in **B**. Ordinate is fraction of the propagated spike. [Hodgkin, 1938.]

membrane potential), but do not necessarily have the same amplitude because some membranes have considerably higher resistance in one direction than in the other. There is still no general picture of the distribution among neurons or of the correlates of this property of **rectification** but it is found in some degree in axons, cell bodies, and synaptic regions.

The **significance of the decremental potentials** has come to be stressed in recent years. They are one of the principal, even if not the only, modes of mixing signals and achieving integration; this role is developed in the next two chapters. We may anticipate so far as to say that the receptive parts of the neuron—a variable fraction but in general the dendrites—act only by means of graded potentials which spread to influence each other or the rest of the neuron electrotonically. Sometimes there is a sequence of one or more kinds of synaptic potentials, mixing with one or more pacemaker potentials arising in autoactive loci and spreading to an adjoining electrically excitable membrane where a local potential amplifies the resultant and sums with it. The graded potentials lack an absolute refractory period and therefore can build up and maintain a summed response. It is clear that the all-or-none impulse arises after a good deal has already transpired and does so in a circumscribed locus. This will be at or beyond the boundary between different kinds of cell membrane: that which is capable of only graded activity and that which has a critical state beyond which it detonates and propagates a spike without decrement throughout the extent of such a membrane—in general, along the length of the axon. The location of this crucial boundary varies with the type of cell, to be discussed in Chapter 4. Moreover, the threshold for spike initiation is sometimes higher in the more proximal parts of the conductile membrane (for example, in the soma) than in somewhat more distal parts (for example, in the initial segment of axon); the difference in excitability may be greater than the decrement in electrotonically spread prepotentials, so that the spike actually arises in the more distal locus. Impulse conduction begins then, significantly downstream from the origins of natural excitation and at a sharply defined place, which however may shift and is not the same for every neuron. Here graded activity has played a decisive role in determining and timing the firing, in combination with the local threshold.

Elsewhere **along the axon** the local potential can be demonstrated by subthreshold stimuli. This has been studied carefully in a few preparations, notably crab leg motor axons, squid giant axons, and frog sciatic axons. As shown in Fig. 3.3, a detectable active response begins to be visible, superimposed on the electrotonic potential of the stimulus, only when the latter exceeds 40–60% of the threshold. With stronger stimuli the local response is disproportionately stronger, increasing very rapidly as the stimuli approach threshold. In form the local potential is typically almost as brief as a spike, with about the same crest time but having a passive or even slightly longer than passive decay instead of the rapid repolarization of the spike. This means that the rate of rise of the local potential is proportionate to its amplitude and therefore that local potentials have no threshold but give origin to the spike at an earlier phase the stronger the stimulus. When the stimulus is several times threshold—as when the stimulus is the approaching spike in an axon with a good safety factor—the spike in the region under observation rises from such an early phase and the local potential is so steep that there is no visible distinction between them. In this case the conduction velocity of the spike is maximal. If the safety factor is not high, the local potential has time to rise to a greater fractional height and is less steep, so that the spike rises with a distinct notch or shoulder and is delayed. The rate of development of the local potential, then, plays a role in the velocity of impulse conduction, in combination with the spike threshold. The safety factor may be low in places, especially at branching points, so that a spike may fail at such places. Both rate of change and threshold are locally determined and are quite independent from place to place along the axon. But they are not properties of infinitely small points; the passive electrical characteristics of the membrane ensure that any change will involve an area. Certain experimental evidence, which space precludes giving here, indicates that a minimum area must be involved to reach threshold. Evidences of a microscopic mosaic in the axon membrane are important in interpretations of miniature subthreshold potentials and the mechanism of response (Tasaki et al., 1958; Grundfest, 1959).

In the ramifying **terminals of axons** it is possible that in some cases the safety factor falls so low that only local potentials are generated and conduction to the tips becomes decremental. The dorsal horn of the vertebrate spinal cord gives evidence of lability of the amplitude of potentials in presynaptic terminals, depending on the events just past in the same fibers and on the activity of neighboring fibers.

We see that conduction does not involve solely nerve impulses but in important roles also graded, decrementing activity. Only nerve impulses can carry signals for distances of many millimeters or more. It is possible to conceive of a system based on a graded event in a region of axon stimulating the adjoining region to active but graded response and thus propagating without decrement but without all-or-none detonation; but since each region experiences only the amplitude of the event immediately preceding, there would be large drifts of amplitude without correction. Some signals would be lost. Others would fluctuate widely in velocity. Those that grew to maximal intensity and velocity might overcome the incomplete refractoriness in the wake of weaker signals ahead and fuse with them. The all-or-none system is not the only possible long-distance propagation principle, but it partly overcomes the problem of slight **differences in excitability** along the fiber. The amplitude, because of the large safety factor, allows a wide variation in local condition of the membrane; the signal is conducted faithfully over long distances. But the information flow, since it is dependent not simply on the presence of the pulses but also on their spacing, is not necessarily faithful. Local variation in excitability or responsiveness can alter the intervals and we are rightly impressed by the evidence that conduction is on the whole without serious distortion. This fact means that the time periods employed in information flow by intervals between impulses are on the whole shorter than the rates of drift of excitability of local areas of axonal membrane.

In the middle 1950's the appreciation of two general conclusions about signals in neurons led to an interpretation of their **possible evolutionary relationship** (Table 3.1). The first general conclusion was the flexibility, variety, and primacy of graded responses in the normal sequence of activation of neurons. The second was the specialized character of the impulse, confined in function to the faithful propagation of a pulsed code determined by integrative processes preceding it, and confined in space to a special and

Table 3.1. A HIERARCHY OF CELLS ARRANGED WITH RESPECT TO DEGREE OF SPECIALIZATION OF MEMBRANES FOR PRODUCING ACTION POTENTIALS WHEN ACTIVE (Bishop, 1956).

Type of Cell	Type of Potential of Activity	Excitability
Embryonic cells	None (?)	
Fat cells, erythrocytes	None (?)	
Gland; liver, kidney, salivary cells	Slow depolarization accompanies functioning	Hormones or nerve impulses
Protozoan, *Plasmodium*	Graded	Electrical, mechanical
Smooth muscle cells	Slow, persists during contraction	Electrical, hormonal
Adductor muscle of clam	Graded, decremental conduction	Electrical, chemical (?)
Invertebrate and some vertebrate skeletal muscle	Graded, like endplate, decremental conduction, or none	Some electrical, others by chemical transmitter only
Skeletal muscle endplate	Graded	Chemical transmitter
Electroplaque, *Torpedo*, skate (modified endplate)	Graded	Chemical transmitter
Muscle spindle, mammalian intrafusal muscle cell	Graded	Chemical transmitter (?)
Electroplaque of eel, *Electrophorus*	Graded plus all-or-none spike	Graded, chemical transmitter; spike, electrical
Dendrite of pyramidal cell	Graded	Electrical, chemical transmitter (?)
Vertebrate heart muscle cell	All-or-none; duration of systole	Electrical, spontaneous
Nitella (unicellular alga)	All-or-none; graded near threshold (?)	Electrical, other (?)
Vertebrate skeletal muscle, "twitch" type	All-or-none spike briefer than contraction	Electrical
Cell body of spinal motoneuron	Reacts like mixture of dendrite and axon	Electrical, chemical (?)
Cell body of lobster cardiac ganglion	Reacts like dendrite; no spike	Electrical, chemical transmitter (?)
Axon	Graded local response plus spike	Electrical

monotonous though extended part of the neuron. The suggestion is inevitable that while the impulse can hardly exist without graded prepotentials, the latter can not only exist, but within limits—purely spatial—can function entirely without impulses. Since amplification of the energy content, measured by current, occurs even in graded responses, a certain amount of decrement can be tolerated; the boosting to restore signals can be imagined to occur in successive neurons with short processes. As outlined in the next chapter, there is some evidence for the already reasonable proposition that neurons might influence each other in some cases without the intervention of impulses, even over specific paths. Bishop, who formulated a definite suggestion that the graded responses are more primitive and may have existed at first without all-or-none impulses (1956), reminded us that nonnervous cells of many kinds have local and graded responses which could have formed the first stages in an evolution of responses specialized for conduction over long distances, at high velocity and at high repetition rate. The question whether the development of such responses ("impulses") coincided with the differentiation of nerve cells from nonnervous cells, or whether there are still some neurons employing only graded activity, cannot be answered yet; up to now no such neurons are known unequivocally, though rods, cones, retinal bipolars, and *Limulus* retinula cells perhaps function this way. An expectation in favor of their existence—if not in primitive nervous systems, at least in certain short-process cells, and if not as exclusive mode of functioning, at least at times—is inherent in the hypothesis.

A special class of responses with obvious interest is that of **active hyperpolarization.** Most preparations have shown only passive, electrotonic membrane changes when anodal or inward current is imposed. But a number of workers have observed active response tending to increase internal negativity. Arvanitaki (1941, 1943) described anodal responses in resting axons (see also Chalazonitis and Arvanitaki, 1957). True spikelike regenerative hyperpolarizing responses to strong inward currents have been seen in several types of preparations, for example an axon partly depolarized by high external potassium. Some agents prolong the ordinary (inside positive) spike greatly; during this phase an active, regenerative repolarization (to inside negative) can be induced, either briefly or persisting (Teorell, 1953; Segal, 1958; Stämpfli, 1958; Tasaki, 1959b, 1959c; Adrian, 1960; Chang and Schmidt, 1960a, 1960b; Reuben et al., 1960).

B. Dendrite and Axon

Current terminology reflects the inadequacy of our knowledge. It cannot be excluded that **dendrites** in some situations carry impulses, but they are not dendrites just because they conduct impulses toward the cell body. At least many dendrites probably are incapable of supporting an all-or-none event; if they can, it is likely to be only at the base, with the impulse normally occurring first in the cell body, spreading outward rather than cellulipetally (Fig. 3.4). Remembering the most common type of invertebrate neuron—the unipolar with a long stem process which can join the main pathway of nervous signals at one end or in the middle—dendrites must be defined as receptive processes with respect to the neuron, rather than the cell body. The **axon,** for its part, is not the only structure supporting an impulse (the soma often can), but it is the long conducting pathway of the neuron between its receptive and its efferent terminals. In the instances examined so far, the axon only functions by carrying impulses, but the possibility has been mentioned that in neurons with short axons electrotonic conduction of graded signals may play a role. In special cases neurosecretory axons function by transporting materials and there are some axons histologically known which have so far not been found to conduct impulses (the majority of fibers in the optic nerve of *Limulus*). These criteria for axon and dendrite are not identical with the classical ones but are not in conflict with them. The newer functional criteria have developed from vertebrate neurophysiology but they serve to extend the usefulness of the terms to invertebrates. No doubt such criteria are temporary usages in the development of a more natural set of concepts; at present, the direction in which this is going seems to be the recognition of relatively sharply differentiated types of cell surface membranes, each with its own functional capacities. The classical cytological differences between vertebrate dendrites and axons are on this basis meaningful since it is not surprising that differentiation in functional capacity goes with differentiation in structure. There is probably more differentiation in structure yet to be recognized, at finer levels (see Chapter 2).

The problem of the **orthodromic direction** of conduction requires attention, particularly in typical invertebrate neurons which are unipolar. The all-or-none process, once started, can propagate in either direction, and the term orthodromic refers to the direction which it takes in the normally or physiologically initiated reaction. The term antidromic is used when, by artificial stimulation downstream from its normal origin, an impulse is caused to conduct in the opposite direction, toward its normal site of initiation. The "normal" direction can readily be inferred when it is obvious which end or which processes of the neuron are normally afferent or efferent for that cell; this is often not obvious in unipolar ganglion cells. Nor is the difference obvious between dendritic and axonal ramifications, as explained further in Chapter 2. Even when there is a long process which can safely be regarded for that reason as an axon, the orthodromic direction is not at once clear, for there is no absolute rule that the axon normally conducts toward or away from the cell body; and in addition, many unipolar cells are situated about midway along the long process. It is necessary to ascertain which of the ramifications at the ends of the axon, or along its course, are receptive and which efferent. We have not learned to do this histologically in every case. Confusion due to similarity of appearance of some of the branchings at the two ends in some common types of intersegmental interneurons in invertebrates is alleviated by the physiological finding that there are neurons which normally conduct now upward, now downward. Evidently some of the tufts which sprout from the axon are receptive and some efferent at each end.

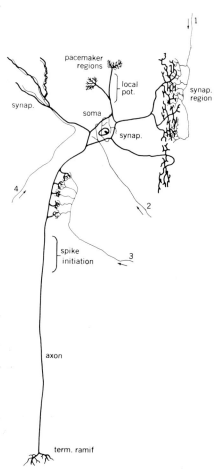

Figure 3.4. Schematic neuron. Based on a cell in the cardiac ganglion of a crab after Alexandrowicz (1932). Several presynaptic pathways converge from diverse sources: inhibiting, driving, accelerating (1, 2, 3, 4). The functionally known pathways cannot yet be identified with anatomically distinct endings, several of which actually can be made by branches of the same presynaptic axon. These produce synaptic potentials of various forms, each in its limited locus. Restricted regions also initiate spontaneous activity ("pacemaker," located arbitrarily in the diagram), local potentials (labeled in one place only but perhaps repeated elsewhere), and propagated impulses ("spike initiation"). Only the axon supports all-or-none activity. Terminal ramifications are presumed to act by graded local potentials, at least in many neurons. Integration occurs at each site of transition or confluence.

III. EXCITABILITY AND ITS PRIMARY PARAMETERS

The excitability of a neuron is not a single variable. The synaptic regions, which may be several and of more than one kind, need not bear a constant relation to each other in their individual sensitivities. Similarly, the endogenic parts of the neuron (Figs. 3.1, 3.4), which respond to antecedent activity of other parts of the same cell, may have quite different excitabilities. This is especially documented for the difference between spike thresholds of soma and initial axon segment in vertebrate spinal motoneurons. But even at one and the same spot on a simple axon there are two separate modes of excitability: that measured by local response and that measured by spike response. The latter is adequately expressed by the value of the threshold, but the local response has no sharp threshold and must be described by a curve relating its relative amplitude to the subthreshold stimulus intensity (Fig. 3.3). Finally, whenever repetitive firing is generated in response to maintained depolarization (and not to discrete arriving volleys), the cell so firing can be said to have a pacemaker process that determines its frequency as a function of this maintained depolarizing condition; this is another form of sensitivity.

It is important to distinguish between **sensitivity and responsiveness** or responsivity. The difference is clear even though it is usually difficult to show which one is responsible when there has been a change. To explain by analogy, I may be very sensitive to insult but weak or unwilling to respond or, conversely, thick-skinned but terrible when aroused! Stated generally, sensitivity is measured by the intensity of stimulus necessary to elicit a given percentage response (the chosen response may be that which is just detectable or that which is just maximal or some fraction of the latter, as distinct from absolute magnitude). Responsiveness is measured by the absolute magnitude of the response to a stimulus of a given fractional intensity (whether just threshold or just maximal or some proportion of the latter).

A. Local Response Excitability

The sensitivity of graded responses has been little studied, but it should be quite significant what changes in the curve of fractional response against stimulus occur with time, with repetition, and with predepolarization. In contrast, an extensive literature deals with spike threshold and its determinants. Historically, it was but slowly admitted that the signs of a graded potential in the neighborhood of the stimulated locus point to an active local response; it was not realized that this conflicts in no way with the all-or-none concept of the impulse arising therefrom. Early workers observed that subthreshold shocks leave behind them a state of heightened excitability so that a second stimulus, by itself subthreshold, suffices to reach threshold and fire the impulse. The state of sensitization rises to an early maximum and then decays (Rushton, 1935), and the rate of decay is characteristic of the nerve. This measurement is a composite; it certainly includes the spike threshold, but it also includes the local response.

Direct demonstration of the active local potential rising out of the passive electrotonic potential at the site of stimulation dates from Katz (1937) on frog nerve, and Hodgkin (1938) on crab nerve; soon thereafter squid, earthworm, and other cases were added. It is still not known whether the excitability as manifested by local response is bound to that manifested by threshold, because changes in the two necessarily go together. But the so-called sodium theory of Hodgkin and Huxley (1952) assumes that this is so and accounts remarkably well for actual values measured on squid axon under certain conditions. Tests of this point under a variety of conditions or in different axons would be technically laborious and have not been done. The unification of spike and graded excitability and of sensitivity and responsiveness or, better, the accounting for these properties by relations between a more fundamental set of independent variables, is as secure at present as that theory is satisfactory on other tests (see "Sodium theory," below).

Differences in local potential excitability between different types of neurons are not well known but clearly occur. Wright (1959) shows (Fig. 3.5) that the claw opener axon in *Homarus* has a very small (1–2 mv), long (15 msec) local response, that is, its spike threshold is low. The slow-closer axon has an intermediate (2–5 mv, 5–10 msec) response and the fast-closer and also the medial giant fiber in the cord have large (15–20 mv), brief (2–5 msec) local responses that are actually followed by a refractory period.

The spatial extent of graded responses is mainly, but not entirely, a function of the space constant of the membranes through which the local circuits flow. That is, a certain amount of active propagation can be seen for a short distance—a fraction of a millimeter—but suffers a decrement; and since the curve of active response against intensity of exciting current falls steeply with falling intensity, soon nothing but electrotonic or passive spread remains (Hodgkin, 1937). This spread is given by the **space constant**—which is the distance to which an electrotonic pulse will be seen, having declined to $1/e$ or about 37% of its initial height. The space constant is determined in turn by internal, external, and membrane resistances, and membrane capacity. It therefore varies with fiber size (being smaller in smaller fibers), though not in a simple manner because of other variables. Representative values may be obtained from these examples: *Sepia*, 200 μ diameter fibers: space constant 5.7 mm; *Loligo*, 500 μ fibers: 2.3 mm; *Carcinus*, 30 μ fibers: 2.0 mm; *Homarus*, 75 μ fibers: 0.81 to 2.95 mm, mean: 1.6 mm; *Rana*, sciatic nerve: 1.3 to 2.9 mm, mean: 2.0 mm. The variation among measurements of the same kind of fiber is often several fold. In very fine fibers such as dendritic processes and axonal terminals, the possibilities of electrotonic spread are of great importance in present views about the role of graded activity. Unfortunately, reliable values for the space constant in such elements are not yet at hand. Eccles gives 0.3 mm as a value for a "standard" dendrite. Ito (1957) gives 0.6 mm for the initial unmyelinated segment of axon in toad dorsal root ganglion cells where the diameter is 5 μ; Ottoson (1959) reports an electrotonic decay to one-half in 1 mm in frog olfactory nerves, $1/e$ would be slightly larger. Measurements made in air or oil or with a limited saline surround are lower than those made in a saline bath; a squid fiber which in a high-resistance medium gives a value of 2.3 mm, gives 6.0 in sea water.

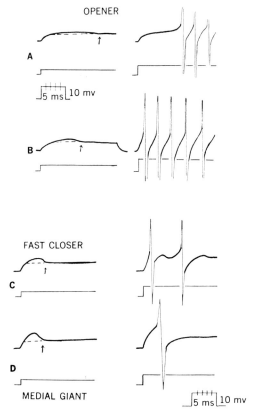

Figure 3.5. Differences in form of local potential in different axons. Subthreshold and suprathreshold responses of three crustacean motor axons. **A.** The left trace shows the subthreshold response of a fiber to the opener muscle superimposed on a shock artifact; the right trace shows the suprathreshold response. The D. C. stimulus is drawn in below, illustrating relative strength of stimulation. **B.** Subthreshold and suprathreshold responses from the same fiber two hours later. The *arrows* indicate termination of subthreshold response. **C.** Subthreshold and suprathreshold responses from an axon causing a fast response of the closer muscle. **D.** Subthreshold and suprathreshold responses from a medial giant fiber. Extracellular recording. [Wright, 1959.]

B. Spike Excitability

Efforts to distill the **primary parameters** of excitation out of the observed variables have a long and complex history and have chiefly dealt with aspects of the spike: form (amplitude and time course), velocity, recovery cycle (refractory periods), and threshold current as function of duration of stimulus (in the case of abruptly applied shocks) and of slope (in the case of slowly rising stimulating currents). Earlier textbooks, reviews in the *Annual Reviews of Physiology* through the 1940's, and the monograph of Katz (1939) may be consulted for accounts of these efforts. A comprehensive review is out of place here but is much needed. The recent reviews have not embraced the older literature, especially in regard to excitability, and it is just here that the quantitative adequacy of the newer theories has been least carefully examined.

1. Sodium theory

The theory put forward by Hodgkin and Huxley (see Hodgkin, 1951; Hodgkin and Huxley, 1952; Cole, 1957, 1958, 1961; Cole et al., 1955; Tasaki, 1959), often called the sodium theory, is the only modern general theory to be formulated quantitatively which essays to account for most of the observed variables just listed. The three main parameters, according to these authors, are (a) the membrane conductance to sodium as a function of membrane potential, (b) the membrane conductance to potassium as a function of membrane potential, and (c) the inactivation of the transport mechanism by which sodium is carried into the cell, as a function of membrane potential. Each of these includes a time constant. Based on measurements of these functions, Hodgkin and Huxley were able to develop equations, relating the functions and the passive electrical characters of the membrane, that gave astonishingly good numerical agreement with directly observed quantities for the form of the subthreshold response, the threshold, the form of the spike, the refractory period, velocity of propagation, and the effect of a number of alterations of the external medium. Furthermore, the theory agrees with observation at least qualitatively in predicting that there will be a minimal rate of rise of a stimulating current below which no response occurs, that something equivalent to accommodation will occur, and that damped oscillatory potentials will occur after a subthreshold stimulus or a spike when the membrane level is not too high. Its adequacy to account for late afterpotentials has not been tested; and it cannot account for any very slow changes in the state of the membrane. The detailed quantitative support for the theory comes entirely from the third-order giant axons of squid, but a large body of experimental findings on other nerve and muscle fibers in vertebrates and invertebrates is in agreement with it. Nevertheless, a number of discrepancies have been reported and the theory cannot be said to be entirely satisfactory (Teorell, 1953; Grundfest et al., 1954; Grundfest, 1955, 1960; Shaw et al., 1956; Simon et al., 1957; Gaffey and Mullins, 1958; Shanes, 1958; Tasaki and Bak, 1958a, 1958b; Mullins, 1959, 1960; Tasaki, 1959; Tasaki et al., 1961).

2. Strength-duration relation

For the present discussion, two of the observed variables—which have received the least quantitative testing under the sodium theory—are important. They are the time factor involved in the relation between threshold current and duration of abruptly applied shocks and the time factor involved in the relation between threshold current and the slope of slowly rising stimuli.

The first of these two relations is empirically given by the strength-duration curve (Fig. 3.6), perhaps the most studied feature of excitable tissues. Katz (1939) and earlier authors referred to by him should be consulted for the numerous problems of accurately determining this relation and of interpreting it. Though little used today as a description of the state of excitability of a neuron, there is no sound reason for its disuse; we do not have more satisfactory methods of comparing the same fundamental properties in a wide range of preparations. The two landmarks of the curve classically employed, after Lapicque, are the rheobase and the chronaxie. The former is the minimal intensity (current or voltage) which

Table 3.2. PROPERTIES OF EXCITATION IN NERVE CELLS AND FIBERS. Most determinations were made with intracellular electrodes. The values are selected as being probably most representative of tissues in good condition; they are not always means or extremes. Accommodation is used as rise of threshold depolarization (change from resting polarization) under maintained or slowly rising predepolarization, therefore only indirectly related to λ. Pacemaker sensitivity is expressed as (i) impulses per second at twice rheobasic maintained current, (ii) impulses per second per 10^{-9} A measured during the low current part of the frequency/current curve. Other quantities are conventionally defined.

	A	B	C	D	E	F	G	H	I	J	K	L
		Rheobase		Chron-axie[a] (msec)	Absolute Refractory Period[a] (msec)	Accommo-dation	Capacity (μf/cm^b)	Resistance of Resting Membrane (ohm-cm^b)	Time Constant of Membrane (msec)	Pacemaker Sensitivity	Velocity[b] (m/sec)	Dia-meter (μ)
	Preparation	A × 10^{-9}	mv									

NERVE CELL BODIES

	Preparation	A × 10^{-9}	mv	Chronaxie (msec)	Abs. Refr. Period (msec)	Accommodation	Capacity	Resistance	Time Const.	Pacemaker Sens.	Velocity	Diameter
1	*Aplysia*, giant cell, visceral ganglion	10–20 (50–200/cm²)	2–30	20–100	5–10	none for 200 msec	11	4000	10–80	(i) 7.5–45 (ii) 0.7–4.5	—	178
2	*Panulirus*, large cell in cardiac ganglion[c]	4–100	2.5–11				(0.003–0.03 per cell)	(0.14–1.4 × 10⁵Ω per cell)	3–16	(i) 15 (ii) 3	—	50
3	*Limulus*, eccentric cell in retina	0.5	4–9	< 6		slight in 1 sec		(6 × 10⁶Ω per cell)		(i) 7.7 (ii) 4–26	—	25
4	*Sphaeroides* (puffer fish) supramedullary ganglion cell	220 (Bennett et al.)	18–25		3	none	5–15 (30, Bennett et al.)	500–1000 (0.6–2.5 × 10⁵Ω per cell)	4–6 (10–20, Bennett et al.)		—	250
5	*Bufo*, dorsal root ganglion cell	1.2	17	2			1.1	2200–4000[d]	2–5		—	90
6	*Bufo*, spinal motoneuron soma	1.4	8–11	4.6		none for 10 msec; 25% at 25 msec;	18	270[e] (4.5 × 10⁶Ω per cell)	4.3		—	30
	Initial segment of axon	1.3	6.5–8.5	2.0		begins at 8 msec, 250% at 25 msec						
7	*Felis*, spinal motoneuron	2–18 (initial axon)	30 (soma) 10	0.76	about 1.5	none for 200 msec, then small from 200–1000 msec	5	600[f] (1.2 × 10⁵Ω per cell) 1000–8000[g]	3.1 4[g]	(ii) 2.5	—	70

NERVE FIBERS

#											
8	Loligo, giant axon	8–10	1.5		none for about 5 msec, then considerable	1.1	1500	1.6	iteration brief or absent	33 (23° C)	500
9	Sepia, giant axon					1.2	9200	14		7 (16° C)	200
10	Carcinus, leg nerve isolated axons, Types I, II, and III				I, none for 17 + sec; II, slow; III, fast	1.1	8000	9	I(i) 33–105, (ii) 5–144; II(i) 150–225, (ii) 30–60; III, no iteration	3–4 (21° C)	30
11	Cambarus, claw nerve										
	fast-closer axon:	81	0.2	2.1	rapid ($\lambda = 8$ msec)				no iteration	20	58
	slow-closer axon:	70	0.46	1.6	intermediate ($\lambda = 15$ msec)				brief iteration	10	41
	opener axon:	41	0.65	2.2	slow or small ($\lambda = 48$ msec)				long iteration	8	36
		(relative values only – external electrodes)									
12	Rana, sciatic nerve, single nodal fiber	0.6	10–15	0.1 (at node) 0.2 max (internode)	1.7	none for 25–40 msec, then slight	3.7 (at node = 1.5×10^{-6} μf per node) 5×10^{-3} (internode)	8–20 (at node = 40×10^6 per node) 100,000 (internode)	0.06 (per node)	31	16
13	Rana, sciatic nerve										
	A fibers		0.05–0.3	0.8–0.9							14–50
	B_1 fibers		0.35–0.45	0.9–1.1							8–16
	B_2 fibers		3.0–4.0	3.5–3.7							3–4.5
	C fibers		3.5–5.0	4.5–10							0.3–0.8 (20–25° C)

a. Values are not strictly comparable; obtained with internal electrodes (except 11, 12 and 13) of various exposed lengths. Values not corrected for differences in temperature of measurement. Higher temperature within a physiological range decreases chronaxie and refractory period, increases accommodation and velocity, does not change others, generally. For additional values, with external electrodes, see Schaefer (1940), Table 18; Lullies (1932); Rosenberg (1925).

b. Values given have usually been measured with very little saline shunt and would be appreciably higher—up to 50%—in sea water or in the animal.

c. The spike neither arises in nor invades the cell; therefore the stimuli injected into it are attenuated by some unknown amount before acting at the spike-initiating locus.

d. Surface area of soma = 1.1×10^{-4} to 2.5×10^{-4} cm²; values given are for larger cells; lower resistance is for cathodal current, higher for anodal.

e. Assuming that surface area of soma and large dendrites = 6,000 μ² (6×10^{-5} cm²).

f. Assuming that surface area of soma and large dendrites = 50,000 μ² (5×10^{-4} cm²).

g. According to Rall (1959).

References: 1. Tauc (1956), personal communication (1957). 2. Otani and Bullock (1959). 3. Fuortes (1958, 1959). 4. Hagiwara and Saito (1957); Bennett, Craig, and Grundfest (1959). 5. Ito (1957, 1959). 6. Araki and Otani (1955, 1959). 7. Frank and Fuortes (1960); Coombs, Eccles, and Curtis (1959). 8. Hodgkin, Huxley, and Katz (1952); Hagiwara and Oomura (1958, 1959). 9. Weidmann (1951). 10. Hodgkin (1947, 1948). 11. Wright et al. (1954, 1955). 12. Hodler, Stämpfli, and Tasaki (1952); Tasaki (1955, 1959). 13. Schaefer (1940).

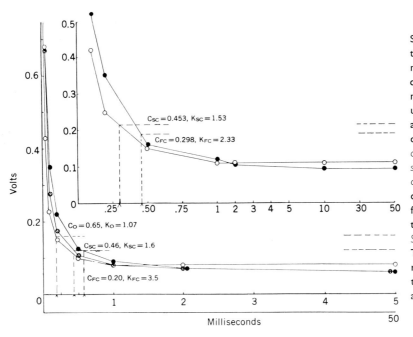

Figure 3.6.

Strength-duration curves of three types of crustacean limb motor axons. The lower curves are drawn through the mean values of all results, the upper curves are results from a crayfish fast- and slow-closer double-fiber preparation. *Open circles* indicate fast fiber closer; *solid circles* show closer, *shaded circles* opener fiber data. The chronaxies *(C)* and k *(K)* values for each fiber are shown with the subletters. F.C., fast closer. S. C., slow closer; O, opener, The *dashed lines* show twice rheobase values used for determining chronaxies. [Wright and Coleman, 1954.]

will reach threshold with a stimulus of long duration. The latter is the threshold duration of a stimulus which is twice the intensity of the rheobase. These have been somewhat discredited but only because they can be misused, as in comparing different preparations in which conditions affecting the strength-duration curve are not quite comparable, for example, electrode spacing and size. The fact remains that a variable not yet shown to be accounted for with more elementary parameters is accessible, is altered by conditions of milieu and past treatment and is different in different axons as well as nerve cell bodies and muscle cells; all of these opportunities deserve systematic treatment.

The most nearly satisfactory theory dealing with this relation is that of Hill (see Katz, 1939), but virtually equivalent formulations are given by Monnier (1934) and by Rashevsky (1936; see also Blair, 1932, 1934, 1936). Except for a small number of assumptions, a single time factor is the essential parameter—Hill's k. This is a brief time, very nearly the same as the time constant of the membrane for electrotonic potentials and therefore approximately one or at most a few milliseconds. In Hill's equation k gives a good fit for many reported data, but discrepancies have been encountered and the interpretation of utilization time and the constant quantity relation with short stimuli are not satisfactory (Katz, 1939; Tasaki, 1953; Tasaki and Sato, 1951; Easton, 1952). Utilization time is the latency of the spike at the site of stimulation. The constant quantity relation is the constancy of the product of current times a function of duration at threshold in a certain range of stimulus durations.

Rheobase values vary widely (Table 3.2). Because of differences in size and effective resistance of different preparations, the current density (doubtless the significant aspect of a stimulus) varies for the same current. Values from less than 10^{-9} A to 10^{-7} A have been reported for transmembrane current in both somata and axons. According to present theory it is the voltage change in the membrane potential which determines stimulus effect, and this has been measured in many cases: values from less than 2 mv (lobster cardiac ganglion cell, Otani and Bullock, 1959) to more than 30 mv (cat spinal motoneuron, Eccles, 1957) have been obtained. General comparisons are difficult but it can be said that the threshold membrane change is less in cells or axons which are close to autoactivity. Because of accommodation, there is no absolute level of membrane polarization which represents a threshold. Significant differences do exist, such as that between initial segment and soma of spinal motoneurons in cat and toad; the initial segment of the axon has a threshold for depolarization of about 10 mv, the soma of about 30 mv (Eccles, 1957).

It is important to note that the spike threshold is normally an influential variable only at one place in the neuron (Fig. 3.4). This is the locus of initiation of spikes, probably generally in a neurite but not out of reach of electrotonic influence from synaptic, pacemaker, and locally responding regions. Downstream in the axon, the threshold can vary widely without consequence except perhaps at certain forks, because the safety factor is usually high.

The **safety factor** varies not only with the tissue and its intrinsic properties, but with the stage of accommodation or refractoriness—the amount of calcium, sodium, and potassium in the medium, and other factors. Since the amplitude of the spike is generally in the range 70–100 mv and the rheobase 10–20 mv, the safety factor is ordinarily between 3.5 and 10. Local regions of low safety factor are known in single-fiber preparations of many animals and probably occur commonly at branching points (possibly also at other points, especially after a period of high-frequency activity). These points may be important, for experimentally produced local regions of low safety factor acquire integrative properties. Impulses may approach and, either after passing through or failing, may leave behind a local condition which some milliseconds later generates one or a series of new impulses which may proceed in the original direction or in the opposite direction (reflection) or both. One-way block may develop. Impulses may jump a depressed region and in so doing may incur a delay of several milliseconds. These phenomena do not require nodes but are described in lobster, polychaete, and earthworm axons (Bullock and Turner, 1950).

The **chronaxie**, or some ratio of minimum short-shock depolarization to the threshold for a long depolarization, would appear to be a reasonable way to compare the excitability of different tissues. The demonstration by Rushton that a minimal area or length of axon has to be changed in potential, means that the geometry of the electrodes and the time for electrotonic distribution of the change affect the values measured. Measurements based on actual depolarization over an isopotential length of membrane have not been made on any number of tissues. The large literature of empirical values of chronaxie is impossible to interpret except in limited comparisons (for collections of values see the tables of Rosenberg in 1925 and Lullies in 1932 in *Tabulae Biologicae*). Changes in chronaxie, like the shortening seen by Golikov et al. (see Koshtoyants, 1957) during the relatively refractory and supernormal periods after a conditioning response, are potentially of interest if the same fibers are used for test and conditioning shocks. It is likely that considerable differences in excitation time exist among different neurons and parts of neurons. For example, large motor fibers in the frog sciatic nerve give values under certain conditions of 0.15 msec; preganglionic parasympathetic fibers inhibiting the heart, in the vagus nerve, 2.0 msec; postganglionic sympathetic fibers to the stomach, 20 msec. Smaller fibers tend to have larger values. Table 3.1 gives values for several modern measurements with intracellular electrodes: axons vary from 0.1 msec for nodal fibers in the frog sciatic to 1.5 msec for the squid giant; cells vary from 0.76 msec for cat motoneurons to more than 20 msec for *Aplysia* giant ganglion cells. The thinly or loosely myelinated but large motor fibers in the crab leg have small values, 0.2–0.65 msec, varying systematically among the three types of axons measured. Crab fibers giving smaller (externally recorded) spikes are characteristically excited by pulses of smaller energy content but they require longer duration for minimum energy pulse than fibers giving large spikes (Easton, 1952). Higher values of chronaxie do not necessarily mean higher rheobase; sometimes the strength-duration curves of different elements intersect. Thus, the lateral giant fibers of the earthworm have a lower threshold than the median for long pulses but have a higher threshold for short pulses (Niki et al., 1953).

Excitation time may be a sensitive indicator of

the effects of chemicals and drugs. Koshtoyants and Treshchalin (1940) reported some that influence rheobase but not chronaxie and others that change both in parallel. Nasonov compared the temperature dependence of excitation parameters in various animals; he found a relative independence of temperature in poikilotherms compared to a dependence in homoiotherms.

Table 3.2 also presents for comparison the main **passive electrical properties** or cable properties of some nerve cells and fibers. Differences in these values, both as a result of diameter of the element and as resistance or capacity per unit area of membrane, have a direct bearing on the activity of the element—for example rate of rise and fall of potential change and velocity of propagation. Values for some of these quantities are known for other preparations not in the Table, for example insect fibers (Boistel, 1959, 1960) and frog sympathetic nerve cells (Nishi and Koketsu, 1960). The relation between current imposed through an internal electrode and the resulting change in steady voltage has been studied for a variety of cells, revealing an asymmetry between the effects of inward and outward current. Membrane resistance falls with outward current above some level, whereas inward current causes no change or sometimes an actual rise in resistance; this has been called delayed rectification (Narahashi, 1960; Hagiwara, 1960).

3. Strength-slope relation

The relation between spike excitability and rate of application of stimulating current is empirically given by a curve of threshold strength against slope (Fig. 3.7). This curve also varies widely from one tissue to another and is even more labile with conditions of the medium and with previous treatment of the preparation. Generally the threshold strength shows a plateau (if it is expressed as threshold depolarization) over some range of slopes, and then rises slightly for very gradually rising stimuli. There is always a **minimal slope;** still more gradual stimuli will not excite at any strength. The rise in threshold with decreasing slopes (or, more generally, with predepolarization) is known as **accommodation.** Classically this term is defined by the strength of

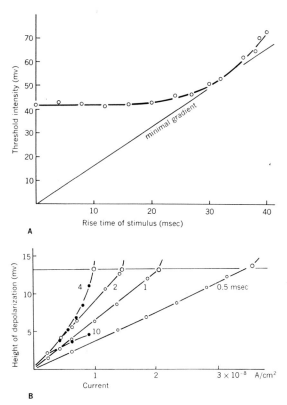

Figure 3.7. Strength-slope curves; the dependence of threshold upon rate of rise of stimulus: accommodation. **A.** Motor axon in the frog sciatic nerve with the sheath removed. The straight line gives the minimal slope that will elicit a response. Extracellular stimulation and recording, so the ordinate values are not absolute values of the depolarization of the membrane achieved by the stimulus. Note that there is no accommodation for the first 20 msec; there is a limited rise in threshold with just supraliminal gradient. [Diecke, 1954.] **B.** Squid giant axon. The same phenomenon can be seen on a different form of plot in which the voltage change caused by internally imposed current is plotted at different moments after the onset of a constant current. The first spike arises 0.5 msec after the start of the current only with very large currents; very small current produces a spike after latency of 4 msec. But the actual depolarization at the time of spike initiation is the same for long latency and short latency spikes. [Hagiwara and Oomura, 1958.]

an applied stimulus adequate to cause an impulse, while varying the slope (usually by charging condensers of different time constants). The result can be expressed by a value λ, in milliseconds (see Solandt, 1936), which is high when accommodation is low or slow. However, it is often overlooked that the latency of the spike may not increase in proportion to the stimulus time constant, so the actual voltage of the stimulus when spikes occur may be rising little or not at all with smaller slopes. More recent intracellular measurements, stimulating over some length of squid axon uniformly (Hagiwara and Oomura, 1958), show the necessity for considering the threshold depolarization actually produced in the membrane rather than applied voltage as the measure of true changes in excitability with slowly rising stimuli. Throughout a phase of many milliseconds there is no critical depolarization rise in squid axon; minimal slope is due to setting in of delayed rectification which reduces the depolarization caused by a given current. At a late phase there is a true increase in the critical depolarization, which coincides with an accommodative rise in the membrane potential (Tasaki, 1950). We may call this a true accommodation and the classical form, which may or may not represent such a change, apparent accommodation.

According to the sodium theory, true accommodation should be bound to the **refractory period** and both are accounted for by a combination of increased potassium conductance (probably equal to the delayed rectification) and sodium inactivation due to the time of subthreshold depolarization of the gradually rising stimulus. This quantitative prediction has not been tested. Hill's theory requires a single time factor to account fairly well for apparent accommodation curves, except in their earliest parts. This is his λ, a time which is usually at least ten times and often several hundred times as long as k, the primary time constant of excitation with brief shocks; λ is therefore from a few tens to some hundred of milliseconds. Measured in nerves or bundles with external electrodes, λ is about 35 msec in winter frog's sciatic nerves; 12–140 msec in ray (*Raja*) motor nerves; 30–60 msec in motor nerves of the cat and 100–1000 msec in sensory nerves. The threshold under these conditions rises approximately linearly during 30–100 msec of a linearly rising current but then gradually stabilizes at 3–4 times rheobase in motor, 1.5–2 times rheobase in sensory nerves (cat); the apparent accommodation is said to break down and further current elicits repetitive firing (Skoglund, 1942; Bernhard, Granit, and Skoglund, 1942).

Lambda is reduced by many agents: increased external potassium or calcium or temperature, manipulation of the preparation, reduced blood supply, time after dissection, and other deteriorating factors—all these increase apparent accommodation.

It has been found that determination of **apparent accommodation** is profoundly influenced by the presence or absence of the connective tissue sheath, at least in frog nerve (Frankenhaeuser, 1952; Tasaki, 1950; Diecke, 1954); whereas intact nerve exhibits a distinct rise in threshold beginning with slopes which reach threshold in less than a millisecond, single fibers free of this sheath exhibit no threshold rise even with slowly rising stimuli. Only just before the minimal slope is a small rise found. Similar behavior has been found in mammalian sensory nerve fibers in situ with circulation intact (Gray and Matthews, 1951). It is highly desirable that a body of comparative data be assembled, preferably using internal electrodes.

Incomplete accommodation is probably the rule, where there is any; that is, an indefinitely maintained current does not witness a complete return of the threshold depolarization to its initial value (think of the height of a test pulse given on top of a predepolarization). There is a residual change which is in the excitatory direction with cathodal or depolarizing currents and depressing after anodal or hyperpolarizing currents. Contrary to the classical dictum then, steady flow of current can exert an excitatory action.

Accommodation alters the effectiveness of synaptic potentials. Just as in anode break excitation, after a period of hyperpolarization, the absolute level of membrane potential at which a spike will be initiated is elevated. In this way an inhibitory synaptic potential, reversed by raising the membrane polarization, can excite. By the same token, during a series of excitatory postsynaptic potentials (e.p.s.p.'s) that maintain a considerable subthreshold depolarization, humps that rise above a critical level initiate spikes at first, but later they must rise materially higher to effect this. During block induced by excess depolarization, anodal pulses in the crayfish stretch receptor cell restore firing only if they are long enough (150 msec) and repeating them shows cumulative effects—a kind of accommodation.

The **range of variation** among nerve fibers and cells with respect to accommodation is large; only a few values are given in Table 3.1, obtained with intracellular electrodes or isolated elements. Most published figures cannot strictly be compared because of such factors as the sheaths, which distort the stimulus. It is clear that some cells and fibers do not accommodate at all for at least many milliseconds, for example, *Aplysia* ganglion cells, puffer supramedullary cells, cat motoneurons (some do and others do not, depending on site of spike origin), crab leg fibers of Hodgkin's type I (1948), the crayfish claw opener fiber. The frog sciatic nerve at least includes such nonaccommodating fibers but it seems likely that there are fibers differing in this respect. Other cells and fibers do exhibit a rise in threshold. Even though the cruder, whole-nerve measurements with slowly rising stimuli are suspected, there are internal electrode or isolated fiber measurements giving a rise in threshold of earlier onset and greater extent than the elements just named. For example, the initial segment of the axon of toad motoneurons begins to show accommodation after about 8 msec of predepolarization and its threshold rises 250% within 25 msec, ten times more than that of the soma, as shown in Fig. 3.8 (Araki and Otani, 1959). Mammalian motoneurons behave in either of two sharply contrasting ways, one with and one without a considerable rise in threshold (Bradley and Somjen, 1961). The fast-closer fiber of the crayfish claw has a value for λ about one-sixth of the opener's and one-half of the slow-closer's (Wright et al., 1954). Single dorsal root fibers of the cat from a cutaneous sensory (saphenous) nerve have little if any accommodation—λ approaches infinity—whereas those from muscular twigs of the popliteal nerve have values of λ between 150–200 msec, representing approximately an increase of 50% in threshold in 50 msec (Granit and Skoglund, 1943). The differences among cells and fibers revealed by this variable have significance in helping or hindering repetitive discharge to steadily maintained stimuli, although the relation is not simple or direct ("Pacemaker Excitability," below).

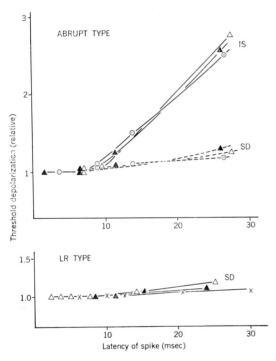

Figure 3.8. Strength-slope curve in cell bodies. The rise of threshold depolarization measured intracellularly as a function of latent time, using exponentially increasing currents in spinal motoneurons of two types. Ordinates are the threshold depolarizations relative to that of responses with the shortest latency (steepest stimuli); the abscissa, latency, measures the rate of rise of the stimulus. The abrupt type are spikes that appear to arise in the initial segment of the axon; the *LR* type are spikes that appear to arise in the soma; *IS*, initial segment spikes; *SD*, soma dendrite spikes that arise later at the same threshold in the abrupt type and are the only response seen in the *LR* type. Note that there is no accommodation for the first several milliseconds and then there is very little for one and much more for the other type of spike. [Araki and Otani, 1959.]

4. Recovery cycle

After an active response the recovery cycle (Fig. 3.9) is possibly related to accommodation, although the latter can vary widely without a striking change in the refractory period. The sodium theory explains them both on the same grounds but no exact application has been

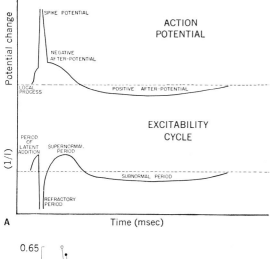
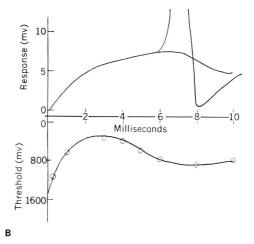
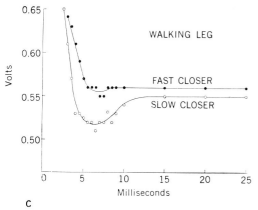

Figure 3.9. Recovery cycle. **A.** Diagrammatic presentation of the common phases; there is not always such a close parallel between the afterpotential and the late excitability. The time scale has been expanded at the left and compressed at the right. Ordinate $1/\iota$ means reciprocal of intensity of stimulation. [Morgan, 1943, *Physiological Psychology*, McGraw Hill, New York.] **B.** The local subthreshold depolarization deflection, upper trace, and dual-shock excitability curve, lower trace, of a medial giant axon in the crayfish. The excitability curve has been inverted to make comparison easier. [Wright, 1959.] **C.** Excitability recovery curves of the fast-closer fiber *(solid circles)* and slow-closer fibers *(open circles)* of the walking leg of the lobster. [Wright and Reuben, 1958.]

made. It is one of the cardinal features of the pulse-coded principle of signal propagation that after each impulse there is a period of complete refractoriness. This gives way to a gradual recovery of the initial excitability, the relatively refractory period. In some preparations there follows a period of supernormality, possibly but not certainly associated with a damped oscillation of the membrane potential. The concept of the **functional refractory period** (Rosenblueth et al., 1949) points up the fact that a given part of the axon is excited normally by the local circuit current of an approaching impulse which is several times larger than necessary due to the safety factor. Therefore the spacing of successive impulses as determined by refractoriness will be that which expresses a safety factor of one; as soon as the recovery of excitability has proceeded far enough that the approaching impulse can just reach the threshold, the functional refractory period is over. Velocity of propagation will be still subnormal.

Absolutely refractory periods vary over a range (Table 3.2) from about 0.4 msec for mammalian axons of the class Aα to 2.0 msec for mammalian C fibers, 3.0 msec for squid giant fibers, and 5–10 msec for ganglion cell bodies of *Aplysia*. The correlation is very close between this quantity and the duration of the spike in each case, but exceptions are known (see Shanes, 1958b). As a rule, both quantities are greater in thinner or less myelinated fibers. Periods even shorter than 0.4 msec are indicated by a number of cases of axons carrying normal signals at repetition rates exceeding 1000 per second—in the mammalian auditory nerve and cortex, for example, and in certain gymnotid fish (Grundfest, personal communication). **Relatively refractory periods** always follow and usually are longer than the absolutely refractory period. Representative values are 3 msec for mammalian A fibers and 12 msec for large frog sciatic axons, but it is difficult to ascertain comparable figures in different nerves because the end point is not sharp and the effect of treatment of the preparation is great. Rate of recovery of normal excitability is said not to vary with diameter in vertebrate fibers. Still more variable is the **supernormal period,** not known in most nerves and developing only with time after dissection in others. It is large and long (5–10 msec) in the small claw-opener fiber of *Homarus*, which is a fiber with a strong iterative tendency, but small or absent in the large, non-oscillatory fast-closer fiber and in the giant fiber of the cord. A still later period of subnormality is related to the positive afterpotential (treated below) and may last many tens or hundreds of milliseconds, especially in smaller fibers.

5. Form of spike

There is no basis for supposing that any large, systematic variation in spike amplitude exists from species to species or between different nerves, measured across the neuronal membrane. But it must also be stated that the observed variation (Table 3.3), well within a factor of three (50–120 mv) for most axons and spike-invaded cells, cannot be entirely attributed to

Table 3.3. RESTING AND SPIKE POTENTIALS.

Values given are judged to be representative of tissues in good condition; they are not always means or extremes but weighted selections from those reported. For additional values see *Handbook of Biological Data*, 1956, Table 280.

	Preparation	Resting Potential (mv)	Spike Potential Peak (mv)	Overshoot Maximum (mv)	Spike Duration (msec)
	A	B	C	D	E
	Fibers				
1	*Loligo*, giant axon	60	120	60	0.75 (18°)
2	*Lumbricus*, median giant	70	100	30	1.0 (20°)
3	*Cambarus*, median giant	90	145	55	2 (18°)
4	*Periplaneta*, giant fibers	70	80–104	26	0.4 (26°)
5	*Carcinus*, 30 μ leg axon	71–94	116–153	60	1.0 (21°)
6	*Rana*, sciatic nerve axon	60–80	100–130	50	1.0 (20°)
	Cells				
7	*Aplysia*, visceral ganglion	40–60	80–120	60	10 (21°)
8	*Onchidium*, visceral ganglion	60–70	80–100	30	9 (21°)
9	*Cambarus*, stretch receptor	70–80	80–90	20	2.5 (21°)
10	*Sphaeroides*, supramedullary	50–80	80–110	40	3 (26°)
11	*Bufo*, dorsal root ganglion	50–80	80–125	57	2.8 (17°)
12	*Bufo*, spinal motoneuron	40–60	40–84	25	2 (17°)
13	*Oryctolagus*, sympathetic	65–82	75–103	25	4–7 (37°)
14	*Felis*, spinal motoneuron	55–80	80–110	40	1–1.5 (37°)

References

(1) Weidmann (1951); Hodgkin and Huxley (1952); Grundfest et al. (1954); Hodgkin (1958).
(2) Kao and Grundfest (1957; see Annelida).
(3) Kao and Grundfest (1956); Watanabe (1958; see Arthropoda).
(4) Yamasaki and Narahashi (1959).
(5) Hodgkin (1951).
(6) Tasaki (1959).
(7) Tauc (1955).
(8) Hagiwara and Saito (1959).
(9) Eyzaguirre and Kuffler (1955).
(10) Hagiwara and Saito (1957); Bennett et al. (1959).
(11) Ito (1957).
(12) Araki and Otani (1953, 1955).
(13) Eccles, R. (1955).
(14) Frank and Fuortes (1956, 1961); Coombs et al. (1959).

poor condition in all the submaximal cases. Injury as well as high frequency discharge do reduce spike height. But in instances where these can reasonably be excluded there is still a variation, doubtless due to "condition" of the material but possibly within a "normal" range. In insects with high potassium in the blood, muscle cells and probably some nerve fibers and cells, at least in the periphery, have relatively low resting and action potentials (see p. 1133). Some cell bodies have very small spikes due to the inability of the membrane to support regenerative action, that is, spikes do not invade (p. 994).

Spikes **recorded by external electrodes** vary in height systematically with fiber diameter, as expected from the smaller core conductance. This relation between diameter and externally recorded spike height is usually taken to be one of simple proportionality since such has been shown in vertebrate myelinated fibers in agreement with the simple proportionality between diameter and velocity. But in other animals or fiber types it is quite obvious that spike height increases much more than in proportion to diameter. For example, in polychaetes having several giant fibers of different sizes (as in *Lumbrineris*, Fig. 14.21), the spikes increase approximately as the square of the diameter, while the velocity is only increasing approximately as the square root of the diameter (see p. 698). Since so little gain in velocity and so much in amplitude is achieved—as when a fourfold increase in diameter gives about twice the velocity and 16 times the spike height, recorded externally—it is tempting to speculate whether the latter can have some adaptive value as such. Giant fibers represent a large cost in volume for a small gain in velocity, if that is the pertinent gain. Externally felt spike height might be significant if electrical transmission is involved, as it is in crayfish at least. This would give an additional meaning to the **evolution of giant fiber systems.**

The **duration of the spike** on the contrary varies systematically with the fiber type. But the range is almost as great among fibers of the same species as among the whole animal kingdom, from jellyfish to mammal. The fastest spikes in the latter group rise to a crest in little more than 0.1 msec and fall to a transition into afterpotential in a total time of about 0.4 msec; slow C fiber spikes in the same animals last 2 msec, with approximately the same proportion of time in rising and falling. Jellyfish nerve net neurons and gastropod ganglion cells fall in the range of 5–10 msec. Large axons of squid, earthworm, crab, and insect have action potentials of 1–3 msec. Even the small axons of these animals produce spikes which do not exceed 3 msec as a rule. In spite of some characteristic differences then, it is impressive how similar the electrical sign of the nerve impulse is throughout the great diversity of animal types and level of organization. The spike which is produced in nerve cell bodies is typically slower than that in the axon by a third or a half.

Since duration varies much less than velocity, there is a large difference between nerve fibers in the **wavelength of the spike,** the length of axon occupied at any moment by the nerve impulse. This is given by multiplying the velocity by the duration. *Carcinus* leg axons of 30 μ conduct at about 4 m/sec (= 4 mm/msec); the spike lasts about 1 msec at 21° C. The wavelength is hence 4 mm. Lower values are given by such slow fibers as those in the nerve net for the swimming beat in *Aurelia* (about 2.3 mm) and the nongiant fibers in worms, arthropods, and molluscs, where velocity is not known accurately but probably lies below 1 m/sec while duration does not usually exceed 2 or 3 msec. The slowest fibers in vertebrates—C fibers—yield values of about 1–2 mm. Longer wavelengths are better known: squid giants, 9–17 mm; earthworm median giant, 30 mm; large motor fibers in lobster, 20 mm; large frog sciatic axon, 40 mm; cat A fiber, 50 mm. Note that many smaller neurons or those with short axons are completely occupied by an impulse at one moment. Only the successive phases —foot, rise, peak, and recovery—move along the fiber.

6. Afterpotentials

The afterpotentials vary more than the other aspects heretofore discussed because each of them may be relatively large or small or absent. Much of this variation depends on the condition of the preparation, the amount of stimulation it has recently undergone, and the membrane potential; nevertheless, there are some characteristic differences between fiber types. For details of the better-known vertebrate fiber types, Erlanger and Gasser (1937) and the textbooks should be consulted; an extensive review emphasizing the underlying mechanisms and the effects of various agents is provided by Shanes (1958).

The rapid repolarization or falling phase of the spike sometimes "undershoots" by passing over into a short phase of **early hyperpolarization** or early positive afterpotential (Fig. 3.10). The term "undershoot" is occasionally used because the potential passes a reference level (here the resting level) in the opposite direction from the overshoot or peak of the spike. The term "positive afterpotential" became established before the days of internal electrodes and refers to the fact that an external electrode on the active region is now positive to a distant reference electrode; hyperpolarization refers only to the membrane potential's being higher at this time than the resting potential prevailing immediately before the last spike or burst, not to its being abnormally high. This phase is absent if the resting potential is high and is increased if it is lowered by moderate depolarization. Squid giant axons exhibit up to 10 mv or more of undershoot, lasting 4–6 msec or more unless the resting potential is artificially raised or is exceptionally high (close to the potassium equilibrium potential). This phase is accompanied by a continued high membrane conductance, now subsiding from the peak value during the spike; potassium permeability is rapidly rising and sodium permeability has not yet fallen too far. As this phase of hyperpolarization declines toward the resting level the conductance reverses for 3 msec to a value lower than at rest, returning again to a small excess conductance which slowly diminishes during the negative afterpotential. Curiously, this undershoot is not found in most types of fibers; it has been called a signature of the squid giant. Crab, crayfish, earthworm, frog, and mammal—at least in their largest fibers—lack such a phase under ordinary conditions; presumably their fibers are usually found closer to the potassium equilibrium level. Cockroach giant fibers may have none or may show up to 7.5 mv of afterpositivity. External potassium increase has a much stronger effect in reducing the undershoot than it has on the spike or resting potential. Cell bodies which produce full spikes are much like axons in that most show an initial phase of afterdepolarization ("negative afterpotential," discussed below) and only a few have a phase of hyperpolarization prior to this, unless they have been preset at a low membrane potential. Toad motoneurons are said to be exceptional in showing a brief hyperpolarizing dip or undershoot even when believed to be in good condition, with high resting and action potentials. But cat motoneurons, spinal ganglion cells, sympathetic ganglion cells, crayfish stretch receptor cells, and some gastropod ganglion cells only show such an early hyperpolarization if they are somewhat depolarized.

The next, and in most elements the first phase, is an **afterdepolarization or negative afterpotential,** so called because an external electrode is still negative to a distant reference electrode, as it was

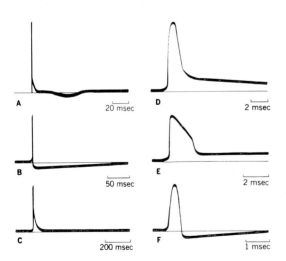

Figure 3.10. Form of the spike and afterpotentials in nerve trunks (left) and in single fibers (right). **A.** Mammalian A fibers recorded extracellularly from a nerve trunk. **B.** Mammalian B fibers. **C.** Mammalian C fibers. **D.** Twitch muscle fiber of a frog, recorded intracellularly. **E.** Single myelinated nerve fiber of a frog recorded extracellularly from a single node. **F.** Squid giant axon recorded intracellularly. [Modified from Tasaki, 1959a.]

during the spike. This is a period of slowed repolarization and is distinct in all fibers with the exception of mammalian B fibers, in which an early and long-lasting positivity obscures it. Many squid giant fibers exhibit only a slowly declining positivity and no negativity. The early phase of the negative afterpotential is part of the spike. The duration varies from a few milliseconds (7 msec in earthworm giant, 12 msec in mammalian A fibers, longer in crayfish giants) to many (50–200 msec in squid giant, 50–80 msec in mammalian C fibers), declining exponentially. Its size varies from zero, as already indicated, to small (less than a millivolt as in large motor fibers in the crab and in squid giants), to several per cent of spike (3–5% in mammalian A and C fibers, 10% or more in crayfish giants). Under given conditions of preparation, these characteristics are fairly typical of the respective fiber types, although there is considerable individual variation. But it is not known what the significant correlates are, for the most part; for example, is it relevant that crab nerve survives 540 min of anoxia but rabbit nerve only 24 min? Many agents alter the size or duration greatly: most notable are veratrine and tetanization (prolonged high-frequency stimulation), which increase and prolong the negative afterpotential to a second or more, and metabolic inhibitors, which reduce it. Tetraethylammonium ions delay repolarization in squid axon, causing long flat-topped spikes. The negative afterpotential may not represent equivalent processes in all fibers. It adds linearly with repetitive stimulation in the squid giant, with the same time constant as it decays (30–100 msec), but not in mammalian A fibers. Nerve cell bodies also exhibit such a phase, with about the same range of variation (toad motoneurons, 5–10 msec, up to 13%; cat motoneurons, 2–6 msec, 10%; crayfish stretch receptor cells, 10–20 msec, about 14% of the spike height).

A long, slow **afterhyperpolarization** or positive afterpotential is general but, under the experimental conditions that have been used, not universal. Squid giant axons normally show no afterpositivity beyond the relatively brief initial undershoot already discussed. Earthworm giant fibers and probably crayfish show no afterhyperpolarization at all, although it has not been stated whether presetting the membrane at a low polarization brings it in. Crab motor axons exhibit a positive afterpotential after a period of high-frequency stimulation and it subsides only in some seconds. Mammalian A fibers, recorded extracellularly and in multifiber bundles, develop such a potential to the extent of 0.2% of the spike height, lasting 40–60 msec; B fibers, 1.5–4.0% for 100–300 msec; C fibers, 1.5% for 300 and more msec. Nerve cells may show a similar phase and it may be larger than in the axons of the same cells. Toad motoneurons develop an afterhyperpolarization of 10% of the spike height compared to 2.6% in their axons. Supramedullary cells in the puffer fish, *Sphaeroides*, develop an 18 mv positivity lasting 20 msec. Cat motoneurons have been most carefully studied (see Eccles, 1957) and the amplitude of this potential, measured as excess polarization over the membrane level prevailing before the spike, has been shown to decrease with increase in prevailing polarization until—when the latter is 90 mv—there is no such excess; at higher levels it reverses.

Positive afterpotential thus tends to compensate for any deviation of the membrane level from the equilibrium level of 90 mv; the **compensation** reaches 30–40%. At usual resting potential the maximum of the afterhyperpolarization is 5 mv and occurs at 10–15 msec, declining slowly over about 100 msec. Perhaps the equilibrium level is not always so high, for in crayfish stretch receptor and gastropod (*Aplysia*) ganglion cells—at high but not artificially raised membrane levels (60–70 mv)—there is no afterhyperpolarization. This is looked for in antidromically invading or directly evoked spikes, not in those normally evoked, because stretch in the one or synaptic activity in the other produces a lowering of the membrane polarization. Under the latter conditions, especially when the cell fires repetitively as is normal, the spike gives way to an immediate hyperpolarization without intervening negative afterpotential. The compensation for the membrane displacement is up to 70%. This hyperpolarization is so prompt that it recalls the undershoot of the squid giant axon but it may last much longer, depending on the prevailing membrane level due to the generator potential imposed by stretch or synaptic bombardment or by an intracellular anode. The declining phase of the hyperpolarization is in fact the gradual depolarization leading to the next spike, which we have heretofore called a pacemaker potential.

It is not possible at present to say what relation this pacemaker form of hyperpolarization has to the early undershoot and the later positive afterpotential of other cells and fibers; the possibility that they are all one process, sometimes interrupted by a temporary afterdepolarization or negative afterpotential, has not been excluded. But there are reasons for suggesting that deflections of potential of the same sign and sequential position are not always the same kind of underlying process in different cells or fibers (Shanes, 1958). Clearly the complex subject of afterpotentials must wait for rationalization upon an understanding of what fundamental events are occurring and what their permutations are in different classes of neurons; a good deal concerning these events is already known (see Shanes) but not enough to separate the independent parallel processes from those which are causally sequential.

Many **agents influence afterhyperpolarization**; it is enhanced notably by yohimbine, cocaine, low Ca^{++}, and tetanization and may last several seconds. During this time excitability is decreased, perhaps not exactly as much as in an electrotonus; K^+ ions continue to flow out long after Na^+ permeability has recovered.

Oscillations are common in some cells and fibers and not in others (Fig. 3.11). Squid giant axons, for example, are prone to give several cycles of damped oscillations after a spike or even after a subthreshold stimulus, at least under some conditions which cannot be specified. Shanes (1949) observed the tendency in all his *Loligo* preparations but other workers have seen it only in some; Shanes saw it in none of his *Sepioteuthis* preparations; Arvanitaki saw it in 25% of her *Sepia* fibers. It is labile; the amplitude is enhanced and the damping reduced by lowered calcium in the medium, by raised potassium, by prior subthreshold depolarization, and by certain drugs like veratrine. For further details the reviews of Shanes (1958), Bonnet (1941), Arvanitaki (1939, 1943), Fessard (1936), and the original papers referred to by these authors may be consulted.

7. *Velocity and its correlates*

Table 3.4 and Fig. 3.12 show selected data on the velocity of conduction in a wide variety of preparations. This property of nerve fibers varies over a range of several hundred to one, and almost this whole range can be found among the nerve fibers of one animal. These statements cannot be made more precise because our knowledge of the lower limits of velocity is based on but a few cases and not an adequate sample. The upper limit, given by the fastest fibers in each species, rises from lower to higher groups: it is about 1 m/sec in coelenterates, 10–20 m/sec in arthropods and molluscs, 40–60 m/sec in fish, and 120 m/sec in mammals. The last figure is not materially higher than the next to last if it is

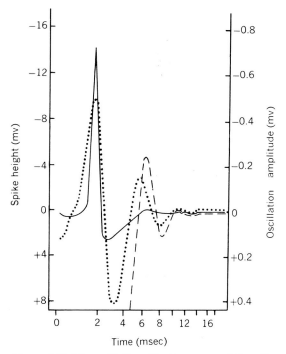

Figure 3.11. Afteroscillations. The time course of the spike *(solid line)* at low amplification and its following oscillations at high amplification *(dashed line)*, compared with that of the slightly subthreshold response also observed with high amplification *(dotted line)* from crab leg motor axons. [Shanes, 1949.]

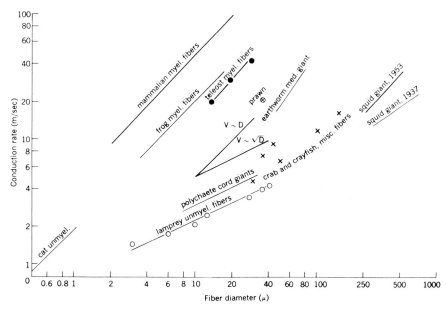

Figure 3.12. Velocity of nerve impulse conduction as a function of fiber diameter. A wide assortment of animals and preparations is compared. Where lines are not drawn the data for fibers of the same type but different diameter are not available.

given a reasonable correction for the temperature difference. Evolution is more conspicuous when the **cost measured in sizes of the fibers** which are required to achieve a given velocity is compared. In the cat, fibers of 3.5 μ conduct at 25 m/sec (37°); velocity in m/sec is about 7 times the diameter in microns. In the frog, 12 μ fibers conduct at this speed (20°); the ratio is about 2, partly due to temperature. Earthworm fibers of 75 μ have this rate (20°); the ratio is now 0.33. Squid giant fibers of 350 μ are needed to achieve the same speed (20°); the ratio becomes 0.07—one hundred times smaller than in the mammal, or, in terms of volume of protoplasm assigned to the task, 10,000 times more expensive.

This is not the place to develop the question of the **role of velocity** in adaptiveness or the intensity of the selective pressure for improvement in this performance. Cases which probably approach the extreme are insect reaction times that have been measured as short as 25 msec (cockroach response to air-puff); of this time about 10% is consumed in conduction so that if the fibers were half as fast the reaction time would be lengthened by 2.5 msec. The differential in time would be considerably more in longer animals and in slower actions.

The consequence of velocity given by the wavelength of the impulse is discussed on p. 145.

A significance of velocity that has been little exploited in studies up till now is the opportunity to make measurements day after day on the same intact animal, as a tool in general physiological research, for example, upon adaptation and acclimation or the effects of various agents or stresses. Many animals permit precise measurement of velocity, at least of the fastest fibers in a bundle, without breaking the skin. Some (earthworms, crayfish, and polychaetes) with giant fibers having unique functions would allow the assurance that the same unit was being measured every day. The advantages of velocity include the precision of its measurement and its lability or sensitivity to conditions.

Besides the factors enumerated below which are primary velocity correlates, it is important to be aware of some factors which can operate to make measured values not comparable. Stretch, taper, external medium, and temperature, will each be mentioned briefly.

Nerves vary widely in **tolerance of stretch.** Those in animals of fixed dimensions, like vertebrates and arthropods, typically tolerate only a few per cent increase of length although they

Table 3.4. VELOCITY OF CONDUCTION OF NERVE IMPULSES IN SELECTED CASES. (Additional values may be found in the respective systematic chapters. See also Schaefer, 1940; *Handbook of Biological Data*, 1956.)

	Animal	Nerve Fiber	Velocity[1] (m/sec)	Diameter incl. Sheath μ	Temp. (°C)	v/d[1]
	A	B	C	D	E	F
		VERTEBRATES				
1	*Felis*	Pyramidal tract (max)	164	12.5	37	13
2		Dorsal spinocerebellar tract (max)	120	16–18		7
3		Dorsal columns, spinal cord (max)	67	10–14		5.5
4		A fibers[2] in peripheral nerves	110	20		5.5
5			80	15		5.3
6			50	10		5
7			70–81	13–15	35	5.3
8			60	7–9		7.5
9			40	8		5
10			30	6		5
11			20	4		5
12			10	2		5
13		C fibers[2], sympathetic	1–2	<1–5		<1
14	*Rana*	A fibers[2], sciatic	40	20	24	2
15			30	15		
16			25	12		
17			20	10		
18			15	8		
19			10	5		
20			7	4		
21		C fibers[2]	0.2–0.6	<1		
22	*Ameiurus* (catfish)	Giant fiber (Mauthner's)	50–60	22–43	10–15	appr. 2
23	*Cyprinus*	Giant fiber (Mauthner's)	55–63	55–65	20–25	1.0
24	*Protopterus* (lungfish)	Giant fiber (Mauthner's)	19	45	20	0.42
25	*Raja*	Dorsal roots	8–36	2–17		appr. 3
26	*Entosphenus* (lamprey)	Giant fiber (Müller's)	5	50	20	0.1
		INVERTEBRATES				
	Arthropoda					
27	*Carcinus*	Leg	4.4(3.9–5.5)	30	21	0.14
28	*Munida*	Leg	6.4	50	17	0.13
29	*Cambarus*	Lateral giant fibers	10–15	70–150	20	appr. 0.12
30		Medial giant fibers	15–20	100–250[3]	20	appr. 0.12
31		Chela, fast-closer	20	58		0.34
32		slow-closer	10	41		0.25
33		opener	8	36		0.22
34	*Homarus*	Medial giant fiber	18	125		0.14
35		Large fiber in cord	7	70		0.10
36		Chela, fast-closer	18–20	80–125		appr. 0.2
37		opener	14–18	70–90		0.2
38		Leg, fast-closer	12–18	70–100		0.17

([1]) Velocity (v) in m/sec (in myelinated fibers) in vertebrates is directly proportional to outside diameter (d) in μ; v = kd. K ≈ 6 in mammals; K ≈ 2 in frogs. Temperature coefficient = 1.8 for 10°.

([2]) A fibers = myelinated fibers of somatic system, sometimes subdivided into α, β, γ and δ in order of descending velocity in ratio of approximately 100:60:40:25; B fibers = myelinated (usually preganglionic) fibers of autonomic nervous system; C = non-myelinated fibers.

(Table continues on next page)

III. EXCITABILITY B. Spike 7. Velocity

(Continuation of Table 3.4)

	Animal	Nerve Fiber	Velocity[1] (m/sec)	Diameter incl. Sheath μ	Temp. (°C)	v/d[1]
	A	B	C	D	E	F
	INVERTEBRATES (Continued)					
39	*Homarus*	Leg opener	7–13	40–80		0.17
40	*Callianassa*	Giant fibers	6.0–7.5	35–40	20–22	0.18
41	*Leander*	Giant fibers	18–23	35	17	0.6
42	*Periplaneta*	Giant fibers	9–12	10–40		0.3
43	*Limulus*	Optic nerve	2	6		0.3
	Mollusca					
44	*Loligo forbesi*	Stellar n., incl. giant fibers	5, 7.5, 11, 16	50, 100, 200, 400	20	0.04–0.1
45	*Loligo pealii*	Giant fibers	18	260	23	0.09
46		Giant fibers	23.5 (21.5–25)	350	23	0.067
47		Giant fibers	30 (27.5–32)	450	23	0.067
48		Giant fibers	35	520	23	0.067
49	*Sepia*	Giant fibers	3, 8	35, 150	20	0.05, 0.09
50		Giant fibers	7.05	231	17	0.03
51		Giant fibers	6.90	168	15	0.04
52		Giant fibers	6.17	126	16	0.05
53	*Aplysia*	Pleurovisceral connective, max.	1	35–50	23	0.03
54	*Ariolimax*	Pedal n. (fastest wave)	0.83	35	21.8	0.024
55	*Helix*	Visceral n. (fastest wave)	0.6–0.7			
56	*Mya*	Pallial n. (fastest wave)	1			
57		Cerebrovisceral connective	0.2–0.5	>4	17	
	Annelida					
58	*Neanthes*	Lateral giant fibers	5	30–37	24	0.15
59		Median giant fiber	4.5	15–18	24	0.27
60		Medial giant fibers	2.5	7–9	24	0.31
61	*Lumbrineris*	Median dorsal giant fiber	10	130	24	0.07
62		Median ventral giant fiber	4.5	27	24	0.17
63	*Harmothoë*	Lateral giant fiber	0.93	13	18	0.07
64	*Diopatra*	Giant fiber	10	130	24	0.08
65	*Myxicola*	Giant fiber	6–20	100–1000[3]	24	0.02–0.06
66	*Arenicola*	Giant fiber	2	25	24	0.08
67	*Aphrodita*	Ventral cord (no giants)	0.05			
68	*Lumbricus*	Median giant fiber	30 (15–45)	50–90[3]	22	0.3–0.6
69		Lateral giant fiber	11.3 (7.5–15)	40–60[3]	22	0.19–0.28
70	*Hirudo*	Ventral cord (no giants)	0.025			
	Coelenterata					
71	*Aurelia*	Nerve net, swimming beat	0.5	6–12[3]		0.05
72	*Calliactis*	Column nerve net, longit.	0.1			
73		circ.	0.15			
74		radial	0.04			
75		Mesentery, longit. conduction	1.2			
76	*Metridium canum*	Mesenteric nerve net		5–13		
77	*Metridium senile*	Mesenteric nerve net		1–5		

([3]) These fibers taper within the range of diameters given.

may withstand considerable stretching force. Squid giant fibers are intermediate and continue to conduct when stretched to about 120% of their initial length; reliable values based on length in situ are not available. Many animals, however, normally extend and contract greatly and it has been found that their nerve fibers partake in some measure of this property. This means that only some of the changes in length of the whole animal are reflected in coiling or zig-zagging of the nerve fibers. Such looping is prominent in unstretched nerves and even in those mildly stretched, for example in earthworms, gastropods, nemertineans, and others. The effect of this gradual straightening out with stretch is to make the apparent velocity of conduction higher as the animal is stretched—obviously an artifact due to lengthening the measured path (without lengthening the actual path). This may perhaps explain the unusual results of Dittmar (1954), on leeches (p. 705). In a few cases, particularly the earthworm (Bullock, Cohen, and Faulstick, 1950), it has been found possible to stretch considerably after all kinks are straightened out and still obtain conduction. Direct observation of the giant fibers in vivo while measuring the velocity of the same fiber shows that speed of conduction (per meter) is usually constant even while diameter is being reduced to one-half by stretch. Jenkins and Carlson (1904) and Carlson (1905) first observed this behavior—though without diameter control—and concluded it was evidence of the fluidity of the conducting substance in nerve! This conclusion is not in accord with contemporary theory but we are not in a position to explain the phenomenon yet. Constant surface area in microscopic dimensions does not appear to explain it. It would be worthwhile to determine the passive membrane properties (p. 674), especially the specific capacitance under conditions of stretch (Martin, 1954). Similar independence of maximum velocity and stretch, once the stretch has proceeded far enough to remove sinuousness, has been shown in several polychaetes and in the gastropods *Ariolimax* and *Aplysia* (Fig. 3.13; Turner, 1951; Goldman, 1961). Values for these animals can therefore be included in comparative

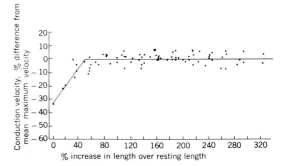

Figure 3.13. The relation of velocity of conduction to stretch in a fiber tolerant of great stretch, the single fiber represented by the R2 response in the right pleural-visceral connective in *Aplysia californica*. Each point represents a single velocity measurement; data from seven specimens. Between 40 and 320% stretch, 86% of the points fall within ± 5% of the horizontal line. [Goldman, 1961.]

tables even though the species has no fixed length, but many values in the literature are of limited use because the stretch was not controlled.

Nerve fibers commonly taper, sometimes so rapidly that a length just sufficient to make accurate velocity determinations changes significantly in diameter. Not only smooth taper but irregular fluctuation in diameter is known in some instances (especially annelids), so without a number of diameter measurements along the length of fiber whose over-all velocity is being determined, it is impossible accurately to place some cases into a comparative scheme such as that of Fig. 3.12.

The **conductivity of the medium** around the nerve fiber affects its velocity, presumably in simple proportion. Most determinations are made upon nerves lifted from the animal into moist air or oil, in order to make the action potential large. Deliberate shunting with external saline solutions or metal increases the velocity by 12–140% (Hodgkin, 1939); the variation probably reflects differences in the residual shunting due to connective tissue and aqueous film when the preparation was lifted into the insulating medium. Published values are rarely controlled with respect to this factor. Shunting by a longitudinal intra-axonal wire increases the velocity up to 250 times (Del Castillo and Moore, 1959). The three foregoing factors doubtless account for a considerable part of the variation among animals, even in the selected data of Table 3.4 and Fig. 3.12. More importantly, these factors may influence the regression coefficients calculated for a given species and nerve fiber type because of systematic differences between specimens; for example, the larger fibers may come from preparations with more shunting tissue adhering, therefore exaggerating their higher velocity.

Temperature influences velocity with a coefficient for $10°C$ of somewhat less than two. The Q_{10} over rather wide ranges of temperature does not differ greatly from about 1.8 in a number of cases studied so far, including frog, crustacean, gastropod, and annelid; it is sometimes even lower (Gasser, 1931; Tasaki and Fujita, 1948; Hodgkin and Katz, 1949b; Engelhardt, 1951; Turner, 1955; Wright, 1958).

A small **acceleration** has been found in the second and subsequent impulses conducted 5 or 10 msec apart in earthworm, polychaete, and frog fibers, but not in other polychaete species and several crustaceans, shown in Fig. 3.14 (Bullock, 1951). But excessive stimulation or other unfavorable conditions cause considerable slowing.

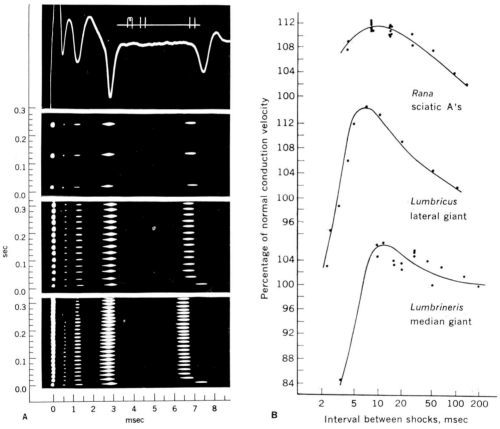

Figure 3.14. Facilitation of conduction velocity. **A.** *Lumbricus*, isolated cord, the time course of facilitation at three stimulus frequencies. Top: conventional picture showing stimulus escape, proximal pickup of median giant, proximal lateral, distal median, and distal lateral spikes, in that order. Note the diagram of the electrodes. The other three frames were made by imposing the spike potentials on the brightening circuit of the cathode ray tube and a slow vertical sweep giving the order and spacing of shocks. Latency on the abscissa measures conduction velocity between the recording electrodes. The lower two frames show marked decrease in latency after the first and second shocks, thereafter constant velocity. **B.** Facilitation as a function of interval between shocks in three different preparations. Each point represents one determination; the curves are fitted by eye and represent preparations typical of their kind. [Bullock, 1951.]

Diameter of the nerve fiber is the most obvious correlate of velocity. Examination of Table 3.4 and Fig. 3.12 discloses not only the general tendency but something of the nature of the relation. In myelinated fibers of vertebrates (and according to Hodes, in squid giant fibers also), the slope of the regression is of the first power; velocity equals a constant times diameter. The constant varies significantly; it is about 7 in the cat, 2 in the frog, and 0.07 in the squid. This indicates the greater diameter needed to achieve the same velocity in the species with the lower constant. Intermediate values occur; earthworm giants (about 0.4) are much more efficient by this standard than squid giants or polychaete giants (less than 0.1). Crustaceans are not particularly efficient even compared with annelids and molluscs except for the extraordinary report on the prawn *Leander*, which has giant fibers said to exhibit the relatively high constant of 0.6 (Holmes, Pumphrey, and Young, 1941).

It should not be assumed, as is commonly done in texts, that the first power relation established for vertebrate myelinated fibers is general. Careful work has not been done on many invertebrates but the data available on some polychaetes suggest that velocity may be proportional to diameter raised to some exponent closer to 0.5 (square root). *Lumbrineris*, for example, has two giant fibers with a diameter ratio of about 5:1, a velocity ratio of about 2:1 (and an externally recorded spike amplitude ratio of about 30:1; see p. 698). *Neanthes* has three sets of fibers with ratios of diameter of approximately 4:2:1, of velocity 2:1.8:1 (and of spike height 12:4:1). Of course, these are not necessarily of the same fiber type; there may be characteristic differences in morphology or intrinsic properties apart from diameter. Nicol and Whitteridge (1955) examined the same, single giant fiber in a polychaete (*Myxicola*) over a diameter range of 100–1000 μ and obtained velocity dependence much better fitted by the square root than the first power function. Two identified single fibers, the inhibitor and the excitor of the opener muscle of the claw of *Astacus*, differ as follows (Eckert and Zacharová, 1957): their diameters are respectively 18 μ and 14 μ (1.29:1), velocities are 3.8 m/sec and 3.1 m/sec (1.22:1), spike amplitudes are as 1.69:1 in a certain case; v/d for both fibers is close to 0.21. A square root relation of velocity to diameter was originally proposed by Pumphrey and Young (1938) for squid giants but was not corroborated by later work of Hodes (1953), who claimed a simple proportionality, as in vertebrates. Hodes may be right but at present it appears likely that there are animals or fiber types with **different velocity-diameter regressions,** including some which approach a square root function. A low v/d ratio makes it somewhat to be expected that a relatively smaller gain in velocity will be achieved by a given size increase than in an animal or fiber type having a high v/d ratio. A square root relation means that to increase velocity 3-fold the diameter must be 9 times greater and the volume of protoplasm 81 times greater. Obviously there is a great advantage in any other factor that can improve velocity and save size. Most of the evolution of velocity has in fact involved other factors, as shown by the displacement of the separate slopes of Fig. 3.12 along the abscissa.

Adey's (1951) graph for giant fibers in a species of earthworm, taken at face value, points to an exponent materially higher than one, but this is unique and needs confirmation, with control of stretch.

Curiously enough this relation is not known in arthropods; the difficulties of determining a valid figure for diameter, of getting a sufficient range of sizes, and the scatter of values even in the best studied material, have prevented a satisfactory set of data in any species. In addition, the demonstration of Hodgkin (1948), Easton (1952), and Wright et al. (1954)—that even among the few large motor fibers of crustacean legs there are quite distinct types based on excitability parameters—make it necessary that future work assure the homogeneity of excitability type or of function.

The elementary dependence of velocity upon diameter calls for new and careful observations in several **lower groups**. Among others, a particular opportunity is afforded by the lampreys and hagfish whose large, unmyelinated Müller's fibers (Schulz, Berkowitz, and Pease, 1956) have a wide diameter spread easily penetrated by microelectrodes. As shown in Table 3.4 they are quite out of line with teleosts and elasmobranchs in the v/d ratio. The paucity of cases in pelecypods, gastropods, insects, nongiant fibers in general, and the many groups still unrepresented, is conspicuous. For the lowest groups the only figures come from conduction pathways which are quite likely to be interrupted by synapses—nongiant propagation in the central nerve cord or spread in through-conducting nerve nets. The values for velocity in nerve nets may therefore be low but, surprisingly, the v/d ratios give no sign of it and are quite comparable to other slower invertebrate fibers, assuming the estimated diameters are not seriously low (in which event the fibers in some nerve nets would be approaching some annelid giants in size). We may conclude from this that the large number of synapses in an anemone mesentery or jellyfish bell probably introduce very little or no delay, and hence are quite likely to be of the electrically transmitting kind like the giant-to-motor synapse in crayfish (p. 215).

There is a general correlation, although it cannot at present be made at all precise, between velocity and **sheath thickness.** Probably it is not simply thickness that is pertinent but some other variable such as compactness times thickness or lipid density or ion impermeability. Understanding of the role of sheaths is poor; it is one of the areas where combined anatomical and physiological work could yield particularly large dividends even with present techniques. What is the significance of increasing numbers of layers and of compactness of layers of sheath cell membrane in invertebrate, presumably nonnodal fibers? What is the path of current through the external circuit in the better sheathed invertebrate fibers? Is the absence of myelin in cyclostomes an indication of loss or an indication that vertebrates developed multilayered Schwann cell membranes around their axons independently and in parallel with higher invertebrates?

Vertebrate A fibers have a myelin sheath which is 20–70% of the diameter of the whole fiber; that is, the radial thickness of the myelin can be as great as the diameter of the axon, on each side of the latter. As between thicker and thinner A fibers, the latter generally have the relatively thicker sheath. The B fibers are not well known in this respect. The C fibers are spoken of as unmyelinated, as are most invertebrate nerve fibers (see Glossary and p. 105 for definition of "myelinated"). The sheaths on C fibers and most invertebrate fibers are either very thin (less than one-tenth the axon diameter) or absent. Giant fibers of crustaceans, squid, polychaetes, and oligochaetes often have quite appreciable sheaths but generally only a few per cent of fiber diameter. Squid giants at less than 1% and earthworm at 10% **correlate with the greater v/d ratio** of the latter. The single and an outstanding exception is the report on the prawn *Leander*, which as we noted above has a strikingly small fiber diameter for its velocity (high v/d ratio); the sheath is said to be from one-fourth to one-half the fiber diameter. Further examination of this point in prawns and other forms is obviously needed. The larger fibers in *Leander* have relatively less sheath: 23% of total fiber diameter in 20–50 μ fibers, 31% in 10–20 μ fibers, and 47% in fibers less than 10 μ thick. Nongiant invertebrate fibers of several up to about 20 μ diameter may have distinct sheaths; these are often indefinite in outer boundary in electron microscope sections and vary in apparent thickness at different points around the fiber, but it is perhaps fair to say that commonly the specialized sheath is but a few per cent of fiber diameter. Possibly of great importance as a correlate of velocity are the different types of specialized sheaths distinguished by structural features (described on pp. 102 and 903). The smaller nerve fibers, comprising the great majority of all fibers and usually less than 2 μ thick, are either quite naked or are buried in the cytoplasm of a simple sheath cell without multilayered double membranes around them. These facts, gross and without quantitative relations, are the basis of the accepted generalization that the sheath is a major correlate of velocity.

In a long series of studies, among which the name of Tasaki should be particularly mentioned, **nodes in the myelin sheath** have been shown to result in saltatory conduction, as seen in Fig.

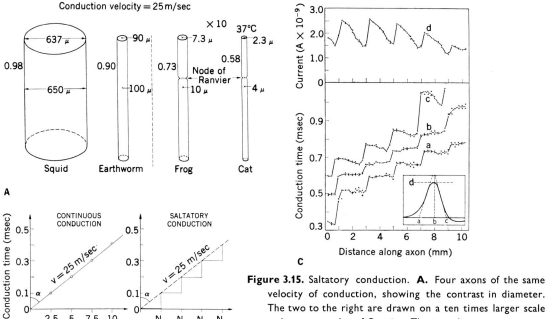

Figure 3.15. Saltatory conduction. **A.** Four axons of the same velocity of conduction, showing the contrast in diameter. The two to the right are drawn on a ten times larger scale and possess nodes of Ranvier. The two diameters given are of the axon itself and of the fiber, including sheaths; the ratio of these two is given at the left. The velocity for the cold-blooded forms is that at about 20°C, in the cat at 37°C. **B.** The latency of the spike at points along the axon. The curve on the left applies to the squid and earthworm, the curve on the right applies to the frog and cat and shows rapid jumps separated by pauses coinciding with the nodes. [**A** and **B**, Muralt, 1958.] **C.** Below: the evidence for the diagram at the right in **B**. The times of arrival at points along the nerve fiber of the rising phase, summit, and falling phase of the spike are plotted. Above: the amplitude of the action potential is plotted at different points along the fiber. The intensity diminishes in the internode and recovers at each node. [Stämpfli, 1958a.]

3.15 (see Tasaki, 1959a). At each node there is a delay of approximately 0.06 msec in frog fibers in good condition. Since the nodes in a 10 μ fiber are spaced about 1.6 mm apart and the velocity in such a fiber is about 20 m/sec or 1.6 mm/0.08 msec the conduction time is virtually accounted for in the node and the long internode does not consume time. Very long internodes are associated with diminishing returns in respect to velocity and very short internodes do not realize a great advantage of the nodal principle. The relation between internode length and velocity is said to have a rather gently rounded maximum and most vertebrate nodal fibers in adults have internodal lengths falling in the region of this maximum, so that length of internode is not a primary variable as between fibers of different velocities. But the fact of nodes has been regarded as a major development in the evolution of conduction because they have been classically associated with vertebrates and the much higher velocity for any given diameter characteristic of this group, at least for A and B fibers. Indeed, nodes have been almost the only distinctively vertebrate feature of neurons, as are v/d ratios above 1. The phylogenetic contrast became all the more interesting when it was shown that a representative cyclostome (*Entosphenus*, a lamprey, Berkowitz, 1955; Schultz, Berkowitz, and Pease, 1956) has large fibers with a low v/d ratio of 0.1, like most invertebrate fibers, and no myelin sheaths, hence no nodes. It appears that high velocity in relation to diameter, and presumably nodes, are first found in elasmobranchs.

However, the report of McAlear et al. (p. 941) alters this picture. They reported (see more detail on pp. 103 and 902) that among several other types of sheaths, there are actually some closely resembling characteristic vertebrate, tightly packed, many-layered myelin sheaths, and which display **fully formed nodes, in crustaceans.** Assuming that the nodes, which were seen in good electron microscope sections, are confirmed, and that the fibers involved are not greatly different from known crustacean fibers in v/d ratio, the correlation between nodes and high velocity is no more an automatic but only a permissive one. The delay at each node in these crustacean fibers must be presumed to be of the order of ten times that in frogs. The evolution of nodes in these two phyla can hardly be thought of as anything but independent, an astonishing example of convergence at the cellular level. But the wonder does not stop here. If the nodal condition in crustacean fibers is of functional significance it means that the myelin is of low conductance and so relatively impermeable to ions. Are there, then, all degrees of permeability? They are perhaps correlated with the number and compactness of the layers of sheath cell membrane, since the well-studied giant fibers of invertebrates have high conductance, even though some (like the earthworm) have twenty or more layers of myelin wrapping, apparently without nodes.

Periodic constrictions in the unusually thick sheath of the giant fibers in *Leander* have been described, although nothing similar has been seen in other crustaceans or invertebrates. These large gaps are of no obvious relation to true nodes of Ranvier. The latter have such a characteristic electron microscope structure that it should be possible to establish or rule out such a relation by suitable re-examination of *Leander*. This is the more interesting because this prawn is reported to have a uniquely high proportion of sheath and of velocity to fiber diameter.

Residual intrinsic factors clearly are the most important in explaining the observed velocities of spike propagation. The delay at nodes, like the rate of rise of active response at a locus in the squid axon, depends on a combination of excitability and responsiveness of the membrane, to the rising intensity of local current flow from preceding regions of activity. The principal variations in velocity among animals are not yet attributable to nodes, sheath thickness, or fiber diameter, but to these dynamic properties.

C. Rhythmicity and Pacemaker Excitability

Many nerve fibers but especially nerve cells and receptive regions are able to fire repetitively to a constant stimulus, that is, to iterate. They initiate a series of more or less rhythmically spaced impulses during an imposed steady current of equivalent depolarizing stimulus. This presents several **new parameters** or properties of excitation. One is the form of sensitivity given by the curve of frequency of discharge against strength of stimulus. A second is adaptation in frequency during the maintained stimulus. A third is the degree of rhythmicity, measured for example by the standard deviation of intervals between spikes. Knowledge is fragmentary on all these points but it is clear that significant differences exist among neurons in each respect. It will also be convenient to mention here the influence of factors impinging on the neuron or its pacemaker region from sources other than specific pathways, for these have their most subtle effects by modifying ongoing frequency.

The form of sensitivity given by the curve of strength of applied current against frequency of repetitive discharge has not yet been shown to be a predictable result of the primary parameters. Since most normal signaling involves iterative firing of impulses, this form of sensitivity is **perhaps the most important** of all. It need not be invoked when the timing of impulses is predetermined by some impinging cue such as a series of very discrete synaptic potentials. But generally a neuron initiates its own rhythmic series of impulses as some function of a general level of excitation resulting from its input; this is shown by the lack of a simple relation, such as one to one, between input and output, or by the fact of a rhythmic series under steady-state stimulation as in incompletely adapting receptor neurons. Whenever several sources of presynaptic influx converge on a given neuron, it is likely that the arriving impulses are in sum nonrhythmic and more numerous than the output, which is generally more or less rhythmic. In all such cases the sensitivity that determines output frequency can be called a **pacemaker sensitivity.**

It is readily shown that frequency is not

determined by the refractory period, except in extreme conditions. Frequency is often very low and smoothly graded. The latency of the first impulse after the onset of a maintained stimulating current tends to be the same as the intervals between impulses, barring adaptation that makes the intervals longer; by measuring the first intervals, or choosing slowly adapting elements, the tendency is clear. This means that we can think of the frequency determination as

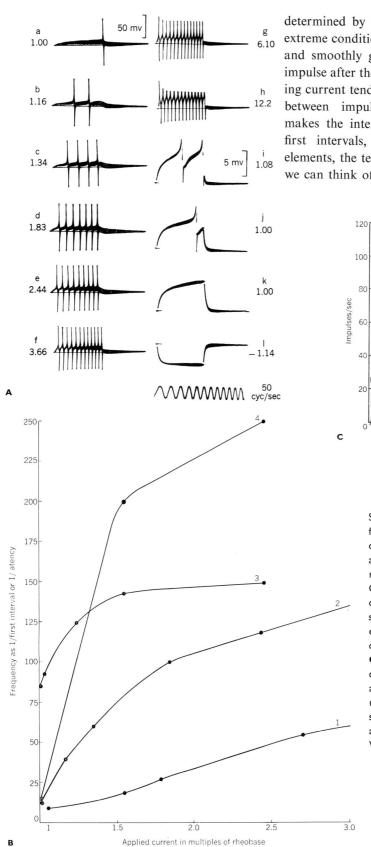

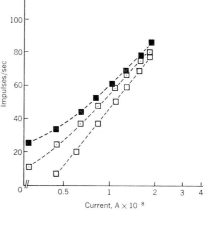

Figure 3.16.

Strength-frequency relation of repetitive firing in axons and receptors. **A.** Records of three types of fibers (records a–h, i–j, and k–l, respectively) which respond differently to constant current. Crab leg nerve. **B.** Curves of the relation of initial frequency to applied current in similar fibers. The current is applied externally and is expressed in multiples of rheobase. [**A** and **B**, Hodgkin, 1948.] **C.** Intracellularly applied current in crayfish stretch receptor of the fast adapting type. In order, from below upward, the curves represent the unstretched condition, medium stretch, and strong stretch. [Terzuolo and Washizu, 1962.]

similar to the unknown processes determining utilization time or latency; something in the level of membrane potential, set by the maintained stimulating current or by the naturally occurring generator potential, influences the rate of rise of the gradual depolarization which we call the pacemaker potential.

Figure 3.16 shows examples of the strength-frequency curve in several cells and fibers. There is a wide range in pacemaker sensitivity. As yet there is no generally accepted convention for expressing this variable (Strumwasser and Rosenthal, 1960). Since it is a curve of gradually saturating frequency, any single value will apply

Table 3.5. COLLECTED VALUES FOR SENSITIVITY OF PACEMAKERS. Three expressions of this form of sensitivity (derived from the curve of frequency of repetitively firing neurons as a function of current imposed) are compared in various preparations.

	Source and Preparation		$\dfrac{\Delta F}{\Delta I}$ Imp./sec $\overline{A \times 10^{-9}}$	$\dfrac{\Delta F}{\Delta V}$ Imp./sec mv	F at 2 × rheobase
	A		B	C	D
	Carcinus, leg nerve				
1	(Hodgkin, 1948)	Fig. 3B	—	—	93
2		Fig. 3A	—	—	105.0
3		Pl. I	—	—	105.0
4		Pl. II	5.2 (8.5–17)[1]	—	33.0
5		Fig. 1 (low I)	25.0	—	105.0
6		Fig. 1 (interm. I)	12.0	—	105.0
7		Fig. 4	144.0	—	>150.0
8	Type II Fig. 6	(1st interval)	29.0	—	147.0
9	Type II Fig. 6	(latency)	58.0	—	225.0
	Limulus eye, eccentric cell[2,3]				
10	(Fuortes, 1958)		4–26	0.75–4	7.7
11	(Fuortes, 1959)	Fig. 3	4.4	0.77	—
12		Fig. 6	23.5	5.0	—
13	dark	Fig. 9	16.7	2.2	7.0
14	light	Fig. 9	12.2	2.2	—
	Frog brain, interneurons				
15	(Strumwasser and Rosenthal, 1960)		2–20	—	—
	Cat, spinal motoneurons[3] (Frank				
16	vide Strumwasser and Rosenthal, 1960)		4–13.6	—	—
	Aplysia, visc. gang. cell[3]				
17	(Tauc, 1955)	Fig. 4	4.5	—	45.0
18		Fig. 5	0.7	—	7.5
	Crayfish, stretch receptor cell				
19	(Terzuolo and Bullock, 1956)		0.1 (large shunt)	—	15.0
	Lobster, cardiac ganglion cell[3]				
20	(Otani and Bullock, 1959)		3.0	—	15.0

([1]) During last second of I; higher value allows × 2 for shunting.

([2]) Measured during steady-state firing, after partial adaptation. The latency of the first spike is 26 times shorter than the intervals after steady state is achieved, which are shown here; latency would therefore give higher values in this table.

([3]) Current delivered through intracellular electrode; other cases all employed external electrodes.

only to a segment of the relation; the low current–low frequency portion is generally linear and being the steepest portion has attracted attention. The simplest measurements are those giving the increment of **impulses per second for each unit of current** imposed; representative values are 2 to 20 impulses per second per millimicroampere but published curves of less than 1.0 and more than 100 are included in Table 3.5. This measurement, based on total current, prevents the comparison of elements having different internal or external resistances; that is, cells and fibers of quite different diameters or with different amounts of external shunting. According to present theory the best measurement would be the **transmembrane potential change** in the relevant part of the neuron, caused by the stimulating current. The membrane change in the soma has been recorded in a few instances; representative values are those of 0.8 to 2 impulses per second per millivolt depolarization in the eccentric cell of the eye of *Limulus* (Fuortes, 1959). Commonly, however, this measurement is not available. In some experiments no voltage change above the noise level has been seen even when current adequate to cause a detectable change in frequency is flowing (Terzuolo and Bullock, 1956, in crayfish stretch receptor). The basic difficulty is that the evidence clearly points to a lack of equipotentiality of the parts of the neuron; a change in membrane potential at some localized region is necessary and sufficient to determine the frequency of impulses generated. This region may be at some distance from the soma and not easily penetrated with intracellular electrodes.

Another means of expressing pacemaker sensitivity is simply the **frequency at some multiple of rheobasic** current; values are shown in Table 3.5—ranging from 7 to 225, for twice rheobase. Even twice rheobase is sometimes so strong as to be nearly saturating and lower multiples exaggerate the error from the uncertainty of the rheobase. Moreover, this method does not permit comparison with the common elements that have a background discharge before intentionally imposed voltage change is applied.

Hodgkin (1948) distinguished **three types of crab fibers:** the first will fire at a wide range of frequencies from 5 to 150 per second according to the current imposed; the second fires only at high frequency (75–225 per second) and is relatively insensitive to current measured by the above ratio; the third will not fire repetitively or requires very high currents to do so. The first shows no phase of supernormal excitability after each impulse, the second has a pronounced supernormal phase. Both of the iterative types can begin after a long latency (up to 2 seconds) and the first can fire as few as 5 times per second, although the relatively refractory period is over in 10 msec. An instructive comparison of the properties of a certain fiber of type one upon standing, is provided by the following data:

	Three hours after isolation	*Six hours after isolation*
Critical depolarization	7.4 mv	12.9 mv
Safety factor	7.15	2.84
Rheobase	9.3×10^{-9}A	24.8×10^{-9}A
Longest response time, at rheobase	160 msec	5.9 msec
Longest time to crest of local response	120 msec	7.1 msec
Lowest repetition rate	9 per second	—
Number of impulses at $2 \times$ rheobase	70	1

Note that on the measure of impulses per second per unit current as well as the measure of frequency at twice rheobase, type two will appear similar to type one because of the high frequency of whatever firing the fibers do, whereas they are actually quite insensitive to change in current (ratio measure). Deterioration changes types one and two into type three. The data of Wright and Adelman (1954) on crayfish fibers appear to show three entirely corresponding categories that are functionally distinct fibers: the opener axon of the claw, the slow-closer and the fast-closer axons resembling respectively types 1, 2, and 3 (Table 3.1).

III. EXCITABILITY C. Pacemakers

Differences in pacemaker sensitivity are as great between fibers of the same animal and nerve as between animals. They are probably at least as important as the more familiar forms of spike sensitivity in central and receptor structures in determining the frequency of nerve impulses generated, the change in this frequency with change in stimulus, and the adaptation rate.

Neuronal elements vary widely in the **degree and in the rate of adaptation** of frequency under constant stimulation. Some adapt so rapidly that they cannot iterate and give only one impulse to a long current pulse. Others show an extremely small and slow adaptation, iterating for long periods at a constant frequency. Tonically firing elements like cardiac pacemakers in *Limulus* and lobsters or position receptors and others with a background discharge, alter their frequency under imposed current without any adaptation. Other elements exhibit adaptation but it is only partial and ends in a **nonadapting fraction** of the response; the nonadapting fraction, measured in absolute frequency of firing, is always less sensitive than the initial or adapting fraction. Intermediate cases show an initial more rapidly adapting phase, followed by a slowly adapting phase which may require minutes. Elements with a strong iterative tendency generally have a low rheobase, slow adaptation, low maximum frequency, and a flatter curve of frequency against stimulus strength, and are more common among thinner and unmyelinated fibers.

Adaptation is distinct from accommodation; at least they cannot yet be identified with each

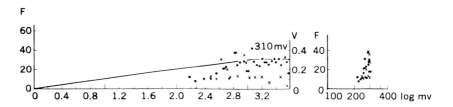

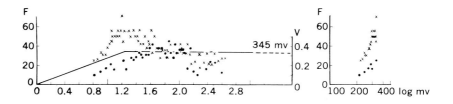

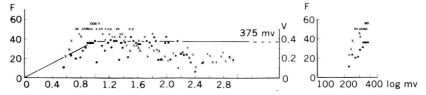

Figure 3.17. The strength-frequency relation in the medial and lateral giant fibers of an earthworm. Slowly rising currents are applied externally to the ventral nerve cord of *Pheretima*, and the latency of the first spike as well as the time of appearance of subsequent spikes is plotted in the left-hand graph while the current is rising and after it plateaus. The right-hand graph shows the frequency as function of voltage before the plateau is reached. The *crosses* represent the median giant fiber, the *circles* represent the lateral giant fibers. Characteristically the lateral giant fires first for slowly rising currents and reaches lower frequencies of firing for any given current. [Katsuki, Chen, and Takeda, 1954.]

other. As shown above, it is usual for true accommodation or rise in threshold depolarization to begin, if at all, only after some period of no accommodation. Adaptation on the contrary is maximal in rate of change at the beginning of a maintained stimulus. Crab leg axons which show no accommodative rise in threshold for some seconds nevertheless show a considerable adaptation, generally partial. The median and lateral giant fibers of the earthworm *Pheretima* (Katsuki et al., 1954) show no detectable accommodation of threshold, and their rheobases are very close together, even to stimuli rising to rheobase in 2.4 sec. They do show an initial higher frequency repetitive discharge and partial adaptation to a steady level after about half a second. The two units show interesting differences in excitability (Fig. 3.17). At any given current the median fires more frequently than the lateral and has thus a higher frequency per unit of current and per multiple of rheobase, but it fires later to a slowly rising current. The strength duration curves cross, that of the lateral giants having the lower rheobase. (Compare Glossary, Adaptation and Accommodation.)

The properties of repetitiousness and adaptation and the dependence upon slope or rate of rise of stimulus in axons imitate remarkably well the behavior of receptor endings. Nerve is a model of sense organs probably because they share fundamentally common parameters of excitability (Skoglund, 1942; Bernhard, Granit, and Skoglund, 1942). It is significant and no coincidence, on this view, that vertebrate sensory nerve fibers are much more repetitious than motor fibers and crab opener fibers than fast-closers—the first of each pair no doubt normally functions more tonically; the second more phasically.

The **perfection of rhythmicity** is a property of interest. Factors still unknown influence the uniformity of the successive intervals; some neurons are characterized by an extreme regularity, others by a wide variance (Fig. 3.18). Typically, the higher the average frequency of discharge, the smaller the relative as well as absolute variation among intervals. This curve in the more perfectly rhythmic units is displaced,

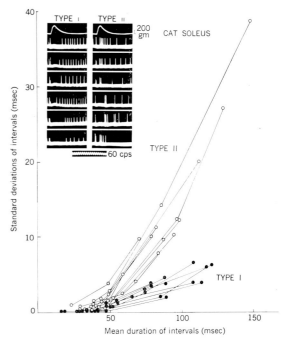

Figure 3.18. Differences in regularity of rhythm. A sample of 18 muscle-spindle receptor units from the soleus muscle of the cat has been examined in teased dorsal root filaments; the ventral roots are cut and the leg denervated except for the soleus nerve. With the tendon severed, the muscle was stretched to different extents to get different mean rates of firing for each unit. At each mean rate the standard deviation of intervals was estimated from plots. Insets show 12 examples of responses during a twitch induced by stimulation of the muscle nerve and illustrate units having more (type 2) and less (type 1) variability; type 1 are probably flower-spray endings, type 2 annulospiral. Both types improve in rhythmicity with higher mean frequency. [Tokizane and Eldred, unpublished.]

such that a given scatter of intervals occurs at lower frequencies. The curve is probably steeper in some and flatter in other instances but quantitative work is still needed.

The question of **how low the frequency can be** has a general interest in connection with the problem of the origin of biological rhythms of activity. We can ascribe heart beats, jellyfish pulsations, and similar events recurrent every few seconds to the elementary rhythmicity of single-celled neuronal pacemakers. It is a challenging problem still in the future whether there is a complete dichotomy or a continuum between these and rhythms with periods of hours and

III. EXCITABILITY C. Pacemakers

days. Evidence implicating the subesophageal ganglion in insects in the diurnal rhythms of activity (Harker, 1956, 1960) does not as yet show that the pace is set by a nerve cell discharging impulses at long intervals.

Low-frequency **rhythmic changes in excitability** are known but not provided for in any of the theories of excitation. Fessard (1936) and Bonnet (1941) in particular have studied these phenomena. In crustacean nerve it has been observed frequently that a delayed response starts after a long latency and often exhibits slow rhythms, as though spontaneous changes in the membrane potential were recurring (Arvanitaki, 1938). A difficulty in the study of excitation manifested by repetitive firing is that it verges with a scarcely perceptible transition into truly spontaneous firing. Many axons and, more important under natural conditions, many spike-initiating regions are depolarized sufficiently by conditions within the range they normally encounter to commence sustained spontaneous discharge. Since this is preceded by an extreme reduction of accommodation and of the rheobase, it is virtually impossible to determine clearly what the effect of controlled stimulating current is; even a single impulse which is really precipitated by the stimulus may inaugurate a train of spontaneous impulses. While inconvenient for the purpose of studying controlled stimuli, these phenomena are instructive in understanding afterdischarge and rhythmic activity in synaptic and integrative structures (Chapters 4 and 5).

Space prevents a full discussion of the literature on examples of rhythmicity and the **influence of various agents** upon it. It is a property basic to nervous function and one whose variations among neurons play an important role in the organization of useful nervous systems. Extended accounts have been provided by Fessard (1936), Arvanitaki (1939, 1943), and Bonnet (1941).

We return to closely related aspects (in particular to spontaneity) in Chapter 5, and there distinguish between sinusoidal oscillations and relaxation oscillations. Little is known concerning the physiological importance of influences known to affect iterative discharge (temperature, osmotic stress, electrical fields, calcium, magnesium, and other ions) or concerning the effectiveness of physiologically important agents such as hormones and nutrients.

Temperature can act with a Q_{10} at least as high as five (position receptor background discharge in lobsters, Cohen, Katsuki, and Bullock, 1953) or virtually one (stretch receptor, crayfish, Burkhardt, 1959). The latter value is significant because analysis shows that the near-independence of temperature is due to two equal and opposite temperature-sensitive aspects of the spike discharge sequence within the neuron (see p. 1023). It is thus more understandable that other neurons should show a negative coefficient and become more active with lowering temperature, at least transitorily (touch receptors, insect, slug, and crayfish central ganglia, Kerkut and Taylor, 1956, 1958). As shown in Fig. 3.19, **osmotic dilution** in one case increased repetition

Figure 3.19.

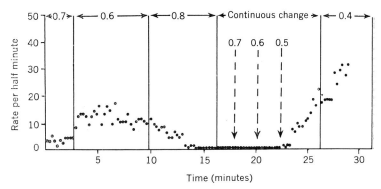

Effects of dilution of bathing medium on spontaneous discharge rate. Isolated ganglion of the gastropod *Agriolimax* in Locke's solution; extracellular recording with fine wire electrodes in the ganglion. Spontaneous activity of units totalled for each 30 sec that increase markedly with dilution from 0.8 Locke to 0.4 Locke over 10 min. The change is significant over the naturally occurring range of osmotic pressure of bodyfluids with desiccation and hydration. [Kerkut and Taylor, 1956.]

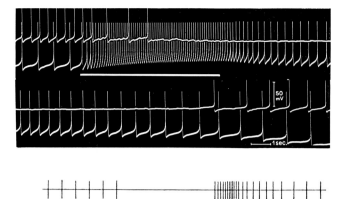

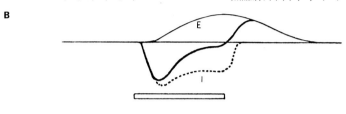

Figure 3.20.

Modulation by light of ongoing activity in nerve cells. **A.** Autorhythmic firing in two ganglion cells of the visceral ganglion of *Aplysia* is influenced in opposite directions by illumination during the white bar. The lower cell is promptly excited to higher frequency; the upper cell is inhibited, probably as the result of inhibitory synaptic bombardment in this case. After the illumination, a slow recovery ensues. In similar experiments, monochromatic light of long wavelengths has been found to inhibit the same cell as is excited by short wave-lengths. [Courtesy A. Arvanitaki and N. Chalazonitis.] **B.** Similar behavior has been recorded from a neuron in the clam *Spisula*. This is a postulated scheme for the contributions of the inhibitory (*I*) and excitatory (*E*) processes to the discharge pattern of the receptor unit. [Kennedy, 1960.]

rate of neurons in a biologically important range and with adaptive results (slug, pedal ganglion, Hughes and Kerkut, 1956).

Light modulates ongoing discharge in visceral ganglion cells of *Aplysia*, as seen in Fig. 3.20 (Arvanitaki and Chalazonitis, 1947, 1949a, 1949b, 1957b, 1958c, 1961; Chalazonitis, 1957). These cells cannot on present evidence be called real photoreceptors, but a similarly responding neuron at a spot in the course of the large pallial nerve of the clam *Spisula* is so called because of its uniqueness and high sensitivity (Kennedy, 1960, p. 1414). In both animals there are two pigments with different action spectra; one pigment mediates excitation, the other, inhibition of a background or dark discharge.

Classified References

General

Adrian, 1932; Arvanitaki, 1938; Bartlett, 1948; Bernstein, 1912; Biedermann, 1895; Bishop, 1951, 1956; Blinks, 1936; Brazier, 1960; Bullock, 1957; Chagas and Paes de Carvalho, 1961; Cole, 1957; Curtis and Cole, 1944; Diecke, 1958; Eccles, 1957; Erlanger and Gasser, 1937; Frank and Fuortes, 1961; Furness, 1961; Grundfest, 1940, 1955, 1956, 1959, 1960a, 1961a, 1961b; Hermann, 1879; Hill, 1932; Höber, 1946; Hodgkin, 1937a, 1937b, 1951; Jasper, 1956; Kan, 1937; Katsuki, 1960; Katz, 1939; Koshtoyants, 1957; Lapique, 1926; Lorente de Nó, 1946, 1947; Lucas, 1917; Lullies, 1932, 1952; McAlear et al., 1958; Merritt, 1952; Monnier, 1934; Muralt, 1946, 1958; Nachmansohn, 1946, 1951, 1952, 1955, 1959; Nachmansohn and Merritt, 1954; Nernst, 1908; Pease, 1955; Rosenberg, 1925; Schaefer, 1940; Schultz et al., 1956; Segal, 1958; Shedlovsky, 1955; Tasaki, 1953, 1958a, 1958b, 1959; Voronin, 1953; Wedensky, 1903; Zhukov, 1946.

Excitability, including Velocity of Conduction *(see also Conductance; Effects of Altering Ion Concentrations; Electrical Characteristics of Axons; Electrical Characteristics of Soma; Spontaneity)*

Abbott, 1961; Adrian, 1920, 1932; Adrian and Lucas, 1912; Arvanitaki and Chalazonitis, 1961; Barnes, 1930, 1931, 1932a, 1932b; Bear et al., 1937a; Berkowitz, 1955; Bernhard et al., 1942; Blair, 1936; Bonnet, 1956a, 1956b; Bradley and Somjen, 1961; Bugnard and Hill, 1935; Bullock, 1950, 1951; Bullock and Turner, 1950; Bullock et al., 1950; Carlson, 1905; Case et al., 1957; Castillo and

Classified References

Moore, 1959; Chalazonitis, 1961; Curtis and Cole, 1944; Davis and Forbes, 1936; Diecke, 1954; Dzhavrishvili, 1959; Easton, 1952; Edwards and Ottoson, 1958; Engelhardt, 1951b; Fabre, 1956; Fleischhacker, 1925; Florey, 1955; Frankenhaeuser, 1952; Frankenhaeuser and Hodgkin, 1956; Gasser and Grundfest, 1939; Gerard, 1931; Goldman, 1961; Gray, 1951; Grundfest, 1961; Hagiwara and Oomura, 1958; Hertz, 1947; Hess and Young, 1949; Hill, 1910, 1936a, 1936b; Hodes, 1953; Hodgkin, 1939; Hodgkin and Huxley, 1952; Holmes et al., 1941; Horridge, 1955; Hughes, 1952; Hursh, 1939; Huxley and Stämpfli, 1949, 1950; Jenkins and Carlson, 1902, 1904a, 1904b; Jordan and Lullies, 1933; Kan and Kusnezov, 1938; Katz, 1939, 1947b; Kugelberg, 1944; Lapicque, 1903a, 1903b, 1926; Lapicque and Veil, 1925; Le Fevre, 1948, 1950; Levin, 1927; Lorente de Nó, 1946, 1947; Lucas, 1907; Lullies, 1934; Martin, 1954; Monnier, 1934, 1950, 1952; Nachmansohn, 1950; Niki et al., 1953; Pumphrey and Young, 1938; Rosenberg, 1934; Rosenblueth et al., 1949; Rushton, 1927, 1935, 1951; Schlote, 1955; Skoglund, 1942; Solandt, 1936; Stämpfli, 1954, 1958a, 1958b; Tasaki, 1950; Tasaki and Sato, 1951; Turner, 1951; Wright and Adelman, 1954; Wright and Coleman, 1954; Wright and Reuben, 1958; Young, 1938.

Electrical Characteristics of Axon *(including both passive and active; see also General; Conductance of the Cell Membrane; Excitability; Spontaneity)*

Adrian, 1932, 1960; Arvanitaki, 1938, 1939a, 1939b, 1940, 1941–1943, 1942; Arvanitaki, Auger, and Fessard, 1936; Arvanitaki, Fessard, and Kruta, 1936; Bishop, 1928, 1937; Bishop and Heinbecker, 1930; Blair, 1932, 1934; Blair and Erlanger, 1933; Bogue and Rosenberg, 1934a, 1934b, 1936; Boistel, 1959, 1960; Boistel and Coraboeuf, 1954b; Buchtal, 1941; Case et al., 1957; Chang and Schmidt, 1960a, 1960b; Cole, 1941, 1955, 1957; Cole and Curtis, 1939, 1941; Cole and Hodgkin, 1939; Coombs et al., 1959; Coraboeuf and Boistel, 1957; Curtis and Cole, 1940, 1942, 1944; Eccles et al., 1933; Eccles and Krnjević, 1959; Eckert and Zacharová, 1957; Edwards and Ottoson, 1958; Findlay, 1959; Florey, 1955; Frank and Fuortes, 1956; Graham and Gerard, 1946; Grundfest, 1961; Hill, 1932; Hodgkin, 1938b, 1947a; Hodgkin and Huxley, 1939, 1945, 1952b, 1952d; Hodgkin and Rushton, 1946; Hodgkin et al., 1952; Horridge, 1953, 1954; Kao and Grundfest, 1955; Karaev, 1938; Katz, 1937, 1947a; Le Fevre, 1950; Lev et al., 1959a, 1959b, 1960; Lorente de Nó, 1947; Lucas, 1917; Marmont, 1940, 1949; Meves, 1960; Monnier and Dubuisson, 1931; Mozhaeva, 1958; Narahashi, 1960; Narahashi and Yamasaki, 1960a, 1960b, 1960c; Nicol and Whitteridge, 1955; Prosser and Chambers, 1938; Pumphrey et al., 1940; Rashbass and Rushton, 1949; Reuben et al., 1960; Rosenblueth et al., 1948; Rosenblueth and Luco, 1950; Rushton and Barlow, 1943; Sasaki and Otani, 1961; Schmitt, 1950; Schmitt and Schmitt, 1940; Segal, 1958; Shanes, 1949a, 1949b, 1958; Shaw et al., 1956; Spyropoulos, 1959; Stämpfli, 1959; Tasaki, 1953; Tasaki and Bak, 1958a, 1958b; Tasaki and Freygang, 1955; Tasaki and Hagiwara, 1957; Tasaki and Spyropoulos, 1958a, 1958b; Tauc, 1962a, 1962b; Tauc and Hughes, 1961; Thies, 1957; Uchizono, 1957, 1959; Ulbricht, 1958; Weidmann, 1951; Wright, 1959; Yamagiwa, 1960; Yamasaki and Narahashi, 1957c, 1959b; Zhukov, 1937.

Electrical Characteristics of Soma, Dendrites, and Nerve Endings *(see also Spontaneity; Effects of Ions; Excitability)*

Albe-Fessard and Buser, 1953, 1955; Arvanitaki and Chalazonitis, 1949d, 1955a, 1955e, 1956a, 1956b, 1958a, 1958b, 1960; Bennett et al., 1959a, 1959b, 1959c; Bishop, 1941, 1958; Bronk, 1936; Buchtal, 1941; Buser and Rougeul, 1955; Cardot and Arvanitaki, 1941; Cardot et al., 1942; Case et al., 1957; Chalazonitis, 1959; Chalazonitis and Arvanitaki, 1957a, 1958; Clare and Bishop, 1955; Davis, 1957; Eccles, 1955, 1957, 1959; Edwards and Ottoson, 1958; Eyzaguirre and Kuffler, 1955, 1956; Fessard and Tauc, 1956; Florey, 1955; Frank, 1959; Fuortes, 1958a, 1958b, 1958c, 1959; Gray and Matthews, 1951; Grundfest, 1958; Grundfest et al., 1954; Hagiwara, 1960, 1961; Hagiwara and Saito, 1958, 1959; Hagiwara et al., 1961; Ito, 1957; Katsuki et al., 1954; Kerkut and Walker, 1961; Kikuchi et al., 1960; Loewenstein, 1959; Loewenstein and Cohen, 1959a, 1959b; Loewenstein and Rathkamp, 1958; Morita, 1959; Morita and Takeda, 1959; Morita and Yamashita, 1959; Morita et al., 1957; Nishi and Koketsu, 1960; Otani, 1960; Otani and Bullock, 1959; Ottoson, 1959; Rall, 1959a, 1959b; Strumwasser and Rosenthal, 1960; Tasaki, 1959c; Tateda and Morita, 1959; Tauc, 1952, 1954, 1955b, 1955c, 1955d, 1955e, 1956a, 1958d, 1960, 1962a, 1962b; Terzuolo and Araki, 1961.

Spontaneity, Autorhythmicity, and Iterative Activity of Nerve Cells and Fibers

Arvanitaki, 1938, 1939a, 1939c, 1939d, 1943a, 1943b; Arvanitaki and Cardot, 1939, 1941a, 1941b, 1941c; Arvanitaki and Chalazonitis, 1955c, 1955d, 1957a; Arvanitaki et al., 1936; Bonnet, 1941, 1956a, 1956b; Cardot and Arvanitaki, 1941; Chalazonitis and Sugaya, 1958; Dudel and Trautwein, 1958; Erlanger and Blair, 1938; Fessard, 1936a, 1936b; Harker, 1951, 1960; Hodgkin, 1948; LeFevre, 1948; Sjodin and Mullins, 1958; Tasaki and Bak, 1957, 1958b; Terzuolo and Washizu, 1962.

Effects of Altering Ion Concentrations *(this bibliography is only a sample since this topic is peripheral to the scope of the book; see also General; Conductance; Effects of Drugs)*

Adelman, 1956; Adelman and Adams, 1959; Adelman and Dalton, 1960; Adrian, 1960; Amatniek et al., 1957; Dalton, 1959; Dalton and Adelman, 1960; Dalton and FitzHugh, 1960; Grundfest et al., 1954; Hagiwara et al.,

1961; Hodgkin, 1947b; Hodgkin and Katz, 1949a, 1949c; Hodgkin and Keynes, 1956; Katz, 1947b; LeFevre, 1948; Narahashi, 1961; Schallek, 1945; Schmidt and Stämpfli, 1957; Shanes and Hopkins, 1948; Shanes et al., 1959; Shaw et al., 1956; Steinbach, 1940; Steinbach et al., 1944; Tasaki, 1959b; Tasaki and Hagiwara, 1957; Waterman, 1941; Werman et al., 1960; Wright et al., 1955; Yamasaki and Narahashi, 1959a.

Effects of Drugs *(Since this topic is peripheral to the scope of the book, this list is only a sample; see also Effects of Altering Ion Concentrations)*

Arvanitaki and Chalazonitis, 1949, 1951, 1954a, 1955b, 1955f; Arvanitaki et al., 1956; Bacq, 1947; Bayliss et al., 1935; Boistel and Coraboeuf, 1954a, 1955; Bonnet, 1937; Boruttau, 1905; Boyarsky et al., 1949; Brady et al., 1958; Brink et al., 1946; Bullock, Nachmansohn, and Rothenberg, 1946; Bullock, Grundfest, Nachmansohn, and Rothenberg, 1947a, 1947b; Bullock, Grundfest, Nachmansohn, Rothenberg, and Sterling, 1946; Chalazonitis, 1959, 1961; Chalazonitis and Arvanitaki, 1957a, 1957b; Coraboeuf, 1954; Gerard, 1946, 1950; Grundfest, 1958; Kitamura, 1960; Lavigne and Coraboeuf, 1955; Meves, 1955; Moore and Turner, 1948; Nachmansohn, 1946, 1952, 1954, 1955, 1959; Nachmansohn and Wilson, 1955; Schmidt and Stämpfli, 1959; Shanes, et al., 1953, 1959; Taylor, 1959; Tobias, 1955; Tobias and Bryant, 1955; Wilson and Cohen, 1953; Wright, 1946; Yamasaki and Ishii, 1952; Yamasaki and Narahashi, 1957d.

Effects of Temperature, Light, and Mechanical Forces *(This list is a sample only)*

Amatniek et al., 1957; Arvanitaki, 1961; Arvanitaki and Chalazonitis, 1947, 1949a, 1949b, 1954b, 1957b, 1958c, 1960, 1961; Cerf, 1957; Chalazonitis, 1957; Cohen et al., 1953; Engelhardt, 1951a; Goldman, 1961; Hodgkin and Katz, 1949b; Ishiko and Loewenstein, 1960; Kennedy, 1960; Martin, 1954; Shanes, 1954; Spyropoulos, 1957; Turner, 1953, 1955; Wright, 1958.

Conductance of the Cell Membrane *(Since this topic is peripheral to the scope of the book, this list is a sample only; see also General; Effects of Altering Ion Concentrations)*

Amatniek et al., 1957; Boistel and Coraboeuf, 1958; Cole, 1955; Gaffey and Mullins, 1958; Grundfest, 1960b, 1961b; Grundfest et al., 1953; Hodgkin, 1951, 1958; Hodgkin and Huxley, 1947, 1952a, 1952c, 1953; Hodgkin and Katz, 1949a; Hodgkin and Keynes, 1953, 1954, 1955a, 1955b; Keynes, 1951a, 1951b; Keynes and Lewis, 1951a; Mullins, 1959, 1960, 1961; Nevis, 1958; Rothenberg, 1950; Shanes, 1950, 1951, 1954; Shanes and Berman, 1955; Stämpfli, 1959; Tasaki et al., 1961; Teorell, 1950, 1953; Treherne, 1961; Ulbricht, 1958; Van Harreveld and Russell, 1954; Wilson, 1956.

Models and Theoretical Studies of Conduction

Akiyama, 1955; Akiyama et al., 1958, 1959; Bishop, 1927; Carricaburu, 1959, 1960; Cole, 1955, 1957, 1958, 1961; Cole et al., 1955; Dalton and FitzHugh, 1960; FitzHugh, 1955; Franck, 1956; Grundfest, 1960b; Lillie, 1920, 1936; Nernst, 1908; Rall, 1959a, 1959b; Rashevsky, 1936, 1960; Rogers, 1954; Tasaki and Bak, 1959; Teorell, 1953.

Chemical Assay of Nervous Tissue *(Since this topic is peripheral to the scope of the book, this list is only a sample)*

Bear and Schmitt, 1939; Bear et al., 1937b; Boell and Nachmansohn, 1940; Caldwell, 1958; Easton, 1950; Keynes and Lewis, 1951b, 1956; Koechlin, 1954, 1955; Lewis, 1952; Libet, 1947, 1948a, 1948b; Maxfield, 1951, 1953; Maxfield and Hartley, 1955, 1957; Nachmansohn, 1946, 1952, 1954, 1955, 1959; Richards and Cutkomp, 1945; Schmitt et al., 1939; Shaw et al., 1956; Silber, 1941; Silber and Schmitt, 1940; Simon et al., 1957; Skou, 1957; Steinbach, 1940, 1941; Steinbach and Spiegelman, 1943; Webb and Young, 1940.

Bibliography

In this list selected references are assembled, representing the more recent general accounts on conduction of nerve impulses in axons and contributions of special interest in connection with comparative neurology. With respect to molecular mechanisms, only a limited selection is included. Some literature, especially on invertebrates, which goes beyond the scope just stated, is given for convenience, but neurochemistry, the effects of drugs, and metabolic aspects of nerve function are barely touched.

ABBOTT, B. C. 1961. Energetics of nerve activity. In: *Biophysics of Physiological and Pharmacological Actions.* A. M. Shanes (ed.). Amer. Ass. Adv. Sci., Washington, D.C.

ADELMAN, W. J., JR. 1956. The effect of external calcium and magnesium depletion on single nerve fibers. *J. gen. Physiol.*, 39:753-772.

ADELMAN W. J., JR. and ADAMS, J. 1959. Effects of calcium lack on action potential of motor axons of the lobster limb. *J. gen. Physiol.*, 42:655-664.

ADELMAN, W. J., JR. and DALTON, J. C. 1960. Interactions of calcium with sodium and potassium in membrane potentials of the lobster giant axon. *J. gen. Physiol.*, 43:609-619.

ADEY, W. R. 1951. The nervous system of the earthworm *Megascolex. J. Comp. Neurol.*, 94:57-103.

Bibliography

ADRIAN, E. D. 1920. The recovery process in excitable tissues. *J. Physiol.* 54:1-31 and 55:193-225.

ADRIAN, E. D. 1932. *The Mechanism of Nervous Action.* Univ. of Pennsylvania Press, Philadelphia.

ADRIAN, E. D. and LUCAS, K. 1912. On the summation of propagated disturbances in nerve and muscle. *J. Physiol.*, 44:68-124.

ADRIAN, R. H. 1960. Potassium chloride movement and the membrane potential of frog muscle. *J. Physiol.*, 151:154-185.

AKIYAMA, I. 1955. The silver nitrate and iron system as an electrochemical model of nervous conduction. *Gunma J. med. Sci.*, 4:41-46.

AKIYAMA, I., NOSE, R., and NOMACHI, T. 1958. Contribution to the properties of the so-called Akiyama model consisting of iron wire and silver nitrate. *Gunma J. med. Sci.*, 7:77-84.

AKIYAMA, I., ISHIKAWA, I., ISHIDA, M., and MORIKAWA, J. 1959. Studies on the stimulating effect of alternating current with electrochemical model of excitation. *Jap. J. Physiol.*, 9:266-273.

ALBE-FESSARD, D. and BUSER, P. 1953. Explorations de certaines activités du cortex moteur du chat par microélectrodes: dérivations endo-somatiques. *J. Physiol., Paris* 45:14-16.

ALBE-FESSARD, D. and BUSER, P. 1955. Activités intracellulaires recueillies dans le cortex sigmoïde du chat: participation des neurones pyramidaux au "potentiel évoqué" somesthésique. *J. Physiol., Paris*, 47:67-69.

AMATNIEK, E., FREYGANG, W., GRUNDFEST, H., KIEBEL, G., and SHANES, A. 1957. The effect of temperature, potassium, and sodium on the conductance change accompanying the action potential in the squid giant axon. *J. gen. Physiol.* 41:333-342.

ARAKI, T. and OTANI, T. 1959. Accommodation and local response in motoneurons of toad's spinal cord. *Jap. J. Physiol.*, 9:69-83.

ARVANITAKI, A. 1938. *Les Variations Graduées de la Polarisation des Systèmes Excitables.* Univ. of Lyons, Thesis. Hermann et Cie., Paris.

ARVANITAKI, A. 1939a. Recherches sur la réponse oscillatoire locale de l'axone géant isolé de *Sepia*. *Arch. int. Physiol.*, 49:209-256.

ARVANITAKI, A. 1939b. Caractères de l'activité électrique graduée et locale de l'axone isolé de *Sepia*. *C. R. Soc. Biol., Paris*, 130:424-428.

ARVANITAKI, A. 1939c. Caractères de l'activité électrique graduée et locale de l'axone isolé de *Sepia*. Réponses réitérées. *C. R. Soc. Biol., Paris*, 130:545-552.

ARVANITAKI, A. 1939d. Contributions à l'étude analytique de la réponse électrique oscillatoire locale de l'axone isolé de *Sepia*. *C. R. Soc. Biol., Paris*, 131:1117-1120.

ARVANITAKI. A. 1939-1940. L'activité électrique sous-liminaire et locale de l'axone normal isolé de *Sepia*, *J. Physiol. path. gen.*, 37:895-912.

ARVANITAKI, A. 1941-1943. Réactions au stimulus anodique. Étude de la réponse électrique locale de signe positif. Observations sur l'axone isolé de *Sepia*. *J. Physiol. path. gen.*, 38:147-170.

ARVANITAKI, A. 1942. Phase de dépression relative de l'irritabilité, consécutive à la réponse négative locale sur l'axone isolé de Seiche. *C. R. Soc. Biol., Paris*, 136:6-11.

ARVANITAKI, A. 1943a. Variations de l'excitabilité locale et activité autorythmique sous-liminaire et liminaire. Observations sur l'axone isolé de *Sepia*. I. Effets immédiats du stimulus test: extra-réponse. *Arch. int. Physiol.* 53:508-532.

ARVANITAKI, A. 1943b. Variations de l'excitabilité locale et activité autorythmique sous-liminaire et liminaire. Observations sur l'axone isolé de *Sepia*. II. Effets secondaires, consécutifs à l'extra-réponse. *Arch. int. Physiol.*, 53:533-559.

ARVANITAKI, A. 1961. Excitation and inhibition of nerve cells by visible and infra-red radiation. In: *Nervous Inhibition.* E. Florey (ed.). Pergamon Press, Oxford.

ARVANITAKI, A. and CARDOT, H. 1939. Les quatre cas possibles de réponses électriques oscillatoires locales aux stimulations galvaniques sur l'axone géant isolé de *Sepia*, *C.R. Soc. Biol., Paris*, 131:1112-1116.

ARVANITAKI, A. and CARDOT, H. 1941a. Les caractéristiques de l'activité rythmique ganglionnaire "spontanée" chez l'aplysie. *C.R. Soc. Biol., Paris*, 135:1207-1211.

ARVANITAKI, A. and CARDOT, H. 1941b. Réponses rythmiques ganglionnaires, graduées en fonction de la polarisation appliquée. Lois des latences et des fréquences. *C.R. Soc. Biol., Paris*, 135:1211-1216.

ARVANITAKI, A. and CARDOT, H. 1941c. Réponses autonomes ganglionnaires à une polarisation appliquée. *C.R. Soc. Biol., Paris*, 135:1216-1221.

ARVANITAKI, A. and CHALAZONITIS, N. 1947. Réactions bioélectriques à la photoactivation des cytochromes. *Arch. Sci. physiol.* 1:385-405.

ARVANITAKI, A. and CHALAZONITIS, N. 1949a. Réactions bioélectriques neuroniques à la photoactivation spécifique d'une héme-protéine et d'une carotène-protéine. *Arch. Sci. physiol.*, 3:27-44.

ARVANITAKI, A. and CHALAZONITIS, N. 1949b. Inhibition ou excitation des potentiels neuroniques à la photoactivation distincte de deux chromoprotéides (caroténoïde et chlorophyllien). *Arch. Sci. physiol.*, 3:45-60.

ARVANITAKI. A. and CHALAZONITIS, N. 1949c. Catalyse respiratoire et potentiels bioélectriques. *Arch. Sci. physiol.* 3:303-338.

ARVANITAKI, A. and CHALAZONITIS, N. 1949d. Prototypes d'interactions neuroniques et transmissions synaptiques. Données bioélectriques de préparations cellulaires. *Arch. Sci. physiol.*, 3:547-565.

ARVANITAKI, A. and CHALAZONITIS, N. 1951. Effets narcotiques sur les biopotentiels neuroniques et sur la catalyse respiratoire. In: *Mécanisme de la Narcose.* C.N.R.S., Paris.

ARVANITAKI, A. and CHALAZONITIS, N. 1954a. Diffusibilité de l'anhydride carbonique dans l'axone géant, ses effets sur les vitesses de l'activité bioélectrique. *C.R. Soc. Biol., Paris*, 148:952-954.

ARVANITAKI, A. and CHALAZONITIS, N. 1954b. Réponses bioélectriques de l'axone géant à l'accélération thermique de sa respiration. *C.R. Soc. Biol., Paris*, 148:1027-1029.

ARVANITAKI, A. and CHALAZONITIS, N. 1955a. Potentiels d'activité du soma neuronique géant *(Aplysia)*. *Arch. Sci. physiol.*, 9:115-144.

ARVANITAKI, A. and CHALAZONITIS, N. 1955b. Les phosphates de sodium sur la respiration et l'activité électrique des neurones. *J. Physiol., Paris*, 47:645-650.

ARVANITAKI, A. and CHALAZONITIS, N. 1955c. Les potentiels bioélectriques endocytaires du neurone géant d'*Aplysia* en activité autorythmique. *C.R. Acad. Sci., Paris*, 240:349-351.

ARVANITAKI, A. and CHALAZONITIS, N. 1955d. Variations lentes et périodiques du potentiel de membrane associées à des groupes de pointes (neurone géant d'*Aplysia*). *C.R. Acad. Sci., Paris*, 240:462-464.

ARVANITAKI, A. and CHALAZONITIS, N. 1955e. Biopotentiels du soma neuronique géant étudiés par dérivation intranucléaire. *C.R. Acad. Sci., Paris*, 240:2016-2018.

ARVANITAKI, A. and CHALAZONITIS, N. 1955f. Evolution décroissante de l'intensité respiratoire et des pentes des potentiels d'activité des neurones, durant l'action narcotique. *C.R. Soc. Biol., Paris*, 149:228-233.

ARVANITAKI, A. and CHALAZONITIS, N. 1956a. Biopotentiels neuroniques à l'échelle infracellulaire. Stimulation mécanique graduée dans le soma géant d'*Aplysia*. *Bull. Inst. océanogr. Monaco*, 53:1-37.

ARVANITAKI, A. and CHALAZONITIS, N. 1956b. Surexcitabilité prénarcotique sur le soma neuronique d'*Aplysia*. (Dérivation endocytaire.) *J. Physiol., Paris*, 48:374-376.

ARVANITAKI, A. and CHALAZONITIS, N. 1957a. Réfractivités introduites par les potentiels positifs du soma neuronique, en activité autoentretenue. *C.R. Acad. Sci., Paris*, 245:445-447.

ARVANITAKI, A. and CHALAZONITIS, N. 1957b. Réponses du soma neuronique à la photoactivation des grains pigmentés de son cytoplasme. *J. Physiol.*, Paris, 49:9-12.

ARVANITAKI, A. and CHALAZONITIS, N. 1958a. Configurations modales de l'activité, propres à différents neurones d'un même centre. *J. Physiol.*, Paris, 50:122-125.

ARVANITAKI, A. and CHALAZONITIS, N. 1958b. Réactions électriques distinctes de deux neurones définis, à un même modificateur. *J. Physiol.*, Paris, 50:125-128.

ARVANITAKI, A. and CHALAZONITIS, N. 1958c. Activation par la lumière des neurones pigmentés. *Arch. Sci. physiol.*, 12:73-106.

ARVANITAKI, A. and CHALAZONITIS, N. 1960. Photopotentiels d'excitation et d'inhibition de différents somata identifiables *(Aplysia)*. Activations monochromatiques. *Bull. Inst. océanogr. Monaco*, 57(1164):1-83.

ARVANITAKI. A. and CHALAZONITIS, N. 1961. Excitatory and inhibitory processes initiated by light and infrared radiations in single identifiable nerve cells (giant ganglion cells of *Aplysia*). In: *Nervous Inhibition.* E. Florey, ed. Pergamon Press, Oxford.

ARVANITAKI, A., AUGER, D., and FESSARD, A. 1936. Existence de longues durées d'utilisation dans le déclenchement galvanique des pulsations nerveuses chez les crustacés. *C.R. Soc. Biol.*, Paris, 121:638-640.

ARVANITAKI, A., CHALAZONITIS N., and OTSUKA, M. 1956. Activité paroxystique du soma neuronique d'*Aplysia* sous l'effet de la strychnine. (Dérivation endosomatique des potentiels.) *C.R. Acad. Sci.*, Paris, 243:307-309.

ARVANITAKI, A., FESSARD A., and KRUTA, V. 1936a. Analyse du potentiel d'action des nerfs viscéraux chez *Sepia officinalis*. *C.R. Soc. Biol.*, Paris, 122:1204-1206.

ARVANITAKI, A., FESSARD, A., and KRUTA. V. 1936b. Mode répétitif de la réponse électrique des nerfs viscéraux et étoilés chez *Sepia officinalis*. *C.R. Soc. Biol.*, Paris, 122:1203-1204.

BACQ, Z. M. 1947. L'acétylcholine et l'adrénaline chez les invertébrés. *Biol. Rev.*, 22:73-91.

BARNES, T. C. 1930. Le diamètre des fibres nerveuses des crabes en rapport avec leurs propriétés fonctionelles. *C.R. Soc. Biol.*, Paris, 105:385-387.

BARNES, T. C. 1931. Impulses in crustacean nerve associated with pressure stimulation. *J. Physiol.*, 71:xii-xiii.

BARNES, T. C. 1932a. Responses in the isolated limbs of Crustacea and associated nervous discharges. *Amer. J. Physiol.*, 99:321-331.

BARNES, T. C. 1932b. The significance of fibre diameter in the sensory nerves of the crustacean limb. *Amer. J. Physiol.*, 100:481-486.

BARTLETT, J. H. 1948. Comparison of transients in inorganic systems with those in plant and nerve cells. *J. cell. comp. Physiol.*, 32:1-29.

BAYLISS, L. E., COWAN, S. L., and SCOTT, D., JR. 1935. The action potentials in *Maia* nerve before and after poisoning with veratrine and yohimbine hydrochlorides. *J. Physiol.*, 83:439-454.

BEAR, R. S. and SCHMITT, F. O. 1939. Electrolytes in the axoplasm of the giant nerve fibers of the squid. *J. cell. comp. Physiol.*, 14:205-215.

BEAR, R. S., SCHMITT, F. O., and YOUNG, J. Z. 1937a. The sheath components of the giant nerve fibres of the squid. *Proc. roy. Soc. (B.)*, 123:496-504.

BEAR, R. S., SCHMITT, F. O., and YOUNG, J. Z. 1937b. Investigations on the protein constituents of nerve axoplasm. *Proc. roy. Soc. (B.)*, 123:520-529.

BENNETT, M. V. L., CRAIN, S. M., and GRUNDFEST, H. 1959a. Electrophysiology of supramedullary neurons in *Spheroides maculatus*. I. Orthodromic and antidromic responses. *J. gen. Physiol.*, 43:159-188.

BENNETT, M. V. L., CRAIN, S. M., and GRUNDFEST, H. 1959b. Electrophysiology of supramedullary neurons in *Spheroides maculatus*. II. Properties of the electrically excitable membrane. *J. gen. Physiol.*, 43:189-219.

BENNETT, M. V. L., CRAIN, S. M., and GRUNDFEST, H. 1959c. Electrophysiology of supramedullary neurons in *Spheroides maculatus*. III. Organization of the supramedullary neurons. *J. gen. Physiol.*, 43:221-250.

BERKOWITZ, E. C. 1955. Anatomical correlates of conduction velocity in the cord of a teleost and a cyclostome. *Anat. Rec.*, 121:264.

BERNHARD, C. G., GRANIT, R., and SKOGLUND, C. R. 1942. The breakdown of accommodation—nerve as model sense-organ. *J. Neurophysiol.*, 5:55-68.

BERNSTEIN, J. 1912. *Elektrobiologie.* Vieweg, Brunswick.

BIEDERMANN, W. 1895. *Elektrophysiologie.* Fischer. Jena. (English translation by F. A. Welby, Macmillan, London, 1896, 1898.)

BISHOP, G. H. 1927. The effects of polarization upon steel wire-nitric acid model of nerve activity. *J. gen. Physiol.*, 11:159-174.

BISHOP, G. H. 1928. The effect of nerve reactance on the threshold of nerve during galvanic current flow. *Amer. J. Physiol.*, 84:417-436.

BISHOP, G. H. 1937. La théorie des circuits locaux, permet-elle de prévoir la forme du potentiel d'action? *Arch. int. Physiol.*, 45:273.

BISHOP, G. H. 1941. The relation of bioelectric potentials to cell functioning. *Annu. Rev. Physiol.*, 3:1-20.

BISHOP, G. H. 1951. Excitability. In: *The Nerve Impulse.* No. 2. Josiah Macy, Jr. Foundation, New York.

BISHOP, G. H. 1956. Natural history of the nerve impulse. *Physiol. Rev.*, 36:376-399.

BISHOP, G. H. 1958. The dendrite: receptive pole of the neurone. *Electroenceph. clin. Neurophysiol.*, 10(Suppl.):12-21.

BISHOP, G. H. and HEINBECKER, P. 1930. Differentiation of axon types in visceral nerves by means of the potential record. *Amer. J. Physiol.*, 94:170-200.

BLAIR, E. A. and ERLANGER, J. 1933. A comparison of the characteristics of axons through their individual electrical responses. *Amer. J. Physiol.*, 106:524-564.

BLAIR, H. A. 1932. On the intensity-time relation for stimulation by electric currents. I, II. *J. gen. Physiol.*, 15:709-755.

BLAIR, H. A. 1934. Conduction in nerve fibres. *J. gen. Physiol.*, 18:125-142.

BLAIR, H. A. 1936. The kinetics of the excitatory process. *Cold Spr. Harb. Symp. quant. Biol.*, 4:63-72.

BLINKS, L. R. 1936. Effects of current flow on bioelectric potential. III. *Nitella*. *J. gen. Physiol.*, 20:229-265.

BOELL, E. J. and NACHMANSOHN, D. 1940. Choline esterase in nerve fibers. *Biol. Bull. Woods Hole*, 79:357.

BOGUE. J. Y. and ROSENBERG, H. 1934a. The rate of development and spread of electrotonus. *J. Physiol.*, 82:353-368.

BOGUE, J. Y. and ROSENBERG, H. 1934b. Action potentials in nerve of *Sepia*. *J. Physiol.*, 83:21P-23P.

BOGUE, J. Y. and ROSENBERG, H. 1936. Electrical responses of *Maia* nerve to single and repeated stimuli. *J. Physiol.*, 87:158-180.

BOISTEL, J. 1959. Quelques caractéristiques électriques de la membrane de la fibre nerveuse au repos d'un insecte *(Periplaneta americana)*. *C.R. Soc. Biol.*, Paris, 153:1009-1113.

BOISTEL, J. 1960. *Caractéristiques Fonctionnelles des Fibres Nerveuses et des Récepteurs Tactiles et Olfactifs des Insectes.* Librairie Arnette, Paris.

BOISTEL, J. and CORABOEUF, E. 1954a. Action de l'anhydride carbonique sur l'activité électrique du nerf isolé d'insecte. *J. Physiol.*, Paris, 46:258-261.

BOISTEL, J. and CORABOEUF, E. 1954b. Potentiel de membrane et potentiels d'action de nerf d'insecte recueillis à l'aide de microélectrodes intracellulaires. *C. R. Acad. Sci.*, Paris, 238:2116-2118.

BOISTEL, J. and CORABOEUF, E. 1955. Etude de quelques facteurs modifiant

Bibliography

l'action de l'anhydride carbonique sur le nerf isolé d'insecte. *J. Physiol., Paris*, 47:102-104.

BOISTEL, J. and CORABOEUF, E. 1958. Rôle joué par les ions Na$^+$ dans la genèse de l'activité électrique du tissu nerveux d'insecte. *C. R. Acad. Sci., Paris*, 247:1781-1783.

BONNET, V. 1937. Action paralysante de la strychnine chez l'écrevisse. Son influence sur l'excitabilité des nerfs moteurs. *C. R. Soc. Biol., Paris*, 124:993-995.

BONNET, V. 1941. *L'activité rythmique de la cellule nerveuse et ses modifications*. Soc. Anonyme de l'Imprimerie A. Rey, Lyon.

BONNET, V. 1956a. Étude de l'effet des modifications de la polarisation membranaire neuronique sur les réactions réflexes de la grenouille spinale. I. Polarisation par courant constant déductions relatives au déterminisme de la post-décharge. *Arch. int. Physiol.*, 64:141-167.

BONNET, V. 1956b. Étude de l'effet des modifications de la polarisation membranaire neuronique sur les réactions réflexes de la grenouille spinale. II. Effet d'une dépolarisation physiologique de longue durée des neurones spinaux sur les caractères de leur décharge réflexe. *Arch. int. Physiol.*, 64:168-191.

BORUTTAU, H. 1905. Elektropathologische Untersuchungen. II. Zur Elektropathologie der marklosen Kephalopodennerven. *Pflüg. Arch. ges. Physiol.* 107:193-206.

BOYARSKY, L. L., ROSENBLATT, A. D., POSTEL, S., and GERARD, R. W. 1949. Action of methylfluoroacetate on respiration and potential of nerve. *Amer. J. Physiol.*, 157:291-298.

BRADLEY, K. and SOMJEN, G. G. 1961. Accomodation in motoneurones of the rat and the cat. *J. Physiol.*, 156:75-92.

BRADY, R. O., SPYROPOULOS, C. S., and TASAKI, I. 1958. Intra-axonal injection of biologically active materials. *Amer. J. Physiol.*, 194:207-213.

BRAZIER, M. A. B. 1960. *The Electrical Activity of the Nervous System*. 2nd ed. Macmillan, New York.

BRINK, F., BRONK, D. W., and LARRABEE. M. G. 1946. Chemical excitation of nerve. *Ann. N.Y. Acad. Sci.*, 47:457-485.

BRONK, D. W. 1936. The activity of nerve cells. *Cold Spr. Harb. Symp. Quant. Biol.*, 4:170-178.

BUCHTAL, F. 1941. Messungen von Potential-Differenzen an einzelnen Zellen. *Tabul. biol. Hague*, 19(2):28-75.

BUGNARD, L. and HILL, A. V. 1935. Electric excitation of the fin nerve of *Sepia*. *J. Physiol.*, 83:425-438.

BULLOCK, T. H. 1950. The course of fatigue in single nerve fibers and junctions. *Int. Physiol. Congr., Abstr. XVIII.* 1950:134-135.

BULLOCK, T. H. 1951. Facilitation of conduction rate in nerve fibres. *J. Physiol.*, 114:89-97.

BULLOCK, T. H. 1957. The trigger concept in biology. In: *Physiological Triggers*. T. H. Bullock (ed.). Amer. Physiol. Soc., Washington.

BULLOCK, T. H. and TERZUOLO, C. A. 1957. Diverse forms of activity in the somata of spontaneous and integrating ganglion cells. *J. Physiol.*, 138:341-364.

BULLOCK, T. H. and TURNER, R. S. 1950. Events associated with conduction failure in nerve fibers. *J. cell. comp. Physiol.*, 36:59-82.

BULLOCK, T. H., COHEN, M. J., and FAULSTICK, D. 1950. Effect of stretch on conduction in single nerve fibers. *Biol. Bull., Woods Hole*, 99:320.

BULLOCK, T. H., NACHMANSOHN, D., and ROTHENBERG, M. 1946. Effects of inhibitors of choline esterase on the nerve action potential. *J. Neurophysiol.*, 9:9-22.

BULLOCK, T. H., GRUNDFEST, H., NACHMANSOHN, D., ROTHENBERG, M. A., and STERLING, K. 1946. Effect of di-isopropyl fluorophosphate (DFP) on action potential and choline esterase of nerve. *J. Neurophysiol.*, 9:253-260.

BULLOCK, T. H., GRUNDFEST, H., NACHMANSOHN, D., and ROTHENBERG, M. A. 1947a. Generality of the role of acetylcholine in nerve and muscle conduction. *J. Neurophysiol.*, 10:11-22.

BULLOCK, T. H., GRUNDFEST, H., NACHMANSOHN, D., and ROTHENBERG, M. A. 1947b. Effect of di-isopropyl fluorophosphate (DFP) on action potential and cholinesterase of nerve. II. *J. Neurophysiol.*, 10:63-78.

BURKHARDT, D. 1959. Effect of temperature on isolated stretch-receptor organ of the crayfish. *Science*, 129:392-393.

BUSER, P. and ROUGEUL, A. 1955. Réception intracellulaire au niveau des cellules pyramidales de la corne d'ammon du chat. *J. Physiol., Paris*, 47:121-123.

CAJAL, S. R. Y. 1954. *Neuron Theory or Reticular Theory?* (Transl. by M. Ubeda Purkiss and C. A. Fox.) Inst. Ramón y Cajal, C.S.I.S., Madrid.

CALDWELL, P. C. 1958. Studies on the internal pH of large muscle and nerve fibres. *J. Physiol.*, 142:22-62.

CARDOT, H. and ARVANITAKI, A. 1941a. Les incréments thermiques critiques relatifs aux phases composantes de la réponse électrique oscillatoire locale. Axone isolé de *Sepia*. *J. Physiol. path. gén.*, 38:9-16.

CARDOT, H. and ARVANITAKI, A. 1941b. Donneés sur les caractéristiques de l'activité électrique du soma neuronique. *Schweiz Med. Wschr.*, 71:395-397.

CARDOT, H., ARVANITAKI, A., and TCHOU, S. H. 1942. Exploration de l'activité électrique sur une cellule nerveuse isolée. *C. R. Soc. Biol., Paris*, 136:367-369.

CARLSON, A. J. 1905. Further evidence of the fluidity of the conducting substance in nerve. *Amer. J. Physiol.*, 13:351-357.

CARRICABURU, P. 1959. Sur quelques propriétés du nerf d'Akiyama. *C. R. Soc. Biol., Paris*, 153:2048-2053.

CARRICABURU, P. 1960a. Oscillations de relaxation du nerf d'Akiyama. *C. R. Acad. Sci., Paris*, 251:906-907.

CARRICABURU, P. 1960b. Fonctionnement auto-rhythmique du nerf d'Akiyama. *C. R. Soc. Biol., Paris*, 154:761-762.

CASE, J. F., EDWARDS, C., GESTELAND, R., and OTTOSON, D. 1957. The site of origin of the nerve impulse in the lobster stretch receptor. *Biol. Bull., Woods Hole*, 113:360.

CERF, J. 1957. Narcose thermique et polarisation membranaire du nerf de la grenouille. *Arch. int. Pharmacodyn.*, 109:300-333.

CHAGAS, C. and PAES DE CARVALHO, A. (eds.). 1961. *Bioelectrogenesis*. Elsevier, Amsterdam.

CHALAZONITIS, N. 1957. *Effets de la lumière sur l'évolution des potentiels cellulaires et sur quelques vitesses d'oxydoréduction dans les neurones*. Bosc, Lyon.

CHALAZONITIS, N. 1959. Chémopotentiels des neurones géants fonctionnellement différenciés. *Arch. Sci. Physiol.*, 13:1-38.

CHALAZONITIS, N. 1961. Chemopotentials in giant nerve cells *(Aplysia fasciata)*. In: *Nervous Inhibition*. E. Florey (ed.). Pergamon Press, Oxford.

CHALAZONITIS, N. and ARVANITAKI, A. 1957a. Pointes et potentiels positifs du soma neuronique en fonction de la température. *C. R. Acad. Sci., Paris*, 245:1079-1081.

CHALAZONITIS, N. and ARVANITAKI, A. 1957b. Évolutions des pentes des biopotentiels et des vitesses de tranfert de l'hydrogène (Axone de *Sepia*). *Colloq. int. Cent. nat. Rech. sci.*, 1955(67):203-240.

CHALAZONITIS, N. and ARVANITAKI, A. 1958. Dérivation endocytaire simultanée de l'activité de différents neurones, in situ. *C. R. Acad. Sci., Paris*, 246:161-163.

CHALAZONITIS, N. and SUGAYA, E. 1958. Effets anoxiques sur l'auto-activité électrique des neutrones [sic] géants d'*Aplysia*. *C. R. Acad. Sci.*, 247:1495-1497.

CHANG, J. J. and SCHMIDT, R. F. 1960a. Action potentials of reversed polarity in Purkinje fibers of dog heart. *Naturwissenschaften*, 47:259.

CHANG, J. J. and SCHMIDT, R. F. 1960b. Prolonged action potentials and regenerative hyperpolarizing responses in Purkinje fibers of mammalian heart. *Pflüg. Arch. ges. Physiol.*, 272: 127-141.

CLARE, M. M. and BISHOP, G. H. 1955. Dendritic circuits: the properties of cortical paths involving dendrites. *Amer. J. Psychol.*, 111:818-825.

COHEN, M. J., KATSUKI, Y., and BULLOCK, T. H. 1953. Oscillographic analysis of equilibrium receptors in crustacea. *Experientia*, 9:434-435.

COLE, K. S. 1941. Rectification and inductance in the squid giant axon. *J. gen. Physiol.*, 25:29-51.

COLE, K. S. 1955. Ions, potentials, and the nerve impulse. In: *Electrochemistry in Biology and Medicine*. T. Shedlovsky (ed.). Wiley, New York.

COLE, K. S. 1957. The nerve trigger. In: *Physiological Triggers*. T. H. Bullock (ed.). Amer. Physiol. Soc., Washington.

COLE, K. S. 1958. Membrane excitation of the Hodgkin-Huxley axon. Preliminary corrections. *J. appl. Physiol.*, 12:129-130.

COLE, K. S. 1961. The advance of electrical models for cells and axons. Proc. Intern. Congr. Biophysics, Stockholm (in press).

COLE, K. S. and CURTIS, H. J. 1939. Electric impedance of the squid giant axon during activity. *J. gen. Physiol.*, 22:649-670.

COLE, K. S. and CURTIS, H. J. 1941. Membrane potential of the squid giant axon during current flow. *J. gen. Physiol.*, 24:551-563.

COLE, K. S. and HODGKIN, A. L. 1939. Membrane and protoplasm resistance in the squid giant axon. *J. gen. Physiol.*, 22:671-687.

COLE, K. S., ANTOSIEWICZ, H. A., and RABINOWITZ, P. 1955. Automatic computation of nerve excitation. *J. Soc. Indust. Appl. Math.*, 3:153-172.

COOMBS, J. S., CURTIS, D. R., and ECCLES, J. C. 1959. The electrical constants of the motoneurone membrane. *J. Physiol.*, 145:505-528.

CORABOEUF, E. 1954. Analyse électrophysiologique de l'action de l'anhydride carbonique sur le nerf isolé. *J. Physiol., Paris*, 46:745-775.

CORABOEUF, E. and BOISTEL, J. 1957. Quelques aspects de la microphysiologie nerveuse chez les insectes. *Colloq. int. Cent. nat. Rech. sci.*, 1955(67):57-72.

CURTIS, H. J. and COLE, K. S. 1940. Membrane action potentials from the squid giant axon. *J. cell. comp. Physiol.*, 15:147-157.

CURTIS, H. J. and COLE, K. S. 1942. Membrane resting and action potentials from the squid giant axon. *J. cell. comp. Physiol.*, 19:135-144.

CURTIS, H. J. and COLE, K. S. 1944. Nerve: excitation and propagation. In: *Medical Physics*. Vol. I. O. Glasser (ed.). Yearbook, Chicago.

DALTON, J. C. 1959. Effects of external ions on membrane potentials of a crayfish giant axon. *J. gen. Physiol.*, 42:971-982.

DALTON, J. C. and ADELMAN, W. J. 1960. Some relations between action potential and resting potential of the lobster giant axon. *J. gen. Physiol.*, 43:597-607.

DALTON, J. C. and FITZHUGH, R. 1960. Applicability of Hodgkin-Huxley model to experimental data from the giant axon of lobster. *Science*, 131:1533-1534.

DAVIS, H. 1957. Initiation of nerve impulses in cochlea and other mechano-receptors. In: *Physiological Triggers*. T. H. Bullock (ed.). Amer. Physiol. Soc., Washington.

DAVIS, H. and FORBES, A. 1936. Chronaxie. *Physiol. Rev.*, 16:407-441.

DEL CASTILLO, J. and MOORE, J. W. 1959. On increasing the velocity of a nerve impulse *J. Physiol.*, 148:665-670.

DIECKE, F. P. J. 1954. Die "Akkomodation" des Nervenstammes und des isolierten Ranvierschen Schnürringes. *Z. Naturf.*, 98:713-729.

DIECKE, F. P. J. 1958. Nervenphysiologie. *Fortschr. Zool.*, 11:208-244.

DUDEL, J. and TRAUTWEIN, W. 1958. Der Mechanismus der automatischen rhythmischen Impulsbildung der Herzmuskelfasern. *Pflüg. Arch. ges, Physiol.*, 267:553-565.

DZHAVRISHVILI, T. D. 1959. Oscillographic analysis of the interaction between nerve fibres. *Sechenov. J. Physiol.*, 45(2):169-175.

EASTON, D. M. 1950. Synthesis of acetylcholine in crustacean nerve and nerve extract. *J. biol. Chem.*, 185(2):813-816.

EASTON, D. M. 1952. Excitability related to spike size in crab nerve fibers. *J. cell. comp. Physiol.*, 40:303-315.

ECCLES, J. C. 1957. *The Physiology of Nerve Cells*. Johns Hopkins Press, Baltimore.

ECCLES, J. C. 1959. Neuron physiology—introduction. In: *Handbook of Physiology*, Sect. 1. Neurophysiology, 1:59-74.

ECCLES, J. C. and KRNJEVIĆ, K. 1959. Potential changes recorded inside primary afferent fibres within the spinal cord. *J. Physiol.*, 149:250-273.

ECCLES, J. C., GRANIT, R., and YOUNG, J. Z. 1933. Impulses in the giant fibres of earthworms. *J. Physiol.* 77:23P-24P.

ECCLES, R. M. 1955. Intracellular potentials recorded from a mammalian sympathetic ganglion. *J. Physiol.*, 130:572-584.

ECKERT, B. and ZACHAROVÁ, D. 1957. Unterschiede zwischen dem erregenden und dem hemmenden Axon des Öffnermuskels der Krebsschere. *Physiol. bohemoslov.*, 6:39-48.

EDWARDS, C. and OTTOSON, D. 1958. The site of impulse initiation in a nerve cell of a crustacean stretch receptor. *J. Physiol.*, 143:138-148.

ENGELHARDT, A. 1951a. Die Temperaturabhängigkeit der Erregungsleitungsgeschwindigkeit im Kalt- und Warmblüternerven. *Z. vergl. Physiol.*, 33:125-128.

ENGELHARDT, A. 1951b. Abhängigkeit der Erregungsleitungsgeschwindigkeit vom Bau der Fasern des peripheren Nerven. *Z. vergl. Physiol.*, 33:378-386.

ERLANGER, J. and BLAIR. E. A. 1938. Comparative observations on motor and sensory fibers with special reference to repetitiousness. *Amer. J. Physiol.*, 121:431-453.

ERLANGER, J. and GASSER, H. S. 1937. *Electrical signs of nervous activity*. Univ. Penn., Philadelphia.

EYZAGUIRRE, C. and KUFFLER, S. W. 1955a. Processes of excitation in the dendrites and in the soma of single isolated sensory nerve cells of the lobster and crayfish. *J. gen. Physiol.*, 39:87-119.

EYZAGUIRRE, C. and KUFFLER, S. W. 1955b. Further study of soma, dendrite, and axon excitation in single neurons. *J. gen. Physiol.* 39:121-153.

FABRE, P. 1956. Variations de la conductibilité des nerfs non myélinisés pendant le passage le l'influx et la période de restauration. *C. R. Soc. Biol., Paris*, 150:177-179.

FESSARD, A. 1936a. Propriétés rythmiques de la matière vivante. Nerfs isolés. Vol. 1: Nerfs myélinisés Vol. 2: Nerfs non myélinisés. Hermann, Paris.

FESSARD, A. 1936b. *Recherches sur l'Activité Rythmique des Nerfs Isolés*. Hermann, Paris.

FESSARD, A. and TAUC, L. 1956. Capacité, résistance et variations actives d'impédance d'un soma neuronique. *J. Physiol., Paris*, 48: 541-544.

FINDLAY, G. P. 1959. Studies of action potentials in the vacuole and cytoplasm of *Nitella*. *Austr. J. Biol. Sci.*, 12:412-426.

FITZHUGH, R. 1955. Mathematical models of threshold phenomena in the nerve membrane. *Bull. math. Biophysics*, 17:257-278.

FLEISCHHACKER, H. 1925. Die Chronaxie beim Menschen. *Tabul. biol.*, Hague, 2:429-436.

FLOREY, E. 1955. Untersuchungen über die Impuls-Entstehung in den Streckrecceptoren des Flusskrebses. *Z. Naturf.*, 108:591-597.

FLOREY, E. 1962. Comparative neurochemistry: inorganic ions, amino acids and possible transmitter substances of invertebrates. In: *Neurochemistry*. 2nd ed. K. A. C. Elliott, I. H. Page, and J. H. Quastel (eds.). Thomas, Springfield, Ill.

Bibliography

FRANCK, U. F. 1956. Models for biological excitation processes. *Prog. Biophysics*, 6:171-206.

FRANK, K. 1959. Identification and analysis of single unit activity in the central nervous system. In: *Handbook of Physiology*, Sect. 1. *Neurophysiology*, 1:261-277.

FRANK, K. and FUORTES, M. G. F. 1956. Stimulation of spinal motoneurones with intracellular electrodes. *J. Physiol.*, 134:451-470.

FRANK, K. and FUORTES, M. G. F. 1961. Excitation and conduction. *Annu. Rev. Physiol.*, 23:357-386.

FRANKENHAEUSER, B. 1952. The hypothesis of saltatory conduction. *Cold Spring Harbor Symp. quant. Biol.*, 17:27-36.

FRANKENHAEUSER, B. and HODGKIN, A. L. 1956. The after-effects of impulses in the giant nerve fibres of *Loligo. J. Physiol.*, 131:341-376.

FUORTES, M. G. F. 1958a. Electric activity of cells in the eye of *Limulus. Amer. J. Opthal.*, 46(5):2:210-223.

FUORTES, M. G. F. 1958b. Generation, conduction and transmission of nerve impulses. *Arch. ital. Biol.*, 96:285-293.

FUORTES, M. G. F. 1958c. Generation of nerve impulses in receptor organs. *Electroenceph. clin. Neurophysiol.*, 10(suppl.): 71-73.

FUORTES, M. G. F. 1959. Initiation of impulses in visual cells of *Limulus. J. Physiol.*, 148:14-28.

FURNESS, F. N. (ed.). 1961. Current problems in electrobiology. *Ann. N. Y. Acad. Sci.*, 94:337-654.

GAFFEY, C. T. and MULLINS, L. J. 1958. Ion fluxes during the action potential in *Chara. J. Physiol.*, 144:505-524.

GASSER, H. S. 1931. Nerve activity as modified by temperature. *Amer. J. Physiol.*, 97:254-270.

GASSER, H. and GRUNDFEST, H. 1939. Axon diameters in relation to spike dimensions and conduction velocity in mammalian fibres. *Amer. J. Physiol.*, 127:393-414.

GERARD, R. W. 1931. Nerve conduction in relation to nerve structure. *Quart. Rev. Biol.*, 6:59-83.

GERARD, R. W. 1946. Nerve metabolism and function. A critique of the role of acetylcholine. *Ann. N. Y. Acad. Sci.*, 47:575-600.

GERARD, R. W. 1950. The acetylcholine system in neural function. *Recent Progr. Hormone Res.*, 5:37-61.

GOLDMAN, L. 1961. The effect of stretch on the conduction velocity of single nerve fibers in *Aplysia. J. cell. comp. Physiol.*, 57:185-191.

GRAHAM, J. and GERARD, R. W. 1946. Membrane potentials and excitation of impaled single muscle fibers. *J. cell. comp. Physiol.*, 28:99-117.

GRANIT, R. and SKOGLUND, C. R. 1943. Accommodation and autorhythmic mechanism in single sensory fibres. *J. Neurophysiol.*, 6:337-348.

GRAY, J. A. B. and MATTHEWS, P. B. C. 1951. A comparison of the adaptation of the Pacinian corpuscle with the accommodation of its own axon. *J. Physiol.*, 114:454-464.

GRUNDFEST, H. 1940. Bioelectric potentials. *Annu. Rev. Physiol.*, 2:213-242.

GRUNDFEST, H. 1955. The nature of the electrochemical potentials of bioelectric tissues. In: *Electrochemistry in Biology and Medicine.*, T. Shedlovsky (ed.). Wiley, N.Y.

GRUNDFEST, H. 1956. Some properties of excitable tissue. In: *Nerve Impulse.* D. Nachmansohn and H. H. Merritt (eds.). Josiah Macy, Jr. Foundation, New York.

GRUNDFEST, H. 1958. Electrophysiology and pharmacology of dendrites. In: *A Symposium on Dendrites. Electroenceph. clin. Neurophysiol.*, Suppl. 10:22-41.

GRUNDFEST, H. 1959. Evolution of conduction in the nervous system. In: *Evolution of Nervous Control.* A. D. Bass (ed.). Amer. Assoc. Adv. Sci., Washington.

GRUNDFEST, H. 1960a. Comparative studies on electrogenic membrane. In: *Inhibitions of the Nervous System and γ-Aminobutyric Acid.* Pergamon Press, Oxford.

GRUNDFEST, H. 1960b. A four-factor ionic hypothesis of spike electrogenesis. *Biol. Bull, Woods Hole*, 119:284.

GRUNDFEST, H. 1961a. Functional specifications for membranes in excitable cells. In: *Regional Neurochemistry.* S. S. Kety and J. Elkes (eds.). Pergamon Press, New York.

GRUNDFEST, H. 1961b. Ionic mechanisms in electrogenesis. *Ann. N. Y. Acad. Sci.*, 94:405-457.

GRUNDFEST, H. 1961c. Excitation by hyperpolarizing potentials. A general theory of receptor activities. In: *Nervous Inhibition.* E. Florey (ed.). Pergamon Press, Oxford.

GRUNDFEST, H., ALTAMIRANO, M., and KAO, C. Y. 1954. Local independence of bioelectric generator. *Fed. Proc.*, 13(1):208-209.

GRUNDFEST, H., KAO, C. Y., and ALTAMIRANO, M. 1954. Bioelectric effects of ions microinjected into the giant axon of *Loligo. J. gen. Physiol.*, 38:245-282.

GRUNDFEST, H., SHANES, A. M., and FREYGANG, W. 1953. The effect of sodium and potassium ions on the impedance change accompanying the spike in the squid giant axon. *J. gen. Physiol.*, 37:25-37.

HAGIWARA, S. 1960. Current-voltage relations of nerve cell membrane. In: *Electrical Activity of Single Cells.* Y. Katsuki (ed.). Igaku Shoin, Tokyo.

HAGIWARA, S. 1961. Nervous activities of the heart in crustacea. *Ergebn. Biol.*, 24:287-311.

HAGIWARA, S. and OOMURA, Y. 1958. The critical depolarization for the spike in the squid giant axon. *Jap. J. Physiol.*, 8:234-245.

HAGIWARA, S. and SAITO, N. 1959a. Voltage-current relations in nerve cell membrane of *Onchidium verruculatum. J. Physiol.*, 148:161-179.

HAGIWARA, S. and SAITO, N. 1959b. Membrane potential change and membrane current in supramedullary nerve cell of puffer. *J. Neurophysiol.*, 22:204-221.

HAGIWARA, S., KUSANO, K., and SAITO, N. 1961. Membrane changes of *Onchidium* nerve cell in potassium-rich media. *J. Physiol.*, 155:470-489.

HARKER, J. E. 1956. Factors controlling the diurnal rhythm of activity of *Periplaneta americana* L. *J. exp. Biol.*, 33:224-234.

HARKER, J. E. 1960. Internal factors controlling the suboesophageal ganglion neurosecretory cycle in *Periplaneta americana* L. *J. exp. Biol.*, 37:164-170.

HERMANN, L. 1879. Physiologie des Nervensystems. I. Allgemeine Nervenphysiologie. *Handb. Physiol.*, 2(1):1-196.

HERTZ, H. 1947. Action potential and diameter of isolated nerve fibres under various conditions. *Acta physiol. scand.*, 13(suppl. 43):1-91.

HESS, A. and YOUNG, J. Z. 1949. Correlation of internodal length and fibre diameter in the central nervous system. *Nature, Lond.*, 164:490-491.

HILL, A. V. 1910. A new mathematical treatment of changes of ionic concentration in muscle and nerve under the action of electric currents, with a theory as to their mode of excitation. *J. Physiol.*, 40:190-224.

HILL, A. V. 1932. *Chemical Wave Transmission in Nerve.* Macmillan, New York.

HILL, A. V. 1936a. Excitation and accommodation in nerve. *Proc. roy. Soc.*, (B) 119:305-355.

HILL, A. V. 1936b. The strength-duration relation for electric excitation of medullated nerve. *Proc. roy. Soc.*, (B), 119:440-453.

HÖBER, R. 1946. The membrane theory. *Ann. N. Y. Acad. Sci.*, 47:373-394.

HODES, R. 1953. Linear relationship between fiber diameter and velocity of conduction in giant axon of squid. *J. Neurophysiol.*, 16:145-154.

HODGKIN, A. L. 1937a. Evidence for electrical transmission in nerve. Part. I. *J. Physiol.*, 90:183-210.

HODGKIN, A. L. 1937b. Evidence for electrical transmission in nerve. Part. II. *J. Physiol.*, 90:211-232.

HODGKIN, A. L. 1938. The subthreshold potentials in a crustacean nerve fibre. *Proc. roy. Soc. (B)*, 126:247-285.

HODGKIN, A. L. 1939. The relation between conduction velocity and the electrical resistance outside a nerve fibre. *J. Physiol.*, 94:560-570.

HODGKIN, A. L. 1947a. The membrane resistance of a non-medullated nerve fibre. *J. Physiol.*, 106:305-318.

HODGKIN, A. L. 1947b. The effect of potassium on the surface membrane of an isolated axon. *J. Physiol.*, 106: 319-340.

HODGKIN, A. L. 1948. The local electric changes associated with repetitive action in a non-medullated axon. *J. Physiol.*, 107:165-181.

HODGKIN, A. L. 1951. The ionic basis of electrical activity in nerve and muscle. *Biol. Rev.*, 26:339-401.

HODGKIN, A. L. 1958. Ionic movements and electrical activity in nerve fibres. *Proc. roy. Soc. (B)*, 148:1-37.

HODGKIN, A. L. and HUXLEY, A. F. 1939. Action potentials recorded from inside a nerve fibre. *Nature, Lond.*, 144:710-711.

HODGKIN, A. L. and HUXLEY, A. F. 1945. Resting and action potentials in single nerve fibres. *J. Physiol.*, 104: 176-195.

HODGKIN, A. L. and HUXLEY, A. F. 1947. Potassium leakage from an active nerve fibre. *J. Physiol.*, 106:341-366.

HODGKIN, A. L. and HUXLEY, A. F. 1952a. Currents carried by sodium and potassium ions through the membrane of the giant axon of *Loligo*. *J. Physiol.*, 116:449-472.

HODGKIN, A. L. and HUXLEY, A. F. 1952b. The components of membrane conductance in the giant axon of *Loligo*. *J. Physiol.*, 116:473-496.

HODGKIN, A. L. and HUXLEY, A. F. 1952c. The dual effect of membrane potential on sodium conductance in the giant axon of *Loligo*. *J. Physiol.*, 116:497-506.

HODGKIN, A. L. and HUXLEY, A. F. 1952d. A quantitative description of membrane current and its application to conduction and excitation in nerve. *J. Physiol.*, 117:500-544.

HODGKIN, A. L. and HUXLEY, A. F. 1952e. Propagation of electrical signals along giant nerve fibres. *Proc. roy. Soc. (B)*, 140:177-183.

HODGKIN, A. L. and HUXLEY, A. F. 1953. Movement of radioactive potassium and membrane current in a giant axon. *J. Physiol.*, 121:403-414.

HODGKIN, A. L. and KATZ, B. 1949a. The effect of sodium ions on the electrical activity of the giant axon of the squid. *J. Physiol.*, 108:37-77.

HODGKIN, A. L. and KATZ, B. 1949b. The effect of temperature on the electrical activity of the giant axon of the squid. *J. Physiol.*, 109:240-249.

HODGKIN, A. L. and KATZ, B. 1949c. The effect of calcium on the axoplasm of giant nerve fibres. *J. exp. Biol.*, 26:292-294.

HODGKIN, A. L. and KEYNES, R. D. 1953. The mobility and diffusion coefficient of potassium in giant axons from *Sepia*. *J. Physiol.*, 119: 513-528.

HODGKIN, A. L. and KEYNES, R. D. 1954. Movements of cations during recovery in nerve. *Symp. Soc. exp. Biol.*, 8:423-437.

HODGKIN, A. L. and KEYNES, R. D. 1955a. Active transport of cations in giant axons from *Sepia* and *Loligo*. *J. Physiol.*, 128:28-60.

HODGKIN, A. L. and KEYNES, R. D. 1955b. The potassium permeability of a giant nerve fibre. *J. Physiol.*, 128: 61-88.

HODGKIN, A. L. and KEYNES, R. D. 1956. Experiments on the injection of substances into squid giant axons by means of a microsyringe. *J. Physiol.*, 131:592-616.

HODGKIN, A. L. and RUSHTON, W. A. H. 1946. The electrical constants of a crustacean nerve fibre. *Proc. roy. Soc. (B)*, 133:444-479.

HODGKIN, A. L., HUXLEY, A. F., and KATZ, B. 1952. Measurement of current-voltage relations in the membrane of the giant axon of *Loligo*. *J. Physiol.*, 116:424-448.

HOLMES, W., PUMPHREY, R. J., and YOUNG, J. Z. 1941. The structure and conduction velocity of the medullated nerve fibres of prawns. *J. exp. Biol.*, 18:50-54.

HORRIDGE, G. A. 1953. An action potential from the motor nerves of the jellyfish *Aurellia aurita* Lamarck. *Nature, Lond.*, 171:400.

HORRIDGE, G. A. 1954. The nerves and muscles of medusae. I. Conduction in the nervous system of *Aurellia aurita* Lamarck. *J. exp. Biol.*, 31:594-600.

HORRIDGE, G. A. 1955. The nerves and muscles of medusae. III. A decrease in the refractory period following repeated stimulation of the muscle of *Rhizostoma pulmo*. *J. exp. Biol.*, 32:636-641.

HUGHES, G. M. 1952. Differential effects of direct current on insect ganglia. *J. exp. Biol.*, 29:387-402.

HUGHES, G. M. and KERKUT, G. A. 1956. Electrical activity in a slug ganglion in relation to the concentration of Locke solution. *J. exp. Biol.*, 33:282-294.

HURSH, J. B. 1939. Conduction velocity and diameter of nerve fibres. *Amer. J. Physiol.*, 127:131-139.

HUXLEY, A. F. and STÄMPFLI, R. 1949. Evidence for saltatory conduction in peripheral myelinated nerve fibers. *J. Physiol.*, 108:315-339.

HUXLEY, A. F. and STÄMPFLI, R. 1950. Saltatory transmission of the nervous impulse. In: *Électrophysiologie, Colloq. int. Cent. nat. Rech. sci.*, 22:309-322.

ISHIKO, N. and LOEWENSTEIN, W. R. 1960. Temperature and charge transfer in a receptor membrane. *Science*, 132:1841-1842.

ITO, M. 1957. The electrical activity of spinal ganglion cells investigated with intracellular microelectrodes. *Jap. J. Physiol.*, 7:297-323.

JASPER, H. H. 1956. Properties of nerve impulses. In: *Nerve Impulse*. D. Nachmansohn and H. H. Merritt (eds.). Josiah Macy, Jr. Foundation, New York.

JENKINS, O. P. and CARLSON, A. J. 1902. The rate of nervous impulse in certain molluscs. *Amer. J. Physiol.*, 8:251-268.

JENKINS, O. P. and CARLSON, A. J. 1904a. The rate of the nervous impulse in the ventral nerve-cord of certain worms. *J. comp. Neurol.*, 13:259-289.

JENKINS, O. P. and CARLSON, A. J. 1904b. Physiological evidence of the fluidity of the conducting substance in the pedal nerves of the slug—*Ariolimax columbianus*. *J. comp. Neurol.*, 14:85-92.

JORDAN, H. J. and LULLIES, H. 1933. Leitung und refraktäre Periode bei den Fussnerven von *Aplysia limacina*. *Z. vergl. Physiol.*, 19:648-665.

KAN, I. L. 1937. The problem of the evolution of conduction in the nervous system. (The VIth All-Union Congress of Physiologists, Biochemists, and Pharmacologists; Collection of reports, 82. In Russian.)

KAN, I. L. and KUSNEZOV, D. P. 1938. Decrement conduction of excitation in non-medullated nerve commissure of *Anodonta*. *Bull. Biol. Med. exp. URSS.* (*Biull. eksp. Biol. Med.*), 6:3-5. (In Russian.)

KAO, C. Y. and GRUNDFEST, H. 1955. Graded response in squid giant axon. *Biol. Bull. Woods Hole*, 109:348.

KARAEV, A. 1938. Range of propagation of local subliminal changes in the nerve. *Bull. Biol. Med. exp. URSS.* (*Biull. eksp. Biol. Med.*) 6:7-9. (In Russian.)

KATSUKI, Y. (ed.). 1960. *Electrical Activity of Single Cells*. Igaku Shoin, Tokyo.

KATSUKI, Y., CHEN, J., and TAKEDA, H. 1954. Fundamental neural mechanism of the sense organ. *Bull. Tokyo med. dent. Univ.*, 1:21-31.

KATZ, B. 1937. Experimental evidence for a non-conducted response of nerve to subthreshold stimulation. *Proc. roy. Soc.* (B), 124:244-276.

KATZ, B. 1939. *Electric Excitation of Nerve*. Oxford Univ. Press, London.

KATZ, B. 1947a. Subthreshold potentials in medullated nerve. *J. Physiol.*, 106:66-79.

KATZ, B. 1947b. The effect of electrolyte deficiency on the rate of conduction in a single nerve fibre. *J. Physiol.*, 106:411-417.

KENNEDY, D. 1960. Neural photoreception in a lamellibranch mollusc. *J. gen. Physiol.*, 44:277-299.

KERKUT, G. A. and TAYLOR, J. R. 1956. The sensitivity of the pedal ganglion of the slug to osmotic pressure changes. *J. exp. Biol.*, 33:493-501.

Bibliography

KERKUT, G. A. and TAYLOR, B. J. R. 1958. The effect of temperature changes on the activity of poikilotherms. *Behaviour*, 13:259-279.

KERKUT, G. A. and WALKER, R. J. 1961. The resting potential and potassium levels of cells from active and inactive snails. *Comp. Biochem. Physiol.*, 2:76-79.

KEYNES, R. D. 1951a. The leakage of radioactive potassium from stimulated nerve. *J. Physiol.*, 113:99-114.

KEYNES, R. D. 1951b. The ionic movements during nervous activity. *J. Physiol.*, 114:119-150.

KEYNES, R. D. and LEWIS, P. R. 1951a. The resting exchange of radioactive potassium in crab nerve. *J. Physiol.*, 113:73-98.

KEYNES, R. D. and LEWIS, P. R. 1951b. The sodium and potassium content of cephalopod nerve fibres. *J. Physiol.*, 114:151-182.

KEYNES, R. D. and LEWIS, P. R. 1956. The intracellular calcium contents of some invertebrate nerves. *J. Physiol.*, 134:399-407.

KIKUCHI R., NAITO, K., and MINAGAWA, S. 1960. Summative action of acetylcholine with physiological stimulus on the generator potential in the lateral eye of the horseshoe crab. *Nature, Lond.*, 187:1118-1119.

KITAMURA, S. 1960. Osmotic pressure and electrical activity. *Jap. J. Physiol.*, 10:51-63.

KOECHLIN, B. A. 1954. The isolation and identification of the major anion fraction of the axoplasm of squid giant nerve fibers. *Proc. nat. Acad. Sci., Wash.*, 40:60-62.

KOECHLIN, B. A. 1955. On the chemical composition of the axoplasm of squid giant nerve fibers with particular reference to its ion pattern. *J. biophys. biochem. Cytol.*, 1:511-530.

KOSHTOYANTS, KH. S. 1957. *Comparative Physiology of the Nervous System*. Vol. II. Publ. House of the Academy of Sciences, Moscow. (In Russian.)

KUGELBERG, E. 1944. Accommodation in human nerves and its significance for the symptoms in circulatory disturbances and tetany. *Acta physiol. scand.*, 8(suppl.24):1-105.

LAPICQUE, L. 1903a. Sur la loi d'excitation électrique chez quelques invertébrés. *C. R. Acad. Sci., Paris*, 136:1147-1148.

LAPICQUE, L. 1903b. Expression nouvelle de la loi d'excitation électrique. *C. R. Acad. Sci., Paris*, 136:1477-1479.

LAPICQUE, L. 1926. *L'Excitabilité en Fonction du Temps*. Hermann, Les Presses Universitaires de France, Paris.

LAPICQUE, M. and VEIL, C. 1925. Vitesse de conduction nerveuse et musculaire comparée à la chronaxie chez la sangsue et le ver de terre. *C. R. Soc. Biol., Paris*, 93:1590-1591.

LAVIGNE, S. and CORABOEUF, E. 1955. Action comparée du gaz carbonique sur les fibres nerveuses motrices et sensitives des crustacés. *J. Physiol., Paris*, 47:209-211.

LE FEVRE, P. G. 1948. Comparison of frog nerve and squid axon with respect to the measurement of accommodation. *Biol. Bull., Woods Hole*, 95:256-257.

LE FEVRE, P. G. 1950. Excitation characteristics of the squid giant axon: a test of excitation theory in a case of rapid accommodation. *J. gen. Physiol.*, 34:19-36.

LEV, A. A., NIKOLSKY, N. N., ROSENTAL, D. L., and SHAPIRO, E. A. 1959a. Dependence between intensity of stimulation and amount of local electrical response of a single nerve fiber of squid. *Tsitologiia (Akad. Nauk SSSR) Moscva*, 1:94-104. (In Russian.)

LEV, A. A., NIKOLSKY, N. N., ROSENTAL, D. L., and SHAPIRO, E. A. 1959b. Propagation of excitation in giant nerve fiber of Pacific squid. *Tsitologiia (Akad. Nauk SSSR) Moscva*, 1:665-671. (In Russian.)

LEV, A. A., NIKOLSKY, N. N., ROSENTAL, D. L., SVINKIN, V. B., and SHAPIRO, E. A. 1960. Single giant fiber of Pacific squid as an object of electrophysiological investigations. In: *Viprosy tsitologii i protistologii Sbornik rabot*. V.I. Vorob'ev (ed.). Akad. Nauk SSSR, Moscow. (In Russian.)

LEVIN, A. 1927. Fatigue, retention of action current and recovery in crustacean nerve. *J. Physiol.*, 63:113-129.

LEWIS, P. R. 1952. The free amino-acids of invertebrate nerve. *Biochem. J.*, 52:330-338.

LIBET, B. 1947. Localization of adenosinetriphosphatase (ATP-ase) in the giant nerve fiber of the squid. *Biol. Bull., Woods Hole*. 93:219.

LIBET, B. 1948a. Enzyme localization in the giant nerve fiber of the squid. *Biol. Bull., Woods Hole*, 95:277-278.

LIBET, B. 1948b. Adenosinetriphosphatase (ATP-ase) in nerve. *Fed. Proc.*, 7:72.

LILLIE, R. S. 1920. The recovery of transmissivity in passive iron wires as a model of recovery processes in irritable living systems. Part I. *J. gen. Physiol.*, 3:107-128.

LILLIE, R. S. 1936. The passive iron wire model of protoplasmic and nervous transmission and its physiological analogues. *Biol. Rev.*, 11:181-209.

LOEWENSTEIN, W. R. 1959. The generation of electric activity in a nerve ending. *Ann. N. Y. Acad. Sci.*, 81:367-387.

LOEWENSTEIN, W. R. and COHEN, S. 1959a. I. After-effects of repetitive activity in a nerve ending. *J. gen. Physiol.*, 43:335-345.

LOEWENSTEIN, W. R. and COHEN, S. 1959b. II. Post-tetanic potentiation and depression of generator potential in a single non-myelinated nerve ending. *J. gen. Physiol.*, 43:347-376.

LOEWENSTEIN, W. R. and RATHKAMP, R. 1958. Localization of generator structures of electric activity in a Pacinian corpuscle. *Science*, 127:341.

LORENTE DE NÓ, R. 1946-1947. Correlation of nerve activity with polarization phenomena. *Harvey Lect.*, 42:43-105.

LORENTE DE NÓ, R. 1947. *A Study of Nerve Physiology*. Rockefeller Institute, New York.

LUCAS, K. 1907. On the rate of variation of the exciting current as a factor in electric excitation. *J. Physiol.*, 36:253-274.

LUCAS, K. 1917. *The Conduction of the Nervous Impulse*. Longmans, Green, London. (Rev. by E. D. Adrian.)

LULLIES, H. 1932. Allgemeine Nervenphysiologie. 1925-1930. *Tabul. biol., Hague*, 8:56-107.

LULLIES, H. 1934. Aktionsströme und Fasergruppen im Extremitätennerven von *Maja squinado*. *Pflüg. Arch. ges. Physiol.*, 233:584-606.

LULLIES, H. 1952. Über "Reizgesetze" und unsere Vorstellungen von den Vorgängen bei der Erregung des Nerven. *Ergebn. Physiol.*, 47:1-23.

MARMONT, G. 1940. Action potential artifacts from single nerve fibers. *Amer. J. Physiol.*, 130:392-402.

MARMONT, G. 1949. Studies on the axon membrane. I. A new method. *J. cell. comp. Physiol.*, 34:351-382.

MARTIN, A. R. 1954. Effect of change in length on conduction velocity in muscle. *J. Physiol.*, 125:215-220.

MAXFIELD, M. 1951. Studies of nerve proteins. Isolation and physicochemical characterization of a protein from lobster nerve. *J. gen. Physiol.*, 34:853-863.

MAXFIELD, M. 1953. Axoplasmic proteins of the squid giant nerve fiber with particular reference to the fibrous protein. *J. gen. Physiol.*, 37:201-216.

MAXFIELD, M. and HARTLEY, R. W., JR. 1955. Two proteins purified from lobster nerve extract. Isolation and physiochemical characterization. *J. biophys. biochem. Cytol.*, 1:279-286.

MAXFIELD, M. and HARTLEY, R. W., JR. 1957. Dissociation of the fibrous protein of nerve. *Biochem. biophys. Acta*, 24:83-87.

MCALEAR, J. H., MILBURN, N. S. and CHAPMAN, G. B. 1958. The fine structure of Schwann cells, nodes of Ranvier and Schmidt-Lanterman incisures in the central nervous system of the crab, *Cancer irroratus*. *J. Ultrastruct. Res.*, 2:171-176.

MERRITT, H. H. (ed.). 1952. *Nerve Impulse*. Josiah Macy, Jr. Foundation, New York.

MEVES, H. 1955. Die Wirkung der Wasserstoffionen und der Kohlensäure auf die Nervenleitungsgeschwindigkeit. *Pflüg. Arch. ges. Physiol.*, 261:249-263.

MEVES, H. 1960. Die Nachpotentiale isolierter markhaltiger Nervenfasern des Frosches bei Einzelreizung. *Pflüg. Arch. ges. Physiol.*, 271:655-679.

MONNIER, A. M. 1934. *L'Excitation Électrique des Tissus.* Hermann, Paris.

MONNIER, A. M. 1950. Le facteur d'amortissement des processus d'excitation. Sa mesure. Sa signification fonctionnelle. In: *Électrophysiologie, Colloq. int. Cent. nat. Rech. sci.*, 22:247-272.

MONNIER, A. M. 1952. The damping factor as a functional criterion in nerve physiology. *Cold Spr. Harb. Symp. quant. Biol.*, 17:69-92.

MONNIER, A. M. and DUBUISSON, M. 1931. Étude à l'oscillographe cathodique des nerfs pédieux de quelques arthropodes. *Arch. int. Physiol.*, 34:25-57.

MOORE, A. R. and TURNER, R. S. 1948. Some effects of narcosis on the nerve impulse in the squid, *Loligo opalescens*. *Physiol. Zoöl.*, 21:224-231.

MORGAN, T. 1943. *Physiological Psychology.* McGraw-Hill, New York.

MORITA, H. 1959. Initiation of spike potentials in contact chemosensory hairs of insects. III. D. C. stimulation and generator potential of labellar chemoreceptor of *Calliphora*. *J. cell. comp. Physiol.*, 54:189-204.

MORITA, H. and TAKEDA, K. 1959. Initiation of spike potentials in contact chemosensory hairs of insects. II. The effect of electric current on tarsal chemosensory hairs of *Vanessa*. *J. cell. comp. Physiol.*, 54:177-187.

MORITA, H. and YAMASHITA, S. 1959. Generator potential of insect chemoreceptor. *Science*, 130:922.

MORITA, H., DOIRA, S., TAKEDA, K., and KUWABARA, M. 1957. Electrical response of contact chemoreceptor on tarsus of the butterfly, *Vanessa indica*. *Mem. Fac. Sci. Kyūshū Univ. (E) (Biol.)*, 2:119-139.

MOZHAEVA, G. N. 1958. The influence of the strength of stimulation on the magnitude of local electric reaction in a nerve. *Biofizika*, 3:286-293.

MULLINS, L. J. 1959. An analysis of conductance changes in squid axon. *J. gen. Physiol.*, 42:1013-1035.

MULLINS, L. J. 1960. An analysis of pore size in excitable membranes. *J. gen. Physiol.*, 43:105-117.

MULLINS, L. J. 1961. The macromolecular properties of excitable membranes. *Ann. N. Y. Acad. Sci.*, 94:390-404.

MURALT, A. VON. 1946. *Die Signalübermittlung im Nerven.* Birkhäuser, Basel.

MURALT, A. VON. 1958. *Neue Ergebnisse der Nervenphysiologie.* Springer, Berlin.

NACHMANSOHN, D. 1946. Chemical mechanism of nerve activity. *Ann. N. Y. Acad. Sci.*, 47:395-428.

NACHMANSOHN, D. 1950. Studies on permeability in relation to nerve function. I. Axonal conduction and synaptic transmission. *Biochim. biophys. Acta*, 4:78-95.

NACHMANSOHN, D. (ed.). 1951. *Nerve Impulse.* Josiah Macy, Jr. Foundation, New York.

NACHMANSOHN, D. 1952. Chemical mechanisms of nerve activity. In: *Modern Trends in Physiology and Biochemistry.* E. S. G. Barron (ed.). Academic Press, New York.

NACHMANSOHN, D. 1955. Metabolism and function of the nerve cell. *Harvey Lect.*, 49:57-99.

NACHMANSOHN, D. 1959. *Chemical and Molecular Basis of Nerve Activity.* Academic Press, N.Y.

NACHMANSOHN, D. and MERRITT, H. H. (eds.). 1954. *Nerve Impulse.* Josiah Macy, Jr. Foundation, New York.

NACHMANSOHN, D. and WILSON, I. B. 1955. Molecular basis for generation of bioelectric potentials. In: *Electrochemistry in Biology and Medicine.*, T. Shedlovsky (ed.). Wiley, New York.

NARAHASHI, T. 1960. Excitation and electrical properties of giant axon of cockroaches. In: *Electrical Activity of Single Cells.* Y. Katsuki (ed.). Igaku Shoin, Tokyo.

NARAHASHI, T. 1961. Effect of barium ions on membrane potentials of cockroach giant axons. *J. Physiol.*, 156:389-414.

NARAHASHI, T. and YAMASAKI, T. 1960a. Behaviors of membrane potential in the cockroach giant axons poisoned by DDT. *J. cell. comp. Physiol.*, 55:131-142.

NARAHASHI, T. and YAMASAKI, T. 1960b. Mechanism of the afterpotential production in the giant axons of the cockroach. *J. Physiol.*, 151:75-88.

NARAHASHI, T. and YAMASAKI, T. 1960c. Mechanism of increase in negative after-potential by dicophanum (DDT) in the giant axons of the cockroach. *J. Physiol.*, 152:122-140.

NERNST, W. 1908. Zur Theorie des elektrischen Reizes. *Pflüg. Arch. ges. Physiol.*, 122:275-314.

NEVIS, A. H. 1958. Water transport in invertebrate peripheral nerve fibers. *J. gen. Physiol.*, 41:927-958.

NICOL, J. A. C. and WHITTERIDGE, D. 1955. Conduction in the giant axon of *Myxicola infundibulum*. *Physiol. comp.*, 4:101-117.

NIKI, I., SONE, T., and HOSOI, E. 1953. Study on the strength-duration curve of *Pheretima* nerve. *Zool. Mag., Tokyo.*, 62:253-256. (In Japanese.)

NISHI, S. and KOKETSU, K. 1960. Electrical properties and activities of single sympathetic neurons in frogs. *J. cell. comp. Physiol.*, 55:15-30.

OTANI, T. 1960. Excitation and accommodation in toad's spinal motoneuron. In: *Electrical Activity of Single Cells.* Y. Katsuki (ed.). Igaku Shoin, Tokyo.

OTANI, T. and BULLOCK, T. H. 1959. Effects of presetting the membrane potential of the soma of spontaneous and integrating ganglion cells. *Physiol. Zoöl.*, 32:104-114.

OTTOSON, D. 1959. Olfactory bulb potentials induced by electrical stimulation of the nasal mucosa in the frog. *Acta physiol. scand.*, 47:160-172.

PEASE, D. C. 1955. Nodes of Ranvier in the central nervous system. *J. comp. Neurol.*, 103:11-16.

PROSSER, C. L. and CHAMBERS, A. H. 1938. Excitation of nerve fibers in the squid *(Loligo pealii)*. *J. gen. Physiol.*, 21:781-794.

PUMPHREY, R. J. and YOUNG, J. Z. 1938. The rates of conduction of nerve fibres of various diameters in cephalopods. *J. exp. Biol.*, 15:453-466.

PUMPHREY, R. J., SCHMITT, O. H., and YOUNG, J. Z. 1940. Correlation of local excitability with local physiological response in the giant axon of the squid *(Loligo)*. *J. Physiol.*, 98:47-72.

RALL, W. 1959a. Dendritic current distribution and whole neuron properties. *Naval Med. Res. Inst., Res. Report* 17:479-525.

RALL, W. 1959b. Branching dendritic trees and motoneuron membrane resistivity. *Exp. Neurol.*, 1:491-527.

RASHBASS, C. and RUSHTON, W. A. H. 1949. The relation of structure to the spread of excitation in the frog's sciatic trunk. *J. Physiol.*, 110:110-135.

RASHEVSKY, N. 1936. Physico-mathematical aspects of excitation and conduction in nerves. *Cold Spr. Harb. Symp.*, 4:90-97.

RASHEVSKY, N. 1960. *Mathematical Biophysics.* 3rd rev. ed. Dover, New York. 2v.

REUBEN, J. P., WERMAN, R., and GRUNDFEST, H. 1960. Properties of indefinitely prolonged spikes of lobster muscle fibers. *Biol. Bull., Woods Hole*, 119:336.

RICHARDS, A. G., Jr. and CUTKOMP, L. K. 1945. The cholinesterase of insect nerves. *J. cell. comp. Physiol.*, 26(1):57-61.

ROGERS, W. A. 1954. *Introduction to Electric Fields.* McGraw-Hill, New York.

ROSENBERG, H. 1925. Allgemeine Nerven-Physiologie. Aufbau, Physikalische Konstanten. *Tabul. biol.*, Hague, 2:384-429.

Bibliography

ROSENBERG, H. 1934. The strength-duration curves of two non-medullated nerves. *J. Physiol.*, 83:23P-24P.

ROSENBLUETH, A. and LUCO, J. V. 1950. The local responses of myelinated mammalian axons. *J. cell. comp. Physiol.*, 36:289-331.

ROSENBLUETH, A., ALANIS, J., and MANDOKI, J. 1949. The functional refractory period of axons. *J. cell. comp. Physiol.*, 33:405-440.

ROSENBLUETH, A., WIENER, N., PITTS, W., and GARCIA RAMOS, J. 1948. An account of the spike potential of axons. *J. cell. comp. Physiol.*, 32:272-317.

ROTHENBERG, M. A. 1950. Studies on permeability in relation to nerve function. II. Ionic movements across axonal membranes. *Biochim. biophys. Acta.* 4:96-114.

RUSHTON, W. A. H. 1927. Effect upon the threshold for nervous excitation of the length of nerve exposed and the angle between current and nerve. *J. Physiol.*, 63:357-377.

RUSHTON, W. A. H. 1935. The time factor in electrical excitation. *Biol. Rev.*, 10:1-17.

RUSHTON, W. A. H. 1951. A theory of the effects of fibre size in medullated nerve. *J. Physiol.*, 115:101-122.

RUSHTON, W. A. H. and BARLOW, H. B. 1943. Single-fibre response from an intact animal. *Nature, Lond.*, 152:597-598.

SASAKI, K. and OTANI, T. 1961. Accommodation in spinal motoneurons of the cat. *Jap. J. Physiol.*, 11:443-456.

SCHAEFER, H. 1940. *Elektrophysiologie.* Vol. I and II. Deuticke, Wien. (Edwards Bros., Ann Arbor, 1944).

SCHALLEK, W. 1945. Action of potassium on bound acetylcholine in lobster nerve cord. *J. cell. comp. Physiol.*, 26:15-24.

SCHLOTE, F. W. 1955. Die Erregungsleitung im Gastropodennerven und ihr histologisches Substrat. *Z. Vergl. Physiol.*, 37:373-415.

SCHMIDT, H. and STÄMPFLI, R. 1957. Die Depolarisation durch Calcium-Mangel und ihre Abhängigkeit von der Kalium-Konzentration. *Helv. physiol. acta*, 15:200-211.

SCHMIDT, H. and STÄMPFLI, R. 1959. Der Einfluss aniso-osmotischer Ringerlösungen auf das Membranpotential markhaltiger Nervenfasern. *Helv. physiol. acta*, 17:219-235.

SCHMITT, F. O. and SCHMITT, O. H. 1940. Partial excitation and variable conduction in the squid giant axon. *J. Physiol.*, 98:26-46.

SCHMITT, F. O., BEAR, R. S., and SILBER, R. H. 1939. Organic and inorganic electrolytes in lobster nerves. *J. cell. comp. Physiol.*, 14:351-356.

SCHMITT, O. H. 1950. Some low frequency characteristics of axoplasm and the nerve membrane. *Biol. Bull.*, 99:344.

SCHULTZ, R., BERKOWITZ, E. C., and PEASE, D. C. 1956. The electron microscopy of the lamprey spinal cord. *J. Morph.*, 98:251-274.

SEGAL, J. 1958. *Die Erregbarkeit der Lebenden Materie.* Fischer, Jena.

SEGAL, J. R. 1958. An anodal threshold phenomenon in the squid giant axon. *Nature, Lond.*, 182:1370.

SHANES, A. M. 1949a. Electrical phenomena in nerve. I. Squid giant axon. *J. gen. Physiol.*, 33:57-73.

SHANES, A. M. 1949b. Electrical phenomena in nerve. II. Crab nerve. *J. gen. Physiol.*, 33:75-102.

SHANES, A. M. 1950. Potassium retention in crab nerve. *J. gen. Physiol.*, 33:643-649.

SHANES, A. M. 1951. Potassium movement in relation to nerve activity. *J. gen. Physiol.*, 34:795-807.

SHANES, A. M. 1954. Effect of temperature on potassium liberation during nerve activity. *Amer. J. Physiol.*, 177:377-382.

SHANES, A. M. 1958. Electrochemical aspects of physiological and pharmacological action in excitable cells. I. The resting cell and its alteration by extrinsic factors. II. The action potential and excitation. *Pharmacol. Rev.*, 10:59-273.

SHANES, A. M. and BERMAN, M. D. 1955. Kinetics of ion movement in the squid giant axon. *J. gen. Physiol.*, 39:279-300.

SHANES, A. M. and HOPKINS, H. S. 1948. Effect of potassium on "resting" potential and respiration of crab nerve. *J. Neurophysiol.*, 11:331-342.

SHANES, A. M., GRUNDFEST, H., and FREYGANG, W. 1953. Low level impedance changes following the spike in the squid giant axon before and after treatment with "veratrine" alkaloids. *J. gen. Physiol.*, 37:39-51.

SHANES, A. M., FREYGANG, W. H., GRUNDFEST, H., and AMATNIEK, E. 1959. Anesthetic and calcium action in the voltage clamped squid giant axon. *J. gen. Physiol.*, 42:793-802.

SHAW, F. H., SIMON, S. E., and JOHNSTONE, B. M. 1956. The non-correlation of bioelectric potentials with ionic gradients. *J. gen. Physiol.*, 40:1-17.

SHAW, F. H., SIMON, S. E., JOHNSTONE, B. M., and HOLMAN, M. E. 1956. The effect of changes of environment on the electrical and ionic pattern of muscle. *J. gen. Physiol.*, 40:263-288.

SHEDLOVSKY, T. (ed.). 1955. *Electrochemistry in Biology and Medicine.* Wiley, New York.

SILBER, R. H. 1941. The free amino acids of lobster nerve. *J. cell. comp. Physiol.*, 18:21-30.

SILBER, R. H. and SCHMITT, F. O. 1940. The role of free amino acids in the electrolyte balance of nerve. *J. cell. comp. Physiol.*, 16:247-254.

SIMON, S. E., SHAW, F. H., BENNETT, S., and MULLER, M. 1957. The relationship between sodium, potassium, and chloride in amphibian muscle. *J. gen. Physiol.*, 40:753-777.

SJODIN, R. A. and MULLINS, L. J. 1958. Oscillatory behavior of the squid axon membrane potential. *J. gen. Physiol.*, 42:39-47.

SKOGLUND, C. R. 1942. The response to linearly increasing currents in mammalian motor and sensory nerves. *Acta physiol. scand.*, 4 (suppl. 12):1-75.

SKOU, J. C. 1957. The influence of some cations on an adenosine triphosphatase from peripheral nerves. *Biochem. Biophys. Acta*, 23:394-401.

SOLANDT, D. Y. 1936. The measurement of "accommodation" in nerve. *Proc. roy. Soc. (B)*, 119:355-379.

SPYROPOULOS, C. S. 1957. The effects of hydrostatic pressure upon the normal and narcotized nerve fiber. *J. gen. Physiol.*, 40:849-857.

SPYROPOULOS, C. S. 1959. Miniature responses under "voltage-clamp." *Amer. J. Physiol.*, 196:783-790.

STÄMPFLI, R. 1954. Saltatory conduction in nerve. *Physiol. Rev.*, 34:101-112.

STÄMPFLI, R. 1958a. La conduction saltatoire de l'influx nerveux. *Pathol. Biol.*, 6:1909-1918.

STÄMPFLI, R. 1958b. Die Strom-Spannungs-Charakteristik der erregbaren Membran eines einzelnen Schnürrings und ihre Abhängigkeit von der Ionenkonzentration. *Helv. physiol. acta*, 16:127-145.

STÄMPFLI, R. 1959a. Is the resting potential of Ranvier nodes a potassium potential? *Ann. N. Y. Acad. Sci.*, 81:265-284.

STÄMPFLI, R. 1959b. Current-voltage curves of Ranvier nodes. In: *La Méthode des Indicateurs Nucléaires dans l'Étude des Transports Actifs d'Ions.* Pergamon Press, London.

STEINBACH, H. 1940. Chemical and concentration potentials in the giant fibers of squid nerves. *J. cell. comp. Physiol.*, 15:373-386.

STEINBACH, H. 1941. Chloride in the giant axons of the squid. *J. cell. comp. Physiol.*, 17:57-64.

STEINBACH, H. B. and SPIEGELMAN, S. 1943. The sodium and potassium balance in squid nerve axoplasm. *J. cell. comp. Physiol.*, 22:187-196.

STEINBACH, H. B., SPIEGELMAN, S. and KAWATA, N. 1944. The effects of potassium and calcium on the electrical properties of squid axons. *J. cell. comp. Physiol.*, 24:147-154.

STRUMWASSER, F. and ROSENTHAL, S. 1960. Prolonged and patterned direct extracellular stimulation of single neurons. *Amer. J. Physiol.*, 198:405-413.

TASAKI, I. 1950. The threshold conditions in electrical excitation of the nerve fiber. II. *Cytologia, Tokyo.* 15:219-263.

TASAKI, I. 1953. *Nervous Transmission.* Thomas, Springfield.

TASAKI, I. 1958a. Electric response of glia cells. *Science,* 128:1209-1210.

TASAKI, I. 1958b. Electrical responses of astrocytic glia from the mammalian central nervous system cultivated in vitro. *Experientia,* 14:220-226.

TASAKI, I. 1959a. Conduction of the nerve impulse. In: *Handbook of Physiology.* Sect. 1. *Neurophysiology.* 1:75-121.

TASAKI, I. 1959b. Demonstration of two stable states of the nerve membrane in potassium-rich media. *J. Physiol.,* 148:306-331.

TASAKI, I. 1959c. Resting and action potentials of reversed polarity in frog nerve cells. *Nature, Lond.,* 184:1574-1575.

TASAKI, I. and BAK, A. F. 1957. Oscillatory membrane currents of squid giant axon under voltage-clamp. *Science,* 126:696-697.

TASAKI, I. and BAK, A. F. 1958a. Current-voltage relations of single nodes of Ranvier as examined by voltage-clamp technique. *J. Neurophysiol.,* 21:124-137.

TASAKI, I. and BAK, A. F. 1958b. Discrete threshold and repetitive responses in the squid axon under "voltage-clamp". *Amer. J. Physiol.,* 193:301-308.

TASAKI, I. and BAK, A. F. 1959. Voltage clamp behavior of iron-nitric acid system as compared with that of nerve membrane. *J. gen. Physiol.,* 42:899-915.

TASAKI, I. and FREYGANG, W. H. 1955. The parallelism between the action potential, action current, and membrane resistance at a node of Ranvier. *J. gen. Physiol.,* 39:211-223.

TASAKI, I. and FUJITA, M. 1948. Action currents of single nerve fibers as modified by temperature changes. *J. Neurophysiol.,* 11:311-315.

TASAKI, I. and HAGIWARA, S. 1957. Demonstration of two stable potential states in the squid giant axon under tatraethylammonium chloride. *J. gen. Physiol.,* 40:859-885.

TASAKI, I. and SATO, M. 1951. On the relation of the strength-frequency curve in excitation by alternating current to the strength duration and latent addition curves of the nerve fiber. *J. gen. Physiol.,* 34:373-388.

TASAKI, I. and SPYROPOULOS, C. S. 1958a. Nonuniform response in the squid axon membrane under "voltage-clamp". *Amer. J. Physiol.,* 193:309-317.

TASAKI, I. and SPYROPOULOS, C. S. 1958b. Membrane conductance and current-voltage relation in the squid axon under "voltage-clamp." *Amer. J. Physiol.,* 193:318-327.

TASAKI, I., TEORELL, T., and SPYROPOULOS, C. S. 1961. Movement of radioactive tracers across squid axon membrane. *Amer. J. Physiol.,* 200:11-22.

TATEDA, H. and MORITA, H. 1959. Initiation of spike potentials in contact chemosensory hairs of insects. I. The generation site of the recorded spike potentials. *J. cell. comp. Physiol.,* 54:171-176.

TAUC, L. 1952. Potentiels de repos et l'action des cellules ganglionnaires de l'écrevisse étudiés à l'aide de microélectrodes capillaires. *C. R. Soc. Biol., Paris,* 146:1668-1670.

TAUC, L. 1954. Réponse de la cellule nerveuse du ganglion abdominal de *Aplysia depilans* à la stimulation directe intracellulaire. *C.R. Acad. Sci. Paris,* 239:1537-1539.

TAUC, L. 1955a. Étude de l'activité élémentaire des cellules du ganglion abdominal de l'aplysie. *J. Physiol., Paris,* 47:769-792.

TAUC, L. 1955b. Divers aspects de l'activité électrique spontanée de la cellule nerveuse du ganglion abdominal de l'aplysie. *C.R. Acad. Sci., Paris,* 240:672-674.

TAUC, L. 1955c. Activités électriques fractionnées observées dans les cellules ganglionnaires de l'escargot (*Helix pomatia*). *C.R. Acad. Sci., Paris,* 241:1070-1073.

TAUC, L. 1955d. Les divers modes d'activité du soma neuronique ganglionnaire de l'aplysie et de l'escargot. In: *Microphysiologie comparée des éléments excitables.* C.N.R.S. (ed.). 67th Colloque International, Paris.

TAUC, L. 1956. Potentiels sous-liminaires dans le soma neuronique de l'aplysie et de l'escargot. *J. Physiol., Paris,* 48:715-718.

TAUC, L. 1958a. Processus post-synaptiques d'excitation et d'inhibition dans le soma neuronique de l'aplysie et de l'escargot. *Arch. ital. Biol.,* 96:78-110.

TAUC, L. 1958b. Action d'un choc hyperpolarisant sur le potentiel d'action et le stade réfractaire du soma neuronique ganglionnaire de l'aplysie. *C.R. Acad. Sci., Paris,* 246:2045-2048.

TAUC, L. 1960. The site of origin of the efferent action potentials in the giant nerve cell of *Aplysia. J. Physiol.,* 152:36P-37P.

TAUC, L. 1962a. Identification of active membrane areas in the giant neuron of *Aplysia. J. gen. Physiol.,* 45:1099-1115.

TAUC, L. 1962b. Site of origin and propagation of spike in the giant neuron of *Aplysia. J. gen. Physiol.,* 45:1077-1097.

TAUC, L. and HUGHES, G. 1961. Sur la distribution partielle des influx efférents dans les ramifications d'un axone non myélinisé. *J. gen. Physiol.,* 53:483.

TAYLOR, R. E. 1959. Effect of procaine on electrical properties of squid axon membrane. *Amer. J. Physiol.,* 196:1071-1078.

TEORELL, T. 1950. Membrane electrophoresis in relation to bio-electrical polarization effects. In: *Électrophysiologie, Colloq. int. Cent. nat. Rech. sci.,* 22:83-97.

TEORELL, T. 1953. Transport processes and electrical phenomena in ionic membranes. *Progr. Biophys.,* 3:305-369.

TERZUOLO, C. A. and ARAKI, T. 1961. An analysis of intra- versus extracellular potential changes associated with activity of single spinal motoneurons. *Ann. N. Y. Acad. Sci.,* 94:547-558.

TERZUOLO, C. A. and BULLOCK, T. H. 1956. Measurement of imposed voltage gradient adequate to modulate neuronal firing. *Proc. nat. Acad. Sci., Wash.,* 42:687-694.

TERZUOLO, C. A. and WASHIZU, Y. 1962. Relation between stimulus strength, generator potential and impulse frequency in stretch receptor of Crustacea. *J. Neurophysiol.,* 25:56-66.

THIES, R. E. 1957. Electrical recording in the living squid. *Biol. Bull., Woods Hole,* 113:333-334.

TOBIAS, J. M. 1955. Effects of phospholipases, collagenase and chymotrypsin on impulse conduction and resting potential in the lobster axon with parallel experiments on frog muscle. *J. cell. comp. Physiol.,* 46:183-207.

TOBIAS, J. M. and BRYANT, S. H. 1955. An isolated giant axon preparation from the lobster nerve cord. Dissection, physical structure, transsurface potentials and microinjection. *J. cell. comp. Physiol.,* 46:163-182.

TREHERNE, J. E. 1961. The movements of sodium ions in the isolated abdominal nerve cord of the cockroach, *Periplaneta americana. J. exp. Biol.,* 38:629-636.

TURNER, R. S. 1951. The rate of conduction in stretched and unstretched nerves. *Physiol. Zoöl.,* 24:323-329.

TURNER, R. S. 1953. Modification by temperature of conduction and ganglionic transmission in the gastropod nervous system. *J. gen. Physiol.,* 36:463-471.

TURNER, R. S. 1955. Relation between temperature and conduction in nerve fibers of different sizes. *Physiol. Zoöl.,* 28:55-61.

UCHIZONO, K. 1957. Further studies on the wave-length of impulse in relation to some physiological constants. *Jap. J. Physiol.,* 7:172-180.

UCHIZONO, K. 1959. Graded spike-height of single nodes of Ranvier. *Acta. Med. Biol., Niigata, Jap.,* 7:11-18.

Bibliography

ULBRICHT, W. 1958. Zustand des Na-Transportsystems und elektrotonische Schwellenänderungen an markthaltigen Einzelfasern. *Pflüg. Arch. ges. Physiol.*, 267:478-490.

VAN HARREVELD, A., and RUSSELL, F. E. 1954. Ionic migration in isolated nerves. *J. cell. comp. Physiol.*, 43:335-346.

VORONIN, L. G. 1953. The evolution of properties of nervous processes. *Rep. 19th Int. Congr., Moscow.*

WATERMAN, T. H. 1941. A comparative study of the effects of ions on whole nerve and isolated single nerve fiber preparations of crustacean neuromuscular systems. *J. cell. comp. Physiol.*, 18:109-126.

WEBB, D. A., and YOUNG, J. Z. 1940. Electrolyte content and action potential of the giant nerve fibres of *Loligo*. *J. Physiol.*, 98:299-313.

WEIDMANN, S. 1951. Electrical characteristics of *Sepia* axons. *J. Physiol.*, 114:372-381.

WERMAN, R., REUBEN, J. P., and GRUNDFEST, H. 1960. Effects of environmental changes on indefinitely prolonged action potentials of lobster muscle fibers. *Biol. Bull., Woods Hole*, 119:347.

WILSON, I. B. 1956. Chemical control of ion movements during nerve activity. *Int. Congr. Biochem., III. Brussels.* 1955:440-444.

WILSON, I. B. and COHEN, M. 1953. The essentiality of acetylcholinesterase in conduction. *Biochem. biophys. Acta*, 11:147-156.

WRIGHT, E. B. 1946. A comparative study of the effects of oxygen lack on peripheral nerve. *Amer. J. Physiol.*, 147:78-89.

WRIGHT, E. B. 1958. The effect of low temperatures on single crustacean motor nerve fibers. *J. cell. comp. Physiol.* 51:29-65.

WRIGHT, E. B. 1959. The subthreshold response of the single crustacean motor axon. *J. cell. comp. Physiol.*, 53:349-375.

WRIGHT, E. B. and ADELMAN, W. J. 1954. Accommodation in three single motor axons of the crayfish claw. *J. cell. comp. Physiol.*, 43:119-132.

WRIGHT, E. B. and COLEMAN, P. D. 1954. Excitation and conduction in crustacean single motor axons. *J. cell. comp. Physiol.*, 43:133-164.

WRIGHT, E. B. and REUBEN, J. P. 1958. A comparative study of some excitability properties of the giant axons of the ventral nerve cord of the lobster, including the recovery of excitability following an impulse. *J. cell. comp. Physiol.*, 51:13-28.

WRIGHT, E. B., COLEMAN, P., and ADELMAN, W. J. 1955. The effect of potassium chloride on the excitability and conduction of the lobster single nerve fiber. *J. cell. comp. Physiol.*, 45:273-308.

YAMAGIWA, K. 1960. Again on the local response as a small area activity. *Jap. J. Physiol.*, 10:456-470.

YAMASAKI, T. and ISHII, T. 1952. Studies on the mechanism of action of insecticides. IV. The effects of insecticides on the nerve conduction of insect. *Oyo-Konchu (J. Nippon Soc. Appl. Ent.)*, 7:157-164. (Japanese with English summary.)

YAMASAKI, T. and NARAHASHI, T. 1957a. Studies on the mechanism of action of insecticides. XIV. Intracellular microelectrode recordings of resting and action potentials from the insect axon and the effects of DDT on the action potential. *Botyu-Kagaku (Sci. Insect Control)*, 22:305-313.

YAMASAKI, T. and NARAHASHI, T. 1957b. Studies on the mechanism of action of insecticides. XV. Effects of metabolic inhibiters, potassium ions and DDT on some electrical properties of insect nerve. *Botyu-Kagaku (Sci. Insect Control)*, 22:354-367.

YAMASAKI, T. and NARAHASHI, T. 1959a. The effects of potassium and sodium ions on the resting and action potentials of the cockroach giant axons. *J. Insect Physiol.*, 3:146-158.

YAMASAKI, T. and NARAHASHI, T. 1959b. Electrical properties of the cockroach giant axon. *J. Insect Physiol.*, 3:230-242.

YOUNG, J. Z. 1938. The functioning of the giant nerve fibres of the squid. *J. exp. Biol.*, 15:170-185.

ZHUKOV, E. K. 1937. After-positivity in the limb nerve of the crab. *Bull. Biol. Med. exp. URSS (Biull. eksp. Biol. Med. Moskva)*, 3(3):297-299.

ZHUKOV, E. K. 1946. Some laws of the evolution of excitation. *J. gen. Biol. (Zh. obshch. Biol. Akad. Nauk SSSR)*, 7:435-454. (In Russian with English summary.)

CHAPTER 4

Comparative Neurology of Transmission

Summary	180
I. The Concept of the Synapse	**181**
A. Definition	181
B. Problems Raised by the Present Concept	182
C. Common Properties of Synapses	183
D. Differences Between Synaptic and Nonsynaptic Membranes	189
II. Nonsynaptic Interaction Between Neurons	**191**
A. Specific Interactions	191
B. Synchronization of Masses of Cells	194
C. Experimentally Imposed Electric Fields	195
D. Chemical Interaction	195
E. Interaction Between Neurons and Glia	195
F. Ephaptic Interaction	196
III. Differences Among Synapses	**197**
A. Differences Based on Dendrite Arrangement	198
1. The mammalian skeletal muscle cell	198
2. The motoneuron in the ventral horn	199
3. The stretch receptor neuron	199
4. The large cells in the cardiac ganglion	200
5. The vertebrate dorsal root ganglion	200
B. Differences Based on Functional Properties	200
1. Excitatory and inhibitory junctions	200
2. Polarized and unpolarized junctions	202
3. Relay and integrative junctions	203
4. Facilitating and nonfacilitating junctions	203
5. Accommodation and iterative response	204
6. Aftereffects; fast- and slow-following	207
7. Long and short delays	207
8. Chemical versus electrical transmission	207
IV. Examples of Better-Known Junctions	**211**
A. Invertebrate Preparations	211
1. Squid giant synapses in the stellate ganglion	211
2. Crayfish giant fiber-to-motoneuron synapses	214
3. Crayfish stretch receptor modulation	216
4. Crustacean neuromuscular junctions	218
5. Lobster cardiac ganglion cells	223
6. *Aplysia* giant ganglion cells	225
7. Other invertebrate preparations	227
B. Vertebrate Preparations	231
1. Neuromuscular junctions	231
2. Electric fish neuroelectroplaque synapses	233
3. Spinal motoneurons	234
4. Autonomic ganglia	235
5. Interneurons in brain and spinal cord	236
V. Evolution of Nervous Transmission	**237**
Classified References	238
Bibliography	240

Summary

Interneuronal spread of nervous influence is called transmission; it occurs in two general forms, synaptic and nonsynaptic. A synapse is a functional connection between two neurons accomplished by contact or near contact of their membranes. The problem cases on this **definition** are the following. Separateness of neurons known to have functional communication may not have been established—for example electrical connections between cardiac ganglion cells in lobster; communication exists between cells of some syncytial units and is not considered synaptic—for example third-order giant cells in squid. Contact or near contact may be doubtful; influences exerted by an electric field that cannot reasonably be called specific for selected postsynaptic cells or exerted by neurosecretory or other chemical agents acting at a distance are not considered synaptic. Functional relations may be unknown or doubtful between elements known to be in contact; probably many good synapses mediate very weak, modulating influence.

Several properties of transmission are common though not defining or universal: integration and lability, special drug and fatigue susceptibility, polarization, delay, and production of special excitatory or inhibitory postsynaptic potentials.

Synaptic membranes are not all alike and are often different in excitable properties from nonsynaptic membrane. Least distinctive from nonsynaptic, axonal membrane is the class of **electrically transmitting** synapses; here current is shown to be present and sufficient, across the postsynaptic membrane, to excite it. In contrast, **chemically transmitting** synapses are most clearly indicated simply by the virtual absence of current through postsynaptic membrane as a result of presynaptic activity.

Excitatory and inhibitory postsynaptic potentials (**e.p.s.p.'s and i.p.s.p.'s**) are characteristic features of junctions, at least of those that transduce (that is, of chemically transmitting junctions) and, by analogy with receptor potentials, of some if not all sensory endings as well. E.p.s.p.'s and i.p.s.p.'s have properties that distinguish them from local endogenic potentials. E.p.s.p.'s contribute to subsequent events mainly by their depolarization; this is greater when the standing level of membrane potential is high and falls to zero at a level near zero. I.p.s.p.'s contribute in most cases mainly by the associated conductance surge that tends to short-circuit excitatory potentials; the hyperpolarization is greater at low levels of membrane potential, zero at a definite high level (the turnover or reversal potential level), and becomes depolarizing at still higher levels.

Dendrites in general, and somata in some cases (especially vertebrates), are a patchwork of differently effective excitatory and inhibitory junctional loci; sometimes there is so little endogenic response (p. 126) that spikes can not arise in or invade these regions.

Several kinds of **nonsynaptic interaction** between neurons are indicated by the evidence available. Electrical connections between specific neurons are known in various forms. Some can carry only slowly changing potentials and not spikes, others can conduct spikes. The possibility that a low-resistance membrane separates the cells cannot be excluded, in which case an unusual form of synapse would be involved. Presynaptic inhibition and presynaptic facilitation are known but it is not yet clear whether a synapse upon a synapse is involved. The synchronization of slow activity in masses of neurons, as in brain waves, possibly represents nonsynaptic interaction. This suggestion is reinforced by the effects of experimentally imposed electrical fields of presumably physiological strength. Interaction by chemical signals of generalized or diffuse or blood-borne nature is another form. There are suggestions of effects mediated by neuroglia, though at present these are extremely vague. Ephaptic transmission is treated in this section; artificial junctions have been useful in the history of synaptology.

Differentiation among synapses is based partly on dendrite arrangement and form of axonal terminals. Additional variables are the amplitude of the presynaptic event at the terminals, the amount of transmitter released (when that is a separate quantity), and the postsynaptic excitability as a function of locus, chemical milieu, electric fields, and past use. Others are the spatial interaction of junctional loci on the same neuron, the proximity or remoteness of loci of endogenic excitability, and the relative importance of temporal summation, facilitation, repetitiveness, and aftereffects. Excitatory and inhibitory postsynaptic potentials differ in duration, amplitude, and the alternatives of afterfacilitation or relative refractoriness. Some junctions exhibit miniature, quantal potentials in a random sequence, apparently representing spontaneous events whose probability and hence average frequency is altered by presynaptic impulses. There are several forms of inhibition not dependent on i.p.s.p.'s, notably refractoriness after an e.p.s.p., excess depolarization from summated e.p.s.p.'s, slowly cumulative hyperpolarization, and presynaptic inhibition, already mentioned. Most synapses are polarized but a number are known in which transmission is unpolarized. A distinction is useful between relay and integrative junctions. Both electrically transmitting and chemically transmitting synapses are recognized; several different chemical transmitters are probable at different synapses.

A series of examples of better-known junctions, as listed in the table of contents, are treated one by one to illustrate the differentiation on the basis of the properties just mentioned.

An attempt is made to discern **differences of evolutionary significance** or general comparisons of major groups with respect to properties of transmission. As great a range of complexity and specialization can be found among the synapses within a single species as between lower and higher animals. Both relay and integrative junctions occur in the nerve net of medusae, as well as inhibitory and excitatory junctions. Probably the variety of specialized junctions is greater in cephalopods, higher crustaceans, and insects—and especially in higher vertebrates—than in lower groups. Synapses on somata are perhaps relatively recent. Large fiber systems and their synapses are apparently derived and have evolved independently many times; primitive systems are thought to have been small-fiber and more integrative than relay. Many of the highest centers, however, employ small-fiber components with short axons, integrative properties, and quite possibly mainly or exclusively graded activity rather than all-or-none spikes—as in presumed ancestral neuroid cells. Some arguments favor the supposition that electrical excitability and transmission as seen in higher animals are derived; on this view, primitive excitability relevant to nervous evolution would be mainly chemical. Two-way transmission can be regarded as more primitive than polarized transmission in coelenterate nerve nets but is probably secondary in giant synapses of annelids and arthropods. Since the main evolution has been not in properties of individual neurons or junctions, but in organized masses of neurons, it seems reasonable to believe that there has been progressively more importance and variety of nonsynaptic interactions in the higher animal groups.

Since the nervous system—in all animals which have one—is apparently made up of cellular units, called neurons, specialized for the generation and propagation of signals, it follows that one of the steps in nervous functioning will be the transfer of messages from one neuron to another. This process we will call transmission, a term which we reserve for intercellular spread as distinct from intracellular spread, or conduction.

Most intercellular interaction, at least that for which most information is available, involves synapses, but some forms of nonsynaptic interaction are known. These will be dealt with in turn; the emphasis of this chapter is as an introduction to the concepts, properties, and different kinds of interactions. In the next chapter the role of these processes, among the others which participate in nervous integration, is discussed.

I. THE CONCEPT OF THE SYNAPSE

A. Definition

A synapse is a functional connection between distinct neurons accomplished through contact or near contact of their membranes. This is a more general definition than some which use one or more of the common but not necessarily universal properties as defining characteristics—for example, chemical transmission, delay, or polarization. These narrower concepts are likely to exclude some normal junctions such as that between giant and motor neurons in the crayfish. The concept adopted here is neither purely structural nor purely functional but requires criteria in both spheres; often evidence is not available for one or another criterion. The following considerations are raised by the wide variety of junctions known, especially in invertebrates. Since our information is based on a small number of particularly favorable cases (see Table 4.1), it is likely that we do not yet have a representative picture of functional connections and hence that the concept of the synapse will evolve still more in the future. By extension, the term is often applied to neuromuscular junctions (but rarely to other neuroeffector junctions) and to junctions between nonnervous sense cells and afferent neurons.

B. Problems Raised by the Present Concept

The problems raised by the present concept include the following. "When do we have two different neurons?" "How close a relation is contact?" "How much influence can be regarded as functional?" "What is *a* synapse as opposed to a synaptic region or a convergence of pathways?"

The first question, concerning the **distinctness of neurons,** is apparently merely one of anatomical information. But each generation of scientists demands more anatomical information so that, as in the septa of the earthworm giant fibers, a clear demonstration of hematoxylin-staining membranes in the twenties (Stough, 1926) did not convince many neurologists in the forties. Finally, electron microscopy appears to provide satisfactory evidence of cell membranes, in this case (Hama, 1959) and others, although this method does not show the molecular relations and, as presently developed, is surely not the last word on synaptic details. Moreover, another difficulty arises due to secondary fusion between cells. Where large processes flow together, a syncytium can easily be recognized and the large fiber regarded as a single unit of compound origin (squid and many annelid giant fibers). But there is evidence for fine anastomosing processes in some cases, as in lobster cardiac ganglia (see section II A, below) indicated by a pathway of low electrical resistance between cells; whether there is a bounding membrane of low resistance separating them is quite unknown. These connections are incapable of conducting nerve impulses but spread slow potentials electrotonically, hence decrementally, between cells several millimeters apart. These may be regarded as separate—they have distinct functions, fire impulses at different times, and even have conventional synaptic connections in addition. For the present we regard these as nonsynaptic relations between separate neurons. There do not appear to be many intergrades as yet between these two types of protoplasmic continuity.

The second question, **"How close a relation is contact?"** may require an arbitrary answer. Not every functional connection between neurons is a synapse. It is possible that some neurosecretory cells influence other neurons at some distance. The anatomical basis for presynaptic inhibition (p. 78) is not yet known and it is therefore uncertain whether we have here a synapse. Masses of neurons beating in synchrony, as in brain waves, bespeak communication which is generally believed not to require impulses but subthreshold, slow rhythms. Direct current or standing potentials of unknown origin have been measured between pairs of points in nervous tissue and as these potentials in the field around neurons change they can modulate ongoing activity. These examples of interactions show why the clause in the definition which specifies contact by specific processes is necessary. However, immediate apposition would not be necessary in the presence of other evidence that proximity is specific with respect to transmission to a certain postjunctional unit. We do not as yet recognize synapses between cell bodies, without the intervention of any processes.

The question **"How much influence can be regarded as functional?"** refers to the difficult border between clearly functionally related neurons and those which, although adjacent or near by, are not obviously functionally connected. It is known that nerve fibers in a nerve exert a weak effect on others in the same nerve, altering their excitability by the electric field surrounding an impulse. Such effects may be, however, insignificant in the face of a large safety factor in the axon. In a neuropile there must be all degrees of influence and it is only our ignorance that permits at present a relatively ready recognition

I. CONCEPT OF SYNAPSE C. Common Properties

of those cases among the few tested where plainly synaptic relations are involved. Probably a large percentage of the cells exerting functional influence on others do so only as modulators or predisposers and cannot by themselves fire the others, so that this borderland between weak influence and none must be extensive.

Ephapses or false synapses are readily distinguished from synapses when they require artificial juxtaposition of isolated nerve fibers or the creation of a cut end of a nerve. But synapses are basically distinguished by the normality of the transmission of excitation which they bring about or, more precisely, by whether transmission is the normal functional significance of the spatial proximity of the neuronal elements concerned. This is not easily decided in every case and we may perhaps expect more rather than fewer difficult instances.

In a different category is the question **"What is a synapse?"** since it does not demand an answer. By extension of the principle that a minimum critical area is necessary in excitation of a nerve fiber, it is probable that to fire a neuron transsynaptically a minimum area must be involved in transmission. But this is surely not a fixed quantity because it will depend on how close to firing the cell is already. Moreover, many synapses are only modulators or predisposers and no basis for a unit of action can be formulated. Anatomically, it is true that in electron micrographs synapses are commonly seen as discretely circumscribed patches of contact through windows in the glia and each one is naturally spoken of as *a* synapse. However, it would be premature to make this the basis for recognizing a unit synapse. The complex ramifications of both efferent and afferent processes in neuropiles, the great range of contacts—between bare meeting in passing and spiral windings of axons ending about a soma—emphasize the synaptic region rather than the discreteness of synapses. The same conclusion is indicated by the evidence from electrophysiology and from electron microscopy that even when only one presynaptic fiber is involved, many loci of contact separated by areas without contact are typical, forming a patchwork of graded spot size.

C. Common Properties of Synapses

Passing from considerations of a definition and its qualifications, we may enumerate the common, though not necessarily universal, properties of synapses as known so far. Schäfer in 1879 clearly stated the idea and the importance of synapses, reasoning from the evidence on coelenterate nerve nets (quoted on p. 471), but it was Sherrington in 1898 who both suggested the term and set forth the properties in detail from evidence in mammalian physiology. The most important properties–of **integration** and lability–are not at all general, for some junctions are normally 1:1 relays. Pharmacological susceptibility is often useful but often quite unreliable and extremely variable from case to case. The most common physiological properties are polarization, delay, and the postsynaptic potentials. The only frequent anatomical property, besides intimacy of contact, is the occurrence of synaptic vesicles in many cases.

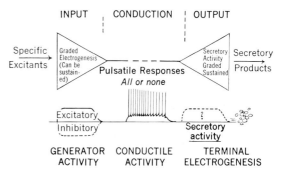

Figure 4.1. Diagrammatic representation of functional components and electrical responses of a neuron. The input or receptive mechanism, which is commonly specialized for a particular form of excitant, produces an electrical response graded in proportion to its stimulus, and which may be brief or sustained. The electrical response may be depolarizing or hyperpolarizing in direction. In the case of depolarization a sufficient level will trigger the pulse-coded conduction component, which propagates the signals without decrement to the output mechanism; there, at least in some neurons, a graded secretory activity proportional to the information included in the arriving pulses is induced and can be sustained. The transmitter released can then be a specific chemical, which operates upon the input mechanism of another cell, or a nonspecific electrical response if the next cell has an electrically excitable input. [Modified from Grundfest, 1957a.]

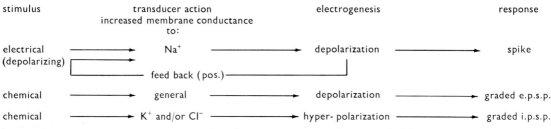

Figure 4.2. A diagram to indicate the differences in ionic mechanism evoked by transducer actions in spike and synaptic membranes. The depolarization caused by electrical stimulus is regenerative in the electrically excitable membrane and produces, therefore, an all-or-none spike, which propagates without decrement. Electrically inexcitable synaptic membrane can produce either depolarizing or hyperpolarizing postsynaptic potentials (excitatory equals e. p. s. p., inhibitory equals i. p. s. p.); these cannot regeneratively feed back on the transducer actions. Therefore the electrical effects are graded in proportion to the stimulus. [Modified from Grundfest, 1959.]

Polarization is only partly due to a characteristic of the transmission mechanism. Where there is only a single presynaptic fiber, as at some nerve-muscle junctions and at giant synapses, one-way transmission is most nearly an intrinsic functional property of the junction. But at many synaptic regions there is convergence of several presynaptic paths, each contributing to the probability of firing of the postsynaptic unit; in these cases irreciprocity can be largely attributed to the organizational feature of convergence. Even where convergence is not obvious, it is possible that geometric asymmetry plays a role (McCulloch, 1938). Several cases are cited below (and see Table 4.1) where unpolarized transmission occurs, and each is apparently associated with histological symmetry (nerve net in coelenterates, giant fibers in serpulids, earthworms, and crayfish). Apart from geometry, the most direct evidence of intrinsic polarization of junctions is in the electrical rectification of applied current in the giant-to-motor synapse in crayfish and in the prevailing presynaptic localization of synaptic vesicles.

Delay at junctions varies between about 0.3 and 10 msec. In many measurements it is probably not minimal synaptic delay proper that is determined but a latency which includes two additional periods, especially in the longer measurements. Conduction time in fine presynaptic terminals is the first; it has in most cases not been estimated. The second is the time between the first appearance of the postsynaptic potential and the beginning of the postspike. This is known and eliminated in the better studied cases. Some presumably monosynaptic delays are a number of milliseconds long—over ten times the minimum figure. In the best known case of electrical transmission, the giant-to-motor junction in the crayfish, the presynaptic action current is visible in the postsynaptic membrane as the first part of the postresponse and hence there is no delay in the transmission process proper. (The value given in the table is measured from extrapolated slopes because the foot of the potentials is not sharp.) The same is probably the case in the septa of the giant fibers of earthworm, crayfish, and polychaete. While delay is then a treacherous criterion, particularly of the number of synapses in a pathway, it is a highly characteristic feature, is often labile, and certainly contributes to integration as well as limiting nervous performance. A summary of some published values appears in Table 4.2 (p. 208), including the effect of temperature and, for comparison, the latencies of some receptor processes.

More general than either delay or polarization is the synaptic property of responding to excitatory input by a depolarization of a few millivolts, the **e.p.s.p.** or excitatory postsynaptic potential (Fig. 4.3) and to inhibitory input by an increase in membrane permeability, usually causing a shift in potential (the **i.p.s.p.** or inhibitory postsynaptic potential), which can be either hyperpolarizing or depolarizing.

Excitation is probably always accompanied by a graded, local, depolarizing prepotential, though it is obscured if the spike starts early due to high excitability. There is no evidence that

Table 4.1. SUMMARY OF PROPERTIES OF REPRESENTATIVE JUNCTIONS

	A	B	C	D	E	F	G	H	I	J	K
	Preparation	Polarized transmission	Facilitation important(¹)	Spatial summation important	Both excitatory and inhibitory endings	Cell can maintain iterative response	Follows input 1:1 only at low frequency	Aftereffects prominent	Delay, minimum, msec	Duration of p.s.p., rise time + decay time to ½, msec	Special Features

RELAY JUNCTIONS

1	Earthworm giants and crayfish lateral giants, segmental septa (Bullock, 1952; Kao and Grundfest, 1957)	0	0	0	0	+	—	0	<0.05	?	macrosynapse, symmetrical low resistance to current electrically transmitting
2	Crayfish, sabellid and serpulid commissural junctions (Bullock, 1952; Furshpan and Potter, 1959)	0	0	0	0	+	+	0	0.5	?	high resistance annelid not electrically transmitting crayfish electrically transmitting, with attenuation
3	Squid giant synapse in stellate ganglion (Bullock and Hagiwara, 1957; Hagiwara and Tasaki, 1958)	+	0	0	0	0	—	0	0.5	1	no electrotonic spread across synapse; cannot be electrically transmitting an accessory prefiber makes synapse with similar properties
4	Crayfish giant to motor synapse (Furshpan and Potter, 1959)	+	0	0 (f+)	+	+?	—	0	<0.1	1.5	electrotonic spread, rectifying electrically transmitting
5	Frog nerve-muscle, fast (twitch) system (Del Castillo and Katz, 1956)	+ (²)	0 (f+)	0	0	0	—	0	0.8	1–2	no electrotonic spread across junction; cannot be electrical "spontaneous" miniature potential, and quantal increments to p.s.p.
6	Nereis and Loligo, nerve-muscle, fast system (Wilson, 1960)	+	0 or —	0	0	0	—	0	<2?	?	rapid adaptation or fatigue; in N., probably graded, with antifacilitation

SIMPLE INTEGRATIVE JUNCTIONS

7	Nereis and Loligo, nerve-muscle, slow system (Wilson, 1960)	?	+	?	0	?	—	0	>3?	?	not as different from fast as in Crustacea

(Table continues on next page)

(Continuation of Table 4.1) Summary of properties of representative junctions.

	A	B	C	D	E	F	G	H	I	J	K
8	Frog, nerve-muscle, slow (small motor) system (Kuffler and V. Williams, 1953) (Burke and Ginsborg, 1956)	+	0	0	0	+	—	+	?	25–40	graded contraction to e.p.s.p.; no electrical excitability, no spike or local response; multiple endings and multiple innervation; great dependence on temporal summation
9	Anemone, medusa and hydroids, neuroneural junctions (Pantin, Horridge, Josephson) in nerve net	0	+++	0	0	?	+	0 (or +)	?	?	functional frequency range about 0.1–1 per second; recruitment of units with different facilitation thresholds
10	Crayfish, stretch-receptor inhibitor (Kuffler, 1958)	—	+	0	0	0	+	?	+	?	15–30	potential of reversal of i.p.s.p. is below maximum resting potential; i.p.s.p.'s temporally summate at functional frequency; no electrotonic spread across synapse
11	Electric eel, nerve-electroplaque (Grundfest, 1957)	—	+	++	+	0	+?	—	0	1–2.4	1	both e.p.s.p. and spike patches intermingled; multiple innervation; early and late facilitation
12	Roach, primary afferent to ascending pseudo-giants (Roeder et al., 1947)	—	+	0 (f+)	0 to +	0	0?	+	++	0.6	?	emphasis on spatial summation for recruiting and for overcoming accommodation but main lability is later in pathway
13	Rabbit, cervical ganglion (R. Eccles, 1955; J. C. Eccles, 1943, 1944)	—	+	— (or +)	++	0 (³)	0	—	+ (⁴)	3.5	>10	spike can arise in initial segment or in soma

COMPLEX INTEGRATIVE JUNCTIONS

	A	B	C	D	E	F	G	H	I	J	K
14	Crayfish, neuromuscular, slow, fast and inhibitory (Hoyle and Wiersma, 1958)	—	+	+++	0 (⁵)	+	?	—	0	?	8;20 (⁶)	multiple innervation; muscle membrane mainly synaptic often no spike involved in response inhibition even at level of zero i.p.s.p. can maintain plateau depolarization
15	Crayfish, primary afferent to ascending interneurons (Prosser, 1935, 1940; Kennedy, 1958; Watanabe, 1958)	+	0	++	+	0	+	0	2 to >6	7 to 20 +	presynaptic impulses (separate fibers): postspikes = 4 : 1; 100 msec functional recovery time of synapses; shows subliminal fringe; either long delays (30 msec) or interneurons involved
16	Lobster, cardiac ganglion cells, followers (Bullock, 1958)	+	++ and —	0 or +	+	+	++ (⁷)	++	?	30	spontaneity and patterned bursts; spike does not invade soma; several synaptic loci; electrotonic spread of slow potentials, not spikes; facilitation and antifacilitation; + and — aftereffects

17	Aplysia, visceral ganglion cells (Tauc, 1958)	+	+	+	+	+(7)	++	?	50 to 200	many converging presynaptic paths; usual p.s.p. is compound often spontaneously firing fractionation of neuron: local potentials fast and slow phases of facilitation e.p.s.p. can block by excess depolarization
18	Cat, motoneuron in spinal cord (J. C. Eccles, 1957)	+	0 (or +)	0? to +	+	0?	— to +(7)	+(8)	0.4	5	spike arises in initial segment of axon; invades soma secondarily single spike for single e.p.s.p.
19	Cat, interneurons in cord, geniculate, cerebral cortex and reticular formation (see text)	+	0, +, —	0, +	0, +	+	+, —	++	lo to hi	lo to hi to >50	repetitive firing to high frequency on one e.p.s.p. many endings only modulate and predispose

ARTIFICIAL JUNCTIONS

20	Squid giant fibers in contact (Arvanitaki, 1942)	0 to +	+	0	0	0	+ or —	0	1.5	1	one-way transmission in certain geometric arrangements
21	Crab leg nerves in contact (Jasper and Monnier, 1938)	0	+	0	0	+?	?	+	?	?	must be hyperexcitable to the point of rhythmic spontaneous firing
22	Frog, cut end of sciatic nerve Cat, cut end of dorsal columns of cord (Granit and Skoglund, 1945; Renshaw and Therman, 1941)	0	+	+	0	0	0	0 to +	0.1	?	preferential transmission in one direction

+, ++, +++: the property is present and, if graded, is of the relative importance indicated.
0: the property is absent, or "not so."
—: the opposite of statement of heading is true.
f: in fatigued condition.

(1) Facilitation is here confined to responses which are each greater than the last, not merely temporal summation adding to a residuum. Facilitation is also distinguished from posttetanic potentiation, which has a special definition and is measured after a prolonged high-frequency (tetanic) stimulation.
(2) Under some conditions in cat, antidromic impulses are reflected up ventral root fibers following a twitch, with less than 0.1 msec latency, showing temporal summation, spatial summation, and facilitation.
(3) After depression of e.p.s.p. by curare a postsynaptic hyperpolarization remains, but whether this is due to separate prefibers or is inhibitory is not known.
(4) It is not clear how far afterdischarge is normal.
(5) It is not necessary to excite more than one nerve fiber but each of these has many endings per muscle fiber. There may be addition of effects of exciting slow and fast nerve fibers simultaneously.
(6) E.p.s.p. 8 ± msec, i.p.s.p. 20 ± msec.
(7) Some presynaptic pathways cause smoothly maintained shift of membrane potential or such small p.s.p.'s that there is no 1 : 1 response, even of subthreshold humps, but only predisposing or modulating influence.
(8) Limited to afterdepression and posttetanic potentiation; not prominent, especially positive aftereffects.

Figure 4.3. Active versus passive (electrotonic) spread. **A.** The spike is the most active of responses; visceral ganglion cell of *Aplysia* excited indirectly through a presynaptic pathway. **B.** The e.p.s.p. alone; the spike has been prevented by artificially raising the resting membrane potential, so this e.p.s.p. cannot reach threshold. **C.** The same, now larger because the membrane potential has been further increased. **D.** A purely passive electrotonic potential resulting from the imposition of a subthreshold pulse of current that dissipates exponentially with a time constant which measures the resistance and capacity of the cell membrane. The scheme at the top shows separate polarizing (p.) and recording (r.) electrodes inserted in the unipolar soma; the presynaptic pathway influenced by the stimulus (st.) probably terminates actually upon branches of the stem process rather than on the soma directly. **E.** Recording the amplitude of an e.p.s.p. at different points along the post-synaptic cell (in this case a muscle fiber), an exponential decay with distance on both sides of the point of maximum amplitude is found, which measures the space constant of the membrane. **F.** Plotting the duration of the rising phase in the same way reveals an increasing delay of the peak as the distance is increased from the junction in either direction. **G.** The same is even more true for the duration of the falling phase. These are likewise dependent on the distributed resistance and capacity in the elongated cylindrical membranes. [**A–D** from Tauc, 1958, 1959; **E–G** from Boyd and Martin, 1956.]

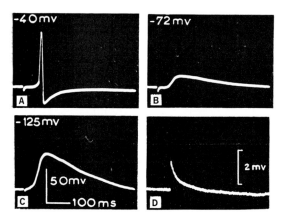
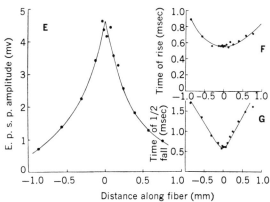

synaptic response exerts its effect immediately on subsequent spike initiation by any means other than the voltage change of the e.p.s.p.; longer term effects occur without sign in membrane potential. The impedance change is not the significant concomitant in excitation as it is in some inhibitory synapses. The e.p.s.p. is graded and depolarizes to a critical level at which the spike takes off explosively, sometimes with the intermediate stage of a local response. Typically this level is about 10 mv less polarized than the resting potential. However, much smaller e.p.s.p.'s are common and they may be influential—in modulating or predisposing neurons to other input and in some cases in controlling the magnitude of a graded response, as in crustacean muscle contraction.

Inhibition is usually but not invariably accompanied by a potential change and, if it is, the direction is usually but not invariably an increase in membrane polarization. In some synapses the potential change itself appears to be the significant event; in others it is less important than the simultaneous rise in membrane conductance or permeability. The potential change can exert an effect on another part of the cell, if not too far away, by its electrotonic spread, whereas the impedance change is more local. The i.p.s.p. is usually smaller than the e.p.s.p. as seen by an electrode in the soma; it rarely exceeds a few millivolts. This is sufficient because the voltage required to be effective is only that which will oppose the passing of the critical level of depolarization for spike initiation. For further differences between excitatory and inhibitory potentials, see pp. 265 and 978.

But even these postsynaptic potential properties are not universal or defining features of all normal junctions. The electrically transmitting synapses do not behave in the way just described; an electrotonic spread of current from presynaptic fiber to postsynaptic membrane accounts for the initial potential change in the latter and in turn stimulates a local potential and, if above the critical amplitude, a spike. The rectifying feature which guarantees polarized transmission in the crayfish giant-to-motor synapse is probably not general and is absent in cases of unpolarized electrical transmission, as in the segmental septal junctions in giant fibers of crayfish and earthworms.

In conclusion it is apparent that no single common property or mechanism is universal even among the limited samples of synapses known. These properties aid in recognition when present but when they are not we must simply look for synapses as functional contacts between neurons.

D. Differences Between Synaptic and Nonsynaptic Membranes

It is generally believed that one or more types of synaptically activated cell surface membranes are quite distinct from other cell surface membranes, in particular that involved in conduction of impulses (Figs. 4.1, 4.2). There may be many basic similarities between the mechanisms of conduction and transmission but these are beyond our present scope. What is germane here is that some apparent differences in properties, whatever be their mechanisms, define these types of membrane and that the different properties appear to be distributed in a fairly sharp, often patchy manner over the cell.

Synaptic membranes are not all alike; one class known from a small number of examples is electrically transmitting but some, possibly most, are **electrically inexcitable**, that is, cannot be made to respond in the same way to electrical stimuli as they do to normal presynaptic input (Fig. 4.4). Furthermore, the latter does not contribute any significant quantity of electric current across the postsynaptic membrane. In contrast, conductile membrane such as that of long axons is electrically excitable; indeed, this is its essential feature. Its main function, to propagate without decrement, depends on the capacity of each segment to be excited by the local electric currents of the all-or-none action potential or spike of the adjacent region. Adequate excitation of most synaptic membranes is accompanied by a considerable irreducible **delay**, whereas conductile membrane can be made to respond with virtually no latency. The class of electrical junctions fits the definition of synapse given above and transmits by flow of current; the membranes are electrically excitable. The distinction of synaptic from conductile membrane in such synapses is a low resistance to flow of current in the direction of transmission. These junctions may have a delay which is vanishingly small but it is entirely to be expected that they

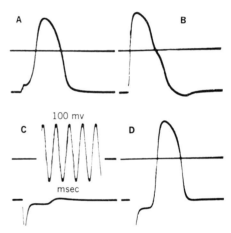

Figure 4.4. The contrast between electrically excitable and inexcitable responses, as seen in the electroplaque of the electric eel. **A, B.** Weak and strong depolarizing electrical stimuli delivered to the electroplaque excite the cell directly; strong stimuli with almost no latency. **C, D.** Stimuli applied in the reverse direction are ineffective directly but stimulate the cell indirectly by way of nerve terminals that are activated by the current and supply the synaptic fraction of the membrane of the electroplaque cell. The weak indirect stimulus evoked only an e.p.s.p. after a latency of almost 2 msec (**C**). The very strong stimulus (**D**) shortened the latency to about 1.7 msec, and the larger e.p.s.p. evoked a spike with brief delay. No synaptic potentials were produced by the direct stimuli (**A, B**), but the strong direct stimulus (**B**) excited nerve fibers that evoked an e.p.s.p. (occurring with the same latency as in **C** and **D**), therefore appearing on the falling phase of the directly elicited spike, while the electrically excitable membrane was absolutely refractory. [Altamirano et al., after Grundfest, 1959.]

can also have a considerable delay like that between nodes in a myelinated vertebrate axon which has been depressed in one way or another. Even long delays before the beginning of active, amplifying postsynaptic responses (as distinct from the electrotonic potential attendant on the presynaptic action current) can be anticipated with electrical transmission in certain conditions.

Synaptic regions are capable of normal response in a **graded** manner, in either depolarizing or hyperpolarizing direction; the latter at least is reversible. They are not capable of response in an all-or-none, overshooting, rapidly repolarizing and then undershooting manner, which are distinctive characters of conductile membrane. Synaptic response can be prolonged, summating, and without refractoriness (Fig. 4.5), but it does not propagate; it spreads decrementally by passive electrotonic current flow, declining to a small percentage at distances of the order of a few millimeters or less. These characteristics are similar to those of the local response in conductile membrane stimulated below spike threshold. But at least the class of e.p.s.p.'s that is not electrically elicitable is different, not only in being incapable of growing up into a regenerative or explosive process, but also in their alterability by imposed conditions.

Based upon differences in the specificity of the ion permeabilities and the electrochemical gradients responsible, the conductile and the electrically inexcitable synaptic potentials behave differently to various preset **levels of the membrane potential** and to changes in the ionic composition of the external fluid. The e.p.s.p. increases in voltage swing with increasing membrane polarization, whereas in conductile membrane such increasing polarization blocks the spike and then depresses the local response to a given stimulus. Lowering the membrane potential too much blocks (inactivates) the spike mechanism but not the e.p.s.p. However, it reduces the size of the e.p.s.p. even more than it does that of the spike. The slope of this dependency would give zero spike or local potential at a highly reverse-polarized membrane potential (about 50 mv inside positive) whereas e.p.s.p. reaches zero at about zero membrane potential. In the few cases tested, the e.p.s.p. persists in the absence of external sodium, while conductile potentials, both spike and local, are dependent on it or certain substitutes in the best-known examples.

Many **drugs** influence synaptic areas but have little or no effect on conductile membrane. There is great variation between preparations and it is not always unequivocal where the agent is in fact acting. Nevertheless, with caution and qualifications, drugs are extremely useful in experimentally demonstrating synaptic and nonsynaptic membranes in particular cases. For example, the effector cell of the electric organ of the electric eel (see p. 233) is a mosaic or patchwork of conductile and synaptic membrane. Eserine, procaine, and d-tubocurarine block the development of the

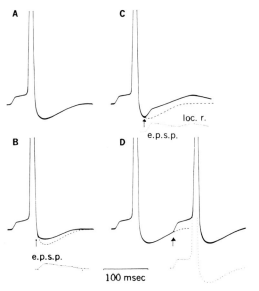

Figure 4.5. Absence of refractoriness in postsynaptic responses in a giant cell of *Aplysia*. **A.** A single shock to the presynaptic nerve first evokes a long-lasting e.p.s.p. out of which rises the spike in the giant cell. **B.** A second stimulus exciting the cell during its refractory period adds a potential (*solid line* beginning at arrow) to the initial response (*broken line*). The difference (*dotted line* below) is due to the second e.p.s.p. **C.** The second stimulus was delivered somewhat later. The added potential also shows a local response (*loc. r.*), which was initiated by the e.p.s.p. in the electrically excitable membranes during the relatively refractory period. **D.** At a longer interval a second stimulus evokes the full response, as in **A**. [Arvanitaki and Chalazonitis, 1956c. Reproduced with the permission of Centre National de la Recherche Scientifique, Paris.]

e.p.s.p. without depolarizing or preventing electric excitation of the conductile mechanism; acetylcholine, carbamylcholine, or decamethonium block the conductile mechanism by excessively depolarizing through enhancing the e.p.s.p.

Conductile membrane responds monotonously to one nonspecific form of input, moving charged particles. Synaptic membranes on the contrary probably have specific and diverse input requirements. Knowledge of chemical transmitters is too scanty to demonstrate this clearly, but certainly suggests that different substances are used. Moreover, receptor endings of the great variety of sensory structures are quite analogous and their characteristic electrical response to the various highly specific adequate stimuli may be subsumed under a common term with synapses: exogenic or transducer potentials (Fig. 4.1). Besides diverse sensibilities, synaptic membranes differ among themselves in the type of response they manifest. These differences are mainly in the ion permeabilities and hence equilibrium potentials. As the present treatment is restricted by space to the operational aspects, the reader must be referred elsewhere for information on mechanisms at a lower level (Eccles, 1957, Grundfest, 1958a, 1958b, 1959).

Some parts of some neurons are probably purely one or another type of membrane; the axon is commonly without synaptic regions, dendrites are perhaps often without spike-supporting ability. But it is not uncommon for a part of a postjunctional cell to behave as though it had two or more **types of membrane intermingled.** Their distribution is one of the chief integrative variables, to anticipate the next chapter, not only because the accessibility of electrically excitable membrane to the synaptic membrane influences the chain of events from e.p.s.p. to local potential and then spike. It is also important because in some cases the spike initiation occurs beyond the soma in the base of the axon and if the soma includes electrically excitable membrane, the spike can "backfire" as it were, discharging the soma secondarily. This invasion of all-or-none activity into a region where graded events from different dendrites and from inhibitory and excitatory synapses are pooled will profoundly affect subsequent arriving events. Some cells do not permit such invasion, presumably because they lack that type of membrane.

II. NONSYNAPTIC INTERACTION BETWEEN NEURONS

Several kinds of evidence indicate that neurons can interact directly and physiologically, other than by classical synaptic contacts. We will here consider electrical evidence of protoplasmic bridges or other specific connections, evidence from synchronization of slow activity of masses of neurons and from the effects of experimentally imposed electrical fields of presumably physiological strength, chemical interactions, and effects mediated by glia. It will be useful to refer to ephapses also under this heading.

A. Specific Interactions

Protoplasmic bridges between neurons are inferred from the fact that subthreshold or even hyperpolarizing electrical pulses put into one neuron through an internal electrode can, in certain cases like lobster cardiac ganglion cells, be seen, attenuated and distorted, by an internal electrode in another cell. Withdrawing either electrode until it is just external to the cell abolishes the electrotonic sign, even if the pulse is made much stronger. Between the large anterior cells in the lobster cardiac ganglion the attenuation has been measured; it would correspond to a bridge about 1 μ in diameter, assuming a resistance of axoplasm and capacity of membrane within reasonable limits (Watanabe, 1958; Hagiwara, Watanabe, and Saito, 1959). Between the same cells and the small posterior ones which normally pace the heart beat there is also clear indication of a connection (Figs. 4.6, 4.7) and in this case it can be seen that the electrotonic spread of potential, although attenuated greatly by a separation of 5 mm, is physiologically effective in altering the frequency of the heart beat. Impulses set up in the large anterior cells have

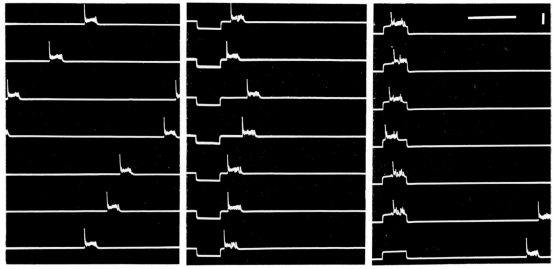

Figure 4.6. Direct electrical interaction by subthreshold currents between separated cells. Synchronization of the pacemaker cell of a lobster cardiac ganglion with repetitive subthreshold pulses delivered inside a distant follower cell. The record shows the burst of synaptic activity in the follower cell driven by the pacemaker cell several millimeters away. In the left column the heartbeat bursts are spontaneously arriving, shown in the continuous record to be slightly out of step with the successive sweeps of the time base. In the middle column long pulses of hyperpolarizing current are delivered intracellularly to the follower neuron and there is a strong tendency for the pacemaker to drive the next heartbeat burst shortly after the end of this current. In the third column a subthreshold depolarizing pulse tends to synchronize (commanded by the pacemaker neuron) with the onset of the current. Calibration marks: 10 mv, 1 sec. [Watanabe and Bullock, 1960.]

no such influence. Because of their brevity they are so attenuated that they are without effect on the other cells; only long pulses can survive the capacitative shunting. The neurons involved have independent impulse initiation, different functions, and even ordinary synaptic contacts, so that it is still justified to speak of them as separate neurons but with some kind of low resistance bridge between them (Watanabe and Bullock, 1960).

One may be prepared to find **various sizes of bridges** and hence degrees of independence of the entities they connect. The paired lateral giant fibers of earthworms normally act quite synchronously but there is a delay of about 0.8 msec for each experimentally compelled crossing and the crossing can be differentially fatigued by repetitive stimulation so that a synaptic connection is suggested between them in each segment. However, using two penetrating electrodes, a pulse injected into one can be seen, attenuated to about a third, in the other (Wilson, 1961; see Annelid Bibliography).

Either Stough (1926) was right in describing transverse anastomoses, or there is a membrane of exceedingly low resistance in both directions. If the latter proves to be true, we are dealing with an electrical synapse. An interpretation consistent with the facts is that normally these bridges conduct so that an impulse initiated in one lateral fiber propagates into the other as well, by way of the next segmental bridge, being thereby delayed only a tenth of a millisecond or so and giving the well-known large, externally recorded, synchronous lateral spike. But the bridge is susceptible to fatigue and has perhaps a low safety factor; the familiar dissociation of the laterals into a variably double-peaked spike and the 0.8 msec delay found by Rushton (1945) would then represent excitation of one lateral by the other across a passive bridge with a delay comparable to that in depressed nodes of Ranvier. Each lateral segmental unit, bounded by its septa anteriorly and posteriorly and having its own cell bodies, could then be regarded as an independent neuron—somewhat more closely

II. NONSYNAPTIC INTERACTION   A. Specific Interactions

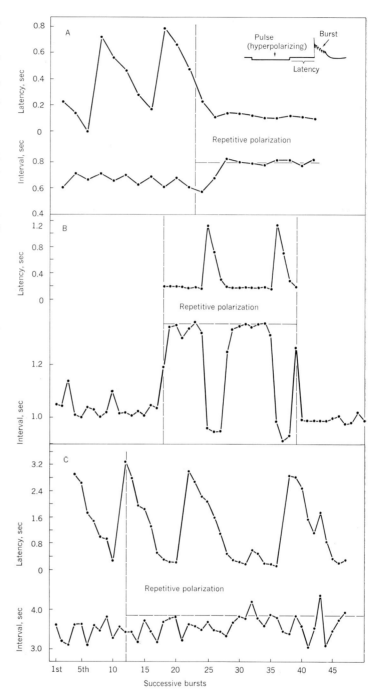

Figure 4.7.

Direct electrical interaction between cells by subthreshold currents. The effect of repetitive hyperpolarization of a follower cell upon the rhythm of the pacemaker in the lobster cardiac ganglion, shown by latency diagrams and interval diagrams. The serial number of the successive bursts is on the abscissa. In each frame (**A, B, C** representing three different cells) the upper trace indicates the latency (see inset, upper right) and the lower trace indicates the interval between successive bursts. The *vertical broken lines* indicate "on" or "off" of the repetitive polarization and the duration of the applied current is shown by *horizontal broken lines*. **A** shows a strong effect; the latency before polarization rises and falls as the phase relation between spontaneous bursts and not-yet-imposed repetitive pulses, at close to the frequency of the spontaneous bursts, drifts through all possible values. When the polarizing pulses are imposed, there is abruptly synchronization, with a minimum latency. **B** shows the same with occasional escape from synchronization. **C** shows no effect except in the 30th to 35th burst. [Watanabe and Bullock, 1960.]

tied to the other lateral than are the units in the lobster heart—because impulses can jump the bridge. But normally the two laterals must be regarded as a singly syncytial, bilateral neuron in each segment, unless a low resistance membrane between them is found.

Exceptionally, two adjacent large cells in the visceral ganglion of *Aplysia* show a specific electrotonic connection. A spike in one cell causes a passive deflection of 2–3 mv with a similar time course and no delay in the neighbor, whereas such a potential is not seen upon withdrawal of the penetrating electrode from the latter. The electrotonic sign is therefore not simply the field of the spike in the conducting fluids. A low resistance bridge of some sort is required (Tauc, 1959). Arvanitaki saw entrainment of one iterative cell by another in 1942. Other examples of low resistance connections have recently been found in supramedullary cells of fish (Bennett, 1960), ventral cord cells of the leech (Hagiwara and Morita, 1962), and between muscle fibers of lobsters (Reuben, 1960).

In each of these cases, if it should be found that there is a membrane of low resistance, representing contact between the two cells, rather than anastomosis, our definition of a synapse would be satisfied and we would have another example of an electrical synapse.

The **dorsal root reflex** and dorsal root potential illustrate another form of relatively specific interaction which is not due to synapses as presently conceived. Inflow of impulses through certain functional classes of fibers in the dorsal roots of the cat somehow alters the membrane potential and the excitability in certain other afferent fibers near by. Thus a volley in a skin nerve produces a large depolarization lasting 100–200 msec in other cutaneous afferents in the dorsal horn. This is probably similar to the classical dorsal root potential (Wall, 1959; Bernhard, 1958; Katz and Miledi, 1962a). In addition there may arise, following a single afferent volley, a prolonged high-frequency antidromic discharge of impulses in both the originally stimulated and in neighboring passive fibers; this is the dorsal root reflex. Both phenomena can be localized in or near the central endings of the sensory fibers and cannot be ascribed to any known synaptic contacts. Evidently there is some kind of "cross talk" between afferent fibers or from interneurons back to afferent axon terminals (Eccles and Krnjević, 1959; Wall, 1959; see also p. 202). As yet there is no indication of a normal functional meaning and these phenomena might be regarded as ephaptic (see below, p. 196).

Presynaptic inhibition has been demonstrated in crustacean neuromuscular junctions by Dudel and Kuffler (1961a). This means that in alpha inhibition inhibitory axons exert an effect directly upon excitatory axons just before the transmitter is released that will excite the muscle. This discovery was made possible by finding that minute quantal events occur in the postsynaptic (muscle) membrane in a random sequence. Properly timed inhibitory impulses reduce the probability of occurrence of such events but not their size. A similar mechanism is suggested for the so-called remote inhibition seen by Frank and Fuortes (1957) in the spinal cord of cats (Eccles, 1961b, 1961c). Facilitation in the crayfish neuromuscular junction also works presynaptically. Until anatomical evidence indicates that a recognizable synapse is involved, this phenomenon may be treated in the present section, as nonsynaptic interaction. Possibly the serial synapses mentioned on p. 78 provide the needed anatomical basis, but this has yet to be shown.

Another example of nonsynaptic but relatively specific action of one set of nerve fibers on another is the alteration in tactile fibers from the skin following **stimulation of sympathetic** fibers to the skin. In noncirculated, isolated frog skin a single strong shock to the sympathetic supply can evoke a brief repetitive discharge in tactile afferents. Weaker shocks lower the threshold to touch and delay adaptation. No such effects occur in smaller afferent fibers excited by skin stretch (Loewenstein, p. 347).

B. Synchronization of Masses of Cells

The synchronization of **slow potential changes** in masses of neurons, as in brain waves (see pp. 31, 318), is probably at least sometimes independent of synaptic pathways and dependent on proximity and geometry. The sinusoidal waves of about 15–40 per second in insect optic ganglia under certain conditions may be examples (see p. 1092). The body of evidence that brain waves are a distinct form of activity not dependent on spikes (though they influence each other) is a strong argument for mass synchronization. Gerard (1941a, 1941b) observed traveling waves induced by caffeine to cross complete anatomical transections, provided the cut surfaces were closely opposed. Bremer (1941) noted the maintenance of synchrony of electrical beating of adjacent segments of the transected spinal cord. Another argument is the anatomical consideration that, whereas specific synaptic connections do not inherently call for any particular architecture of the masses and layers of neurons—since the fidelity of the axon in preserving and conducting messages regardless of distance makes it unnecessary—the evolution of the nervous system is an eloquent display of progressively more elaborate, specific, and circumscribed architecture (Gerard, p. 344). The interaction among neurons which these arguments indicate may be similar to the experimental mechanism next described, but it cannot be taken for granted; there are possibly mechanisms of transmission without impulses which are not so gross but utilize processes and arborizations, all within a distance such that decrementally spreading events are still effective.

C. Experimentally Imposed Electric Fields

The effects of passing current through whole masses of tissue have been repeatedly reported and are not without significance. One reason is that it has been estimated that the voltages and current densities which are sufficient to cause changes in frequency of firing of active neurons are within the range of those known to be present in intact tissues (see further, pp. 157–163). The sensitivity involved must be looked upon as quite a different form of neuronal sensitivity from the usually measured forms. Changes in membrane potential which are far below threshold are adequate to modulate already ongoing discharge. This means that the small departures from internal isopotentiality which are permitted by the low resistance of neuronal cytoplasm—or to put it in another way, the differences in standing transmembrane potential between one part of the neuron and another, even without any change in the average membrane potential—are influential at critical loci in predisposing the cell to fire. Neurons are therefore reciprocally affected by imposed current of opposite polarities; one direction of flow increases frequency of firing, the other decreases it. This is evidence that the firing frequency is determined at a limited locus, asymmetrically placed with respect to current entering and leaving the neuron. How far this available means of interaction actually exists normally is difficult to assess. It is treated with more detail in the next chapter (p. 262). Another reason that findings with imposed current are significant is that several signs point to some unexpected organization of whole masses of cells. Crayfish ganglia generally show an increase in spontaneous firing to one polarity of current and a decrease to the opposite polarity; insects may show a coordinated posture which is opposite for the two polarities.

D. Chemical Interaction

Effects of chemicals elaborated by some cells upon the activity of other cells, other than by synaptic transmitters, are not well known but certainly occur and are only relatively nonspecific. Neurosecretion is perhaps the principal mechanism of concern here and is treated in Chapter 6. Stimulation of the reticular system in the brain stem can activate the electroencephalogram of a slab of cortex isolated but for its blood supply; so can stimulation of the peripheral stump of a divided splanchnic nerve. Sympathetic stimulation enhances the response of tactile afferents in isolated frog skin apparently by release of adrenalin (p. 308). Sperry (1958) reviews evidence of quite specific morphogenetic influences which are believed to be chemical.

E. Interaction Between Neurons and Glia

The neuron-glia interaction has been suggested by several types of evidence (see Windle, p. 121; Galambos, 1961). It is not at all clear what kind of effects are involved, but the possibility must be kept in mind that these traditionally nonnervous constituents may play a role more direct than merely supporting and nourishing neurons, in certain meanings of those terms. Some glia actively change shape in tissue culture. Changes of membrane potential have been recorded with an intracellular electrode in glia cells (Tasaki and Chang, 1958; Hild et al., 1958). Electron micrographs show such intimacy between glia and nerve cell membranes that virtually no intercellular space remains; the glia are apparently the external medium through which local circuits are completed and must presumably be the source of the ions which move in and out of neurons (but see Chapters 2 and 5).

F. Ephaptic Interaction

The term ephapse was introduced by Arvanitaki in 1942, from the Greek for "the action of touching," to distinguish certain contacts from synapses, which are loci for the "action of joining or linking." Synapses have come to mean surfaces of contact which are anatomically and functionally designed for the purpose (a shorthand expression which can as well be stated in nonteleological terms: surfaces of contact specially arranged or physiologically specialized and normally acting as funtional links between neurons). By the term ephapse, Arvanitaki designated those surfaces of contact which achieve transmission but which are artificial or not normally functioning for transmission, whether these surfaces are brought together experimentally or are naturally in contact. Grundfest (1959) has aptly termed them "false synapses." He has, however, used the term in a different way from its original and present usage, to mean any electrically transmitting junction—even when it is a normal one, as in the segmental septal contacts in earthworm and crayfish lateral giant fibers and the crustacean giant-to-third root motoneuron junctions. This is consistent with his definition of synapse as a chemically transmitting junction, but present usage is consistent with the general definition of a synapse as a natural junction which achieves transmission, without regard to the mechanism.

A number of **examples of false synapses** or ephapses have been studied, beginning with the demonstration of Galvani that a frog muscle could by its action current stimulate the nerve of another preparation lying in contact with it. Hering in 1882 showed that if one distal branch of a nerve is stimulated, fibers in another branch could be excited by transmission at a freshly cut end of the common trunk, proximally. Careful analyses of similar cut-end ephapses have been made in another preparation where the cut is distal and where the dorsal and ventral roots, divided from the spinal cord, are used as stimulating and recording sites (Granit and Skoglund, 1945a, 1945b). This preparation is important among other reasons for showing that a purely electrical junction can be in different degrees polarized; the transmission from some kinds of fibers (acting as pre-ephaptic) to others (acting as postephaptic) is easier than the reciprocal transmission. Cut-end ephapses also occur at the site of transection of fiber tracts in the cord (Renshaw and Therman, 1941). The most detailed information, however, con-

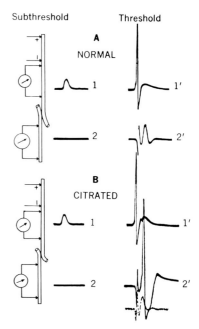

Figure 4.8. Ephaptic interaction. **A.** Two isolated giant axons from the squid are arranged as shown in the diagrams at the left, with contact between the fibers completed by sea water. A weak stimulus (left) evokes a local response in the pre-ephaptic fiber (seen in trace 1). This is not propagated to the ephapse and has no effect on the postephaptic fiber (trace 2). When a stronger stimulus evokes a spike in the pre-ephaptic fiber (1′) the postephaptic fiber generated a local response (2′); ahead of it is seen the electrotonic pickup of the pre-ephaptic spike. **B.** Excitability of the axons was increased by removing calcium ions from the medium with citrate. The weak stimulus still could not evoke activity in the postephaptic fiber (2). When a stronger stimulus evoked a spike (1′) the postephatic fiber also produced a spike (2′), arising on a step that is a local response, as seen in **A**, and without a spike in the lowest trace. [Arvanitaki, 1942.]

III. DIFFERENCES AMONG SYNAPSES

Figure 4.9.

Excitability changes caused by field currents. *Upper left.* A spike was produced by a stimulus to one of a pair of crab nerve fibers arranged as in the diagram at upper right. The electrical excitability of the second fiber is shown *(lower left)* in relation to the time the spike passed the testing region. In the interval before the spike had reached that site the excitability of the fiber was depressed. During the time that activity resided at the tested level the excitability was augmented. This was followed by a second depressed phase as the activity propagated out of the tested site. [Katz and Schmitt, 1940.] *Right.* Diagrams of the anodal, cathodal, and anodal-polarizing sequence generated in the inactive fiber by the spike in an adjoining fiber *(top)* and of different field current conditions produced by different geometrical arrangements *(bottom)*. [Eccles, after Grundfest, 1959.]

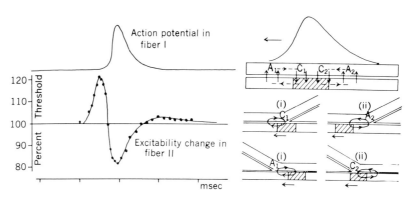

cerns the interaction between isolated nerves and even single fibers brought into artificial contact (Katz and Schmitt, 1940, 1942; Jasper and Monnier, 1938; Arvanitaki, 1942; Nagahama, 1950; Fig. 4.8). In all these cases the pre-ephaptic action current is adequate to excite some response in the postephaptic fibers only under specially favorable conditions—absence of shunting, hyperexcitability of the fibers, or even autorhythmicity. This is not surprising in view of the absence of specializations to channel the current such as exist in natural, electrically transmitting synapses. It is instructive that even among such unspecialized false synapses there can be great variation in synaptic delay (apparent delay ranges up to 27 msec!), polarization, facilitation, spatial summation, and afterdischarge, due to the geometry of the contact and the properties of the postjunctional membrane.

Short of transmission, or of enough excitation to effect active response, nerve fibers in trunks and even cell bodies in ganglia are demonstrably influenced by the activity of neighbors, in their **thresholds to test shocks** or in their phase relations in rhythmic firing (Fig. 4.9). In nerve trunks with a high safety factor such changes in excitability are probably not significant, that is, normally produce no changes in activity. In other cases it may be conjectured that although actual changes in impulse timing occur, they may play no role or are equivalent to meaningless noise. But it is difficult to rule out the possibility that there is a gradation from slight to strong electrotonic interaction and from incidental and valueless to adapted and valuable—that is to say, synaptic—interaction.

III. DIFFERENCES AMONG SYNAPSES

Junctional regions and neuronal terminations differ fantastically in finer anatomy and this suggests functional differentiation. Many cases are illustrated in detail in Chapter 2, and in the several systematic accounts of the groups. We can expect contrasts in properties and mechanisms when some cells receive but one presynaptic fiber (as in the mammalian skeletal muscle cell) and others receive several fibers from each of several sources (as in the crustacean muscle cell or Mauthner's cell of fishes). Presynaptic fibers which climb up postsynaptic dendrites like vines must operate somewhat differently from those ending like bottle brushes or like leafless bushes or blunt, thorny blades of cactus (Figs. 2.18–20, 19.15). There are endings winding around the soma or simply abutting on it, others making contact with the axon itself or with shorter or longer branches of it, and others related to purely receptive processes or dendrites.

These are referred to as axosomatic, axoaxonal, and axodendritic synapses, respectively. Besides their shape and relation to each other, these structures, both terminal and receptive, are often disposed in characteristic and hence presumably meaningful ways in the tissue architecture as noted in the foregoing section. The most extreme and suggestive examples are perhaps those in the vertebrate cerebellar cortex, where dendrites of each Purkinje cell are shaped like a fan; the planes of these fans for all the Purkinje cells are parallel, while the axons of granule cells run through many fans at right angles to their plane. Stratification of synaptic regions and glomerulus formation are further examples of the same principle. A recently discovered arrangement, not yet evaluated fully but of great potential significance, is the serial synapse (p. 78), where one axon ends on the terminals of another, perhaps explaining the physiologically observed presynaptic excitation and inhibition (see further, Chapter 2).

The role of the components of the neuron in transferring information from one unit to another is so bound up with transmission that we must consider here some features of intraneuronal organization, though these are further developed in Chapter 5.

Dendrites (see pp. 40, 131 for definition) are the component of the neuron most important in adaptability, gradability, evaluation of different inputs, and probably in persistent change—in short, in integration. The terminal ramifications of the axon share in this and are at least sometimes active in graded rather than spike form. The soma is important in the vertebrates, where it is commonly a multipolar cell with presynaptic fibers ending on it. But in invertebrates in general, barring special exceptions, the soma is believed to have no synaptic endings upon it and is often electrically inactive; when it is capable of antidromic activity, it may exert some secondary influence on the output of the neuron. The region of transition between graded, receptive response and all-or-none, propagating activity is an intraneuronal boundary of profound importance; it may lie in different places in different cells and may not be immovable but somewhat labile. Less obvious boundaries occur in some cells between synaptic potential and local potential regions of dendrites or soma and between spike propagating and graded response terminals of axons. These components and their boundaries differ widely in properties in different nerve cells.

The **main variables** for transmission can be summarized as (a) the presynaptic spike height (or graded potential amplitude if the effective terminal action of the presynaptic fiber is graded), (b) the amount of transmitter released (when that is a separate quantity, as it probably is in many cases), (c) the postsynaptic excitability as a function of locus on the neuron and as influenced by the chemical milieu, electric fields, other physical conditions and past history, (d) the spatial interaction of junctional regions on the same neuron, (e) the proximity or remoteness of areas of cell membrane with an excitability to the local currents flowing due to the synaptic potentials, and whether this response is local or regenerative. Additional variables having to do with the temporal changes in the transmission mechanism under continued input are (f) temporal summation, (g) facilitation or antifacilitation, (h) repetitiveness, and (i) aftereffects (persistence or rebound). These are treated below as differences among synapses and are compared in a series of examples of transmission; they are reconsidered as bases of integration in the next chapter.

A. Differences Based on Dendrite Arrangement

It will be useful to point out several contrasting cells representing different types of arrangements of the parts of the neuron and hence of junctional regions (Fig. 4.10).

1. The mammalian skeletal muscle cell

In the mammalian skeletal muscle cell a single presynaptic fiber ends on a postjunctional unit and the circumscribed area of the contact is dendritelike in being the receptive region of the muscle cell and responding with a graded (endplate) potential equivalent to an e.p.s.p. All the rest of the muscle surface is axonlike in showing

III. DIFFERENCES AMONG SYNAPSES A. Dendrite Arrangement

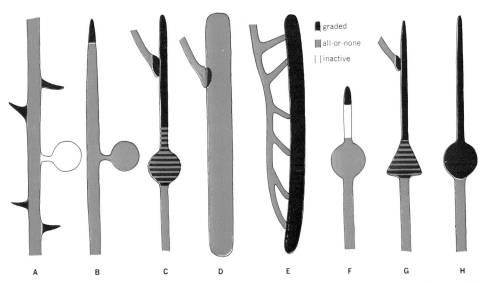

Figure 4.10. The distribution of graded and all-or-none conducting membrane in different excitable cells. All-or-none firing membrane is capable of graded, subthreshold activity, but the membrane here called graded is in general capable only of such activity, and with the possible exception of some axonal terminals, which may not be essentially different, all cases indicated are receptive (synaptic, dendritic type of membrane). **A.** Invertebrate unipolar interneuron, like those described in crayfish, with inactive somata. **B.** Invertebrate unipolar with active soma, like the large cells in *Aplysia*; also vertebrate dorsal root ganglion cell. **C.** Vertebrate spinal motoneuron with synaptic and spike supporting membrane in the soma and with some presynaptic terminals, probably graded and perhaps equivalent to synaptic membrane in receiving presynaptic inhibition (not shown). **D.** Vertebrate uniterminal twitch-type muscle. **E.** Invertebrate multiterminal graded type of muscle. **F.** Crayfish stretch receptor; the presumed inactive stretch of dendrite is not certain. **G.** Pyramidal cell in the cerebral cortex. **H.** Lobster cardiac ganglion cell. The exact position of the boundaries between types of membrane is both uncertain and labile; in several there is evidence that it is normally downstream from the position shown, sometimes because the upstream regions have a higher threshold but are still capable of all-or-none firing. [Modified from Bishop, 1958.]

all-or-none spikes (muscle action potentials), which cannot excite the endplate membrane; that is, to put it the other way around, the endplate cannot support a regenerative electrical process, although, because of its size and proximity, it feels the full spike electrotonically.

2. The motoneuron in the ventral horn

In the cat spinal cord this motoneuron has its site of spike initiation separated from the nearest concentration of synaptically activated membrane by some distance, even though a short one; the first-named site is normally in or distal to the initial segment of the axon; the second, in dendrites and possibly also in the soma. Once initiated, the spike can propagate back into the soma and basal dendrites, where, however, the threshold is high. The distal dendrites are considered to be capable only of graded response though the evidence for this comes mainly from cortical pyramidal cells. Decremental, electrotonic spread must be an important part of the mechanism in these cells.

3. The stretch receptor neuron

The stretch receptor neuron in the abdominal muscles of crayfish is similar in locus of spike origin and in the secondary invasion of the spike into the soma. But apparently only the distal dendrites are transducers, producing a receptor potential, and the noninvasion of spikes into this region is strongly indicated. The requirement for electrotonic current flow between the receptive region and the spike-initiating region in order to trigger spikes is even greater than in (2), since the two regions are so far apart.

4. The large cells in the cardiac ganglion

These cells, in lobsters and in abdominal ganglia of crayfish, exhibit normal functioning without spikes in the soma; thus they are more dendritelike than the preceding cases. Dendrites branch both from the soma and from the axon (or axons) and receive different kinds of presynaptic endings in each place. Synaptic potentials seem to arise both near and far from the spike-initiating region. This arrangement is highly integrative in offering possibilities for complex weighting of input, and for mixing of the graded responses. In addition, there are one or more loci of true spontaneity which interact with input according to their distance apart in the neuron.

5. The vertebrate dorsal root ganglion

This ganglion is an extreme case of removal of the receptive processes from the soma. Only the peripheral terminations which are actually sensory are dendrites and they connect with the soma by a long axon, as defined by both histological and functional criteria. The impulse can enter the soma in this instance but this happens after the central branch of the T-shaped axon has propagated the impulse beyond the bifurcation, so that it has little functional significance, at least on a short term basis. Spontaneous firing, often in bursts preceded by growing sinusoidal oscillations, has been seen in dorsal root cells, but at present we cannot evaluate the significance of this for the normal physiology of these neurons.

The dorsal root ganglion cell suggests that the neuron soma is not essential for conduction or transmission and the abundance of unipolars in invertebrates points in the same direction. Direct experimental proof that transmission from neuron to neuron can continue **after removal of the soma** has been achieved by Bethe (1897) in the crab *Carcinus*, and confirmed by Young (1938a) in *Loligo* and Tauc (1960c) in *Aplysia*. The first two were done by surgically severing the stem processes of a group of neurons; Tauc simply hyperpolarized the soma so that it was inactivated. In each case afferent stimulation caused a normal and successful transmission.

B. Differences Based on Functional Properties

A distinction between the last group of differences and the following can hardly be maintained; the separation is essentially one of convenience.

1. Excitatory and inhibitory junctions

In contrast with those transmitted events which increase the likelihood of the postsynaptic neuron's becoming active, certain events appear to be inhibitory, having a depressing effect on the production of impulses by the postsynaptic cell. This dichotomy is certainly one of the most important known, for a useful nervous system can hardly be imagined without "the sculpturing effect of inhibition" (as Eccles has called it) acting on what would otherwise be the massive incoordinate activity of a convulsing excitatory system. We cannot agree that the study of the transmission of information between neurons can be reduced to an analysis of the elementary differences between these two classes, for phenomena such as facilitation, afterdischarge, rebound, the weighting of input, and spontaneity have historically been neglected in such analyses. The term **inhibition** is confined by some workers to those cases which are both specific and direct and where no conditioning excitatory influence is involved. We will use the unqualified term, phenomenologically and regardless of mechanism. So long as an active input causes a reduced output or probability of output, within the structure to which the term is applied, inhibition is an appropriate description. If the decrease in output can be attributed to a reduction in input, inhibition does not apply.

Excitation is always accompanied by an initial depolarization which may, if the explosive or regenerative, all-or-none impulse follows, turn into a brief reverse polarization of the membrane, the "overshoot." In the ordinary, presumably chemically transmitting, synapses, this excitatory prepotential is called an e.p.s.p. (excitatory postsynaptic potential). In electrically transmitting synapses the initial potential is simply the electrotonic spread of the presynaptic action current and this may cause a local response. In any event, the prepotentials must depolarize to a critical level to initiate a spike. This level is not fixed but varies with the milieu, recent events, and possibly more remote events. Values given for "normal" conditions range from 2 and 40 mv of depolarization, usually around 10 to 15 mv. Variation is systematic, at least sometimes; for example, the threshold depolarization of the motorneuron soma in the cat under given conditions is about 30 mv, whereas that of the initial segment of axon is about 10 mv. In nearly spontaneous follower cells of the lobster cardiac ganglion at a certain phase of the cardiac cycle it is less than 1.3 mv.

Specific, direct **inhibitory** synaptic transmission has been examined with the intracellular electrode in several cases, such as the spinal motoneuron in mammals, the muscle receptor organ in crustaceans, cardiac ganglion cells and skeletal muscle cells in crustaceans, and visceral ganglion cells in gastropods. It is usually associated in all these cases with a hyperpolarizing inhibitory synaptic potential or i.p.s.p. and this hyperpolarization begins immediately as the effective transmission takes place. In spinal motoneurons and similar cases the inhibition is a function of the size and sign of the i.p.s.p., which algebraically sums with any e.p.s.p. at the time. In contrast, in crustacean neuromuscular transmission the so called alpha inhibitory effect is not given by the size or direction of the i.p.s.p. but is a function of the large conductance increase which does not diminish and reverse with changing membrane potential. There can be strong inhibition when the membrane potential is at such a level that the i.p.s.p. is zero or even reversed and depolarizing. The impedance change may be effective more locally than the potential change, which has an electrotonic spread. (See pp. 221, 975 for contrast with beta inhibition, the more common type.)

In the present work the considerable information available on the chemical mechanism of e.p.s.p.'s and i.p.s.p.'s cannot be reviewed (see Eccles, 1961b). But their performance under different physiological conditions is relevant and the most conspicuous is the **alteration with changed membrane potential.** The two kinds of synaptic potential differ (Fig. 4.11). The e.p.s.p.'s increase in amplitude with the membrane potential; the dependency varies but it is not known whether this is considerable or significant. In the better known cases the slope of the dependency crosses the line of zero e.p.s.p. at about zero membrane potential. The i.p.s.p.'s, in each of the several cases studied, exhibit a dependency in the same direction but passing through zero i.p.s.p. at about 5 mv above the normal resting membrane potential—for example, at about 60 mv if

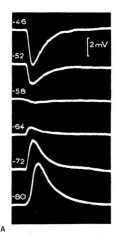
A

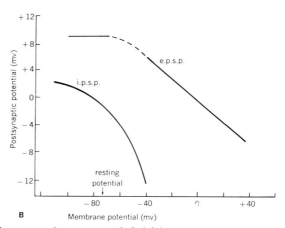
B

Figure 4.11. The dependence of synaptic potential upon membrane potential. **A.** Inhibitory postsynaptic potential in a visceral ganglion cell of *Aplysia* at various membrane potentials, which are set by a separate polarizing electrode inserted into the soma; the normal resting potential is lower than the 60 mv reversal point. [Tauc, 1959.] **B.** Comparison of this dependence for i.p.s.p. and e.p.s.p. in spinal motoneurons. [Eccles, 1957.]

the resting potential is 55 mv. The consequence is that the i.p.s.p. is normally a small increase in potential, not exceeding 5 mv, but if the membrane has been preset at a higher level, say 60 mv, there will be no i.p.s.p.; if preset at a still higher level, there will be a small depolarization —toward but not passing the 60 mv level. This level is called the reversal point or equilibrium level. The concomitant increase in permeability does not pass through zero and reverse but is effective even at the reversal point when there is no i.p.s.p.; in some synapses it is so influential as a shunt, reducing the size of a simultaneous e.p.s.p., that inhibitory action is pronounced regardless of i.p.s.p. magnitude or sign. The inhibitory transmitter tends to clamp the membrane at the equilibrium level. In other cases the potential change is more influential and a reversed i.p.s.p. actually excites or facilitates excitation. These cases are given in more detail below (Section IV).

Besides such specific inhibitory synapses, several other types result in inhibition (Tauc, 1960d). Some e.p.s.p.'s summate to such a considerable depolarization that the spike is blocked; "excess" depolarization block is a familiar result of artificially imposed current but in *Aplysia* at least it happens as a result of natural presynaptic bombardment. In lobster cardiac ganglion cells and in *Aplysia* visceral ganglion cells, some depolarizing synaptic potentials are followed by an apparently secondary phenomenon, a slow hyperpolarization which inhibits. This can take the form of a slowly accumulating increase in polarization with repetition of e.p.s.p.'s or a long-lasting (50 msec) and slowly developing polarization following a single e.p.s.p. Some e.p.s.p.'s are followed by a relative refractoriness, although most are not; this can result in inhibition, by a normally excitatory path, without the conditioning excitation's reaching the level of firing.

Presynaptic excitation and inhibition are well demonstrated in crustacean neuromuscular control, and strong evidence exists also for the mammalian spinal cord. This localization, perhaps based on serial synapses (p. 78) offers a wide potential as an additional mechanism of integrative control (see further, pp. 194, 979). Other forms of excitation and inhibition not yet identified with anatomical synapses are treated above under nonsynaptic interactions.

A feature of the mechanism of transmitter action that has now been found in several excitatory junctions in crustaceans and vertebrates is revealed by **miniature quantal postsynaptic potentials.** These may occur spontaneously, in the absence of presynaptic impulses, and do so in a random sequence; presynaptic impulses increase the probability and hence the average frequency but not the size of the unit events. It is believed that these quantal events reflect the release of packets of transmitter by the presynaptic endings and that this release is an independently determined event in numerous spots of the junctional area (Del Castillo and Katz, 1953, 1954a, 1954b, 1956; Boyd and Martin, 1956; Liley, 1956; Dudel and Kuffler, 1961a, 1961b, 1961c; Katz and Miledi, 1962b).

2. Polarized and unpolarized junctions

Most junctions are polarized or irreciprocally transmitting. This means not only that no impulse can be produced in the presynaptic neuron by stimulation of the postsynaptic neuron, but that in those preparations permitting an internal electrode in the presynaptic terminal region (as in the giant synapse of the squid stellate ganglion and the giant-to-motor synapse in crayfish abdominal ganglia) neither can a sign of a junctional potential be seen when an impulse is elicited in the postsynaptic fiber. However, in the latter synapse a hyperpolarization from any cause in the postfiber does produce a similar potential electrotonically in the prefiber. The normal significance of this, if any, is unknown.

In several cases there are junctions which are apparently unpolarized, as indicated by reciprocal transmission. These are all relay or 1:1 junctions. (a) The paired giant fibers of the serpulid *Protula* can each excite the other at a circumscribed locus in the head. Histologically, a single symmetrical contact between them is found in the brain. Similar reciprocal junctions in the brain connect the paired medial giant fibers in crayfish, in the ghost shrimp *Callianassa*, and others. These are probably all decussating giant fibers. (b) Similar in connecting symmetrical neuronal units are the commissural (or "collateral") synapses between the lateral giant fibers of crayfish. These are transverse connections repeated in successive segments. The possibility of separate, reciprocal, polarized contacts in the two directions cannot be ruled out, but it cannot be supported either. Unlike the corresponding structures in earthworms, which are apparently anastomoses with no more resistance than the axoplasm, the commissural connections in crayfish provide a high resistance between the two lateral giants. (c) The giant fibers of earthworms and the lateral giant fibers of crayfish are divided into segmental units by apparently complete septa, which under the electron microscope look like ordinary axon surface membranes (Hama, 1959); on present evidence they may be regarded as chains of distinct neurons. They transmit impulses in both directions—hence the septa represent unpolarized synapses. A low resistance to electrotonic current is reported as in electrically transmitting junctions. (d) Nerve nets in coelenterates propagate in all directions. Physiological as well as histological evidence (see Chapter 8) is strongly in favor of synapses, at least in most cases. The possibility of paired junctions, one polarized in each direction, is contradicted by the excellent silver and methylene blue stains of Pantin (1952), Batham, Pantin, and Robson (1960), Horridge (1956a), and Mackie (1960). An a priori more probable alternative is that polarized junctions are randomly distributed so that spread can occur in either direction; however, the path taken would always have to be quite zigzag. Reciprocal propagation in narrow bridges of tissue of approximately known neuron density should permit or exclude this according to the frequency of one-way block. Such experiments have not been systematically done but many casual observations have been reported; block of only one direction has been seen repeatedly but is nevertheless rare. On balance it is most likely that the junctions in nerve nets are usually unpolarized. Certain polarized junctions are known in these systems in definite places, as detailed in Chapter 8.

3. *Relay and integrative junctions*

A number of the best studied cases are relay junctions. In addition to those mentioned just above, examples are the giant synapse in the stellate ganglion of squid, the giant-to-motor synapses in crayfish, twitch-type neuromotor junctions like the all-or-none vertebrate skeletal muscle junctions, and polychaete, squid, and crustacean giant systems. In relay junctions neither temporal nor spatial summation is required and a single impulse is delivered for each one received. They are simple in these respects but are more likely to be derived than primitive.

All degrees of integrative junctions occur. In some of the junctions in coelenterates, particularly at the endings of nerve fiber on muscle, repetition of impulses is essential and facilitation is a prominent feature determining the consequence of stimulation. The same is true—on a vastly shorter time scale—at most crustacean neuromuscular endings. In the synapses between caudal hair receptor fibers and ascending second-order neurons located in the last abdominal ganglion of crayfish, spatial summation is necessary—several hair receptors must be simultaneously fired. The same is true of vertebrate autonomic ganglia. More complex are the systems in which one incoming impulse evokes several outgoing, as in the crayfish abdominal ganglia (Wiersma, 1952a; Watanabe, 1958). Finally, the most complex systems are those in which several different inputs influence the output, and each input is weighted differently in effectiveness. Such systems include the convergence of sensory inputs on giant fibers, the motion-detecting cells in optic ganglia, many motoneurons, and most higher internuncial neurons.

4. *Facilitating and nonfacilitating junctions*

Among integrative junctions an important variable is the effect of repetition. Some synaptic potentials are larger the shorter the interval since the last, within limits. If this is due only to the

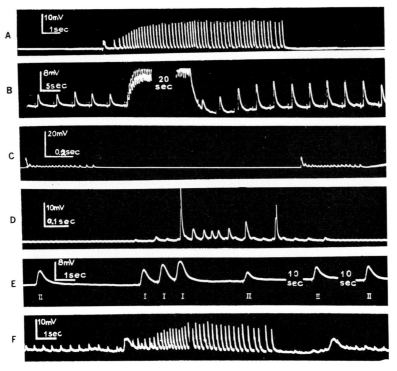

Figure 4.12.

Facilitation and antifacilitation. **A.** Facilitation of e.p.s.p.'s in *Aplysia*. [Tauc, 1958.] **B.** Early and late phases of the same, due to a prolonged high-frequency stimulation. [Fessard and Tauc, 1958.] **C.** Antifacilitation in a follower cell in the lobster cardiac ganglion. [Terzuolo and Bullock, original.] **D.** The same in a crab; the larger e.p.s.p. varies in amplitude with the interval since the last; the small bumps are e.p.s.p.'s from another presynaptic path and do not exhibit antifacilitation. [Terzuolo and Bullock, original.] **E.** Heterologous antifacilitation in *Helix*, interaction of synaptic deflections from different pathways (*I, II*) such as to depress *II* following *I*. [Fessard and Tauc, 1958.] **F.** The same in *Aplysia* but arising within the ganglion, not due to applied stimuli. [Tauc, 1958.]

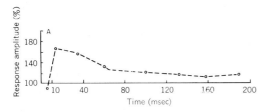

Figure 4.13. Time course of facilitation. The electroplaque of the electric eel develops an e.p.s.p. of a height indicated here as 100% following a single presynaptic volley. If this testing shock is preceded at the intervals shown on the abscissa by a similar conditioning volley, the e.p.s.p. is enhanced except at the shortest intervals. This dependence upon interval is very subject to alteration with drugs and other conditions. [Altamirano et al., after Grundfest, 1959.]

circumstance that the new potential change rises from a residuum of the last one, it is properly called summation; but if the successive increments grow it is facilitation (see further, p. 269). Some junctions show no facilitation. Others show a diminution of successive p.s.p.'s even within a normal range of frequency. This has been called defacilitation but it is more appropriately termed antifacilitation (Fig. 4.12). These effects are manifested promptly and with few arriving impulses; they reach maximum within a few impulses or seconds and they decline usually in tenths of seconds (Fig. 4.13). (Compare fatigue, adaptation; see Glossary.)

Under prolonged repetitive stimulation alterations in response are sometimes brought out, even in preparations that show no effect of brief barrages. Best known is a long-lasting and slowly developing enhancement called posttetanic (or postactivation) potentiation. The growth in effectiveness of an arriving test impulse continues for thousands of impulses and lasts for minutes after input ceases (see further, Eccles, 1957).

5. Accommodation and iterative response

Accommodation (see Glossary and previous chapter for definition) has seldom been measured in structures other than axons. By the nature of the operation used to measure it, the results apply to whatever membrane happens to respond to the applied shock. Since this membrane must be electrically excitable, the accommodation of synaptic membrane cannot be estimated with present forms of controlled stimuli. (The amplitude of a p.s.p. to a constant presynaptic event does not measure excitability as distinct from responsiveness; pp. 133, 259.) At present the soma is the only region other than axon accessible to internal electrodes. Here the measurement is feasible but it is not clear whether corresponding structures are being compared in different preparations. However, the techniques now in use determine the influence of slope of slowly rising current on the spike threshold and not on the excitability of the graded local potential, which is the significant form of response in this region intervening between synapses and spike initiation. Comparisons of the accommodation of the soma are then only partly satisfactory as an indication of the difference between axon and preaxonal regions in respect to this property, but comparison is desirable because it is just in these preaxonal regions that slowly changing potentials and subthreshold depolarization are normally important. In the few cases studied, cell bodies show less accommodation than axons of the same cells (see Chapter 3). Because comparable data are too sparse, this parameter is not included in Table 4.1.

Cell bodies are more able to fire repeatedly under maintained depolarization than axons, in general (Fig. 4.14). This property of **iterativeness** is distinct from low accommodation although of course depending on it; high accommodation cuts short iterated activity. Pacemaker loci in neurons presumably have little or no accommodation.

Actually the accommodation may not begin for some milliseconds of maintained or slowly increasing depolarization, as is true also for some axons. The threshold in spinal motoneurons of the toad does not rise for 10 msec in either soma or initial segment of axon but then rises more steeply for the latter. Measuring the amount of depolarization required to reach threshold with different rates of exponentially increasing stimuli delivered intracellularly, there is only a 25% increase after 20–30 msec in the soma while a 50–300% increase has occurred in the initial segment (Araki and Otani, 1959). In spinal interneuron somata in the cat a test pulse of 14 msec duration evoked more than 10 impulses at a current strength which produced only 2 or 3 in axons of motoneurons; this is due not to a different rheobase but primarily to a steep rise of the number of impulses with the strength of the current just above rheobase in the cell somata (see Hunt and Kuno, 1959). The frequency of a repetitive discharge to a long stimulating pulse usually falls even if it does not stop. This is taken to be a sign of accommodation. The same neuron may sustain a long, high-frequency discharge to certain natural synaptic inputs and this can be explained as due to the rise and fall of the e.p.s.p.'s under the bombardment.

Besides an increase in threshold which may develop during a long depolarization, there has been found a decrease in the threshold **during a long hyperpolarization.** At first, a given hyperpolarization adds just that much to the shock required to reach firing level. But after some time the firing level shifts toward a higher membrane potential. Stimulation becomes easier and a certain e.p.s.p. is more likely to elicit a spike.

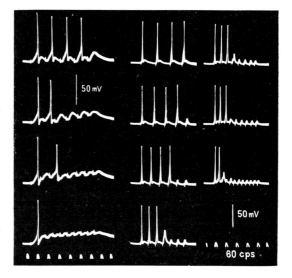

Figure 4.14. Repetitive activation of a neuron and accommodation. Orthodromic and antidromic repetitive activation of a motoneuron with increasing frequencies, showing the accommodation taking place during e.p.s.p. accumulation; the cell is capable of more repetitive firing when driven at the same frequency antidromically without the sustained depolarization of the e.p.s.p. Left column: orthodromic activation with decreasing stimulus intervals from above downward. Middle and right columns: antidromic activation with the same intervals in the middle column and still shorter intervals in the right column. [Araki and Otani, 1959.]

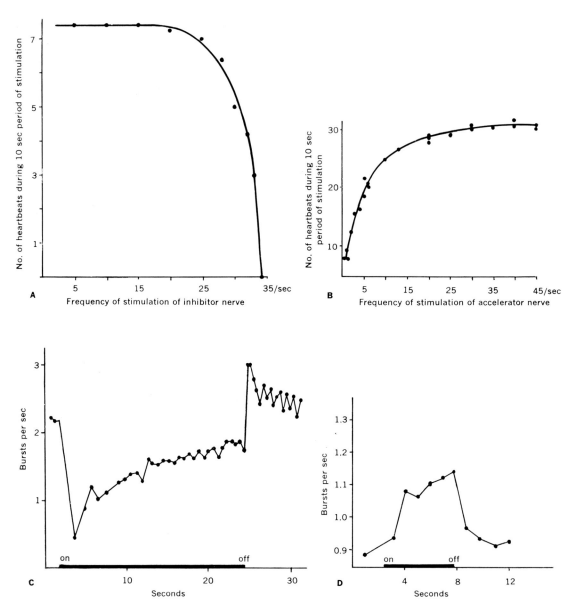

Figure 4.15. Frequency dependence and time course of acceleration and inhibition in the cardiac ganglion of Crustacea. **A.** Dependence of heart rate upon frequency of stimulation of the left cardioinhibitor nerve fiber in a crayfish. **B.** Dependence of heart rate upon frequency of stimulation of right cardioaccelerator nerve fibers in a crayfish. [**A** and **B**, Florey, 1960.] **C.** The time course of inhibition of spontaneous ganglion activity during 49 per second firing of the cardioinhibitor axon. **D.** Time course of acceleration of spontaneous ganglion activity during 22 per second firing in a single cardioaccelerator fiber in the dorsal nerve. Note the slow growth of acceleration and the slow decay to normal activity, in contrast to inhibition. [**C** and **D**, Maynard, 1961; see p. 348.]

This is illustrated well in lobster cardiac ganglion cells and in spinal interneurons (Kolmodin, 1957).

During a maintained or slowly rising depolarizing current there is often seen a **subthreshold oscillation,** for example in toad motoneurons and in insect muscle fibers. This is an important sign of the tendency to iterativeness because it is not due to repolarization following a spike and restimulation when recovery has advanced. The differences which have been noted but not quantified in this tendency (Table 4.1) suggest an important degree of variation among junctions.

6. Aftereffects; fast- and slow-following

Afterpotentials are commonly more pronounced and long-lasting when recorded in cells than in axons, but neurons vary widely in this respect. It is not clear whether these potentials have anything to do with synaptic membrane but they are certainly influential in determining the excitability of some part of the sequence of signal transmission, and therefore the output as function of input arriving at different phases of afterpotential (following antecedent activity). Cells differ in the sequence and magnitude of activity after cessation of input (Fig. 4.15): some persist in the discharge or the inhibition which had been underway (= positive aftereffect; see Fig. 4.16). Others promptly rebound into a postexcitatory depression or postinhibitory discharge (= negative aftereffect; see Fig. 4.17). Sometimes one of these succeeds the other and several cycles of firing and silence may transpire before dying out.

Junctions differ in the frequency to which they can follow incoming rhythmic impulses. Some relay systems can deliver output impulses 1:1 with input impulses up to 100 or even several hundred per second, as in squid giant, some insect muscle, some electric organs in fish. Others can follow only to frequencies of the order of 10 per second, as in primary afferent synapses in the cockroach last abdominal ganglion. This distinction of course does not apply to the most integrative synapses, where output is not 1:1 with input even at the lowest frequencies and where several unsynchronized inputs necessarily converge.

7. Long and short delays

Again, we cannot fairly compare delays at 1:1 with those at many:1 junctions. But even in the former category it is clear that some have a minimum delay of a few tenths of a millisecond and others ten or more milliseconds. Only in those where we have records of the synaptic potential rather than simply the first postspike, can the delay here discussed be measured. In only a few cases is the presumption strong that we have to do with a single junction and no

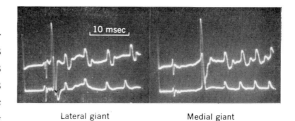

Lateral giant   Medial giant

Figure 4.16. Transmission with repetitive firing to a single presynaptic impulse. Discharges recorded externally with microelectrodes on the second roots of the third abdominal ganglion of a crayfish, left (lower traces) and on stimulation of a central giant fiber, right (upper traces). On the left, stimulation of the left lateral giant fiber alone. On the right, stimulation of the right medial giant fiber alone. After the beginning of the sweep the stimulus artifact is seen; then the large spike in the giant fiber is recorded simultaneously in the left and right roots and—after a short delay in the case of the lateral giant and a longer delay in the case of the medial giant—a burst of impulses, higher in frequency after the medial giant. [Wiersma, 1952.]

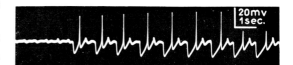

Figure 4.17. Rebound discharge after i.p.s.p. Intracellular records from a large ganglion cell in *Helix*. Note that spikes arise on the recovery phase of the i.p.s.p. at a level of membrane potential higher than the resting level. [Tauc, 1960.]

intercalated neurons. The selected cases in Table 4.2 show the wide range of values. They are nevertheless always in the range of milliseconds except for the electrotonically transmitting junction, where there is virtually or actually no delay.

8. Chemical versus electrical transmission

The discussion which has been waged for decades over these alternatives is far from finished. The adequacy of evidence to indicate one or the other in particular instances will continue to be a lively subject. But we are now in a position to say—as many predicted years ago—that each mechanism in all probability actually prevails in different cases and that therefore synaptic transmission is not monistic; by the same token, a synapse is not definable by a certain mechanism. The two cases known

Table 4.2. JUNCTIONAL DELAYS AND RECEPTOR LATENCIES. This table gives minimum delays except when otherwise indicated. Synaptic delay is defined for this purpose as the time transpiring between first electrical sign of a single presynaptic impulse or synchronized volley occurring in presynaptic terminals and the first electrical sign of response in the postsynaptic unit. It is not always possible to bring published values to a common basis because different interpretations prevail as to the moment in a presynaptic spike when activity can be said to have entered the terminals, or, in the other cases, different bases have been used for correcting for conduction time between presynaptic electrodes and the terminals. All figures except those footnoted include the original author's corrections. Note that time is measured to the beginning of postsynaptic response, often a synaptic potential and not a spike. Since the spike may be initiated any time up to or slightly beyond the crest of the junctional potential, a separate section gives values for crest time of such potentials. Central, peripheral, and artificial synapses are included, and response times (to first sign, slow wave or spike) of some primary sensory neurons to abruptly applied adequate stimuli are given in a separate section.

	Group and Species*	Preparation, Junction and Condition	Corrected Delay (msec)*	Temp. (°C)
		Synapses, including Neuromuscular Junctions		
	Coelenterata			
1	Sea anemone, *Metridium senile*	Nerve net; mesentery; through-conduction pathway	<2.5[1]	—
	Annelida			
2	Plume worm, *Protula intestinum*	Natural synapse between giants, in brain; fresh	0.8	16
3	*Protula intestinum*	Natural synapse between giants, in brain; fatigued	6.5	16
4	*Protula intestinum*	"Quasi-artificial" jumping between giant fibers; fresh	0.6	16
5	*Protula intestinum*	"Quasi-artificial" jumping between giant fibers; fatigued	7+	16
6	Earthworm, *Lumbricus terrestris*	Oblique septum between segmental giant fiber units	<0.1[1]	24
	Arthropoda			
7	Crayfish, *Cambarus clarkii*	Lateral giant fibers, segmental, septal synapse	<0.1	—
8	*Cambarus clarkii*	Lateral giant fibers, commissural synapse	0.5	—
9	*Cambarus clarkii*	Ipsilateral lateral giant to third root motor fiber	0.1	20
10	*Cambarus* sp.	First synapse in proprioceptive pathway; exposed ventral ganglia	3.5–4.5[1]	—
11	Ghost shrimp, *Callianassa*	Last abdominal ganglion; giant central fiber to motor fibers in telson	0.25	20–22
12	Cockroach, *Periplaneta americana*	Last abdominal ganglion; cercal afferents to ascending pseudogiant fibers	0.6–1.5	—
	Mollusca			
13	Slug, *Ariolimax columbianus*	Pedal ganglion, isolated	33	7.6
14	*Ariolimax columbianus*	Pedal ganglion	19	21.8
15	Squid, *Loligo opalescens*	Synapse in stellate ganglion, 2nd to 3rd order giant fibers	0.5	24
	Chordata			
16	Frog, *Rana* sp.	Neuromuscular junction; semitendinosus	0.8	22.5
17	Cat, *Felis cattus*[4 or 5]	Neuromuscular junction; soleus and other muscles	0.55–0.65	37–39
18	*Felis cattus*[5]	Monosynaptic reflex, ventral horn motoneurons, wire electrodes	0.3–0.45[13]	37
19	Rabbit, *Oryctolagus cuniculus*[3]	Trochlear motoneurons, stimulating superior colliculus	0.7 aver.	37
20	*Oryctolagus cuniculus*[3]	Trochlear motoneurons, weak stimulus	0.9 max.	37
21	*Oryctolagus cuniculus*[3]	Trochlear motoneurons, facilitated	0.5 min.	37
22	Cat, *Felis cattus*[10]	Cochlear nucleus in medulla (trapezoid fibers)	0.8	—
23	Turtle, *Pseudemys* sp	Sympathetic ganglion, superior cervical, B-fibers	8[14]	—
24	*Pseudemys* sp.	Sympathetic ganglion, superior cervical, C-fibers	25[14]	—
25	Cat, *Felis cattus*[4]	Sympathetic ganglion, superior cervical, synapse facilitated	2	35
26	*Felis cattus*[5]	Stellate ganglion, in situ	3–4	37–39

Artificial Synapses

27	Earthworm, *Lumbricus terrestris*	Single giant fiber, after-discharge arising near anodally depressed locus	5+	24
28	Crabs, several species	Two isolated nerves or fibers in contact	7+	—
29	Cuttlefish, *Sepia officinalis*	Two isolated giant fibers in contact; normal or citrated	2.5^2–5	—
30	Cuttlefish, *Sepia officinalis*	The same; citrated; rhythmic subthreshold activity	>40^2	—
31	Cat, *Felis cattus*[7] or [4]	Cut end of nerve; A-fibers; motor to sensory	0.1–0.3	—
32	*Felis cattus*[9]	Cut end of dorsal columns of spinal cord; dorsal root to dorsal root	0.1–0.3	—

Response Time of Sense Organs to Abruptly Applied Physiological Stimuli

Mechanoreceptors

33	Crayfish, *Cambarus* sp.	Tactile hairs on telson	0.5–1.5	—
34	Frog, *Rana temporaria*	Touch receptors, dorsal skin	0.7–14.6 max.[1]	20.6–29.6
35	Cat, *Felis cattus*[7] or [4]	Pacinian corpuscle, single, mesenteric	0.5–1.5	—
36	*Felis cattus*[8]	Baroceptors in carotid body, single fibers	<$10^{1,2}$	—
37	*Felis cattus*	Auditory nerve spikes, click stimulus (incl. 0.1 msec latency of microphonic)	0.6^1min.–0.8^1max.	—
38	Guinea pig, *Cavia porcellus*	Cochlear microphonic to action potential	0.15^1	—

Radiation receptors

39	Clam, *Mya arenaria*	Photoreceptors in siphon; electrical response at "on."	720–16,000	20
40	Horseshoe crab: *Limulus*	Spikes in optic nerve; near-maximal and near-threshold stimuli	77–750	—
41	*Limulus*	ERG,[12] intracellular electrode in ommatidial receptor	70	—
42	*Limulus*	ERG, near-maximal light flash and 5 log units weaker, respectively	10;55	25–28
43	Isopoda, *Ligia occidentalis*	ERG, near-maximal intensity and 10^{-5} of this	6–20	22
44	Grasshopper, *Melanoplus*	ERG, near-maximal and 10^{-6} of this	9.3–59.6	—
45	Fly, *Calliphora*	ERG, near-maximal	6^2	—
46	Frog, *Rana* sp.	ERG, a-wave; 10^6 and 10^2 times threshold	28–120	18
47	Cat, *Felis cattus*	ERG, a-wave	4	—
48	*Felis cattus*[4]	ERG; b-wave, near-maximal and very weak stimuli, respectively	25;80	—
49	Rattlesnake, *Crotalus viridis*[11]	Infrared receptor in facial pit organ	15–50+	23

Rise Time (foot to summit) of Junctional Potentials

50	Plume worm, *Protula*	Giant synapse in brain, in situ	1	16
51	Various crabs, crayfish	Various leg muscles; end-plate potential	3	17
52	Squid, *Loligo pealii*	Giant synapse in stellate ganglion	0.3–1.5+	23
53	Frog, *Rana temporaria*[11]	Neuromuscular junction; isolated skeletal muscle; internal electrode	1.2	20
54	*Hyla aurea*	Isolated muscle fiber; external microelectrode on single endplate	0.5	—
55	Cat, *Felis cattus*[11]	Neuromuscular junction; soleus strip with circulation intact	0.8	37–39
56	*Felis cattus*[6]	Spinal cord; ventral horn motoneuron; internal electrode	0.6–1.0	36–38
57	*Felis cattus*	Sympathetic ganglion, stellate, in situ	10–20	37–39
58	Rabbit, *Oryctolagus cuniculus*[11]	Sympathetic ganglion, superior cervical, isolated	25–35	35

*([1]) Uncorrected for (i.e. includes) conduction time. ([2]) Measured from illustrations. electroretinogram. ([13]) Measured from peak of positive wave of presynaptic volley. ([3]) Decorticate. ([4]) Decerebrate. ([5]) Nembutal. ([6]) Pentobarbitone. ([7]) Chloralose. ([14]) Measured from first reversal of sign of diphasic prespike; conduction-time ([8]) Chloralose-urethane. ([9]) Dial. ([10]) Avertin anesthesia. ([11]) Curarized. ([12]) ERG = correction not stated.

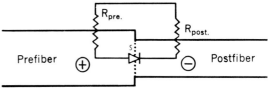
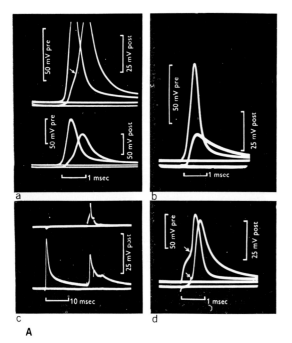

Figure 4.18. An electrically transmitting synapse. The central giant-to-motoneuron synapse in the crayfish; internal microelectrodes have been placed inside the presynaptic and inside the postsynaptic axons, permitting both recording and polarizing, to change the membrane potential level. **A.** Presynaptic potential on the upper trace and postsynaptic on the lower trace. Note the near simultaneity of takeoff of the potentials in pre- and postfibers. The second p.s.p. in c and that in d were evoked by direct stimulation of the prefiber with an intracellular electrode; otherwise the stimulus was applied externally to the dorsal surface of the cord. The two records in a were recorded from the same synapse at different amplifications. The p.s.p. at about the point indicated by the arrow in a exceeded threshold and evoked a spike. Subthreshold p.s.p.'s are shown in b–d. In c, the first p.s.p. evoked by external stimulation was apparently due to firing of one of the medial giant fibers, for no spike was seen in the impaled lateral giant fiber (upper trace). The second p.s.p. accompanied a lateral giant spike (85 mv) evoked by intracellular stimulation. The small hump on its falling phase was probably associated with firing of the other lateral giant fiber. In d the arrows indicate the end of the depolarizing current pulse supplied to the fiber. **B.** The proposed equivalent circuit for the giant-to-motor synapse. Pre- and postfibers are shown as though terminating end to end. The "input resistances" of the two fibers are represented by the elements, R_{pre} and R_{post} and the synaptic resistance by S. The plus and minus signs indicate the direction of the potential difference across the junction for which S is small. Inasmuch as R_{pre} and R_{post} are shown as constant resistor elements, the diagram only holds for small steady potential changes. [Furshpan and Potter, 1959a.]

most intimately from the standpoint of recording inside presynaptic fiber and postsynaptic fiber, close to the junction, are the squid stellate ganglion giant synapse and the crayfish giant-to-motoneuron synapse (Fig. 4.18). These are described in detail in the next section but, in brief, the former proves not to employ intercellular electric current whereas the latter does.

The evidence of electrotonic current or its absence, providing the electrode is not only internal but in the synaptic region, can exclude or permit a major role of current flow from presynaptic events. In contrast, pharmacological evidence has been able only to build up a presumption on the basis of several converging types of observation. That some acetylcholine appears in bathing fluids after junctional activity in certain preparations is clear. The principal quantitative evidence comes from preparations in which it is protected from enzymatic destruction, by inhibiting cholinesterase with eserine and stimulating for a long time. It is not known how much of the acetylcholine recovered comes from presynaptic endings, or is only released because of the presence of eserine or is released after transmission is complete. The potency of acetylcholine topically applied to the synaptic region of a nerve muscle preparation in causing a response of the muscle is clear. Electrophoretic ejection of acetylcholine from a micropipette in the neighborhood of a muscle endplate, during a brief interval of time, has shown its capacity to set up artificial endplate potentials differing little except in their longer time course from normal ones. Doses of the order of 10^8 to 10^9 molecules are effective; how much of this dose is wasted due to the distance between pipette and target is unknown. The ratio between the amount recovered under eserine (divided by the number of impulses during the long stimulation) and the amount adequate to cause a response has varied widely in the literature; Katz, in his 1959 review, gives a ratio of 100 for the excess of the latter over the former. Whether such a figure is significant or impressive depends on the weighting given to the questions above. Most workers find the suggestion convincing that this argues in favor of chemical transmission. However, in fairness, there has been little serious consideration by most authors of the suggestion that acetylcholine plays its main role, not intercellularly but intramembranously in both pre- and postsynaptic membranes. The pharmacological, histochemical, and biochemical evidence is actually compatible with either view, although it is usually held to support the former on balance. Only the electrophysiological evidence of lack of current flow in the postsynaptic membrane during the presynaptic impulse is selective. It is clear that the issue is not passé just because it is old and that new evidence will be forthcoming which will change our present picture. It is here accepted that the vertebrate nerve muscle junction, electric organ, spinal motoneuron, and squid giant synapses—and by analogy many other synapses which share postsynaptic potential properties with these—are chemically transmitting. The crayfish giant-to-motoneuron synapse, the septal synapses of the lateral giant fibers of crayfish and earthworms, the commissural synapses between right and left giants, the inhibitory synapse upon Mauthner's cell in fish, from the contralateral VIIIth nerve, the calyciform synapses in the ciliary ganglion in the chick, and quite possibly the junctions in the large bipolar nerve net in medusae are all electrically transmitting.

A proper review of the literature on mechanisms of transmission is beyond the scope of this work. It is in line with the present level of treatment, however, to state the conclusion from this body of information: transmission in different situations and animals is not always accomplished by one and the same mechanism, nor are a large number of different mechanisms employed. Rather, it appears likely that there have been developed several closely related and several quite different means of transmitting excitation from one neuron to another or to an effector. What these means are, apart from the broad generalities in the last paragraph, can hardly be described even in the most tentative terms. Florey (1961b) has provided an excellent review of the comparative physiology of transmitter substances.

IV. EXAMPLES OF BETTER-KNOWN JUNCTIONS

The better-known junctions are for the most part those where it has been possible to insert an intracellular recording electrode into the postsynaptic region, permitting visualization of subthreshold events, particularly small synaptic potentials. Some are known only from external electrode recording—which does permit in favorable situations, a view of larger synaptic potentials. A few have been successfully impaled with two and even more ultramicroelectrodes, for instance two in the postsynaptic fibers close to the junction and one or two in the prefiber; when a second electrode is inserted in a given neuron it permits imposing changes in the membrane potential or, in special cases, injection or iontophoretic addition of drugs or ions.

In this section a brief characterization of nearly all the better-known synaptic preparations is provided (Table 4.1). Emphasis is placed on the properties and performance and less on the mechanism. Since a general view together with comparisons is important, the main vertebrate cases are necessarily included, as well as instances of artificial synapses.

A. Invertebrate Preparations

1. Squid giant synapses in the stellate ganglion

One of the best known in many respects, this relatively simple junction, more properly known

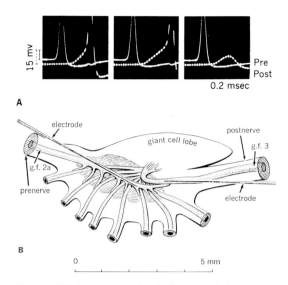

Figure 4.19. A presumed chemically transmitting synapse. The giant synapse of the squid between the second- and third-order giant fibers cannot be electrically transmitting and therefore is probably chemically transmitting. Microelectrodes have been inserted inside the presynaptic and inside the postsynaptic axons, permitting recording and also polarizing to alter the membrane potential. **A.** Presynaptic spikes *(continuous line)* and postsynaptic responses *(broken line)* recorded as transmission is failing after a prolonged repetitious stimulation. Note the long delay between the onset of pre- and postactivity. [Hagiwara and Tasaki, 1958.] **B.** Slightly diagrammatic representation of the preparation used, drawn to scale; in some cases a second electrode was inserted in either the pre- or postfiber and in the later experiments of Hagiwara and Tasaki the tips of the electrodes were pushed closer to the actual area of synaptic contact. [Bullock and Hagiwara, 1957.]

as the distal giant synapse, is known anatomically (Figs. 4.19, 4.20) from the work of Young (1939), Robertson (p. 118), and Hama (p. 114). A giant, hardly tapering presynaptic fiber representing the second-order neuron in the giant system comes from the pleurovisceral ganglion in the head, through the mantle connective, and divides in the stellate ganglion into about ten branches; each ends blindly applied to the side of one of the third-order giant fibers. These are motor axons arising in cell bodies in a special lobe of the ganglion (see p. 1483); in squid (*Loligo*) there is one in each of the ten or so stellar nerves supplying the mantle. The last stellar nerve has the largest of these axons and

this is often referred to as "the" giant fiber of the squid. Its synapse can be studied without interference from the others of the same kind in the ganglion, either by penetration with intracellular electrodes or by cutting the other stellar nerves close to the ganglion and thus causing prompt failure of the damaged giant stumps remaining in the ganglion. Electrophysiological analysis of this preparation has been reported by Bullock (1948), Bullock and Hagiwara (1957), Bryant (1958), and Hagiwara and Tasaki (1958). Because of the anatomy, it is easy to prove that the large electrical spikes recorded come from the giant fibers and that the region where pre- and postsynaptic fibers lie side by side is the site of impulse transmission. The electrical evidence confirms the isolation of the pre- from the post-fiber and of the several postfibers in the respective stellar nerves from each other.

The giant synapses in the squid are found to be polarized, short delay 1:1 relay junctions, in the fresh condition. No summation, facilitation, or afterdischarge occurs normally and transmission can follow incoming impulses at least as high as 475 per second. Minimum **synaptic delay,** measured to the start of the e.p.s.p., is given as 0.5 msec or less at 24°C, and the Q_{10} (9–24°) is very low, about 1.3 (Bullock, 1956b). (The temperature dependence of the latency of the postspike is much higher and even conduction in the third-order axon appears to be more sensitive to temperature than is synaptic delay.) These measurements of delay are significant because they are almost the only ones where conduction time in fine presynaptic terminals is not a complication, since the arrival time of the prespike at the junction can be confirmed with an internal electrode in the prefiber right at the synaptic region. A further finding permitted by this arrangement is that fatigue from repeated stimulation does not in this case affect the delay but only the rate of rise of the e.p.s.p. and hence the postspike latency. This may be taken to mean that lability is more fundamentally a postsynaptic property; in more integrative junctions there is evidence of lability in the presynaptic terminals as well. After the postspike has failed, continued presynaptic impulses hold the junction in a fatigued state; therefore fatigue does not depend on postfiber firing and the e.p.s.p. causes refractoriness which prevents recovery.

The **time course of the e.p.s.p.** in the squid is brief (Fig. 4.20), crest time is about 0.5 msec and it falls to 37% amplitude in about the same time. But these values are labile and increase with fatigue; the decay is therefore not entirely passive. The e.p.s.p. may reach 20% of the spike height, which is a fairly high spike threshold. The e.p.s.p. amplitude declines to half about 3 mm from the synapse but this again is labile, showing that some active though decremental propagation is involved. There is sometimes a small undershoot (hyperpolarizing afterpotential) after the e.p.s.p., but usually there is none. Commonly a second synaptic potential can be elicited immediately after or even on top of the first, showing slight facilitation; but this varies among preparations and often there is strong refractoriness which may last more than five msec. A synaptic potential can be superimposed upon the falling phase of a postsynaptic spike during a time when that membrane is probably refractory to direct electrical stimulation, suggesting that the synaptic excitation process is something distinct from electrical excitation.

Presetting the membrane potential of the postfiber to values higher than the resting potential— that is, to greater polarization, by increasing the negativity of an internal electrode—increases the e.p.s.p. as well as the spike, but to quite different degrees. The slope of dependence of e.p.s.p. amplitude on membrane potential extrapolates to zero synaptic potential at about zero membrane potential, suggesting as one possible mechanism that the synaptic potential results from an increase in permeability to several kinds of ions; but the spike dependence extrapolates to zero at a highly reversed membrane potential of high internal positivity, close to the value expected for a purely sodium permeability increase.

Of great importance for understanding integration in other synapses is the demonstration of dependence of e.p.s.p. on **presynaptic spike amplitude** (Fig. 4.21). In the present preparation it is possible to alter the prespike height by inserting an electrode into the prefiber and

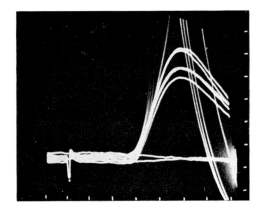

Figure 4.20. Intracellular record from the postunit of the squid giant synapse at high amplification to show the usual finding that there is no detectable deflection attributable to the field potential of the arriving presynaptic spike. Calibration in millivolts and milliseconds. [Bullock and Hagiwara, 1957.]

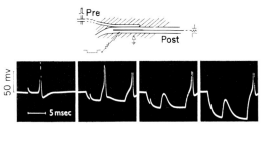

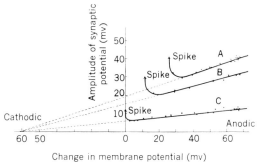

Figure 4.21. Effect of hyperpolarization of the postsynaptic membrane upon its response. An inward current pulse is delivered through a separate microelectrode inserted in the postfiber and stimulating shocks are delivered to the presynaptic axon during this current pulse. The spike is blocked and the e.p.s.p. grows larger with hyperpolarization. The three curves in the graph were obtained at different rates of repetition of presynaptic stimulation: **A**, at 1 per second; **B**, at 1.5 per second; **C**, at 10 per second. [Hagiwara and Tasaki, 1958.]

hyperpolarizing or depolarizing it, as described in the preceding paragraph for the postfiber. The significant finding is that a small change in the presynaptic spike amplitude has a large effect and a nonlinear effect on the postsynaptic potential amplitude and hence on the occurrence or latency of a spike. In the case illustrated in Fig. 4.22, a 30% reduction of the prespike almost completely prevented a synaptic potential and, just below the critical height for a postspike, a 10% change in prespike caused a twofold change in e.p.s.p. This is compatible with the assumptions that any lability in the presynaptic spike alters the amount or effectiveness of the synaptic transmitter produced at the terminals and that either this dependence or that of the e.p.s.p. upon the transmitter is steep and nonlinear, with what amounts to a threshold.

Another feature of consequence is the absence of **spread of current** across the junction. No appreciable potential is developed across the postsynaptic membrane as a result of large currents passing across the presynaptic membrane, whether from normal prespikes or imposed pulses in either polarity. This forces us to conclude that the transmission cannot be electrical.

A second or **accessory presynaptic fiber,** also of large size, makes synaptic contact with the same postfiber more proximally, or closer to the cell bodies (Fig. 25.38). Bryant has succeeded in stimulating this prefiber while recording from the third-order giant internally and found that it is not inhibitory or essentially different from the main giant synapse; it appears to be an alternate pathway for exciting the same motor giants. Again, the proximal synapses may be used for unilateral and local control of the mantle, a suggestion based only on permissive anatomy. Possibly under conditions of high-frequency prolonged activity, when one synapse may fail, the other could act as a booster to insure continued transmission. Unless one or other of these possibilities occurs it is difficult at present to discern any functional significance for these junctions, main or accessory. They would in that event appear to be merely of morphogenetic or trophic convenience in providing an avalanching pathway to a large musculature, without integration but only high-fidelity relay of signals, just as in conduction in the axon.

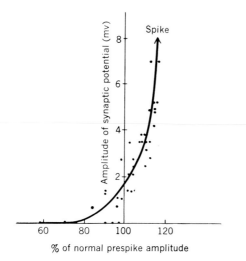

Figure 4.22. The dependence of amplitude of the e.p.s.p. upon amplitude of the presynaptic spike. The presynaptic spike height is controlled by polarizing the membrane of the prefiber and the resulting effect on the e.p.s.p. amplitude is recorded with an electrode inside the postfiber. Unless the prespike is more than 80% of normal amplitude there is no postsynaptic response; above that minimum the effect is quite nonlinear and steep. [Hagiwara and Tasaki, 1958.]

2. Crayfish giant fiber-to-motoneuron synapses

The giant system of crayfish provides a number of synapses of special interest, including the (a) segmental (or septal), (b) the commissural (or collateral), both belonging to the lateral giant fibers, (c) little-known junctions on the afferent side of the giants, and (d) several better-known junctions of the efferent side. The large motor axons emerging in the third roots in each abdominal segment form synapses on their way across the midline and out, with (α) the ipsilateral lateral or segmental giant fiber (designated by the side of emergence of the motor axon), (β) the ipsilateral medial giant fiber, (γ) the contralateral medial giant fiber, (δ) the commissural collateral of the contralateral lateral giant fiber, and (ε) the contralateral symmetrical motor axon, at the point where they both cross the midline; in addition, (ζ) physiological evidence points to a small-fiber, inhibitory input from the central neuropile. The most detailed

IV. EXAMPLES OF JUNCTIONS   A. Invertebrate   2. Crayfish giant motor

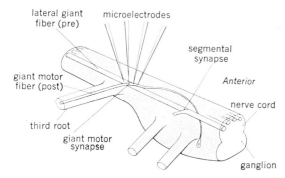

Figure 4.23. A diagram of the preparation for study of the giant motor synapse of a crayfish. The course of one motor axon is shown in a portion of the abdominal nerve cord; of the several junctions made with central giant fibers en route, only its junction with the ipsilateral giant fiber is shown. A polarizing and recording microelectrode are inserted in the prefiber and the postfiber. [Furshpan and Potter, 1959.]

information is available for (α), including electron microscopy (Fig. 2.27). The anatomy is due to Johnson (see p. 867), Robertson (p. 906), and Hama (p. 74), and the properties revealed by external electrodes to Wiersma (1947) and by internal electrodes to Furshpan and Potter (1959a, 1959b). Again the structure is relatively simple (Fig. 4.23) and the prefiber large, even larger than the postfiber. Contrary to expectation from earlier theoretical volume-conductor considerations as to electrical transmission, this junction works in spite of the prefiber's continuing past the synapse.

The crayfish giant-to-motor synapse is polarized and normally 1:1. Its e.p.s.p. is quite like that of the squid in time course, and size. A striking feature however is the extremely short delay—about 0.1 msec from beginning of prespike to beginning of e.p.s.p. This is more significant taken together with the other remarkable property: either normal impulses or subthreshold pulses spread across the synapse and are visible as appreciable potentials in the postsynaptic membrane—quite in contrast to the squid and indeed most other normal junctions so far examined. Presynaptic hyperpolarization does not so spread; further, postsynaptic hyperpolarization does cause appreciable hyperpolarization in the prefiber. These results justify the hypothesis of Furshpan and Potter that the "synaptic membrane" in this junction is a **rectifier** of low resistance to positive current flowing from pre- to postfiber but of high resistance to electrotonic flow in the opposite direction (Fig. 4.24). The "synaptic rectifier" is oriented in the right direction to allow presynaptic action currents to stimulate the postfiber

Figure 4.24. The polarized electrically conducting synapse acts as a rectifier. *Above*, the complete current voltage characteristic of the giant-to-motor synapse of the crayfish; positive values of V_s signify that the prefiber side of the junction was electrically positive with respect to the postfiber side. *Below*, the synaptic rectifier hypothesis in the four possible situations, imposing current through a microelectrode inside the pre- or inside the postfiber in the depolarizing or in the hyperpolarizing direction. The junction is indicated by a *dotted line* or a *heavy bar*, representing a low or high synaptic resistance, respectively. The arrows give the direction of (positive) current entering or leaving the current-passing microelectrode; dashed lines indicate negligible current flow due to high synaptic resistance. Diagrams *a* and *d* correspond to the two situations in which transsynaptic effects have been observed; in both cases current would cross the junction in the same direction. [Furshpan and Potter, 1959a.]

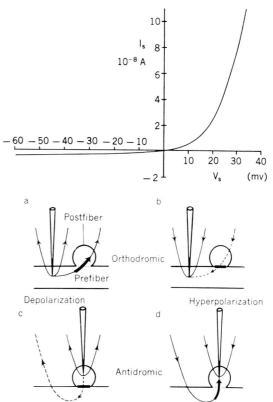

but to prevent antidromic transmission. There is evidence that the electrotonic current permitted with a normal prespike is adequate to account for normal transmission. On the other hand it would be difficult on a chemical theory of transmission to account for the antidromically produced hyperpolarization of the prefiber by a hyperpolarizing pulse applied to the postsynaptic membrane. A strong argument is thus adduced for a type of electrical transmission. A number of quantitative and qualitative tests of the hypothesis have been carried out with results agreeing with expectation. This is a highly significant discovery for it can hardly be supposed that the mechanism of these synapses is entirely peculiar to the crustacean abdomen. However, apart from the near absence of delay there is little difference in properties or possibilities between these and chemically transmitting synapses like that in the squid, unless it happens that the time constants—insufficiently known as yet—permit

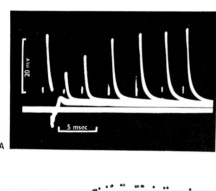

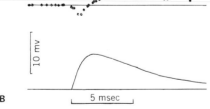

Figure 4.25. Small fiber inhibitory transmission to the motor neurons of the crayfish. **A.** Seven e.p.s.p.'s produced by stimulation of the lateral giant fiber were successively superimposed on a slow potential produced via small fibers. **B.** Tracing of a similar experiment; the peaks of the giant e.p.s.p.'s are indicated by dots; in the absence of a slow potential these were 20 mv in amplitude; during the active phase of the slow potential the giant e.p.s.p.'s are depressed, during the falling phase of the slow potential they are enhanced. [Furshpan and Potter, 1959b.]

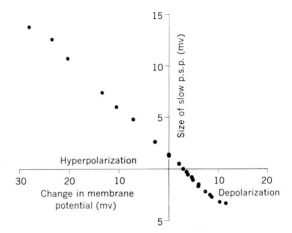

Figure 4.26. Dependence of the slow p.s.p.'s on change in postfiber membrane potential. Depolarizing slow p.s.p.'s are plotted above the horizontal axis. At resting potential the p.s.p.'s are 1.4 mv in size; they reverse sign about 3 mv below resting potential. [Furshpan and Potter, 1959b.]

postfiber undershoot or afterpotential to modify slightly the prefiber membrane potential. This would have little effect unless synaptic potentials in this unit, the central giant fiber, were close to critical amplitude as a result of antecedent small-fiber activity such as that described by Kao and Grundfest (1957) in earthworms.

The same postsynaptic motor axons exhibiting rapid, excitatory p.s.p.'s from central giant fiber activity also exhibit **slow p.s.p.'s** with an inhibitory effect from activity presumably of small fibers in the cord; this is junction (ε), above (Fig. 4.25). Transmission across these synapses appears to be chemical. Like i.p.s.p.'s elsewhere they reverse in sign at a membrane level not far from the resting potential and tend to drive the membrane toward this level—that is, to inhibit by opposing large depolarizing e.p.s.p.'s. But curiously, these slow p.s.p.'s are always seen as small depolarizations at resting membrane level and their reversal level is at a membrane polarization a few millivolts lower than the resting level (Fig. 4.26).

3. Crayfish stretch receptor modulation

One of the most valuable preparations in neuronal physiology is the muscle receptor organ

(MRO) in the dorsal trunk musculature of crayfish and some allies (see Chapter 18). In each segment, on each side lies an organ consisting of two thin bundles of muscle fibers and two large, multipolar primary sensory neurons, each with dendritic sensory terminals in a muscle bundle and an axon proceeding centrally. Besides motor axons to the muscle bundles, one (crayfish) or two (lobster) inhibitory axons travel in the same nerve with the afferent fibers, ending among the dendrites of the neurons (Fig. 4.27). The neurons are excited by stretch or by contraction of the muscle bundle; one neuron is phasic and the other tonic in response to a maintained stretch. Both fire very rhythmically so that small changes in frequency are noticeable. By stimulating an inhibitory axon during controlled stretch, the sensory discharge can be modulated. This makes a useful object of study since it is clear that a single presynaptic fiber is involved and no interneuron. Details revealed by intracellular recording may be found in the papers of Kuffler, Eyzaguirre, and Edwards.

The resting potential in the unstretched or relaxed state is 70–80 mv. Graded stretch produces graded depolarization; this is the receptor potential which must be 10–20 mv of depolarization from maximum resting potential in order to initiate impulses. These are fired repetitively as long as the receptor potential does not decline below the critical level. As an intermediate step in the firing we can recognize the progressive depolarizing potential change exactly similar to that called a pacemaker potential in spontaneous cells. The rate of depolarization or slope of this potential change is a function of the general level of membrane potential as determined by stretch, and in turn determines the frequency of firing. The impulses have been shown to be initiated some distance out in the axon and to propagate both away from the cell and into it, but it is supposed that the spikes do not invade the region of the dendrites where the receptor potential is maximal.

A train of inhibitory impulses elicits 1:1 a succession of **i.p.s.p.'s** of long duration (20–40 msec) which do not facilitate but summate at sufficient frequency. Their sign and size depend

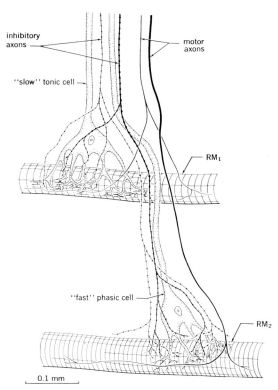

Figure 4.27. Diagram of the crayfish stretch receptor preparation. This represents the pair of organs present on each side of each segment of the abdomen. RM_1, receptor muscle 1; RM_2, receptor muscle 2. [Modified from Alexandrowicz, Florey and Florey, and Kuffler and Eyzaguirre.] See also Fig. 18.14.

on the prevailing membrane potential, just as for other i.p.s.p.'s; the reversal level is usually a few millivolts on the depolarized side of the resting potential in the unstretched state, which is the maximal resting level. Occasionally the reversal level is much further below the resting level; the i.p.s.p. is then not a minute potential but is a large depolarizing hump and actually precipitates an impulse. This instance shows that at the spike-initiating site it is the potential change that matters rather than the impedance change, which is always a drop regardless of prevailing membrane potential. Compared to other inhibitory junctions, the i.p.s.p.'s are an unusually large fraction—as much as 90%—of the difference between the starting level of membrane potential and the reversal level; 20% is given as normal value in cat motoneurons and even smaller values in crustacean neuromuscular junctions. This explains the lack of facilitation for the humps are already almost as large as they can become. When a series of inhibitory impulses arrives during a sensory discharge, they find the prevailing membrane level on the depolarized side of the reversal point so that the i.p.s.p.'s act to repolarize. They reduce the receptor potential to a given stretch, partly by short circuiting, and thus, providing the inhibitory frequency is high enough, prevent the pacemaker potentials from being initiated.

Additional information is obtained by sending **antidromic impulses** into the cell from stimulating electrodes on the afferent axon, during conditions when endogenous impulses will not arise. At very high levels of membrane polarization there is no undershoot or afterhyperpolarization, but instead, a slow repolarization. Curiously, the membrane level when this reverses into an afterhyperpolarization is different from that for i.p.s.p. reversal, sometimes higher, sometimes lower. An antidromic impulse may be blocked before entering the soma, in which case either an electrotonic ripple or a local response is recorded there; either will facilitate a second antidromic impulse which therefore is likely to grow up into a spike. Occasionally an antidromic spike leaves behind it a condition such that, some milliseconds after it has passed off, a new impulse arises in or near the cell and propagates orthodromically. This "reflection" is doubtless quite abnormal but is possibly a valuable clue, as is often true of abnormal behavior, to inherent capacities which may in other situations be utilized normally.

A discovery of quite general importance made on this preparation is that lowered temperature increases the receptor potential to a given stretch and at the same time raises the threshold of spike initiation, that is, both are shifted toward further depolarization. The net effect is a near **temperature independence** of the frequency of impulses caused by stretch—an elegant example of the way two constituent processes with ordinary Q_{10} values can be combined to stabilize the final result against temperature. Slight differences in the dependence of Q_{10} upon degree of stretch and upon time cause an "anomalous" transient increased frequency with lowered temperature, and some change in the final steady-state effect according to whether the stretch is strong or weak (Burkhardt, 1959).

4. Crustacean neuromuscular junctions

In the series of progressively more integrative junctions we come abruptly to a much higher degree of integration in these peripheral synapses. This is associated with (a) several presynaptic axons ending on the same postsynaptic cell and having different effects, (b) a high development of facilitation, and (c) the presence of inhibitory endings. The flexibility and complexity inherent in these properties chiefly account for the fine gradation of both speed and strength of muscle contraction and the reciprocal inhibition of antagonists, accomplishments which in animals depending on recruitment of motor units, as in vertebrates, are entirely centrally determined.

Our information is based almost entirely upon the higher crustaceans but it has proven of real importance that a variety of species among the decapods has been examined, for there are significant differences between them. Key references are Katz (1949), Hoyle (1957), Hoyle and Wiersma (1958), Furshpan (1959), Fatt (1961), Dudel and Kuffler (1961d); see further, Chapter 17.

Crustacean muscles are innervated by a very small number of nerve fibers—from two to five. These nerve fibers branch many times and each one innervates at least a large proportion of the total number of muscle fibers in the muscle and, in addition, supplies many endings to each of these muscle fibers. Therefore each muscle fiber receives not only many endings but in most instances endings from two, three, four, or five different efferent axons. There is a considerable percentage of muscle fibers which receive only one axon, but they apparently never dominate a muscle. The endings are of the order of tenths of a millimeter apart and most of the membrane of the muscle fiber is synaptic or endplate-like in being excited by the transmitter or transmitters. There is electrically excitable membrane and at least in some muscles a spike can be elicited by an electrical stimulus and will propagate without decrement; but normally if a spike is produced at all, by arriving nerve impulses, it arises at many sites on the muscle fiber at once. Many muscles probably rarely or never produce spikes. However, it is difficult to say how much of the graded response may be like local potentials, that is, a secondary response of electrically excitable membrane to the antecedent junctional potential. There is apparently a contribution from such a mechanism so the whole of the observed graded muscle potential cannot be called an e.p.p. or p.s.p.; but there is no reason to avoid these terms as some authors have done. The spacing of the endings in relation to the spatial decrement of graded potential results in only a small inequality along the muscle fiber in the intensity of response—a ratio of about 1.4 to 1 between maximum and minimum transmembrane response potential has been found, the points being a few millimeters apart. Presumably mechanical tension follows a somewhat similar course, that is contraction develops in a graded manner fairly uniformly along the fiber.

The e.p.s.p.'s characteristically **facilitate** and the frequency range from no contraction to maximal contraction, may be typically from 10 or 20 to 200 or 300 per second. Facilitation continues to develop for many impulses, again varying widely, but comes to an equilibrium for each frequency. There are second-order effects comparable to posttetanic potentiation, that is, in a certain range of time the repetition of a facilitating series of stimuli shows "facilitation of facilitation."

Different kinds of response are produced by the different nerve fibers supplying the same muscle fiber. There is always one (but there may be two) which is inhibitory. The others are primarily distinguished by the speed of the muscle contraction which they cause—fast, slow, or intermediate. They also differ, but not consistently, in the size, rate of development, and rate of facilitation of the junctional potentials they evoke, and in the maximum response possible and in ease of fatigue (Fig. 4.28). Each of these characteristics varies from cell to cell as recorded with an intracellular electrode. The contraction is not measured for individual muscle fibers and possibly varies also. On the average, as seen in multiunit recording, the "slow"

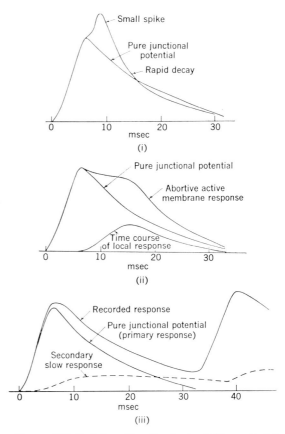

Figure 4.28. Varieties of junctional potential in crustacean nerve muscle systems. The tracings illustrate three different types of departure of the response from the "pure" junctional potential; fast response of the closer of *Panulirus*. (i) Comparison of "pure" junctional potential and j.p. giving rise to a small spike. (ii) Comparison of j.p. giving rise to an abortive secondary response, and the time course of the latter alone when the j.p. is subtracted. (iii) Comparison of "pure" j.p. with that when secondary slow response obtains. Pictures taken at a frequency of 30 per second. [Hoyle and Wiersma, 1958a.]

nerve fiber causes smaller, more facilitating potentials (Fig. 4.29), which are likely to achieve a lower maximum height, to decay more slowly, and to fatigue more slowly than p.s.p.'s caused by the "fast" nerve fiber. These quantitative relations, however, vary greatly from muscle to muscle. The variation in time of decay of individual potentials is taken to mean that some—the longer ones—include a component of active but decremental contribution superimposed on the passive decay due to membrane time constant; others—the shortest ones—include a

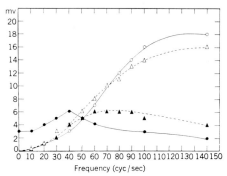

Figure 4.29. The relationship between the height of junctional potentials and the maintained depolarization produced by different frequencies of excitation. *Open symbols,* fast fiber; *filled symbols,* slow fiber responses. The *circles* give the final j.p. magnitudes for each and the *triangles* the corresponding maintained depolarizations, measured to the lowest point on the trace. Each point is the average of readings from 16 muscle fibers of the closer of *Panulirus*. [Hoyle and Wiersma, 1958a.]

spike with its characteristic rapid repolarization attributed to "sodium inactivation." The nerve fiber causing fast contraction is more likely to produce spikes, but these are not usual in crustacean muscles. Local, abortive spikes are more common than full, all-or-nothing overshooting spikes; the former are in a continuum with local potentials but show some sign of the beginnings of regenerative action and sometimes of accelerated repolarization but abort before reaching large amplitude. A few muscles do give full spikes to single nerve impulses and corresponding tension twitches which are very fast; in most muscles, when spikes occur it is at high-frequency stimulation and only a small additional hump appears on the already high level of tension.

The evidence indicates a **chain of several processes** at different sites between nerve impulse and muscle contraction (Fig. 4.30). These can differ between fast and slow systems and between muscles or species. Besides quantitative variation, Hoyle and Wiersma (1958c) argue for qualitative differences in the sense that at least two different transmitters must be invoked, one for fast and one for slow junctions. The transfer functions between membrane potential and contraction are also unequal for slow and fast contraction and therefore plural. Whereas, in many instances there is a relation over a wide range between tension of the whole muscle and membrane potential plateau of typical component muscle fibers during various states of excitation and inhibition, this relation is not always found. In certain muscles stimulation of the "fast" axon at low frequencies (10–15 per second) produces large junctional potentials but no contraction, while stimulation of the "slow" axon at the same frequency produces much smaller potentials and a considerable contraction. This is the so-called "paradox" phenomenon; it is not yet explained.

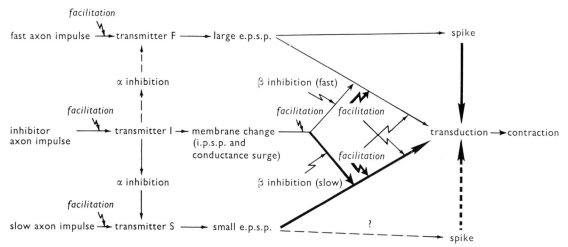

Figure 4.30. The nerve muscle transmission system in Crustacea. The diagram illustrates the case where three nerve fibers—a slow exciter, a fast exciter, and an inhibitor—innervate a muscle fiber. The points in the chain of effect where facilitation occurs and where the different types of inhibition occur are indicated; heavy *arrows* indicate greater importance.

Even in cases where there is no paradox, a contraction is sometimes seen when intracellular electrodes record no potential change above the noise level of about 0.1 mv. No theory which makes muscle contraction dependent on membrane potential as such can therefore be general.

Slow and fast excitatory systems controlling muscle contraction are not confined to crustaceans. Insects, polychaetes, oligochaetes, hirudineans, pelecypods, gastropods, and cephalopods also manifest the distinction; only in the insects is there direct intracellular evidence that the same muscle fiber is capable of both kinds of response according to which nerve fiber excites it but this seems likely to be more common.

Inhibitory postsynaptic potentials in crustacean muscle are remarkable for the unimportance of the magnitude and even the sign of the potential in relation to the capacity to cancel excitation. Polarizing i.p.s.p.'s are found, but only when the resting potential is low; more often there is no potential change, that is the resting potential coincides with the reversal potential for the i.p.s.p. By setting the membrane polarization higher, depolarizing i.p.s.p.'s are obtained but they are still associated with an inhibitory effect. Under each of these conditions there is a uniform membrane impedance drop on inhibitory nerve stimulation and it is supposed that this rather than the potential is significant at least insofar as reducing e.p.s.p.'s is concerned (Fig. 4.31).

But a curious complication is that inhibition of contraction can be produced either with or without a reduction in the excitatory junction potentials. Most commonly it occurs without reduction of e.p.s.p.'s (simple or **beta inhibition** of Katz, 1949); this must mean that the inhibitory transmitter attacks a link in the contraction release process either after or quite apart from the membrane potential mechanisms. Reduction of e.p.s.p by inhibitor impulses (supplemented or **alpha inhibition** of Katz, 1949), when it can be obtained at all, requires that the inhibiting impulses arrive, within narrow limits, simultaneously with the excitatory. This is not important for simple inhibition. There is always in inhibition a reduction of the maintained plateau of depolarization produced by excitatory stimulation, whether

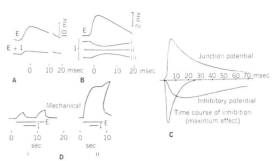

Figure 4.31. Inhibition in crustacean muscle fibers. **A.** Upper trace: intracellular record of junction potential during stimulation of the slow nerve fiber at 33 per second. Lower trace: combined stimulation of motor axon and inhibitor axon, showing continuation of j.p. by inhibitor action. **B.** Time courses of single exciter potential and three single inhibitor potentials: (i) at 48 mv resting potential, which is slightly depolarized; (ii) at 73 mv (normal) resting potential; (iii) at 95 mv (hyperpolarized) resting potential. **C.** The time relations of the junction potential, the inhibitory potential, and the inhibitory effect. **D.** The mechanical effect of an inhibitory action: (i) during stimulation of the slow nerve fiber, inhibition is complete; (ii) during stimulation of the fast nerve fiber, very slight inhibition; postinhibitory facilitation is evident. [**A, B, C**, opener of *Eupagurus*, Fatt and Katz, 1953; **D**, Cambarus claw, Van Harreveld and Wiersma, 1937; all from Hoyle, 1957.]

the individual inhibitory potentials are depolarizing or hyperpolarizing; but we have already seen that there is no simple relation between this membrane level and muscle tension. Inhibition builds up with frequency of stimulation of the inhibitory nerve fiber and with time, and when i.p.s.p.'s can be seen they facilitate and summate to a plateau. The efficacy of a given frequency is different against slow and fast excitatory transmission; the latter is more resistant to inhibition, but this varies among muscles quantitatively. There are several muscles which receive two different inhibitor fibers, one in common with several other muscles and one specific to itself; these have appreciably different characteristics but it is too early to say that they must have different transmitters.

Recently Reuben and Grundfest (1960) and Dudel and Kuffler (1961a, 1961b, 1961c) have discovered that **miniature, quantal potentials** lie at the basis of crustacean muscle e.p.s.p.'s very much as in frog and cat (pp. 231, 974). A random sequence of spontaneous miniature potentials occurs; presynaptic impulses simply modify the probability of their occurrence and hence the average frequency.

Analysis of the probability of these potentials, thought to represent release of quanta of transmitter, has led to the conclusion that both facilitation and inhibition are primarily taking place **presynaptically** (Dudel and Kuffler, 1961b, 1961c, 1961d).

Insect neuromuscular junctions have been carefully examined, with the aid of intracellular electrodes; the monograph of Hoyle (1957a) and Chapter 17 of the present work should be consulted for details. Judging from the instances so far known, the main features are quite similar to those of crustaceans but do not reach as complex a development. Not more than three types of nerve fibers supply a muscle (or muscle flag), causing respectively (a) fast and (b) slow e.p.s.p.'s with contractions and (c) a hyperpolarizing postsynaptic potential unaccompanied by contraction when alone. Many muscle fibers receive all three types but not all muscle fibers do; many receive only fast nerve fibers and possibly some receive only slow fibers. Fast nerve fibers give an e.p.s.p. and on top of it a spike, without facilitation; the mechanical response is a large twitch and only at high frequency (about 50 per second) do these fuse into a smooth tetanus. Slow fibers produce only small junctional potentials which facilitate, summate, and accompany a slow, smooth contraction without refractory period (Fig. 4.32). The e.p.s.p.'s may be from 1 mv up to 60 mv in height, in the latter event not facilitating. If fast and slow impulses arrive at the same time, the electrical response is not the sum but is only a slightly augmented fast impulse, whereas if the slow impulse just precedes the fast, the response to the latter is reduced. The third type of nerve fiber would be expected to inhibit but it has not been observed to do so. It does yield an increase in membrane polarization which summates as much as 20 mv but which in no case passes a total of 70 mv membrane potential. Since the spike is not blocked, it takes off from a higher base line and is actually greater in amplitude than it seems against this background; correspondingly, the twitch is stronger. Hoyle gives reasons for believing that this is an adaptation to overcome the tendency for the blood to accumulate excessive potassium concentrations—a result of the vegetable diet of locusts and some allies—which in turn tends to lower the muscle resting potential.

Species vary widely but a good many have at times or chronically a **high blood potassium, a**

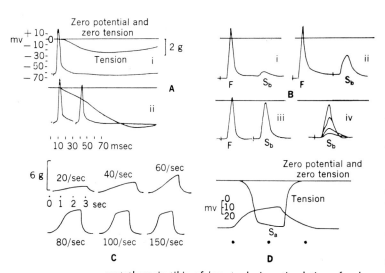

Figure 4.32.

Fast and slow fiber activity in insect muscle. **A.** Single and paired fast fiber responses of the metathoracic extensor tibialis muscle of *Locusta*. The twitch tension is recorded at the same time on the upper beam of the oscillograph. **B.** (i)–(iii) Some typical responses in fibers of the metathoracic extensor tibialis of *Schistocerca*. The fast fiber response (F) appears first and is followed by the slow fiber response (S). (iv) The extent of facilitation occurring at different frequencies of stimulation in an initially small slow fiber response. **C.** Tension records from the metathoracic tibia of *Locusta* during stimulation of a slow nerve fiber at increasing frequencies. **D.** Tension correlated with extent of depolarization in a fiber of the metathoracic extensor tibialis muscle of *Locusta*, which demonstrated the slow (S_a) response. Tension measured at tibial tip: muscle tension is 40 times this. [Hoyle, 1957.]

low sodium, and a high magnesium, extremely unfavorable conditions for maintaining a high resting potential, a large action potential, and effective transmission. The nerves appear to be like those of other groups at least with respect to the first two of these ions; the muscles may be quite different. The problem is dealt with on pp. 982 and 1133 in its general aspects but in the present connection it should be pointed out that part of the tolerance of these conditions lies in the provision for neuromuscular transmission to be distributed at numerous, closely spaced sites along the muscle fiber, making a propagated spike unnecessary, whereas even a considerable depressed e.p.s.p. can still cause some contraction. There is probably a significantly higher specific tolerance to magnesium, at least in the periphery, than in junctions in other phyla. As regards transmission at central synapses where all-or-none spikes have to be generated, there should be and there is greater protection, in the form of a sheath around the central nervous system, which maintains a more equable ionic balance in the intercellular medium inside it; the subject is treated fully in Chapter 20.

Another special problem brought out by certain insects is the existence of extremely fast, **fibrillar muscles** which contract and relax repeatedly at frequencies far higher than those of their action potentials. Examples are the flight muscles in fast-flying flies, bees, and beetles, haltere muscles in dipterans, and sound-producing muscles in some cicadas. Like the muscles of crustaceans, but in quite a different way, these show the indirect relation between muscle membrane potential change and contraction. Evidently some depolarization is necessary, but only a few action potentials each second put the fast insect muscles in a condition permitting a high-frequency cycle of activation and deactivation of contraction, determined peripherally.

5. Lobster cardiac ganglion cells

Still more integrative is the transmission in the nine-celled cardiac ganglion of lobsters and crabs. The five follower neurons receive input from the pacemakers, from each other, and from fibers from the central nervous system, some of

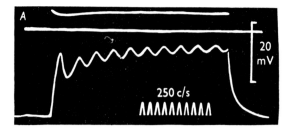

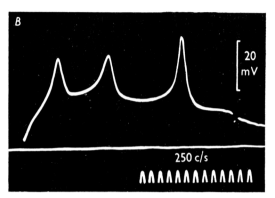

Figure 4.33. Oscillatory potential changes evoked in a locust muscle fiber by applying a steady outward current. The upper record in **A** shows the voltage drop produced by the current flowing through the membrane across a monitor resistance. [Del Castillo, Hoyle, and Machne, 1953.]

which accelerate and others of which inhibit the heart rate. They are unusual invertebrate neurons in being multipolar and with synapses upon the soma as well as on dendritic processes, which come off both the soma and the main neurites. This makes it possible for an electrode inside the soma to detect a good deal of the relevant activity, including prepotentials, whereas there is reason to doubt that much of the activity potentials are appreciably felt in some typical invertebrate cell bodies. The heart beat in decapods and many other crustaceans is driven by a burst of impulses from the cardiac ganglion and this burst recurs with the same temporal patterning of several to many impulses from each of the nine cells for hundreds of beats. This preparation has therefore been studied from the point of view of parameters of integration and of the origin of temporal pattern. Although it has been possible to insert two electrode tips in one cell and in some cases a third in another cell, the results are not easy to interpret because there are evidently synaptic loci closer and farther from the electrode, not to speak of spike-initiating loci and pacemaking loci in the same cell, any of which may be multiple and at some distance from the soma. Details may be found in the papers of Maynard, Matsui, Hagiwara, Watanabe, Otani, Terzuolo and Bullock.

E.p.s.p.'s of long duration are evoked 1:1 by impulses from the posterior small cells which normally pace the heart (see Table 4.1). These show a new property for our series—**antifacilitation**. The amplitude is a function of interval; it is therefore large for the first arriving impulse after an interval between bursts, minimal immediately following, and slowly grows again because the pacemaker fires a burst of declining frequency. This basic pattern is characteristic of the follower cells of *Panulirus* (Fig. 4.34) and is a primary determinant of its spike output. But superimposed are other, later, and smaller e.p.s.p.'s from other posterior cells or followers. Spikes arise not simply from the peaks of highest depolarization, since some other parameters seem to influence the firing level of the spike-initiating locus out in the axon—probably distortion during electrotonic spread from synaptic loci, the recovery cycle after the preceding impulse, and the like. Consequently the number and pattern of spikes from a follower will be quite different from that which impinged upon it. The **spikes cannot invade the soma**, presumably for lack of sufficient electrically excitable membrane (or too low a safety factor), and are seen by the electrode in the soma much attenuated, and only slightly repolarizing the accumulated synaptic depolarization.

Even follower cells have a tendency to **spontaneous discharge** which is expressed toward the end of the interburst interval if the tendency is high or the driving from presynaptic sources is low. Pacemaker potentials arising in circumscribed loci are then added to the several different e.p.s.p.'s and attenuated spikes in the record. Furthermore, local potentials are clearly recognizable at a critical level of pacemaker potentials and in turn either rise to spike firing level or fail to do so. Local potentials may be caused by synaptic potentials but it is not easy to be sure when one is seeing a local potential, as it is when there is no presynaptic activity.

The several inputs have **different transmission characteristics**. Some e.p.s.p.'s show marked facilitation instead of the antifacilitation already mentioned, and others show neither. Certain cases show no individual e.p.s.p. with each arriving impulse but only a smooth, slow shift of the membrane level; possibly a slow component like that seen in the crustacean neuromuscular junction potential is here relatively large and the usual initial hump is attenuated by distance from the synaptic site. Finally, the so-called inhibitory

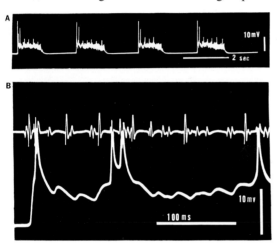

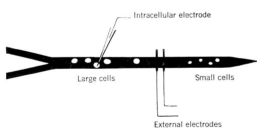

Figure 4.34. Activity in the lobster cardiac ganglion. Intracellular and external recording electrodes, arranged as shown in the diagram; the former sees only the activity of one follower neuron, and the latter sees the activity in axons of many large and small cells. **A.** On a slow time scale the successive bursts of synaptic activity and spikes seen in the soma of a follower neuron represent successive heartbeats. **B.** On a fast time scale the lower trace shows the intracellular record in which the bumps represent e.p.s.p.'s and occasional spikes rise to a height that is very limited because the impulse originates in the axon at some distance and does not invade the soma; the upper trace indicates the activity seen by the external electrodes—spikes of complex form from smaller and larger axons show the activity of others of the nine cells. [Watanabe, 1958.]

presynaptic fiber—fortunately for analysis there is but one on each side—causes facilitating p.s.p.'s which are hyperpolarizing in some cells and depolarizing in others at the same time. This may be dependent on the membrane level although it is not obvious. What is clearer is that there can be a summing hyperpolarization, even when the individual humps are depolarizing, or there can be a summing depolarization; the membrane level for reversal of sign is not the same for the initial hump and the late, accumulating component.

The present example differs from most of those preceding in its ability to maintain a repetitive discharge during a maintained depolarization, its inability to follow high frequency input, and the importance of aftereffects. The last are both positive and negative; that is, persistence of effect and rebound reversal of effect. Afterdischarge and afterinhibition represent the positive; postinhibitory discharge and postexcitatory depression, the negative. These aftereffects are differently developed for different inputs. Thus the small posterior pacemaker cells cause postexcitatory depression initially in the large anterior followers, but the accelerator input from the central nervous system shows a long positive aftereffect. The inhibitor fiber from the central nervous system has a short afterinhibition, but has a more pronounced rebound discharge which can be used to drive the ganglion faster than its spontaneous pace if one gives recurrent bursts of inhibitory impulses.

6. *Aplysia giant ganglion cells*

The large ganglion cells of many gastropods, especially pulmonates, have been classical objects of cytological research. *Aplysia*, like other opisthobranchs, not only has large cells but offers other advantages for experimental work. A dozen or more of the cells of the visceral ganglion are exceptionally favorable for intracellular electrophysiology. This is due not only to their size but to their location and properties as well. They are parts of discrete, integrative ganglia of the central nervous system, mediating reflexes and influences from other ganglia. They not only exhibit excitatory and inhibitory synaptic responses—each in several forms—but they have an enduring ability to fire repetitively even at extremely low frequencies and an unusual degree of fractionation of the neuron during activity.

The duration of each of the activity potentials is very long and is a direction function of cell size; size becomes an important integrative parameter since large cells can summate lower frequency input than small cells. The cells mainly studied are the largest (300–800 μ in diameter) of a continuous size spectrum but commonly one stands out in size and can be called giant. The giant cell is apparently an interneuron projecting forward through the long connectives to the pleural ganglia. It receives presynaptic fibers from the pleurovisceral connectives of both sides, from peripheral nerves, and from interneurons in the visceral ganglia. The cells are unipolar but the anatomy is not adequately known, particularly in regard to the location of the synapses relative to the cell body. An extreme development of connective tissue trabeculae penetrating the cytoplasm on the inner or axonal side of the soma and continued in the axon as a series of longitudinal folds may have some functional significance (see pp. 98 and 1302). Most of the knowledge of the physiology of this preparation is due to Tauc (see 1958); Arvanitaki and Chalazonitis (1949, 1956, 1957) have contributed significantly, and Hagiwara and Kusano (1961) have studied a similar preparation in the marine pulmonate *Onchidium*.

Figure 4.3 shows the response of a typical cell to a single presynaptic impulse, first at the normal resting potential of 40 mv, then at increased membrane polarization imposed by a second intracellular electrode. At first a spike rises rapidly, overshooting by 35 mv, lasting 17 msec, and undershooting 20 mv. Raising the membrane level greatly delays the spike origin and uncovers the e.p.s.p., which grows larger as the polarization is further increased. The duration of the e.p.s.p. is 50–70 msec in cells of 70 μ diameter but 200–300 msec in cells of 400 μ. This represents a greater **excess of e.p.s.p.** over a purely electrotonic decay-time constant in the smaller cells (Fig. 4.35). The cells of the snail *Helix* do not achieve such dimensions but cells of a given

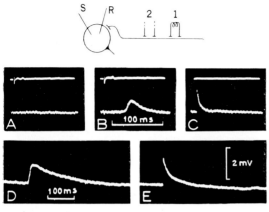

Figure 4.35. Comparison of e.p.s.p. and electrotonic potentials. Intracellular records from a large visceral ganglion cell of *Aplysia* that can be activated orthodromically through a presynaptic fiber or stimulated directly by an intracellular current electrode in addition to the recording electrode. **A, B, C,** cell of 70 μ diameter; **D, E,** a cell of 350 μ diameter. **A.** Subthreshold stimulation of the presynaptic nerve, whose total multifiber activity is recorded with external electrodes on the upper trace; no response in the cell. **B.** A just liminal stimulation produces in an all-or-none manner a deflection in the lower trace, which is a synaptic potential. It is all-or-none because the presynaptic fiber is all-or-none; a stronger stimulus would have excited more presynaptic fibers and caused a larger e.p.s.p., which could have grown large enough to cause a postsynaptic spike. **C.** In the same cell a brief pulse of current is applied directly through the intracellular stimulating electrode and the dissipation of this current is shown on the lower trace, representing the electrotonic potential and its decay with the time constant of the soma membrane. **D** and **E** are identical to **B** and **C** but in a large cell; note that the time constant of passive decay is longer and that of the decay of e.p.s.p. is much longer. [Tauc, 1959.]

size may have even longer lasting potentials. Tauc encountered no heterofacilitation—facilitation of p.s.p. evoked through one pathway by activity in another; he did observe homofacilitation and heterodepression. The spike invades the soma but it does not wipe out or prevent an e.p.s.p., which points to the separateness of the membranes supporting the two and the electrical inexcitability of the synaptic membrane. Spikes may arise apparently directly and this is taken to mean that the prepotentials at that time have a rapid rate of rise and the spike arises far from the soma where the threshold is low; this is characteristic of the undepressed state. In some cases a synaptic potential may be visible as a shoulder preceding the spike. In other cases another step is visible which is identified with the local potentials that have already been described. A rhythmic succession of spikes arises spontaneously if the cell is intentionally depolarized slightly—and sometimes without such intervention; the spikes maintain a steady frequency indefinitely, sometimes one in many seconds. In this event a gradually rising depolarizing potential, corresponding to the pacemaker potential described before, occupies the interval between spikes, which may then rise without any abrupt discontinuity.

Local potentials of a particular behavior have been called **pseudospikes.** Appearing like miniature spikes and hence much briefer, rapidly rising and sharper crested than synaptic potentials, these are clearly of endogenous origin rather than 1:1 with any external arriving events. This is shown by their abolition with increasing polarization and their greater probability of occurrence with decreasing membrane polarization. They are only found in some cells; in others they cannot be produced even by depolarization. They may be rhythmic or nonrhythmic, are superficially of a few invariant sizes in a given cell, summating, and therefore spike-engendering at a critical height. Although they are not adequately understood, it seems safe to conclude that some local inequalities are involved in the autoactivity determined by membrane potential; perhaps they represent quantal activity confined to distant branches.

Typical **inhibitory postsynaptic** potentials are encountered both as a consequence of stimulation of a preganglionic nerve and as a part of intrinsic spontaneous activity of presynaptic elements in the isolated ganglion (Fig. 4.36). At a normal resting potential of 40 mv the i.p.s.p. may be 15 mv; it reverses polarity at 60 mv. Often the amplitude is a much smaller fraction of the difference between prevailing and reversal levels. The form and duration are similar to those of the e.p.s.p. A reversed, depolarizing i.p.s.p. is excitatory, facilitating spike initiation, and indicates that impedance drop is not the only significance of the response. Unlike the preceding

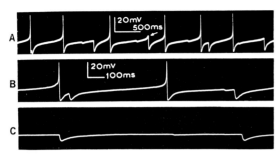

Figure 4.36. Inhibitory postsynaptic potentials caused by spontaneous activity in isolated ganglion. Activity arising without stimulation in a large visceral ganglion cell of *Aplysia*. This cell shows spontaneous firing preceded by typical pacemaker potentials. These are interrupted at intervals by hyperpolarizing deflections interpreted as i.p.s.p.'s due to activity of spontaneous interneurons elsewhere in the ganglion. The i.p.s.p. can cancel the developing pacemaker potential *(arrow)*. In **C** the i.p.s.p.'s appear alone because the spontaneous activity has been suppressed by an applied hyperpolarization. [Tauc, 1959.]

example, the afterhyperpolarization following a spike does exert a strong effect on the height of i.p.s.p.'s which fall during this time (Fig. 4.36).

Tauc (1960d) has underlined the existence of several **other forms of inhibitory phenomena**, not requiring i.p.s.p.'s. One is excess depolarization due to e.p.s.p.'s which can completely block the production of spikes (see Fig. 5.5, p. 267). Another results from the accumulation of afterhyperpolarization; this is remarkable in that it builds up after repeated e.p.s.p.'s even when no spikes have been fired (see Fig. 5.6, p. 267). This can reach such an extreme form that a single shock to the preganglionic connective in a snail, *Helix*, often causes a dispersed barrage of presynaptic activity which elicits e.p.s.p.'s in the cell under observation and, after a brief augmentation of excitability (0.5 sec), a slowly growing and long lasting hyperpolarization (20–30 sec) with inhibition. Long lasting inhibition is said to vary seasonally. Other integrative possibilities are discussed in Chapter 5, pp. 265-269.

7. Other invertebrate preparations

In a number of the following cases transmission has been studied with somewhat less intimacy. For the most part the properties, as far as they are known, are given only in the table and are not discussed here. Two groups of examples of reciprocal transmission, which in the absence of special explanation are treated as unpolarized junctions, fall into the class of simple relays. The **segmental, septal synapses** of earthworm giant fibers and of crayfish lateral giants probably have their equivalents in many species related to these favorite laboratory subjects. The discontinuity of the segmental units is attested by rigorous anatomical criteria, some by electron microscopy, and accords with the distribution of cell bodies which recur in every segment. Although equally rigorous electrical measurements have not been made, it is probable that the resistance across the septal synapses is low, as in the crayfish giant-to-motor synapse, but unlike that instance, not rectifying; that is, the resistance to flow of electrotonic current from one segmental giant unit to the next is presumably equally low in both directions. This would explain the absence of appreciable delay and would mean that we do not expect a synaptic potential as distinct from the electrically excitable response of ordinary conducting membrane; the latter can be local and graded, imitating an e.p.s.p., if the spike is blocked or delayed as by a rise in its threshold or a slowing of the prepotential rise. **Commissural connections** between right and left crayfish lateral giants and between the paired giant fibers of sabellid polychaetes and the junctions in the brain between right and left median giants in crayfish, ghost shrimp (*Callianassa*), and serpulid (*Protula*, Fig. 4.37) are probably not electrically transmitting since they have a delay of half a millisecond or more and, in the first case at least, no electrotonic spread from one neuron to the other. Except for the feature of two-way transmission, they resemble the squid giant synapse.

A curious category of junctions has been encountered between the paired giant fibers of serpulid and sabellid polychaetes (especially *Protula* and *Spirographis*) after fatigue or ablation of the connections in the anterior end just mentioned. In mid-body, where no anatomical connections can be found in good silver impregnations or are usually manifest functionally, cross-over of impulses occurs in a synaptic

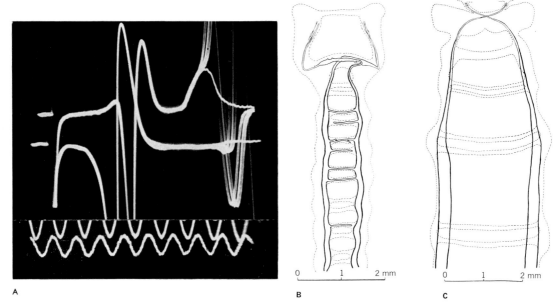

Figure 4.37. The giant synapse in the serpulid *Protula* and the sabellid *Spirographis*. **A.** Synaptic transmission in the anterior end. The upper beam gives pre- and postsynaptic potentials, led off by electrodes 3 and 6 mm back of the extreme anterior end. The lower beam shows the ascending and descending spikes, led off 9 and 12 mm from this end. Repetitive stimulation (27 mm from the anterior end) has brought the synapse to the point of failure of spike transmission, without increasing synaptic delay measured to the start of e.p.s.p. Time in milliseconds. **B.** Diagrammatic plan of the giant fibers and central nervous system in the anterior end of the sabellid *Spirographis*. **C.** Diagrammatic plan of the giant fibers in the central nervous system in the anterior end of *Protula*. All dimensions to scale. [Bullock, 1953.]

manner, delaying and labile but unpolarized. The cross takes place at a localizable spot but this spot may shift smoothly along the body, millimeters in seconds. Such facultative crosses in mid-body occur sporadically from time to time in intact specimens. Lacking fixed position and anatomical basis they act as though artificial; hence they have been called **quasi-artificial**. The giants are far apart and the communication must be by processes; the anatomical results can only exclude large processes. But it is strange that the large number of transverse processes called for by the shifting of locus should be exerting no detectable influence most of the time. Another interest of these crossovers is that sometimes the presynaptic impulse propagates right past the point of cross; sometimes it goes only one way or the other. Intracellular recording and impedance measurements are needed to help in the interpretation of these junctions.

Giant interneuron-to-motoneuron synapses, strongly resembling those which in the crayfish are of unique importance as the best established electrically transmitting junctions, are found in *Nereis* and *Harmothoë* among polychaetes. As in the crustacean, here also several central giants make synaptic contact with the same final common motor axon and any one of them can fire the muscle, which is a twitch type appropriate to a startle response.

Fast and slow nerve-muscle systems are widespread. In coelenterates, polychaetes, sipunculids, gastropods, pelecypods, cephalopods, crustaceans, insects, holothuroids, tunicates, and vertebrates apparently equivalent differences have been found between responses of the same muscle (sometimes of the same muscle fiber, sometimes of separate sets of muscle fibers) to two different types of nerve fiber. One motor axon (or group of axons) brings about twitches of greater rate of rise of contraction and muscle action potential, less summation and less facilitation or none; another fiber causes slow, markedly summating or facilitating tonic responses upon repetitive stimulation. The latter (and sometimes the former) employs graded rather than all-or-none unit responses; the fast system may show refractoriness after each unit response, but the slow transmission does not. The distinction between responses is not great in some, such as *Neanthes* and *Octopus*, but it is highly developed in others, especially crustaceans, where there may be several intermediate responses of the same muscle fibers brought about by a third and fourth motor nerve fiber. Transmission in slow systems (and in some fast systems) is integrating rather than simply relaying and in all invertebrates so far studied this aspect is a prominent part of the control of motor activity; conversely, in all vertebrates—even the frog, where slow muscle fibers are widespread—integrative transmission is much less prominent. It is not lacking, however; a significant percentage of mammalian neuromuscular junctions fail to transmit 1:1 in slack muscle. Graded and local response is an important character of the intrafusal muscle fibers of the intramuscular stretch sense organs called muscle spindles. Many muscles like these in vertebrates apparently include only the fast or only the slow type of response.

Still more integrative are the **neuromuscular junctions of coelenterates.** Although none are known as intimately as intracellular electrodes would permit, quite a variety are indicated by the fact of separate fast and slow muscles of actinians, fast and slow responses of one muscle in ceriantharians, and of medusae, and luminescence in soft corals. These are highly dependent on the frequency and number of nerve impulses reaching the neuroeffector junction; some muscles do not even exhibit a visible response to the first or to the first few arriving impulses. It is now believed that graded response of each postjunctional cell is responsible, rather than recruitment of all-or-none units. At least in some muscles there is clearly a duality of motor response systems, fast and slow, as is so common in higher groups.

Interneural synapses in the **nerve net** in coelenterates behave in the same way. In addition, the over-all transmission is two-way. The possibility of a distribution of individual junctions, each of which is polarized, is discussed on page 464. At present there is some slight reason to believe this is not the case and that we have to do with unpolarized synapses. Although synapses, as opposed to anastomoses, may be more widespread, they can be studied physiologically only in those cases of non-through-conducting nets or through-conducting nets which have been converted into the former, as by drugs. Recently a number of diverse examples of nets which require repeated stimuli to propagate excitation to a distance and in which the distance depends on the number and spacing of the shocks, have come to light. The first was the oral disc edge-raising mechanism in certain anemones. Others include the retraction of individual polyps in colonies of alcyonarians, madreporarians, and hydroids, and tentacle movements in many of these and in medusae. These examples differ among the species in ways that indicate significant quantitative differences in the properties and arrangement of the junctions. Unlike the neuromuscular junctions, we must believe that recruitment of units which have varying thresholds with respect to amount of facilitation required is the basis of grading the extent of response; independent evidence points to the presence already in these systems of all-or-none nerve impulses. The time scale of the build-up and decay of facilitation, which otherwise is so similar to that in crustacean neuromuscular and other junctions, is here extremely slow; some junctions operate at intervals between impulses of many seconds though not at still lower rates.

Hardly more integrative, but emphasizing gradation by spatial summation of converging presynaptic lines on each postsynaptic unit, is the **first sensory synapse** in the last abdominal ganglion of the cockroach, between mechanoreceptor afferents and ascending pseudogiant fibers. Transmission can follow 1:1 only when incoming volleys are at low frequency. Since normal input is unsynchronized and several afferent fibers converge on each postsynaptic unit, the output frequency must normally be no simple function but an integrated result of arriving impulses. After-discharge is a conspicuous feature but is abol-

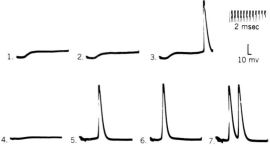
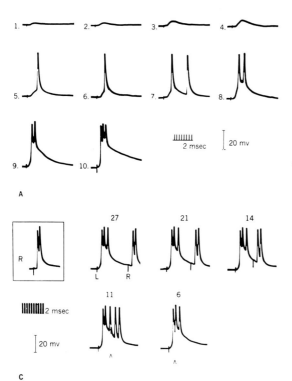

Figure 4.38. Some properties of transmission in crayfish abdominal ganglion. **A.** Intensity series, electrical stimulation of one root pair, recording of a multiply discharging unit. Synaptic potential becomes extremely large in 9 and 10. Note that synaptic potential is not repolarized by spikes, suggesting noninvasion of the junctional membrane. **B.** 1, 2, and 3: electrical stimuli to one pair of roots; maximum intensity produces a synaptic potential which only occasionally generates a spike. 4 to 7: input from the opposite side produces single and then multiple firing at low stimulus intensity and with very low amplitude synaptic potentials. The properties of transmission and spike initiation can be very different even in the same cell according to the source of presynaptic activation. **C.** Interactions of high intensity input producing multiple firing. Synaptic activity remains after a multiple discharge in which the spikes apparently do not invade the synaptic region. This allows the two inputs to add even when very close together; occlusion occurs only when the shocks are but 6 msec apart. Numbers give the intervals in milliseconds; the inset shows the response to the R pathway alone. [Courtesy D. Kennedy.]

ished by decapitation, revealing the presence of descending influence predisposing the abdominal ganglion cells. This influence discriminates between predisposing to afterdischarge and sensitizing the cell to afferent input; the latter is not obviously affected but the former is.

Similar but perhaps still more complex are the synapses in the **last abdominal ganglion** of the crayfish between incoming mechanoreceptor afferents and large ascending interneurons. This preparation, introduced by Prosser in 1935, was the first monosynaptic transmission studied. In a typical case one impulse in each of four afferent fibers, produced by slight bending of as many caudal hairs, is necessary to elicit one postsynaptic impulse. There is no facilitation but temporal summation is important. The transmitted event can follow arriving volleys only up to about 10 per second and adapts fairly soon. Although the functional recovery time is fairly long, the cell can discharge at high frequency briefly, to a single strong stimulating volley. Ganglionic delays are widely variable, from as short as 2 msec to over 30 msec; the longer values suggest the mediation of interneurons. As is true throughout the summary in Table 4.1, no absolute reliance should be placed on those symbols which represent the lack of inhibition or of afterdischarge or repetitive firing; they may actually be absent in given cases or only relatively less conspicuous, or it may be that the proper conditions for bringing out one or another of these features have not been utilized.

The listed characters apply to large postsynaptic fibers. Kennedy and Preston (1960) have also recorded from **small interneurons** (Figs. 4.38, 20.1, 20.2). These show a variety of spontaneous activity patterns, rhythmic and less rhythmic. They often respond both to tactile and to photic input (via intrinsic photoreceptor cells in the ganglion, see p. 1097); touch causes transient bursts and light a tonic frequency increase, or occasionally one inhibits the other. The circumstantial evidence as to the location of the electrode tip in these experiments strongly suggests that the cell bodies of the crayfish ventral cord ganglia are not invaded by spikes and, consistent with their known anatomy, synaptic potentials are not seen in them either. Both of these signs of activity are recorded only when the electrode is in a process in the region of the neuropile; spikes alone may be seen in the connectives. Interestingly, some of the smaller interneurons with rhythmic spontaneous firing cannot be made to respond to photic, tactile, proprioceptive, or even temperature stimulation. Another finding which is understandable on the basis of the anatomy is that more than one spike form or amplitude can be recorded from inside the same neuron. They may have different firing patterns or different thresholds for the same sensory input. The same feature was earlier reported in lobster cardiac ganglion cells and was interpreted as reflecting separate spike-initiating loci in different processes or branches, that is, functionally more than one axon. Separate loci may also account for the occasional "piggy-back" spikes during brief high-frequency bursts: a second spike occurs on the falling phase of the first, when the membrane at the locus of origin of the first should be not only excessively depolarized but refractory. Similar behavior is known in some vertebrate neurons, for example, Renshaw cells; and here also it may be attributed to a shifting locus.

Watanabe (1958) distinguished three **types of transmission** through the isolated abdominal ganglion of crayfish: (a) one in which output follows input impulses 1:1, (b) one in which output frequency usually begins higher than the input, at a level proportional to the input in the range 50–150 per second and then decreases gradually, and (c) one in which output varied in successive trials and an afterburst occurred following the end of the input. The output frequency acts as though determined by competition between the synaptic potential and the residual afterhyperpolarization.

B. Vertebrate Preparations

1. Neuromuscular junctions

There is another 1:1 relay junction which is known in detail and this again provides some features of unique interest. Most vertebrate striated muscle fibers answer to arriving nerve impulses in an all-or-none manner, showing under usual conditions no temporal summation; this is termed the twitch or fast muscle system as distinct from the slow system found in the frog. The special features to be mentioned have been demonstrated in the twitch muscles which are unlike arthropod and probably most muscle in being innervated, not at many but at a very few, usually one point. Reference to the relevant literature may be found in Del Castillo and Katz (1956) and Fatt (1959). The twitch junctions are like the squid example above in most respects, including the absence of appreciable electrotonic spread from pre- to postsynaptic membrane but the specialized synaptic membrane is more clearly set off from the surrounding nonsynaptic, propagating membrane. It is easier to see the lack of invasion of the synaptic (end plate) region by an antidromic action current, since an orthodromically arriving nerve impulse can cause a conspicuous potential change in the end plate even during the peak of the action current.

A remarkable finding, first made here but now extended to *Limulus* eye and crustacean muscle, and possibly frog motoneurons, is that the postsynaptic or end-plate potential is **made up of small quanta** of nearly uniform, all-or-none potentials (Fig. 4.39). There occur in these end plates "spontaneous" potential changes at random intervals, each having the same form and time course as an end-plate potential but miniature in size. The evidence is that the quanta of which the e.p.s.p. is composed are identical with these spontaneous, miniature potentials. Facilitation and depression are notable when the full propagated muscle action current has been abolished by drugs in order to reveal the e.p.s.p. and they appear to involve an increased or decreased statistical probability of individual quanta responding to a nerve impulse, rather than primarily a change in the amplitude of a given number

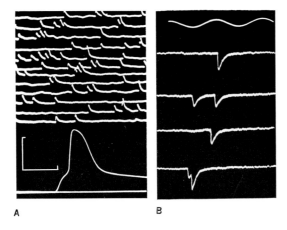
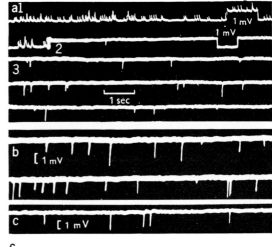
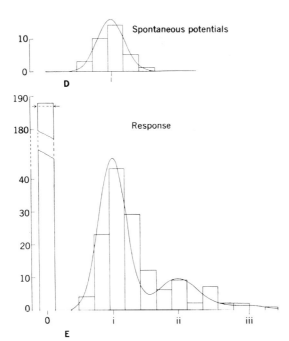

Figure 4.39. Spontaneous miniature junction potentials. **A.** Intracellular recording from a muscle fiber of the frog's toe; electrode inserted at the end plate of the nerve fiber. Upper portion shows spontaneous miniature deflections of quantal sizes and random succession of intervals. Amplification is high (calibration = 3.6 mv) and time scale slow (47 msec). The lower part, taken at low amplification (50 mv) and high speed (2 msec), shows the response of this fiber to a nerve impulse; 2 mm away from the end plate the neurally evoked spike is similar but the junction potentials are attenuated greatly. **B.** Extracellular recording reveals the junction potentials are inverted and fewer because the electrode "sees" a smaller region of the muscle fiber; high amplification (scale in millivolts) and fast sweep (top, 50 cps). **C.** a, b, c, from different end plates, prostigmine-treated muscle. a1, internal recording, 1 mv calibration; a2, a3, external recording at two different spots on the same end plate—such an electrode records from a fraction of an end plate. b, c, similar external records from other end plates. **D.** Histogram of the distribution of amplitudes of spontaneous miniature potentials in a certain experiment. The units on the abscissa = mean amplitude of spontaneous potentials = 0.875 mv; the smooth curve is drawn to fit. **E.** In the same experiment—calcium-deficient muscle, in which transmission gives only e.p.s.p.'s—end plate responses (e.p.s.p.'s) and failures (bar at zero) are given by the histogram; the smooth curve is a test of the hypothesis that e.p.s.p.'s are made up statistically of units like the spontaneous potentials and is the expected distribution on this basis. Arrows show the expected number of failures. [**A, B, C,** Fatt and Katz, 1952b; **D,** Del Castillo and Katz, 1954a.]

of quanta. It has been suggested that the spontaneous miniature potentials are true e.p.s.p.'s resulting from random release of small packets of acetylcholine of uniform size, which are perhaps the synaptic vesicles seen with the electron microscope. Whether or not this hypothesis is true, it is difficult to understand the relation between the release of endplate potentials by "spontaneity" and by nerve impulses. Agents such as excess magnesium, lowered calcium or sodium, and strong depolarization block nerve impulses without stopping spontaneous miniature potentials.

Other muscle fibers are scattered among the

preceding types in frog muscles but in the cat are apparently confined to the intrafusal fibers of the muscle spindle, which respond normally with a **graded and summating potential** and a contraction that falls off with distance from a local focus of maximum intensity. This is called a slow muscle system. Apparently the whole muscle membrane is like an endplate or synaptic membrane for not only is no potential superimposed on the junctional potential but no response to electrical stimulation can be aroused from direct, intracellular application of current; this accords with properties of chemically excited synapses.

2. Electric fish neuroelectroplaque synapses

Since there are technical limitations upon the use of the same intracellular electrode for recording the potential of the neuroplasm and for passing currents of more than a certain magnitude into the cell to set its membrane potential at a desired value, it is a distinct advantage to be able to place more than one electrode tip inside the same cell. In order to learn something about propagation or attenuation, still another electrode inserted at a distance in the same cell is valuable. It is understandable, therefore, that each of the preparations so far discussed has been one permitting an array of impaling electrodes, sometimes in both pre- and postsynaptic elements.

The effector cells or electroplaques of electric organs in several different teleosts and elasmobranches are large and flattened, reaching several millimeters in diameter; they have proven extraordinarily useful in clarifying the relations between graded synaptic potentials, graded nonsynaptic ("local") potentials, and all-or-none spikes. In addition they introduce some simple integrative properties. There are many important details of anatomy and function concerned with orientation, timing, summing, and with the fundamental mechanism of elementary discharges which are beyond the scope of this book. Many authors have contributed to the literature on the control of electric organ discharge; among them, Grundfest may be specially mentioned and his review (1957c) will serve to introduce the earlier literature while Bennett et al., in a series of papers since 1959, have added greatly to

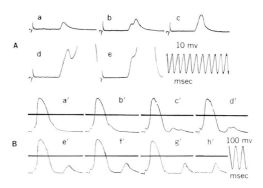

Figure 4.40. Postsynaptic potentials in the electric eel electroplaque. **A.** With an intracellular recording electrode in the electric cell, increasing stimuli to a nerve produce a stepwise increase of the e.p.s.p. (a to c). A still larger stimulus evokes a spike (d and e); the e.p.s.p. first generates a local, graded response of the electrically excitable spike-generating membrane. **B.** When the neural stimulus evokes a p.s.p. during the absolute refractory period (a' and b') the response lacks this component of graded activity of the electrically excitable membrane. Later (c' to g') the local response develops, grows, arises earlier, and fuses with the p.s.p. The combined response is seen in isolation in h'. This series of records was taken at approximately one tenth the amplification of sets a to e. The base line denotes the zero for the resting potential. [Altamirano et al., after Grundfest, 1959.]

what was known and have compared a wide variety of species.

The electric organ synapse in the main organ of the electric eel is a polarized, 1:1 junction with a brief e.p.s.p. and delay (Fig. 4.40). It is like the muscle and the squid giant synapses in these respects, as it is also in the lack of spread of electrotonic current from prespike to postsynaptic membrane. The distinctive new feature is that a single presynaptic impulse evokes only a small e.p.s.p., far below threshold for spike initiation. Repetitive stimulation at suitable frequencies builds up the response to the critical level; the facilitation has a large early component lasting up to 80 msec and a small late component detectable for more than a second. In addition to this homosynaptic temporal summation there is spatial summation. Many presynaptic nerve fibers terminate on each electroplaque and their e.p.s.p.'s sum if they are nearly synchronous. If they are more than about 2 msec apart there is no summation, so that temporal summation is practically confined to successive impulses in the same prefiber.

The contrast between **electrically excitable and synaptic membrane** is particularly obvious in the electric cell. Either a critical height of e.p.s.p. or a direct electric shock can evoke an all-or-none, regenerative, and propagating spike. If the nerve is stimulated at such a time as to deliver an impulse during the refractory period, an e.p.s.p. can be seen on top of a directly evoked spike. No direct shock can elicit this e.p.s.p. component. Hyperpolarization blocks the spike but increases the e.p.s.p. to a nerve impulse, without blocking it. Even then, a direct shock does not elicit the e.p.s.p. The e.p.s.p. cannot be elicited by electrical stimulation but only by the transmitter provided by nerve endings. The two kinds of membrane are thus quite different, but in this case are intermingled, like patches, over the innervated side of the electroplaque. The other side of the flat, disc-shaped cell is perhaps made up of inert membrane; at least it does not discharge. This must mean that it is not electrically excitable. This silence is, in *Electrophorus*, the secret of the series addition of the voltage of the many active cells comprising the organ. The synaptic potential is not the only graded and decremental potential; just below spike threshold so-called local responses are obtained which differ from the e.p.s.p.'s (in the several ways enunciated in Chaps. 3 and 4) and resemble spikes except in falling short of explosive regeneration.

The distinction between the types of membrane is not apparent in some other families of electric fish where there is no overshooting, electrically excitable spike (*Raja, Torpedo, Narcine,* and *Astroscopus*); the entire active membrane appears to be equivalent to a muscle end plate, fantastically densely innervated, and the discharge is probably a pure e.p.s.p.

3. Spinal motoneurons

Much of the most detailed information has come from study of these cells, especially in the cat and with intracellular electrodes. The summaries of Eccles (1957, 1959) and Grundfest (1959) should be consulted. As indicated in Table 4.1, the more familiar junctions upon these cells, in particular those of stretch receptor afferents, are relatively uncomplicated in integrative properties. Facilitation and aftereffects are not prominent, and transmission is mainly 1:1. Attention has been focused upon the elementary biophysical characteristics; less is known about the transmission from higher centers or local, nonsynchronized and internuncial presynaptic elements. It is known that cerebellar stimulation at moderate frequency causes a smooth shift of membrane potential without sign of individual p.s.p.'s and this seems probably to be an event of very general importance in normal central function wherever highly synchronized volleys at low frequency and producing powerful driving action are not involved.

To summarize a somewhat controversial literature on the **localization of events** in the neuron, it appears that under usual conditions spikes arise in the initial segment of the axon when that region is brought to its threshold by the antecedent potential changes in dendrites and soma. Sometimes a local potential is separately visible and the spike grows out of it. The spike then propagates distally down the axon and at the same time proximally and invades the soma where, during its repolarizing phase, it cancels much of the residual e.p.s.p.—more than in the case of *Aplysia* ganglion cells or lobster cardiac ganglion cells. In some conditions, as in toad spinal motoneurons when the resting potential is low, the spike appears to arise in the same region as the local response, namely in the soma (Araki and Otani, 1959, Fig. 4.41). Recent evidence indicates the occurrence of spontaneous synaptic potentials in frog motoneurons reminiscent of the miniature end-plate potentials in muscle (p. 231) (Katz and Miledi, 1962b).

Accommodation has been examined with exponentially rising stimulating currents introduced into the soma. As in the axon, there is no rise in the critical level for spike initiation—that is, the absolute membrane level—during several milliseconds of maintained and rising subthreshold depolarization. But after 10 msec the threshold does begin to rise, more steeply for the ini-

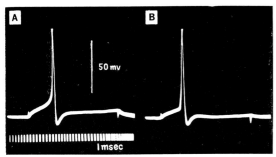

Figure 4.41. Two modes of spike iniation: (a) the so-called LR type, arising in the soma with a low resting potential; (b) the so-called abrupt type, arising in the initial segment of the axon with a high resting potential. [Araki and Otani, 1959.]

tial segment than for the soma; the soma has somewhat less accommodation, reaching only 25% above normal threshold after 20–30 msec, as compared to 50–300% in the initial segment. The chronaxie of the cell is also longer. Hill's λ comes out to 25–70 msec, which is small and therefore probably belongs to the initial segment. Accommodation probably plays a role when incoming dorsal root impulses are too high in frequency. The motoneuron fails to follow and indeed fails to initiate spikes at all although the e.p.s.p.'s are not depressed. Apparently the maintained depolarization of the summed e.p.s.p.'s causes the threshold of the initial segment to rise until they are subthreshold in height. There is some tendency, although not in every cell, for periodic undulations of soma potential during a slowly rising depolarization; this seems to be a low-grade tendency to iterative activity under maintained stimulation, a tendency which is so highly developed in some other cells.

4. Autonomic ganglia

The stellate ganglion of the cat and the superior cervical sympathetic ganglion of the rabbit have been favorite objects, even long before

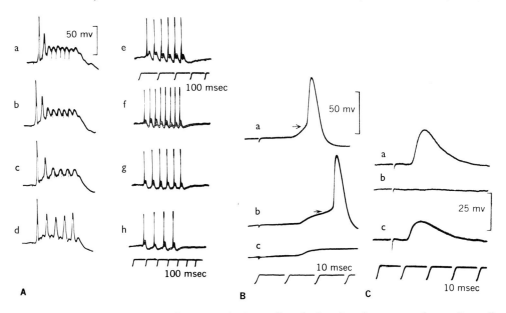

Figure 4.42. Transmission in the mammalian sympathetic ganglion. **A.** A series of responses of a ganglion cell to repetitive stimulation recorded intracellularly: a, b, c, d are at frequencies of approximately 160 per second, 125 per second, 100 per second, and 80 per second; e is the response to 35 per second followed by its time scale; f, g, h are responses to frequencies of 20 per second, 15 per second, and 10 per second, respectively, with time scale following. The cell follows only low frequencies and at high frequencies builds up accumulative synaptic potential. **B.** a is the action potential recorded from a ganglion cell in response to a single preganglionic volley; b, c are the responses when the stimulus strength is reduced. Arrows mark approximately the commencement of the spike. **C.** Synaptic potentials recorded in a cell with a resting potential of 65 mv: a is the response soon after penetration of the ganglion cell; b shows that a small reduction of stimulus strength produced no response; c, stimulus strength as in a, the record taken several minutes later. [R. Eccles, 1955.]

microelectrodes came into use. There is no evidence clearly indicating interneurons within the ganglia, so that in a simple preparation one has a single synapse in series, many in parallel. Intracellular microelectrodes have clarified the picture of what one of these postganglionic cells can do under controlled, but still multiunit, preganglionic stimulation. Eccles' (1957) summary and the literature cited there may be consulted for details. Apparently single presynaptic fibers have a feeble influence and spatial summation is important, as also are temporal summation and long afterpotentials. The e.p.s.p. takes several milliseconds to rise and many to fall. By abolishing the spikes and depressing the e.p.s.p. with curare, a postsynaptic hyperpolarization is revealed to an incoming volley but it is not definitely known whether this is due to special prefibers or even whether it is accompanied by inhibition. A further and still longer, late depolarizing wave lasts for seconds. The e.p.s.p.'s can follow synchronized incoming volleys 1:1 up to fairly high frequency and they sum in the process, but postspikes are not produced after the first few. Spikes can arise either in the soma or in the initial segment of the axon. These cells are unusual in exhibiting a considerable degree of potentiation after a single presynaptic volley but, as in motoneurons the potentiation is only maximal with prolonged repetitive stimulation. Little is known of the output generated by normal, asynchronous input but it would be expected to be only moderately integrative—removed from the relay category mainly because each cell receives input from many prefibers out of phase, and hence cannot under active bombardment relay one output spike for each incoming one.

5. Interneurons in brain and spinal cord

It might be expected that in this great and anatomically diverse population of cells the most complex and integrative properties with respect to interaction on neurons would be found, and there is considerable indication that this is the case. Single unit records of central interneurons have been studied from the dorsal and ventrolateral horns of the spinal cord, the reticular formation in various regions, several thalamic nuclei, cerebellar and cerebral cortices (Albe-Fessard and Buser, 1953; Tasaki et al., 1954; Phillips, 1956; Frank and Fuortes, 1956; Eccles, 1957; Kolmodin, 1957; Hunt and Kuno, 1959).

Details of the experiences with different types of neurons cannot be given here. It suffices in the present connection to point out that at least some instances have been recorded of significant differences from typical motoneuron transmission. Single arriving impulses in certain cases have powerful excitatory action and produce large e.p.s.p.'s; in some cases, repetitive discharge of the postsynaptic cell. In other situations arriving input can only weakly modify ongoing activity or predispose the cell to other input. Many types of cells are prone to high frequency iterative firing. Often it has been noted that there is much less afterhyperpolarization than in motoneurons. The p.s.p.'s are usually compound and are not wiped out by the occurrence of spikes. Clear i.p.s.p.'s have rarely been seen, but because of the nature of the experiments the pathway between stimulating and recording sites is often not well enough known

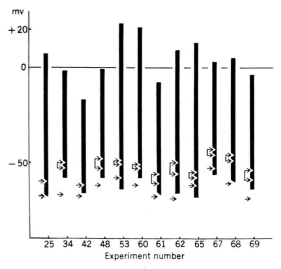

Figure 4.43. Activity of interneurons in the mammalian cerebral cortex. *Vertical bars* show peak and tail voltages of spikes (both natural and antidromic) in 12 experiments. *Arrows* show range of fluctuation in potential observed in each experiment. Bracketed pairs of arrows show range of firing level; lowermost arrows show maximum polarization. [Phillips, 1956.]

to distinguish between positive inhibition and the withdrawal of excitation. Offdischarges, afterdischarges, postexcitatory rebound depression and postinhibitory rebound excitation, accumulation of inhibition (summation), and escape from inhibition (adaptation) have all been described from central interneurons (Strumwasser and Rosenthal, 1960). Certainly this is one of the most fruitful frontiers and a great deal more can and must be learned before a systematization of the properties and degrees of freedom of interneurons will be at all complete.

V. EVOLUTION OF NERVOUS TRANSMISSION

No definite trends are evident by comparison of the properties of synaptic transmission in lower and higher groups of animals. The main impression is that the physiology of the neuron is essentially the same. The range of variation of properties is as great among the junctions in a given animal as it is between widely unrelated species. The characteristics of the junctions in jellyfish and worms are not at an extreme end of the neural spectrum nor are those of mammals. There is certainly a **greater differentiation into kinds** of junctions in the higher vertebrates, and also in the higher invertebrates, than in simpler invertebrates. Based mainly on histological information but also on considerable physiological sampling, specific types of connection and combinations of integrative properties appear to characterize many regions, nuclear masses, and strata in animals with more complex nervous systems. Even in worms and medusae transmission in some pathways is different from that in others—in degree of facilitation, relative synaptic power of a single impulse, time course of aftereffects, and the like. Both excitatory and inhibitory pathways are known even in medusae. Both relay and integrative transmission occur in nerve nets. Slow and fast muscle systems with their corresponding junctions are found in coelenterates as well as in higher forms, but the variety of specialized junctions is not as great as in cephalopods, higher crustaceans, insects, and vertebrates. It would be more difficult to assert confidently that vertebrates have a greater variety than cephalopods or insects. Perhaps lower vertebrates have less and higher vertebrates more variety. Bishop suggests that synapses on the soma are recent in vertebrates, and related to the evolution of large-fiber systems; these he proposes have been superimposed upon older, small-fiber systems.

There is a certain reasonableness to the proposition that in general **relay junctions are more derived** than primitive. Integrative properties are not only useful for complex transfer functions but also resemble the graded behavior of simple and nonnervous cells. We have already concluded in the preceding chapter that the more primitive form of response is the graded rather than the all-or-none impulse. We have seen that already in protozoans both depolarization and hyperpolarization occurs. It is certainly primitive for each cell to be sensitive to certain forms of impinging events and to respond. Furthermore the response may take place in a different part of the cell and therefore give opportunity for summation of arriving events in different places as well as at successive moments, with some kind of integration of these in determining the response. The **primitive graded and local character has been retained**—and possibly, though we cannot be sure, further perfected or differentiated—in the mechanism of synaptic transmission. Some cells exhibit only this type and no all-or-none response —slow muscle fibers in the frog and many arthropods, and electroplaques in *Torpedo* and other marine electric fish. No nerve cells are known to react only with local, graded responses. But it seems highly probably that this is true of some of the smaller intrinsic cells of specialized masses. The total extent of the axons of many such neurons is far less than the space constant measured in ordinary nerve fibers; that is, decremental spread of electrotonic potentials in a wide size range of fibers is such that only half of the effectiveness of a brief pulse, and less than that for a slower event, has been lost in a distance of a millimeter or so. Even if the distance is considerably less in extremely small fibers, there would appear to be little advantage in the decrementless impulse, for example, in small neurons in the optic ganglia of many insects.

Since the essence of the regenerative impulse is the ability of the membrane in which it occurs to respond to the local electric current from preceding regions, one is tempted to propose that **electrical excitability is derived** and electrical inexcitability is more primitive. There is no decisive evidence for or against this supposition. If it is true, the electrically transmitting synapses would be specialized; even more so would be the rectifying or one-way electrically transmitting synapses. Chemical excitability, often with a high degree of chemical specificity, is surely a property of single-celled organisms.

Two-way transmission can be regarded as primitive in the nerve nets of coelenterates, polarized junctions as specialized, though no strong case can be made for this. But the two-way commissural synapses in annelids and arthropods are doubtless specialized.

What is the basis for polarized transmission? Several distinct areas are known, both morphological and functional. The former refers mainly to the convergence of presynaptic branches on a limited postsynaptic region, permitting spatial summation, which would not be available in the reverse direction (McCulloch, 1938). It seems likely that in addition the polarized, chemically transmitting synapses are composed of membranes presynaptically and postsynaptically so specialized physiologically that they could not exchange roles, the one releasing transmitter and the other transducing this into a response; such specialization is suggested by electron microscopy. A potent force for polarization is the general property of dendrites which prevents antidromic impulses from invading them.

In spite of a considerable body of knowledge and refinement of technic in the study of transmission, there is room in our area of ignorance or uncertainty for the occurrence among synapses of a good deal of specialization which may be the **main evolutionary achievement** at this level. Particular forms of nonlinearity between input and output may be the specialized character of certain junctions in higher animals. Permutations of faster and slower temporal summation, weighting of input from different presynaptic sources, repetitiveness or accommodation (perhaps different for different inputs), and aftereffects of various degrees and sign are in all likelihood diagnostic of functional categories evolved from the highly plastic primitive forms of interaction. To this should be added the further degrees of freedom available in the nonsynaptic modes of interaction, about which still less is known in comparative terms.

Classified References

General

Bacq, 1941, 1947; Bremer, 1951a, 1951b, 1953, 1956, 1961; Bremer and Bonnet, 1950; Bullock, 1940, 1947, 1950b, 1950c, 1952a, 1952b, 1956a, 1958, 1959a; 1959b; Burgen and MacIntosh, 1955; Dale, 1937; Del Castillo and Katz, 1956; Eccles, 1950, 1952, 1957, 1959, 1961; Elliott et al., 1955; Eyzaguirre, 1961; Fatt, 1959; Fatt and Katz, 1952c; Feldberg, 1945, 1951; Fessard and Posternak, 1950; Florey, 1951c, 1951d, 1957a, 1961; Forbes, 1922; Furshpan 1959; Grundfest, 1957a, 1957b, 1958a, 1958b, 1959; Hebb, 1957a; Hoyle, 1957a, 1957b; Katz, 1949; Koelle, 1959; Koppanyi, 1948; Korey and Nurnberger, 1956; Kuffler, 1952, 1958, 1960; Lloyd, 1944; Lorente de Nó, 1946–1947; McCulloch, 1938; McLennan, 1961; Minz, 1955; Nachmansohn, 1948, 1959; Pantin, 1937, 1952; Richards, 1955; Rosenblueth, 1950; Ross, 1956; Seite, 1961; Sperry, 1958; Terzuolo and Bullock, 1956; Welsh, 1948, 1957a, 1957b; Welsh and Schallek, 1946; Wiersma, 1941, 1952b, 1953; Wilson, 1952.

Nonsynaptic Interactions of Neurons

Arvanitaki, 1940, 1942a, 1942b; Arvanitaki and Chalazonitis, 1949d, 1956a, 1956c; Bennett, 1960; Dudel and Kuffler, 1961b, 1961c, 1961d; Dzhavrishvili, 1959; Eccles and Krnjević, 1959; Eccles et al., 1960; Galambos, 1961; Gerard, 1941a, 1941b; Gerard and Libet, 1939, 1940; Granit and Skoglund, 1945a, 1945b; Hagiwara et al., 1959; Hagiwara and Morita, 1962; Hild et al., 1958; Jasper and Monnier, 1938; Katz and Schmitt, 1940, 1942; Krnjevic and Miledi, 1959; Libet and Gerard, 1939, 1941; Marrazzi and Lorente de Nó, 1944; Nagahama, 1950; Renshaw and Therman, 1941; Reuben, 1960; Rosenblueth, 1941; Tasaki and Chang, 1958; Tauc, 1959a, 1959c; Wall, 1958, 1959; Wall et al., 1956; Watanabe, 1958a; Watanabe and Bullock, 1960.

Miniature Quantal Events at Junctions

Del Castillo and Katz, 1953, 1954a, 1954b; Dudel and Kuffler, 1961a; Fatt and Katz, 1952; Ishiko and Loewenstein, 1959; Reuben and Grundfest, 1960; Wakabayashi, and Ikeda, 1957; Yeandle, 1958.

Conductance Changes at Junctions

Coombs et al., 1953, 1955a, 1955b, 1960; Del Castillo and Katz, 1954c; Edwards and Hagiwara, 1959; Fatt, 1961; Werman, 1960b.

Studies on Arthropod Junctions

Bethe, 1897; Boistel and Fatt, 1958; Bullock and Terzuolo 1957; Bullock et al., 1943; Burgen and Kuffler, 1957; Cerf et al., 1957a, 1957b; Del Castillo et al., 1953; Dudel and Kuffler, 1961b, 1961c, 1961d; Easton, 1957; Ellis and Hoyle, 1954; Eyzaguirre and Kuffler, 1954b; Fatt and Katz, 1952a, 1953a, 1953b; Fielden, 1960; Florey, 1956; Florey and Hoyle, 1961; Furshpan and Potter, 1957, 1959a, 1959b; Furshpan and Wiersma 1954; Grundfest and Reuben, 1961; Grundfest et al., 1959; Hagiwara, 1953, 1958; Hagiwara and Bullock, 1957; Hagiwara and Watanabe, 1954, 1956; Hagiwara et al., 1960; Hoyle, 1955, 1958a, 1958b, 1959; Hoyle and Wiersma, 1958a, 1958b, 1958c; Kao, 1960; Katz, 1936; Katz and Kuffler, 1946; Kennedy and Preston, 1960; Kuffler and Eyzaguirre, 1955; Kuffler and Katz, 1946; Marmont and Wiersma, 1938; Maynard, 1953a, 1953b, 1953c, 1958; Nagahama, 1950, 1954; Otani and Bullock, 1959; Pantin, 1934, 1936a, 1936b, 1936c; Preston and Kennedy, 1960; Prosser, 1935a. 1935b, 1937, 1943a; Pumphrey and Rawdon-Smith, 1937; Ripley and Wiersma, 1953; Roeder and Weiant, 1950; Roeder et al., 1947; Ruck, 1958, 1961; Terzuolo and Bullock, 1958; Van Harreveld and Wiersma, 1937, 1939; Watanabe, 1958; Watanabe and Grundfest, 1961; Watanabe et al., 1960; Werman and Grundfest, 1959; Wiersma, 1933, 1937, 1938, 1947, 1949, 1952a; Wiersma and Adams, 1950; Wiersma and Ellis, 1942; Wiersma and Helfer, 1941; Wiersma and Marmont, 1936; Wiersma and Ripley, 1954; Wiersma and Schallek, 1947; Wiersma and Turner, 1950; Wiersma and Van Harreveld, 1934, 1935, 1938, 1939; Wilson, 1954; Yamasaki and Narahashi, 1960.

Studies on Molluscan Junctions

Arvanitaki and Chalazonitis, 1956e, 1957b; Bryant, 1958b, 1959; Bullock, 1946, 1948a; Bullock and Hagiwara, 1957; Fessard and Tauc, 1957, 1958, 1960; Fröhlich and Loewi, 1907; Hagiwara and Kusano, 1961; Hagiwara and Tasaki, 1958; Kusano and Hagiwara, 1961; Prosser and Young, 1937; Richards, 1929; Tauc, 1955, 1956, 1957b, 1957c, 1957d, 1958a, 1958c, 1958e, 1959a, 1959b, 1959d, 1960b, 1960c, 1960d; Tauc and Gerschenfeld, 1961; Young, 1938a, 1938b.

Studies on Vertebrate Junctions and Cells

Albe-Fessard and Buser, 1953; Araki and Otani, 1959; Barron, 1940; Barron and Matthews, 1938; Bennett, Crain, and Grundfest, 1959; Bennett and Grundfest, 1959, 1961a, 1961b, 1961c; Bennett, Wurzel, and Grundfest, 1961; Bonnet, 1957; Bonnet and Bremer, 1938; Boyd and Martin, 1956; Bremer, 1941a, 1941b, 1953; Bronk, 1939; Coombs et al., 1953, 1955a, 1955b; Curtis and Eccles, 1960; Del Castillo and Katz, 1954c; Desmedt and Monaco, 1961; Easton, 1955, 1956; Eccles, J. C., 1935, 1943, 1961c; Eccles and Krnjević, 1959; Eccles, Eccles, and Magni, 1960; Eccles, Fatt, and Koketsu, 1953, 1954; Eccles, R. M., 1955; Fatt and Katz, 1951, 1952d, 1953c; Frank and Fuortes, 1955, 1956, 1957; Fuortes et al., 1957; Grundfest, 1958a, 1958b; Grundfest and Bennett, 1961; Hubbard, 1959; Hunt and Kuno, 1959; Hutter and Loewenstein, 1955; Ito, 1959; Ito and Saiga, 1959; Keynes et al., 1961; Kolmodin, 1957; Kuffler, 1942a, 1942b, 1949, 1953; Kuffler and Vaughan Williams, 1953; Laporte and Lorente de Nó, 1950a, 1950b, 1950c; Larrabee and Bronk, 1947; Liley, 1956; Lloyd, 1946, 1949, 1951a, 1951b; Loewenstein, 1959a, 1959b; Lorente de Nó, 1938; Lorente de Nó and Laporte, 1950; Nastuk, 1953; Nishi and Koketsu, 1960; Phillips, 1956; Renshaw, 1946; Sechenov, 1881; Sperry, 1958; Tasaki and Chang, 1958; Tasaki et al., 1954.

Studies on Transmission in other Groups

Batham et al., 1960; Bullock, 1943, 1945, 1953; Bullock and Turner, 1950; Hama, 1959; Horridge, 1955, 1956b; Kao and Grundfest, 1957; Mackie, 1960; Pantin, 1935a, 1935b, 1952; Prosser, 1954; Prosser and Melton, 1954; Ross, 1952, 1955, 1957; Ross and Pantin, 1940; Rushton, 1945; Stough, 1926.

Effects of Temperature on Junctions

Benthe, 1954; Bullock, 1956b; Burkhardt, 1959a, 1959b; Kerkut and Taylor, 1956, 1958; Schenck et al., 1956; Turner, 1953.

Assay of Nervous Tissue *(since this topic is peripheral to the scope of this book, this list is only a sample)*

Artemov and Mitropolitanskaja, 1938; Augustinsson, 1948; Augustinsson and Grahn, 1954; Bacq, 1935, 1937, 1947; Bacq and Mazza, 1935; Bazemore et al., 1956, 1957; Bullock and Nachmansohn, 1942; Burgen and MacIntosh, 1955; Colhoun, 1959; Dale et al., 1936; Defretin and Riff, 1948; Easton, 1950; Elliott and Florey, 1956; Elliott et al., 1955; Florey, 1951b, 1951c, 1951d, 1953a, 1954c, 1957b; 1961; Florey and Biederman, 1960; Florey and Chapman, 1961; Florey and Florey, 1954, 1958; Florey and McLennan, 1955c; Gerebtzoff, 1956; Griffith, 1892; Hoyle, 1954; Jullien and Vincent, 1938b; Koshtoyants, 1936; Lewis and Smallman, 1956; Loewenstein and Molina, 1958; Maynard and Maynard, 1960a, 1960b; Maynard, D. M. and Welsh, 1959; McLennan, 1961; Means, 1942; Mikalonis and Brown, 1941; Minz, 1955; Nachmansohn, 1937, 1950, 1959; Parrot, 1941; Persky and Gold, 1948; Richard, 1955; Smallman, 1956; Smallman and Wolfe, 1956; Smith, 1939; Stedman and Stedman, 1938; Tobias et al., 1946; Van der Kloot, 1960; Vincent and Jullien, 1938; Walop, 1951; Welsh, 1939a, 1957a, 1957b; Wolfe and Smallman, 1956.

Effects of Drugs *(since this topic is peripheral to the scope of the book, this list is only a sample)*

Bacq, 1932a, 1932b, 1933a, 1933b, 1934, 1941, 1947; Bacq and Coppée, 1937; Barry et al., 1937; Bellemare and Belcourt, 1955; Boardman and Collier, 1946; Bonnet, 1938; Botsford, 1941; Bryant, 1958b; Buck et al., 1952; Burgen and MacIntosh, 1955; Cardot and Jullien, 1940; Cate, 1933; Chauchard, A. B. and Chauchard, P., 1933; Chauchard, B. and Chauchard, P., 1950, 1952; Chauchard, P. and Chauchard, J., 1942; Colhoun, 1958; Davenport, 1941, 1942; Del Castillo and Katz, 1954d; Dresden, 1949; Edwards and Kuffler, 1959; Ellis et al., 1942; Ewer and Berg, 1954; Feldberg, 1945, 1951; Florey, 1951a, 1954a, 1954b, 1954c, 1956, 1957a, 1957b, 1961; Florey and Florey, 1953, 1954a, 1954b; Florey and McLennan, 1955a, 1955b, 1955c, 1955d, 1955e; Florey and Merwin, 1961; Garrey, 1942; Gerschenfeld and Tauc, 1961; Grundfest, 1958b; Hebb, 1957a; Horridge, 1956; Hoyle, 1955; Hughes and Kerkut, 1956; Jullien and Vincent, 1938a; Kerkut and Taylor, 1956; Knowlton, 1942; Koelle, 1959; Koppanyi, 1948; Koshtoyants, 1936, 1957; Kruta, 1935, 1936; Kuffler and Edwards, 1958; Kuperman et al., 1959; Loewi, 1921; Lorente de Nó, 1948; Lowenstein, 1942; McLennan, 1961; Michaelis et al., 1949; Milburn et al., 1960; Minz, 1955; Moore, 1918; Nachmansohn, 1948, 1950, 1959; Narahashi and Yamasaki, 1960; Obreshkove, 1941; Pantin, 1935; Prosser, 1938, 1940, 1942; Prosser and Buehl, 1939; Purpura et al., 1959; Reid and Vaughan Williams, 1949; Reiter, 1957; Richards, 1929; Roaf, 1935; Robbins and Van der Kloot, 1958; Roeder, 1948a, 1948b; Roeder et al., 1947; Ross, 1945a, 1945b, 1957; Schallek, 1945; Schallek and Wiersma, 1948a, 1948b; Schallek et al., 1948; Seite, 1961; Smith, 1947; Tauc, 1958b; Tauc and Gerschenfeld, 1960a, 1960b, 1961; Tobias and Kollros, 1946; Van der Kloot, 1960; Van der Kloot et al., 1958; Wait, 1943; Waterman, 1941; Welsh, 1939b, 1942, 1948, 1957a, 1957b; Welsh and Gordon, 1947; Welsh and Haskin, 1939; Welsh and Moorhead, 1960; Whitteridge, 1948; Wiersma and Schallek, 1948; Wiersma and Zawadzki, 1948; Wiersma et al., 1953; Wright, 1949; Wu, 1939a, 1939b; Yamasaki and Ishii, 1952, 1954a, 1954b; Yamasaki and Narahashi, 1958.

Anatomical References Cited in this Chapter *(see also Chapter 2)*

Batham et al., 1960; Dun, 1951; Florey and Florey, 1955; Hama, 1959; Horridge, 1956b; Mackie, 1960; Richards, 1955; Scharf, 1953; Stough, 1926; Young, 1936, 1937, 1939.

Bibliography

In this list a selection is given of citations on the more important modern and certain older works on junctional transmission. Papers on the effects of drugs and other chemicals and on the assay of active compounds in nervous tissue are represented only by samples, since this field is beyond the scope of the present work.

ALBE-FESSARD, D. and BUSER, P. 1953. Explorations de certaines activités du cortex moteur du chat par microélectrodes: dérivations endosomatiques. *J. Physiol., Paris*, 45: 14-16.

ARAKI, T. and OTANI, T. 1959. Accommodation and local response in motoneurons of toad's spinal cord. *Jap. J. Physiol.*, 9:69-83.

ARTEMOV, N. M. and MITROPOLITANSKAJA, R. L. 1938. Content of acetylcholine-like substances in the nerve tissue and of choline esterase in the hemolymph of crustaceans. *Bull. Biol. Med. exp. URSS. (Biull. eksp. Biol. Med. Moskva)*, 5:378-381.

ARVANITAKI, A. 1940. Temps courts et définis dans la transmission de l'excitation au niveau de la synapse expérimentale axono-axonique. *C. R. Soc. Biol., Paris*, 133:208-211.

ARVANITAKI, A. 1942a. Interactions électriques entre deux cellules nerveuses contiguës. *Arch. int. Physiol.*, 52:381-407.

ARVANITAKI, A. 1942b. Effects evoked in an axon by the activity of a contiguous one. *J. Neurophysiol.*, 5: 89-108.

ARVANITAKI, A. and CHALAZONITIS, N. 1949. Prototypes d'interactions neuroniques et transmissions synaptiques. Données bioélectriques de préparations cellulaires. *Arch. Sci. physiol.*, 3:547-565.

ARVANITAKI, A. and CHALAZONITIS, N. 1956a. Stimulation d'une aire de la membrane somatique par l'activité d'un autre lieu cellulaire contigu (soma géant d'*Aplysia*). *C. R. Acad. Sci., Paris*, 242:1814-1816.

ARVANITAKI, A. and CHALAZONITIS, N. 1956b. Interactions bioélectriques entre aires somatiques autoactives contiguës (soma géant d'*Aplysia*). *C. R. Soc. Biol., Paris*, 150:700.

ARVANITAKI, A. and CHALAZONITIS, N. 1956c. Activations du soma géant d'*Aplysia* par voie orthodrome et par voie antidrome (dérivation endocytaire). *Arch. Sci. physiol.*, 10:95-128.

ARVANITAKI, A. and CHALAZONITIS, N. 1957. Actions inhibitrices sur la genèse des potentiels positifs du soma neuronique. *C. R. Acad. Sci., Paris*, 245:1029-1032.

AUGUSTINSSON, K. 1948. Cholinesterases. A study in comparative enzymology. *Acta physiol. scand.*, 15:1-182.

AUGUSTINSSON, K. and GRAHN, M. 1954. The occurrence of choline esters in the honeybee. *Acta physiol. scand.*, 32:174-190.

BACQ, Z. M. 1932a. Action des ions potassium sur la musculature des chromatophores des céphalopodes. *C. R. Soc. Biol., Paris*, 111:220-222.

BACQ, Z. M. 1932b. Action de l'ergotamine sur les muscles des chromatophores des céphalopodes. *C. R. Soc. Biol., Paris*, 111:223-224.

BACQ, Z. M. 1933a. Action de l'adrénaline, de l'ergotamine et de la tyramine sur le ventricule médian isolé de *Loligo pealii*. *C. R. Soc. Biol., Paris*, 114:1358-1360.

BACQ, Z. M. 1933b. Réactions du ventricule médian isolé de *Loligo pealii* à l'acétylcholine, à l'atropine et aux ions K, Ca. et Mg. *C. R. Soc. Biol., Paris*, 114:1360-1361.

BACQ, Z. M. 1934. Réactions de divers tissus isolés du calmar (*Loligo pealii*) à l'adrénaline, à l'acétylcholine, à l'ergotamine et aux ions. *C. R. Soc. Biol., Paris*, 115:716-717.

BACQ, Z. M. 1935. Occurrence of unstable choline esters in invertebrates. *Nature, Lond.*, 136:30-31.

BACQ, Z. M. 1937. Nouvelles observations sur l'acétylcholine et la cholineestérase chez les invertébrés. *Arch. int. Physiol.*, 44:174-189.

BACQ, Z. M. 1941. Physiologie comparée de la transmission chimique des excitations nerveuses. *Ann. Soc. zool. Belg.*, 72:181-203.

BACQ, Z. M. 1947. L'acétylcholine et l'adrénaline chez les invertébrés. *Biol. Rev.*, 22:73-91.

BACQ, Z. M. and COPPÉE, G. 1937. Contraste entre les vers et les mollusques en ce qui concerne leur réaction à l'ésérine. *C. R. Soc. Biol., Paris*, 125:1059-1060.

BACQ, Z. M. and MAZZA, F. 1935. Identification d'acétylcholine extraite des cellules ganglionnaires d'*Octopus*. *C. R. Soc. Biol., Paris*, 120:246-247.

BARRON, D. H. 1940. Central course of "recurrent sensory discharges." *J. Neurophysiol.*, 3:403-406.

BARRON, D. H. and MATTHEWS, B. H. C. 1938. The interpretation of potential changes in the spinal cord. *J. Physiol.*, 92:276-321.

BARRY, D.-T., CHAUCHARD, A., and CHAUCHARD, B. 1937. Action de la nicotine sur l'excitabilité de l'appareil neuromoteur chez le crabe. *C. R. Soc. Biol., Paris*, 126:574-576.

BATHAM, E. J., PANTIN, C. F. A., and ROBSON, E. A. 1960. The nerve-net of the sea-anemone *Metridium senile*: the mesenteries and the column. *Quart. J. micr. Sci.*, 101:487-510.

BAZEMORE, A. W., ELLIOTT, K. A. C., and FLOREY, E. 1956. Factor I and γ-aminobutyric acid. *Nature, Lond.*, 178:1052-1053.

BAZEMORE, A. W., ELLIOTT, K. A. C., and FLOREY, E. 1957. Isolation of Factor I. *J. Neurochem.*, 1:334-339.

BELLEMARE, E. R. and BELCOURT, J. 1955. Action du dérivé cyanuré du DDT sur le système nerveux de *Periplaneta americana* (L.). *Rev. canad. Biol.*, 14:95-107.

BENNETT, M. V. L. 1960. Electrical connections between supramedullary neurons. *Fed. Proc.*, 19:282.

BENNETT, M. V. L. and GRUNDFEST, H. 1959. Electrophysiology of electric organ in *Gymnotus carapo*. *J. gen. Physiol.*, 42:1067-1104.

BENNETT, M. V. L. and GRUNDFEST, H. 1961a. The electrophysiology of electric organs of marine electric fishes. II. The electroplaques of main and accessory organs of *Narcine brasiliensis*. *J. gen. Physiol.*, 44:805-818.

BENNETT, M. V. L. and GRUNDFEST, H. 1961b. The electrophysiology of electric organs of marine electric fishes. III. The electroplaques of the stargazer, *Astroscopus y-graecum*. *J. gen. Physiol.*, 44:819-843.

BENNETT, M. V. L. and GRUNDFEST, H. 1961c. Studies on the morphology and electrophysiology of electric organs. III. Electrophysiology of electric organs in mormyrids. In: *Bioelectrogenesis*. C. Chagas and A. Paes de Carvalho (eds.). Elsevier, Amsterdam.

BENNETT, M. V. L., CRAIN, S. M., and GRUNDFEST, H. 1959. Electrophysiology of supramedullary neurons in *Spheroides maculatus*. I, II, and III. *J. gen. Physiol.*, 43:159-250.

BENNETT, M. V. L., WURZEL, M., and GRUNDFEST, H. 1961. The electrophysiology of electric organs of marine electric fishes. I. Properties of electroplaques of *Torpedo nobiliana*. *J. gen. Physiol.*, 44:757-804.

BENTHE, H. F. 1954. Über die Temperaturabhängigkeit neuromuskulärer Vorgänge. *Z. vergl. Physiol.*, 36:327-351.

BERNHARD, C. G. 1958. On undifferentiated neuronal spread of excitation. In: *Submicroscopic Organization and Function of Nerve Cells*. H. Fernández-Morán and R. Brown (eds.). Academic Press, New York. (*Exp. Cell Res.*, Suppl. 5:201-220.)

BETHE, A. 1897. Das Centralnervensystem von *Carcinus maenas*. *Arch. mikr. Anat.*, 50:589-639.

BISHOP, G. H. 1958. The dendrite: receptive pole of the neurone. *Electroenceph. clin. Neurophysiol.*, Suppl. 10:12-21.

BOARDMAN, D. L. and COLLIER, H. O. J. 1946. Effect of magnesium deficiency on neuromuscular transmission in the shore crab, *Carcinus maenus*. *J. Physiol.*, 104:377-383.

BOISTEL, J. and FATT, P. 1958. Membrane permeability change during inhibitory transmitter action in crustacean muscle. *J. Physiol.*, 144:176-191.

BONNET, V. 1938. Action de la strychnine et de l'acétylcholine sur la rythmicité neuronique chez les crustacés. *C. R. Soc. Biol., Paris*, 127:804-806.

BONNET, V. 1957. La transmission synaptique d'influx au niveau des dendrites superficiels de l'écorce cérébrale. *Arch. int. Physiol.*, 65:506-511.

BONNET, V. and BREMER, F. 1938. Relation des potentiels réactionnels spinaux avec les processus d'inhibition et de sommation centrale. *C. R. Soc. Biol., Paris*, 127:812.

BOTSFORD, E. F. 1941. The effect of physostigmine on the responses of earthworm body wall preparations to successive stimuli. *Biol. Bull., Woods Hole*, 80:299-313.

BOYD, I. A. and MARTIN, A. R. 1956. The end plate potential in mammalian muscle. *J. Physiol.*, 132:74-91.

BREMER, F. 1941a. L'activité électrique "spontanée" de la moëlle épinière. *Arch. int. Physiol.*, 51:51-84.

BREMER, F. 1941b. Le tétanos strychnique et le mécanisme de la synchronisation neuronique. *Arch. int. Physiol.*, 51:211-260.

BREMER, F. 1951a. Transmissions synaptiques ganglionnaire et centrale. *Arch. int. Physiol.*, 59:475-626.

BREMER, F. 1951b. Aspects électrophysiologiques de la transmission synaptique. *Arch. int. Physiol.*, 59:588-602.

BREMER, F. 1953a. Aspects physiologiques de la mémoire. *Centre Belge de Navigation*, 9:1-16.

BREMER, F. 1953b. Strychnine tetanus of the spinal cord. In: *The Spinal Cord*. G. E. W. Wolstenholme (ed.). (Ciba Foundation Symposium.) Churchill, London.

BREMER, F. 1956. Convergences neuroniques et facilitations centrales. In: *Problems of the Modern Physiology of the Nervous and Muscle Systems*. S. P. Narikashvili (ed.). Acad. Sci. Georgian SSR, Tbilisi.

BREMER, F. 1961. L'interprétation des potentiels électriques de l'écorce cérébrale. In: *Structure and Function of the Cerebral Cortex.* D. B. Tower and J. P. Schadé (eds.). Elsevier, Amsterdam.

BREMER, F. and BONNET, V. 1950. Les potentiels synaptiques et leur interprétation. In: *Électrophysiologie, Colloq. int. Cent. nat. Rech. sci.,* 22: 363-393.

BRONK, D. 1939. Synaptic mechanisms in sympathetic ganglia. *J. Neurophysiol.,* 2:380-401.

BRYANT, S. H. 1958. Transmission in squid giant synapses. The importance of oxygen supply and the effects of drugs. *J. gen. Physiol.,* 41:473-484.

BRYANT, S. H. 1959. The function of the proximal synapses of the squid stellate ganglion. *J. gen. Physiol.,* 42:609-616.

BUCK, J. B., KEISTER, M. L., and POSNER, I. 1952. Physiological effects of DDT on *Phormia* larvae. *Ann. ent. Soc. Amer.,* 45:369-384.

BULLOCK, T. H. 1940. The existence of unpolarized synapses. *Anat. Rec.,* 78(suppl.):67.

BULLOCK, T. H. 1943. Neuromuscular facilitation in scyphomedusae. *J. cell. comp. Physiol.,* 22:251-272.

BULLOCK, T. H. 1945. Functional organization of the giant fiber system of *Lumbricus. J. Neurophysiol.,* 8:55-72.

BULLOCK, T. H. 1946. A preparation for the physiological study of the unit synapse. *Nature, Lond.,* 158: 555-556.

BULLOCK, T. H. 1947. Problems in invertebrate electrophysiology. *Physiol. Rev.,* 27:643-664.

BULLOCK, T. H. 1948. Properties of a single synapse in the stellate ganglion of squid. *J. Neurophysiol.,* 11: 343-364.

BULLOCK, T. H. 1950a. The course of fatigue in single nerve fibers and junctions. *Int. Physiol. Congr., XVIII,* Copenhagen, 1950:134-135. (Abs.).

BULLOCK, T. H. 1950b. Separation of excitation and response of the postsynaptic membrane. *Anat. Rec.,* 108:607.

BULLOCK, T. H. 1952a. The invertebrate neuron junction. *Cold Spr. Harb. Symp. quant. Biol.,* 17:267-273.

BULLOCK, T. H. 1952b. Electrical similarities and differences between synaptic transmission and axonal conduction. In: *Nerve Impulse.* H. H. Merritt, (ed.). Josiah Macy, Jr. Foundation, New York.

BULLOCK, T. H. 1953. Properties of some natural and quasi-artificial synapses in polychaetes. *J. comp. Neurol.,* 98:37-68.

BULLOCK, T. H. 1956a. Delays at synapses and peripheral junctions. In: *Handbook of Biological Data.* Saunders, Philadelphia.

BULLOCK, T. H. 1956b. Temperature sensitivity of some unit synapses. *Pubbl. Staz. zool. Napoli,* 28:305-314.

BULLOCK, T. H. 1958. Parameters of integrative action of the nervous system at the neuronal level. In: *Symposium on Ultramicroscopic Structure and Function of Nerve Cells. Exper. Cell Res.,* Suppl. 5:303-337.

BULLOCK, T. H. 1959a. Initiation of nerve impulses in receptor and central neurons. *Rev. Modern Physics,* 31:504-514; and in: *Biophysical Science—A Study Program.* J. L. Oncley (ed.). Wiley, New York.

BULLOCK, T. H. 1959b. The neuron doctrine and electrophysiology. *Science,* 129:997-1002.

BULLOCK, T. H. and HAGIWARA, S. 1957. Intracellular recording from the giant synapse of the squid. *J. gen. Physiol.,* 40:565-577.

BULLOCK, T. H. and NACHMANSOHN, D. 1942. Choline esterase in primitive nervous systems. *J. cell. comp. Physiol.,* 20:239-242.

BULLOCK, T. H. and TERZUOLO, C. A. 1957. Diverse forms of activity in the somata of spontaneous and integrating ganglion cells. *J. Physiol.,* 138:341-364.

BULLOCK, T. H. and TURNER, R. S. 1950. Events associated with conduction failure in nerve fibers. *J. cell. comp. Physiol.,* 36:59-82.

BULLOCK, T. H., BURR, H. S., and NIMS, L. F. 1943. Electrical polarization of pacemaker neurons. *J. Neurophysiol.,* 6:85-98.

BURGEN, A. S. V. and KUFFLER, S. W. 1957. Two inhibitory fibres forming synapses with a single nerve cell in the lobster. *Nature, Lond.,* 180: 1490-1491.

BURGEN, A. S. V. and MACINTOSH, F. C. 1955. The physiological significance of acetylcholine. In: *Neurochemistry,* K. A. C. Elliott et al. (eds.). Thomas, Springfield, Ill.

BURKE, W. and GINSBORG, B. L. 1956. The electrical properties of the slow muscle fibre membrane. *J. Physiol.,* 132:586-598.

BURKHARDT, D. 1959a. Effect of temperature on isolated stretch-receptor organ of crayfish. *Science,* 129: 392-393.

BURKHARDT, D. 1959b. Die Erregungsvorgänge sensibler Ganglienzellen in Abhängigkeit von der Temperatur. *Biol. Zbl.,* 78:22-62.

CARDOT, H. and JULLIEN, A. 1940. Action de la pourpre sur l'excitabilité du nerf et du muscle. *C. R. Soc. Biol., Paris,* 133:521-523.

CATE, J. TEN. 1933. L'action de quelques substances pharmacologiques sur le ganglion stellaire des céphalopodes. *Arch. néerl. Physiol.,* 18:1-14.

CERF, J., GRUNDFEST, H., HOYLE, G., and MCCANN, F. V. 1957a. The nature of electrical responses of doubly-innervated insect muscle fibers. *Biol. Bull., Woods Hole,* 113:337-338.

CERF, J., GRUNDFEST, H., HOYLE, G., and MCCANN, F. V. 1957b. Neuromuscular transmission in the grasshopper *Romalea microptera. Biol. Bull., Woods Hole,* 113:338.

CHAUCHARD, A. B. and CHAUCHARD, P. 1933. Influence du chloroforme sur l'excitabilité de l'appareil neuromotor chez les crustacés. *C. R. Soc. Biol., Paris,* 113:136-138.

CHAUCHARD, B. and CHAUCHARD, P. 1950. Action du curare sur l'excitabilité neuromusculaire des crustacés. *Bull. Lab. Marit. Dinard,* 33:29-30.

CHAUCHARD, B. and CHAUCHARD, P. 1952. Action de l'histamine sur le système nerveux des pagures. *Bull. Lab. marit. Dinard,* 36:1-2.

CHAUCHARD, P. and CHAUCHARD, J. 1942. Action des pyréthrines sur l'excitabilité nerveuse chez les crustacés. *Bull. Lab. marit. Dinard,* 24:72-76.

COLHOUN, E. H. 1958. Acetylcholine in *Periplaneta americana* L. II. Acetylcholine and nervous activity. *J. Insect Physiol.,* 2:117-127.

COLHOUN, E. H. 1959. Acetylcholine in *Periplaneta americana* L. III. Acetylcholine in roaches treated with tetraethyl phosphate and 2,2-bis (p-chlorophenyl)-1,1,1-trichloroethane. *Canad. J. Biochem. Physiol.,* 37:259-272.

COOMBS, J. S., ECCLES, J. C., and FATT, P. 1953. The action of the inhibitory synaptic transmitter. *Aust. J. Sci.,* 16:1-5.

COOMBS, J. S., ECCLES, J. C., and FATT, P. 1955a. The specific ionic conductances and the ionic movements across the motoneural membrane that produce the inhibitory post-synaptic potential. *J. Physiol.,* 130:326-373.

COOMBS, J. S., ECCLES, J. C., and FATT, P. 1955b. Excitatory synaptic action in motoneurones. *J. Physiol.,* 130:374-395.

CURTIS, D. R. and ECCLES, J. C. 1960. Synaptic action during and after repetitive stimulation. *J. Physiol.,* 150:374-398.

DALE, H. H. 1937. Transmission of nervous effects by acetylcholine. *Harvey Lect.,* 32:229-244.

DALE, H. H., FELDBERG, W., and VOGT, M. 1936. Release of acetylcholine at voluntary motor nerve endings. *J. Physiol.,* 86:353-380.

DAVENPORT, D. 1941. The effects of acetylcholine, atropine and nicotine on the isolated heart of the commercial crab, *Cancer magister* Dana. *Physiol. Zoöl.,* 14:178-185.

DAVENPORT, D. 1942. Further studies in the pharmacology of the heart of *Cancer magister* Dana. *Biol. Bull., Woods Hole,* 82:255-260.

Bibliography

DEFRETIN, R. and RIFF, T. 1948. La glycogène du tissu nerveux chez *Anodonta cygnea* L. *C. R. Soc. Biol., Paris*, 142:1108-1110.

DEL CASTILLO, J. and KATZ, B. 1953. Statistical nature of "facilitation" at a single nerve-muscle junction. *Nature, Lond.*, 171:1016-1017.

DEL CASTILLO, J. and KATZ, B. 1954a. Quantal components of the end-plate potential. *J. Physiol.*, 124:560-573.

DEL CASTILLO, J. and KATZ, B. 1954b. Statistical factors involved in neuromuscular facilitation and depression. *J. Physiol.*, 124:574-585.

DEL CASTILLO, J. and KATZ, B. 1955a. Electrophoretic application of acetylcholine to the two sides of the end-plate membrane. *J. Physiol.*, 125:16-17P.

DEL CASTILLO, J. and KATZ, B. 1955b. The membrane change produced by the neuromuscular transmitter. *J. Physiol.*, 125:546-565.

DEL CASTILLO, J. and KATZ, B. 1956. Biophysical aspects of neuro-muscular transmission. *Progr. Biophys.*, 6:121-170.

DEL CASTILLO, J., HOYLE, G., and MACHNE, X. 1953. Neuromuscular transmission in a locust. *J. Physiol.*, 121:539-547.

DESMEDT, J. E. and MONACO, P. 1961. Mode of action of the efferent olivocochlear bundle on the inner ear. *Nature, Lond.*, 192:1263-1265.

DRESDEN, D. 1949. *Physiological Investigation into the Action of DDT.* (Thesis, University of Utrecht) Van der Wiel, Arnhem.

DUDEL, J. and KUFFLER, S. W. 1961a. The quantal nature of transmission and spontaneous miniature potentials at the crayfish neuromuscular junction. *J. Physiol.*, 155:514-529.

DUDEL, J. and KUFFLER, S. W. 1961b. Mechanism of facilitation at the crayfish neuromuscular junction. *J. Physiol.*, 155:530-542.

DUDEL, J. and KUFFLER, S. W. 1961c. Presynaptic inhibition at the crayfish neuromuscular junction. *J. Physiol.*, 155:543-562.

DUDEL, J. and KUFFLER, S. W. 1961d. Presynaptic inhibition at the neuromuscular junction in crayfish. In: *Nervous Inhibition.* E. Florey (ed.). Pergamon Press, Oxford.

DUN, F. T. 1951. The terminal arborization of nerve fibres as an important factor in synaptic and neuromuscular transmission. *J. cell. comp. Physiol.*, 38:133-135.

DZHAVRISHVILI, T. D. 1959. Oscillographic analysis of the interaction between nerve fibres. *Sechenov. J. Physiol.*, 45(2):169-175.

EASTON, D. M. 1950. Synthesis of acetylcholine in crustacean nerve and nerve extract. *J. biol. Chem.*, 185:813-816.

EASTON, D. M. 1955. Intracellular action potentials modified at muscle end-plate by adjacent fiber activity. *J. Neurophysiol.* 18:375-387.

EASTON, D. M. 1956. Dependence of intracellular end-plate action potential on adjacent fiber activity. *Amer. J. Physiol.* 187:199-202.

EASTON, D. M. 1957. Facilitation and inhibition in crustacean neuromuscular system. *Physiol. comp.*, 14(4):415-428.

ECCLES, J. C. 1935. The action potential of the superior cervical ganglion. *J. Physiol.*, 85:179-206.

ECCLES, J. C. 1943. Synaptic potentials and transmission in sympathetic ganglion. *J. Physiol.*, 101:465-483.

ECCLES, J. C. 1950. A review and restatement of the electrical hypotheses of synaptic excitatory and inhibitory action. In: *Électrophysiologie, Colloq. int. Cent. nat. Rech. sci.*, 22:441-458.

ECCLES, J. C. 1952. The electrophysiological properties of the motoneurone. *Cold Spr. Harb. Symp. quant. Biol.*, 17:175-183.

ECCLES, J. C. 1957. *The Physiology of Nerve Cells.* Johns Hopkins, Baltimore.

ECCLES, J. C. 1959. Neuron physiology —introduction. In: *Handbook of Physiology*, Sect. 1. *Neurophysiology*, 1:59-74.

ECCLES, J. C. 1961a. The synaptic mechanism for postsynaptic inhibition. In: *Nervous Inhibition.* E. Florey (ed.). Pergamon Press, Oxford.

ECCLES, J. C. 1961b. The mechanism of synaptic transmission. *Ergebn. Physiol.*, 51:299-430.

ECCLES, J. C. 1961c. The nature of central inhibition. *Proc. roy. Soc.* (B), 153:445-476.

ECCLES, J. C. and KRNJEVIĆ, K. 1959. Presynaptic changes associated with post-tetanic potentiation in the spinal cord. *J. Physiol.*, 149:274-287.

ECCLES, J. C., ECCLES, R. M., and MAGNI, F. 1960. Presynaptic inhibition in the spinal cord. *J. Physiol.*, 154:28P.

ECCLES, J. C., FATT, P., and KOKETSU, K. 1953. Cholinergic and inhibitory synapses in a central nervous pathway. *Aust. J. Sci.*, 16:50-54.

ECCLES, J. C., FATT, P., and KOKETSU, K. 1954. Cholinergic and inhibitory synapses in a pathway from motor-axon collaterals to motoneurones. *J. Physiol.*, 126:524-562.

ECCLES, R. M. 1955. Intracellular potentials recorded from a mammalian sympathetic ganglion. *J. Physiol.*, 130:572-584.

EDWARDS, C. and HAGIWARA, S. 1959. Potassium ions and the inhibitory process in the crayfish stretch receptor. *J. gen. Physiol.*, 43:315-321.

EDWARDS, C. and KUFFLER, S. W. 1959. The blocking effect of γ-aminobutyric acid (GABA) and the action of related compounds on single nerve cells. *J. Neurochem.*, 4:19-30.

ELLIOTT, K. A. C. and FLOREY, E. 1956. Factor I—inhibitory factor from brain. Assay. Conditions in brain. Stimulating and antagonizing substances. *J. Neurochem.*, 1:181-192.

ELLIOTT, K. A. C., PAGE, I. H., and QUASTEL, J. H. 1962. *Neurochemistry: the Chemical Dynamics of Brain and Nerve.* 2nd ed. Thomas, Springfield, Ill.

ELLIS, C. H., THIENES, C. H., and WIERSMA, C. A. G. 1942. The influence of certain drugs on the crustacean nerve-muscle system. *Biol. Bull., Woods Hole*, 83:334-352.

ELLIS, P. E. and HOYLE, G. 1954. A physiological interpretation of the marching of hoppers of the African migratory locust (*Locusta migratoria migratorioides* R. & F.). *J. exp. Biol.*, 31:271-279.

EWER, D. W. and BERG, R. VAN DEN. 1954. A note on the pharmacology of the dorsal musculature of *Peripatopsis. J. exp. Biol.*, 31:497-500.

EYZAGUIRRE, C. 1961. Excitatory and inhibitory processes in crustacean sensory nerve cells. In: *Nervous Inhibition.* E. Florey (ed.). Pergamon Press, Oxford.

EYZAGUIRRE, C. and KUFFLER, S. W. 1954b. Inhibitory activity in single cell synapses. *Biol. Bull., Woods Hole*, 107:310-311.

FATT, P. 1954. Biophysics of junctional transmission. *Physiol. Rev.*, 34:674-710.

FATT, P. 1959. Skeletal neuromuscular transmission. In: *Handbook of Physiology.* Sect. 1. *Neurophysiology*, 1:199-213.

FATT, P. 1961. The change in membrane permeability during the inhibitory process. In: *Nervous Inhibition.* E. Florey (ed.). Pergamon Press, Oxford.

FATT, P. and KATZ, B. 1951. An analysis of the end-plate potential recorded with an intra-cellular electrode. *J. Physiol.*, 115:320-370.

FATT, P. and KATZ, B. 1952a. Electric responses of single crustacean muscle fibres. *J. Physiol.*, 117:158-168.

FATT, P. and KATZ, B. 1952b. Spontaneous subthreshold activity at motor nerve endings. *J. Physiol.*, 117:109-128.

FATT, P. and KATZ, B. 1952c. Some problems of neuro-muscular transmission. *Cold Spr. Harb. Symp. quant. Biol.*, 17:275-280.

FATT, P. and KATZ, B. 1952d. The effect of sodium ions on neuromuscular transmission. *J. Physiol.*, 118:73-87.

FATT, P. and KATZ, B. 1953a. Distributed "end-plate potentials" of crustacean muscle fibres. *J. exp. Biol.*, 30:433-439.

FATT, P. and KATZ, B. 1953b. The effect of inhibitory nerve impulses on a crustacean muscle fibre. *J. Physiol.*, 121:374-389.

FATT, P. and KATZ, B. 1953c. Chemoreceptor activity at the motor endplate. *Acta physiol. scand.*, 29:117-125.

FELDBERG, W. 1945. Present views on the mode of action of acetylcholine in the central nervous system. *Physiol. Rev.*, 25:596-642.

FELDBERG, W. 1951. Some aspects in pharmacology of central synaptic transmission. *Arch. int. Physiol.*, 59:544-560.

FESSARD, A. and POSTERNAK, J. 1950. Les mécanismes élémentaires de la transmission synaptique. *J. Physiol., Paris*, 42:319-445.

FESSARD, A. and TAUC, L. 1957. Comparaison entre la dissipation des potentiels postsynaptiques et électrotoniques dans le soma neuronique de l'aplysia. *J. Physiol., Paris*, 49:162-164.

FESSARD, A. and TAUC, L. 1958. Effets de répétition sur l'amplitude des potentiels postsynaptiques d'un soma neuronique. *J. Physiol., Paris*, 50:277-281.

FESSARD, A. and TAUC, L. 1960. Variations prolongées du rythme de neurones autoactifs provoquées par la stimulation synaptique. *J. Physiol., Paris*, 52:101.

FIELDEN, A. 1960. Patterns of conduction in the caudal ganglion of the crayfish. *Physiol. Zoöl.*, 33:161-169.

FLOREY, E. 1951a. Reizphysiologische Untersuchungen an der Ascidie *Ciona intestinalis* L. *Biol. Zbl.*, 70:523-530.

FLOREY, E. 1951b. Vorkommen und Funktion sensibler Erregungssubstanzen und sie abbauender Fermente im Tierreich. *Z. vergl. Physiol.*, 33:327-377.

FLOREY, E. 1951c. Neurohormone und ihre Funktion bei Arthropoden. *Verh. deutsch. zool. Ges.*, 1951:199-206.

FLOREY, E. 1951d. Neurohormone und Pharmakologie der Arthropoden. *PflSchBer.*, 7:81-141.

FLOREY, E. 1953. Über einen nervösen Hemmfactor in Gehirn und Rückenmark. *Naturwissenschaften*, 40:295-296.

FLOREY, E. 1954a. Über die Wirkung von Acetylcholin, Adrenalin, Nor-Adrenalin, Faktor I und anderen Substanzen auf den isolierten Enddarm des Flusskrebses, *Cambarus clarkii* Girard. *Z. vergl. Physiol.*, 36:1-8.

FLOREY, E. 1954b. Experimentelle Erzeugung einer "Neurose" bei der Honigbiene. *Naturwissenschaften*, 41:171.

FLOREY, E. 1954c. An inhibitory and an excitatory factor of mammalian central nervous system, and their action on a single sensory neuron. *Arch. int. Physiol.* 62:33-53.

FLOREY, E. 1956a. The action of Factor I on certain invertebrate organs. *Canad. J. Biochem. Physiol.*, 334:669-681.

FLOREY, E. 1956b. Adaptationserscheinungen in den sensiblen Neuronen der Streckreceptoren des Flusskrebses. *Z. Naturf.*, 118:504-513.

FLOREY, E. 1957a. Chemical transmission and adaptation. *J. gen. Physiol.*, 40:533-545.

FLOREY, E. 1957b. Further evidence for the transmitterfunction of Factor I. *Naturwissenschaften*, 44:424-425.

FLOREY, E. 1960. Studies on the nervous regulation of the heart beat in decapod Crustacea. *J. gen. Physiol.*, 43:1061-1081.

FLOREY, E. 1961a. A new test preparation for bio-assay of Factor I and gamma-aminobutyric acid. *J. Physiol.*, 156:1-7.

FLOREY, E. 1961b. Comparative physiology: transmitter substances. *Ann. Rev. Physiol.*, 23:501-528.

FLOREY, E. 1961c. Excitation, inhibition and the concept of the stimulus. In: *Nervous Inhibition*. E. Florey (ed.). Pergamon Press, Oxford.

FLOREY, E. and BIEDERMAN, M. A. 1960. Studies on the distribution of Factor I and acetylcholine in crustacean peripheral nerve. *J. gen. Physiol.*, 43:509-522.

FLOREY, E. and CHAPMAN, D. D. 1961. The non-identity of the transmitter substance of crustacean inhibitory neurons and gamma-aminobutyric acid. *Comp. Biochem. Physiol.*, 3:92-98.

FLOREY, E. and FLOREY, E. 1953. Über die Bedeutung von 5-Hydroxytryptamin als nervöser Aktionssubstanz bei Cephalopoden und dekapoden Crustaceen. *Naturwissenschaften*, 40:413-414.

FLOREY, E. and FLOREY, E. 1954a. Über die mögliche Bedeutung von Enteramin (5-Oxy-Tryptamin) als nervöser Aktionssubstanz bei Cephalopoden und dekapoden Crustaceen. *Z. Naturf.*, 9B:58-68.

FLOREY, E. and FLOREY, E. 1954b. Über die Wirkung von 5-Oxytryptamin (Enteramin) in der Krebsschere. *Z. Naturf.*, 9B:540-547.

FLOREY, E. and FLOREY, E. 1955. Microanatomy of the abdominal stretchreceptor of the crayfish *Astacus fluviatilis* L. *J. gen. Physiol.*, 39:69-85.

FLOREY, E. and FLOREY, E. 1958. Studies on the distribution of Factor I in mammalian brain. *J. Physiol.*, 144:220-228.

FLOREY, E. and HOYLE, G. 1961. Neuromuscular synaptic activity in the crab (*Cancer magister*). In: *Nervous Inhibition*. E. Florey (ed.). Pergamon Press, Oxford.

FLOREY, E. and MCLENNAN, H. 1955a. Die Wirkung des Hemmungsfaktors aus Gehirn und Rückenmark auf periphere und zentrale synaptische Übertragung. *Naturwissenschaften*, 42:51-52.

FLOREY, E. and MCLENNAN, H. 1955b. Is ATP the sensory transmitter substance? *Naturwissenschaften*, 42:561.

FLOREY, E. and MCLENNAN, H. 1955c. The release of an inhibitory substance from mammalian brain and its action on peripheral synaptic transmission. *J. Physiol.*, 129:384-392.

FLOREY, E. and MCLENNAN, H. 1955d. Effects of an inhibitory factor (Factor I) of brain on central synaptic transmission. *J. Physiol.*, 130:446-455.

FLOREY, E. and MCLENNAN, H. 1959. The effects of Factor I and of gamma-aminobutyric acid on smooth muscle preparations. *J. Physiol.*, 145:66-76.

FLOREY, E. and MERWIN, H. J. 1961. Inhibition in molluscan hearts and the role of acetylcholine. In: *Nervous Inhibition*. E. Florey (ed.). Pergamon Press, Oxford.

FORBES, A. 1922. Interpretation of spinal reflexes in terms of present knowledge of nerve conduction. *Physiol. Rev.*, 2:361-414.

FRANK, K. and FUORTES, M. G. F. 1955. Potentials recorded from the spinal cord with microelectrodes. *J. Physiol.*, 130:625-654.

FRANK, K. and FUORTES, M. G. F. 1956. Unitary activity of spinal interneurones of cats. *J. Physiol.*, 131:424-435.

FRANK, K. and FUORTES, M. G. F. 1957. Presynaptic and post-synaptic inhibition of monosynaptic reflexes. *Fed. Proc.*, 16:39-40.

FRÖHLICH, A. and LOEWI, O. 1907. Scheinbare Speisung der Nervenfaser mit mechanischer Erregbarkeit seitens ihrer Nervenzelle. (Nach Versuchen an *Eledone moschata*.). *Zbl. Physiol.*, 21:273-276.

FUORTES, M. G. F., FRANK, K., and BECKER, M. C. 1957. Steps in the production of motoneuron spikes. *J. gen. Physiol.*, 40:735-752.

FURSHPAN, E. J. 1959. Neuromuscular transmission in invertebrates. In: *Handbook of Physiology*. Sect. 1. *Neurophysiology*, 1:239-254.

FURSHPAN, E. J. and POTTER, D. D. 1957. Mechanism of nerve-impulse transmission at a crayfish synapse. *Nature, Lond.*, 180:342-343.

FURSHPAN, E. J. and POTTER, D. D. 1959a. Transmission at the giant motor synapses of the crayfish. *J. Physiol.*, 145:289-325.

FURSHPAN, E. J. and POTTER, D. D. 1959b. Slow post-synaptic potentials recorded from the giant motor fibre of the crayfish. *J. Physiol.*, 145:326-335.

FURSHPAN, E. J. and WIERSMA, C. A. G. 1954. Local and spike potentials of impaled crustacean muscle fibers on

stimulation of single axons. *Fed. Proc.*, 13:51.
GALAMBOS, R. 1961. A glia-neural theory of brain function. *Proc. nat. Acad. Sci., Wash.*, 47:129-136.
GARREY, W. E. 1942. An analysis of the action of acetylcholine on the cardiac ganglion of *Limulus polyphemus*. *Amer. J. Physiol.*, 136:182-193.
GERARD, R. W. 1941a. The interaction of neurones. *Ohio J. Sci.*, 41:160-172.
GERARD, R. W. 1941b. Intercellular electric fields and brain function. *Schweiz. Med. Wochenschr.*, 12:555-559.
GERARD, R. W. and LIBET, B. 1939. On the unison of neurone beats. In: *Livro de Homenagem.* A. M. Ozorio de Almeida (ed.). Rio de Janeiro.
GERARD, R. W. and LIBET, B. 1940. The control of normal and "convulsive" brain potentials. *Amer. J. Psychiat.*, 96:1125-1151.
GEREBTZOFF, M. A. 1956. Conditions d'existence des céphalopodes et localisation de l'acétylcholinestérase au niveau de leurs fibres nerveuses. *C. R. Soc. Biol., Paris*, 150:1815-1817.
GERSCHENFELD, H. and TAUC, L. 1691. Pharmacological specificities of neurones in an elementary central nervous system. *Nature, Lond.*, 189:924-925.
GRANIT, R. and SKOGLUND, C. R. 1945a. Facilitation, inhibition and depression at the "artificial synapse" formed by the cut end of a mammalian nerve. *J. Physiol.*, 103:435-448.
GRANIT, R. and SKOGLUND, C. R. 1945b. The effect of temperature on the artificial synapse formed by the cut end of the mammalian nerve. *J. Neurophysiol.*, 8:211-217.
GRIFFITH, A. B. 1892. Sur les tissus nerveux de quelques invertébrés. *C. R. Acad. Sci., Paris*, 115:562-563.
GRUNDFEST, H. 1957a. Electrical inexcitability of synapses and some consequences in the central nervous system. *Physiol. Rev.*, 37:337-361.
GRUNDFEST, H. 1957b. Excitation triggers in post-junctional cells. In: *Physiological Triggers.* T. H. Bullock (ed.). Amer. Physiol. Soc., Washington.
GRUNDFEST, H. 1957c. The mechanisms of discharge of the electric organs in relation to general and comparative electrophysiology. *Progr. Biophys.*, 7:1-85.
GRUNDFEST, H. 1958a. Electrophysiology and pharmacology of dendrites. *Electroenceph. clin. Neurophysiol.*, 10 (Suppl.):22-41.
GRUNDFEST, H. 1958b. An electrophysiological basis for neuropharmacology. *Fed. Proc.*, 17:1006-1018.
GRUNDFEST, H. 1959. General physiology and pharmacology of synapses and some implications for the mammalian central nervous system. *J. nerv. ment. Dis.*, 128:473-496.

GRUNDFEST, H. and BENNETT, M. V. L. 1961. Studies on the morphology and electrophysiology of electric organs. I. Electrophysiology of marine electric fishes. In: *Bioelectrogenesis.* C. Chagas and A. Paes de Carvalho (eds.). Elsevier, Amsterdam.
GRUNDFEST, H. and REUBEN, J. P. 1961. Neuromuscular synaptic activity in lobster. In: *Nervous Inhibition.* E. Florey (ed.). Pergamon Press, Oxford.
GRUNDFEST, H., REUBEN, J. P., and RICKLES, W. H., Jr. 1959. The electrophysiology and pharmacology of lobster neuromuscular synapses. *J. gen. Physiol.*, 42:1301-1323.
HAGIWARA, S. 1953. Neuro-muscular transmission in insects. *Jap. J. Physiol.*, 3:284-296.
HAGIWARA, S. 1958. Synaptic potential in the motor giant axon of the crayfish. *J. gen. Physiol.*, 41:1119-1128.
HAGIWARA, S. and BULLOCK, T. H. 1957. Intracellular potentials in pacemaker and integrative neurons of the lobster cardiac ganglion. *J. cell. comp. Physiol.*, 50:25-47.
HAGIWARA, S. and KUSANO, K. 1961. Synaptic inhibition in giant nerve cell of *Onchidium verruculatum*. *J. Neurophysiol.*, 24:167-175.
HAGIWARA, S. and MORITA, H. 1962. Electrotonic transmission between two nerve cells in leech ganglion. *J. Neurophysiol.*, 25:721-731.
HAGIWARA, S. and TASAKI, I. 1958. A study on the mechanism of impulse transmission across the giant synapse of the squid. *J. Physiol.*, 143:114-137.
HAGIWARA, S. and WATANABE, A. 1954. Action potential of insect muscle examined with intracellular electrode. *Jap. J. Physiol.*, 4:65-78.
HAGIWARA, S. and WATANABE, A. 1956. Discharge in motoneurons of cicada. *J. cell. comp. Physiol.*, 47:415-428.
HAGIWARA, S., KUSANO, K., and SAITO, S. 1960. Membrane changes in crayfish stretch receptor neuron during synaptic inhibition and under action of gamma-aminobutyric acid. *J. Neurophysiol.*, 23:505-515.
HAGIWARA, S., WATANABE, A., and SAITO, N. 1959. Potential changes in syncytial neurons of a lobster cardiac ganglion. *J. Neurophysiol.*, 22:554-572.
HAMA, K. 1959. Some observations on the fine structure of the giant nerve fibers of the earthworm, *Eisenia foetida*. *J. biophys. biochem. Cytol.*, 6:61-66.
HEBB, C. O. 1957. Biochemical evidence for the neural function of acetylcholine. *Physiol. Rev.*, 37:196-220.
HILD, W., CHANG, J. J., and TASAKI, I. 1958. Electrical responses of astrocytic glia from the mammalian central nervous system cultivated in vitro. *Experientia*, 14:220-226.
HORRIDGE, G. A. 1955. The nerves and muscles of medusae. IV. Inhibition in *Aequorea forskalea*. *J. exp. Biol.*, 32:642-648.
HORRIDGE, G. A. 1956a. The responses of *Heteroxenia* (Alcyonaria) to stimulation and to some inorganic ions. *J. exp. Biol.*, 33:604-614.
HORRIDGE, G. A. 1956b. The nervous system of the ephyra larva of *Aurellia aurita*. *Quart. J. micr. Sci.*, 97:59-74.
HOYLE, G. 1954. Changes in the blood potassium concentration of the African migratory locust (*Locusta migratoria migratorioides* R. & F.) during food deprivation, and the effect on neuromuscular activity. *J. exp. Biol.*, 31:260-270.
HOYLE, G. 1955a. The effects of some common cations on neuromuscular transmission in insects. *J. Physiol.*, 127:90-103.
HOYLE, G. 1955b. Neuromuscular mechanisms of a locust skeletal muscle. *Proc. roy. Soc. (B)*, 143:343-367.
HOYLE, G. 1957a. *Comparative Physiology of the Nervous Control of Muscular Contraction.* Cambridge University Press, Cambridge.
HOYLE, G. 1957b. Nervous control of insect muscles. In: *Recent Advances in Invertebrate Physiology.* B. T. Scheer (ed.). Univ. of Oregon Publ., Eugene.
HOYLE, G. 1958a. Two inhibitory fibres forming synapses with a single cell. *Nature, Lond.*, 181:1134.
HOYLE, G. 1958b. Studies on neuromuscular transmission in *Limulus*. *Biol. Bull., Lond.*, 115:209-218.
HOYLE, G. 1959. The meaning and significance of neuromuscular facilitation. *Int. Congr. Zool., Proc. XV*, 1958:459-461.
HOYLE, G. and WIERSMA, C. A. G. 1958a. Excitation at neuromuscular junctions in Crustacea. *J. Physiol.*, 143:403-425.
HOYLE, G. and WIERSMA, C. A. G. 1958b. Inhibition at neuromuscular junctions in Crustacea. *J. Physiol.*, 143:426-440.
HOYLE, G. and WIERSMA, C. A. G. 1958c. Coupling of membrane potential to contraction in crustacean muscles. *J. Physiol.*, 143:441-453.
HUBBARD, J. I. 1959. Post-activation changes at the mammalian neuromuscular junction. *Nature, Lond.*, 184:1945-1947.
HUGHES, G. M. and KERKUT, G. A. 1956. Electrical activity in a slug ganglion in relation to the concentration of Locke solution. *J. exp. Biol.*, 33:282-294.
HUNT, C. C. and KUNO, M. 1959. Properties of spinal interneurones. *J. Physiol.*, 147:346-363.
HUTTER, O. F. and LOEWENSTEIN, W. R. 1955. Nature of neuromuscular facilitation by sympathetic stimulation in the frog. *J. Physiol.*, 130:559-571.

ISHIKO, N. and LOEWENSTEIN, W. R. 1959. Spontaneous fluctuations in generator potential in a receptor membrane. *Nature, Lond.*, 183:1724-1726.

ITO, M. 1959. An analysis of potentials recorded intracellularly from the spinal ganglion cell. *Jap. J. Physiol.*, 9:20-32.

ITO, M. and SAIGA, M. 1959. The mode of impulse conduction through the spinal ganglion. *Jap. J. Physiol.*, 9:33-42.

JASPER, H. W. and MONNIER, A. M. 1938. Transmission of excitation between excised non-myelinated nerves. An artificial synapse. *J. cell. comp. Physiol.*, 11:259-277.

JULLIEN, A. and VINCENT, D. 1938a. Sur l'action de l'acétylcholine sur le coeur des mollusques. L'antagonisme curare-acétylcholine. *C. R. Acad. Sci., Paris*, 206:209-211.

JULLIEN, A. and VINCENT, D. 1938b. Les esters de choline dans quelques organes des mollusques. *C. R. Acad. Sci., Paris*, 206:1145-1147.

KAO, C. Y. 1960. Postsynaptic electrogenesis in septate giant axons. II. Comparison of medial and lateral giant axons of crayfish. *J. Neurophysiol.*, 23:618-635.

KAO, C. Y. and GRUNDFEST, H. 1957. Postsynaptic electrogenesis in septate giant axons. I. Earthworm median giant axon. *J. Neurophysiol.*, 20:553-573.

KATZ, B. 1936. Neuro-muscular transmission in crabs. *J. Physiol.*, 87:199-221.

KATZ, B. 1949. Neuro-muscular transmission in invertebrates. *Biol. Rev.*, 24:1-20.

KATZ, B. 1959. Mechanisms of synaptic transmission. In: *Biophysical Science—A Study Program*. J. L. Oncley (ed.). Wiley, New York.

KATZ, B. and KUFFLER, S. W. 1946. Excitation of the nerve-muscle system in Crustacea. *Proc. roy. Soc. (B)*, 133:374-389.

KATZ, B. and MILEDI, R. 1962a. An "antidromic reflex" in the frog's spinal cord, and its abolition by curare. *J. Physiol.*, 162:42P (demonstr.).

KATZ, B. and MILEDI, R. 1962b. The nature of spontaneous synaptic potentials in motoneurones of the frog. *J. Physiol.*, 162:51P-52P.

KATZ, B. and SCHMITT, O. H. 1940. Electrical interaction between two adjacent nerve fibres. *J. Physiol.*, 97:471-488.

KATZ, B. and SCHMITT, O. H. 1942. A note on interaction between nerve fibres. *J. Physiol.*, 100:369-371.

KENNEDY, D. and PRESTON, J. B. 1960. Activity patterns of interneurons in the caudal ganglion of the crayfish. *J. gen. Physiol.*, 43:655-670.

KERKUT, G. A. and TAYLOR, B., Jr. 1956a. The sensitivity of the pedal ganglion of the slug to osmotic pressure changes. *J. exp. Biol.*, 33:493-501.

KERKUT, G. A. and TAYLOR, B., Jr. 1956b. Effect of temperature on the spontaneous activity from the isolated ganglia of the slug, cockroach and crayfish. *Nature, Lond.*, 178:426.

KERKUT, G. A. and TAYLOR, B., Jr. 1958. The effect of temperature changes on the activity of poikilotherms. *Behaviour*, 13:259-279.

KEYNES, R. D., BENNETT, M. V. L., and GRUNDFEST, H. 1961. Studies on the morphology and electrophysiology of electric organs. II. Electrophysiology of the electric organ of *Malapterurus electricus*. In: *Bioelectrogenesis*. C. Chagas and A. Paes de Carvalho (eds.). Elsevier, Amsterdam.

KNOWLTON, F. P. 1942. The action of certain drugs on crustacean muscle. *J. Pharm. & exp. Ther.*, 75:154-160.

KOELLE, G. B. 1959. Neurohumoral agents as a mechanism of nervous integration. In: *Evolution of Nervous Control*, Amer. Assoc. Adv. Sci., Washington, D.C.

KOLMODIN, G. M. 1957. Integrative processes in single spinal interneurones with proprioceptive connections. *Acta physiol. scand.*, 40 (Suppl. 139):1-89.

KOPPANYI, T. 1948. Acetylcholine as a pharmacological agent. *Johns Hopk. Hosp. Bull.*, 83:532-561.

KOREY, S. R. and NURNBERGER, J. I. 1956. *Progress in Neurobiology: I Neurochemistry*. Hoeber-Harper, New York.

KOSHTOYANTS, KH. S. 1936a. On methods of action of acetylcholine, tested by the new biological indicator and on cholinestarases of invertebrates. *Bull. Ehk.*, 2:37-40. (In Russian.)

KOSHTOYANTS, KH. S. 1936b. On adrenaline-like substances in the organs of invertebrates. (Ganglia of mollusks.) *Bull. Ehk.*, 2:41-43. (In Russian.)

KOSHTOYANTS, KH. S. 1957. Peculiarities of nervous regulation and action of mediators in mollusks. *Izv. Akad. Nauk Armyan. S. S. R., Biol.*, 10(7):13-16. (In Russian.)

KRNJEVIĆ, K. and MILEDI, R. 1959. Presynaptic failure of neuromuscular propagation in rats. *J. Physiol.*, 149:1-2.

KRUTA, V. 1935. Sur l'action de l'acétylcholine et de l'atropine sur le coeur de *Sepia officinalis*. *C. R. Soc. Biol., Paris*, 119:608-610.

KRUTA, V. 1936. Action de quelques alcaloïdes sur les nerfs cardiaques chez les céphalopodes. *C. R. Soc. Biol. Paris*, 122:585-586.

KUFFLER, S. W. 1942a. Electric potential changes at an isolated nerve-muscle junction. *J. Neurophysiol.*, 5:18-26.

KUFFLER, S. W. 1942b. Responses during refractory period at myoneural junction in isolated nerve-muscle fibre preparation. *J. Neurophysiol.*, 5:199-209.

KUFFLER, S. W. 1949. Transmitter mechanism at the nerve-muscle junction. *Arch. Sci. physiol.*, 3:585-601.

KUFFLER, S. W. 1952. Transmission processes at nerve-muscle junctions. In: *Modern Trends in Physiology and Biochemistry*. E. S. G. Barron (ed.). Academic Press, New York.

KUFFLER, S. W. 1953. The two skeletal nerve-muscle systems in frog. *Arch. exp. Path. Pharmak.*, 220:116-135.

KUFFLER, S. W. 1958. Synaptic inhibitory mechanisms. Properties of dendrites and problems of excitation in isolated sensory nerve cells. *Exp. Cell Res. (Suppl.)* 5:493-519.

KUFFLER, S. W. 1960. Excitation and inhibition in single nerve cells. *Harv. Lect.*, 54:176-218.

KUFFLER, S. W. and EDWARDS, C. 1958. Mechanism of gamma aminobutyric acid (GABA) action and its relation to synaptic inhibition. *J. Neurophysiol.*, 21:589-610.

KUFFLER, S. W. and EYZAGUIRRE, C. 1955. Synaptic inhibition in an isolated nerve cell. *J. gen. Physiol.*, 39:155-184.

KUFFLER, S. W. and KATZ, B. 1946. Inhibition at the nerve muscle junction in Crustacea. *J. Neurophysiol.*, 9:337-346.

KUFFLER, S. W. and VAUGHAN WILLIAMS, E. M. 1953. Small-nerve junctional potentials. *J. Physiol.*, 121:289-317.

KUPERMAN, A. S., WERNER, G., and GILL, E. W. 1959. Presynaptic activity in drug-induced neuromuscular facilitation. *Biol. Bull., Woods Hole*, 117:390-391.

KUSANO, K. and HAGIWARA, S. 1961. On the integrative synaptic potentials of *Onchidium* nerve cell. *Jap. J. Physiol.*, 11:96-101.

LAPORTE, Y. and LORENTE DE NÓ, R. 1950a. Properties of sympathetic B ganglion cells. *J. cell. comp. Physiol.*, 35 (Suppl. 2):41-60.

LAPORTE, Y. and LORENTE DE NÓ, R. 1950b. Potential changes evoked in a curarized sympathetic ganglion by presynaptic volleys of impulses. *J. cell. comp. Physiol.*, 35 (Suppl. 2):61-106.

LAPORTE, Y. and LORENTE DE NÓ, R. 1950c. Dual mechanism of synaptic transmission through a sympathetic ganglion. *J. cell. comp. Physiol.*, 35 (Suppl. 2):107-153.

LARRABEE, M. and BRONK, D. W. 1947. Prolonged facilitation of synaptic excitation in sympathetic ganglia. *J. Neurophysiol.*, 10:139-154.

LEWIS, S. E. and SMALLMAN, B. N. 1956. The estimation of acetyl-

choline in insects. *J. Physiol.*, 134:241-256.
LIBET, B. and GERARD, R. W. 1939. Control of the potential rhythm of the isolated frog brain. *J. Neurophysiol.*, 2:153-169.
LIBET, B. and GERARD, R. W. 1941. Steady potential fields and neurone activity. *J. Neurophysiol.*, 4:438-455.
LILEY, A. W. 1956. The quantal components of the mammalian end-plate potential. *J. Physiol.*, 133:571-587.
LLOYD, D. P. C. 1944. Functional organisation of the spinal cord. *Physiol. Rev.*, 24:1-17.
LLOYD, D. P. C. 1946. Facilitation and inhibition of spinal motorneurons. *J. Neurophysiol.*, 9:421-438.
LLOYD, D. P. C. 1949. Post-tetanic potentiation of response in monosynaptic reflex pathways of the spinal cord. *J. gen. Physiol.*, 33:147-170.
LLOYD, D. P. C. 1951a. Electrical signs of impulse conduction in spinal motoneurons. *J. gen. Physiol.*, 35:255-288.
LLOYD, D. P. C. 1951b. After-currents, after-potentials, excitability, and ventral root electrotonus in spinal motoneurons. *J. gen. Physiol.*, 35:289-321.
LOEWENSTEIN, W. R. 1959a. The generation of electric activity in a nerve ending. *Ann. N. Y. Acad. Sci.*, 81:367-387.
LOEWENSTEIN, W. R. 1959b. Properties of a receptor membrane. Spatial summation of electric activity in a non-myelinated nerve ending. *Nature, Lond.*, 183:1724.
LOEWENSTEIN, W. R. and MOLINA, D. 1958. Cholinesterase in a receptor. *Science*, 128:1284.
LOEWI, O. 1921. Über humorale Übertragbarkeit der Herznervenwirkung. *Pflüg. Arch. ges. Physiol.*, 189:239.
LORENTE DE NÓ, R. 1938. Limits of variation of the synaptic delay of motoneurons. *J. Neurophysiol.*, 1:187-244.
LORENTE DE NÓ, R. 1946-1947. Correlation of nerve activity with polarization phenomena. *Harvey Lect.*, 42:43-105.
LORENTE DE NÓ, R. 1948. Quaternary ammonium ions and sodium ions in nerve physiology. *Johns Hopk. Hosp. Bull.*, 83:497-529.
LORENTE DE NÓ, R. and LAPORTE, Y. 1950. Refractoriness, facilitation and inhibition in a sympathetic ganglion. *J. cell. comp. Physiol.*, 35 (Suppl.2):155-192.
LOWENSTEIN, O. 1942. A method of physiological assay of pyrethrum extracts. *Nature, Lond.*, 150:760-762.
MACKIE, G. O. 1960. The structure of the nervous system in *Velella*. *Quart. J. micr. Sci.*, 101:119-131.
MARMONT, G. and WIERSMA, C. A. G. 1938. On the mechanism of inhibition and excitation of crayfish muscle. *J. Physiol.*, 93:173-193.
MARRAZZI, A. S. and LORENTE DE NÓ, R. 1944. Interaction of neighbouring fibres in myelinated nerve. *J. Neurophysiol.*, 7:83-101.
MAYNARD, D. M., Jr. 1953a. Inhibition in a simple ganglion. *Fed. Proc.*, 12:95.
MAYNARD, D. M., Jr. 1953b. Activity in a crustacean ganglion. I. Cardio-inhibition and acceleration in *Panulirus argus*. *Biol. Bull., Woods Hole*. 104:156-170.
MAYNARD, D. M., Jr. 1953c. Integration in the cardiac ganglion of *Homarus*. *Biol. Bull., Woods Hole*, 105:367.
MAYNARD, D. M., Jr. 1958. Correlations between heart rate and size in the lobster. *Anat. Rec.*, 132:475.
MAYNARD, D. M. and WELSH, J. H. 1959. Neurohormones of the pericardial organs of brachyuran Crustacea. *J. Physiol.*, 149:215-227.
MAYNARD, E. A. and MAYNARD, D. M. 1960a. Cholinesterase in the crustacean muscle receptor organ. *J. Histochem. Cytochem.*, 8:376-379.
MAYNARD, E. A. and MAYNARD, D. M. 1960b. Cholinesterases in the nervous system of the lobster, *Homarus americanus*. *Anat. Rec.*, 137:380.
MCCULLOCH, W. S. 1938. Irreversibility of conduction in the reflex arc. *Science*, 87:65-66.
MCLENNAN, H. 1961. Inhibitory transmitters—a review. In: *Nervous Inhibition*. E. Florey (ed.). Pergamon Press, Oxford.
MEANS, O. W., Jr. 1942. Cholinesterase activity of tissues of adult *Melanoplus differentialis* (Orthoptera, Acrididae). *J. cell. comp. Physiol.*, 20:319-324.
MICHAELIS, M., ARANYO, N. I., and GERARD, R. W. 1949. Inhibition of brain dehydrogenases by "anti-cholinesterases." *Amer. J. Physiol.*, 157:463-467.
MIKALONIS, S. J. and BROWN, R. H. 1941. Acetylcholine and cholinesterase in the insect central nervous system. *J. cell. comp. Physiol.*, 18:401-403.
MILBURN, N., WEIANT, E. A., and ROEDER, K. D. 1960. The release of efferent nerve activity in the roach, *Periplaneta americana*, by extracts of the corpus cardiacum. *Biol. Bull., Woods Hole*, 118:111-119.
MINZ, B. 1955. *The Role of Humoral Agents in Nervous Activity*. Thomas, Springfield, Ill.
MOORE, A. R. 1918. Reversal of reaction by means of strychnine in planarians and starfish. *J. gen. Physiol.*, 1:97-100.
NACHMANSOHN, D. 1937. Cholinesterase in the central nervous system. *Nature, Lond.*, 140:427.
NACHMANSOHN, D. 1948. The role of acetylcholine in conduction. *Johns Hopk. Hosp. Bull.*, 83:463-493
NACHMANSOHN, D. 1950. Studies on permeability in relation to nerve function. I. Axonal conduction and synaptic transmission. *Biochim. biophys. Acta*, 4:78-95.
NACHMANSOHN, D. 1959. *Chemical and Molecular Basis of Nerve Activity*. Academic Press, New York.
NAGAHAMA, H. 1950. Axon-axon transmission of nerve impulses, as tested by motor axons of the cheliped of the crayfish. *Annot. zool. jap.*, 24:29-37.
NAGAHAMA, H. 1954. Influence of inhibitory nerve impulse on the contraction of the leg muscle of the crayfish. *J. Fac. Sci. Tokyo Univ.*, (4) 7:15-30.
NARAHASHI, T. and YAMASAKI, T. 1960. Studies on the mechanism of action of insecticides. XVIII. Nervous and cholinesterase activities in the cockroach as affected by demeton and methyldemeton. *Jap. J. Appl. Ent. Zool.* 4:64-69.
NASTUK, W. M. 1953. The electrical activity of the muscle cell membrane at the neuromuscular junction. *J. cell. comp. Physiol.*, 42:249-272.
NISHI, S. and KOKETSU, K. 1960. Electrical properties and activities of single sympathetic neurons in frogs. *J. cell. comp. Physiol.*, 55:15-30.
OBRESHKOVE, V. 1941. The action of acetylcholine, atropine and physostigmine on the intestine of *Daphnia magna*. *Biol. Bull., Woods Hole*, 81:105-113.
OTANI, T. and BULLOCK, T. H. 1959. Effects of presetting the membrane potential of the soma of spontaneous and integrating ganglion cells. *Physiol. Zoöl.*, 32:104-114.
PANTIN, C. F. A. 1934. On excitation of crustacean muscle I. *J. exp. Biol.*, 11:11-27.
PANTIN, C. F. A. 1935a. Response of the leech to acetylcholine 1935. *Nature, Lond.*, 135:875.
PANTIN, C. F. A. 1935b. The nerve net of the Actinozoa I. Facilitation. *J. exp. Biol.*, 12:119-138.
PANTIN, C. F. A. 1935c. The nerve net of the Actinozoa IV. Facilitation and the "staircase". *J. exp. Biol.*, 12:389-396.
PANTIN, C. F. A. 1936a. On excitation of crustacean muscle II. Neuromuscular facilitation. *J. exp. Biol.*, 13:111-130.
PANTIN, C. F. A. 1936b. On the excitation of crustacean muscle III. Quick and slow responses. *J. exp. Biol.*, 13:148-158.
PANTIN, C. F. A. 1936c. On the excitation of crustacean muscle. IV. Inhibition. *J. exp. Biol.*, 13:159-169.
PANTIN, C. F. A. 1937. Junctional transmission of stimuli in the lower animals. *Proc. roy. Soc.* (B), 123:397-399.
PANTIN, C. F. A. 1952. The elementary nervous system. *Proc. roy. Soc.*, (B)140:147-168.

PARROT, J.-L. 1941. Recherches sur la transmission chimique de l'influx nerveux chez les crustacés. Libération d'une substance active sur l'intestin de *Maia squinado* par l'excitation des nerfs cardio-inhibiteurs. *C. R. Soc. Biol., Paris*, 135:929-933.

PERSKY, H. and GOLD, M. 1948. The choline acetylase and choline esterase content of some invertebrate tissues. *Biol. Bull., Woods Hole*, 95:278.

PHILLIPS, C. G. 1956. Intracellular records from Betz cells in the cat. *Quart. J. exp. Physiol.*, 41:58-69.

PRESTON, J. B. and KENNEDY, D. 1960. Integrative synaptic mechanisms in the caudal ganglion of the crayfish. *J. gen. Physiol.*, 43:671-681.

PROSSER, C. L. 1935a. A preparation for the study of single synaptic junctions. *Amer. J. Physiol.*, 113:108.

PROSSER, C. L. 1935b. Action potentials in the nervous system of the crayfish. V. Temporal relations in presynaptic and postsynaptic responses. *J. cell. comp. Physiol.*, 7:95-111.

PROSSER, C. L. 1937. Synaptic transmission in the sixth abdominal ganglion of the crayfish. *Biol. Bull., Woods Hole*, 73:346.

PROSSER, C. L. 1938. Evidence for chemical control of "spontaneous" activity of isolated ganglia. *Amer. J. Physiol.*, 123:165.

PROSSER, C. L. 1940. Action potentials in the nervous system of the crayfish. Effects of drugs and salts upon synaptic transmission. *J. cell. comp. Physiol.*, 16:25-38.

PROSSER, C. L. 1942. An analysis of the action of acetylcholine on hearts, particularly in arthropods. *Biol. Bull., Woods Hole*, 83:145-164.

PROSSER, C. L. 1943. Single unit analysis of the heart ganglion discharge in *Limulus polyphemus*. *J. cell. comp. Physiol.*, 21:295-305.

PROSSER, C. L. 1954. Activation of non-propagating muscle in *Thyone*. *J. cell. comp. Physiol.*, 44:247-253.

PROSSER, C. L. and BUEHL, C. C. 1939. Oxidative control of "spontaneous" activity in the nervous system of the crayfish. *J. cell. comp. Physiol.*, 14:287-297.

PROSSER, C. L. and MELTON, C. E., Jr. 1954. Nervous conduction in smooth muscle of *Phascolosoma* proboscis retractors. *J. cell. comp. Physiol.*, 44:255-275.

PROSSER, C. L. and YOUNG, J. Z. 1937. Responses of muscles of the squid to repetitive stimulation of the giant nerve fibers. *Biol. Bull., Woods Hole*, 73:237-241.

PUMPHREY, R. J. and RAWDON-SMITH, A. F. 1937. Synaptic transmission of nervous impulses through the last abdominal ganglion of the cockroach. *Proc. roy. Soc. (B)*, 122:106-118.

PURPURA, D. P., GIRADO, M., SMITH, T. G., CALLAN, D. A., and GRUNDFEST, H. 1959. Structure-activity determinants of pharmacological effects of amino acids and related compounds on central synapses. *J. Neurochem.*, 3:238-268.

REID, G. and VAUGHAN WILLIAMS, E. M. 1949. The development of sensitivity to acetylcholine in denervated muscle. *J. Physiol.*, 109:25-31.

REITER, M. 1957. Die Wirkung von Acetylcholine auf das isolierte Herz von *Aplysia limacina*. *Pubbl. Staz. zool. Napoli*, 29:226-228.

RENSHAW, B. 1946. Central effects of centripetal impulses in axons of spinal ventral roots. *J. Neurophysiol.*, 9:191-204.

RENSHAW, B. and THERMAN, P. O. 1941. Excitation of intraspinal mammalian axons by nerve impulses in adjacent axons. *Amer. J. Physiol.*, 133:96-105.

REUBEN, J. P. 1960. Electrotonic connections between lobster muscle fibers. *Biol. Bull., Woods Hole*, 119:334.

REUBEN, J. P. and GRUNDFEST, H. 1960. Inhibitory and excitatory miniature postsynaptic potentials in lobster muscle fibers. *Biol. Bull., Woods Hole*, 119:335-336.

RICHARDS, A. G. 1955. Structure, chemistry and pathology of the central nervous system of arthropods. In: *Neurochemistry*. K. A. C. Elliott et al. (eds.). Thomas, Springfield, Ill.

RICHARDS, O. W. 1929a. The effect of neurophil drugs on the fiddler crab, *Uca pugnax*. *Biol. Bull., Woods Hole*, 56:28-31.

RICHARDS, O. W. 1929b. The conduction of the nervous impulse through the pedal ganglion of *Mytilus*. *Biol. Bull., Woods Hole*, 56:32-40.

RIPLEY, S. H. and WIERSMA, C. A. G. 1953. The effect of spaced stimulation of excitatory and inhibitory axons of the crayfish *Physiol. comp.*, 3:1-17.

ROAF, H. E. 1935. The effect on invertebrates of drugs which act upon the vertebrate autonomic nervous system. *J. Physiol.*, 86:19P-20P.

ROBBINS, J., and VAN DER KLOOT, W. G. 1958. The effect of picrotoxin on peripheral inhibition in the crayfish. *J. Physiol.*, 143:541-552.

ROEDER, K. D. 1948a. The effect of anticholinesterases and related substances on nervous activity in the cockroach. *Johns Hopk. Hosp. Bull.*, 83:587-600.

ROEDER, K. D. 1948b. The effects of potassium and calcium on the nervous system of the cockroach *Periplaneta americana*. *J. cell. comp. Physiol.*, 31:327-338.

ROEDER, K. D. and WEIANT, E. A. 1950. The electrical and mechanical events of neuro-muscular transmission in the cockroach. *Periplaneta americana* (L.). *J. exp. Biol.*, 27:1-13.

ROEDER, K. D., KENNEDY, N. K., and SAMSON, E. A. 1947. Synaptic conduction to giant fibers of the cockroach and the action of anticholinesterases. *J. Neurophysiol.*, 10:1-10.

ROSENBLUETH, A. 1941. The stimulation of myelinated axons by nerve impulses in adjacent myelinated axons. *Amer. J. Physiol.*, 132:119-128.

ROSENBLUETH, A. 1950. *The Transmission of Nerve Impulses at Neuroeffector Junctions and Peripheral Synapses*. MIT and Wiley, New York.

Ross, D. M. 1945a. Facilitation in sea anemones. I. The action of drugs. *J. exp. Biol.*, 22:21-31.

Ross, D. M. 1945b. Facilitation in sea anemones. II. Tests on extracts. *J. exp. Biol.*, 22:32-36.

Ross, D. M. 1952. Facilitation in sea anemones. III. Quick responses to single stimuli in *Metridium senile*. *J. exp. Biol.*, 29:235-254.

Ross, D. M. 1955. Facilitation in sea anemones. IV. The quick response of *Calliactis parasitica* at high temperatures. *J. exp. Biol.*, 32:815-821.

Ross, D. M. 1956. Neuromuscular transmission in sea animals. *Int. Physiol. Congr., Abstr. XX*, 1956:780.

Ross, D. M. 1957a. The action of tryptamine and 5-hydroxytryptamine on muscles of sea anemones. *Experientia*, 13:192-194.

Ross, D. M. 1957b. Quick and slow contractions in the isolated sphincter of the sea anemone, *Calliactis parasitica*. *J. exp. Biol.*, 34:11-28.

Ross, D. M. and PANTIN, C. F. A. 1940. Factors influencing facilitation in Actinozoa. The action of certain ions. *J. exp. Biol.*, 17:61-73.

RUCK, P. 1958. Postsynaptic giant fiber response in ocellar nerves of dragonflies. *Anat. Rec.*, 132:499.

RUCK, P. 1961. Electrophysiology of the insect dorsal ocellus. II. Mechanisms of generation and inhibition of impulses in the ocellar nerve of dragonflies. *J. gen. Physiol.*, 44:629-639.

RUSHTON, W. A. H. 1945. Action potentials from the isolated nerve cord of the earthworm. *Proc. roy. Soc., (B)*132:423-437.

SCHALLEK, W. 1945. Action of potassium on bound acetylcholine in lobster nerve cord. *J. cell. comp. Physiol.*, 26:15-24.

SCHALLEK, W. and WIERSMA, C. A. G. 1948a. The influence of various drugs on a crustacean synapse. *J. cell. comp. Physiol.*, 31:35-47.

SCHALLEK, W. and WIERSMA, C. A. G. 1948b. Effects of anti-cholinesterases on synaptic transmission in the crayfish. *Physiol. comp.*, 1:63-67.

SCHALLEK, W., WIERSMA, C. A. G., and ALLES, G. A. 1948. Blocking and protecting actions of amines and ammonium compounds on a crustacean synapse. *Proc. Soc. exp. Biol., N. Y.*, 68:174-178.

Bibliography

SCHARF, J.-H. 1953. Die Beziehungen der Lipoide zu den perizellulären Strukturen der Ganglienzellen bei einigen Wirbellosen im Vergleich zu Wirbeltieren. *Z. Zellforsch.*, 38: 526-570.

SCHENCK, E., LUSCHNAT, K., and BRUNE, H. F. 1956. Einfluss der Temperatur auf neuromuskuläre Bahnung und Hemmung beim Flusskrebs. *Pflüg. Arch. ges. Physiol.*, 263:476-491.

SECHENOV, I. M. 1881. Galvanische Erscheinungen an der cerebrospinalen Axe des Frosches. *Pflüg. Arch. ges. Physiol.*, 25:281-284.

SEITE, R. 1961. Données récentes de cytophysiologie nerveuse. La transmission synaptique de l'excitation et le problème de l'acétylcholine liée (essai sur certaines données biochimiques, physiologiques et morphologiques) *Année biol.*, 37:217-253.

SMALLMAN, B. N. 1956. Mechanisms of acetylcholine synthesis in the blowfly. *J. Physiol.*, 132:343-357.

SMALLMAN, B. N. and WOLFE, L. S. 1956. Soluble and particulate cholinesterase in insects. *J. cell. comp. Physiol.*, 48:197-213.

SMITH, R. I. 1939. Acetylcholine in the nervous tissues and blood of crayfish. *J. cell. comp. Physiol.*, 13:335-344.

SMITH, R. I. 1947. The action of electrical stimulation and of certain drugs on cardiac nerves of the crab, *Cancer irroratus*. *Biol. Bull., Woods Hole*, 93:72-88.

SPERRY, R. W. 1958. Physiological plasticity and brain circuit theory. In: *Biological and Biochemical Bases of Behavior*. H. R. Harlow and C. N. Woolsey (eds.). Univ. of Wisconsin Press, Madison.

STEDMAN, E. and STEDMAN, E. 1938. Mechanism of the biological synthesis of acetylcholine. *Nature, Lond.*, 141:39-40.

STOUGH, H. B. 1926. Giant nerve fibers of the earthworm. *J. comp. Neurol.*, 40:409-463.

STRUMWASSER, F. and ROSENTHAL, S. 1960. Prolonged and patterned direct extracellular stimulation of single neurons. *Amer. J. Physiol.*, 198:405-413.

TASAKI, I. and CHANG, J. J. 1958. Electric response of glia cells in cat brain. *Science*, 128:1209-1210.

TASAKI, I., POLLEY, E. H., and ORREGO, F. 1954. Action potentials from individual elements in cat geniculate and striate cortex. *J. Neurophysiol.*, 17:454-474.

TAUC, L. 1955a. Réponse de la cellule nerveuse du ganglion abdominal d'*Aplysia punctata* activée par voie synaptique. *J. Physiol., Paris*, 47:286-287.

TAUC, L. 1956b. Potentiels post-synaptiques inhibiteurs obtenus dans les cellules nerveuses de ganglion abdominal de l'aplysie. *C. R. Acad. Sci., Paris*, 242:676-678.

TAUC, L. 1957a. Potentiels postsynaptiques d'inhibition obtenus dans les somas neuroniques des ganglions de l'aplysie et de l'escargot. *J. Physiol., Paris*, 49:396-399.

TAUC, L. 1957b. Stimulation du soma neuronique de l'aplysie par voie antidromique. *J. Physiol., Paris*, 49:973-986.

TAUC, L. 1957c. Développement du potentiel post-synaptique en présence du potentiel d'action dans le soma neuronique du ganglion d'escargot (*Helix pomatia*). *C. R. Acad. Sci., Paris*, 245:570-573.

TAUC, L. 1958a. Processus postsynaptiques d'excitation et d'inhibition dans le soma neuronique de l'aplysie et de l'escargot. *Arch. ital. Biol.*, 96:78-110.

TAUC, L. 1958b. Potentiel de repos d'un soma neuronique (ganglion d'aplysie) soumis à l'action de solution d'alcool éthylique, d'aldéhyde formique ou d'acétone. *Arch. Sci. physiol.*, 12: 31-36.

TAUC, L. 1958c. Analyses unitaires d'activités synaptiques chez l'aplysie, révélant la mise en jeu de neurones intermédiaires dans le ganglion abdominal. *J. Physiol., Paris*, 50: 541-544.

TAUC, L. 1958d. Quelques précisions sur l'origine du potentiel postsynaptique d'inhibition dans la préparation ganglionnaire de l'aplysie. *J. Physiol., Paris*, 50:1107-1116.

TAUC, L. 1959a. Interactions neuronales synaptiques et non synaptiques dans le ganglion abdominal de l'aplysie. *J. Physiol., Paris*, 51:570-571.

TAUC, L. 1959b. Preuve expérimentale de l'existence de neurones intermédiaires dans le ganglion abdominal de l'aplysie. *C. R. Acad. Sci., Paris*, 248:853-856.

TAUC, L. 1959c. Interaction non synaptique entre deux neurones adjacents du ganglion abdominal de l'aplysie. *C. R. Acad. Sci., Paris*, 248:1857-1859.

TAUC, L. 1959d. Sur la nature de l'onde de surpolarisation de longue durée observée parfois après l'excitation synaptique de certaines cellules ganglionnaires de mollusques. *C. R. Acad. Sci., Paris*, 249:318-320.

TAUC, L. 1960a. Diversité des modes d'activité des cellules nerveuses du ganglion déconnecté de l'aplysie. *C. R. Soc. Biol., Paris*, 154:17-21.

TAUC, L. 1960b. The site of origin of the efferent action potentials in the giant nerve cell of *Aplysia*. *J. Physiol.*, 152:36P-37P.

TAUC, L. 1960c. Maintien de la transmission synaptique dans le neurone géant d'aplysie sans activation du soma ou en l'absence du soma. *C. R. Acad. Sci., Paris*, 250:1560-1562.

TAUC, L. 1960d. Evidence of synaptic inhibitory actions not conveyed by inhibitory post-synaptic potentials. In: *Inhibition in the Nervous System and Gamma-aminobutyric Acid*. Pergamon Press, New York.

TAUC, L. and GERSCHENFELD, H. 1960a. L'acétylcholine comme transmetteur possible de l'inhibition synaptique chez l'aplysie. *C. R. Acad. Sci., Paris*, 251:3076-3078.

TAUC, L. and GERSCHENFELD, H. M. 1960b. Effet inhibiteur ou excitateur du chlorure d'acetylcholine sur le neurone d'escargot. *J. Physiol., Paris*, 52:236.

TAUC, L. and GERSCHENFELD, H. M. 1961. Cholinergic transmission mechanisms for both excitation and inhibition in molluscan central synapses. *Nature, Lond.*, 192:366-367.

TERZUOLO, C. A. and BULLOCK, T. H. 1956. Measurement of imposed voltage gradient adequate to modulate neuronal firing. *Proc. nat. Acad. Sci., Wash.*, 42:687-694.

TERZUOLO, C. A. and BULLOCK, T. H. 1958. Acceleration and inhibition in crustacean ganglion cells. *Arch. ital. Biol.*, 96:117-134.

TOBIAS, J. M. and KOLLROS, J. J. 1946. Loci of action of DDT in the cockroach (*Periplaneta americana*). *Biol. Bull., Woods Hole*, 91:247-255.

TOBIAS, J. M., KOLLROS, J. J., and SAVIT, J. 1946. Acetylcholine and related substances in the cockroach, fly and crayfish and the effect of DDT. *J. cell. comp. Physiol.*, 28:159-182.

TURNER, R. S. 1953. Modification by temperature of conduction and ganglionic transmission in the gastropod nervous system. *J. gen. Physiol.*, 36:463-471.

VAN DER KLOOT, W. G. 1960. Factor S —a substance which excites crustacean muscle. *J. Neurochem.*, 5: 245-252.

VAN DER KLOOT, W. G., ROBBINS, J., and COOKE, I. M. 1958. Blocking by picrotoxin of peripheral inhibition of crayfish. *Science*, 127:521-522.

VAN HARREVELD, A. and WIERSMA, C. A. G. 1937. The triple innervation of crayfish muscle and its function in contraction and inhibition. *J. exp. Biol.*, 14:448-461.

VAN HARREVELD, A. and WIERSMA, C. A. G. 1939. The function of the quintuple innervation of a crustacean muscle. *J. exp. Biol.*, 16: 121-133.

VINCENT, D. and JULLIEN, A. 1938. De la teneur des principaux organes de *Murex* en esters de la choline. *C. R. Soc. Biol., Paris*, 129:602-603.

WAIT, R. B. 1943. The action of acetylcholine on the isolated heart of *Venus mercenaria*. *Biol. Bull., Woods Hole*, 85:79-85.

WAKABAYASHI, T. and IKEDA, K. 1957.

Phylogenetic studies on miniature electrical oscillation in insect. *Jap. J. Physiol.*, 7:222-231.

WALL, P. D. 1958. Excitability changes in afferent fibre terminations and their relation to slow potentials. *J. Physiol.*, 142:1-21.

WALL, P. D. 1959. Repetitive discharge of neurons. *J. Neurophysiol.*, 22: 305-320.

WALL, P. D., LETTVIN, J. Y., MCCULLOCH, W. S., and PITTS, W. H. 1956. The nature and origin of prolonged events in the terminal arborisations of spinal afferent fibres. *Int. Physiol. Congr., Abstr. XX*, 1956:941-942.

WALOP, J. N. 1951. Studies on acetylcholine in the crustacean central nervous system. *Arch. int. Physiol.*, 59:145-156.

WATANABE, A. 1958. The interaction of electrical activity among neurons of lobster cardiac ganglion. *Jap. J. Physiol.*, 8:305-318.

WATANABE, A. and BULLOCK, T. H. 1960. Modulation of activity of one neuron by subthreshold slow potentials in another in lobster cardiac ganglion. *J. gen. Physiol.*, 43:1031-1045.

WATANABE, A. and GRUNDFEST, H. 1962. Impulse propagation at the septal and commissural junctions of crayfish lateral giant axons. *J. gen. Physiol.*, 45:267-308.

WATANABE, A., SMITH, T. G., and GRUNDFEST, H. 1960. Segmental and crossed ephaptic transmission in crayfish lateral giant axons. *Fed. Proc.*, 19:298.

WATANABE, Y. 1958. Transmission of impulses through abdominal ganglia in the crayfish, *Cambarus clarkii*. *J. Fac. Sci., Hokkaido Univ.*, (6)14: 17-29.

WATERMAN, T. H. 1941. A comparative study of the effects of ions on whole nerve and isolated single nerve fiber preparations of crustacean neuromuscular systems. *J. cell. comp. Physiol.*, 18:109-126.

WELSH, J. H. 1939a. Chemical mediation in crustaceans. I. The occurrence of acetylcholine in nervous tissues and its action on the decapod heart. *J. exp. Biol.*, 16:198-219.

WELSH, J. H. 1939b. Chemical mediation in crustaceans. II. The action of acetylcholine and adrenalin on the isolated heart of *Panulirus argus*. *Physiol. Zoöl.*, 12:231-237.

WELSH, J. H. 1942. Chemical mediation in crustaceans. IV. The action of acetylcholine on isolated hearts of *Homarus* and *Carcinides*. *J. cell. comp. Physiol.*, 19:271-279.

WELSH, J. H. 1948. Concerning the mode of action of acetylcholine. *Johns Hopk. Hosp. Bull.*, 83:568-579.

WELSH, J. H. 1957a. Serotonin as a possible neurohumoral agent: evidence obtained in lower animals. *Ann. N. Y. Acad. Sci.*, 66:618-630.

WELSH, J. H. 1957b. Neurohormones or transmitter agents. In: *Recent Advances in Invertebrate Physiology*. B. T. Scheer (ed.). Univ. Oregon, Eugene.

WELSH, J. H. and GORDON, H. T. 1947. The mode of action of certain insecticides on the arthropod nerve axon. *J. cell. comp. Physiol.*, 30: 147-171.

WELSH, J. H. and HASKIN, H. H. 1939. Chemical mediation in crustaceans. III. Acetylcholine and autotomy in *Petrolisthes armatus* (Gibbes). *Biol. Bull., Woods Hole*, 76:405-415.

WELSH, J. H. and MOORHEAD, M. 1960. The quantitative distribution of 5-hydroxytryptamine in the invertebrates, especially in their nervous systems. *J. Neurochem.*, 6:146-169.

WELSH, J. H. and SCHALLEK, W. 1946. Arthropod nervous systems: a review of their structure and function. *Physiol. Rev.*, 26:447-478.

WERMAN, R. 1960. Electrical inexcitability of the synaptic membrane in the frog skeletal muscle fibre. *Nature, Lond.*, 188:149-150.

WERMAN, R. and GRUNDFEST, H. 1959. Properties of prolonged action potentials in insect muscle. *Fed. Proc.*, 18:169.

WHITTERIDGE, D. 1948. The role of acetylcholine in synaptic transmission: a critical review. *J. Neurol. Neurosurg Psychiat.*, 11:134-140.

WIERSMA, C. A. G. 1933. Vergleichende Untersuchungen über das periphere Nerven-muskelsystem von Crustaceans. *Z. vergl. Physiol.*, 19:349-385.

WIERSMA, C. A. G. 1937. Die doppelte motorische Innervation des Scherenschliessers von *Astacus fluviatilis*. *Z. vergl. Physiol.*, 24:381-386.

WIERSMA, C. A. G. 1938. Function of the giant fibers of the central nervous system of the crayfish. *Proc. Soc. exp. Biol. N. Y.*, 38:661-662.

WIERSMA, C. A. G. 1941. The efferent innervation of muscle. *Biol. Symp.*, 3:259-291.

WIERSMA, C. A. G. 1947. Giant nerve fiber system of the crayfish. A contribution of comparative physiology of synapse. *J. Neurophysiol.*, 10:23-38.

WIERSMA, C. A. G. 1949. Synaptic facilitation in the crayfish. *J. Neurophysiol.*, 4:267-275.

WIERSMA, C. A. G. 1952a. Repetitive discharges of motor fibers caused by a single impulse in giant fibers of the crayfish. *J. cell. comp. Physiol.*, 46:399-419.

WIERSMA, C. A. G. 1952b. Neurons of arthropods. *Cold Spr. Harb. Symp. quant. Biol.*, 17:155-163.

WIERSMA, C. A. G. 1953. Neural transmission in invertebrates. *Physiol. Rev.*, 33:326-355.

WIERSMA, C. A. G. and ADAMS, R. T. 1950. The influence of nerve impulse sequence on the contractions of different crustacean muscles. *Physiol. comp.*, 3:20-33.

WIERSMA, C. A. G. and ELLIS, C. H. 1942. A comparative study of peripheral inhibition in decapod crustaceans. *J. exp. Biol.*, 18:233-236.

WIERSMA, C. A. G. and HELFER, R. G. 1941. The effects of peripheral inhibition on the muscle action potentials of the crab. *Physiol. Zoöl.*, 14:296-304.

WIERSMA, C. A. G. and MARMONT, G. 1936. On the mechanism of inhibition of crayfish muscle. *Proc. nat. Acad. Sci., Wash.*, 22:502-504.

WIERSMA, C. A. G. and RIPLEY, S. H. 1954. Further functional differences between fast and slow contractions in certain crustacean muscles. *Physiol. comp.*, 3:327-336.

WIERSMA, C. A. G. and SCHALLEK, W. 1947. Potentials from motor roots of the crustacean central nervous system. *J. Neurophysiol.*, 10:323-330.

WIERSMA, C. A. G. and SCHALLEK, W. 1948. Influence of drugs on response of a crustacean synapse to preganglionic stimulation. *J. Neurophysiol.*, 11:491-496.

WIERSMA, C. A. G. and TURNER, R. S. 1950. The interactions between the synapses of a single motor fiber. *J. gen. Physiol.*, 34:137-145.

WIERSMA, C. A. G. and VAN HARREVELD, A. 1934. On the nerve-muscle system of the hermit crab (*Eupagurus bernhardus*). Inhibition of the contraction of the abductor of the claw. *Arch. néerl. Physiol.*, 19:459-468.

WIERSMA, C. A. G. and VAN HARREVELD, A. 1935. On the nerve-muscle system of the hermit crab (*Eupagurus bernhardus*). The action currents of the muscles of the claw in contraction and inhibition. *Arch. néerl. Sci. (3C)*, 20:89-102.

WIERSMA, C. A. G. and VAN HARREVELD, A. 1938. The influence of the frequency of stimulation on the slow and fast contraction in crustacean muscle. *Physiol. Zoöl.*, 11:75-81.

WIERSMA, C. A. G. and VAN HARREVELD, A. 1939. The interactions of the slow and the fast contraction of crustacean muscle. *Physiol. Zoöl.*, 12: 43-49.

WIERSMA, C. A. G. and ZAWADZKI, B. 1948. On the relation between different ions and peripheral inhibition in crustacean muscle. *J. cell. comp. Physiol.*, 32:101-103.

WIERSMA, C. A. G., FURSHPAN, E., and FLOREY, E. 1953. Physiological and pharmacological observations on muscle receptor organs of the crayfish. *Cambarus clarkii*. Girard. *J. exp. Biol.*, 30:136-150.

WILSON, D. M. 1960a. Nervous control of movement in annelids. *J. exp. Biol.*, 37:46-56.

WILSON, D. M. 1960b. Nervous control of movement in cephalopods. *J. exp. Biol.*, 37:57-72.

WILSON, D. M. 1961. "The connections between the lateral giant fibers of earthworms," *Comp. Biochem. Physiol.*, 3:274-284.

WILSON, I. B. 1952. Biochemical similarities and differences between synaptic transmission and axonal conduction. In: *Nerve Impulse*. H. H. Merritt (ed.). Josiah Macy, Jr. Foundation, New York.

WILSON, V. J. 1954. Slow and fast responses in cockroach leg muscles. *J. exp. Biol.*, 31:280-290.

WOLFE, L. S. and SMALLMAN, B. N. 1956. The properties of cholinesterase from insects. *J. cell. comp. Physiol.*, 48:215-235.

WRIGHT, E. B. 1949. The action of erythroidin, curare, and chlorobutanol in the crayfish. *J. cell. comp. Physiol.*, 33:301-332.

WU, K. S. 1939a. On the physiology and pharmacology of the earthworm gut. *J. exp. Biol.*, 16:184-197.

WU, K. S. 1939b. The action of drugs, especially acetylcholine, on the annelid body wall (*Lumbricus, Arenicola*). *J. exp. Biol.*, 16:251-257.

YAMASAKI, T. and ISHII, T. 1952. Studies on the mechanism of action of insecticides. V. The effects of DDT on the synaptic transmission in the cockroach. *Oyō-Konchū (J. Nippon. Soc. Appl. Ent.)*, 8:111-118. (In Japanese with English summary.)

YAMASAKI, T. and ISHII, T. 1954a. Studies on the mechanism of action of insecticides. VII. Activity of neuron soma as a factor of development of DDT symptoms in the cockroach (*Periplaneta americana*). *Botyu-Kagaku (Sci. Insect Control)*, 19:1-14. (In Japanese with English summary.)

YAMASAKI, T. and ISHII, T. 1954b. Studies on the mechanism of action of insecticides (IX). Repetitive excitation of the insect neurone soma by direct current stimulation and effects of DDT. *Jap. J. Appl. Zool.*, 19:16-28. (In Japanese with English summary.)

YAMASAKI, T. and NARAHASHI, T. 1958. Synaptic transmission in the cockroach. *Nature, Lond.*, 182:1805-1806.

YAMASAKI, T. and NARAHASHI, T. 1960. Synaptic transmission in the last abdominal ganglion of the cockroach. *J. Insect Physiol.*, 4:1-13.

YEANDLE, S. 1958. Evidence of quantized slow potentials in the eye of *Limulus*. *Amer. J. Ophthal.*, 46:82-87.

YOUNG, J. Z. 1936. Structure of nerve fibres and synapses in some invertebrates. *Cold Spr. Harb. Symp. quant. Biol.*, 4:1-5.

YOUNG, J. Z. 1937. The physical and chemical properties of nerve fibres and the nature of synaptic contacts. *Trans. Faraday Soc.*, 33:1035-1040.

YOUNG, J. Z. 1938a. Synaptic transmission in the absence of nerve cell bodies. *J. Physiol.*, 93:43P-45P.

YOUNG, J. Z. 1938b. The functioning of the giant nerve fibres of the squid. *J. exp. Biol.*, 15:170-185.

YOUNG, J. Z. 1939. Fused neurons and synaptic contacts in the giant nerve fibres of cephalopods. *Philos. Trans. (B)*, 229:465-503.

CHAPTER 5

Mechanisms of Integration

Summary	254
I. Introduction: The Levels of Integration	257
II. The Unit Level: Integrative Properties of Neurons	258
A. A Frame of Reference: Loci and Modes of Lability	259
1. Eight forms of responsiveness	259
2. Four forms of sensitivity	261
B. The Degrees of Freedom: Intracellular Permutations	264
1. The distribution over the neuron of types of cell membrane	264
2. Excitation and inhibition	265
3. Facilitation and antifacilitation	269
4. Negative and positive aftereffects	270
5. The alternative effects of milieu	271
6. Movements of parts of neurons	271
III. Integration at the Level of Organized Groups of Neurons	272
A. Input and its Interpretation	273
1. The parameters of the code	273
2. Analysis by labeled lines and spatial representation	274
3. Differentiation among modalities in central effect	276
4. The requirements of an analyzer	277
5. The recognition of predetermined stimulus pattern	279
6. Thresholds to integrated input; "decision units"	281
7. Integrative possibilities of neuronal circuits	285
B. Output and its Control	287
1. Neuroeffector integration	287
2. Nerve nets and other uncentralized control systems	288
3. Reflex integration	289
4. The origin of patterned discharge	295
C. Associative Levels of Moderate Complexity	299
1. Degrees of coordination; "relative coordination"	299
2. Functions tending to topographic segregation	302
3. Specific and nonspecific systems	306
4. Central control of input	307
5. The weighting of influx, mood, or set	309
6. Recovery from damage; "plasticity"	309
7. The role of numbers of neurons	312
8. Structural specializations among associative neurons	312
IV. Spontaneity, Its Sources and Consequences	314
A. Spontaneity at the Neuronal Level	314
1. Neuronal pacemakers	314
2. "Brain waves"	318
3. Cellular spontaneity which does not pace other cells	319
B. Spontaneity at the Behavioral Level	319
1. The central origin of certain rhythms	319
2. Phases of behavior	320
3. Tonic input maintaining central excitatory state; "Stimulationsorgane"	322

V. Neurological Deductions from the Study of Behavior	323
A. The Analysis of Behavior as a Control System	323
B. Elementary Fixed Action Patterns	328
C. Kineses, Taxes, and Reversal of Sign	329
D. Instinctive Behavior	331
E. The Release and Probability of Specific Behavior Patterns	333
F. Learning	337
G. Unity of Action, Attention, Arousal	338
Classified References	339
Bibliography	341

Summary

Having treated elsewhere the details of conduction, transmission, neuronal organization, and the peculiarities of the several groups of invertebrates, this chapter undertakes a survey of the principles and mechanisms available to the nervous system with which it can integrate—that is, **determine the output** of any cell or group of cells as some function of the input. Integrative mechanisms are considered at three **levels:** the unit neuron, the group of neurons comprising a ganglion or organized central mass, and the whole animal as deduced from behavior.

INTEGRATIVE PROPERTIES OF NEURONS

At the neuronal level responsiveness is to be distinguished from sensitivity and at present eight **forms of responsiveness** can be recognized: (1) generator potentials, (2) pacemaker potentials, (3) synaptic, (4) local, (5) spike, and (6) afterpotentials, (7) potentials associated with activity of terminal arborizations, and (8) release of transmitter substance in chemically transmitting endings. These occur in different kinds of cell surface membrane. Activity is usually a sequence of several of these linked by transfer functions which represent integrations of the impinging activity with conditions of the milieu and the history of the cell. Four **forms of sensitivity** can be recognized: (1) spike threshold at the spike initiating locus, (2) synaptic and (3) local potential excitability at more distributed loci and (4) modifiability of on-going activity by weak electric fields. Like the forms of responsiveness, these also represent integrations of the cell milieu and history.

Within this framework the main alternative properties or degrees of freedom which can contribute to integration are as follows: (1) Different distributions of the several kinds of cell membrane over the soma, axon, and dendrites can be profoundly influential. Dendrites are the most important structures for the integrative functions of the nervous system. Intracellular sites are more important than synapses as loci of labile and evaluating processes. (2) Specific transmitters and nonspecific electrically charged particles can cause either excitation or inhibition, and each of these is shown by a complex of measured signs which generally are nonlinear with respect to each other. Inhibition can occur in at least six ways: (a) refractoriness following a response, (b) refractoriness following a subthreshold excitation, (c) specific synaptic pathways that selectively increase membrane permeability and usually membrane potential, (d) excessive depolarization, (e) electric potential gradient, and (f) presynaptically, by unknown means. Some of these are accompanied by an appreciable hyperpolarization. (3) Repetitive input may cause facilitation or antifacilitation. (4) Aftereffects may follow the end of influx and may be negative or positive. (5) A great variety of chemical and physical agents in the environment of the neuron can cause either stimulatory or depressing effects. (6) Parts of neurons are believed to move or grow during functional life in ways that have integrative significance.

INTEGRATION AT THE LEVEL OF ORGANIZED GROUPS OF NEURONS

The best known form of communication between neurons is the **pulsed code,** in which frequency and number of impulses are the main parameters of each line or axon. Temporal patterning, within a given average frequency, is probably another part of the code.

The interpretation of input is considered first, starting with the principle of **analysis by labeled lines.** Differences among modalities depend not only on the adequate stimulus they are sensitive to, but on the **imperativeness** of their message as measured by their weight in the central processes determining output. Subdivisions within a general modality,

Summary

thick and thin fibers serving the same modality, and the problems of the evolution of modalities are emphasized. The problem of distinguishing signals in the presence of fluctuating background activity dictates some conclusions about the **requirements of a central analyzer.** The commonplace neural property of **recognition** of predetermined pattern is an important integrative achievement and is understandable as the accomplishment either of a single neuron (or a hierarchy which stepwise decodes the more complex patterns) or of certain kinds of networks. For each parallel and each successively higher central integration (comparison or decision) there must be **decisive recognition units** (cell or network), each one competent to receive all relevant inputs, to determine whether a criterion has been reached and, if so, to trigger the whole machine normally dependent on that decision—for example, to choose between competing instinctive acts. Novel principles of integrative action are still to be expected. The enormous possibilities of specific networks or circuits of neurons for performing complex integrative processing are being investigated chiefly in models; these show great power with relatively simple circuits of neuronlike analog elements.

Next, **output and its control** is surveyed. Labile, integrative relations at the neuroeffector junction have been found to be important in all groups carefully studied, from coelenterates to vertebrates. The simplest organized systems for controlling effectors are the **nerve nets,** as found in coelenterates; here a small number of integrative variables permit remarkably diversified behavior.

A condensed review of **reflexology** emphasizes the properties and rules of integrative action. The most remarkable is the lack of algebraic summation of simultaneously stimulated antagonistic reflexes; instead of interfering, all but one are precluded. During maintained stimulation adequate for both, they may alternate but at any moment it is either-or. A switching mechanism is indicated which distinguishes between antagonistic and allied reflexes, the latter freely coexisting to give characteristic reflex figures. The adaptive unity of the individual is already enforced at this level, no matter what constellation of simultaneous stimuli impinges. The factors that determine the selection and timing of reflexes elicited successively are found to be reminiscent of neuronal ones. There is a built-in hierarchical order of dominance, there is a weighted evaluation of input strength, and there is a strong rebound especially important in chain reflexes.

Patterned discharge may be due to (a) peripheral cues, with centrally determined timing elements and distribution, as in a cough or a step, (b) proprioceptively fed-back rhythms, or (c) complex central pacemakers, as in neurogenic hearts, "vacuum" actions, and others. Feedback may importantly shape the details in the last case.

Some additional basic principles are enumerated under "Associative levels." The perfect coordination of familiar reciprocal inhibition of antagonistic movements, as in eye muscles and respiration, is complemented by the less well-known **relative coordination** exemplified in rhythmic fin movements in fish. Here all degrees of coordination can be seen. Two general phenomena recognized are superposition and the magnet effect. In superposition a fin shows amplitude modulation of its own rhythm by a different one belonging to another, "dominant" fin. In the magnet effect one rhythm attracts another toward its frequency. The two effects may interplay in a complex way like coupled nonlinear oscillators. Instead of a purely triggered reflex view of central coordination, this evidence indicates intrinsic automaticities and built-in, but labile, coupling.

As in an industrial organization, we can find **two categories of functions,** those which are segregated into special departments and those which are diffused or repeated in many places. Much of the classical work and much of the future of neurology consists in the effort to define these two categories in the central nervous system.

Parallel specific and nonspecific systems, known in vertebrates, are fundamental respectively to the information-discriminating and the general arousing or galvanizing spheres: without the former the animal is blind, deaf, or anesthetic; without the latter it cannot be roused from a sleeplike state. **Central control of input** is widespread and occurs in the periphery as well as after one or more relays. Affect, set, or **disposition** is not a uniquely human trait, but deeply influences the weighting of influx and the determination of response in many lower animals which go through moods as though experiencing a shifting pleasure-pain balance.

Plasticity in the sense of functional regulation in adults is believed to be poorly developed in all animals except man and some primates. Recovery after severe peripheral disarrangement in young amphibians and fish (for example, crossing flexor and extensor or right and left optic nerves) manifests a profound specificity and regulatory power of developing or regenerating central connections, which has also been called plasticity. The so-called plasticity of insects, which determines specific locomotor patterns characteristic for the number of appendages whose joints remain movable after amputation or immobilization, is probably a different phenomenon. It results from built-in pathways and functional gradients—partly sensory feedback, partly central patterning.

A discussion of the dependence of evolutionary advance upon **numbers of neurons** and interconnections leads to the conclusion that while they do, roughly, increase with functional complexity, something else must be increasing also—some integrative property associated with **quality of connections** as distinct from quantity alone. Functional differentiation is reflected in anatomical specializations which might in a measure correspond with such a development.

SPONTANEITY, ITS SOURCES AND CONSEQUENCES

A special section is devoted to manifestations of spontaneity. By mixing response to stimulation through complex circuits with spontaneous activity in selected loci, the possibilities for determining output by some indirect function of input are enormously increased. Spontaneous activity means change of state of neurons without change of state of the impinging environment. It is probably common. There has been significant recent advance in the intimate description of neuronal pacemaker action. **"Brain waves"** apparently represent a fluctuating state of many neurons, more or less synchronized, separate from spikes though interacting with them. This is another expression of graded cellular activity, perhaps especially developed in dendritic processes. Meager evidence indicates that brain waves may not be merely by-products but causes as well. Invertebrate brain waves appear consistently different in character from vertebrate brain waves, which are remarkably uniform from fish to man in spite of the development of cortex and associated structures.

Rhythmic movements of hearts, fins, parapods, and gut segments are often due to intrinsic spontaneity, more or less modified by sensory feedback. They may resemble built-in reflex patterns but with a trigger from an internal clock. Intrinsically competent rhythmic mechanisms may be profoundly shaped by phasic peripheral input and by experience and by volition. Slow successions of phases of activity, as in burrowing polychaetes and actinians, seem not attributable to environmental triggers or obvious biological need thresholds but to intrinsic and possibly nervous "clocks."

The idea that some **receptors have a noninformational tonic function** (besides their sensory one) in maintaining a central excitatory state is regarded as likely and significant but the evidence available is ambiguous.

DEDUCTIONS FROM THE STUDY OF BEHAVIOR

The final section presents deductions about physiological mechanisms which must be presumed from studies on behavior. In the first part, analyses of certain cases treated as **control systems** are shown to be capable of specifying connections, amplifying factors, and dynamic features of components and hence accuracy, phase-amplitude relations, and final steady state. The question can be answered—in the negative—Is a basic reflex such as the optomotor interrupted, when in voluntary movements it seems to be circumvented? The reflex tending to prevent departure from a given state can be shown in several ingenious ways to be still operating but counteracted by a superimposed higher command. In certain situations it has been found that a voluntary command must leave behind itself in the center some trace, or coded **state of expectation** (von Holst's efference copy), with which the resulting change in afferent flux is compared; normally matching in some way (that is, the expected change in influx being in fact realized), the two sets of messages result in normal perception. The efference copy is an example of an integrative device which must have physiological reality, inferred from experiments on behavior. One other example is the **feed-forward** connection as applied to receptor-effector relations, and another is the very **brief use of high amplification** feedback to gain sensitivity without instability. Similar methods show the probable circuit equivalent for visual detection of movement in an insect to be one achieving an **autocorrelation** in time, of specified channels, with multiplication of both magnitude and sign of stimuli.

Indicative of built-in patterns of connections and dynamic relations, of different degrees of complexity, are the **elementary fixed action patterns, kineses, taxes, and instincts.** Kineses and instincts express grades of spontaneity which are possibly compatible with the pacemaker actions known physiologically. The element of randomness is interesting in the kineses; it can be ascribed to interaction of pacemakers without appealing to molecular indeterminancy. Simple and complex instances of spatial and temporal integration of afferent information—such as the filtering and summation of background and one or more releasing and steering stimuli—are seen in taxes and the release of instinctive acts. The integrative constants can be altered by milieu and experience. It is concluded that features such as the hierarchical sequence of different actions, the typically alternative character of instinctive acts within a hierarchical level, and the specific readiness or probability of appearance (accumulating "action-specific potential") bespeak processes and centers of exactly the kinds already inferred in earlier sections of the chapter.

It is emphasized that basically the problem presented by **learning** is neurophysiologically similar to that of instincts: namely, how a given set of constants and relations among dynamic properties of junctions and graded membrane actions can be so stable. What it is that changes during learning is of profound interest, but it is secondary to **what stabilizes** a persistent pattern. The belief is expressed that presently known principles of integrative action do not suffice to explain the highest forms of behavior and that yet to be discovered are properties which may be called **emergent mechanisms** that inhere in the organized masses of neurons.

The supreme integration, for which we also assume a physiological mechanism, is the achievement of adaptive **unity of action** by the normal organism. In addition to other factors this must be based upon arousal, attention, and the mutual exclusion of conflicting directions of attention and action, therefore upon high-level, either-or switching, or decision units.

I. INTRODUCTION: THE LEVELS OF INTEGRATION

This chapter takes stock of the presently known or reasonably indicated modes of integration actually employed by the nervous systems of animals, including man. This is simply a way of looking at the physiology of neural systems, to emphasize the operations accomplished as opposed to describing their structural organization in one group of animals, such as laboratory mammals, or to analyzing their mechanisms in molecular or biochemical terms. A survey of modes of integration really embraces all the adaptive "nervous" processes of the nervous system—that is, those processes other than mere conduction and one-to-one transmission—whether based on functional properties or structural relations, and is therefore central to a proper understanding of this most complex of the organ systems. The present knowledge of these processes is both vast and fragmentary. An effort to systematize it today is bound to be incomplete, not to say naive, even in respect to some of the indications already in the literature and—it is to be hoped—soon obsolete!

Paraphrasing the dictionary, integration is a process of putting parts together into a whole. An equivalent **definition** in terms more useful neurophysiologically, would be that integration is a process or set of processes resulting in an output not identical to input but still some function of the input. This definition is very broad, making integration in respect to the neural systems the general term for the "nervous" achievements of those systems (not to speak of other systems), as opposed to the achievements of distributing information by conduction over distances. It appears to be the most useful definition and focuses attention on the principles involved which are often identical in a wide range of anatomically distinct entities and often plural within a given anatomical entity. Needless to say, receptors and effectors both perform integrative functions, and other organ systems do likewise—notably the endocrine glands.

The definition applies not only to cases of converging input determining an output but also to the single line interrupted by an element with some transfer function, because here information is being mixed from the input with that stored in the transformer or filter; furthermore, the latter is likely to be subject to change with physiological condition.

Some years ago a symposium, entitled "Levels of Integration in Biological and Social Systems" (Redfield, 1942; see especially Gerard, 1942), expounded the history of thinking on a much abused subject, and made the argument that we cannot expect the most rapid progress in understanding if we investigate only at the most basic level permitted by the techniques available. With this I agree. Insistence upon studying society only by observation of individuals, individual behavior only by research on brains and glands, the nervous system only by unit neuron analysis, the neuron only by biochemical dissection, and so on to ultimate irreducibles, is abuse of a worthy objective. For the present and the forseeable future, we must treat integration as **occurring at many levels,** each doubtlessly based upon those below it but, for some time at least, not fully explained by study of the next or the next more fundamental level. Each level deserves attention even without reference to the techniques or concepts developed by the lower-level disciplines.

The **scope** of the present survey requires consideration of integrative performance of major parts of nerve cells but not of the prenervous events accounting for excitability and response. It includes synaptic events but not their mechanisms of transmitters and conductance changes; it includes local and spike potentials but not their bases in ionic movements or permeability. The higher levels of integration among cellular units in functional groups and among organized groups making up whole nervous systems belong here. At the other end of the continuum, it goes only as far as some of the more obvious inferences about neural substrates that are permitted by elementary observations of behavior.

Illustrations and conclusions come from all kinds of animals. Other things being equal, examples are chosen from the lower animals out of a personal bias that they have been neglected relative to their potential value for revealing secrets of nature. Literature citations are illustrative and by no means comprehensive.

II. THE UNIT LEVEL: INTEGRATIVE PROPERTIES OF NEURONS

Why do we begin with the neuron? There are complex levels below it. But on significant functional as well as anatomical, embryological, and trophic grounds, the cell theory is probably nowhere more apparent and entrenched than in neurology. We deal with the neuron as a uniquely significant integrative level because it is a unit like a person in society—it generally speaks with one voice. At least with respect to the impulse-coded output which is the main output for long-distance propagation of specific signals, most neurons, having but one axon, speak with one voice after integrating all that they receive.

The neuron doctrine can no longer be construed to mean that, when any part of the neuron is excited, an all-or-none wave spreads to all the rest. Nor is it possible to construe Cajal's ancillary doctrine, that neurons are dynamically polarized, to mean that in order to influence the cell body or impulse initiation, dendrites necessarily conduct impulses. And similarly, the synapse requires a new perspective. Beginning with the early workers who applied the cell theory to the nervous system—notably Romanes, the physiologist of medusae, and later Cajal and Sherrington—it has been doctrine to the present day that neuronal integrations are the result of the properties of synapses, a magnificent induction (Romanes, 1885, p.79) made even before the neuron doctrine was formulated. These properties were elaborated in prescient detail by Foster and Sherrington when they introduced the term synapse in 1897 (see Fulton, 1949, p. 55). But, while the synapse as we understand it today (Chapter 4) is no less complex or integrative than we have been taught all these years, we have seen that at least four other places in the neuron are also integrative—the extra-synaptic contiguous membrane, the soma or confluence(s) of receptive processes, the impulse-initiating locus, and the subterminal axonal arborization (Figs. 3.1, 3.4). But while the current of recent discovery emphasizes the fractionation of the neuron and the differences in function among its parts, nevertheless the neuron doctrine is compatible with these findings and still forms the synthesis of facts from anatomy, embryology, pathology, and physiology.

The **neuron doctrine may be stated** in the light of contemporary histology and physiology in this form. The structures principally involved in the more immediate and "nervous" activities of the nervous system are all part of cellular units called neurons. These usually consist of one nerve cell body with one nucleus and one or more processes (prolongations) variously branched;

they are separated from other such units by a cell membrane visible with the electron microscope. They are derived, in the typical case, each from a single embryonic cell called a neuroblast. All parts of a neuron depend in the first place on the integrity of the nucleated cell body for trophic maintenance, whatever else may also play a role; so that in a mass of fibers only those which have suffered damage to or severance from their cell bodies degenerate, while neighbors may survive. The neuron is commonly a functional unit by reason of convergence of its inputs to determine a single, pulsed output of propagated all-or-none events. But it need not act as a whole and its parts may act in different ways. It is commonly dynamically polarized by irreciprocity of spread from one part of the neuron to another or across the junctions between neurons. This is not different in essentials from Cajal's doctrine but recognizes exceptional cases and functional qualifications. It permits for example, a significant role for interaction between cells as well as within them by nonimpulse, hence gradable, forms of activity. It leaves room, vaguely, for important functions of glia, increasingly suspected to be more than passive, supporting, and nutritive.

The **synapse** (recapitulating Chapter 4) is still any functional contact between two neurons or by extension neuroeffector junctions. This disregards differences in mechanism of transmission, which become categories of synapses. (The word ephapse, introduced in 1942 by Arvanitaki, is reserved for artificial synapses, contra Grundfest, 1958, who uses it for electrically transmitting junctions.) The functional aspect of the definition makes it impossible to generalize how circumscribed or extensive a structure constitutes a synapse; usually a region of some histological complexity is minimally involved. The "contact" qualification excludes any interactions which may occur between neurons separated by some distance; the latter interactions are called "field effects." The "two neurons" requirement excludes any junction where it is not believed that a cell membrane separates cellular units. The definition does not depend on the properties of one-way transmission, delay, electrical inexcitability, fatigability, and drug susceptibility, but these properties are very common.

There being few if any purely relay neurons, the vast majority either (1) receive convergent input from several sources, or (2) receive input from only one source but temporally integrate that input, or (3) initiate activity as a result of some nonnervous input, as in receptors or "spontaneous" central neurons. These are the **three general classes** that comprise neuronal integration and it will be obvious that these classes embrace the length and breadth of the nervous system considered from the standpoint of what makes the neurons fire.

A. A Frame of Reference: Loci and Modes of Lability

The activity of individual neurons is determined by their sensitivity and their responsiveness. These are distinguishable properties: sensitivity measures the intensity of stimulation necessary to produce a threshold response; responsiveness measures the amplitude of response resulting from a threshold stimulus. More generally stated—independent of working only at threshold—sensitivity is measured by the intensity of stimulus necessary to elicit a given percentage of the maximal graded response. Responsiveness is measured by the amplitude of response for a given percentage of maximal stimulus (defined as that giving the largest possible graded response). We now turn to the main aspects of each, to establish a frame of reference within which we may examine the degrees of freedom permitting neuronal integration.

1. Eight forms of responsiveness

The absolute magnitude of response is in most cases measured as a change in electrical potential difference across the nerve cell membrane. Grundfest (1957a, 1957b, 1959) has reintroduced the useful term electrogenesis, meaning simply the production of an electrical sign (Fig. 3.1). But it should not be forgotten that response has other aspects, and certain forms of response—such as the production of specific transmitter substances at nerve endings—may have no electrical sign detectable by methods in use. We may divide the electrogenic forms of neuronal activity into seven and consider the release of transmitter substances as an eighth. The first four categories here distinguished are **prepotentials** and they are not all successive; any of the first three may inaugurate activity. But it seems probable that normally there is always a prepotential preceding and itself causing the spike or spikes. (See further, Chapter 3.)

(a) Generator potentials or receptor potentials are the transduced response potentials of sensory structures. They may be rapidly adapting or maintained (Fig. 3.2), according to the receptor. They are graded, and while in most cases known are depolarizing in sign, they are probably in some cases hyperpolarizing. When large enough and depolarizing, a generator potential brings on firing through the intermediation of **(b), (d),** and **(e).**

(b) Pacemaker potentials are gradually increasing depolarizations leading, at some critical threshold, to the triggering of a local potential or spike (Fig. 3.2). Either of these can bring on a restoration process which repolarizes the membrane to a high level, whence a new pacemaker potential can begin and a rhythm in the nature of a relaxation oscillation can result. Alternatively, the pacemaker potential is a smooth undulation, almost sinusoidal, and gives rise to local or spike potentials near its peak of depolarization but is not rapidly repolarized or indeed influenced appreciably by them. These forms of activity are descriptively distinct from **(a).** Pacemaker potentials occur in spontaneous, pacemaker centers like those in hearts (both neurogenic and myogenic) and in central neurons (for example, in the cortex, Phillips, 1956) as well as in receptors (for example, crayfish stretch receptor, Eyzaguirre and Kuffler, 1955a).

The frequency of a repetitive, spontaneous discharge is determined by the rate of pacemaker depolarization and the spike threshold. Each of these variables is modifiable by natural or artificial influences and the frequency alone is no measure of either. Frequency represents an integration of the various steady-state factors that act on the rate of depolarization and the threshold.

(c) Synaptic potentials are depolarizing or hyperpolarizing transient shifts in membrane potential which result directly from the arrival of impulses in a competent presynaptic pathway. They are thus transducer potentials of events in the specifically nervous environment of the neuron. Like all the others, except **(e),** they are graded. Synaptic potentials of usefully different amplitude, form, and sign occur in the same neuron, both as a result of different inputs and in consequence of lability of a given transmitter locus or response locus. (Lability does not deny reproducibility under some conditions, but asserts significant variability under other conditions, known or unknown.)

(d) Local potentials are graded internal responses which decrement with distance; they are apparently due to antecedent activity (generator, synaptic, or pacemaker) in the same neuron. They are labile and commonly trigger a spike at a critical amplitude. On these grounds they should be recognized as a distinct category even though the membrane mechanisms may be found to be the same as those producing spikes but without the regenerative, explosive character.

(e) Spike potentials are the all-or-none, propagated signs of nerve impulses. Being quantal and usually with a good factor of safety, they are not significantly labile normally and are not involved in integration except where they arise and terminate. They represent a specialization for conveying information over long distances with minimum loss and noise. Spikes are not properly regarded as digital signals because the information is at least ordinarily conveyed by the intervals between them; these are continuously graded and potentially noise-introducing. In practically all cases spikes are in the direction of internal positivity of the cell membrane, therefore initially depolarizing and then briefly reverse-polarizing. In a few instances (for example lobster muscle fibers, Reuben et al., 1960) explosive, all-or-none hyperpolarization has been seen but this is not yet known to be used as a propagated signal.

(f) Afterpotentials may be regarded as part of the spike-initiated event but because they commonly last long after the impulse has completed its function with respect to the next neuron (usually a good part of a second, sometimes minutes) and are highly labile and graded, they are better considered as separate, though dependent. They integrate past spike activity and milieu but with what significance is not well understood—perhaps chiefly in determining the next two events.

(g) The **activity of terminal arborizations** is perhaps commonly graded, so that the spike does not reach the endings unmodified. Little is known definitely and it seems likely that the opposite is true in many neurons. But the form of response in these portions of the neuron, as a function of all that has gone before, will be an additional variable of crucial importance in determining output (Wall, 1958; Eccles and Krnjević, 1959a, 1959b; Dudel and Kuffler, 1961).

(h) Release of transmitter substance in the case of chemically transmitting endings must be recognized as a process which could follow the preceding events with some graded or labile function and hence could contribute to integration (for example, see Thesleff, 1959).

Several of these forms of response characteristically occur in separate parts of the neuron— this is the **doctrine of differentiated loci.** Generally discontinuous forms of response, they are best regarded for the present as properties of distinct regions of the cell surface representing types of membrane. It is believed that the different kinds of membrane are sometimes distributed like a mosaic. The properties of these different response mechanisms are different, for example, with respect to the nature of the adequate stimulus. At least some of the synaptic responses are not excitable by direct electrical stimulation; all the others are normally activated by local current flow. The refractoriness following activity is quite different, in the cases of synaptic, local, and spike potentials and the conductance changes in the membrane involve differention specificities (see Grundfest, 1958).

The important feature at this point is that the neuron has several forms of activity in its several parts, permitting output to be a complex function of input because the coupling or **transfer functions** between successive forms of activity are often complex—for example, nonlinear but continuous, or discontinuous with a threshold (see Bullock, 1957a). Certain loci, notably that of spike initiation but also of pacemaker and local potentials, have been thought to shift sites in a given neuron under certain conditions—the intracellular localization is dynamic.

Each form of responsiveness can be regarded as **integrating at any moment the existing conditions** of many factors, including the milieu, both specific and nonspecific chemicals, external electrical fields, and preceding activity; other long-term factors must exist but their nature and locus are as yet obscure.

2. Four forms of sensitivity

One aspect of excitability, by which we mean sensitivity or the stimulus necessary to produce a given relative response, is the classical form measured by **spike threshold.** This is the critical level of membrane potential change necessary to trigger the initiation of a propagating impulse. It is a parameter which is concerned, in normal activity, only with a limited part of the neuron— namely, the locus of origin of the conducted impulse. At least for some neurons this is at or near the base of the axon; in others the spike origin appears to occur farther from the cell body; in vertebrate dorsal root ganglion cells it is in the periphery; possibly in some neurons it is in the soma. This parameter may also be of importance at certain branchings of the axon where the safety factor is low. The threshold is neither a fixed level of membrane potential nor a fixed change in membrane potential, though commonly it lies at about 50–60 mv inside negative, equivalent to a reduction of about 10–20 mv below resting level. It is subject to change with various conditions including fatigue, accommodation, excess depolarization, hyperpolarization, and unknown factors (see Chapter 3).

Figure 5.1. Nonlinear input-output curves for single units. **A.** The potential close to a stimulating electrode in a crab axon, expressed as a fraction of the spike potential (ordinate), measured at an arbitrary moment after application of a shock of strength given as fraction of threshold (abscissa). [Hodgkin, 1938.] **B.** Amplitude of the excitatory postsynaptic potential at different heights of presynaptic spike, altered at will by controlling the presynaptic membrane potential. [Hagiwara and Tasaki, 1958.]

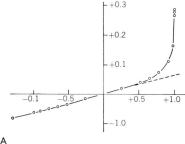
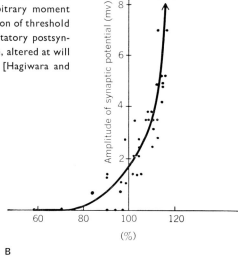

A second form of sensitivity, the first in the sequence of activation, may be called the **synaptic excitability.** This probably does not have a sharp threshold but a continuous curve relating amplitude of postsynaptic potential to amplitude of arriving presynaptic input. Hagiwara and Tasaki (1958) have recorded these quantities simultaneously with intracellular electrodes in the squid axon (Fig. 5.1). We may expect this curve to be typically nonlinear, as well as labile, according to the conditions and the preceding activity. In most neurons this excitability is not confined to so restricted a region of the neuronal surface as the preceding but is distributed over the whole region of the surface involved in synapses—in vertebrates, both dendrites and soma; in invertebrates, an as yet undelimited part of the neuronal shrubbery. The input to which the synapse is excitable may be electrical or chemical in different cases. Some synaptic membranes are inexcitable electrically; that is, they do not manifest any sensitivity leading to active response with electrical stimuli but are only displaced passively, as though current were an inadequate stimulus (see Grundfest, 1957a). Other synapses, like the first and third forms of sensitivity, are probably normally activated by electric current (Furshpan and Potter, 1959).

Third, if, as we concluded above, **local potentials** are a separate form of response, there must be a sensitivity thereof to the preceding potential, for example synaptic or pacemaker potential. In the branching terminals of a type of axon where only decrementing activity occupies the ending, we may count an equivalent form of excitability; the normal stimulus is the local current flow from the spike. There is little information on the behavior of this sensitivity but presumably it is labile and nonlinear.

Each of these three sensitivities can be considered a transfer function between normally successive, causally connected events. The transfer functions between the other phases of activity are too little known to permit significant statement.

A fourth form of sensitivity—and one still less understood—is that to **nonspecific fields of electric current,** steady or fluctuating, passing through the tissue. These can exert marked physiological effects, excitatory or inhibitory according to the polarity of the current flow. Quantitatively, the significant feature is that extremely weak fields are effective in modulating already on-going activity. The voltage drop of an effective field for the crayfish stretch receptor neuron, measured in the external medium across 100 μ, including the whole soma of a neuron (Fig. 5.2), can be 100 μv. This is smaller—by more than an order of magnitude—than the threshold voltage change that must be applied across the membrane of the cell with an internal electrode in order to stimulate. The voltage gradient across the soma membrane produced by this field is too small to measure with the methods used. Since externally imposed current that enters the cell must leave it, there can be no average membrane potential change. This leads to the important conclusion that the membrane **potential across a limited region of the neuron is critical** in controlling firing. The critical region may be extremely sensitive (some small fraction of a millivolt) and its excitability extremely stable in the case of highly rhythmic firing modulated by a steady field. Currents adequate to produce distinct acceleration or deceleration of background are apparently in the same range as physiologically available currents within nervous tissue, for example those due to brain waves. Gross fields of current through central nervous masses may cause coordinated, reciprocal postural movements, so there must be assumed an orientation of current sensitive neuronal structures, in opposite directions for antagonistic muscles (Hughes, 1952). Whether such electrical fields play a significant role in normal function cannot confidently be stated on the evidence available.

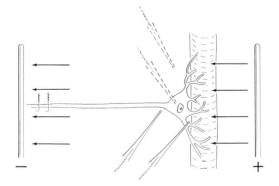

Figure 5.2. Diagram of the experiment to measure the voltage gradient near a neuron soma adequate to alter the frequency of firing. The cell is a crayfish stretch receptor. The strength of an imposed field is measured with microelectrodes on the dendritic and axonal sides, in the extracellular space. An internal electrode, referred to an external one *(dashed lines)* does not detect a measureable effect of the imposed current. But electrodes on the axon record a modulation of the maintained discharge.

In harmony with this idea and with the notions expressed by a number of authors (Bremer, 1951; Gerard, 1953; Fessard, 1956; Bullock, 1953a) is the pregnant possibility that, **even without impulses, one neuron can influence another.** Strong suggestions from study of masses of neurons (Libet and Gerard, 1941) and definite evidence at the unit level (Watanabe and Bullock, 1959) require that we reckon with communication among nerve cells, not only by pulsed forms of input and output with conversion to continuously graded forms between, but by the graded form of activity without the intervention of all-or-none events. The graded and low-amplitude forms of decrementally propagated activity may, even in the absence of spikes, modulate on-going activity in adjacent units. This possibility runs through all the sections above and below and increases by a large factor the possibilities for complexity of integrative relations among neurons.

The mechanism of subthreshold interactions may not be a field effect, for example, in the case analyzed by Watanabe and Bullock. They passed repetitive subthreshold current pulses through a follower cell in the lobster cardiac ganglion and observed synchronization of the pacemaker cell, whose soma is several millimeters distant. The pulses must be longer than about 50 msec and close to the natural frequency of the pacemaker. Withdrawal of the current-carrying electrode from the follower cell abolishes the effect. It seems likely that a cell prolongation carrying slow potentials but not brief ones is involved (see further, Chapter 4).

We can regard each steady-state sensitivity and **each labile transfer function as integrating** at any moment the present conditions of many factors: the milieu, specific and nonspecific chemicals, membrane potential, electrical fields, preceding activity (which may have left refractoriness or facilitation behind it), and longer-term factors in the genetic and developmental background (instincts, learning, "plastic" compensation— each of which see below). This integration is not necessarily the same as that which we noted above for responsiveness; the two cannot be identified as the same process, just as sensitivity and responsiveness cannot be equated.

B. The Degrees of Freedom: Intracellular Permutations

Each of the following sections presents alternative possibilities of response of single neurons. Together they represent the main degrees of freedom within the foregoing general framework.

1. The distribution over the neuron of types of cell membrane

If, as is probable in some neurons, the spike is initiated in the proximal segment of the axon and if synapses with excitatory input are confined to distal dendrites, that input will have only a modulating effect on the frequency of spike initiation and may be unable to fire a spike by itself. But if the spike originates in proximal dendrites or if the synaptic membrane is on the soma, input will more directly control output. Again, if there is a separation between spike-initiating locus and synaptic membrane—the one in the proximal axon, the other in several dendrites—there will then be opportunity for much more integration of different inputs, weighting them according to built-in or transient coupling functions between synaptic potential and spike, than if the two are close together as in the squid giant synapse.

Where **dendrites or receptive processes** are able to conduct excitation only by electrotonic spread, their role will be quite different from that where they actively propagate. Many dendrites apparently spread excitation only electrotonically, hence decrementally, even toward the soma. The spread depends on the duration and rate of rise of the dendritic activity and the distribution of membrane capacity (Rall, 1959). Thus dendrites acquire a highly special significance, far from that classically assigned to them as the processes carrying impulses toward the cell body. On this view they become the integrative structures par excellence, enormously hypertrophied synaptic areas, perhaps interspersed with nonsynaptic but local potential-generating membrane. The algebraic addition of the events in various branches depends greatly upon the geometry, the tapering, the twigs, the area of membrane whose time constant is changed by active response, the change in responsiveness with change in membrane potential. Dendrites behaving in this way are probably impotent to fire explosively and cannot be invaded by an impulse from the axon. Our understanding of dendritic function is certainly just beginning to emerge, as witness the recent "Symposium on Dendrites" (Ward, 1958). Bishop declares that "the dendrite is the most flexible, variable and adaptable of the three neuronal segments" (dendrite, soma, and axon). It seems probable that dendrites are the **most important structures** for the typically nervous functions of the nervous system. Historically the synapse has been regarded as the seat of labile nervous functions—fatigue, facilitation, evaluation, valving, learning—for the good reason that it was supposed the neuron responds throughout its extent whenever an impulse is initiated in it. Now, however, it seems more probable that intracellular sites are at least as important in integrating various converging inputs and auto-activity. Especially must the transition zones between different types of responding membrane be crucial, whether synaptic, local, spike, or auto-active.

Remembering the conspicuous histological differences among animals, particularly between higher invertebrates and vertebrates, with respect to the identification of axon and dendrites, we may expect a comparative physiology of these functions in the future (see Chapter 3). Indeed, within each animal we know such a diversity of form and arrangement of synaptic endings and dendrites as is unparalleled in other aspects of neuronal configuration and we may therefore look for important differentiation of function—a **comparative microphysiology,** as Fessard (1956) has called it. Indeed there is already much evidence of differentiation among dendrites and synapses in electrical and pharmacological properties, of patchiness of the cell membrane in respect of these properties, and of different incoming paths impinging on different parts of neurons (see Clare and Bishop, 1955; Tasaki et al., 1954; Purpura and Grundfest, 1956; Grundfest, 1957a, 1958, 1959a; Li et al., 1951, 1956a, 1956b; Bishop, 1958; Eccles, 1951; Cragg and Hamlyn, 1955; Tauc, 1957, 1958).

To return to specific alternatives, the boundary between spike-supporting and solely graded response-supporting regions of the neuron is now on one side of the **soma,** now on the other. Most cells studied heretofore permit the spike to enter the soma if not to be initiated there (for example trunk-muscle stretch receptors in crustacea, giant ganglion cells in gastropods, mammalian motoneurons). Others, like the large anterior cells in the lobster cardiac ganglion, do not, and the spike appears to an electrode in the soma as a 5–10 mv miniature deflection, smaller than some synaptic potentials; it is seen only as an electrotonically conducted fraction of the full spike, which originates out in a process and propagates cellulifugally (Fig. 5.3). The **relation between regions of the neuron is irreciprocal,** such that graded response regions can determine the initiation of impulses in other regions without themselves being fired. This preserves the membrane not only from the intense depolarization and overshooting reverse polarization of the spike, but also from the strong recovery repolarization which may tend to wipe out accumulated and integrated prepotentials (although not necessarily—some synaptic potentials last long after a spike has passed near by, Tauc, 1957; Hagiwara, 1958). Thus influx of various sorts, requiring time and repetition to build up, can control a repetitive output of quite different frequency and pattern.

Size of the soma or other region summating graded potentials is an integrative variable because it is one of the factors determining time constant; large cells can sum electrotonically and give a spike even after several hundred milliseconds, whereas small cells receiving input from the same source and stimulus may not be able to sum at all, as Tauc has shown in *Aplysia*.

A corollary is the possibility of two or more independent impulses in the same neuron. This has been observed (Fig. 5.3) and is believed to depend on the presence of two or more processes capable of acting like axons, each initiating its own spike, which cannot invade the soma and clash with the others.

The neuron is far from a perfect functional unit fired throughout its extent by each impulse arising in it. The classical view—that an impulse spreads throughout all processes—must yield in the face of evidence of regional activity. The firing of impulses is to be regarded as a specialization of one portion of the neuron: that mediating simple conduction. Great variety in integrative possibilities is available according to the spatial distribution of the several sorts of surface membrane over the neuron.

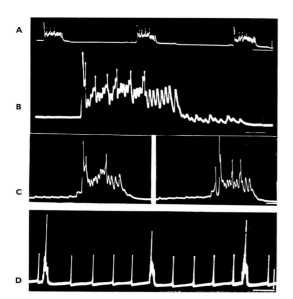

Figure 5.3. Small amplitude spikes and two independent spikes, seen in the soma. A microelectrode is inserted into cells of the crustacean cardiac ganglion of the large anterior follower type. Each heartbeat is represented by a burst of activity (3 such in **A**, 1 in **B**, 2 in **C**, 3 in **D**). Most of the activity consists of excitatory postsynaptic potentials, but small, 5 to 10 mv spikes rise from these. **A** and **B** show a single size of spike, **C** and **D** two different sizes. In **D** the smaller spike rises from spontaneous slow pacemaker potentials between the bursts. Calibration marks represent 500 msec and 18 mv for **A**, 110 msec and 4 mv for **B**, 200 msec and 7 mv for **C**, and 200 msec and 20 mv for **D**. [Bullock and Terzuolo, 1957.]

2. Excitation and inhibition

The consequence of input may be either the increase or decrease in the probability of output. Each may be brought about in more than one way. Either may result from blockade of the effect of the other input whenever that is present. Both inhibition and excitation can be the result of active responses of the synaptic membranes.

Inhibition may result from (a) some degree of refractoriness following a suprathreshold excitatory response or (b) following a subthreshold excitatory postsynaptic potential. The latter alternative provides a direct depression which is graded, has a short latency, can last several milliseconds, and does not require the discharge of an initial impulse from the inhibited cell or a special, purely inhibitory pathway. (c) Inhibition may result from weak electrical fields acting to reduce on-going activity already present or the likelihood of its beginning (see pp. 157, 262). (d) Excessive depolarization blocks impulses. This is not solely an abnormal result of experiment. Suppression of on-going repetitive discharge has been observed as a result of arriving depolarizing synaptic action—that is, ordinary excitatory postsynaptic potentials, when the summed depolarization is large (Fig. 5.5). (e) Inhibition sometimes occurs presynaptically; for example, crustacean excitatory neuromuscular junctions liberate fewer quanta of transmitter substance when the inhibitory impulse arrives at the right moment (Dudel and Kuffler, 1961). (f) Inhibition may result from impulses in specific fibers that cause an active synaptic response of increased conductance to potassium, chloride, or all ions, and hence usually a hyperpolarization (i.p.s.p.); this is the so-called direct inhibition. It is regarded by some authors as due to the specificity of the transmitter agent; by others as due to specificity of the responding patch of synaptic membrane. (g) Hyperpolarization is observed of a character not readily attributable to i.p.s.p.'s; perhaps there are several distinct phenomena in this category. Tauc in *Aplysia* (see 1960), Holmgren and Frenk (1961) in *Helix*, and others have noted large, slow, often cumulative hyperpolarization during or after stimulating a nerve (p. 227); changes with time and repetition can be complex and suggest long-term effects.

It is important to note that the observed effect of a given input **depends on the parameter used to measure it.** In the moderately complex case of a neuron firing in intermittent bursts, we may, for example, measure the interval between bursts, the number of spikes in a burst, the maximum frequency, the average frequency, the membrane potential, or other signs. These may not all change in the same proportion when a simply inhibitory input is imposed through a single axon; indeed some of these measures may change in the direction usually associated with excitation (Maynard, 1955), as seen in Fig. 5.4.

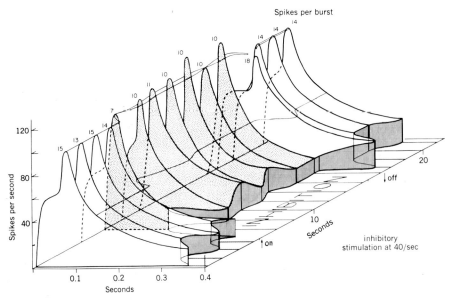

Figure 5.4. The disproportionate effect of inhibition on different measures of unit activity. Responses of a single cell of the lobster cardiac ganglion are plotted, showing each burst (representing one heartbeat) as a plane. Note the disproportionate effect of the single inhibitory axon on duration of burst, spikes per burst, the shoulder of the ascending phase, and even a reverse effect on peak frequency. Adaptation to inhibition and rebound are also shown. [Maynard, 1961.]

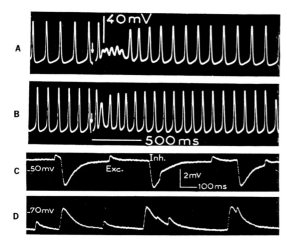

Figure 5.5. Some effects of the level of the membrane potential. Ongoing activity seen with an electrode inside a large cell of the isolated visceral ganglion of *Aplysia*. In **A** and **B** the cell is autorhythmic; at the *arrow* stimulation of a preganglionic nerve results in strong synaptic bombardment of the cell under observation. This is sufficient to block the firing by excess depolarization in **A** but only to accelerate briefly and almost block in **B. C.** Without applied stimulation this cell exhibits e.p.s.p.'s and large i.p.s.p.'s resulting presumably from spontaneous activity in other presynaptic interneurons. **D.** Using a separate polarizing electrode inserted into the same soma, the membrane potential is raised to 70 mv and the i.p.s.p.'s now reverse to depolarizing deflections. [Tauc, 1958.]

As already set forth in Chapter 4, excitatory input usually produces a depolarizing synaptic potential (e.p.s.p.) and inhibitory input a hyperpolarizing synaptic potential (i.p.s.p.); see Fig. 5.5. Successive arriving impulses may progressively modify the membrane potential and favor activity or silence, respectively. The amplitude of the synaptic potential can be changed by presetting the **membrane potential**: depolarizing synaptic potentials become smaller as the membrane level is present at lower potentials (less internal negativity) and hyperpolarizing synaptic potentials are reduced by raising the membrane potential. This effect is much greater for inhibitory potentials. The latter are reduced to zero at a membrane potential somewhat above the usual resting level and at still higher levels are actually reversed in sign (Figs. 5.5 and 4.11). The reversal level for excitatory input is much lower than the resting level. In some cases reversing the sign of inhibitory synaptic potentials also reverses the sign of the effect, that is, actually excites, but in other cases this is not so (Hoyle and Wiersma, 1958). There is evidence that the membrane potential change may in some cases not be paralleled by changes in **excitability**; for example, hyperpolarizing end plate potentials in insect muscle do not apparently inhibit (Hoyle, 1957). The reverse is also true: in some cases excitability can change without any change in membrane potential (see "Negative and positive aftereffects," below).

Repetitive synaptic activation manifests additional variable properties. Where there is accumulation of potential change by summating

Figure 5.6. Cumulative and long-lasting forms of inhibition. **A.** Inhibitory hyperpolarization results from repetitive impulses arriving at a lobster cardiac ganglion cell through a specific pathway, even though individual deflections are depolarizing. [Terzuolo and Bullock, 1958.] **B.** e.p.s.p.'s summate and then, owing to increase in their repolarizing phase, accumulate a hyperpolarization—although again the individual deflections are initially depolarizing. Visceral ganglion cell of *Aplysia*. [Tauc, unpublished.] **C.** A slow inhibitory hyperpolarization continues to grow even after cessation of the causal input in a ganglion cell of *Aplysia*. [Tauc, unpublished.] **D.** A single shock to a certain presynaptic path in *Helix* results in interruption of spontaneous spiking and hyperpolarization for several seconds. [Tauc, 1955.] **E.** Long, slow hyperpolarization can be evoked in *Helix* even without discrete synaptic potentials. [Tauc, 1955.]

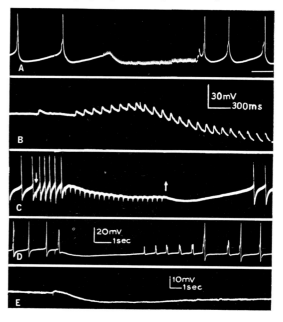

synaptic potentials, the initial phase of an inhibitory synaptic potential may be depolarizing while the slower and accumulating falling phases shift the membrane level inexorably toward a hyperpolarized potential (Fig. 5.6). The membrane level at which they reverse is different for the two phases (Otani and Bullock, 1959). Tauc (1958, 1960) has shown that a burst of excitatory synaptic potentials after a period of augmenting depolarizing and excitatory effect similarly can develop an accumulating hyperpolarization and consequent inhibition. This case is remarkable furthermore in that hyperpolarization continues to grow for several tenths of a second after the presynaptic input has ceased, only slowly yielding to a reduction in the polarization toward the previous level (Fig. 5.6).

There is a wide difference among neurons in the **tendency to repetitive discharge.** This is most obvious under stimulation by maintained current but is also apparent with transient prepotentials (Fig. 5.7). The factors responsible include refractory period, accommodation, and the time constant of the system; no one of these alone can account for the observed frequency (Fuortes, 1958).

Interaction of synaptic potentials from different presynaptic fibers can result in effects which are not the simple algebraic sum. There is limited information on this phenomenon but it provides the possibilities of additional complexity in integration (Tauc, 1958 in *Aplysia* cells); see also Fig. 4.12. Acceleratory presynaptic impulses in the lobster cardiac ganglion can not cancel the effect of inhibitory presynaptic impulses to the same ganglion; even maximal stimulation frequency of acceleratory fibers does not shift the threshold frequency for an inhibitory effect of stimulating the inhibitory fiber or accelerate more than trivially a partially inhibited heart beat. Clearly the interaction is highly asymmetrical. Another aspect of the same feature is the slower build up, slower adaptation, and the longer aftereffect of acceleration (Maynard, 1955; Florey, 1960).

Presynaptic inhibition may perhaps be the same as the phenomenon called remote inhibition seen in the depression of effectiveness of particular dorsal root inflows in the mammalian spinal cord by the presence of certain parallel inflows. It is believed to be exerted on or near the terminals of the sensory axons and may represent therefore a kind of axo-axonal influence or a reverse synaptic transmission from dorsal horn cells to afferent fibers (Eccles, 1959; Eccles and Krnjević, 1959a; Frank, 1959; Dudel and Kuffler, 1961). The excitability of certain sensory terminals is known to be selectively altered by specific input in the sensory nerves (Wall et al., 1956).

A principle has been enunciated by Eccles (1957) to the effect that a given incoming fiber which is excitatory to some cells must relay through an additional, **intercalated neuron in order to exert an inhibitory effect** upon others. This is based on several cases in the spinal cord (contra Retzlaff, 1957). We may question the generalization of this to a principle, even for mammals. In several suitable cases analyzed in the crustaceans and gastropods, it could be shown that one and the same incoming fiber, called inhibitory because of its effect on the heart rate, inhibits only certain cells while exciting others. The intervention of internuncial cells could be excluded on the evidence (Terzuolo and Bullock, 1958; Tauc, 1958, 1960).

Inhibition has been found clearly manifested

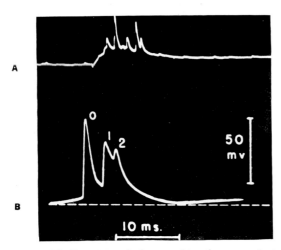

Figure 5.7. Repetitive discharge arising from transient prepotentials. **A.** A single long synaptic deflection in *Helix* causes several spikelike or local potentials. [Tauc, unpublished.] **B.** A crayfish stretch receptor occasionally shows multiple deflections with short intervals. [Eyzaguirre and Kuffler, 1955.]

as low as the coelenterate nervous system (see Chapter 8). The suggestions that it becomes relatively more important with cephalization and that in its lower forms it is likely to be a total, nonspecific action, becoming differentiated with further evolution, remain for future evaluation.

3. Facilitation and antifacilitation

When arriving input is in the form of repetitive nerve impulses, the result may be facilitation or antifacilitation or neither. It is important to recognize that these processes, like those in the next section (4), manifest integrative events of great flexibility which are **not indicated by the membrane potential.** They are events in the domain of excitability or responsiveness and are invisible until the response to a test stimulus shows where they stood at that moment.

Facilitation is a name for an unseen process as a result of which successive visible responses or increments of response are larger. Antifacilitation is said to occur when they are smaller. **Temporal summation** is the cumulative increase in height of responses, visible or assumed, whether the successive increments are facilitating or the contrary or equal. It is thus a more general term and can include facilitation on the showing or the assumption that the process underlying the augmenting responses is an accumulation of something at a lower level. At each level of observation—intracellular potentials, brain potentials, muscle tension, and behavior—augmenting successive responses may be called facilitation, though the basis may be a nonfacilitating summation of an antecedent, not visible at that level. For example, diminishing but summating synaptic potentials might reach threshold and cause a spike muscle twitch. With grosser recording this appearance of response on the second or later stimulus would be called facilitation. Temporal summation requires that the successive responses last longer than the intervals between them; facilitation may describe either records of responses which are at that level of observation completed, the record having returned to its base line, or response increments arising out of their uncompleted predecessors. These definitions are more restrictive than some usage common in the literature, but follow well-established precedent and make useful distinctions. Antifacilitation is operationally equivalent to a relative refractoriness or depression but the term is used (Hagiwara and Bullock, 1957) to indicate that it occurs under conditions which cannot be regarded as excessively stimulating, either by abnormally high frequency or by too long a duration. It is encountered in rapid response mechanisms which must not dominate the activity for too long.

The same neuron can exhibit facilitating synaptic potentials to some presynaptic influx, antifacilitating synaptic potentials to other influx and synaptic potentials which neither augment nor decline to still other influx (Bullock and Terzuolo, 1957). Of importance is the **separate course of change in different places** or processes; for example, besides the change in synaptic potential amplitude, which is different for different incoming pathways, the firing level may change so that larger or smaller prepotentials are required to trigger a spike. According to current interpretations, a presynaptic locus in axonal terminals is involved in at least some cases, but the postsynaptic membrane is no doubt capable of this change in some cases also. It is not easy to say whether the change is in excitability or responsiveness.

Facilitation due to prior activity in the same presynaptic pathway as the test impulse may be called **homologous. Heterologous facilitation** and depression are also known, in which activity in one presynaptic pathway alters the size of a postsynaptic potential evoked by another (Wiersma and van Harreveld, 1939; Fessard and Tauc, 1958).

Like other measures of input-output relations in nerve cells, facilitation is commonly **nonlinear** with respect to different frequencies of input. Separate processes of different time scale falling under the heading of facilitation may be observed, even in the same cell—for example, **early and late facilitation.** The augmenting response may be brought on after a period of intense activity and is then called **posttetanic potentiation;** this also may have separate early and late forms. Little studied, but perhaps of importance, is the phenomenon of **facilitation of facilitation,** where the course and degree of augmentation is itself enhanced by repetition. While some forms of facilitation last only milliseconds, others are reported in simple preparations lasting hours and in one case seven days (Wilson, 1959). An upper limit for neurons cannot be given, for the preparations and reponses grade into complex and behavioral ones and hence into learning and memory.

If facilitation rises rapidly to an early maximum and then decays rapidly, as in some crustacean neuromuscular junctions, an interesting property results—a strong **sensitivity to temporal pattern** of incoming impulses (see "Input and its interpretation," p. 273). Thus the response to 10 impulses per second arriving in five closely spaced pairs may be up to 5000% greater than that to the same number per second, evenly spaced. In other nearby junctions on the same muscle fiber, facilitation rises and decays slowly and there is no such pattern sensitivity. The consequences of varying the amounts and time courses of facilitation and antifacilitation can thus be dramatic. Temporal pattern probably is important in many central synapses but direct experimental demonstration is still needed.

As a consequence of differing balances between the processes tending toward excitation and those tending toward depression, there are some neurons whose **maximum freqency of firing** is very high—well above 1000 per second—and others which can only fire at low frequencies.

Accommodation is used here as a hysteresis of excitability, a gradual return to initial excitability during maintained depolarization or hyperpolarization. Possibly related to these other processes, at least phenomenologically, it is highly likely that the underlying causes are multiple and not uniform for all cases. Accommodation at junctional regions has received little attention, but is probably more important than in nerve fibers, where a true rise in firing level of the membrane with maintained or slowly rising depolarization has been denied in single fibers (see Chapter 3). A number of observations indicate that the threshold change for initiation of spikes by controlling the membrane level of the soma is at first lowered with a maintained depolarization and then after many milliseconds gradually rises.

Similarly the threshold is at first raised by hyperpolarization but recovers while the membrane level is maintained (Kolmodin and Skoglund, 1958; Tauc, 1958; Otani and Bullock, 1959). Tauc found the impedance of the whole soma slowly recovers to its original level during hyperpolarization which at first increased it.

Excessive depolarization—several tens of millivolts—can block firing and this type of suppression can even be induced by synaptic potentials of the "excitatory" type (Tauc, 1958); see Fig. 5.5.

4. Negative and positive aftereffects

There may be no appreciable aftereffects or there may be a negative or a positive aftereffect or both in sequence. A positive aftereffect is represented by afterdischarge following the cessation of an excitatory input or by afterinhibition following cessation of inhibitory input. Negative aftereffect is represented by rebound inhibition after excitatory input or by rebound discharge after inhibitory input (Fig. 5.8).

These tendencies vary greatly from cell to cell; they can be quite strong. Like the preceding group of phenomena, these show the complex and integrative changes in probability of firing that are not reflected necessarily in the membrane

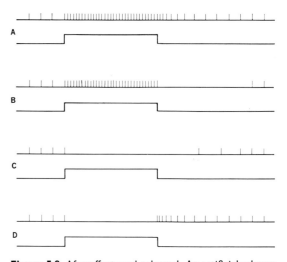

Figure 5.8. Aftereffects and rebound. An artificial scheme showing positive aftereffects **(A, C)** and negative aftereffects **(B, D**; rebound) following excitatory and inhibitory input. Duration of input marked by the lower trace.

potential but are owing to "invisible" changes in excitability and tendency to auto-activity. (Often aftereffects are correlated with slow membrane potentials.) A strong positive aftereffect may bring on a period of rebound, so that one or more complete cycles of oscillation are sometimes seen.

Even when there is little or no on-going activity to inhibit, the arrival of a short burst of inhibitory impulses may in certain preparations be followed by rebound discharge, and repetition of the inhibitory burst can elicit periodic rebound activity far greater than the activity that was going on before inhibition. This has been called **paradoxical driving** (Maynard, 1955). The bursts of rebound activity may themselves be sufficient to bring on compensatory silence and this in turn another phase of activity, as though a damped oscillation had been set up; it is possible to observe three or four well-formed bursts of activity in succession.

5. The alternative effects of milieu

Central neurons have been found which respond to or are influenced in excitability by pressure (Alanis and Matthews, 1952), oxygen, carbon dioxide, hypo- and hypertonicity (Andersson, 1953; Kerkut and Taylor, 1956; Cross and Green, 1959), temperature and light (see p. 163), and a number of hormones (Faure, 1957; Ozbas and Hodgson, 1958; Sawyer, 1959), to mention agencies which may normally act in vivo. In addition, changes in ionic environment, in pH, and in a wide variety of organic compounds exert stimulatory or depressing effects and may in some cases be of significance in life. Many of these actions are probably equivalent to sensory reception but they can be regarded as integrative if the same neuron adds this type of exciting or inhibiting effect to other forms of input to determine its output. At the neuronal level little is known about this type of interaction, but it is probably an area in which important advances will soon be made.

6. Movements of parts of neurons

It has frequently been proposed that neurons respond to some normal events by meaningful movements of axonal or dendritic terminals or even the cell body. This certainly occurs in phylogeny, in ontogeny, and in tissue culture, but the mechanisms are not necessarily similar. Certain kinds of movement, especially of the cell body in the direction toward which the principal dendrites extend, have been described under the name neurobiotaxis by Ariëns Kappers (1929). Other authors have suggested a phenomenon of chemotropism to embrace instances in which it is supposed that the nerve fiber grows along a chemical gradient. Embryological processes do not concern us in the present connection, but the persistent suggestion that terminal processes or synaptic connections are **anatomically labile during normal function** must be mentioned (Wiedersheim, 1890; Weber, 1949; J. C. Eccles, 1958; R. M. Eccles and Westerman, 1959; Dercum, 1922). Speidel (1941) demonstrated all stages of terminal degeneration, growth, and regeneration in the normal adult central nervous system, using silver impregnation methods. Clear evidence that such movement is functionally meaningful in the normal experience of the cell has yet to be obtained, but it would appear that minute alterations in topographic relations, dimensions, distances, or numbers of twigs would have pronounced effects, so labile have been the transfer functions so far considered.

It is clear that already at the neuronal level there are many properties and degrees of freedom which in combination and permutation can provide for an almost unlimited degree of complexity. The evidence detailed in the systematic chapters permits the conclusion that in respect of the existence of these integrative parameters of single nerve cells, processes, and junctions, the invertebrates, even **lower phyla, are on an equal plane with the nervously most advanced animals,** though perhaps not in their utilization in masses of cells. There is a wider spectrum of development of integrative properties among the neurons of a given species than we can with assurance recognize between major groups. So great are the possibilities for lability, variability, and complexity of transfer functions, even at the neuronal level, the problem of extrapolating upward changes its complexion from that of a few years ago—"Can we get really complex behavior from large numbers of all-or-none units?"—to quite a different one. This is: What constraints on the possible permutations are there and how is consistent performance ensured in face of the possibilities for inconsistency?

III. INTEGRATION AT THE LEVEL OF ORGANIZED GROUPS OF NEURONS

Turning from the integrative properties of neurons as units to the naturally occurring organized groups of neurons in ganglia and central nervous systems, we recognize immediately an enormously greater complexity. Here are ranged functional clusters of a very few cells, a spectrum of more and more elaborate integrative masses and the most complex systems we know of in nature. This great spread embraces, we believe, the remaining secrets of the physiological basis of behavior, an awe-inspiring problem, only slightly ameliorated by the article of faith that the study of systems in various animals and situations will shed light on the most complex instances. In this and the next section but one we will draw a blurred line between two levels of function of organized groups of neurons: first, those which have been studied by more analytical, physiological means concerned with manipulating and measuring the activity of components; and second, the higher levels of achievement of nervous systems which have been studied, sometimes no less analytically or experimentally but by manipulating the whole animal and observing its behavior. These latter levels are dealt with in "Deductions from the study of behavior," p. 323.

One possible order in which we might approach the **principles of organization of systems of neurons** is the following. First we might compare sequences of neurons on the basis of the plan or structure. The simplest (a) are chains of elements that accomplish a **transformation**; one input line leads into the integrative element and one output line leads away, after alteration in the signal dependent on the transfer function. (b) Then there are systems showing **convergence** of two or more input lines; the output is determined by weighting and mixing inputs, by summation or multiplication, linearly or nonlinearly. (c) The case of **divergence** may be trivial, as when several output lines carry identical signals, or it may be a significant new class, as when output lines from the integrative element, whether single neuron or loop, carry different signals. (d) Mixtures of the above can be called **networks**. (We reserve the term nerve nets for a special class, defined on p. 462.) Second, we might contrast neuron circuits on the basis of dynamic properties. One dichotomy really connected by a continuous spectrum, distinguishes (e) systems that passively wait for input—complete **reaction systems**; these comprise an input filter of greater or less complexity, some sort of central, internuncial connections, and a constellation of efferent elements. The range is enormous—from simple independent effectors to complex, hierarchically inhibited reflexes. (f) In contrast are systems that include inherent automatic activity or **spontaneous temporal pattern** (other than random). These range from heart beats to locomotion of central origin and perhaps much more complex levels. (g) Combinations of the last two are probably the rule in higher systems. The continuity between them lies in considering more and more simplified inputs adequate to set off reaction systems and more and more specified steady-state conditions necessary to maintain spontaneity. The two are also similar in that the central nervous system, with its complex circuitry and transfer functions at each neuronal integrating locus, supplies the principal body of information for the behavior observed.

A third alternative and the one chosen for this section is to subdivide the problems into afferent, efferent, and associative components. The first are concerned with the input and its interpretation, the second with the output and its control. The associative are those which cannot be readily designated as either parts of afferent pathways or centers. Note that internuncial neurons or interneurons may be clearly in an afferent system or in an efferent system and are not all associative in the present meaning.

A. Input and its Interpretation

The following principles apply to the input to any functional group of neurons; most of our relevant information, however, comes from study of the input from sensory receptors to the central nervous system. The principles discovered by sensory physiology that are most basic to the general problem are: (a) information comes to any center via neurons, therefore discrete fibers, which may be considered as individual lines of communication, and (b) apparently all this information is carried by nerve impulses which, being discrete all-or-none unit events, may be considered as individual elements of a code.

1. The parameters of the code

As a consequence of the second principle above, the code is pulsed (by quantal events of insignificant duration relative to ordinary messages). The evidence at present is against meaningfully different kinds of impulses in the same nerve fiber. The code is sometimes said to be digital and binary; strictly it is not, since the interval between impulses is graded and a source of noise. The **number** of impulses (duration of a series) and their **frequency** comprise the major parameters within each line; these are analog variables. Occasionally it seems that frequency is not used to convey any significant information; some afferent fibers merely discharge at a fixed frequency as though signaling only "the stimulus is present" and not its intensity. It is found in certain afferent fibers that a given stimulus adds a constant number of spikes per second regardless of the level of background frequency and hence of the percentage change in frequency.

Temporal relations between lines convey important information in many cases; for example in distinguishing signals from noise, in marking the time of onset, in analyzing frequency of vibration, flicker or low-pitched sounds.

The question whether the nervous system employs **temporal pattern** in its code, defined as any

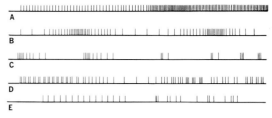

Figure 5.9. Temporal patterns. Diagrams of types of sequences available for pattern coding of impulses in single lines. **A.** Long maintained, constant, or slowly changing frequencies can signal steady or slowly changing states, not divisible into messages. Drawn to emphasize that this grades into **B**, in which slowly changing frequency signals events rather than states and hence can be regarded as divided into messages. This again grades into **C**, in which the messages are still more discrete and the code may utilize both the number of impulses in a message and the general form of the frequency change during a message. This grades into **D**, in which the microstructure is considered to be significant. Unequal but regular and meaningful intervals such as doublets or triplets of spikes may carry information different from the same average frequency with a different ratio of intervals. This is drawn to show different ratios and the gradation into regular spacing. **E.** Not a distinct category, but a feature of importance in each category, is the tolerance of irregularity or random fluctuations of interval or the significance of regularity of rhythm as such; here extremes are drawn, such as are commonly seen. The random spacing at right suggests that only average frequency, integrated over some time, is utilized; if true, the high regularity at left has no significance.

nonrandom, specified sequence of pulses, may be subdivided as follows (Fig. 5.9). (i) One class of patterned spikes has considerable sequences with unchanging or slowly changing intervals (pure or gradually changing frequencies; "pure tone patterns"). (ii) A second class has recurring abrupt changes in interval (on, off, increase, or decrease) occurring within a period of the order of an interval. The second class can be further divided into (iiα) a subclass in which the pattern occurs once for each recurrence of a given discrete message (burst; "word or phoneme pattern") and (iiβ) a subclass in which specified spike sequences recur frequently during a unit message (for example, series of spike doublets or triplets; "harsh tone pattern"). The first class and the subclass iiα ("pure tone patterns" and bursts or "phoneme patterns") are used commonly by nervous systems (for example, see Wall and Cronly-Dillon, 1961). The second subclass (iiβ; series of doublets or the like) would add greatly to the capacity of each line to carry information; evidence is still meager as to its employment by animals. Wiersma and Adams (1950) found a pronounced difference of effect between doublets and evenly spaced stimuli of the same total number per second in certain neuromuscular junctions of crustaceans and little or no difference in others. Modalities with high temporal resolution (optic, acoustic) follow stimuli up to moderately high frequencies and, in some special cases, to quite high frequencies (for example dipteran optic, homoiotherm acoustic systems), and for this reason probably do not ordinarily use this type of pattern as a code up to these frequencies. However, in the case found by Arden in rabbit retina (personal communication), where there is a background discharge of the same average frequency as that during a series of light flashes which evoke grouped bursts, possibly the pattern is useful; certainly the fact that separate lines come into phase must be a prime factor. Refractory periods set an upper limit, and play or unreliability of intervals between impulses sets a lower limit to the number of impulses that can make up a significant, coded bit of information. On the present evidence it must be said that pattern coding, in the sense of subclass (iiβ), is probably not a significant feature of afferent signals. But the **number of lines and labels on the lines** or nerve fibers activated are clearly significant parameters.

Signals of the sort used by the nervous system encounter several **sources of ambiguity.** As an example, reference may be made to the discussion of the crustacean statocyst on p. 1027.

2. Analysis by labeled lines and spatial representation.

As a consequence of the first principle, (a) above, and of the stable anatomical connections of the organized central organ, we obtain the possibility of analysis by labeled lines (Fig. 5.10). That is, the central organ distinguishes quality

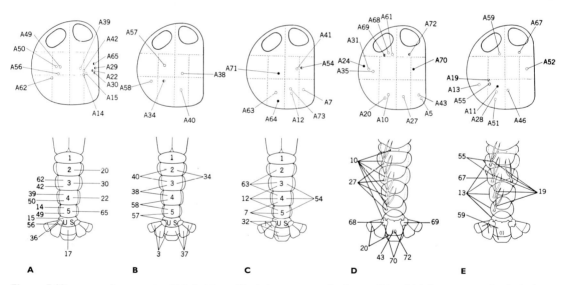

Figure 5.10. An example of labeled lines. Single interneurons in the crayfish, which fire upon mechanical stimulation in the periphery, are mapped for the area of the skin to which they are sensitive. The upper diagrams show the positions of these units in a cross section of the circumesophageal connective; the lower diagram shows the zones of the periphery whose stimulation each neuron responds to. **A.** Dorsal side of abdomen with areas innervated by interneurons that cover only small dorsal parts, limited to single segments. **B.** The same, indicating areas innervated by interneurons covering areas of two segments. **C.** Interneurons responding to touch of dorsal hairs of three adjacent abdominal segments. **D.** Ventral view, indicating joints to which different interneurons respond. **E.** Ventral view, indicating areas of some of the interneurons responsive to touch of hairs. [Wiersma and Hughes, 1961.]

of stimulus by source of impulses, this line or group of lines having the meaning "optic stimuli," that one "acoustic stimuli," and so on. (The phrase "having the meaning" is an abbreviated form of the conclusion from experiment that this line mediates responses caused by the respective form of stimulus.) Grossly different types of stimulation, often called **modalities** after the different qualities of sensation they produce in man, are certainly distinguished in this way, although it is possible that this is not universally so. Some modalities in some animals appear to be normally carried by mixed lines, or with an ambiguous label ("cold or touch"; see p. 867). An increased number of modalities implies that an increased diversity of specific sensory mechanisms must have evolved. An important task for the future is the unraveling of the labels as a function of the evolutionary stage.

Even **within a modality,** labeled lines play a large role, for example, in the analysis of pitch, color, specific chemicals, and locus of mechanical stimulation. Ambiguity exists in varying degree but, according to the usual interpretation, is compensated for by central integrative processes which compare many lines of different, overlapping ranges. Wiersma (1952, 1961a) has found afferent interneurons excited by touch to a small spot on a leg of a crayfish, others to the general region of the leg, others to the whole leg, to several legs, to all legs of that side, even to all legs of both sides. Kolmodin (1957) found second-order central neurons in the cat which respond to proprioceptive stimuli from one muscle group, others to synergists from several neighboring joints, to parallel antagonists, to antagonists at several joints, even to both proprioceptors and exteroceptors; among each type of second-order neuron some were excited, others excited and inhibited according to the source. Such a series of **overlapping lines** can signal messages of different degrees of specificity and, by combinations, of some degree of complexity and presumably with great economy in number of lines (compare Williams, 1958; Landahl and Williams, 1958). This is probably a highly developed feature of the arthropod nervous system.

Lines for different modalities and also lines with local sign within modalities can be followed to at least third order, and in the case of the optic system in both invertebrates and vertebrates, to fourth-order neurons before losing their labels. Nevertheless, at some of the relays between orders of neurons **integrative processes** take place, so that even before a line reaches a level where it is integrated with other modalities or where output is formulated, and it therefore ceases to be afferent, its label changes—that is, its meaning or representation is altered. Just what alteration or integrative achievement occurs at the several relays is one of the relatively nearer frontiers of investigation and one where lower animals are likely to be peculiarly valuable.

One sort of afferent processing is inferred in cases like the single modality, where both locus and intensity of stimulation of the sensory surface (for example skin or retina or cochlea) are signaled in a set of lines. The central analyzer can not depend on labeled lines to tell locus and on impulse frequency to tell intensity, because of overlapping receptive fields of individual lines, each having a more sensitive central region and less sensitive—or quite differently sensitive—peripheral region. Each line can not tell whether it carries signals from a weak central stimulation or a strong peripheral one within its receptive field. We assume that the analyzer must depend on a **formula of labeled lines and a profile of their frequencies** to extract locus and intensity (Fig. 5.11). This is accomplished in some systems with high resolution relative to the field of and the discrimination of any one line.

Some systems go beyond this to achieve **contrast enhancement of edges** or discontinuities by using the principle of providing an excitatory focus with an inhibitory surrounding. Second- or higher-order neurons are excited by input from a central region of its whole receptive field but inhibited by input from the peripheral regions of the field (for example touch areas of units in postcentral cortex of monkey, retinal areas of certain retinal ganglion cells, and units in striate cortex of cat). Two-point tactile discrimination seems to involve a temporal alternation in activity of two overlapping groups of cortical cells which accentuates, not a spatial summation, but a reciprocal refractoriness, similarly steepening the gradient between closely adjacent areas. Mountcastle et al. (1957) discovered many adjacent microcolumns in the somatic sensory cortex, each column of cells representing a locality and a modality. The intermingled but distinct modalities are defined by electrical response to joint movement or to skin pressure or to hair deflection. Our ability to localize somatic stimulation rests on a map of excited and surrounding inhibited clusters of these specific columns (see also Barlow et al., 1957; Powell and Mountcastle, 1958; Ratliff, 1961).

More complex changes in the labels on lines occur as biologically meaningful aspects of the sensory influx are extracted; we may say that messages are being transformed or abstracted (see below, "Pattern recognition," p. 279).

The peculiar architecture of the cortex of the cerebellum strongly suggests another type of spatial analysis of input: a kind of **toposcopic display** based on graded, decrementally spread activity in complex, interlacing arborizations oriented at precise angles to each other.

3. Differentiation among modalities in central effect

Besides the peripheral differentiation between modalities such that each has its meaning for the kind of stimulus which is adequate to initiate its signals, there is a differentiation in central effect. It is as though the different signals have different values with respect to their insistency or their **compellingness**. In mammals it is possible to define a system of neurons representing labeled lines for a variety of harmful environmental events—the nociceptive system, or the pain system—after the sensation known to man. This form of input has a dominance over others in central effect. In many vertebrates it is thought that olfactory influx is also peculiarly compelling, as measured not only by immediate dominance, but, for example, by the ease and longevity of the associations it can establish. At the other end of the scale it is not so obvious which modalities have least effect. In the terms of this hierarchy it seems likely that the steady-state receptors—of posture, position, temperature, pressure, to name the better known—belong at the bottom, though by another measure they outdistance the intense but briefly acting alarm modalities—as the tortoise did the hare—namely in over-all control, day in and day out.

At the present stage of knowledge very little can be said in the way of generalizations about the **evolution of the modalities** and their differen-

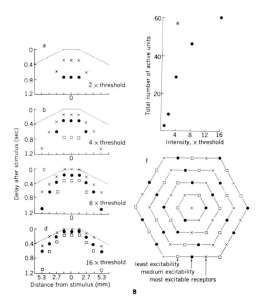

Figure 5.11. Coding intensity and locus of a stimulus. **A.** Two types of "coding": frequency (left) and spatial (right) representations. Impulses are shown traveling upward from receptor ends to central nervous system. **B.** The spatial representation as found in tactile afferents from the cat's foot pad when excited by mechanical pulses at one point; diagram based on quantitative experimental data. Assuming a number of equidistant receptors of 3 levels of excitability (f), a stimulus twice threshold will give the number of impulses at the distances from the stimulated point and at the time after stimulus shown in (a). The dotted line is the latency due to the traveling time of the mechanical wave in the pad. Stronger stimuli bring in responses shown in (b) to (d). The total number of impulses at each stimulus strength is given in (e). [Courtesy J. A. B. Gray.]

tiation. The direction has been from few to many, from general to subdivided, and probably from mixed to pure. Already in the coelenterates there are several clearly distinct types of receptors, and we can not place an upper limit on the number owing to the difficulty of recognizing functionally differentiated but not anatomically distinct receptors. This is all the more important because, while at first glance one might expect that the evolutionary level would also be indicated by total number or density of receptors, the more recent histologic and physiological findings —for example in coelenterates and annelids— show a very large population of primary sensory neurons. In man the total number of afferent nerve fibers entering the central nervous system has been estimated to be about five times that of the efferent fibers leaving. We have no figures for most groups of animals, but it seems safe to estimate from a variety of data that in annelids and arthropods the figure is higher by at least a factor of one hundred. This probably means not a relative decrease in number of afferents in man, but an increase in number of efferents. Multiplication of sense cells appears to be simple and primitive but establishing labeled lines is more difficult and apparently only slowly emerges in evolution.

There is another interesting form of differentiation in meaning and central effectiveness— within a modality. Drawing again upon vertebrate experience, in a number of cases it has been found that there are **thick and thin afferent** nerve fibers serving the same sense organ. The thin fibers come from receptor units of lower threshold to physiological stimuli, lower rate of adaptation, lower maximum frequency of discharge; the thicker fibers come from elements of higher threshold, higher rate of adaptation, and higher maximum frequency of discharge under strong stimulation. Further, the thin fiber receptors are more prone to continuous, background discharge and often they are more anatomically ramified so a larger number of sensory terminals converges into one fiber (Bullock, 1953b). Thus in the lateral line of fish, stretch receptors in the lung in mammals, touch receptors around the cat's whiskers, and in the trunk muscle stretch receptors in crustaceans, thick and thin afferent fibers travel side by side and carry signals of different meaning. Zotterman (1939) has expressed the opinion that the thin fibers, at least with respect to certain kinds of afferents, are disproportionately more influential in determining reflex response than the thick. This needs to be confirmed in other situations for it is highly significant as an example of another integrative mechanism, providing as it does over and beyond the code common to all lines, for an **evaluation of input** in determining output, an evaluation which is not only different according to the stimulus, but even for the same stimulus, according to the line—thick or thin—over which it arrives.

4. The requirements of an analyzer

These considerations dictate already much of the design of the central integrative receiving stations which, as noted, are distributed along the afferent pathway. Since an analyzer may receive a large number of impulses, from many fibers which are not in phase, and may pass on an output which is rhythmic but some function of the input, it must (a) begin with a **pulse-to-continuous-signal converter**; that is, it must translate the code into another form. (b) It must end by converting the integrated result back into pulse-coded form for propagation over nerve fibers to the next destination. (This is the second such sequence: at the receptor a similar conversion generated impulses out of the uncoded stimulus.) (c) The analyzer must be able to distinguish **signals** representing stimuli from fluctuations representing **noise** in the system. In the case of afferent lines which are always silent until stimulated, this places no demands on the central organ. But in the common case of receptors with a continual background discharge whose frequency is increased or decreased by stimuli (Fig. 5.12), a serious problem is raised because of the fluctuations in background frequency (Fitzhugh, 1957; Bullock, 1957b) and is only serious when the fluctuations approach or exceed weak signals. It is similar to the problem in communication engineering of signal detection in a noisy channel. As already emphasized, the nervous system—at least in some animals but perhaps much less in others

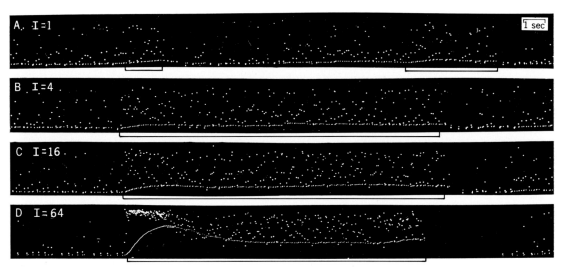

Figure 5.12. Example of weak signals in the presence of noise. Response of a single infrared sensitive receptor in the facial pit of the rattlesnake to successively stronger stimuli, consisting of radiant heat from a lamp. These units typically have a background activity and the successive intervals are very irregular; it is therefore difficult to detect a small response. In the method of recording illustrated each spike causes a bright spot on a slowly moving film at a vertical position that is higher when the interval is shorter since the last spike (pulse interval plotting). The same spike also causes another bright spot by way of an integrating circuit with a time constant of 2.5 sec, seen as the lower smooth line of dots. This kind of simple integration can detect a weaker stimulus, but as seen at high intensities it is slow in exhibiting a deflection. The figures show the limitations of detecting weak and brief stimuli by a frequency code on a nonrhythmic background. [Bullock and Diecke, original.]

(Wiersma, 1952)—uses the principle of redundancy and **compares the activity in many parallel lines.** Even though all are noisy, if they carry the same stimulus message, the information transfer compared to a single line is improved and the threshold is lowered by this means. The principle is widely used in instrumentation, as in coincidence counters for cosmic rays and in detection of evoked potentials on the brain.

But **with respect to each line** we may consider the question, what change in the input to a central analyzer constitutes a signal of stimulation and not a random fluctuation in background discharge? Experimental physiology has not yet provided an answer, but since in face of technical developments pertinent findings appear imminent, we may attempt to examine some of the alternatives. (i) The actual instantaneous frequency or reciprocal of the interval since the last impulse is not likely to be the signal at which the central analyzer looks, since it would have to rise to relatively high frequency to be reliably different from a chance fluctuation in background and the threshold would therefore be high. (ii) Frequency averaged over some period offers the possibility of useful compromise: If the period is short there is not much gain in sensitivity; if long, reaction time, detection of brief stimuli, flicker resolution, and spatial localization of moving stimuli will suffer. It seems probable that analyzers answering to steady-state input from nonadapting receptors, like some position, temperature, and pressure receptors, use such a simple process. But in the case of phasic input (due either to transient stimuli or to adaptation), more information can be preserved by a more complex processing. Among the possible formulations we may choose a simple one, stated in two steps. (iii) The ratio, frequency-averaged-over-some-short-time-just-past (F_{t_1}, t_1 being approximately a minimum expected signal duration) to frequency-averaged-over-some-longer-period (F_{t_2}, approximate background frequency) provides higher sensitivity for brief signals—on the order of t_1. For the same reasons as given just above, t_1 cannot be too long. Within certain limits (Bullock, 1957b), the longer t_2 is, the better. (iv) If there is drift of background frequency, still better extraction of signal from noise can occur if the receiver places a higher value upon changes in this ratio which occur rapidly since these are more likely to be real stimulus signals: several short intervals in succession are highly improbable as spontaneous fluctuations of background. We may, for example, propose $[F_{t_1}/F_{t_2}] \times$ [rate of change of this ratio]. (d) A somewhat similar treatment can be applied to the requirements which determine theoretically the analyzer properties with respect to **temporal and spatial resolution** of incoming signals and then to higher-level afferent centers analyzing the pattern. But it is not known whether these specifications approximate those found in actual nervous systems.

Barlow (1961) argues that neural systems not only abstract desired information and discard the undesired but also **reversibly compress information** while reducing redundancy, sometimes close to the input. There is adaptive value in doing so and he believes that some existing evidence can be viewed as supporting the reality of such a process.

5. The recognition of predetermined stimulus pattern

Normal stimuli impinge on several to many receptor units. Afferent centers are probably innately, or by learning, "tuned" to respond selectively to certain combinations of activity in the labeled lines and to certain temporal sequences. It is difficult to induce normal movements by electrical stimulation of large sensory nerves, presumably because this does not reproduce essential features of the spatial and temporal patterning.

Examples of pattern recognition abound in the literature of animal behavior. Birds react to the specific calls of their own kind, fur seals recognize their mates and pups in a crowd, courtship movements in many fish depend on a combination of stimuli—visual, chemical, thermal, and auditory. Some birds give an alarm reaction to a cruciform silhouette moving overhead with the short end forward, when it resembles a hawk, but not to the same silhouette moving long end forward, when it resembles a goose. Many hunting wasps seek out particular species of prey insects in which to lay their eggs.

Platt (1960) discusses the simple recognition of nonstraightness of a line due to a jog; this may be taken to mean that a predetermined pattern of neuronal activity represents "straightness". He suggests it is a certain movement of the eyes which, for a straight line, causes no change in input signal in the receptor channels and accounts for the extremely high resolution—equivalent to 1/30 the diameter of a retinal cone. If this is the mechanism, then the recognition of the direction of the line depends in turn on the central sensory analyzer "knowing" the direction of the motor movement. Sutherland (1957) states that an octopus can be taught to distinguish between vertical and horizontal but not between 90°-separated oblique patterns of parallel lines. Even at the reflex level, as Sherrington emphasized (1906), normally adequate stimuli are patterned. The pre-existing posture or movement is often a decisive part of the input determining the next movement.

In many cases a constellation, including several modalities or several different receptor types within one modality (color, pitch of sound, sweetness of taste, quality of odor), plus a specific topographic distribution and a particular time course, seems to be recognized and required to trigger a given response. The complexity of the formula of modalities, submodalities, and spatial and temporal sequence of stimuli appears staggering. Detailed information on how sensory influx is processed to accomplish recognition is fragmentary (but see Rosenblith, 1961). Evidence for distinct operations including filtering, linear transformation, auto- and cross-correlation, multiplication, higher-order nonlinear processes, and template matching (or "input A is big enough and B is small enough") has been adduced. A few examples from the fascinating insights accumulating in this field follow.

Contrast enhancement at discontinuities simultaneously presented in spatially patterned visual stimuli is produced by inhibition of receptor channels around a stimulated one; this occurs already in the first-order neuron in the eye of *Limulus* (Hartline, 1959; Ratliff, 1961) and is widely found more centrally. Sharpening of response area is evident in the successive neurons of the auditory path; threshold curves of sound frequency against intensity are broad in afferent cochlear fibers and narrower in third- or fourth-order units (Katsuki, 1961).

Filtering is shown by the behavioral evidence that the beetle *Chlorophanus* discards information about phase in the detection of movement of regular patterns of mixed frequencies in its visual field. Cats are said to waste some ability in their visual receptor mechanism to distinguish colors. These are nervous filters; limitations of the sensory transducers of course provide extensive selection of stimuli at an earlier level (Marler, 1961). Extraction of particular aspects of stimuli obviously must occur to account for observed behavior. But to see it in single units early in the afferent pathway is impressive. Maturana et al. (1960) provide clear instances in the optic nerve fibers of frogs (Fig. 5.13). One type of fiber, for example, fires on relative movement between small objects and background and not upon movement of a large object, a large field of stripes, or a small object together with its background; ambient light level is unimportant. This type of fiber is collected into a certain stratum of the optic tectum while the other strata collect other types. Higher-order cells in the optic tectum exhibit still higher-order abstraction: some, for example, fire to any small movement occurring in a large visual field but having done so will only fire to further small movements of the same object, within certain limitations of time and of large, fast movements. The result is a unit that signals every step of a walking insect throughout a large angle, and if it freezes it remembers its whereabouts for some seconds, ignoring distracting stimuli (Lettvin et al., 1961). Hubel (1959; Hubel and Wiesel, 1959; see also Jung, 1961) encountered cells in a mammalian cerebral cortex which respond weakly to diffused light but strongly to movement in one direction!

These cases come from electrophysiological recording of single neuron activity. The mathematical forms of the operations behind them are not known. In other cases the latter has been analyzed in whole animals by manipulating the stimulus-response sequence, but the neuronal mediation has not been disclosed. The detection of movement of the visual field in the beetle *Chlorophanus* and in the bee *Apis* has been found to require specified forms of linear and of non-

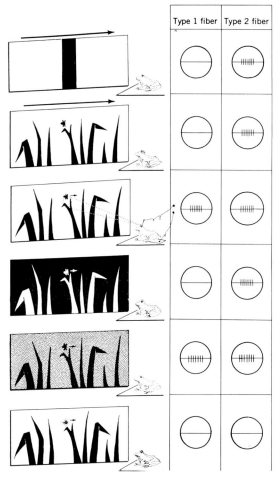

Figure 5.13. Abstraction at an early stage in the afferent pathway. Functional categories of fibers in the optic nerve of the frog illustrate the integrative processing that must occur in the retina. The diagram, based on data of Maturana et al. (1960) indicates—for two of the fiber types in the optic nerve—the presence or absence of response in the form of impulse bursts to various forms of stimulation in the visual field. Fiber type **1** does not respond to on and off of light, to movement of the whole visual field or of large objects but only to movement of small objects relative to the background (if they are darker than the background), and the response is independent of the general level of illumination. Fiber type **2** does not respond to on and off or to movement of the visual field or of objects in it that have only fuzzy edges, but to any movement of the whole field or objects within it where there are sharp boundaries.

linear **transformation,** multiplication, and auto- and cross-correlation (Reichardt, 1957, 1961). This study is further detailed on p. 1094. **Exclusion** as well as inclusion is necessary in many specific stimulus patterns, that is, "A *and* not B." Since sensory physiology does not give much grounds to believe in narrow gates or windows of unit response to varying intensity, but only in a threshold, it is probable that inhibition of some kind is required for exclusion. Possibly inhibitory input plays a role equal in general to excitatory input so far as pattern recognition is concerned. Another operation can be inferred from anatomical considerations: **spatial representation** of some graded resultant of input, involving many neurons. The cerebellar cortex of vertebrates suggests a primarily two-dimensional display of an analog function of input more derived than simply topographic mapping. The fanning of Purkinje cell dendrites in a consistent plane and the passage of granular cell axons at right angles to this plane would be unnecessary if all processing of information were accomplished by impulses and specific connections.

Shape recognition has attracted attention as a problem amenable to precise treatment. When a visual system is presented with a shape we may assume some kind of analog representation of the shape is set up among the neurons and, further, that abstracting takes place, that is, discarding of information. Some pre-set criterion is met or is not met and within limits that vary among species, shapes can be recognized regardless of tilt, position in the field, size, and even certain missing parts. Some common output is triggered for a certain class of stimuli; the main efforts of experimenters have so far been directed toward discerning the class characters. Concerning the large amount of discarded information in the input, other abstractions performed by the visual system doubtless use some of it (movement detection, ambient light level estimation, and so on). Much is known about the shape discrimination abilities of *Octopus* and of a few mammals; tests have excluded some hypotheses such as distinguishing between shapes by total perimeter, total area, ratio of circumference to longest axis, presence or absence of straight lines, presence or absence of angles (see p. 1504). Deutsch (1960a) proposed a system of connected neurons of very modest specification based on the performance of *Octopus* and reasonable neuron geometry; although it is probably not the system used by *Octopus* (Sutherland, 1959, 1960), it calls attention to simple possibilities for extraction of desired information.

A special problem in recognition is presented by the commonplace but mysterious feats of **detection of absolute** as distinct from relative quantities in stimuli. Within a certain accuracy, most of us recognize absolute light intensity, color, time, saltiness, even sound frequency (pitch) and, more clearly, because we regulate them reflexly, body temperature, blood pressure, blood oxygen concentration. Both receptors and higher-level analyzers must be stabilized against slow drift. Stated as a problem of calibration against some standard, the phenomena are in the same class with change detection, where also stability of the physiological constants relating stimulus magnitude to response magnitude is important.

6. Thresholds to integrated input; "decision units"

Wherever several incoming lines are compared, pooled, or integrated to read information about several aspects of the stimulating situation, the analyzer must consist of functional units each of which performs this integrating operation on a number of input lines or on the already partly integrated intermediate stages in reading these lines.

Consider any moderately complex central integration such as the release of chirping in a cricket or flight in a bird by a certain stimulus configuration—time of day and year, suitable temperature, sight, and sounds. For simplicity, confine attention to such either-or acts—a large class of behavior. The "decision" to initiate the act depends on a criterion of quantity and quality of impulses (labeled lines) being met. If the integration is performed in the nervous system then there must be, before the command enters the effectors, a final **determination by some competent functional unit** whether the criterion has been met. All relevant input and central predisposition must be read and finally evaluated by a single integrator after all the partial evaluations. Whatever its composition, it is a unit because it is necessary and sufficient to perform this act.

If averaging of many penultimate integrators is involved, something must perform the averaging and deliver an unequivocal answer—or else the action is not centrally determined. No doubt some actions are integrated finally only by mechanical averaging of the activity of motor units. But it seems likely there are many that are integrated centrally, such as initiation of chirping or of flight, recognition of faces or of voices. For each of these, or more precisely for each pair or set of alternative events, there must be a single final functional unit (or a number of equivalent units) upon each of which all preceding lines converge.

Since the final functional unit is, like a military general, competent to render an answer, it may be called a "decision" unit or recognition-of-criterion unit or high-level trigger unit. The term "decision" may suggest a lack of determination by input. We assume determination for all actions and use "decision" precisely because it is those actions which have been said to be "decided" that we wish to cover—that is, to include with all other actions determined by input and existing biases and patterns. Not only are there many such units—one for each pair or alternative set of recognitions which the organism has acquired by birth and training—but there is physiological evidence to support the theoretical expectation that many **successive levels of recognitions** to subcriteria occur, which gradually converge on the narrowest bottleneck in the stimulus-response sequence. The analogy to an army suggests the hierarchy of convergences and integrations of lines of incoming information, the encoding, decoding, and re-encoding, the predetermined differences in the value and the meaning of the same message depending on what line it comes on, the processing of data from lower levels in order to make interpretation possible, and other similarities. All these are features of our present picture of the nervous system.

Further to consider the **characteristics of decision making** in a nervous system, it is reasonable to expect in different cases greater or less redundancy—that is, duplicate and equivalent recognition-of-criterion units with the same input. This would explain tolerance of injury and provide for greater uniformity of threshold. Duplicate units would each have to receive from all the inputs and lower integrative levels involved and command the same output paths—though they would not necessarily be identical in their answers or even uninfluenced by each other. An uncritical number of units must closely agree in threshold to give the sharp onset characteristic of the all-or-none or either-or actions such as those we chose for illustration. If a certain threshold number or proportion of units were required, the structure at the next level which would have to measure this number or proportion would be the interesting and relevant unit in the present context. The only requirement for the decision units is that each of them incorporate in its answer all the preceding lines and stages of integration so that it is competent to determine the next stage, whether it be an effector or a higher level. Each action, recognition, or sequential integrative level must finally be triggered by a single unit or an uncritical number of units with the same input requirement, within normal play. It is permitted and probable that in the nervous system some of the same input that influences the decision unit influences units downstream from it, predisposing them or setting the stage for the trigger message from the decision unit. The downstream units are not equivalent to it insofar as they do not receive all the relevant incoming sources of information. Stimulation of a decision unit alone may not precipitate the normal behavior because the stage is not set. There are a number of cases, however, where focal stimulation has released characteristic either-or complex acts (p. 1143) such as chirping in a cricket (see Fig. 5.23) and vivid recall in man. It is precisely the cases where a preformed output pattern is triggered that are under consideration here.

Decision units can be significantly labile in their judgments. As a result of internal state, of the time elapsed since some previous action, of bias due to prevailing input or feedback (experience), or as a result of "spontaneous" central initiative, a meaningful change may take place in the weight assigned to certain inputs or in overall threshold.

Decision units may be said to depend on maintained activity because they must be able to signal both increase and decrease in frequency; otherwise they would be only unidirectional and a final balance between opposing messages would be deferred, as may be the case in symmetrical so-called "half-centers" (Gesell, Brassfield, and Lillie, 1954; Retzlaff, 1957). The final integration may then be between opposing muscles.

Decision units are not required for cases where the final integration of messages in different lines occurs mechanically in the joints. The output of decision units must not be continuously graded in the cases of either-or behavior. We may reasonably anticipate that the output will be an **abrupt change from one range of frequency** of firing to a different range.

What are these units which, like miniature sentient beings, receive all that goes before, act adaptively upon it, and pass on the result to the next level? They must be numerous, for many integrative levels in the path from receptor to effector may have a decision-making character, simply by having a sharply discontinuous relation between input and output. And there are many discrete aspects of behavior in which choices are made. Both lower animals and higher must possess such elements, since either-or behavior is a prominent part of the repertoire in both. It is not of course necessary that the units be everywhere the same in composition or mechanism.

(a) One possibility is that certain **single neurons are decision units.** We know of no smaller element than the neuron which can act in so unitary a way on the basis of converging nerve impulses. Every neuron is a decision-making element when it changes from a silent to an impulse-firing state. Only those which integrate a complex input or control a large output are interesting in this context. The capacity of a single neuron to integrate a large number of incoming impulses and to give a single-impulse answer, as in giant neurons like Mauthner's, demonstrates its suitability for a decision unit. This is true at least of those neurons whose response is go–no go; but it is not at all limited to them. The single neuron's possibilities for complex evaluation and interaction of inputs and for abrupt change from low- to high-frequency firing fit it for higher-level and responsible decisions. In fact we know experimentally of no larger element which can compare, evaluate, add, and multiply inputs in the central nervous system.

(b) A theoretical alternative is that a **mass of randomly connected neurons** constitutes a trigger unit. Beurle (1956) has pointed out that activity with a sharp threshold can be propagated in a mass of a large number of cells, each assumed to have only some of the properties known for real neurons and connected without specification except that the probability of interconnection falls with distance. As an amplitude-sensitive switch, the mass can integrate by permitting subthreshold excitatory and inhibitory input to influence its critical level. Several loops leaving and reentering the mass, also uncritical in connections, can stabilize the amplitude of the propagated waves of excitation. The waves can terminate on one cell or a specific link. In order to be useful the mass must connect specifically at both input and output. If use were to produce a change in the connections, size, or constants of some of the cells, then conditioning, trial and error learning, and recall could be explained. By postulating many local masses of this sort—each performing a limited integration—something resembling the brain may be imagined. Storage of species-characteristic knowledge and of individually learned information is in many cells, each a part of the storage of many items.

This concept can provide the requirements under consideration and calls for a modest degree of specific organization relative to the number of cells. There is rather sharp distinction between the class of connections which is specific and critical—that is, inputs and outputs—and the class which is random within the given functionally defined mass. A useful nervous system would consist of a large number of such unit masses.

The redundancy and freedom from a critical number of neurons resembles the same features mentioned in the previous alternative, but only partly. In Beurle's model the proportion of active neurons is critical, though the mass is the decision unit that measures this, not a single neuron or an uncritical number of equivalent independent neurons. The redundancy is therefore of a different kind. The absolute amount of organization is the same, being the connections into and out of and between unit masses of single cells. But the redundancy in Beurle's model does not involve duplication of the specific pathways. The physical basis of stored information is not different at the neuronal level. Neither model has been formally developed sufficiently to compare the complexity of achievements they permit or to evaluate their relative advantages.

(c) The third alternative is probably more attractive to most contemporary neurophysiologists. This is a **multiple-input, metastable feedback loop,** consisting of a definitely specified meshwork of mutually interacting neurons. In order that the meshwork possess a threshold, a regenerative chain reaction is assumed to take place by positive feedback. The principal difference between the feedback loop and the single-cell model is that the constellation of inputs converging to provide the criterion of recognition or triggering does not need to converge on one cell; some of the necessary inputs may arrive at one cell of the network and some at another, up to any degree of complexity. Each cell then contributes an output to the others, but only if all of a certain predetermined set receive input, in a predetermined amount and temporal pattern, will the network constants provide the regeneration for self-re-excitation, amounting to the passing of a threshold. The system resembles a flip-flop or a multivibrator circuit with two (or more) stable states. It may be self-restoring or a specific input may be required to return it to one of its stable states.

The output may be from one cell or from several or many of the cells; hence several output fibers may go to different fractions of the whole effector mechanism. In the simple form of this mechanism, based on component cells having one axon, the cells in the efferent pathway next downstream will feel the impulses that are a part of the loop feedback before a loop threshold has been reached. In this case, since there must be a clear difference between outputs before and after threshold, we must suppose that a significant jump in frequency of impulses is the normal signal of triggering. Otherwise different forms of output must be invoked—for example, dendritic potentials for the interactions within the loop. The impulse output may now change from zero to a finite value or, as before, from one frequency to a considerably higher frequency. If the output frequency is continuously graded, then some later junctional unit must read this and provide the sharp threshold required in a decision unit.

This multiple-input, metastable feedback loop points to no one cell but to the whole loop as responsible for the recognition of input criterion and provision of a threshold. It can at the same time be the mechanism for many alternative actions by having more than two stable states. This feature might provide a mechanism for recognizing individuals or symbols and to that extent it is possibly more important in higher vertebrates. More complexity and specification of connections is necessary than in the other alternatives. Like them, it can of course employ redundant cells.

Another class of decisions is considered by Sholl and Uttley (1953), who advocate pattern recognition by a statistical process among many units which provides a continuous measure of pattern difference, an analog rather than a digital answer. Instead of "A differs from B" or "A does not differ from B," the continuous measure has the form "item A differs from ensemble C by more than does item B." Such a process is not at all excluded by the foregoing argument, but for decisions different than those leading to recognition of familiar faces or to the takeoff of a fly. Analog events are fundamental to both classes of decisions, but in the latter they eventuate in either-or answers.

Any or all three of the decision mechanisms discussed is consistent with present neurophysiology. The single-cell type is known to exist, for example, in giant fiber startle response systems. Randomly connected masses are suggested by many neuropiles. The specific network with positive feedback is the most appealing, although it does not inherently offer any greater flexibility or complexity of achievement, economy of cells, or economy of specific or predetermined connections and transfer functions. All three types of mechanism can provide the same features of performance, stability, redundancy, tolerance of injury, and modifiability. I see no reason to propose at present any real dichotomy in the ways in which triggered actions are set off, as between lower animals and higher, for example, or between instinctive and learned acts.

7. Integrative possibilities of neuronal circuits

Much has been written about the superficially limitless potentialities of prearranged circuits of neurons for accomplishing nervous functions by processing signals in specified ways (see Lorente de Nó, 1934, 1938; McCulloch and Pitts, 1943; McCulloch, 1949; Jeffress, 1951; Von Neumann, 1958; Ashby, 1960; Rashevsky, 1960; Reiss, 1962; Fessard, 1961). Little is actually known about specific neuronal connections that exist in nervous systems beyond the fact that some types of connectivity occur: converging inputs, recurrent collaterals from the output axon, recurrent (feedback) neurons from output back to input, pathways that diverge and eventually rejoin, and a few similar elementary types. (See for example, Scheibel and Scheibel, 1958a.) This type of evidence together with the form of neurons and their pulse-coded activity, as well as evidence of specific connections of regions down to minute subdivisions of the brain, understandably encourage the speculation that specified circuits of some complexity are a major principle of neural function. For the most part, this is still a theoretical area, exploring the properties of known combinations and searching for the simplest circuits that can perform certain operations. This approach is important because it is too much to expect that finer anatomy will gradually describe the actual connections in adequate detail; plausible circuits may point to experimental tests of their actuality (but see Horridge, 1961).

The simplifying limitations known are few. Unlike radios, only a few kinds of elements and a few kinds of connections between them occur over and over. The possibility is usually passed over that small circuits operate without all-or-none impulses, with only graded activity—useful in amplitude for a few hundreds of microns; such signals probably do play a role even between units and along specific cell prolongations. Considering only the pulse-coded neurons, the **major principle of their dynamic organization** is that they interact by opposing excitatory and inhibitory tendencies. But the complicating feature is that we have to deal not with passive input-output functions but with the modulation of autogenic activity, and not with one-to-one events but with trains of impulses coming in and going out, each more or less rhythmic. Probabilistic descriptions of the changing tendencies to fire have become attractive but must reckon with the commonly quite rhythmic spacing of impulses. Because the long-distance code is pulsed, earlier models treated the entire system in the same way, but more realistic models include some degree of simulation of the complex sequence of graded events known in synaptic, dendritic, and other loci (Harmon, 1959).

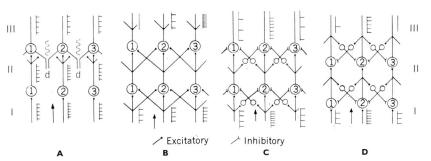

Figure 5.14. Circuits of neuron networks. Four diagrams showing some typical connection patterns that are supposed to exist and most probably to coexist (together with others) in real networklike structures of the brain. Trains of impulses showing relative frequencies at successive levels are represented and in one place slow waves, synchronized by hypothetical dendrite interaction, d. Further explanation in text. [Fessard, 1961.]

Figure 5.14 shows four elementary forms of connection that probably exist—in various mixtures and together with more complex forms—in real nervous systems. Figure 5.14, A represents independent lines and hence no network, except that it includes a symbol for possible dendrodendritic interaction, for example, of slow waves. Figure 5.14, B is a purely excitatory scheme, emphasizing that convergence upon any cell is the essence of circuitry and hence that graded events are basic to the integration. Alteration of the frequencies of impulses in the lines can occur as a result of transmission- or transfer-functions that give a sigmoid input-output curve (Fig. 5.15); cascading such curves can provide abrupt step functions or thresholds. Figure 5.14, C is the classical case of reciprocal inhibition, here done by an interpolated inhibitory neuron, as is known in the spinal cord (Eccles, 1959). Figure 5.14, D is retroactive inhibition, also known in the spinal cord, and here generalized to an array of cells.

The frequencies of impulses shown are only one of many possibilities inherent in such networks; here the central lines are assumed to receive the highest input frequencies and the result of the circuit is a focal sharpening—that is, contrast or **contour enhancement**. Given certain transfer functions, the networks of C or D can act like a multivibrator; that is, the networks can **gate or switch** very rapidly from a maximal frequency in one output line and minimal in another, to the opposite (Reiss, 1962). This can recur at a steady rate, providing a **pacemaker role** in which no single cell is the essential locus. Other dynamic properties of the elements convert this simple array into a **comparison circuit**, the two outputs reflecting in their durations the relative magnitudes of the input frequencies. Reiss has set out exactly the conditions for each of these consequences of reciprocal inhibition and in addition the slight changes that can provide **sensory scanning**, periodically or irregularly sampling each of a number of input lines. If a number of input lines converge on one cell, a drastic increase in one line may be lost in the murmur of others, but the reciprocal inhibition network allows one line to dominate and thus affords an **alarm system**. An array with inhibitory cross connections can be viewed as an arena with competition between parallel streams of signals; when one wins control, it is a dictatorial network. A spectrum of properties is possible from democratic to dictatorial; oligarchy is the compromise that results from not having all possible reciprocal inhibitory connections and probably prevails in nervous systems. **Reliability** can be achieved with unreliable elements by employing reciprocal inhibition, so that the fatigue or adaptation in a single cell need not reduce the system output; elements can be automatically shifted from active to standby or recuperating status. **Redundancy** can be built in without a proportionate "cost" in energy required to maintain activity in many cells. Time and **load sharing** are related uses of the same principles. By adding inhibition of the inhibitor neurons, new possibilities accrue—for example, use of the same network for either antagonistic or synergic action of muscles, according to whether input arrives over a single line or both lines (Fig. 5.16). **Delay lines** are readily provided and could be useful in measuring velocity, in discriminating certain events, or for reducing random spurious counts, for example. **Null detection**, multiplication of input frequencies, and filtering are similarly inherent in simple networks as available properties. Whether the nervous system uses all the operations such circuits can perform remains to be answered.

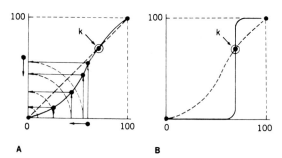

Figure 5.15. Input-output curves. **A.** Theoretical S-shaped input-output curve relative to the transmission of groups of impulses through one barrier of monosynaptic junctions. The intersection with the bisector k is a critical point. This figure is intended to show how transmission to several successive synaptic barriers, such as are present in networks, is necessarily decremental, as shown here, if the representative point starts below k. **B.** The resulting steplike transmission curve of a network containing a sufficient number of successive synaptic barriers. [Fessard, 1961.]

III. GANGLIONIC INTEGRATION B. Output Control 1. Neuroeffector integration

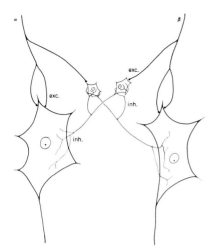

Figure 5.16. A hypothetical neuron circuit that would give reciprocal inhibition when either α or β brings input alone, or synergic output when α and β both carry input. It therefore provides a flexible control permitting either antagonistic or mutual effector action. [Szentagothai, unpublished.]

Sustained reverberation of signals, recirculating in loops with regeneration to prevent fading, is an oft-discussed possibility. The anatomical basis is clearly present but definite transfer functions are necessary in addition. Self-sustained activity apart from autorhythmicity arising in small foci has not been found as yet. It is expected by those who see some advantages for explaining learning (Young, 1951) and is discounted by others who see a smearing of information and prefer other means of persistence (Fessard, 1961).

Opposed to specific circuitry are the two possibilities: (a) that neurons are connected randomly, within limits, and (b) that neurons interact nonspecifically, by other than axon connections. A well-formulated exposition of the potentialities of randomly connected arrays was given by Beurle (1956); the achievements on reasonable assumptions of propagating waves of activity are impressive. Quite possibly such arrays exist here and there in nervous systems. But their inputs and outputs must have specified connections and a moderate amount of specification within the array could give richer attributes with fewer cells (see also p. 281, Decision units). Interaction of neurons in fields or masses other than by fiber-to-fiber contacts is a much mooted possibility. It would greatly smear specific signal transmission but may nevertheless exist. It is thought to be the basis of some synchronization of slow potential changes in many cells, for example in brain waves (see p. 318), but the evidence is not decisive.

B. Output and its Control

1. Neuroeffector integration

There is no integration in the neuromuscular relations as classically described in mammalian skeletal muscle. This is because every nerve impulse arriving was supposed to be transmitted successfully and to cause an all-or-none muscle twitch; the input-output relation would be one-to-one. Actually it is now known that even in this type of muscle the percentage of one-to-one transmitting junctions depends on the state of stretch. And it must be concluded now that this kind of junction is a special case rather than being a satisfactory paradigm of neuroeffector relations in general. Most kinds of muscle in most animals —and quite likely other kinds of effectors as well —are controlled by **junctions with graded response** (see Hoyle, 1957). This means output is some function of input other than one; that is, there is integration.

In the better-studied animals—for example the arthropods—most junctions between nerve and skeletal muscle are strongly **dependent on facilitation;** impulses arriving far apart elicit only small contraction or none. The gradation of mechanical response is mainly by frequency of arriving impulses. The same is true of certain muscle fibers in vertebrates—for example the small motor system scattered through muscles in amphibians and the intrafusal fibers of muscle spindles in mammals.

In arthropods and probably elsewhere there is a further elaboration of the possibilities for peripheral integration. Two or even three or **four motor nerve fibers may end on the same muscle fiber,** each capable of activating a contraction. Each contraction has its own characteristic relation between tension and impulse frequency, its own speed of development of tension, rate of build-up and decay of facilitation, and plateau contraction for a given frequency. Based on the mechanical effect, the motor nerve fibers are called slow, fast, or intermediate. In a few muscles (for example crayfish claw closer) there occur nerve fibers at whose junctions a single impulse is capable of causing a maximal, all-or-nothing twitch.

Another degree of freedom is introduced in most of the same cases (arthropods, and some indications in annelids and molluscs) by the possession of one or two nerve fibers ending on the same muscle fiber with the motor axons just mentioned, which cause **inhibition** of contraction, or at least hyperpolarization of the muscle membrane potential. In the insects studied there is hyperpolarization without frank inhibition (p. 982), but in the crustaceans studied inhibition is pronounced even when there is virtually no hyperpolarization as well as when there is (p. 975). The inhibition requires facilitation and is graded with frequency, so that algebraic addition of motor and inhibitory effects is finely graded over a wide range.

Actually, there is more than one stage of labile and graded transmission of effect between nerve impulse and mechanical event. The **coupling between muscle action potential and contraction** is evidently quite indirect (Hoyle and Wiersma, 1958) but at present we know little about the integrating mechanisms connected with this coupling in the crustaceans. In the special case of high-frequency, insect fibrillar muscle, however, we see an instructive extreme. Here normal contraction occurs at high frequency—up to several hundred per second, for example, in dipteran flight—while muscle action potentials occur at five to ten per second and not in a fixed ratio. The action potentials are essential to the occurrence of the contraction but clearly control them only indirectly (Pringle, Roeder, and Boettiger; p. 983).

Second-order motor neurons are known in the periphery in several phyla. The vertebrate autonomic is the best known case. But in polychaetes and in echinoderms many motoneurons are supposed to occur in or near the muscles and to receive excitation from ordinary central (first-order) motoneurons.

Little is known of the **control of other effectors** in neural terms. Luminescent organs in soft corals, ctenophores, polychaetes, insects, and enteropneusts are apparently controlled by facilitating neuroeffector junctions (see the respective systematic chapters). The same is true of mammalian salivary gland cells (Lundberg, 1958). Peripheral integrative mechanisms are more probably primitive than derived and in any case are widespread, from coelenterates to vertebrates.

2. Nerve nets and other uncentralized control systems

We have briefly considered integrative principles operating in afferent systems and at neuroeffector junctions. It remains to consider the ways in which the two are coupled. This section and the next examine simple systems, first the uncentralized and then the lower centralized ones.

Nerve nets are treated in detail in Chapter 8 because most of our knowledge of them derives from coelenterates. They are defined as anatomically dispersed systems of neurons, so connected that excitation can spread through some considerable number of neurons in any direction and diffusely, bypassing incomplete cuts. (Protoplasmic continuity is irrelevant; it is probably sometimes present and sometimes absent.) Inherent in these properties is a distribution system that can control the areal extent and the decrement in intensity of response, according to the intensity, duration, and character of the stimulus and species-characteristic features of the net. Examples are given in Chapter 8 of limited spread and of through-conduction; sometimes these are two separate systems, slow and fast, in the same epithelium. Interaction between the two can be irreciprocal and localized in nervous concentrations which represent the first ganglia in our sequence of phyla. The other properties that permit variation and adaptation include (a) a wide-ranging dependence upon facilitation with respect to number as well as interval between impulses, (b) different degrees of spatial summation of input, (c) varying tendencies to fire repetitively as a function of input (the first nervous integration in our sequence of phyla), (d) different numbers of neurons with which each neuron makes functional contact, (e) the proportion of the units in the net ready to be excited or to transmit at any moment, (f) the development of specialized

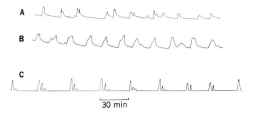

Figure 5.17. Example of complexity arising in simple nerve net systems. Spontaneous activity of isolated rings of the body wall of the sea anemone *Stomphia*. **A.** Oral disc region. **B, C.** Basal disc region. **A** and **B** were from the same animal. Note the very regular periodicity in **C**. [Hoyle, 1960.]

tracts, polarization of conduction, and regional differentiation, and (g) several forms of shorter- and longer-term spontaneous patterns (Fig. 5.17).

Information concerning **nerve nets in higher phyla** is a most unsatisfactory matter. Although probably developed locally in many animals, good evidence is available for very few except coelenterates. In ctenophores a nerve net is most probable but has not actually been shown. In platyhelminths it is possible but the evidence mainly shows its unimportance if present; in nemertineans it is still less important or likely. Among pseudocoelomates only entoprocts have given some evidence for a nerve net, while the nematodes and rotifers quite probably lack a nerve net in the skin. Phoronids seem likely to possess one and ectoprocts have yielded a little positive evidence for colonial coordination by nerve net. There is no good reason to propose such a system in annelids, onychophorans, or arthropods at present. In molluscs it is at most rare; only the palps of some clams have been shown to conduct diffusely around partial cuts. The foot of gastropods has a plexus and local autonomy but diffuse spread is not definitely known. Cephalopod skin has been suspected of having a nerve net but present evidence is against it. In echinoderms the most conspicuous situation is the epidermal system controlling echinoid spine movement; but here the criterion of diffuse conduction is not met (discussed in the next paragraph) and we must conclude it is not a nerve net. There does appear to be a true nerve net of very limited distance of spread in asteroid skin, controlling the papulae, which are respiratory evaginations of the body wall. Hemichordates exhibit a good epidermal nerve net physiologically; tunicates probably but not certainly have one; cephalochordates show one histologically. Vertebrates have dispersed plexuses, notably in the wall of the gut; but we do not really know whether there is nervous conduction of excitation in a diffuse manner, able to bypass cuts and not dependent on mechanical restimulation or muscle propagation.

Anatomically dispersed systems other than nerve nets must be reckoned with. There may be a significant chapter in comparative neurology yet to be written on forms of organization of uncentralized systems or peripheral plexuses that do not conform to the definition of nerve net. For example, the leaning and waving of spines of sea urchins, often rapid, conspicuous, and apparently purposive, is controlled by a superficial, dispersed conduction system, probably nervous but only spreading excitation in straight lines, not bending around corners or bypassing cuts (Kinosita, p. 1546; Bullock, Chernetski, Biederman, and Thorson, unpublished results). It probably involves relays—labile junctions that determine the distance of spread and the decrement in intensity. Sites initiating spontaneous, apparently haphazard waving may be independent for each spine or in some cases may transiently cause a number of neighboring spines to wave in phase. The snail foot is another dispersed, autonomous system of nerve cells and fibers controlling muscles, according to rules of organization not yet worked out as to types of spread of excitation, but evidently not a nerve net. This should be a profitable field for further work and for insight into diverse forms of communication systems in animals.

3. Reflex integration

Sherrington started the concept of the reflex, one of the truly great and fruitful abstractions in biology, with these words: "**The unit reaction in nervous integration is the reflex,** because every reflex is an integrative reaction and no nervous action short of a reflex is a complete act of integration" (1906, emphasis his). While there have been many who have pointed out the limitations of the concept and Sherrington as clearly as anyone else emphasized the nonexistence of a discrete circuit insulated from others, the fundamental usefulness of recognizing this category of responses has been amply proven by the insight which experiments based upon it have provided. The reflex is a useful abstraction but we qualify the broad statement that it is the unit of all nervous integration: (a) the arousal functions of the reticular activating system in mammals cannot be resolved into reflexes, nor can (b) mere sensing, (c) autochthonous actions, or (d) many instincts.

Familiar examples of the phenomena under consideration are the stretch reflex, the flexion, crossed extension, and the scratch (dog) reflexes, the reflexes of micturition and defecation, the pinna, swallowing, stepping, sneezing, salivation, blink, accommodation, and a host of other reflexes familiar to the clinical neurologist. They are significant as ready-made, unlearned, adaptive movements, prompt and coordinated. The **coordination** is not only within each reflex, which we might explain as a fixed pattern, but even when conflicting reflexes are stimulated simultaneously there is almost invariably a resolution in favor of adaptive coordination among them. The present account of reflexology necessarily draws mainly from mammalian literature, but it is believed that the principles enunciated are probably general. The integrative principles rather than the characteristically vertebrate features of organization are the present concern. Analysis of invertebrates along these lines has lagged.

There are **simple and complex reflexes**. At the former extreme, stretch reflexes may be mentioned; at the latter, copulation. Stretch, at least in mammalian skeletal muscle, induces contraction of the same muscle, reflexly. But in extensors the contraction can be tonic with maintained stretch, while in flexors it is generally strictly phasic. The one is normally a postural reflex, the other a protective withdrawal from an injury. Copulation, though successful after transection of the spinal cord and normally a self-completing act difficult to interrupt, can in mammals at least justifiably be considered a chain of reflexes, some of which are themselves complex. Implicit in the concept of reflexes—since there is almost invariably more than one reflex utilizing a given muscle, and hence more than one central mechanism converging on the same motoneurons—is the concept of the **final common path** (see Glossary). The very expression emphasizes the integrative function.

What are the **properties of reflexes** which manifest integration? (a) The threshold stimulus is very much dependent on conditions. (b) Gradation of response does not closely correspond with gradation of stimulus above threshold. Indeed, reflexes are in general difficult to grade, some being almost all-or-none, like the extensor thrust in the dog. With increasing stimulation, there is often no change until there occurs an abrupt spread to a wider area of movement or an abrupt increase in duration. (c) If the stimulus is repetitive, there is usually a poor correspondence between its rhythm and that of the reflex response. (d) Single afferent impulses are usually not adequate; temporal summation is usually necessary to elicit a response. For each animal, certain topographic and modality combinations reinforce each other. The best-known instances are in the dog, for which Sherrington's papers and bibliography may be consulted (1947). (e) A depressed excitability typically follows a reflex and is often quite long. With frequent stimulation, the depression determines the frequency of the response—for example in the scratching or the stepping reflex in the dog and in the abnormal clonus of rhythmically stimulated reflexes which are normally smoothly tetanic.

(f) The phenomenon known as **afterdischarge** or the prolongation of the motoneuron activity after the cessation of the stimulus is a prominent feature of many reflexes. It is as though the mechanism is organized to complete a certain movement in a controlled way. It used to be thought that the only available explanation of afterdischarge was sustained activity in reverberating circuits, but other mechanisms are now becoming known. It seems more than likely that the tendency of some individual neurons to afterdischarge is involved (see preceding section). Continuously discharging ("spontaneous") units are common and their rhythm may be re-established after a reflex change only slowly.

(g) Spatially and temporally **patterned control**

of several muscles is probably involved in all reflexes. The extreme case of minimal involvement is perhaps the stretch reflex in mammals. The activated musculature in this instance is highly localized—restricted even to that portion of a single muscle from which the afferent stretch excitation comes. (This is adaptively important because the slightly different time constants of neighboring parts of the muscle prevent oscillation.) But the stretch reflex not only involves activation of certain motoneurons; it involves inhibition of others. The ones inhibited commonly embrace the motoneurons of muscles whose action would be antagonistic to that of the activated muscle. Most if not all reflexes involve both excitation and inhibition, showing thereby an innately organized, nicely adaptive reciprocal innervation. Not only spatially but temporally patterned, the timing of the inhibition is coordinated with that of the excitation, as is clearly seen in the alternating reflexes already referred to—stepping and scratching. Even simple flexion and crossed extension reflexes show characteristic temporal patterning: the simple flexion reflex is dramatically sudden in onset, the full number of motor units being activated nearly synchronously; the crossed extension is a recruiting reflex, slowly adding motor units and muscle tension for some time, even though the stimuli be given with the same time course. (h) **Irradiation** with increasing intensity of stimulus occurs in some reflexes—for example the protective flexion reflex. At threshold a response may involve a limited part of a synergic muscle group across one joint, but with irradiation it may spread to other joints of the same appendage, to other appendages and segmental levels, to the head and neck. The spread is generally saltatory and is confined strictly to certain lines or muscle groups. However, apparently uninvolved muscles may in fact be involved as objects of inhibition. The possible movements are thus circumscribed in a characteristic pattern. Pflüger formalized certain rules, including the following, restated and qualified. If a response is unilateral, it is usually ipsilateral. If a response is bilateral and not quite equal on the two sides, the stronger response is usually ipsilateral. Irradiation is generally polarized—spread is easier in some reflexes cephalad, in others caudad.

The result of patterned irradiation is that for each main form and locus of stimulation there is adopted by the animal a certain set of muscle contractions and inhibitions, presenting to view a **"reflex figure."** We shall return in a moment to this concept when we consider the results of simultaneous stimulation in two distinctly separate places.

The following additional rules have been recognized for mammals. More intense stimulation is required to bring in segmentally more distant parts. Each dorsal root has in its own segment or near it a motor path of as low a threshold as any available to it. The motor mechanisms of a given segment are of very unequal accessibility to any given afferent path even of the same segment. The reflex movement elicited from one spinal region exhibits much uniformity despite shift of stimulating locus within that region; that is, the receptive field of a type reflex is usually plurisegmental. The principle emerges inescapably that—in spite of its morphological conservatism—the nervous system is innately **organized functionally in terms of movement patterns,** not in terms of morphological segments.

The most remarkable integrative features of this level of organization are to be seen when two or more separate reflexogenic sites are stimulated simultaneously. If the individual reflexes involve compatible or coordinative movements, a compound reflex results. If they involve antagonistic movements, they are usually not both released, one cancelling the other, but one is precluded and the other acts without interference. At any moment, it is either-or, like the visual impression of a drawing of steps or cubes in perspective (Fig. 5.18). In either allied or antagonistic reflexes simultaneously stimulated, there is typically **no algebraic summation of the component simple reflexes** and we are at present without an explanation of the switching mechanism. This is a prime example of the new integrative features of the ganglionic level, because at the neuronal level algebraic summation of excitatory and inhibitory inputs seems to be the usual mechanism.

Summation does occur between allied reflexes

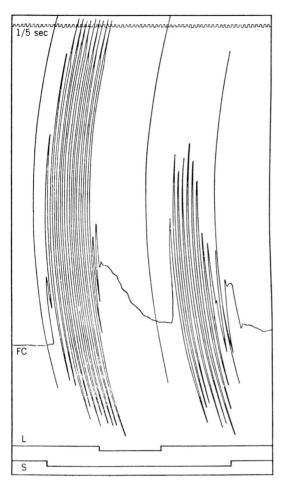

Figure 5.18. The either-or property of antagonistic reflexes. Simultaneous adequate stimulation for two incompatible reflexes does not result in intermediate posture or action but one dominates over the other (even though sometimes briefly, resulting in vacillation). This example shows the contractions (downward on the record) of the left hip flexor *(F C)* of a dog when, to stimulation adequate for the scratch reflex *(S)*, is added stimulation of the foot *(L)*. This interrupts the clonic scratching completely, though not instantly, and its effect outlasts for a short time the foot stimulus. (The vertical arcs mark the moments of onset of *S*, cessation of *L*, and cessation of *S*.) [Sherrington, 1906.]

for a short time after any of them, there is a summation period, shorter for weaker reflexes and for those farther apart. Subthreshold summation is particularly important in releasing reflexes as a result of ineffective stimulation to different areas simultaneously. Interestingly, phasic reflexes, such as scratching and stepping, are not influenced either in frequency or phase by summation of separate stimuli, though they may be increased in amplitude. Successive stimuli (to be considered further below) are even more effective than simultaneous ones, at least in the case of subthreshold tactile stimuli (compare a rolling disc with the edge of a card). The term reinforcement has been applied to the lowering of the threshold of one reflex by an allied one, which is a corollary of subthreshold summation. Sherrington also used the term facilitation for this phenomenon—quite within the present strict sense, although he did not explicitly limit it to this case (see Glossary).

It is said that the pool of available motoneurons is fractionated, since each reflex uses a part of it. If we add the separate outputs of several reflexes using the same pool—by adding muscle tension or motor nerve action potential—we find the total is larger than the maximum possible at any moment; therefore the several reflexes must be sharing some of the motoneurons; that is, there is **convergence.** Two consequences of convergence are **occlusion** and **facilitation:** occlusion is the preoccupation of some neurons by one input so that they are not available to another and is more obvious with strong stimulation, since it involves more of the pool; facilitation is more obvious with weak stimulation because then a larger proportion of the neurons influenced by either stimulus alone are influenced only to a subthreshold degree, making up the so-called **subliminal fringe** (Fig. 5.19). The effect of these two processes, other things being equal, is to narrow the range of response between liminal and maximal stimuli. The two are not symmetrical in causation, occlusion being the result of the anatomical relations, facilitation being a physiological property, resulting in recruitment into action of hitherto silent units. Another consequence of convergence is the comparison and evaluation of different inputs,

in many cases, sometimes in intensity, sometimes in area, but if summation is not adaptive it does not occur. Actually many of the familiar reflexes—flexion, extensor thrust, pinna, crossed extension—are type reflexes, there being a number of separate reflexes possible from different receptive zones which overlap and are plurisegmental. Not only during these reflexes but also

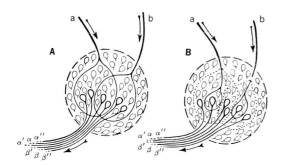

Figure 5.19. Occlusion and the subliminal fringe. **A.** Occlusion; two excitatory afferents, a and b, with their fields of supraliminal effect in the motoneuron pool of a muscle. By itself a activates four units ($\alpha, \alpha', \alpha'', \beta'$); by itself b activates four units ($\beta, \beta', \beta'', \alpha'$). Concurrently, they activate not eight but six; the contraction deficit by occlusion is equivalent to two units (α' and β'). **B.** Subliminal fringe. Weaker stimulation of a and b restricts their supraliminal fields of effect in the pool as shown by the *continuous line* limit. By itself a activates one unit; b similarly; concurrently they activate four units ($\alpha, \alpha', \beta, \beta'$), owing to summation of subliminal effect in the overlap in the subliminal fields outlined by *dots*. (Subliminal fields of effect are not indicated in **A**.) [Sherrington, 1929.]

including the opportunity for inhibition to play a significant role.

At any one moment, normally, stimulation is impinging on a constellation of loci and modalities, so that compound reflexes are the rule. Certain of the stimuli will be pre-eminent and therefore reinforcing allied reflexes will determine the posture and movements of the animal, preoccupying the final common paths to the exclusion of antagonistic reflexes. In the intact organism there are probably even fewer reflexes indifferent to the presence of others than in the usually studied preparation (for example spinalized or decerebrate). The knitting together of all the simultaneously combined reactions, far from being a simple sum of individual reflexes, is a complex and highly adaptive achievement, the more so because, preserving its adaptive character, the result is exceedingly variable from time to time and from specimen to specimen.

Turning to the case of **different reflexes elicited successively,** it is understandable that antagonistic as well as synergic movements can be employed. As the constellation just referred to changes like a kaleidoscope, the normal succession of stimuli inevitably results in conflicting stimuli overlapping in time, so that the shift from one reflex to another actually involves the already mentioned dominance of one, and exclusion of the other of two competing reflexes or groups of allied reflexes. There may be an alternation back and forth between them if both stimuli are maintained. But we gain the impression inescapably that the whole animal acts as though, like our minds, it has a single track, however often it may vacillate! (See further, p. 338.)

Relative intensities of competing stimuli represent possibly the most powerful of the **determinants of the succession** of winning reflexes. Consistent with what has been said, there is generally no intermediate intensity which gives an intermediate response, except in the case of allied reactions which can be graded. While there is usually no confusion, that is, mixing, of antagonistic reflexes, there may be hesitation—that is, the latency of the second reflex may be lengthened by the presence of the first. Furthermore, if the stimulus for the first response is still present, that response may reappear in altered form after the second has ceased; that is to say, successive reflexes may influence each other, for each is a part of the stimulus constellation determining the next.

A general property of reflexes which finds its greatest role in succession is rebound or **negative aftereffect** (Fig. 5.20). Although foreshadowed in the properties of unit neurons, we must admit that this is another of the major mysteries of coordinated nervous activity. Its consequence is to help in stopping one action and starting another, for not only are the active muscles of a given movement inhibited by the rebound but the reciprocally inhibited ones are excited. Alternating reflexes, such as locomotion, are probably significantly aided by this mechanism.

Perhaps related is an interesting class known as **compensatory reflexes.** These are illustrated by an active reflex return of a leg to a pre-existing condition after a brief interruption brought about by either a passive or an active movement. Once again we are reminded how "knowingly" regulatory the reflex level can be.

Chain reflexes are the important class where the

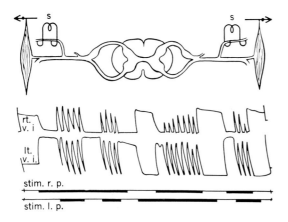

Figure 5.20. Production of rhythmic reflexes by continuous stimulation of two inputs. Stepping in the hind leg of the dog during concurrent stimulation of right and left peroneal nerves. The diagram at the top represents the experimental arrangement. The upper two traces are myographic records of tension in the right and left vastus intermedius muscles *(rt. v. i., lt. v. i.,)*, which are knee extensors. The lower two traces signal repetitive stimulation of afferent nerve fibers of the right *(r. p.)* and left peroneal *(l. p.)* nerves. Stepping ceases when either nerve is stimulated alone. s, stimulating electrodes. [Creed et al., 1932.]

adequate stimulus for each reflex is provided by the execution of the preceding one. It may be a proprioceptive input due to the position of a member, or a tactile input due to contact accomplished by the last movement, a change in photic input or some other. Micturition and defecation in mammals and the walking of leeches, with its succession of sucker attachments, sucker release, shortening, and elongation, are good examples. The timing is determined less by central time constants, which play the leading role in most of the cases so far discussed, and more by the peripheral events themselves. The whole sequence has a unity as real as that of a simple reflex and we cannot but conclude that the possession of a single afferent and a single efferent limb or a **single "arc" should not be a defining feature** of the category of nervous mechanisms called reflexes. Even in simple reflexes, in mammals, it is now well demonstrated that there is often a primary effect upon the motoneurons which supply only the intrinsic muscle fibers of the stretch-receptive muscle spindles (gamma efferents) and this change in stretch receptors is responsible for a second input signal which now controls the main muscle motoneurons. Since stretch receptors have been discovered recently in one invertebrate group after another, it will be interesting to see whether such servo-loop mechanisms occur elsewhere.

A **clear hierarchical character of reflexes** is observed and indicates a built-in value system basic to their integration. Some reflexes are said to be more potent than others; they take precedence or dominate when simultaneously stimulated. The nociceptive or protective reflexes are the most potent, the postural are least potent, and the others are intermediate in approximately the same order as the sensation in man produced by that form of stimulation is affective. For example, the clasping reflex in the frog—a sexual, not a postural reflex—is, as expected from the affective strength of the sensations in the sexual sphere in man, very strong. There is a difference also in fatigability, the postural being lower, the protective reflexes higher in this respect, perhaps correlated with their relative tonicity versus phasicity. These properties are little known in invertebrates but should be of fundamental interest.

Also of fundamental interest is the distinction in **ontogeny** which has attracted very wide notice since it was elaborated by Coghill, between the earlier- and the later-developing movements. Broadly speaking, it is the contention that the first to appear in lower vertebrates (mammalian development is interpreted differently; see Barcroft and Barron, 1939; Windle, 1940) are mass or total reactions and at this stage modifiability of response is primarily by nonnervous means. Later in the embryonic lower vertebrate, more local responses appear, new systems of neurons are ready to function for reflexes of individual limbs. Lastly, there is a synthesis permitting the coordination of local reflexes and the appearance of initiative and spontaneous actions, which show that the behavior of the whole organism is not only a sum of its reflexes, secondarily assembled, but is fundamentally a unitary, integrated whole.

Further consideration of these aspects would take us afield, but it is well to emphasize here that, in the intact animal, reflexes are heavily dependent on **descending** influences from higher levels

III. GANGLIONIC INTEGRATION B. Output Control 4. Pattern formulation

of the central nervous system and, at least in vertebrates, on "horizontal" influence from the nonspecific facilitating and inhibiting systems in the reticular substance. In widely varying degree for different species of animal and of reflex, there is a dependence upon some tonic influence to permit the reflex to occur at all. And in addition there is still more complex dependence on phasic influences which modulate and can suppress or reverse reflex action. One influence recently studied is called **recurrent inhibition**; this is the effect exerted upon ventral horn motoneurons of the spinal cord by antidromic stimulation of ventral root fibers that conduct impulses via recurrent collaterals back into internuncials in the cord. Granit (1961) and his collaborators in other papers have shown how this inhibition acts under various conditions of excitatory drive to the motoneuron. It appears to be a rather complex mechanism for controlling frequency of firing, especially stabilizing or limiting its range, suppressing feebly excited fringe cells, and preventing afterdischarge from lingering on.

4. The origin of patterned discharge

The output of single neurons and of groups of neurons is normally probably always patterned (p. 273)—that is, temporally and spatially distributed in a meaningful, nonrandom way. One way of stating the function of the nervous system or of any significant part of it is that it formulates appropriately patterned messages in code. The question how this formulation takes place is surely one of the core questions of general neurology. Curiously, it has received little direct attention, although a great body of related information is known.

Temporal pattern can be regarded as the more basic aspect. Theoretically, it can be expected to arise in one of two general ways: by following (a) timing cues from peripheral causes or (b) timing cues from central pacemakers.

(a) **Timing cues from peripheral causes** are exemplified by ordinary reflexes. Except in a few special cases, the input pattern of impulses is greatly "distorted" in the output. Commonly a large number of unsynchronized input impulses arriving at a given neuron cause a smaller number of rhythmic output impulses. The peripheral cause which confers the timing may be an environmental event—as in an eye-blink to an approaching object (Fig. 5.21, A).

Probably the most common peripheral cause of *rhythmic* patterns in *terrestrial* animals (note the two qualifiers) is **sensory feedback from proprioceptors** (Fig. 5.21, B). Any case belongs in this category where a feedback loop exists which actually starts the next cycle of events before some central pacemaker does (Fig. 5.21, C). The difficulty of distinguishing categories in practice is obvious from this requirement but the distinctions are nevertheless real.

A subcategory of some interest must be those patterns derived from receptors with **maintained**

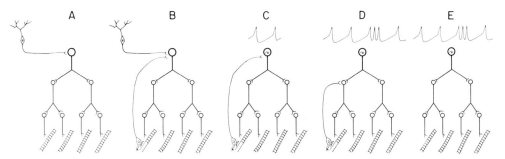

Figure 5.21.   Five mechanisms of pattern formulation. The three levels of neurons are understood to represent branching chains in whose junctions integrative properties may alter the actual impulses and deliver them spatially as well as temporally distributed, to the effectors *(bottom)*. **A** and **B** are shown with receptors, **C, D,** and **E** with spontaneous pacemakers giving simple or group discharges. **B** and **C** have proprioceptive feedback acting on the trigger neuron, **D** only on the shaping of the pattern. [Bullock, 1961.]

background discharge. The function connecting input and output may become complex. This is probably not the unusual but the usual case, especially if we take into account maintained background discharge of some of the central neurons in the pathway. A related class (discussed with examples on p. 322) consists of actions which depend on the presence of a tonic input from certain receptors to sustain the central excitatory state, even when those receptors are presumably not necessary for their sensory-informational contribution or for timing the pattern ("Stimulationsorgane" of von Buddenbrock and others).

Although in this general class the rhythmically recurring patterns are peripherally determined by feedback which is essential to the rhythmicity, the diagram (Fig. 5.21, B) shows that there must usually be some **triggering signal to start the rhythm** from a resting condition. This may be a sensory input other than feedback or a centrally arising command.

Most actions triggered by input signals contain elements of pattern not in the input. Consider eye movements, swallowing, coughing, a cricket's chirp, a grasshopper's hop, a squid's color change, taxes, and instincts, not to speak of more complex behavior. From the feeding movements of a sea anemone to formation of a word in human speech, the predetermined **central contribution to the pattern** is enormous even if the initiation follows a peripheral cue. Our examples have been confined to motor acts, but is it not just as likely that the output of most central masses of organized neurons constituting an integrating level is also patterned in a way not contained in its input? A pre-existing pattern, awaiting permissive input and then triggering input to release it, may reasonably be inferred to be, not universal, but frequent.

(b) **Timing cues from central pacemakers** are another way of triggering patterned output. In principle it should be no surprise to find that a perfectly coordinated sequence of reciprocal activation of antagonistic muscles forming an adaptive action can arise purely centrally. For in only one uncomplicated feature does this differ from the picture just drawn for reflexes—namely, that a central clock is provided which rings an alarm at intervals, triggering a predetermined mechanism.

The clock or pacemaker or central automaticity is by definition not dependent on phasic peripheral cues but it always depends on certain **steady-state conditions**—for example, temperature and ionic and organic milieu. It is highly likely that the automaticity often depends also on steady-state or background impulses, as it were, to keep it awake. Under certain steady conditions, an isolated heart beats, a respiratory center discharges in an isolated brain stem (Salmoiraghi and Burns, 1958), an embryo moves its limbs in a coordinated way before reflex arcs are formed, a resting millipede starts to walk, a ctenophore abruptly changes the direction of its ciliary metachronal wave, a hovering fish moves now this fin, now that. Wells (1950) concludes that the rhythmic movements of *Arenicola* are neither responses to environmental cues nor to the internal needs they normally satisfy. Still one can be sure some state or substance has accumulated to a threshold and caused the alarm to ring. But this is exactly what we mean by spontaneity: the determination by internal factors of when the alarm shall ring, given a steady application of energy or accumulation of something. Some examples of electrically recorded rhythms in the absence of sensory nerves are cited on pp. 320 and 1155.

The theoretical alternative of a **spontaneous rhythm modulated in frequency by feedback**—that is, by phasic input—is probably common (Fig. 5.21, C). It can properly be called a central automatism, since removal of the phasic input can alter but not stop the rhythm. But the feedback might also be competent to maintain the rhythm if it were possible to stop the spontaneity, and might in some cases take control by having time constants for a higher frequency than the autochthonous one.

The **feedback may only modulate the form** but not the frequency of the rhythm by affecting not the pacemaker but its followers (Fig. 5.21, D). The disturbances of walking in a tabetic or a blinded man show the effect of feedback on the form as distinct from the occurrence of a pattern.

The **feedback in any of these cases may be**

from central follower neurons and not only from the periphery. It may be a complex mixture including positive and negative, specific (from one muscle) and less specific (from many muscles), fast- and slow-adapting influx and may be more or less dependent on higher central influences (Eccles and Lundberg, 1958a).

Central automatism without feedback is not unreasonable (Fig. 5.21, E), at least feedback of immediate role in timing. Of course an eventual feedback is inevitable—the success or failure of the action for the biological welfare of the organism will influence the evolution of the mechanism. We know of such purely central cases; in fact, several kinds of cases have been discovered.

Some pacemakers are **fixed-frequency alarm clocks** so far as normal sensory influence is concerned. The weak electric organs of gymnotid fish like *Eigenmannia* discharge at a constant frequency of about 350 per second, day and night, uninfluenced by excitement, food, or other fish; only temperature influences it, among the factors so far tested (Lissmann, 1958). The electric discharge of *Torpedo* (Albe-Fessard and Szabo, 1954) and the sound production in the cicada, *Graptosaltria* (Hagiwara and Watanabe, 1956), involve fixed-frequency motoneuron firing; sensory input can determine only the duration of a burst.

More commonly, the neurons between pacemaker and effector are integrative followers, **modifying the details of the temporal pattern** within each cycle. In each heartbeat of a lobster a complex burst of several dozen impulses arises in a pacemaker neuron of the cardiac ganglion, even when the ganglion is isolated from the heart. It fires many times during a burst, and the other neurons—followers—fire repeatedly, each in a different pattern; for example, starting at high frequency and declining. The actual frequencies, time courses, durations, and numbers of impulses are individually characteristic and recur consistently for hundreds of heartbeats. This has been analyzed in some detail and the conclusion reached that a primary patterned burst arises in a single cell, the pacemaker, not dependent on feedback of spikes from other cells to formulate it. Intracellular mechanisms, presumably two or more interacting processes, must be postulated (see Sections II and IV of this chapter).

These, then, are the main alternative sources of initiation of pattern. It is probable that each of them actually occurs in certain cases. In most cases it has not been unequivocally determined which category is involved, for example, in locomotion in insects, fish, and terrestrial vertebrates. The methodology necessary is similar to that used in the analysis of control systems (see p. 323), namely, the interruption of, or injection of a spurious signal into, selected parts of the presumptive loop. To accomplish this in a known manner—for example, to cut all the sensory feedback without disturbing other elements, such as reducing some essential tonic input—is technically difficult because most nerves are mixed. As a result the literature, though abundant, is conflicting and in most instances inadequate for a satisfactory evaluation.

Some statements are readily established, however. When the action never occurs until released by external stimuli, it can be said to be a **pure case of peripheral timing.** Although "never" is a hard requirement to satisfy, possibly many reflexes such as the lateral giant fiber startle responses in earthworm and crayfish are examples and, barring the discovery of very complete "vacuum" mating and feeding, without mate or prey, the finer movements of copulatory organs or mouthparts may be. But in view of the known occurrence of vacuum activity (see Verplanck, 1957), this cannot be assumed without careful examination in any given case. The importance of the central elements in determining the shape and course of such responses is clear upon consideration of the simplicity of the timing trigger in the stimulus situation. Wiersma (1952) speaks of push-button responses exemplified by giant fiber-mediated escape movements. These are peripherally timed but it is not correct to speak of peripherally "controlled" movements. If my enemy strikes, he may precipitate my action but he does not control it!

When the action is obviously disturbed in pattern by cutting some sensory pathways, it can be said that **peripheral elements** at least play a role in shaping the pattern. The hind limbs of a frog, if all dorsal roots contributing to them are severed, are somewhat abnormal in details of righting and hopping, though these movements can be performed. A mantid makes errors in striking at prey after neck proprioceptors are cut. The immediate change of gait of insects upon amputation of some of the legs and the restoration of normal gait, in some experiments, by gluing pegs onto the stumps, points to the same conclusion in these instances (Hughes, 1958).

But it cannot be said that the rhythm itself, apart from its detailed form, is peripheral or feedback-determined unless rigorous experiments show that the rhythm fails when nothing but timing cues in the input are cut off. The experiments of Gray and Lissmann (1940, 1946) on ambulation in toads appear to have shown just this. A toad continued to show stepping movements with all dorsal roots cut but one, the minor cutaneous tenth dorsal root, but stopped when that was cut. The probability that the minimum essential input was phasic, providing a timing cue, rather than merely tonic, was indicated by the behavior of preparations with three of the limbs plus the dorsal musculature immobilized by section of their motor roots and the fourth limb left at first intact, but with its nerve supply exposed. Relatively weak stimuli to any limb sufficed to elicit a clear ambulatory rhythm. But when the dorsal roots of the intact limb were severed, all rhythmic response disappeared, even to strong stimuli. The previously intact limb still gave specific reflex responses but only monophasically. "It is difficult to draw any conclusion other than that when [phasic] proprioceptor impulses from other limbs are effectively excluded, the impulses arising in the proprioceptor endings of the intact limb are essential for the maintenance of the ambulatory rhythm" (Gray, 1950, p. 117).

Similar experiments on other species are extremely desirable, for both the results just mentioned and the opposite results (listed on p. 319) seem clear; some relation to species or type of movement seems required. The use of additional techniques is also desirable—for example, interrupting ventral roots and looking for rhythmic discharge in their central stumps by electrical recording. Still another technique of servoloop analysis, that of injecting artificial signals, has also been little used. Von Holst manipulated the fins of a fish and adduced evidence that this did not determine or even seriously influence the rhythm normally controlling that fin, since the rhythm could still be seen as usual superimposed on the separate rhythm of another fin.

Another type of evidence pointing to pattern formulation by central neurons, even without any reafference (feedback), is the literature on **brain stimulation in unanesthetized animals,** a technique now in use by many authors. In predisposed humans, Penfield has found that certain regions of the cortex can be crudely stimulated electrically and complex, vivid audiovisual experiences triggered. The subject reports a scene as though reliving it, each time and only while the electrical stimulus is applied to that spot. Less complex but normally coordinated movements occur in lower forms. Although many regions of the brain are "silent" or yield simple jerks or twitches, particular loci call up entirely normal-appearing sequences such as chirping or antenna-cleaning in crickets (Huber, 1960), compulsive drinking in a goat (Andersson, 1953), stuffing imaginary seeds into the mouth in a pocket mouse (Strumwasser and Cade, 1957), crowing and many other actions in chickens (Hunsperger, 1956; Molina et al., 1959; von Holst and St. Paul, 1960). Intensity of electric shocks has in some instances little effect beyond changing the probability of occurrence of the characteristic response; localization within the brain is sometimes not narrow and is sometimes multiple (see also pp. 1142, 1498).

The category of patterned movements represented by most of these responses ("fixed action patterns") is of special interest for our problem. Other familiar examples are sneezing, jumping, spitting, stinging, swallowing, seizing, displaying, and the like. They are in general not rhythmic, like heartbeats and locomotion, but are brief and episodic and suggest in certain cases rather less **opportunity for feedback control** of the pattern than in others. This could well be studied with the techniques already mentioned. Mittelstaedt (1960) provides a review of a few cases already examined from the point of view of control circuits. The central origin, which is the aspect of interest here, is found in such diagrams as "higher command" or an equivalent term. It seems at present likely that for many relatively complex behavioral actions the nervous system contains not only genetically determined circuits but also genetically determined physiological properties of their components, so the complete act is represented in coded form and awaits only an adequate trigger (Bullock, 1957a), either internal or external. In the former event we have a built-in tendency to spontaneous discharge under suitable steady-state conditions, which may include deprivation of external releasers and gradual increase of probability of release, by weaker stimuli or eventually "in a vacuum," without external trigger.

Boycott and Young (1950) wrote a relevant passage in discussing the comparative study of learning:

From such experiments we get the impression that an animal, whether it is *Octopus*, *Sepia* or any other, is provided with a delicate system of springs or drives that very readily set it into motion. The effect of a "stimulus" is to allow a complicated set of nervous actions to begin, so that the movement which results continues for a considerable time, even after cessation of the original stimulus. During this time, therefore, the action is being produced, as it were, from within the organism. It is the function of the more elaborate nervous systems to allow such behaviour to be long and complicated. The large masses of nervous tissue found in higher animals have in part this very function of allowing sustained action under internal control.

C. Associative Levels of Moderate Complexity

Our principles of integration have become gradually more difficult to specify. Some cannot be confined to the level under which they are listed. As the higher levels are considered it is still harder to discern new mechanisms as distinct from the extensions of the ones recognized before. In fact, the prime question cannot yet be answered: do the highest achievements of the nervous system rest on emergent mechanisms or on the quantitative addition of units and interconnections with the same properties as found in levels or nervous systems of less complex behavioral achievement? This distinction may not be familiar but it is quite fundamental (Redfield, 1942). We are used to the fact that a relatively complete knowledge of one level, say the properties of hydrogen and of oxygen, do not yet remove the surprise from the discovery of the properties of the next, in this instance of water. The question just posed may be equivalent to asking whether the highest nervous systems are in fact a different level in this meaning than the nervous systems in lower animals and at lower anatomical levels. It is the working hypothesis of the writer that they are, that there are emergent mechanisms which remain to be discovered before we can explain the more complex forms of behavior, especially that of man, in physiological terms, and that these discoveries will solve the brain-mind problem, removing the temptation to place the mind in the dualist's position as having liaison with the brain but being itself a nonmaterial separate entity (Eccles, 1953). Until these discoveries, however, this remains more an article of faith than a hypothesis.

The **associative level** is clearly real, though not easily distinguished from the levels of analysis of input and formulation of output. This rests, to be sure, on vertebrate physiology—indeed, chiefly upon clinical neurology. But there are tell-tale signs of the same distinction in bee, lobster, and octopus. It is rather unsatisfactorily defined as the level, region, or function not mainly concerned with either one of the other spheres—sensory processing or motor control—but with both. In the following paragraphs we shall not find the answer to the question whether new integrative principles are at work in the associative level, but we shall find properties and specializations which could not be seen at lower levels.

1. Degrees of coordination; "relative coordination"

Phasic events in nerve cells or effectors may be either coordinated or uncoordinated. The familiar cases of the former, like all those considered above under reflexology, are absolutely coordinated; we see a certain play of phase relations but there are not degrees of coordination. Of course, one can distinguish the obligately from the facul-

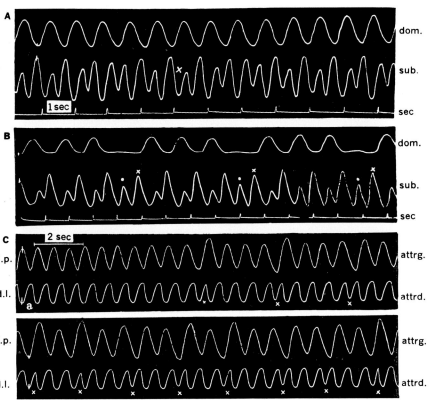

Figure 5.22. The interaction of rhythms; superposition and the magnet effect. **A.** Mechanogram of the movements of the left pectoral fin *(upper)* and the dorsal *(lower)* of the fish *Labrus*, spontaneously moving after transection of the medulla. The upper rhythm is "dominant" and irreciprocally influences the amplitude of the lower "subordinate" rhythm. The effect is independent of holding

tatively coordinated instances. In normal humans the movements of the extrinsic eye muscles and of the respiratory muscles are in the first category, those of the fingers and limbs in the second because they can under different stimuli or voluntarily be activated in a coordinated or an uncoordinated manner. But there are well-studied examples of a less familiar class—that of relative coordination, a term and a class whose recognition is chiefly due to von Holst (see 1943, 1939, and earlier); all degrees of coordination from perfect to zero can be found and the partial coordination is not coincidence but is highly systematic.

Little enough is understood of the mechanisms of absolute coordination, beyond the assumption that it depends on fixed circuits with the timing built in by the properties of the elements of the pathways. But still less appreciated are the interesting problems provided by the phenomena of relative coordination. The problems of absolute coordination have been touched upon in the preceding sections; we may examine the problems of relative coordination superficially here, although space precludes an explanation of the ingenious methods of analysis which we owe to von Holst. These methods were first applied to the peculiarly favorable movements executed by the fins of certain fishes. Rhythmic events offer in a given time far more opportunity to establish and measure coordination than aperiodic events (like jumping in a grasshopper). In many fish the several fins are not always perfectly coordinated but evidence **separate rhythms,** which, without losing their individuality, interact in law-abiding ways.

Two principal processes have been invoked: superposition and magnet effects. **Superposition** is shown when the mechanogram of a fin, for example, the dorsal fin of *Labrus*, contains systematic deviations from a single frequency and am-

or manipulating the pectoral fin and therefore is not due to proprioception. The vertical marks at the left are coincidence marks for the levers. Time is indicated in seconds. **B.** The same in a case of regular spontaneous block or skipping of one cycle of the dominant rhythm. The two fins are for the moment locked in a 2:1 frequency relation, but besides this an amplitude modulation is prominent. The dorsal fin beats, which are in phase with the pectoral, are small but become larger when the latter is blocked. Those out of phase are largest but become significantly reduced following a missing pectoral beat (x). **C.** Magnet effect in the fish *Sargus* between the left pectoral fin (l. p.) and the dorsal lobe of the tail fin (d. l.). Beginning in perfect syn-

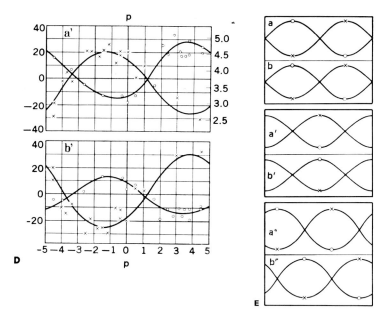

chrony the frequencies gradually move apart during the record; the tail rhythm becomes more frequent but is partially held by the pectoral, more and more often escaping (x). The ventral lobe shows the same rhythm as the dorsal but a stronger superposition effect on its amplitude. **D.** Analysis of **C.** (a') The deviations of duration of individual beats of the dorsal lobe of the tail fin from the average are plotted against time, expressed as tenths of the pectoral fin cycle, before and after an arbitrary reference point in each cycle. *Circles* are the peaks (movement of the tail fin to the left), *crosses* the valleys (movement to the right). (b') Deviations of velocity of the beat (angle of slope between peak and valley, from the horizontal) from the average are plotted on the same abscissa. **E.** Scheme of the results obtained by analysis as in **D.** *Top.* Pure superposition or amplitude effect of one rhythm upon the other. *Middle.* Pure magnet or frequency attracting effect. *Bottom.* Equal mixture of the two. The significant feature of each case is the comparison of the relative position of the (a) and (b) or duration and velocity curves. The scheme holds for all cases where the attracted or the subordinate rhythm is higher than the attracting or the dominant rhythm. [From von Holst, 1935, 1936b.]

plitude and hence speed of movement. These are regarded as resulting from the superimposition of two rhythms. One of the two can be identified as belonging to that fin, the other as belonging to another fin; these can be called the dependent and the dominant rhythms, respectively. Typically the pectoral fin is dominant and shows only its own rhythm; the dorsal is dependent and though it shows its own rhythm at times in pure form, at other times it is influenced in some degree by the dominant rhythm (Fig. 5.22). The interaction is such that amplitude is increased when the rhythms are 180° out of phase; variations in frequency and in strength of the dominant—which can be simultaneously recorded in pure form in the other fin—produce precisely the expected effects on a simple theory of superimposed rhythms in which the imposed modulates the amplitude and frequency of the dependent rhythm periodically about its own natural value as an average. Cases have been recorded of three and more rhythms in the same spinal fish, mixing in the dependent fins in systematic fashion. Von Holst emphasized that the strengthening and weakening of nervous action by superpositions of the same and of opposite sign are separate and distinct from the more familiar forms of enhanced excitability and inhibition.

The fin **rhythms are generated** in cells other than the motoneurons and in all probability not in receptors but within the central nervous system (see further under "Spontaneity," p. 319). The motoneurons mix and preserve the separate rhythms and in addition integrate other central and peripheral influences. Reflexly evinced heightening or diminishing of the fin movements occurs without altering the automatic rhythmic generators or even the degree of interaction of two rhythms. The rhythmic automatisms can be suppressed without cutting off the fins from excitation from reflex or from other central sources, including other fainter rhythms. Von Holst imagines that the automatic cells do not directly drive the motor cells but—under proper conditions of activity of the latter, determined by various central and peripheral influences—that they displace the excitation equilibrium, pushing it rhythmically in reciprocal directions in the antagonistic motor cell group.

The **magnet effect** is regarded as a different process. Here, one rhythm tends to pull the other into its frequency (Fig. 5.22). The attracted rhythm shows periodic weakening of its susceptibility and the attracting rhythm shows periodic fluctuation of its strength and direction of action —retarding or accelerating. A dominant, superposing rhythm can be attracted by the dependent but magnetically attracting rhythm; or the same rhythm can be both dominant and attracting. The combined result of the two processes leads to complicated forms of periodicity. Peripheral stimuli may influence the coupling functions, heightening or diminishing them quite independently of the excitatory or inhibitory actions these stimuli otherwise exert (see further "Plasticity," pp. 309, 1146). Some influences can destroy the synergy of the cells within one ordinarily perfectly coordinated group—itself probably a magnet effect— so, for example, the individual rays of a fin exhibit separate rhythms and the whole fin an irregular flutter recalling fibrillation. Many special cases resembling phenomena in the vertebrate heart are seen—for example amplitude and frequency alternation, block, Wenckebach's period, and escaped beats.

Here is certainly a set of integrative mechanisms of extreme interest, deserving more attention. At the lower levels of cellular analysis, internal events during these interactions would be showpieces of neuronal integration. Models reproducing these properties are readily available and are in the class of **coupled, nonlinear oscillators** (von Holst, 1939; Pringle, 1951). Especially during the lag period, before synchronization of two oscillators is established, the two frequencies can be different though influencing each other. The influence can be quite asymmetrical, favoring one of the two frequencies. Pringle (1951) has based an ingenious hypothesis of brain function, including learning, on this asymmetry of coupling in a large population of oscillators where cells are often shared by two or more loops.

It seems extremely likely that interactions like the magnet and superposition effects play a role in a wide category of events where more than one source of timing is involved—where two or more elementary fixed patterns of action are coordinated, as well as in entirely central processes not directly expressed in motor output. They have been found in man—for example in arm and leg phase relations in swimming the "crawl"—in insects and centipedes, and in head and leg coordination in the chicken (Bangert, 1960). All these involve locomotion or rhythmically repeated movements, and this merely because analysis is easier.

Von Holst underlined the support these phenomena provide for the concept that **coordination** is basically the result of centrally provided relations and automaticities which are only modified and steered by peripheral stimuli, as opposed to a purely "reflex" theory of coordination, meaning one based on stimulus-response relations. It is encouraging to find that essentially new principles and mechanisms at the level of the spinal animal are accessible to discovery and analysis. (See further, p. 319.)

2. Functions tending to topographic segregation

One of the continual sources of surprise and fascination in functional neurology is the kind of activity, process, or sign found to have some degree of anatomical segregation—in the classical phrase, functional localization. The chief means of analysis are the signs resulting from ablations or lesions and the responses to stimulation, either artificial or pathological. Both are subject to **difficulties of interpretation,** even aside from artifacts. A symptom appearing after ablation may not mean that the region removed has a center for that function; this is clear in the extreme case where the ablation merely interrupts a tract. Some symptoms prove to be temporary though lasting for minutes to months, and are ascribed to "shock"; this appears to be important only in mammals. The absence of effect of ablation has time and again proved to be due simply to the absence of suitable observation or functional tests. The effects of electrical stimulation may be due to current spread, especially to nearby tracts, and hence involving distant centers. Absence of effect may mean simply that this neuron aggregate is not organized to respond to the nonspecific stimulation of our lightning bolt-like electric shocks or may be due to depressing effects of the anesthetic. While these remarks are not exhaustive they suggest that caution, convergence of methods, and the repetition of classical experiments are all needed. This field is today actively advancing and rich harvests are to be expected, particularly in comparative studies among lower forms, both vertebrate and invertebrate.

A new technique of great power is to record from many sites in the brain while delivering a stimulus (such as a light flashing 10 times per second) that elicits distinctive and easily recognized potentials. The sites showing this marked response are at first very widespread, then greatly narrowed as habituation to the flashing takes place; again it is widespread—but not in all the same places—when flashing is paired with punishment, and then advances through phases to still another pattern as learning and correct response to avoid punishment is developed (John and Killam, 1961).

In a complex organization such as an industrial factory there can be recognized **two categories of functions,** those which are segregated into special buildings, offices, or departments and those which are diffused or repeated in many of these places. Divisions of engineering, production, advertising, and sales illustrate the first category. Analyzing, record keeping, learning from experience, and decision making are in various degrees diffused, decentralized, or repeated in each division, although perhaps localized within the division. Other functions which are diffused—such as reading, subtracting, multiplying, and comparing—are properties of the individual units of the system. The higher functions may also be carried out by individuals but whether this is an instructive feature or a flaw in the analogy with nervous systems cannot be decided at present. There are other attractive aspects of the analogy but the only purpose here is to clarify a distinction between types of function. Great progress in neurophysiology will have been made when we can construct the corresponding lists of localized and diffused functions in the nervous system. This, of course, has been a major effort for more than a century and a great body of knowledge exists for the mammals, much of it undoubtedly highly relevant to a general consideration of lower animals. The following represents a selection of information, especially from higher primates.

So-called **primary sensory areas** of the cerebral cortex are easily defined in man; they mediate simple, crude sensation for a given modality, corresponding to a specific afferent pathway. The best examples are the areas for vision, hearing, and skin senses, and these are subdivided into representation for parts of the peripheral field. Not all modalities have specific areas at the cortical level: pain is usually said not to be produced by any stimulation of the cortex, nor prevented by any ablation of it. Temperature, position, touch, and pressure are not grossly separate on the cortex, but Mountcastle (see 1961) found a clear segregation of certain submodalities among them into adjacent microcolumns of the somatic sensory cortex. The extent of the area representing a given receptive field in the cortex is larger or smaller, not in correlation with the hierarchical position of the modality in the scale of dominance of reflexes, but in a scale which appears to be the importance of fine discriminations. Thus distance receptors are assigned more space than proprioceptors or nociceptors or interoceptors, and within the area of touch response more cortical space is given to parts of the body surface like the hand in man and the snout in the pig. Something of the same kind is true of the cortical area more concerned with motor functions. Details of the localization of each sensory modality cannot be given here; the point at issue is the segregation in the cortex of primary from secondary receiving functions. **Secondary sensory areas** are indicated by lesions which cause loss of ability to recognize familiar objects, without loss of crude sensation. Some authors recognize tertiary sensory areas in which lesions cause disturbances of associations such as danger, though both sensation and simple recognition are spared.

Electrical stimulation elicits **motor acts** from a limited group of brain regions; some give only simple movements such as opening and closing the hand or turning the head. From a different group of structures more complex but still quite reproducible actions are produced, such as vocalizations, eating, or sucking motions with salivation and swallowing. Other circumscribed loci yield postural and attitudinal responses involving many muscles. Some of these responses are complex and have the appearance of moods such as fright or aggression or feeding and may appear to involve "hallucinated" objects (Hess, 1956, 1957; Huber, 1953, 1955, 1960; Strumwasser and Cade, 1957; von Holst and St. Paul, 1960). These findings are of great interest both from the point of view of assigning anatomical locus to differences in function and as a distinct type of opening or handhold in the problem of the dynamic organization of central representation. The value for anatomy is clear from the fact that such normal, complex, and coordinated movements are generally not elicited by stimulating tracts, white matter, or nerves. It is not easy to establish "the" locus of a function or "the" function of a locus but methods and sophistication in the area are improving. The problem of dynamics has its own interest; little is known quantitatively of the ways in which behavior control is fractionated, graded, summed, and timed. We have still to learn the influence of stimulus parameters in complex responses, exemplified by clucking, crowing, pecking, and standing in chickens (von Holst and St. Paul) or by still more complex responses such as reaction-to-enemy-on-ground. The authors named show the opportunities for study of interplay between loci of different levels of integration —for example, they find a locus evoking only neck-twitch, a locus for peering-movement including neck-twitch, and a locus for the combination: peering plus vocalization plus attack on foreign object. Stimulating in two places at once adds to the possibilities of detecting dynamic relations, as does the manipulation of environment—for example introducing a stuffed weasel.

Learning and **memory** cannot apparently be localized narrowly, but certain learned items are —speech and the visual, tactile, and auditory recognitions of familiar experiences. Speech and the symbolization associated with it have more than one area, and complex dissociations of parts of the total function can be caused by restricted lesions in the dominant hemisphere. Something called short-term memory can be differentially damaged by restricted lesions (Sanders and Young, 1940; Nielsen, 1958). Memory of events in one's personal life can be lost without affecting memory of impersonal "common" knowledge such as the date 1492. In patients predisposed by an epileptogenic process, Penfield could reproducibly evoke complex, vivid recall of specific scenes and sounds by local electrical stimulation of the temporal lobe. Circumscribed lesions in man have been followed by loss of "body image" on one side; such patients are unaware of one side of their body or do not regard it as belonging to them. More bizarre are the rare cases of visual agnosia for animate or for inanimate objects (Nielsen, 1946). Such patients, in consequence of localized cortical lesions, overlook all objects in their visual field which share the quality of animateness (or in other patients, of inanimateness) by their own standards; thus, one may correctly identify several kinds of flowers, familiar people, fingers, arms, and even artificial teeth but not notice a hat, telephone, automobile, or food.

Many localized lesions in man cause **subtle deficits in behavior;** the proper designation of the functions represented is a long, slow task. Lesions confined to the cingulate cortex are said to produce severe loss of volition including will to respond, although sensation and consciousness are apparently present. Small losses in the mesencephalic central gray cause enduring coma. Consciousness is not lost with any cortical excision but is vulnerable to restricted deeper insults. The deficits following loss of extensive prefrontal cortex, the most recently acquired structures in evolution, are subtle and not in the sphere of memory or intellectual skills but in the realm of social responsibility and personality.

Overt **behavior associated with particular "emotions"** and drives (that is, behavior suggestive of

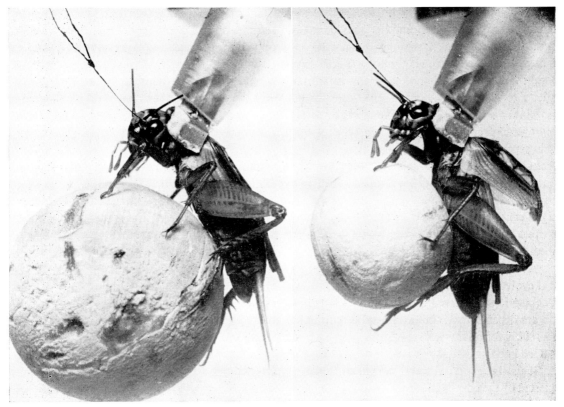

Figure 5.23. Focal brain stimulation producing behavioral signs in the unanesthetized insect. The mole cricket *Acheta domestica* is fixed so that head, antennae, legs, and wings are movable; it holds a cork ball that turns when it runs. Electrodes of fine tungsten wire are inserted into the brain. *Left:* before stimulation. *Right:* during stimulation of a sharply defined region. The cricket sings in a normal manner and moves its antennae and wings to a typical posture. [Courtesy F. Huber.]

emotions in man) is precipitated by lesions and stimulation in unanesthetized animals in definite circumscribed areas. Marked signs of docility or tameness—or, on the other hand, of wildness—can be produced at will. Actions indicating appetite, thirst, or mating drive have likewise been exaggerated following specified local interference. In connection with psychological discussions over the concepts of emotion and drive, it is significant that such clear-cut syndromes are reproducibly evoked by lesion or stimulus, and we may expect knowledge to accumulate rapidly in this sphere not only for vertebrates (Hess, 1956; Molina et al., 1959) but also invertebrates where a beginning is at hand (Huber, 1960); see also Fig. 5.23.

Lower associative functions which are segregated in the vertebrate nervous system are extensively known. Only a few examples will be mentioned. Several vegetative functions are represented in limited regions of the brain—for example thermoregulation, respiration, and some of the functions controlled by the pituitary gland. The central mechanisms, though they are localized, are not entirely concentrated into one center but may be spread out over some distance or successively higher centers may be superimposed. Nor are the different functional centers entirely discrete anatomically; they overlap, and apparently quite unrelated functions may coexist in the same mass of cells. Some of the same cells are probably involved in widely different functions, perhaps having a common denominator such as sympathetic excitation.

Another extraordinarily large mass of nervous tissue devoted to lower associative functions (corpus striatum, cerebellum, and so on) has to do with skeletal **muscle movements,** both tonic and phasic. These masses are not essential for reflex or voluntary use of muscle but if they are damaged, symptoms in the motor sphere result—compulsive movements, loss of associated movements, awkwardness, or tremor. Within the cerebellum, coordination of equilibratory movements required as a result of influx from labyrinthine receptors of position and motion may be segregated from that of postural contractions using proprioceptive and exteroceptive influx from all parts of the body, and this in turn may be segregated from coordination of commands coming from cortical levels. The full functional significance of many structures (for example the corpus striatum) cannot accurately be stated. It would be of real interest for comparative neurophysiology because there has been great anatomical evolution of these masses in the vertebrates. Lacking homologous structures in invertebrates we are dependent on the functional definition in searching for analogous structures.

One important feature which seems to have an analogue in invertebrates is the **tonic inhibition** exerted by the corpus striatum upon spontaneous movement. Lesions of sufficient extent in this mass in mammals release various kinds of forced or uncontrollable, compulsive movements. One attempt to refer in a few words to the important principle involved in the integrated initiation and suppression of phasic and tonic contractions is that of Hughlings Jackson, who spoke of "the cooperative antagonism of the cerebrum and cerebellum." A possibly analogous relation obtains between brain and ventral ganglia in insects and others (see Huber, 1960; Prosser and Brown, 1961). The ganglia of the segments bearing locomotor appendages include mechanisms coordinating the movements within and between them. The subesophageal ganglion regulates the state of excitation and spontaneity in these thoracic centers and the brain governs the directions of locomotion, body attitude, and hence tonus. (See further under the several systematic chapters.)

3. Specific and nonspecific systems

A discovery of sweeping importance in the coordinative functions of the nervous system is that, parallel with the system of labeled lines or specific pathways for the several modalities of input, there is a system of neurons and pathways not specific for any modality but profoundly influential upon the interpretation of input and determination of output. Here again, we borrow a general principle known thus far only in mammals and underline the opportunity and significance of ascertaining the nature of its evolution in lower vertebrates and invertebrates.

In mammals, fibers leave the classical specific sensory pathways and enter the **reticular formation,** where interneurons can be found which are activated by widely disparate modalities and regions of the body, hence are "nonspecific." This formation is not only considerable in mass and in extent along the neuraxis, but by its activity or quiescence permits or prevents the specific pathways from exerting any activating effect upon the motor command systems. If even a considerable fraction of it is removed or silenced, the brain and the animal remain in a sleeplike state (Fig. 5.24). Smaller lesions have related effects—for example interfering with learning. Before the functional appreciation of the reticular substance, the framework of our understanding of the central nervous system was by levels, the higher superimposed on the lower, the suprasegmental upon the segmental. Now it is clear that we must recognize at the same time a subdivision at right angles to the levels—from the cortex to the caudal end of the cord there is a nonspecific thicket side by side with the specific fascicles, with a functional distinction of a new type. In contrast to the hierarchy of information-extracting levels, superimposing ever more complex associations of specific messages, the relatively nonspecific reticular system, broadly speaking, is necessary for and acts to galvanize the specific system into action, unifying its efforts by confining the direction of **attention,** thus **arousing, alerting,** and focusing action. It can suppress input already arrived, by inhibiting—at the first relay as well as later! It can generally facilitate or inhibit the motor response by acting to "set the stage" at the lower motor level, that is, influence the result of later arriving corticospinal command. And it can awaken the cortex and deeper gray masses, permitting their full expression. If deprived of this influence, the animal tends

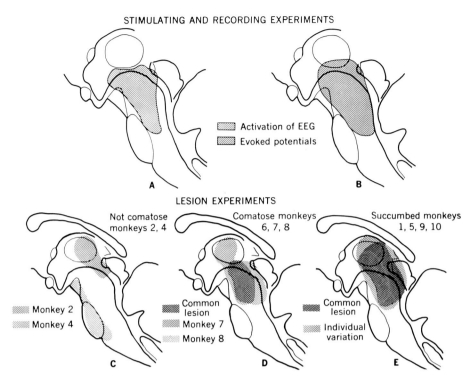

Figure 5.24. Reticular activating system. Midsagittal reconstructions of the brain stem of the monkey, showing **(A)** area from which activation of the electroencephalogram is obtained on direct stimulation; **(B)** area from which evoked potentials are recorded on peripheral stimulation; **(C)** area whose destruction does not produce comatose animals; **(D)** area whose destruction produces comatose monkeys, **(E)** area whose destruction the animals do not survive. [French and Magoun, 1952.]

toward a passively vegetable existence even though the specific systems are intact. Whether this is an ancient heritage or is a novelty associated with the mammalian habit of life becomes a general neurological question. In any case it is an integrative mechanism of overriding importance (Magoun, 1958; Pribram, 1958; Jasper et al., 1958).

Other evidence of parallel systems is at hand: one for more specific and another for relatively unspecific information. The great ascending somatic sensory systems of the mammalian spinal cord and brain stem are the lemniscal and the spinothalamic pathways. The lemniscal has physiological properties suggestive of specific information carriage: high ratio of activity following stimulation to background activity, absence of randomness in the distribution of intervals between spikes in a unit, high resistance to changes in condition of the animal. The spinothalamic system conveys information into the brain of a more diffuse and unspecific type as to the locus and form of the stimulation to the body surface (Poggio and Mountcastle, 1960).

The diffuse activating system might be likened to the labor relations department of an industrial organization; it parallels all other departments and, though nonspecific with reference to the particular function of each unit, nevertheless plays a vital role in keeping all kinds of specific functions active toward a unified end.

4. Central control of input

The preceding paragraph has introduced us to the principle that the central organ may listen to or may shut off or turn down its receivers. This occurs not only as a relatively nonspecific event, it is provided for in many sense organs as a specific, local action. We shut our eyes and cover our ears; the pupils and the tensor tympani reflexly contract. There is a great deal of movement of antennae, raising of eyestalks, dabbing with taste-receptive surfaces, turning of auditory

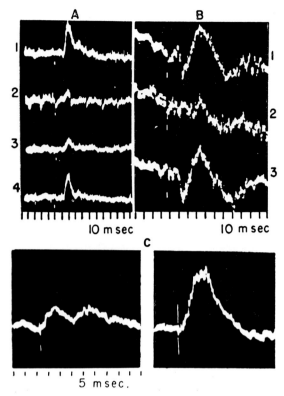

Figure 5.25. Central control of afferent input. **A.** The influence of hypothalamic and toe pad stimulation on responses to sciatic nerve stimulation, recorded within the brain stem reticular formation. Recording from the right side of the midbrain reticular formation in a curarized cat without central anesthesia: (1) before, (2) during, (3) 8 sec after, and (4) 20 sec after repetitive stimulation in the right hypothalamic region. **B.** The same: (1) before, (2) during, and (3) 10 sec after pinching the toe pads of the right hind limb. [Hernández-Peón and Hagbarth, 1955.] **C.** Relief of tonic descending inhibitory influences by cord transection; curarized cats without central anesthesia. Response recorded in the left ventral column to feeble L_7 dorsal root stimulation: (left) before and (right) 1 hour after high cord transection. In each experiment the stimulus intensity and location were kept constant; the dorsal columns had been transected at level L_4. [Hagbarth and Kerr, 1954.]

organs, and the like. Also muscular is the resetting of the sensitivity of stretch receptors by control of the tension of the intrinsic muscle fibers in both vertebrate and crustacean muscle. In some instances there is **direct nervous control of the receptor** neuron in the periphery—for example, in the crustacean muscle receptor organ, where an inhibitor axon ends on and about the dendrites of the primary sensory neuron just where

they transduce mechanical into electrical events (Eyzaguirre and Kuffler, 1955). In the mammalian ear, the cochlear microphonic potential is slightly potentiated and the action potentials in the auditory nerve strongly reduced by stimulation of a central tract, the crossed olivocochlear bundle (Desmedt and Monaco, 1961); controls indicate a direct effect on the sense organ in the organ of Corti, possibly upon both the hair cell and the afferent nerve endings (p. 72).

Another form of modulation of receptor activity has been discovered in the isolated frog skin, prepared to permit recording from tactile afferents while stimulating sympathetic efferents (Loewenstein, 1956). Here a pronounced **autonomic enhancement of afferent activity** takes place, manifested as a prolonging of adaptation and a lowering of threshold. The effect is specific for this type of afferent and does not involve pressure-sensitive receptors in the skin.

Editing of input occurs at more central stages as well (Fig. 5.25)—for example, at a **second- or third-order neuron** in the cochlear nucleus of the cat (Hernández-Peón, 1959; Ruben and Sekula, 1960) and probably in the retina. Both local and more distant pathways may be involved. In the vertebrate retina centrifugal fibers have been described which pass out the optic nerve to modify the response (Dodt, 1956; Granit, 1956; Maturana et al., 1960). In addition there are intraretinal circuits which depress the elements around a strongly stimulated spot (Hartline et al., 1960). Higher-level editing in successive stages probably continues all the way to perception or motor command (Bruner, 1957).

The advantages of being able to control the range of receptors and to permit in this way some input to dominate attention are obvious. But the accuracy with which the central nervous system is informed about the world depends on whether it "knows" by how much it has changed the sensitivity. In some cases it is quite reasonable to assume that it does, and that therefore no essential ambiguity results. In other cases, however, it is difficult to suppose that the central nervous system can but roughly correct for the change in sensitivity, and therefore that these forms of influx can have a precise quantitative meaning.

Besides sensitivity and attention, it is altogether likely that central influences on input are exerted selectively with respect to its information content. It has already been noted that the process of analysis and integration of input by central receivers involves convergence and successive selective manipulations (p. 279). These events overlap with central influences on input. Both lead to **selective sacrifice of information** but to an adaptively simplified final answer.

5. *The weighting of influx, mood, or set*

The integrative mechanisms of the nervous system determine not only quantities but qualities as distinct as pleasure is from pain. (We reject the ancient proposal that the one is really the absence of the other.) In turn, the pleasure-pain balance, the affect, set, or "Stimmung" (see von Holst, 1950; these terms are not necessarily equivalent) decisively **colors the interpretation of input** and the formulation of output (Fig. 5.26; 5.39). At first glance this may seem anthropomorphic. But there is abundant objective evidence that lower animals—octopods, insects, even medusae and sea anemones—have lasting phases of quite different responsiveness from other periods. One cannot avoid thinking of moods. Either integrated sensory input or spontaneous changes can set up a disposition to react in a certain limited direction and to place more weight upon certain types of stimulation than otherwise. A "sulking" octopus which has been punished in a learning trial, a wasp in the act of stuffing food into a hole in the ground, an onuphid polychaete which has been removed from its tube, or a satiated sea anemone—all place an altered value upon many forms of sensory influx, and do so, we assume, because a definite pattern of nervous activity has so predisposed them. These patterns are important parts of both innate and learned behavior (Bruner, 1957; Pribram, 1959).

Sperry (1958) has developed the argument that much of the flexibility due to such alternatives is **mediated by sets or biases comparable to the facilitation** of certain pathways rather than others. If a spinal salamander is tapped on the tail, the hind leg retracts or protracts depending on which pathway is favored by the proprioceptive inflow

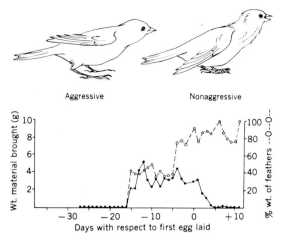

Figure 5.26. Nervous bias. Certain states, which have been called moods or set, are manifested by posture but include a complex of changes in readiness and receptivity, and therefore involving integrative functions in the nervous system. The illustration is from the canary; in the upper left is the aggressive form of the raised wing posture, in the upper right is the nonaggressive form of the raised wing posture. [Hinde, 1955.] The graph shows an example of a change in response, in this case to nesting material provided continuously to a naive female. Until two weeks before the first egg was laid no nest-building activity was evident; after starting abruptly it changed in character so that the percentage of feathers used increased greatly even after the first egg, when tendency to gather material (measured by weight) fell off. The nests built were removed daily. [Hinde, 1958.]

due to posture. The central circuits are so constructed that numerous alternative pathways of discharge are available to each integrating center. Without any alteration in its wiring plan, fleeting shifts in central state can link the same input signal to any of various responses and simultaneously inhibit each of the alternatives. Normally the nervous system does not get confused by activating inappropriate mixtures. Once more we are confronted by an adaptive unity in decisions of either-or character, an integrative phenomenon par excellence.

6. *Recovery from damage; "plasticity"*

The phenomenon of plasticity in the central nervous system refers to **functional regulation** after damage, in the young or in the adult. The subject cannot be treated at length here. It is indicative of higher-order integrative mechanisms, not at all understood. Extensive regulation is possible in some species and ages. Primates and especially man recover considerably from initial deficit after many types of lesions. These include peripheral nerves cross-sutured into the wrong peripheral field, as well as central damage. Not all central recovery is attributed to recovery from shock. Younger animals show less shock and less permanent deficit from the same lesion than older animals. In man, **relearning** can be increased with motivation. Adult lower forms, including rats, frogs, and fish, relearn or recover from functional deficit less easily. Thus disarrangements—reinnervation of the triceps muscle from the antagonistic biceps nerve, contralateral crossed innervation of the right foot from nerves from the left, or inverted eyes—result in permanently maladaptive behavior according to some reports or in significant readjustment of central nervous connections only after 8–12 months (rabbit, Shamarina, 1958).

The most impressive evidence of regulatory central rearrangement comes from experiments imposing severe peripheral disarrangement in young amphibians (Fig. 5.27) and fishes. Complete recovery of adaptive reflexes and normally coordinated movements occurs following section of optic, vestibular, trigeminal, or spinal nerves or roots. This means that orderly and appropriate central connections are established by **regenerating** fibers growing into the higher centers, as though they can find their way precisely to the same connections they had before, despite chaotic interweaving in the scar. Pattern and color vision, skin area localization, and modality are all restored. Relearning is not involved; the reflexes develop just the same even though maladaptive, as when the eye is rotated 180° or a regenerating nerve is compelled to enter the wrong side of the brain or cord. Actual **alteration of central connections** can result from the same developmental process of seeking synaptic contact specific for the peripheral termination of the neuron. Crossunion of flexor and extensor nerves in young amphibians, which fails to lead to adaptive recovery in later stages or in higher animals (but see Shamarina, 1958), is followed by appropriate muscle-specific rearrangement of central relations and adaptive recovery. A supernumerary limb grafted into the trunk region becomes innervated by nerves not normally supplying limb muscles and skin. Central connections are established appropriate to the new periphery, providing coordinated movements in phase with the normal limbs and cutaneous reflexes like those elicited by stimulating corresponding points on normal limbs. The plasticity involved in these phenomena is one of embryonic development and becomes progressively less as determination proceeds from broad to narrow categories (Weiss, 1936, 1950a, 1950b; Sperry, 1945, 1951, 1958).

Physiological alterations which may play some role in functional plasticity have been noted in several conditions. Disuse, retrograde chromatolysis from interruption of an axon, and longlasting posttetanic potentiation have each been shown to produce characteristic changes in the physiological properties of neurons (see Eccles, 1958). Growth of nerve fiber terminals as tentative, sprouting, and retracting buds (Speidel, 1941) has already been referred to. Young has proposed a principle of double dependence—that neurons are only maintained in a normal state if they both receive input and deliver output; thus cells of the lateral geniculate nucleus of mammals atrophy either as a result of retinal or of striate lesions.

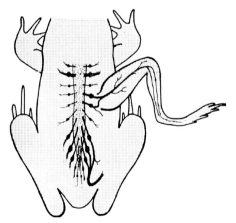

Figure 5.27. Developmental reorganization of central synaptic connections according to peripheral termination of nerve fibers. By the transplantation of an extra limb to the back, cutaneous fibers, destined normally to innervate trunk skin, can be made to connect with limb skin. The central processes thereupon form synaptic relations appropriate for the limb instead of the trunk. [Miner from Sperry, 1951a.]

There is a considerable literature on so-called **plasticity in arthropod locomotor movements,** originally regarded as akin to learning. Bethe (see 1930) described how crabs which normally right themselves with the last (4th) pair of walking legs, after their loss use the new last (3rd) pair. Von Buddenbrock (1953; and see Hughes, 1957) has assembled cases in terrestrial arthropods in which the normal sequence of leg movements in locomotion is altered in a fixed and, within limits, predictable manner to a new one for each possible amputation or combination of amputations. It is as though the various gaits were predetermined and the necessary coordination and interaction of centers of timing ready for the eventuality (Fig. 5.28). And it suggests that the gait used depends on information as to which legs are available or, more specifically, which joints are subject to stress from pushing against the substratum. In *Dixippus*, either immobilizing the joints or shortening the leg so that it does not reach the ground induces a gait as though the leg were missing. The normal gait is restored by freeing the joints or by artificially lengthening the leg or bringing a small platform up into contact with the shortened member. Von Holst (1943) has found that the coupling constant which measures the closeness of coordination between legs in centipedes alters in a characteristic way with amputations. In *Lithobius* the phase difference between legs five segments apart progressively increases with loss of legs between; if all but two pairs of legs of this 30-legged animal are removed, it assumes a typical lizard gait regardless of the distance between the pairs; within limits, if any three pairs are left, it assumes an insect gait. This species is said to be **"endoplastic,"** relatively more dependent on internal factors for determination of its coordination. Another chilopod, *Geophilus*, is called **"exoplastic,"** emphasizing a greater influence of peripheral input; this is most conspicuously manifest on rough ground, where every foot steps exactly into the place left by its predecessor.

Examples can be multiplied of the **substitution of another fixed pattern for one prevented,** especially in locomotion of annelids and anthropods. When the interference takes the form of amputation of a leg, the pattern-determining system presumably is influenced by sensory feedback caused by the altered mechanical conditions. It is also presumably influenced by the altered tonic input which has no particular position-information content. In any event, while a primary role can be assigned to peripheral factors—and only these have received attention in many recent discussions (see references in Hughes, 1957)—central factors are at least equally important, for a chain-reflex theory depends on central connections to give an adaptive result. Beyond this the possibility remains that the feedback, while essential, determines the patterning only very indirectly. The timing of successive phases may not be chiefly peripheral loop time. And the determination of which leg shall move next may not be entirely, though it is probably very largely, peripheral. When the interference is by immobilization or cutting connectives between ganglia, the results are still harder to interpret; different species yield different results (compare annelids, chilopods, diplopods, and aquatic and terrestrial insects; Gray, Lissmann, and Pumphrey, 1938;

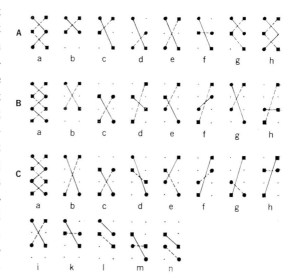

Figure 5.28. Predetermined alternatives in the gait of arthropods subject to various amputations. The legs that move nearly synchronously in normal running are represented by a common symbol—*squares* or *circles*. The legs that move nearly synchronously after various amputations are connected by *solid* or *dashed lines*. The *small dots* represent amputated legs. **A.** *Dixippus*. **B.** *Opilio*. **C.** *Carcinus*. [von Holst, 1935.]

Gray, 1939; von Holst, 1939a, 1943; von Buddenbrock, 1953; Hughes, 1952, 1957, 1958); see also pp. 319, 1146.

The relevance of these phenomena in the present context is that an **apparent plasticity may in fact be its opposite**—a stereotypy in a complex and adaptable form. On either central or peripheral emphasis, predetermined, fixed, alternative patterns can be instantly substituted by altering components of a complex of interlocked circuits and loops and hence changing certain input magnitudes, coupling functions, or time constants. Certainly it is not necessary to consider with von Buddenbrock (1953) that immediate change in gait, in many sequences with many alternative amputations, argues against the possibility of centers of coordination and fixed pathways because of the unreasonable number of preformed discrete circuits supposedly required. A theory with any mixture of peripheral and central factors could account for such substitution with a considerable but limited number of hierarchically arranged centers and coupling functions. Insects do show gradual recovery as well, resembling the plasticity of mammals, for example, after circling or rolling resulting from unilateral blinding (see Carthy, 1958; Jahn, 1960).

7. The role of numbers of neurons

Many authors have been concerned about the numbers of neurons and of connections between neurons, which according to estimates for various gray masses in mammals are astronomically high (see Sholl, 1956). The contribution of afferent and efferent neurons does not account for most of the number in higher invertebrates and vertebrates. Association neurons and their interconnections form the vast majority. But the association areas of the brain, cortical and subcortical, are tolerant of considerable loss of nervous tissue in mammals. No signs whatever have been detected after small lesions and sometimes fairly sizable lesions in certain areas. As a first approximation it is assumed that there is great **redundancy,** especially in higher centers and in higher vertebrates.

The assumption is often made that the first and principal answer to the problem of accounting for **increasing complexity of nervous achievement** is increase in numbers of neurons and their interconnections in the nervous system. There is certainly some correlation between these, though it is far from regular, and a greater number of neurons is neither a necessary nor a sufficient difference between two species to account for their behavioral differences. Behavior is probably not greatly disparate in complexity—comparing ant and lobster, small and giant squid, mouse, and elephant—despite some suggestions to the contrary (Rensch, 1956). On the other hand, mouse behavior is probably more complex than that of a large shark, under whose midbrain alone a mouse could hide. The motor and sensory centers do not entirely account for the difference in brain size, nor does difference in cell size—there must be a large discrepancy in numbers of cells in associative centers, in favor of the large shark.

The conclusion is compelling that while numbers of cells and connections do in general increase with functional complexity, **something else must be increasing also.** As a general statement, independent of gross anatomical novelties like the cortex in the higher vertebrates, it is difficult to say what this is. For the time being one may have to be content with the indefinite—but emphatic—reference to a complexity of quality in cell types and connections and to the superposition of new levels of integrative machinery, as in the sequence: archicortex, paleocortex, neocortex.

8. Structural specializations among associative neurons

The anatomical evidences of special mechanisms for integration at the level of many neurons are not, in general, confined to the associative level, strictly defined, but they become more conspicuous as one moves farther from the first-order sensory and the final motor neurons. They are dealt with in a different connection in Chapter 2.

Avalanche spread is provided for in many variants, each neuron sending axonal terminals to many neurons which in turn have ramifying connections with many more. **Recurrent collaterals** are highly developed in some places—a branch of the axon returning to end among other cells of the same kind as the parent cell (Fig. 5.29). If these have an excitatory effect they will greatly

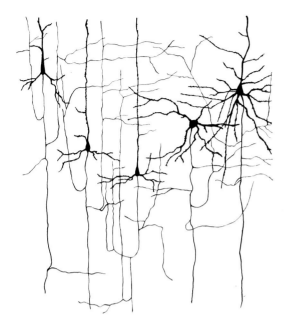

Figure 5.29. Recurrent collaterals. Neurons with long axons from the fourth layer of the cortex in the motor area of man. Golgi method. The axons as they descend toward the white matter give off collaterals that run back into the cortex and end among the dendrites of their own and similar cells. [Cajal, 1912; see Bibliography of Chapter 2.]

enhance avalanche spread, but they are more probably generally inhibitory, acting to intensify the contrast between the active cell and its neighbors. It is the simplest of an ever more complex spectrum of **reverberating circuits** which could mediate positive or negative feedback in any of a variety of combinations, with amplifying and integrating stages. Enhancing the possibilities of complex circuits are the **compact short axon cells** so characteristic of and abundant in higher central masses; these are variously designated Golgi type II, globuli, and granule cells.

The concentration of endings and synapses in such centers signifies some functional meaning for the **diversity of neuropile texture** of invertebrates and vertebrates and of the gray matter peculiar to the latter. The gradual unraveling of various neuropiles is certainly one of the concrete objectives of the neuroanatomy of the future, for in this incredibly dense tangle must surely lie the substratum of the highest forms of behavior at least of invertebrates, and according to Herrick (1948), of lower vertebrates as well.

Differentiation among neuropiles is marked—we may simply recall the elaborate optic masses of arthropods and cephalopods, the corpora pedunculata and various glomeruli treated elsewhere in this volume. These will serve also to call to mind the extreme differentiation among neurons in types of endings and synaptic arrangements and pose the question whether these are all accounted for by the variety of functionally specialized synapses recognized above (in Chapter 4 and in section II of this chapter). Higher centers in higher vertebrates do not depend on neuropiles but, beginning with the tectum of the midbrain and later the cerebral cortex, have evolved characteristic superficial gray matter and laminar organization. The cerebellar cortex has developed these features and in addition a striking rectilinear pattern; certain important dendrites are disposed in parallel planes and a class of numerous axons, carrying input to these dendrites, pass in straight lines through the dendritic arbors at right angles to this plane (Fig. 5.30).

Figure 5.30. Anatomical specialization. The cerebellar cortex illustrates the development of superficial gray matter, laminar organization, and a striking rectilinear pattern. Dendrites of Purkinje neurons branch in a plane and axons of the granule cells rise into this layer, bifurcate, and run for long distances at right angles to this plane. gran., granule cell axons; Purk. c., Purkinje cell layer; Purk. den., Purkinje dendrites. [Scheibel and Scheibel in Möllendorff-Bargmann, *Handbuch der mikroskopischen Anatomie des Menschen*, vol. 4, pt. 8; Springer.]

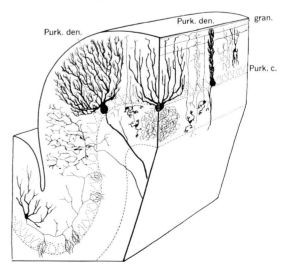

Even so simple a feature as **fiber size** is thought to correlate with fairly subtle functional significance; at least the correlates are still speculative or vague. Bishop (1959) proposes that large axons in somesthetic systems have evolved with certain modalities and with projection to the cerebral cortex; small fibers are supposed to be more primitive and used in more generalized responses, and large fibers to be more recent and discriminating.

Many other topics relevant to associative levels cannot be developed here for want of space. Especially rich in suggestive detail is the literature of human and higher mammalian neurology. The topics treated will serve as examples of the integrative features at this level and point to opportunities for significant comparative investigation of lower vertebrate and invertebrate groups.

IV. SPONTANEITY, ITS SOURCES AND CONSEQUENCES

A. Spontaneity at the Neuronal Level

By mixing response to stimulation through complex circuits with spontaneous activity in selected loci, the possibilities for determining output by some indirect function of input are enormously increased. The term spontaneous means repetitive **change of state of neurons without change of state of the effective environment**—that is, activity without stimulation other than the standing conditions. Of course the activity occurs only if many aspects of the milieu remain within certain limits—for example, the temperature and the ionic balance. These could be thought of as steady-state stimuli; but unless there is evidence of physiologically significant control of the activity by normal changes in these aspects of mileu, the term stimulus is not appropriate. In the sense of ongoing activity under normal conditions, many neurons are spontaneous and many are not, and the distribution is probably highly significant for integrative function. Of the spontaneous cells, some are arranged to drive or pace other cells in a specific pattern, anatomically or temporally or both; some are apparently not normally pacers of other cells; and some seem to exert a physiologically essential influence of an unspecific, unpatterned sort upon other cells. These are discussed briefly in turn.

1. Neuronal pacemakers

Theoretically, two kinds of endogenous rhythm-determining mechanisms are possible—one dependent on impulses circulating in a reverberating circuit and the other resting on autorhythmicity in individual cells. The latter is known to exist—for example in the cardiac ganglion of decapod crustaceans. Reverberating circuits cannot be excluded but are not experimentally demonstrated as pacemakers.

Another basic distinction can be made between two types of pacemakers. There are those so communicating with the conducting system into which their activity is delivered that they can be fired and hence their rhythm reset by antidromic impulses in the system; that is, they are **accessible** to impulses introduced from other sources. An example is the pacemakers for swimming pulsation in medusae. Access and resetting do not mean that the system is a simple relaxation oscillator; there may be, for example, an extra long pause after two firings close together, because the effects of preceding activity have not been wiped out at each firing. The vertebrate heart also belongs to this class, although the interjected impulse must be initiated in the region of the sinus in order to reset the rhythm. If the extra impulse originates in the ventricle, it always finds the conducting path back to the pacemaker region refractory, owing to the slow conduction, and can therefore never reach that region. Theoretically there is another class in which the pacemaker is **inaccessible** to impulses in the system which it drives. This would be the case where a pacemaker cell had only one-way synaptic connection and no feedback connections with downstream neurons. In multicellular systems, it therefore represents the simpler case. No unequivocal example, however, has been thoroughly analyzed. The cardiac ganglion of crustaceans

IV. SPONTANEITY A. Neuronal Level 1. Pacemakers

approaches this condition, since the pacemaker neuron can usually not be influenced by impulses alone in the follower cells; but slow potentials in the latter do feed back to the pacemaker and affect it. Rhythmic strychnine spikes in the mammalian cortex and spinal cord cannot be reset by indirect stimuli. A general discussion is provided by Pantin and Vianna Dias (1952).

The principal recent advances concern the **intimate description of the pacemakers** based on autorhythmicity in single cells where synaptic driving can be excluded (Eyzaguirre and Kuffler, 1955a, 1955b; Weidmann, 1956; Trautwein and Dudel, 1958; Dudel and Trautwein, 1958; Bullock and Terzuolo, 1957). It has been found that there is an electrical indication of the progress of the intracellular clock as it steadily advances between impulse discharges, permitting studies of the origin and the nature of the rhythm. Two types of electrical changes have been seen. In the more common, a gradually increasing depolarization, called the pacemaker potential eventually leads to a critical level of membrane potential which fires first a local response, and this in turn fires an all-or-nothing propagating impulse (Fig. 5.31). The recovery phase of the impulse in-

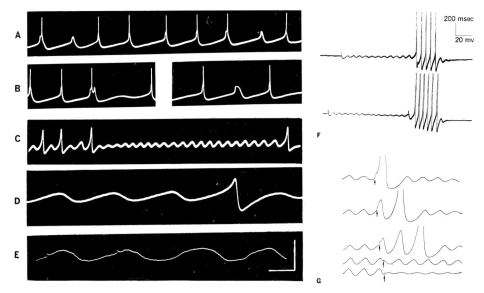

Figure 5.31. Pacemaker action; two types. Records showing the subthreshold ongoing change between intermittent discharges. **A** and **B** (continuous record). Cardiac ganglion cell of the lobster *Panulirus*, showing the relaxation oscillation type with progressively depolarizing pacemaker potential, leading to local and spike or only to local potentials, which in turn can repolarize. The time mark in **E** represents 350 msec. [Bullock and Terzuolo, 1957.] **C.** Pleurovisceral ganglion cell of *Helix*, showing sinusoidal type of oscillation, which does not require intervention of any new event. Spikes are superimposed from time to time. Time mark equals 1 sec. [Tauc, unpublished.] **D.** The same at higher film speed. Time equals 0.1 sec; vertical calibration equals 20 mv (see calibration marks for **E**). **E.** External electrodes on the myxomycete *Physarum*. Time equals 1.87 min; vertical calibration equals 20 mv. [Tauc, 1954.] **F.** Visceral ganglion cell of *Aplysia* after treatment with pyruvate, acting "undamped"; an orthodromic activation causes an e.p.s.p. that initiates a growing series of oscillations of membrane potential, leading eventually to a burst of spikes. [Arvanitaki and Chalazonitis, 1956. Reproduced with the permission of Centre National de la Recherche Scientifique, Paris.] **G.** Isolated axon of *Sepia* showing subthreshold local oscillation of membrane potential following subthreshold shocks. At different phases of the oscillation a test shock is applied *(arrow)* and its effect shows the differences in excitability at these times. During the rising (depolarizing) phase of oscillation a spike may be triggered; during the falling phase the oscillation may be delayed or stopped. Time scale of the order of 10 msec per oscillation cycle. [Arvanitaki and Chalazonitis, 1949. Reproduced with the permission of Centre National de la Recherche Scientifique, Paris.]

cludes a strong repolarization which carries the membrane potential back to a high level from which another gradual depolarization can begin, starting the next cycle. In the less commonly encountered type, an undulatory or sinusoidal fluctuation of membrane potential occurs and does not require any new form of activity to complete the cycle. Usually it leads, at each maximum of depolarization, to local potentials and thence to spikes, but the repolarization is not due to the recovery phase of the latter; it is the continuation of the sine wave and therefore is as slow as the depolarization (Fig. 5.31).

The first type is a **relaxation oscillation** and as such depends on two processes and a threshold to determine the rhythm. There must be a cellular mechanism capable of slowly increasing depolarization; there must be a threshold for the next process; and there must be a mechanism capable–whatever else it does–of restoration of the initial condition. The "whatever else" normally includes initiation of an overshooting, propagating spike, but it is possible at least in some cells (Bullock and Terzuolo, 1957) for the local response to repolarize the membrane, and the spike may therefore be regarded as an optional epiphenomenon. In view of the regularity of many rhythms, that is, the small variation of intervals between spikes, each of these requirements—the rate of gradual depolarization, the level of critical potential, and the rate and extent of repolarization—must be capable of maintaining a high degree of constancy over many cycles. Considering the low frequencies of some rhythms—one cycle in some seconds or more—the rate of change of membrane potential is low so a small deviation in terms of millivolts per second or a small fluctuation in threshold in terms of percentage would mean a large change in frequency. The form of the pacemaker potential as recorded from the soma varies in different cells; it is commonly linear but often accelerating in slope. The evidence indicates that the pacemaker potential arises in a fraction of the pacemaker neuron—for example in the cardiac ganglion of crustaceans in a locus at some distance from the soma, out in the processes. Moreover, this locus is separate from that of spike initiation, though the two must of course be within interacting distance. The pacemaker locus may not even be invaded by the spike, that is, may never be depolarized completely; there may be more than one such locus in a neuron at one time. Whatever the nature of activity at such loci, it appears probable that the pacemaker potential is not simply an afterpotential of a spike or the passive recovery of potential owing to the membrane time constant, although it cannot be excluded that these factors have an influence in some cases. The pacemaker loci not only interact with spike-initiating loci, some distance away, but also with synaptic loci in cases where the spontaneous cell is also subject to being driven synaptically.

This picture not only serves for the neurons which are strictly spontaneous but also for those which are sustaining a rhythm in response to a stimulus—for example primary sensory neurons, to judge from the most intimately analyzed receptor process available, the trunk muscle stretch receptor of decapod crustaceans (Burkhardt, 1958); see also pp. 216, 1019. Here, however, a crucial additional process makes itself known, a potential representing the transducing action of the sensory structures, a function of stretch in this case, and called the **generator potential.** It differs from the pacemaker potential in that it can be maintained steadily by a subthreshold stretch, like a DC potential (see Fig. 4.2). Above threshold, when the two potentials presumably coexist, it is not clear whether the rhythmic process is determined simply by passive time constants and the level of the generator or whether the relation between the two is more complex, for example an active pacemaker potential whose rate of change is influenced by the generator level.

In the class of pacemakers under consideration it is a defining feature that a single cell can pace the whole system. But, to judge from the medusae swimming mechanism and the lobster heartbeat mechanisms, there are likely to be provided **duplicate pacemaker cells,** which represent, at any one moment, standbys without function. In the medusae it has been shown that the redundancy serves to keep the frequency higher and more uniform than one pacemaker alone will do; the role of actual pacemaker for the moment is rotated among them, the cell with the fastest rhythm always having that role. The pacemaker integrates various factors in the steady-state milieu and also may be subject to sensory influences of several kinds. The internal organization of a pacemaker in *Aurelia* provides self-regulation of the rhythm against deviation (Horridge, 1958). An overcompensating regulation could explain series of bursts, such as the patterned discharges of the lobster cardiac ganglion (see also pp. 223, 993).

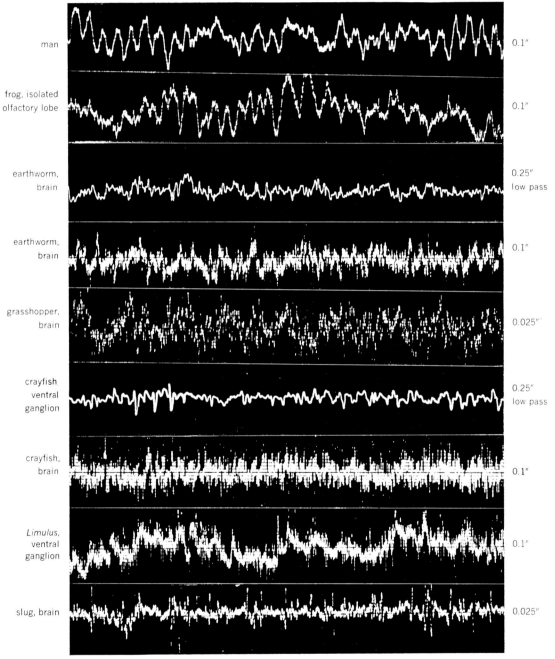

Figure 5.32. Brain waves, invertebrate and vertebrate. Note the differences in time scale; *low pass* signifies that the filters were set to attenuate high frequencies severely; the other records employ wideband amplifiers. [Bullock, 1945.] Time in seconds, on the right, gives the duration of each whole strip.

2. "Brain waves"

Since the brain in invertebrates is not called an encephalon, the formal term electroencephalogram is not quite appropriate and no other has been introduced! But electrical activity is recorded from the brain and other ganglia of invertebrates. And although it has been little studied, it is of great comparative interest because there appears to be a general **difference between the brain waves in invertebrates and vertebrates** (Bullock, 1945). From fish to man—therefore with or without a cortex, not to speak of most of the structures of the thalamus and basal ganglia of mammals—the brain exhibits smooth, low-frequency, sinusoidal waves dominated by rhythms of less than 50 per second and mostly less than 10 per second (Gerard, 1941; Enger, 1956). But in all invertebrates examined, including annelids, arthropods, and molluscs, the dominant character of the electrical activity recorded by surface electrodes is spiky (Fig. 5.32). The frequencies of maximum energy would be in the hundreds of cycles per second. The spikes look like single or well-synchronized nerve impulses. Spikes are certainly taking place in the vertebrate brain under normal conditions but it requires special means of recording to observe them. Comparing the large cephalothoracic ganglionic mass of *Limulus* and the brain of a small frog, it can be concluded that size of ganglion is not the critical difference, and on similar grounds it seems possible to rule out size of axons. Apart from the spikes, invertebrate ganglia do show slow waves, and in some cases these appear to demand something beyond an envelope of individual spikes to account for them (see, for example, Gogava, p. 1376). Thus the large, sinusoidal, regular waves of 40 per second and thereabouts shown by Jahn and Crescitelli (1938, 1939, 1941) and Burkhardt (1954) in the brain and optic ganglia of insects after certain regimes of light and dark (Fig. 5.33) do mobilize spikes around one phase but suggest a separate underlying rhythm (Adrian, 1931; Bullock, 1945). Perhaps it is because such rhythms are commonly unsynchronized in separate cells that large slow waves are not generally prominent in invertebrates; synchronized slow waves may be somehow dependent on the vertebrate type of dendrite, cytologically quite different from those in invertebrates, or on the proximity of the soma or the character of the neuroglia.

Slow changes of state as a separate form of activity from spikes are now generally accepted for vertebrate neurons (Fig. 5.33). It is a common opinion that slow cortical potentials are largely or exclusively dendritic in origin (Clare and Bishop, 1955, 1956; Purpura and Grundfest, 1956; Tasaki et al., 1954; Bremer, 1960). The slow waves have been recorded from microelectrodes in single cells and the question that remains uppermost is how a large mass of cells is **synchronized** with respect to this subthreshold fluctuation. Second only to this question is whether the synchronized waves themselves exert some **intercellular influence,** that is, any effect apart from the spikes which are influenced in each cell by its own subthreshold change. There are suggestions that rhythmically firing cells have been shown to be so sensitive to fields of electric potential in the tissue that voltage gradients of

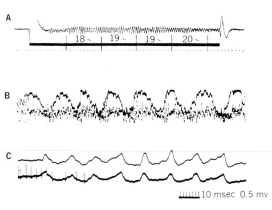

Figure 5.33. Independence of slow waves and spikes; insect optic system and cat brain. **A.** Oscillatory potentials from the optic system of the grasshopper *Melanoplus*. These waves are characteristic of a certain stage of light adaptation. The *dark bar* indicates the duration of a light stimulus. Time is 0.1 sec. [Crescitelli and Jahn, 1942.] **B.** Similar waves in the locust *Schistocerca*. The same electrodes go to two amplifiers, one passing slow waves, the other spikes. Spikes tend to occur at one phase of the slow wave but the latter can hardly be an envelope of the former. Time, 0.025 sec. [Original.] **C.** Microelectrode 1700 μ deep in the cortex of the cat (lower) and surface macroelectrode (upper), showing the same feature. [Li and Jasper, 1951.]

the order of those in brain waves can modulate their frequency (Terzuolo and Bullock, 1956). Spreading waves such as those induced by caffeine or occurring autochthonously will continue uninterrupted across transections of the brain or cord (Gerard and Libet, 1940; Bremer, 1941). Therefore, in addition to the indication of spontaneity, the behavior of brain waves suggests the operation of special integrating mechanisms. They are probably diverse and complex. Contrary to a common assumption, an electrogram dominated by fast, "desynchronized" waves does not necessarily represent an excited state and a slow, "synchronized" record an inhibited state; a desynchronization may go with excitation or with inhibition, even with a lowered excitability (Kogan, 1958; Grundfest, 1958, 1961c). A full discussion cannot be given here of the hitherto nearly exclusively mammalian literature. Suffice it to suggest that significant findings await further attention devoted to the nature and effect of brain waves and their comparative study (Rijlant, 1931; Prosser, 1936; Bonnet, 1938, 1941; Roeder, 1939, 1941; Bullock, 1945). Much data of interest is already at hand from mammalian studies concerning regional interactions, synchronization, frequency peaking ("transformer action"), propagations, and especially empirical correlations of brain waves with other events and states (Jasper and Shagass, 1941; Gibbs and Gibbs, 1950; Li et al., 1952; Sato et al., 1957; Fessard, 1959; Walter, 1959; Jasper, 1936, 1960; Aladzhalova, 1958; Brazier, 1960).

3. Cellular spontaneity which does not pace other cells

The follower cells in the lobster cardiac ganglion, when cut off from the pacemakers and thus released from regular driving with its rebound depression, show a spontaneity of their own (Maynard, 1954; Matsui, 1955). In many ganglia this spontaneity never shows itself as long as the ganglion is intact; in some, where the frequency of the pacemaker is low, there is enough time for recovery from the depression and one or a few spontaneous spikes may arise in the last fraction of a second before the next heartbeat burst is precipitated by the pacemakers.

In other words, in most ganglia most of the time this spontaneity of the followers is not permitted to express itself by reaching the level of impulse initiation. This is another and curious category of spontaneous neurons, like the muscle fibers of the vertebrate heart. They could be common and remain unknown because the conditions are not as favorable as in the lobster preparation for uncovering the facts. The functional meaning of such a category invites speculation. It is possible for these cells to maintain a heartbeat in the absence of the pacemakers, but as an explanation of the possession of spontaneity this is not attractive. It seems likely that this **represents a way of poising** the cells so close to threshold that they will follow with (a) a small synaptic input or (b) with a short delay or (c) in conjunction with the property of noninvasion of the soma and pacemaker potential regions by the spikes (pp. 223, 264) with a high-frequency, repetitive burst or (d) any combination of the three. If this is a reasonable hypothesis, then it is probable that there are indeed many such cells, normally allowed to express their spontaneity only to a subthreshold degree as preparation for synaptic response.

B. Spontaneity at the Behavioral Level

Since spontaneity is so potent a factor in nervous integration, we cannot leave it without mentioning a few other types of observation in which it makes itself evident and in complex form. These are based upon intact animals or animals with much of the nervous system intact. In the presence of the peripheral receptors we cannot pretend there is spontaneity in the sense of no stimulation, and the evidence mentioned therefore rests on the occurrence of phasic behavior when it cannot be ascribed to phasic stimulation, that is, in the presence of steady-state stimulation only. This conforms to our definition of spontaneity: change of nervous activity without change in the effective environment.

1. The central origin of certain rhythms

Besides the clear cases of neurogenic cardiac rhythms there are other rhythmic processes which have been recorded electrically from the central nervous system after experimentally isolating it from the periphery. The isolated ventral nerve cord of the beetle *Dytiscus* is a classical case (Adrian, 1931). The respiratory neurons in mammals are **not dependent on phasic input** (Salmoiraghi and Burns, 1958; Oberholzer and Tofani, 1960). Hughes and Wiersma (1960) have found persistence of the rhythmic discharge of motoneurons to a swimmeret in crayfish after isolation of the abdominal cord from the periphery. Miller (1960) and Wilson (1961) find the breathing and flying rhythms of a locust to be central and not dependent on phasic input. The evidence is strong that the rhythm of swimming pulsations in medusae is of nervous origin (p. 502) and it may be supposed that the rhythm of tentacle activity of certain alcyonarian polyps is also (p. 519).

Peristaltic creeping in the earthworm probably arises as a central rhythm and is not dependent on phasic input although its spread along the animal depends on tonic input from tactile receptors or proprioceptors. A denervated length of ventral nerve cord, isolated except for its connection with intact segments at one end, can pace those segments in regular reciprocal contraction of circular and longitudinal muscles and shows electrical bursts at the same time. Other evidence is reviewed below in Chapter 14. Parapodial creeping in *Nereis* is even more dependent on intrinsic nervous automaticity. Swimming in *Hirudo* seems likely also to come from a neurogenic rhythm rather than from chain reflexes, as is clearly the case for the walking movements of this animal.

There has been a discussion in the literature, especially with respect to locomotion, as to the adequacy of explanation by centrogenic rhythms. Von Holst (1939a) supported the centrogenic interpretation. Gray (1936, 1939; Gray et al., 1936, 1938) inclined to a similar view in the 1930's but later (see 1950) argued for the essential role of feedback and reflexes in calling up the successive phases. Wells (1955) is impressed by the evidence of intrinsic spontaneity in his material. It seems on present evidence that among the diverse activities of animals some are very little dependent and others heavily dependent on sensory feedback, at least for their shaping (p. 296). It is difficult to adjudicate unequivocally on the proper subjective emphasis and this is not the place to review specific instances. For the purposes of this section it is sufficient to recognize that **evidence exists for central rhythms.** Weiss (1950a) describes tadpoles swimming after deafferentation and axolotl spinal cord implants rhythmically controlling implanted limbs long before afferents are differentiated. A strong indication comes also from the movements of fins of teleosts such as *Labrus* (von Holst, 1939). A "dependent" fin (p. 300) shows the influence of a "dominant" fin's rhythm and this rhythm can be shown not to be due to proprioceptive timing, for it is not altered by passive manipulations of the dominant fin or by denervation and removal of its muscles. There are in fact many potentially independent rhythms, as when the rays of a fin under certain conditions move separately; these manifest desynchronized central rhythms existing in spite of the mechanical and presumably proprioceptive factors tending to synchronize them (see also p. 1155).

These considerations also prepare us to find that many cases will be based upon such an **intrinsically competent mechanism plus superimposed modifications** as a result of the normally available phasic afferent signals—for example adjusting the gait to the terrain. At another level they prepare us to find that such mechanisms, entirely competent to carry out the function independently, may be innate, may be modified in an enduring way by experience, and may be modified again by superimposed "voluntary" control. These remain expectations for the present since no case is sufficiently well analyzed to demonstrate unequivocally all these stages in the same piece of behavior.

2. Phases of behavior

Closely related to these locomotor activities are the rhythmic movements of the polychaete *Arenicola marina*, which lives in a U-shaped burrow (see Wells, 1949, 1950, 1955). Under stable conditions there may occur every 6–7 minutes a burst of feeding movements, every 20–60 minutes a sequence of tailward creeping, forward irrigation with headward creeping, tailward irrigation, and then defecation. In Wells' words (1949, p. 476), "Such an integrated pattern [of cyclic activity] could be achieved by means of a hierarchy of reflexes with appropriate responses to the worm's various needs—to empty gullet, to a full rectum, and to oxygen lack or CO_2 excess. But all the evidence suggests that in *Arenicola* the organization is based quite differently, on spontaneously active 'clocks' in its oesophagus and nerve cord. The rhythmic outbursts begin spontaneously, without any external stimulus or any biological need; they subside without any kind of satisfaction; and normally their rhythm plays an important role in patterning the life of the worm as a whole." The brain is unnecessary for these forms of activity but the ventral cord is necessary; at least some of the spontaneity is intrinsic in muscle in the wall of the esophagus (Fig. 5.34). The internal clocks of the two principal rhythms mentioned do not directly influence each other, but when they conflict one wins; there is **no summation** or cancellation although one may dominate in one part of the body and the other in another.

In the last section we noted the severe requirements in stability of gradual depolarization rate and in constancy of threshold for a rhythm of several seconds. Here we have regular rhythms of many minutes, apparently arising in nerve cells. It seems quite possible that at least some of the host of diurnal, tidal, and possibly even slower clocks known (see Pittendrigh, 1958, 1961) will be found to be nervous. Certainly they **can occur in the absence of a nervous system,** as in *Euglena*. And, although Harker (1956, 1958, 1960) has located the clock in an insect in the subesophageal ganglion, there is no assurance that it is nervous or in a nerve cell. Temporal patterns of growth and activity are among the basic properties of living material. But it is at least as likely that cellular clocks will occur in the nervous system as elsewhere, especially for those activities requiring coordination of an extensive musculature. The analysis of *Arenicola* lends support to this possibility.

Long-term recording of the activity of a number of polychaetes and of sea anemones (see Batham and Pantin, 1950) shows that over and above short-term rhythms, these animals are creatures of moods! Periods of feeding may persist for hours only to give way to a period of some other form of activity or a prolonged period of complete quiescence. Some details and illustrations for sea anemones are given below, in Chapter 8, and in Fig. 8.36. *Chaetopterus*, several sabellids, and *Arenicola* have similarly written the complex record of their own activities for days without giving a hint of the reason for abruptly shifting from one mood to another. The shifts can be precipitated by stimulation and the spontaneous activity in each phase can be modified by the environment or by obvious needs, but in the absence of these imposed triggers the system will **not remain quiescent or monotonously continue** the same activity for very long. And its changes in mood at the same time alter its reaction to sensory stimulation and hence its integration of all the factors determining output.

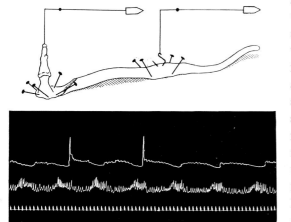

Figure 5.34. Spontaneous phases of activity. Mechanograms of contractions of parts of the lug worm *Arenicola*. The lower line records esophageal outbursts from the extrovert, the upper from body wall of a midbody region. Experiments show that these phases are not brought on by lack of food or oxygen or excess carbon dioxide. Time in minutes. [Wells, 1949.]

Many other aspects of behavior indicate central automaticities. At a higher level of complexity, for example, reference may be made to autogenic aspects of perception in man, tending to inject a normalizing distortion into input that is noisy or ambiguous (Bruner, 1957). Still higher examples may be represented by our abilities in imagery.

3. Tonic input maintaining central excitatory state; "Stimulationsorgane"

There is an interesting theory which states that in addition to their sensory functions some receptors have an essential function in maintaining the central excitatory state or central tonus. In reference to this role such sense organs have been called, by the German advocates of the idea, "Stimulationsorgane"—that is, they stimulate the central nervous system (see von Buddenbrock, 1952).

Among the instances which have suggested this provocative but still unproven idea, the following examples may be mentioned. Muscle tone in vertebrates has been found in various species to be severely decreased upon loss of input from the equilibratory fibers of the eighth cranial nerve. Gray and Lissmann (1946a, 1946b) found that a toad with the spinal cord deafferented would still swim, but upon the further loss of the labyrinths the animal became completely immobile. Similarly, in the fly *Sarcophaga* loss of the six legs does not stop flight, but the additional loss of the halteres abolishes the ability to fly. These organs are equilibrium receptors—vibrating gyroscopes—of varying importance in different species of dipterans. In addition they are responsible for the tone in many muscles, especially the legs. After unilateral removal in *Tipula* posture is abnormal, and after bilateral removal a general weakness so overcomes the animal that it is only able to drag its body along, resting on the ground, with difficulty. In species which will still vibrate the wings, the frequency and the amplitude of beat are decreased (see Faust, 1952). The fly *Philonicus* is reported to be unable to walk after removal of wings and halteres. A frog from which the skin has been completely removed lies motionless and flaccid; but any small piece of skin remaining suffices to keep the tone and spontaneity of movement at a much higher level (Ozorio de Almeida and Piéron, 1924).

Besides mechanoreceptors optic receptors are the most important in this sphere. The serpulid polychaete *Hydroides* and the barnacle *Balanus*, which normally respond quickly to slight increase in light and to shadow, respectively, are unresponsive for a few to many minutes after some hours in the dark, when the receptors should be most sensitive (von Buddenbrock, 1930).

On this theory, the widespread phenomenon of photokinesis can be interpreted as simply an increase in central excitability as a result of increase in light and the protagonists cite **kineses** as evidence for the theory. Although quite compatible with the idea, it cannot be admitted that the mere fact of a kinesis argues for the general stimulating role as opposed to or in addition to a sensory role. If it could be shown that the responsiveness to all kinds of environmental stimuli was lower in the dark and higher in the light, the case would be stronger.

The theory has suffered from ambiguity of the cases cited as evidence. It is not correct to say that the clearest proof results from the immobility following extirpation of a sense organ. Blindfolding a man greatly reduces his movements. What is necessary is the finding of a nonspecific loss of excitability to various forms of stimulation after silencing the one afferent nerve. What would also provide telling evidence would be the amelioration of this deficit by allowing the sense organ to continue sending impulses to the central nervous system but somehow preventing it from conveying specific sensory information. Thus if cutting the optic nerve resulted in generalized, nonspecific lack of response, whereas covering the eyes with a frosted plastic film did not, the usual criticisms would be dispelled. But this makes it clear that the crux of the problem is the subjective question, what is specific response and what is specific or sensory input? For the human it is easy to agree that frosted glass destroys the essential sensory value of the eyes, but this is a conclusion based on a great body of observation about humans; in other cases it must be decided de novo whether light versus dark is an insignificant part of the sensory role. Similarly, an important element of judgment is involved in how many diverse tests are required to show that unresponsiveness is generalized and nonspecific. The cases mentioned provide some of the desired evidence and are certainly suggestive. The idea is not unattractive and finds support, though not decisive evidence, in the great reduction of brain waves with deafferentation in mammals (Bremer, 1935, 1938) and in the prolonged increase in general electrical activity of all ganglia in *Cambarus* after a brief light stimulus to the eyes (Prosser, 1936).

The part played by the sense organ in the normal life of the animal needs to be adequately determined. If it is concluded that in fact an unspecific role in tonic stimulation is a significant part of the role played by a sense organ, we may draw an analogy with the relatively unspecific reticular activating system in mammalian brains (see pp. 306-307) and ascribe an integrative function hitherto identified with a large mass of central tissue to peripheral neuron activity, at least to that degree.

V. NEUROLOGICAL DEDUCTIONS FROM THE STUDY OF BEHAVIOR

In the search for clues and signs of integrating mechanisms, methods of observation utilizing the behavior of the whole animal have proved revealing. Not only at the highest levels where this is the only method open, but even with respect to some relatively simple levels such methods have great power. They have shown the existence of phenomena calling for physiological mechanisms which would not be known or studied if the nervous system were approached only by purely physiological methods as these are usually construed. Our objectives in this section will have to be limited to a cursory introduction to some of the approaches and findings, with most attention to lower forms of behavior.

A. The Analysis of Behavior as a Control System

Interest has developed recently in the analysis of selected cases of behavior in the manner of a systems engineer, treating them as examples of control systems. Given information about some of the components, especially the receptors and the effectors that are involved and the way in which the actions of those effectors in turn alter the input to the receptors, certain deductions can be made. These are in the nature of "at least such and such relations must obtain and mechanisms for them must be present" (or more accurately "such and such relations provide a model with the same characteristic behavior"). Further, certain predictions can be made about the accuracy, stability, final state, and time course of the behavior, in the form of "these characteristics or limitations must obtain unless additional receptors and interrelations are in operation." Besides the special and comparative interest in ascertaining the class of control systems to which given examples of animal behavior belong, certain basic discoveries of general interest can be attributed to this approach.

Let us consider a series of instances, of roughly increasing complexity (Fig. 5.35; Mittelstaedt, 1957, 1960, 1961).

Case I: In a constant and horizontally nondirectional environment, a centipede can exhibit a sudden change in gait or in direction of locomotion; this is apparently a random spontaneous event—better, a trivial event, because it doesn't matter whether it occurs at that moment or not. It is possibly triggered centrally or peripherally by some random event or some minute inhomogeneity; the cognomen "spontaneous central steering" in the figure represents no more than a hypothesis.

In *case II* external events and directional signs take control, as when a firefly turns toward a source of flashing light ("open loop system"). But here the signal is so brief that there is no opportunity for the animal's turning to be corrected by renewed comparison with the stimulus. The output does not influence the

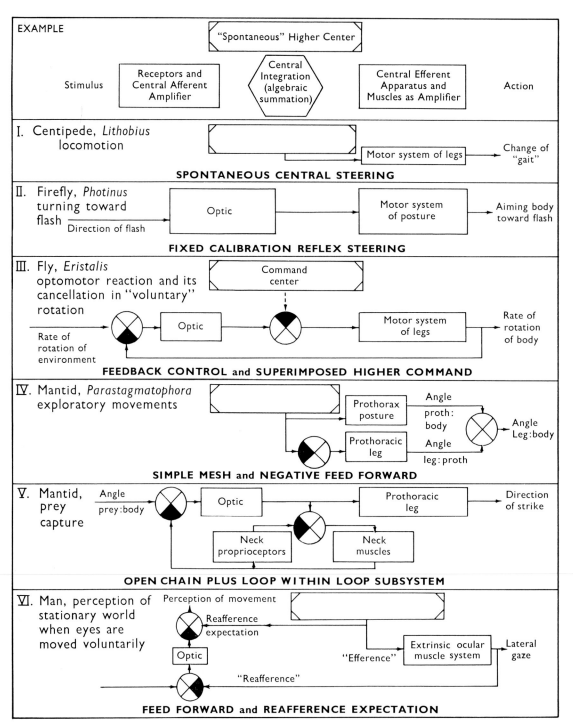

Figure 5.35. Six examples of pieces of behavior analyzed as control systems. *Arrows* signify that information must be transmitted at least in the manner shown, but not to exclude more complex pathways and relays en route. The *boxes* signify that various kinds of processing must take place, but not to exclude other kinds and sites of processing. Simple *rectangles* act as amplifiers, motors, or transducers of information. Sensory systems on the left and motor systems on the right. *Circles* represent algebraic summation of converging inputs; the *blackened triangle* simply means that the sign of the information transmitted is reversed befor the summation. The symbol for spontaneous higher center or command center represents the requirement for systems that can internally initiate activity.

input and depends for its fitness entirely upon the accuracy of the built-in performance characteristics or **calibration** of the amplifiers (transformations or transfer functions) in sensory and motor systems. Perturbations in these or in transmission links will harm performance and cannot be compensated for. Dynamically the only speed limitations are the transit time, external loads, and inertia of the components.

Case III is the **feedback** circuit ("closed loop system") with negative feedback which partially compensates for any extraneous disturbance affecting aim; sensitivity to amplifier drifts is much reduced, especially at high gain. A significant finding is that when higher commands take control, this automatic balancing against extraneous disturbances continues to operate.

Thus we can answer the old question: in view of the strong optomotor reflex (to choose a particular case), which forces an animal to turn with the rotation of a striped cylinder around it, how can it walk and turn voluntarily in stationary surroundings? **Is the reflex interrupted?** No; apparently it is not but is counteracted by a superimposed higher command. That the reflex is still acting can be seen by either of two methods. (a) Reversing the input resulting from the output—by twisting the head of a fly 180° and fastening it so. Thus the result of the command does not cancel but adds to the reflex, and causes the fly to circle compulsively as soon as it starts a voluntary movement (see Braemer, 1957; also von Holst and Mittelstaedt, 1950; Sperry, 1950 and earlier, and Szentágothai and Székely, 1956, on fish and other vertebrates). (b) Changing the strength of the input resulting from the output—for example by putting a fish in a centrifuge, so that a given tilt results in stronger stimulation of labyrinthine position receptors—changes the magnitude of voluntary departures from the reflex-stable state (that is, voluntary tilting) in the expected way, namely, fish spontaneously tilt themselves to a smaller degree.

The control system of case III is limited in speed of operation by well-defined phase-amplitude relations inherent in the circuit; deviating from these it becomes unstable and may go into uncontrolled oscillation. In addition, compensation is delayed because there must be appreciable deviation before the system can act against it. The control system of case III belongs to the **class of constant-error control systems** ("proportional control") in which the operation is such that the controlled variable—the motor output measured as rate of rotation of the body relative to the substratum—is kept proportional to the signal going through the system, that is, to the net input. Such systems can never achieve perfect compensation but will come to a steady state with deviation (of angular velocity, in the case of a rotating cylinder) which is a simple function of the total amplification—that is, the fly rotates slightly slower than the cylinder; reciprocally, the actual amplification can be computed from the observed deviation from perfect steering. Changes in amplification factor produce less than the reciprocal change in deviation; amplification cannot reach high values relative to inertia, or uncontrolled oscillation will result. Another class, not illustrated, is that of **zero-error control systems** ("integrating control") in which the operation is such that the controlled variable represents at any time the time integral of the past amplified signal going through the system. Common examples are gun pointing and prey capture—considered as a whole—and therefore any process in which the controlled action ceases when the objective is achieved. But in one carefully analyzed case of prey capture (by the praying mantis), it is found that the basic process is in fact a proportional control system in which the residual error is generally smaller than the prey (case V).

Case IV is a simple mesh of parallel paths with negative **feed-forward**. For simplicity the centrally initiated exploratory movements of the prothorax and its legs in the mantid are chosen, in order to consider its properties without receptor input. Again the accuracy—here the agreement between actual and expected output—is entirely dependent on the calibration of the transfer functions or amplification factors; it shares this and the stability and time characters with case II. But the result of feeding forward a portion of the input negatively through a parallel path is to eliminate system-inherent speed limitations and to permit, in a closed loop system, perfect following of input by output. Combined with sensory feedback, "ideal" compensation can be maintained (within the validity of the calibration of feed-forward). That is, there need be no residual deviation in phase or amplitude if the input changes do not exceed certain limits. In fact error can be "anticipated" since the counteracting process starts before the disturbance reaches the output.

Case V illustrates two parallel feedback **subcircuits which act as rivals;** each prevents the other from quite reaching its target but permits accurate localization of the prey of the mantis and following it, though only partially with the approximate center of the eye, located on a movable member, the head, relative to the effector, the foreleg. This system has the limitations of feedback systems, already mentioned, but it means that the steering of the strike is **unaffected by loading** the neck muscles, or by changes in their strength, and is determined only by the amplification factors of the optic and proprioceptive mechanisms. The output of each of the latter, at equilibrium, contains the information necessary to steer the foreleg strike; that is, for any given angle between the prey and the body axis, the mechanism will turn the head to a certain unique degree so that the remaining angle between prey and optic axis or the angle between head and prothorax can inform the foreleg exactly the direction in which to strike. It turns out that in this animal the optic source is used more than the proprioceptive, but this only matters in experimentally imposed damage. The prothoracic leg amplifier is in an open loop of the system, as figured here. This means that its calibration is—to the extent that the system works—compensating for the inherent residual error between head-to-body angle and prey-to-body angle, as though it "knows" this steady-state error. We will refer to the case again below.

Case VI is taken from a segment of human behavior but the mechanism it reveals is so interesting that it becomes important to look for it elsewhere. The surprising experimental result is that when a shifting of gaze is willed, but is prevented by local drug paralysis of the extrinsic ocular muscles, a distinct perception of movement of the surroundings is experienced—an illusion which is equal and opposite to the one occurring upon manually moving the passive eyeball. Von Holst, who is responsible for this analysis of the classical problem (see Duke-Elder, 1949, pp. 3858-3860) of the stationary appearance of the surroundings with voluntary but not with imposed movements of the eye, has applied the term "efference copy" to the link which the analysis requires. The essential feature is not a duplicate of the command but a preservation of some coded message that is a function of the command until there has been time for the change in input caused by the command to return to same region and be "compared." If some criterion of expected correspondence is met in the integrating center, the perception "no movement" results. Sokoloff also emphasizes the opposite effects of accordance and discord-

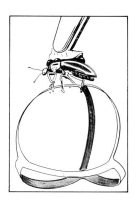
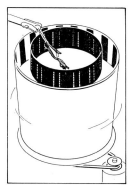

Figure 5.36. Experimental situation for analysis of the mechanism of detection of movement in the visual field. A beetle is fixed to a support but carries a light grass globe which turns as it walks. At the intersection it chooses right or left according to the impinging visual stimuli. These are delivered in various patterns, by rotating stripes seen through stationary slots. The four examples above show some possible sequences. The upper, interrupted lines are the stationary slots, the lower lines represent the moving stripes. Seen through the slots, *a* appears as a steady forward-moving pattern, *b* gives a fluctuating backward-moving pattern, *c* a fluctuating forward-moving pattern, and *d* a steady backward-moving pattern. [Hassenstein, 1951.]

ance between input and a central expectancy in higher levels of behavior. This link, "efference copy," or **coded central expectation is a prime example** of a physiologically required mechanism discovered by experimental means, using behavior as the measured variant and treating the whole as a control system.

Besides the discovery of links, integrating centers, and the relative roles of component mechanisms, this approach can also **reveal dynamic features** or properties. In the prey capture by mantids, Mittelstaedt (1957) found, in addition to the relationships just reviewed above, some special features which illustrate this point. Mantids follow the movements of a fly by two kinds of head movements: quick, saccadic jerks or slow, continuous turning, according to the distance from the prey. The former are less than 80 msec in duration so it seemed possible they were in the class of movements which, like our case II, are not controlled during their execution, while the slow, continuous turnings would be. An elegant analysis of performance by measuring the exact position of the freely moving head, showed that the jerks are so well steered and are so affected by removal of the proprioceptors as to require the assumption that they too are controlled by a feedback process. But the end points of the jerks do not correspond with the final steady state of the feedback process, and it must hence be inferred that the feedback is blocked in its action upon the jerks before it has come to equilibrium. An important principle may be involved here: it is inherent in simple feedback circuits that either amplification must be sacrificed, or speed of centering on target or **the circuit must be used briefly**, intermittently resampling the error, and opened between times to prevent instability. It seems possible that we have here an example of the brief use of a circuit and its opening before it has time to come to rest —a trick of great power where high amplification control is needed; some high-gain neuronal network may be supposed to operate intermittently for short periods.

Similar methods were used to show that one of the amplifiers—the neck muscle control unit— has the property of being sensitive only to the time integral of its input. Another example is the demonstration by Autrum (1955) that the central nervous mechanisms involved in the **stroboscopic illusion** in flies have temporal limits very much shorter than those in man, in agreement with the properties of the peripheral receptor mechanisms which have much higher flicker fusion frequencies (see also Küpfmüller and Poklekowsky, 1956; Stark and Sherman, 1957; Küpfmüller, 1961; Stegemann, 1961; Keidel, 1961).

A further example will serve to show another kind of finding concerning integrative mechanisms, by related methods. Analysis of the tendency of the bettle *Chlorophanus* to turn with rotating stripes has permitted a number of conclusions about the mechanism of detection of and response to movement (Fig. 5.36, B). The comparison performed by the beetle's visual system of the moments when a visible discontinuity strikes successive receptor elements takes place between adjacent ommatidia (shown to be unit sensory elements for movement perception) and also between adjacent-but-one ommatidia, but not between those farther separated. The significant comparison of moment of stimulation involves pairs of events, not longer series; that is, the reaction to a complex succession of stimuli is the **sum of reactions to all pairs** of successive stimuli in adjacent and adjacent-but-one ommatidia, of which the stimulus situation consists.

Let us represent stimulation of ommatidium A by an increase in light intensity by A^+ and a temporal succession—increase in light in ommatidium A, then decrease in light in ommatidium B —by A^+B^-. Then we can represent any situation by a formula compounded of A, B, and C. The analysis shows that the reaction to $A^+B^+C^+$ is in fact the sum of reactions, $A^+B^+ + B^+C^+ + A^+C^+$, and the same is true for each other possible formula. Now, the reaction to A^+B^- or A^-B^+ is a negative optomotor movement—the beetle turns in the opposite direction to that of the movement. But the reaction to A^-B^- is positive, the same as that to A^+B^+. The intensity of the reaction (strength of tendency to turn again and again in the same direction under experimental conditions forcing a choice between left and right every few seconds) is a direct quadratic function of the stimulus intensities. Hence, we must infer that the sensory input and motor output are linked by a process which works not as an algebraic summation but as **a multiplier, both of intensities and signs.** In a theory of summation of inputs of separate ommatidia, no optomotor response is expected to a rotating cylinder bearing stripes of random sequences of white, black, and grays. But a turning tendency occurs and follows expectation for a multiplying **correlation process.** The same methods have permitted determination of the time constants of the filters within and between ommatidial channels, the nonlinearity of the filters at very low velocities of rotation (due to partial adaptation during the light changes in the single ommatidia); the linearity of the scale of grays in the case of medium and high velocities; the blindness to phase in the beetle but sensitivity to phase in the bee (Hassenstein, 1951, 1958, 1961; Hassenstein and Reichardt, 1956; Reichardt, 1957, 1961; Kunze, 1961; see p. 1094).

B. Elementary Fixed Action Patterns

In the report on the round-table conference on nomenclature in animal behavior, held at Cambridge, England in 1949 (Thorpe, 1951), there is distinguished a category of behavior patterns within instinctive behavior called "fixed action patterns." These are inherited, relatively complex movement patterns, compared to the simpler reflexes, as characteristic of the species or group as gross structural features, and are little modifiable by external stimuli except in intensity. The acts of biting, antenna cleaning, drinking, scratching, jumping, defecating, coition, beckoning (as in fiddler crabs), threatening, chirping, and others are examples (Fig. 5.37). They overlap with or may be considered as complex reflexes. They are components of instincts and often the final consummatory act. Physiologically and anatomically they present the same problem as reflexes. We may confidently deduce that the nervous system contains **genetically determined circuits and physiological properties** of integration—timing, thresholds, frequencies, aftereffects—such that the complete act is represented in coded form and awaits only the adequate trigger. The circuits include provision for altering the probability of occurrence of the event under tonic influence from higher centers, for either-or resolution of conflicting action tendencies, for refractoriness on the one hand and, on the other, for decreasing thresholds leading to true spontaneous discharge of the action in the proper stimulus situation.

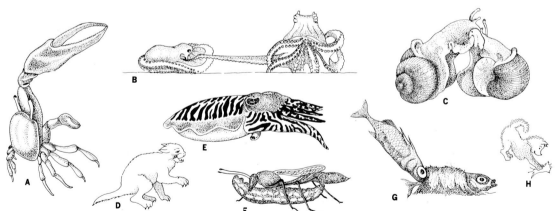

Figure 5.37. Elementary fixed action patterns. Examples of the species-characteristic components of higher behavioral sequences. Learning may modify the use of these actions but very little their actual pattern. **A.** Fiddler crab, *Uca pugilator*, waving its claw. [Tinbergen, 1951, after Meisenheimer.] **B.** Mating in *Octopus*. [Buddenbrock, 1956.] **C.** Mating in *Helix pomatia*. [Tinbergen, 1951, after Meisenheimer.] **D.** European wildcat striking with its paw. [Lindemann, 1955.] **E.** Male *Sepia officinalis* in sexual display. [Tinbergen, 1951.] **F.** The digger wasp *Ammophila campestris* with prey. [Tinbergen, 1951, after Baerends.] **G.** Male three-spine stickleback stimulating the female to spawn by quivering. [Tinbergen, 1951, after Ter Pelkwijk and Tinbergen.] **H.** The mouse jump in the domestic dog. [Lorenz, 1954.]

C. Kineses, Taxes, and Reversal of Sign

A kinesis has been defined as locomotor behavior not involving a steering reaction but in which turning, when it occurs, is random in direction. Some function of the turning—either its rate or amount—is related to the intensity of a stimulus. The onset or incidence of turning in a steady-stimulus situation can be considered an interesting special case of spontaneity and its control. Locomotion is maintained by the steady stimulus; in the class of orthokineses the rate of locomotion varies with the strength of the stimulation. So far the problem is one of spontaneity of this action pattern and, at least by analogy with some familiar systems like that governing the beat of neurogenic hearts, would seem to require no new or complex integrative mechanisms. But the specification that turning occurs repeatedly and that its direction is random introduces a fundamental property of which nothing has been said thus far. True randomness of individual events has often been invoked in speculative treatments of the physical basis of behavior. One interest is in the possibility that inherently unpredictable events, originating in noise at the level of very small particles, may determine macroscopic events through the enormous amplification supplied by the nervous system. Some encouragement for such speculations has been found in the occurrence of apparently random events at the level of subthreshold junction potentials in certain nerve-muscle endings (pp. 231, 974), miniature end plate potentials arising spontaneously, with the frequency character of noise. But the reality of "indeterminate" events at the level of the organism has been questioned. Insofar as kineses really obey their definition, they have this special interest. It is a consequence of our knowledge of the code used by neurons that a **source of randomness** is available without appeal to molecular events, in any case where several active neurons converge on one. Even if the frequencies of the individual cells are not random, but are fluctuating in a way determined by systematic variations in electrogenesis and in threshold, a modest number of independent cells will together make up an input experimentally indistinguishable from true noise. But while there is no difficulty making models which would provide randomness, how and where it is actually done is quite unknown. The additional specification for the class of **klinokineses,** which is defined by the frequency of the random turning being a function of stimulus intensity, makes this question more pertinent as well as offering a valuable clue in the search.

The sensory input maintaining locomotion in a kinesis may be the same as that used in a quite opposite way for other aspects of behavior. Locusts are kept in flight by the stimulation of hairs on the frons and vertex by the air stream— a case of positive feedback acting in a dynamic or driving function. The same input is used as negative feedback in the controlling function that affects flight direction.

One approach to the evaluation of alternative hypotheses for locomotor orientation is the application of probability theory and other mathematical techniques (for example Patlak, 1953a, 1953b); within the limitations of the biological adequacy of the assertions and postulates, these methods offer powerful means of analyzing potential adaptive value of candidate mechanisms.

The word **taxis** has been used in at least three different senses in modern writings: for locomotion either directly toward or away from a source of stimulation, for any locomotor behavior involving a steering, and for the spatial correction movement or turning component resulting in orientation. These phenomena present no new problems from the physiological point of view. This is not to say that they are understood or that they are all alike, even according to any one of the definitions. The classification of subtypes in older schemes has been regarded as unsatisfactory and no new system has attained stability. The categories defined behaviorally may turn out to comprise different combinations of a small number of physiological mechanisms such as increased turning tendency or increased rate with stronger stimuli.

The several kinds of oriented locomotion present exceptionally clear illustrations of the problems already before us. Without undertaking an exhaustive survey, a few examples will be recounted. One of the simple kinds of reaction—**klinotaxis** (see Fraenkel and Gunn, 1940)—requires a comparison of intensity of stimulation at successive moments in time. The blowfly larva, *Calliphora*, crawls away from light by turning its head alternately to right and to left and swinging the body away from the side toward which the head is turned when a stronger light is received. It is when rather than where the light falls that matters; if the light is directly overhead but is switched on only when the maggot turns its head leftward, it crawls continually to the right. This could be a **simple temporal integration** in a few neurons, systematically linked with the alternate left and right phases: body flexor motoneurons on the side opposite that to which the head is turned may be differentially sensitive to the light receptor signals for "light-increasing."

Another common reaction requires simultaneous comparison of intensities on two receptors, on the right and left of the midline—the **tropotaxis** of Kühn (see Fraenkel and Gunn, 1940). Without turning to this side and that, *Daphnia*, *Armadillidium*, *Planaria*, and other animals can move directly toward or away from a source of stimuli and, if one receptor is removed, circle even in a uniform field of stimulation, as when a light is overhead or diffuse. (It is assumed, in these cases, that sense organ removal sets up no message itself, therefore that the sense organ has no resting frequency.) There **must be at some point a comparison**—unless the neurons for the two sides are calibrated for absolute intensity. Here we recognize the general problem already enunciated (p. 281) of decision making. The decision on which way and how far to turn may be merely a result of competition between muscles or between bilateral neurons and hence subject to the wandering of their amplification factors, or there may be a control unit—probably a cell—which compares and decides. There may be several cells but they must each be equivalent in receiving input from both left and right, initiating output according to which of the two is stronger.

Still another taxis has been distinguished, again by the apparent common denominators of physiological mechanism. The **telotaxis** of Kühn (see Fraenkel and Gunn, 1940) achieves directedness of locomotion even when there is asymmetry of stimulation, by seemingly fixating an image of the source in one part of the retina. This category was created for optic reactions of the type of the dragonfly larva pursuing its prey, where unilateral blinding or simultaneous stimulation from two sources does not interfere with the ability to steer toward a source. In the test with two sources, where the tropotactic reaction results in steering between them, the telotactic animal may vacillate—approaching in a direct line toward one of the sources only to switch to a direct line for the other. But it is a representation of an object—and therefore a certain **anatomically defined group of neurons**—which determines the correction movements if a balance between stimulated and unstimulated neurons is involved; it is within the afferent system of each side and not between sides of the body. This is the class examined more closely as a control system in Section A., above.

Physiologically, the **light compass reaction** is similar, except that the angle between the object and the locomotion is kept at a constant value other than zero. Tinbergen (1951) has given reasons for regarding the learned homing of bees and certain wasps as basically similar—the **menotaxis** of Kühn. Here simple tests show that the orientation is to a configuration of several objects. If a wasp is trained to use a circle of pine cones around its nest, and some of the cones are moved or taken away, the wasp's choices show that it reacts not to any particular part of the total situation but to a combination of landmarks, even up to the last moment. Perhaps only a slightly more complex integration of sensory information is involved, but otherwise the problem appears to be of the same class physiologically. In ants a menotaxis has been shown to consist of a simple phototaxis plus a continuous tendency to turn by a definite, learned amount, in one direction on the way out from the nest, in the reverse direction on the way home (Jander, 1957). In the same study it was found that ants have the remarkable ability to integrate many successively experienced angles: after learning an indirect route home, removal of previous obstacles is followed by the ants taking a straight path directly home!

The simpler taxes may have a **positive or negative sign.** Baylor and Smith (1957) find that *Daphnia* exhibits negative geotaxis under conditions of falling temperature (within narrow limits), and given certain conditions of the other variables. Likewise, with rising hydrostatic pressure, high pH, and low redox potential, as when bacteria-fed, they are negatively geotactic, whereas under the opposite conditions of each parameter they are geopositive. They are photopositive with rising temperature, pressure, pH, and redox potential—as when algae-fed. Schneirla (1953) mentions reversal of sign in insects after feeding, mating, and agents which alter metabolic rate. Permutations are to be expected and have been found, as in the lowering of the point of temperature reversal of phototaxis, with age in droneflies. The common phenomenon of reversal of sign of a taxis, of decisive importance in the life of countless creatures, only represents for us another and more striking case of the alteration of an integrative function by superimposed input. In this instance there is a threshold and an either-or effect, since it is found that all intermediates between positive and negative taxes can generally not be obtained. At present it seems reasonable to assume that single cells or their equivalent are performing the integration and determining the direction.

D. Instinctive Behavior

In spite of vicissitudes, twists, and turns in the construction of its meaning, and a long period of neglect in the first decades of this century, the term instinct has been found useful for what appears on modern reexamination to be a natural category of behavior. While it has long stood for behavior which is relatively complex (compared to fixed action patterns), markedly stereotyped, and related to a situation rather than to a local stimulus, its use by comparative ethologists has resulted in considerable refinement, especially in interpretation of the organization of instinct. A recent definition is that of Thorpe (1951). "Instinct = an inherited and adapted system of coordination within the nervous system as a whole, which when activated finds expression in behaviour culminating in a fixed action pattern. It is organized on a hierarchical basis, both on the afferent and efferent sides. When charged, it shows evidence of action-specific-potential and a readiness for release by an environmental releaser." The wording is meant to imply that some alteration with experience is possible but that the instincts are genetically determined in a form which is **self-differentiating** or endogenous. They thus do not require instruction from another member of the species or conditioning by external reward or punishment except what may be imposed by a component of the environment essential to normal development. The **hierarchical** clause indicates that no single act or invariable sequence of acts comprises the instinct, but a system of adaptively coordinated higher and lower brackets of actions (Fig. 5.38). The wording of the third sentence of Thorpe's definition emphasizes a difference between instincts and reflexes and fixed action patterns; indeed, in the view of some it is the most fundamental part of the concept—namely, the dependence of instincts upon the organism being primed or in a state of readiness to perform the behavior patterns of one instinct in preference to all other behavior patterns. This **specific readiness** diminishes or disappears when the consummatory act of the charged instinct has been performed. This is the concept of the "action-specific-potential." The name has attracted criticism but we may think of a specific probability; it is equivalent to an internal drive in the sense that it is not simply or usually an expression of any primary physiological need (Lorenz, 1950; see also Tinbergen, 1951; Thorpe, 1956; Hinde, 1960). The requirement of many instinctive acts for a **specified releaser** and the internal mechanism this presupposes are discussed in the next section.

The following points relevant to the present work are tenable even if some instincts do not satisfy all parts of Thorpe's definition; certain cases appear not to require an environmental

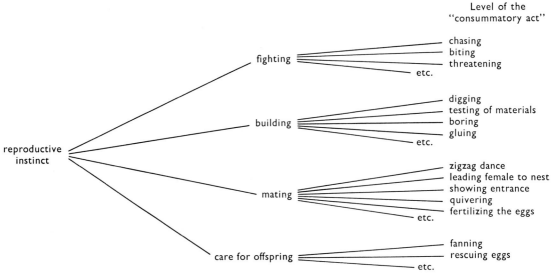

Figure 5.38. The principle of hierarchical organization. The reproductive instinct of the male three-spine stickleback. [Tinbergen, 1951.]

releaser; some are uniformly probable and hence show no evidence of action specific potential; some do not always culminate in a consummatory act. The first inference which the facts point to is that there must be laid down in the inherited and self-differentiating structure of the nervous system complex patterns of efferent commands, sufficient to account for all the instinctive acts of the species. This conclusion is similar to that reached with respect to the elementary fixed motor patterns, kineses, and taxes, but the patterns of connections now under consideration must be more elaborate and indeed largely made up of combinations of these simpler ones. Further, the evidence indicates that the motor patterns involved often have afferent control by servo loops during the performance of the instinctive action; hence the inherited pattern includes the constants of amplification of these loops. Many instances permit the additional qualification that the inherited self-differentiating pattern can be modified by learning, and, in fact, may require this in order to be steered in an adaptive direction.

The complexities which must be inferred are greater when we look more closely at actual cases. Preceding the final and often simple fixed motor pattern called the consummatory act, there may be a sequential **hierarchy of so-called appetitive actions.** These begin with apparently aimless searching movements and progress through stages of narrower search, each stage ending when a certain stimulus-constellation is encountered, constituting a releaser for the next. Thus the appetitive phases appear complex, variable, and plastic, though they are essentially repetitive re-enactments of a few types of oriented locomotion, modified within strict limits by afferent input. They are reminiscent of the forced or compulsive movements the clinical neurologist sees in certain patients. Under given conditions and until an exhaustion or reduction of some central state, repetition occurs without new stimuli being required. Some appetitive actions are variable according to the environmental features they bring the animal into, others are quite stereotyped. In the natural setting they are, of course, highly adaptive and appear purposeful. The result of ethological research has been to show that it is not the filling of a basic biological need but simply the release of the next stage and finally of the consummatory act that satisfies the drive behind such activity. The central ethological experiment is perhaps that which makes the instinct miscarry—that is, perform normally although under conditions where it cannot satisfy the biological need. For each stage in a hierarchic sequence there may be postulated a

system of connections leading to the corresponding efferent signals, with a releaser or initial afferent signal pattern adequate to start activity in this system, a limiting set of steady afferent signals necessary to permit this activity, and some device for providing recurrence in the cases of repetitive appetitive behavior—for example, a spontaneous pacemaker or a feedback loop restimulating the same act by afferent impulses due to the completion of the preceding occurrence, or a responsiveness to steady-state afferent signals in the right combination. Shutting off a released fixed action may be done by inhibition due to the releasing situation for the next action or by central or peripheral feedback (Prechtl, 1956). This brings us to a closer inspection of the mechanisms releasing instinctive acts and the probability of those acts.

E. The Release and Probability of Specific Behavior Patterns

The term **releaser** has come to be applied to that specific feature or complex of features in a situation which elicits an instinctive activity or a mood; a synonym is "sign stimulus" (sometimes "releaser" is confined to synonymy with "social releaser"). It differs from the term stimulus as used by physiologists only in complexity, and often not even in that. Examples range from the vibration of the water surface, releasing approach in the hunting *Notonecta*, to the learned configuration of an artificial flower, releasing approach in a bee. Both require other background input simultaneously such as daylight, season, calm weather, and a certain internal state. According to the internal state there may be no response at all or any one of a variety of responses, even to the same object; for example, a caterpillar may release catching and stinging acts in a wasp of the genus *Ammophila* which is hunting, or drawing-in movements in one which has just opened a nest, or stuffing movements in one which is filling a nest entrance, or even throwing-away movements in one which is digging out a nest (Baerends, 1941, 1958). The odor is probably the stimulus in hunting, and optical stimuli are effective if the caterpillar is dropped in transport;

perhaps a different aspect of the object is effective for each releasing action. One is reminded of the growing evidence of a neurophysiological basis of directed attention with suppression of competing input signals (pp. 306-309).

Experiments with models of various degrees of distortion and simplification have shown that the **actual stimulus is much less than the potential** in typical releasers; the animal does not use all that its sense organs can perceive. For example, the apparent recognition of sex may be simply response to a single aspect such as crouching posture. Commonly the essential quality is relative—the nearer or the higher or the smaller of two objects is the more likely to release. The configurational sign may be extracted not only as a static relation but dynamically, as in the well-known difference in reaction of certain young birds to a moving silhouette that suggests a goose if moved with long neck forward and a hawk if moved in the opposite direction. Some learned releasers may be fairly detailed; others, effective the first time they are encountered and without opportunity for instruction, may be simple or complex but are often highly specific, as in the case of odors. Social releasers (those performed by animals, releasing actions by other animals) are apt to be little more than the minimum movement required to form the sign stimulus. A distinction can be made between purely **releasing** stimuli and **steering** stimuli which direct the distinctive movement, as when butterflies respond to scent from a spray by alighting on colored papers.

Most of the selection and recognition of releasing stimuli—the **filtering**—is probably central and demands neurophysiological mechanisms, many of which are unlearned and species-constant. In some cases the filtering of effective aspects of sign stimuli is probably a result of the properties of the receptors (Marler, 1961). The term **innate releasing mechanism**—IRM—has been used for the whole internal mechanism of reception and recognition, from receptors to central circuits. It would appear to be a special case of the class of preset recognition circuits which includes the simple requirements noted above for reflexes and the most complex learned symbols in man.

Many instinctive acts respond to a single sign stimulus, especially acts to social releasers. But most **respond to the sum of several features** of the releasing situation. Certain male butterflies will begin the pursuit of a moving object—normally the sexual pursuit of a female—if a sufficient sum of three graded factors is reached: the light intensity, the nearness, and the zigzaggedness of the movement. The separate aspects display the interesting relation to each other of being additive and equivalent—in the sense that a certain strength of one (nearness, for example) is completely interchangeable with a certain intensity of another (brightness, for example). This has been called the law of heterogeneous summation (Tinbergen, 1951). Occasionally there is interaction as when thinness of models of a mother gull's bill is more important when it is held vertically than horizontally, in eliciting pecking by a nestling gull, perhaps because shadows are relatively more important for the thin bill (Weidmann, personal communication). These additional properties of releasing mechanisms bespeak a physiological integration of exactly the kind we seek in this chapter, namely, a process and a locus where several coded inputs come together and are evaluated, algebraically added, and together determine an output which is a function of input but a very indirect one.

Tinbergen hypothesized **centers for the main instincts** and then for the successive stages in the hierarchy of a normal sequential performance. Although it was not an essential feature of his "centers" that they should each occupy an anatomically discrete region, as opposed for example, to being a widespread network of specific pathways, it is significant that we can now deduce the strong likelihood that they are, in fact, anatomically concentrated. Not without overlap and not without widespread circuits of essential connections, nevertheless some sort of "heart" of the system for a given instinct or component behavior pattern is probable; here drives and relevant external and internal states—including weather, hormones, moods, and so on—can exert background, threshold, and stage-setting effects. The reasons are twofold: the a priori expectation from neurophysiological experience that at least many of the integrated factors will converge on a small locus in order to interact, and the experimental finding that crude, local brain stimulation can precipitate specific, normal, complex behavior patterns. Electrical stimulation in the brain of waking vertebrates causing sleep, drinking, feeding, and other complex acts has already been referred to (p. 298). Huber (1960) has inaugurated the local stimulation of brain structures in insects and reported discrete but complex effects in the behavior sphere, not due to paralysis or sensory deficit. In a certain part of the corpora pedunculata electrical stimulation elicits antenna cleaning or leg cleaning movements or chirping. (Of course, the mere fact of a center or of the release of complex behavior by local brain stimulation is not sufficient reason to call that action instinctive.)

But the most important characteristic of instinctive behavior and function of its nervous centers, in the view of some (Thorpe, 1956; Hinde, 1953, 1959a, 1959b, 1960), remains to be mentioned. This may be stated as the greatly increased **readiness to occur** or probability of occurrence of a given instinctive act as time goes on under suitable conditions but withholding the adequate releasers, and the decreased probability after several performances of the act. The increased probability occurs during the appetitive behavior or at least the mood specific to and normally preceding the action, and is manifested by a gradual lowering of threshold so that fewer sign stimuli suffice, or signs of lower stimulating value (nearness, brightness, and so on); see Fig. 5.39. The extreme is sometimes seen in explosion or **spontaneous discharge** ("vacuum activity") in the absence of any releaser (not however, independent of all receptors, as sometimes stated, because a certain combination of steady-state stimuli must be constantly received). Some birds are said to have a tonic hypothalamic sex center because they reproduce continuously unless stopped by unfavorable temperatures or water availability. Decreased probability develops when an action is repeatedly elicited (or even when the releaser is perceived but the action is not performed), the requirement for releaser intensity goes up, fewer trials result in reaction

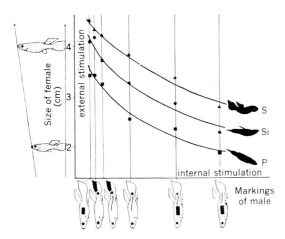

Figure 5.39. Probability of release of a specific act as function of intensity of internal state and external stimulus. Male guppies, *Lebistes reticulatus* show by certain of their color markings the intensity of an internal state making for readiness to mate. As this state increases the smaller females are adequate to release fixed postures (*P, Si, S*), representing approaches to copulation. [Baerends, 1958.]

and the extreme of complete refractoriness can be attained. Quite naturally these facts led to the postulation of an **accumulating potential** energy as time goes on without release of the action, and an expenditure or exhaustion of this potential with each performance. A basic property of this change in probability or potential is that it is commonly specific to a given instinct or even component phase, namely, the one normally released by the sign stimuli withheld (but not always—see Prechtl, 1956). Hence the term "action-specific potential" (earlier, "action-specific energy"). Obviously, the more a.s.p., the less the sum of releaser stimuli necessary to trigger—since the latter is the measure of the former. A formulation of the accumulating readiness and its increase and decrease in terms of feedback is perfectly compatible with the foregoing.

Under certain conditions it is possible to speak of **afterdischarge,** as when fighting movements continue after disappearance of a rival. Other conditions have led to use of the term **accommodation,** as when a sign stimulus creeps into range and, before eliciting the reaction, reaches an intensity above that which would trigger reaction if suddenly introduced. It is believed that the inertia manifested in accommodation is different from that in afterdischarge. Some cases of **rhythmic behavior** are ascribed to the exhausting of a.s.p. following each performance, its slow buildup in presence of a maintained stimulus to a level above that for sudden stimulation, then a burst of action leading to discharge of the accumulated a.s.p.

Unquestionably, this is one of the most fundamental contributions of the study of behavior to the study of physiology. The evidence that **some factor accumulates and dissipates**, specific for given patterns of efferent signals, is in fact quite physiological in character, methodology, and units. But physiologists might have been a long time in recognizing the existence of such phenomena and initiating the search for their basis. At the present stage we are far enough along to see the question, "What is accumulating and where?" The first hypotheses grow naturally out of physiology already known. Ewer (1957) has suggested two alternatives. (a) Diverse excitatory substances may be assumed to accumulate in the blood, having been produced anywhere. Each center must be assumed to be specifically sensitive to its own particular excitant. The nature of the excitant produced therefore determines what center is activated and consequently what behavior results. (b) The production and accumulation of excitant may be assumed to occur in the centers themselves. In this case a single substance would suffice. The specificity is topographic, the behavior being determined by the locus in which accumulation takes place.

By pointing to the CO_2-sensitive cells in the vertebrate respiratory center, the osmosensitive cells in the drinking center, and the heat-sensitive cells which must exist in the temperature-regulating center, Ewer suggests the operation of alternative (a) in these better-studied cases. But he points to automatisms in the invertebrates, such as the rhythm in the isolated earthworm nerve cord, the irrigation cycle in *Arenicola*—and we might add the cardiac ganglion of neurogenic hearts—to suggest the existence of centers "which are rhythmically active without external excitation, either in the form of nervous input or alterations in blood composition." We may waive discussion of whether the invertebrate cases are really different from the vertebrate, because the point of interest is that both depend on the physiological property of **neuronal pacemakers under control of steady-state conditions.** If a certain one of these conditions turns activity on and off in normal functioning, we tend to speak of a receptor; if normally the conditions permit rhythmic activity to continue for long periods, we speak of a spontaneous central rhythm. But in both, in all probability, we are dealing with the same mechanism. In the section above, where spontaneity at the neuronal level is discussed, we saw that the most common form is a relaxation oscillation dependent on a slowly developing depolarization (the pacemaker potential), a critical point (the threshold), and a discharge which, whatever else it does, must carry the system back to the initial condition, permitting another cycle to begin. There is no lack of precedent or acceptable physiological models capable of explaining in principle the observed facts of behavior, but whether it is actually anything like the known mechanisms of single neurons which is responsible is the question that presents such a challenge.

There is one more concept growing out of the same facts—that of **block.** Ethologists have tended to think of the accumulated action-specific potential as being blocked from discharge into the expression of the appropriate action until the specific releaser is encountered which acts on the releasing mechanism of that action. The releasing mechanism was thought to remove the block (but see van Iersel and Bol, 1958, for a new view). There is little independent evidence of the block; it is thought to be required. Certainly it presents no insuperable physiological difficulty, if as before we may fall back on known mechanisms at lower levels. A block could be an inhibition and the releasing mechanism could be focused on this rather than on the accumulating probability mechanism. But a consequence of comparing the latter with a pacemaker neuron is that the threshold stimulus in this neuron is changing pari passu with the gradual depolarization, so unless there is positive evidence of block it is not necessary to invoke it. Some kind of **inhibition** does seem more than likely in the mutual exclusion of instincts and even in component actions within one instinct—for example, digging out a nest, placing food in a nest, and closing the nest in the digger wasp. This has the general appearance of the reciprocal inhibition familiar to reflexology.

The **main deductions from behavior** with respect to the release and probability of performance of instinctive actions are then the following. (a) The preset circuits which comprise the several instincts, whether innate or modifiable-innate, must include on their afferent side sometimes complex recognition mechanisms actuated by relations between a few selected aspects of the total environmental situation. The separate aspects are summed (have a certain interchangeable equivalence). A critical threshold is reached only in the presence of steady-state permissive influx of several kinds, external and internal. There may be several of these releasing mechanisms (RM's) for each of the several major instincts. There may be groups of alternative RM's having reciprocally inhibiting connections (leading, for example, within the major instinct of reproductive activities to different kinds of fighting, to display, or to threat). Furthermore, groups of hierarchically related RM's occur, having sequential and progressively less variable character (for example, food-getting in *Notonecta*: hunting movements encounter the releaser for approach, approach leads to the releaser for pounce, which in turn brings receptors into the range of releasers for piercing, and, if the last releaser is encountered, consummation is by sucking). (b) Each releasing mechanism must include or be connected with a specific mechanism having the property of gradually increasing the probability of occurrence of a given action or lowering its threshold during periods of preparedness for that action but absence of the releaser ("hunger"), and the further property of decreasing the probability or raising the threshold as the action is performed (eating—but not necessarily nourishing). (c) There must be a higher-level integrating mechanism which, on the basis of internal and external states and mutual inhibition, permits or triggers one or another of the several major moods or central states of readiness to discharge a given complex of behavior patterns.

These are not physiologically incredible requirements. They are staggeringly complex, but of a quite different order from those called for by the higher types of behavior in vertebrates and especially man. They seem amenable to explanation in terms of the integrative properties at the neuronal level, each separate requirement and characteristic having some precedent in known cellular physiology. The question is, however, very much open whether actually these already recognized integrative parameters, as listed earlier in this chapter, account for the observed instinctive behavior, or whether yet-to-be-discovered parameters operate, as we conclude must be true for higher-level phenomena.

F. Learning

It has already been proposed that learning is a function diffused through the nervous system and operating separately in different levels and spheres. The analogy was drawn with functions in an industrial concern such as computation, as opposed to departmentalized functions such as advertising. By the same token, one may be prepared to find that there are simple and complex forms of learning and even specialized aspects of memory, like a departmentalized record storage.

Changes in the parameters of single neuronal units enduring for many minutes have been recorded in acute preparations (for example posttetanic potentiation, Eccles, 1958). Facilitation attributable to the neuromuscular junction has been claimed to last at least 7 days in *Stomphia*, a swimming sea anemone (Wilson, 1959). Pathologic states have been shown to alter properties of the simplest reflex arcs in the cat spinal cord (Eccles, Libet, and Young, 1958). Disuse caused by dorsal root section altered reflex response of motoneurons (Eccles, 1958). Asymmetrical posture of limbs in the rat induced by unilateral brain lesions persists if the cord is transected after 45 minutes (Chamberlain et al., 1961). Cytoplasmic ribonucleic acid has been found to change with nerve cell activity, and this has led to the speculation that the contents of the cell play a role in alteration with experience, the membrane only expressing the result. Whether these cases are germane to memory is uncertain; at least partly it is a matter of definition.

With reference to higher forms of persistent change, there is no real insight into mechanism as yet. Theories embodying physiological explanations of conditioning have been reviewed elsewhere (Thorpe, 1956; Deutsch, 1960b; Morrell, 1961). Even models are lacking for those examples which, like Lashley's trained rats, continue to manifest the learning after any of a variety of lesions and also after deep anesthesia and convulsive stimulation. It is understandable therefore that research in the physiological, as distinct from the experimental psychological, analysis of learning is chiefly concerned with these objectives: (1) the anatomical localization of **structures involved** in some way in a given learning process, and (2) the observation by electrical recording of localized **events correlated** with a given learning process. The almost insuperable difficulty in both approaches is to distinguish those loci or correlated events which are, like muscle action potentials, after the fact, or, like retinal action potentials, only partly integrated preconditions.

The results of these studies are nevertheless interesting and often unexpected. Certain kinds of conditioning are found to have different topographic vulnerabilities to lesion from others. Animals seemingly capable of little learning by some tests may perform apparent feats if the task is in the right phase of normal activities and especially if connected with learned steering of instinctive actions. Learning may be so little generalized that an ant taught a visual T-maze with one eye covered must relearn when the cover is switched to the other eye. Circumscribed regions are found to which electrical shock acts as a reward: mammals can quickly learn that pressing a bar delivers a shock and will press at high repetition rates (Olds, 1958). Loci acting as punishment are similarly demonstrated. Distinctive potentials evoked by a flickering light as the sign of punishment to come (conditioned stimulus) change their distribution in the brain during learning and even appear spontaneously. After learning has reached a certain perfection, they are narrowly localized and never appear spontaneously (John and Killam, 1961).

A promising preparation for the study of operant conditioning in a simple form, in the isolated insect thorax, has been recently discovered by Horridge (1962).

A healthy result of the effort to comprehend the nature of persistent change in nervous connections and in weighting of different inputs is the realization that this is really a special case of the general problem of the **maintenance of stability** among a given set of relations between dynamic properties of junctions and graded membrane actions. Since this is one formulation of the problem of the neurological basis of stereotyped behavior, we see that basically innate and acquired patterns of behavior present the same problem. How or what changes in learning is of secondary interest to how or what stabilizes the amplification factors and mixing constants in any persistent pattern.

G. Unity of Action, Attention, Arousal

The reciprocal inhibition of conflicting reflexes and instinctive acts has already been alluded to. The principle involved has the widest application and is one of the outstanding achievements of the nervous system, as well as one of the least understood. Not only is an antagonistic reflex or fixed pattern inhibited, but also each one of a long list of alternative courses of action, extending to the highest levels of behavior. Mutual exclusion is, of course, a necessary facet of behavior which includes pushbutton responses, like giant fiber startle reflexes —a necessary facet, that is, if frequent jams and stalls from coincidence of stimuli are to be avoided. The most careful observation of animal behavior has repeatedly emphasized the lack of such maladaptive cancellation and the reality of a unity of action at any moment. Sherrington (1947, p. 235) put it this way: "The resultant singleness of action from moment to moment is a keystone in the construction of the individual whose unity it is the specific office of the nervous system to perfect. The interference of unlike reflexes and the alliance of like reflexes in their action upon their common paths seem to lie at the very root of the great psychical process of 'attention.'"

Little can be said about the physiological mechanism of this important function. It must take place at a fairly high level since the actions inhibited are so complex that they share doubtless many neurons above the level of the final common path. Unlike the inhibitions that have become familiar through direct electrical recording, this type must not be graded, since even strong competitive stimulation either does not have any effect or breaks through and usurps the whole system, extinguishing the previous activity. This does not mean the elementary cellular mechanism of inhibition needs to be a new kind, but only that this application of it is so powerful as to be all-or-nothing. These considerations in turn support the argument already referred to that the separate stereotyped behavior patterns have centers where an integrative executive receives decisive input and controls his particular switch.

This kind of functional relation can be supposed to be one of the most important in evolution. When our information is more complete it may be expected that lower and higher groups will differ markedly along a quantitative scale in the development of mutual exclusion, of concentration and scope of control of executive centers. Perhaps it is a characteristic of human behavior that it is less replete with either-or switching than that of lower animals; in our most human moments we are torn between this and that and a good deal of our activity is in fact an attempt to compromise between conflicting aims. Has evolution proceeded toward more frustration, more balancing; does this represent more freedom?

An additional facet in the over-all concentration of actions in adaptive behavior has been demonstrated in higher vertebrates: the requirement of activity of an arousal system or alerting system of neurons. This widespread, nonspecific activating system represented by reticular substance at all levels, has already been discussed (p. 306).

It is indicative of the promise of the future that such basic and seemingly intangible behavior phenomena as general alertness, focused attention, and singleness of action should now be the subject of physiological and anatomical analysis.

Classified References

General

Adrian et al., 1954; Brazier, 1960; Bremer, 1960; Buddenbrock, 1953; Bullock, 1957a; Fulton, 1949; Gerard, 1942, 1953, 1960; Harlow and Woolsey, 1958; Herrick, 1924, 1948, 1949; Jeffress, 1951; Jennings, 1906; Kappers, 1929; Koshtoyants, 1957b; Prosser and Brown, 1961; Redfield, 1942; Sherrington, 1906.

Integration at the Neuronal Level

Adelman et al., 1960; Alanis and Matthews, 1952; Albe-Fessard and Szabo, 1954; Arvanitaki, 1942; Bennett, 1959; Bennett et al., 1959; Bernhard, 1950, 1953; Bishop, 1958; Boistel and Fatt, 1958; Bremer, 1951, 1956b; Bullock, 1953a, 1953b, 1957b, 1958, 1959; Bullock and Terzuolo, 1957; Burgen and Kuffler, 1957; Burkhardt, 1959a, 1959b; Clare and Bishop, 1955; Cross and Green, 1959; Davis, 1961; Dudel and Kuffler, 1961a, 1961b; Dudel and Orkand, 1960; Eccles, 1951, 1957, 1959, 1960, 1961a, 1961b; Eccles and Krnjević, 1959a, 1959b; Eccles and Westerman, 1959; Eccles, Eccles, and Magni, 1960a, 1960b; Eccles, Libet, and Young, 1958; Eyzaguirre and Kuffler, 1955a, 1955b; Fatt, 1961; Faure, 1957; Fessard, 1956; Fessard and Tauc, 1958; Florey, 1960, 1961; Frank, G. B., 1958; Frank, K., 1959; Frank and Fuortes, 1960; Freygang, 1958; Fuortes, 1958, 1959; Furshpan and Potter, 1957, 1959; Galambos, 1961; Gellhorn, 1953, 1957; Gerard, 1941; Gerard and Libet, 1940; Graham and Gerard, 1946; Granit and Phillips, 1956, 1957; Gray, 1959; Grundfest, 1957a, 1957b, 1958, 1959a, 1959b, 1961a, 1961b; Hagiwara, 1958, 1961; Hagiwara and Bullock, 1957; Hagiwara and Morita, 1962; Hagiwara and Saito, 1959; Hagiwara and Tasaki, 1958; Harris et al., 1958; Hodgkin, 1938; Hoyle and Wiersma, 1958; Hughes, 1952; Katz, 1947; Kennedy, 1958, 1960; Kennedy and Preston, 1960; Kerkut and Taylor, 1956, 1958; Kolmodin, 1957; Kolmodin and Skoglund, 1958; Krnjević and Miledi, 1959; Libet and Gerard, 1941; Lloyd, 1961; Lundberg, 1958; Matsui, 1955; Maynard, 1953, 1954, 1955, 1960, 1961; Morita, 1959; Morita and Takeda, 1959; Morita and Yamashita, 1959a, 1959b; Morita et al., 1957; Otani and Bullock, 1959; Ozbas and Hodgson, 1958; Phillips, 1956; Preston and Kennedy, 1960; Purpura and Grundfest, 1956; Retzlaff, 1957; Retzlaff and Fontaine, 1960; Reuben, et al., 1960; Rosenblueth, 1950; Sawyer, 1959; Speidel, 1941; Strumwasser, 1961; Tasaki et al., 1954; Tateda and Morita, 1959; Tauc, 1955, 1956, 1957, 1958, 1960; Terzuolo, 1959; Terzuolo and Bullock, 1956, 1958; Thesleff, 1959; Van der Kloot, 1961; Wall, 1958; Wall et al., 1956; Ward, 1958; Watanabe, 1958; Watanabe and Bullock, 1959; Weber, 1949; Wiedersheim, 1890; Wiersma, 1952, 1958; Wiersma and Adams, 1950; Wiersma and Van Harreveld, 1939; Wilson, 1959; Wolbarsht, 1960; Young, 1946.

Integration at the Level of Groups of Neurons

Adrian, 1959; Auger and Fessard, 1928, 1929; Autrum, 1955; Ayrapet'yants, 1952; Baglioni, 1907; Barlow et al., 1957; Baylor and Smith, 1957; Békésy, 1958, 1959, 1960; Biryukov, 1952; Bishop, 1959; Bonnet, 1938; Broadbent, 1961; Bruner, 1957; Buller et al., 1960a, 1960b; Bullock, 1961; Burkhardt, 1958; Burns, 1958; Buser, 1955; Bykov, 1942; Clare and Bishop, 1956; Cohen, 1960; Creed et al., 1932; Dercum, 1922; Deutsch, 1960a, 1960b; Dijkgraaf, 1955, 1961; Dodwell and Sutherland, 1961; Dow and Moruzzi, 1958; Duke-Elder, 1949; Eccles and Lundberg, 1958a; Eckert, B., 1959; Eckert, R. O., 1960; Faust, 1952; Fessard, 1961; FitzHugh, 1957; Furshpan, 1959; Gesell et al., 1954; Granit, 1961; Granit and Phillips, 1956, 1957; Granit and Rutledge, 1960; Granit, Haase, and Rutledge, 1960; Granit, Pascoe, and Steg, 1957; Granit, Pompeiano, and Waltman, 1959; Hagiwara and Watanabe, 1956; Hartline, 1959; Hartline, Ratliff, and Miller, 1961; Hartline, Wagner, and MacNichol, 1952; Holmgren and Frenk, 1961; Holst, 1935, 1936a, 1936b, 1939a, 1939b, 1943, 1950, 1954, 1955; Holst and Mittelstaedt, 1950; Horridge, 1961; Hoyle, 1957; Hoyle and Wiersma, 1958; Hubel, 1959; Hubel and Wiesel, 1959; Hughes, 1952, 1958; Hughes and Wiersma, 1960; Jansen and Brodal, 1958; Jasper et al., 1958; Jung, 1961; Katsuki, 1961; Katsuki et al., 1954; Kuffler, 1953; Lavín et al., 1959; Lettvin et al., 1961; Lippold et al., 1960; Lissmann and Machin, 1958; Livingston, 1959; Lorente de Nó, 1934, 1938; Machin and Lissmann, 1960; Magoun, 1953, 1958; Maturana et al., 1960; Mountcastle, 1961; Mountcastle et al., 1957; Nakao, 1958; Neff, 1960; Nielsen, 1946, 1958; Oberholzer and Tofani, 1960; Olds, 1958; Poggio and Mountcastle, 1960; Penfield, 1960; Platt, 1960; Powell and Mountcastle, 1958; Ratliff, 1961; Ratliff and Hartline, 1959; Ratliff, Floyd, and Mueller, 1957; Ratliff, Miller, and Hartline, 1958; Rensch, 1956; Rosenblith, 1961; Ross, 1957; Rushton, 1959; Sawyer, 1959; Scheibel and Scheibel, 1958a; Sherrington, 1906, 1929; Sholl, 1956; Sholl and Uttley, 1953; Sprague et al., 1960; Sutherland, 1957, 1959, 1960; Tauc and Hughes, 1961; Tomita, 1958; Voronin, 1957; Wall and Cronly-Dillon, 1961; Wiersma, 1957, 1961a, 1961b; Zotterman, 1939.

Central Control of Afferent Input

Chambers et al., 1960; Desmedt, 1960; Desmedt and Monaco, 1961; Dodt, 1956; Eccles and Lundberg, 1958b; Granit, 1956; Hagbarth and Kerr, 1954; Hernández-Peón, 1959, 1961; Hernández-Peón and Hagbarth, 1955; Loewenstein, 1956; Marler, 1961; Molina and Hunsperger, 1959; Ruben and Sekula, 1960.

Plasticity of Function

Barcroft and Barron, 1939; Bethe, 1930; Bethe and Woitas, 1930; Coghill, 1929; Eccles, 1958; Hughes, 1957; Shamarina, 1958; Sperry, 1945, 1951a, 1951b, 1958; Székely, 1958; Szentágothai and Székely, 1956; Weiss, 1936, 1941, 1950a, 1950b, 1960b, 1961; Windle, 1950; Young, 1951.

Brain Stimulation

Andersson, 1953; Andersson et al., 1958; Grimm, 1960; Hess, 1956, 1957; Holst and St. Paul, 1960; Huber, 1953, 1955, 1960; Hunsperger, 1956; Penfield, 1958; Penfield and Rasmussen, 1950; Strumwasser and Cade, 1957.

Theoretical Models and Computer Studies

Adolph, 1959; Ashby, 1960; Barlow, 1961; Bergeijk and Harmon, 1960; Bernard and Kare, 1962; Beurle, 1956; Cragg and Temperley, 1954; Crane, 1961; Farley, 1962; Farley and Clark, 1960; Fessard, 1961; Harmon, 1959, 1961; Harmon and Wolfe, 1959; Hassenstein, 1951, 1958, 1961; Hassenstein and Reichardt, 1956; Hick, 1952; Hick and Crossman, 1952, 1955; Householder, 1946; Householder and Landahl, 1945; Keidel, 1961a, 1961b; Küpfmüller, 1961; Küpfmüller and Poklekowsky, 1956; Landahl, 1961a, 1961b; Landahl and Williams, 1958; Levinson and Harmon, 1961; McCulloch, 1949; McCulloch and Pitts, 1943; Mittelstaedt, 1954, 1957, 1960, 1961; Mittelstaedt and Holst, 1953; Patlak, 1953a, 1953b; Rall, 1959; Rashevsky, 1956, 1960; Reichardt, 1957, 1961; Reiss, 1961; Rochester et al., 1956; Stark and Cornsweet, 1958; Stark and Sherman, 1957; Stark et al., 1958; Stegemann, 1961; Van Bergeijk, 1961; Von Neumann, 1958; Williams, 1958.

Spontaneity, Brain Waves, Phases of Behavior, Tonic Input:

Adrian, 1931; Aladzhalova, 1958; Albe-Fessard and Szabo, 1954; Arvanitaki, 1938; Barcroft and Barron, 1939; Batham and Pantin, 1950; Bonnet, 1941; Bremer, 1935, 1938, 1941, 1960; Buddenbrock, 1930, 1952; Bullock, 1945, 1957b, 1958, 1959, 1961; Bullock and Terzuolo, 1957; Burkhardt, 1954; Burns, 1958; Coghill, 1929; Cragg and Hamlyn, 1955; Crescitelli and Jahn, 1939, 1942; Dudel and Kuffler, 1961a, 1961b; Dudel and Orkand, 1960; Dudel and Trautwein, 1958; Enger, 1956; Fessard, 1959; Gerard and Young, 1937; Gibbs and Gibbs, 1950; Gray, 1936, 1938, 1939, 1950; Gray and Lissmann, 1940a, 1940b, 1946a, 1946b; Gray and Sand, 1936; Gray et al., 1938; Grundfest, 1961c; Hagiwara and Bullock, 1957; Hagiwara and Watanabe, 1956; Harker, 1956, 1958, 1960; Horridge, 1958; Hughes, 1958; Hughes and Wiersma, 1960; Jahn and Crescitelli, 1938, 1939, 1941; Jasper, 1936, 1960; Jasper and Shagass, 1941; Kogan, 1958; Li and Jasper, 1951; Li, Cullen, and Jasper, 1956a, 1956b; Li, Jasper, and Henderson, 1952; Lilly, 1960; Matsui, 1955; Matsui and Shibuya, 1958; Maynard, 1954, 1961; Miller, 1960; Olds, 1960; Ozorio de Almeida and Piéron, 1924; Pantin and Vianna Dias, 1952; Pittendrigh, 1958, 1961; Prosser, 1936, 1940, 1943; Rijlant, 1931; Roeder, 1939, 1941, 1955a, 1955b; Roeder et al., 1960; Romanes, 1885; Salmoiraghi and Burns, 1958; Sato et al., 1957; Tasaki et al., 1954; Trautwein and Dudel, 1958; Walter, 1959; Weiant, 1958; Weidmann, 1956; Weiss, 1936, 1941, 1950a, 1950b; Wells, 1949, 1950, 1955; Wilson, 1961; Windle, 1940; Wolsky, 1933.

Analysis of Behavior, Control Systems, Learning (see also Spontaneity)

Adolph, 1959; Baerends, 1941, 1958; Bangert, 1960; Baylor and Smith, 1957; Boycott and Young, 1950, 1958, 1950–1961; Brady, 1960; Braemer, 1957; Broadbent, 1961; Buddenbrock, 1956; Burkhardt, 1958; Bykov and Slonim, 1960; Carthy, 1958; Chamberlain et al., 1961; Child, 1924; Coghill, 1929; Deutsch, 1960b; Eccles et al., 1958; Eccles and Westerman, 1959; Eckert, 1960; Ewer, 1957; Fraenkel and Gunn, 1940; Galambos, 1960; Galambos and Morgan, 1960; Gray, 1936, 1939, 1950; Gray and Lissmann, 1938, 1940a, 1940b, 1946a, 1946b; Gray and Sand, 1936; Gray, Lissmann, and Pumphrey, 1938; Harlow and Woolsey, 1958; Hassenstein, 1951, 1958, 1961; Hassenstein and Reichardt, 1956; Hebb, 1949; Hernández-Peón, 1960; Hernández-Peón et al., 1958; Hinde, 1953, 1955, 1958, 1959a, 1959b, 1960; Holst, 1935a, 1935b, 1936a, 1936b, 1939a, 1939b, 1943, 1950, 1954, 1955; Holst and Mittelstaedt, 1950; Holst and St. Paul, 1960; Horridge, 1962; Huber, 1953, 1955, 1960; Hunsperger, 1956; Iersel and Bol, 1958; Jahn, 1960; Jander, 1957; Jasper and Shagass, 1941; Jeffress, 1951; Jennings, 1906; John and Killam, 1961; Konorski, 1948;

Kunze, 1961; Küpfmüller and Poklekowsky, 1956; Lindemann, 1955; Lissmann and Machin, 1958; Lorenz, 1950, 1954; Machin and Lissmann, 1960; Mittelstaedt, 1957, 1960, 1961, 1962; Mittelstaedt and Holst, 1953; Molina and Hunsperger, 1959; Morrell, 1961; Nakao, 1958; Olds, 1958; Penfield, 1958; Prechtl, 1956; Pribram, 1958, 1959; Pringle, 1951; Reichardt, 1957, 1961; Sanders and Young, 1940; Schneirla, 1953; Schöne, 1959; Sperry, 1950; Strumwasser and Cade, 1957; Sutherland, 1957, 1959, 1960; Thorpe, 1951, 1956; Tinbergen, 1951; Varjú, 1959; Wagner, et al., 1961; Weiss, 1959; Wells, 1958; Wendler, 1961; Windle, 1940; Wolda, 1961; Wolstenholme and O'Connor, 1958; Wright Air Dev. Div., 1960; Young, 1946.

Bibliography

A limited selection from the large relevant literature is cited. The reader should depend heavily on the bibliographies of the papers and books given here and on those of other chapters, particularly Chapter 3, on Excitability and Conduction, and Chapter 4, on Transmission.

ADELMAN, W. J., PAUTLER, E., and EPSTEIN, S. 1960. Analysis of repetitive transient response of lobster motor axons. *Amer. J. Physiol.*, 199:367-372.

ADOLPH, A. R. 1959. Feedback in physiological systems: an application of feedback analysis and stochastic models to neurophysiology. *Bull. math. Biophys.*, 21:195-216.

ADRIAN, E. D. 1931. Potential changes in the isolated nervous system of *Dytiscus marginalis. J. Physiol.*, 72:132-151.

ADRIAN, E. D. 1959. Sensory mechanisms—Introduction. In: *Handbook of Physiology*, Sect. 1. *Neurophysiology*, 1:365-367.

ADRIAN, E. D., BREMER, F., JASPER, H. H., and DELAFRESNAYE, I. F. (eds.). 1954. *Brain Mechanisms and Consciousness.* Blackwell, Oxford.

ALADZHALOVA, N. A. 1958. Electrographic study of pharmacologically induced changes in the ultra-slow waves of the cortical potentials. *Sechenov J. Physiol. (Fiziol Zh.)* 44:757-764. (English Translation.)

ALANIS, J. and MATTHEWS, B. H. C. 1952. The mechano-receptor properties of central neurones. *J. Physiol.*, 117:P59-P60.

ALBE-FESSARD, D. and SZABO, TH. 1954. Étude microphysiologique du neurone intermédiaire d'une chaîne réflexe disynaptique. *C. R. Soc. Biol., Paris,* 148:281-284.

ANDERSSON, B. 1953. The effect of injections of hypertonic NaCl-solutions into different parts of the hypothalamus of goats. *Acta physiol. scand.*, 28:188-201.

ANDERSSON, B., JEWELL, P. A., and LARSSON, S. 1958. An appraisal of the effects of diencephalic stimulation of conscious animals in terms of normal behaviour. In: *Neurological Basis of Behaviour.* G. E. W. Wolstenholme and C. M. O'Connor (eds.). J. & A. Churchill, London.

ARVANITAKI, A. 1938. Propriétés rythmiques de la matière vivante. II. *Etude expérimentale sur le myocarde d'Helix.* Hermann, Paris.

ARVANITAKI, A. 1942. Effects evoked in an axon by the activity of a contiguous one. *J. Neurophysiol.*, 5:89-108.

ARVANITAKI, A. and CHALAZONITIS, N. 1949. Prototypes d'interactions neuroniques et transmissions synaptiques. Données bioélectriques de préparations cellulaires. *Arch. Sci. physiol.*, 3:547-565.

ARVANITAKI, A. and CHALAZONITIS, N. 1956. Activations du soma géant d'*Aplysia* par voie orthodrome et par voie antidrome (dérivation endocytaire). *Arch. Sci. physiol.*, 10:95-128.

ASHBY, W. R. 1960. *Design for a Brain.* Wiley, New York.

AUGER, D. and FESSARD, A. 1928. Recherches sur l'excitabilité du système nerveux des insectes. *C. R. Soc. Biol., Paris,* 99:305-307.

AUGER, D. and FESSARD, A. 1929. Observations complémentaires sur un phénomène de contractions rythmées provoquées par excitation galvanique chez certains insectes. *C. R. Soc. Biol., Paris,* 101:897-899.

AUTRUM, H. 1955. Die Zeit als physiologische Grundlage des Formensehens. *Stud. Gener.*, 8:526-530.

AYRAPET'YANTS, E. SH. 1952. *Higher Nervous Activity and Receptors of Internal Organs.* Acad. Sci. USSR, Leningrad. (In Russian.)

BAERENDS, G. P. 1941. Fortpflanzungsverhalten und Orientierung der Grabwespe *Ammophila campestris. Jur. Tijd. Entom.*, 84:71-275.

BAERENDS, G. P. 1958. The contribution of ethology to the study of the causation of behaviour. *Acta physiol. pharmacol. néerl.*, 7:466-499.

BAGLIONI, S. 1907. *Zur Analyse der Reflexfunktion.* Bergmann, Wiesbaden.

BANGERT, H. 1960. Untersuchungen zur Koordination der Kopf- und Beinbewegungen beim Haushuhn. *Z. Tierpsychol.*, 17:143-164.

BARCROFT, J. and BARRON, D. H. 1939. Movement in the mammalian foetus. *Ergebn. Physiol.*, 42:107-152.

BARLOW, H. B. 1961. The coding of sensory messages. In: *Current Problems in Animal Behaviour.* W. H. Thorpe and O. L. Zangwill (eds.). Cambridge Univ. Press, Cambridge.

BARLOW, H. B., FITZHUGH, R., and KUFFLER, S. W. 1957. Change of organization in the receptive fields of the cat's retina during dark adaptation. *J. Physiol.*, 137:338-354.

BATHAM, E. J. and PANTIN, C. F. A. 1950. Phases of activity in the sea-anemone, *Metridium senile* (L.) and their relation to external stimuli. *J. exp. Biol.*, 27:377-399.

BAYLOR, E. R. and SMITH, F. E. 1957. Diurnal migration of plankton crustaceans. In: *Recent Advances in Invertebrate Physiology.* B. T. Scheer (ed.). Univ. of Oregon Press, Eugene.

BÉKÉSY, G. VON. 1958. Funneling in the nervous system and its role in loudness and sensation intensity on the skin. *J. acoust. Soc. Amer.*, 30:399-412.

BÉKÉSY, G. VON. 1959. Neural funneling along the skin and between the inner and outer hair cells of the cochlea. *J. acoust. Soc. Amer.*, 31:1236-1249.

BÉKÉSY, G. VON. 1960. Neural inhibitory units of the eye and skin. Quantitative description of contrast phenomena. *J. opt. Soc. Amer.*, 50:1060-1070.

BENNETT, M. V. L. 1959. Neurite and soma spikes in a monopolar neuron. *Fed. Proc.*, 18:10. (See Bibliography in Chapter 4 for further references.)

BENNETT, M. V. L., CRAIN, S. M., and GRUNDFEST, H. 1959. Electrophysiology of supramedullary neurons in *Spheroides maculatus*. III. Organization of the supramedullary neurons. *J. gen. Physiol.*, 43:221-250.

BERGEIJK, W. A. VAN and HARMON, L. D. 1960. What good are artificial neurons? *Wadd Tech. Rep.*, 60-600: 395-406. (Bell Tel. Lab., Bionics Symp.)

BERNARD, E. E. and KARE, M. R. (eds.).

1962. *Biological Prototypes and Synthetic Systems*. Vol. 1. Plenum Press, New York.

BERNHARD, C. G. 1950. Potentiels lents et modifications d'excitabilité dans des systèmes réciproques de la moelle épinière. In: *Électrophysiologie, Colloq. int. Cent. nat. Rech. sci.*, 22:395-406.

BERNHARD, C. G. 1953. Analysis of the spinal cord potentials in leads from the cord dorsum. In: *The Spinal Cord*. (CiBa Foundation Symposium). G. E. W. Wolstenholme (ed.). Churchill, London.

BETHE, A. 1930. Studien über die Plastizität des Nervensystems. I. Mitteilung. Arachnoideen und Crustaceen. *Pflüg. Arch. ges. Physiol.*, 224:793-820.

BETHE, A. and WOITAS, E. 1930. Studien über die Plastizität des Nervensystems. II. Mitteilung. Coleopteren, Käfer. *Pflüg. Arch. ges. Physiol.*, 224:821-835.

BEURLE, R. L. 1956. Properties of a mass of cells capable of regenerating pulses. *Philos. Trans.* (B), 240:55-94.

BIRYUKOV, D. A. 1952. Comparative physiology and pathology of conditioned reflexes. *J. Higher Nervous Activity (Zh. vyssh. nervn. Deyat.)* 2(4). (In Russian.)

BISHOP, G. H. 1958. The dendrite: receptive pole of the neurone. *Electroenceph. clin. Neurophysiol.*, 10(Suppl.):12-21.

BISHOP, G. H. 1959. The relation between nerve fiber size and sensory modality: phylogenetic implications of the afferent innervation of cortex. *J. nerv. ment. Dis.*, 128:89-114.

BOISTEL, J. and FATT, P. 1958. Membrane permeability changes during inhibitory transmitter action in crustacean muscle. *J. Physiol.*, 144:176-191.

BONNET, V. 1938. Contribution à l'étude du système nerveux ganglionnaire des crustacés. *Arch. int. Physiol.*, 47:397-433.

BONNET, V. 1941. L'activité rythmique de la cellule nerveuse et ses modifications. Soc. Anonyme de l'imprimerie A. Rey, Lyon.

BOYCOTT, B. B. and YOUNG, J. Z. 1950. The comparative study of learning. *Symp. Soc. exp. Biol.*, 4:432-453.

BOYCOTT, B. B. and YOUNG, J. Z. 1950-1961. (Other relevant citations are given in the Bibliography in Chapter 25.)

BOYCOTT, B. B. and YOUNG, J. Z. 1958. Reversal of learned responses in *Octopus vulgaris* Lamarck. *Anim. Behav.*, 6:45-52.

BRADY, J. V. 1960. Temporal and emotional effects related to intracranial electrical self-stimulation. In: *Electrical Studies on the Unanesthetized Brain*. E. R. Ramey and D. S. O'Doherty (eds.). Hoeber, New York.

BRAEMER, W. 1957. Verhaltensphysiologische Untersuchungen am optischen Apparat bei Fischen. *Z. vergl. Physiol.*, 39:374-398.

BRAZIER, M. A. B. 1960. *The Electrical Activity of the Nervous System*. 2nd ed. Macmillan, New York.

BREMER, F. 1935. Quelques propriétés de l'activité électrique du cortex cérébral "isolé". *C. R. Soc. Biol., Paris*, 118:1241-1244.

BREMER, F. 1938. Effets de la déafférentation complète d'une région de l'écorce cérébrale sur son activité électrique spontanée. *C. R. Soc. Biol., Paris*, 127:355-359.

BREMER, F. 1941. L'activité électrique "spontanée" de la moëlle épinière. *Arch. int. Physiol.*, 51:51-84.

BREMER, F. 1951. Aspects électrophysiologiques de la transmission synaptique. *Arch. int. Physiol.*, 59:588-626.

BREMER, F. 1956. Convergences neuroniques et facilitation centrales. In: *Problems of the Modern Physiology of the Nervous and Muscle Systems*, Akad. Nauk, Georgian SSR. Tbilisi.

BREMER, F. 1960. L'Interprétation des potentiels électriques de l'écorce cérébrale. In: *Structure and Function of the Cerebral Cortex*. D. B. Tower and J. P. Schadé (eds.). Elsevier, Amsterdam.

BROADBENT, D. E. 1961. Human perception and animal learning. In: *Current Problems in Behaviour*. W. H. Thorpe and O. L. Zangwill (eds.). Cambridge Univ. Press, Cambridge.

BRUNER, J. S. 1957. Neural mechanisms in perception. *Psychol. Rev.*, 64:340-358.

BUDDENBROCK, W. VON. 1930. Untersuchungen über den Schattenreflex. *Z. vergl. Physiol.*, 13:164-213.

BUDDENBROCK, W. VON. 1952. *Vergleichende Physiologie*. Vol. I. Sinnesphysiologie. Birkhäuser, Basel.

BUDDENBROCK, W. VON. 1953. *Vergleichende Physiologie*. Vol. II. Nervenphysiologie. Birkhäuser, Basel.

BUDDENBROCK, W. VON. 1956. *The Love Life of Animals*. Muller, London.

BULLER, A. J., ECCLES, J. C., and ECCLES, R. M. 1960a. Differentiation of fast and slow muscles in the cat hind limb. *J. Physiol.*, 150:399-416.

BULLER, A. J., ECCLES, J. C., and ECCLES, R. M. 1960b. Interactions between motoneurones and muscles in respect of the characteristic speeds of their responses. *J. Physiol.*, 150:417-439.

BULLOCK, T. H. 1945. Problems in the comparative study of brain waves. *Yale J. Biol. Med.*, 17:657-679.

BULLOCK, T. H. 1953a. A contribution from the study of cords in lower forms. In: *The Spinal Cord*, G. E. W. Wolstenholme (ed.). Churchill, London.

BULLOCK, T. H. 1953b. Comparative aspects of some biological transducers. *Fed. Proc.*, 12:666-672.

BULLOCK, T. H. 1957a. The trigger concept in biology. In: *Physiological Triggers and Discontinuous Rate Processes*. T. H. Bullock (ed.). Amer. Physiol. Soc., Washington.

BULLOCK, T. H. 1957b. Neuronal integrative mechanisms. In: *Recent Advances in Invertebrate Physiology*. B. T. Scheer (ed.). Univ. Oregon Press, Eugene.

BULLOCK, T. H. 1958. Parameters of integrative action of the nervous system at the neuronal level. In: *Symposium on Ultramicroscopic Structure and Function of Nerve Cells*. R. Brown (ed.). *Exp. Cell Res. Suppl.*, 5:323-337.

BULLOCK, T. H. 1959. Initation of nerve impulses in receptor and central neurones. *Rev. Mod. Physics.* 31:504-514, and in *Biophysical Science—A Study Program*. J. L. Oncley (ed.). Wiley, New York.

BULLOCK, T. H. 1961. The origins of patterned nervous discharge. *Behaviour*, 17:48-59.

BULLOCK, T. H. and TERZUOLO, C. A. 1957. Diverse forms of activity in the somata of spontaneous and integrating ganglion cells. *J. Physiol.*, 138:341-364.

BURGEN, A. S. V. and KUFFLER, S. W. 1957. Two inhibitory fibres forming synapses with a single nerve cell in the lobster. *Nature, Lond.*, 180:1490-1491.

BURKHARDT, D. 1954. Rhythmische Erregungen in den optischen Zentren von *Calliphora erythrocephala*. *Z. vergl. Physiol.*, 36:595-630.

BURKHARDT, D. 1958. Die Sinnesorgane des Skeletmuskels und die nervöse Steuerung der Muskeltätigkeit. *Ergebn. Biol.*, 20:27-66.

BURKHARDT, D. 1959a. Effect of temperature on isolated stretch-receptor organ of crayfish. *Science*, 129:392-393.

BURKHARDT, D. 1959b. Die Erregungsvorgänge sensibler Ganglienzellen in Abhängigkeit von der Temperatur (Untersuchungen an den abdominalen Streckrezeptoren des Sumpfkrebses, *Astacus leptodactylus*). *Biol. Zbl.*, 78:22-62.

BURNS, B. D. 1958. *The mammalian cerebral cortex*. Arnold, London.

BUSER, P. 1955. Analyse des réponses électriques du lobe optique à la stimulation de la voie visuelle chez quelques vertébrés inférieurs. Masson, Paris.

BYKOV, K. M. 1942. *Cerebral Cortex and Internal Organs*. Kirov. (In Russian.)

BYKOV, K. M. and SLONIM, A. D. 1960. *Investigation of the Complex Reflex Activity of Animals and Man under Natural Conditions*. Akad. Nauk,

Bibliography

SSSR, Moscow. (Transl. by U.S. Joint Publications Research Service. Distr. by Office of Technical Services U. S. Dept. of Commerce.)

CARTHY, J. D. 1958. *An Introduction to the Behaviour of Invertebrates.* Allen & Unwin, London.

CHAMBERLAIN, T. J., HALICK, P., and GERARD, R. W. 1961. Fixation of experience in the rat spinal cord. *Physiologist*, 4(3):17.

CHAMBERS, W. W., LEVITT, M., CARRERAS, M., and LIU, C. N. 1960. Central determination of sensory processes. *Science*, 132:1489.

CHILD, C. M. 1924 *Physiological Foundations of Behavior.* Holt, New York.

CIBA FOUNDATION SYMPOSIUM. 1959. *Pain and Itch, Nervous Mechanisms; in honour of Y. Zotterman.* Churchill, London.

CLARE, M. H. and BISHOP, G. H. 1955. Properties of dendrites; apical dendrites of the cat cortex. *Electroenceph. clin. Neurophysiol.*, 7:85-98.

CLARE, M. H. and BISHOP, G. H. 1956. Potential wave mechanisms in cat cortex. *Electroenceph. clin. Neurophysiol.*, 8:583-602.

COGHILL, G. E. 1929. *Anatomy and the problem of behaviour.* Macmillan, New York.

COHEN, M. J. 1960. A proprioceptive system in the legs of the crab *Cancer magister*. *Anat. Rec.*, 137:346.

CRAGG, B. G. and HAMLYN, L. H. 1955. Action potentials of the pyramidal neurones in the hippocampus of the rabbit. *J. Physiol.*, 129:608-627.

CRAGG, B. G. and TEMPERLEY, H. N. V. 1954. The organization of neurones: a co-operative analogy. *Electroenceph. clin. Neurophysiol.*, 6:85-92.

CRANE, H. D. 1961. On the complete logic capability and realizability of trigger-coupled neuristors. Stanford Res. Inst., Project No. 3286, Interim Report 4.

CREED, R. S., DENNY-BROWN, D., ECCLES, J C., LIDDELL, E. G. T., and SHERRINGTON, C. S. 1932. *Reflex Activity of the Spinal Cord.* Clarendon Press, Oxford.

CRESCITELLI, F. and JAHN, T. L. 1939. The electrical response of the dark-adapted grasshopper eye to various intensities of illumination and to different qualities of light. *J. cell. comp. Physiol.*, 13:105-112.

CRESCITELLI, F. and JAHN, T. L. 1942. Oscillatory electrical activity from the insect compound eye. *J. cell. comp. Physiol.*, 19:47-66.

CROSS, B. A. and GREEN, J. D. 1959. Activity of single neurones in the hypothalamus: effect of osmotic and other stimuli. *J. Physiol.*, 148:554-569.

DAVIS, H. 1961. Some principles of sensory receptor action. *Physiol. Rev.*, 41:391-416.

DERCUM, F. X. 1922. *An essay on the physiology of mind.* Saunders, Philadelphia.

DESMEDT, J. E. 1960. Neurophysiological mechanisms controlling acoustic input. In: *Neural Mechanisms of the Auditory and Vestibular Systems.* G. L. Rasmussen and W. Windle (eds.). Thomas, Springfield.

DESMEDT, J. E. and MONACO, P. 1961. Mode of action of the efferent olivocochlear bundle on the inner ear. *Nature, Lond.*, 192:1263-1265.

DEUTSCH, J. A. 1960a. The plexiform zone and shape recognition in the octopus. *Nature, Lond.*, 185:443-446.

DEUTSCH, J. A. 1960b. *The structural Basis of Behavior.* Univ. Chicago Press, Chicago.

DEUTSCH, J. A. 1962. Higher nervous function: the physiological bases of memory. *Annu. Rev. Physiol.*, 24:259-286.

DIJKGRAAF, S. 1955. The physiological significance of the so-called proprioceptors. *Acta physiol. pharm. néerl.*, 4:123-126.

DIJKGRAAF, S. 1961. Wahrnehmung und Beantwortung des Erdschwerereizes bei Tieren. *Studium gen.*, 14:479-494.

DODT, E. 1956. Centrifugal impulses in rabbit's retina. *J. Neurophysiol.*, 19:301-307.

DODWELL, P. C. and SUTHERLAND, N. S. 1961. Facts and theories of shape discrimination. *Nature, Lond.*, 191:578-583.

DOW, R. S. and MORUZZI, G. 1958. *The Physiology and Pathology of the Cerebellum.* Univ. Minn. Press, Minneapolis.

DUDEL, J. and KUFFLER, S. W. 1961a. The quantal nature of transmission and spontaneous miniature potentials at the crayfish neuromuscular junction. *J. Physiol.*, 155:514-529.

DUDEL, J. and KUFFLER, S. W. 1961b. Mechanism of facilitation at the crayfish neuromuscular junction. *J. Physiol.*, 155:530-542.

DUDEL, J. and ORKAND, R. K. 1960. Spontaneous potential changes at crayfish neuromuscular junctions. *Nature, Lond.*, 186:476-477.

DUDEL, J. and TRAUTWEIN, W. 1958. Der Mechanismus der automatischen rhythmischen Impulsbildung der Herzmuskelfaser. *Pflüg. Arch. ges. Physiol.*, 267:553-565.

DUKE-ELDER, W. S. 1949. *Text-book of Ophthalmology.* Vol. 4. Mosby, St. Louis.

ECCLES, J. C. 1951. Interpretation of action potentials evoked in the cerebral cortex. *Electroenceph. clin. Neurophysiol.*, 3:449-464.

ECCLES, J. C. 1953. *The Neurophysiological Basis of Mind.* Oxford University Press, London.

ECCLES, J. C. 1957. *The Physiology of Nerve Cells.* Johns Hopkins Press, Baltimore.

ECCLES, J. C. 1958. Problems of plasticity and organization at simplest levels of mammalian central nervous system. *Perspectives Biol. Med.*, 1:379-396.

ECCLES, J. C. 1959. Neuron physiology —introduction. In: *Handbook of Physiology*, Sect. 1. *Neurophysiology*, 1:59-74.

ECCLES, J. C. 1960. The properties of the dendrites. In: *Structure and Function of the Cerebral Cortex.* D. B. Tower and J. P. Schadé (eds.). Elsevier, Amsterdam.

ECCLES, J. C. 1961a. The nature of central inhibition. *Proc. roy. Soc.*, (B) 153:445-476.

ECCLES, J. C. 1961b. Inhibitory pathways to motoneurons. In: *Nervous Inhibition.* E. Florey (ed.). Pergamon Press, Oxford.

ECCLES, J. C. and KRNJEVIĆ, K. 1959a. Potential changes recorded inside primary afferent fibres within the spinal cord. *J. Physiol.*, 149:250-273.

ECCLES, J. C. and KRNJEVIĆ, K. 1959b. Presynaptic changes associated with posttetanic potentiation in the spinal cord. *J. Physiol.*, 149:274-287.

ECCLES, J. C., ECCLES, R. M., and MAGNI, F. 1960a. Presynaptic inhibition in the spinal cord. *J. Physiol.*, 154:28P.

ECCLES, J. C., ECCLES, R. M., and MAGNI, F. 1960b. Monosynaptic excitatory action on motoneurones regenerated to antagonistic muscles. *J. Physiol.*, 154:68-88.

ECCLES, J. C., LIBET, B., and YOUNG, R. R. 1958. The behaviour of chromatolysed motoneurones studied by intracellular recording. *J. Physiol.*, 143:11-40.

ECCLES, R. M. and LUNDBERG, A. 1958a. Integrative pattern of Ia synaptic actions on motoneurones of hip and knee muscles. *J. Physiol.*, 144:271-298.

ECCLES, R. M. and LUNDBERG, A. 1958b. Significance of supraspinal control of reflex actions by impulses in muscle afferents. *Experientia*, 14:197-199.

ECCLES, R. M. and WESTERMAN, R. A. 1959. Enhanced synaptic function due to excess use. *Nature, Lond.*, 184:460-461.

ECKERT, B. 1959. Über das Zusammenwirken des erregenden und des hemmenden Neurons des M. Abductor der Krebsschere beim Ablauf von Reflexen des myotatischen Typus. *Z. vergl. Physiol.*, 41:500-526.

ECKERT, R. O. 1960. Feedback in the crayfish stretch receptor system. *Anat. Rec.*, 137:351-352.

ENGER, P. S. 1956. The electroencephalogram of the codfish (*Gadus callarias*). Spontaneous electrical activity and reaction to photic and acoustic stimulation. *Acta physiol. scand.*, 39:55-72.

EWER, R. F. 1957. Ethological concepts. *Science*, 126:599-603.

EYZAGUIRRE, C. and KUFFLER, S. W. 1955a. Processes of excitation in the dendrites and in the soma of single isolated sensory nerve cells of the lobster and crayfish. *J. gen. Physiol.*, 39:87-119.

EYZAGUIRRE, C. and KUFFLER, S. W. 1955b. Further study of soma, dendrite and axon excitation in single neurons. *J. gen. Physiol.*, 39:121-153.

FARLEY, B. G. 1962. Some results of computer simulation of neuron-like nets. *Fed. Proc.*, 21:92-96.

FARLEY, B. G. and CLARK, W. A. 1960. Activity in networks of neuron-like elements. *Proc. 4th Lond. Symp. Information Theory*. C. Cherry (ed.). (In press.)

FATT, P. 1961. The change in membrane permeability during the inhibitory process. In: *Nervous Inhibition*. E. Florey (ed.), Pergamon Press, Oxford.

FAURE, J. 1957. Activité électrique du cerveau, comportement, fonctions cortico-viscérales en relation avec l'état hormonal chez l'animal et chez l'homme. *Electroenceph. clin. Neurophysiol.* (Suppl.) 6:257-269.

FAUST, R. 1952. Untersuchungen zum Halterenproblem. *Zool. Jb. (allg. Zool.)*, 63:325-366.

FESSARD, A. 1956. Formes et caractères généraux de l'excitation neuronique. *Int. Physiol. Congr., Abstr. XX*. 1956:1:35-58.

FESSARD, A. 1959. Brain potentials and rhythms—introduction. In: *Handbook of Physiology*, Sect. 1. *Neurophysiology*, 1:255-259.

FESSARD, A. 1961. The role of neuronal networks in sensory communications within the brain. In: *Sensory Communication*. W. A. Rosenblith (ed.). MIT Press and Wiley, New York.

FESSARD, A. and TAUC, L. 1958. Effets de répétition sur l'amplitude des potentiels postsynaptiques d'un soma neuronique. *J. Physiol., Paris*, 50:277-281.

FITZHUGH, R. 1957. The statistical detection of threshold signals in the retina. *J. gen. Physiol.*, 40:925-948.

FLOREY, E. 1960. Studies on the nervous regulation of the heart beat in decapod Crustacea. *J. gen. Physiol.*, 43:1061-1081.

FLOREY, E. 1961. Excitation, inhibition and the concept of the stimulus. In: *Nervous Inhibition*. E. Florey (ed.). Pergamon Press, Oxford.

FRAENKEL, G. S. and GUNN, D. L. 1940. *The Orientation of Animals*. Clarendon Press, Oxford.

FRANK, G. B. 1958. Nature of the steady potential across the mammalian cerebral cortex. *Fed. Proc.*, 17:48.

FRANK, K. 1959. Basic mechanisms of synaptic transmission in the central nervous system. *Inst. Radio Eng. Trans. Med. Electr.*, vol. ME-6:85-88.

FRANK, K. and FUORTES, M. G. F. 1960. Accomodation of spinal motoneurones of cats. *Arch. ital. Biol.*, 98:165-170.

FRENCH, J. D. and MAGOUN, H. W. 1952. Effects of chronic lesions in central cephalic brain stem of monkeys. *A. M. A. Arch. Neurol. Psychiat.*, 68:591-604.

FREYGANG, W. H., Jr. 1958. An analysis of extracellular potentials from single neurons in the lateral geniculate nucleus of the cat. *J. gen. Physiol.*, 41:543-564.

FULTON, J. F. 1949. *Physiology of the nervous system*. Oxford Univ. Press, New York.

FUORTES, M. G. F. 1958. Electric activity of cells in the eye of *Limulus Amer. J. Ophthal.*, 46:210-223.

FUORTES, M. G. F. 1959. Integrative mechanisms in the nervous system. *Amer. Nat.*, 93:213-224.

FURSHPAN, E. J. 1959. Neuromuscular transmission in invertebrates. In: *Handbook of Physiology*, Sect. 1. *Neurophysiology*, 1:239-254.

FURSHPAN, E. J. and POTTER, D. D. 1957. Mechanism of nerve-impulse transmission at a crayfish synapse. *Nature, Lond.*, 180:342-343.

FURSHPAN, E. J. and POTTER, D. D. 1959. Transmission at the giant motor synapses of the crayfish. *J. Physiol.*, 145:289-325.

GALAMBOS, R. 1960. Some neural correlates of conditioning and learning. In: *Electrical Studies on the Unanesthetized Brain*. E. R. Ramey and D. S. O'Doherty (eds.). Hoeber, New York.

GALAMBOS, R. 1961. A glial-neural theory of brain function. *Proc. nat. Acad. Sci., Wash.* 47:129-136.

GALAMBOS, R. and MORGAN, C. T. 1960. The neural basis of learning. In: *Handbook of Physiology*, Sect. 1. *Neurophysiology*, 3:1471-1499.

GELLHORN, E. 1953. *Physiological Foundation of Neurology and Psychiatry*. Univ. Minnesota Press, Minneapolis.

GERARD, R. W. 1941. The interaction of neurones. *Ohio J. Sci.*, 41:160-172.

GERARD, R. W. 1942. Higher levels of integration. In: *Levels of integration in biological and social systems*. R. Redfield (ed.). Cattell, Lancaster, Pa.

GERARD, R. W. 1953. Neurophysiology in relation to behavior. In: *Mid-century psychiatry*. R. R. Grinker (ed.). Thomas, Springfield.

GERARD, R. W. 1960. Neurophysiology: an integration (molecules, neurons and behavior). In: *Handbook of Physiology*, Sect. 1. *Neurophysiology*, 3:1919-1965.

GERARD, R. W. and LIBET, B. 1940. The control of normal and "convulsive" brain potentials. *Amer. J. Psychiat.*, 96:1125-1151.

GERARD, R. W. and YOUNG, J. Z. 1937. Electrical activity of the central nervous system of the frog. *Proc. roy. Soc.*, (B) 122:343-352.

GESELL, R., BRASSFIELD, C. R., and LILLIE, R. H. 1954. Implementation of electrical energy by paired half-centers as revealed by structure and function. *J. comp. Neurol.*, 101:331-406.

GIBBS, F. A. and GIBBS, E. L. 1950. *Atlas of electroencephalography*. Addison-Wesley, Cambridge, Mass.

GRAHAM, J. and GERARD, R. W. 1946. Membrane potentials and excitation of impaled single muscle fibers. *J. comp. cell. Physiol.*, 28:99-117.

GRANIT, R. 1956. Centrifugal and antidromic effects on the retina. In: *Problems in Contemporary Optics*. Istituto Nazionale di Ottica, Firenze.

GRANIT, R. 1961. Regulation of discharge rate by inhibition, especially by recurrent inhibition. In: *Nervous Inhibition*. E. Florey (ed.). Pergamon Press, Oxford.

GRANIT, R. and PHILLIPS, C. G. 1956. Excitatory and inhibitory processes acting upon individual Purkinje cells of the cerebellum in cats. *J. Physiol.*, 133:520-547.

GRANIT, R. and PHILLIPS, C. G. 1957. Effects on Purkinje cells of surface stimulation of the cerebellum. *J. Physiol.*, 135:73-92.

GRANIT, R. and RUTLEDGE, L. T. 1960. Surplus excitation in reflex action of motoneurones as measured by recurrent inhibition. *J. Physiol.*, 154:288-307.

GRANIT, R., HAASE, J., and RUTLEDGE, L. T. 1960. Recurrent inhibition in relation to frequency of firing and limitation of discharge rate of extensor motoneurones. *J. Physiol.*, 154:308-328.

GRANIT, R., PASCOE, J. E., and STEG, G. 1957. The behaviour of tonic *a* and *γ* motoneurones during stimulation of recurrent collaterals. *J. Physiol.*, 138:381-400.

GRANIT, R., POMPEIANO, O., and WALTMAN, B. 1959. Fast supraspinal control of mammalian muscle spindles: extra- and intrafusal co-activation. *J. Physiol.*, 147:385-398.

GRAY, J. 1936. Studies in animal locomotion. IV. The neuromuscular mechanism of swimming in the eel. *J. exp. Biol.*, 13:170-180.

GRAY, J. 1939. Studies in animal locomotion. VIII. The kinetics of locomotion of *Nereis diversicolor*. *J. exp. Biol.*, 16:9-17.

GRAY, J. 1950. The role of peripheral sense organs during locomotion in the vertebrates. *Symp. Soc. exp. Biol.*, 4:112-126.

GRAY, J. and LISSMANN, H. W. 1938. Studies in animal locomotion. VII.

Bibliography

Locomotory reflexes in the earthworm. *J. exp. Biol.*, 15:506-517.

GRAY, J. and LISSMANN, H. W. 1940a. The effect of de-afferentiation upon the locomotor activity of amphibian limbs. *J. exp. Biol.*, 17:227-236.

GRAY, J. and LISSMANN, H. W. 1940b. Ambulatory reflexes in spinal amphibians. *J. exp. Biol.*, 17:237-251.

GRAY, J. and LISSMANN, H. W. 1946a. Further observations on the effect of de-afferentiation on the locomotory activity of amphibian limbs. *J. exp. Biol.*, 23:121-132.

GRAY, J. and LISSMANN, H. W. 1946b. The coordination of limb movements in the amphibia. *J. exp. Biol.*, 23:133-142.

GRAY, J. and SAND, A. 1936. The locomotory rhythm of the dogfish (*Scyllium canicula*). *J. exp. Biol.*, 13:200-209.

GRAY, J., LISSMANN, H. W., and PUMPHREY, R. J. 1938. The mechanism of locomotion in the leech (*Hirudo medicinalis* Ray). *J. exp. Biol.*, 15:408-430.

GRAY, J. A. B. 1959. Initiation of impulses at receptors. In: *Handbook of Physiology*, Sect. 1. *Neurophysiology*, 1:123-145.

GRIMM, R. J. 1960. Feeding behavior and electrical stimulation of the brain of *Carassius auratus*. *Science*, 131:162-163.

GRUNDFEST, H. 1957a. Electrical inexcitability of synapses and some consequences in the central nervous system. *Physiol. Rev.*, 37:337-361.

GRUNDFEST, H. 1957b. Excitation triggers in post-junctional cells. In: *Physiological Triggers*. Amer. Physiol. Soc., Washington.

GRUNDFEST, H. 1958. An electrophysiological basis for neuropharmacology. *Fed. Proc.*, 17:1006-1018.

GRUNDFEST, H. 1959a. Synaptic and ephaptic transmission. In: *Handbook of Physiology*, Sect. 1. *Neurophysiology*, 1:147-197.

GRUNDFEST, H. 1959b. Evolution of conduction in the nervous system. In: *Evolution of Nervous Control*. Amer. Assoc. Adv. Sci., Washington.

GRUNDFEST, H. 1961a. Varieties of inhibitory processes. In: *Nervous Inhibition*. E. Florey (ed.). Pergamon Press, Oxford.

GRUNDFEST, H. 1961b. Excitation by hyperpolarizing potentials. A general theory of receptor activities. In: *Nervous Inhibition*. E. Florey (ed.). Pergamon Press, Oxford.

GRUNDFEST, H. 1961c. The interpretation of electrocortical potentials *Ann. N. Y. Acad. Sci.*, 92:877-889.

HAGBARTH, K.-E. and KERR, D. I. B. 1954. Central influences on spinal afferent conduction. *J. Neurophysiol.*, 17:295-307.

HAGIWARA, S. 1958. Synaptic potential in the motor giant axon of the crayfish. *J. gen. Physiol.*, 41:1119-1182.

HAGIWARA, S. 1961. Nervous activities of the heart in crustacea. *Ergebn. Biol.*, 24:287-311.

HAGIWARA, S. and BULLOCK, T. H. 1957. Intracellular potentials in pacemaker and integrative neurons of the lobster cardiac ganglion. *J. cell. comp. Physiol.*, 50:25-47.

HAGIWARA, S. and MORITA, H. 1962. Electrotonic transmission between two nerve cells in leech ganglion. *J. Neurophysiol.* 25:721-731.

HAGIWARA, S. and SAITO, N. 1959. Membrane potential change and membrane current in supramedullary nerve cell of puffer. *J. Neurophysiol.*, 22:204-221.

HAGIWARA, S. and TASAKI, I. 1958. A study on the mechanism of impulse transmission across the giant synapse of the squid. *J. Physiol.*, 143:114-137.

HAGIWARA, S. and WATANABE, A. 1956. Discharges in motoneurons of cicada. *J. cell. comp. Physiol.*, 47:415-428.

HARKER, J. E. 1956. Factors controlling the diurnal rhythm of activity of *Periplaneta americana* L. *J. exp. Biol.*, 33:224-234.

HARKER, J. E. 1958. Diurnal rhythms in the animal kingdom. *Biol. Rev.*, 33:1-52.

HARKER, J. E. 1960. Internal factors controlling the suboesophageal ganglion neurosecretory cycle in *Periplaneta americana* L. *J. exp. Biol.*, 37:164-170.

HARLOW, H. F. and WOOLSEY, C. N. 1958. *Biological and Biochemical Bases of Behavior*. Univ. Wisconsin Press, Madison.

HARMON, L. D. 1959. Artificial neuron. *Science*, 129:962-963.

HARMON, L. D. 1961. Studies with artificial neurons, I. Properties and functions of an artificial neuron. *Kybernetik*, 1:89-101.

HARMON, L. D. and WOLFE, R. M. 1959. An electronic model of a nerve cell. *Semiconductor Products.*, Aug. 1959: 36-40.

HARRIS, G. W., MICHAEL, R. P., and SCOTT, P. P. 1958. Neurological site of action of stilboestrol in eliciting sexual behaviour. In: *Neurological Basis of Behaviour*, G. E. W. Wolstenholme and C. M. O'Connor (eds.). J. & A. Churchill, London.

HARTLINE, H. K. 1959. Receptor mechanisms and the integration of sensory information in the eye. In: *Biophysical Science—A Study Program*, J. L. Oncley (ed.). Wiley, New York.

HARTLINE, H. K., RATLIFF, F., and MILLER, W. H. 1961. Inhibitory interaction in the retina and its significance in vision. In: *Nervous Inhibition*. E. Florey (ed.). Pergamon Press, Oxford.

HARTLINE, H. K., WAGNER, H. G., and MACNICHOL, E. F., Jr. 1952. The peripheral origin of nervous activity in the visual system. *Cold Spr. Harb. Symp. quant. Biol.*, 17:125-142.

HASSENSTEIN, B. 1951. Ommatidienraster und afferente Bewegungsintegration. (Versuche an dem Rüsselkäfer *Chlorophanus viridis*.) *Z. vergl. Physiol.*, 33:301-326.

HASSENSTEIN, B. 1958. Über die Wahrnehmung der Bewegung von Figuren und unregelmässigen Helligkeitsmustern. *Z. vergl. Physiol.*, 40:556-592.

HASSENSTEIN, B. 1961. Wie sehen Insekten Bewegungen? *Naturwissenschaften*, 48:207-214.

HASSENSTEIN, B. and REICHARDT, W. 1956. Systemtheoretische Analyse der Zeit-, Reihenfolgen- und Vorzeichenauswertung bei der Bewegungsperzeption des Rüsselkäfers *Chlorphanus*. *Z. Naturf.*, 11B:513-524.

HEBB, D. O. 1949. *The Organization of Behavior*. Wiley, New York.

HERNÁNDEZ-PEÓN, R. 1959. Centrifugal control of sensory inflow to the brain and sensory perception. *Acta Neurol. Latinoamer.*, 5:279-298.

HERNÁNDEZ-PEÓN, R. 1960. Neurophysiological correlates of habituation and other manifestations of plastic inhibition (internal inhibition). *Electroenceph. clin. Neurophysiol.*, Suppl. 13:101-114. In: Moscow Colloq. on EEG of Higher Nervous Activity. H. H. Jasper and G. D. Smirnov (eds.).

HERNÁNDEZ-PEÓN, R. 1961. Reticular mechanisms of sensory control. In: *Sensory Communication*. W. A. Rosenblith (ed.). MIT and Wiley, New York.

HERNÁNDEZ-PEÓN, R. and HAGBARTH, K.-E. 1955. Interaction between afferent and cortically induced reticular responses. *J. Neurophysiol.*, 18:44-55.

HERNÁNDEZ-PEÓN, R., GUZMÁN-FLORES, C., ALCARAZ, M., and FERNÁNDEZ-GUARDIOLA, A. 1958. Habituation in the visual pathway. *Acta Neurol. Latinoamer.*, 4:121-129.

HERRICK, C. J. 1924. *Neurological Foundations of Animal Behavior*. Holt, New York.

HERRICK, C. J. 1948. *The Brain of the Tiger Salamander Ambystoma tigrinum*. U. Chicago Press, Chicago.

HERRICK, C. J. 1949. A biological survey of integrative levels. In: *Philosophy for the Future*. R. W. Sellars, V. J. McGill, and M. Farber (eds.). Macmillan, New York.

HESS, W. R. 1956. *Hypothalamus und Thalamus*. Thieme, Stuttgart.

HESS, W. R. 1957. *The functional Organization of the Diencephalon*. J. R. Hughes (ed.). Grune and Stratton, New York. (Translation of *Das Zwischenhirn*.)

HICK, W. E. 1952. On the rate of gain of information. *Quart. J. exp. Psychol.*, 4:11-26.

HINDE, R. A. 1953. Appetitive behaviour, consummatory act, and the hierarchical organisation of behaviour—with special reference to the Great Tit (*Parus major*). *Behaviour*, 5:189-224.

HINDE, R. A. 1955. A comparative study of the courtship of certain finches (Fringillidae). *Ibis*, 97:706-745.

HINDE, R. A. 1958. The nest-building behaviour of domesticated canaries. *Proc. zool. Soc. Lond.*, 131:1-48.

HINDE, R. A. 1959a. Unitary drives. *Anim. Behav.*, 7:130-141.

HINDE, R. A. 1959b. Recent trends in ethology. In: *Psychology: a Study of a Science*. S. Koch (ed.). McGraw-Hill, New York.

HINDE, R. A. 1960. Energy models of motivation. *Symp. Soc. exp. Biol. Cambridge*, 14:199-213.

HODGKIN, A. L. 1938. The subthreshold potentials in a crustacean nerve fibre. *Proc. roy. Soc.(B)* 126:247-285.

HOLMGREN, B. and FRENK, S. 1961. Inhibitory phenomena and "habituation" at the neuronal level. *Nature, Lond.*, 192:1294-1295.

HOLST, E. VON. 1935a. Alles oder Nichts, Block, Alternans, Bigemini und verwandte Phänomene als Eigenschaften des Rückenmarks. *Pflüg. Arch. ges. Physiol.*, 236:515-532.

HOLST, E. VON. 1935b. Die Koordination der Bewegung bei den Arthropoden in Abhängigkeit von zentralen und peripheren Bedingungen. *Biol. Rev.*, 10:234-261.

HOLST, E. VON. 1936a. Vom Dualismus der motorischen und der automatisch-rhythmischen Funktion im Rückenmark und vom Wesen des automatischen Rhythmus. *Pflüg. Arch. ges. Physiol.*, 237:356-378.

HOLST, E. VON. 1936b. Über den "Magnet-Effekt" als koordinierendes Prinzip im Rückenmark. *Pflüg. Arch. ges. Physiol.*, 237:655-682.

HOLST, E. VON. 1939a. Die relative Koordination als Phänomen und als Methode Zentralnervöser Funktionsanalyse. *Ergebn. Physiol.*, 42:228-306.

HOLST, E. VON. 1939b. Über die nervöse Funktionsstruktur des rhythmisch tätigen Fischrückenmarks. *Pflüg. Arch. ges. Physiol.*, 241:569-611.

HOLST, E. VON. 1943. Über relative Koordination bei Arthropoden (mit Vergleichsversuchen am Regenwurm.) *Pflüg. Arch. ges. Physiol.*, 246:847-865.

HOLST, E. VON. 1950. Quantitative Messung von Stimmungen im Verhalten der Fische. *Symp. Soc. exp. Biol.*, 4:143-172.

HOLST, E. VON. 1954. Relations between the central nervous system and the peripheral organs. *Brit. J. Anim. Behav.* 2:89-94.

HOLST, E. VON. 1955. Die Beteiligung von Konvergenz und Akkommodation an der wahrgenommenen Grössenkonstanz. *Naturwissenschaften.* 42:444-445.

HOLST, E. VON and MITTELSTAEDT, H. 1950. Das Reafferenzprinzip. (Wechselwirkung zwischen Zentralnervensystem und Peripherie.) *Naturwissenschaften*, 37:464-476.

HOLST, E. VON and ST. PAUL, U. VON. 1960. Vom Wirkungsgefüge der Triebe. *Naturwissenschaften*, 47:409-422.

HORRIDGE, G. A. 1958. The nerves and muscles of medusae. VI. The rhythm. *J. exp. Biol.*, 36:72-91.

HORRIDGE, G. A. 1961. The organization of the primitive central nervous system as suggested by examples of inhibition and the structure of neuropile. In: *Nervous Inhibition*. E. Florey (ed.). Pergamon Press, Oxford.

HORRIDGE, G. A. 1962. Learning of leg position by headless insects. *Nature, Lond.*, 193:697-698.

HOUSEHOLDER, A. S. 1946. Mathematical biophysics and the central nervous system. *Acta biotheor.*, Leiden, 8:67-76.

HOUSEHOLDER, A. S. and LANDAHL, H. D. 1945. *Mathematical Biophysics of the Central Nervous System*. Principia Press, Bloomington.

HOYLE, G. 1957. *Comparative Physiology of the Nervous Control of Muscular Contraction*. University Press, Cambridge.

HOYLE, G. 1960. Neuromuscular activity in the swimming sea anemone, *Stomphia coccinea* (Müller). *J. exp. Biol.*, 37:671-688.

HOYLE, G. and WIERSMA, C. A. G. 1958. Inhibition at neuromuscular junctions in crustacea. *J. Physiol.* 143:426-440.

HUBEL, D. H. 1959. Single unit activity in striate cortex of unrestrained cats. *J. Physiol.*, 147:226-238.

HUBEL, D. H. and WIESEL, T. N. 1959. Receptive fields of single neurones in the cat's striate cortex. *J. Physiol.*, 148:574-591.

HUBER, F. 1953. Verhaltensstudien am Männchen der Feldgrille (*Gryllus campestris* L.) nach Eingriffen am Zentralnervensystem. *Zool. Anz. Suppl.*, 17:138-149.

HUBER, F. 1955. Sitz und Bedeutung nervöser Zentren für Instinkthandlungen beim Männchen von *Gryllus campestris* L. *Z. Tierpsychol.*, 12:12-48.

HUBER, F. 1960. Untersuchungen über die Funktion des Zentralnervensystems und insbesondere des Gehirnes bei der Fortbewegung und der Lauterzeugung der Grillen. *Z. vergl. Physiol.*, 44:60-132.

HUGHES, G. M. 1952. Differential effects of direct current on insect ganglia. *J. exp. Biol.*, 29:387-402.

HUGHES, G. M. 1957. The co-ordination of insect movements. II. The effect of limb amputation and the cutting of commissures in the cockroach (*Blatta orientalis*). *J. exp. Biol.*, 34:306-333.

HUGHES, G. M. 1958. The co-ordination of insect movements. III. Swimming in *Dytiscus*, *Hydrophilus*, and a dragonfly nymph, *J. exp. Biol.*, 35:567-583.

HUGHES, G. M. and WIERSMA, C. A. G. 1960. The co-ordination of swimmeret movements in the crayfish, *Procambarus clarkii* (Girard). *J. exp. Biol.*, 37:657-670.

HUNSPERGER, R. W. 1956. Affektreaktionen auf elektrische Reizung im Hirnstamm der Katze. *Helv. physiol. acta*, 14:70-92.

IERSEL, J. J. A. VAN and BOL, A. C. A. 1958. Preening of two tern species. A study on displacement. *Behaviour*, 13:1-88.

JAHN, T. 1960. Optische Gleichgewichtsregulation und zentrale Kompensation bei Amphibien, insbesondere bei der Erdkröte (*Bufo bufo* L.). *Z. vergl. Physiol.*, 43:119-140.

JAHN, T. L. and CRESCITELLI, F. 1938. The electrical response of the grasshopper eye under conditions of light and dark adaptation. *J. cell. comp. Physiol.*, 12:39-55.

JAHN, T. L. and CRESCITELLI, F. 1939. The electrical response of the Cecropia moth eye. *J. cell. comp. Physiol.*, 13:113-119.

JAHN, T. L. and CRESCITELLI, F. 1941. Electrical oscillations from insect eyes. *Amer. J. Physiol.*, 133:339-340.

JANDER, R. 1957. Die optische Richtungsorientierung der Roten Waldameise (*Formica rufa* L.). *Z. vergl. Physiol.*, 40:162-238.

JANSEN, J. and BRODAL, A. 1958. Das Kleinhirn. In: *Handbuch der Mikroskopischen Anatomie des Menschen.* Bd. IV. Nervensystem. Teil 8. W. v. Möllendorff and W. Bargmann (eds.). Springer, Berlin.

JASPER, H. H. 1936. Cortical excitatory state and synchronism in the control of bioelectric autonomous rhythms. *Cold Spr. Harb. Symp. quant. Biol.*, 4:320-338.

JASPER, H. H. 1960. Unspecific thalamo-cortical relations. In: *Handbook of Physiology*, Sect. 1. Neurophysiology, 2:1307-1321.

JASPER, H. H. and SHAGASS, C. 1941. Conditioning the occipital alpha rhythm in man. *J. exp. Psychol.*, 28:373-388.

JASPER, H. H., PROCTOR, L. D., KNIGHTON, R. S., NOSHAY, W. C., and COSTELLO, R. T. (eds). 1958. *Reticular Formation of the Brain*. Little, Brown, Boston.

JEFFRESS, L. A. (ed.). 1951. *Cerebral Mechanisms in Behavior* (The Hixson Symposium). Wiley, New York.

JENNINGS, H. S. 1906. *Behavior of the Lower Organisms*. Columbia Univ. Press, New York.

Bibliography

JOHN, E. R. and KILLAM, K. F. 1961. Studies of electrical activity of brain during differential conditioning in cats. In: *Recent Advances in Biological Psychiatry*, Vol. III. J. Wortis (ed.). Grune and Stratton, New York.

JUNG, R. 1961. Neuronal integration in the visual cortex and its significance for visual information. In: *Sensory Communication*. W. A. Rosenblith (ed.). MIT Press and Wiley, New York.

KAPPERS, C. U. ARIËNS. 1929. *The Evolution of the Nervous System in Invertebrates, Vertebrates and Man*. De erven F. Bohn, Haarlem.

KATSUKI, Y. 1961. Neural mechanism of auditory sensation in cats. In: *Sensory Communication*. W. A. Rosenblith (ed.). MIT Press and Wiley, New York.

KATSUKI, Y., CHEN, J. and TAKEDA, H. 1954. Fundamental neural mechanism of the sense organ. *Bull. Tokyo med. dent. Univ.*, 1:21-31.

KATZ, B. 1947. Subthreshold potentials in medullated nerve. *J. Physiol.*, 106:66-79.

KEIDEL, W. D. 1961a. Grenzen der Übertragbarkeit der Regelungslehre auf biologische Probleme. *Naturwissenschaften*, 48:264-276.

KEIDEL, W. D. 1961b. Grundprinzipien der akustischen und taktilen Informationsverarbeitung. *Ergebn. Biol.*, 24:213-246.

KENNEDY, D. 1958. Electrical activity of a "primitive" photoreceptor. *Ann. N. Y. Acad. Sci.*, 74:329-336.

KENNEDY, D. 1960. Neural photoreception in a lamellibranch mollusc. *J. gen. Physiol.*, 44:277-299.

KENNEDY, D. and PRESTON, J. B. 1960. Activity patterns of interneurons in the caudal ganglion of the crayfish. *J. gen. Physiol.*, 43:655-670.

KERKUT, G. A. and TAYLOR, J. R. 1956. The sensitivity of the pedal ganglion of the slug to osmotic pressure changes. *J. exp. Biol.*, 33:493-501.

KERKUT, G. A. and TAYLOR, B. J. R. 1958. The effect of temperature changes on the activity of poikilotherms. *Behaviour*, 13:259-279.

KOGAN, A. B. 1958. Electrophysiological indices of excitation and inhibition in the cerebral cortex. *Sechenov J. Physiol.* (Fiziol. Zh.) 44:774-782.

KOLMODIN, G. M. 1957. Integrative processes in single spinal interneurones with proprioceptive connections. *Acta physiol. scand.*, 40:1-89.

KOLMODIN, G. M. and SKOGLUND, C. R. 1958. Slow membrane potential changes accompanying excitation and inhibition in spinal moto- and interneurons in the cat during natural activation. *Acta physiol. scand.*, 44:11-54.

KONORSKI, J. 1948. *Conditioned Reflexes and Neuron Organization*. Cambridge Univ. Press, Cambridge.

KOSHTOYANTS, KH. S. 1957b. *Comparative Physiology of the Nervous System*. Izdatel'stbo Akademii Nauk, SSSR, Moskva. (In Russian.)

KRNJEVIĆ, K. and MILEDI, R. 1959. Presynaptic failure of neuromuscular propagation in rats. *J. Physiol.*, 149:1-22.

KUFFLER, S. W. 1953. Discharge patterns and functional organisation of mammalian retina. *J. Neurophysiol.*, 16:37-68.

KUNZE, P. 1961. Untersuchung des Bewegungssehens fixiert fliegender Bienen. *Z. vergl. Physiol.*, 44:656-684.

KÜPFMÜLLER, K. 1961. Nachrichtenübertragung und Nachrichtenverarbeitung (Neue gedankliche Werkzeuge). *Naturwissenschaften*, 48:177-184.

KÜPFMÜLLER, K. and POKLEKOWSKY, G. 1956. Der Regelmechanismus willkürlicher Bewegungen. *Z. Naturf.*, 11B:1-7.

LANDAHL, H. D. 1961a. A note on mathematical models for the interaction of neural elements. *Bull. math. Biophys.*, 23:91-97.

LANDAHL, H. D. 1961b. Mathematical models of neurone interaction. In: *Switching Circuit Theory and Logical Design*. Amer. Inst. elec. Eng. Special Publication S-134.

LANDAHL, H. D. and WILLIAMS, C. M. 1958. Representation of modality in cutaneous sensibility. *Bull. math. Biophys*, 20:309-315.

LAVÍN, A., ALCOCER-CUARÓN, C., and HERNÁNDEZ-PEÓN, R. 1959. Centrifugal arousal in the olfactory bulb. *Science*, 129:332-333.

LETTVIN, J. Y., MATURANA, H. R., PITTS, W. H., and McCULLOCH, W. S. 1961. Two remarks on the visual system of the frog. In: *Sensory Communication*. W. A. Rosenblith (ed.). MIT Press and Wiley, New York.

LEVINSON, J. and HARMON, L. D. 1961. Studies with artificial neurons, III. Mechanisms of flicker-fusion. *Kybernetik*, 1:107-117.

LI, C.-L. and JASPER, H. H. 1951. Microelectrode studies of the electrical activity of the cerebral cortex in the cat. *J. Physiol.*, 121:117-140.

LI, C.-L., CULLEN, C., and JASPER, H. H. 1956a. Laminar microelectrode studies of specific somatosensory cortical potentials. *J. Neurophysiol.*, 19:111-130.

LI, C.-L., CULLEN, C., and JASPER, H. H. 1956b. Laminar microelectrode analysis of cortical unspecific recruiting response and spontaneous rhythms. *J. Neurophysiol.*, 19:131-143.

LI, C.-L., JASPER, H., and HENDERSON, L. 1952. The effect of arousal mechanisms on various forms of abnormality in the electroencephalogram. *Electroenceph. Clin. Neurophysiol.*, 4:513-526.

LIBET, B. and GERARD, R. W. 1941. Steady potential fields and neurone activity. *J. Neurophysiol.*, 4:438-455.

LILLY, J. C. 1960. Learning motivated by subcortical stimulation: The "start" and the "stop" patterns of behavior. In: *Electrical Studies on the Unasthetized Brain*. E. R. Ramey and D. S. O'Doherty (eds.). Hoeber, New York.

LINDEMANN, W. 1955. Über die Jugendentwicklung beim Luchs (*Lynx l. lynx* Kerr.) und bei der Wildkatze (*Felis s. silvestris* Schreb.). *Behaviour*, 8:1-45.

LIPPOLD, O. C. J., NICHOLLS, J. G., and REDFEARN, J. W. T. 1960. Electrical and mechanical factors in the adaptation of a mammalian muscle spindle. *J. Physiol.*, 153:209-217.

LISSMANN, H. W. 1958. On the function and evolution of electric organs in fish. *J. exp. Biol.*, 35:156-191.

LISSMANN, H. W. and MACHIN, K. E. 1958. The mechanism of object location in *Gymnarchus niloticus* and similar fish. *J. exp. Biol.*, 35:451-486.

LIVINGSTON, R. B. 1959. Central control of receptors and sensory transmission systems. In: *Handbook of Physiology*, Sect. 1. Neurophysiology, 1:741-760.

LLOYD, D. P. C. 1961. A study of some twentieth century thoughts on inhibition in the spinal cord. In: *Nervous Inhibition*. E. Florey (ed.). Pergamon Press, Oxford.

LOEWENSTEIN, W. R. 1956. Modulation of cutaneous mechanoreceptors by sympathetic stimulation. *J. Physiol.*, 132:40-60.

LORENTE DE NÓ, R. 1934. Studies on the structure of the cerebral cortex. *J. Psychol. Neurol.*, 45:381-438.

LORENTE DE NÓ, R. 1938. Analysis of the activity of the chains of internuncial neurons. *J. Neurophysiol.*, 1:207-244.

LORENZ, K. 1950. The comparative method in studying innate behaviour patterns. *Symp. Soc. exp. Biol.*, 4:221-268.

LORENZ, K. Z. 1954. *Man Meets Dog*. (Transl. from the German by M. K. Wilson.) Methuen, London.

LUNDBERG, A. 1958. Electrophysiology of salivary glands. *Physiol. Rev.*, 38:21-40.

MACHIN, K. E. and LISSMANN, H. W. 1960. The mode of operation of the electric receptors in *Gymnarchus niloticus*. *J. exp. Biol.*, 37:801-811.

MAGOUN, H. W. 1953. An ascending reticular activating system in the brain stem. *Harvey Lect.*, 47:53-71.

MAGOUN, H. W. 1958. *The Waking Brain*. Thomas, Springfield.

MARLER, P. 1961. The filtering of external stimuli during instinctive behaviour. In: *Current Problems in Animal Behaviour*. W. H. Thorpe and O. L. Zangwill (eds.). Cambridge Univ. Press, Cambridge.

MATSUI, K. 1955. Spontaneous discharges of the isolated ganglionic trunk of the lobster heart. (*Panulirus japonicus*). *Sci. Rep. Tokyo Kyoiku Daig.* (*B*), 7:257-268.

MATSUI, K. and SHIBUYA, T. 1958. Effects of some drugs on the spontaneous activity of the isolated ganglionic trunk of the lobster heart (*Panulirus japonicus*). *Jap. J. Zool.*, 12:189-201.

MATURANA, H. R., LETTVIN, J. Y., MCCULLOCH, W. S., and PITTS, W. H. 1960. Anatomy and physiology of vision in the frog (*Rana pipiens*). *J. gen. Physiol.*, 43:129-175.

MAYNARD, D. M., Jr. 1953. Activity in a crustacean ganglion. I. Cardio-inhibition and acceleration in *Panulirus argus*. *Biol. Bull., Woods Hole*, 104:156-170.

MAYNARD, D. M. 1954. Activity in a crustacean ganglion. II. Pattern and interaction in burst formation. *Biol. Bull., Woods Hole*, 109:420-436.

MAYNARD, D. M. 1955. *Direct inhibition in the lobster cardiac ganglion*. PhD dissertation, Dept. of Zoology, Univ. of California, Los Angeles.

MAYNARD, D. M. 1960. Electrical activity of single neurons in the lobster brain. *Anat. Rec.*, 137:380.

MAYNARD, D. M. 1961. Cardiac inhibition in decapod crustacea. In: *Nervous Inhibition*. E. Florey (ed.). Pergamon Press, Oxford.

MCCULLOCH, W. S. 1949. The brain as a computing machine. *Electr. Eng.*, 68:492-497.

MCCULLOCH, W. S. and PITTS, W. 1943. A logical calculus for ideas immanent in nervous activity. *Bull. Math. Biophysics*, 5:115-133.

MELZACK, R. and WALL, P. D. 1962. On the nature of cutaneous sensory mechanisms. *Brain*, 85:331-356.

MILLER, G. A., GALANTER, E., and PRIBRAM, K. H. 1960. *Plans and the Structure of Behavior*. Holt, New York.

MILLER, P. L. 1960. Respiration in the desert locust. I. The control of ventilation. *J. exp. Biol.*, 37:224-236.

MITTELSTAEDT, H. 1954. Regelung und Steuerung bei der Orientierung der Lebewesen. *Regelungstechnik*, 2:226-232.

MITTELSTAEDT, H. 1957. Prey capture in mantids. In: *Recent Advances in Invertebrate Physiology*. B. T. Scheer (ed.). Univ. Oregon Publ., Eugene.

MITTELSTAEDT, H. 1960. The analysis of behavior in terms of control systems. In: *Transactions of the Fifth Conference on Group Processes*. B. Schaffner (ed.). Josiah Macy, Jr., Foundation, New York.

MITTELSTAEDT, H. 1961a. Die Regelungstheorie als methodisches Werkzeug der Verhaltensanalyse. *Naturwissenschaften*, 48:246-254.

MITTELSTAEDT, H. 1961b. Probleme der Kursregelung bei frei beweglichen Tieren. In *Aufnahme und Verarbeitung von Nachrichten Durch Organismen*. Hirzel, Stuttgart.

MITTELSTAEDT, H. 1962. Control systems of orientation in insects. *Annu. Rev. Ent.*, 7:177-198.

MITTELSTAEDT, H. and HOLST, E. VON. 1953. Reafferenzprincip und Optomotorik. *Zool. Anz.*, 151:253-259.

MOLINA, A. F. DE and HUNSPERGER, R. W. 1959. Central representation of affective reactions in forebrain and brain stem: electrical stimulation of amygdala, stria terminalis, and adjacent structures. *J. Physiol.*, 145:251-265.

MORITA, H. 1959. Initiation of spike potentials in contact chemosensory hairs of insects. III. D. C. Stimulation and generator potential of labellar chemoreceptor of *Calliphora*. *J. cell. comp. Physiol.*, 54:189-204.

MORITA, H. and TAKEDA, K. 1959. Initiation of spike potentials in contact chemosensory hairs of insects. II. The effect of electric current on tarsal chemosensory hairs of *Vanessa*. *J. cell. comp. Physiol.*, 54:177-187.

MORITA, H. and YAMASHITA, S. 1959a. Generator potential of insect chemoreceptor. *Science*, 130:922.

MORITA, H. and YAMASHITA, S. 1959b. The back-firing of impulses in a labellar chemosensory hair of the fly. *Mem. Fac. Sci., Kyushu Univ.* (*E*.), 3:81-87.

MORITA, H. S., DOIRA, S., TAKEDA, K., and KUWABARA, M. 1957. Electrical response of the contact chemoreceptor on tarsus of the butterfly, *Vanessa indica*. *Mem. Fac. Sci., Kyushu Univ.* (*E*.), 2:119-139.

MORRELL, F. 1961. Electrophysiological contributions to the neural basis of learning. *Physiol. Rev.*, 41:443-494.

MOUNTCASTLE, V. B. 1961. Some functional properties of the somatic afferent system. In: *Sensory Communication*. W. A. Rosenblith (ed.). MIT Press and Wiley, New York.

MOUNTCASTLE, V. B., DAVIES, P. W., and BERMAN, A. L. 1957. Response properties of neurons of cat's somatic sensory cortex to peripheral stimuli. *J. Neurophysiol.*, 20:374-407.

NAKAE, H. 1958. Emotional behavior produced by hypothalamic stimulation. *Amer. J. Physiol.*, 194:411-418.

NEFF, W. D. 1960. Sensory discrimination. In: *Handbook of Physiology*, Sect. 1. *Neurophysiology*, 3:1447-1470.

NIELSEN, J. M. 1946. *Agnosia, apraxia, aphasia*. Hoeber, New York.

NIELSEN, J. M. 1958. *Memory and amnesia*. San Lucas Press, Los Angeles.

OBERHOLZER, R. J. H. and TOFANI, W. O. 1960. The neural control of respiration. In: *Handbook of Physiology*, Sect. 1. *Neurophysiology*, 2:1111-1129.

OLDS, J. 1958. Selective effects of drives and drugs on "reward" systems of the brain. In: *Neurological Basis of Behaviour*. G. E. W. Wolstenholme and C. M. O'Connor (eds.). J. & A. Churchill, London.

OLDS, J. 1960. Differentiation of reward systems in the brain by self-stimulation technics. In: *Electrical Studies on the Unanesthetized Brain*, E. R. Ramey and D. S. O'Doherty (eds.). Hoeber, New York.

OTANI, T. and BULLOCK, T. H. 1959. Effects of presetting the membrane potential of the soma of spontaneous and integrating ganglion cells. *Physiol. Zoöl.*, 32:104-114.

OZBAS, S. and HODGSON, E. S. 1958. Action of insect neurosecretion upon central nervous system in vitro and upon behavior. *Proc. nat. Acad. Sci., Wash.*, 44:825-830.

OZORIO DE ALMEIDA, M. and PIÉRON, H. 1924. Sur les effets de l'extirpation de la peau chez la Grenouille. *C. R. Soc. Biol., Paris*, 90:420-422.

PANTIN C. F. A. and VIANNA DIAS, M. 1952. Rhythm and afterdischarge in medusae. *Ann. Acad. bras. Sci.*, 24:351-364.

PATLAK, C. S. 1953a. Random walk with persistence and external bias. *Bull. math. Biophys.*, 15:311-338.

PATLAK, C. S. 1953b. A mathematical contribution to the study of orientation of organisms. *Bull. math. Biophys.*, 15:431-476.

PENFIELD, W. 1958. The role of the temporal cortex in recall of past experience and interpretation of the present. In: *Neurological Basis of Behaviour*. G. E. W. Wolstenholme and C. M. O'Connor (eds.). J. & A. Churchill, London.

PENFIELD, W. 1960. Neurophysiological basis of the higher functions of the nervous system. In: *Handbook of Physiology*, Sect. 1, *Neurophysiology*, 3:1441-1445.

PENFIELD, W. and RASMUSSEN, T. 1950. *The cerebral cortex of man*. Macmillan, New York.

PHILLIPS, C. G. 1956. Intracellular records from Betz cells in the cat. *Quart. J. exp. Physiol.*, 41:58-69.

PITTENDRIGH, C. S. 1958. Perspectives in the study of biological clocks. In *Perspectives in marine biology*. A. A Buzzati-Traverso (ed.). Univ. of Calif. Press, Berkeley and Los Angeles.

PITTENDRIGH, C. S. 1961. Circadian rhythms and the circadian organization of living systems. *Cold Spr. Harb. Symp. quant. Biol.*, 25:159-184.

PLATT, J. R. 1960. How we see straight lines. *Sci. Amer.*, 202:121-129.

POGGIO, G. F. and MOUNTCASTLE, V. B. 1960. A study of the functional contributions of the lemniscal and spinothalamic systems to somatic sensibility. Central nervous mechanisms in pain. *Johns Hopk. Hosp. Bull.*, 106:266-316.

Bibliography

POWELL, T. P. S. and MOUNTCASTLE, V. B. 1958. Afferent inhibition in somatic sensory system of the monkey. *Fed. Proc.*, 17:126.

PRECHTL, H. F. R. 1956. Neurophysiologische Mechanismen des formstarren Verhaltens. *Behaviour*, 9:243-319.

PRESTON, J. B. and KENNEDY, D. 1960. Integrative synaptic mechanisms in the caudal ganglion of the crayfish. *J. gen. Physiol.*, 43:671-681.

PRIBRAM, K. H. 1958. Comparative neurology and the evolution of behavior. In: *Behavior and evolution*. A. Roe and G. G. Simpson (eds.). Yale Univ. Press, New Haven.

PRIBRAM, K. H. 1959. On the neurology of thinking. *Behav. Sci.*, 4:265-287.

PRINGLE, J. W. S. 1951. On the parallel between learning and evolution. *Behaviour*, 3:174-215.

PROSSER, C. L. 1936. Rhythmic activity in isolated nerve centers. *Cold Spr. Harb. Symp. quant. Biol.*, 4:339-346.

PROSSER, C. L. 1940. Effects of salts upon "spontaneous" activity in the nervous system of the crayfish. *J. cell. comp. Physiol.*, 15:55-65.

PROSSER, C. L. 1943. An analysis of the action of salts upon abdominal ganglia of crayfish. *J. cell. comp. Physiol.*, 22:131-145.

PROSSER, C. L. and BROWN, F. A. 1961. *Comparative Animal Physiology*. 2nd ed. Saunders, Philadelphia.

PURPURA, D. P. and GRUNDFEST, H. 1956. Nature of dendritic potentials and synaptic mechanisms in cerebral cortex of cat. *J. Neurophysiol.* 19:573-595.

RALL, W. 1959. Branching dendritic trees and motoneuron membrane resistivity. *Exp. Neurol.*, 1:491-527.

RASHEVSKY, N. 1956. A neural mechanism for adjustment to optimal conditions, with possible reference to visual accommodation. *Bull. math. Biophysics.*, 18:189-203.

RASHEVSKY, N. 1960. *Mathematical Biophysics*. 3rd ed. Dover, New York. 2 v.

RATLIFF, F. 1961. Inhibitory interaction and the detection and enhancement of contours. In: *Sensory Communication*. W. A. Rosenblith (ed.). MIT Press and Wiley, New York.

RATLIFF, F. and HARTLINE, K. 1959. The responses of *Limulus* optic nerve fibers to patterns of illumination on the receptor mosaic. *J. gen. Physiol.* 42:1241-1255.

RATLIFF, F. and MUELLER, C. G. 1957, Synthesis of "on-off" and "off" responses in a visual-neural system. *Science*, 126:840-841.

RATLIFF, F., MILLER, W. H., and HARTLINE, H. K. 1958. Neural interaction in the eye and the integration of receptor activity. *Annals N. Y. Acad. Sci.*, 74:210-222.

REDFIELD, R. (ed.). 1942. Levels of integration in biological and social systems. *Biological Symposia* vol. VIII. J. Cattell (ed.). Cattell, Lancaster, Pa.

REICHARDT, W. 1957. Autokorrelations-Auswertung als Funktionsprinzip des Zentralnervensystems (bei der optischen Bewegungswahrnehmung eines Insektes). *Z. Naturf.*, 12b:448-457.

REICHARDT, W. 1961. Autocorrelation, a principle for the evaluation of sensory information by the central nervous system. In: *Sensory Communication*. W. A. Rosenblith (ed.). MIT Press and Wiley, New York.

REISS, R. F. 1962. A theory and simulation of rhythmic behavior due to reciprocal inhibition in small nervenets. *Proc. 1962 Spring Joint Computer Conference*, Amer. Fed. Inform. Processing Soc., 21:171-194.

RENSCH, B. 1956. Increase of learning capability with increase of brain-size. *Amer. Nat.*, 90:81-95.

RETZLAFF, E. 1957. A mechanism for excitation and inhibition of the Mauthner's cells in teleost: A histological and neurophysiological study. *J. comp. Neurol.*, 107:209-226.

RETZLAFF, E. and FONTAINE, J. 1960. Reciprocal inhibition as indicated by a differential staining reaction. *Science*, 131:104-105.

REUBEN, J. P., WERMAN, R., and GRUNDFEST, H. 1960. Oscillatory hyperpolarizing responses in lobster muscle fibers. *Fed. Proc.*, 19:298.

RIJLANT, P. 1931. Etude à l'oscillographe cathodique du ganglion cardiaque de la Limule polyphème. *C. R. Soc. Biol., Paris*, 108:1144-1147.

ROCHESTER, N., HOLLAND, J. H., HAIBT, L. H., and DUDA, W. L. 1956. Tests on a cell assembly theory of the action of the brain, using a large digital computer. *Inst. Rad. Eng. Trans. Information Theory*, IT-2(3): 80-93.

ROEDER, K. D. 1939. Synchronized activity in the optic and protocerebral ganglia of the grasshopper, *Melanoplus femur—rubrum*. *J. cell. comp. Physiol.*, 14:299-307.

ROEDER, K. D. 1941. The effect of potassium and calcium on the spontaneous activity of the isolated crayfish nerve cord. *J. cell. comp. Physiol.*, 18:1-13.

ROEDER, K. D. 1955a. Spontaneous activity and behavior. *Sci. Mon., N.Y.*, 80:362-370.

ROEDER, K. D. 1955b. Spontaneous activity in the last abdominal ganglion and sexual activity in insects. *Amer. J. Physiol.*, 183:656.

ROEDER, K. D., TOZIAN, L., and WEIANT, E. A. 1960. Endogenous nerve activity and behaviour in the mantis and cockroach. *J. Insect Physiol.*, 4:45-62.

ROMANES, G. J. 1885. *Jellyfish, starfish and sea urchins, being a research on the primitive nervous systems*. Internat. Sci. Ser., Appleton, New York.

ROSENBLITH, W. A. (ed.). 1961. *Sensory Communication*. MIT Press and Wiley, New York.

ROSENBLUETH, A. 1950. *The Transmission of Nerve Impulses at Neuroeffector Junctions and Peripheral Synapses*. MIT and Wiley, New York.

ROSS, D. M. 1957. Quick and slow contractions in the isolated sphincter of the sea anemone, *Calliactis parasitica*. *J. exp. Biol.*, 34:11-28.

RUBEN, R. J. and SEKULA, J. 1960. Inhibition of central auditory response. *Science*, 131:163.

RUSHTON, W. A. H. 1959. Excitation pools in the frog's retina. *J. Physiol.*, 149:327-345.

SALMOIRAGHI, G. C. and BURNS, B. D. 1958. Rhythmicity of breathing—a study with extracellular microelectrodes. *Fed. Proc.*, 17:139.

SANDERS, F. K. and YOUNG, J. Z. 1940. Learning and other functions of the higher nervous centers of *Sepia*. *J. Neurophysiol.*, 3:501-526.

SATO, K., MIMURA, K., OZAKI, T., YAMAMOTO, Y., MASUYA, S., and HONDA, N. 1957. On the "transforming action" of the brain shown in the brain wave. *Jap. J. Physiol.*, 7:181-189.

SAWYER, C. H. 1959. Nervous control of ovulation. In: *Progress in the Endocrinology of Reproduction*. C. W. Lloyd (ed.). Academic Press, New York.

SCHEIBEL, M. and SCHEIBEL, A. 1958a Structural substrates for integrative patterns in the brain stem reticular core. In: *Reticular Formation of the Brain*. H. H. Jasper et al.(eds.). Little, Brown, Boston.

SCHEIBEL, M. E. and SCHEIBEL, A. B. 1958b. (See Fig. 140 in Jansen and Brodal, 1958.)

SCHNEIRLA, T. C. 1953. Insect behavior in relation to its setting. In: *Insect Physiology*. K. D. Roeder (ed.). Wiley, New York.

SCHÖNE, H. 1959. Die Lageorientierung mit Statolithorganen und Augen. *Ergebn. Biol.*, 21:161-209.

SHAMARINA, N. M. 1958. Readjustment of innervation relationships in the central nervous system resulting from transplantation of antagonist muscles. *Sechenov J. Physiol. (Fiziol. Zh. SSR)*, 44:991-1000.

SHERRINGTON, C. S. 1906. *The integrative action of the nervous system*. Yale Univ. Press, New Haven. 2nd ed. 1947, with bibliography. Cambridge University Press, Cambridge.

SHERRINGTON, C. S. 1929. Some functional problems attaching to convergence. *Proc. roy. Soc.*, (B)105: 332-362.

SHOLL, D. A. 1956. *The Organization of the Cerebral Cortex*. Wiley, New York.

SHOLL, D. A. and UTTLEY, A. M. 1953. Pattern discrimination and the visual cortex. *Nature, Lond.*, 171:387-388.

SPEIDEL, C. C. 1941. Adjustments of nerve endings. *Harvey Lect.*, 36:126-158.

SPERRY, R. W. 1945. The problem of central nervous reorganization after nerve regeneration and muscle transposition: a critical review. *Quart Rev. Biol.*, 20:311-369.

SPERRY, R. W. 1950. Neural basis of the spontaneous optokinetic response produced by visual inversion. *J. comp. physiol. Psychol.*, 43:482-489.

SPERRY, R. W. 1951a. Regulative factors in the orderly growth of neural circuits. *Symp. Soc. Study Devel. Growth*, 10:63-87.

SPERRY, R. W. 1951b. Mechanisms of neural maturation. In: *Handbook* S. *Experimental Psychology*, S. of Stevens (ed.). Wiley, New York.

SPERRY, R. W. 1958. Physiological plasticity and brain circuit theory. In: *Biological and Biochemical Bases of Behavior*, H. R. Harlow and C. N. Woolsey (eds.). Univ. of Wisconsin Press, Madison.

SPRAGUE, J. M., STELLAR, E., and CHAMBERS, W. W. 1960. Neurological basis of behavior in the cat. *Science*, 132:1498.

STARK, L. and CORNSWEET, T. N. 1958. Testing a servoanalytic hypothesis for pupil oscillations. *Science*, 127:588-590.

STARK, L. and SHERMAN, P. M. 1957. A servoanalytic study of consensual pupil reflex to light. *J. Neurophysiol.*, 20:17-26.

STARK, L., CAMPBELL, F. W., and ATWOOD, J. 1958. Pupil unrest: an example of noise in a biological servomechanism. *Nature*, 182:857-858.

STEGEMANN, J. 1961. Die Regelung der retinalen Beleuchtungsstärke. *Naturwissenschaften*, 48:254-258.

STRUMWASSER, F. 1961. Modes of synaptic operation and their relevance for pattern formation in an integrative ganglion. *Fed. Proc.*, 20:338.

STRUMWASSER, F. and CADE, T. J. 1957. Behavior elicited by brain stimulation in freely moving vertebrates. *Anat. Rec.*, 128:630-631.

SUTHERLAND, N. S. 1957. Visual discrimination of orientation and shape by the octopus. *Nature, Lond.*, 179:11-13.

SUTHERLAND, N. S. 1959. A test of a theory of shape discrimination in *Octopus vulgaris* Lamarck. *J. comp. physiol. Psychol.*, 52:135-141.

SUTHERLAND, N. S. 1960. Visual discrimination of shape by *Octopus*: open and closed forms. *J. comp. physiol. Psychol.*, 53:104-112. (Additional references are given in the Bibliography of Chapter 25.)

SZÉKELY, GY. 1958. The role of the specificity of sensory areas in the development of reflexes. *Acta biol. hung. (Suppl.)* 2:38-39.

SZENTÁGOTHAI, J. and SZÉKELY, GY. 1956. Elementary nervous mechanisms underlying optokinetic responses, analyzed by contralateral eye grafts in urodele larvae. *Acta physiol. hung.*, ,10:43-55.

TASAKI, I., POLLEY, E. H., and ORREGO, F. 1954. Action potentials from individual elements in cat geniculate and striate cortex. *J. Neurophysiol.*, 17:454-474.

TATEDA, H. and MORITA, H. 1959. Initiation of spike potentials in contact chemosensory hairs of insects. I. The generation site of the recorded spike potentials. *J. cell. comp. Physiol.*, 54:171-176.

TAUC, L. 1954. Phénomènes bioélectriques observés dans la plasmode d'un myxocète (*Physarum polycephalum*). *J. Physiol., Paris*, 46:659-669.

TAUC, L. 1955. Étude de l'activité élémentaire des cellules du ganglion abdominal de l'aplysie. *J. Physiol., Paris*, 47:769-792.

TAUC, L. 1956. Potentiels sous-limnaires dans le soma neuronique de l'aplysie et de l'escargot. *J. Physiol., Paris*, 48:715-718.

TAUC, L. 1957. Les divers modes d'activité du soma neuronique ganglionnaire de l'aplysie et de l'escargot. *Colloq. int. Cent. nat. Rech. sci.*, 1955(67):91-119.

TAUC, L. 1958. Processus post-synaptiques d'excitation et d'inhibition dans le soma neuronique de l'aplysie et de l'escargot. *Arch. ital. Biol.*, 96:78-110.

TAUC, L. 1960. Evidence of synaptic inhibitory actions not conveyed by inhibitory post-synaptic potentials. In: *Inhibition in the Nervous System and Gamma-aminobutyric acid.* E. Roberts (ed.). Pergamon Press, New York.

TAUC, L. and HUGHES, G. M. 1961. Sur la distribution partielle des influx efférents dans les ramifications d'un axone non myélinisé. *J. Physiol., Paris*, 53:483-484.

TERZUOLO, C. A. 1956. Cerebellar inhibitory and excitatory actions upon spinal extensor motoneurons. *Arch. ital. Biol.*, 97:316-339.

TERZUOLO, C. A. and BULLOCK, T. H. 1956. Measurement of imposed voltage gradient adequate to modulate neuronal firing. *Proc. Nat. Acad. Sci., Wash.*, 42:687-694.

TERZUOLO, C. A. and BULLOCK, T. H. 1958. Acceleration and inhibition in crustacean ganglion cells. *Arch. ital. Biol.*, 96:117-134.

THESLEFF, S. 1959. Motor end-plate "desensitization" by repetitive nerve stimuli. *J. Physiol.*, 148:659-664.

THORPE, W. H. 1951. The definition of some terms used in animal behaviour studies. *Bull. anim. Behav.*, 9:34-49.

THORPE, W. H. 1956. *Learning and Instinct in Animals.* Harvard Univ. Press, Cambridge, Mass.

TINBERGEN, N. 1951. *The Study of Instinct.* Clarendon Press, Oxford.

TOMITA, T. 1958. Mechanism of lateral inhibition in the eye of *Limulus*. *J. Neurophysiol.*, 21:419-429.

TRAUTWEIN, W. and DUDEL, J. 1958. Hemmende und "erregende" Wirkungen des Acetylcholin am Warmblüterherzen. Zur Frage der spontanen Erregungsbildung. *Pflüg. Arch. ges. Physiol.*, 266:653-664.

VAN BERGEIJK, W. A. 1961. Studies with artificial neurons, II. Analog of the external spiral innervation of the cochlea. *Kybernetik*, 1:102-107.

VAN DER KLOOT, W. G. 1961. Inhibition in the neuro-endocrine systems of invertebrates. In: *Nervous Inhibition.* E. Florey (ed.). Pergamon Press, Oxford.

VARJÚ, D. 1959. Optomotorische Reaktionen auf die Bewegung periodischer Helligkeitsmuster. *Z. Naturf.*, 148:724-735.

VERPLANCK, W. S. 1957. A glossary of some terms used in the objective science of behavior. *Psychol. Rev.*, 64(Suppl.):1-42.

VON NEUMANN, J. 1958. *The Computer and the Brain.* Yale Univ. Press, New Haven.

VORONIN, L. G. 1957. *Comparative Physiology of Higher Nervous Activity.* Lectures. Moscow Univ., Moscow. (In Russian.)

WAGNER, R., MITTELSTAEDT, H., STEGEMANN, J., KOHLER, I., and KEIDEL, W. D. 1961. Steuerung und Regelung in der Biologie. In: *Verhandlungen der Gesellschaft Deutscher Naturforscher und Ärzte.* 101. Versammlung zu Hannover. Springer, Berlin. (Also in *Naturwissenschaften*, 48:242-276.)

WALL, P. D. 1958. Excitability changes in afferent fibre terminations and their relation to slow potentials. *J. Physiol.*, 142:1-21.

WALL, P. D. and CRONLY-DILLON, J. R. 1961. Pain, itch, and vibration. *Amer. Med. Ass. Arch. Neurol.*, 2:365-375.

WALL, P. D., LETTVIN, J. Y., MCCULLOCH, W. S., and PITTS, W. H. 1956. The nature and origin of prolonged events in the terminal arborisations of spinal afferent fibres. *Int. Physiol. Congr., Abstr. XX*, 1956:941-942.

WALTER, W. G. 1959. Intrinsic rhythms of the brain. In: *Handbook of Physiology*, Sect. 1. *Neurophysiology*, 1:279-298.

WARD, A. A. 1958. *A Symposium on Dendrites.* (Vol. 10, Suppl. to *Electroenceph. clin. Neurophysiol.*) Int. Fed. Soc. Electroenceph. clin. Neurophysiol., Montreal, Canada.

Bibliography

WATANABE, A. 1958. The interaction of electrical activity among neurons of lobster cardiac ganglion. *Jap. J. Physiol.*, 8:305-318.

WATANABE, A. and BULLOCK, T. H. 1959. Modulation of activity of one neuron by subthreshold slow potentials in another in lobster cardiac ganglion. *J. gen. Physiol.*, 43:1031-1045.

WEBER, A. 1949. Instabilité morphologique de synapses centrales. *Experientia*, 5:461-471.

WEIANT, E. A. 1958. Control of spontaneous activity in certain efferent nerve fibers from the metathoracic ganglion of the cockroach, *Periplaneta americana*. *Int. Congr. Entomol., Proc. X*, 1956 (v.2):81-82.

WEIDMANN, S. 1956. *Electrophysiologie der Herzmuskelfaser*. Huber, Bern.

WEISS, P. 1936. Selectivity controlling the central-peripheral relations in the nervous system. *Biol. Rev.*, 11:494-531.

WEISS, P. 1941. Autonomous versus reflexogenous activity of the central nervous system. *Proc. Amer. phil. Soc.*, 84:53-64.

WEISS, P. 1950a. Experimental analysis of co-ordination by the disarrangement of central-peripheral relations. *Symp. Soc. exp. Biol.*, 4:92-111.

WEISS, P. 1950b. Problems of the development, growth, and regeneration of the nervous system and of its functions. In: *Genetic Neurology*. P. Weiss (ed.). Univ. Chicago Press, Chicago.

WEISS, P. 1959. Animal behavior as system reaction: orientation toward light and gravity in the resting postures of butterflies (*Vanessa*). *Gen. Systems: Ybk. Soc. Gen. Systems Res.*, 4:1-44.

WEISS, P. 1960. Modifiability of the neuron. *Arch. Neurol.*, 2:595-599.

WEISS, P. 1961. The concept of perpetual neuronal growth and proximodistal substance convection. In: *Regional Neurochemistry*. S. S. Kety and J. Elkes (eds.). Pergamon Press, New York.

WELLS, G. P. 1949. The behaviour of *Arenicola marina* L. in sand, and the role of spontaneous activity cycles. *J. Mar. Biol. Ass. U. K.*, 28:465-478.

WELLS, G. P. 1950. Spontaneous activity cycles in polychaete worms. *Symp. Soc. exp. Biol.*, 4:127-142.

WELLS, G. P. 1955. *The Sources of Animal Behaviour*. An inaugural lecture delivered at University College, London, 5 May 1955. H. K. Lewis, London.

WELLS, M. J. 1958. Nerve structure and function. *Advanc. Sci., Lond.*, 1958 (57):449-457.

WENDLER, G. 1961. Die Regelung der Körperhaltung der Stabheuschrecken (*Carausius morosus*). *Naturwissenschaften*, 48:676-677.

WIEDERSHEIM, R. 1890. Bewegungserscheinungen im Gehirn von *Leptodora hyalina*. *Anat. Anz.*, 5:673-679.

WIERSMA, C. A. G. 1952. Neurons of arthropods. *Cold Spr. Harb. Symp. quant. Biol.*, 17:155-163.

WIERSMA, C. A. G. 1957. Neuromuscular mechanisms. In *Recent Advances in Invertebrate Physiology*. B. T. Scheer (ed.). Univ. Oregon Press, Eugene.

WIERSMA, C. A. G. 1958. On the functional connections of single units in the central nervous system of the crayfish *Procambarus clarkii* Girard. *J. comp. Neurol.*, 110:421-471.

WIERSMA, C. A. G. 1961a. Reflexes and the central nervous system. In: *The Physiology of Crustacea*, Vol. II. T. H. Waterman (ed.). Academic Press, New York.

WIERSMA, C. A. G. 1961b. Inhibitory neurons: a survey of the history of their discovery and of their occurrence. In: *Nervous Inhibition*. E. Florey (ed.). Pergamon Press, Oxford.

WIERSMA, C. A. G. and ADAMS, R. T. 1950. The influence of nerve impulse sequence on the contractions of different crustacean muscles. *Physiol. comp.*, 2:20-33.

WIERSMA, C. A. G. and BOBBERT, A. C. 1961. Membrane potential changes on activation in crustacean muscle fibers. *Acta. Physiol. Pharm. Néerl.*, 10:51-72.

WIERSMA, C. A. G. and HUGHES, G. M. 1961. On the functional anatomy of neuronal units in the abdominal cord of the crayfish, *Procambarus clarkii* (Girard). *J. comp. Neurol.*, 116:209-228.

WIERSMA, C. A. G. and VAN HARREVELD, A. 1939. The interactions of the slow and the fast contraction of crustacean muscle. *Physiol. Zoöl.*, 12:43-49.

WIERSMA, C. A. G., WATERMAN, T. H., and BUSH, B. M. H. 1961. The impulse traffic in the optic nerve of decapod Crustacea. *Science*, 134:1435.

WILLIAMS, C. M. 1958. Representation of locality in a biological information system. *Bull. math. Biophys.*, 20:217-230.

WILSON, D. M. 1959. Long-term facilitation in a swimming sea anemone. *J. exp. Biol.*, 36:526-532.

WILSON, D. M. 1961. The central nervous control of flight in a locust. *J. exp. Biol.*, 38:471-490.

WINDLE, W. F. 1940. *Physiology of the Fetus*. Saunders, Philadelphia.

WOLBARSHT, M. L. 1960. Electrical characteristics of insect mechanoreceptors. *J. gen. Physiol.*, 44:105-122.

WOLDA, H. 1961. Response decrement in the prey catching activity of *Notonecta glauca* L. (Hemiptera). *Arch. néerl. Zool.*, 14:61-89.

WOLSKY, A. 1933. Stimulationsorgane. *Biol. Rev.*, 8:370-417.

WOLSTENHOLME, G. E. W. and O'CONNOR, C. M. 1958. *Neurological Basis of Behaviour*. CIBA Foundation Symposium. Churchill, London.

WRIGHT AIR DEVELOPMENT DIVISION. 1960. *Bionics Symposium. Living Prototypes—the Key to New Technology*. WADD Techn. Rep. 60-600. Office of Technical Services, U.S. Dept. Commerce, Washington, D.C.

YOUNG, J. Z. 1946. Effects of use and disuse on nerve and muscle. *Lancet*, 2:109-113.

YOUNG, J. Z. 1951. Growth and plasticity in the central nervous system. *Proc. Roy. Soc. (B)*, 139:18-37.

ZOTTERMAN, Y. 1939. Touch, pain and tickling: an electrophysiological investigation on cutaneous sensory nerves. *J. Physiol.*, 95:1-28.

ZOTTERMAN, Y. 1959. (See CIBA.)

CHAPTER 6

Neurosecretion

by HOWARD A. BERN *and* IRVINE R. HAGADORN

Summary	354	I. Echiuroidea	378
I. Phenomenon and Concept of Neurosecretion	356	J. Onychophora	378
		K. Crustacea Entomostraca	379
II. Nature of Neurosecretion	360	L. Crustacea Malacostraca	379
A. The Neurosecretory Material	360	M. Insecta	388
B. The Neurosecretory Process	363	N. Arthropoda Myriapoda	398
C. The Neuronal Properties of Neurosecretory Cells	367	O. Arthropoda Arachnida	400
		P. Mollusca Amphineura	403
D. The Functions of Neurosecretion	368	Q. Mollusca Gastropoda	403
III. Neurosecretory Systems	369	R. Mollusca Scaphopoda	405
A. Introduction	369	S. Mollusca Pelecypoda	405
B. Platyhelminthes	373	T. Mollusca Cephalopoda	405
C. Nemertinea	373	U. Echinodermata	407
D. Nematoda	373	V. Protochordates	407
E. Annelida Polychaeta	373	W. Brief Summary of Vertebrate Neurosecretory Systems	408
F. Annelida Oligochaeta	374		
G. Annelida Hirudinea	377	Classified References	410
H. Sipunculoidea	378	Bibliography	413

Summary

The phenomenon of neurosecretion has been found in nearly all groups possessing a centralized nervous system. The **generalized neurosecretory system** encountered in metazoan animals consists of a group of neurosecretory cells and a tract of secretion-bearing axons that terminate in close association with a vascular system, forming a "neurohemal organ" for storage and release of the secretory product. Frequently the neurohemal structure lies in close contact with a nonneural endocrine gland (for example, the adenohypophysis of vertebrates and the corpus allatum of insects). The **definition** of neurosecretion rests on the existence of signs of glandular activity in cells otherwise definable as neurons. Cytologic signs of secretion—granules, globules, droplets—present in the cytoplasm of neurons suggest the presence of neurosecretory activity, although other substances, such as neural pigment, may present a similar appearance. The **concept** of neurosecretion attaches functional significance to the products of neuronal glandular activity, although such significance can be ascribed at present with finality only to some vertebrate and arthropod neurosecretory systems. The products of these systems are true hormones. Implicit in the concept of neurosecretion is the confidence that definite hormonal factors also will be found to be produced by neurosecretory systems described in other groups of animals.

The **morphologic nature** of neurosecretion has been investigated by light, phase, dark-field, and electron microscopy. There is no specific stain for neurosecretory material per se; however, certain so-called Gomori combinations are particularly useful. Neurosecretory materials may stain with the major components of these combinations (chrome hematoxylin, paraldehyde fuchsin) or with the acid stains. With the electron microscope, granules in the size range 1000–3000 Å have been observed in a variety of neurohemal structures and neurosecretory cell bodies. These "elementary neurosecretory granules" may aggregate into inclusions of various sizes, which become stainable and visible with the light microscope.

The **chemical nature** of neurosecretion may differ in different neurosecretory systems and in the same system in different groups of animals. Protein, especially sulfur-rich protein, is the ubiquitous component; however, carbohydrate and lipid constituents are frequently reported. The cytochemical variability of neurosecretory material suggests that the stainable materials may represent carrier or parent molecules, rather than the actual hormones, which appear to be of lower molecular weight.

The **locus of synthesis** of neurosecretory material is generally considered to be the cell body. Materials synthesized in the perikaryon are believed to be transported along the axon to the axon ending. A role in the manufacture of neurosecretory material has been claimed for various structures in the perikaryon, including the nucleus, Nissl substance, and mitochondria. The electron microscope strongly suggests the intervention of the Golgi complex in the supramolecular organization into granules of substances which may be formed elsewhere, probably in association with the endoplasmic reticulum. The possibility of synthesis of neurosecretory material throughout the neurosecretory cell and its axon has been suggested; modification or "maturation" of the neurosecretory substance as it passes from the perikaryon along the axon to the end organ seems probable in many cases. However, the apparent absence of the perquisites for protein synthesis in the axon suggests modification of a preformed substrate rather than local synthesis de novo.

Axonal transport of neurosecretory material from the cell body to the axonal ending is supported by three lines of evidence: (a) secretory products may be demonstrated cytologically throughout the neurosecretory cell, including the axon; (b) transection of a neurosecretory fiber tract results in a "pile-up" of secretion proximal to the point of section and a simultaneous depletion of secretion distally; (c) direct observation of movement of secretory materials along axons has been reported. The motive force for this proximodistal transport is assumed to be axoplasmic flow: the generally occurring movement of axoplasm from the perikaryon to the periphery of all nerve cells.

Release of neurosecretory material appears to occur primarily at the axon endings, although some evidence suggests that release may occur from the soma of some neurosecretory cells and conceivably along the axon as well. The finding of synaptic vesicles along with the elementary neurosecretory granules in the axon terminals of many neurosecretory cells suggests that the vesicles may play some role in promoting the release of neurosecretory material.

The **neuronal nature** of neurosecretory cells seems well established. The morphologic attributes are in most cases clear; axons and dendrites, Nissl substance (or equivalent ribonucleoprotein accumulations), neurofibrils, and on the ultrastructural level synaptic vesicles, have been described in neurosecretory cells of a variety of animals. Until recently,

Summary

however, the electrophysiologic attributes have been unexplored. Extracellular recordings from both vertebrate and invertebrate neurosecretory systems have made it seem probable that neurosecretory cells can conduct; intracellular recordings from caudal neurosecretory cells of teleosts support this conclusion. Release of neurosecretory material from the end organs may thus be controlled by the electrical activity of the neurosecretory cells themselves; however, separate conduction by parallel nonneurosecretory fibers also remains a definite possibility. Possible participation of glial elements in the release mechanism or in the transformation of secretory product can also be envisaged.

A **survey of the invertebrate groups** shows that, with the exception of the Coelenterata and Ctenophora, all major phyla have been examined for the presence of neurosecretory cells; such cells are present in almost all cases. Evolutionary convergence has resulted in the occurrence of similar neurosecretory systems in a number of major groups.

In the **Nemertinea** and **Platyhelminthes**, there is cytological evidence for neurosecretion in the cerebral ganglia; in the **Nematoda** neurosecretory cells are present in the cephalic nerve ring.

In the **Annelida** neurosecretory cells are present in all three major classes; a neurohemal organ is present in the brain. In oligochaetes and polychaetes, neurosecretion has been implicated in several physiologic processes, especially reproduction and regeneration.

The **Sipunculoidea** have neurosecretory cells in the brain, which send their axons to a neurohemal structure at the front of the brain. An association of the secretory cycle with reproduction has been reported.

In the **Onychophora**, neurosecretory cells of unknown function are present throughout the nervous system, but no neurohemal organ has been found. The **Arthropoda** possess prominent and complex neurosecretory systems. The Arachnida possess well-developed neurosecretory systems; with the possible exception of the Pantopoda and Xiphosura, all show variations of a common theme. Neurosecretory cells are found in the brain and ventral ganglia; those of the protocerebrum are associated with neurohemal organs which are usually retrocerebral in position. Nonneural organs of suggested endocrine significance are frequently present in the prosoma. Neurosecretory activity in various groups of arachnids has been correlated with molting and with reproduction. In the Myriapoda neurosecretory cells are present throughout the central nervous system; those of the protocerebrum send their secretions via their axons to a neurohemal organ lying in the head. A second neurohemal structure, presumably associated with tritocerebral neurosecretory cells, is also present in the head. The functions of these systems are not known.

The **Crustacea** possess several elaborate neurosecretory systems. The principal systems present in higher Malacostraca are (a) the X-organ-sinus gland system in the eyestalk, (b) the postcommissure organs associated with the tritocerebrum, and (c) the pericardial organ-anterior ramifications system. The eyestalk X-organ consists of a group of neurosecretory cell aggregations, with fibers terminating in the neurohemal sinus gland (which also receives fibers from other areas of the central nervous system). The postcommissure and pericardial organs are largely neurohemal structures. The neurosecretory systems govern a wide variety of functions, including (a) light-adaptive responses of retinal pigments; (b) somatic pigmentation; (c) molting (through the agency of a nonneural gland, the Y-organ), regeneration and growth; (d) reproduction; (e) metabolism; (f) heart rate. Less well developed systems subserving similar functions are present in lower Malacostraca. Among Entomostraca neurosecretory cells have been reported in several groups; their functions have not yet been elucidated.

In the **Insecta**, the neurosecretory systems approach those of the Crustacea in extent of development. The principal system consists of protocerebral neurosecretory cells (medial and lateral), which send their axons to the neurohemal corpus cardiacum. Neurosecretory cells also occur in other portions of the nervous system, and the subesophageal ganglion appears to have both neuroendocrine and neurohemal functions. The corpus cardiacum may possess intrinsic endocrine activity; a second, nonneural, endocrine organ, the corpus allatum, is also present in the head region. The corpus cardiacum and the corpus allatum often form a retrocerebral complex; in higher Diptera a third nonneural endocrine structure, the ecdysial (molting) gland also becomes associated to form the so-called ring gland. The principal neurosecretory product of the protocerebrum-corpus cardiacum system is a tropic hormone which acts upon the ecdysial gland to induce the secretion of ecdyson ("molting hormone"). This molt-favoring hormone acts with the product of the corpus allatum (neotenin) to result in larval, pupal, and imaginal development. Materials obtainable from the corpus cardiacum may show important myotropic and behavioral effects, and in a few species influence pigmentation. The subesophageal ganglion may also be involved in some species in affecting pigmentation and motor activity, as well as egg diapause.

In **Mollusca**, all classes, with the possible exception of the Amphineura, have been found to possess signs of neurosecretory activity. The Gastropoda show particularly prominent indications of glandular activity, although neural pigments may possibly mimic neurosecretion in some species. Several neurohemal and neuroglandular areas have been reported. A relation of neurosecretory activity to reproduction and to osmoregulation may occur in some gastropods. In the Pelecypoda neurosecretory cells are found in the cerebral and visceral ganglia. Again, a correlation with reproduction is suggested. In the Scaphopoda neurosecretory cells are also present. In the Cephalopoda cells found in association with folliclelike structures in both decapods and octopods may not be neurosecretory, as originally thought.

Among the **Protochordates,** cytologic evidence of neurosecretion has been reported in both Urochordata and Cephalochordata.

In the **Vertebrata,** neurosecretory systems closely analogous in format to those of arthropods have been described. The principal system is the hypothalamo-neurohypophysial system. Axons from neurosecretory cells of the hypothalamus end in the neurohypophysis, a neurohemal structure, consisting of the pars nervosa, which releases hormones (oxytocin, vasopressin, vasotocin) into the systemic circulation, and the median eminence, which releases hormones into the hypophysial portal system and thus controls adenohypophysial function. Various other extrahypothalamic nuclei also may show signs of secretory activity. In fishes a second major system, the caudal neurosecretory system, terminating in the neurohemal urohypophysis in teleosts, is seen at the posterior end of the spinal cord; its function is not yet established, but may be related to osmoregulation.

In general, throughout the animal kingdom neurosecretory systems appear to be fundamental to endocrine control over major aspects of the physiology of the organism, including growth and reproduction. Although important nonneural endocrine organs occur among invertebrates and vertebrates, the over-all integration of endocrine control lies within the central nervous system and is in large part accomplished by the secretion of specific hormonal factors by specialized nerve cells—the neurosecretory neurons.

I. PHENOMENON AND CONCEPT OF NEUROSECRETION

The **history of our knowledge** of neurosecretion begins with the description by Speidel in 1919 of secretory-appearing neurons in the caudal region of the spinal cord of elasmobranchs. (More than forty years later, the function of these cells has still not been established.) About ten years after Speidel's largely ignored finding, Ernst Scharrer reported evidence of secretory activity in hypothalamic neurons of the minnow. During the 1930's Scharrer described similar "glandular neurons" as a constant occurrence in the hypothalamus of other vertebrates. In addition, during this period neurosecretion was first described in crustaceans by Hanström, in insects by Weyer, and in several other invertebrate groups by Berta Scharrer. Beginning in 1949 Bargmann and his associates laid a firm basis for understanding the functional significance of hypothalamic neurosecretion in vertebrates. This comprehension was facilitated in large part by their elucidation of the neurosecretory "pathway" by which the secretory product was conveyed to the neurohypophysis—an example of a "neurohemal organ" (a term first introduced by Knowles in the middle 1950's). The modern orientation toward neurosecretion as a general phenomenon is in fact only about ten years old at this writing. During the decade 1950–1960 an enormous literature has accumulated, expanding broadly the information available on the occurrence and the structure of neurosecretory systems and, to a less extent, their functioning.

The **phenomenon of neurosecretion** is nearly ubiquitous among metazoan animals. Almost without exception, wherever a central nervous system exists along with a degree of cephalization, some nerve cells with secretory activity can be found. The occurrence of some neurosecretory cells can be reasonably expected in representatives of all invertebrate and vertebrate groups, although many of the smaller invertebrate phyla have yet to be examined.

The **generalized neurosecretory system** encountered in metazoan animals is composed of a

I. PHENOMENON AND CONCEPT OF NEUROSECRETION

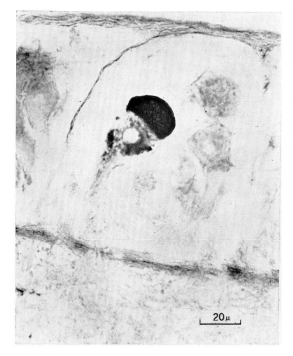

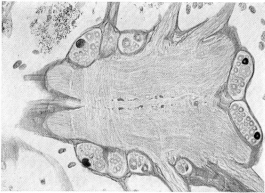

Figure 6.1. Left: Hirudo medicinalis (Hirudinea). Subesophageal ganglion showing a large neurosecretory cell (α cell). Note lightly stained ordinary neurons. Paraldehyde fuchsin stain. Right: Frontal section through subesophageal ganglion of the leech Theromyzon rude, stained with paraldehyde fuchsin and counterstains, to illustrate typical staining reactions, as well as bilateral symmetry in the location of stainable cells. Compartments 21, on the right (anterior), contain a pair of α cells, staining with paraldehyde fuchsin ("Gomori-positive"). Compartments 13, on the left (posterior), contain a pair of prominent β cells, staining with orange G ("Gomori-negative"). The ganglion is about 700 μ long.

source of secretion in the form of groups of neurosecretory cells and a tract of secretion-bearing fibers that terminate in close association with a vascular system, forming a neurohemal organ for storage and release of the secretory product. Evolutionary convergence presumably has resulted in the occurrence of systems that lend themselves to close analogy in several major groups (Hanström, 1947, 1948, 1953, 1957; B. Scharrer, 1959; E. Scharrer, 1959).

The **definition of neurosecretion** rests in part on the existence of signs of secretion in cells that are otherwise definable as neurons (Fig. 6.1). The essential criterion is a cytologic one—to date, mainly on the level of the light microscope. Regardless of the nature of the stain employed, revelation of granules or globules, which in epithelial cells would be referred to as "secretion," marks a nerve cell as *possibly* neurosecretory.

Some confusion has arisen over the so-called "special" **staining methods** for neurosecretion, notably the combinations of chrome hematoxylin and phloxin and of paraldehyde fuchsin, fast green, chromotrope, and orange G. Materials stainable with chrome hematoxylin or paraldehyde fuchsin are often referred to as "Gomori-positive" (the term strictly applies only to chrome hematoxylin staining), and this term regrettably has taken on a quasi-cytochemical significance. In fact, much neurosecretion is stainable with the acid components of these mixtures and is accordingly "Gomori-negative"; see Fig. 6.1. The "special" methods also are anything but specific; nonneurosecretory cells and materials associated with the nervous system—degenerating nerve cells, reduced melanin, neuromelanin, lipofuscins (ceroid), mast cells, insect amebocytes, vertebrate phagocytes, elastic fibers, reticulin, and so on—are all "Gomori-positive" (see, for example, de Groot, 1957). Older staining methods, including the azan or the Masson combinations, are still useful in the detection of secretion in neurons, but in the past they have failed to reveal neurosecretory cells in some forms which now are known to possess them.

The **concept associated with neurosecretion,** as distinguished from the bare phenomenon just defined, involves more than the presence of cells in the nervous system containing stainable secretory substance. Implicit in the concept is attachment of functional significance to these cells and to their products. To these products the generic term "neurohormones" can be applied, although Welsh also covers neurotransmitter substances (neurohumors) by this term. Accordingly, whereas a relation to specific hormonal products can be ascribed with assurance only to vertebrate, insect, and crustacean neurosecretory systems at this date, the concept involves confidence that hormonal factors also will be found associated with parallel systems in other groups of animals. The presence of glandular neurons among other neurons provides a source of chemical mediators, the secretion of which is under the modulative influence of a variety of nervous pathways, themselves responsive to changes in the internal and external environments. Sustained, cyclical, and "emergency" secretory patterns are all conceivable, depending on the nature and intensity of the nervous input (Fig. 6.2).

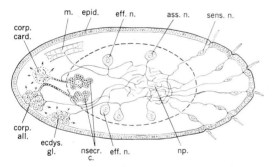

Figure 6.2. Schematic figure illustrating the interactions of neurons, neurosecretory cells, and endocrine organs in an insect. [Modified from Wigglesworth, 1959.] *ass. n.*, interneuron; *corp. all.*, **corpus allatum**; *corp. card.*, **corpus cardiacum**; *ecdys. gl.*, **ecdysial gland**; *eff. n*, **effector neuron**; *epid.*, **epidermis**; *m.*, **muscle**; *np.*, **neuropile**; *nsecr. c.*, **neurosecretory cell**; *sens. n.*, **sensory neuron**.

The term "neurosecretion" has been applied in the past to any stainable materials found in neurons, that are normal cell products. Eventual release of the material from the cell is not always certain; the liberation of material from the secretory granules, droplets, or binary systems (Smith, 1952) into the cell interior may be the ultimate fate. Picard and his collaborators at Marseilles (Picard and Stahl, 1956; Stahl, 1957) have developed the **notion of "neuroelaboration"** to include RNA-associated lipoprotein material which appears often to be a product of "nuclear secretion" and which remains intracytoplasmic in location. This material, they suggest, may be a carrier substance for neurotransmitters. Neurosecretion would be a special type of "neuroelaboration." Moussa and Banhawy (1960) describe the formation of material by Golgi dictyosomes in the motor neurons of the locust *Schistocerca* and refer to this also as "neurosecretion." The dictyosomes produce tubular "neurofibrils" in the cockroach, according to Wigglesworth (1960a).

The relation of the "Gomori-positive" material to "wear-and-tear" **nerve pigments** is unclear, but so-called neurosecretory cells in the mesencephalon of amphibians (Loebel and Guijon, 1957) or throughout the brain of many teleosts (Stahl, 1957; Stolk, 1960; Hagadorn and Bern, unpublished) may be tinctorially reactive because of their content of neuromelanin (E. Scharrer, 1935). In *Rana pipiens*, melanophores may mimic neurosecretory cells, particularly those associated with the ependyma; both nerve and ependymal cells themselves contain stainable melanin (Dawson, 1953). Much of the so-called secretory material demonstrable in gastropod neurons (see especially *Cylichna* and *Aplysia*) may be really pigment of no secretory significance, although secretory processes as such have been described in neurons of *Helix* (Thomas, 1951; Krause, 1960; Dalton, 1960).

The mere presence of **"giant cells" in the central nervous system** is not sufficient basis for the conclusion that they are neurosecretory. Some such confusion on the parts of de Lerma (1949a) in regard to certain Coleoptera and of Hanström (1943) in regard to the hemipteran *Lygaeus* has occurred (compare Weber, 1952). The deutocerebral cells ("Gomori-positive") described by Grandori (1956) in muscid Diptera are difficult to accept as neurosecretory cells and rather resemble hemocytes. In gastropod ganglia some of the largest cells do not appear to be neurosecretory, and this is also true, in the strict sense, for the giant cells in leech ventral cord ganglia (but compare Legendre, 1959b). In vertebrates, Mauthner's neurons and the supramedullary cells of puffers are examples of giant nerve cells that are not neurosecretory. Although it is the magnocellular components of the hypothalamic nuclei that are looked upon as neurosecretory, in the caudal neurosecretory system of some teleosts there appear to be more small neurosecretory (Dahlgren) cells than giant ones, even occasionally to the complete exclusion of the latter (see Enami and Imai, 1955).

Staining reactions alone provide an inadequate basis for conclusions regarding the occurrence of neurosecretion. In this chapter we have tried to employ the following **criteria** in making judgments. (a) The observation of stainable materials in neurons indicates the possibility of neurosecretion. (b) If changes in quantity or quality of the presumed secretory material can be correlated with seasonal change or with physiologic condition, the occurrence of a secretory cycle is indicated and the existence of neurosecretion is probable. (c) If the presumed neurosecretory material can be associated in quantity and in location with the occurrence of a hormonal factor, then neurosecretion can be considered as established.

Even when the staining cell is established as neurosecretory, interpretation of the physiologic significance of a neurosecretory cell (or any gland cell) filled with secretion requires caution. The alternative explanations are: (a) the cell is actively releasing hormone into the vascular system, but is producing secretion even faster than it can release it; or (b) release of hormone is slight or absent, so that even a low rate of synthesis causes accumulation of secretion in the axon and cell body. Stainability as a criterion of activity does appear to have different meanings in different forms. For example, certain neurosecretory cell groups in some insects contain the most secretion during periods when little or no hormone is being released. (See papers by Lea and by Highnam in Heller and Clark, 1962.) By contrast, the greatest stainability of certain cells and axons in polychaetes occurs when the factors they produce are apparently being discharged (Bobin and Durchon, 1952; Durchon, 1952; Clark, 1959; Clark and Clark, 1959).

For many years after the first description of hypothalamic neurosecretion by E. Scharrer (1928), the cytologic data were interpreted by most workers as signifying cell senescence and cell degeneration, and not as indicative of a vital phenomenon. There are still adherents to this school of thought today (Spatz, 1954; Hagen, 1954; Diepen and Engelhardt, 1958), and the term **"physiologic degeneration"** is used to express the idea of apocrine or holocrine secretion resulting in partial or total degeneration of the cells (see also Sano, Tamiya, and Ishizaki, 1957). In this view, beading of axons would represent stages in regeneration, and the Herring bodies would be "sich loslösende gomoripositive Achsenzylinderauftreibungen" (Spatz, 1954). However, owing to the revelations of the electron microscope, the unique staining images of the neurosecretory cell—beaded axons (see Fig. 6.6), clumps of secretory granules, colloid masses, and Herring bodies—are in little further danger of being termed artificial or cytopathologic. Even the massive accumulations of stainable material, the Herring bodies, prove to be membrane-limited aggregates (true axonal enlargements) of enormous numbers of elementary neurosecretory granules interspersed among healthy-appearing mitochondria.

Knowles and Carlisle (1956) have attempted to add to the concept of the neurosecretory cell the fact that its axon neither innervates an organ nor is in synapse with another neuron. Neurosecretion-bearing axons proceed in many cases to areas where they are in close contact with blood vessels or hemocoels. This association of axonal bulbs and vascular structures is defined as a **neurohemal organ,** serving as a storage-release area for the neurosecretory product. However, there are some neurosecretory neurons whose axons do not terminate in known neurohemal organs but accompany ordinary axons toward the periphery, for example, some neurosecretory cells of the ventral ganglia in leeches. In addition, the intimate relation of axonal terminations to elements such as the parenchymatous pituicytes of the vertebrate neurohypophysis or urohypophysis, or the intrinsic cells of the corpus cardiacum, opens the question of possible innervation, especially since synaptic vesicles may appear in large numbers in some of the terminals (Palay,

1957; Gerschenfeld et al., 1960; Holmgren and Chapman, 1960; Willey and Chapman, 1960, their Fig. 2). The adenohypophysial cells of fishes and the corpus allatum cells of insects may be under the direct influence of neurosecretion-bearing nerve fibers. Therefore we cannot include the presence of a neurohemal organ as a necessary component of all functional neurosecretory systems, but the absence of true synaptic terminations may prove to be the only unique feature of neurosecretory neurons, along with their hormonogenic function.

The term neurosecretion could be applied improperly to neurons whose endings **release transmitter substances;** Champy and Champy-Coujard (1957) have attempted the development of special staining methods for adrenergic, cholinergic, and histamine-liberating nerve fibers. Synaptic vesicles (Fernández-Morán, 1957; de Robertis, 1959) can be distinguished from the larger, generally homogeneously electron-dense, elementary neurosecretory granules (see below); nevertheless, they could be looked upon as a variety of secretion granule releasing "neurohormonal" transmitter substances (Gray and Whittaker, 1960). Since it is widely accepted that chemical factors play a role in most nervous transmission, on this basis the definition of the neurosecretory neuron could be broadened to include all neurons (Horridge, 1961). In view of the constant renewal of neuron cytoplasm and the occurrence of axoplasmic flow (apparently unidirectional—Weiss and Hiscoe, 1948), there is some merit to the point of view expressed by Brattgård, Edström, and Hydén (1958) that the neuron is "an enormous gland cell structure whose lively protein metabolism serves the specific nerve function." However, the attempt to be all-inclusive with the term "neurosecretion" is self-defeating and leads to a multiplication of semantic difficulties. Accordingly, it is profitable to refer to the cytologic and functional criteria and to concede that the neuron engages in forms of synthetic activity other than that related to neurosecretion proper.

Concerning neurosecretion per se, however, **major questions still exist** in regard to the nature of the neurosecretory material (structurally and chemically), the nature of the neurosecretory process (synthesis, transport, and release), and the neuronal properties of the neurosecretory cell (the morphologic appurtenances and the electrical properties). These issues are considered in some detail in the following sections.

II. NATURE OF NEUROSECRETION

The nature of the neurosecretory material and of the neurosecretory process—in both morphologic and chemical terms—is considered in this section on the basis of data from invertebrates and vertebrates. Cytologically and cytochemically, similarities are more striking than differences; in the electron microscope, elementary granules in neurohemal areas are proving to be constant features, partially diagnostic of the neurosecretory phenomenon regardless of the animal examined.

A. The Neurosecretory Material

1. Morphologic nature

That the stainable neurosecretory material (NSM) is not a fixation artifact is indicated by ordinary light, phase, and dark-field microscope studies of crustacean, insect, and vertebrate materials (Carlisle, 1953, 1958; Palay and Wissig, 1953; Passano, 1954; E. Thomsen, 1954; Thomsen and Thomsen, 1954; Miyawaki, 1955, 1960; Nayar, 1955; Takahashi et al., 1957; Takahashi, 1957; Maynard, 1961). Large inclusions occur in the X-organ perikarya and axons in *Sesarma*, measuring 7 μ wide and 13 μ long and composed of "granule spheres" 2–4 μ in diameter, which are in turn composed of fine granules 0.3 μ in diameter (Passano, 1954). Still smaller granules (in the 1000 Å range) can be detected with the electron microscope. In the living state, neurosecretory cells and material possess a characteristic bluish-white opacity.

The problem of **defining precisely the secretory granule** visible in the cytoplasm of neurons with the light microscope is repeated when ultrastructural observations are reviewed. A gross distinction can be made between the elementary granules of neurosecretion (Figs. 6.3, 6.9, 6.21, 6.22) and synaptic vesicles in nerve endings on the basis of size (1500 Å, in contrast to 300 Å in diameter) and electron density (often homogeneously dense and surrounded by a membrane, in contrast to being peripherally dense only). However, there is a real gradation in size of the elementary neurosecretory granules; in some neuro-

man, 1960). In both hypothalamic and caudal neurosecretory cells of the goldfish, 1000 Å granules and 1 μ droplets are distinguishable (Palay, 1960; Bern and Nishioka, unpublished).

Neurosecretory granules, or structures simulating them, **sometimes occur where not expected.** For example, the presynaptic axons in the myenteric plexus of the guinea pig regularly show such granules, along with synaptic vesicles (Hager and Tafuri, 1959). Occasional 800 Å particles are seen among the 200 Å vesicles in central nervous sys-

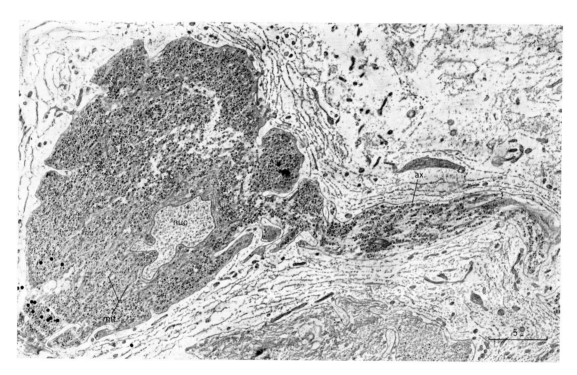

Figure 6.3. Electron micrograph of a neurosecretory cell from the leech *Theromyzon*. Axon containing elongate mitochondria and elementary neurosecretory granules; *mit.*, mitochondria; *nuc.*, nucleus. Note cytoplasm of perikaryon loaded with elementary neurosecretory granules. Cell is surrounded by glial cell membranes, which project into cell body and axon *(ax.)*. [Micrograph by R. S. Nishioka.]

hemal areas (sinus gland—Hodge and Chapman, 1958; Knowles, 1958; neurohypophysis—Duncan, 1956; urohypophysis—Holmgren and Chapman, 1960; Bern and Nishioka, unpublished), at least two kinds of nerve endings appear to be distinguishable: those containing smaller granules (ranging from 500–1000 Å in diameter) and those containing larger ones (from 1500–2000 Å). In the corpus cardiacum the two size classes have means of 1500 Å and 3000 Å (Willey and Chap-

tem neurites in the lepidopteran *Pholus labruscoe* (Trujillo-Cenóz, 1959), and a somewhat similar mixture of larger "droplets" (1200–2500 Å), medium-sized vesicles (1000–1500 Å), and smaller granules (occasionally vesicular, 300–500 Å) occur in nerve endings in the cockroach *Periplaneta* (Hess, 1958). The "droplets" closely resemble known neurosecretory granules. The finding of some cytologic evidence of secretion in sympathetic and other ganglionic neurons and in neurons of the central nervous system not ordinarily considered neurosecretory (Thomas, 1951) is thus repeated on the ultrastructural level. In this connection, the small amount of information available on the ultrastructure of other cytoplasmic inclusions indicates that the "Gomori-positive" neuromelanin granule is considerably larger than the elementary neurosecretory granule and would not be confused with the latter (Bern and Hagadorn on supramedullary and other neurons of the puffer, unpublished). This is also true of the pigment droplets in *Aplysia* neurons (Simpson, Bern, and Nishioka, unpublished).

In representatives of all vertebrate groups, neurohypophysial nerve endings contain **elementary neurosecretory granules** in the 1000–3000 Å range (Green and Van Breemen, 1955; Duncan, 1956; Bargmann and Knoop, 1957, 1960; Bargmann, Knoop, and Thiel, 1957; Fujita, 1957; Palay, 1957; Bargmann, 1958; Hartmann, 1958; Legait and Legait, 1958, 1959; Green and Maxwell, 1959; Howe, 1959; and Gerschenfeld et al., 1960). The elementary granules occasionally coalesce into relatively huge structures (Bargmann, 1958). Elementary granules have also been described in the urohypophysis (neurohemal organ) of the caudal neurosecretory system (Enami and Imai, 1958; Enami, 1959; Sano and Knoop, 1959). The "Gomori-negative" neurosecretion of this system is ultrastructurally indistinguishable from the "Gomori-positive" neurohypophysial material. **Change in size** of the granules en route from the cell body occurs in the caudal system, and also in the hypothalamo-hypophysial tract (Green and Van Breemen, 1955; Gerschenfeld et al., 1960); these observations raise the question of possible continued synthesis within the axons.

Granules in the nervi corporis allati of the moth *Celerio* (see Fig. 6.22) measure 1000–3000 Å (Schultz, 1960), similar to the "droplets" found by Hess (1958) in the cockroach ventral ganglionic chain. Electron-dense neurosecretory granules in *Cambarellus* (Fingerman and Aoto, 1959) range from 300–3000 Å (300–1650 Å in the sinus gland) depending on their location; these granules are not bounded by membranes, as are those described by Meyer and Pflugfelder (1958) in the corpus cardiacum of *Carausius* (500–2000 Å; mean = 1000 Å).

Detailed studies of the elementary neurosecretory granules at very high magnifications have revealed **some internal structure** and tend to point to the so-called homogeneous granule as a vesicle conceivably filled with ultraparticles measuring about 50 Å in diameter (Knowles, 1960, in the pericardial organ of *Squilla*). In 1952, Schiebler commented on the granular internal structure of the pig neurohypophysial granules obtained by differential centrifugation. The limiting membrane of neurosecretory granules has been described by Carlisle and Knowles (1959; see their Plate III) and Knowles (1960) in the sinus gland, pericardial organ, and postcommissure organ of *Squilla* and *Palaemon*, where formation of the granule within tubes of the endoplasmic reticulum was suggested.

The triple limiting membrane (Knowles) or double-layered halo (Wetzstein) is also seen in the case of neurohypophysial granules (Palay, 1957; Bargmann and Knoop, 1957) and of chromaffin ultragranules (about 1800 Å) in the mammalian adrenal medulla (Wetzstein, 1957). In general, the **vesicular nature** of most neurosecretory granules seems established. The electron density of these granules in the mammalian neurohypophysis can be altered by physiologic stimuli such as dehydration (Palay, 1957; Gerschenfeld et al., 1960) or histamine (Hartmann, 1958); under such stimulation the vesicles appear as sacs emptied of their contents but without change in size.

The **relationship of neurosecretory axonal endings to capillaries** or sinuses is still unclear. In most cases the terminal bulbs appear to end at some distance from the capillary endothelium and are often partially enveloped by glial cell processes in vertebrates. Axonal endings are evident at the periphery of parenchymatous pituicytes in the rat (Rennels and Drager, 1955), and axons may project into the pituicytes of teleosts as far as the nuclei of the latter (Legait and Legait, 1958). A similar relationship occurs between axonal endings and corpus cardiacum cells (Meyer and Pflugfelder, 1958). In addition to being ensheathed by glial elements, some vertebrate neurosecretion-bearing fibers may be myelinated (Tamiya et al., 1956; Legait and Legait, 1958, 1959), but most are not.

2. Chemical nature

The **chemical nature** of the neurosecretory material long has eluded investigators. Cytochemical tests at one time allowed the conclusion that neurosecretory material in mammals and fishes was a complex glycolipoprotein (Schiebler, 1951, 1952); there is still good evidence for the occurrence of carbohydrate (Rehm, 1959; Gabe, 1960) and lipid components, especially phospholipid (Hadler et al., 1957; Brousse et al., 1958), along with the protein of neurosecretory material in various sites, especially sulfur-rich protein (Barrnett, 1954; Sloper, 1957, 1958). In the vertebrate hypothalamo-hypophysial system, the active hormonal principles are cystine-containing octapeptides (Du Vigneaud, 1956); the cytochemical finding of **protein-bound disulfide** groups in some (but by no means all) invertebrate neurosecretory material has led Brousse et al. (1958) to refer to a convergent relation ("rapport de convergence") between invertebrate and hypothalamic neurosecretory cells. The somewhat variable morphology and the variable staining reactions of neurosecretory material provide a basis for the expectation that neurosecretory material is not a single chemical entity. The different content of carbohydrate in hypothalamic neurosecretory material in representatives of different vertebrate classes (Gabe, 1960) may indeed reflect a major difference in the parent protein ("protéine-mère"); the oligopeptide hormones themselves may vary in the substitution of one amino acid for another.

The nature of the relation of neurosecretory material to the actual hormones in any neurosecretory locale remains undefined. Neurosecretory material possibly serves as a **carrier substance** ("Trägersubstanz") for hypothalamic neurosecretion. A correspondence between physiologic state and excess or depletion of neurosecretory material was established thoroughly by the Kiel school and others (for example, Kratzsch, 1951; Ortmann, 1951; Andersson and Jewell, 1957). Hild and Zetler (1953) first established a relation between the occurrence of antidiuretic, pressor, and oxytocic activities and of neurosecretory material in the hypothalamic nuclei and supraoptico-hypophysial tract (tuber cinereum) of the dog. Neurohypophysial extracts, synthetic oxytocin, and cystine, imbedded in agar and treated by standard histologic methods, are all "Gomori-positive" (Rodeck, 1959). If a carrier protein itself contains cystine, it will appear as stainable neurosecretion even if there be no neurohypophysial hormone associated with it. Soluble protein decreases along with antidiuretic hormone activity from the rat neurohypophysis (Albers and Brightman, 1959). The stainable protein of neurosecretory material well may be a carrier ("van Dyke protein" or "neurophysine") from which the hormonally active peptides can be dissociated and recombined, rather than a parent compound (Acher and Fromageot, 1957; Chauvet et al., 1960). Paralleling the vertebrate work, Fingerman and Aoto (1958) have extracted the dark-red pigment-concentrating hormone from the stubs of the optic nerves and from the brain after eyestalk removal in *Cambarellus*, and have pointed out that the stubs are engorged also with neurosecretory material. Considerable effort is now being directed toward isolation and characterization of elementary neurosecretory granules from various sources. Chemical studies of such isolated granules should help decide the relation of their major components to the hormones presumably contained within them.

B. The Neurosecretory Process

1. Locus and mode of synthesis of neurosecretory material

The generally accepted picture of the production of secretion by neurosecretory cells is based on the morphologic observations of E. Scharrer and of W. Bargmann and their collaborators. According to their views, neurosecretory material is **synthesized in the perikaryon** and is **transported distally** inside the axon to its point of discharge. Possibly indicative of the true secretory nature of these neurons in vertebrates, the Golgi apparatus is not always a perinuclear network as in ordinary neurons, but often is polarized "apically" between nucleus and axon hillock (Stahl and Seite,

1954). However, neurosecretory material is seen also in the dendrites of some cells (see Wilson et al., 1957).

Cytologic studies have indicated nuclear secretion of neurosecretory material in some forms (Palay, 1943, and Ortmann, 1958, in teleost hypothalamic neurons) and a close association with Nissl substance (Scharrer et al., 1945; Palay and Wissig, 1953) and with cytoplasmic RNA generally (Edström and Eichner, 1958). Association of neurosecretory material with mitochondria has been suggested repeatedly (compare Ortmann, 1958); Pardoe and Weatherall (1955) have reported that oxytocic and pressor activities are present in cytoplasmic particles from the neurohypophysis that behave like mitochondria.

Early **electron microscope studies** of neurosecretory perikarya were not very informative (Fingerman and Aoto, 1959; Enami, 1959; Löblich and Knezevic, 1960) but did reveal granules similar to those seen in the axons. In vertebrate hypothalamic and caudal neurosecretory neurons, the elementary granules are associated with Golgi membranes (Sano and Knoop, 1959; Palay, 1960; Bern et al., 1961; Murakami, 1961, 1962). This association is also seen in cells of the lepidopteran and phasmid pars intercerebralis (Nishiitsutsuji-uwo, 1960, 1961; Stiennon and Drochmans, 1961), the brachyuran thoracic ganglion (Miyawaki, 1960b), the earthworm brain (Scharrer and Brown, 1961), the leech supraesophageal ganglion and cockroach pars intercerebralis (Bern et al., 1961), and certain neurons of *Helix* and *Aplysia* (Dalton, 1960; Simpson, Bern, and Nishioka, unpublished).

A prominent endoplasmic reticulum is also present in cells not loaded with formed granules. The **Golgi complex** may contribute to the supramolecular organization of the neurosecretory substance into detectable granules (see also Naisse, 1961). There is no evidence for involvement of the Golgi membranes in the transport of the granules to the exterior of the cells. Claims have been made that mitochondria may transform (Green and Maxwell, 1959) or fragment (Nishiitsutsuji-uwo, 1960) into neurosecretory granules, or that neurosecretory granules are manufactured along the course of the axon as a result of mitochondrial activity (Knowles, 1958). Gerschenfeld et al. (1960), however, regard the membrane-limited neurosecretory vesicle as a synthetic organelle in its own right.

A **hypothetical scheme** of neurosecretory material production can be presented, based on the ultrastructural evidence. The protein component of neurosecretory material may be synthesized in the endoplasmic reticulum (granular membranes), from which it then passes to the Golgi (agranular) membranes. A portion of the double Golgi membrane then detaches as a vesicle. The contained neurosecretory material may increase in amount and density as the vesicle moves toward the axon hillock to be transported down-axon as a self-contained organelle. Accordingly, modification of the contents may continue up till the time of release.

In addition to the majority of workers who accept the notion of axonal transport (see below), there are some workers (Bodian, 1951; Green and Maxwell, 1959; Knowles, 1959) who suggest the possibility of neurosecretory material **synthesis throughout the neurosecretory neuron** including its axon, as von Euler (1958) feels must occur with catecholamines. The "growth" of elementary neurosecretory granules en route from perikaryon to terminal also implies continued synthesis in the neurite. Some recent work on the uptake of S^{35}-labeled amino acids (especially cystine) shows concentration of the isotope in axons of the neurosecretory cells of rats and rabbits that can be ascribed only to local synthesis of protein (Goslar and Schultze, 1958). However, Sloper (1958) and Sloper et al. (1960) indicate arrival of the S^{35} label in the infundibular process by way of the pituitary stalk, and their findings support the concept of neurosecretion transport. The peripheral axoplasm of ordinary neurons does not appear to have the perquisites for substantial biosynthesis (see Schmitt, 1959) and the electron microscope fails to reveal ribosome concentrations in the axonal cytoplasm. Peripheral synthesis of acetylcholine esterase has been claimed (Koenig and Koelle, 1960); however, peripheral activation of preformed "zymogen" synthesized in the perikaryon provides an alternative explanation.

2. Axonal transport of neurosecretory material

Burgen (1959) has emphasized as a major problematic aspect of the concept of neurosecretion the alleged proximodistal migration of neurosecretory material, although the report of the "Convegno" on neurosecretion (1954) considered the evidence supporting the notion of transport as "adequate" (p. 88). There are three principal lines of **evidence for the existence of axon-borne secretion.**

(a) Stainable materials: droplets, granules, and vesicles along axons of neurosecretory cells are readily demonstrated cytologically and can be interpreted as reflecting the **passage of particulate material.** The "classic" neurite of the neurosecretory cell shows granules arranged in a "string-of-pearls" fashion—so-called beaded axons (Fig. 6.6). However, nodular-appearing fibers are associated not only with neurosecretory cells but with other neurons as well (Knoche, 1958), and beading and vesiculation are a general consequence of nerve constriction (Weiss and Hiscoe, 1948; Weiss, 1955). In the case of neurosecretion, the transported material accumulates in the axonal bulbs, which are specialized terminations of the axons, or in fine axonal branches (as in the pericardial organ); these terminals are generally "in contact" with capillaries, sinusoids, or sinuses of the blood-vascular or hemocoelic system.

(b) Experiments initiated by Hild (1951) establish a double consequence of transection of a tract of neurosecretion-bearing fibers. There is an **accumulation** ("pile-up") of secretory material proximal to the cut, accompanied by a depletion of secretory material distal to the cut, including that ordinarily stored in the neurohemal area. "Pile-up" in the "cardiac-recurrent" nerve of living *Calliphora* has been observed under darkfield illumination by E. Thomsen (1954); see also Fingerman and Aoto (1958), on *Cambarellus*.

(c) In vivo observation of **secretion passing along axons** has been claimed; essentially, there are only two such records, both by Carlisle (1953a, 1953b, 1958a), on neurosecretory tracts in living crustaceans and fish. In addition, Hild (1954) claims to have observed movement distally in the process of a cultured paraventricular neuron.

Axoplasmic flow is a generally accepted phenomenon and can be demonstrated directly or indirectly in a variety of ways. Section, ligature, or constriction of nerves results in damming of the axonal contents, evident morphologically and by virtue of intracellular concentration of enzymes, substrates, and products proximal to the block (catecholamines, Raab and Humphreys, 1947; von Euler, 1950; Tainter and Luduena, 1950; choline acetylase, Hebb and Waites, 1956; succinic dehydrogenase and diaphorases, Friede, 1959; acetylcholine, MacIntosh, 1959). Pressure also causes changes in fiber diameters consistent with the notion of proximodistal (centrifugal) flow (Causey, 1949). Smith (1952) claims that the form of the droplets in toad sympathetic ganglion perikarya can be ascribed to the effect of "moving cytoplasm which accelerates as it enters the narrow axon."

In Friede's studies (1959), the sectioned cat or rat sciatic nerve shows an accumulation of histochemically demonstrated succinic dehydrogenase in the proximal stump, that is reminiscent of neurosecretion "pile-up." P^{32}-labeled material (presumably largely inositol diphosphate as phosphoprotein) has been found to move peripherad in guinea pig sciatic nerves at a rate of about 2.7 mm per day, 1/20 the rate of endoneurial (interaxonic) flow (Gerard, 1950; Samuels et al., 1951). The **rate of movement of material** presumably produced in the cell body is consistent with the rate of fiber growth or regeneration and with the more generalized phenomenon of axoplasmic (intra-axonal) flow (Weiss and Hiscoe, 1948). Weiss (1944) proposed a balance of central synthesis and peripheral destruction to account for the unidirectional flow, based on the changes seen following chronic constriction of mammalian nerve. However, Ochs and Burger (1958) suggest that the data of Samuels et al. (1951) support distal movement of certain cellular constituents rather than the entire axoplasm; their own data support the concept of intra-axonal transport distally, as do those of Weiss (1959) and

Miani (1960). Weiss found that newly synthesized protein (tagged by the administration of radioactive amino acids) moved distally from the perikaryon in normal neurons; Miani found the same movement in regenerating neurons.

Attempts have been made to compare the **rate of movement of secretion** along the axon with the rate of axoplasmic flow. On a physiologic basis, the basal rate of movement of antidiuretic hormone in the dog hypothalamo-hypophysial tract has been computed as being about 0.2 mm per day, which is 1/5 the minimum rate ascribed to axoplasmic flow but is not incompatible with this explanation for transport (Sloper, 1958). The rate of movement of droplets, granules, and mitochondria in the X-organ connective of *Dromia* and *Lysmata* has been estimated at 2–4 μ per minute (about 4.5 mm per day) and at 100–150 μ per hour (Carlisle, 1953a, 1953b). These estimates correspond well with the estimate of the rate of axoplasmic flow of 1–2 mm per day in vertebrates (Weiss and Hiscoe, 1948; Weiss, 1959). In 1958 Carlisle estimated the rate of movement of intra-axonal neurosecretory clumps in the pituitary stalk of the goosefish as being 100–200 μ per minute. This rate corresponds to the rate of endoneurial (interaxonal) flow in mammalian limb nerves (about 150–500 μ per minute, according to Weiss et al., 1945).

This discussion leads one to emphasize that the phenomenon of neurosecretory transport and "pile-up" may represent nothing more than a special case of a phenomenon which occurs in all neurons and which is connected with the transport of essential materials (including the enzymes and substrates involved in the anabolism and catabolism of cholinergic and adrenergic substances) toward the nerve endings. Hanström (1955) definitely conceives of the flow of neurosecretory material as similar to the flow of axoplasm in regenerating nerves.

3. *Release of neurosecretory material*

The **axonal terminal** is the generally accepted point of release of the hormones associated with neurosecretory material; however, release may also occur elsewhere. Release along the axon is conceivable, as well as release from the perikaryon itself, as in certain arthropod neurosecretory cells. The reason "pile-up" does not occur after transection of some crustacean secretory neurons is possibly because such cells are able to release their material from the cell body (Matsumoto, 1958). This also may be the case with the insect pars intercerebralis, where there is some physiologic evidence for the discharge of ecdysiotropic hormone directly from the brain.

Synaptic vesicles are present in the axonal bulbs of Dahlgren cells when neurosecretory granules are scarce (Sano and Knoop, 1959). This observation and similar findings for the pars nervosa (Palay, 1957; Bargmann, Knoop, and Thiel, 1957; Hartmann, 1958; and Gerschenfeld et al., 1960) and for the median eminence (Barry and Cotte, 1961; Kobayashi et al., 1961) are somewhat unexpected, since neurosecretory cells characteristically do not innervate effector organs. If the synaptic vesicles are associated with release of transmitter substances (see discussion in Hess, 1958), the function of these structures in neurosecretion-bearing fibers is uncertain because there is no postsynaptic cell and therefore no synapse. There appears to be an inverse relationship between the occurrence of the smaller synaptic vesicles and the elementary neurosecretory granules in nerve endings; in the chronically dehydrated toad, the synaptic vesicles of neurosecretory axon terminals in the pars nervosa are increased in number over the normal.

If the neurosecretory nerve cell conducts impulses to trigger the release of material from its own neurosecretory granules (see below), the action of the so-called synaptic vesicles may be entirely intracellular in affecting the **permeability properties** of the neurosecretory granule (Gerschenfeld et al., 1960) or of the cell membrane (Koelle, 1961; Koelle and Geesey, 1961). Another possible function of the synaptic vesicles concerns the relationship of neurosecretion-laden nerve endings to the **glialike elements** in the neurohemal organ. In several papers involving cytologic and cytochemical observations, Gabe has emphasized that many neurohemal areas have their own secretory function, possibly in the production of specific materials or in the conversion of neurosecretory material to active principles. The neurosecretion-bearing fibers conceivably could be secretomotor to the "pituicyte" and its invertebrate equivalents. Whether pituicytes actually contain "Gomori-positive" neurosecretory material still is being debated; paraldehyde fuchsinophilic inclusions (gliosomes) are characteristic of some invertebrate glia (Pipa, 1961a). Recently, the equivalence of the vesicular structures in the ferret neurohypophysis to synaptic vesicles has been questioned (Holmes and Knowles, 1960), since size of the vesicles tends to parallel the caliber of the fibers containing them. Kobayashi et al. (1961) consider the possibility that these small vesicles arise from neurosecretory vesicles after they have discharged their contents. In any case, the presence of synaptic vesiclelike granules along with elementary neurosecretory granules in axonal terminals is proving to be a useful criterion in defining diffuse neurohemal areas (Hagadorn and Nishioka, 1961).

C. The Neuronal Properties of Neurosecretory Cells

1. Morphologic features

The possession of neuronal features by neurosecretory cells is assumed by most workers in the field. At least some vertebrate neurosecretory neurons show **dendrites:** bipolar hypothalamic neurons of amphibians (Wilson et al., 1957); supraoptic neurons of dog and monkey (Sano, 1958; Fox and Zabors, 1960). Conventional methods have not yet established dendrites associated with most invertebrate neurosecretory neurons; however, the neurosecretory neurons of *Ascaris* are bipolar (Gersch and Scheffel, 1958) and those of the crustacean thoracic ganglion possess granule-free dendrites (Maynard, 1961). There is good evidence for the existence of **Nissl substance,** at least major cytoplasmic RNA-protein accumulations, in all active neurosecretory cells examined. **Neurofibrils** are present in some (if not all) neurosecretory cells, as demonstrated by conventional methods: Kastl (1954), Sano, Tamiya, and Ishizaki (1957), Ishizaki, Ishida, and Kawakatsu (1959), on dog and human hypothalamic neurons; Hagadorn (1958),

on leech neurosecretory neurons. **Synaptic vesicles** are also often present.

Hanström (1954) considers the secretory attributes of the neurosecretory cell to have been **acquired secondarily** to their neuronal properties and, in some cases, even at the expense of the latter. Certainly the occurrence of complex neurohemal areas, which increase the efficiency of release of glandular products from the nervous system, represents evolutionary specialization. On the other hand, Clark (1956b) considers secretory ability to be a **primitive feature** of neurons—the neurosecretory phenomenon occurs throughout the central nervous system in some lower invertebrate groups. Almost all annelid neurons show secretory granules on the ultrastructural level (Scharrer and Brown, 1961; Hagadorn and Nishioka, 1961). However, there seems to be little basis for a general application of Clark's (1956a) suggestion that some neurosecretory cells may have originated from epidermal mucous cells which have been incorporated into the nervous system and have come to resemble neurons cytologically, although it may be true in special instances.

If a secretory cell in the nervous system has a greatly reduced axon or none at all, and no dendrites, the problem forces itself upon us boldly: **is it a nerve cell** and hence neurosecretory? How is a nerve cell to be recognized aside from its prolongations? At present no definite answer can be given: inclusions are of help but not decisive; cytology comparable to ordinary nerve cells in the same animal and contrasting with the glia is also helpful but not always conclusive. Fortunately, not many actual cases are so equivocal; most neurosecretory cells have processes or are obviously special members of a cluster of ordinary nerve cells.

2. Conduction by neurosecretory cells

Neuroendocrine reflexes, such as those involved in certain pigment changes in crustaceans or in milk let-down in mammals, involve a hormonal efferent pathway that cannot—in view of the rapidity of the response—be ascribed to the passage of neurosecretion from perikaryon to axonal bulb. Some factor must act locally at the point of discharge of the neurohormonal factor (on the axonal bulbs in the sinus gland or in the neurohypophysis) to cause immediate discharge of hormone. The critical issue, not yet resolved, is whether the information responsible for release is conducted to the bulbs by the neurosecretory axons themselves (the "economical" method) or by a parallel set of nonneurosecretory fibers "innervating" the bulb region or changing the bulb microenvironment (possibly through glial elements) so as to facilitate release. The possibility that the glial sheath itself plays a conducting role cannot be discounted (see Galambos, 1961).

The **conduction of impulses** by neurosecretory cells is probable, largely on indirect evidence. Neurohormones are released from neurohemal organs by electrical stimulation of neurosecretory centers (from neurohypophysis, Harris, 1947; from corpus cardiacum, Hodgson and Geldiay, 1959). Physiologically active neurohemal organs (for example, sinus gland) also show electrical activity (Milburn, cited by Bliss, 1956). Neurosecretory centers show electrical activity upon physiologic activation (Van der Kloot, 1955, 1958; Nakayama, 1955; Cross and Green, 1959). Neurosecretory tracts or nerves, which include some nonneurosecretion-bearing fibers, show electrical evidence of impulse conduction: goosefish hypophysial stalk (Potter and Loewenstein, 1955; Carlisle, 1958a); *Limulus* lateral rudimentary eye (Waterman and Enami, 1954). The first direct evidence of electrical properties of neurosecretory neurons similar to those of other neurons comes from intracellular recording from Dahlgren cell bodies of the eel caudal neurosecretory system, which reveals that at least these giant cells can conduct (Morita et al., 1961).

D. The Functions of Neurosecretion

Neurosecretory systems are **central to the operation of endocrine mechanisms** wherever they exist among metazoan organisms. There is no endocrinology in any animal without its neuroendocrinologic facets. There are many endocrine organs which are completely nonneural: almost all endocrine glands of vertebrates; the corpus allatum and ecdysial gland of insects; the Y-organ and androgenic gland of crustaceans; the optic gland of octopods. However, the normal functioning of most of these nonneural glands is dependent upon nervous system regulation through hormonal or nervous pathways or both. In some major invertebrate groups (several classes of molluscs, the annelids, the arachnids) nonneural endocrine organs have yet to be definitely described; however, neurosecretory cell groups associated with regulatory mechanisms are the rule.

The basic system of the neurosecretory nucleus, the neurosecretion-bearing tract, and the neurohemal organ for storage and release is encountered repeatedly throughout the animal kingdom. Closely analogous systems occur among all major metazoan groups. In many species neurohemal areas have not been defined but can be expected to exist; in other well-studied species (for example, crabs—see Maynard, 1961) new neurohemal areas are being discovered as the fibers from neurosecretory cell groups are traced to their terminations. The neurohemal area provides a locus for the storage of preformed hormone; release is subject to delicate modulation by nervous impulses conducted or transmitted to the axonal terminals.

We may briefly state the functions identified with the best-known systems. In oligochaete and polychaete **annelids** the neurosecretory areas appear to be involved in reproductive and regenerative activity. In **crustaceans** the brain–X-organ–sinus gland neurosecretory system is responsible for regulating somatic pigmentation, retinal pigment migration, molting (through hormonal control over the Y-organ), and various aspects of metabolism. Additional crustacean neurosecretory systems: the brain–postcommissure organ system and the thoracic ganglion–pericardial organ system may affect pigmentation and heart rate, respectively. In **insects** the brain–corpus cardiacum system affects growth, differentiation, and molting (including regulation of the pupal diapause where it occurs) through its essential hormonal control over the ecdysial gland. Other neurosecretory areas, especially cell groups of the subesophageal ganglion, appear to regulate cyclic behavior patterns and pigmentation of larvae, pupae, and adults in a few species. In **vertebrates** the principal neurosecretory system originating in the hypothalamus controls adenohypophysial function by the release of hormones in the median eminence region and produces the neurohypophysial hormones that are released from the pars nervosa of the pituitary complex.

In all these instances environmental and endogenous stimuli can trigger **neurohormone release** by activating nervous centers. Thus, as examples, illumination may result in pigmentary changes and in molting in a crustacean; a blood meal may trigger molting in an insect; copulation may result in ovulation and in increased oviducal motility in a vertebrate (see E. Scharrer, 1959).

The **association of neurohemal areas with epithelial organs** of nonneural origin is a frequent morphologic finding. The best-known examples are the relationship between the neurohypophysis (especially the median eminence) and the adenohypophysis in vertebrates and between the corpus cardiacum and the corpus allatum in insects. In molluscs, epithelial vesicles (occasionally transient in nature) are seen in contact with the nervous system: the follicle gland of gastropods, the epistellar organ of octopods, and the parolfactory vesicle of decapods. In crustaceans this relation is represented by the pars distalis of the X-organ (sensory pore X-organ), which appears to have glandular as well as neurohemal features.

Table 6.1 presents a summary of the various neurosecretory systems encountered in invertebrate animals and also lists nonneural structures, both associated with the neurosecretory system and separate from it, which are of established or suggested endocrine significance. Because of the close functional association of neurosecretory cells with endocrine mechanisms generally, this table can serve as a key to what is known about the endocrine system as a totality in the various animal groups.

An **inhibitory relation of the nervous system to gonadal maturation** is of general occurrence and may possibly involve neurosecretion, although neurosecretion is not known to be concerned in many cases. In representatives of many animal groups, removal of the brain or damage to certain cerebral centers results in premature sexual maturity or in stimulation of reproductive function. This appears to be true in polychaetes (where the brain secretes a factor inhibiting gamete maturation and the heteronereid transformation), in lamellibranchs (where neurosecretory material from the cerebral ganglion may be related to inhibition of spawning), in octopods (where the nerve from the brain to the optic gland inhibits its secretion of a gonadotropic factor), in crustaceans (where the eyestalk system produces an ovary-inhibiting hormone, and also a testis-inhibiting hormone in crabs), in insects (where protocerebral innervation of the corpus allatum is inhibitory to its production of a gonadotropic factor in females), and in mammals (where lesions in various parts of the brain will result in precocious physiologic and/or behavioral sexual activity). There are occasional exceptions to this generality: in vertebrates, the normal neurohormonal hypothalamic control over secretion of two gonadotropins (LH and FSH) appears to be stimulatory; in adult oligochaetes the brain is the source of a gonad-stimulating factor.

III. NEUROSECRETORY SYSTEMS

A. Introduction

In this section we shall define the morphology of neurosecretory systems in the various animal groups, as well as review the status of knowledge of their functional import. However, in the cases of crustaceans and insects, the literature on fundamental physiologic investigations of neurosecretion and on the nonneural endocrine organs is so extensive that reference can be made only to critical earlier papers and to reviews which can be used to document more fully the evolution of current concepts in arthropod neuroendocrinology. The later literature has been cited extensively, but nevertheless is only represented by less than 50% of the available papers.

Table 6.1. NEUROSECRETORY SYSTEMS AND POSSIBLE ENDOCRINE ORGANS IN INVERTEBRATES.

Group	Neurosecretory Cell Groups	Neurohemal Organs	Epithelial Organs of Possible Endocrine Significance Anatomically Associated with Neurosecretory System	Possible Endocrine Organs Not Associated with Neurosecretory System	Comments	References (if not given in text)
Platyhelminthes	In brain					
Nemertinea	In brain			Cerebral organ (?) (connected to the brain)	Cerebral organ activity possibly related to spawning	
Nematoda	In ganglion of major lateral papillary nerve of cephalic nerve ring					Gersch (1957) Gersch and Scheffel (1958)
Annelida	In brain, ventral nerve cord, and stomatogastric system; in hypodermis	Base of supra-esophageal ganglion		Gonads (?)		
Sipunculoidea	In brain	Papilliform processes at anterior end of brain (?)		Internephridial organ (?)		Harms (1948)
Echiuroidea	In ventral nerve cord				Secretion-bearing axons in ventral nerve cord	Hagadorn (unpublished)
Onychophora	In cerebral ganglion and ventral nerve chain	Infracerebral organs (?)*				
Arthropoda						
– Crustacea	In brain and ventral ganglia. X-organs (eyestalk groups). In pericardial organs and anterior ramifications	Sinus gland. Pericardial organ and anterior ramifications. Postcommissure organs. Pars distalis of X-organ (?)*	Pars distalis of X-organ (?)	Y-organ. Androgenic gland. Circumorbital gland (?)		
– Insecta	In brain, ventral nerve ganglia, stomatogastric ganglia	Corpus cardiacum* Subesophageal ganglion (?)*	Corpus allatum	Ecdysial gland (prothoracic, thoracic, ventral, peritracheal, and so on). Gonads.		

– Myriapoda	In brain and ventral nerve cord	Cerebral gland* Connective bodies Hypocerebral formations			
– Arachnida					
– – Xiphosura	In circumesophageal nerve ring and ventral nerve cord Lateral rudimentary eye				
– – Pantopoda	In some ganglia (?)	Ventral glands (?)		Ventral glands probably not neurosecretory (Fig. 16.4)	Sanchez (1958, 1959)
– – Araneida	In brain and ventral nerve masses	Primary and secondary Schneider's organs*	Anterior organ (?) Endocrine cells of Millot (?)		
– – Phalangida	In brain and ventral nerve masses	Paraganglionic plaques*	Molting gland (?)		
– – Scorpionida	Protocerebral cells posterior to globuli In subesophageal ganglion	Organs of Police ("stomatogastric ganglia")*		Nerve fibers to organs of Police via lateral and intestinal nerves	Gabe (1955a) Habibulla (1961)
– – Pseudo- scorpionida	In protocerebrum In subesophageal nerve mass	Parapharyngeal organs*		Neurohemal organs precerebral in position	Gabe (1955a)
– – Acarina	Two paired groups in protocerebrum In subesophageal nerve mass	Paraganglionic organs*		Brain required for molting in *Ornithodoros*	Gabe (1955a) Cox (1960)
– – Solifuga	In protocerebrum, tritocerebrum, ventral nerve mass	Perineural capsule (?)	Paired paraldehyde fuchsin-positive vesicles near corpora pedunculata (?)	Neurohemal organs not comparable to retrocerebral organs of other arachnids	Junqua (1957)
Mollusca					
– Amphineura	Absent (?)			Use of newer staining methods needed	B. Scharrer (1937)
– Gastropoda	In all parts of nervous system	Frontal glands	Follicle glands (?) Frontal glands (?) and similar structures	Gonads	

(continued on next page)

(Continuation of Table 6.1)

Group	Neurosecretory Cell Groups	Neurohemal Organs	Epithelial Organs of Possible Endocrine Significance Anatomically Associated with Neurosecretory System	Possible Endocrine Organs Not Associated with Neurosecretory System	Comments	References (if not given in text)
Mollusca (cont.)						
— Pelecypoda	In cerebral and visceral ganglia					
— Scaphopoda	In cerebral, pleural, and anterior buccal ganglia				No secretory cycle	Gabe (1949, 1954b)
— Cephalopoda						
—— Octopoda	In subesophageal ganglionic mass and stomatogastric system. Cells associated with epistellar body (?)		Epistellar body (?) (cells resemble photoreceptors)	Optic gland Branchial bodies (?)		
—— Decapoda	Cells associated with parolfactory vesicles (?)		Parolfactory vesicles (?)	Optic gland (?)		
Echinodermata	In circumoral nerve ring and radial nerves					Unger (1960)
Protochordates						
— Tunicata	In cerebral ganglion	Neural gland (?)*	Neural gland (?) Asymmetric gland (?)	Endostyle (?)		
— Cephalochordata	In brain		Hatschek's pit (?)	Endostyle (?)		

* = intrinsic secretory cells present   ? = questionable evidence

The bibliography on the other invertebrate groups, including the smaller arthropod groups, is essentially complete except for preliminary reports. Most of the work on these last groups is of recent vintage and is suggestive of the great need for experimental investigation of neuroendocrine mechanisms in worms, molluscs, arachnids, and myriapods. Many of the minor groups, of which little or nothing is known, are treated only in Table 6.1. For these groups and for tabulated material not mentioned in the text, references are included in the table.

The vertebrates have been treated at the end of this section in a highly condensed fashion in order to permit comparison of the neurosecretory phenomenon and its operative significance throughout the animal kingdom.

B. Platyhelminthes

Among flatworms, the simplest animals with a centralized nervous system, neurosecretory cells have been found in the brain of the planarian *Polycelis* (Lender and Klein, 1961). In *Dugesia* and *Polycelis* a water-soluble factor is produced by the cerebral ganglion which governs the regeneration of excised eyespots by the neoblasts. The anteroposterior and mediolateral gradients of regenerative ability have been related to gradients in the numbers of neurons present along these two axes (Stéphan-Dubois and Lender, 1956; Török, 1958). In the polyclad *Leptoplana*, possible neurosecretory cells are found in the brain (Turner, 1946).

C. Nemertinea

Neurosecretion was initially considered absent from nemertineans. B. Scharrer (1941b) has discussed the bearing of the nonneurosecretory cerebral organ of the Nemertinea on the concept of neurosecretion. The cerebral organ is generally considered to be a sensory organ (Hyman, 1951, pp. 43-44); in the majority of cases it is a compound structure consisting of glandular cells in intimate association with a ganglion. The close relation to both nervous and vascular systems suggests an endocrine function. In at least some forms the nervous and glandular portions of the cerebral organ arise from the same primordium.

Ablation of the anterior portion of the body of *Lineus* causes premature maturation of the ovules in the isolated posterior portion of the body; the ovules of the anterior portion remain immature. Spawning is correlated with a cycle of activity in the cerebral organs (Gontcharoff and Lechenault, 1958). The cerebral ganglia contain neurosecretory cells (Lechenault, 1962).

D. Nematoda

(*See Table* 6.1 *and Fig.* 11.8)

E. Annelida Polychaeta

The major work with the polychaetes has been done on the Nereidae and Nephtyidae, although data are available for several other groups—Aphroditidae, Arenicolidae, Pectinariidae, Polynoidae, Sabellidae, Serpulidae, Syllidae, Terebellidae (B. Scharrer, 1937; Arvy, 1954; Junqua, 1957; Durchon, 1959; Hauenschild, 1959b; recent review by Durchon, 1960). In nereids, **four types of neurosecretory cells** are found in the supraesophageal ganglion, based on cytology and staining characteristics (B. Scharrer, 1936, 1937; Schaefer, 1939; Bobin and Durchon, 1953). Two of these (the c- and d-cells) are stages in the secretory cycle of a single cell type. In the Nephtyidae (Fig. 6.4), at least three main types have been observed (Clark, 1959). Neurosecretory cells have also been observed in the larvae of several polychaete families (Korn, 1959, 1960).

The **endocrine role of the brain** is similar in the nereids and nephtyids; the brain has been found to produce a factor ("juvenile hormone") which has two effects: (1) inhibition of maturation of gametes, and (2) inhibition of the heteronereid transformation in those forms which normally undergo it. Secretion in a neurohemal structure at the base of the brain is prominent in the atokous stages when the juvenile hormone is present, and disappears when the juvenile hormone is no longer active, as in the heteronereid (Bobin and Durchon, 1952; Durchon, 1952, 1956a, 1956b; Hauenschild, 1956a, 1956b;

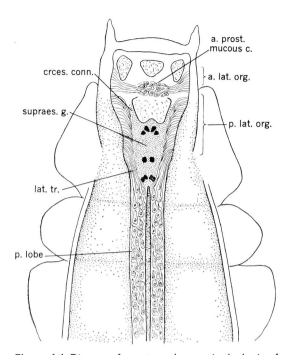

Figure 6.4. Diagram of secretory elements in the brain of *Nephtys* (Polychaeta). *a. lat. org.*, anterior lateral organ; *a. prost. mucous c.*, anterior prostomial mucous cells; *crces. conn.*, circumesophageal connective; *lat. tr.*, lateral tract of posterior lobe cell processes; *p. lobe*, posterior lobe; *p. lat. org.*, posterior lateral organ; *supraes. g.*, supraesophageal ganglion. Dark areas in supraesophageal ganglion represent nuclei showing especially prominent neurosecretory activity. The posterior lobes are composed of "Gomori-positive" mucous gland cells, which in some species (for example *N. cirrosa*) may take on a neuronlike appearance. [Based on Clark, 1955, 1958a, 1958b, 1959.]

Clark, 1959). The secretion originates in neurosecretory cells at the rear of the brain (Clark, 1959; Hauenschild, 1959a). The syllid brain completely lacks juvenile hormone, and the inhibitory influence is exerted by the proventriculus or possibly by neurosecretory cells in the ventral nerve cord in the region of the proventriculus.

The c- and d-cells of nereids show a **cycle of secretion** associated with epitoky; secretion is prominent only in those forms showing the heteronereid change (Bobin and Durchon, 1953; Durchon and Frézal, 1955). There is as yet little evidence to support the suggestion that the secretion is released to the outside environment to act as an attractant to the opposite sex during swarming.

Extensive **accumulation of carbohydrates** occurs in the neurons of the brain and ventral nerve cord prior to swarming. However, this concentration does not occur in the area of the neurohemal organ described above (Defretin, 1955, 1956).

Endocrine control of regeneration by neurosecretory cells of the brain also occurs in both nereids and nephtyids (Harms, 1948; Durchon, 1956c; Clark and Clark, 1959; Clark and Bonney, 1960; Hauenschild, 1960; Clark and Evans, 1961). In *Nereis* the brain factor is required during a three-day critical period after injury. Following the critical period, regeneration can proceed in the absence of the brain, although at a reduced rate.

F. Annelida Oligochaeta

The **oligochaete brain** has been implicated in the same endocrine functions as have been demonstrated for the polychaetes, though strictly comparable work has been done in the two groups only in the case of regeneration. Activity of the neurosecretory cells of the earthworm supra- and subesophageal ganglia has been correlated with the **possible control of four processes:** (a) maturation of gametes, (b) development of somatic sex characters, (c) regeneration, and (d) physiologic color change.

1. Cytology

There are **several types of neurosecretory cells** present in the brain of oligochaetes, although the terminology is somewhat confused: cell types given the same name by different workers sometimes appear to be related to different physiologic conditions. Attempts to equate earthworm neurosecretory cells to cell types in the polychaete brain are as yet unjustified. Cell classes (Table 6.2) have been set up by various workers according to several criteria: tinctorial affinities with various staining mixtures, location in the brain, cell morphology, and response to such factors as ambient light and operative injury.

The **acidophilic cells in the supraesophageal ganglion,** discovered by B. Scharrer (1937), have been studied in several groups of oligochaetes (Lumbricidae, Lumbriculidae, Tubificidae, Nai-

Table 6.2. POSSIBLE EQUIVALENCE OF NEUROSECRETORY CELL TYPES IN OLIGOCHAETA[1].

Herlant-Meewis	Hubl	Harms	Brandenburg	Michon and Alaphilippe	Aros and Vigh	Description	Physiologic Correlates
a-cells	a-cells (in part) b-cells (in part)	a-cells (in part) Blue cells (in part)		a-cells		Located in posterolateral portion of cerebral ganglion; dense cytoplasm; nucleus often eccentric; axons do not leave supraesophageal ganglion	Reproductive cycle (Herlant-Meewis, Hubl)
b-cells	a-cells (in part) b-cells (in part) c-cells	a-cells (in part) Blue cells (in part)		b-cells		Located laterally below a-cell region near point of emergence of anterior connectives; fusiform; radial orientation. Some axons leave supraesophageal ganglia via circumesophageal connectives	Regeneration (Hubl)
Subesophageal neurosecretory cells	u-cells (?)	Red and blue cells	Stage 6 = u-cells of Hubl	d-cells (?) (seen in one species only)		Located in subesophageal ganglion	Regeneration (Hubl)
Large and medium-sized neurons	Large and small neurons			Δ-cells	Dorsomedial cell group	Internal to and larger than a- and b-cells	Color change (Aros and Vigh); egg-laying (Herlant-Meewis, Hubl)

([1]) With special reference to the Lumbricidae.

didae, Enchytraeidae) by Harms (1948), Marapão (1959), Deuse-Zimmermann (1960), and Otremba (1961). The azan technique discloses two secretory cell types in *Lumbricus*, the a-cells and the blue cells, in the supra- and subesophageal ganglia and in the first two ventral segmental ganglia. Glialike cells with a sparsely granular secretion also are seen. In the Lumbricidae the neurosecretory cells form a cortex ("Cerebralorgan") on the posterior surface of the supraesophageal ganglion. Paraldehyde fuchsin-positive cells are seen in the brain and ventral nerve cord of *Enchytraeus* and *Tubifex*; *Nais* possesses them only in the ventral nerve cord. Peculiar associations of neurosecretory cells of the ventral ganglia with the hypodermis occur in *Tubifex* and *Nais*; *Tubifex* also possesses hypodermal sensory cells which show signs of neurosecretory activity. Neurosecretion in earthworms has also been described briefly by Brandenburg (1956a), Michon and Alaphilippe (1959), and Aros and Vigh (1959).

The principal work on neurosecretion in earthworms was done on *Eisenia foetida* by Herlant-Meewis (1955, 1956a, 1956b, 1959) and on several species of *Lumbricus* and of *Allolobophora* by Hubl (1953, 1956). Both workers have described four basic cell types in the brain, although it is probable that their types are not completely equivalent (Table 6.2). Three secretory cell types are present in the supraesophageal ganglion—the a-cells, the b-cells, and the "large and medium-sized neurons" (Fig. 6.5)—and one additional type in the subesophageal ganglion (Herlant-Meewis, 1955, 1956a). Hubl observed Herlant-Meewis's large and medium-sized neurons, as well as a-cells and b-cells, in the supraesophageal ganglion. He observed a third cell type (the c-cell) in the supraesophageal ganglion during regeneration, which he believed to be derived from b-cells; u-cells were also found in the subesophageal ganglion during regeneration.

2. Physiology

(a) Reproduction. The brain of *Lumbricus* is the probable source of a hormonal factor governing development of somatic sex characters (Avel, 1929). It has been found that the a-cells show a cycle of secretion, storing colloid in their perikarya during sexual inactivity and releasing it during the spring and summer reproductive period (Hubl, 1953; Herlant-Meewis, 1955, 1956a, 1956b).

Removal of either supra- or subesophageal ganglion blocks egg-laying until the missing part is regenerated. This blockage is due to regression of the somatic sex characters and suppression of meiosis in the gonads (Herlant-Meewis, 1956a, 1959). The a-cells may be involved in the control of the somatic sex characters; in addition, the gonads may exert some sort of feed-back control on the brain: gonadectomy results in drastic alterations in the a-cells (Hubl, 1953, 1956)

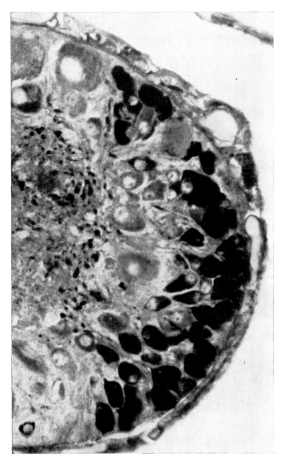

Figure 6.5. Sagittal section of posterior region of cerebral ganglion of *Eisenia foetida* (Oligochaeta) to show paraldehyde fuchsinophilic neurosecretory a-cells and neurosecretory material in neuropile. Lightly stained cells are the "gros et moyens neurones" (see text). Histologic picture is characteristic of worm collected in winter. [Herlant-Meewis, 1955.]

(b) Regeneration. The effects of the nervous system on regeneration in earthworms have been studied by several workers (see Avel, 1947), but only recently has regeneration been studied from the standpoint of neurosecretion (Harms, 1948; Hubl, 1953, 1956). In *Lumbricus* the anterior portion of the nervous system is indispensable for regeneration; the influence is humoral and may be due to the restriction of the neurosecretory cells to this portion of the nervous system in *Lumbricus*.

Following the amputation of the hind end, activation of the a-, b-, and u-cells occurs. The b-cells become vacuolated and are present as aggregates in the dorsolateral portions of the brain in some species. Hubl refers to them as c-cells at this stage. The brain is needed for regeneration only during a critical period, lasting one or two days following the injury.

(c) Color change. Exposed to sunlight or to ultraviolet light *Lumbricus rubellus* darkens, and vacuolation appears in the dorsomedial group of neurosecretory cells of the brain. When the animals are replaced in the dark, paling occurs and the dorsomedial cells become filled with secretion. The other neurosecretory cells in the brain do not respond to this treatment. It is possible that the dorsomedial neurosecretory cells may exert a chromatophorotropic effect on the animal (Aros and Vigh, 1959, 1961). Scharrer and Scharrer (1954a) cite the report of McVay (1942) on the extraction of a chromatophorotropin from the dorsal and posterior portions of the brain of *Lumbricus*; this substance was not effective in *Lumbricus* itself (McVay, unpublished). Dehydration also has effects on the neurosecretory activity of the brain (Aros and Bodnár, 1960).

3. Neurosecretory cell types

Resolution of the neurosecretory cell types present in the oligochaetes is rendered difficult by the variety of stains used and by the almost total lack of overlapping physiologic data. Cell classification based on tinctorial affinities (Harms, 1948; Hubl, 1953) has been compromised by the finding that the cells change their affinities during the course of their secretory cycle. Thus the cell types described by Harms and Hubl, though homogeneous tinctorially, may be mixed populations from the standpoint of function. In addition, there may be real differences in cell morphology and location in different genera. Despite these reservations, we have employed staining affinities, cell morphology and location, destination of axons, and responses to experimental manipulations as bases for the comparisons in Table 6.2.

G. Annelida Hirudinea

At least **two types of neurosecretory cells** occur in the brain of the leech (B. Scharrer, 1937; Hagadorn, 1958; Nambudiri and Vijayakrishnan, 1958; Legendre, 1959b). The two cell types are found in the supra- and subesophageal ganglia (Fig. 6.6) and ventral nerve cord of *Theromyzon*, *Placobdella*, *Erpobdella*, and *Hirudo* on the basis of staining and cytochemical reactions. No intergradation between the two cell types is seen. Neurons specialized for secretion (Fig. 6.1) make up about 5% of the total number of neurons in the brain of *Theromyzon*; with the electron microscope these cells are found to contain enormous numbers of elementary granules (Fig. 6.3). Secretion-bearing fiber tracts occur in all portions of the central nervous system including the ventral nerve cord, where transection experiments suggest that secretion is transported anteriorly toward the brain. A loose network of secretion-bearing axon endings occurs in the connective tissue sheath of the posterior surface of the dorsal commissure, which may function as a primitive neurohemal structure. Many secretion-bearing axons, particularly in the ventral nerve cord, must end elsewhere, presumably diffusely. Some have been observed in the segmental nerves.

In *Hirudo* the neurosecretory cells of the ventral nerve cord, and especially of the subesophageal ganglion, are said to give a positive chromaffin reaction. Legendre believes that neurosecretion may be simply a maximum expression of the neurohumoral activity of the nerve cell. Certain of the chromaffin cells (the "giant" chromaffin cells, of which there are two

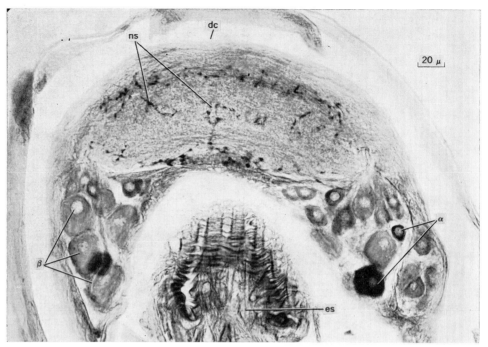

Figure 6.6. *Theromyzon rude* (Hirudinea). Frontal section of the supraesophageal ganglion; anterior at the top. [Hagadorn, 1958, *J. Morph.*, **102**:55.] α, α neurosecretory cell; β, β neurosecretory cell; *dc*, dorsal commissure; *es*, esophagus; *ns*, neurosecretion in α-cell axons.

in the anteromedial compartment of each ventral ganglion) have been demonstrated to produce an adrenalinlike substance, and earlier workers have suggested that these neuronlike cells have developed endocrine attributes at the expense of their neuronal properties (see Perez, 1942).

H. Sipunculoidea

Neurosecretion has been observed in the supraesophageal ganglion in four genera of sipunculids, including *Phascolion strombi*, *Phascolosoma vulgare*, *Golfingia elongata*, and *Sipunculus nudus* (Gabe, 1953d; Carlisle, 1959d; Koller, 1959; Åkesson, 1961). In *Sipunculus* two neurosecretory cell types are seen; only one cell type has been reported in the other three genera. Axonal transport is observed only in *Sipunculus*. These cells are acidophilic and PAS-positive; they react weakly or not at all to the chrome hematoxylin stain of Gomori. In *Phascolion* a secretory cycle is possibly related to the degree of maturation of the gonocytes.

A **neurohemal organ** is present in *Sipunculus* in the form of papilliform processes ("finger organs") on the anterior surface of the cerebral ganglion, projecting into the hemocoel. Before terminating in these finger organs, the secretion-bearing axons loop into a sensory organ associated with a ciliated canal which opens to the outside. Carlisle suggests that this association may influence the release of the neurosecretory substances from the finger organs, but the association is not confirmed by Åkesson (1961). Brain extracts affect motility of the nephridia in *Phascolosoma*; the effects may be due to the secretory cells of the brain (extracts of ventral nerve cords do not have these effects).

I. Echiuroidea

(*See Table* 6.1)

J. Onychophora

Neurosecretory cells are found in the brain, in the ventral cords, and along the pedal nerves (Gabe, 1954a; Sanchez, 1958). In the brain, five groups of chrome hematoxylin-positive cells are found: one paired group bilaterally in the antennal segment, one pair near the posterior segment of the cerebral trabeculae, and a single dorsomedial group near the central body. Occasional large phloxinophil cells also are seen. Axonal transport of the neurosecretory material occurs, but the axons cannot be traced to any neurohemal organ.

Sanchez believes that the **infracerebral organs** are neurohemal in nature. These structures, intimately connected with the brain in the antennal segment, are vesicular in form and contain a lamellated, chrome hematoxylin-positive concretion. Cuénot (1949) suspected that they have an endocrine function; however, neither Sanchez nor Gabe has been able to trace neurosecretion-bearing axons to these organs.

K. Crustacea Entomostraca

Studies on neurosecretion in the Entomostraca have been conducted on the cirripedes (Barnes and Gonor, 1958), cladocerans (Sterba, 1957), anostracans (Lochhead and Resner, 1958), ostracods (Weygoldt, 1961), and copepods (Carlisle and Pitman, 1961).

Among the Cirripedia two types of neurosecretory cells are found in the sub- and supraesophageal ganglia of *Pollicipes*. The secretion appears to be released from the cell body as well as transported along the axons. Neurosecretory cells also were observed in *Chthamalus dalli* and in four species of *Balanus*. They are absent from sexually immature animals. Chromatophorotropic activity has been demonstrated in the central nervous system of *Balanus*, *Chelonibia*, and *Lepas* (Sandeen and Costlow, 1961).

Among the Cladocera neurosecretory cells occur in the brains of *Daphnia* (Fig. 21.3) and *Simocephalus*; cell groups are present in the proto- and deutocerebrum, the second antennal segment, the mandibular segment, and possibly the maxillary segment of the brain. Axonal transport of the secretions is suggested, but no evidence of storage has been seen. An increase in the amount of secretion is observed at the beginning of egg maturation, but the amount of secretion is small at the time of discharge of the eggs into the brood pouch.

In *Artemia* and *Eubranchipus* (Anostraca) eyestalk removal affects neither molting nor reproduction. Neurosecretory cells occur in the supra- and subesophageal ganglia, but not in the eyestalks. No neurohemal area was found.

Among the Ostracoda, both unipolar and bipolar neurosecretory cells occur in the protocerebrum of *Cyprideis*, and in the supraesophageal ganglion, circumesophageal connectives, and subesophageal ganglion of *Cypris*. In *Cyprideis* all secretory neurons stain with chrome hematoxylin; in *Cypris* some of the cells are phloxinophilic.

In the Copepoda neurosecretory cells occur in lateral areas of the brain; their axons carry secretion to the frontal organ. The secretion is prominent in summer and absent or sparse in winter; it may be related to diapause. Extracts from *Calanus* have chromatophorotropic effects on *Leander*.

L. Crustacea Malacostraca

1. General

As has been pointed out already, the extent of the literature on the Crustacea necessitates their incomplete coverage here. Inasmuch as there are several recent reviews extensively devoted to crustacean endocrinology, little mention is made here of the literature prior to 1952, except where essential to our present understanding of neurosecretion in the crustaceans. Some attempt has been made to present a reasonably complete bibliography of the literature since 1956, although only the more indispensable references are cited in the text. For general considerations of the earlier literature and for further information on the nonneural endocrine organs, reference should be made to the following: Hanström (1939, 1948), Kleinholz (1942, 1957, 1961), Brown (1944, 1950, 1952), Panouse (1947), Parker (1948), Gabe (1954b), Knowles and Carlisle (1956), Carlisle and Knowles (1959), Carlisle (1960), Passano (1960, 1961), Scheer (1960), and Welsh (1961a).

Studies of neurosecretion in the Malacostraca have been focused mainly on the decapods, and to a less degree on the isopods and hoplocarids.

The eyestalks have been established as a versatile and important source of hormonal factors, although two other neurosecretory organs of more specialized function (postcommissure and pericardial organs) have been discovered. **Neurosecretory control** is important in at least six areas of crustacean endocrinology: (a) light-adaptive responses of retinal pigments (see Chapter 19 for anatomy of compound eyes); (b) somatic pigmentation; (c) molt, regeneration, and growth; (d) reproduction; (e) metabolism; (f) heart rate.

The neurosecretory systems of the Malacostraca are **complex and widespread** throughout the central nervous system. Neurons suspected of secretory function on cytologic grounds have been found in all ganglia examined (Fig. 6.7), with the exception of the stomatogastric and cardiac ganglia (Matsumoto, 1958). Accordingly, the ganglia of the optic centers, the brain, the connective ganglia, the tritocerebral commissure of natantians, and the thoracic ganglia of brachyurans, natantians, and isopods, all show neurosecretory cells (see, for example, Bliss, Durand, and Welsh, 1954; Knowles, 1955; Knowles and Carlisle, 1956; Matsumoto, 1958, 1959; Fingerman and Aoto, 1959; Duveau, 1961).

Up to eleven different **types of neurosecretory cells** have been recognized cytologically in malacostracans (Enami, 1951; Durand, 1956; Matsumoto, 1958; Potter, 1958; Carlisle, 1959b). The various cell types have yet to be related in most cases to specific physiologic activities. Two **mechanisms of release** of secretion from neurosecretory cells have been postulated. Axonal transport of secretion from the perikaryon to a storage-release organ is accepted generally. In addition, some cell types may release secretions from the perikaryon instead of, or in addition to, release from the axonal endings (Parameswaran, 1956; Matsumoto, 1958). The relative importance of the two modes of release is said to vary with the cell type.

2. The sinus gland

The sinus gland was early recognized as a potent source of **hormone(s) affecting pigmentation** (Hanström, 1935, 1937). The sinus gland has been found in all malacostracans examined with the exception of the Leptostraca, Syncarida, and Cumacea (Hanström, 1948). In its simplest form (mysids and euphausiids) the sinus gland is a thickened disk on the neurilemma of the medulla terminalis, bordering a blood sinus from which it is separated by a thin membrane. The nuclei present are relatively few in number and are similar to those of the neurilemma. In higher forms this basic structure may become more elaborate as the result of invagination to form a cup-shaped structure and of branching along internal blood channels—most complex in the Astacura (Hanström, 1948). In most forms a single pair of sinus glands is present, always in association with the optic centers; in marine isopods (*Idothea* and others) two pairs of sinus glands have been described by Oguro (1959a, 1959b; see below). The sinus glands are said to be absent in the parasitic isopods *Argeia* and *Athelges* (Oguro, 1961). In all cases the sinus glands receive extensive innervation from the medulla

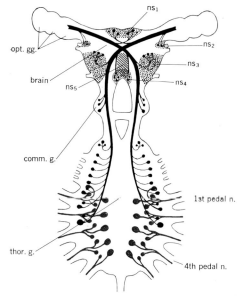

Figure 6.7. Generalized diagram of neurosecretory pathways in crabs. Neurohemal organs and eyestalk system are omitted. In *black* are shown the main routes leading to the optic ganglia; in *red* are shown the fibers leading from the thoracic ganglion into the pedal nerves. ns_1–ns_5, cerebral neurosecretory cell groups B_1–B_5, according to Bliss, Passano, and Welsh (1956)— group ns_5, indicated by *hatching*, is located ventrally; *comm. g.*, commissural ganglion; *opt. gg.*, optic ganglia; *thor. g.*, thoracic ganglion. [Modified from Matsumoto, 1958.]

terminalis and elsewhere. Their palisaded appearance (Fig. 6.8) was ascribed by Hanström to the presence of "secretory canals," which he felt transported the secretions to the blood sinus.

Initially it was believed that the sinus gland was the site of production of the hormones contained therein, and some recent workers (see especially Gabe, 1954b) still consider that the sinus gland possesses some inherent secretory ability. However, following the development of improved surgical methods, it was found that removal of the sinus gland alone did not duplicate many of the physiologic effects of eyestalk ablation (Frost, Saloum, and Kleinholz, 1951; Bliss, 1953; Passano, 1953; and others). This led Bliss and Welsh (1952) to conclude that the sinus gland is little more than an **aggregation of axonal endings** from neurosecretory cells of the eyestalk and elsewhere; the nuclei present are considered to belong to connective tissue cells. This suggestion has been supported strongly by other workers (for example, Passano, 1953; Carlisle and Knowles, 1959), on both morphologic and physiologic grounds. The electron microscope reveals axons filled with secretory granules (Fig. 6.9), which are oriented perpendicularly to the free surface of the organ (thus the palisaded appearance); interspersed are nonglandular cells comparable to pituicytes (Hodge and Chapman, 1958; Knowles, 1959). That the neurosecretory material brought to the sinus gland may undergo modification en route and/or locally (conceivably with the participation of sinus gland cells) is indicated by changes in staining and cytochemical reactions (Knowles and Carlisle, 1956; Carlisle, 1958b, 1959b; Rehm, 1959; see Potter, 1958, for dissent). It also has been reported that while sinus gland extracts are strong chromactivators, X-organ extracts are ineffective unless boiled or treated with alcohol (see Knowles, 1955).

The sources of the **axons forming the sinus gland** are various (Fig. 6.10); the medulla terminalis ganglionic X-organ (see Knowles and Carlisle, 1956, for basis of terminology) is the major contributor, but neurosecretory cells from other optic ganglia, the brain, and the connective and thoracic ganglia also send fibers to the sinus

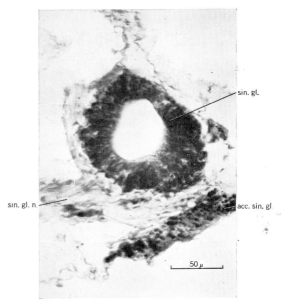

Figure 6.8. Sinus gland from eyestalk of *Crangon*, stained with paraldehyde fuchsin. Note palisaded appearance of sinus gland *(sin. gl.)*. *acc. sin. gl.*, accessory sinus gland; *sin. gl. n.*, sinus gland nerve, showing some neurosecretory material.

gland. The axonal endings in the sinus gland tend to be grouped, at least in some species, by cell type and possibly, therefore, by function (Potter, 1958, and Rehm, 1959, in *Carcinus* and *Callinectes*; Carlisle, 1959b, in *Pandalus*). Potter relates six axonal types observed in the sinus gland with six types of neurosecretory cells occupying characteristic positions in the medulla terminalis ganglionic X-organ. The several chromactivators vary in amount in the different lobes of the sinus gland of *Palaemon* (Knowles, 1955). The hypothesis that each cell type produces its own characteristic secretion is also supported by the electron microscope, which discloses that although the elementary neurosecretory granules may differ considerably in size, each axon in the sinus gland tends to contain granules of only one size class (Hodge and Chapman, 1958; Knowles, 1959).

3. The X-organs

The structures in the eyestalk designated as "X-organs" are of **two essentially different types,** the ganglionic X-organs and the sensory pore X-organ (Carlisle and Passano, 1953). The latter is also called the sensory papilla X-organ or pars

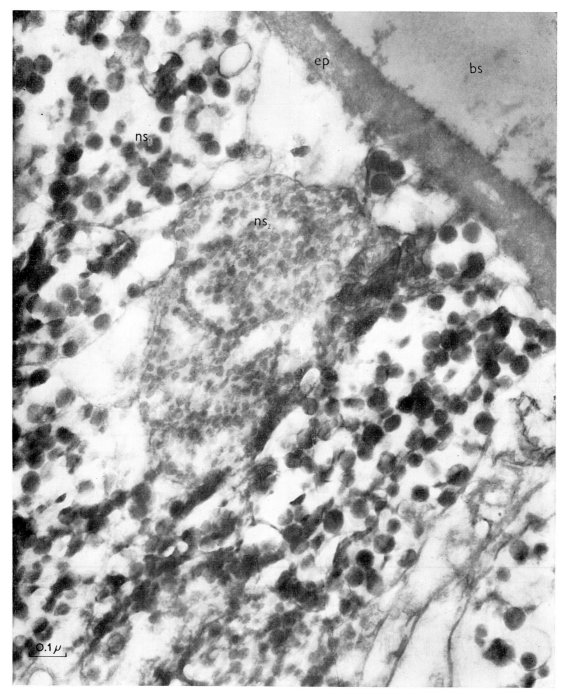

Figure 6.9. Electron micrograph of sinus gland in *Squilla* (Hoplocarida). *bs*, blood sinus; *ep*, epineurium; ns_1, ns_2, axonal endings containing elementary neurosecretory granules of two different mean diameters. [Knowles, 1959. From Gorbman (ed.), *Comparative Endocrinology*, John Wiley and Sons, New York.]

distalis X-organ. These two components may be combined into a single structure in the medulla terminalis; this is the common case in the Reptantia (especially the Brachyura), in which the sensory papilla is reduced or absent. In most of the lower Malacostraca, Hoplocarida, and Decapoda Natantia, where a sensory papilla or pore is commonly present, the sensory pore

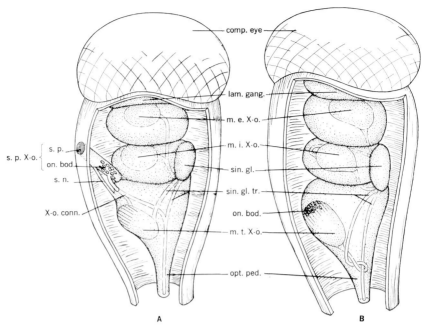

Figure 6.10. Diagrams of optic ganglia in crustacean eyestalks. **A.** With sensory pore, as in most Natantia; **B.** without sensory pore, as in some Brachyura and Astacura. *comp. eye*, compound eye; *lam. gang.*, lamina ganglionaris; *m. e. X-o, m. i. X-o, m. t. X-o*, ganglionic X-organs of medullae interna, externa, terminalis; *on. bod.*, "onion bodies"; *opt. ped.*, optic peduncle; *sin. gl.*, sinus gland; *sin. gl. tr.*, sinus gland tracts; *s.n.* sensory nerve; *s. p. X-o.*, sensory pore X-organ (pars distalis); *X-o. conn.*, X-organ connective. Striped area on medulla terminalis X-organs indicates location of neuron perikarya terminating as "onion bodies."

X-organ is separate from the ganglionic X-organ. Hanström was unable to find the sensory pore X-organ in tanaids, isopods, or amphipods; among the decapods it is lacking in *Astacus*, *Sesarma*, *Aratus*, *Potamon*, and possibly *Uca* (Hanström, 1947, 1948; Matsumoto, 1958).

The **sensory pore X-organ** (Fig. 6.11), where separate from the medulla terminalis, is a compound structure enclosed in a connective tissue sheath and lying in a blood sinus near the sensory pore. Three components can be recognized: (a) typical bipolar sensory cells associated with the sensory pore; (b) round or elongate cells of epithelioid appearance (Carlisle, 1953c, 1959b), derived from neuroblasts of the medulla terminalis (Dahl, 1957)—Matsumoto (1958) considers them neurosecretory despite their apparent lack of axons; (c) characteristic concentrically lamellated concretions referred to as "onion bodies." The "onion bodies" have been interpreted as accumulated secretory products of the sensory pore X-organ, but Carlisle (1953c, 1959b) indicates that they are in reality axon endings of neuro-

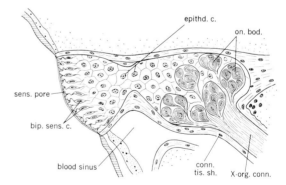

Figure 6.11. Composite diagram of crustacean sensory pore X-organ (as in *Lysmata* or *Pandalus*). [Adapted from Carlisle, 1953c, 1959b.] *bip. sens. c.*, bipolar sensory cell; *conn. tis. sh.*, connective tissue sheath; *epithd. c.*, epithelioid cells; *on. bod.*, "onion bodies" (axonal endings?); *sens. pore*, sensory pore; *X-org. conn.*, X-organ connective.

secretory cells located in the medulla terminalis ganglionic X-organ and in the brain. These axons form the "X-organ connective"; each axon branches to terminate as a series of "onion bodies" (see Knowles and Carlisle, 1956). Signs of cyclic secretory activity have been detected both in the epithelioid cells and in the "onion bodies" (Matsumoto, 1958; Carlisle, 1959b). The bipolar sensory cells show no signs of secretory activity; their axons form a sensory nerve which may run to the medulla terminalis separately or may be joined partially or completely to the X-organ connective.

In the isopods, the situation with respect to the sinus gland and the sensory pore X-organ is somewhat confused. Two structures have been observed in the region of the optic lobes in isopods: the organ of Bellonci (Bellonci, 1881) and the pseudofrontal organ (Gräber, 1933). Amar (1948, 1950, 1953; see also Gabe, 1952b, 1952c, 1954b; Oguro, 1960a) has homologized the pseudofrontal organ to the sinus gland and the organ of Bellonci to "l'organe X," presumably the sensory pore X-organ.

The **ganglionic X-organs** (Fig. 6.10) are composed entirely of the cell bodies of neurosecretory cells. The medulla terminalis ganglionic X-organ is most prominently and consistently present, but ganglionic X-organs may occur also in the medulla interna and medulla externa (Knowles and Carlisle, 1956; Potter, 1958; Carlisle, 1959b), along with solitary neurosecretory cells scattered throughout the optic ganglia. In *Carcinides* and *Callinectes*, where most of the neurosecretory cells are found in the medulla terminalis ganglionic X-organ (Potter, 1958), six types of cells can be distinguished tinctorially; in *Pandalus*, where a medulla interna ganglionic X-organ and a medulla externa ganglionic X-organ also are present (Carlisle, 1959b), only three cell types are seen in the medulla terminalis ganglionic X-organ; two additional types are present in the medulla interna ganglionic X-organ and in the medulla externa ganglionic X-organ. The axons from the medulla terminalis ganglionic X-organ contribute prominently to the sinus gland through the X-organ–sinus gland tract; the medulla terminalis ganglionic X-organ also contributes axons to the sensory pore X-organ (where present) by the X-organ connective. The medulla interna ganglionic X-organ and medulla externa ganglionic X-organ (where present) contribute only to the sinus gland.

4. Physiology of the eyestalk neurosecretory system

The crustacean eyestalk is associated with a variety of neuroendocrine control mechanisms. **Control of somatic pigmentation** is ascribed to a series of chromatophorotropins extractable from the sinus glands and the postcommissure organs (see below). These factors presumably originate in the medulla terminalis ganglionic X-organ and in various regions of the central nervous system (Matsumoto, 1954b; Knowles, 1955; Pasteur, 1958). There is no agreement yet as to the number of factors involved, and there is considerable evidence for some species-specific patterns (Brown, 1952; Knowles, 1956; Fingerman and Aoto, 1958b; Fingerman, Sandeen, and Lowe, 1959). Furthermore, the extraction of effective materials from neurosecretory areas does not mean that these materials are hormones normally released in the organism. An attractive working hypothesis is based on some evidence that the neurosecretory nuclei contain high molecular weight compounds (proteins?) with multiple effects, whereas the neurohemal areas (sinus gland, postcommissure organ) release smaller peptides with more restricted activities (compare Knowles et al., 1955; Knowles, 1956).

Regulation of the distal retinal pigment may involve two factors, one light-adapting and one dark-adapting (see Kleinholz, 1958; Kleinholz et al., 1962; Fingerman et al., 1959a, 1959b). Light-adapting hormone, according to Kleinholz (1958), is not identical with the erythrophore-concentrating hormone (a somatic chromatophorotropin; see Edman et al., 1958). The two factors are extractable from the eyestalk and the central nervous system; light-adapting hormone and dark-adapting hormone are both found in the sinus gland in Macrura, but the brachyuran sinus gland is poor in light-adapting hormone (dark-adapting hormone content has not been measured). Kleinholz (1936) early pointed out that hormonal treatment does not duplicate exactly the effect of light or dark in most species.

Molting of crustaceans is also under hormonal control. In most forms eyestalk removal results in accelerated molting, and extracts from the medulla terminalis ganglionic X-organ–sinus gland complex inhibit molting. These facts and others led to the conclusion that the medulla terminalis ganglionic X-organ neurosecretory cells produce a molt-inhibiting hormone that is stored in and released from the sinus gland (Passano, 1953). In some forms, however, eyestalk removal results in molt inhibition, and extracts may result in molt acceleration. Even in different populations of the same species, the results of eyestalk removal may differ (Carlisle, 1959a). A second factor (molt-accelerating hormone) originating in the brain and medulla terminalis and stored in the sensory pore X-organ has been postulated (Carlisle and Dohrn, 1953; Carlisle, 1953a). Variations in the relative amounts of secretion of molt-accelerating hormone from the brain and eyestalk and of molt-inhibiting hormone from the eyestalk only may explain the conflicting reports from different laboratories on the effects of eyestalk ablation. A further possible explanation is related to the precocious sexual maturation occurring in eyestalkless animals due to the lack of ovary-inhibiting hormone normally released from the sinus gland (see below). The inhibition of molt occurring in ovigerous animals has been used to explain the apparent inhibition of molt occurring in some forms following eyestalk removal (Drach, 1955; Vernet-Cornubert, 1960).

With the discovery (Gabe, 1953c) of a non-neural endocrine gland, the **Y-organ** (see also Echalier, 1959), a direct effect of these neurosecretory products (molt-inhibiting hormone and molt-accelerating hormone) on molting now appears unlikely. The Y-organ is a good candidate for the crustacean analog (and possibly homologue) of the insect ecdysial gland. The Y-organ is located in either the antennary or maxillary segment, depending upon the location of the excretory organ. The product of the Y-organ appears to have somewhat the same effect as insect ecdyson, and pure ecdyson will induce molting in crustaceans (Marcos-Gallego and Stamm, 1959; Karlson and Skinner, 1960). The effect of molt-inhibiting hormone and molt-accelerating hormone upon the Y-organ may parallel the action of the neurosecretory ecdysio-tropic hormone (the "brain hormone") of insects.

The eyestalk (medulla terminalis ganglionic X-organ–sinus gland) is the source of **ovary-inhibiting hormone** in all crustaceans examined (Carlisle, 1953b) and of a **testis-inhibiting hormone** in crabs only (Demeusy, 1953, 1960; Carlisle, 1954b). The Y-organ may be involved in the chain of events regulating gonad development (Arvy et al., 1956), but another newly discovered non-neural endocrine gland, the **androgenic gland** of Charniaux-Cotton, also is involved importantly (Charniaux-Cotton, 1956, 1958; Carlisle, 1959c). The androgenic gland is located near the genital pore in most male malacostracans, but is embedded partially in the testis itself in isopods (Balesdent-Marquet, 1960; Legrand and Juchault, 1960a, 1960b; Katakura, 1960). This organ determines testicular development, along with the formation of male secondary sex characters; female secondary sex characters are under ovarian control. The gonadotropic (inhibitory) principle may act upon the testis in male crabs through the androgenic gland (Charniaux-Cotton, 1960).

Metabolic effects also may result from eyestalk removal and administration of eyestalk extracts (Scheer, 1957). Removal of the eyestalks results in increased O_2 consumption (possibly owing to loss of molt-inhibiting hormone), in derangements of Ca^{++} (Travis, 1951a, 1951b) and water metabolism (Heller and Smith, 1948; Carlisle, 1956a) related to molting, and in alterations of the blood sugar level. Extracts from the eyestalk, the postcommissure organ, and the central nervous system reveal the presence of various **myotropic factors** affecting the contraction of visceral musculature (Gersch, 1959).

5. Cerebral neurosecretory areas

By comparison with the eyestalks, relatively little is known about neurosecretion in the brain. Among the **decapods**, three cell types have been recognized in the **supraesophageal ganglion of Brachyura** (Enami, 1951; Bliss and Welsh, 1952; Matsumoto, 1958; Konok, 1960); only two have been observed in the crayfish (Durand, 1956). Bliss et al. (1954) plotted five general areas of the brain in *Gecarcinus* and *Orconectes* (Fig. 6.7,

B_1 to B_5), which contain neurosecretory cells. The exact outlines of these "maps" vary from species to species owing to fusion or splitting of the various areas.

The only malacostracans other than the decapods in which the neurosecretory cells of the brain have been studied are the **isopods** (Fig. 6.12) (Amar, 1953; Miyawaki, 1958; Matsumoto, 1959; Oguro, 1960a, 1960b, 1961). Up to four cell types have been recognized in the brain and optic lobes, and they have been equated to four of the cell types observed in crabs by Matsumoto.

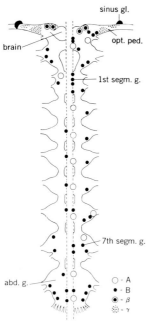

Figure 6.12. Diagram of distribution of neurosecretory cells in the central nervous system of *Armadillidium* (Isopoda). Dorsal view on left; ventral view on right. [Modified from Matsumoto, 1959.] *abd. g.*, abdominal ganglion; *opt. ped.*, optic peduncle; *sinus gl.*, sinus gland; *segm. g.*, segmental ganglia.

The **sites of release** of the secretions of the neurosecretory cells of the brain have been partially elucidated. Axons bear secretory products from the brain to the sinus gland and, to a small extent, to the sensory pore X-organ (Bliss and Welsh, 1952; Carlisle, 1953c, 1959b; Bliss et al., 1954; Potter, 1958); transection of the optic peduncles has resulted in a proximal accumulation of both neurosecretory material and chromatophorotropic activity at the site of injury (Fingerman and Aoto, 1958a; Matsumoto, 1958). Knowles (Carlisle and Knowles, 1959) finds that the neurosecretion-bearing fibers that supply the postcommissure organs (see below) in *Palaemon*, *Penaeus*, and *Squilla*, originate in the brain.

The **circumesophageal connective ganglia** in crabs contain two types of neurosecretory cells (Enami, 1951; Parameswaran, 1956; Inoue, 1957; Matsumoto, 1958); there also has been a suggestion (Knowles, 1953) that some of the cells in the tritocerebral commissure of natantians may be neurosecretory. The cells in crabs probably contribute axons to the tracts leading to the brain and optic ganglia.

6. Neurosecretory areas in the ventral ganglia

The ganglia of the thorax and abdomen have received even less attention than has the brain, except in the Brachyura. Neurosecretory cells occur in the thoracic ganglia of *Penaeus* and *Lysmata* (Knowles and Carlisle, 1956) and *Crangon* (Clausen and Palz, 1959) and in the ventral ganglia of *Cambarellus* (Fingerman and Aoto, 1959) and *Astacus* (Clausen and Palz, 1959). The function of these cells is unknown. Up to five types of neurosecretory cells have been described in the thoracic ganglion of crabs (Enami, 1951; Bliss and Welsh, 1952; Bliss et al., 1954; Matsumoto, 1954a; Parameswaran, 1956; Inoue, 1957). Matsumoto (1958) reports four **routes of discharge** of secretion in the thoracic ganglion: (a) peripheral discharge from the cell body into tissue fluid or capillaries; (b) axonal transport through the circumesophageal connectives to the brain and eyestalks; (c) axonal transport by axon collaterals to tissue spaces in the thoracic ganglion; (d) axonal transport out the pedal nerves. Type C neurosecretory cells (Matsumoto, 1954a) send their axons through the cardiac inhibitory nerves to contribute importantly to the pericardial organs and anterior ramifications (see below); about four type B cells send their axons out each segmental nerve to the pericardial organs (Maynard, 1961b). Matsumoto (1954b, 1958) attempted to correlate the several cell types with various physiologic activities; Nayar and Parameswaran (1955) and Miyawaki (1956a, 1956b) studied their cytochemistry. One of the three cell types may be involved in the control of somatic pigmentation (Matsumoto, 1954b); implants of thoracic ganglion induce precocious ovarian development in *Potamon* (Ōtsu, 1960).

7. The postcommissure organs

The discovery that chromatophorotropins could be obtained from portions of the central nervous system other than the eyestalks (see Brown, 1933; Knowles, 1939) and the pinpointing of the tritocerebral commissure as a highly potent source of these hormones (Brown and Ederstrom, 1940; Brown, 1946) led to the description of the postcommissure organs as **neurohemal structures** by Knowles (1951, 1953). These organs have been observed in hoplocarids and in several Natantia and *Pachygrapsus* among the decapods (see Matsumoto, 1958; Carlisle and Knowles, 1959). In Natantia they are paired swellings or plaques, interconnected in some species (as in *Palaemon*), lying along the course of the postcommissure nerves which take their origin from the tritocerebral commissure. Microscopically each plaque consists of connective tissue of the epineurium surrounding ramifications of nerve fibers entering from the postcommissure nerves. Four or five of these nerve fibers pass down each connective from the brain, enter the commissure on each side, and bifurcate, sending one branch to each of the postcommissure nerves (Fig. 6.13). In the distal portion of the nerve, each fiber splits into numerous tiny branches which remain enclosed in a common sheath until they reach the postcommissure organ proper. There they spread out and ultimately terminate in bulbous endings in the epineurium of the organ (Knowles, 1958).

The **nerve cell bodies** associated with the postcommissure fibers are unknown, though Carlisle and Knowles (1959) believe they are probably in the tritocerebrum. Other cells also may contribute to the postcommissure organs. Certain cells in the tritocerebral commissure show signs of secretion, but their axons have not been traced. Two large cells in each connective ganglion also contribute axons to the postcommissure nerves. Their cell bodies show signs of secretion, but

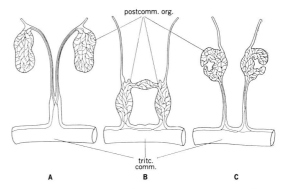

Figure 6.13. Diagrams of postcommissure organs *(postcomm. org.)* of three crustacean species. **A.** *Penaeus.* **B.** *Palaemon (Leander).* **C.** *Squilla.* tritc. comm., tritocerebral commissure. [Carlisle and Knowles, 1959.]

their axons pass through the postcommissure organs, apparently without contributing to them, and innervate muscles immediately behind the organs. In *Pachygrapsus* (Brachyura) a pair of postcommissure nerves arises from the tritocerebral commissure. Each passes dorsally and ultimately divides into two branches. One of these terminates in a network of fibers on the ligamentum ventrale capitis. This network is homologous to the plaques seen in *Palaemon* (Maynard, 1961a). The second branch innervates a muscle possibly homologous to the molting muscle of *Palaemon*.

The postcommissure organs are important in **hormonal control of color change,** being the major source of such hormones in *Squilla*, while sharing equal honors with the sinus gland in the natantians (Knowles, 1954; Carlisle and Knowles, 1956). The physiologic properties of the brachyuran organ have not yet been examined. Hara (1952a, 1952b) has extracted a factor from the commissure in shrimp which depresses heart rate.

8. The pericardial organs and anterior ramifications

The first extensive work on the pericardial organs was done by Alexandrowicz (1952, 1953a, 1953b, 1953c, 1954, 1955; see Alexandrowicz and Carlisle, 1953, for earlier references, and Miyawaki, 1955b, Matsumoto, 1958, and Maynard, 1961b, for recent studies).

The **pericardial organs** consist of the fine branches of nerve fibers surrounded by loose con-

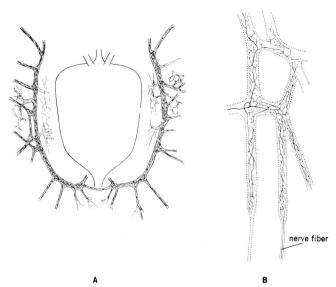

Figure 6.14.

Pericardial organs of decapod crustaceans. **A.** Flanking the heart of *Eupagurus*. **B.** Anterior portion of right organ of *Cancer* to show ramification of a single fiber entering the anterior bar (finer branching not shown). The anterior bar spans the two anterior openings of the branchiocardiac veins and is connected by longitudinal trunks to the posterior bar lying across the third (posterior) opening of the branchiocardiac veins. The labeled nerve fiber runs toward the posterior bar. [Redrawn from Alexandrowicz, 1953a. Reproduced with the permission of the Council of the Marine Biological Association of the United Kingdom.]

nective tissue, in the form either of trunks hanging freely in the pericardial cavity or of plexuses in the membranes lining the pericardial cavity (Fig. 6.14). Both types may coexist in the same animal. In either case they are so arranged as to be bathed in the blood traveling from the gills to the heart. In *Squilla* the fibers derive from two sources: cells lying in the dorsal nerve trunks associated with the pericardial organs and cells (as yet unlocated) lying in the anterior central nervous system, presumably in the subesophageal ganglion. In decapods the peripherally located cells number about 25 and lie in the pericardial organs themselves; a large number of fibers also come from the thoracic ganglion (Maynard, 1961b; see Matsumoto, 1958, for dissent). In addition to the findings in *Squilla* and in a wide variety of decapods, pericardial organs are present in amphipods and possibly in isopods and mysids (Alexandrowicz).

The **anterior ramifications** are formed by a branch of the cardiac inhibitory nerve. They are composed mainly of the terminations of type C neurosecretory cells from the thoracic ganglion, and lie in the region of the ventral respiratory muscles (Maynard, 1961b).

The pericardial organs are neurohemal structures which store and release materials affecting the **frequency and amplitude of the heartbeat** (Alexandrowicz and Carlisle, 1953). The active material may be an orthodihydroxyindole alkylamine, possibly 5,6-dihydroxytryptamine (Carlisle, 1956b; Carlisle and Knowles, 1959). Maynard and Welsh (1959) have isolated 5-hydroxytryptamine from the brachyuran pericardial organ; however, they believe that the major activity of pericardial organ extracts is conferred by another, as yet unidentified, substance which is probably a polypeptide.

While the pericardial organs are effectively situated to help regulate cardiac output, the anterior ramifications are so placed as to discharge their secretions into the venous blood reaching the respiratory muscles and gills. It is thus possible that the secretions of the anterior ramifications affect either the rate of pumping of the respiratory current or the peripheral resistance of the branchial blood vessels. That these two structures (pericardial organs and anterior ramifications) work in close coordination is suggested by the possibility that they may receive axon collaterals from the same secretory neurons. The result may be regulation of the rate of gas transport through the body and its exchange in the gills (Maynard, 1961b).

M. Insecta

1. General

Neurosecretory phenomena in insects are evident in cerebral, subesophageal, and other ganglia. The principal neurohemal organ is the

III. NEUROSECRETORY SYSTEMS M. Insecta

Figure 6.15.
Diagram to show interrelations of brain, stomatogastric nervous system, and retrocerebral endocrine organs in a generalized insect. [Redrawn from Weber, 1952.] *c. a.*, corpus allatum; *c. c.*, corpus cardiacum; *deutc.*, deutocerebrum; *ecd. gl.₁*, ecdysial gland (prothoracic); *ecd. gl.₂*, ecdysial gland (ventral—present in hemimetabolous larvae and apterygotes); *fr. g.*, frontal ganglion; *hypoc. g.*, hypocerebral ganglion; *msth. g.*, mesothoracic ganglion; *n. conn.*, nervus connectivus; *n. c. a.*, nervus corporis allati; *n. c. c.*, nervi corporis cardiaci I and II; *n. rec.*, nervus recurrens, paired at the level of n. rec.'; *protc.*, protocerebrum; *proth. g.*, prothoracic ganglion; *subes. g.*, subesophageal ganglion; *tritc.*, tritocerebrum; *ventr. g.*, ventricular ganglion.

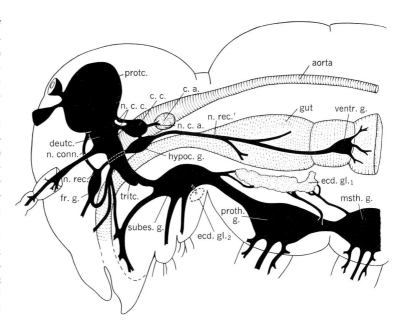

corpus cardiacum; this organ and the nonneural corpus allatum (and occasionally the nonneural ecdysial gland) form a retrocerebral endocrine complex (Fig. 6.15). These glandular structures are intimately involved anatomically with the aorta and the hypocerebral ganglion. Reviews of insect neurosecretory systems have been published recently by Raabe (1959), Van der Kloot (1960), and Scheer (1960); detailed considerations of the structure of various components of the retrocerebral complex can be found in Nabert (1913), Hanström (1940, 1941), Cazal (1948), Williams (1949a), de Lerma (1950), and Pflugfelder (1952).

2. Cerebral neurosecretory systems

Neurosecretory cells were first described in insects in the brain of *Apis* (Weyer, 1935). Since that time it has emerged that the major neurosecretory system in insects is one involving primarily two organs, the **protocerebrum and the corpus cardiacum,** and their connections. The principal neurosecretory nuclei of the protocerebrum in most insects are paired groups of cells, close to the midline in adults and sometimes fused, forming the bulk of the pars intercerebralis (Fig. 6.16); according to Formigoni (1956) 60–65 neurosecretory cells are present on each side in the third larval stage of the *Apis* worker. From these cells, fibers proceed first anteriorly for a short distance, then ventrally to decussate, and finally posteriorly to issue forth as distinct nerves of variable length according to species, to the corpus cardiacum (the internal nervi corporis cardiaci or n.c.c.I; see Fig. 6.17).

In many insects a **second pair of neurosecretory nuclei** is found laterally in the protocerebrum near the corpora pedunculata. From these small

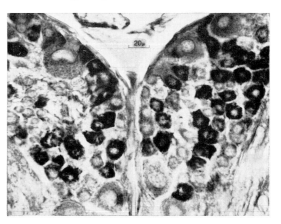

Figure 6.16. *Periplaneta* (Blattodea), adult. Pars intercerebralis of protocerebrum to show large number of fuchsinophilic neurosecretory cells and occasional large acidophilic cell. [From Gorbman and Bern, *A Textbook of Comparative Endocrinology*, John Wiley and Sons, New York, 1962.]

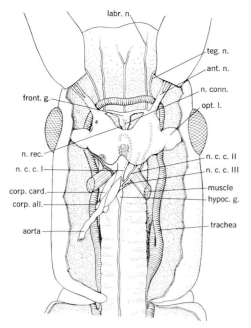

Figure 6.17. *Carausius morosus* (Phasmidae). Dissection of the head from the dorsal side, to show the relative positions of nervous and endocrine structures. The *dotted areas* in the cerebral ganglion indicate neurosecretory regions of the pars intercerebralis, and of the lateral protocerebral group and the tritocerebrum on the left side. [Modified from Raabe, 1959.] *ant. n.,* antennal nerve; *corp. all.,* corpus allatum; *corp. card.,* corpus cardiacum; *front. g.,* frontal ganglion; *hypoc. g.,* hypocerebral ganglion; *labr. n.,* labral nerve; *n. c. c. I, II, III,* nervi corporis cardiaci; *n. conn.,* nervus connectivus; *n. rec.,* nervus recurrens; *opt. l.,* optic lobe; *teg. n.,* tegumentary nerve.

and often indistinct cell groups, fibers proceed posteriorly to form a second pair of nerves to the corpora cardiaca, the external nervi corporis cardiaci or n.c.c.II. These latter nerves are evidently absent from some insects—for example *Andrena* (Hymenoptera) (Brandenburg, 1956b) and dipterous larvae (Fraser, 1957); some structures initially reported as nerves have turned out to be muscle strands.

In general, the lateral neurosecretory cells may not become apparent until late larval and adult life, and the nervi corporis cardiaci II may not develop until late larval stages (see Bounhiol, Gabe, and Arvy, 1953, in *Bombyx*; Arvy and Gabe, 1953a, in the Odonata). However, both medial and lateral protocerebral neurosecretory cell groups are visible in all larval stages of the dermapteran *Forficula* (Lhoste, 1957). In the larval stages of *Drosophila*, 8–10 neurosecretory cells can be seen in each half of the cerebral ganglion; in the adult there are 24–32 medial cells and 10–14 lateral cells in the brain ("Gomori-positive and -negative"), a further indication of change in the extent of the neurosecretory centers with metamorphosis (Köpf, 1957, 1958). Two groups of 3–4 lateral cells and 32 medial cells also characterize the brain of the hemipteran *Iphita* (Nayar, 1955a, 1956a), but only one pair of nervi corporis cardiaci is indicated.

The generalized picture of medial and lateral groups of secretory neurons outlined above holds for many adult insects, but the **larval picture may be much more complicated.** Six pairs of cell groups are present in the larval brain of the fly *Lucilia* (Fig. 6.18, A; Fraser, 1959a), and the distribution of neurosecretory cell groups in caterpillars does not conform to the "simple" medial and lateral designations (Rehm, 1955). A third pair of cell groups is located in the posterior part of the larval brain of the lepidopteran *Chilo*

Figure 6.18. *Lucilia* (Diptera) larva. **A.** Diagram of the distribution of neurosecretory cell groups in the brain, indicated by Roman numerals. **B.** Distribution of five pairs of neurosecretory cells in the first abdominal ganglion, drawn from a transverse section. [Redrawn from Fraser, 1959a, 1959b.]

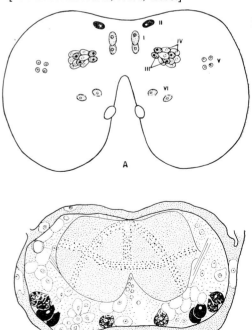

(Mitsuhashi and Fukaya, 1960). In some Diptera the nervi corporis cardiaci II take their origin from cells intermingled with those of the medial group (Cazal, 1948). Paired tritocerebral neurosecretory cell groups occur in *Carausius* (Raabe, 1959) and in the pentatomid *Chrysocoris* (Ganguly and Deb, 1960), and a third group has been described in Neuroptera "planipennes" (Arvy, 1956). A third pair of nervi corporis cardiaci occurs in *Carausius* (Fig. 6.17; Dupont-Raabe, 1957), in *Oncopeltus* (Johansson, 1958), in the thysanuran *Thermobia* (Chaudonneret, 1950), and possibly in other forms as well.

In the Apterygota the **lateral frontal organs** appear to be homologous to the neurosecretory cell groups that are incorporated into the brain of pterygote insects (Gabe, 1953e); nerves lead from these organs to the corpus cardiacum. In the thysanuran lepismatids, the median frontal organ also supposedly contains secretory cells (de Lerma, 1947; Gabe, 1953e), which may in fact be photoreceptors (Pipa, Bern, and Nishioka, unpublished).

3. Other neurosecretory groups

Neurosecretory cells occur also in the **subesophageal ganglion** (Day, 1940; B. Scharrer, 1941a), and these cells may bear an important relation to the humoral functions of this ganglion in the production of a factor responsible for egg-diapause in some lepidopterans (Fukuda, 1952; Hasegawa, 1957) and for embryo-diapause in *Locustana* (Jones, 1956a, 1956b). Secretion of the egg-diapause factor in the silkworm may be dependent on nervous control of the ganglion by the pupal brain through the circumesophageal connectives (Fukuda, 1952), although this is not accepted by Hasegawa (1957). The complexity of the interactions is illustrated well by Morohoshi's study (1959) implicating the corpus allatum as the source of a factor favoring development of nondiapause eggs. The subesophageal ganglion is an inhibitory center for motor activity in *Periplaneta* (Milburn et al., 1960), and a humoral role in the regulation of cyclic activity is indicated (Harker, 1956, 1960a, 1960b). The ganglion also has been described as a possible neurohemal area (Raabe, 1959) for the release of tritocerebral neurosecretory material regulating pigment changes in *Carausius*. In some forms (paleopterans, Coleoptera), the ecdysial (ventral) gland innervation from the subesophageal ganglion supposedly originates from the neurosecretory cells (Arvy and Gabe, 1953a, 1953b).

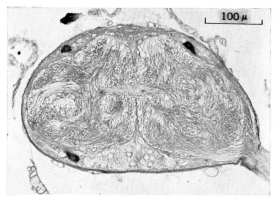

Figure 6.19. Prothoracic ganglion of *Oncopeltus* (Heteroptera). Transverse section to show three of the four paraldehyde fuchsin-staining neurosecretory cells. [Johansson, 1958.]

Other ganglia of the insect nervous system: ventral (Fig. 6.19) and stomatogastric, also show neurosecretory cells (see Bounhiol, Gabe, and Arvy, 1953; Johansson, 1958; Füller, 1960). Three types of neurosecretory cells are identifiable in the thoracic and abdominal ganglia of *Blaberus* (Geldiay, 1959).

Within a single species, the location and number of neurosecretory cells are **relatively constant.** Thus, in an abdominal ganglion of a *Lucilia* larva there are five pairs of stainable cells (Fig. 6.18, B) and there is a total of 40–48 neurosecretory cells in the brain (Fig. 6.18, A) (Fraser, 1959a, 1959b). The prothoracic ganglion of *Oncopeltus* has four large neurosecretory cells (Johansson, 1958; see Fig. 6.19). In the subesophageal ganglion of *Periplaneta* there are consistently four neurosecretory A cells; aberrations ("asymmetry") are infrequent (B. Scharrer, 1941a, 1955a). Their very constancy may bespeak an important role for such cells, even though for many groups of secretory neurons in the insects, as in other groups, the functional significance is completely undetermined as yet.

4. The corpus cardiacum

Extensive considerations of the comparative anatomy of the corpus cardiacum, its nerve

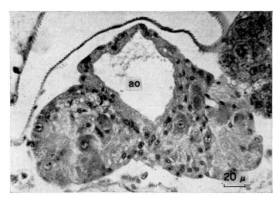

Figure 6.20. Corpora cardiaca of *Oncopeltus* (Heteroptera) Transverse section to show absence of stainable neurosecretion and cellular nature of organ in this species. *ao*, aorta. [Johansson, 1958.]

supply, and more recently, its connections with neurosecretory centers in the brain are available in Hanström (1937b, 1938, 1940, 1941, 1947), Cazal (1948), Weber (1952), Scharrer and Scharrer (1954a, 1954b), Gabe (1953f, 1954b), and Raabe (1959).

In some apterygotes, part or all of the **protocerebral nerve endings** occur in the aortic wall (Gabe, 1953e), even where the corpus cardiacum is well developed (as in lepismatids). A similar situation has been described in the belostomid hemipteran *Hydrocyrius*, where all the neurosecretory material passes into the "cardioglial" tissue of the aortic rudiment, which connects all the retrocerebral elements including the ecdysial gland (Junqua, 1956). Neurosecretory material also passes directly into the aorta in *Iphita* (Nayar, 1956a) and in *Oncopeltus* (Johansson, 1958). The corpora cardiaca-allata are devoid of neurosecretory material in *Oncopeltus* (Fig. 6.20).

The **corpus cardiacum** is the principal terminus (neurohemal organ) for the neurosecretory product of the brain. In most insects neurosecretory material accumulates in this structure, and appears either to be converted into a product that passes into the blood (Wigglesworth, 1954a) or to be passed unchanged as needed into the blood. The corpus cardiacum is a complex structure histologically, to whose organization the adjective "chaotic" has been applied more than once. In addition to its generally accepted role as a depot organ for neurosecretory material (Fig. 6.21), two other functions must be considered.

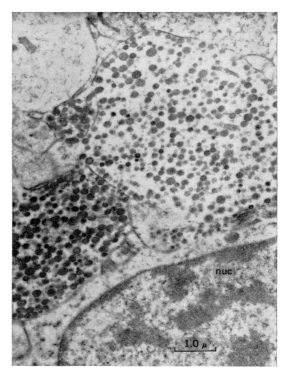

Figure 6.21. Electron micrograph of corpus cardiacum of *Periplaneta*. Note axons containing variously sized elementary neurosecretory granules. Processes at top probably belong to intrinsic cells of the corpus cardiacum (compare B. Scharrer on *Leucophaea*). *nuc*, nucleus of intrinsic cell. [Micrograph by R. S. Nishioka.]

In the first place, there are generally present recognizable epithelioid elements, called **intrinsic cells,** that contain secretory material and appear to undergo a secretory cycle; in the second place, small numbers of large cells, recognizable as **true neurons,** occur occasionally (Joly, 1945a, in *Dytiscus*; Boisson, 1949, in *Bacillus*; Clements, 1956, in *Culex pipiens*). The corpus cardiacum is described as an ordinary ganglion in *Ctenocephalus* (Wenk, 1953). Hanström emphasizes the ganglionic nature of the corpus cardiacum, based on its ontogeny, although in some forms (for example, the Hymenoptera, according to M. Thomsen, 1954a, 1954b) nerve cells are completely absent. Another cell type is also present in the form of small nondescript cells, often syncytial in arrangement, and suggestive of glia. An additional conceivable role of the corpus cardiacum lies in the fact that a nerve from nervi corporis cardiaci I and II, generally including some fibers bearing neurosecretory material, passes along

Figure 6.22.

Celerio (Lepidoptera) adult. Electron micrograph of fibers of the nervus corporis allati, showing neurosecretory granules within the axons. The granules are often unevenly distributed and sometimes seem to be separated by a membrane from the rest of the axon *(arrows). nuc*, nucleus of sheath cell; *tr*, tracheole. [Schultz, 1960.]

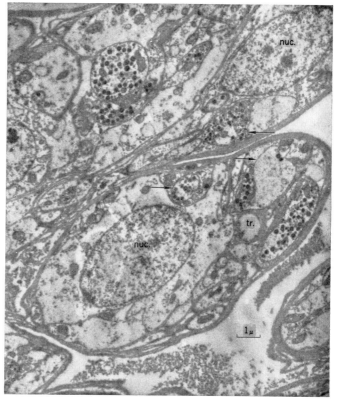

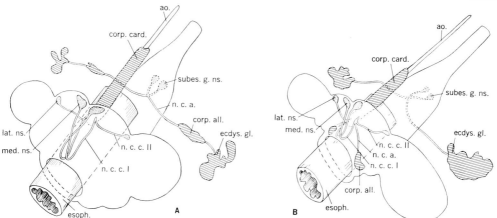

Figure 6.23. The relation of the cephalic endocrine glands *(hatched)* to neurosecretory areas *(stippled)* in **(A)** Ephemeroptera and **(B)** Odonata. [Redrawn from Arvy and Gabe, 1953a.] *ao.*, aorta; *corp. all.*, corpus allatum; *corp. card.*, corpus cardiacum; *ecdys. gl.*, ecdysial (ventral) gland; *esoph.*, esophagus; *lat. ns.*, lateral neurosecretory cells; *med. ns.*, medial neurosecretory cells of the pars intercerebralis; *n. c. a.*, nervus corporis allati; *n. c. c.* I and II, nervi corporis cardiaci; *subes. g. ns.*, neurosecretory cells of the subesophageal ganglion.

the corpus cardiacum and partly through it to the corpus allatum (Arvy and Gabe, 1953a, 1954; Schultz, 1960); see Fig. 6.22. The fate of the fibers from the cardiacal neurons is not known, but they may pass to the corpus allatum also. In the Ephemeroptera the corpora allata receive their innervation from the subesophageal ganglion instead of the nervi corporis cardiaci (Cazal, 1948; Arvy and Gabe, 1953a); see Fig. 6.23.

The relation of the brain to the corpus cardiacum may involve more than the mere utilization of the latter for the storage and discharge of the brain's secretory product. The material borne by the axons also may serve as the **precursor for the synthesis** of specific materials by the corpus cardiacum. Ultrastructural studies (Meyer and Pflugfelder, 1958; Willey and Chapman, 1960) reveal virtual envelopment of neurosecretory axons by the intrinsic cells. Furthermore, the correlation between neurosecretory activity in the brain and secretory activity in the corpus cardiacum suggests the receipt of informational stimuli from the brain; **nonneurosecretory fibers** join the intracerebral tracts of the nervi corporis cardiaci I shortly after their decussation in *Leucophaea* (Engelmann, 1957; see also Strumm-Zollinger, 1957, on *Platysamia*). These last fibers innervate the corpus allatum and inhibit its gonadotropic function in *Leucophaea*.

The problem of the **origin of the secretory product** of the corpus cardiacum is complicated by the tinctorial properties of neurosecretory material (Pipa, 1961b). Not all the pars intercerebralis secretion is "Gomori-positive"; acidophilic material is seen in some of these neurosecretory cells (sometimes in the same cell where "Gomori-positive" material is seen also), as well as in the lateral neurosecretory cells. The secretion of the corpus cardiacum proper is also acidophilic, and in this fact lies much of the confusion as to which cells are producing what. The electron microscope reveals both osmiophil and osmiophobe secretory cells, the latter surrounding the neurosecretory axon terminals (Meyer and Pflugfelder, 1958).

At the present time it is accepted generally (see especially de Lerma, 1954; Highnam, 1961) that the corpus cardiacum has a **secretory function** and secretory cells of its own (Fig. 6.23). Occasionally, the cellular nature of the corpus cardiacum is so evident as to make it difficult to distinguish its cells from those of the corpus allatum fused to it (Risler, 1957, on *Thrips*). In a sense, the epithelioid cells of the corpus cardiacum may be analogized with adrenal medulla cells. The claim that the orthodiphenol heart stimulant (see Davey, 1961) from the corpus cardiacum does not originate in the brain and does not require intact nervi corporis cardiaci (Cameron, 1953) may point to an even greater functional autonomy than that possessed by vertebrate chromaffin tissue, which is dependent upon preganglionic sympathetic innervation.

In **summary,** the status of the corpus cardiacum is complicated by several features: (a) it has a variable anatomy in different insect groups and enjoys a variable degree of intimacy with the aorta; (b) it is neural in origin but includes only occasional neurons in its final makeup; (c) it is innervated by one, two, or three nervi corporis cardiaci from the brain which apparently bring to it important quantities of neurosecretory material, both "Gomori-positive and -negative"; (d) it may have its own cyclic secretory function, in part independent and in part dependent on the brain; (e) it may change in format and histology during development (see Highnam, 1958), and the larval corpus cardiacum does not always have the same histology or structural connections as the adult.

5. *The retrocerebral complex*

The retrocerebral glands of the insect include not only the corpus cardiacum, but also the corpus allatum, and frequently the ecdysial gland (see below). As either single or paired structures, the corpora cardiaca and allata appear to be present in all insect groups, including the Apterygota (with the possible exception of *Campodea*), and are firmly established as parts of the endocrine system. In the past, these structures have been considered largely in relation to the stomatogastric nervous system, including the frontal and hypocerebral ganglia (sympathetic) and the recurrent nerve. However, we have already defined the corpus cardiacum primarily as an extension and terminus of protocerebral tracts. Nervous connection with the hypocerebral ganglion does exist (nervus cardiostomatogastricus of Weber, 1952) and a nervus aorticus goes to the aorta, but these are apparently of secondary importance. Branches also leave the nervi corporis cardiaci and the corpus cardiacum itself to anastomose with cephalic and thoracic nerves. There appears to be a constant nerve supply to the corpus allatum from the protocerebrum (generally through the corpus cardiacum) or from the subesophageal ganglion (paleopterans, Dictyoptera); see Fig. 6.23.

III. NEUROSECRETORY SYSTEMS M. Insecta

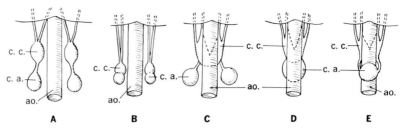

Figure 6.24. Major types of relations of retrocerebral organs, from the dorsal view, according to Cazal (1948). **A.** Simple lateralized; **B.** distal lateralized; **C.** semicentralized; **D.** centralized; **E.** annular. Note the two pairs of nervi corporis cardiaci originating in the protocerebrum and the variable prominence of the nervi corporis allati, depending on the approximation of the corpus cardiacum and the corpus allatum. *ao.*, aorta; *c. a.*, corpus allatum; *c. c.*, corpus cardiacum.

Several **types of organization** of the retrocerebral complex occur (Cazal, 1948; Fig. 6.24), which can be summarized as follows (Weber, 1952): (a) simple bilaterally symmetric with a pair of corpora cardiaca and a pair of corpora allata (Trichoptera, Lepidoptera, Aphaniptera, Homoptera, psyllinids, the coccid *Pulvinaria*); (b) distal bilaterally symmetric with a fused corpus cardiacum–allatum on each side at some distance from the brain (some Diptera and some Coleoptera); (c) semicentralized with the corpora cardiaca partly fused ventral to the aorta and paired corpora allata (most apterygotes, paleopterans, Mallophaga, Orthoptera, Blattodea, Mecoptera, Hymenoptera, some Diptera, some Coleoptera); (d) centralized with the fused corpora cardiaca and allata ventral to the aorta (Plecoptera, Dermaptera, Psocoptera, many Homoptera and Heteroptera including *Rhodnius*, brachyceran Diptera); (e) fused ring type (see Vogt, 1941, 1942b; Possompès, 1954) with a single corpus cardiacum fused to the hypocerebral ganglion ventral to the aorta and a single corpus allatum dorsal to the aorta; the circumaortic (Weismann's) ring is completed by a pair of nervi corporis allati leading from the corpus cardiacum to the corpus allatum and imbedded in the larval ecdysial gland (higher Diptera; see Fig. 6.25). The ring may be formed only by the ecdysial gland in some calliphorines and is not invariably a three-component structure (Fraser, 1955).

The **variations** in the anatomy of the retrocerebral complex are indicated under the various insect orders (Chapter 21). Within a single insect order, bizarre arrangements may be encountered, of no apparent adaptive (or taxonomic) value.

No ordinal consistency is encountered. For example, the Diptera include some forms with individualized organs and others with fusion into Weismann's ring (Fig. 6.25). The parasitic nycteribiid fly *Basilia nana* shows a large solitary corpus allatum in the thorax, innervated by branches of cerebral nerves, but no corpus cardiacum is in evidence (Nussbaum, 1960). Among the Plecoptera the corpora allata may be paired, or asymmetrically or symmetrically fused to the single corpus cardiacum and hypocerebral ganglion (Arvy and Gabe, 1954).

The **corpus allatum** is a nonneural epithelial structure derived from an ectodermal fold between the mandible and the first maxillary segment. The cells so derived migrate dorsally and posteriorly to their final association with the corpus cardiacum (itself derived, as is the hypocere-

Figure 6.25. Section through the "ring gland" (Weismann's ring) of the young pupa of *Eristalis* (Diptera). [Redrawn from Cazal, 1948.] *corp. all.*, corpus allatum; *corp. card.*, corpus cardiacum; *ecdys. gl.*, ecdysial gland; *n. corp. all.*, nervus corporis allati.

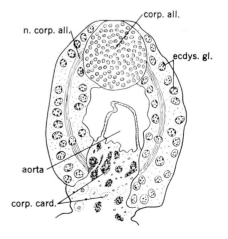

bral ganglion, from the foregut ectoderm of the dorsal esophageal wall).

In addition to the corpus allatum, a **second nonneural epithelial structure,** derived from the ectoderm of the second maxillary segment, becomes located finally in the ventral head region (as in the paleopterans, Plecoptera, and Coleoptera), or in the thoracic region (as in Lepidoptera and Hemiptera) (Fig. 6.15), or retrocerebrally as part of the ring gland in higher Diptera. This structure goes under many names (ventral gland, peritracheal gland, subesophageal body, intersegmental organ, prothoracic gland, thoracic gland, pericardial gland), but it is essentially the "glande de mue" or molting gland. We propose the term **"ecdysial gland"** to emphasize its functional import and to avoid unnecessary synonymy based on a variable anatomic location. This structure is innervated in most forms studied from the subesophageal, prothoracic, or mesothoracic ganglia (Fig. 6.15), or from a combination of these ganglia. In the ring gland of *Calliphora* it is dependent on the pars intercerebralis for maintenance (Possompès, 1958). Four neurons occur in each larval ecdysial gland of the calliphorine fly *Protophormia*, some of whose axons are traceable to the corpus cardiacum (Fraser, 1957).

6. Physiology of the insect neuroendocrine system

The functions of insect "neurohormones" have been reviewed in detail (Raabe, 1959; van der Kloot, 1960; Scheer, 1960). Beginning with Kopeć's pioneer experiments on *Lymantria* in 1917, the importance of the brain for pupation has been generally accepted. (Kopeć's 1912 experiments had led him to conclude that the nervous system had *no* relation to metamorphosis!) This finding was confirmed in Lepidoptera by Caspari and Plagge (1935), Bounhiol (1936), and Kühn and Piepho (1936), and the hormonal nature of the brain influence was established by ablation and replacement technics. Wigglesworth's classical studies on *Rhodnius* (summarized by him in 1954b, 1959), led to an understanding of an interrelation between the growth influence of the brain and that of the corpus allatum in determining normal larval development and molting. The intervention of presumably nonneural factors (Vogt, Bodenstein, Fukuda in the early forties) in the causation of molting complicated the early understanding, until it was established that the brain factor acted through a nonneural structure, the ecdysial gland (Williams, 1947, 1949a), which is recognized now as the source of the directly acting molting hormone (growth and differentiation hormone or ecdyson) (see Wigglesworth, 1939, 1951). The brain hormone thus becomes a **tropic hormone** responsible for the development, maintenance, and functional activity of the ecdysial gland (see also Williams, 1952). Initially, this ecdysiotropic factor was analogized to the tropic hormones of the pars anterior of the vertebrate adenohypophysis. In view of the neurosecretory origin of the ecdysiotropic hormone, a more accurate analogy is to the hypophysiotropic factors produced by hypothalamic nuclei, which themselves regulate adenohypophysial function. The brain and ecdysial hormones are not species-specific (Williams, 1946, 1947).

Some differences exist in regard to the **areas of the brain** required for ecdysial gland activation. In *Rhodnius* and *Calliphora* only the medial neurosecretory cells appear to be required; in *Platysamia* both medial and lateral cell groups are needed (Strumm-Zollinger, 1957). The corpus cardiacum itself generally does not initiate molting in the absence of the neurosecretory cells, an indication of its function as a storage-release organ for ecdysiotropin. On the other hand, removal of the corpus cardiacum can delay molting in some species. In other species (*Platysamia* and *Rhodnius*), intactness of the brain–corpus cardiacum connection is not essential, and the latter organ is dispensable. This indicates direct discharge of ecdysiotropin into the blood, though other experimental data indicate the possibility of a rapid reconstitution of a neurohemal terminus (Strumm-Zollinger, 1957), much as occurs after sinus gland removal from the crustacean eyestalk, or more slowly after neurohypophysectomy or hypophysial stalk transection in vertebrates (Billenstien and Leveque, 1955; Barker-Jørgensen et al., 1956).

In general, larval or nymphal development occurs under the "balanced" direction of ecdyson, the secretory product of the ecdysial gland produced under the tropic influence of the "brain hormone" or ecdysiotropin, and neotenin (the juvenile hormone), the secretory product of the larval or nymphal corpus allatum. Ecdyson favors progressive development in the adult direction including the larval molts and also the pupal molt (the "brain hormone" may act synergistically with ecdyson—Kobayashi and Burdette, 1961); neotenin favors progressive larval development and the maintenance of larval characteristics. The ecdysial gland degenerates after imaginal differentiation is complete, but the corpus allatum may take on a new functional role as a source of gonadotropic hormone in the adult female. The gonadotropin is evidently identical with neotenin (Williams, 1956; Wigglesworth, 1960b).

Neural control over corpus allatum function in the paleopterans presents a nice parallelism between the cytologic indices of activity in neurosecretory cell groups and the presumed activity of organs innervated by these cell groups (Arvy and Gabe, 1953a). The two pairs of nervi corporis cardiaci terminate in the corpus cardiacum in both Ephemeroptera and Odonata (Fig. 6.23). The corpora allata receive their innervation from the nervi corporis cardiaci in the Odonata and from nerves from the subesophageal ganglion in the Ephemeroptera. The subesophageal ganglion also innervates the ecdysial (ventral) gland in both orders. Maximum protocerebral neurosecretory activity is accompanied by corpus cardiacum enlargement in both orders, but by corpus allatum enlargement in the Odonata only. Diminution of subesophageal neurosecretory activity is accompanied by atrophy of both corpus allatum and ecdysial gland in Ephemeroptera, but only of the ecdysial gland in Odonata.

Neural control over corpus allatum function in neopteran insects is complex. When protocerebral innervation of the corpus allatum (through the nervi corporis allati extension of the nervi corporis cardiaci) is destroyed, the corpus allatum enlarges in some forms (*Leucophaea*— Scharrer, 1952a; starved *Oncopeltus*—Johansson, 1958). However, in *Calliphora* destruction of the medial neurosecretory cells results in allatal atrophy (E. Thomsen, 1952). Nervous or hormonal stimulation from the cerebral neurosecretory centers is evidently responsible for corpus allatum–ovary activation in anautogenous mosquitoes (Larsen and Bodenstein, 1959). In *Leucophaea* the corpus allatum appears to receive essential stimulatory innervation from the subesophageal ganglion (for this ganglion's other functions, refer to p. 391) and inhibitory innervation from the protocerebrum, controlling secretion of gonadotropic hormone in the female. At least part of this mechanism is also operative in *Diploptera* and in *Oncopeltus* (Lüscher and Engelmann, 1955; Engelmann, 1957, 1959; Johansson, 1958).

In addition to the "ordinary" innervation of the corpus allatum, it should be remembered that many of the fibers of the nervi corporis allati bring **neurosecretory material to the corpus allatum** (Stutinsky, 1952; Arvy and Gabe, 1953a; Bounhiol, 1957; Schultz, 1960); the significance of this transport is essentially unknown, but Bounhiol states that the neurosecretory material itself is inhibitory to corpus allatum function in *Bombyx*. E. Thomsen and Møller (1959) report ovary retardation in *Calliphora* following severance of the neurosecretion-bearing nervi oesophagei from the corpus cardiacum to the gut, or following removal of the medial neurosecretory cells. In both cases there is a notable decrease in intestinal proteinase.

Raabe (1959) includes a detailed discussion of the relation between the primary neurosecretory apparatus (pars intercerebralis and corpus cardiacum) and **water metabolism** (see Stutinsky, 1953a; Nuñez, 1956; Nayar, 1960). The experimental results are contradictory, and Raabe points out that since different insects have different water problems depending on habitat, it is not surprising that ablation and replacement of pars intercerebralis and/or corpus cardiacum do not always have the same effects. Evidence exists for both diuretic and antidiuretic principles, with the corpus allatum as a further possible source of a water-active principle. How much of the water changes observed are traceable to general metabolic (or pharmacologic) effects of extracts and implants has yet to be determined.

Hormonal control of **adaptive pigmentation changes (physiologic color changes)** is not a constant phenomenon in insects as it is in crustaceans; the well-studied phasmids and the larva of the mosquito *Corethra* represent exceptional, although not unique, cases (see Dupont-Raabe, 1957; L'Hélias, 1957). The tritocerebral neurosecretory origin of the chromatophorotropin regulating physiologic color change in *Carausius* has been established (Dupont-Raabe, 1957). The corpora allata have been suggested as the controllers of green pigment production **(morphologic color change)** in phasmids and acridians (Joly, 1950, 1958; Jones, 1956a, 1956b); the corpus cardiacum, brain, and prothoracic glands have been implicated in other species as regulators of pigment synthesis. Brown pupal color in the lepidopteran *Cerura* is traceable to ecdyson action (Bückmann, 1956) and in *Papilio* and *Pieris* to the presence of the intact supraesophageal–subesophageal–prothoracic ganglionic complex—even the nervous connections must be intact (Hidaka, 1956; Ohtaki, 1960). In other Lepidoptera no dependence of the pupal coloration on cephalic or prothoracic structures can be demonstrated (see Hidaka, 1957).

In addition to the importance of the subesophageal ganglion in influencing activity (Harker, 1956, 1960a, 1960b; Milburn et al., 1960), the corpus cardiacum contains **a material which depresses spontaneous activity** in isolated cockroach nerve cords (Özbas and Hodgson, 1958; Barton-Browne et al., 1961). In the intact animal, Milburn et al. visualize the corpus cardiacum principle as indirectly releasing efferent nerve activity by inhibition of the subesophageal inhibitory center. A variety of **"myotropic" factors** influencing heart rate and smooth muscle contraction (oviduct, gut, and Malpighian tubules) has been isolated from neurosecretory structures, other nervous tissues, and the corpus allatum (see Florey, 1951; Gersch, 1959). Whether such extractable factors play a normal neurohormonal role in insects has yet to be established.

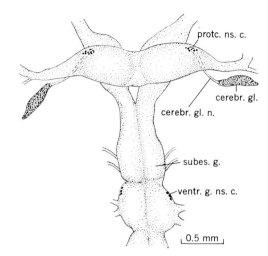

Figure 6.26. Neurosecretory cell groups and cerebral glands of *Lithobius* (Chilopoda). [Redrawn from Palm, 1955.] *cerebr. gl.*, cerebral gland; *cerebr. gl. n.*, cerebral gland nerve; *protc. ns. c.*, protocerebral neurosecretory cells; *subes. g.*, subesophageal ganglion; *ventr. g. ns. c.*, neurosecretory cells of first ventral ganglion.

N. Arthropoda Myriapoda

1. General

The term "Myriapoda," as used here, includes the more or less closely related groups of the Chilopoda, Diplopoda, Symphyla, and Pauropoda. Most of the work on neurosecretion has been done on the chilopods and diplopods; almost nothing is known of the situation in symphylans and pauropods.

The myriapods are found to conform to the general pattern already seen in the insects and crustaceans and also characteristic of arachnids. Neurosecretory cells are present in the central nervous system; those in the protocerebrum send their secretions by their axons to a neurohemal organ lying in the head. As is the case in many other arthropods, this neurohemal organ may possess secretory activity of its own.

Early work led to the description of an organ in the head of *Lithobius*, which was referred to by Holmgren (1916) as the frontal organ. Its glandular nature caused it to be renamed **cerebral gland** (Figs. 6.26, 6.27) (Fahlander, 1938). The cerebral gland is innervated from the protocerebrum; Fahlander traced its nerve to the area of the protocerebral globuli II. De Lerma (1951) tentatively compared the cerebral glands to the

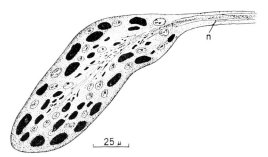

Figure 6.27. Histology of cerebral gland of *Lithobius* (Chilopoda). Note large colloid secretion masses *(black);* branching in center of gland of fibers of nerve from brain, *n.* [Redrawn from Palm, 1955.]

lateral frontal organs of Thysanura. Later workers have described neurosecretory cells in the area mentioned by Fahlander, as well as in other areas of the central nervous system. Neurosecretory material has been traced from the protocerebral cells to the cerebral gland, where it is stored until released. The structure of the cerebral gland is typical of the neurohemal organs observed throughout the arthropods: secretion-laden nerve fibers from the brain are interspersed among glandular cells intrinsic to the organ (Fig. 6.27). The location of the cerebral glands is variable; in some genera they lie immediately under the optic lobes (as in *Thereuopoda*), while in others (for example *Lithobius*) they lie in contact with the hypodermis at some distance from the brain.

2. Chilopoda

Neurosecretion has been found in 18 chilopod genera representing the four orders Geophilomorpha, Scolopendromorpha, Lithobiomorpha, and Scutigeromorpha (Palm, 1955; Gabe, 1956). In all four groups, neurosecretory cells are present in the protocerebrum. Secretory neurons occur in the subesophageal ganglion and possibly in the deutocerebrum and ventral segmental ganglia. Except in *Lithobius*, where they are scattered anterior and medial to the globuli (Fig. 6.26), the protocerebral neurosecretory cells are gathered into clearly defined bilateral islands of cells. The axons lead ventrolaterally to the aboral face of the protocerebrum, where they exit as the cerebral gland nerves. The development of the neurosecretory system in *Lithobius* has been studied by Palm (1955). The cerebral gland is present in the youngest stages after hatching; at this time, the neurosecretory cells of the protocerebrum are indistinguishable from the ordinary neurons of the brain. At about the 13 mm stage (Palm considers 23 mm and above as mature) secretion appears, first in the cerebral glands and then in the neurosecretory cells of the brain. Palm suggests that the secretions of the protocerebral neurosecretory cells probably are not needed for molting and growth, because they are absent from the smallest specimens. Rather, they may be involved in promoting development of adult characteristics. However, Joly (1962) presents evidence that the cerebral glands are involved in molting. The neurosecretory cells of the ventral nerve cord were observed only in adult animals. Scheffel (1961) presents an extensive discussion of neurosecretion in *Lithobius*.

3. Diplopoda

Although B. Scharrer (1937) failed to observe neurosecretion in the brain of *Julus*, newer staining methods have revealed secretory neurons in the brain (medial to globuli II) of five genera, including *Julus* (Gabe, 1959). The secretion-bearing axons terminate in a neurohemal cerebral gland, which also possesses intrinsic secretory activity, as in the chilopods. Tritocerebral and subesophageal neurosecretory cells are also present; neurosecretory fibers of presumed protocerebral origin lead into the ventral nerve cord (Sahli, 1958). Paired neurohemal organs (connective bodies) are found on the circumesophageal connectives near the tritocerebral commissure. These may be storage organs for the tritocerebral neurosecretory cells (Prabhu, 1959). The neurosecretory systems of diplopods have been further clarified by Prabhu (1961) and Sahli (1961, 1962). "Hypocerebral formations" appear to be part of the system.

4. Symphyla

Protocerebral neurosecretory cells and cerebral glands are present in *Scutigerella* (Juberthie-Jupeau, 1961). The paired fusiform organs (Jupeau, 1956) are innervated from the subesophageal ganglion; a homology to the ecdysial (ventral) glands of lower pterygotes has been suggested.

5. Pauropoda

Tiegs's (1947) comparison of the pseudocular glands of *Pauropus* to the cerebral glands of other myriapods is unlikely, owing to the apparent absence of connection with the protocerebrum; no secretory neurons have been reported in the pauropods.

O. Arthropoda Arachnida

1. General

The arachnids appear to have well-developed neurosecretory systems; in the spiders these systems equal in complexity parallel structures in insects and decapod crustaceans. The histology of arachnid neurosecretory areas has been well studied (Brown and Cunningham, 1941; B. Scharrer, 1941c; Legendre, 1953, 1954, 1955, 1959a; Gabe, 1955a; Sanchez, 1954, 1958, 1959; Herlant-Meewis and Naisse, 1957; Junqua, 1957; Kühne, 1959; Naisse, 1959), but little is known of their functional significance.

With the possible exceptions of the Pantopoda (Pycnogonida), the Xiphosura, and the Palpigrada (which have not been studied), the various orders of arachnids display **variations of a single structural theme**: neurosecretory cells are found in the brain, with axons leading to neurohemal organs (usually retrocerebral); nonneural organs of suggested endocrine significance are frequently present in the prosoma.

True neurosecretory cell groups (believed by Legendre, 1959b, to be chromaffin) are located bilaterally in all ganglionic masses, in addition to mesodermal glandular cells of glial nature (Legendre, 1959a). Commonly there are two bilateral groups in the protocerebrum, two in the tritocerebrum (cheliceral ganglion), and one bilateral group in each neuromere of the subesophageal nerve masses. Cyclic secretory activity is seen in the protocerebral cells, primarily associated with molting according to Gabe and with reproduction according to Legendre and Kühne. The subesophageal cells always appear filled with secretion.

The neurosecretory cells show **axonal transport**, but only some cerebral cells can be traced to neurohemal organs. Neurohemal organs have been found in all orders studied except the pantopods and xiphosurans and consist of fiber endings in association with the vascular system. These structures resemble the insect corpus cardiacum more than the crustacean sinus gland, in their possession of possible intrinsic secretory units. **Nonneural glandular structures** (ventral gland, anterior gland, endocrine cells of Millot, molting gland) may correspond to the insect ecdysial gland and crustacean Y-organ (Sanchez, 1954, 1959; Herlant-Meewis and Naisse, 1957; Legendre, 1959a; Naisse, 1959).

2. Araneida

Some 26 genera of spiders have been studied (see Gabe, 1955a; Legendre, 1959a; Kühne, 1959). Neurosecretory cell groups are present throughout the **brain** (Fig. 6.28). Two bilateral pairs of neurosecretory cells occur in the protocerebrum: a pair of oral groups composed of about 8 cells each, situated near the midline of the protocerebrum, and a pair of aboral groups located more laterally, just in front of the central body. These cell groups show evidence of a secretory cycle. Neurosecretory cells also are found in the ganglia of the pedipalps, of the walking legs, and of the abdominal nerve mass. In addition, other cerebral cell groups have been reported (Fig. 6.28).

The araneid **retrocerebral complex** (Fig. 6.29) consists of the primary and secondary organs of Schneider, and in some forms (as in *Tegenaria*) an appendage to the primary organs, termed the "Tropfenkomplex." Histologically Schneider's organs consist of axonal endings, glandular cells (from which secretion leads in some forms to the "Tropfenkomplex"), and infrequent bipolar neurons (absent from the secondary organs in labidognaths). The primary organ consistently receives a principal nerve, which leaves the brain near the tritocerebral commissure; a second (accessory) nerve is evidently present in some forms and may carry fibers from aboral neurosecretory cells. Each secondary organ receives innervation from the brain by the pharyngeal nerve and, according to Legendre, from the prin-

Figure 6.28.

Generalized diagram of the neurosecretory system in the brain of the spider. **A.** Dorsolateral view. **B.** Anteroventral view. *1*, neurosecretory pathways according to Gabe (1955a); *2*, pathways according to Legendre (1959a). *acc. n.*, accessory nerve; *chel. n.*, cheliceral nerve; *esoph.*, esophagus; *leg g. 1–4*, leg ganglia; *opt. n.*, optic nerve; *protc.*, protocerebrum; *pdp. g.*, pedipalp ganglion; *ph. n.*, pharyngeal nerve of Schneider's organ II; *pr. n.*, principal nerve of Schneider's organ I; *subes. g.*, subesophageal ganglion; *Schn. org. I, II*, Schneider's organs; *tritc.*, tritocerebrum; *Tropf.*, "Tropfenkomplex." Neurosecretory cell groups: *ab. pr. ns.*, aboral protocerebral; *lat. ch. ns.*, lateral cheliceral; *leg g. ns.*, leg ganglion; *or. pr. ns.*, oral protocerebral; *pdp. g. ns.*, pedipalp ganglion; *rst. ch.ns.*, rostral (oral) cheliceral. [Modified from Kühne, 1959.]

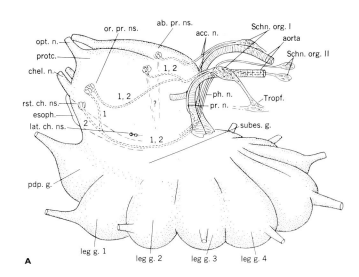

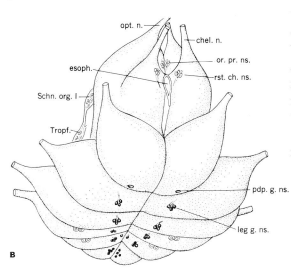

cipal nerve by an interganglionary nerve leading from the primary organ. Secretion has been reported associated with all these nerves; the intracerebral pathways have not been defined completely, but are represented tentatively in Fig. 6.28, A.

3. Phalangida

As in spiders, neurosecretory cells occur in bilateral groups in specific locations in the **brain** (Gabe, 1955a; Herlant-Meewis and Naisse, 1957; Naisse, 1959). Two principal fiber tracts (from aboral and oral cell groups) join on each side to form a pair of **paraganglionic plaques** in the neurilemma of the posterolateral surface of the brain. These plaques are similar to Schneider's organs. The oral neurosecretory cells show activity related to molting (a "molting gland" also has

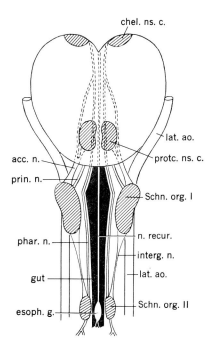

Figure 6.29.

Diagram of the neurosecretory centers and the retrocerebral endocrine complex of the Araneida, dorsal view. [Redrawn from Legendre, 1954.] *acc. n.*, accessory nerve; *chel. ns. c.*, cheliceral neurosecretory cells; *esoph. g.*, esophageal ganglion; *interg. n.*, interganglionic nerve; *lat. ao.*, lateral aorta; *n. recur.*, nervus recurrens; *phar. n.*, pharyngeal nerve; *protc. ns. c.*, protocerebral neurosecretory cells; *prin. n.*, principal nerve; *Schn. org. I, II*, Schneider's organs.

been described in the prosoma of the phalangids) and to reproduction. Other cell groups become increasingly active with the attainment of sexual maturity.

4. Scorpionida, Pseudoscorpionida, Acarina, Solifuga, Pantopoda (Fig. 6.30). (Refer to Table 6.1)

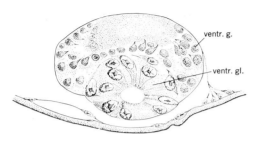

Figure 6.30. Ventral ganglion *(ventr. g.)* of larval *Achelia* (Pantopoda) to show its associated ventral gland *(ventr. gl.)*. [Redrawn from Sanchez, 1959.]

5. Xiphosura

Neurosecretory cells are present in the **circumesophageal nerve ring** and ventral nerve cord of *Limulus* (B. Scharrer, 1941c); see Fig. 6.31. The secretory cells at the end of their cycle contain large vacuoles filled with colloid. Their distribution in the nerve ring shows a gradient of increasing frequency from anterior to posterior, but the actual number of cells observed was variable, from one or two cells to as high as 2500 cells per ring. The number of cells present showed no relation to daily or yearly cycles or to sex; however, larger animals possessed more cells. Despite the prominence of the neurosecretory phenomenon in *Limulus*, neither neurosecretory tracts nor neurohemal organs have yet been described.

A substance has been extracted from the nerve ring of *Limulus*, which causes darkening of eyestalkless *Uca* (Brown and Cunningham, 1941). This substance was present in increased amounts around the circumesophageal ring. A correspondence of chromatophorotropic activity with the sometimes demonstrable gradient in neurosecretory cell number was suggested.

Signs of glandular activity are also present in nerve cells associated with the **lateral rudimentary eye** of *Tachypleus* and *Limulus* (Waterman and Enami, 1954). Secretion is formed in the sensory cells and is discharged into the surrounding connective tissue, where it may be picked up by the blood. Aqueous extracts of these organs from *Tachypleus* were found to have chromatophorotropic effects on juvenile *Sesarma*, though similar extracts from *Limulus* had no effect on *Uca* chromatophores. The function of these chromactivators in the xiphosurans themselves is unknown.

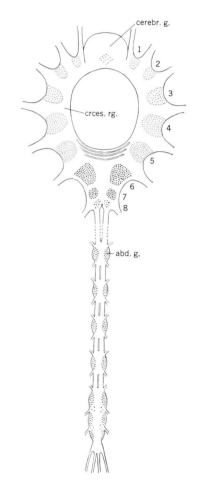

Figure 6.31. Diagram of central nervous system of *Limulus* to show increasing numbers of neurosecretory cells anteroposteriorly around circumesophageal ring *(crces. rg.)*. Stippled areas represent neurosecretory cell groups. Ganglia of circumesophageal ring are numbered 1–8; *abd. g.*, abdominal ganglion of ventral nerve cord; *cerebr. g.*, cerebral ganglion. [Redrawn from B. Scharrer, 1941c.]

P. Mollusca Amphineura

(See Table 6.1)

Q. Mollusca Gastropoda

Neurosecretion has been reported in all orders of gastropods; glandular activity has been observed in nearly every portion of the central nervous system in one species or another, although there is considerable variation from genus to genus in the ganglia involved. In some forms (as in *Cylichna*, Lemche, 1956), most of the nervous system, both central and peripheral, contains chrome hematoxylin-positive or paraldehyde fuchsin-positive materials. Neuroglandular organs—associations of presumed neurosecretory cells with glandular or sensory cells—have been described in several forms: in the distal portions of both sets of tentacles in *Helix*, the frontal organs of nudibranchs, the follicle gland and mediodorsal bodies in Basommatophora and Stylommatophora, and elsewhere. In the nudibranchs intracellular materials are so common that it is difficult to draw a line between ordinary nerve cells and neurons specialized for glandular activities (Tuzet et al., 1957; see note of caution below).

1. Prosobranchia

In the Monotocardia neurosecretory cells are found in various ganglia, especially the pleural and supraintestinal; they are absent from the buccal and pedal ganglia in all genera studied. In the taenioglossids the secretion is acidophilic; in stenoglossids it appears to change its tinctorial affinities according to the stage of the secretory cycle (B. Scharrer, 1937; Gabe, 1951, 1953a). The presence of neurosecretion in the Diotocardia is still uncertain (B. Scharrer, 1937; Gabe, 1954b).

2. Opisthobranchia

Cytologic indicators of neurosecretion are prominent in both tectibranchs and nudibranchs (B. Scharrer, 1935, 1937; Scharrer and Scharrer, 1945; Gabe, 1953b, 1954b; Pavans de Ceccatty and Planta, 1954; Sanchez and Pavans de Ceccatty, 1957; Tuzet et al., 1957). Secretory activity is observed most frequently in the cerebral ganglion (Fig. 6.32), especially in nudibranchs. Signs of axonal transport are said to be rare.

Lemche (1955, 1956) studied the tectibranch *Cylichna*. In addition to finding several possible neurohemal organs, he reported finding **neurosecretory material in most areas of the nervous system** with the exception of the osphradial and branchial ganglia. Chrome hematoxylin-staining granules were found in most peripheral nerves; in neurons, presumably motoneurons, innervating muscle fibers; and in neurons ending on or near unicellular glands. He concludes from this that neurosecretion in gastropods may differ from that in vertebrates in playing a more direct role in the innervation of effectors. While this is possible, a **note of caution** should be observed: the prominence of yellowish or reddish, probably lipoprotein, nerve pigments in the gastropods is well known. In the vertebrates at least, these pigments are presumed to be inert by-products of cellular metabolism, a sign of age. It is also known that chrome hematoxylin stains these "wear-and-tear" pigments intensely—so intensely that Pearse (1954) recommends its use in their detection. Until such time as the role of these products in *Cylichna* is more clearly demonstrated, it would be desirable to treat their presence with the caution shown by Picard, Stahl, and others in their studies of a similar situation in

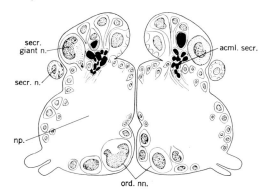

Figure 6.32. Diagram of section through the cerebral ganglion of *Pleurobranchaea* (Opisthobranchia). [Redrawn from B. Scharrer, 1935.] *acml. secr.*, accumulated secretion; *np.*, neuropile; *ord. nn.*, ordinary neurons; *secr. giant n.*, secretory giant neuron; *secr. n.*, secretory neuron.

fishes (see general discussion). The same caution may well be applied to some of the observations in the nudibranchs (see below). Certain neurons in *Aplysia* (and in *Helix*) show granules in the 2000 Å size range (Dalton, 1960; Simpson, Bern and Nishioka, unpublished; Rosenbluth, personal communication).

Several associations of **neurosecretory cells with sensory or glandular structures** have been described in nudibranchs. In *Aeolis*, *Polycera*, and *Doris* an olfactory bulb (accessory ganglion of the rhinophore?) lies at the base of the rhinophore; it contains neurosecretory cells which are associated with a paired frontal organ which lies in front of the cerebral ganglion. The neurohemal organs in *Doris*, *Aeolis*, and *Helix* (see below) appear to produce a phloxinophilic secretion of their own (Sanchez and Pavans de Ceccatty, 1957). In the central nervous system of the nudibranchs, cytologic manifestations of neurosecretion seem widespread, and Tuzet et al. state that it is difficult to distinguish between a "secretory" function common to all neurons and a supposed hormonogenic neurosecretory specialization of certain neurons. It is probable that the same caution should be exercised here as with *Cylichna*.

3. Pulmonata

Neurosecretion has been observed in both Stylommatophora and Basommatophora both in vivo and after fixation (Grzycki, 1951; Gabe, 1954b; Herlant-Meewis and van Mol, 1959; Krause, 1960; van Mol, 1960a; Pelluet and Lane, 1961). Two cell types have been distinguished in *Helix*, five in *Ferrissia*. The sites of greatest activity are the cerebral, pleural, and supraintestinal ganglia.

Two areas of particular interest in the cerebral ganglia of several genera of Basommatophora have been studied: the mediodorsal bodies and the lateral lobes (Lever, 1957, 1958a, 1958b; Lever et al., 1959); see Fig. 6.33. The **mediodorsal bodies** are composed solely of a type C cell; they are small bodies, commonly paired, lying on the mediodorsal side of the cerebral ganglia. Mediodorsal bodies were also observed in *Succinea putris*, a stylommatophoran.

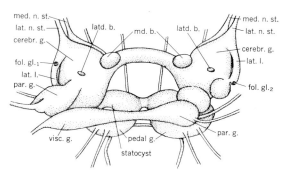

Figure 6.33. Central nervous system of *Ferrissia* (Pulmonata). [Modified from Lever, 1957, 1958a, 1958b.] *cerebr. g.*, cerebral ganglion; *latd. b.*, laterodorsal body; *lat. l.*, lateral lobe; *lat. n. st.*, lateral nerve stem; *md. b.*, mediodorsal body; *med. n. st.*, medial nerve stem; *par. g.*, parietal ganglia; *pedal g.*, pedal ganglia; *visc. g.*, visceral ganglia. *fol. gl.*$_1$, position of follicle gland in *Ferrissia* and *Acroloxus*; *fol. gl.*$_2$, position of follicle gland in *Ancylus*.

The **lateral lobes** arise as ectodermal invaginations called the cerebral tubes. In adult basommatophorans and in the stylommatophoran *Oncidiella*, a portion of these tubes remains patent in each lateral lobe, forming a small follicle termed the **follicle gland** (Fig. 6.34). This follicle gland consists of a single epithelial layer surrounding a lumen which contains a variable amount of a colloid material. Neurosecretory cells, including types A$_3$ and B, send processes to this structure. In one basommatophoran, *Ovatella*, the lumen of the cerebral tube remains open to the outside, and the "Gomori-positive" colloid can be seen within it. With the exception of

Figure 6.34. *Ancylus* (Pulmonata). *fol. gl.*, follicle gland, showing small secretory droplets near the apex of the cells. Paraldehyde fuchsin stain. [Lever, 1958a.]

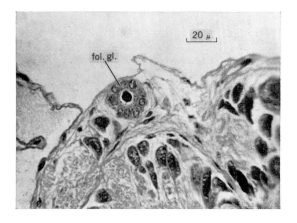

Oncidiella, the gland seems to degenerate in adult stylommatophorans (*Helix*, *Arion*, and others), although it is functional in the embryo until shortly after hatching (van Mol, 1960b). In *Helix* embryos it is filled with colloid secretion even though the neurosecretory cells do not become apparent until after its degeneration (Sanchez and Bord, 1958). Its embryology has been studied in *Arion* by van Mol (1960b). Organs in the tentacles of *Helix* are similar to the olfactory lobes of nudibranchs (Tuzet et al., 1957; Sanchez and Pavans de Ceccatty, 1957).

Secretory activity occurs in the cerebral and buccal ganglia of *Arion rufus*, and in the cerebral ganglion of *A. subfuscus* (Herlant-Meewis and van Mol, 1959; van Mol, 1960a). The cells stain with paraldehyde fuchsin and with alcian blue; they show axonal transport of their material. Part of the secretion from the buccal ganglion enters the posterior gastric nerve; the remainder seems to enter the second pharyngeal nerve. The secretion may serve a function similar to that suggested for neurosecretion in the **digestion of proteins** in *Calliphora* (E. Thomsen and Møller, 1959). The neurosecretory cells of the cerebral ganglia lie in the mesocerebrum and send their axons out the nerve of the cerebral artery, with whose wall they are in intimate contact.

4. Physiology

There is little direct evidence concerning the role of any of the neurosecretory phenomena described above. Correlations have been found between the **state of the gonads** and the cytologic appearance of the neurosecretory cells in some species of all orders (see Pelluet and Lane, 1961). Neurosecretion in the heteropods (Monotocardia) is most prominent at the time when the gonocytes are in an early stage of maturation, and the neurosecretory cells become depleted in these animals at the time of fertilization and spawning. The pleural ganglia may secrete a hormone that affects water balance (Hekstra and Lever, 1960; see also Lever and Joosse, 1961; Lever, Jansen, and De Vlieger, 1961; Lever, Kok, Meuleman, and Joosse, 1961).

Gersch and Deuse (1960) have extracted from most parts of the nervous system of *Aplysia* materials which have a **cardio-accelerator effect** in the same species. Meng (1960) and Kerkut and Laverack (1960) have done similar work on *Helix*.

R. Mollusca Scaphopoda

(See Table 6.1)

S. Mollusca Pelecypoda

Neurosecretory cells have been detected in the cerebral, pedal, and visceral ganglia of the orders Protobranchiata, Filibranchiata, Pseudolamellibranchiata, and Eulamellibranchiata (Gabe, 1955b; Lubet, 1955a; Gabe and Rancurel, 1958; Fährmann, 1961), and their physiology has been studied to some extent (Lubet, 1955b, 1956, 1957). Neurosecretion has yet to be reported in the septibranchs.

A relation between reproduction and the neurosecretory cells in the cerebral and visceral ganglia of *Mytilus* and of *Chlamys* has been suggested (Lubet). In these animals, which may release gametes two or three times during the course of the reproductive period, cerebral neurosecretory material is observed to disappear in some or all of the cells several days prior to each discharge of the gametes. The material is replenished between successive spawnings, and the cells become depleted completely by the end of the reproductive period. Emission of the gametes may depend on the lifting of an internal inhibition correlated with the disappearance of neurosecretory material from the cerebral ganglion. If this inhibitory influence is humoral, and if Lubet's suggestion (1957, p. 28) that removal of the cerebral ganglion may speed maturation of the gametes (especially the ova) is also correct, an interesting parallel with the situation in polychaetes is seen.

T. Mollusca Cephalopoda

Three organs of possible endocrine significance are associated with the nervous system of the cephalopods: the epistellar bodies of octopods, the parolfactory vesicles of decapods, and the optic glands. The first two structures may be neurosecretory; both lie outside the brain. B. Scharrer (1937) failed to find neurosecretory cells in the brain of *Octopus* and *Sepia*, though Hagadorn (unpublished) has observed paraldehyde fuchsin-positive cells in the subesophageal portion of the brain of *Octopus*, and Bogoraze and Cazal (1944) found possible secretory activity in neurons of the stomatogastric system.

The **epistellar body** is a vesicle, occasionally divided or branching, on the stellate ganglion of octopods near the root of the last stellar nerve (Young, 1936). The gland is separated from the main body of the ganglion by a thin connective tissue sheath which is interrupted at one point by the entry of nerve fibers from the mantle connective. It contains two types of cells: a columnar epithelial cell layer lining the vesicle and a subjacent layer of cells, unipolar or bipolar, which send processes between the epithelial cells into the lumen (Fig. 6.35). The lumen is filled with the swollen endings of cell processes, as well as with colloid and cellular debris. The subjacent cells have been described as neurosecretory since they appeared to contain secretion droplets, have a neuronal appearance, and are innervated by the mantle connective. Young believes them to be homologous to the syncytial third-order giant fiber neurons of decapods, since they occupy the same relative position as the cells of the giant fibers in *Loligo*. Cazal and Bogoraze (1944, 1949) have questioned Young's interpretation, stating that while the bipolar cells are nonsecretory nerve cells, the unipolar cells are secretory cells derived from the epithelial layer, all stages in the transition being seen. Boycott and Young (1956) have by implication reaffirmed their belief that the subjacent cells are neurosecretory. The epistellar organ is said to show a secretion cycle, the climactic stage being reached in June to

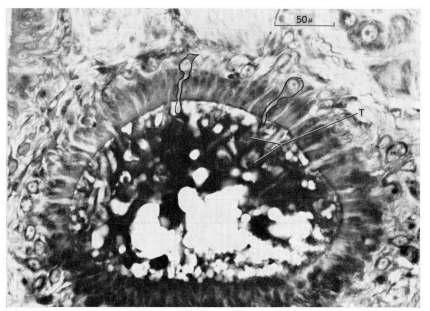

Figure 6.35. *Octopus* (Cephalopoda). Epistellar body. Note cells (see text) basal to palisade cells, with processes projecting between the latter into the lumen. Two cells are outlined in black. T, enlarged terminals of processes. Paraldehyde fuchsin.

August in *Octopus vulgaris* at Naples, but no endocrine function has as yet been ascribed to it. Recent electron microscope studies reveal that the processes of the subjacent neurons resemble the photoreceptive elements of the eye, and there is little evidence of neurosecretory activity. The epistellar organ would appear to be a rudimentary photoreceptor (Nishioka et al., 1962).

The **parolfactory vesicles** occur just below the optic stalk of decapods (Fig. 6.36) (Thore, 1939; Boycott and Young, 1956). These vesicles, similar in structure to the epistellar bodies of octopods, are not found in the latter. As with the epistellar bodies, there may be one to several parolfactory vesicles on each side; occasionally they are tubu-

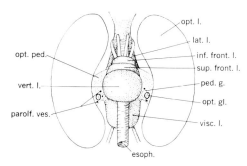

Figure 6.36. Diagram of brain of *Illex* (Decapoda) to show location of optic glands *(opt. gl.)* and parolfactory vesicles *(parolf. ves.)*; esoph., esophagus; inf. front. l., inferior frontal lobe; *lat. l.*, lateral lobe; *opt. l.*, optic lobe; *opt. ped.*, optic peduncle; *ped. g.*, peduncular ganglion *(broken circle)*; *sup. front. l.*, superior frontal lobe; *vert. l.*, vertical lobe; *visc. l.*, visceral lobe. [Redrawn from Haefelfinger, 1954.]

lar rather than vesicular. Boycott and Young (1956) state that "... the walls of the vesicles represent modified nerve cells, the contents modified neuropil." The parolfactory and epistellar vesicles in cephalopods resemble each other and also the follicle glands observed in *Ferrissia* and other gastropods by Lever. A careful embryologic study of these vesicles is needed to resolve the conflicting views of Young and of Cazal and Bogoraze, and also possibly to establish the degree of relationship to the gastropod organs.

The **nonneurosecretory optic glands** (Fig. 6.36) are the best established cephalopod endocrine organs. They lie on the optic stalk and have been found in all cephalopods examined with the exception of *Nautilus*. In *Octopus*, sexual maturation is under the direct stimulatory hormonal control of the optic glands, which are in turn subjected to nervous inhibition from the brain. The inhibitory centers in their turn are activated by visual stimuli (Wells and Wells, 1959; Wells, 1960).

U. Echinodermata

(See Table 6.1)

V. Protochordates

1. Tunicata

The status of neurosecretion in the tunicates is controversial, particularly with respect to the supposed evolutionary relationship of the neural complex to the vertebrate pituitary gland. The neural complex is composed of three structures: the cerebral ganglion, the neural gland, and the ciliated funnel (Fig. 27.20, p. 1580). Paraldehyde fuchsin-positive or chrome hematoxylin-positive materials occur in nerve cells of the cerebral ganglion of *Ciona, Pyura,* and *Styela* (Mazzi, 1952; Dawson and Hisaw, 1956). The controversy arises over the frequent **attempts to homologize the neural gland and ciliated funnel** to either the pars nervosa of the pituitary or to the entire pituitary gland. The arguments for and against this homology center on three lines of evidence: embryologic, histologic, and endocrinologic (Dodd, 1955).

Early workers (such as van Beneden and Julin, 1884) claimed that the neural gland and ciliated funnel were derived from both neurectoderm and a pharyngeal inpocketing, just as is the vertebrate pituitary. This claim seems amply refuted by later workers (see Elwyn, 1937), who find that both structures are derived completely from the neural tube. The embryologic evidence thus tends to discourage a comparison with the vertebrate pars distalis. Histologically, it has been claimed that the mode of secretion in the neural gland approximates that of the adenohypophysis; however, the neural gland has a holocrine mode of secretion, which has no counterpart in the pituitary (Pérès, 1943, 1946).

The **endocrinologic evidence is conflicting.** Claims have been made for the presence of hormones comparable in action to the hormones proper to all three portions of the mammalian pituitary. Oxytocic, pressor, and melanophore-expanding activities have been found associated with the neural complex by earlier workers (Butcher, 1930; Bacq and Florkin, 1935, 1946; Carlisle, 1950); however, more recent work shows that the oxytocic activity is neither specific to the neural complex nor due to oxytocin (Dodd, 1959; Sawyer, 1959). Convincing evidence for melanophore-expanding hormone activity appears to be lacking (Dodd, 1955, p. 187; 1959). Authors also do not agree on the gonadotropic activity claimed for the neural complex (Hogg, 1937; Carlisle, 1951a, 1954a); again, recent work has failed to confirm this (Dodd, 1955, 1959).

An **"asymmetric gland"** is associated with the neural gland. It has the appearance of an endocrine organ, and undergoes periodic fluctuations in activity which may have something to do with the "awakening" of sexual activity (Pérès, 1943).

2. Cephalochordata

Glandular activity is present in the central nervous system of the amphioxus (Pavans de Ceccatty, 1957). Chrome hematoxylin-stainable materials are observable in the "dorsal giant cells of Joseph," in the region of the anterior pigment spot, and in the infundibular organ. In the medulla, stainable materials also may be seen in the giant cells of Rhode; the material is traceable out into the axons. Similar material is seen in other fibers of undetermined origin. Fibers containing a phloxinophil material also are observed.

Two other organs (nonneurosecretory) associated with the nervous system in amphioxus are of possible endocrine significance: Hatschek's pit and the infundibular organ. Hatschek's pit, on the basis of embryologic evidence, may be homologous to the pars distalis of the pituitary, but no physiologic data are available (Goodrich, 1917). The infundibular organ may be analogous, if not homologous, to the subcommissural organ of vertebrates (Hofer, 1959). Olsson and Wingstrand (1954) disagree with this comparison, although they report that Reissner's fiber originates in the infundibular organ in the amphioxus. A homology is suggested between the infundibular organ and the hypothalamo-hypophysial complex of vertebrates. Nothing definite is known concerning the function of either the infundibular organ or the subcommissural organ (see Olsson, 1958).

W. Brief Summary of Vertebrate Neurosecretory Systems

In vertebrates, evidence of neurosecretion can be found in many areas of the central nervous system. The principal neurosecretory system, the hypothalamo-hypophysial system (see general reviews), present in all vertebrates, consists of secretory cells (described first by E. Scharrer in 1928) located in certain **hypothalamic nuclei**, from which axons lead to two neurohemal areas: the **median eminence and the pars nervosa** of the neurohypophysis. In the former area, the axon terminals are associated with capillaries of the hypophysial portal system, and the secretions are evidently involved in the neurohormonal regulation of the pars distalis of the pituitary (see Harris, 1960). The median eminence is especially well developed in birds, and the neurosecretion-bearing axons are said to form loops in this region and thence to proceed to the pars nervosa (Benoit and Assenmacher, 1959). However, recent observations indicate that these loops do not exist in some birds, but that axons terminate in the median eminence as in the pars nervosa (Oksche, personal communication; Kobayashi et al., 1961).

The general picture of hypothalamic hormonal control over adenohypophysial function apparently involves an **excitatory (release) influence** on five hormones (follicle-stimulating, luteinizing, growth, adrenocorticotropic, thyrotropic) and an **inhibitory influence** on two hormones (mammotropic or luteotropic, chromatotropic or intermedin). Evidence is best established for the existence of a specific adrenocorticotropin-release factor.

In the pars nervosa the neurosecretory material is associated with a **family of specific hormonal factors—octapeptides** (du Vigneaud, 1956; Sawyer et al., 1960)—which are released into the blood stream and help regulate such diverse activities as blood pressure, renal excretion of salt and water, milk ejection, uterine contractility, and possibly extrarenal water and salt movements in some anamniotes. Fibers bearing neurosecretion can be seen penetrating the pars intermedia in many vertebrates (see Green and Maxwell, 1959); in fishes the neurohypophysis and pars intermedia often are united to form a neurointermediate lobe. In cyclostomes many neurosecretory fibers terminate on the anteroventral surface of the mesencephalon (Oztan and Gorbman, 1960), not unlike the situation in annelids.

In anamniotes there is supposed to be a single pair of hypothalamic neurosecretory nuclei: the **preoptic nuclei**. In amniotes each preoptic nucleus is represented by two nuclei: the **supraoptic and the paraventricular**. Actually, the single pair of neurosecretory cell groups in anamniotes often can be subdivided, and the two pairs of discrete nuclei in amniotes are not always defined clearly. In addition, another neurosecretory nucleus **(lateralis tuberis)** has been implicated in various higher vertebrates as possibly being related to regulation of gonadotropin release; this nucleus is especially prominent in teleosts.

Despite the huge amount of study directed to the hypothalamo-hypophysial neurosecretory system, several basic issues remain unresolved. The uncertain nature of the relation between neurosecretory material and the neurohypophysial hormones has already been discussed (p. 363). The relation between neurosecretory material and adenohypophysial control factors is even less clear. There are nonneurosecretory hypothalamic and higher nervous centers which also may be involved in the direct and indirect regulation of adenohypophysial function. It appears firmly established that the hypothalamic control over the pars distalis is not nervous, but is certainly neurohormonal.

Additional neurosecretory pathways are present in the brain. Fibers also pass from the hypothalamic neurosecretory nuclei into the telencephalon and to the roof of the diencephalon (Legait, 1958). A prominent hypothalamo-paraphysial tract has been described in snakes (E. Scharrer, 1951). Discharge of neurosecretion into the third ventricle also has been described (see Noda et al., 1955; Wilson et al., 1957; Talanti and Kivalo, 1961). In teleosts especially, but also in amphibians, various extrahypothalamic nuclei in the brain show evidence of secretory activity, including the Purkinje cells of the cerebellum (Thomas, 1951; Shanklin et al., 1957). In birds groups of secretory motor horn cells are reported to occur in the lumbar region (Sano, 1958; see Ghosh et al., 1962, for dissent). No function has been ascribed yet to these various secretory areas.

A second major neurosecretory system of vertebrates is located at the **posterior end of the spinal cord in fishes** (see Enami, 1959; Sano, 1961). In teleosts, neurosecretory Dahlgren cells —often of giant size and equipped with endocellular capillaries (also encountered in hypothalamic neurons of some fish)—have their secretion-bearing fibers terminating in a posterior neurohemal organ generally located in the last vertebral element, or in the urostyle. This structure, the **urophysis** (also called urohypophysis or neurophysis spinalis caudalis), is a counterpart of the neurohypophysis; the hypothalamo-neurohypophysial system and the **caudal neurosecretory system** represent almost quasi-serial homology. In elasmobranchs there is no delineated neurohemal organ, and the short processes of the Dahlgren cells terminate diffusely among capillaries in the ventral part of the spinal cord. Secretory activity of the Dahlgren neurons of rays represents the first description of neurosecretion in any animal (Speidel, 1919). A function has yet to be assigned definitely to this caudal system, but a relation to osmoregulation (Bern and Takasugi, 1962) and to buoyancy regulation has been suggested.

Evidences of secretion have also been reported outside the central nervous system in some spinal ganglion cells (Seite and Chambost, 1958), in various ganglionic neurons of the sympathetic nervous system (Lennette and Scharrer, 1946; Thomas, 1951; Smith, 1952), and in neurons sometimes located in the adrenal medulla (Eichner, 1951; Ito, 1954; Stutinsky, 1959). Changes corresponding to hormonal events have been described in the rodent ganglion cervicalis uteri (Lehmann and Stange, 1953; Takahashi, 1960). The presynaptic axon terminations of the myenteric plexus show numerous elementary neurosecretory granules (Hager and Tafuri, 1959). If nonneuronal tissue is considered, a variety of additional areas of the vertebrate brain also may qualify as secretory. These areas include the pineal–parietal eye complex (Eakin et al., 1961), the subcommissural organ and the cells secreting Reissner's fiber (Olsson, 1958), and the ependyma lining the ventricles, especially in the region of the infundibular recess (Löfgren, 1960). It should be emphasized that none of the secretory areas described in this paragraph is "neurosecretory" in the sense discussed in Section I.

Classified References

Space does not permit an exhaustive bibliography of this area. The following is a limited selection emphasizing lower animals, principles, and recent and physiological works.

General Reviews and Symposia *(the compilations marked by an asterisk were not available by the time of completion of this chapter):*

Bargmann et al., 1958; Barker-Jørgensen and Larsen, 1960; Bern, 1962; Carlisle and Knowles, 1959; Colloq. Int., CNRS, 1949; Convegno sulla Neurosec., 1954a; Curri et al., 1958; Enami, 1957; Florey, 1951; Gabe, 1953f, 1954b; Gersch, 1959; *Gorbman and Barrington, 1962; Green and Maxwell, 1959; Hanström, 1939, 1947, 1953, 1957; Harker, 1958; Harris, 1960; Hebb, 1959; *Heller and Clark, 1962; Hild, 1956; Hydén, 1960; Ortmann, 1960; Parker, 1948; Rothballer, 1957; Roussy and Mosinger, 1946; Scharrer, B., 1955b, 1959; Scharrer, E., 1959; Scharrer and Scharrer, 1954a, 1954b; Scheer, 1960; Schiebler, 1954; Sloper, 1958; Stutinsky, 1953b; *Takewaki, 1962; Van der Kloot, 1961b; Welsh, 1955, 1959, 1961a, 1961b.

Morphology of Neurosecretory Cells

Benoit and Assenmacher, 1959; Bern and Hagadorn, 1959; Bern et al., 1962; Bodian, 1951; Champy and Champy, 1957; Clark, 1956a, 1956b; Dawson, 1953; Diepen and Engelhardt, 1958; Edström and Eichner, 1958; Enami and Imai, 1955; Fernández-Morán, 1957; Fox and Zabors, 1960; Fridberg, 1959; Galambos, 1961; Gray and Whittaker, 1960; Groot, 1957; Hagadorn, 1958, 1962; Hagen, 1954; Hanström, 1954; Horridge, 1961; Ishizaki et al., 1959; Kastl, 1954; Koelle, 1961; Koelle and Geesey, 1961; Lewis and Lever, 1960; Matsumoto, 1958; Maynard, 1961b; Miyawaki, 1955a; Moussa and Banhawy, 1960; Naisse, 1961; Ortmann, 1958; Palay, 1943; Palay and Wissig, 1953; Passano, 1954; Picard and Stahl, 1956; Rennels and Drager, 1955; Robertis, 1959; Sano et al., 1957; Scharrer, E., 1935; Scharrer, E. et al., 1945; Schmitt, 1959; Shanklin et al., 1957; Spatz, 1954; Stahl, 1957; Stahl and Seite, 1954; Stolk, 1960; Tamiya et al., 1956; Thomas, 1951; Thomsen and Thomsen, 1954; Wigglesworth, 1960a; Wilson et al., 1957.

Nature of Neurosecretory Material

Acher and Fromageot, 1957; Albers and Brightman, 1959; Andersson and Jewell, 1957; Barrnett, 1954; Brousse et al., 1958; Chauvet et al., 1960; Du Vigneaud, 1956; Eichner, 1958; Gabe, 1960; Goslar and Schultze, 1958; Hadler et al., 1957; Hild, 1951; Hild and Zetler, 1953; Howe and Pearse, 1956; Kratzsch, 1951; Lasansky and Sabatini, 1957; Loebel and Guijón, 1957; Ortmann, 1951; Pardoe and Weatherall, 1955; Rehm, 1959; Rodeck, 1959; Schiebler, 1951, 1952a, 1952b; Sloper, 1957; Sloper et al., 1960; Takahashi, 1957.

Electron Microscope Studies of Neurosecretion

Bargmann, 1958; Bargmann and Knoop, 1957, 1960; Bargmann et al., 1957; Barry and Cotte, 1961; Bern et al., 1961; Carlisle and Knowles, 1959; Dalton, 1960; Duncan, 1956; Enami and Imai, 1958; Fährmann, 1961; Fingerman and Aoto, 1959; Fujita, 1957; Fujita and Hartmann, 1961; Gerschenfeld et al., 1960; Green and Maxwell, 1959; Green and Van Breemen, 1955; Hagadorn and Nishioka, 1961; Hager and Tafuri, 1959; Hartmann, 1958; Hess, 1958; Hodge and Chapman, 1958; Holmes and Knowles, 1959, 1960; Holmgren and Chapman, 1960; Howe, 1959; Knowles, 1959, 1960; Kobayashi et al., 1961; Legait and Legait, 1958, 1959; Löblich and Knezevic, 1960; Meyer and Pflugfelder, 1958; Miyawaki, 1960b; Murakami, 1961, 1962; Nishiitsutsuji-Uwo, 1960, 1961; Palay, 1957, 1958, 1960; Robertson, 1957; Sano and Knoop, 1959; Scharrer, E. and Brown, 1961; Schiebler; 1952a; Schultz, 1960; Stiennon and Drochmans, 1961; Trujillo-Cenóz, 1959; Wetzstein, 1957; Willey and Chapman, 1960.

Neurosecretory Cells—Conduction

Bliss, 1956; Carlisle, 1958b; Cross and Green, 1959; Harris, 1947; Hodgson and Geldiay, 1959; Morita et al., 1961; Nakayama, 1955; Potter and Loewenstein, 1955; Van der Kloot, 1955a, 1958; Waterman and Enami, 1954.

Neurosecretory Cells—Transport

Burgen, 1959; Carlisle, 1953a, 1953b, 1958a; Causey, 1949; Convegno sulla Neurosec., 1954b; Euler, 1950, 1958; Fingerman and Aoto, 1958a; Friede, 1959; Gerard, 1950; Green and Maxwell, 1959; Hanström, 1955; Hebb and Waites, 1956; Hild, 1954; Knoche, 1958; Knowles, 1958; Koenig and Koelle, 1960; MacIntosh, 1959; Miani, 1960; Ochs and Burger, 1958; Raab and Humphreys, 1947; Samuels et al., 1951; Smith, 1952; Tainter and Luduena, 1950; Thomsen, E., 1954; Weiss, 1944, 1955, 1959; Weiss and Hiscoe, 1948; Weiss et al., 1945.

Platyhelminthes

Lender and Klein, 1961; Stéphan-Dubois and Lender, 1956; Török, 1958; Turner, 1946.

Nemertinea

Gontcharoff and Lechenault, 1958; Hyman, 1951; Scharrer, B., 1941b.

Nematoda

Gersch, 1957; Gersch and Scheffel, 1958.

Polychaeta

Arvy, 1954; Bobin and Durchon, 1952, 1953; Clark, 1955, 1958a, 1958b, 1959; Clark and Bonney, 1960; Clark and Clark, 1959; Clark and Evans, 1961; Defretin, 1955, 1956; Durchon, 1952, 1956a, 1956b, 1956c, 1957, 1959, 1960; Durchon and Frézal, 1955; Harms, 1948; Hauenschild, 1956a, 1956b, 1959a, 1959b, 1960; Korn, 1959, 1960; Schaefer, 1939; Scharrer, B., 1936, 1937.

Oligochaeta

Aros and Bodnár, 1960; Aros and Vigh, 1959; Avel, 1929, 1947; Brandenburg, 1956a; Deuse-Zimmermann, 1960; Harms, 1948; Herlant-Meewis, 1955, 1956a, 1956b, 1959; Hubl, 1953, 1956; Marapão, 1959; Michon and Alaphilippe, 1959; Scharrer, B., 1937; Scharrer and Scharrer, 1954a; Schmidt, 1947.

Hirudinea

Hagadorn, 1958; Legendre, 1959b; Nambudiri and Vijayakrishnan, 1958; Perez, 1942; Scharrer, B., 1937.

Sipunculoidea

Åkesson, 1961; Carlisle, 1959d; Gabe, 1953d; Koller, 1959.

Onychophora

Cuénot, 1949; Gabe, 1954a; Sanchez, 1958.

Crustacea

Alexandrowicz, 1952, 1953a, 1953b, 1953c, 1954, 1955; Alexandrowicz and Carlisle, 1953; Amar, 1948, 1950, 1953; Arvy et al., 1956; Balesdent-Marquet, 1960; Barnes and Gonor, 1958; Bellonci, 1881; Bliss, 1953, 1956; Bliss and Welsh, 1952; Bliss et al., 1954; Brown, 1933, 1944, 1946, 1950, 1952; Brown and Ederstrom, 1940; Carlisle, 1953a, 1953b, 1953c, 1953d, 1954b, 1956a, 1956b, 1956c, 1958b, 1959a, 1959b, 1959c, 1960; Carlisle and Dohrn, 1953; Carlisle and Knowles, 1959; Carlisle and Passano, 1953; Carlisle and Pitman, 1961; Charniaux-Cotton, 1956, 1958; Clausen and Franz-Egon, 1959; Dahl, 1957; Demeusy, 1953, 1960; Drach, 1955; Drach and Gabe, 1962; Durand, 1956; Duveau, 1961; Echalier, 1959; Edman et al., 1958; Enami, 1951; Fingerman and Aoto, 1958a, 1958b, 1959, 1960; Fingerman, Lowe, et al., 1959a, 1959b; Fingerman, Sandeen, et al., 1959; Gabe, 1952a, 1952b, 1952c, 1953c, 1954b; Gersch, 1959; Gräber, 1933; Hanström, 1935, 1937a, 1939, 1947, 1948; Hara, 1952a, 1952b; Heller and Smith, 1948; Hodge and Chapman, 1958; Inoue, 1957; Karlson and Skinner, 1960; Kleinholz, 1936, 1942, 1957, 1958; Kleinholz et al., 1962; Knowles, 1939, 1951, 1953, 1954, 1955, 1956, 1958, 1959; Knowles and Carlisle, 1956; Knowles et al., 1955; Konok, 1960; Legrand and Juchault, 1960a, 1960b; Lochhead and Resner, 1958; Marcos-Gallego and Stamm, 1959; Matsumoto, 1954a, 1954b, 1958, 1959; Maynard, 1961a, 1961b; Maynard and Welsh, 1959; Miyawaki, 1955a, 1955b, 1956a, 1956b, 1956c, 1958, 1960a, 1960b; Nayar and Parameswaran, 1955; Oguro, 1959a, 1959b, 1960a, 1960b, 1961; Otsu, 1960; Panouse, 1947; Parameswaran, 1956; Parker, 1948; Passano, 1953, 1954, 1961; Pasteur, 1958; Pérez-González, 1957; Potter, 1958; Prabhu, 1961; Rehm, 1959; Sandeen and Costlow, 1961; Scheer, 1957; Sterba, 1957; Travis, 1951a, 1951b; Valente, 1959; Vernet-Cornubert, 1961; Welsh, 1961a; Weygoldt, 1961; Yamamoto, 1955, 1961.

Insecta

Arvy, 1956; Arvy and Gabe, 1947, 1953a, 1953b, 1954; Barton-Browne et al., 1961; Bodenstein, 1953a, 1953b, 1954, 1955, 1957; Boisson, 1949; Bounhiol, 1936, 1938, 1957; Bounhiol et al., 1953; Brandenburg, 1956b; Brousse et al., 1958; Brown and Harker, 1960; Bückmann, 1956; Cameron, 1953; Carlisle et al., 1955; Caspari and Plagge, 1935; Cazal, 1946, 1947, 1948; Cazal and Guerrier, 1946; Chaudonneret, 1950; Clements, 1956; Companjen, 1958; Davey, 1961; Day, 1940, 1943; Deroux-Stralla, 1948; Dupont-Raabe, 1952a, 1952b, 1957, 1959; Engelmann, 1957, 1959; Etkin, 1955; Formigoni, 1956; Fraser, 1955, 1957, 1959a, 1959b, 1960; Fukuda, 1952, 1953a, 1953b, 1953c; Füller, 1960; Gabe, 1953e; Ganguly and Banerjee, 1960; Ganguly and Deb, 1960; Geldiay, 1959; Gersch, 1956, 1961; Gersch and Kothes, 1956; Gersch and Unger, 1957; Gersch et al., 1957; Gilbert and Schneiderman, 1961; Grandori, 1954, 1956; Grandori and Carè, 1953; Grison, 1949; Hanström, 1937b, 1938, 1940, 1941, 1943; Harker, 1956, 1960a, 1960b; Hasegawa, 1957; Herlant-Meewis and Paquet, 1956; Hidaka, 1956, 1957; Highnam, 1958, 1961; Ichikawa and Nishiitsutsuji, 1951, 1952; Johansson, 1958; Joly, 1945a, 1945b, 1950, 1958; Jones, 1956a, 1956b; Junqua, 1956; Karlson, 1956, 1958, 1960; Kirchner, 1960; Klug, 1958; Knowles et al., 1955; Kobayashi and Burdette, 1961; Kobayashi and Kirimura, 1958; Koller, 1948; Kopéc, 1912, 1917, 1922; Köpf, 1957,

1958; Kühn and Piepho, 1936; Kuhnen and Beck, 1958; Larsen and Bodenstein, 1959; Lees, 1955, 1959; Lerma, 1937, 1947, 1949a, 1949b, 1950, 1954, 1956; Levinson and Platonova, 1948; L'Hélias, 1950, 1952, 1956, 1957; Lhoste, 1957; Lüscher and Engelmann, 1955; Milburn et al., 1960; Mitsuhashi and Fukaya, 1960; Monro, 1956, 1958; Morohoshi, 1959; Mothes, 1960; Nabert, 1913; Nayar, 1954a, 1954b, 1955a, 1955b, 1955c, 1956a, 1956b, 1956c, 1957, 1958, 1960; Noirot, 1957; Nuñez, 1956; Nussbaum, 1960; Nyst, 1942; Ohnishi, 1957; Ohtaki, 1960; Ozbas and Hodgson, 1958; Perez, 1940; Pflugfelder, 1937, 1952; Piepho, 1946; Pipa, 1961; Possompès, 1946, 1947, 1948, 1953a, 1953b, 1954, 1958; Power, 1948; Raabe, 1959; Rahm, 1952; Rehm, 1951, 1955; Risler, 1957; Schaller, 1959; Scharrer, B., 1941a, 1952a, 1952b, 1952c, 1955a, 1956, 1958; Scharrer and Scharrer, 1944; Scharrer and Harnack, 1961; Schneiderman, 1957; Schneiderman and Gilbert, 1959; Sellier, 1956; Strangways-Dixon, 1959; Strich-Halbwachs, 1959; Stumm-Zollinger, 1957; Stutinsky, 1952, 1953a; Thomsen, E., 1952, 1954, 1956; Thomsen and Møller, 1959; Thomsen, M., 1951, 1954a, 1954b; Unger, 1957; Van der Kloot, 1955a, 1955b, 1960, 1961a; Vogt, 1941, 1942a, 1942b, 1942c, 1946; Watterson, 1959; Weber, 1952; Wenk, 1953; Weyer, 1935; Wigglesworth, 1939, 1951, 1954a, 1954b, 1959, 1960a, 1960b; Williams, 1946, 1947, 1948, 1949a, 1949b, 1952, 1953, 1956, 1960; Zee and Pai, 1944.

Arachnida

Brown and Cunningham, 1941; Cox, 1960; Gabe, 1955a; Habibulla, 1961; Herlant-Meewis and Naisse, 1957; Junqua, 1957; Kühne, 1959; Legendre, 1953, 1954, 1955, 1959a, 1959b; Naisse, 1959; Sanchez, 1954, 1958, 1959; Scharrer, B., 1941c; Waterman and Enami, 1954.

Myriapoda

Fahlander, 1938; Gabe, 1956, 1959; Holmgren, 1916; Joly, 1962; Jupeau, 1956; Juberthie-Jupeau, 1961; Lerma, 1951; Palm, 1955; Prabhu, 1959; Sahli, 1958, 1961, 1962; Scharrer, B., 1937; Scheffel, 1961; Tiegs, 1947.

Amphineura

Scharrer, B., 1937.

Gastropoda

Gabe, 1951, 1953a, 1953b, 1954b; Gabe and Prenant, 1952; Gersch and Deuse, 1960; Grzycki, 1951; Hekstra and Lever, 1960; Herlant-Meewis and Mol, 1959; Kerkut and Laverack, 1960; Krause, 1960; Lemche, 1955, 1956; Lever, 1957, 1958a, 1958b; Lever and Boer, 1959; Lever and Joosse, 1961; Lever, Jansen, and De Vlieger, 1961; Lever, Kok, Meuleman, and Joosse, 1961; Meng, 1960; Mol, 1960a, 1960b; Pavans de Ceccatty and Planta, 1954; Pearse, 1954; Pelluet and Lane, 1961; Sanchez and Bord, 1958; Sanchez and Pavans de Ceccatty, 1957; Scharrer, B., 1935, 1937; Scharrer and Scharrer, 1937, 1945; Tuzet et al., 1957.

Scaphopoda

Gabe, 1949, 1954b.

Pelecypoda

Fährmann, 1961; Gabe, 1955b; Gabe and Rancurel, 1958; Lubet, 1955a, 1955b, 1956, 1957.

Cephalopoda

Bogoraze and Cazal, 1944; Boycott and Young, 1956; Cazal and Bogoraze, 1944, 1949; Haefelfinger, 1954; Hutchinson, 1928; Nishioka et al., 1962; Scharrer, B., 1937; Thore, 1939; Wells, 1960; Wells and Wells, 1959; Young, 1936.

Echinodermata

Unger, 1960.

Protochordates

Bacq and Florkin, 1935, 1946; Beneden and Julin, 1884; Brien, 1948; Butcher, 1930; Carlisle, 1950, 1951a, 1951b, 1953e, 1954a; Dawson and Hisaw, 1956; Dodd, 1955, 1959; Elwyn, 1937; Goodrich, 1917; Hofer, 1959; Hogg, 1937; Mazzi, 1952; Olsson, 1958; Olsson and Wingstrand, 1954; Pavans de Ceccatty, 1957; Pérès, 1943, 1946; Sawyer, 1959; Sengel and Kieny, 1962.

Vertebrata—General Reviews and Symposia

Bargmann et al., 1958; Barker-Jørgensen and Larsen, 1960; Convegno sulla Neurosec., 1954a; Curri et al., 1958; Enami, 1957; Gorbman and Barrington, 1962; Green and Maxwell, 1959; Hanström, 1947, 1957; Harris, 1960; Hebb, 1959; Heller and Clark, 1962; Hild, 1956; Ortmann, 1960; Rothballer, 1957; Roussy and Mosinger, 1946; Scharrer, B., 1959; Scharrer, E., 1959; Scharrer and Scharrer, 1954a, 1954b; Schiebler, 1954; Sloper, 1958; Stutinsky, 1953b; Takewaki, 1962; Welsh, 1955, 1959.

Bibliography

ACHER, R. and FROMAGEOT, C. 1957. The relationship of oxytocin and vasopressin to active proteins of posterior pituitary origin. Studies concerning the existence or non-existence of a single neurohypophysial hormone. In: *The Neurohypophysis.* H. Heller (ed.). Butterworths, London.

ÅKESSON, B. 1961. The development of *Golfingia elongata* Keferstein (Sipunculidea) with some remarks on the development of neurosecretory cells in sipunculids. *Ark. Zool. Kungl. Svens. Vetens. Akad.,* 13:511-531.

ALBERS, R. W. and BRIGHTMAN, M. W. 1959. A major component of neurohypophysial tissue associated with antidiuretic activity. *J. Neurochem.,* 3:269-276.

ALEXANDROWICZ, J. S. 1952. Notes on the nervous system in the Stomatopoda. I. The system of median connectives. *Pubbl. Staz. zool. Napoli,* 23:201-214.

ALEXANDROWICZ, J. S. 1953a. Nervous organs in the pericardial cavity of the decapod Crustacea. *J. Mar. biol. Ass. U.K.,* 31:563-580.

ALEXANDROWICZ, J. S. 1953b. Notes on the nervous system in the Stomatopoda, II. The system of dorsal trunks. *Pubbl. Staz. zool. Napoli,* 24:29-39.

ALEXANDROWICZ, J. S. 1953c. Notes on the nervous system in the Stomatopoda. III. Small nerve cells in motor nerves. *Pubbl. Staz. zool. Napoli,* 24:39-45.

ALEXANDROWICZ, J. S. 1954. Innervation of an amphipod heart. *J. Mar. biol. Ass. U.K.,* 33:709-719.

ALEXANDROWICZ, J. S. 1955. Innervation of the heart of *Praunus flexuosus* (Mysidacea). *J. Mar. biol. Ass. U.K.,* 34:47-53.

ALEXANDROWICZ, J. S. and CARLISLE, D. B. 1953. Some experiments on the function of the pericardial organs in Crustacea. *J. Mar. biol. Ass. U.K.,* 32:175-192.

AMAR, R. 1948. Un organe endocrine chez *Idotea. C. R. Acad. Sci., Paris,* 227:301-303.

AMAR, R. 1950. Les formations endocrines cérébrales des Isopodes marins. *C. R. Acad. Sci., Paris,* 230:407-409.

AMAR, R. 1953. Sur l'existence de cellules neurosécrétrices dans le cerveau de *Rocinela* (Crustacea Isopoda). *Bull. Soc. zool. Fr.,* 78:171-173.

ANDERSSON, B. and JEWELL, P. A. 1957. The effect of long periods of continuous hydration on the neurosecretory material in the hypothalamus of the dog. *J. Endocrin.,* 15:332-338.

AROS, B. and BODNÁR, E. 1960. Histologische Untersuchungen über die neurohumorale Funktion von *Eisenia rosea* (Oligochaeta). *Symp. Biol. hung.,* 1:191-202.

AROS, B. and VIGH, B. 1959. Changes in the neurosecretion of the nervous system of earthworm under various external conditions. *Acta biol. hung.,* suppl. 3:47.

AROS, B. and VIGH, B. 1961. Neurosecretory changes in the nervous system of *Lumbricus rubellus* Hoffm. provoked by various experimental influences. *Acta biol. Acad. Sci. hung.,* 12:87-98.

ARVY, L. 1954. Sur l'existence de cellules neurosécrétrices chez quelques Annélides polychètes sédentaires. *C. R. Acad. Sci., Paris,* 238:511-513.

ARVY, L. 1956. Le système pars intercerebralis - corpus cardiacum - corpus allatum chez *Euroleon nostras* Fourcroy (Névroptéroïde Planipenne). In: *Bertil Hanström: Zoological Papers in Honour of his Sixty-fifth Birthday.* K. G. Wingstrand (ed.). Zool. Inst., Lund.

ARVY, L. and GABE, M. 1947. Contribution à l'étude cytologique et histochimique des formations endocrines rétrocérébrales de la larve de *Chironomus plumosus. Rev. canad. Biol.,* 6:777-796.

ARVY, L. and GABE, M. 1953a. Données histo-physiologiques sur la neurosécrétion chez les Paléoptères (Ephéméroptères et Odonates). *Z. Zellforsch.,* 38:591-610.

ARVY, L. and GABE, M. 1953b. Particularités histophysiologiques des glandes endocrines céphaliques chez *Tenebrio molitor* L. *C. R. Acad. Sci., Paris,* 237:844-846.

ARVY, L. and GABE, M. 1954. The intercerebralis - cardiacum - allatum system of some Plecoptera. *Biol. Bull., Woods Hole,* 106:1-14.

ARVY, L., ECHALIER, G., and GABE, M. 1956. Organe Y et gonade chez *Carcinides maenas* L. *Ann. Sci. nat. (Zool.),* 18:263-268.

AVEL, M. 1929. Recherches expérimentales sur les caractères sexuels somatiques des Lumbriciens. *Bull. biol.,* 63:149-318.

AVEL, M. 1947. Les facteurs de la régénération chez les annélides. *Rev. suisse Zool.,* 54:219-235.

BACQ, Z. M. and FLORKIN, M. 1935. Mise en évidence, dans le complexe "ganglion nerveux-glande neurale" d'une Ascidie (*Ciona intestinalis*), de principes pharmacologiquement analogues à ceux du lobe postérieur de l'hypophyse des Vertébrés. *Arch. int. Physiol.,* 40:422-428.

BACQ, Z. M. and FLORKIN, M. 1946. Sur la spécificité des principes extraits de la région neuro-glandulaire de l'Ascidie *Ciona intestinalis. Experientia,* 2:451.

BALESDENT-MARQUET, M. L. 1960. Disposition, structure et mode d'action de la glande androgène d'*Asellus aquaticus* L. (Crustacé Isopode). *C. R. Acad. Sci., Paris,* 251:803-805.

BARGMANN, W. 1958. Elektronenmikroskopische Untersuchungen an der Neurohypophyse. In: *II Internat. Symp. Neurosekretion.* W. Bargmann et al. (eds.). Springer, Berlin.

BARGMANN, W. and KNOOP, A. 1957. Elektronenmikroskopische Beobachtungen an der Neurohypophyse. *Z. Zellforsch.,* 46:242-251.

BARGMANN, W. and KNOOP, A. 1960. Über die morphologischen Beziehungen des neurosekretorischen Zwischenhirnsystems zum Zwischenlappen der Hypophyse (licht- und elektronen-mikroskopische Untersuchungen). *Z. Zellforsch.,* 52:256-277.

BARGMANN, W., HANSTRÖM, B., and SCHARRER, E. (eds.). 1958. *II Internationales Symposium über Neurosekretion.* Springer, Berlin.

BARGMANN, W., KNOOP, A., and THIEL, A. 1957. Elektronenmikroskopische Studie an der Neurohypophyse von *Tropidonotus natrix* (mit Berücksichtigung der Pars intermedia). *Z. Zellforsch.,* 47:114-126.

BARKER-JØRGENSEN, C. and LARSEN, L. 1960. Comparative aspects of hypothalamic-hypophyseal relationships. *Ergebn. Biol.,* 22:1-29.

BARKER-JØRGENSEN, C., ROSENKILDE, P., and WINGSTRAND, K. G. 1956. Regeneration of the neural lobe of the pituitary gland in the toad, *Bufo bufo* L. In: *Bertil Hanström: Zoological Papers in Honour of his Sixty-fifth Birthday.* K. G. Wingstrand (ed.). Zool. Inst., Lund.

BARNES, H. and GONOR, J. J. 1958. Neurosecretory cells in the cirripede *Pollicipes polymerus,* J. B. Sowerby. *J. Mar. Res.,* 17:81-102.

BARRNETT, R. J. 1954. Histochemical demonstration of disulfide groups in the neurohypophysis under normal and experimental conditions. *Endocrinology,* 55:484-501.

BARRY, J. and COTTE, G. 1961. Etude préliminaire au microscope électronique de l'éminence médiane du cobaye. *Z. Zellforsch.,* 53:714-724.

BARTON-BROWNE, L., DODSON, L. F., HODGSON, E. S., and KIRALY, J. K. 1961. Adrenergic properties of the cockroach corpus cardiacum. *Gen. comp. Endocrin.* 1:232-236.

BELLONCI, G. 1881. Sistema nervoso e organi dei sensi della *Sphaeroma*

serratum. Mem. Accad. Lincei (3), 10:91-104.
BENEDEN, F. VAN and JULIN, C. 1884. Le système nerveux central des ascidies adultes et ses rapports avec celui des larves urodèles. *Arch. Biol., Paris*, 5:317-367.
BENOIT, J. and ASSENMACHER, I. 1959. The control by visible radiations of the gonadotropic activity of the duck hypophysis. *Recent Progr. Hormone Res.*, 15:143-164.
BERN, H. A. 1962. The properties of neurosecretory cells. *Gen. comp. Endocrin.*, suppl. 1:117-132.
BERN, H. A. and HAGADORN, I. R. 1959. A comment on the elasmobranch caudal neurosecretory system. In: *Comparative Endocrinology*. A. Gorbman (ed.). Wiley, New York.
BERN, H. A. and TAKASUGI, N. 1962. The caudal neurosecretory system of fishes. *Gen. comp. Endocrin.*, 2:96-110.
BERN, H. A., NISHIOKA, R. S., and HAGADORN, I. R. 1961. Association of elementary neurosecretory granules with the Golgi complex. *J. Ultrastruct. Res.*, 5:311-320.
BERN, H. A., NISHIOKA, R. S., and HAGADORN, I. R. 1962. Neurosecretory granules and the organelles of neurosecretory cells. *Mem. Soc. Endocr.*, 12:21-34.
BILLENSTIEN, D. C. and LEVEQUE, T. F. 1955. The reorganization of the neurohypophyseal stalk following hypophysectomy in the rat. *Endocrinology*, 56:704-717.
BLISS, D. E. 1953. Endocrine control of metabolism in the land crab, *Gecarcinus lateralis* (Fréminville). I. Differences in the respiratory metabolism of sinusglandless and eyestalkless crabs. *Biol. Bull., Woods Hole*, 104:275-296.
BLISS, D. E. 1956. Neurosecretion and the control of growth in a decapod crustacean. In: *Bertil Hanström: Zoological Papers in Honour of his Sixty-fifth Birthday*. K. G. Wingstrand (ed.). Zool. Inst., Lund.
BLISS, D. E. and WELSH, J. H. 1952. The neurosecretory system of brachyuran Crustacea. *Biol. Bull., Woods Hole*, 103:157-169.
BLISS, D. E., DURAND, J. B., and WELSH, J. H. 1954. Neurosecretory systems in decapod Crustacea. *Z. Zellforsch.*, 39:520-536.
BOBIN, G. and DURCHON, M. 1952. Étude histologique du cerveau de *Perinereis cultrifera* Grube (annélide polychète.) Mise en évidence d'un complexe cérébro-vasculaire. *Arch. Anat. micr.*, 41:25-40.
BOBIN, G. and DURCHON, M. 1953. Sur le cerveau d'une annélide en voie de transformation hétéronéréidienne(*Perinereis cultrifera* Grube) et sur le déroulement d'un phénomène de neurosécrétion. *Arch. Anat. micr.*, 42:112-126.

BODENSTEIN, D. 1953a. Studies on the humoral mechanisms in growth and metamorphosis of the cockroach, *Periplaneta americana*. II. The function of the prothoracic gland and the corpus cardiacum. *J. exp. Zool.*, 123:413-434.
BODENSTEIN, D. 1953b. Studies on the humoral mechanisms in growth and metamorphosis of the cockroach, *Periplaneta americana*. III. Humoral effects on metabolism. *J. exp. Zool.*, 124:105-115.
BODENSTEIN, D. 1954. Humoral agents in insect morphogenesis. In: *Aspects of Synthesis and Order in Growth*. D. Rudnick (ed.). Princeton Univ. Press, Princeton.
BODENSTEIN, D. 1955. Endocrine mechanisms in the life of insects. *Recent Progr. Hormone Res.*, 10:157-182.
BODENSTEIN, D. 1957. Studies on nerve regeneration in *Periplaneta americana*. *J. exp. Zool.*, 136:89-116.
BODIAN, D. 1951. Nerve endings, neurosecretory substance and lobular organization of the neurohypophysis. *Johns Hopk. Hosp. Bull.*, 89:354-376.
BOGORAZE, D. and CAZAL, P. 1944. Remarques sur le système stomatogastrique du Poulpe (*Octopus vulgaris* Lamarck). Le complexe rétro-buccal. *Arch. Zool. Exp. Gén.*, 84:115-131.
BOISSON, C.-J. 1949. Recherches histologiques sur le complexe allatocardiaque de *Bacillus rossii* Fabr. *Bull. biol.*, Suppl. 34:1-92.
BOUNHIOL, J.-J. 1936. Dans quelles limites l'écérébration des larves de Lépidoptères est-elle compatible avec leur nymphose? *C. R. Acad. Sci., Paris*, 203:1182-1184.
BOUNHIOL, J.-J. 1938. Rôle possible du ganglion frontal dans la métamorphose de *Bombyx mori* L. *C. R. Acad. Sci., Paris*, 206:773-774.
BOUNHIOL, J.-J. 1957. La métamorphose se produit, chez *Bombyx mori*, après suppression, au dernier stade larvaire, des rélations nerveuses entre cérébroïdes et corps allates, ceux-ci restant longtemps imprégnés de neurosécrétion. *C. R. Acad. Sci., Paris*, 245:1087-1089.
BOUNHIOL, J.-J., GABE, M., and ARVY, L. 1953. Données histophysiologiques sur la neuro-sécrétion chez *Bombyx mori* L., et sur ses rapports avec les glandes endocrines. *Bull. biol.*, 87: 323-333.
BOYCOTT, B. B. and YOUNG, J. Z. 1956. The subpedunculate body and nerve and other organs associated with the optic tract of cephalopods. In: *Bertil Hanström: Zoological Papers in Honour of his Sixty-fifth Birthday*. K. G. Wingstrand (ed.). Zool. Inst., Lund.
BRANDENBURG, J. 1956a. Neurosekretorische Zellen des Regenwurms. *Naturwissenschaften*, 43:453.
BRANDENBURG, J. 1956b. Das endo-

krine System des Kopfes von *Andrena vaga* Pz. (Ins. Hymenopt.) und Wirkung der Stylopisation (*Stylops*, Ins. Strepsipt.). *Z. Morph. Ökol. Tiere*, 45:343-364.
BRATTGÅRD, S.-O., EDSTRÖM, J. E., and HYDÉN, H. 1958. The productive capacity of the neuron in retrograde reaction. *Exp. Cell Res.*, Suppl. 5:185-200.
BRIEN, P. 1948. Embranchement des Tuniciers. Morphologie et réproduction. *Grassé's Traité de Zoologie*, 11:553-894.
BROUSSE, P., IDELMAN, S., and ZAGURY, D. 1958. Mise en évidence de lipoprotéines à groupements-SH au niveau des grains de sécrétion des cellules neurosécrétrices de la blatte, *Blabera fusca* Br. *C. R. Acad. Sci., Paris*, 246:3106-3108.
BROWN, F. A., JR. 1933. The controlling mechanism of chromatophores in *Palaemonetes*. *Proc. nat. Acad. Sci., Wash.*, 19:327-329.
BROWN, F. A., JR. 1944. Hormones in the Crustacea. Their sources and activities. *Quart. Rev. Biol.*, 19:118-143.
BROWN, F. A., JR. 1946. The source and activity of *Crago*-darkening hormone (CDH). *Physiol. Zoöl.*, 19:215-223.
BROWN, F. A., JR. 1950. Endocrine mechanisms. In: *Comparative Animal Physiology*. C. L. Prosser (ed.). Saunders, Philadelphia.
BROWN, F. A., JR. 1952. Hormones in crustaceans. In: *The Actions of Hormones in Plants and Invertebrates*. K. V. Thimann (ed.). Academic Press, New York.
BROWN, F. A. and CUNNINGHAM, U. 1941. Upon the presence and distribution of a chromatophorotropic principle in the central nervous system of *Limulus*. *Biol. Bull., Woods Hole*, 81:80-95.
BROWN, F. A. and EDERSTROM, H. E. 1940. Dual control of certain black chromatophores of *Crago*. *J. exp. Zool.*, 86:53-69.
BROWN, R. H. J. and HARKER, J. E. 1960. A method of controlling the temperature of insect neurosecretory cells in situ. *Nature, Lond.*, 185:392.
BÜCKMANN, D. 1956. Die Umfärbung der Raupen von *Cerura vinula* unter verschiedenen experimentellen Bedingungen. *Naturwissenschaften*, 43: 43-44.
BURGEN, A. S. V. 1959. Introduction to symposium on neurosecretion. *Canad. J. Biochem. Physiol.*, 37:307-308.
BUTCHER, E. O. 1930. The pituitary in the ascidians (*Molgula manhattensis*). *J. exp. Zool.*, 57:1-11.
CAMERON, M. L. 1953. Secretion of an orthodiphenol in the corpus cardiacum of the insect. *Nature, Lond.*, 172:349-350.
CARLISLE, D. B. 1950. Una localizzazione più esatta del principio cro-

Bibliography

matoforotropico della regione neurale della *Ciona intestinalis*. *Pubbl. Staz. zool. Napoli*, 22:192-199.

CARLISLE, D. B. 1951a. On the hormonal and neural control of the release of gametes in ascidians. *J. exp. Biol.*, 28:463-472.

CARLISLE, D. B. 1951b. Corpora lutea in an ascidian, *Ciona intestinalis*. *Quart. J. micr. Sci.*, 92:201-203.

CARLISLE, D. B. 1953a. Studies on *Lysmata seticaudata* Risso (Crustacea Decapoda). IV. On the site of origin of the moult-accelerating principle—experimental evidence. *Pubbl. Staz. zool. Napoli*, 24:285-292.

CARLISLE, D. B. 1953b. Studies on *Lysmata seticaudata* Risso (Crustacea Decapoda). V. The ovarian inhibiting hormone and the hormonal inhibition of sex-reversal. *Pubbl. Staz. zool. Napoli*, 24:355-372.

CARLISLE, D. B. 1953c. Studies on *Lysmata seticaudata* Risso (Crustacea Decapoda). VI. Notes on the structure of the neurosecretory system of the eyestalk. *Pubbl. Staz. zool. Napoli*, 24:435-447.

CARLISLE, D. B. 1953d. Note préliminaire sur la structure du système neurosécréteur du pédoncle oculaire de *Lysmata seticaudata* Risso (Crustacea). *C. R. Acad. Sci., Paris*, 236:2541-2542.

CARLISLE, D. B. 1953e. Origin of the pituitary body of chordates. *Nature, Lond.*, 172:1098.

CARLISLE, D. B. 1954a. The effect of mammalian lactogenic hormone on lower chordates. *J. Mar. biol. Ass. U. K.*, 33:65-68.

CARLISLE, D. B. 1954b. Studies on *Lysmata seticaudata* Risso (Crustacea Decapoda). VIII. The lack of influence of eyestalk ablation and of injection of eyestalk extracts on testicular weight and degree of development of the male genital ducts. *Pubbl. Staz. zool. Napoli*, 25:241-245.

CARLISLE, D. B. 1956a. On the hormonal control of water balance in *Carcinus*. *Pubbl. Staz. zool., Napoli*, 27:227-231.

CARLISLE, D. B. 1956b. An indolealkylamine regulating heart-beat in Crustacea. *Biochem. J.*, 63:32P-33P.

CARLISLE, D. B. 1956c. Studies on the endocrinology of isopod crustaceans. Moulting in *Regia oceanica* L. *J. Mar. biol. Ass. U. K.*, 35:515-520.

CARLISLE, D. B. 1958a. Neurosecretory transport in the pituitary stalk of *Lophius piscatorius*. In: *II Internat. Symp. Neurosekretion*. W. Bargmann et al. (eds.). Springer, Berlin.

CARLISLE, D. B. 1958b. Activation of hormonal secretions. A crustacean chromactivator. *Nature, Lond.*, 182:33-34.

CARLISLE, D. B. 1959a. Moulting hormones in *Palaemon* (*Leander*) (Crustacea Decapoda). II. Differences between populations. *J. Mar. biol. Ass. U. K.*, 38:351-359.

CARLISLE, D. B. 1959b. On the sexual biology of *Pandalus borealis* (Crustacea Decapoda). I. Histology of incretory elements. *J. Mar. biol. Ass. U. K.*, 38:381-394.

CARLISLE, D. B. 1959c. Sexual differentiation in Crustacea Malacostraca. *Mem. Soc. Endocrin.*, 7:9-16.

CARLISLE, D. B. 1959d. On the neurosecretory system of the brain and associated structures in *Sipunculus nudus*, with a note on the cuticle. *Gunma J. med. Sci.*, 8:183-194.

CARLISLE, D. B. 1960. Moulting cycles in Crustacea. *Symp. zool. Soc. London*, 2:109-120.

CARLISLE, D. B. and DOHRN, P. F. R. 1953. Studies on *Lysmata seticaudata* Risso (Crustacea Decapoda). II. Experimental evidence for a growth and moult-accelerating factor obtainable from eyestalks. *Pubbl. Staz. zool. Napoli*, 24:69-83.

CARLISLE, D. B. and KNOWLES, F. G. W. 1959. *Endocrine Control in Crustaceans*. Cambridge Univ. Press, Cambridge.

CARLISLE, D. B. and PASSANO, L. M. 1953. The X-organ of Crustacea. *Nature, Lond.*, 171:1070-1071.

CARLISLE, D. B. and PITMAN, W. J. 1961. Diapause, neurosecretion and hormones in Copepoda. *Nature, Lond.*, 190:827-828.

CARLISLE, D. B., DUPONT-RAABE, M., and KNOWLES, F. 1955. Recherches préliminaires relatives à la séparation et à la comparaison des substances chromactives des Crustacés et des Insectes. *C. R. Acad. Sci., Paris*, 240:665-667.

CASPARI, E. and PLAGGE, E. 1935. Versuche zur Physiologie der Verpuppung von Schmetterlingsraupen. *Naturwissenschaften*, 23:751-752.

CAUSEY, G. 1949. The effect of pressure on nerve-fibre size. *J. Anat., Lond.*, 83:75.

CAZAL, P. 1946. Corps paracardiaques et corps allates chez les Japygidae. *Bull. biol.*, 80:477-482.

CAZAL, P. 1947. Recherches sur les glandes endocrines rétrocérébrales des insectes. 2. Odonates. *Arch. Zool. exp. gén.*, 85:55-82.

CAZAL, P. 1948. Les glandes endocrines rétro-cérébrales des Insectes (étude morphologique). *Bull. biol.*, Suppl. 32:1-227.

CAZAL, P. and BOGORAZE, D. 1944. La glande épistellaire du poulpe *Octopus vulgaris* Lam., organe neuricrine. *Arch. Zool. exp. gén.*, 84:10-22.

CAZAL, P. and BOGORAZE, D. 1949. Les glandes neurocrines des Céphalopodes. *Année biol.*, 25:225-238.

CAZAL, P. and GUERRIER, Y. 1946. Recherches sur les glandes endocrines rétro-cérébroïdiennes des Insectes. I. Etude morphologique chez les Orthoptères. *Arch. Zool. exp. gén.*, 84:303-334.

CHAMPY, C. and CHAMPY-COUJARD, C. 1957. L'étude histochimique des sécrétions des neurones. *Acta anat.*, 30:169-174.

CHARNIAUX-COTTON, H. 1956. Déterminisme hormonal de la différenciation sexuelle chez les Crustacés. *Année biol.*, 32:371-399.

CHARNIAUX-COTTON, H. 1958. Contrôle hormonal de la différenciation du sexe et de la réproduction chez les Crustacés supérieurs. *Bull. Soc. zool. Fr.*, 83:314-336.

CHARNIAUX-COTTON, H. 1960. Sex determination. In: *The Physiology of Crustacea*. Vol. 1. T. H. Waterman (ed.). Academic Press, New York.

CHAUDONNERET, J. 1950. La morphologie céphalique de *Thermobia domestica* (Packard) (Insecte aptérygote thysanoure). *Ann. Sci. nat. (Zool.)*, 12:145-302.

CHAUVET, J., LENCI, M. T., and ACHER, R. 1960. L'ocytocine et la vasopressine du Mouton: réconstitution d'un complexe hormonal actif. *Biochim. biophys. Acta*, 38:266-272.

CLARK, R. B. 1955. Caractères histologiques des cellules neurosécrétrices de *Nephthys*. *C. R. Soc. Biol., Paris*, 241:1171.

CLARK, R. B. 1956a. On the origin of neurosecretory cells. *Ann. Sci. nat. (Zool.)*, 18:199-207.

CLARK, R. B. 1956b. On the transformation of neurosecretory cells into ordinary nerve cells. *Förh. K. fysiogr. Sällsk. Lund*, 26:1-8.

CLARK, R. B. 1958a. The micromorphology of the supraesophageal ganglion of *Nephtys*. *Zool. Jb. (Zool.)*, 68:261-296.

CLARK, R. B. 1958b. The posterior lobes of *Nephtys*: observations on three New England species. *Quart. J. micr. Sci.*, 99:505-510.

CLARK, R. B. 1959. The neurosecretory system of the supra-oesophageal ganglion of *Nephtys* (Annelida; Polychaeta). *Zool. Jb. (Zool.)*, 68:395-424.

CLARK, R. B. and BONNEY, D. G. 1960. Influence of the supra-esophageal ganglion on posterior regeneration in *Nereis diversicolor*. *J. Embryol. exp. Morph.*, 8:112-118.

CLARK, R. B. and CLARK, M. E. 1959. Role of supraesophageal ganglion during the early stages of caudal regeneration in some errant polychaetes. *Nature, Lond.*, 183:1834-1835.

CLARK, R. B. and EVANS, S. M. 1961. The effect of delayed brain extirpation and replacement on caudal regeneration in *Nereis diversicolor*. *J. Embryol. exp. Morphol.*, 9:97-105.

CLAUSEN, D.-M. and PALZ, F.-E. 1959. Das Problem der Sekretion im Zentralnervensystem der Crustaceen. *Zool. Anz.*, 22 (suppl.):92-101.

CLEMENTS, A. N. 1956. Hormonal control of ovary development in mosquitoes. *J. exp. Biol.*, 33:211-223.

COLLOQUES INTERNATIONAUX DU C.N.R.S. 1949. IV. *Endocrinologie des Arthropodes.* Paris, 1947. (See also *Bull. biol. France Belg.*, Suppl. 33, 1949.)

COMPANJEN, W. 1958. Neurosecretory activity in the cerebral ganglion of *Bupulis piniarius* L. (Geometridae, Lepidoptera), during winter. *Acta physiol. pharm. néerl.*, 7:513-514.

CONVEGNO SULLA NEUROSECREZIONE. 1954a. *Pubbl. Staz. zool. Napoli*, 24 (suppl.):1-98.

CONVEGNO SULLA NEUROSECREZIONE 1954b. Summary of the symposium. *Pubbl. Staz. Zool. Napoli*, 24(suppl.): 87-90.

COX, B. L. 1960. Hormonal involvement in the molting process in the soft tick, *Ornithodoros turicata* Dugès. *Anat. Rec.*, 137:347.

CROSS, B. A. and GREEN, J. D. 1959. Activity of single neurones in the hypothalamus: effect of osmotic and other stimuli. *J. Physiol.*, 148:554-569.

CUÉNOT, L. 1949. Les Onychophores. *Grassé's Traité de Zoologie*, 6:3-37.

CURRI, S. B., MARTINI, L., and KOVAC, W. 1958. *Pathophysiologia Diencephalica.* Springer, Vienna.

DAHL, E. 1957. Embryology of X organs in *Crangon allmanni. Nature, Lond.*, 179:482.

DALTON, A. J. 1960. Morphology and physiology of the Golgi apparatus. In: *Cell Physiology of Neoplasia.* M. D. Anderson Hospital and Tumor Institute, Univ. of Texas Press, Austin.

DAVEY, K. G. 1961. The mode of action of the heart accelerating factor from the corpus cardiacum of insects. *Gen. comp. Endocrin.*, 1:24-29.

DAWSON, A. B. 1953. The occurrence of regional distribution of perivascular melanophores within the optic lobes of the frog, *Rana pipiens. Anat. Rec.*, 117:37-47.

DAWSON, A. B. and HISAW, F. L., Jr. 1956. The occurrence of neurosecretory cells in the "cerebral" ganglion of tunicates. *Anat. Rec.*, 125:582. (Abs.)

DAY, M. F. 1940. Neurosecretory cells in the ganglia of Lepidoptera. *Nature, Lond.*, 145:264.

DAY, M. F. 1943. The homologies of the ring gland of Diptera Brachycera. *Ann. ent. Soc. Amer.*, 36:1-10.

DEFRETIN, R. 1955. Recherches cytologiques et histochimiques sur le système nerveux des Néréidiens. *Arch. Zool. exp. gén.*, 92:73-140.

DEFRETIN, R. 1956. Les sécrétions des polyosides et ses rapports avec l'épitoquie chez les Néréidiens. *Ann. Sci. nat.* (Zool.), 18:209-222.

DÉMEUSY, N. 1953. Effets de l'ablation des pédoncules oculaires sur le développement de l'appareil génital mâle de *Carcinus maenas* Pennant. *C. R. Acad. Sci., Paris*, 236:974-975.

DÉMEUSY, N. 1960. Différenciation des voies génitales mâles du crabe *Carcinus maenas* Linné. Rôle des pédoncules oculaires. *Cahiers Biol. mar.* (Sta. biol. Roscoff), 1:259-277.

DEROUX-STRALLA, D. 1948. Recherches anatomo-histologiques préliminaires à une étude des mécanismes endocrines chez les Odonates. *Bull. Soc. zool. Fr.*, 73:31-36.

DEUSE-ZIMMERMANN, R. 1960. Vergleichende Untersuchungen über Neurosekretion bei Enchytraeidae, Tubificidae, und Naididae. *Z. Zellforsch.*, 52:801-816.

DIEPEN, R. and ENGELHARDT, F. 1958. Neuronale Phänomene im Hypothalamus-Hinterlappen System. In: *Pathophysiologia Diencephalica.* S. B. Curri, L. Martini, and W. Kovac (eds.). Springer, Vienna.

DODD, J. M. 1955. The hormones of sex and reproduction and their effects in fish and lower chordates. *Mem. Soc. Endocrin.*, 4:166-187.

DODD, J. M. 1959. Discussion. In: *Comparative Endocrinology.* A. Gorbman (ed.). Wiley, New York. (pp. 262-263.)

DRACH, P. 1955. Système endocrinien pédonculaire, durée d'intermue et vitellogénèse chez *Leander serratus* (Pennant) crustacé décapode. *C. R. Soc. Biol., Paris*, 149:2079-2083.

DRACH, P. and GABE, M. 1962. Evolution cyclique de la neurosécrétion de l'organe X des Carididés au cours de l'intermue. *C. R. Acad. Sci., Paris*, 254:165-167.

DUNCAN, D. 1956. An electron microscope study of the neurohypophysis of a bird, *Gallus domesticus. Anat. Rec.*, 125:457-472.

DUPONT-RAABE, M. 1952a. Étude morphologique et cytologique du cerveau de quelques phasmides. *Bull. Soc. zool. Fr.*, 76:386-397.

DUPONT-RAABE, M. 1952b. Substances chromactives de crustacés et d'insectes. Activité réciproque, répartition, différences qualitatives. *Arch. Zool. exp. gén.*, 89:102-112.

DUPONT-RAABE, M., 1957. Les mécanismes de l'adaptation chromatique chez les insectes. *Arch. Zool. exp. gén.*, 94:61-293.

DUPONT-RAABE, M. 1959. Recherches relatives à la régulation des mouvements pigmentaires chez les insectes. *Gunma J. med. Sci.*, 8:291-300.

DURAND, J. B. 1956. Neurosecretory cell types and their secretory activity in the crayfish. *Biol. Bull., Woods Hole*, 111:62-76.

DURCHON, M. 1952. Recherches expérimentales sur deux aspects de la reproduction chez les annélides polychètes; l'épitoquie et la stolonisation. *Ann. Sci. nat.* (Zool.), 14:117-206.

DURCHON, M. 1956a. Nouvelles recherches expérimentales sur l'épitoquie des néréidiens (Annélides polychètes). *Ann. Sci. nat.* (Zool.) 18:1-13.

DURCHON, M. 1956b. Rôle du cerveau dans la maturation génitale et le déclenchement de l'épitoquie chez les néréidiens. *Ann. Sci. nat.* (Zool.), 18:269-274.

DURCHON, M. 1956c. Influence du cerveau sur les processus de régénération caudale chez les néréidiens (Annélides polychètes). *Arch. Zool. exp. gén.*, 94:1-9.

DURCHON, M. 1957. Problèmes posés par le comportement des néréidiens au moment de leur reproduction. *Année biol.*, 33:31-42.

DURCHON, M. 1959. Contribution à l'étude de la stolonisation chez les syllidiens (Annélides polychètes): I. Syllinae. *Bull. biol.*, 93:155-219.

DURCHON, M. 1960. L'endocrinologie chez les annélides polychètes. *Bull. Soc. zool. Fr.*, 85:275-301.

DURCHON, M. and FRÉZAL, J. 1955. Étude comparée d'un phénomène de neurosécrétion observé dans le cerveau des néréidiens (Annélides polychètes), au moment de la maturité génitale. *C. R. Acad. Sci., Paris*, 241:445-447.

DUVEAU, J. 1961. Données morphologiques sur la voie neurosécrétrice protocéphalique de *Nebalia geoffroyi* Leach. *Bull. Soc. zool. France*, 86:51-58.

DU VIGNEAUD, V. 1956. Hormones of the posterior pituitary gland: oxytocin and vasopressin. *Harvey Lect.*, 1954-1955, 50:1-26.

EAKIN, R. M., QUAY, W. B., and WESTFALL, J. A. 1961. Cytochemical and cytological studies of the parietal eye of the lizard, *Sceloporus occidentalis. Z. Zellforsch.*, 53:449-470.

ECHALIER, G. 1959. L'organe Y et le déterminisme de la croissance et de la mue chez *Carcinus maenas* (L.), crustacé décapode. *Ann. Sci. nat.* (Zool.), (12)1:1-60.

EDMAN, P., FÄNGE, R., and ÖSTLUND, E. 1958. Isolation of the red pigment concentrating hormone of the crustacean eyestalk. In: *II Internat. Symp. Neurosekretion.* W. Bargmann et al. (eds.). Springer, Berlin.

EDSTRÖM, J. E. and EICHNER, D. 1958. Quantitative Ribonukleinsäure-Untersuchungen an den Ganglienzellen des Nucleus supraopticus der Albino-Ratten unter experimentellen Bedingungen (Kochsalz-Belastung). *Z. Zellforsch.*, 48:187-200.

EICHNER, D. 1951. Zur Frage der Neurosekretion der Ganglienzellen des Nebennierenmarkes. *Z. Zellforsch.*, 36:293-297.

EICHNER, D. 1958. Topochemische Untersuchungen am neurosekreto-

Bibliography

rischen Zwischenhirn-Hypophysensystem der Albino-Ratte unter normalen und experimentellen Bedingungen. *Z. Zellforsch.*, 48:402-428.

ELWYN, A. 1937. Some stages in the development of the neural complex in *Ecteinascidia turbinata*. *Bull. neurol. Inst. N. Y.*, 6:163-177.

ENAMI, M. 1951. The sources and activities of two chromatophorotropic hormones in crabs of the genus *Sesarma*. II. Histology of incretory elements. *Biol. Bull., Woods Hole*, 101:241-258.

ENAMI, M. 1957. *An Introduction to the Study of Neurosecretion.* Kyodo Ishyo Shuppanshya, Tokyo. (In Japanese.)

ENAMI, M. 1959. The morphology and functional significance of the caudal neurosecretory system of fishes. In: *Comparative Endocrinology.* A. Gorbman (ed.). Wiley, New York.

ENAMI, M. and IMAI, K. 1955. Studies in neurosecretion. V. Caudal neurosecretory system in several freshwater teleosts. *Endocr. jap.*, 2:107-116.

ENAMI, M. and IMAI, K. 1958. Studies in neurosecretion. XII. Electron microscopy of the secrete granules in the caudal neurosecretory system of the eel. *Proc. Jap. Acad.*, 34:164-168.

ENGELMANN, F. 1957. Die Steuerung der Ovarfunktion bei der ovoviviparen Schabe *Leucophaea maderae* (Fabr.). *J. Insect Physiol.*, 1:257-278.

ENGELMANN, F. 1959. The control of reproduction in *Diploptera punctata* (Blattaria). *Biol. Bull., Woods Hole*, 116:406-419.

ETKIN, W. 1955. Metamorphosis. In: *Analysis of Development.* B. H. Willier, P. A. Weiss, and V. Hamburger (eds.). Saunders, Philadelphia.

EULER, U. S. VON. 1950. Noradrenaline (arterenol), adrenal medullary hormone and chemical transmitter of adrenergic nerves. *Ergebn. Physiol.*, 46:261-307.

EULER, U. S. VON. 1958. Distribution and metabolism of catechol hormones in tissues and axons. *Recent Progr. Hormone Res.*, 14:483-512.

FAHLANDER, K. 1938. Beiträge zur Anatomie und systematischen Einteilung der Chilopoden. *Zool. Bidr. Uppsala*, 17:1-149.

FÄHRMANN, W. 1961. Licht- und elektronenmikroskopische Untersuchungen des Nervensystems von *Unio tumidus* (Philipsson) unter besonderer Berücksichtigung der Neurosekretion. *Z. Zellforsch.*, 54:689-716.

FERNÁNDEZ-MORÁN, H. 1957. Electron microscopy of nervous tissue. In: *Metabolism of the Nervous System.* D. Richter, (ed.). Pergamon Press, London.

FINGERMAN, M. and AOTO, T. 1958a. Evidence for axonal transport of neurosecretory material from the supraesophageal ganglia into the eyestalk of the dwarf crayfish, *Cambarellus shufeldti*. *Anat. Rec.*, 131:552-553.

FINGERMAN, M. and AOTO, T. 1958b. Electrophoretic analysis of chromatophorotropins in the dwarf crayfish, *Cambarellus shufeldti*. *J. exp. Zool.*, 138:25-50.

FINGERMAN, M. and AOTO, T. 1959. The neurosecretory system of the dwarf crayfish *Cambarellus schufeldti*, revealed by electron and light microscopy. *Trans. Amer. micr. Soc.*, 78:305-317.

FINGERMAN, M. and AOTO, T. 1960. Effects of eyestalk ablation upon neurosecretion in the supraesophageal ganglia of the dwarf crayfish, *Cambarellus shufeldti*. *Trans. Amer. micr. Soc.*, 79:68-74.

FINGERMAN, M., LOWE, M. E., and SUNDARARAJ, B. I. 1959a. Darkadapting and light-adapting hormones controlling the distal retinal pigment of the prawn, *Palaemonetes vulgaris*. *Biol. Bull., Woods Hole*, 116:30-36.

FINGERMAN, M., LOWE, M. E., and SUNDARARAJ, B. I. 1959b. Hormones controlling the distal retinal pigment of the crayfish *Orconectes clypeatus*. *Amer. Midl. Nat.*, 62:167-173.

FINGERMAN, M., SANDEEN, M. I., and LOWE, M. E. 1959. Experimental analysis of the red chromatophore system of the prawn, *Palaemonetes vulgaris*. *Physiol. Zoöl.*, 32:128-149.

FLOREY, E. 1951. Neurohormone und Pharmakologie der Arthropoden. *PflSchBer.*, 7:81-141.

FORMIGONI, A. 1956. Neurosécrétion et organes endocrines chez *Apis mellifica* L. *Ann. Sci. nat. (Zool.)*, 18:283-291.

FOX, C. A. and ZABORS, T. E. 1960. Neurons of the supra-optic nucleus in Golgi preparations. *Anat. Rec.*, 136:335.

FRASER, A. 1955. Location and innervation of the corpus cardiacum in the larvae of some Calliphorinae (Diptera). *Nature, Lond.*, 175:817-818.

FRASER, A. 1957. The retrocerebral endocrine organs of the larva of *Protophormia terrae-novae* Robineau Desvoidy (Diptera: Cyclorrhapha). *Proc. R. ent. Soc., London* (A), 32:40-46.

FRASER, A. 1959a. Neurosecretion in the brain of the larva of the sheep blowfly, *Lucilia caesar*. *Quart. J. micr. Sci.*, 100:377-394.

FRASER, A. 1959b. Neurosecretory cells in the abdominal ganglia of larvae of *Lucilia caesar* (Diptera). *Quart. J. micr. Sci.*, 100:395-399.

FRASER, A. 1960. Humoral control of metamorphosis and diapause in the larvae of certain Calliphoridae (Diptera: Cyclorrhapha.) *Proc. roy. Soc. Edinb.*, 67:127-140.

FRIDBERG G. 1959. A histological evidence of the homology between Dahlgren's cells in rays and teleosts. *Acta zool., Stockh.*, 40:101-104.

FRIEDE, R. L. 1959. Transport of oxidative enzymes in nerve fibers; a histochemical investigation of the regenerative cycle in neurons. *Exp. Neurol.*, 1:441-466.

FROST, R., SALOUM, R., and KLEINHOLZ, L. H. 1951. Effect of sinus gland and of eyestalk removal on rate of oxygen consumption in *Astacus*. *Anat. Rec.*, 111:572.

FUJITA, H. 1957. Electron microscopic observation on the neurosecretory granules in the pituitary posterior lobe of dog. *Arch. histol. jap.*, 12:165-172.

FUJITA, H. and HARTMANN, J. F. 1961. Electron microscopy of neurohypophysis in normal, adrenaline-treated and pilocarpine-treated rabbits. *Z. Zellforsch.*, 54:734-763.

FUKUDA, S. 1952. Function of the pupal brain and suboesophageal ganglion in the production of non-diapause and diapause eggs in the silkworm. *Annot. zool. jap.*, 25:149-155.

FUKUDA, S. 1953a. Determination of voltinism in the univoltine silkworm. *Proc. Jap. Acad.*, 29:381-384.

FUKUDA, S. 1953b. Determination of voltinism in the multi-voltine silkworm. *Proc. Jap. Acad.*, 29:385-388.

FUKUDA, S. 1953c. Alteration of voltinism in the silkworm following transection of pupal oesophageal connectives. *Proc. Jap. Acad.*, 29:389-391.

FÜLLER, H. B. 1960. Morphologische und experimentelle Untersuchungen über die neurosekretorischen Verhältnisse im Zentralnervensystem von Blattiden und Culiciden. *Zool. Jb. (Zool.)*, 69:223-250.

GABE, M. 1949. Sur la présence de cellules neurosécrétrices chez *Dentalium entale* Deshayes. *C. R. Acad. Sci., Paris*, 229:1172-1173.

GABE, M. 1951. Donnés histologiques sur la neurosécrétion chez les Ptérotracheidae (Hétéropodes). *Rev. canad. Biol.*, 10:391-410.

GABE, M. 1952a. Particularités histochimiques de l'organe de Hanström (organe X) et de la glande du sinus chez quelques crustacés décapodes. *C. R. Acad. Sci., Paris*, 235:90-92.

GABE, M. 1952b. Sur l'existence d'un cycle sécrétoire dans la glande du sinus (organe pseudofrontal) chez *Oniscus asellus* L. *C. R. Acad. Sci., Paris*, 235:900-902.

GABE, M. 1952c. Particularités histologiques de la glande du sinus et de l'organe X (organe de Bellonci) chez *Sphaeroma serratum* Fabr. *C. R. Acad. Sci., Paris*, 235:973-975.

GABE, M. 1953a. Particularités morphologiques des cellules neurosécrétrices chez quelques Proso-

branches monotocardes. *C. R. Acad. Sci., Paris,* 236:323-325.
GABE, M. 1953b. Peculiarités histologiques des cellules neurosécrétrices chez quelques Gastéropodes opisthobranches. *C. R. Acad. Sci., Paris,* 236:2166-2168.
GABE, M. 1953c. Sur l'existence, chez quelques crustacés malacostracés, d'un organe comparable à la glande de la mue des insectes. *C. R. Acad. Sci., Paris,* 237:1111-1113.
GABE, M. 1953d. Données histologiques sur la neuro-sécrétion chez quelques Sipunculiens. *Bull. Lab. marit. Dinard,* 38:3-15.
GABE M. 1953e. Données histologiques sur les glandes endocrines céphaliques de quelques thysanoures. *Bull. Soc. zool. Fr.,* 78:177-193.
GABE, M. 1953f. Quelques acquisitions récentes sur les glandes endocrines des Arthropodes. *Experientia,* 9:352-356.
GABE, M. 1954a. Sur l'existence de cellules neurosécrétrices chez quelques Onychophores. *C. R. Acad. Sci., Paris,* 238:272-274.
GABE, M. 1954b. La neuro-sécrétion chez les invertébrés. *Année biol.,* 30:5-62.
GABE, M. 1955a. Données histologiques sur la neurosécrétion chez les arachnides. *Arch. Anat. micr.,* 44:351-383.
GABE, M. 1955b. Particularités histologiques des cellules neuro-sécrétrices chez quelques lamellibranches. *C. R. Acad. Sci., Paris,* 240:1810-1812.
GABE, M. 1956. Contribution à l'histologie de la neuro-sécrétion chez les chilopodes. In: *Bertil Hanström: Zoological Papers in Honour of his Sixty-fifth Birthday.* K. G. Wingstrand (ed.). Zool. Inst., Lund.
GABE, M. 1959. Emplacement et connexions des cellules neuro-sécrétrices chez quelques diplopodes. *C. R. Acad. Sci., Paris,* 239:828-830.
GABE, M. 1960. Présence de composés décélable par la réaction à l'acide périodique-Schiff dans le produit de neurosécrétion hypothalamique chez quelques Vertébrés. *C. R. Acad. Sci., Paris,* 250:937-939.
GABE, M. and PRENANT, M. 1952. Quelques particularités histologiques d'*Acteon tornatilis* L. *Bull. Soc. zool. Fr.,* 77:220-228.
GABE, M. and RANCUREL, P. 1958. Caractères histologiques des cellules neurosécrétrices chez quelques *Teredo* (Mollusques lamellibranches). *Bull. Inst. franç. d'Afrique noire,* (A)20:73-78.
GALAMBOS, R. 1961. A glia-neural theory of brain function. *Proc. nat. Acad. Sci., Wash.,* 47:129-136.
GANGULY, D. N. and BANERJEE, M. 1960. Morphological study of the neurosecretory system of the plant bug *Macroceroea grandis* (Gray), Family: Pyrrhocoridae; Hemiptera. *Proc. zool. Soc. Bengal,* 13:71-90.
GANGULY, D. N. and DEB, D. C. 1960. Studies on the cephalic incretory pathways of *Chrysocoris stolli* Wolff (Pentatomidae; Hemiptera). *Anat. Anz.,* 109:28-35.
GELDIAY, S. 1959. Neurosecretory cells in ganglia of the roach, *Blaberus cranifer.* *Biol. Bull., Woods Hole,* 117:267-274.
GERARD, R. W. 1950. Some aspects of neural growth, regeneration, and function. In: *Genetic Neurology.* P. Weiss (ed.). Univ. Chicago Press.
GERSCH, M. 1956. Untersuchungen zur Frage der hormonalen Beeinflussung der Melanophoren bei der *Corethra*-Larve. *Z. vergl. Physiol.,* 39:190-208.
GERSCH, M. 1957. Wesen und Wirkungsweise von Neurohormonen im Tierreich. *Naturwissenschaften,* 44:525-532.
GERSCH, M. 1959. Neurohormone bei wirbellosen Tieren. *Zool. Anz.,* Suppl. 22:40-76.
GERSCH, M. 1961. Insect metamorphosis and the activation hormone. *Amer. Zool.,* 1:53-57.
GERSCH, M. and DEUSE, R. 1960. Über herzaktive Faktoren aus dem Nervensystem von *Aplysia.* *Zool. Jb. (Zool.),* 68:519-534.
GERSCH, M. and KOTHES, G. 1956. Neurohormonalen Wirkungsantagonismus beim Farbwechsel von *Dixippus morosus.* *Naturwissenschaften,* 43:542.
GERSCH, M. and SCHEFFEL, H. 1958. Sekretorisch tätige Zellen im Nervensystem von *Ascaris.* *Naturwissenschaften,* 45:345-346.
GERSCH, M. and UNGER, H. 1957. Nachweis von Neurohormonen aus dem Nervensystem von *Dixippus morosus* mit Hilfe papierchromatographischer Trennung. *Naturwissenschaften,* 44:117.
GERSCH, M., UNGER, H., and FISCHER, F. 1957. Die Isolierung eines Neurohormons aus dem Nervensystem von *Periplaneta americana* und einige biologische Testverfahren. *Wiss. Z. Fr.-Schiller Univ. Jena (Math. Naturwiss. Reihe),* 6:125-129.
GERSCHENFELD, H. M., TRAMEZZANI, J. H., and ROBERTIS, E. DE. 1960. Ultrastructure and function in neurohypophysis of the toad. *Endocrinology,* 66:741-762.
GHOSH, A., BERN, H. A., GHOSH, I., and NISHIOKA, R. S. 1962. Nature of the inclusions in the lumbosacral neurons of birds. *Anat. Rec.,* 143:195-217.
GILBERT, L. I. and SCHNEIDERMAN, H. A. 1961. Some biochemical aspects of insect metamorphosis. *Amer. Zool.,* 1:11-51.
GONTCHAROFF, M. and LECHENAULT, H. 1958. Sur le déterminisme de la ponte chez *Lineus lacteus.* *C. R. Acad. Sci., Paris,* 246:1929-1930.
GOODRICH, E. S. 1917. "Proboscis pores" in craniate vertebrates, a suggestion concerning the premandibular somites and hypophysis. *Quart. J. micr. Sci.,* 62:539-553.
GORBMAN, A. and BARRINGTON, E. J. W. (eds.). 1962. Symposium on Neurosecretion (Denver, 1961). *General and Comparative Endocrinology,* 2:1-169.
GOSLAR, H. G. and SCHULTZE, B. 1958. Autoradiographische Untersuchungen über den Einbau von S^{35}-Thioaminosäuren im Zwischenhirn von Kaninchen und Ratte. *Z. mikr.-anat. Forsch.,* 64:556-574.
GRÄBER, H. 1933. Über die Gehirne der Amphipoden und Isopoden. *Z. Morph. Ökol. Tiere,* 26:334-371.
GRANDORI, L. 1954. Anello di Weismann e neurosecrezioni in *Calliphora erythrocephala* Meig. e *Musca domestica* L. *Boll. Zool. agr. Bachic., Univ. Milano,* 20:51-59.
GRANDORI, L. 1956. Anello di Weismann, metamorfosi e neurosecrezioni in *Calliphora erythrocephala* Meig. e *Musca domestica* L. *Boll. Lab. Zool. Portici,* 33:198-244.
GRANDORI, L. and CARÈ, E. 1953. I corpi faringei (corpora cardiaca) in *Calliphora erythrocephala* Mg. adulta. *Boll. Zool. agr. Bachic., Univ. Milano,* 19:3-10.
GRAY, E. G. and WHITTAKER, V. P. 1960. The isolation of synaptic vesicles from the central nervous system. *J. Physiol.,* 153:35P-37P.
GREEN, J. D. and MAXWELL, D. S. 1959. Comparative anatomy of the hypophysis and observations on the mechanism of neurosecretion. In: *Comparative Endocrinology.* A. Gorbman (ed.). Wiley, New York.
GREEN, J. D. and VAN BREEMEN, V. L. 1955. Electron microscopy of the pituitary and observations on neurosecretion. *Amer. J. Anat.,* 97:177-227.
GRISON, P. 1949. Effets d'implantation de cerveaux chez le doriphore *Leptinotarsa decemlineata* Say. en diapause. *C. R. Acad. Sci., Paris,* 228:428-430.
GROOT, J. DE. 1957. Neurosecretion in experimental conditions. *Anat. Rec.,* 127:201-217.
GRZYCKI, M. S. 1951. Topography and structure of the probable neurosecretory material in the ganglion cells of the snails (*Limnaea stagnalis* L., *Planorbis corneus* L. and *Paludina vivipara* L.). *C. R. Cl. Sc. Math. Nat. Acad. Pol.,* 1-2, No. 5.
HABIBULLA, M. 1961. Secretory structures associated with the neurosecretory system of the immature scorpion, *Heterometrus swammerdami.* *Quart. J. micr. Sci.,* 102:475-480.
HADLER, W. A., TRAMEZZANI, J. B., BEREZIN, A., SESSON, A., and LISON, L. 1957. Action des lécithinases sur le

Bibliography

matériel neurosécrétoire de l'hypophyse postérieure et l'hypothalamus. *C. R. Acad. Sci., Paris*, 245:2095-2097.

HAEFELFINGER, H. R. 1954. Inkretorische Drüsenkomplexe im Gehirn decapoder Cephalopoden (Untersuchungen an *Illex coindeti*). *Rev. suisse Zool.*, 61:153-162.

HAGADORN, I. R. 1958. Neurosecretion and the brain of the rhynchobdellid leech, *Theromyzon rude* (Baird, 1869). *J. Morph.*, 102:55-90.

HAGADORN, I. R. 1962. Neurosecretory phenomena in the leech *Theromyzon rude*. *Mem. Soc. Endocrin.*, 12:313-321.

HAGADORN, I. R. and NISHIOKA, R. S. 1961. Neurosecretion and granules in neurones of the leech brain. *Nature, Lond.*, 191:1013-1014.

HAGEN, E. 1954. Morphologische und experimentelle Untersuchungen am Hypophysen-Zwischenhirnsystem. *Verh. anat. Ges. Jena*, 51:93-95.

HAGER, H. and TAFURI, W. L. 1959. Elektronenoptische Untersuchungen über die Feinstruktur des Plexus myentericus (Auerbach) im Colon des Meerschweinchens *(Cavia cobaya)*. *Arch. Psychiat. Nervenkr.*, 199:427-471.

HANSTRÖM, B. 1935. Preliminary report on the probable connection between the blood gland and the chromatophore activator in decapod crustaceans. *Proc. nat. Acad. Sci., Wash.*, 21:584-585.

HANSTRÖM, B. 1937a. Die Sinusdrüse und der hormonal bedingte Farbwechsel der Crustaceen. *K. svenska VetenskAkad. Handl.*, 16:1-99.

HANSTRÖM, B. 1937b. Vermischte Beobachtungen über die chromatophoraktivierenden Substanzen der Augenstiele der Crustaceen und des Kopfes der Insekten. *Acta Univ. Lund, Avd. 2*, 32:1-11.

HANSTRÖM, B. 1938. Untersuchungen aus dem Oeresund XXVI. Zwei Probleme betreffs der hormonalen Lokalisation im Insektenkopf. *Acta Univ. Lund, Avd. 2*, 34:17.

HANSTRÖM, B. 1939. *Hormones in Invertebrates*. Oxford Univ. Press, Oxford.

HANSTRÖM, B. 1940. Inkretorische Organe, Sinnesorgane und Nervensystem des Kopfes einiger niederer Insektenordnungen. *K. svenska VetenskAkad. Handl.*, (3)18:1-265.

HANSTRÖM, B. 1941. Die Corpora cardiaca und Corpora allata der Insekten. *Biol. gen.*, 15:485-531.

HANSTRÖM, B. 1943. Ergänzende Beobachtungen über das Corpus cardiacum und das Stirnauge der Machiliden und das Gehirn der Campodeiden. *Förh. K. fysiogr. Sällsk. Lund*, 13:215-219.

HANSTRÖM, B. 1947. Three Principal Incretory Organs in the Animal Kingdom. Einar Munksgaard, Copenhagen.

HANSTRÖM, B. 1948. The brain, the sense organs, and the incretory organs of the head in the Crustacea Malacostraca. *Bull. biol., Suppl.* 33:98-126.

HANSTRÖM, B. 1953. Neurosecretory pathways in the head of crustaceans, insects and vertebrates. *Nature, Lond.*, 171:72-73.

HANSTRÖM, B. 1954. On the transformation of ordinary nerve cells into neurosecretory cells. *Förh. K. fysiogr. Sällsk. Lund*, 24:75-82.

HANSTRÖM, B. 1955. Notes on the hypothalamic neurosecretion in the wolf. *Förh. K. fysiogr. Sällsk. Lund*, 25:89-100.

HANSTRÖM, B. 1957. The comparative aspect of neurosecretion with special reference to the hypothalamo-hypophysial system. In: *The Neurohypophysis*. H. Heller (ed.). Butterworths, London.

HARA, J. 1952a. On the hormones regulating the frequency of the heartbeat in the shrimp *Paratya compressa*. *Annot. zool. jap.*, 25:162-171.

HARA, J. 1952b. On the effects of extracts of the grapsoid crab, *Sesarma picta*, and of the head of the pill bug, *Armadillidium vulgare*, upon the heartbeat in the shrimp, *Paratya compressa*. *Annot. zool. jap.*, 25:411-414.

HARKER, J. E. 1956. Factors controlling the diurnal rhythm of activity of *Periplaneta americana* L. *J. exp. Biol.*, 33:224-234.

HARKER, J. E. 1958. Diurnal rhythms in the animal kingdom. *Biol. Rev.*, 33:1-52.

HARKER, J. E. 1960a. The effect of perturbations in the environmental cycle on the diurnal rhythm of activity of *Periplaneta americana* L. *J. exp. Biol.*, 37:154-163.

HARKER, J. E. 1960b. Internal factors controlling the suboesophageal ganglion neurosecretory cycle in *Periplaneta americana* L. *J. exp. Biol.*, 37:164-170.

HARMS, J. W. 1948. Über ein inkretorisches Cerebralorgan bei Lumbriciden, sowie Beschreibung eines verwandten Organs bei drei neuen *Lycastis*-Arten. *Arch. EntwMech. Org.*, 143:332-346.

HARRIS, G. W. 1947. The innervation and actions of the neurohypophysis; an investigation using the method of remote-control stimulation. *Philos. Trans.* (B), 232:385-441.

HARRIS, G. W. 1960. Central control of pituitary secretion. In: *Handbook of Physiology*. Sect. 1. *Neurophysiology*, 2:1007-1038.

HARTMANN, J. F. 1958. Electron microscopy of the neurohypophysis in normal and histamine-treated rats. *Z. Zellforsch.*, 48:291-308.

HASEGAWA, K. 1957. The diapause hormone of the silkworm, *Bombyx mori*. *Nature. Lond.*, 179:1300-1301.

HAUENSCHILD, C. 1956a. Hormonale Hemmung der Geschlechtsreife und Metamorphose bei dem Polychaeten *Platynereis dumerilii*. *Z. Naturf.*, 11B:125-132.

HAUENSCHILD, C. 1956b. Weitere Versuche zur Frage des Juvenilhormons bei *Platynereis*. *Z. Naturf.*, 11B:610-612.

HAUENSCHILD, C. 1959a. Zyklische Veränderungen an den inkretorischen Drüsenzellen im Prostomium des Polychaeten *Platynereis dumerilii* als Grundlage der Schwarmperiodizität. *Z. Naturf.*, 14B:81-87.

HAUENSCHILD, C. 1959b. Hemmender Einfluss der Proventrikelregion auf Stolonisation und Oocyten-Entwicklung bei dem Polychaeten *Autolytus prolifer*. *Z. Naturf.*, 14B:87-89.

HAUENSCHILD, C. 1960. Abhängigkeit der Regenerationsleistung von der inneren Sekretion im Prostomium bei *Platynereis dumerilii*. *Z. Naturf.*, 15B:52-59.

HEBB, C. O. 1959. Chemical agents of the nervous system. *Int. Rev. Neurobiol.*, 1:165-193.

HEBB, C. O. and WAITES, G. M. H. 1956. Choline acetylase in antero- and retro-grade degeneration of a cholinergic nerve. *J. Physiol.*, 132:667-671.

HEKSTRA, G. P. and LEVER, J. 1960. Some effects of ganglion extirpations in *Limnaea stagnalis*. *Proc. Kon. Ned. Akad. Wet.*, (C) 63:271-282.

HELLER, H. and CLARK, R. B. (eds.). 1962. *Proceedings of the Third International Symposium on Neurosecretion*. Memoir No. 12, Soc. for Endocrinology. Academic Press, New York.

HELLER, H. and SMITH, B. 1948. The water-balance principle of crustacean eyestalk extracts. *J. exp. Biol.*, 25:388-394.

HERLANT-MEEWIS, H. 1955. Neurosécrétion chez les Oligochètes. *Bull. Acad. Belg. Cl. Sci.*, 41:500-508.

HERLANT-MEEWIS, H. 1956a. Reproduction et neurosécrétion chez *Eisenia foetida* (Sav.) *Ann. Soc. zool. Belg.*, 87:151-183.

HERLANT-MEEWIS, H. 1956b. Croissance et neurosécrétion chez *Eisenia foetida* Sav. *Ann. Sci. nat. (Zool.)*, 18:185-198.

HERLANT-MEEWIS, H. 1959. Phénomènes neuro-sécrétoires et sexualité chez *Eisenia foetida*. *C. R. Acad. Sci., Paris*, 248:1405-1406.

HERLANT-MEEWIS, H. and MOL, J.-J. VAN. 1959. Phénomènes neurosécrétoires chez *Arion rufus* et *Arion subfuscus*. *C. R. Acad. Sci., Paris*, 249:321-322.

HERLANT-MEEWIS, H. and NAISSE, J. 1957. Phénomènes neurosécrétoires et glandes endocrines chez les opilions. *C. R. Acad. Sci., Paris*, 245:858-860.

HERLANT-MEEWIS, H. and PAQUET, L. 1956. Neurosécrétion et mue chez *Carausius morosus* Brdt. *Ann. Sci. nat. (Zool.)*, 18:163-169.

HESS, A. 1958. The fine structure of nerve cells and fibers, neuroglia, and sheaths of the ganglion chain in the cockroach *(Periplaneta americana)*. *J. biophys. biochem. Cytol.*, 4:731-742.

HIDAKA, T. 1956. Recherches sur le déterminisme hormonal de la coloration pupale chez lépidoptères I. Les effets de la ligature, de l'ablation des ganglions et de l'incision des nerfs chez prépupes et larves âgées de quelque papilionides. *Annot. zool. jap.*, 29:69-74.

HIDAKA, T. 1957. Recherches sur le déterminisme hormonal de la coloration chez lépidoptères II. Sur le cas de deux nymphalides. *Annot. zool. jap.*, 30:83-85.

HIGHNAM, K. C. 1958. Activity of the brain/corpora cardiaca system during pupal diapause "break" in *Mimas tiliae* (Lepidoptera). *Quart. J. micr. Sci.*, 99:73-88.

HIGHNAM, K. C. 1961. The histology of the neurosecretory system of the adult female desert locust, *Schistocerca gregaria*. *Quart. J. micr. Sci.*, 102:27-38.

HILD, W. 1951. Experimentell-morphologische Untersuchungen über das Verhalten der "neurosekretorischen Bahn" nach Hypophysenstieldurchtrennungen, Eingriffen in den Wasserhaushalt und Belastung der Osmoregulation. *Virchows Arch.*, 319:526-546.

HILD, W. 1954. Das morphologische, kinetische und endokrinologische Verhalten von hypothalamischem und neurohypophysärem Gewebe in vitro *Z. Zellforsch.*, 40:257-312.

HILD, W. 1956. Neurosecretion in the central nervous system. In: *Hypothalamic-Hypophysial Interrelationships*. W. S. Fields et al. (eds.). Thomas, Springfield.

HILD, W. and ZETLER, G. 1953. Über die Funktion des Neurosekrets im Zwischenhirn - Neurohypophysensystem als Trägersubstanz für Vasopressin, Adiuretin und Oxytocin. *Z. ges. exp. Med.*, 120:236-243.

HODGE, M. H. and CHAPMAN, G. B. 1958. Some observations on the fine structure of the sinus gland of a land crab, *Gecarcinus lateralis*. *J. biophys. biochem. Cytol.*, 4:571-574.

HODGSON, E. S. and GELDIAY, S. 1959. Experimentally induced release of neurosecretory materials from roach corpora cardiaca. *Biol. Bull., Woods Hole*, 117:275-283.

HOFER, H. 1959. Über das Infundibularorgan und den Reissnerschen Faden von *Branchiostoma lanceolatum*. *Zool. Jb. (Anat.)*, 77:465-490.

HOGG, B. M. 1937. Subneural gland of ascidian *(Polycarpa tecta)*: an ovarian stimulating action in immature mice. *Proc. Soc. exp. Biol., N. Y.*, 35:616-618.

HOLMES, R. L. and KNOWLES, F. G. W. 1959. Electron microscope observations on the neurohypophysis of the ferret. *Nature, Lond.*, 183:1745.

HOLMES, R. L. and KNOWLES, F. G. W. 1960. "Synaptic vesicles" in the neurohypophysis. *Nature, Lond.*, 185:710-711.

HOLMGREN, N. 1916. Zur vergleichenden Anatomie des Gehirns bei Polychäten, Onychophoren, Xyphosuren, Arachniden, Crustaceen, Myriapoden und Insekten. *K. svenska VetenskAkad. Handl.*, 56:1-303.

HOLMGREN, U. and CHAPMAN, G. B. 1960. The fine structure of the urophysis spinalis of the teleost fish, *Fundulus heteroclitus* L. *J. Ultrastruct. Res.*, 4:15-25.

HORRIDGE, G. A. 1961. The organization of the primitive central nervous system as suggested by examples of inhibition and the structure of the neuropile. In: *Nervous Inhibition*. E. Florey (ed.). Pergamon Press, New York.

HOWE, A. 1959. A combined light and electron microscopic examination of the neural lobe of the pituitary of the rat. *J. Anat., Lond.*, 93:572.

HOWE, A. and PEARSE, A. G. E. 1956. A histochemical investigation of neurosecretory substance in the rat. *J. Histochem. Cytochem.*, 4:561-569.

HUBL, H. 1953. Die inkretorischen Zellelemente im Gehirn der Lumbriciden. *Arch. EntwMech. Org.*, 146:421.

HUBL, H. 1956. Über die Beziehungen der Neurosekretion zum Regenerationsgeschehen bei Lumbriciden nebst Beschreibung eines neuartigen neurosekretorischen Zelltyps im Unterschlundganglion. *Arch. EntwMech. Org.*, 149:73-87.

HUTCHINSON, G. E. 1928. The branchial gland of the Cephalopoda: a possible endocrine organ. *Nature, Lond.*, 121:674-675.

HYDÉN, H. 1960. The neuron. In: *The Cell*. Vol. 4. J. Brachet and A. E. Mirsky (eds.). Academic Press, New York.

HYMAN, L. H. 1951. *The Invertebrates: Platyhelminthes and Rhynchocoela*. Vol. 2. McGraw-Hill, New York.

ICHIKAWA, M. and NISHIITSUTSUJI, J. 1951. Studies on insect metamorphosis. I. Role of the brain in the imaginal differentiation of lepidopterans. *Annot. zool. jap.*, 24:205-211.

ICHIKAWA, M. and NISHIITSUTSUJI, J. 1952. Studies on insect metamorphosis. II. Determination of the critical period for pupation in the Erisilkworm, *Philosamia cynthia ricini*. *Annot. zool. jap.*, 25:143-148.

INOUE, H. 1957. On the neurosecretory cells of *Pachygrapsus crassipes*. *Mem. Gakugei Fac. Akita Univ. nat. Sci.*, 7:84-92. (In Japanese with English summary.)

ISHIZAKI, N., ISHIDA, Y., and KAWAKATSU, Y. 1959. Histological studies on the hypothalamic neurosecretory nucleus. VII. Histological findings of intracellular neurofibrils in the hypothalamic neurosecretory cell. *Arch. histol. jap.*, 18:429-438.

ITO, T. 1954. Neurosecretory phenomenon of the ganglion cells in the adrenal medulla of the golden hamster. *Okajimas Folia Anat. Jap.*, 26:221-226.

JOHANSSON, A. S. 1958. Relation of nutrition to endocrine-reproductive functions in the milkweed bug *Oncopeltus fasciatus* (Dallas) (Heteroptera: Lygaeidae). *Nytt Mag. Zool., Oslo*, 7:1-132.

JOLY, P. 1945a. La fonction ovarienne et son contrôle humoral chez les Dytiscidés. *Arch. Zool. exp. gén.*, 84:49-164.

JOLY, P. 1945b. Les corrélations humorales chez les insectes. *Année biol.*, 21:1-34.

JOLY, P. 1950. Les hormones sexuelles des insectes. *Bull. biol.* Suppl. 33:81-86.

JOLY, P. 1958. Les corrélations humorales chez les Acridiens. *Année biol.*, 34:97-118.

JOLY, R. 1962. Les glandes cérébrales, organes inhibiteurs de la mue chez les Myriapodes Chilopodes. *C. R. Acad. Sci., Paris*, 254:1679-1681.

JONES, B. M. 1956a. Endocrine activity during insect embryogenesis. Function of the ventral head glands in locust embryos (*Locustana pardalina* and *Locusta migratoria*, Orthoptera). *J. exp. Biol.*, 33:174-185.

JONES, B. M. 1956b. Endocrine activity during insect embryogenesis. Control of events in development following the embryonic moult (*Locusta migratoria* and *Locustana pardalina*, Orthoptera). *J. exp. Biol.*, 33:685-696.

JUBERTHIE-JUPEAU, L. 1961. Données sur la neurosécrétion protocérébrale et mise en évidence de glandes céphaliques chez *Scutigerella pagesi* Jupeau (Myriapode, Symphyle). *C. R. Acad. Sci., Paris*, 253:3081-3083.

JUNQUA, C. 1956. Étude morphologique et histophysiogique des organes endocrines de l'*Hydrocirius columbiae* Spin (Hémiptères bélostomidés). *Bull. biol.*, 90:154-162.

JUNQUA, C. 1957. Aspects histologiques du système nerveux d'un Solifuge. *Bull. Soc. zool. Fr.*, 82:136-138.

JUPEAU, L. 1956. Présence d'organes glandulaires céphaliques chez *Scutigerella immaculata* Newport (Symphyles). *C. R. Acad. Sci., Paris,* 243:96-98.

KARLSON, P. 1956. Biochemical studies on insect hormones. *Vitamins and Hormones,* 14:228-266.

KARLSON, P. 1958. Zur Chemie und Wirkungsweise der Insektenhormone. *Proc. 4th Int. Cong. Biochem.* XII. *Biochemistry of Insects,* pp. 37-47.

KARLSON, P. 1960. Pheromones. *Ergebn. Biol.,* 22:212-225.

KARLSON, P. and SKINNER, D. 1960. Attempted extraction of crustacean moulting hormone from isolated Y-organs. *Nature, Lond.,* 185:543-544.

KASTL, E. 1954. Über die intracytären Neurofibrillen in den Ganglienzellen des Nucleus supraopticus bei Mensch und Hund. *Acta neuroveg.,* 8:437-445.

KATAKURA, Y. 1960. Transformation of ovary into testis following implantation of androgenous glands in *Armadillidium vulgare,* an isopod crustacean. *Annot. zool. jap.,* 33:241-244.

KERKUT, G. A. and LAVERACK, M. S. 1960. A cardio-accelerator present in tissue extracts of the snail *Helix aspera. Comp. Biochem. Physiol.,* 1:62-71.

KIRCHNER, E. 1960. Untersuchungen über neurohormonale Faktoren bei *Melolontha vulgaris. Zool. Jb. (Zool.),* 69:43-62.

KLEINHOLZ, L. H. 1936. Crustacean eyestalk hormone and retinal pigment migration. *Biol. Bull., Woods Hole,* 70:159-184.

KLEINHOLZ, L. H. 1942. Hormones in Crustacea. *Biol. Rev.,* 17:91-119.

KLEINHOLZ, L. H. 1957. Endocrinology of invertebrates, particularly crustaceans. In: *Recent Advances in Invertebrate Physiology.* B. T. Scheer (ed.). Univ. of Oregon Press, Eugene.

KLEINHOLZ, L. H. 1958. Neurosecretion and retinal pigment movement in crustaceans. In: *II Internat. Symp. Neurosekretion.* W. Bargmann et al. (eds.). Springer, Berlin.

KLEINHOLZ, L. H. 1961. Pigmentary effectors. In: *The Physiology of Crustacea.* Vol. 2. T. H. Waterman (ed.). Academic Press, New York.

KLEINHOLZ, L. H., BURGESS, P. R., CARLISLE, D. B., and PFLUEGER, O. 1962. Neurosecretion and crustacean retinal pigment hormone: distribution of the light-adapting hormone. *Biol. Bull.* 122:73-85.

KLUG, H. 1958. Neurosekretion und Aktivitätsperiodik bei Carabiden. *Naturwissenschaften,* 45:141-142.

KNOCHE, H. 1958. Über die Ausbreitung und Herkunft der nervösen Nodulusfasern in Hypothalamus und Retina. *Z. Zellforsch.,* 48:602-616.

KNOWLES, F. G. W. 1939. The control of the white reflecting chromatophores in Crustacea. *Pubbl. Staz. zool. Napoli,* 17:174-182.

KNOWLES, F. G. W. 1951. Hormone production within the nervous system of a crustacean. *Nature, Lond.,* 167:564.

KNOWLES, F. G. W. 1953. Endocrine activity in the crustacean nervous system. *Proc. roy. Soc.,* (B) 141:248-267.

KNOWLES, F. G. W. 1954. Neurosecretion in the tritocerebral complex of crustaceans. *Pubbl. Staz. zool. Napoli,* 24 (Suppl.):74-78.

KNOWLES, F. G. W. 1955. Crustacean colour change and neurosecretion. *Endeavour,* 14:95-104.

KNOWLES, F. G. W. 1956. Some problems in the study of colour-change in crustaceans. *Ann. Sci. nat. (Zool.),* 18:315-324.

KNOWLES, F. G. W. 1958. Electron microscopy of a crustacean neurosecretory organ. In: *II Internat. Symp. Neurosekretion.* W. Bargmann et al. (eds.). Springer, Berlin.

KNOWLES, F. G. W. 1959. The control of pigmentary effectors. In: *Comparative Endocrinology.* A. Gorbman (ed.). Wiley, New York.

KNOWLES, F. G. W. 1960. A highly organized structure within a neurosecretory vesicle. *Nature, Lond.,* 185:709-710.

KNOWLES, F. G. W. and CARLISLE, D. B. 1956. Endocrine control in the Crustacea. *Biol. Rev.,* 31:396-473.

KNOWLES, F. G. W., CARLISLE, D. B., and DUPONT-RAABE, M. 1955. Studies on pigment activating substances in animals. I. The separation by paper electrophoresis of chromactivating substances in arthropods. *J. Mar. biol. Ass. U. K.,* 34:611-635.

KOBAYASHI, M. and BURDETTE, W. J 1961. Effect of brain hormone from *Bombyx mori* on metamorphosis of *Calliphora erythrocephala. Proc. Soc. exp. Biol., N.Y.,* 107:240-242.

KOBAYASHI, M. and KIRIMURA, J. 1958. The "brain" hormone in the silkworm, *Bombyx mori* L. *Nature, Lond.,* 181:1217.

KOBAYASHI, H., BERN, H. A., NISHIOKA, R. S., and HYODO, Y. 1961. The hypothalamo-hypophyseal neurosecretory system of the parakeet, *Melopsittacus undulatus. Gen. comp. Endocrin.,* 1:545-564.

KOELLE, G. B. 1961. A proposed dual neurohumoral role of acetylcholine: its functions at the pre- and postsynaptic sites. *Nature, Lond.,* 190:208-211.

KOELLE, G. B. and GEESEY, C. N. 1961. Localization of acetylcholinesterase in the neurohypophysis and its functional implications. *Proc. Soc. exp. Biol., N.Y.,* 106:625-628.

KOENIG, E. and KOELLE, G. B. 1960. Acetylcholinesterase regeneration in peripheral nerve after irreversible inactivation. *Science,* 132:1249-1250.

KOLLER, G. 1948. Rhythmische Bewegung und hormonale Steuerung bei den Malpighischen Gefässen der Insekten. *Biol. Zbl.,* 67:201-211.

KOLLER, G. 1959. Hormonale Regulation bei *Phascolosoma vulgare. Zool. Anz.,* Suppl. 22:84-91.

KONOK, I. 1960. Studies on the neurosecretory activity of the brain in the fresh water crustacean, *Astacus leptodactylus* Eschscholz (Decapoda). *Ann. Inst. Biol. hung. Acad. Sci.,* 27:15-28.

KOPEĆ, S. 1912. Über die Funktionen des Nervensystems der Schmetterlinge während der successiven Stadien ihrer Metamorphose. *Zool. Anz.,* 40:353-360.

KOPEĆ, S. 1917. Experiments on metamorphosis of insects. *Bull. int. Acad. Cracovie (Acad. pol. Sci.),* 1917B:57-60.

KOPEĆ, S. 1922. Studies on the necessity of the brain for the inception of insect metamorphosis. *Biol. Bull., Woods Hole,* 42:323-342.

KÖPF, H. 1957. Über Neurosekretion bei Drosophila. *Biol. Zbl.,* 76:28-42.

KÖPF, H. 1958. Beitrag zur Topographie und Histologie neurosekretorischer Zentren bei Drosophila. II. Larven- und Puppenstadien. *Zool. Anz.,* 21 (Suppl.):439-443. *(Verh. dtsch. zool. Ges.)*

KORN, H. 1959. Vergleichend-embryologische Untersuchungen an *Harmothoë* Kinberg 1857. (Polychaeta, Annelida). Organogenese und Neurosekretion. *Z. wiss. Zool.,* 161:3-4.

KORN, H. 1960. Das larvale Nervensystem von *Pectinaria* Lamarck und *Nephthys* Cuvier (Annelida, Polychaeta). *Zool. Jb. (Anat.),* 78:427-456.

KRATZSCH, E. 1951. Experimentell-morphologische Untersuchungen am Zwischenhirn-Hypophysensystem der Ratte bei Polyurie infolge Alloxanvergiftung (mit besonderer Berücksichtigung der Pituizyten). *Z. Zellforsch.,* 36:371-380.

KRAUSE, E. 1960. Untersuchungen über die Neurosekretion im Schlundring von *Helix pomatia* L. *Z. Zellforsch.,* 51:748-776.

KÜHN, A. and PIEPHO, H. 1936. Über hormonale Wirkungen bei der Verpuppung der Schmetterlinge. *Nachr. Akad. Wiss. Göttingen, math.-phys. Kl.,* (6) 2:141-154.

KÜHNE, H. 1959. Die neurosekretorischen Zellen und der retrocerebrale neuro-endokrine Komplex von Spinnen (Araneae, Labidognatha) unter Berücksichtigung einiger histologisch erkennbarer Veränderungen während des postembryonalen Lebensablaufes. *Zool. Jb. (Anat.),* 77:527-600.

KUHNEN, H. and BECK, H. 1958. Versuche über myotrope Wirkungen von Arthropoden-Hormonen. *Zool. Anz.,* 22 (suppl.):101-108. *(Verh. dtsch. zool. Ges.)*

LARSEN, J. R. and BODENSTEIN, D. 1959. The humoral control of egg matu-

ration in the mosquito. *J. exp. Zool.*, 140:343-381.
LASANSKY, A. and SABATINI, D. D. 1957. Distribution des groupes sulfhydrile et disulfure dans la neurohypophyse et l'hypothalamus du Crapaud. *C. R. Soc. Biol., Paris*, 151:1755.
LECHENAULT, H. 1962. Sur l'existence de cellules neurosécrétrices dans les ganglions cérébroïdes des Lineidae (Hétéronémertes). *C. R. Acad. Sci., Paris*, 255:194-196.
LEES, A. D. 1955. *The Physiology of Diapause in Arthropods*. Cambridge Univ. Press, Cambridge.
LEES, A. D. 1959. Photoperiodism in insects and mites. In: *Photoperiodism and Related Phenomena in Plants and Animals*. R. B. Withrow (ed.). Amer. Ass. Adv. Sci., Wash., D. C.
LEGAIT, E. and LEGAIT, H. 1958. Recherches sur l'ultrastructure de l'hypophyse de quelques Téléostéens. *C. R. Soc. Biol., Paris*, 152:130-133.
LEGAIT, E. and LEGAIT, H. 1959. Recherches sur l'ultrastructure de la neurohypophyse. *C. R. Ass. Anat.*, 45:514-518.
LEGAIT, H. 1958. Les voies extra-hypothalamo-neurohypophysaires de la neurosécrétion diencéphalique dans la série des vertébrés. In: *II Internat. Symp. Neurosekretion*. W. Bargmann et al. (eds.). Springer, Berlin.
LEGENDRE, R. 1953. Recherches sur les glandes prosomatiques des araignées du genre: *Tegenaria*. *Ann. Univ. Saraviensis*, 2:305-333.
LEGENDRE, R. 1954. Données anatomiques sur le complexe neuro-endocrine rétrocérébral des aranéides. *Ann. Sci. nat. (Zool.)*, 16:420-426.
LEGENDRE, R. 1955. L'organe pariétal des aranéides. *Ann. Univ. Saraviensis*, 4:145-150.
LEGENDRE, R. 1959a. Contribution à l'étude du système nerveux des aranéides. *Ann. Sci. nat. (Zool.)*, 1:339-473.
LEGENDRE, R. 1959b. Sur la présence de cellules neurosécrétrices dans les ganglions sus-oesophagiens de la sangsue médicinale (*Hirudo medicinalis* L.), suivie de quelques considérations sur la neurosécrétion. *Bull. biol.*, 93:462-471.
LEGRAND, J.-J. and JUCHAULT, P. 1960a. Disposition métamérique du tissu sécréteur de l'hormone mâle chez les différents types d'Oniscoïdes. *C. R. Acad. Sci., Paris*, 250:764-766.
LEGRAND, J.-J. and JUCHAULT, P. 1960b. Mise en évidence anatomique et expérimentale des glandes androgènes de *Sphaeroma serratum* Fabricius (Isopode, Flabellifère). *C. R. Acad. Sci., Paris*, 250:3401-3402.
LEHMANN, H. J. and STANGE, H. H. 1953. Über das Vorkommen vakuolenhaltiger Ganglienzellen im Ganglion cervicale Uteri trächtiger und nichtträchtiger Ratten. *Z. Zellforsch.*, 38:230-236.

LEMCHE, H. 1955. Neurosecretion and incretory glands in a tectibranch mollusc. *Experientia*, 11:320-322.
LEMCHE, H. 1956. The anatomy and histology of *Cylichna*. *Spolia zool. Mus. Hauniensis*, 16:1-278.
LENDER, T. and KLEIN, N. 1961. Mise en évidence de cellules sécrétrices dans le cerveau de la Planaire *Polycelis nigra*. Variation de leur nombre au cours de la régénération postérieure. *C. R. Acad. Sci., Paris*, 253:331-333.
LENNETTE, E. H. and SCHARRER, E. 1946 Neurosecretion. IX. Cytoplasmic inclusions in peripheral autonomic ganglion cells of the monkey. *Anat. Rec.*, 94:85-92.
LERMA, B. DE. 1937. Osservazioni sul sistema endocrino degli insetti ("corpora allata" e "corpi faringei"). *Arch. zool. (ital.), Napoli*, 24:339-368.
LERMA, B. DE. 1947. L'organo frontale mediale di Ctenolepisma Targionii (Grassi e Rov.): suo valore di organo endocrino. *Arch. zool. (ital.), Napoli*, 32:1-18.
LERMA, B. DE. 1949a. Sulla presenza di cellule giganti nel cerebron e nella massa gangliare sottoesofagea dei Coleotteri. *Boll. Zool.*, 16:169-177.
LERMA, B. DE. 1949b. Gli organi frontali degli Insetti apterigoti. *Ann. Ist. sup. Sc. Lett. S. Chiara, Napoli*, 1:153-157.
LERMA, B. DE. 1950. Endocrinologia degli Insetti. *Boll. Zool.*, 17(suppl.):67-192.
LERMA, B. DE. 1951. Note originali e critiche sulla morfologia comparata degli organi frontali degli Artropodi. *Ann. Ist. Mus. zool. Napoli*, 3:1-26.
LERMA, B. DE. 1954. Osservazioni sulla neurosecrezione in *Hydrous piceus* L. (Coleotteri). *Pubbl. Staz. zool. Napoli*, 24 (suppl.):56-58.
LERMA, B. DE. 1956. Corpora cardiaca et neurosécrétion protocérébrale chez le coléoptère *Hydrous piceus* L. *Ann. Sci. nat. (Zool.)*, 18:235-250.
LEVER, J. 1957. Some remarks on neurosecretory phenomenon in *Ferrissia* sp. (Gastropoda Pulmonata). *Proc. Kon. Ned. Akad. Wet.*,(C)60:510-536.
LEVER, J. 1958a. On the occurrence of a paired follicle gland in the lateral lobes of the cerebral ganglia of some Ancylidae. *Proc. Kon. Ned. Akad. Wet.*, (C)61:235-242.
LEVER, J. 1958b. On the relation between the medio-dorsal bodies and the cerebral ganglia in some pulmonates. *Arch. néerl. Zool.*, 13 (suppl.):194-201.
LEVER, J. and JOOSSE, J. 1961. On the influence of the salt content of the medium on some special neurosecretory cells in the lateral lobes of the cerebral ganglia of *Lymnaea stagnalis*. *Proc. Kon. Ned. Akad. Wet.*, (C) 64:630-639.
LEVER, J., JANSEN, J., and DE VLIEGER, T. A. 1961. Pleural ganglia and water balance in the fresh water pulmonate *Limnaea stagnalis*. *Proc. Kon. Ned. Akad. Wet.*, (C) 64:531-542.
LEVER, J., KOK, M., MEULEMAN, E. A., and JOOSSE, J. 1961. On the location of Gomori-positive neurosecretory cells in the central ganglia of *Lymnaea stagnalis*. *Proc. Kon. Ned. Akad. Wet.*, (C) 64:240-247.
LEVER, J., BOER, H. H., DUIVEN, R. J. T., LAMMENS, J. J., and WATTEL, J. 1959. Some observations on follicle glands in pulmonates. *Proc. Kon. Ned. Akad. Wet.*, (C) 62:139-144.
LEVINSON, L. B. and PLATONOVA, G. N. 1948. Neurosecretory cells of the honey bee. *C. R. Acad. Sci. URSS.*, 60:129-132.
LEWIS, P. R. and LEVER, J. D. 1960. The association of certain chemical activities with intracellular structure. *J. Roy. micr. Soc.*, 78:104-110.
L'HÉLIAS, C. 1950. Étude des glandes endocrines post-cérébrales de la larve d'*Apis mellifica* (Hyménoptère). *Bull. Soc. zool. Fr.*, 75:70-74.
L'HÉLIAS, C. 1952. Étude des glandes endocrines postcérébrales et du cerveau de la larve des *Lophyrus pini* L. et *rufus* André (Hyménoptères). *Bull. Soc. zool. Fr.*, 77:106-112.
L'HÉLIAS, C. 1956. Les hormones du complexe rétrocérébral du phasme *Carausius morosus*, action chimique et identification du squelette commun de ces hormones. *Année biol.*, 60:203-219.
L'HÉLIAS, C. 1957. Isolement de substances pré- ou co-hormonales du complexe postcérébral de *Carausius morosus* et de *Clitumnus extradentatus* (Ins. chéleutoptères). *Bull. biol.*, 91:241-263.
LHOSTE, J. 1957. Données anatomiques et histophysiologiques sur *Forficula auricularia* L. (Dermoptère). *Arch. Zool. exp. gén.*, 95:75-252.
LÖBLICH, H. J. and KNEZEVIC, M. 1960. Elektronenoptische Untersuchungen nach akuter Schädigung des Hypophysen-Zwischenhirnsystems. *Beitr. path. Anat.*, 122:1-30.
LOCHHEAD, J. H. and RESNER, R. 1958. Functions of the eyes and neurosecretion in Crustacea Anostraca. *Int. Congr. Zool., Proc. XV.*, 397-399.
LOEBEL, S. F. and GUIJON, K. P. 1957. Estudios histoquímicos sobre neurosecreción en el cerebro de *Bufo spinolosus*. *Biológica, Santiago*, 24:15-30.
LÖFGREN, F. 1960. The infundibular recess, a component in the hypothalamo-adenohypophyseal system. *Acta morph. néerl.-scand.*, 3:55-78.
LUBET, P. 1955a. Cycle neurosécrétoire chez *Chlamys varia* L. et *Mytilus edulis* L. (mollusques lamellibranches). *C. R. Acad. Sci., Paris*, 241:119-121.
LUBET, P. 1955b. Le déterminisme de la ponte chez les lamellibranches (*My-

tilus edulis L.). Intervention des ganglions nerveux. *C. R. Acad. Sci., Paris,* 241:254-256.

LUBET, P. 1956. Effets de l'ablation des centres nerveux sur l'émission des gamètes chez *Mytilus edulis* L. et *Chlamys varia* L. (mollusques lamellibranches). *Ann. Sci. nat. (Zool.),* 18:175-184.

LUBET, P. 1957. Cycle sexuel de *Mytilus edulis* L. et de *Mytilus galloprovincialis* LMK. dans le bassin d'Arcachon (Gironde). *Année biol.,* 33:19-29.

LÜSCHER, M. and ENGELMANN, F. 1955. Über die Steuerung der Corpora allata-Funktion bei der Schabe *Leucophaea maderae. Rev. suisse Zool.,* 62:649-657.

MACINTOSH, F. C. 1959. Formation, storage, and release of acetylcholine at nerve endings. *Canad. J. Biochem. Physiol.,* 37:343-356.

MARAPÃO, B. P. 1959. The effect of nervous tissue extracts on neurosecretion in the earthworm, *Lumbricus terrestris. Catholic Univ. Amer. (Biol. Studies),* 55:1-37.

MARCOS-GALLEGO, P. and STAMM, M. D. 1959. Hormonas en crustáceos. *Rev. esp. Fisiol.,* 15:263-268.

MATSUMOTO, K. 1954a. Neurosecretion in the thoracic ganglion of the crab, *Eriocheir japonicus. Biol. Bull., Woods Hole,* 106:60-68.

MATSUMOTO, K. 1954b. Chromatophorotropic activity of the neurosecretory cells in the thoracic ganglion of *Eriocheir japonicus. Biol. J. Okayama Univ.,* 4:234-248.

MATSUMOTO, K. 1958. Morphological studies on the neurosecretion in crabs. *Biol. J. Okayama Univ.,* 4:103-176.

MATSUMOTO, K. 1959. Neurosecretory cells of an isopod, *Armadillidium vulgare* Latreille. *Biol. J. Okayama Univ.,* 5:43-50.

MAYNARD, D. M. 1961a. Thoracic neurosecretory structures in Brachyura. I. Gross anatomy. *Biol. Bull., Woods Hole,* 121:316-329.

MAYNARD, D. M. 1961b. Thoracic neurosecretory structures in Brachyura. II. Secretory neurons. *Gen. comp. Endocrin.,* 1:237-263.

MAYNARD, D. M. and WELSH, J. H. 1959. Neurohormones of the pericardial organs of brachyuran Crustacea. *J. Physiol.,* 149:215-227.

MAZZI, V. 1952. Esistono fenomeni neurosecretori nelle ascidie? *Boll. Zool.,* 19:161-162.

MENG, K. 1960. Die Beeinflussung der Tätigkeit des *Helix*-Herzens durch das extrakardiale Nervensystem. *Zool. Jb. (Zool.),* 68:567-576.

MEYER, G. F. and PFLUGFELDER, O. 1958. Elektronenmikroskopische Untersuchungen an den Corpora cardiaca von *Carausius morosus* Br. *Z. Zellforsch.,* 48:556-564.

MIANI, N. 1960. Proximo-distal movement along the axon of protein synthesized in the perikaryon of regenerating neurons. *Nature, Lond.,* 185:541.

MICHON, J. and ALAPHILIPPE, F. 1959. Contribution à l'étude de la neurosécrétion chez les Lumbricinae. *C. R. Acad. Sci., Paris.* 249:835-837.

MILBURN, N., WEIANT, E. A., and ROEDER, K. D. 1960. The release of efferent nerve activity in the roach, *Periplaneta americana,* by extracts of the corpus cardiacum. *Biol. Bull., Woods Hole,* 118:111-119.

MITSUHASHI, J. and FUKAYA, M. 1960, The hormonal control of larval diapause in the rice stem borer, *Chilo suppressalis.* III. Histological studies on the neurosecretory cells of the brain and the secretory cells of the corpora allata during diapause and post diapause. *Jap. J. appl. Ent. Zool.,* 4:127-134.

MIYAWAKI, M. 1955a. Neurosecretory cells of the crab *Telmessus cheiragonus* (Tilesius) in the living condition. *Annot. zool. jap.,* 28:163-166.

MIYAWAKI, M. 1955b. Observations on the pericardial organs in two kinds of crab, *Paralithodes brevipes* (Brandt) and *Telmessus cheiragonus* (Tilesius). *Zool. Mag., Tokyo,* 64:137-140.

MIYAWAKI, M. 1956a. PAS-positive material in the neurosecretory cells of the crab, *Telmessus cheiragonus* (Tilesius). *Annot. zool. jap.,* 29:151-154.

MIYAWAKI, M. 1956b. Cytological and cytochemical studies on the neurosecretory cells of a Brachyura, *Telmessus cheiragonus* (Tilesius). *J. Fac. Sci. Hokkaido Univ.,* (6)12: 516-520.

MIYAWAKI, M. 1956c. Histological observations on the incretory elements in the eyestalk of a Brachyura, *Telmessus cheiragonus* (Tilesius). *J. Fac. Sci. Hokkaido Univ.,* (6) 12:325-332.

MIYAWAKI, M. 1958. On the neurosecretory system of the isopod, *Idotea japonica. Annot. zool. jap.,* 31:216-221.

MIYAWAKI, M. 1960a. On the neurosecretory cells of some decapod Crustacea. *Kumamoto J. Sci.,* 5:1-20.

MIYAWAKI, M. 1960b. Studies on the cytoplasmic globules in the nerve cells of the crabs, *Gaetice depressus* and *Potamon dehaani.* II. Observations by conventional, phase-contrast and electron microscopes. *Kumamoto J. Sci.,* 5:29-40.

MOL, J.-J. VAN. 1960a. Phénomènes neurosécrétoires dans les ganglions cérébroïdes d'*Arion rufus. C. R. Acad. Sci., Paris,* 250:2280-2281.

MOL, J.-J. VAN. 1960b. Etude histologique de la glande céphalique au cours de la croissance chez *Arion rufus* (Linné). *Ann. Soc. Roy. Zool. Belg.,* 91:45-55.

MONRO, J. 1956. A humoral stimulus to the secretion of the brain-hormone in Lepidoptera. *Nature, Lond.,* 178: 213-214.

MONRO, J. 1958. Cholinesterase and the secretion of the brain hormone in insects. *Aust. J. Sci.,* 2:399-406.

MORITA, H., ISHIBASHI, T., and YAMASHITA, S. 1961. Synaptic transmission in neurosecretory cells. *Nature, Lond.,* 191:183.

MOROHOSHI, S. 1959. Hormonal studies on the diapause and non-diapause eggs of the silkworm, *Bombyx mori* L. *J. Insect Physiol.,* 3:28-40.

MOTHES, G. 1960. Weitere Untersuchungen über den physiologischen Farbwechsel von *Carausius morosus* Br. *Zool. Jb. (Zool.),* 69:133-162.

MOUSSA, T. A. and BANHAWY, M. 1960. The Golgi dictyosomes during the differentiation and growth of the nerve cells of *Schistocerca gregaria* with special reference to the problem of neurosecretion. *J. Roy. micr. Soc.,* 79:19-36.

MURAKAMI, M. 1961. Elektronenmikroskopische Untersuchungen über die neurosekretorischen Zellen im Hypothalamus von *Gecko japonicus. Arch. histol. jap.,* 21:323-337.

MURAKAMI, M. 1962. Elektronenmikroskopische Untersuchung der neurosekretorischen Zellen im Hypothalamus der Maus. *Z. Zellforsch.,* 56:277-299.

NABERT, A. 1913. Die Corpora allata der Insekten. *Z. wiss. Zool.,* 104: 181-358.

NAÏSSE, J. 1959. Neurosécrétion et glandes endocrines chez les opilions. *Arch. Biol. Belg.,* 70:217-264.

NAÏSSE, J. 1961. Élaboration de la neurosécrétion au niveau de vésicules golgi-ergastoplasmiques chez l'opilion. *C. R. Acad. Sci., Paris,* 252: 185-186.

NAKAYAMA, T. 1955. Hypothalamic electrical activities produced by factors causing discharge of pituitary hormones. *Jap. J. Physiol.,* 5:311-316.

NAMBUDIRI, P. N. and VIJAYAKRISHNAN, K. P. 1958. Neurosecretory cells of the brain of the leech *Hirudinaria granulosa* (Sav.) *Curr. Sci.,* 27:350-351.

NAYAR, K. K. 1954a. The structure of the corpus cardiacum of *Locusta migratoria. Quart. J. micr. Sci.,* 95:245-250.

NAYAR, K. K. 1954b. The neurosecretory system of the fruit fly *Chaetodacus cucurbitae* Coq. I. Distribution and description of the neurosecretory cells in the adult fly. *Proc. Indian Acad. Sci. (B),* 40:138-144.

NAYAR, K. K. 1955a. Studies on the neurosecretory system of *Iphita limbata* Stal. I. Distribution and structure of the neurosecretory cells of the nerve ring. *Biol. Bull., Woods Hole,* 108:296-307.

NAYAR, K. K. 1955b. Neurosecretory cells in larvae of gall midges. *Curr. Sci.*, 24:90-91.

NAYAR, K. K. 1955c. Studies on the neurosecretory system of *Iphita limbata* Stal. II. Acid phosphatase and cholinesterase in the neurosecretory cells. *Proc. Indian Acad. Sci. (B)*, 42:27-30.

NAYAR K. K. 1956a. Studies on the neurosecretory system of *Iphita limbata* Stal. (Hemiptera). III. The endocrine glands and the neurosecretory pathways in the adult. *Z. Zellforsch.* 44:697-705.

NAYAR, K. K. 1956b. Studies on the neurosecretory system of *Iphita limbata* Stal. (Pyrrhocoridae: Hemiptera). IV. Observations on the structure and functions of the corpora cardiaca of the adult insect. *Proc. nat. Inst. Sci. India (B)*, 22:171-184.

NAYAR, K. K. 1956c. Effect of extirpation of neurosecretory cells on the metamorphosis of *Iphita limbata* Stal. *Curr. Sci.*, 25:192-193.

NAYAR, K. K. 1957. Water content and release of neurosecretory products in *Iphita limbata* Stal. *Curr. Sci.*, 26:25.

NAYAR, K. K. 1958. Studies on the neurosecretory system of *Iphita limbata* Stal. V. Probable endocrine basis of oviposition in the female insect. *Proc. Indian Acad. Sci. (B)*, 47:233-251.

NAYAR, K. K. 1960. Studies on the neurosecretory system of *Iphita limbata* Stal. VI. Structural changes in the neurosecretory cells induced by changes in water content. *Z. Zellforsch.*, 51:320-324.

NAYAR, K. K. and PARAMESWARAN, R. 1955. Succinic dehydrogenase in the neurosecretory cells of the thoracic ganglion of the crab. *Curr. Sci.*, 24:341.

NISHIITSUTSUJI-UWO, J. 1960. Fine structure of the neurosecretory system in Lepidoptera. *Nature, Lond.*, 183:953-954.

NISHIITSUTSUJI-UWO, J. 1961. Electron microscopic studies on the neurosecretory system in Lepidoptera. *Z. Zellforsch.*, 54:613-630.

NISHIOKA, R. S., HAGADORN, I. R., and BERN, H. A. 1962. The ultrastructure of the epistellar body of the octopus. *Z. Zellforsch.*, 27:406-421.

NODA, H., SANO, Y., and NAKAMOTO, T. 1955. Über den Eintritt des hypothalamischen Neurosekrets in den dritten Ventrikel. *Arch. histol. jap.* 8:355-360.

NOIROT, C. 1957. Neurosécrétion et sexualité chez le termite à cou jaune *Calotermes flavicollis* F. *C. R. Acad. Sci., Paris*, 245:743-745.

NUÑEZ, J. A. 1956. Untersuchungen über die Regelung des Wasserhaushaltes bei *Anisotarsus cupripennis* Germ. *Z. vergl. Physiol.*, 38:341-354.

NUSSBAUM, R. 1960. Der Thorax von *Basilia nana* (Diptera, Nycteribiidae). *Zool. Jb. (Anat.)*, 78:313-368.

NYST, R. H. 1942. Structure et rapports du système nerveux, du vaisseau dorsal et des annexes cardiaques chez *Dixippus morosus* Br. *Ann. Soc. zool. Belg.*, 73:150-164.

OCHS, S. and BURGER, E. 1958. Movement of substance proximo-distally in nerve axons as studied with spinal cord injections of radioactive phosphorus. *Amer. J. Physiol.*, 194:499-506.

OGURO, C. 1959a. On the sinus gland in four species belonging to the Idoteidae (Crustacea, Isopoda). *J. Fac. Sci. Hokkaido Univ.*, 14:260-285.

OGURO, C. 1959b. Occurence of accessory sinus gland in the isopod, *Idotea japonica*. *Annot. zool. jap.*, 32:71-73.

OGURO, C. 1960a. On the neurosecretory system in the cephalic region of the isopod, *Tecticeps japonicus*. *Endocrin. Jap.*, 7:137-145.

OGURO, C. 1960b. Effects of castration on the neurosecretory system in the marine isopod, *Tecticeps japonicus*. I. Effects on the sinus gland of male animals. *Annot. zool. jap.*, 33:37-41.

OGURO, C. 1961. On the neurosecretory system of two parasitic isopods, *Argeia pugettensis* Dana and *Athelges japonicus* Shiino. *Annot. zool. jap.*, 34:43.

OHNISHI, E. 1957. Sur la nature de la coloration des chrysalides de *Papilio xuthus* et de *P. protenor*. *Bull. Soc. Hist. nat. Toulouse*, 92:181-185.

OHTAKI, T. 1960. Humoral control of pupal coloration in the cabbage white butterfly, *Pieris rapae crucivora*. *Annot. zool. jap.*, 33:97-103.

OLSSON, R. 1958. *The Subcommissural Organ*. Haeggström, Stockholm.

OLSSON, R. and WINGSTRAND, K. G. 1954. Reissner's fibre and the infundibular organ in *Amphioxus*—results obtained with Gomori's chrome alum haematoxylin. *Univ. Bergen Årb. Naturv. R. (Publ. Biol. Sta.)*, 14:1-15.

ORTMANN, R. 1951. Über experimentelle Veränderungen der Morphologie des Hypophysenzwischenhirnsystems und die Beziehung der sog. "Gomorisubstanz" zum Adiuretin. *Z. Zellforsch.*, 36:92-140.

ORTMANN, R. 1958. Neurosekretion und Proteinsynthese. *Z. mikr.-anat. Forsch.*, 64:215-227.

ORTMANN, R. 1960. Neurosecretion. In: *Handbook of Physiology*. Sect. 1. *Neurophysiology*, 2:1039-1067.

OTREMBA, P. 1961. Beobachtungen an neurosekretorischen Zellen des Regenwurmes (*Lumbricus* spec.). *Z. Zellforsch.*, 54:421-436.

ŌTSU, T. 1960. Precocious development of the ovaries in the crab, *Potamon dehaani*, following implantation of the thoracic ganglion. *Annot. zool. jap.*, 33:90-96.

OZBAS, S. and HODGSON, E. S. 1958. Action of insect neurosecretion upon central nervous system in vitro and upon behavior. *Proc. nat. Acad. Sci., Wash.*, 44:825-830.

ÖZTAN, N. and GORBMAN, A. 1960. Responsiveness of the neurosecretory system of larval lampreys (*Petromyzon marinus*) to light. *Nature, Lond.*, 186:167-168.

PALAY, S. L. 1943. Neurosecretion. V. The origin of neurosecretory granules from the nuclei of nerve cells in fishes. *J. comp. Neurol.*, 79:247-275.

PALAY, S. L. 1957. The fine structure of the neurohypophysis. In: *Ultrastructure and Cellular Chemistry of Neural Tissue*. H. Waelsch (ed.). Hoeber-Harper, New York.

PALAY, S. L. 1958. The morphology of secretion. In: *Frontiers in Cytology*. S. L. Palay (ed.). Yale Univ. Press, New Haven.

PALAY, S. L. 1960. The fine structure of secretory neurons in the preoptic nucleus of the goldfish (*Carassius auratus*). *Anat. Rec.*, 138:417-459.

PALAY, S. L. and WISSIG, S. L. 1953. Secretory granules and Nissl substance in fresh supraoptic neurones of the rabbit. *Anat. Rec.*, 116:301-313.

PALM, N.-B. 1955. Neurosecretory cells and associated structures in *Lithobius forficatus* L. *Ark. Zool.*, (2)9:115-129.

PANOUSE, J. B. 1947. Les corrélations humorales chez les crustacés. *Année biol.*, 23:33-70.

PARAMESWARAN, R. 1956. Neurosecretory cells of the central nervous system of the crab, *Paratelphusa hydrodromous*. *Quart. J. micr. Sci.*, 97:75-82.

PARDOE, A. U. and WEATHERALL, M. 1955. The intracellular localization of oxytocic and vasopressor substances in the pituitary gland of rats. *J. Physiol.*, 127:201-212.

PARKER, G. H. 1948. *Animal Colour Changes and their Neurohumours*. Cambridge Univ. Press, Cambridge.

PASSANO, L. M. 1953. Neurosecretory control of molting in crabs by the X-organ sinus gland complex. *Physiol. comp.*, 3:155-189.

PASSANO, L. M. 1954. Phase microscopic observations of the neurosecretory product of the crustacean X-organ. *Pubbl. Staz. zool. Napoli*, 24(suppl):72-73.

PASSANO, L. M. 1960. Molting and its control. In: *The Physiology of Crustacea*. Vol. 1. T. H. Waterman (ed.). Academic Press, New York.

PASSANO, L. M. 1961. The regulation of crustacean metamorphosis. *Amer. Zoologist*, 1:89-95.

PASTEUR, C. 1958. Influence de l'ablation de l'organe X sur le comportement chromatique de *Leander serratus* (Pennant). *C. R. Acad. Sci., Paris*, 246:320-322.

PAVANS DE CECCATTY, M. 1957. Les

Bibliography

activités sécrétrices du système nerveux central de l'amphioxus. *C. R. Acad. Sci., Paris*, 244:2645-2647.

PAVANS DE CECCATTY, M. and PLANTA, O. VON. 1954. Note sur le système nerveux central des éolidiens (Mollusques Nudibranches). *Bull. Soc. zool. Fr.*, 79:152-158.

PEARSE, A. G. E. 1954. *Histochemistry, Theoretical and Applied.* Little, Brown, Boston.

PELLUET, D. and LANE, N. J. 1961. The relation between neurosecretion and cell differentiation in the ovotestis of slugs (Gasteropoda: Pulmonata). *Can. J. Zool.*, 39:789-805.

PÉRÈS, J. M. 1943. Recherches sur le sang et les organes neuraux des tuniciers. *(Thèse Paris) Ann. Inst. Océanogr., Paris*, 21:229-359.

PÉRÈS, J. M. 1946. L'organe neural des Polyclinidae. *Bull. Mus. Hist. nat., Paris*, (21)18:69-79.

PEREZ, H. V. Z. 1942. On the chromaffin cells of the nerve ganglia of *Hirudo medicinalis* L. *J. comp. Neurol.*, 76:367-401.

PEREZ, Z. 1940. Les cellules sécrétrices du cerveau de quelques lépidoptères. *Ann. Fac. Sci. Porto* (Acad. Polytechn., Oporto, Portugal), 25:92-99.

PÉREZ-GONZÁLEZ, M. D. 1957. Evidence for hormone-containing granules in sinus glands of the fiddler crab, *Uca pugilator*. *Biol. Bull., Woods Hole*, 113:426-441.

PFLUGFELDER, O. 1937. Bau, Entwicklung und Funktion der Corpora allata und cardiaca von *Dixippus morosus* Br. *Z. wiss. Zool.*, 149:477-512.

PFLUGFELDER, O. 1952. *Entwicklungsphysiologie der Insekten.* Akad. Verlagsges., Leipzig.

PICARD, D. and STAHL, A. 1956. Signification fondamentale de certaines activités élaboratrices des cellules nerveuses. Étude critique de la notion actuelle de neurosécrétion. *J. Physiol., Paris*, 48:73-95.

PIEPHO, H. 1946. Versuche über die Rolle von Wirkstoffen in der Metamorphose der Schmetterlinge. *Biol. Zbl.*, 65:141-148.

PIPA, R. L. 1961a. Studies on the hexapod nervous system. III. Histology and histochemistry of cockroach neuroglia. *J. comp. Neurol.*, 116:15-26.

PIPA, R. L. 1961b. Studies on the hexapod nervous system. IV. A cytological and cytochemical study of neurons and their inclusions in the brain of a cockroach, *Periplaneta americana* L. *Biol. Bull., Woods Hole*, 121:521-534.

POSSOMPÈS, B. 1946. Les glandes endocrines post-cérébrales des Diptères. I. Étude chez la larve de *Chironomus plumosus* L. *Bull. Soc. zool. Fr.*, 71:99-109.

POSSOMPÈS, B. 1947. Les glandes endocrines post-cérébrales des Diptères. II. Étude sommaire des corpora allata et des corpora cardiaca chez la larve de *Tipula* sp. *Bull. Soc. zool. Fr.*, 72:57-62.

POSSOMPÈS, B. 1948. Les glandes endocrines post-cérébrales des Diptères. III. Étude chez la larve de *Tabanus* sp. *Bull. Soc. zool. Fr.*, 73:228-235.

POSSOMPÈS, B. 1953a. Recherches expérimentales sur le déterminisme de la métamorphose de *Calliphora erythrocephala* Meig. *Arch. Zool. exp. gén.*, 89:203-364.

POSSOMPÈS, B. 1953b. Les données expérimentales sur le déterminisme endocrine de la croissance des insectes. *Bull. Soc. zool. Fr.*, 78:240-275.

POSSOMPÈS, B. 1954. Données expérimentales sur l'activation de l'anneau de Weismann par le cerveau chez *Calliphora erythrocephala* Meig. *Pubbl. Staz. zool. Napoli*, 24(suppl.): 59-62.

POSSOMPÈS, B. 1958. Evolution des cellules neuro-sécrétrices protocérébrales et de la glande péritrachéenne de *Calliphora erythrocephala* Meig (Diptère) après section des connexions nerveuses entre le cerveau et l'anneau de Weismann. In: *II Internat. Symp. Neurosekretion.* W. Bargmann et al. (eds.). Springer, Berlin.

POTTER, D. D. 1958. Observations on the neurosecretory system of portunid crabs. In: *II Internat. Symp. Neurosekretion.* W. Bargmann et al. (eds.). Springer, Berlin.

POTTER, D. D. and LOEWENSTEIN, W. R. 1955. Electrical activity of neurosecretory cells. *Amer. J. Physiol.*, 183:652.

POWER, M. E. 1948. The thoracicoabdominal nervous system of an adult insect, *Drosophila melanogaster*. *J. comp. Neurol.*, 88:347-409.

PRABHU, V. K. K. 1959. Note on the cerebral glands and a hitherto unknown connective body in *Jonespeltis splendidus* Verhoeff (Myriapoda, Diplopoda). *Curr. Sci.*, 28:330-331.

PRABHU, V. K. K. 1961. The structure of the cerebral glands and connective bodies of *Jonespeltis splendidus* Verhoeff (Myriapoda, Diplopoda). *Z. Zellforsch.*, 54:717-733.

RAAB, W. and HUMPHREYS, R. J. 1947. Secretory function of sympathetic neurones and sympathin formation in effector cells. *Amer. J. Physiol.*, 148:460-469.

RAABE, M. 1959. Neurohormones chez les insectes. *Bull. Soc. zool. Fr.* 84:272-316.

RAHM, U. H. 1952. Die innersekretorische Steuerung der postembryonalen Entwicklung von *Sialis lutaria* L. (Megaloptera). *Rev. suisse Zool.*, 59:173-237.

REHM, M. 1951. Die zeitliche Folge der Tätigkeitsrhythmen inkretorischer Organe von *Ephestia kühniella* während der Metamorphose und des Imaginallebens. *Arch. EntwMech. Org.*, 145:205-248.

REHM, M. 1955. Morphologische und histochemische Untersuchungen an neurosekretorischen Zellen von Schmetterlingen. *Z. Zellforsch.*, 42:19-58.

REHM, M. 1959. Observations on the localisation and chemical constitution of neurosecretory material in nerve terminals in *Carcinus maenas*. *Acta histochem.*, 7:88-106.

RENNELS, E. G. and DRAGER, G. A. 1955. The relationship of pituicytes to neurosecretion. *Anat. Rec.*, 122:193-203.

RISLER, H. 1957. Der Kopf von *Thrips physapus* L. (Thysanoptera, Terebrantia). *Zool. Jb. (Anat.)*, 76:251-302.

ROBERTIS, E. DE. 1959. Submicroscopic morphology of the synapse. *Int. Rev. Cytol.*, 8:61-96.

ROBERTSON, J. D. 1957. The ultrastructure of nodes of Ranvier. *J. Physiol.*, 137:8P-9P.

RODECK, H. 1959. Zusammenhänge zwischen Neurosekret und den sogenannten Hypophysenhinterlappenhormonen. II. Untersuchungen an handelsüblichen Hypophysenhinterlappenextrakten. III. Untersuchungen zur färberischen Darstellung von synthetischem Oxytocin. IV. Untersuchungen an schwefelhaltigen Aminosäuren. *Z. ges. exp. Med.*, 132:113-121; 122-135; 225-233.

ROTHBALLER, A. B. 1957. Neuroendocrinology. *Exc. Med. (III. Endocr.)*, 11(11):i-xii.

ROUSSY, G. and MOSINGER, M. 1946. *Traité de Neuro-endocrinologie.* Masson, Paris.

SAHLI, F. 1958. Quelques données sur la neurosécrétion chez le diplopode *Tachypodoiulus albipes* C. L. Koch. *C. R. Acad. Sci., Paris*, 246:470-472.

SAHLI, F. 1961. Sur une formation hpocérébrale chez les Diplopodes Iulides. *C. R. Acad. Sci., Paris*, 252:2443-2444.

SAHLI, F. 1962. Sur le système neurosécréteur de Polydesmoïde *Orthomorpha gracilis* C. L. Koch (Myriapoda, Diplopoda). *C. R. Acad. Sci., Paris*, 254:1498-1500.

SAMUELS, A. J., BOYARSKY, L. L., GERARD, R. W., LIBET, B., and BRUST, M. 1951. Distribution, exchange and migration of phosphate compounds in the nervous system. *Amer. J. Physiol.*, 164:1-12.

SANCHEZ, S. 1954. Les glandes neurosécrétrices des Pycnogonides. *C. R. Acad. Sci., Paris*, 239:1078-1080.

SANCHEZ, S. 1958. Cellules neurosécrétrices et organes infracérébraux de *Peripatopsis moseleyi* Wood (Onychophores) et neurosécrétion chez *Nymphon gracile* Leach (Pycnogoni-

des). *Arch. Zool. exp. gén.*, 96:57-62.
SANCHEZ, S. 1959. Le développement des Pycnogonides et leurs affinités avec les Arachnides. *Arch. Zool. exp. gén.*, 98:1-101.
SANCHEZ, S. and BORD, C. 1958. Origine des cellules neurosécrétrices chez *Helix aspera* Mull. *C. R. Acad. Sci., Paris*, 246:845-847.
SANCHEZ, S. and PAVANS DE CECCATTY, M. 1957. Neurosécrétion et fonction élaboratrice des neurones chez quelques Mollusques Gastéropodes. *C. R. Soc. Biol., Paris*, 151:2172-2173.
SANDEEN, M. I. and COSTLOW, J. D., JR. 1961. The presence of decapod-pigment-activating substances in the central nervous system of representative Cirripedia. *Biol. Bull., Woods Hole*, 120:192-205.
SANO, Y. 1958. Beobachtungen zur Morphologie der Neurosekretion bei Wirbeltieren. In: *II Internat. Symp. Neurosekretion.* W. Bargmann et al. (eds.). Springer, Berlin.
SANO, Y. 1961. Das caudale neurosekretorische System bei Fischen. *Ergebn. Biol.*, 24:191-212.
SANO, Y. and KNOOP, A. 1959. Elektronenmikroskopische Untersuchungen am kaudalen neurosekretorischen System von *Tinca vulgaris*. *Z. Zellforsch.*, 49:464-492.
SANO, Y., TAMIYA, M., and ISHIZAKI, N. 1957. Histological studies of the neurosecretory nuclei. IV. On the amitosis like findings in the hypothalamic neurosecretory nuclei of dog. *Arch. histol. jap.*, 12:173-183.
SAWYER, W. H. 1959. Oxytocic activity in the neural complex of two ascidians, *Chelyosoma productum* and *Pyura haustor*. *Endocrinology*, 65:520-522.
SAWYER, W. H., MUNSICK, R. A., and VAN DYKE, H. B. 1960. Antidiuretic hormones. *Circulation*, 21(suppl.):1027-1037.
SCHAEFER, K. 1939. Lage und Sekretion der Drüsennervenzellen von *Nereis diversicolor* Müll. *Zool. Anz.*, 125:195-202.
SCHALLER, F. 1959. Contrôle humoral du développement postembryonnaire d'*Aeshna cyanea* Müll. (Insecta odonate). *C. R. Acad. Sci., Paris*, 248:2525-2527.
SCHARRER, B. 1935. Ueber das Hanströmsche Organ X bei Opisthobranchiern. *Pubbl. Staz. zool. Napoli*, 15:132-142.
SCHARRER, B. 1936. Über Drüsen-Nervenzellen im Gehirn von *Nereis virens* Sars. *Zool. Anz.*, 113:299-302.
SCHARRER, B. 1937. Über sekretorisch tätige Nervenzellen bei wirbellosen Tieren. *Naturwissenschaften*, 25:131-138.
SCHARRER, B. 1941a. Neurosecretion. II. Neurosecretory cells in the central nervous system of cockroaches. *J. comp. Neurol.* 74:93-108.
SCHARRER, B. 1941b. Neurosecretion. III. The cerebral organ of the nemerteans. *J. comp. Neurol.*, 74:109-130.
SCHARRER, B. 1941c. Neurosecretion IV: Localization of neurosecretory cells in the central nervous system of *Limulus*. *Biol. Bull., Woods Hole*, 81:96-104.
SCHARRER, B. 1952a. Neurosecretion XI. The effects of nerve section on the intercerebralis-cardiacum-allatum system of the insect *Leucophaea maderae*. *Biol. Bull., Woods Hole*, 102:261-272.
SCHARRER, B. 1952b. Hormones in insects. In: *The Action of Hormones in Plants and Invertebrates.* K. V. Thimann (ed.). Academic Press, New York.
SCHARRER, B. 1952c. Über neuroendokrine Vorgänge bei Insekten. *Pflüg. Arch. ges. Physiol.*, 255:154-163.
SCHARRER, B. 1955a. "Castration cells" in the central nervous system of an insect (*Leucophaea maderae* Blattaria). *Trans. N.Y. Acad. Sci.*, 17:520-525.
SCHARRER, B. 1955b. Hormones in invertebrates. In: *The Hormones.* Vol. 3. G. Pincus and K. V. Thimann (eds.). Academic Press, New York.
SCHARRER, B. 1956. Corrélations endocrines dans la réproduction des Insectes. *Ann. Sci. nat. (Zool.)*, 18:231-234.
SCHARRER, B. 1958. Neuro-endocrine mechanisms in insects. In: *II Internat. Symp. Neurosekretion.* W. Bargmann et al. (eds.). Springer, Berlin.
SCHARRER, B. 1959. The role of neurosecretion in neuroendocrine integration. In: *Comparative Endocrinology.* A. Gorbman (ed.). Wiley, New York.
SCHARRER, B. and HARNACK, M. VON. 1961. Histophysiological studies on the corpus allatum of *Leucophaea maderae*. III. The effect of castration. *Biol. Bull., Woods Hole*, 121:193-208.
SCHARRER, B. and SCHARRER, E. 1944. Neurosecretion. VI. A comparison between the intercerebralis-cardiacum-allatum system of the insects and the hypothalamo-hypophyseal system of the vertebrates. *Biol. Bull., Woods Hole*, 87:242-251.
SCHARRER, E. 1928. Die Lichtempfindlichkeit blinder Elritzen (Untersuchungen über das Zwischenhirn der Fische). *Z. vergl. Physiol.*, 7:1-38.
SCHARRER, E. 1935. Über das Pigment im Amphibiengehirn. *Zool. Anz.*, 109:304-307.
SCHARRER, E. 1951. Neurosecretion. X. A relationship between the paraphysis and the paraventricular nucleus in the garter snake *Thamnophis* sp. *Biol. Bull., Woods Hole*, 101:106-113.
SCHARRER, E. 1959. General and phylogenetic interpretations of neuroendocrine interrelations. In: *Comparative Endocrinology.* A. Gorbman (ed.). Wiley, New York.
SCHARRER, E. and BROWN, S. 1961. Neurosecretion. XII. The formation of neurosecretory granules in the earthworm, *Lumbricus terrestris* L. *Z. Zellforsch.*, 54:530-540.
SCHARRER, E. and SCHARRER, B. 1937. Über Drüsen-Nervenzellen und neurosekretorische Organe bei Wirbellosen und Wirbeltieren. *Biol. Rev.*, 12:185-216.
SCHARRER, E. and SCHARRER, B. 1945. Neurosecretion. *Physiol. Rev.*, 25:171-181.
SCHARRER, E. and SCHARRER, B. 1954a. Hormones produced by neurosecretory cells. *Recent Progr. Hormone Res.*, 10:183-240.
SCHARRER, E. and SCHARRER, B. 1954b. Neurosekretion. *Handb. mikro. Anat. Menschen*, 6(5):953-1066.
SCHARRER, E., PALAY, S. L., and NILGES, R. G. 1945. Neurosecretion. VIII. The Nissl substance in secreting nerve cells. *Anat. Rec.*, 92:23-31.
SCHEER, B. T. 1957. The hormonal control of metabolism in decapod crustaceans. In: *Recent Advances in Invertebrate Physiology.* B. T. Scheer (ed.). Univ. of Oregon Press, Eugene.
SCHEER, B. T. 1960. The neuroendocrine system of arthropods. *Vitamins and Hormones*, 18:141-204.
SCHEFFEL, H. 1961. Untersuchungen zur Neurosekretion bei *Lithobius forficatus* L. (Chilopoda). *Zool. Jb. (Anat.)*, 79:529-556.
SCHIEBLER, T. H. 1951. Zur Histochemie des neurosekretorischen hypothalamisch-neurohypophysären Systems. I. *Acta anat.*, 13:233-255.
SCHIEBLER, T. H. 1952a. Cytochemische und elektronenmikroskopische Untersuchungen an granulären Fraktionen der Neurohypophyse des Rindes. *Z. Zellforsch.*, 36:563-576.
SCHIEBLER, T. H. 1952b. Zur Histochemie des neurosekretorischen hypothalamisch-neurohypophysären Systems. II. *Acta anat.*, 15:393-416.
SCHIEBLER, T. H. 1954. Morphologie und Funktion neurosekretorischer Zellgruppen, insbesondere des hypothalamisch-neurohypophysären Systems. *Endokrinologie*, 31:1-16.
SCHMIDT, L. A. 1947. Induced neurosecretion in *Lumbricus terrestris*. *J. exp. Zool.*, 104:365-373.
SCHMITT, F. O. 1959. Molecular organization of the nerve fiber. In: *Biophysical Science.* J. L. Oncley (ed.) Wiley, New York.
SCHNEIDERMAN, H. A. 1957. Onset and termination of insect diapause. In: *Physiological Triggers.* T. H. Bullock (ed.). Amer. Physiol. Soc., Washington, D. C.
SCHNEIDERMAN, H. A. and GILBERT, L. I. 1959. The chemistry and physiology of insect growth hor-

mones. In: *Cells, Organism and Milieu.* D. Rudnick (ed.). Ronald Press, New York.

SCHULTZ, R. L. 1960. Electron microscopic observations of the corpora allata and associated nerves in the moth, *Celerio lineata. J. Ultrastruct. Res.,* 3:320-327.

SEITE, R. and CHAMBOST, G. 1958. Élaborations figurées de la cellule nerveuse: Étude du ganglion rachidien chez le chat. *Z. Zellforsch.,* 47: 498-506.

SELLIER, R. 1956. Gonades, cerveau et morphogénèse alaire chez les grillons. *Ann. Sci. nat. (Zool.),* 18:251-255.

SENGEL, P. and KIENY, M. 1962. Action de divers liquides nutritifs et du complexe "glande neurale-ganglion nerveux-organe vibratile" sur les gonades de *Molgula manhattensis* (Tunicier Ascidiacé) cultivées in vitro. *C. R. Acad. Sci., Paris,* 254: 1682.

SHANKLIN, W. M., ISSIDORIDES, M., and NASSAR, T. K. 1957. Neurosecretion in the human cerebellum. *J. comp. Neurol.,* 107:315-338.

SLOPER, J. C. 1957. Presence of a substance rich in protein-bound cystine or cysteine in the neurosecretory system of an insect. *Nature, Lond.,* 179:148-149.

SLOPER, J. C. 1958. Hypothalamoneurohypophysial neurosecretion. *Int. Rev. Cytol.,* 7:337-389.

SLOPER, J. C., ARNOTT, D. J., and KING, B. C. 1960. Sulphur metabolism in the pituitary and hypothalamus of the rat: A study of radioisotope-uptake after the injection of ^{35}S DL-cysteine, methionine, and sodium sulphate. *J. Endocrin.,* 20:9-23.

SMITH, S. W. 1952. Neurosecretory phenomena in sympathetic ganglion cells of *Bufo marinus* with particular reference to their significance for Weiss's theory of proximo-distal movement of axoplasm. *Anat. Rec.,* 112:390. (Abs.)

SPATZ, H. 1954. Das Hypophysen-Hypothalamus-System und seine Bedeutung für die Fortpflanzung. *Verh. anat. Ges. Jena,* 51:46-86.

SPEIDEL, C. G. 1919. Gland-cells of internal secretion in the spinal cord of the skates. *Carnegie Inst., Wash.,* 13(281):1-31.

STAHL, A. 1957. Recherches sur les élaborations cellulaires et la neurosécrétion dans l'encéphale des poissons téléostéens. *Acta anat.,* Suppl. 28:1-158.

STAHL, A. and SEITE, R. 1954. Cytologie des cellules nerveuses de l'hypothalamus. *Pubbl. Staz. zool. Napoli,* 24 (suppl.):24.

STÉPHAN-DUBOIS, F. and LENDER, T. 1956. Corrélations humorales dans la régénération des Planaires paludicoles. *Ann. Sci. nat. (Zool.),* 12:223-230.

STERBA, G. 1957. Die neurosekretorischen Zellgruppen einiger Cladoceren (*Daphnia pulex* und *magna, Simocephalus vetulus*). *Zool. Jb. (Anat.),* 76:303-310.

STIENNON, J. A. and DROCHMANS, P. 1961. Electron microscope study of neurosecretory cells in Phasmidae. *Gen. comp. Endocrin.,* 1:286-294.

STOLK, A. 1960. Cytoplasmic inclusions of the ageing neurones of some teleosts. *Nature, Lond.,* 186:332-333.

STRANGWAYS-DIXON, J. 1959. Hormonal control of selective feeding in female *Calliphora erythrocephala* Meig. *Nature, Lond.,* 184(suppl. 26): 2040-2041.

STRICH-HALBWACHS, M.-C. 1959. Contrôle de la mue chez *Locusta migratoria. Ann. Sci. nat. (Zool.)* (12)1: 483-570.

STRUMM-ZOLLINGER, E. 1957. Histological study of regenerative processes after transection of the nervi corporis cardiaci in transplanted brains of the Cecropia silkworm (*Platysamia cecropia* L.). *J. exp. Zool.,* 134:315-326.

STUTINSKY, F. 1952. Étude du complexe rétro-cérébral de quelques insectes avec l'hématoxyline chromique. *Bull. Soc. zool. Fr.,* 77:61-67.

STUTINSKY, F. 1953a. Mise en évidence d'une substance antidiurétique dans le cerveau et la complexe rétrocérébrale d'une blatte *Blabera fusca. Bull. Soc. zool. Fr.,* 78:202-204.

STUTINSKY, F. 1953b. La neurosécrétion chez les Vertébrés. *Année biol.,* 29:487-516.

STUTINSKY, F. 1959. Sur l'aspect de certaines cellules nerveuses de la médullo-surrénale du rat. *Gunma J. med. Sci.,* 8:195-198.

TAINTER, M. L. and LUDUENA, F. P. 1950. Sympathetic hormonal transmission. *Recent Progr. Hormone Res.,* 5:3-35.

TAKAHASHI, O. 1960. On the formation of vacuoles in the nerve cells of the ganglion cervicalis uteri in the rat and mouse. *Okajimas Folia Anat. Jap.,* 34:189-205.

TAKAHASHI, S. 1957. Phase contrast microscopic observations on the posterior pituitary of dog, with special reference to the difference between the neurosecretory substance in the hypothalamic neurosecretory nucleus and in the posterior lobe. *Arch. histol. jap.,* 12:311-316.

TAKAHASHI, S., IWASA, N., and MAEDA, T. 1957. On the neurosecretory anterior horn cells in the lumbosacral portion of the avian spinal cord. III. Phase contrast microscopic observations. *Arch. histol. jap.,* 12:457-463.

TAKEWAKI, K. (ed.). 1962. Progress in Comparative Endocrinology. *General and Comparative Endocrinology,* suppl. 1.

TALANTI, S. and KIVALO, E. 1961. The infundibular recess in the brain of *Camelus dromedarius* with particular reference to its neurosecretory pathways into the third ventricle. *Experientia* 17:470-471.

TAMIYA, M., IMOTO, T., and OKA, M. 1956. Histologische Untersuchung über den Tractus hypothalamo-hypophyseus. II. Über die markhaltigen Nervenfasern im Tractus hypothalamo-hypophyseus. *Arch. histol. jap.,* 11:393-406.

THOMAS, O. L. 1951. A comparative study of the cytology of the nerve cell with reference to the problem of neurosecretion. *J. comp. Neurol.,* 95: 73-101.

THOMSEN, E. 1952. Functional significance of the neurosecretory cells and the corpus cardiacum in the female blowfly, *Calliphora erythrocephala* Meig. *J. exp. Biol.,* 29:137-172.

THOMSEN, E. 1954. Studies on the transport of neurosecretory material in *Calliphora erythrocephala* by means of ligaturing experiments. *J. exp. Biol.,* 31:322-330.

THOMSEN, E. 1956. Observations on the oenocytes of the adult *Calliphora erythrocephala* Meig. In: *Bertil Hanström: Zoological Papers in Honour of his Sixty-fifth Birthday.* K. G. Wingstrand (ed.). Zool. Inst., Lund.

THOMSEN, E. and MØLLER, I. 1959. Neurosecretion and intestinal proteinase activity in an insect, *Calliphora erythrocephala* Meig. *Nature, Lond.,* 183:1401-1402.

THOMSEN, E. and THOMSEN, M. 1954. Darkfield microscopy of living neurosecretory cells. *Experientia,* 10:206-207.

THOMSEN, M. 1951. Weismanns ring and related organs in larvae of Diptera. *Danske Selsk. Biol. Skr.,* 6: 1-32.

THOMSEN, M. 1954a. Observations on the cytology of neurosecretion in various insects (Diptera and Hymenoptera). *Pubbl. Staz. zool. Napoli,* 24(suppl.):46-47.

THOMSEN, M. 1954b. Neurosecretion in some Hymenoptera. *K. Danske Vidensk. Selskab.,* 7:1-24.

THORE, S. 1939. Über ein neues Organ bei den decapoden Cephalopoden. *K. fysiogr. Sällsk. Lund Förh.,* 9:105-111.

TIEGS, O. W. 1947. The development and affinities of the Pauropoda, based on a study of *Pauropus silvaticus.* Part I. *Quart. J. micr. Sci.,* 88:165-268.

TÖRÖK, L. V. 1958. Experimental contributions to the regenerative capacity of *Dugesia* (= *Euplanaria*) *lugubris* O. Schm. *Acta Biol. hung.,* 9:79-98.

TRAVIS, D. F. 1951a. Physiological changes which occur in the blood and urine of *Panulirus argus* Latreille during the molting cycle. *Anat. Rec.*, 111:157.

TRAVIS, D. F. 1951b. The control of the sinus glands over certain aspects of calcium metabolism in *Panulirus argus* Latreille. *Anat. Rec.*, 111:503.

TRUJILLO-CENÓZ, O. 1959. Study on the fine structure of the central nervous system of *Pholus labruscoe* L. (Lepidoptera). *Z. Zellforsch.*, 49:432-446.

TURNER, R. S. 1946. Observations on the central nervous system of *Leptoplana acticola*. *J. comp. Neurol.*, 85:53-65.

TUZET, O., SANCHEZ, S., and PAVANS DE CECCATTY, M. 1957. Données histologiques sur l'organisation neuroendocrine de quelques Mollusques Gastéropodes. *C. R. Acad. Sci., Paris*, 244:2962-2964.

UNGER, H. 1957. Untersuchungen zur neurohormonalen Steuerung der Herztätigkeit bei Schaben. *Biol. Zbl.* 76:204-225.

UNGER, H. 1960. Neurohormone bei Seesternen (*Marthasterias glacialis*). *Symposia Biol. Hung.*, 1:203-207.

VALENTE, D. 1959. Contribuição para o estudo da neurosecreção nos crustáceos. *Bol. Fac. Filos. Ciênc. S. Paulo*, 232 (Zool. No. 22):5-75.

VAN DER KLOOT, W. G. 1955a. The control of neurosecretion and diapause by physiological changes in the brain of the Cecropia silkworm. *Biol. Bull., Woods Hole*, 109:276-294.

VAN DER KLOOT, W. G. 1955b. Neurosecretion and the physiology of the brain of the Cecropia silkworm. *J. cell. comp. Physiol.*, 46:359.

VAN DER KLOOT, W. G. 1958. The resting potentials of neurons in the Cecropia brain during diapause and development. *Proc. 10th. Int. Congr. Ento.*, 2:79.

VAN DER KLOOT, W. G. 1960. Neurosecretion in insects. *Ann. Rev. Ent.*, 5:35-52.

VAN DER KLOOT, W. G. 1961a. Insect metamorphosis and its endocrine control. *Amer. Zool.*, 1:3-9.

VAN DER KLOOT, W. G. 1961b. Inhibition in the neuro-endocrine systems of invertebrates. In: *Nervous Inhibition*. E. Florey (ed.). Pergamon Press, New York.

VERNET-CORNUBERT, G. 1961. Connaissances actuelles sur le déterminisme hormonal de la mue chez les Décapodes et étude de quelques phénomènes qui lui sont liés. *Arch. Zool. exp. gén.*, 99:57-76.

VOGT, M. 1941. Anatomie der pupalen *Drosophila*-Ringdrüse und ihre mutmassliche Bedeutung als imaginales Metamorphosezentrum. *Biol. Zbl.*, 61:148-158.

VOGT, M. 1942a. Die "Puparisierung" als Ringdrüsenwirkung. *Biol. Zbl.*, 62:149-154.

VOGT, M. 1942b. Ein drittes Organ in der larvalen Ringdrüse von *Drosophila*. *Naturwissenschaften*, 30:66-67.

VOGT, M. 1942c. Zur hormonalen Bedeutung des *Drosophila*-Gehirnes und seiner hormonal bedingten imaginalen Entwicklung. *Naturwissenschaften*, 30:470-471.

VOGT, M. 1946. Inhibitory effects of the corpora cardiaca and of the corpus allatum in *Drosophila*. *Nature, Lond.*, 157:512.

WATERMAN, T. H. and ENAMI, M. 1954. Neurosecretion in the lateral rudimentary eye of *Tachypleus*, a xiphosuran. *Pubbl. Staz. zool. Napoli*, 24 (suppl.):81-82.

WATTERSON, R. L. (ed.). 1959. Endocrine organs of arthropods, with special emphasis on insects. In: *Endocrines in Development*. Univ. Chicago Press.

WEBER, H. 1952. Morphologie, Histologie und Entwicklungsgeschichte der Articulaten. *Fortschr. Zool.*, 9:18-231.

WEISS, P. 1944. Evidence of perpetual proximo-distal growth of nerve fibers. *Biol. Bull., Woods Hole*, 87:160. (Abs.)

WEISS, P. 1955. Nervous system (neurogenesis). In: *Analysis of Development*. B. H. Willier, P. A. Weiss, and V. Hamburger (eds.). Saunders, Philadelphia.

WEISS, P. 1959. Evidence by isotope tracers of perpetual replacement of mature nerve fibers from their cell bodies. *Science*, 129:1290.

WEISS, P. and HISCOE, H. B. 1948. Experiments on the mechanism of nerve growth. *J. exper. Zool.*, 107:315-395.

WEISS, P., WANG, H., TAYLOR, A. C., and EDDS, M. V., JR. 1945. Proximodistal fluid convection in the endoneurial spaces of peripheral nerves, demonstrated by colored and radioactive (isotope) tracers. *Amer. J. Physiol.*, 143:521-540.

WELLS, M. J. 1960. Optic glands and the ovary of *Octopus*. *Symp. zool. Soc. London*, 2:87-108.

WELLS, M. J. and WELLS, J. 1959. Hormonal control of sexual maturity in *Octopus*. *J. exp. Biol.*, 36:1-33.

WELSH, J. H. 1955. Neurohormones. In: *The Hormones*. Vol. 3. G. Pincus and K. V. Thimann (eds.). Academic Press, New York.

WELSH, J. H 1959. Neuroendocrine substances. In: *Comparative Endocrinology*. A. Gorbman (ed.). Wiley, New York.

WELSH, J. H. 1961a. Neurohumors and neurosecretion. In: *The Physiology of Crustacea*. Vol 2. T. H. Waterman (ed.). Academic Press, New York.

WELSH, J. H. 1961b. Neurohormones of Mollusca. *Amer. Zool.*, 1:267-272.

WENK, P. 1953. Der Kopf von *Ctenocephalus canis* Curt. (Aphaniptera). *Zool. Jb. (Anat.)*, 73:103-164.

WETZSTEIN, R. 1957. Elektronenmikroskopische Untersuchungen am Nebennierenmark von Maus, Meerschweinchen und Katze. *Z. Zellforsch.*, 46:517-576.

WEYER, F. 1935. Ueber drüsenartige Nervenzellen im Gehirn der Honigbiene, *Apis mellifica* L. *Zool. Anz.*, 112:137-141.

WEYGOLDT, P. 1961. Zur Kenntnis der Sekretion im Zentralnervensystem der Ostrakoden *Cyprideis littoralis* (G. S. Brady) (Podocopa Cytheridae) und *Cypris pubera* (O. F. M.) (Podocopa Cypridae). Neurosekretion und Sekretzellen im Perineurium. *Zool. Anz.*, 166:69-79.

WIGGLESWORTH, V. B. 1939. Source of the moulting hormone in *Rhodnius*. *Nature, Lond.*, 144:753.

WIGGLESWORTH, V. B. 1951. Source of moulting hormone in *Rhodnius*. *Nature, Lond.*, 168:558.

WIGGLESWORTH, V. B. 1954a. Neurosecretion and the corpus cardiacum of insects. *Pubbl. Staz. zool. Napoli*, 24(suppl.):41-45.

WIGGLESWORTH, V. B. 1954b. *The Physiology of Insect Metamorphosis*. Cambridge Univ. Press, Cambridge.

WIGGLESWORTH, V. B. 1959. *The Control of Growth and Form: A Study of the Epidermal Cell in an Insect*. Cornell Univ. Press, Ithaca, N.Y.

WIGGLESWORTH, V. B. 1960a. Axon structure and the dictyosomes (Golgi bodies) in the neurones of the cockroach, *Periplaneta americana*. *Quart. J. micr. Sci.*, 101:391-398.

WIGGLESWORTH, V. B. 1960b. Insect metamorphosis and the juvenile hormone. *Proc. Acad. Sci. Amst.* (C), 63:286-290. (*Proc. Kon. Ned. Akad. Wet.*)

WILLEY, R. B. and CHAPMAN, G. B. 1960. The ultrastructure of certain components of the corpora cardiaca in orthopteroid insects. *J. Ultrastruct. Res.*, 4:1-14.

WILLIAMS, C. M. 1946. Physiology of insect diapause: The role of the brain in the production and termination of pupal dormancy in the giant silkworm, *Platysamia cecropia*. *Biol. Bull., Woods Hole*, 90:234-243.

WILLIAMS, C. M. 1947. Physiology of insect diapause. II. Interaction between the pupal brain and prothoracic gland in metamorphosis of the giant silkworm, *Platysamia cecropia*. *Biol. Bull., Woods Hole*, 98:89-98.

WILLIAMS, C. M. 1948. Extrinsic control of morphogenesis as illustrated in the metamorphosis of insects. *Growth*, 12:61-74.

WILLIAMS, C. M. 1949a. The prothoracic glands of insects in retrospect and in prospect. *Biol. Bull., Woods Hole*, 97:111-114.

WILLIAMS, C. M. 1949b. The endocrinology of diapause. *Bull. biol.*, Suppl. 33:52-56.

Bibliography

WILLIAMS, C. M. 1952. Physiology of insect diapause: IV. The brain and prothoracic glands as an endocrine system in the Cecropia silkworm. *Biol. Bull., Woods Hole*, 103:120-138.

WILLIAMS, C. M. 1953. Morphogenesis and the metamorphosis of insects. *Harvey Lect.*, 47:126-155.

WILLIAMS, C. M. 1956. The juvenile hormone of insects. *Nature, Lond.*, 178:212-213.

WILLIAMS, C. M. 1960. The juvenile hormone. *Acta endocr.*, Suppl. 34: 189-191.

WILSON, L. D., WEINBERG, J. A., and BERN, H. A. 1957. The hypothalamic neurosecretory system of the tree frog, *Hyla regilla*. *J. comp. Neurol.*, 107:253-272.

YAMAMOTO, Y. 1955. Effect of ovariotomy on the sinus gland of the isopod crustacean, *Armadillidium vulgare*. *Annot. zool. jap.*, 28:92-99.

YAMAMOTO, Y. 1961. Hormonal control of gastrolith in the crayfish, *Procambarus clarkii*. *Annot. zool. jap.*, 34:38-42.

YOUNG, J. Z. 1936. The giant nerve fibre and epistellar body of cephalopods. *Quart. J. micr. Sci.*, 78:367-386.

ZEE, H. C. and PAI, S. 1944. Corpus allatum and corpus cardiacum in *Chironomus* sp. *Amer. Nat.*, 78:472-477.

PART II The Lower Phyla

CHAPTER **7**

Protozoa, Mesozoa, and Porifera

Summary	434	B. Anatomy of Fibrillar Systems	442
I. Simpler Responses of Protozoa	**435**		
A. Overt Reactions to Stimuli	435	**III. Lasting Changes of State**	**449**
B. Action Potentials and Electrical Stimulation	436	**IV. Mesozoa**	**449**
II. Coordination of Separate Regions	**437**	**V. Porifera**	**450**
A. Experimental Evidence	437		
1. Changes in direction of the waves	437	Classified References	453
2. The mechanism of metachronal wave coordination	439	Bibliography	454

Summary

The prompt, vigorous, and adaptive changes in movements of some ciliates suggest the presence of specialized structures for conducting excitation, but it is not immediately clear whether this is so or whether the cell membrane or the effector organelles themselves can conduct in the pattern necessary for coordination. Electrical properties of the cell membrane show many similarities to those of nerve cells, including potential changes associated with activity and a threshold current for just perceptible change in ciliary beat. Change in direction of ciliary beat or stopping or starting are nearly synchronous over the animal and appear to require no special conducting elements; electrotonically spread membrane changes, together with a built-in gradient of membrane properties and of spontaneous frequency of ciliary beat, would seem to account for the facts. The mechanism of coordination of the metachronal wave is more difficult. Several findings, especially the possibility of altering the frequency and wave length without altering the velocity of spread of the ciliary wave, speak against a theory wherein each cilium mechanically stimulates the next. But evidence does support the idea that each cilium or local region responds actively and with time loss to some stimulus received from the preceding—like a chain reflex or a nerve impulse. The time consumed appears to vary with the number of cilia rather than with the distance traversed, suggesting a cilium-to-cilium reflex rather than a simple neuroid impulse in the cell membrane. Such a picture may employ, but does not need, specialized conducting structures between cilia. The impetus handed from region to region, regeneratively, may be extremely weak and local, for each cilium has a spontaneity and needs only a phasing signal.

The anatomical evidence is chiefly from electron microscopic findings and indicates that in generalized ciliated areas there is no continuous system of subsurface elements that offer the variety of directions possible to the ciliary wave, that is, no continuous meshwork. Kinetodesmal fibrils are too restricted in cross connections to suggest a role in this function. Other fibrils (for example, finer ones, of the order of 20 mμ) are present in the ectoplasm but their distribution and connections are not well enough known to assign such a role to them. The classical silver line systems have all been identified with pellicular or surface configurations that do not represent continuous conducting surfaces; in some species these configurations form polygons but in others they are mainly longitudinal and again would not provide a basis for metachronal waves that can take any direction.

In special cases there are subsurface fibrils connecting distant cirri and membranelles. The cutting experiments of Taylor in 1920 have been the best evidence for some essential role of these fibrils other than contractile or supporting. But recent reexamination fails to confirm his conclusions. At present we are without evidence of formed elements functioning as conducting or coordinating structures in Protozoa.

There is no suggestion of specialized nervous structures in the Mesozoa.

Porifera exhibit several forms of response. Rarely is there any sign of conduction beyond the probable area of action of the stimulus. One such rare sign is the report of very slow general body contraction to pricking the base of a colony of *Tethya*. Electrical stimulation has in a few cases caused response, but not more than 3 mm away. There is no correlation apparent between responses of pores and oscula or between either of these and the beating of choanocyte flagella. Histological claims of nerve cells must be considered as premature on the standards accepted in other phyla.

Whether the protozoans have some kind of specialized system that conducts excitation is a question that has interested zoologists for many years and that has attracted new interest and activity in recent years. We shall see that the answer for the moment is "no," but only on balance of evidence; its unequivocal demonstration is not yet attained. The question we have raised is the presence of a system that is specialized. Certainly there is conduction from place to place—so the mere presence of coordination is not in question. Certainly there are not nerve cells connected into a system—so the presence of a nervous system as defined for metazoans is not at issue. But if there is some specialization of organelles, then its possible relation to metazoan systems should be examined. We shall see that the inquiry is still at the stage of asking, "Of what meaning for the normal conduction of excitation in the protozoan are the molecular and macromolecular patterns and organelles that are known?"

Among the several classes of Protozoa, the ciliates are by far the most likely to provide such evidence on the basis both of anatomical differentiation and of functional elaboration. Even here the question is still far from settled, and it is practically only with the ciliates that we have to deal. Even within this class, the structures and experiments described from one species are not yet comparable to those from other species so that a contemporary account must be fragmented and not a general picture. By the same token, the importance of studying a number of representative forms is clearly emphasized.

I. SIMPLER RESPONSES OF PROTOZOA

A. Overt Reactions to Stimuli

Many reactions of protozoans resemble strikingly those of small metazoans living beside them—rotifers, turbellarians, nematodes, oligochaetes. Prompt, vigorous, and adaptive movements or changes in direction of movement bespeak sensitivity, specialized effector mechanisms, and an appropriate coupling between them (for examples, see the following, among many other accounts: Jennings, 1906; Danisch, 1921; Hempelmann, 1926; Mast, 1941; Biriukov, 1957; Nicol, 1958). The question at hand is not disposed of by saying that the reaction is automatic. Rather, we are concerned to know whether it is the same specialized structure that both receives and responds or whether there is specialized conduction between separate structures.

The fact of **specificity of response** is suggestive. Predatory ciliates, *Didinium* for example, use the same effector organelles—cilia—to respond differently to prey species, indifferent species, and noxious stimuli. The receptors are unknown but the character of the responses suggests that they can hardly be the same structures as the effectors. *Paramecium* chooses between food and other objects, even selecting certain species of bacteria for ingestion, rejecting others; here it is possible that the receptor could be the effector. More indicative of conduction are the responses of *Vorticella*, which withdraws abruptly by coiling its long stalk in response to stimulation of the "head."

Conduction of excitation is believed to be independent of contraction. It is of interest in connection with the **evolution of all-or-none** response (see p. 130) that this dramatically sudden movement is reported variously as all-or-none (*Carchesium*, Sugi, 1959) and as graded (*Vorticella*, Danisch, 1921); the myonemes in the "head" alone may contract or those in the proximal part of the stalk, in the distal part, or in the whole stalk. Other forms of stimuli may cause reversal of the normal direction of ciliary beat or separation of the "head" so that it swims away. Complete gradation is also reported for *Paramecium* (Párducz, 1956), from slight change in beating direction of cilia on a limited region to complete reversal. Still more indicative of conduction are the specific and coordinated types of swimming and creeping in forms like *Euplotes*, discussed below.

Effectors are obviously localized; **receptors are shown to be localized** in some cases (Horton, 1935; Jennings and Jamieson, 1902; Alverdes, 1922; Mast, 1941). The cytopharynx region is specially receptive, and the single or grouped, nonmotile, stiff cilia of certain ciliates are believed to be. The most remarkable organelles for reception are the eyespots of flagellates like *Euglena*, *Volvox*, and *Pouchetia;* these may have a light-sensitive structure contained in a pigment cup and a simple pellicular thickening or a large spherical lens. The principal effector in these forms, however—one or more flagella—is contiguous with the eyespot, so that although the two structures are differentiated, there is not an obvious requirement for conduction between them except at the molecular level. Pacemaker regions are also probably localized (see below); Jensen (1959) proposes that there are three in *Paramecium*.

B. Action Potentials and Electrical Stimulation

Membrane potentials have been measured by inserting electrodes into protozoan cells by a number of workers (see Buchtal, 1941, for a summary of older values). **Resting potentials** vary widely but are generally low compared to metazoan nerve and muscle cells. Kamada (1934) recorded a maximum of 29 mv in *Paramecium*, lower with increased external cations. This investigation, incidentally, used 2–3 μ quartz capillary electrodes, long antedating the modern ultramicroelectrode (usually called the Ling-Gerard electrode). Popova et al. (unpubl.) gave values of 25–35 mv in *Opalina* and Kinosita (1954) reported values from 10–50 mv; Umrath (1956) found a maximum of 65 mv and a mean of 18 mv in *Amoeba;* Hisada (1957) arrived at a mean of 48 mv in *Noctiluca*. As in familiar cells, the interior is negative to the exterior and increased potassium ion concentration in the medium lowers the potential, linearly with log [K] except that from zero to 1.4 mM there is no effect (*Opalina*, Naitoh). A similar potential apparently exists at the protoplasm-vacuole interface in *Noctiluca*, with the protoplasm negative. Yamaguchi (1960a) found changes of potential difference synchronous with pulsation of the contractile vacuole in *Paramecium* by means of the microelectrode inserted into the region closely adjacent to the contractile vacuole. The resistance between an internal electrode and one in the medium—and therefore the resistance for the whole cell at rest—is given as 2.7×10^6 ohms for *Opalina*, with inward current; outward current from 0.5×10^{-9} A upward encounters a lowering resistance which plateaus at about 0.3×10^6 ohms with a current ten times this level. These values are quite similar to those found in metazoan nerve cells, which, however, are much smaller in size. Most interestingly, there are topographic differences in the membrane resistance and these form a consistent pattern as a gradient across the cell. There is a corresponding gradient of effectiveness of electric shocks delivered through a local external microelectrode, perhaps due to the difference in membrane resistance (see further below).

Action potentials have been recorded in a few studies. Umrath (1956) observed graded depolarizing waves in *Amoeba* which at maximum exceeded the resting potential and reversed by as much as 11 mv, inside positive. The action potentials exhibit facilitation, that is, successive responses grow and presumably spread farther. Apparently spontaneous fluctuations in membrane potential occur in *Opalina* (about 1 sec in duration, about 10 mv), sometimes almost sinusoidal in the vicinity of ten c.p.s. (Koshtoyants and Kokina, 1958). Surprisingly, in *Noctiluca* such spontaneous waves are clearly hyperpolarizing and seemingly all-or-none (Hisada). They are invariably associated with a beat of the flagellar tentacle and are probably not artifacts of the movement. On top of a resting potential of about 48 mv they add enough to bring the total potential to 75 mv. Often occurring in groups up to two or three per second, a representative case figured had 16 spikes in 10 seconds. A prepotential inaugurates the burst and precedes each spike in it. Such a burst corresponds to a deflection of the tentacle consisting of many small beats. A depolarizing pulse applied through a second internal electrode elicits a spike and a beat and if it is a long pulse, so that spontaneous spikes occur, adds height to them by shifting their base line, though they rise only to the same peak level. Applied current in the opposite, inward direction abolishes spontaneous spikes, but may provoke a movement of the tentacle in the opposite direction (Watanabe and Hagiwara, personal communication).

Electrical pulses also evoke brief luminescent flashes, one for each pulse and graded in brightness with the intensity of the current. Repetitive pulses at intervals up to 1.4 sec produce summation and facilitation and, at higher frequencies, fusion of flashes all occurring within the photogenic cell (Nicol, 1958). These properties resemble those of metazoans with nervous control of luminescence. The natural stimulus is mechanical and it seems likely that this sets up trains of action potentials.

Except for the spontaneous hyperpolarization, protozoan cells apparently show **electrophysiological behavior quite similar to metazoan cells,** at least to the extent of graded responses and possibly, in the case of *Noctiluca*, something like all-or-none activity as well. Further underlining the resemblance, *Paramecium* has been stimulated with linearly rising current and found to manifest a reproducible threshold in the just perceptible change in ciliary beat; this threshold does not rise with more gradual slopes lasting up to ten or more seconds, but then it rises considerably just before the minimum effective slope is reached—demeanor typical of squid giant and other axons but with about a thousandfold difference in time scale (Kinosita, 1939). Unquestionably there is a great deal yet to be learned by careful study of the electrical activity of protozoans, relevant to the physiology of the metazoan nerve cell, whether they have differentiated conducting organelles or not.

II. COORDINATION OF SEPARATE REGIONS

Virtually all the evidence favoring specialized conduction paths concerns the rapid coordination of cilia over widely separated areas, often with intervening areas devoid of cilia. The best indications come from experimental manipulation; the assignment of function to anatomically defined fibrils or networks of lines is much more uncertain.

A. Experimental Evidence

Depressant chemicals sequentially abolish three aspects of ciliary activity; the ability to change the direction of the metachronal wave, the coordination among cilia—that is, the wave itself—and finally, the beat itself (Okajima, 1954). The **separability of these aspects** is also shown by cutting and by electrical stimulation, which can affect either of the first two alone: in *Opalina* a stimulus causes a change in direction; a transverse cut interrupts the metachronal wave but not the simultaneous change in direction of waves, on the two sides of the cut. Apparently we have two mechanisms to look for—that mediating change in ciliary beat and that mediating the metachronal wave itself.

1. Changes in direction of the waves

The most detailed information comes from *Opalina*, an oval, flattened form of uncertain affinities but behaving in these respects like a simple ciliate. The body is completely clothed with cilia and the metachronal waves move across the surface sometimes in one pattern, sometimes in another. Okajima, who has applied an ingenious technique of **local electrical stimulation,** gives numbers to these patterns. Threshold stimulation causes a local change in the wave pattern (see also Kanno, 1958). Threshold varies systematically and a map of isothreshold lines (Fig. 7.1) resembles a map of the lines of equal intensity needed to change the waves in a certain number of square microns of area. There is, therefore, a permanent gradient in the cell which is also expressed in the topographic differences in velocity of propagation. The correspondence is not always good, so that one may say velocity involves more than excitability; other separate variables, perhaps latency and rate of movement of each cilium, may be factors. Velocity differs by a factor of two in different directions of the metachronal wave, for example, 100 μ/sec and 200 μ/sec.

There is also a permanent gradient in the cell which may be expressed in the topographical differences in magnitude of change in beating direction in response to a given increase of current intensity. Hence, the various patterns of the metachronal wave observed in the organism may be explained in connection with the intensity of autonomous stimulation and the local differences in sensitivities and reactivities of the cilia on the cell surface. The usual global changes in direc-

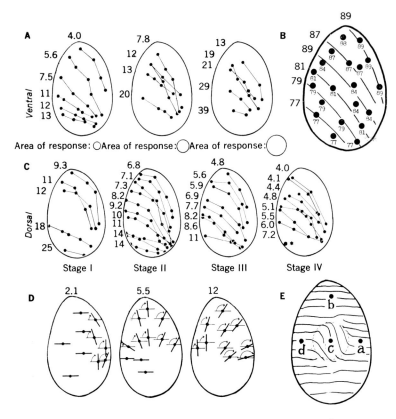

Figure 7.1.

A. The "equivalent lines" required to stimulate a definite area (shown just below as a *small circle*) on the cell surface of *Opalina*, which is in the state of stage I. Numerals represent relative values of the current intensity, when minimal threshold at fixed point* (see stage IV) is taken as 10 (12–18°C). **B.** Topographical difference in electrical conductivity of the membrane. Numerals represent the current intensities expressed in percentage of that measured without the organism. The applied voltage and the negative hydrostatic pressure at the electrode tip were kept constant throughout the measurement (26.4°C). **C.** The "equithreshold lines" on the cell surface of *Opalina* in each stage of the cell. The numerals indicate relative values when the threshold value at a fixed point* in stage IV is taken as 10 (15–24°C). **D.** Local change in direction of the metachronal wave when points on the ventral surface of *Opalina* are stimulated by direct current of intensities represented in relative values by the numerals. Thin and thick *straight lines* respectively represent the direction of the metachronal wave before and after stimulation. *Arrows* indicate the changes in direction of the wave. **E.** An example of local response appearing within the area of dotted circle when *Opalina* in stage I is stimulated by direct current at a fixed point *c* on the ventral side. Note marked change in direction of metachronal wave in a restricted area. [**A, C, D, E,** Okajima, 1953; **B,** Naitoh, 1958.]

tion of metachronal waves are almost simultaneous all over the cell and are therefore independent of the prevailing pattern of waves; in other words, the new pattern of waves is set up before a metachronal wave is conducted.

Kinosita (1954) found a **change in membrane potential** synchronous with spontaneous alternations in direction of beat of the cilia (see also Kokina, 1960); moreover, he was able to plot a systematic decrease in internal negativity of the potential with the increase in angular deviation in direction of beat from normal (Fig. 7.2). Reversible change in direction of beat can be produced by microinjecting Ca^{++} precipitants—which, however, lower the electrical excitability (Ueda, 1956)—or by injecting isotonic solutions enough to cause a local **swelling**, in which event the change in beating direction is also local. Local stretching with needles can do the same. Generalized swelling by sudden decrease in tonicity of the medium, regardless of which electrolytes are used, causes a general reversal of beat, whereas any shrinking change in medium causes an augmentation in the beating in its normal direction (Naitoh, 1959). Okajima carefully pressed a 10–15 μ glass rod against the cilia and those immediately affected stopped activity. The patterns of coordinated waves were nevertheless sometimes maintained, while at other times the wave pattern was different on the two sides (Fig. 7.3); sudden changes in pattern occurred nearly synchronously on the two sides. The evidence is compatible with the notion that there is a spontaneous tendency to beat rhythmically or cyclically in every region, but with a gradient of inherent frequency-tendency falling from anterior left to posterior right.

II. COORDINATION OF SEPARATE REGIONS A. Experimental Evidence

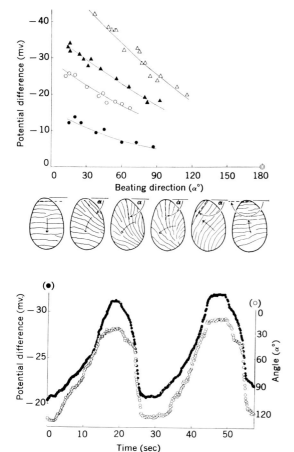

Figure 7.2. *Upper.* Relation between angle α as shown in the diagrams below and the electric potential of the cell interior with reference to the outside in four individuals of *Opalina* during the spontaneous repetition of ciliary reversal in Ringer's solution. The diagrams show a series of patterns of metachronal waves on the ventral surface with increasing degrees of excitation from left to right. The waves are observed from the dorsal side through the transparent protoplasm. The *lower arrows* indicate the directions of wave propagation. *Lower.* A typical result of simultaneous determinations of the time change of potential difference *(solid circles)* and of beating direction of cilia (angle α; *hollow circles*) during the spontaneous activity of *Opalina* in Ringer's solution; obtained from microcinematographic record. The sign of potential difference is that of the cell interior with reference to the external surface. [Kinosita, 1954.]

Any localized or generalized change in the membrane reflected in its potential alters this; the change is graded and usually global and rapidly spread, but can be local as under a microelectrode or local stretch. This hypothesis of the mechanism of spread of change in direction of beat therefore utilizes electrotonically spread, graded changes in the cell membrane.

2. The mechanism of metachronal wave coordination

Such a mechanism is more easily assigned to the cell membrane or a generalized constituent than to any connected system of formed elements. Since the wave can be in any direction, any formed structures necessary for conduction must be in a fine mesh network. The glass rod experiment is equivocal as to the role of mechanical stimulation of one cilium by the preceding. The experiment of cutting off a swathe of cilia (or uprooting them) has not been done. The same things are true in other species, even in metazoans such as ctenophores (Verworn, p. 534; Parker, p. 532). It might be said that the fact that coordination can sometimes be blocked speaks for an essential role of some form of mechanical transmission, either by contact or by water displacement, unless it be postulated that the glass rod pressure was strong enough to block conduction in surface or subsurface structures. Preservation of coordination when holding some cilia quiet does not indicate subsurface conduction because of the possibility that the cilia are nevertheless struggling; it only indicates that any me-

Figure 7.3. Patterns of the metachronal wave on the ventral surface of *Opalina* on the two sides of a compressing glass needle during spontaneous alternation of excitation. *Thin arrows* indicate the direction of wave conduction; *thick arrows* show the sequence of spontaneous alternation of the wave pattern. [Okajima, 1954.]

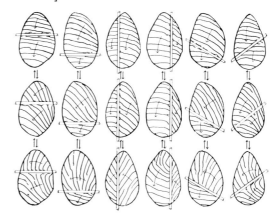

chanical transmission does not require a full beat or water current. It would be important in this connection if it were found that the angle between the direction of effective stroke and that of the metachronal wave can freely change in the individual, from time to time. This does not seem to have been found, although it is known that this angle differs in different organisms (Párducz, Worley, Knight-Jones). If it can change independently of metachronal wave direction, it would be evidence that the spread of excitation is more than a purely mechanical process.

An important additional **limitation on the requirements of any mechanical process** which may mediate the metachronal wave is at least true of the peristomial membranelles of *Stentor* (Sleigh, 1956, 1957). Increased viscosity of the medium (over a certain range) and other agents act to alter the frequency and wave length but not the velocity of the wave. In order to permit a mechanical transmission, it must be assumed that the more sluggish beat is accompanied by a simultaneous drop by just the right amount in the threshold mechanical event required to stimulate the next cilium and brought about by the same agent, for example, a viscous medium. Frequency and velocity have different temperature coefficients. The inherent frequency of different parts of the peristomial row of ciliary plates is different, as shown by isolating them with cuts, but the velocity of the wave in the isolated parts is the same. In the cytopharynx region the wave length is shorter although the frequency is the same as in the rest of the peristome in the intact animal; this means that the velocity is less, per unit distance. But the number of cilia is greater per unit distance and it turns out that the number of cilia per wave (6.4) is the same as elsewhere; since the average frequency is 27.7 waves/sec, the velocity is 178 cilia/sec or 5.6 msec/cilium. As opposed to a theory of continuous, neuroid conduction at less than a millimeter per second, these facts point to a step-by-step or **chain reflex theory,** in which each cilium requires time to respond and by its response somehow sends a stimulus, rapidly propagated, to trigger the next one; the whole cycle takes place in 5.6 msec. Most workers have followed Gray (1930), at least to the extent of supporting the primacy of some mechanical consequence of beat as supplying the trigger; but the evidence of Sleigh makes this theory unlikely since it would require, in *Stentor* at least, a quantitative coincidence of separate effects on excitability and mechanical beat as a result of diverse agents like methyl cellulose and reduced magnesium. Apparently the event in each cilium which mediates the trigger for the next should not be temporally tied to the mechanical. It certainly is labile, for other agents do affect velocity; when they do, frequency is also altered, suggesting that the unquestionably internal pacemaker process is of the same nature as, or indeed identical with, the excitation process.

The medium through which transmission takes place may be the cell surface or the ectoplasm, at least in holotrichs like *Paramecium*. Several workers have noted that pieces of the animal without nucleus or endoplasm still show metachronal waves.

Cutting just into the ectoplasm interrupts the coordinated wave. On a mechanical theory, in view of the occasional successful passage of the wave under a glass rod holding the cilia, it must be supposed that a reduced and abnormal mechanical event suffices for transmission. But in the cutting experiments, transmission does not occur across the apposed edges; *this* sort of reduced and abnormal mechanical event does not suffice (Verworn, 1889; MacDougall, 1928; Worley, 1934; Okajima, 1954; Sleigh, 1957). The metachronal wave in *Paramecium* and *Opalina* cannot circumvent the incisions by conducting around them, whereas the change of direction can. In *Spirostomum* the metachronal wave can to some extent spread laterally and so bypass partial incisions.

The facts of coordination of metachronal waves are not compatible directly with a propagation through the cell surface like that of nerve impulses, but a significant degree of similarity and a cell membrane locale are still possible. The most important similarity between wave coordination and nerve impulse is that each region responds, according to its local responsiveness, to the stimulus of the preceding region and in turn excites the next—each step being a labile, time-consuming process. The present meager evidence suggests that the time spent is a function of the number of cilia rather than the distance traversed, pointing to the saltatory rather than the simple neuroid alternative. A further difference from the simple nerve impulse picture is that each cilium has a spontaneous tendency to beat and only requires a phasing signal to coordinate it with the others; this signal may be extremely weak and local, which would account for the absence of an action potential visible to an electrode in the cytoplasm, which averages the potential of a large area of membrane.

After forty years, still the best known and almost the only direct evidence of **formed structures mediating the coordination** between parts of a protozoan body is that of Taylor (1920). He performed cutting experiments on *Euplotes*, a strategically favorable organism; it has widely separated tufts of cilia (the cirri) in addition to the specialized band of ciliary rows (the membranelles), all stopping and starting and reversing in a highly coordinated manner. As a result of properly timed and directed beating of these effectors there are three creeping movements: (a) straight ahead, (b) a quick backward movement, (c) a turn to the right (aborally); and six swimming movements: (a) forward without spiral revolutions, (b) forward in spiral revolutions, (c) a circus movement to the right, (d) a circus movement to the left (orally), (e) a sharp turn to the right, and (f) a rapid backward movement without revolutions. As we shall see below, there are **anal cirri fibers** connecting each of those cirri with a region of convergence called the **motorium**. Severing the anal cirri fibers was said to destroy coordination between these cirri and the membranelles: creeping movement (b) is infrequent, swimming movement (e) is seldom seen, and (f) is never seen. There is also a fiber under the mem-

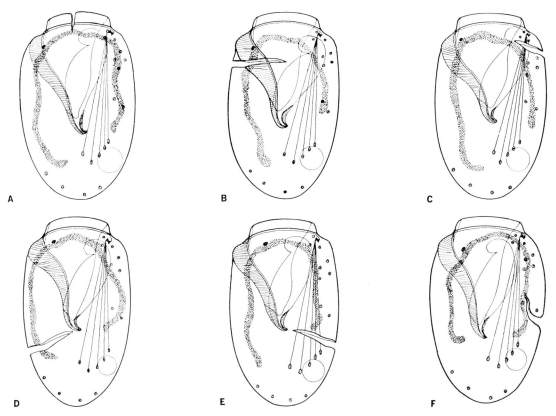

Figure 7.4. Microsurgical mutilation of *Euplotes* used in the testing of presumptive coordinating function of fibrils and membranelles. Lower left and lower right are controls reported to have no serious effect upon coordination; the other incisions are described as causing specific deficits. [Taylor, 1920.]

branelles and severing this results in conspicuous differences in the activity of the membranelles on the two sides of the cut and in abnormal spiral revolutions in swimming. Cuts which invade the region of the motorium or the sides from which fibers converge on it interrupt coordination of the membranelles and anal cirri in the same way as more distal cuts that sever the anal cirri fibrils.

None of the cuts incapacitates any of the cirri in the performance of the various movements proper to it; by chance they will from time to time be moving in synchrony with the membranelles so as to produce a normal creeping or swimming movement. It is the infrequency of these occurrences that suggested to Taylor the lack of coordination and the conclusion that the fibers mediate this normally. A number of types of incisions were performed as controls; any which did not sever either membranelle fiber or anal cirri fibers did not impair normal creeping or swimming movements. It is highly improbable that the fibers are either supporting or contractile. The movements of the anal cirri are in four directions and all four continue after severing the fibers. The fibers are feebly attached to the ectoplasm, not directly to the basal structures of the cilia; to displacement by the microneedle they show no stretch or resilience; they do not kink or curl when cut. It is extremely desirable that these critical experiments be repeated, for at their face value—and the thoroughness and corroborative detail give them weight—they strongly suggest that the complex system of fibers in *Euplotes* are specialized for mediating rapid and specific coordination among its locomotor organelles. Perhaps this system is superimposed on a local metachronal wave-producing system in each cirrus, much like the nervous system in some metazoans superimposes a control over cilia which have their own local autonomy. In the pedal cilia of the gastropod *Alectrion*, for example, the central nervous system can stop, start, and grade the speed of movement (Copeland, 1919). Similar experiments on *Euplotes* have been carried out by Okajima; these have not been published, but I am kindly permitted to refer to them.

He could not find a difference in coordination of the anal cirri between animals with cuts of the same size, some with the anal cirri fibers interrupted and others not. The anterior part of the normal organism is more sensitive than the posterior, and stimulation of the anterior tip sometimes fails to induce reaction in the anal cirri. However, such a tendency is almost unaffected by a partial incision of the cell body, even if it cuts the anal cirri fibers. Thus he is inclined to put less weight on the significance of fiber structures mediating the coordination and more on the importance of the anteroposterior gradient of sensitivity.

Acetylcholinesterase has been found in ciliates and acetylcholine and anticholinesterases have effects for example on the threshold of galvanotaxis and on summation of stimuli (Popova et al., unpubl.; Seaman and Houlihan, 1951; Koshtoyants and Kokina, 1957). Seaman located the entire acetylcholinesterase activity of the cell in the "fibrillar system" of the pellicle of *Tetrahymena* (1951); recently he has isolated large amounts of pure kinetosome and found it highly active for specific cholinesterase. Just how much specialization of response mechanisms this means, cannot, of course, be estimated without much more information. The specific cholinergic enzymes are not confined to nervous tissue in higher animals but neither are they very widespread among tissues (Burgen and MacIntosh, 1955; Nachmansohn, 1959).

B. Anatomy of Fibrillar Systems

The literature on microscopically recognized systems of structures believed to play some role in coordination of locomotor organelles in protozoans is formidable; by no means all references describing such structures are listed here, but more than 70 are mentioned. They appear to fall into an earlier, exuberant period of description based on cytological stains (for example, Yocum, 1918; MacDonald, 1922; Pickard, 1927; ten Kate, 1927; Lund, 1933; Hammond, 1937), overlapping with a period, in the twenties and thirties, of diminishing activity except for a school exploiting special silver stains (J. von Gelei, Klein, G. Gelei), a period of scepticism (ten Kate, 1927; Jacobson, 1931; Rees, 1931), and a

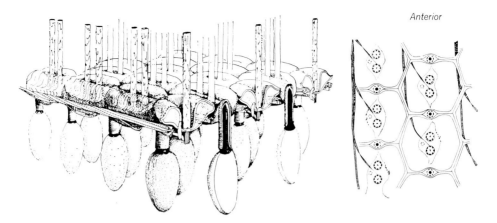

Figure 7.5. Reconstruction of the pellicle system of *Paramecium* as it would appear in perspective view. The relationships of the kinetodesma and trichocysts to the hexagonally packed ciliary corpuscles that form the outer surface are also shown. The plan view is an interpretive diagram of a ciliary corpuscle in idealized section in the tangential plane, including portions of three kineties. Kinetodesmal fibrils are shown leaving the anterior region of the posterior cilium in each corpuscle; this point of departure is below the level of the inner peribasal membrane and would not be visible in an actual thin section of so superficial a slice. Cross sections of trichocyst tips appear sandwiched between adjacent ciliary corpuscles and in a line parallel to and between the kinetodesmal bundles. These interpretations are the result of electron microscope studies. [Ehret and Powers, 1959.]

recent surge of activity based on the use of the electron microscope (see classified list of authors, p. 453). While a large body of new information has been added to the old, and some clarifications provided, the subject is more fluid than ever and there is major disagreement as to whether any of the known fibrous structures are relevant at all directly to the problem of conduction of excitation.

Fibrillar or linear structures in various species have been described in the belief that they may have something to do with conduction. But since their number is large and not reducible to a few classes, and since there is no clear evidence for most of them to implicate them in this function, the following account must be only a summary of the better known examples. Reviews have been written by Taylor (1941), Klein (1955), Párducz (1958a, 1958b), and Ehret and Powers (1959); see also Ehret (1960).

By far the most attention has been devoted to *Paramecium* (Figs. 7.5, 7.6). Ehret and Powers have offered a reconciliation on the basis of the several electron microscope studies which takes into account in a plausible way all of the light microscope and silver-staining structures of the cell surface. These had accumulated entities and terminologies to the point where Párducz (1958a), in a thorough review had to recognize four different systems of ostensibly fibrillar nature, based on the partially overlapping descriptions of many observers. It would be out of place to review the confused history of the terms here; the authors just mentioned provide this perspective. The finding of **electron microscopy, which has been the key** to reinterpretation, is that the surface of some ciliates is raised in a pattern of blisters whose margins correspond to silver lines. In *Paramecium* the surface is composed of a mosaic of basically hexagonally packed pellicular blisters that Ehret and Powers call ciliary corpuscles. In several other ciliates the blisters are longitudinal, meridionial, and not cut up into polygons (*Tetrahymena, Colpidium, Glaucoma, Stentor*).

In *Paramecium* the center of each polygon is depressed, defining a recessed bit of outside space; the cilium, or in many cases two cilia, penetrates the pellicle here. Bradfield (1955) proposed that the two central fibrils of each cilium are conducting and that the ring of nine outer paired fibrillar elements are contractile. The cilium is rooted in a cilium base—kinetosome or "Basalkörperchen"

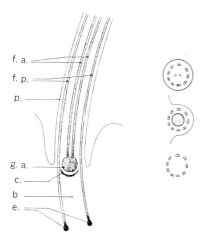

Figure 7.6. Diagram of the ultrastructure of the cilium and the blepharoplast of an isotrich ciliate, seen in longitudinal section. At the right transverse sections at the levels of the free part of the cilium (above), the axial granule (middle), and of the blepharoplast (below). [Noirot-Timothée, 1958.] *b.*, blepharoplast; *c.*, granular cap; *e.*, terminal thickenings of the peripheral fibers; *f. a.*, axial fibers of the cilium; *f. p.*, peripheral fibers of the cilium; *g. a.*, axial granule; *p.*, pellicle.

(Fig. 7.6). Bradfield also thought that rhythmic impulses arise in this basal body and spread up the central ciliary fibrils. The sequence of activation of the outer fibrils would then determine the direction of beat. Sleigh (1960) shows that there need be no set path determining the sequence.

The only fibrillar constituent below the surface so far recognized as a regularly patterned feature, is the **kinetodesmal fibril**; bundles of finer fibrils (20 mμ) are said to be associated with ciliary bases also but their pattern has not been established (Roth, 1958a; Schneider, 1960). The kinetodesmal fibril arises at the anterior side of the base of the cilium (the posterior cilium if there are two) and runs anteriorly, swerving slightly to the left. Each kinetodesmal fibril runs the length of a small number of the hexagonal ciliary corpuscles, joining a strand of several others from the adjacent cilia in the same longitudinal row. These short, overlapping fibrils are not hollow but have a transversely striated dense composition. It seems to us that they **can be ruled out** as adequate conducting paths for the metachronal waves, except in some indirect role, because the waves can progress longitudinally in straight lines, whereas the rows of cilia and the axis of the ciliary corpuscles exhibit a sharp deflection, as pointed out by Párducz (Fig. 7.7). Ehret and Powers warn against faulty functional speculations deduced from structural appearances, and this may be one. Fragments of fibrillar structures have been seen in electron micrographs which do not belong to this system and it is possible that there are still undetected patterned systems.

The membranous junctions of the hexagonal ciliary corpuscles account for the silver-staining "outer lattice system" or "pellicular ridge pattern" ("Indirekt verbindendes System" of Klein; "Äusseres Stützgerüstsystem" of Gelei; "Äussere Gittersystem" of Párducz; "argyrome" of French authors). The trichocysts open to the outside through pores located in the junction between the posterior border of one polygon and the anterior border of the next; thus they are in line with the cilia longitudinally. The double membranes forming the inner and outer walls of the circumciliary depression are believed by Ehret and Powers to account for several of the

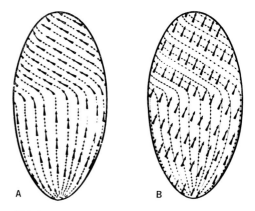

Figure 7.7. Scheme of the two possible variants of excitatory conduction in *Paramecium*. **A.** The coordinating impulses (designated by *arrows*) run in an isolated manner along the interciliary fibers, with some delay toward the right. **B.** The waves of excitation are propagated toward the front in the surface layer of the cell plasma itself, with their front edges parallel to the course of the metachronal ciliary wave. In the posterior half of the body the arrangement of the metachronal waves is identical in both cases, but in the anterior end—where the rows of cilia are bent sideways from their meridional course—this sort of characteristic difference appears. [Párducz, 1958.]

systems of lines seen in microscopic preparations especially stained with silver. (The outer peribasal membrane gives the "Direkt verbindendes System," "Zirkulärfibrille" or "Nebenfibrillen," "Interziliärfasern," "Neuroneme System," "Infraziliäre Gittersystem," depending on the level of the optical or actual section; the inner peribasal membrane gives the "infraciliary lattice system" or "Innere Gittersystem" or "Infraziliäre Gittersystem.") Thus, not only are there now supposed to be **no fibril systems forming the light microscopist's networks,** but the membranes which account for the silver lines are the bounding membranes of independent cavities, the ciliary corpuscles.

The picture which emerges of the relatively undifferentiated ectoplasmic apparatus in *Paramecium* is not without its **possible functional meaning.** The experimental evidence has led to the notion that spread of ciliary excitation is a "chain reflex," or step-by-step process in which each step takes time and requires the active response of something associated with the cilium in order to stimulate the next one (Fig. 7.8). This concept does not call for a continuous conducting system but for excitable units each capable of reaching and influencing the next after its own active state has built up to a sufficient degree. There must be some determination of the direction of effective stroke and this might be the resultant of predisposing factors and the direction from which the excitation arrives. The cilium itself or its kinetosome is differentiated around its axis so that by exciting one side it may be provoked to beat in one direction. The membrane of the cell between cilia may be the conducting medium, carrying the wave of excitation, perhaps very fast and perhaps decrementally, for the few microns necessary. It may be convenient for the membrane to be divided into contiguous units, one for each cilium (or pair); this might represent in fact a kind of unit organelle capable of exciting the next—with a delay, like a miniature subcellular synapse. However, this surface mosaic is known only in some species; others are only longitudinally subdivided. How the spread is kept in a straight line in *Paramecium* instead of a given unit exciting the one obliquely ahead and to the left or the right may be merely a matter of timing: those two latter will have been excited earlier by the patches of membrane directly behind them. There must, though, be something

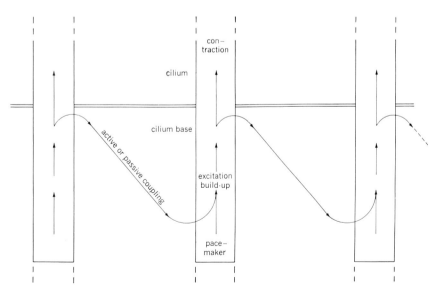

Figure 7.8. A diagrammatic representation of the chain reflex theory of metachronal coordination. Each cilium has a pacemaker but is also coupled to neighboring cilia in such a manner that excitation in one can trigger excitation in another with a delay. The essential event need not necessarily be the mechanical contraction of one cilium; the excited state at its base may, without morphologically differentiated structures, excite a neighbor. [Modified from Sleigh, 1957.]

more to it than this, to explain, for example, why the metachronal wave does not readily spread around a partial incision.

In a number of other species polygons have been seen in the pellicle or ectoplasm apparently corresponding to those of *Paramecium*, but in *Opalina*, *Stentor*, *Tetrahymena*, *Colpidium*, and *Glaucoma* they are not developed, so they are at least not general or necessary to ciliary coordination. A variety of fibrous and ridgelike strengthenings of the surface is known but does not suggest any direct relevance to the coordination function. A kinetosome or basal body (blepharoplast) at the root of both the cilium and the flagellum is a feature in common and possibly is of vital importance in the coordination function. There is variation in the detailed composition of the kinetosome which cannot be assigned a meaning at present. Various kinds of fibers attach to or originate from the kinetosome. In some forms, such as *Tetrahymena*, they resemble the kinetodesma of *Paramecium*, the length of the individual fibrils varying with the species. In some other holotrichs and in *Opalina* the orientation of the fibrils differs and they cannot readily be said to be homologous. Sometimes they appear to connect adjacent kinetosomes but even then with no or few cross connections between longitudinal rows (kineties) such as would be required for a structural basis of metachronal wave coordination, since those waves can change direction not only by 180° but also by 90° and intermediate values. In *Stentor*, fibrils from the kinetosomes gather in sheets and these sheets into lamellae passing for long distances posteriorly. In other ciliates examined with the electron microscope, fibrils or bundles of filaments whose course is unknown arise from some or all kinetosomes (see, for example, Noirot-Timothée, 1958). In some instances skeletal fibers are intimately associated with kinetosomes. Bundles of 20 mμ fibrils—about one-third the size of kinetodesmal fibers—are loosely associated with ciliary bases, but without special pattern.

Euplotes is a **more specialized case**, as it was in connection with experimental evidence and in agreement with other evidence that its group, the hypotrichs, is highly specialized. According to the electron microscope work of Roth (1957), there are fine filaments 20 mμ in diameter closely associated with the basal portion (kinetosome) of most active cilia, including those in the cirri and in the membranelles. These often gather in bundles which correspond with the fibers reported by Yocum (1918), Taylor (1920), Turner (1933), and Hammond (1937), seen with the light microscope in both fixed and stained and unstained living material. Besides the fibril bundles, individual filaments occur. Roth agrees with the previously claimed connections of fibrils and adds others, so that, taken at face value, there are now described connections by means of these filaments or bundles of them between: (a) cirri and a special region called the motorium, mentioned further below; (b) membranelles and motorium; (c) cirrus and adjacent cirrus; (d) membranelle and adjacent membranelle; (e) cirrus and membranelle directly; (f) cirrus and a system of subpellicular filaments running predominantly antero-posteriorly and also from side to side; (g) motorium and subpellicular filament system; (h) cilium and adjacent cilium; and (i) the region of the micronucleus and unknown structures. Not every cirrus is connected with every other cirrus or every membranelle. The "bristle" cilia on the body, which are nonmotile, lack these filaments. Some of the filaments connect with a superficial dense zone of the kinetosome but apparently most of them arise close to the more basal dense zone; although these zones are only 170 mμ apart, they are believed to correspond to the "basal plate" (superficial) and "fiber plate" (deep) of Taylor.

Filamentous structures in micrographs are abundant, and the reality of specific connections —as distinct from a fairly diffuse or inconstant system of fibrous inclusions parallel to and close under the cell surface—calls for additional work and, preferably, reconstruction. Nevertheless there is support for and no reason here to doubt the fiber system of the light microscopists. On a few occasions Roth saw a close grouping of filaments forming a fibrous mass never more than 4 μ across, in the region of the cytostome. Upon this mass converge fibril bundles from different directions. The facts that this is not a commonplace structure (most of the cytoplasm in this region not being fibrous) and that it has relations of a motorium led Roth to postulate this identification with the body of that name described by Yocum (1918) and Hammond (1937). Turner (1933) denied a special body but also saw a convergence of main fibers in this region. Proof for this identification would require much better three-dimensional localization and tracing of connections, but it is significant that circumscribed, densely filamentous masses receiving bundles can be found just beneath the pellicle.

Nothing has caused more scepticism than the **claim of a structure called the motorium** or neuromotorium. First by Sharp (1914) in *Epidinium* (= *Diplodinium*) and later by some 25 other authors in nearly as many genera of ciliates, there has been described a region, not over a few microns wide but sometimes rather elongate, which receives fibers from two or more of the main parts of the subpellicular fiber system called the neuromotor system. Its only real diagnostic feature is that it represents a convergence of principal fibers; generally it is said to be located anteriorly and often near the oral apparatus. The neuromotor fibers, which are not questioned except as to function and hence name, are not easy to stain and to distinguish from other structures, pellicular folds, and discharged trichocysts; the motorium is still more difficult to distinguish, for the same reasons. In addition, because of its position and size it is likely to be obscured by heavily staining membranelle structures or to be destained before them. Many unpublished failures to find it have doubtless contributed to the healthy scepticism. There are only two published denials of a motorium in genera which had been claimed to possess one (Rees, 1930, 1931; Turner, 1933) and in both cases subsequent work has reaffirmed its existence. Turner in fact only denied a swollen, bilobed mass, while confirming the confluence of the membranelle and the anal cirri fibers in *Euplotes*. Rees had earlier made a classical error in describing as neuromotor fibrils what later were identified as probably patterns of discharged trichocysts, thereby undermining somewhat the confidence of other authors in his work.

As we have remarked elsewhere, it is a telling historical fact that most of this kind of critical descriptive detail has been left to one-shot authors who have not had a long seasoning in microscopy and its pitfalls. Nevertheless, many of the reports are rich in corroborative detail, have been written after the sceptical reaction set in, and cannot be dismissed without equally careful work. The situation at present cannot last, with the powerful tool of electron microscopy available. Using this tool, Roth has briefly reported the existence of a local dense mass of fine fibrils, which he identifies with the motorium, in *Paramecium* and *Euplotes*. It would seem that there is little question that subsurface fibers of some sort (Fig. 7.9), distinct from the pellicular blister patterns, form an elaborate system, characteristic

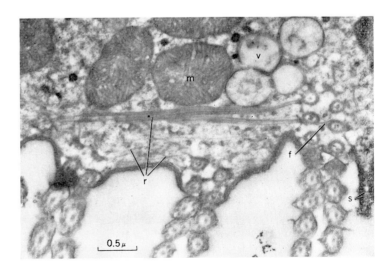

Figure 7.9.

A section through the pellicle of *Euplotes*, including portions of three membranelles. Rootlet filaments (r) both separately and in bundles are shown interconnecting the membranelles. Between the bases of a few cilia, rootlet filaments (f) are shown interconnecting adjacent cilia. Filaments of the subpellicular system (s) are cut across just below the pellicular membrane. m, mitochondria; v, ciliary vesicles. [Roth, 1958.]

of the species, especially among the more highly differentiated locomotor organelles such as membranelles and cirri and in the cytopharynx or its equivalent, and that there is often a convergence of principal fibers in an anterior or paraoral position. However, it should not be forgotten that many of the fibrils do not appear to connect the cirri with anything, merely radiating into neighboring cytoplasm.

Details of the so-called neuromotor fibers, in some cases including their formation in regeneration or behavior in reproduction, and in every case including a motorium, are given by Sharp (1914) for *Diplodinium;* Yocum (1918) for *Euplotes;* McDonald (1922) for *Balantidium;* Campbell (1926, 1927) for *Tintinnopsis* and *Favella;* Pickard (1927) for *Boveria;* Visscher (1927) for *Dileptus;* MacDougall (1928) for *Chlamydodon;* Hilton (1928) for *Lichnophora;* Calkins and Bowling (1929) for *Dallasia;* Lynch (1929, 1930) for *Entorhipidium* and *Lechriopyla;* Calkins (1930) for *Uroleptus;* Brown (1930) for *Paramecium;* MacLennan and Campbell (1931) for *Eupoterion;* Kofoid and MacLennan (1932) for *Diplodinium;* Lund (1933) for *Paramecium;* Powers (1933) for *Entodiscus;* Kidder (1933) for *Conchophtherius;* Worley (1933) for *Paramecium;* Lucas (1934) for *Metopus;* Bush (1934) for *Haptophrya;* MacLennan (1935) for *Ichthyophtherius;* Lund (1935) for *Oxytricha;* Hammond (1937) for *Euplotes;* Ellis (1937) for *Fabrea;* Rosenberg (1937) for *Nyctotherus.* Taylor (1941) provides a useful review and should be consulted for the references.

Much more questionable than the mere existence of something in the positions described is the identification of that thing—its **meaning and relation to other structures,** and its function. The first few electron microscope studies of the more complex ciliates give promise of answers to some of these questions; strikingly clear bundles—which have the relations of the light microscope fibers—have been seen, as has already been mentioned for *Euplotes* and also in *Stentor* (Randall and Jackson, 1958); see Fig. 7.10. But some doubt will remain concerning the reality of the structure called "motorium," especially in simpler forms like *Paramecium,* until electron microscope evi-

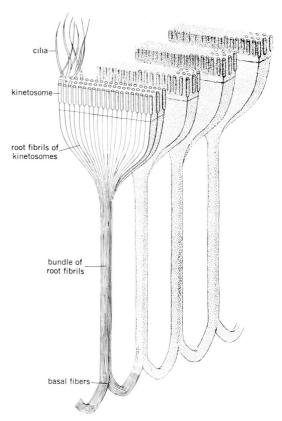

Figure 7.10. Schematic diagram of the organization of the adoral membranelles of *Stentor*. [Redrawn from Randall and Jackson, 1958.]

dence is offered with some pretensions of corroborative detail "to lend verisimilitude to an otherwise bald and unconvincing narrative"! Functionally, evidence is but meager for ascribing a coordinating role to it and hence for its name. Rees (1922) for *Paramecium* and MacDougall (1928) for *Chlamydodon* state that injuries inflicted anywhere other than the region of the motorium have little or no effect on ciliary activity whereas injuries made with microneedles in that region have a pronounced effect, in particular destroying the ability to reverse the beat, as is true also of fragments of the animal. MacLennan (1935) studying dedifferentiation and redifferentiation in *Ichthyophtherius* found that coordinated action of the cilia permitting normal swimming does not begin until the morphological linkage of the neuromotor fibrils has taken place. Taylor's experiments on *Euplotes* remain the most convincing (Fig. 7.4) but clearly more is needed to establish unequivocally a role in coordination. None of the evidence indicates that the motorium region is anything more than a convergence which therefore permits more consequence from a small lesion. It is difficult to imagine solid fibers or bundles of fibrils conducting impulses.

Since it is generally accepted that true nervous conduction is a phenomenon of the surface membrane, the fibrillae of ciliates cannot be regarded as direct precursors of nerve fibers even if they do conduct excitation. There is no evidence for a mechanism by which solid intracellular fibrils might conduct signals; actually we can only exclude the possibility that it is done in the same way as in nerves. But the evidence is still indirect as well as scanty for a specifically conducting role as distinct from some unspecified role in coordination.

If it is difficult to say anything definite about ciliates with respect to the reality or the structural basis of specialized coordination mechanisms, it is much more difficult for **flagellates and other protozoan groups.** Flagella do not exhibit the type of coordination necessary in ciliary activity for they do not beat with an effective stroke in one direction and a slack return but with traveling undulations. A basal body or kinetosome for each flagellum is regarded as homologous with that of the cilium and one or more fibrils generally occur, associated with it. But no pattern of connections suggesting a coordinating function among the flagella, when there are several or many, has emerged (see for example Grassé, 1956; Rouiller and Fauré-Fremiet, 1958; Pitelka and Schooley, 1958; Roth, 1958; Anderson and Beams, 1959). There is probably no special conducting function mediated by structures connecting individual cells of **colonial forms** like *Volvox*.

III. LASTING CHANGES OF STATE

Protozoa illustrate in a simple form some of the features usually associated with the more complex achievements of the nervous system. Apparently internally initiated changes in activity contribute to the appearance of variability in behavior—especially among forms living in contact with surfaces, like *Stentor*, in comparison with more free-swimming forms (see in particular Jennings, 1906; Hempelmann, 1926). Variation in state—of either activity or readiness to respond to stimuli—has been regarded as a fundamental aspect of nervous and prenervous behavior (Batham and Pantin, 1950), as distinct from a concept of purely stimulus-response behavior. Rhythms of altered responsiveness or spontaneity are well established (for example Pittendrigh, 1960, 1961; Ehret, 1959; Withrow, 1959; Hastings, 1959). More or less lasting changes can be provoked by repeated stimulation as well as by what has been called acclimation. There is some evidence of learning of a sort, both in avoidance and approach, and some contrary evidence (Bramstedt, 1935; French, 1940; Dembowski, 1950; Gelber, 1952, 1956, 1957, 1958; Jensen, 1957). Reviews such as that of Warden, Jenkins, and Warner (1940) should be consulted for the earlier literature. Even if learning or improvement with practice occurs it is not to be taken for granted that they involve the same formed structures, if any, that intervene between stimulus and unconditioned response.

IV. MESOZOA

The Mesozoa comprise a small group of extremely simple multicellular endoparasites. During one stage in a complex life cycle they have the characteristic form of an outer, syncytial layer enclosing one or more inner reproductive cells (Lameere, 1922; Nouvel, 1933; Hyman, 1940; McConnaughey, 1949). Although little is known of their activities and of any possible coordination of parts, there is no suggestion in their anatomy or in superficial observation of their movements of any specialized nervous structures.

V. PORIFERA

The two main features of interest in connection with the nervous system are the paucity of response at any considerable distance from a stimulated region and the cellular or very loose tissue grade of construction. The general conclusion of most authors is that no true sensory cells or nerve cells exist—hence no classical nervous system—and that the observed responses do not yet call for any specialized conducting system (Hyman, 1940, 1959).

The **activities and responses of sponges,** so far as they are grossly visible, consist in alterations in shape, closing and opening of dermal pores (the incurrent openings), closing and opening of oscula (excurrent), and propulsion of water by the flagellated choanocytes. These are all slow, and reactions to stimuli take many seconds or minutes. Most species can undergo localized or general changes in shape, attributed to contractions of the pinacocytes, collencytes, or myocytes. Of the three classes, the Hyalospongiae (hexactinellids) are usually thought to have the least such ability but are almost unknown; the Calcispongiae generally exhibit little or no contractility except for some ascons (Wintermann, 1951; Jones, 1957); the Demospongiae are generally able to contract their cortex which is well developed in the subclass Tetractinellida; possibly the layers of elongate cells under the surface are responsible. **General body contractions** may result from handling, injury, removal from the water, or spontaneous action. Wintermann (1951) has recorded both responses and spontaneous contractions dramatically, in time-lapse cinephotomicrographs of cultures between slides of well-organized colonies of the fresh-water *Ephydatia* and *Spongilla*. A complete contraction involves total collapse of canals and takes 10–20 min, with as much again for relaxation. Contraction, which spreads from the region stimulated, can be provoked by heat, certain poisons, or light. Cell-to-cell conduction of excitation is considered but Wintermann favors the possibility of coordination by substances carried in the water currents. Other examples of coordinated contraction are known. Pavans de Ceccatty et al. (1960) refer to the expulsion of larvae by *Reniera* as a violent ejection. Although the mechanism of spread has not been studied, it recalls the contractions of the amnion and other cases of muscle cell to muscle cell spread.

Closure of oscula is active and opening is passive; oscular closing is the commonest response of sponges, especially Demospongiae. It occurs in one or another species in response to injury, high temperatures, certain harmful chemicals, CO_2, oxygen lack, still water, exposure to air, and to touch. Jones (1957) found the osculum of *Leucosolenia* contracts in bright light but concluded this is not under nervous control. A loose sphincterlike ring of elongate cells, recalling some smooth muscle cells, is the effector and is believed to react directly to the stimulating agent. Parker (1909, 1910a, 1910b, 1919) developed this concept under the name of **independent effector,** a term embracing coelenterate nematocysts, cilia of all animals apart from the occasional superimposed nervous control, the isolated iris of some lower vertebrates, myogenic hearts, and various other cases.

Only rarely is there any **sign of conduction** to oscula beyond the probable area of effectiveness of the stimulus. The difficulties of such a demonstration are obvious; the stimulus must be adequate but local. Electrical currents were ineffective in the species studied by Prosser (1960, several fresh-water and marine forms, unnamed) but Pavans de Ceccatty et al. (1960) obtained a brisk general contraction, a spurt of water, and then oscular closure, in response to electrical current in *Tethya;* needles were inserted some centimeters away and then time was allowed for recovery from the mechanical stimulation. Stretch or other quick mechanical deformation (Prosser), local stagnation by carefully placing a glass tube over an osculum (Parker, loc. cit.), and needle pricks (Pavans de Ceccatty et al.) are effective. Response may be graded. In Prosser's species it showed temporal but not spatial summation. In *Tethya* Pavans de Ceccatty et al. found spatial summation and the longest distance of spread of excitation yet reported. Pricking the base of a colony brought about a pronounced general body contraction with a large loss of volume, beginning after several seconds and culminating in 3–7 min, accompanied by closure of the most distant oscula—4 to 8 cm away—beginning in about 30 sec and completed after some minutes. Successive pricks in different places about the base were summed and became the most efficacious form of stimulation. In the same animal CO_2 bubbled into the water provokes a general retraction in several minutes; this is not due to the bubbling for air has no such effect. Oxygen causes a restoration of the volume, distension of the surface, and opening of the oscula. Parker reported closure of an osculum of *Hymeniacadon* (*Stylotella*) 10 min after a pin was stuck into the sponge up to 15 mm away. McNair (1923) stimulated the rim of an osculum of the fresh-water form *Ephydatia*, which has oscula on the tips of chimneylike tubes of dermal membrane held up like tents by the spicules (Fig. 7.11). He described a contraction wave passing down the 1–2 mm chimney, implying a conduction at 0.25 mm/sec; however, in view of the short distance the possibility that contraction of one region pulls on another cannot be ruled out. Electrical stimuli are effective and response is graded with current strength. Conduction is extremely limited at best; none occurs to neighboring oscula even 2–3 mm away, after cutting or pricking the body of the sponge with a needle. A 15–60 sec electrical stimulation 2–4 mm from an osculum can cause increased flow of water through that osculum for 10–15 min; it is not known whether this is an action on the propelling cells or on the resistance to flow.

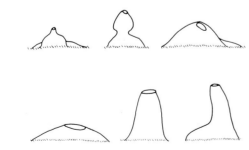

Figure 7.11. Motor reactions of the fresh water sponge *Ephydatia*. Various forms in which the same oscular chimney was observed from time to time. [McNair, 1923.]

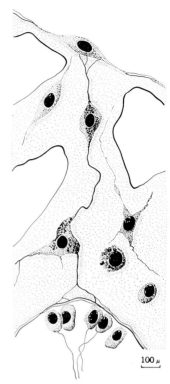

Figure 7.12. Cells in the sponge *Sycon raphanus* stained with ammoniacal silver. Connecting a pinacocyte of the outer surface with choanocytes of the inner surface are two cells with long fiber processes; these are regarded as nerve cells by Pavans de Ceccatty (1955), who kindly provides the figure. Note in addition the amoebocytes and collencytes.

The **dermal pores** are not as reactive as oscula; in *Hymeniacadon* they close to injury, ether, chloroform, strychnine, or cocaine in the water. The complex mechanism of opening and closing in *Hymeniacadon* is described by Wilson (1910); it is said to involve a syncytial dermal membrane with a hole through it that can enlarge or close completely, but this view has been challenged.

There is **no correlation apparent** between the responses of pores and of oscula or between either and the beating of the choanocyte flagella; the latter are themselves not coordinated.

Prosser (1960) looked for bioelectric potentials associated with response or spread of excitation. As yet no definite potentials could be identified with such events. Mitropolitanskaya (1941) could not find cholinesterase in the sponges she studied.

Histological evidence of nerve cells is also equivocal. Von Lendenfeld (1885, 1887) described cells having expansions by the slightly cautious term "nervous type cells" (Fig. 7.12). Their evidence is an argentophilia in Cajal reduced silver impregnation and a general resemblance to ganglion cells in coelenterates. Three types are distinguished: basic, spider, and bladder cells. Intracellular fibrils are said to suggest neurofibrils. They report the appearance of endings of the processes on muscle type cells in the walls of canals, on fibers of the mesenchyme thought to be supporting, and on choanocytes, as well as freely in the large intercellular spaces. Junctions with each other are seen now and then, and are described as protoplasmic fusions; the other endings occur only by contiguity. Two kinds of prolongations are distinguished: numerous fine, short ones homologized with dendrites and, usually, a single stout, long one called an axon. Pavans de Ceccatty (1955) felt that there is a tendency in the more advanced sponges to become organized into a coherent system, though the coherence and organization are so slight as to be far simpler than on the coelenterate level. The same author in 1959 supports von Lendenfeld's old claim of sensory cells. In *Hippospongia*, but apparently not in some other species, in very definite places in the walls of canals—and not in the general epidermis—there are cells oriented radially, reaching the surface and bearing a projecting "palpocil" (Fig. 7.13).

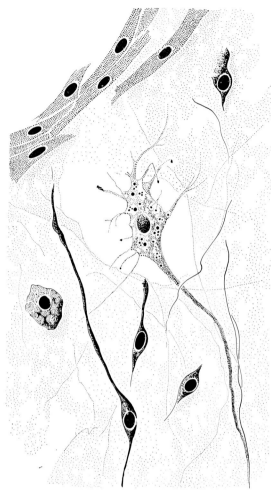

Figure 7.13. Cells of the sponge *Hippospongia* impregnated with silver. The central multipolar cell is regarded as a nerve cell centrifugal with respect to the cavity of a canal defined by a sphincter. [Pavans de Ceccatty, 1959.]

sensory cells and ganglion cells in *Sycandra* and other heterocoel but not in homocoel sponges, but virtually all later writers have been unconvinced of his identifications. The lophocytes discovered by Ankel and Wintermann-Kilian (1952) at first suggested to these authors a sort of nerve net, but they and others do not now regard these cells as in any way nervous. After many decades of the doctrine that sponges have no nerve cells, the problem has been reopened by Tuzet and Pavans de Ceccatty (1952). In genera including *Sycon, Grantia, Leucandra, Leucosolenia, Cliona, Halicondria, Pachymatisma, Halisarca, Hippospongia,* and *Tethya*, these authors identify certain

These claims are of fundamental importance, affecting our views of the most primitive nervous system. The wealth of detail and apparent care with which the observations were made, especially by the French school, cannot be dismissed lightly. Roskin (1957) accepts their views without contributing any original observations. However, owing to the inherent limitations of the methods and the animals used, it must be concluded that the identification of nerve cells in sponges is premature on the standards acknowledged for other Metazoa. Largely based on silver impregnation—itself no argument for such identification and a notoriously poor and distorting cytological procedure—and limited to light microscopy of fixed material, the interpretations are nevertheless vulnerable in their own context. For the cells do not clearly differ from many amoeboid mesenchymatous cells; and they are said to be as plastic and reversible as other types, such as collencytes. Whether other careful observers will agree that these cells are a distinct category will be important to see. Eberl-Rothe (1960) gives no opinion. It is the view of Pavans de Ceccaty that all cells, even choanocytes, are plastic and engaged in a constant differentiation and dedifferentiation about a common polyvalent type, and therefore that it is only to be expected that nerve cells present will do the same and exhibit all intergrades.

The meagerness of the evidence for a general interconnected system and for a physiological conducting system weakens the probability of cell identification. Endings on choanocytes, which show no coordination with each other or with body contractions or oscular closure, endings on scleroblasts and mesenchyme fibers, and—mostly —endings on apparently nothing, also weaken rather than strengthen the argument. One wonders whether observations in life would show these cells to be freely wandering and maintaining no regular connections as a system. The French authors recognize the limited need, in known responses, for a conducting system and ascribe a primarily trophic coordinating function to their system which would be essentially vegetative, but the evidence for this is extremely tenuous.

Jones (1962) has recently provided a full and detailed review of the literature relevant to the question, "is there a nervous system in sponges?" He finds that none of the evidence requires an affirmative answer and in the main is better explained otherwise, in agreement with the conclusions stated here.

Classified References

PROTOZOA

Simpler Responses

Alverdes, 1922; Biriukov, 1957; Buchtal, 1941; Danisch, 1921; Hisada, 1957; Horton, 1935; Kamada, 1934; Kinosita, 1939; Mast, 1941; Nicol, 1958; Sugi, 1959; Umrath, 1956; Viaud and Schwab, 1953.

Ciliary Coordination—Experimental

Bonner, 1954; Bullock and Nachmansohn, 1942; Burgen and MacIntosh, 1955; Copeland, 1919; Doroszewski, 1958; Gray, 1930; Jennings and Jamieson, 1902; Jensen, 1959; Kamada, 1934; Kanno, 1958; Kinosita, 1936, 1938, 1954; Knight-Jones, 1954; Kokina, 1960; Koshtoyants, 1958; Koshtoyants and Kokina, 1957; Kraft, 1890; MacDougall, 1928; Milicer, 1935; Mitropolitanskaya, 1941; Nachmansohn, 1959; Naitoh, 1958, 1959; Okajima, 1953, 1954a, 1954b; Párducz, 1954, 1956a, 1956b, 1958; Popova, Koshtoyants, et al., 1950; Seaman and Houlihan, 1951; Sleigh, 1956, 1957; Steinach, 1908; Taylor, 1920; Ueda, 1956; Verworn, 1889; Worley, 1934; Yow, 1958.

Anatomy of Fibrillar Systems

Anderson and Beams, 1959; Blanckart, 1957; Bradfield, 1955; Brown, 1930; Calkins, 1930; Calkins and Bowling, 1929; Campbell, 1926, 1927; Ehret and Powers, 1959; Gelei, G. 1937, 1940; Gelei, J., 1926a, 1926b, 1926c, 1929, 1932, 1934a, 1934b, 1935, 1936, 1939, 1940; Gelei and Horváth, 1931; Grassé, 1956; Hammond, 1937; Hammond and Kofoid, 1937; Hilton, 1928; Jacobson, 1931; Katen, 1927; Klein, 1926, 1928, 1930, 1931, 1932, 1937, 1941, 1942a, 1942b, 1943, 1955; Kofoid and MacLennan, 1932; Lund, 1933, 1935; MacDonald, 1922; MacDougall, 1928; MacLennan, 1935; Metz et al., 1953; Neresheimer, 1903; Noirot-Timothée, 1958a, 1958b, 1958c; Párducz, 1957, 1958a, 1958b; Pensa, 1926; Pickard, 1927; Pitelka, 1956, 1959; Pitelka and Schooley, 1958; Randall and Jackson, 1958; Rees, 1922, 1930, 1931; Roth, 1956, 1957, 1958a, 1958b; Rouiller and Fauré-Fremiet, 1958a, 1958b; Rouiller et al., 1956; Schneider, 1959; Schuberg, 1905; Sedar and Porter, 1955; Sharp, 1914; Turner, 1930, 1933; Visscher, 1927; Worley, 1933; Worley et al., 1953; Yocum, 1918.

Lasting States

Alverdes, 1923; Batham and Pantin, 1950; Bramstedt, 1935; Dembowski, 1950; Ehret, 1959; French, 1940; Gelber, 1952, 1956a, 1956b, 1956c, 1957a, 1957b, 1957c, 1958; Hastings, 1959; Hempelmann, 1926; Jennings, 1906; Jensen, 1957; Pittendrigh, 1958, 1959; Warden et al., 1940; Withrow and Withrow, 1959.

MESOZOA

Hyman, 1940; Lameere, 1922; McConnaughey, 1949; Nouvel, 1933.

PORIFERA

Ankel and Wintermann-Kilian, 1952; Bacq, 1941, 1947; Eberl-Rothe, 1960; Hilton, 1920b; Hyman, 1940, 1959; Jones, 1962; Lendenfeld, 1885a, 1885b, 1887, 1889; McNair, 1923; Mitropolitanskaya, 1941; Parker, G., 1909; Parker, G. H., 1910a, 1910b, 1919; Pavans de Ceccatty, 1955, 1959, 1960; Roskin, 1957; Tuzet et al., 1952; Tuzet and Pavans de Ceccatty, 1952, 1953a, 1953b, 1953c, 1953d; Wierzejski, 1935; Wilson, 1910; Wintermann, 1951.

Bibliography

ALVERDES. F. 1922. Zur Lokalisation des chemischen und thermischen Sinnes bei *Paramaecium* und *Stentor*. *Zool. Anz.*, 55:19-21.

ALVERDES. F. 1923. Neue Bahnen in der Lehre vom Verhalten der niederen Organismen. Springer, Berlin.

ANDERSON, E. and BEAMS, H. W. 1959. The cytology of *Tritrichomonas* as revealed by the electron microscope. *J. Morph.*, 104:205-235.

ANKEL, W. E. and WINTERMANN-KILIAN, G. 1952. Eine bei *Ephydatia fluviatilis* neu gefundene hochdifferenzierte Zellart und die Struktur der Doppelepithelien. *Z. Naturf.*, 7B: 475-481.

BACQ, Z. M. 1941. Physiologie comparée de la transmission chimique des excitations nerveuses. *Ann. Soc. zool. Belg.*, 72:181-203.

BACQ, Z. M. 1947. L'acétylcholine et l'adrénaline chez les invertébrés. *Biol. Rev.*, 22:73-91.

BATHAM, E. J. and PANTIN, C. F. A. 1950. Phases of activity in the seaanemone, *Metridium senile* (L.), and their relation to external stimuli. *J. exp. Biol.*, 27:377-399.

BIRIUKOV, D. A. 1957. Notes on the evolution of excitation. *Sechenov J. Physiol. (Fiziol. Zh.)*, 43:447-452.

BLANCKART, S. 1957. Die Oberflächenstrukturen von *Paramecium* spec. and *Opalina ranarum*. *Z. wiss. Mikr.*, 63: 276-287.

BONNER, J. T. 1954. The development of cirri and bristles during binary fission in the ciliate *Euplotes eurystomus*. *J. Morph.*, 95:95-107.

BRADFIELD, J. R. G. 1955. Fibre patterns in animal flagella and cilia. *Symp. Soc. exp. Biol.*, 9:306-334.

BRAMSTEDT, F. 1935. Dressurversuche mit *Paramecium caudatum* und *Stylonychia mytilus*. *Z. vergl. Physiol.*, 22:490-516.

BROWN, V. E. 1930. The neuromotor apparatus of *Paramecium*. *Arch. Zool. exp. gen.*, 70:469-481.

BUCHTAL, F. 1941. Messungen von Potential-Differenzen an einzelnen Zellen. *Tabul. biol.*, Hague, 19 (2): 28-75.

BULLOCK, T. H. and NACHMANSOHN, D. 1942. Choline esterase in primitive nervous systems. *J. cell. comp. Physiol.*, 20:239-242.

BURGEN, A. S. V. and MACINTOSH, F. C. 1955. The physiological significance of acetylcholine. in: *Neurochemistry*, K. A. C. Elliott, I. H. Page, and J. H. Quastel (eds.). Thomas, Springfield, Ill.

CALKINS, G. N. 1930. *Uroleptus halseyi* Calkins, III. The kinetic elements and the micronucleus. *Arch. Protistenk.*, 72:49-70.

CALKINS, G. N. and BOWLING, R. 1929. Studies on *Dallasia frontata* Stokes. II. Cytology, gametogamy, and conjugation. *Arch. Protistenk.*, 66:11-32.

CAMPBELL, A. S. 1926. The cytology of *Tintinnopsis nucula* (Fol) Laackmann, with an account of its neuromotor apparatus, division, and a new intranuclear parasite. *Univ. Calif. Publ. Zool.*, 29:179-236.

CAMPBELL, A. S. 1927. Studies on the marine ciliate *Favella* (Jörgensen), with special regard to the neuromotor apparatus and its rôle in the formation of the lorica. *Univ. Calif. Publ. Zool.*, 29:429-452.

COPELAND, M. 1919. Locomotion in two species of the gastropod genus *Alectrion* with observations on the behavior of pedal cilia. *Biol. Bull., Woods Hole*, 37:126-138.

DANISCH, F. 1921. Über Reizbiologie und Reizempfindlichkeit von *Vorticella nebulifera*. *Z. allg. Physiol.*, 19:133-190.

DEMBOWSKI, J. 1950. On conditioned reactions of *Paramaecium caudatum* towards light. *Acta Biol. exp., Varsovie*, 15:5-17.

DOROSZEWSKI, M. 1958. Experimental studies on the conductive role of ectoplasm and the silverline system in ciliates. *Acta Biol. exp. Varsovie*, 18:69-88.

EBERL-ROTHE, G. 1960. Über das Zwischengewebe der wirbellosen Tiere. I. Spongien. *Thalassia Jugoslav.*, 2:5-32.

EHRET, C. F. 1958. Information content and biotopology of the cell in terms of cell organelles. *Symposium on Information Theory in Biology* (Gatlinburg), 1956:218-229.

EHRET, C. F. 1959. Photobiology and biochemistry of circadian rhythms in non-photosynthesizing cells. *Fed. Proc.*, 18:1232-1240.

EHRET, C. F. 1960. Organelle systems and biological organization. *Science*, 132:115-123.

EHRET, C. F. and POWERS, E. L. 1959. The cell surface of *Paramecium*. *Int. Rev. Cytol.*, 8:97-133.

FRENCH, J. W. 1940. Trial and error learning in *Paramecium*. *J. exp. Psychol.*, 26:609-613.

GELBER, B. 1952. Investigations of the behavior of *Paramecium aurelia*: I. Modification of behavior after training with reinforcement. *J. comp. physiol. Psychol.*, 45:58-65.

GELBER, B. 1956a. Investigations of the behavior of *Paramecium aurelia*: II. Modification of a response in successive generations of both mating types. *J. comp. physiol. Psychol.*, 49:590-593.

GELBER, B. 1956b. Investigations of the behavior of *Paramecium aurelia*: III. The effect of the presence and absence of light on the occurrence of a response. *J. genet. Psychol.*, 88:31-36.

GELBER, B. 1957a. A trigger for behavioral change? (*Paramecium aurelia*) *J. Protozool.*, 4:16.

GELBER, B. 1957b. Investigations of the behavior of *Paramecium aurelia*: VI. Reinforcement with three values of training. *Amer. Psychologist*, 12:428.

GELBER, B. 1957c. Food or training in *Paramecium*. *Science*, 126:1340-1341.

GELBER, B. 1958. Retention in *Paramecium aurelia*. *J. comp. physiol. Psychol.*, 51:110-115.

GELBER, B. and RASCH, E. 1956. Investigations of the behavior of *Paramecium aurelia*: V. The effects of autogamy (nuclear reorganization). *J. comp. physiol. Psychol.*, 49:594-599.

GELEI, J. VON. 1926a. Sind die Neurophane von Neresheimer neuroide Elemente? *Arch. Protistenk.*, 56: 232-242.

GELEI, J. VON. 1926b. Zur Kenntnis des Wimperapparates. *Z. ges. Anat. I. Z. Anat. EntwGesch.*, 81:530-553.

GELEI, J. VON. 1929. Über das Nerven-

Bibliography

system der Protozoen. *Állatte. Közl.*, 26:164-190. (In Hungarian with German summary.)

GELEI, J. VON. 1932. Die reizleitenden Elemente der Ciliaten in nass hergestellten Silber- bzw. Goldpräparaten. *Arch. Protistenk.*, 77:152-174.

GELEI, J. VON. 1934a. Der feinere Bau des Cytopharynx von *Paramecium* und seine systematische Bedeutung. *Arch. Protistenk.*, 82:331-362.

GELEI, J. VON. 1934b. Das Verhalten der ectoplasmatischen Elemente des Parameciums während der Teilung. *Zool. Anz.*, 107:161-177.

GELEI, J. VON. 1935. Historisches und neues über die interciliaren Fasern und ihre morphologische Bedeutung. *Z. Zellforsch.*, 22:244-254.

GELEI, J. VON. 1936. Das erregungsleitende System der Ciliaten. *Internat. Congr. Zool., XII, Lisbon*, 1936: 174-209.

GELEI, J. VON. 1937. Ein neues Fibrillensystem im Ectoplasma von *Paramecium;* zugleich ein Vergleich zwischen dem neuen und dem alten Gittersystem. *Arch. Protistenk.*, 89:133-162.

GELEI, J. VON. 1939. Das äussere Stützgerüstsystem des *Parameciumk*örpers. *Arch. Protistenk.*, 92:245-272.

GELEI, J. VON. 1940a. *Cinetochilum* und sein Neuronemensystem. *Arch. Protistenk.*, 94:57-79.

GELEI, J. VON. 1940b. Körperbau und Erregungsleitung bei den Ciliaten. Eine Studie an *Loxocephalus* und einigen anderen Ciliaten. *Arch. Protistenk.*, 93:273-316.

GELEI, J. VON and HORVÁTH, P. 1931. Die Bewegungs- und reizleitenden Elemente bei *Glaucoma* und *Colpidium*, bearbeitet mit der Sublimat-Silbermethode. *Magyar Biol. Kut. Munk.*, 4:40-58.

GRASSÉ, P.-P. 1956. L'ultrastructure de *Pyrsonympha vertens* (Zooflagellata Pyrsonymphina): les flagelles et leur coaptation avec le corps, l'axostyle contractile, le paraxostyle, le cytoplasme. *Arch. Biol., Paris*, 67:595-611.

GRAY, J. 1930. The mechanism of ciliary movement. VI. Photographic and stroboscopic analysis of ciliary movement. *Proc. roy. Soc. (B)*, 107:313-332.

HAMMOND, D. M. 1937. The neuromotor system of *Euplotes patella* during binary fission and conjugation. *Quart. J. micr. Sci.*, 79:507-557.

HAMMOND, D. M. and KOFOID, C. A. 1937. The continuity of structure and function in the neuromotor system of *Euplotes patella* during its life cycle. *Proc. Am. phil. Soc.*, 77:207-218.

HARTMANN, M. 1925. Mesozoa. *Kükenthal's Handb. Zool.*, 1:996-1014.

HASTINGS, J. W. 1959. Unicellular clocks. *Annu. Rev. Microbiol.*, 13:297-312.

HEMPELMANN, F. 1926. Tierpsychologie vom Standpunkte des Biologen. Akademische Verlagsgesellschaft, Leipzig.

HILTON, W. A. 1920a. Nervous system and sense organs. II. Protozoa. *J. Ent. Zool.*, 12 (suppl.):6-12.

HILTON, W. A. 1920b. Nervous system and sense organs. III. The sponges. *J. Ent. Zool.*, 12(suppl.):13-14.

HILTON, W. A. 1928. Nervous system and sense organs. XXX. The nervous system in certain protozoans. *J. Ent. Zool.*, 20:19-24.

HISADA, M. 1957. Membrane resting and action potentials from a protozoan, *Noctiluca scintillans*. *J. cell. comp. Physiol.*, 50:57-71.

HORTON, F. M. 1935. On the reactions of isolated parts of *Paramecium caudatum*. *J. exp. Biol.*, 12:13-16.

HYMAN, L. H. 1940. *The Invertebrates: Protozoa through Ctenophora.* McCraw-Hill, New York.

HYMAN, L. H. 1959. *The Invertebrates.* Vol. V. *Smaller Coelomate Groups.* McGraw-Hill, New York.

JACOBSON, I. 1931. Fibrilläre Differenzierungen bei Ciliaten. *Arch. Protistenk.*, 75:31-100.

JENNINGS, H. S. 1906. *Behavior of the Lower Organisms.* Macmillan, New York.

JENNINGS, H. S. and JAMIESON, C. 1902. Studies on reactions to stimuli in unicellular organisms. X. The movements and reactions of pieces of ciliate infusoria. *Biol. Bull., Woods Hole*, 3:225-234.

JENSEN, D. D. 1957. More on "learning" in Paramecia. *Science*, 126:1341-1342. (See also ibid., 125:191.)

JENSEN, D. D. 1959. A theory of the behavior of *Paramecium aurelia* and behavioral effects of feeding, fission, and ultraviolet microbeam irradiation. *Behaviour*, 15:82-122.

JONES, W. C. 1957. The contractility and healing behaviour of pieces of *Leucosolenia complicata*. *Quart. J. micr. Sci.*, (3) 98:203-217.

JONES, W. C. 1962. Is there a nervous system in sponges? *Biol. Rev.*, 37:1-50.

KAMADA, T. 1934. Some observations on potential differences across the ectoplasm membrane of *Paramecium*. *J. exp. Biol.*, 11:94-102.

KANNO, F. 1958. Galvanic stimulation and ciliary response in *Opalina*. *Zool. Mag. Tokyo* 67:165-168. (In Japanese with English summary.)

KATE, C. G. B. TEN. 1927. Über das Fibrillensystem der Ciliaten. *Arch. Protistenk.* 57:362-426.

KIDDER, G. W. 1933. On the genus *Ancistruma* Strand (*Ancistrum* Maupas). I. The structure and division of *A. mytili* Quenn. and *A. isseli* Kahl. *Biol. Bull., Woods Hole*, 44:1-20.

KINOSITA, H. 1936. Effect of change in orientation on the electrical excitability in *Paramecium*. *J. Fac. Sci. Tokyo Univ.*, 4:189.

KINOSITA, H. 1938. Electrical stimulation of *Paramecium* with two successive subliminal current pulses. *J. cell. comp. Physiol.*, 12:103-117.

KINOSITA, H. 1939. Electrical stimulation of *Paramecium* with linearly increasing current. *J. cell. comp. Physiol.*, 13:253-261.

KINOSITA, H. 1954. Electric potentials and ciliary response in *Opalina*. *J. Fac. Sci. Univ. Tokyo*, (4) 7:1-14.

KLEIN, B. M. 1926. Ergebnisse mit einer Silbermethode bei Ciliaten. *Arch. Protistenk.*, 56:243-279.

KLEIN, B. M. 1928. Die Silberliniensysteme der Ciliaten. Weitere Resultate. *Arch. Protistenk.*, 62:177-260.

KLEIN, B. M. 1930. Die Silberliniensystem einiger Ciliaten. Weitere Ergebnisse IV. *Arch. Protistenk.* 69:235-326.

KLEIN, B. M. 1931. Über die Zugehörigkeit gewisser Fibrillen bzw. Fibrillenkomplexe zum Silberliniensystem. *Arch. Protistenk.*, 74:401-416.

KLEIN, B. M. 1932. Das Ciliensystem in seiner Bedeutung für Lokomotion. Koordination und Formbildung mit besonderer Berücksichtigung der Ciliaten. *Ergebn. Biol.*, 8:75-179.

KLEIN, B. M. 1937. Regionäre Reaktionen im Silberlinien- oder neuroformativen System der Ciliaten. *Arch. Protistenk.*, 88:192-210.

KLEIN, B. M. 1941. Äusseres Stützgerüst und neuroformatives System der Ciliaten. Eine grundsätzliche Betrachtung und Auseinandersetzung. *Ann. Naturh. (Mus.) Hofmus., Wien*, 52:20-53.

KLEIN, B. M. 1942a. Differenzierungsstufen des Silberlinien- oder neuroformativen Systems. *Arch. Protistenk.*, 96:1-30.

KLEIN, B. M. 1942b. Das Silberlinien- oder neuroformative System der Ciliaten. *Ann. Naturh. (Mus.) Hofmus, Wien*, 53:156-336.

KLEIN, B. M. 1955. Potenzen erster nervlicher Differenzierungen. *Acta neuroveg.*, 12:1-24.

KNIGHT-JONES, E. W. 1954. Relations between metachronism and the direction of ciliary beat in Metazoa. *Quart. J. micr. Sci., N. S.*, 95:503-521.

KOFOID, C. A. and MACLENNAN, R. F. 1932. Ciliates from *Bos indicus* Linn. II. A revision of *Diplodinium* Schuberg. *Univ. Calif. Publ. Zool.*, 37:53-152.

KOKINA, N. N. 1960. Ionic interrelations and the role of potassium in the rhythmic variations in the intracellular potential of Infusoria. *Biophysics* 5:159-168.

KOSHTOYANTS, K. S. and KOKINA, N. N. 1957. On the role of the acetylcholinecholinesterase system in the phenomena of galvanotaxis and summation of stimuli in *Paramecium*. *Biofizika*, 2:47-52.

KOSHTOYANTS, KH. S. and KOKINA,

N. N. 1958. Rhythmic bioelectric phenomena in the single-celled organism *Opalina ranarum*. *Biofizika*, 3:422-425.

KRAFT, H. 1890. Zur Physiologie des Flimmerepithels bei Wirbelthieren. *Pflüg. Arch. ges. Physiol.*, 47:196-235.

LAMEERE, A. 1922. L'histoire naturelle des Dicyemides. *Bull. Acad. Belg., Cl. Sci.*, (5) 8:779-792.

LENDENFELD, R. VON. 1885a. Das Nervensystem der Spongien. *Zool. Anz.*, 8:47-50.

LENDENFELD, R. VON. 1885b. The histology and nervous system of the calcareous sponges. *Proc. Linn. Soc. N. S. Wales*, 9:977-983.

LENDENFELD, R. VON. 1887. Synocils, Sinnesorgane der Spongien. *Zool. Anz.* 10:142-145.

LENDENFELD, R. VON. 1889. Experimentelle Untersuchungen über die Physiologie der Spongien. *Z. wiss. Zool.* 48:406-700.

LUND, E. E. 1933. A correlation of the silverline and neuromotor systems of *Paramecium*. *Univ. Calif. Publ. Zool.*, 39:35-76.

LUND, E. E. 1935. The neuromotor system of *Oxytricha*. *J. Morph.*, 58:257-277.

MACDOUGALL, M. S. 1928. The neuromotor apparatus of *Chlamydodon* sp. *Biol. Bull., Woods Hole*, 54:471-484.

MACLENNAN, R. F. 1935. Dedifferentiation and redifferentiation in *Ichthyophthirius*. I. Neuromotor system. *Arch. Protistenk.*, 86:191-210.

MAST, S. O. 1941. Motor responses in unicellular animals. In: *Protozoa in Biological Research*, G. N. Calkins and F. M. Summers (eds.). Columbia Univ. Press, New York.

MCCONNAUGHEY, B. H. 1949. Mesozoa of the family Dicyemidae from California. *Univ. Calif. Publ. Zool.*, 55:1-34.

MCDONALD, J. D. 1922. On *Balantidium coli* (Malmsten) and *Balantidium suis* (sp. nov.), with an account of their neuromotor apparatus. *Univ. Calif. Publ. Zool.*, 20:243-300.

MCNAIR, G. T. 1923. Motor reactions of the fresh water sponge, *Ephydatia fluviatilis*. *Biol. Bull., Woods Hole*, 44:153-166.

METZ, C. B., PITELKA, D. R., and WESTFALL, J. A. 1953. The fibrillar systems of ciliates as revealed by the electron microscope. I. *Paramecium*. *Biol. Bull., Woods Hole*, 104:408-425.

MILICER, W. 1935. Recherches expérimentales sur le système neuromoteur de *Paramecium caudatum*. *Acta Biol. exp., Varsovie*, 9:174-194. (In Polish, with French summary.)

MITROPOLITANSKAYA, R. L. 1941. On the presence of acetylcholine and cholinesterase in the Protozoa, Spongia and Coelenterata. *C. R. Acad. Sci. (Dokl. Akad. Nack. SSSR)*, 31:717-718.

NACHMANSOHN, D. 1959. *Chemical and Molecular Basis of Nerve Activity*. Academic Press, New York.

NAITOH, Y. 1958. Direct current stimulation of *Opalina* with intracellular microelectrode. *Annot. zool. jap.*, 31:59-73.

NAITOH, Y. 1959. Relation between the deformation of the cell membrane and the change in beating direction of cilia in *Opalina*. *J. Fac. Sci. Univ. Tokyo*, (4), 8:357-369.

NERESHEIMER, E. R. 1903. Über die Höhe histologischer Differenzierung bei heterotrchien Ciliaten. *Arch. Protistenk.*, 2:305-324.

NICOL, J. A. C. 1958. Observations on luminescence in *Noctiluca*. *J. Mar. biol. Ass. U. K.*, 37:535-549.

NOIROT-TIMOTHÉE, C. 1958a. Étude au microscope électronique des fibres rétrociliaires des Ophryoscolecidae: leur ultrastructure, leur insertion, leur rôle possible. *C. R. Acad. Sci., Paris*, 246:1286-1289.

NOIROT-TIMOTHÉE, C. 1958b. L'ultrastructure du blépharoplaste des infusoires ciliés. *C. R. Acad. Sci., Paris*, 246:2293-2295.

NOIROT-TIMOTHÉE, C. 1958c. Quelques particularités de l'ultrastructure d'*Opalina ranarum* (Protozoa, Flagellata). *C. R. Acad. Sci., Paris*, 247:2445-2447.

NOIROT-TIMOTHÉE, C. 1959. Recherches sur l'ultrastructure d'*Opalina ranarum*. *Ann. Sci. nat. (Zool.)*, (12) 1:265-281.

NOUVEL, H. 1933. Recherches sur la cytologie, la physiologie et la biologie des dicyémides. *Ann. Inst. océanogr. Monaco*, 13:167-255.

OKAJIMA, A. 1953. Studies on the metachronal wave in *Opalina*. I. Electrical stimulation with the microelectrode. *Jap. J. Zool.*, 11:87-100.

OKAJIMA, A. 1954a. Studies on the metachronal wave in *Opalina*. II. The regulating mechanism of ciliary metachronism and of ciliary reversal. *Annot. zool. Jap.*, 27:40-45.

OKAJIMA, A. 1954b. Studies on the metachronal wave in *Opalina*. III. Time-change of effectiveness of chemical and electrical stimuli during adaptation in various media. *Annot. jap. zool.*, 27:46-51.

PÁRDUCZ, B. 1954. Reizphysiologische Untersuchungen an Ziliaten. I. Über das Aktionssystem von *Paramecium*. *Acta microbiol.*, 1:175-221.

PÁRDUCZ, B. 1956a. Reizphysiologische Untersuchungen an Ziliaten. V. Zum physiologischen Mechanismus der sog. Fluchtreaktion und der Raumorientierung. *Acta biol. hung.*, 7:73-99.

PÁRDUCZ, B. 1956b. Reizphysiologische Untersuchungen an Ziliaten. VI. Eine interessante Variante der Fluchtreaktion bei *Paramecium*. *Ann. hist.-nat. Mus. hung.*, 7:363-369.

PÁRDUCZ, B. 1957. Über den feineren Bau des Neuronemensystems der Ziliaten. *Ann. hist.-nat. Mus. hung.* 8:231-246.

PÁRDUCZ, B. 1958a. Das interziliäre Fasersystem in seiner Beziehung zu gewissen Fibrillenkomplexen der Infusorien. *Acta. biol. hung.* 8:191-218.

PÁRDUCZ, B. 1958b. Reizphysiologische Untersuchungen an Ziliaten. VII. Das Problem der vorbestimmten Leitungsbahnen. *Acta biol. hung.*, 8:219-252.

PARKER, G. 1909. The origin of the nervous system and its appropriation of effectors. I. Independent effectors. *Pop. Sci. Mon.*, 75:56-64.

PARKER, G. H. 1910a. The reactions of sponges, with a consideration of the origin of the nervous system. *J. exp. Zool.*, 8:1-41.

PARKER, G. H. 1910b. The phylogenetic origin of the nervous system. *Anat. Rec.*, 4:51-58.

PARKER, G. H. 1919. *The Elementary Nervous System*. Lippincott, Philadelphia.

PAVANS DE CECCATTY, M. 1955. Le système nerveux des éponges calcaires et siliceuses. *Ann. Sci. nat. (Zool.)*, 17:203-290.

PAVANS DE CECCATTY, M. 1959. Les structures cellulaires de type nerveux chez *Hippospongia communis* Lmk. *Ann. Sci. nat. (Zool.)*, (12) 1:105-112.

PAVANS DE CECCATTY, M. 1960. Les structures cellulaires de type nerveux et de type musculaire de l'éponge siliceuse *Tethya lyncurium* Lmk. *C. R. Acad. Sci., Paris*, 251:1818-1819.

PAVANS DE CECCATTY, M., GARGOUIL, M., and CORABOEUF, E. 1960. Les réactions motrices de l'éponge *Tethya lyncurium* à quelques stimulations expérimentales. *Vie et Milieu*, 11:594-600.

PENSA, A. 1926. Particolarità strutturale di alcuni protozoi cigliati in rapporto con la contrattilità. *Monit. zool. ital.*, 37:165-173.

PICKARD, E. A. 1927. The neuromotor apparatus of *Boveria teredinidi* Nelson, a ciliate from the gills of *Teredo navalis*. *Univ. Calif. Publ. Zool.*, 29:405-428.

PITELKA, D. R. 1956. An electron microscope study of cortical structures of *Opalina obtrigonoidea*. *J. biophys. biochem. Cytol.*, 2:423-432.

PITELKA, D. R. 1959. Ultrastructure of the silver-line system in three tetrahymenid ciliates. *J. Protozool.*, 6 (suppl.):22.

PITELKA, D. R. and SCHOOLEY, C. N. 1958. The fine structure of the flagellar apparatus in *Trichonympha*. *J. Morph.*, 102:199-245.

PITTENDRIGH, C. S. 1960. Circadian rhythms and the circadian organization of living systems. *Cold Spr. Harbor Symp. quant. Biol.*, 25:159-184.

PITTENDRIGH, C. S. 1961. On temporal

organization in living systems. *Harvey Lect.*, 56:93-125.
PROSSER, C. L. 1960. Mechanical responses of sponges. *Anat. Rec.*, 138:377.
RANDALL, J. T. and JACKSON, S. F. 1958. Fine structure and function in *Stentor polymorphus. J. biophys. biochem. Cytol.*, 4:807-829.
REES, C. W. 1922. The neuromotor apparatus of *Paramaecium. Univ. Calif. Publ., Zool.*, 20:333-364.
REES, C. W. 1930. Is there a neuromotor apparatus in *Diplodinium ecaudatum*? *Science*, 71:369-370.
REES, C. W. 1931. The anatomy of *Diplodinium medium. J. Morph.*, 52:195-215.
ROSKIN, G. I. 1957. Nervous system of sponges. *Adv. mod. Biol., Moscow (Usp. Sovr. Biol.)*, 43:199-207. (In Russian.)
ROTH, L. E. 1956. Aspects of ciliary fine structure in *Euplotes patella. J. biophys. biochem. Cytol.*, 2 (4, suppl.): 235-240.
ROTH, L. E. 1957. An electron microscope study of the cytology of the protozoan *Euplotes patella. J. biophys. biochem. Cytol.*, 3:985-1000.
ROTH, L. E. 1958a. A filamentous component of protozoan fibrillar systems. *J. Ultrastruct. Res.*, 1:223-234.
ROTH, L. E. 1958b. Ciliary coordination in the Protozoa. *Exp. Cell Res.*, 5(suppl.):573-585.
ROUILLER, C. and FAURÉ-FREMIET, E. 1958a. Structure fine d'un flagelle chrysomonadien: *Chromulina psammobia. Exp. Cell. Res.*, 14:47-67.
ROUILLER, C. and FAURÉ-FREMIET, E. 1958b. Ultrastructure des cinétosomes à l'état de repos et à l'état cilifère chez un cilié péritriche. *J. Ultrastruc. Res.*, 1:289-294.
ROUILLER, C., FAURÉ-FREMIET. E., and GAUCHERY, M. 1956. Origin ciliaire des fibrilles scléroprotéiques pédonculaires chez les ciliés péritriches. *Exp. Cell. Res.*, 11:527-541.
SCHNEIDER, L. 1959. Neue Befunde über den Feinbau des Cytoplasmas von *Paramecium* nach Einbettung in Vestopal W. *Z. Zellforsch.*, 50:61-77.
SCHNEIDER, L. 1960. Elektronenmikroskopische Untersuchungen über das Nephridialsystem von *Paramaecium. J. Protozool.*, 7:75-90.
SCHUBERG, A. 1905. Über Cilien und Trichocysten einiger Infusorien. *Arch. Protistenk.*, 6:61-110.
SEAMAN, G. R. and HOULIHAN, R. K. 1951. Enzyme systems in *Tetrahymena geleii* S. II. Acetylcholinesterase activity. Its relation to motility of the organism and to coordinated ciliary action in general. *J. cell. comp. Physiol.*, 37:309-321.
SEDAR, A. W. and PORTER, K. R. 1955. The fine structure of cortical components of *Paramecium multimicronucleatum. J. biophys. biochem. Cytol.*, 1:583-604.
SHARP, R. G. 1914. *Diplodinium ecaudatum* with an account of its neuromotor apparatus. *Univ. Calif. Publ. Zool.*, 13:42-122.
SLEIGH, M. A. 1956. Metachronism and frequency of beat in the peristomial cilia of *Stentor. J. exp. Biol.*, 33:15-28.
SLEIGH, M. A. 1957. Further observations on co-ordination and the determination of frequency in the peristomial cilia of *Stentor. J. exp. Biol.*, 34:106-115.
SLEIGH, M. A. 1960. The form of beat in cilia of *Stentor* and *Opalina. J. exp. Biol.*, 37:1-10.
STEINACH, E. 1908. Die Summation einzeln unwirksamer Reize als allgemeine Lebenserscheinung. I. *Pflüg. Arch. ges. Physiol.*, 125:239-289.
SUGI, H. 1959. Contraction and relaxation in the stalk muscle of *Carchesium. Annot. zool. jap.*, 32:163-169.
SUGI, H. 1960. Propagation of contraction in the stalk muscle of *Carchesium. J. Fac. Sci. Tokyo Univ. (IV)*, 8:603-615.
TAYLOR, C. V. 1920. Demonstration of the function of the neuromotor apparatus in *Euplotes* by the method of microdissection. *Univ. Calif. Publ. Zool.*, 19:403-471.
TAYLOR, C. V. 1941. Fibrillar systems in ciliates. In: *Protozoa in Biological Research*. G. N. Calkins and F. M. Summers (eds.). Columbia University Press, New York.
TURNER, J. P. 1930. Division and conjugation in *Euplotes patella* Ehrenberg with special reference to the nuclear phenomena. *Univ. Calif. Publ. Zool.*, 33:193-258.
TURNER, J. P. 1933. The external fibrillar system of *Euplotes* with notes on the neuromotor apparatus. *Biol. Bull., Woods Hole*, 64:53-66.
TUZET, O. and PAVANS DE CECCATTY, M. 1952. Les cellules nerveuses de *Grantia compressa pennigera* Haeckel (éponge calcaire heterocoele). *C. R. Acad. Sci., Paris*, 235:1541-1543.
TUZET, O. and PAVANS DE CECCATTY, M. 1953a. Les cellules nerveuses de l'éponge calcaire homocoele *Leucandra johnstoni* Cart. *C. R. Acad. Sci., Paris*, 236:130-133.
TUZET, O. and PAVANS DE CECCATTY, M. 1953b. Les cellules nerveuses et neuro-musculaires de l'éponge: *Cliona celata* Grant. *C. R. Acad. Sci., Paris*, 236:2342-2344.
TUZET, O. and PAVANS DE CECCATTY, M. 1953c. Les lophocytes de l'éponge *Pachymatisma johnstonni* Bow. *C. R. Acad. Sci., Paris*, 237:1447-1449.
TUZET, O. and PAVANS DE CECCATTY, M. 1953d. Les cellules nerveuses de l'éponge *Pachymatisma johnstonni* Bow. *C. R. Acad. Sci., Paris*, 237:1559-1561.
TUZET, O., LOUBATIÈRES, R., and PAVANS DE CECCATTY, M. 1952. Les cellules nerveuses de l'éponge *Sycon raphanus* O. S. *C. R. Acad. Sci., Paris*, 234:1394-1396.
UEDA, K. 1956. Intracellular calcium and ciliary reversal in *Opalina. Jap. J. Zool.*, 12:1-10.
UEDA, K. 1961. Electrical properties of *Opalina.* I. Factors affecting the membrane potential. *Annot. zool. jap.*, 34:99-110.
UMRATH, K. 1956. Elektrische Messungen und Reizversuche an *Amoeba proteus. Protoplasma*, 47:347-358.
VERWORN, M. 1889. *Psychophysiologische Protistenstudien. Experimentelle Untersuchungen.* Fischer, Jena.
VISSCHER, J. P. 1927. A neuromotor apparatus in the ciliate *Dileptus gigas. J. Morph.*, 44:373-381.
WARDEN, C. J., JENKINS, T. N., and WARNER, L. H. 1940. *Comparative Psychology*. Vol. 2. Ronald Press, New York.
WIERZEJSKI, A. 1935. Süsswasserspongien. *Mem. Acad. Sci., Cracovie (B)*, 9:1-242.
WILSON, H. V. 1910. A study of some epithelioid membranes in monaxonid sponges. *J. exp. Zool.*, 9:537-577.
WINTERMANN, G. 1951. Entwicklungsphysiologische Untersuchungen an Süsswasserschwämmen. *Zool. Jb. (Anat.)*, 71:427-486.
WITHROW, R. B. 1959. *Photoperiodism and Related Phenomena in Plants and Animals*. Am. Assoc. Adv. Sci., Washington.
WORLEY, L. G. 1933. The intracellular fibre systems of *Paramecium. Proc. nat. Acad. Sci., Wash.*, 19:323-326.
WORLEY, L. G. 1934. Ciliary metachronism and reversal in *Paramecium, Spirostomum* and *Stentor. J. cell. comp. Physiol.*, 5:53-72.
YAGIU, R. and SHIGENAKA, Y. 1959. Electron microscopical observation of *Condylostoma spatiosum* Ozaki & Yagiu, in ultrathin section IV. The fibrils between the basal granule and the longitudinal fibrillar bundle. *Zool. Mag., Tokyo*, 68:414-418. (In Japanese, with English summary.)
YAMAGUCHI, T. 1960a. Studies on the modes of ionic behavior across the ectoplasmic membrane of *Paramecium*. I. Electric potential differences measured by the intracellular microelectrode. *J. Fac. Sci. Tokyo Univ. (IV)*, 8:573-591.
YAMAGUCHI, T. 1960b. Studies on the modes of ionic behavior across the ectoplasmic membrane of *Paramecium*. II. In- and outfluxes of radioactive calcium. *J. Fac. Sci. Tokyo Univ. (IV)*, 8:593:601.
YOCOM, H. B. 1918. The neuromotor apparatus of *Euplotes patella. Univ. Calif. Publ. Zool.*, 18:337-396.
YOW, F. W. 1958. A study of the regeneration pattern of *Euplotes eurystomus. J. Protozool.*, 5:84-88.

CHAPTER 8

Coelenterata and Ctenophora

Summary	460
COELENTERATA	
I. Evidence of a Nerve Net and its Properties	462
II. Anatomy	466
A. Large Bipolar Cell Nets	467
B. Small Multipolar Cell Nets	469
C. Cytology	470
D. Synapses and Neuromuscular Junctions	471
E. Regional Differentiation	473
F. Nerve Rings and Marginal Ganglia	475
G. Receptor Cells and Sense Organs	478
III. Physiology	481
A. Properties of the Nervous Elements	481
B. Differentiation of Nerve Net Systems in Polyps	486
1. Fast specific nerve nets	486
2. Slow diffused nerve nets	487
3. Responses to mechanical stimulation	490
4. Responses to chemical and other stimuli	491
5. Regional specialization	492
C. Differentiation of Nerve Net Systems in Medusae	493
1. Fast specific nerve nets	493
2. Slow diffused nerve nets	495
3. Interaction of nerve nets	499
IV. Spontaneity, Internal State, and Phases of Activity	502
V. Details of the Groups of Coelenterates	506
A. Hydrozoa	506
1. Hydra	506
2. Corymorpha	509
3. Colonial hydroids	510
4. Siphonophora and Chondrophora	512
B. Anthozoa	514
1. Actiniaria	514
2. Ceriantharia	515
3. Madreporaria	515
4. Alcyonaria	517
C. Scyphozoa	519
CTENOPHORA	
I. Anatomy	520
II. Physiology	523
Classified References	526
Bibliography	527

Summary

The structural and functional **elements** of the nervous system of coelenterates are not fundamentally different from those in higher animals.

Many of the neurons are organized into nervous networks. Others are not so arranged and the organization in many cases is not at all understood. Among the variety of conducting systems certain cases form a natural group to which the term nerve net is applied. A **nerve net** is a system of functionally connected nerve cells and fibers anatomically dispersed through some considerable portion of an animal and so arranged as to permit diffuse conduction of nervous excitation, that is, in relatively direct paths between many points. The paths, as opposed to indirect routing through a distant ganglion or central structure, are multiple and confer a tolerance of incomplete cuts. The net can be either quite unoriented or somewhat preferentially oriented but not really tractlike. Nothing is specified about protoplasmic continuity or synapses; either may obtain.

Synapses in the nerve nets are the rule, on both anatomical and physiological evidence. The nerve net in hydras appears to be an exception and others are not excluded. Here nerve cells arise separately and appear to fuse secondarily during development. **Most synapses are unpolarized** but some are polarized, for example those in marginal ganglia mediating the influence of the slow nerve net upon the faster system. Many synapses normally transmit one-to-one but others require facilitation by successive impulses or spatial summation. Anatomically they are simple crossings, intertwinings, or contacts in passing.

Graded and decrementally propagated activity has been suggested as a possible stage in the **evolution of the all-or-none impulse.** But there is no evidence of such stages and the only reason to believe in them is the small size of some hydrozoans, where electrotonic spread could possibly suffice. The large size (up to meters) of closely related solitary hydroids and the coordination over considerable distance in colonial hydroids makes this unlikely. In addition, typical nerve impulses have been recorded from several coelenterate nerve nets, including large medusae and small hydroid polyps. Nonnervous types of spread of excitation probably occur here and there; proof that spread is nervous has been offered in only a few cases but the presumption is strong in many.

An independence of intensity of direct electrical stimuli above threshold and a **strong dependence on facilitation,** that is, on frequency and number of shocks, is a general characteristic of nerve net systems, although not a universal one. Many muscles have different time scales and facilitation dependence so that frequency of stimulation can control character of response. These principles obtain both in phasic (fast relaxing) and in tonic muscles (slowly relaxing), for example in anemones and in medusae where the presence or absence of mechanical summation makes the response superficially different. Both have relatively fast and slow components of contraction, under the control of separate conduction systems, suggesting double innervation of the muscle.

Normal stimuli—mechanical, chemical, and so on —operate by evoking a **certain number and temporal pattern of impulses** in a net. Sense cells probably act only in a graded fashion and their actions sum to determine firing in the net. Adaptation is important; mechanical stimuli which cause high frequencies of firing also cause rapid adaptation so that the number of impulses is small. Since clear responses with low-frequency thresholds require higher numbers of impulses, this helps in selectivity of response. The formation of the impulse group or burst by the sense cell–nerve net junction represents probably the **first nervous integration in evolution.** It is decisive in determining among several normal behavioral alternatives, that is, no other integrative process occurs before the nerve net-muscle junction in the common case of through-conducting pathways, which are essentially single motor units excitable in many places and controlling several muscles. Here, then, is a strategically simple case for the study of the relation between neurophysiology and behavior.

Chemical stimuli, at least food, apparently never excite sense cells capable of **firing a through-conducting net.** Mechanical stimuli in some species can do so only if injurious; there is a whole conducting system solely developed for response to frank injury. This is particularly true in some colonial forms.

Physiological and anatomical evidence makes it clear that there must be in many places **at least two nerve nets** coexisting in the same epithelium, making only indirect functional contact with each other, responding to different kinds of stimuli and eliciting different forms of response. There may be **one or more through-conducting pathways** coexisting and these may be extensive or confined; they may rule many muscles widespread through the anemone, jellyfish, colonial hydroid, or coral, or they may be restricted to the symmetrical responses of single coral

Summary

polyps or but few muscles in the solitary forms. Through-conducting nets generally conduct faster and their facilitation is only at the neuro effector junction. **Non-through-conducting systems** vary widely in extent, properties and relations with the through-conducting systems. Some are highly local and cannot be made to spread excitation to any considerable distance. Others can be converted into through-conducting by repetitive stimulation, that is, some kind of integrative process occurs in the net itself, presumably at neuro-neural junctions. Some spread excitation with successive stimuli in approximately uniform increments of radius, whereas others spread with smaller and smaller increments and still others with increasing increments of radius. These facts have stimulated new hypotheses of spread in non-through-conducting nerve nets. One is based on the assumption that only a certain proportion of the available neurons in a given area are excited by a stimulus and that the number originally excited is a prime factor in determining spread. Another is based on interneural facilitation and in certain instances repetitive discharge.

Interaction between conducting systems may occur (a) only at the level of the muscle, (b) within the nervous system, or (c) not at all. Exemplifying the second case, it occurs in medusae in the marginal ganglia or ring nerves, which therefore represent the **first integrating concentrations of nervous tissue** in the animal kingdom. No corresponding locus is known in polyps. Typically the interaction is quite irreciprocal: the slow system influences the occurrence or manifestation of activity in the fast. The latter is regarded as derived, the former as more primitive. Thus the most primitive system appears to be one with integrative processes at its junctions with receptors, within its own junctions and at neuroeffector junctions. The differentiation of nets and long pathways exists (for example, tentacle to manubrium) representing the beginning of local sign or **labeled lines** in nerve fibers.

True nervous spontaneity occurs in many places—that is, intermittent discharge in the absence of any known stimuli except the steady-state milieu. In pulsating jellyfish there are multiple potential pacemakers in the marginal ganglia or nerve rings; they are within the conducting system, accessible to impulses therein, which hence fire them and reset the rhythm. Pacemaking is handed around from one to the other, assuring the fastest rhythm available. The pacemakers are also, irreciprocally, accessible to influence from other conducting systems and from sense organs; they are internal clocks subject to modulation by adequate phasic and tonic stimuli.

Medusae and anemones show a succession of phases of activity of a variety of forms, both spontaneously and as sequelae of stimuli. In contrast to the view of Parker, the present view upholds Jennings' theory that fluctuations in internal state are of profound importance in these simple nervous systems. The **release of a complex, predetermined pattern** of movements by any of several simple stimuli is suggested—and the same pattern can be initiated spontaneously. A number of examples of behavioral sequences have been analyzed in terms of nerve nets, facilitation, and nervous organization, including escape-swimming of an anemone, feeding in siphonophores, alcyonarians, hydromedusae, and others.

The nerve cells are bipolar and multipolar, never unipolar; they are typically isopolar. Sensory nerve cells probably far outnumber motoneurons, as in higher groups. In different places the epidermal or the gastrodermal subepithelial plexus may dominate. Differentiation takes several forms: preferred orientation of conduction, preferred polarity, nearly 100:1 differences in velocity, occasional pure tracts without lateral spread, and the development of simple ganglia in medusae. The **differences between groups**—different genera of anemones, hydroids, corals, and medusae, as well as the three classes—are only differences in the proportions of these several specializations. The differences between genera of the same class are greater than any general differences between the classes.

Further work is particularly needed on the finer anatomy, on the interactions between parts of the nervous system, on the loci of such interactions in polyps, on the intimate organization of the marginal ganglia and ring nerves, and on the details of spread of excitation in nerve nets.

Ctenophores have a nervous system much like coelenterates. A single, synaptic net seems indicated anatomically. Cordlike condensations of the net under the eight comb plate rows are further differentiated under each comb plate. Here unipolar neurons occur for the first time in our sequence of phyla. The apical complex includes a statocyst, four possible ganglia and connecting nerve strands to the comb plate rows. Several presumptive special sensory regions are insufficiently described. The main sign of nervous function is control of the comb plate cilia. Slower local pacemakers are normally dominated by faster apical ones. There are probably several pathways in the plate row nerves; these can inhibit ciliary beat, accelerate or reverse metachronal waves, retract comb plates, and cause luminescence. Some of these actions can be differential on different rows, under apical control. There is no evidence of diffuse conduction in the general plexus. The meridional cords are physiologically much more than simply concentrations of a nerve net.

COELENTERATA

The nervous system of coelenterates occupies a strategic position in our understanding of neurological evolution. This phylum we consider to be the most simply organized of which we have a record, which possesses a true nervous system. The system appears here fully formed with a diverse set of connected neurons. There is coordination of behavior in still simpler phyla, and conduction of excitation from one place to another. But we do not find transitions or intergrades between animals without a nervous system—that is to say, Protozoa, Porifera, and Mesozoa—and the simplest animals with one—the Coelenterata.

Elsewhere we have examined the evolution of types of neurons anatomically and functionally and have concluded that, although there is considerable evolution of nervous units, this does not account for the difference in achievements between the nervous systems of the major groups of animals. These achievements in more complex behavior must then rest on something in the pattern of the units, something in the interaction of groups of units. The functional organization of the system in coelenterates is therefore in a real sense the **beginning of our picture of the evolution of the nervous system.**

An additional point of interest is that a proportionally large amount of new work has been concentrated on this group in recent years and an unusually large fraction of our information about the nervous system of a major group is physiological. The field is moving rapidly and important changes can be expected since new tools—electrical recording, electron microscopy, and computer simulation—have already been brought to bear on the problem.

I. EVIDENCE OF A NERVE NET AND ITS PROPERTIES

Nature presents for our consideration two main forms of coelenterates: **polyps,** generally attached or sedentary, and **medusae,** commonly free-swimming. (We will not here give special consideration to those medusae that remain attached to the polyp as specialized reproductive stages.) Considering first the scyphomedusae, we note that these have generally eight marginal bodies and that any one is capable of initiating the normal rhythmic swimming contractions of the bell. By careful examination or by recording or by cooling down the preparation, especially after removal of seven of the marginal bodies, it can be seen that the normal contractions originate in the region of a marginal body and spread rapidly over the whole bell, to excite the musculature nearly simultaneously. Not only is the character of the contraction quite normal after removal of seven of the marginal bodies, but even after removing the last one we can substitute a single electrical shock and obtain a contraction which appears normal. The significant fact is that this shock can be applied at any point on the margin or subumbrellar surface of the bell. There must be an elaborate **system of pathways to mediate this total reflex,** involving almost simultaneous contraction of a large musculature. But unlike the system of pathways in higher animals, where also a stimulus applied at any of widely separated parts of the body can elicit a general response, in the medusa we can remove any part of the animal and the response is not interrupted. As Romanes in England and Eimer in Germany showed in classical demonstrations simultaneously in 1874 (see their later, full papers for citations), not only large ablations of any region whatsoever but also deep incisions in a complex pattern, calculated to interrupt any direct or even very indirect pathway, fail to interrupt the propagation of the wave of contraction (Fig. 8.1). The only condition is that the narrowest strip of tissue must not be less than about 1 mm in width. We can conclude that the pathways of conduction are diffuse or meshlike and are everywhere present in the subumbrella, although the minimum width of the strip which is necessary tells us something

I. EVIDENCE AND PROPERTIES OF A NERVE NET

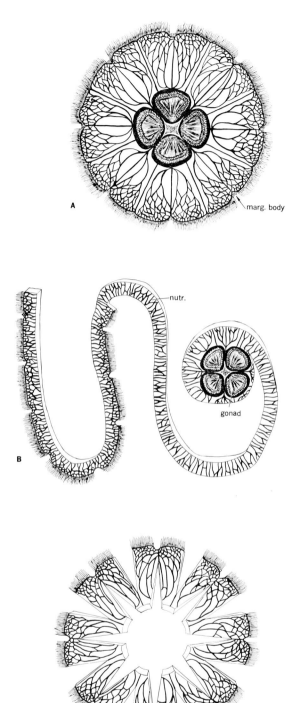

Figure 8.1. *Aurelia aurita.* **A.** Intact adult seen from below. **B** and **C.** Types of mutilation to demonstrate diffuse conduction of excitation. *marg. body*, marginal body—note that all but one have been removed in **B** and **C**; *nutr.*, nutritive tubes of the gastrovascular cavity. [Romanes, 1885.]

about the density of the elements of the net functional at the moment.

The pathways of conduction have been **shown to be nervous** in various ways, as opposed to the possibility of propagation through muscle or through mechanical restimulation. (a) Some agents, such as Mg^{++} ion, in a critical dose depress muscular response completely without preventing conduction, through a local area of application. (b) A second shock shortly after a first can cause a propagated wave of excitation through the conduction system even though the muscle is still refractory. Such a wave is most easily shown by stimulating first one end of a long strip preparation and then the other, at the critical moment after arrival of the first wave: the second wave, traveling back along the strip, goes through a certain length which is refractory for motor response; then it comes to recovered muscle and causes visible contraction. Occasionally entrapped circuit waves suddenly halve in frequency or miss a beat; clearly the excitation wave in the conducting system is on schedule, but muscle refractoriness has set in. (c) Finally, some species have muscle-free regions of the subumbrella, through which conduction occurs just as it does through the muscular regions. These tests for the nervous character of conduction pathways have not been applied to every situation studied but it is usually assumed that physiological observations about conducting systems are observations about nerve cell and fiber systems. This assumption is not entirely safe in coelenterates, where some very slow responses occur. Careful tests should be done in more cases and exact anatomical studies made to see whether a reasonable substratum for the conduction occurs. (See also Parker, 1919b, pp. 109–111.)

The essence of a nerve net is the **relative diffuseness** of the available conducting pathways. Nothing need be specified about the intimate relations between the structural elements of the pathways, in particular about the continuity or discontinuity of the neurons. Conduction must be continuous from one part of the net to another, but the neurons do not have to be continuous any more than in any other nervous system; however, in certain cases they may be.

Although the concept of a nerve net is anatomical in depending on a widespread system of many available pathways and not on the protoplasmic continuity thereof, it is apparent that a physiological demonstration is necessary to show that the neurons are actually a conducting system rather than a plexus of separate elements, as in the skin of higher animals.

The diffused character of the nervous elements is also indicated by the functional property of a **high relative autonomy.** Small pieces of the animal are capable of mediating complete responses nervously. The receptors are scattered widely through the epidermis, conducting elements are scattered widely, and muscles are scattered widely. At least certain kinds of response occur in isolated pieces to physiological stimuli of normal intensity and are susceptible to the same drugs that affect nervous conduction in the whole animal.

The nerve net is suggested by the phenomenon of the **entrapped circuit wave.** If a medusa is prepared by cutting out the center of the bell, a ring-shaped preparation results. Stimulation at one point in this ring-shaped strip usually initiates two waves of contraction which pass around the ring in opposite directions to meet and cancel each other at the opposite side. But occasionally one of these two waves fails to propagate and the other continues around the ring. If the preparation is large enough that the conduction time around the ring exceeds the refractory period of the muscle, the contraction wave on its return to the point of stimulation finds the muscle recovered, with nothing to prevent its proceeding around a second time. Once started, such an entrapped circuit wave will proceed for hundreds of revolutions; in one case, reported by Harvey in *Cassiopea*, the wave continued for over 11 days, having traveled a total of 457 miles at 77.5 cm/sec. Presumably the wave will stop only when a new impulse arises spontaneously from a pacemaker locus or is elicited by stimulation, or when regeneration of nervous tissue across the center of the preparation shortens the conduction path.

Conduction occurs in either direction through all these preparations and across any bridge wide enough to permit any conduction, whatever its axis of orientation.

Experiments such as conduction around partial cuts suffice to indicate the presence of a nerve net, that is, a diffusely conducting system of nervous elements (see definition, p. 1603). The additional properties described below serve to substantiate and in some respects to qualify the diffuseness, as well as to characterize the nerve net, as that form of nervous organization is manifested in the species of coelenterates so far studied.

Horridge (1953) first demonstrated unitary action potentials in the nerve net of *Aurelia* (see also Yamashita, 1957; Passano, 1958; and Josephson, 1961c). Pantin (1935a) had already shown that in the anemone a single electric shock of short duration initiates a widely conducted event which has the character of a unitary event, that is, it is easily distinguished from two such events initiated by two shocks, however spaced. Bullock (1943) found the same in medusae. From these results we may conclude that the nerve net in coelenterates is traversed by **all-or-none nerve impulses** of the same general character as those in higher animals. Other forms of response—for example, graded and decrementally propagated events such as those postulated by Bishop (p. 130) and Grundfest (p. 171) as antedating all-or-none impulses in evolution—are not excluded by this evidence. But we have no reason at present to believe that such events play any role in conduction over considerable distances in these animals.

Unpolarized synapses may be inferred from the foregoing data and from evidence given below that there are in fact anatomical discontinuities in the nerve net. The alternative, of course, is to believe that at each point in the net there are two reciprocally polarized synapses, that is to say, duplicate synapses between each pair of nerve cells. If these are about as far apart as the ordinary meshes of the network, a narrow bridge, just wide enough to conduct (that is, to include continuity of the meshwork), should commonly be polarized. If paired reciprocally polarized connections between neurons occur at a microscopic level, the light microscope gives no evidence of it in recent excellent silver impregnations and methylene blue stains (Fig. 8.2). Possibly there are polarized patches at an electron microscope level but at present it is justified to speak of un-

I. EVIDENCE AND PROPERTIES OF A NERVE NET

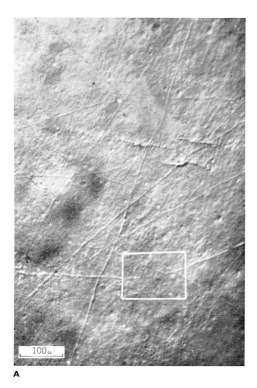

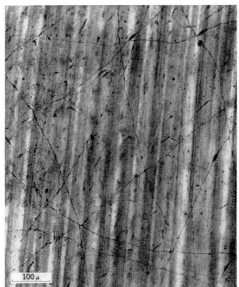

Figure 8.2. Plan of large bipolar celled nets. **A.** *Aurelia*. The living nerve fibers seen by oblique illumination of the subumbrella; within the rectangle, the apparent gap in the course of the fiber shows the position of a cell body. [Horridge, 1954.] **B.** *Aurelia*. Subumbrellar sheet, showing nerve cells and fibers crisscrossing the circular muscle fibers; drawn by Schäfer (1879) after gold chloride impregnation. **C.** Whole mount of a perfect mesentery, showing the nerve net and scattered sense cells underlain by muscle fibers. Methylene blue, fixed with Susa/phosphomolybdic acid. [Batham, Pantin, and Robson, 1960.]

polarized junctions. All the synapses are not unpolarized. In significant places there is found a greater or lesser degree of irreciprocity—transmission is more readily produced one way than the other. This is itself a strong argument for discontinuity.

A **dependence on facilitation** seems to be a general characteristic associated with nerve nets.

This means that responses mediated by nerve nets generally require repetitive stimulation and the amplitude is a function of frequency. This is due, in some places, to a property of the neuroeffector junction and in other places to a property of the synapses in the net. It is of interest because this is really an **integrative property** and we are led to believe that such integration is characteristic of the very first forms of nervous organization. One-to-one responses, such as those occurring in many nerve-muscle junctions and relay nuclei of higher animals, although perhaps simpler, may be more derived in evolutionary terms.

Multiple nerve nets may be a common feature of

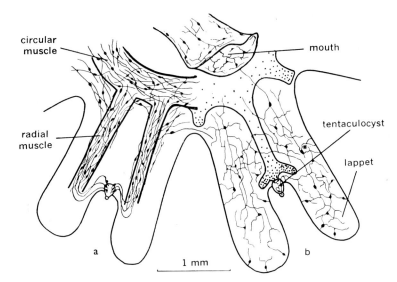

Figure 8.3.

Aurelia aurita ephyra. Two arms of the bell, showing the main structures related to the nerve nets. *a*, the muscle strips and bipolar nerve net system. *b*, cells of the multipolar nerve net and the underlying gastric cavity. [Horridge, 1956a.]

coelenterates (Fig. 8.3). It will become apparent why this is necessary for organisms with some diversity of response as we examine the relatively limited numbers of parameters available to a nerve net with which to determine various forms of response. The demonstration is now clear at least in several cases that discrete networks overlie each other, apparently on the same side of the mesoglea, connected separately to effectors and communicating with each other only at limited regions.

II. ANATOMY

The history of coelenterate anatomical research is remarkable in that the early work (Schäfer, 1879; the Hertwig brothers, 1878, 1879) has come back into favor after many years during which its results were believed to have been superseded (Bethe, 1903; Parker, 1919b). With the present perspective we can see the importance of the most exacting standards in evaluating evidence concerning the **identification of nerve cells** and their relations in coelenterates—and indeed in other animals and parts of animals where nervous tissue is diffusely distributed among other tissues. Identifying isolated peripheral nerve cells is at best difficult, but it is made easier when one can follow the process of a cell to a nerve or to connection with an obvious central nervous organ. Lacking this possibility in coelenterates it is necessary to fall back upon a combination of circumstantial evidence from as many sources as possible, as Pantin eloquently set forth in his Croonian lecture (1952). When we cannot obtain direct experimental evidence, such as the production of physiological signs by cytologically identified lesion or recording or stimulation, we need both of the following types of evidence. (a) The cells and fibers should be anatomically consistent with the identification of nerve cells, for example by staining like nerve cells in higher animals and also not staining in ways characteristic of nonnervous structures, such as collagen. (b) The distribution of the cells and the connections of their processes should be consistent with some independent physiological evidence of nervous functions. Together, these tests can give powerful evidence in favor of the identification of nerve cells and their relations.

But these criteria depend upon visualizing the neurons, and this has been the historical difficulty. In the hands of some workers (for example O. and R. Hertwig, Schneider, Zoja, Horridge) exceedingly simple **methods** such as observation of living tissue (Fig. 8.2) and dissociation of dilute osmic-fixed tissue have yielded valuable evidence. Methylene blue-stained preparations

II. ANATOMY A. Large Bipolar Cell Nets

(Bethe, 1895, 1903; Grošelj, 1909; Hadzi, 1909; Bozler, 1927; Woollard and Harpman, 1939; Komai, 1942; Pantin, 1946; Horridge, 1956a; Spangenberg and Ham, 1960; Batham, Pantin, and Robson, 1960) and metallic impregnation methods (Schäfer, 1879; Havet, 1901; Wolff, 1904; Niedermeyer, 1919; Woollard and Harpman, 1939; Leghissa, 1949, 1950; Pantin, 1952; Horridge, 1954a; Batham, Pantin, and Robson, 1960; Batham, 1956; Mackie, 1960) have provided the main outlines and many of the details of our picture of nerve nets (Fig. 8.2). These same methods have failed in the hands of many workers or have given results which must be regarded, in the light of later evidence, as contaminated by staining of nonnervous elements to a serious degree (Havet, 1901, 1922; Leghissa, 1949). It is only because of unusually exacting standards and patient work that Pantin, Batham, Robson, Horridge, and Mackie have after many years of uncertainty provided relatively complete pictures of the anatomy of the nerve net in certain places. In significant respects these can be correlated with physiological facts.

A. Large Bipolar Cell Nets

The best anatomical descriptions refer to a wide-meshed **lattice of long straight elements** running in different directions at the base of the ectodermal epithelium of the subumbrella of medusae such as *Aurelia* and *Rhizostoma* and in the mesenteries of anemones such as *Metridium* and hydrozoans such as *Velella*. Already in 1879, Schäfer (Fig. 8.2) described this quite accurately from gold impregnations. Hesse (1895), Bethe, (1903), Bozler (1927), Woollard and Harpman (1939), Komai (1942), Pantin (1952), Horridge (1954a), Batham, Pantin, and Robson (1960), Batham (1956), and Mackie (1960) have confirmed and extended the description. The elements of the lattice are bipolar and occasionally tripolar nerve cells, all isopolar, whose processes can be followed for lengths up to 8 mm. They maintain a fairly straight course and only occasionally branch. Bozler distinguished in the medusan *Rhizostoma* three size classes whose cell bodies are respectively 40, 75, and 110 μ in modal length and 7, 11, and 15 μ wide, and whose axons are 2, 4, and 7 μ in diameter. In *Aurelia* the axons may reach 12 μ in diameter. The general aspect of this lattice is well shown by several methods in Figs. 8.2, 8.4, 8.5, 8.7.

This kind of net is very likely to show **preferential orientation** in certain axes. In the subumbrellar net of *Aurelia* there is no preferred orientation except in two regions. Along the edge of the bell most fibers run tangentially or at a small angle to the tangent, whereas just central to each of the marginal bodies the radial direction strongly predominates (Fig. 8.4). Bozler reported similar local preferences in *Rhizostoma*. Horridge calls this net the giant fiber net in the ephyra larva of *Aurelia*; it is here almost confined to the epithelium overlying the muscle strips and is strongly oriented radially over the radial muscle and circularly over the circular muscle. It will hereafter be termed physiologically a fast, specific nerve net.

In anemones of the genus *Metridium* bipolar nets of very elongate neurons are especially developed on the **mesenteries**. The face of the

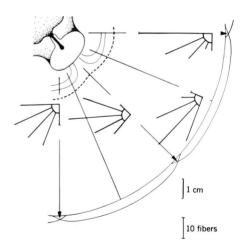

Figure 8.4. The angular distribution of large nerve fibers at four places on the concave surface of the bell of *Aurelia*. At each place 100 fibers were examined by using an ordinary binocular microscope and oblique illumination. A histogram with one angular coordinate has been drawn at each place. Orientation is marked only in two regions: at the edge of the bell, where only a few of the axons run radially, and just central to the marginal body, where they appear to radiate from the stalk. [Horridge, 1954a.]

mesentery bearing the retractor muscle has a dense lattice of relatively thick fibers (average, 1.4 μ; maximum, 4–6 μ); the opposite, radial muscle face has a sparse lattice of finer fibers (Fig. 8.5). Values for fiber diameter have as yet not been correlated with the great stretch which these structures normally tolerate. There is a preferential orientation in the vertical axis on the retractor face. Nearly all the bipolar cells terminate at both ends in expanded endings on the underlying muscle. They are therefore **motoneurons.** But they also form the pathway for through-conduction at relatively high velocity (1–2 m/sec) by way of numerous relays—a synapse at least every 3 mm, probably much more often. They are therefore interneurons at the same time.

The dense network of bipolars is roughly correlated in its occurrence with fast muscle lying below it, comparing regions of mesentery and species of anemones. The processes run for an average of 4–8 mm end to end; the course is strikingly straight but occasionally abrupt changes in direction occur. The main input to these mesenteric nets is from sense cells in the adjacent entoderm; the main output is to the subjacent muscles, which are thereby activated nearly simultaneously all over the mesentery and in all the mesenteries. This through-conduction system is therefore not so much a pathway between two places as a fast distribution system lying over an extended effector. Communication with other parts of the nervous system has been seen but little studied. Some cells send processes into the nervous network over the parietal muscles; others into the oral disc. The latter is ectodermal; presumably nerve fibers follow the epithelium of pores through the body wall.

There is as yet no evidence of double innervation of muscle fibers in *Metridium*, although two distinct kinds of contraction, fast and slow, are caused apparently in the same muscle fibers, for example the retractors in the mesenteries, by the same through-conduction nerve net, according to frequency and number of impulses (Batham et al., 1960). The only other places in polyps where well-developed networks of large nerve cells have been found are the ectoderm of the oral disc of certain anemones (*Metridium*), as hinted at by Hilton already in 1927, and the general ectoderm of the floating chondrophoran *Velella* (see further, p. 512).

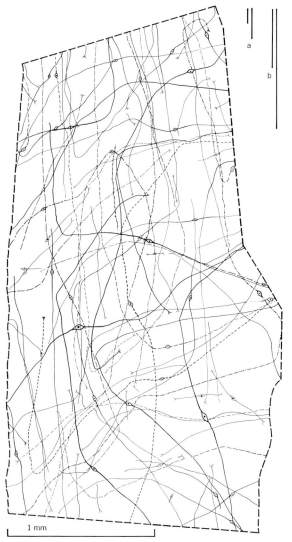

Figure 8.5. Plan of the nerve net on one face (retractor muscle face) of a mesentery of *Metridium senile*. A whole mount of the mesentery was stained with silver and every nerve cell and fiber traced with a camera lucida. *a* and *b* show respectively the ranges of lengths of the off-retractor and of the retractor muscle fibers underlying the nerve net, for comparison with the lengths of nerve fibers. Natural ends of nerve fibers shown by forked tips; these all dip down into the muscle layer. Dots and dashes only for convenience in following elements; relative thickness of fibers indicated but not to scale. [Batham, Pantin, and Robson, 1960.]

B. Small Multipolar Cell Nets

Physiological evidence cited below makes it clear that there must be in many places at least **two nerve nets coexisting** in the same epithelium, making only indirect functional contact one with the other, responding to different kinds of stimuli, and eliciting different forms of response. This has been elegantly demonstrated functionally and anatomically by Horridge (1956a) in the ephyra larva of *Aurelia* (Fig. 8.3). Besides the large bipolar cell net is a network of smaller, mostly multipolar, cells, hereafter called physiologically the slow, diffused nerve net.

The latter lies in the same epithelium and extends beyond, into epithelia not occupied by the large bipolar net. It is less homogeneous in structure, consisting of elements of several kinds: sense cells and bipolar and multipolar ganglion cells, some of which are presumed to be motor. This net extends over the whole aboral surface of the larva around the sides of the arms and the disc and occurs over the whole oral surface. Bipolar cells with long processes are arranged radially along the arms and up the sides of the mouth, with a few extending around the mouth. The subumbrellar portion of the net is put into connection with a network spreading over the folds of the mouth and extending down the inside toward the stomach. When the rudiments of marginal tentacles have appeared their nerve cells are characteristic of this small multipolar net. The **axons are extremely thin** in this net and have many sharp angles and twists, contrasting with the broader straighter fibers of the large bipolar net. Because the fibers are at the limit of resolution, structure cannot be made out at the endings; although they apparently connect with each other or with nerve cells it is impossible to give evidence on the question of continuity or contiguity. The evidence that this net is distinct from the large fiber bipolar net rests therefore on physiological experiment rather than histological observation (see below, p. 495).

A multipolar net has been described in several actinians and in *Tubularia* by Leghissa (1949a, 1949b, 1950). Relatively condensed areas in the latter genus occur at the base of the distal circle of tentacles and at the junction of the crown and stalk, but there is nothing to provide a basis for the two distinct forms of electrical activity seen by Josephson (1962). A concentration of sense cells and ganglion cells and preferential circular orientation in the oral disc of anemones is important as this area is often implicated in discussion of the higher levels of integration of behavior.

A convincing histological description confirming the existence of a **distinct, small multipolar net** in the chondrophoran *Velella* (Fig. 8.6)

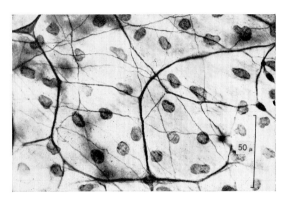

Figure 8.6. Two nerve nets in the aboral ectoderm of *Velella*. The "open" (small fiber) and the "closed" (large fiber) systems are essentially separate. The junctions of fibers of the closed system are said to be anastomoses. Silver impregnation. Nuclei belong to ectoderm cells. [Mackie, 1960.]

in the same epithelia as the large-celled network, is given by Mackie (1960); the same situation appears to exist in *Porpita* (Mackie, in MS; see further, p. 512).

In the subumbrellar sheet of the scyphozoan *Rhizostoma* abundant small multipolar isopolar nerve cells are scattered among the less numerous large bipolars and usually lie up against the axons of the latter (Fig. 8.7). The processes of these small nerve cells are very fine and possibly taper beyond the limit of resolution of the light microscope. Bozler (1927) described them as characteristically running alongside the bipolar cell axons just before disappearing and he believed that they were in functional contact but without protoplasmic continuity. We can only raise the question, without answering it, whether in fact these small multipolars are in functional connection with the large bipolars or form a

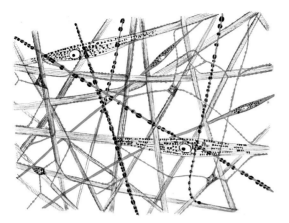

Figure 8.7. Ectodermal nerve net (or nets) of the medusa *Rhizostoma*, combined from several preparations. Note large bipolar cells and their fibers *(gray)*, and small, multipolar cells *(blue)*, possibly forming separate nets. Although Bozler regarded as synapses the contact zones between thick and thin fibers, this is now moot, as is the question of which group the small bipolars belong to. Rongalit white variation of methylene blue. [Bozler, 1927.]

distinct nerve net, corresponding to the diffuse net in the ephyra of *Aurelia*. The latter is more likely since in the related *Cassiopea* there are physiologically two nerve nets.

In the epidermis of the column of actinians sensory neurons are seen but ganglion cells are not. The sensory neurons may directly innervate muscle or penetrate the mesoglea to supply the gastrodermal net; we do not know. If there is a net in the epidermis, it could conceivably be composed solely of **interconnected primary sensory neurons.** This conception is based on the possibility that at least a good proportion of the sensory neurons possess more than one neurite, which make functional connection with the neurites of other sense cells; excitation could thereby spread from one place to another through sense cells acting as interneurons and finally get directly to effectors through motor endings of the same cells. But the hypothesis can not be really supported or excluded on the evidence available.

C. Cytology

Cytological details have been supplied by the authors mentioned above and, in addition, by Grošelj (1909), Hadži (1909), Wolff (1904), O. and R. Hertwig (1880), McConnell (1932), Leghissa (1949a, 1949b), Spangenberg and Ham (1960). Their accounts deal chiefly with nerve cells in regions other than those just described. It is interesting that Schäfer, the first of these cytologists, was able to say:

These appearances are so obvious as to allow of no question that we have before us undoubted nerve-fibers and bipolar-ganglion cells. The tissue which they underlie being just as clearly muscular, with well-characterised cross-striae, it is interesting to observe, even so low down in the metazoic scale as the medusae, that the textures, which in the higher animals are generally looked upon as the most highly differentiated, should have already attained a degree of structural complexity and of functional activity in respects scarcely inferior to the nervous and muscular tissues of vertebrates.

The forms of the nerve cells are bipolar and multipolar, never unipolar; all but sensory cells are typically isopolar, though some short and some long processes are described on multipolars. Sizes vary widely. Small nerve cells are about 3–10 μ in diameter; large ones, 6–25 μ. The "giant" bipolars of *Aurelia* are 10–20 μ wide (presumably fair-sized adults); those of ephyra larvae of this form are 6–8 μ wide (Horridge, 1954a, 1956a). Nerve cells in *Actinia* are given as 5–30 μ in diameter (Leghissa, 1949a). Nerve fibers range down below the limit of the light microscope; as recently shown in the electron microscope they are often 0.25 μ (Fig. 2.45). Large fibers ("giant bipolars") are 1 μ in ephyra stages of *Aurelia* and up to 12 μ but rarely above 8 μ, in the adult. The "thick" through-conduction fibers of the mesentery of *Metridium* reach 4–6 μ at a cell length from the cell body.

Space does not permit a detailed description of the **cytoplasmic inclusions** (see Chapter 2), but we may refer to the large clear nucleus with a conspicuous nucleolus, characteristic of some of the larger nerve cells. There may be a considerable number of refractile droplets of various sizes (up to 3 μ) arranged in rows, especially in the cell bodies and occasionally along large axons. Spangenberg and Ham (1960) mention small heterochromatic granules in cells and fibers. These or other inclusions may be equivalent to the Nissl substance of other animals, but we lack sufficient information on their nature to permit a decision. Locally **neurofibrils** have been repeatedly described from the living fibers, as refractile dark straight lines. According to the more recent descriptions they do not form a net in the cell body. They show their integrity as well as their soft inelastic character in hypertonic solutions; gross coiling of the whole bundle or herniation of axoplasm, with the fibrils looping through the hernia (Fig. 8.8), is reported.

An important feature is the **absence of differentiation of processes.** We cannot distinguish axons and dendrites and when the former term is used it implies only a long process, presumably specialized for conduction and without numerous branches or endings. Another feature of nerve cells and fibers in coelenterate nerve nets is the absence of any sheath cells. The appearance is that of growing neurites in tissue cultures.

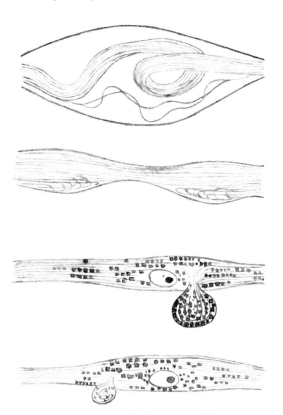

Figure 8.8. Bipolar cells of *Rhizostoma* in hypertonic sea water, to illustrate the reality of neurofibrils seen in the still living cell. In the lower two examples the plasma has flowed out through the ruptured membrane to form a sacklike swelling, and fibrils are continuous into and out of the sac. [Bozler, 1927.]

D. Synapses and Neuromuscular Junctions

The axons of large bipolar cells frequently make obvious contact where they cross or meet. They may intertwine once or twice, or may simply adhere for a short distance (Fig. 8.9). Synapses with multiple contacts occur, fibers running alongside each other for as much as 30 μ. The contact is simple, symmetrical, and made en passant. Pantin counted a minimum of 20 junctions in the most direct route vertically through a 10 cm mesentery of a sea anemone. He measured the over-all conduction time (85 msec) and then could calculate that the synaptic delay at each junction must be well below 2.5 msec, because even at that value the velocity of conduction in the fibers would be an unlikely 3 m/sec (see further, p. 481).

It is particularly interesting in view of the subsequent acceptance of syncytial continuity as a characteristic of nerve nets that Schäfer, in 1879, before the formulation of the neuron doctrine, had already taken such a clear position with respect to these junctions in the jellyfish bell:

It seemed at first sight almost incredible that with such a prodigious number of nerve-fibres, exhibiting so close an interlacement, there should be no actual junctions of the intercrossing nerves. And it was especially difficult of credence because some of the experiments of Mr. Romanes, performed with the view of testing the amount of section which the tissue could endure without loss of nervous (or excitational) continuity, seemed to point to the existence of a structurally continuous network of nerve-fibres. Nevertheless, there can be no doubt that the fibres do not come into anatomical continuity. On the other hand, it can readily be seen that each nerve-fibre comes at one or more points of its course into very close relations with other nerve-fibres. Two fibres, for example, may sometimes be observed to bend towards each other out of their previous course, in order to run closely side by side for a greater or less distance, and in such cases one fibre may hook around the other or they may even be two or three times intertwined. At other places a number of fibres come together from different parts and join in a very close entanglement, the fibres in which run for the most part parallel, and it is only with difficulty that the individual fibres can be followed. So that although there is no actual anatomical continuity abundant opportunity is afforded for inductive action, whether electrical or of some other

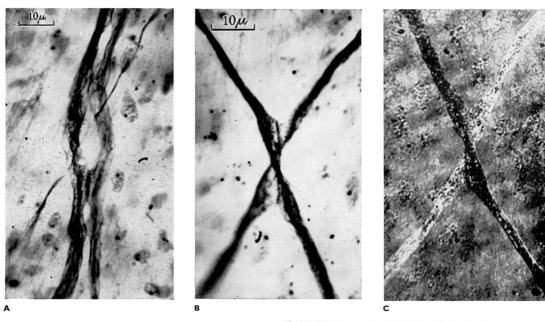

Figure 8.9. Axons and contacts in *Metridium* mesentery prepared as a whole mount, stained with silver. **A.** Parallel axons with four synaptic contacts. **B.** Crossed axons in synaptic contact. **C.** The same chiasma as **B** under crossed polaroids, with mica plate compensator. **D.** Similar to **B.** In **A**, **B** and **D** note black line of demarcation. [Pantin, 1952.]

kind. That physiological continuity is thus maintained it seems as yet premature to conjecture.

The Hertwigs, Hesse, Havet, and Grošelj came to the same conclusion; but after Bethe, using questionable techniques and following Zoja and Schneider, had reported protoplasmic continuity, the latter concept took root and remained essentially unchallenged for a quarter of a century. Bozler (1927a, 1927b) addressed himself particularly to this question and with the advantage of beautiful preparations by the Rongalit white modification of methylene blue (Fig. 8.7) confirmed the older conclusion of discontinuity in abundant detail. On present views his relevant evidence is the lack of fusion of large bipolars with each other and of small multipolars with each other; the reality of functional contacts between fine fibers and thick fibers is at the moment in question, judging from electrophysiology of other medusae. Most recent authors on forms other than *Hydra* (Niedermeyer, Woollard and Harpman, Pantin et al., Horridge, Mackie,

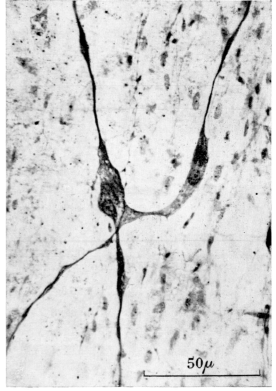

contra Leghissa) have come to the same conclusion as Bozler and we may say for the present that whereas anastomoses probably occur in some other sites, in the large-celled network of the subumbrellar sheet of medusae and in the mesentery of actinians **discontinuity between neurons** is clearly the rule. Nerve nets in other situations may or may not conform to this finding and each should be examined for itself. Mackie (1960) believes there is an "open," synaptic nerve net and a distinct "closed" syncytial net in *Velella* (Fig. 8.6; see also p. 512). Leghissa's (1949, 1950) claim for continuity in actinians is not considered to be adequately founded by later authors. Several workers agree that *Hydra* shows anastomosis (see p. 507). The electron microscope is of course needed but the serious problem is to find the relevant spot to focus upon and to trace the fibers to two distinct cells to prove they belong to different ones (see Fig. 2.26).

Endings upon muscle have been seen by several authors. They may be rather simple or may take the form of specialized expansions, sometimes with numerous short branches (Fig. 8.10). Batham, Pantin, and Robson (1960) have obtained apparently quite complete silver impregnations showing expanded endings on both ends of nearly every bipolar cell in *Metridium* mesenteries. There are very many muscle cells per ending. The muscle fibers are much finer (less than 1 μ) than the larger nerve fibers supplying them (5–12 μ) as well as much shorter.

E. Regional Differentiation

Other nerve nets have been reported upon briefly. In the endoderm of *Rhizostoma* a strong net of bipolar elements of uniform size occurs. A weak, small-celled net together with radial muscle fibers occurs on the opposite face of the mesenteries of anemones from that bearing the through-conducting large-celled net. But for the most part our information about regional differences is based upon the identification of nervous elements and the thickness of the nerve fiber layer in sections, rather than on the actual demonstration of connected sets of nerve cells; most data on this topic is from *Metridium* and a few other actinians (Fig. 8.11; compare the general anatomy, Fig. 8.12).

The **oral disc** represents the greatest concentration of nervous elements in anemones. Sense cells and large ganglion cells are abundant, especially around the bases of tentacles. Bipolar cells with long axons occur, preferentially oriented in the radial direction and more numerous in radial tracts. The nerve fiber layer is at its thickest in the disc, both between the radial tracts, where the cells are few, and in the tracts, where they are more numerous.

The **tentacles** of actinians are the next most heavily supplied with nerve cells, containing both sensory cells and ganglion cells, chiefly longitudinal bipolars.

The wall of the **pharynx** has a thick nerve fiber layer but ganglion cells are sparse. It has many sense cells but seemingly not enough to account for the thickness of the layer. The meaning of this concentration is not clear. It cannot readily be regarded as a pathway between epidermal and gastrodermal nerve nets as some authors have proposed, since there is little evidence of communication, on the basis of the properties observed, between oral disc and mesenteries or from epidermal column receptors to mesenteries.

The **gastrodermis of the mesenteries** on the

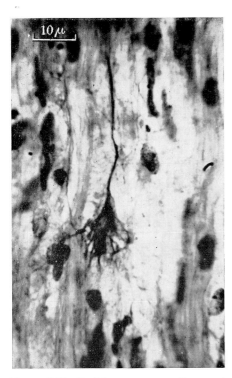

Figure 8.10. *Metridium* mesentery. Presumed end plate of axon on retractor muscle field. [Pantin, 1952.]

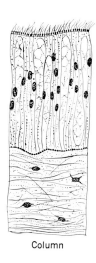
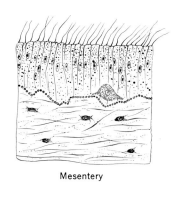

Oral disc

Column

Mesentery

Figure 8.11. Cross sections through oral disc, column, and mesentery of the anemone *Sagartia parasitica.* Nerve cells shown arbitrarily as *blue.* [Redrawn from Hertwig and Hertwig, 1879.]

retractor muscle face presents the previously referred to, well-developed, specialized through-conduction pathways, on both histological and physiological evidence. Compared to the epidermis of column and pedal disc, nervous elements are relatively abundant in the gastrodermis generally. In *Metridium*, with a retractor muscle 2 mm wide, there is a band of longitudinally oriented bipolar cells, over and parallel to the muscle about 50–100 neurons across. This is an extremely dense network for these animals. On each side of the muscle, the net becomes sparse, and is especially so on the radial muscle face; but

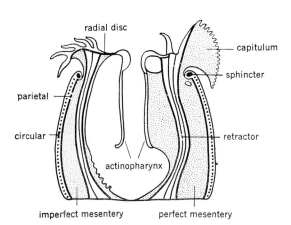

Figure 8.12. Diagram of the anemone *Metridium*, in vertical section to show the musculature. [Batham and Pantin, 1954.]

at the attachment of mesentery and body wall it is again dense and almost tractlike. In the endodermal epithelium of the body wall there are a good many nerve fibers parallel to the circular muscle and a few diagonal to it. We lack an adequate descriptive basis for the stretch sensitivity of the column and pedal disc, now believed to be due to endodermal sense cells, and also for the non-through-conducting net which mediates local response of the column musculature and for the fast circular conduction in the sphincter muscle of some actinians. Havet described some sense cells and ganglion cells in the **acontia,** but Parker considers these to be virtually independent effectors with hardly any nervous system.

The **epidermis of the column** appears to have an exceedingly sparse net. The physiological experiments of Parker, which seemed to indicate conduction in this net, can be reinterpreted as due to gastrodermal sensory and conducting elements excited by mechanical stimuli through the body wall. However, sensory neurons have been seen in the column epidermis and we are uncertain whether to regard these as sending fibers through the mesoglea to the gastrodermis or innervating epidermal musculature directly or contributing to a net in the epidermis. In certain anemones that are regarded as primitive (*Pro-*

tanthea and *Gonactinia*), the nerve fiber layer in the epidermis is well developed throughout the column and includes interneurons (Carlgren, (1893).

The **pedal disc** is probably very different in different species: in some it apparently contains a well-developed network of nerve fibers; in others, an exceedingly sparse one.

A vexatious question has been the presence or absence of **nervous connections through the mesoglea**, especially in actinians. The Hertwigs (1880), Wolff (1904), and Grošelj (1909) denied the appearance of such nerve cells or fibers. Von Heider (1879, 1895, 1899), Havet (1901, 1922), Parker (1912), and Parker and Titus (1916) believed there must be such connections and reported seeing such fibers and cells. We conclude, with Pantin (1952), that there is no unequivocal demonstration of such cells in the actinian mesoglea and that the physiological evidence for such communication is also equivocal in view of the discovery of the abundance and sensitivity of sense cells in the gastrodermis which are probably excited normally by mechanical conduction through the body wall. There is, of course, communication between the gastrodermal and epidermal nerve nets where they meet in the stomodeum and it is quite possible that other more localized communication exists—for example, through the epithelium of the pores, which in many anemones put the gastrovascular cavity into connection with the outside. The possibility that direct connections through the mesoglea are present but simple and widely scattered, would of course be very difficult to exclude. Nerve fibers are said to cross the mesoglea commonly in the chondrophoran *Velella* (Mackie, 1960) and between the nerve rings in hydromedusae (see next section).

F. Nerve Rings and Marginal Ganglia

The highest level of differentiation of nervous organization in the coelenterates is achieved in the active medusae by the formation of local concentrations of nerve cells and fibers into nerve rings or marginal ganglia. **Hydrozoan medusae** characteristically display two circular thickenings of the subepidermal nerve fiber layer close to the margin of the bell, one just above and one just below the point of attachment of the velum (Figs. 8.13 and 8.14). Each ring consists of a relative concentration of parallel, circularly running nerve fibers, and in the overlying epidermis a greater density than elsewhere of nerve cell bodies. Presumably there is also a **concentration of endings and interconnections** between neurons. This structure does not properly deserve the name nerve, because it is not a distinctly demarcated cable of nerve fibers and because it contains cell bodies and connections. It comes closer to meriting the name ring ganglion or nerve cord; the name is not applied to it, but only because the degree of specialization and distinctness from the rest of the nervous system is, on present evidence, apparently not quite in a class with the usual structures to which these terms are applied.

The **two nerve rings** are different in several respects (Hertwig and Hertwig, 1878; Hyde, 1902). The upper is said to consist of very fine fibers only; the lower, to contain many large (2 μ) fibers. The lower has many more ganglion cells (inter- and motoneurons), receives afferent fibers

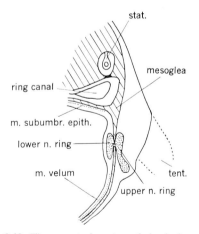

Figure 8.13. The marginal region of the hydromedusan *Geryonia*, showing the position and relations of the two nerve rings and the statocyst. Hertwig and Hertwig are the main authority for the connections between nerve rings. [Horridge, 1955a.] *lower n. ring*, lower nerve ring; *m. subumbr. epith.*, muscle fibers of subumbrellar epithelium; *m. velum*, muscle fibers of velum; *ring canal*, ring canal of gastrovascular system; *stat.*, statocyst; *tent.*, tentacle; *upper n. ring*, upper nerve ring.

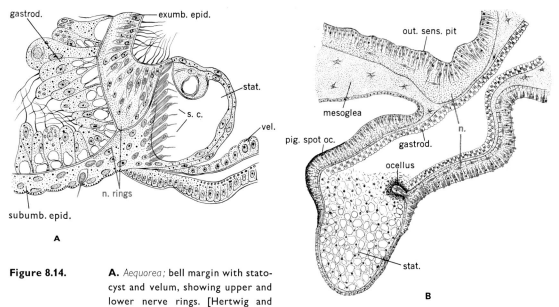

Figure 8.14. **A.** *Aequorea;* bell margin with statocyst and velum, showing upper and lower nerve rings. [Hertwig and Hertwig, 1878.] **B.** *Aurelia;* section through marginal body and sensory pit. [Schewiakoff, 1889.] *exumb. epid.,* exumbrellar epidermis; *gastrod.,* gastrodermis; *n.,* nerve tissue; *n. rings,* upper and lower nerve rings; *ocellus,* pigment cup ocellus; *out. sens. pit,* outer sensory pit; *pig. spot oc.,* pigment spot ocellus; *s. c.,* sensory nerve cells; *subumb. epid.,* subumbrellar epidermis; *stat.,* statocyst; *vel.,* velum.

from the statocysts (Horridge, 1955a, stated that these go to the upper ring), and sends efferents to the velar and subumbrellar musculature, thereby controlling the rhythmic pulsations of locomotion. The upper ring receives afferents from scattered sensory cells in the epithelium of the bell margin, from the tentacles and from ocelli when present. The two are in communication by many fibers which pass back and forth through the mesoglea of the velar attachment. The resemblance of the nerve rings to ganglia is heightened by the fact that the velum with its striated circular muscles of locomotion is apparently devoid of ganglion cells and is innervated from the lower ring. Functional properties of the rings, to be treated below, further emphasize the same point. The rings are in general connection with the subepidermal plexus of the subumbrella and of the exumbrella. Thickenings of this plexus are commonly seen along the radial gastrovascular canals, possibly providing the connection with the nerve net of the manubrium which experiments require. The histological picture is still unsatisfactory, lacking as it does an adequate basis for the functionally demonstrated independence of two conducting systems and of the differentiation of function within the two nerve rings, which goes beyond an emphasis on sensory connections in the upper and motor in the lower ring (see below, p. 499).

In the scyphozoan medusae, nerve rings are absent except in the order Cubomedusae. But a concentration of ganglion cells, fibers, and endings occurs near each marginal body, of which there are usually four (perradial or interradial), eight (both), or occasionally more (perradial, interradial, and adradial). The marginal bodies are called rhopalia (rhopalioids in Stauromedusae) and are complex and highly varying structures (Fig. 8.14, B); of principal interest here is the occurrence of at least two, commonly three or four, different specialized sensory epithelia. These include one or more sensory pits commonly called olfactory pits, a statocyst, and frequently ocelli of simple or complex form. In addition, there may be sensory lappets or tentacles. This **concentration of sensory structures,** together with the physiological evidence that the initiation of the rhythmic pulsation is localized in these regions and that separate nerve nets come into physiolog-

II. ANATOMY F. Nerve Rings and Ganglia

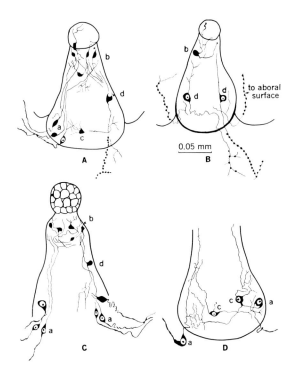
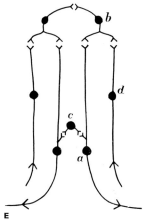

Figure 8.15. Neuron structure of coelenterate ganglion. **A.** The four types of cells, a–d. **B.** The connections with the multipolar nerve net. **C.** The connections with the bipolar cell net. **D.** Neurons of type c make a connection between the cells of type a on the two sides. **E.** Suggested functional arrangement of neurons in the ganglion with input from the multipolar nerve net and output to the bipolar cell net. This is hypothetical and derived from physiological results. [Horridge, 1956a.]

ical contact here, has focused attention upon this concentration of nervous tissue. But before the work of Horridge (1956a) almost nothing had been made out concerning its special character.

Horridge has distinguished **four kinds of nerve cells,** using reduced methylene blue on the ephyra larva of *Aurelia* (Fig. 8.15). (a) Bipolar cells with processes of the type seen in the large fibered net form a group at each side of the base. One process of each cell sweeps out of the ganglion to make synaptic contact with the large bipolar net. The other process dives into the ganglion and ends in fine ramifications among cells and fibers of the next type. (b) Bipolar and multipolar cells with short branching processes form a cluster at the apex of the ganglion and contribute to a dense network of fibers suggestive of the neuropile of higher forms. This and the third type are intrinsic to the ganglion. (c) A pair of cells, one on each side at or near the base of the ganglion, connects the two clusters of type (a) cells of the two sides. Each cell has one long process and one or more short branching processes. (d) At the sides of the ganglion, bipolar cells send a short process toward the apex to end in the network near type (b) and a long process out of the ganglion to connect with fibers of the small multipolar cell net on the oral surface. The scheme (Fig. 8.15) is derived from the physiological evidence cited below and shows how the available anatomy **provides the essentials of a center of integration** between two nerve nets, as well as of automatism, subject to modulation. Possibly in other species and in adults, the picture will be more complex and will include separate pathways of input from the different sense organs.

The concentrations of nerve cells in the nerve rings and marginal bodies of jellyfish are so specialized, particularly on the functional evidence and in the intimacy of contact with the sense organs (which are also novel in medusae and have been regarded as the first true organs in the animal scale), that they have frequently been **regarded as a central nervous system.** This is a matter of degree and therefore of taste. Certainly, the recent evidence has emphasized the complexity and the incipient centralization. However, only a few of the more integrative functions require these structures and the latter are located and replicated wherever these functions arise, not requiring conduction into a distant central organ and back out again.

G. Receptor Cells and Sense Organs

Most of the afferent input of the nervous system of coelenterates comes from **scattered primary sensory neurons,** but there are developed in the medusae true sense **organs,** the first we know of in the animal kingdom (organized groups of several kinds of cells). The scattered sensory neurons are of different form in different animals and sometimes in different parts of the same animal. Typically spindle shaped, with a distal flagellum, they produce proximally one, two, or three long neurites (Batham et al., 1960). Krasinska (1914) describes one kind of superficial sensory neuron in the scyphomedusan *Pelagia,* which has numerous short stiff bristlelike projections. In addition she reports deeper cell bodies which send a branching process toward the surface, a type of sensory neuron found scattered through higher groups and commonly spoken of as sensory cells with free nerve endings. Most other authors on coelenterates have not mentioned such cells. Krasinska further believes that occasional ganglion cells of the nerve net send a peripheral process toward the surface in the hydromedusan *Geryonia.* She distinguishes two kinds of superficial sensory neurons on the basis of the form of the sensory hairs and the presence of basal granules.

Grošelj (1909), Havet (1901, 1922), and others have also described **different forms of sensory cells** (Fig. 8.16), some having extreme attenuation, some with a cap on the end or with cilia or bristlelike projections. In some anemones, cells with two proximal processes passing into the subepithelial plexus of nerve fibers are the commonest. In the tentacles of *Cerianthus,* most cells are provided with a single proximal process and most of these

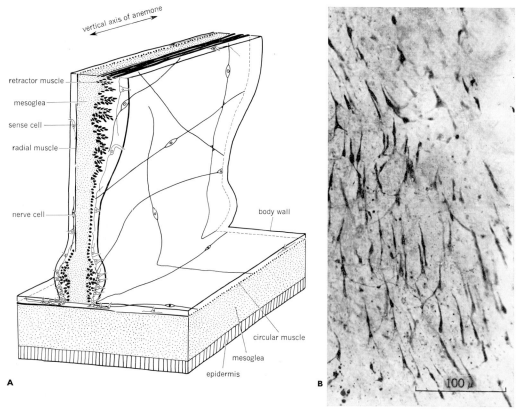

Figure 8.16. Concentration of sense cells and the nervous composition of the actinian body wall and mesentery. **A.** Diagram of the outer body wall and a mesentery, with the sense cells and long bipolar nerve cells. Note the scarcity of sense cells in the body wall and their abundance in the base of the mesentery. **B.** The sense cell concentration seen in a methylene blue preparation. [Batham, Pantin, and Robson, 1960.]

pass toward the base of the tentacle. In other anemones such as *Bunodes* cells with two proximal processes occur but they are less common. In this form it is remarked that in the pedal disc there are many long, threadlike sensory neurons.

The evidence is indicative that the proximal processes of sensory neurons make **synaptic contact with ganglion cells** in the nerve nets and with each other (Batham et al., 1960). Terminations of these processes on muscle fibers or elsewhere have not been certainly seen; they do not cross the mesoglea, at least in *Metridium* (contra Leghissa 1949a, 1949b, 1950). The great muscle and nerve sheet on the mesenteries is supplied by sensory neurons in the gastrodermis of the mesenteries immediately overlying it and of the adjacent inside of the body wall. There is a high correlation between the abundance of sense cells in the epithelium, the concentration of the nerve fiber layer, and, with some exceptions, the abundance of ganglion cells in subepithelial plexuses. Details for *Metridium* can be found in Batham et al. (1960). It seems probable that sensory nerve cells far outnumber motoneurons—as in higher animals; the sense cells of coelenterates probably also outnumber interneurons. Nothing can be said as yet concerning the **differentiation of function** of these cells, that is, whether they are all normally activated by several forms of environmental change including mechanical, photic, and chemical, for example, or whether some or all of them are less sensitive to some forms of stimuli than are others.

The term **sense organs** may be applied to three groups of structures: ocelli, static organs, and dense patches of sensory epithelium, often in pits. These are all developed nearly exclusively in medusae, as opposed to polyps. In medusae they are practically confined to special regions of the margin of the bell, but may be derived from either gastrodermis or epidermis. In hydrozoan medusae there is a band of concentrated sensory epithelium overlying each of the two nerve rings but ocelli and statocysts are usually confined to the **tentacular bulbs** at the base of the tentacles. These bulbs have other roles, in nematocyst formation and digestion, but usually bear, according to the species, an ocellus or a statocyst, uncommonly both, plus patches of sensory epithelium. Scyphomedusae lack nerve rings (except Cubomedusae) but have more elaborate marginal bodies—**rhopalia.** Usually in multiples of four, the rhopalia are composite bodies set in a niche with a hoodlike roof, on the sides of the bell or on pedalia or between a pair of lappets. Basically the rhopalium is a static or equilibrium sense organ in the form of a hollow club, but in some species it bears eyes and it or its base or hood has patches or pits of sensory epithelium.

Photoreceptor organs or **ocelli** are of scattered occurrence in hydromedusae (typically absent in Trachylina) and less common in scyphomedusae (present in *Aurelia*, some Coronatae, best in Cubomedusae, lacking in general in Semaeostomae and Rhizostomae). Ocelli appear as colored spots, usually one on each tentacle bulb (although sometimes on the bell margin in hydromedusae) and one or a group on each rhopalium in scyphomedusae. The simplest are merely a patch of epithelium flush with the rest, with pigment cells and sense cells. The pigment cells are presumably not sensory, nor the pigment photochemical. The role of such cells in these diffuse patches is obscure, and proof that they are specialized organs for photoreception is hardly adequate. More strongly suggestive of this function are the cup ocelli, in which the epithelium with pigment cells and sense cells is enfolded. Many transitional forms are found, up to cups filled with a lenslike cuticular mass (*Sarsia*). The most complex ocelli are those of hydromedusae like *Tiaropsis*; here the pigment cup is gastrodermal and the sense cells are inside the pigment layer, with their axons emerging through the aperture of the cup; that is, the eye is said to be inverted. The axons pass to the upper nerve ring. A similar series of degrees of complexity is found in the scyphomedusae. In the best-developed examples, among cubomedusans, each rhopalium may bear one or two large, complex eyes (*Charybdea*, Berger, 1900) and four smaller simpler ones. The large eyes have a biconvex cellular lens, a precisely organized and stratified grouping of pigment cells and packed retinal sensory cells of two kinds, and a vitreous mass (Fig. 8.17). It is presumed that the axons go to the adjacent rhopalial ganglia. Dem-

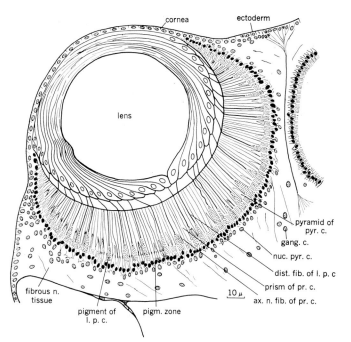

Figure 8.17.
The distal, complex eye of *Charybdea* in sagittal section. [Berger, 1900.] *ax. n. fib. of pr. c.*, axial nerve fiber of prism cell; *dist. fib. of l. p. c.*, distal fiber of long pigment cell; *fibrous n. tissue*, fibrous nerve tissue; *gang. c.*, ganglion cell; *nuc. pyr. c.*, nucleus of pyramid cell; *pigm. zone*, pigment zone (note that this is omitted in a sector to bring out other structures); *pigment of l.p.c.*, pigment of long pigment cell; *prism of pr. c.*, prism of prism cell; *pyramid of pyr. c.*, pyramid of pyramid cell.

onstration of a light-perceiving function of ocelli in *Aurelia* was provided by Horstmann (1934); light or shadow has an effect on the behavior of many forms, but little or no effect on others.

Equilibrium or **static sense organs** are of two types, statocysts and lithostyles; each occurs in both classes of medusae but many hydromedusae lack either type. **Statocysts** are found in all grades from simple epidermal depressions to closed vesicles. In hydromedusae they are in the base of the velum or on the subumbrellar side (often between tentacle bases) and in scyphomedusae in the rhopalia; rarely they may be displaced to the aboral pole (*Hydroctena* and other hydromedusae, Dawydoff, 1953). Their special cell type is the lithocyte, which has a rounded movable concretion or statolith in it, composed largely of calcium carbonate or sulphate plus organic material. These cells presumably give weight to a mass which hangs, bends, or otherwise places stress on the receptors. The latter are long, haired sensory nerve cells variously situated near the lithocytes. There may be, in hydrozoans, fewer than 10 lithocytes and a similar number of sense cells, especially in the closed vesicle type (Fig. 8.14), or the lithocytes may be in regular rows of 10 to 20 or in irregular masses, numbering up to hundreds, with several sense cells for each lithocyte.

Lithostyles, sense clubs, tentaculocysts, and cordyli are all essentially similar, though ranging widely in size and complexity (Fig. 8.18). In some Leptomedusae and many Narcomedusae, often small and simple hydrozoans, they supplement statocysts and therefore suggest a different kind of sensibility. In the scyphomedusae, often large and relatively advanced, they are the prin-

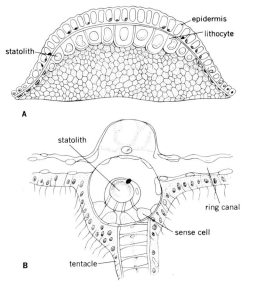

Figure 8.18. Statocysts. **A.** The open type of statocyst as seen in *Mitrocoma*, in optical section. **B.** The closed type as seen in *Obelia*. [Hertwig and Hertwig, 1878.]

cipal static organs and compose most of the rhopalium. Experimental evidences of their function in orientation has been obtained by Fränkel (1925) and Bozler (1926). In these complex forms there are several circumscribed areas of sensory epithelium in a basal cushion and around the sides and base of the large mass of polygonal gastrodermal cells, each with its statolith. A distinct layer of nerve fibers with ganglion cells underlies some of these areas. Modern functional-anatomical study is needed of the relations between these concentrations and the rhopalial ganglion, the location of the pacemakers, the connections with the separate nerve nets, and the afferents from the ocelli in some fully formed adult scyphozoan; this study deserves the effort because here is the most ganglionic, centralized, and differentiated part of the nervous system of coelenterates. Some forms, such as the sessile *Haliclystus*, have lost all special sensory structures on the marginal rhopalioid bodies.

Other sense organs are less clear. Pits or simple patches of specialized epithelium with abundant nerve cells are known in both hydro- and scyphomedusae. The term olfactory pit has been used because of some resemblance to structures so designated in higher invertebrates. Tactile combs are small clusters of cells with very long hair believed to be sensory.

Epitheliomuscular cells are repeatedly described in coelenterates (Hess et al., 1957) but their actual abundance is not yet understood. Grimstone et al. (1958) imply that every muscle fiber is part of such a cell in the mesentery of *Metridium*. From the point of view of this book the question of interest is whether the epithelial portion acts as a receptor or is a precursor of sense cells. At present no answer can be given. Batham, Pantin, and Robson (1960) point out that the nerve net forms a stratum at the subepithelial level where the nerve fibers must cross a forest of radial strands connecting muscle fibers and their epithelial portions, suggesting the possibility of neuromotor junctions here.

III. PHYSIOLOGY

A. Properties of the Nervous Elements

Abundant evidence testifies that the nerve fibers are basically similar to those of higher animals in essential neurological properties. **Conduction in the axon** is nondecremental, all-or-none, and followed by a refractory period. Action potentials have been recorded from several medusae and hydroids (Horridge, 1954b; Passano, 1958; Yamashita, 1957; Josephson, 1961c; see also Fig. 8.19,

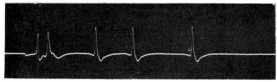

A. *Cordylophora*: Mechanical stimulation

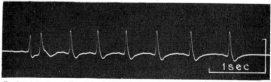

B. *Cordylophora*: Electrical stimulation

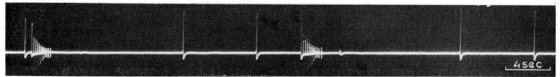

C. *Tubularia*: Spontaneous discharge

Figure 8.19. Electrical discharges recorded from hydroids. Metal macroelectrodes pick up activity in the stalk below the hydranth. **A.** A mechanical prod elsewhere on the colony may cause a burst of activity. **B.** An electric shock elsewhere on the colony may do the same. **C.** *Tubularia* has a complex spontaneous rhythm. Evidence from varying stimulus intensity indicates the single events are all-or-none; they are perhaps nerve action potentials in many parallel fibers of an elongated nerve net. The form and height variations may possibly be due to differences in synchrony or number of active fibers in the net. [Josephson, 1962.]

Table 8.1. VELOCITIES OF PROPAGATION OF EXCITATION IN SOME COELENTERATES. (These are reported figures of over-all apparent velocity and include any junctional delays. It should be realized that various factors influence velocity; in particular, it increases with facilitation and decreases with fatigue; effects of stretch are uncertain.)

Genus	Pathway	Velocity (m/sec)	Temp. (°C)	Ref.
Hydrozoa				
Cordylophora	Electric potential in stolon	0.027	22	1
Tubularia	Electric potential in stem	0.16	19	2
Pennaria	Colonial, local, graded	0.01	26	1
Hydractinia	Colonial, local, graded	0.025	15	1
Aequorea	Radial conduction system	0.002–0.02	20	3
Aequorea	Swimming beat system	0.90	20	3
Physalia	Fishing filament (stretched ?)	0.12	26	4
Scyphozoa				
Haliclystus	Through-conduction, calyx	0.07–0.15	12	5
Pelagia	Radial, centrifugal, subumbrella	0.01		6
Pelagia	Swimming beat	0.24		7
Cotylorhiza	Swimming beat interval 5 sec	0.62		8
Cotylorhiza	Swimming beat interval 3.4 sec	0.71		8
Cotylorhiza	Swimming beat interval 3.2 sec	0.83		8
Aurelia	Swimming beat	0.50		9
Aurelia	Tentacle contraction wave	0.25		9
Cassiopea	Swimming beat: fast action potential	0.45	28	10
Cassiopea	Swimming beat: circuit wave	0.77	29	11
Cassiopea	Slow through-conducting net in subumbrella	0.18	28	10
Cassiopea	Slow wave of contraction after marginal stimulation	0.08	25	12
Cyanea	Spontaneous waves crossing muscle	0.02		12
Anthozoa				
Calliactis	Pedal disc; circumferential	0.15	19	13
Calliactis	Column; vertical	0.10	19	13
Calliactis	Column; radial	0.04	19	13
Calliactis	Mesentery; vertical	1.20	19	13
Calliactis	Oral disc; margin, circular	1.00	19	13
Calliactis	Oral disc; radial	0.60	19	13
Cerianthus	Column through-conduction	1.3	22	14
Pennatula	Luminescent wave, colonial	0.05	15	15
Leioptilus	Luminescent wave, colonial	0.26	20.5	16
Cavernularia	Luminescent wave, colonial			
	(weak glow, from any part)	0.10		17
	(strong glow, from peduncle)	0.075		17
Renilla	Luminescent wave, colonial	0.04–0.08	21	18
Tubipora	Polyp retraction, colonial	0.15–0.20	20	19
Plexaurid gorgonian	Colonial, through-conducting	0.40	29	20
Fungia	Oral disc, through-conducting	0.4–0.5		19
Acropora	Colonial, through-conducting	0.02		19

(1) Josephson, 1961; (2) Josephson, 1962; (3) Horridge, 1955c; (4) Parker, 1932; (5) Gwilliam, 1960; (6) Bozler, 1926b; (7) Horstmann, 1934; (8) Bethe, 1935; (9) Romanes, 1878; (10) Passano, personal communication; (11) Harvey, 1912; (12) Horridge, 1956b; (13) Pantin, 1935; (14) Horridge, 1958; (15) Panceri, 1872; (16) Davenport and Nicol, 1956; (17) Honjo, 1944; (18) Parker, 1920; Nicol, 1955; Buck, 1954; (19) Horridge, 1957; (20) original.

and p. 494). Potentials recorded from single units are spikelike, of the order of 10 msec in duration at habitat temperature. It is argued with some force that these quantal events account for the nervous conduction of excitation in these animals. In *Aurelia* and *Cassiopea* there are two kinds of action potentials with different form and properties. Chronaxies of the mesenteric nerve fibers of an anemone have been recorded as 2–4 msec and absolute refractoriness apparently lasts 20–200 msec in several preparations (anemone mesentery, Pantin, 1935a; medusa swimming net, Bullock, 1943; *Renilla* colonial net, Nicol, 1955; alcyonarian colonial through-conduction, Horridge, 1956c). Velocity of conduction is given for a number of cases in Table 8.1; note that some values presumably include synapses. Mayer (1914) used the trapped circus wave preparation to test various agents on velocity, claiming a direct proportionality between this measure and the concentration of sodium plus potassium plus calcium—whether he diluted sea water with isotonic sugar or $MgCl_2$ or distilled water! M. A. Biederman and I (unpublished) could not find this effect in *Rhizostoma* and question its reality. We did note that although velocity was sometimes constant within a few percent during many measurements, at other times it varied sporadically by as much as 20–30%; the variation could be assigned to the conduction system rather than to neuromuscular delay and was not a function of the interval since the last response or of stimulation strength. Whether synaptic delays, axonal conductions, alternate routes, or slow net influence is involved remain open questions.

Evidence of **synaptic transmission** comes from the anatomy, treated above (p. 472), the cases of interneural facilitation, treated below (p. 487), and in special places the existence of irreciprocal transmission (p. 499), as well as the presumption created by finding nondecremental conduction in the axon. In view of the diffuse spread in a nerve net, in any direction, interaction between neurons can be said to be reciprocal. The argument for the legitimacy of unpolarized junctions is summarized on p. 464.

Delay at junctions is poorly known and values cannot be given for long delays. There is some evidence that delay can be very short. In the mesenteric net of *Calliactis*, a fast, through-conducting system, Pantin estimated a minimum of 20 synapses in the most direct route vertically through a 10 cm stretch of a silver preparation; possibly the number is considerably higher (Fig. 8.5). Since the over-all time of conduction of excitation across this distance is about 85 msec, an apparent velocity of 1–2 m/sec is achieved; this is a very high value for coelenterates—and indeed all lower metazoans—and equals that in some polychaete giant fibers (see Table 3.4, p. 150). If we are not to assign an even higher value to the true fiber velocity, there is no time left for 20 synaptic delays. This evidence then permits the speculation that in these junctions we may have already the electrically transmitting, low-resistance synaptic membranes known in some crayfish and earthworm junctions, where there is no delay (see Chapter 4, p. 214).

When, for the first time, coelenterates were stimulated with single shocks in experiments by Pantin in 1935, a previously confused picture of the functional organization of the nervous system of these animals began rapidly to clarify. Mechanical stimuli evidently result in trains of impulses initiated by receptors and spreading through the nerve nets (Passano and Pantin, 1955). Since they vary in number and spacing, they give responses which are difficult to analyze. Electrical stimulation by single shocks, which probably give rise to individual impulses in the conducting system, shows that response is **independent of the strength** but entirely **dependent on the frequency** and number of individual shocks. A single shock of any strength delivered anywhere on an anemone (for example *Calliactis*, *Metridium*) usually elicits no visible response. If another shock is delivered sufficiently soon, there is a small response. Successive stimuli, spaced neither too far apart nor too close together, elicit successive increments of contraction of increasing amplitude until a plateau is reached. The corresponding increments are higher, the higher the frequency (Fig. 8.20). Some invisible process evidently grows rapidly and declines slowly after each impulse, and has the effect of enhancing the response to the next. It makes no

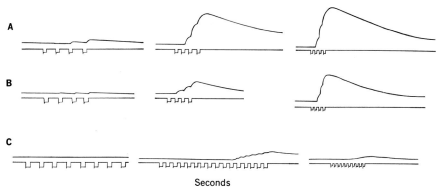

Figure 8.20. Mechanograms of contraction in the sphincter muscle of the sea anemone *Calliactis*, showing facilitation and its dependence on interval between stimuli and on certain ions. **A.** In sea water; stimuli every 2 sec, every 1 sec, and every 0.5 sec. **B.** The same after 28 min in 50% 0.4M MgCl$_2$ + 50% sea water; facilitation is depressed. **C.** The same after 55 min; depression is much greater. Replacing the muscle in sea water caused good recovery. [Ross and Pantin, 1940.]

difference whether successive stimuli to the column of *Metridium*, at moderate to high frequencies (0.2–3 per second) are delivered to the same or to different points, far apart; the conducting system into which such impulses enter is through-conducting and assures symmetrical contraction. Following Sherringtonian usage, Pantin called this process **facilitation.** It is important to distinguish it from summation, which occurs at the same time in anemones due to the slow relaxation and which simply means that each response adds to a persisting fraction of the preceding response. Facilitation is said to occur when the successive increments, or whole contractions in the case of rapidly relaxing, nonsummating responses of medusae, are larger (see further, Chapter 5). Facilitation can occur at the neuromuscular junction and it can occur between elements of a nerve net, at neuro-neural junctions.

A special kind of facilitation, probably in the muscle, is **heterofacilitation;** this is shown by enhanced contraction of swimming muscles in medusae (*Cassiopea*, Passano et al., 1961), when impulses in the slow nerve net have primed the fast response. Contraction follows an impulse directly only in the fast nerve net and in *Cassiopea* none is seen following slow net impulses alone. Facilitation has also been used by Passano to describe a small increase in velocity of propagation of impulses in the slow nerve net of medusae upon repetition, a phenomenon seen also in higher animals (Bullock, 1951).

Neuromuscular facilitation has been examined in some detail, particularly in actinians, by Pantin and Ross and collaborators. With rising temperatures, facilitation decays more rapidly and hence becomes less conspicuous at any given frequency of stimulation (Fig. 8.21). According to the interpretation of Ross (1955), facilitation develops more rapidly as well, until at 25° to 30°C in *Calliactis* (from Plymouth) it develops fast enough during the excitation time of the muscle that even a single shock or the first of a series now produces a contraction, larger with higher temperatures but at best very small. Thus we have a transition in the same animal between nonfacilitating transmission, which is effective beginning with the first stimulus, and the highly facilitating transmission, which exhibits no response to the first shock and which is characteristic of these animals under their normal conditions. The **effects of ions and drugs** on facilitation have been the subject of several papers (Ross and Pantin, 1940; Ross, 1952, 1955, 1960a, 1960b).

Varying the frequency of stimulation over a wide range (a) causes different types of contraction in the same muscle and (b) brings in different muscles. The first has been shown by Ross (1957b) in the isolated sphincter of *Calliactis*. In the range 0.2–3 per second the result is a series of quick facilitating steplike responses, beginning with the

III. PHYSIOLOGY   A. Properties of Elements

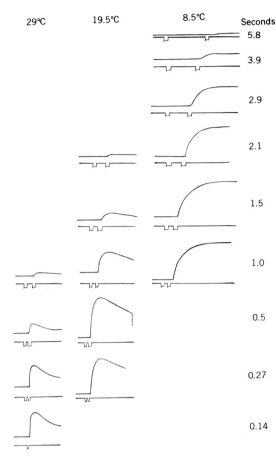

Figure 8.21. Effect of interval and temperature on facilitation in the anemone *Bunodactis*. Mechanograms of contraction to pairs of electrical stimuli at different intervals and temperatures. [Pantin and Vianna Dias, 1952.]

second shock. At one shock per 2–15 sec, a slow smooth contraction occurs with a minimum of several shocks (shortest latency is at 10 shocks in 70 sec at 18°C). It is possible that double innervation is involved as well as separate conducting systems. Ross believes that slow rhythmical movements are the basic primitive form of neuromuscular activity in these animals. Isolated strips or rings of some muscles are spontaneously active; rings of others are not (Ross, 1957b; Batham and Pantin, 1954; Hoyle, 1960).

Different muscles are brought in at different frequencies, all at the same threshold intensity and all symmetrically, in response to electrical stimulation of the column. The mechanism by which a variety of possible responses—of the circular muscles, the parietal longitudinals, the mesenteric longitudinals, and the sphincter—can be evoked is thus elucidated on the basis of different overlapping ranges of facilitating frequencies (Fig. 8.22). In one case, a shock every 10 sec called forth circular muscle contraction after many shocks; one every 3 sec was followed by slow parietal shortening; one every 2 sec, by more parietal contraction, but even earlier by a modest mesenteric longitudinal response; one shock every 1.5 sec gave an enormous mesenteric contraction and a small, slow sphincter action; at one shock per 0.6 sec the sphincter predominated and only a small mesenteric longitudinal contribution followed it. In the same order can be arranged the interval threshold, the threshold number of stimuli, and the time to full contraction (many minutes for circular, 1–2 min for parietals, 0.5–1 min for mesenteric, and some seconds for sphincter muscle). All muscular responses contract the anemone; expanding is accomplished by fluid pressure created by ciliary inflow (Chapman, 1949).

Inhibition—in the proper sense of a positive event or stimulation acting to suppress response—is clearly developed in coelenterate neuromuscular systems (Horridge 1955a, 1955c, 1956b;

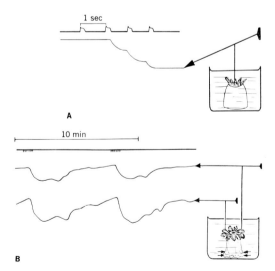

Figure 8.22. Contrast in time scale of adequate stimuli to different muscles. **A.** A moderate response is given by the retractor muscle of *Metridium* to 4 shocks, at 1 per second. **B.** A moderate response is given by the parietal muscles of the same animal to 10 shocks at 1 per 5 sec. [Redrawn from Batham and Pantin, 1954.]

Ewer, 1960; Moore, 1926a). Stimulation of a pedal disc ring preparation of the anemone *Calliactis* at the onset of a spontaneous contraction prevents the further development of this contraction; relaxation is hastened if stimulation comes just after the peak of contraction. The very long latent periods characteristic of pedal ring preparations are interpreted as interactions of excitation and inhibition. In *Aequorea* and other hydromedusae the spontaneous nervously paced beat is inhibited, as is also the rapid transmission of this wave around the margin by nervous excitation conducted in the separate radial conduction system; presumably the inhibition takes place in the marginal nerve rings where the two systems come into functional contact. The action is one-way; the slow system acts on the fast but the reverse action does not occur. A similar modulation of fast nerve net activity by the slow net appears to occur in the scyphomedusae, *Cyanea* and *Cassiopea*, perhaps localized at the neuromuscular junction.

Sensory adaptation to mechanical stimuli is pronounced, so that at high intensities, when high frequencies of impulses are sent into the net, only a few impulses occur. If high frequencies are associated with few impulses and low frequencies with protracted discharge, the selection of different groups of responding muscles will be aided, since they vary widely in their neuromuscular demands in just these respects. We can see that the formation of this impulse group or train represents perhaps the **first nervous integration in evolution**. It presumably occurs at the sense cell-nerve net junction, and in the anemone through-conducting net the train of impulses enters a single motor unit of symmetrical conducting system and four muscles. The mechanism of this integration is the more interesting because (a) it is the decisive physiological process determining several normal behavioral alternatives, and (b) it is peripheral and reduplicated all over the column and pedal disc. Evidence given below places it in the gastrodermis and strengthens the supposition that the sense cells, by graded, summating responses, trigger impulses which actually are initiated in the through-conducting net.

Evidence of **chemicals affecting neural activity** is meager. (Bacq, 1941, 1947; Bullock and Nachmansohn, 1942; Florey, 1951; Ross, 1957a, 1960a, 1960b; Mathias et al., 1957). Acetylcholine is said to be absent and when applied ineffective. Cholinesterase is said to be present in *Tubularia* but not in *Cyanea;* drugs which usually affect this enzyme have no action, according to modern accounts. In contrast to their effects in higher animals, tryptamine is more effective than 5-hydroxytryptamine; it is excitatory. The latter substance is present in high level (500 γ/g of freeze-dried tissue) in coelenteric tissues of *Calliactis* and low in its tentacles; it is not detectable at all in *Metridium* or *Actinia*. More significantly, it is not present in jellyfish ganglia, free from tentacles (unpublished).

B. Differentiation of Nerve Net Systems in Polyps

Possibly *Hydra* has a nearly uniformly developed net (but see p. 506). Most other forms studied, including minute colonial hydroids, floating chondrophorans, anemones, and corals show evidence of differentiation of the nervous system at least regionally, with usually two (sometimes three) contrasting types of nerve net, largely coexisting in the same region of the body. These differ in anatomy as already described (p. 467). One is commonly faster in conducting, recalling the frequent occurrence in higher phyla of parallel fast and slow neuromuscular systems. The most conspicuous difference in physiology in the present animals, however, is that one is through-conducting for a specific reaction system and the other mediates variable, local movements.

1. Fast specific nerve nets

These propagate a single impulse throughout the extent of the system relatively rapidly and control some characteristic muscular response adaptive for symmetrical, general, and also rapid contraction, such as jellyfish swimming pulsations or polyp withdrawal. Most of the preceding section on facilitation is relevant to the functioning of such a system in the behavior of anemones. Further examples are given in the systematic account. It should be realized at this point that among the unanswered questions on the nervous basis of observed responses in these animals, some apply to this simplest part of the system. At present there is no satisfactory explanation of the sharp distinction between quite fast and very slow contractions in the same neuromuscular system. Further, Batham and Pantin (1954) found reciprocal inhibition in actinians but no closer study of its basis has yet been made (but see p. 499 for inhibition in medusae).

2. Slow diffused nerve nets

The fast through-conducting system cannot bring into action all the musculature. In *Metridium*, for example, radial muscles are not excited by the mesenteric-retractor system, but only by adequate stimuli that arouse a separate system.

Stimulation of the margin of the oral disc in *Calliactis* causes a localized edge-raising, which extends farther around the margin with successive shocks. Each shock after the first brings in a definite segment and apparently also invisibly facilitates for a time, at the barrier to conduction into the next region. These barriers are reasonably interpreted as synapses in the nerve net and this facilitation is taken to be interneural. Circumscribed responses which extend farther with stronger mechanical stimuli are not now explained by assuming that decremental conduction in nerve nets is a property different from conduction in higher nerve fibers; they are regarded as a function of junctions requiring facilitation and of the number and frequency of impulses initiated in the nerve net by the receptors.

These labile systems for propagating excitation are of fundamental neurological interest as **primitive integrative mechanisms**—meaning that their output is different from, but some function of, their input (see Chapter 5). They are temporospatial converters, integrating an input—whose main parameters are temporal—to determine an output—whose main parameter is anatomical spread and sometimes also decrement of response with distance. The current explanations of spread with decrement represent perhaps the most obvious departure of recent thinking about nerve nets from the classical. We are indebted to Horridge (1957) and Josephson (1961c) for the bases of a new view on this significant problem.

Working on a variety of species of Red Sea corals (see below, p. 515), Horridge found **five types of spread** of the polyp retraction response over the colony. (a) There may be immediate through-conduction. (b) Some species can not be made to respond more than locally, no matter how strong or numerous the stimuli. (c) In the coral *Palythoa*, successive shocks produce excitation which spreads progressively farther across the colony in equal increments of radius at each shock, for as many as fifty shocks at two-second intervals. (d) Species like *Porites* respond over an extensive area to the first shock and then spread the excitation very little more with successive shocks (Fig. 8.23). (e) Conversely, *Sarcophyton*

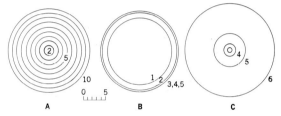

Figure 8.23. Diagrammatic representation of types of spread of excitation across coral colonies. A succession of electric shocks is given in the center; the *circles* show the distance of spread of response after each shock. The serial number of the shock is printed beside some of the circles. **A.** *Palythoa* shows uniform increments of radius of spread. **B.** *Porites* shows decreasing increments, soon reaching a ceiling less than the size of large colonies. **C.** *Sarcophyton* appears to show increasing increments of radius, soon involving the whole of a large colony. Apart from through-conduction systems, these are the principal types of spread so far observed. The scale is in units corresponding to the suggested size of the conducting elements [Horridge, 1957.]

responds with slowly incrementing area to the first several shocks and then with much larger increments until the whole colony is involved. The last three cases are the significant ones for the present discussion.

Pantin provided the first satisfactory explanation of increments of spread with repeated stimuli, in the terms given in the first paragraph above. The model assumed is a meshwork of units connected by junctions, each of which is in one of two states: ready to transmit or not ready to transmit. The "not ready" state is converted into the "ready" by facilitation induced by impulses arriving at the junction, the so-called **interneural facilitation**. This model fits the observations in the edge-raising response of *Calliactis* and in the coral *Palythoa*. The units between junctions must be coextensive with the increments of spread, in a first approximation.

But this model is inadequate to explain the last two types of spread, (d) and (e). Although a diffuse, somewhat randomly arranged network of neurons in one plane is perhaps the simplest assemblage of interacting neurons that can be imagined, activity in such a system could involve limitations that are by no means obvious. Given the common properties of neurons that are demonstrated for coelenterates, a certain degree of interconnection (average number of other neurons making synaptic contact with a given neuron, plus distribution about the average), and a certain percentage of junctions ready to transmit, it is neither obvious nor analytically feasible to predict spread as a function of the number of neurons stimulated or of frequency of stimuli. Simulation by working models is necessary and, because of the assumption of random distribution of transmitting junctions and of neurons with different numbers of synapses, repeated simulation runs are needed to assess the variability. In order to interpret experiments on living systems believed to possess such networks, we must know **what properties can be expected** from them as presently conceived.

Horridge (1957) studied the behavior of two models, one mechanical, the other mathematical. He rejected the former because repeated experiments "stimulating" at one locus produced spread too variable in extent. Actually his discussion serves the useful purpose of focusing attention on **variability as a property** worth observing, since it was found to be systematically greater as the percentage of transmitting junctions increased. The **mathematical model** introduced the assumption that many units are initially stimulated simultaneously: it thereby attempted to reduce the variability. In Horridge's formulation a certain proportion ("density," d_0) of the units in a given area is excited, according to stimulus strength. This value can be multiplied by the probability, P, that an impulse in any unit will succeed in transmitting into a neighboring unit; the product, d_1, will be the new density (actually, number) after impulses of the first generation have all crossed the first set of junctions. In the non-through-conducting cases $P < 1.0$ and therefore $d_1 < d_0$; the number of active units falls exponentially as junctions are crossed, or $d_n = d_0(P)^n$. The probability, P, expresses the labile, graded, and integrative properties of the nerve cell and can be assumed to change with successive stimuli and spatial summation, that is, with density of active units. This adds a new property to the model. If these effects are positive, then $d_1 = d_0 (P + d_0)$ and the shape of the curves of number of active units with successive junctions crossed will change. At a critical initial (stimulated) number the probability will rapidly rise to 1 and through-conduction will supervene.

Repetitive stimulation was supposed to activate an increasing number of units at each stimulus, a proposition for which there is little evidence. Interneural facilitation is hardly recog-

---

Figure 8.24. Experiments with a model of the nerve net. A large digital computer creates simulated networks of elements like bipolar neurons, distributing at random the specified lengths of neurons and types of connections (with respect to requirement for facilitation), each at a specified abundance. **A.** Diagram of a portion of one of these artificial nets. Each straight line simulates a neuron; its length is defined by the number of crossings of other neurons. Every crossing represents a synapse and may be ready to transmit or may require facilitation of a chosen amount; facilitation dies away after arrival of an impulse with one of several time constants. If an impulse crosses a junction it promptly occupies the whole "neuron" and confronts the new set of junctions. **B.** Example of part of the output of the computer after a "stimulus" has been given to a whole region of the net, defined by 16 junctions designated x. The excitation promptly spread throughout the elements stimulated, across the junctions which were ready to transmit—in this case 25% of the junctions—and on until blocked by junctions requiring facilitation. All junctions reached and

nized in the model, as developed, though evidence of its reality is rather compelling. A **limitation of the model** is that knowing the number of active units after some junctions have been crossed is not equivalent to knowing the density or the distance to which excitation has spread, since the percentage of units active in a subarea may fall in an unspecified manner. Also, the equations give maximum likelihood estimates for a series of trials—with each trial being made up of a number of individual events—and demand independence between the individual events forming a single

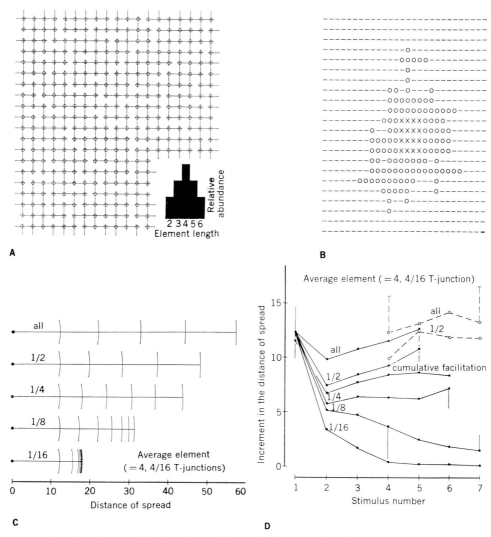

excited, the computer typed o; the remaining junctions, still unexcited, are typed –. **C.** Spread following repetitive stimulation in nets with average element lengths of 4 and with 25% of the junctions ready to transmit. Each point is the average of 15 different nets. The *fractions* refer to the proportion of junctions which would, if reached by an impulse, remain facilitated until the arrival of the next impulse at the frequency of stimulation used. Facilitation was unidirectional and the properties of the junctions at each crossing were independently determined. This figure shows the total distance of spread following successive stimuli, by *arcs* of concentric circles. The unit of distance is the length between adjacent crossings. Note that a limit is soon reached (*Porites*-type spread) when adequate facilitation is scarce. **D.** Plot of results like the preceding. *Open circles* joined by *dashed lines* were determinations made by stimulating in the corners of nets to give more radius for spread. *Vertical bars* give standard deviations for representative points, sometimes shown on one side only. [Josephson et al., 1961.]

trial. In actual nerve nets, when the density of active units is high, occlusion of available pathways probably becomes important.

The **digital simulation model** of Josephson, Reiss, and Worthy (1961) was designed to permit many trials on a large-capacity computer. It incorporates the main features of Horridge's: the initial mixed population of junctions ready to transmit and junctions not transmissive at the moment, and the initial stimulation of a variable number of elements. It adopts a topology, based on the bipolar cell nets, of straight line segments (neurons) making synaptic contact where they cross, but distorts the network deliberately by making it rectilinear (Fig. 8.24). Distance between junctions is thus constant and becomes the unit of length. A frequency of elements of different lengths is chosen, as is the percentage of junctions assumed to be transmissive; the computer then constructs a particular net by distributing such elements at random in a plane. All-or-none activity, uniform conduction velocity and junctional delay, and relatively long refractory period are assumed. Nontransmissive junctions are facilitated, that is, become transmissive for a predetermined period of time. At the choice of the experimenter, crossings can be treated as unpolarized synapses, facilitating in both directions at once, or as double, reciprocally polarized synapses with independent facilitation.

The results of repeated tests with this model show what can be expected, although a given net may deviate by chance. Disregarding for the moment the possibility of repetitive firing to single stimuli, nets which spread excitation to a considerable distance will give quite variable distances: factors leading to greater spread, such as a larger proportion of transmissive junctions or a larger number of junctions per "neuron," lead also to greater variability. Increasing the "strength" of stimulus (number of elements stimulated) increases the spread only within a limited range, soon reaching a limit as occlusion of possible pathways becomes important. Spread of excitation with repeated stimulation can occur in equal increments of distance for each stimulus or, with certain permutations of the chosen properties, in decreasing increments. Spread in increasing increments, as in *Sarcophyton*, requires postulates not in the model, such as repetitive firing to single shocks or recruitment of a new, through-conducting net. The increments of spread can be considerably greater than the average length of a "neuron." The model permits more realism by adding complexity; it might contribute to the design of laboratory experiments— selecting for example between one-way and two-way facilitation or assessing the possible role of subthreshold summation and electrotonic spread. Repetitive firing to single shocks can completely alter the properties of the model. Such firing has been found to occur in some animals (p. 511); but most corals and other coelenterates respond to brief shocks as though with a single impulse.

3. Responses to mechanical stimulation

The responses occurring after mechanical stimulation have been carefully studied by Passano and Pantin (1955) in *Calliactis*. By carefully controlling the mechanical stimulator it was possible to obtain liminal stimuli which appeared to produce the same effect as a single electrical shock. Stronger mechanical stimulation produced more and more impulses in the nerve net, evidenced by discrete increments of contraction at increasing frequency, and hence increasing facilitation and contraction. Normal mechanical stimuli therefore probably elicit in the nerve net a train of impulses whose frequency and duration largely determine the character of the response. Passano and Pantin conclude that there are grave difficulties in assuming that impulses arise in the receptors themselves and favor the possibility that the receptors by a graded response initiate impulses in the nerve net. There is clear summation of subliminal mechanical stimuli. Interestingly enough, the excitable elements in the column wall are tangential and hence are sensitive to stretch, not pressure. The receptors are probably purely endodermal and thus stimulated indirectly, through the body wall.

Much of the determination of behavior in these animals can be ascribed to the **distribution of sensitivities of receptors** and their connections with the nerve net, just as in higher animals. Thus it requires strong stimulation of the column to elicit the closure reflex, whereas weak stimulation (4,000 times lower threshold) of the oral disc elicits movements of feeding or rejection. These differences in response may only sometimes be ascribed to differences in facilitation threshold of different muscles to activity in a common nerve net. It seems likely that there are in fact **several distinct conducting systems** which may be spoken of as separate nerve nets, such that impulses traveling in one may be unable, or only with difficulty, to get into another. This conclusion is supported by work described below on medusae but it cannot be readily adopted for any given case without additional evidence to support it against the alternative of **frequency discrimination by effectors** of impulses in a common conducting system. Thus von Uexküll's (1921) report that the tentacles of *Anemonia* respond to certain chemical stimuli by contraction of the circular muscles and to mechanical stimuli by contraction of longitudinal muscles may mean that there are separate receptors, each connected to a distinct nerve net distributed to the effectors differently, or we may have the same receptors responding with a different frequency or duration of discharge into the common nerve net, the effectors discriminating by their respective facilitation thresholds.

Josephson (1961a) discovered a high **sensitivity to minute disturbances in the water** on the part of a small colonial hydroid, *Syncoryne* (Fig. 8.25). A copepod brought near a polyp causes no response until the copepod moves but this immediately releases an accurate bending and grasping by the polyp. A glass prod near the polyp, actuated by a controlled electromechanical transducer, releases the same response following a single-step displacement, forward or backward, of only 2 μ! The response shows attenuation with distance, and Josephson explains the direction of bending on the basis of differential stimulation of the two sides of the animal by the mechanical event.

4. Responses to chemical and other stimuli

The chemical stimuli normally encountered **never activate the through-conducting system,** nor evoke a symmetrical response, unless the stimulus is by chance symmetrical. From the results of the electrical analysis we must suppose that this means a separate conducting system with its own set of sense cells. The necessary condition seems to be sufficiently prolonged excitation. Adaptation ordinarily prevents mechanical stimuli from initiating a feeding response, although starved anemones have a lowered threshold, and a fully patterned feeding sequence may even occur spontaneously (see p. 504). Well-fed anemones commonly fail to respond positively to food even when so fed as to prevent direct sensory adaptation. A series of electrical stimuli can produce feeding or rejection according to whether it is prolonged or brief (Pantin and Pantin, 1943).

The **feeding response** is a highly adaptive and well-coordinated sequence of events. Normally the nematocysts discharge first; this is usually regarded as completely independent of the nervous system but some evidence suggests that it depends on the physiological state of that system in *Hydra* (Glumac, 1953). As independent effectors, nematocysts require highly specific chemical priming and mechanical triggering (Pantin, 1942). The discharge plays a role in subsequent neuromuscular feeding movements. As first shown in *Hydra* (Loomis, 1955) and later confirmed in marine species (*Physalia*, *Aurelia* scyphistomae, and *Campanularia*, Lenhoff and Schneiderman, 1959), the muscular phases of feeding are extraordinarily sensitive to glutathione (10^{-4}–10^{-8} M) and this is evidently released from prey tissues by the penetrant nematocyst threads. It seems possible that it acts on chemoreceptor sense cells to initiate the feeding movements reflexly. The effect is especially pronounced upon muscles of the mouth,

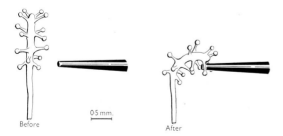

Figure 8.25. The response of a hydroid *(Syncoryne)* when the manipulator holding the glass rod is lightly tapped. [Josephson, 1961.]

which expands to incredible dimensions and applies itself to the food—or to the glass bottom of a dish—as an enormous disc. Pantin and Pantin (1943) showed that *Anemonia* can give a feeding response both to non-diffusible, water-insoluble substances and to amino acids, so this mechanism is not a necessary one in all species. Passano reports no response to glutathione in *Cyanea* or *Cassiopea*.

A rather **complex sequence of movements** may be involved, especially in anthozoans. Movements of the tentacles, the margin of the oral disc, the disc itself, the mouth, and sometimes the pharynx but rarely the column, occur in appropriate degree and timing. Much of this can be shown to be due to conduction of excitation from the stimulated tentacles but direct stimulation of other tentacles and lips acts also in regulating the intensity and timing. **Reversal of ciliary beat** is found in many species and is presumed to be under nervous control. Even in isolated pieces of the oral disc of *Metridium*, food juice causes a reversal of the direction of both the cilia of the ridges and the grooves, which beat in opposite senses (Parker and Marks, 1928).

The main feature of these responses is their **local nature.** They can be less extensive or more extensive, but unless the food directly stimulates tentacles—or, in colonial forms, polyps which are far apart—the spread is always quite limited. This is really an essential condition for the local depression of the oral disc and the bending of the lips toward the food. It has not been analyzed as carefully as the non-through-conducting systems treated above (p. 490) and below under medusae (p. 495) but it seems likely that interneural facilitation is a primary phenomenon.

Properties of the photic sensibility of *Metridium* were analyzed by North (1957), who concludes that the high variability in reaction time is perhaps due to the uncertainty of quantum capture caused by low concentration of the photosensitive pigment; on this assumption, calculations suggest that anemones respond to less than 10 quanta of absorbed light and can integrate low intensities over a minute or more of latent period. No photoreceptor structures have been found in this animal.

5. Regional specialization

The conducting system is physiologically different in the several parts of anemones. A vertical strip of the column conducts rapidly if it includes some mesenteries, but if it does not, conduction takes place slowly. The mesenteries contain high-speed pathways with a **preferential orientation** in the vertical axis.

A **favored polarity** of conduction was shown long ago in the tentacles of anemones (see Grošelj, 1909; Rand, 1909; Chester, 1912; Parker, 1917c; von Uexküll, 1921; Pantin, 1935c). Stimulating a tentacle locally in the middle of its length produces a contraction of its longitudinal muscles which is stronger on the proximal than on the distal side; cutting across a tentacle results in closure of the wound of the central stump only. Certain chemical stimuli cause contraction of the circular muscle distally; mechanical stimuli can do the same if long maintained or elicit only local longitudinal shortening. The tentacles can be brought to response by stimuli to the column wall, and even the first shock causes some of them to shorten slightly, but, in the opposite direction it is only with difficulty that tentacle stimulation spreads to the through-conduction system. After 10 to 20 shocks the sphincter of the oral disc may contract in *Calliactis*. Since this represents spread through junctions between parts of the nervous system, Pantin speaks of the centrifugal polarity of interneural facilitation in this case. Polarity is also commonly evident in the oral disc: in some species mouth stimulation readily conducts radially outwards, whereas edge stimulation requires a long period of facilitation before spread to the mouth will take place. In other species, for example fungid corals, the opposite is true. It seems clear that at least a large part of the explanation of these favored polarities lies in (a) the sensitivity, direction of main process, and connections of the sense cells, and (b) the existence of separate conducting systems with connections between them which are not fully reciprocal.

Differentiation of velocity of conduction in different regions and directions is pronounced. Figure 8.26 shows this for a species of anemone.

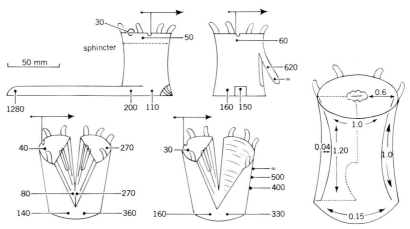

Figure 8.26. Latent periods of contraction of the sphincter muscle in *Calliactis* after adequate stimulation to various points in operated animals. Figures are milliseconds; temperature 18–20°C. At the right a summary scheme of the velocities of conduction in different directions estimated in this way; figures in m/sec. [Pantin, 1935.]

We have then, differentiation of favored axis, of preferred polarity, and of velocity of conduction, differentiation between through-conducting and non-through-conducting regions, and the occurrence in many places of more than one overlying net—besides differences in requirements of neuromuscular junctions.

C. Differentiation of Nerve Net Systems in Medusae

The most conspicuous activity of free medusae is a **rhythmic pulsation** of the whole bell, serving as a locomotor movement (except in Stauromedusae, which see, p. 519). Based on independent experiments of Eimer and Romanes, published nearly simultaneously in 1877, we know that this is a true nervously paced rhythm, originating in any one of the marginal bodies of scyphomedusae or in any part of the margin of hydromedusae. Ablation of all but one of the marginal bodies in scyphomedusae or the whole margin but for one small piece in hydromedusae still permits entirely normal swimming beats, though the frequency is lowered. Removing the last marginal body or piece of margin stops the rhythm—with but rare single beats appearing in scyphozoan medusae. The pacemakers are not confined to certain radii in the nerve rings of hydromedusae but there is less tendency to spontaneity in the rest of the nervous system. It is clear that there is a great **redundancy of pacemakers.** They are all accessible to—or rather, are invaded by—impulses in the swimming beat conducting system. Therefore they fire and reset each other, and after each beat all begin anew the process leading to discharge. Whichever reaches that state first, paces the system, and as the various loci of spontaneity wax and wane in the rate of this process, the pace will be determined now by this one, now by that one, at the highest frequency and best rhythmicity available. The pacemakers are also modulated irreciprocally by other conducting systems, that is, those into which they do not fire. Light, among other external conditions, can also influence them—presumably by sense cell connections with the pacemakers. (See further pp. 499 and 519.) Other activities of jellyfish contrast with this symmetrical, synchronous contraction; they are the **slow and local movements** of tentacles, mouth parts, and even of the swimming muscles themselves.

1. Fast specific nerve nets

Figure 8.27 shows the results of controlled single-shock stimulation to a quiescent medusa bell. The rhythmic pulsation has been stopped by removing the pacemakers in the marginal bodies. There is a **through-conducting system** which brings the whole bell into symmetrical contraction

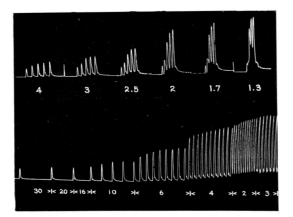

Figure 8.27. Facilitation in the jellyfish *Rhopilema*. Mechanogram of responses of a strip preparation without marginal bodies, stimulated at the intervals shown (seconds). [Bullock, 1943.]

nearly simultaneously. Contraction and relaxation are relatively rapid; the muscle is phasic in contrast to that of anemones. The movement is precisely like the normal swimming pulsation and we are permitted to believe that each of those pulsations can be caused by a single impulse spreading over the subumbrellar sheet.

This supposition has been directly confirmed by Horridge (1954b; see also Passano, 1958 et seq.), who succeeded in recording a typical **action potential** from a fiber of a large bipolar cell of the subumbrellar net of *Aurelia* under direct visual control, at the same time recording the movement resulting from it. A single impulse suffices to produce a contraction of the type of the swimming beat. A narrow bridge of tissue with a single axon can conduct a contraction wave in either direction and the wave is not conducted through the muscle (see also other evidence for this conclusion in Parker, 1919b, pp. 109-111).

In *Cassiopea* Passano found the **fast net spikes** to be composite and jagged, and therefore produced by a small number of imperfectly synchronized units, in contrast to the slow system described below (p. 497). The neuromuscular delay (fast spike to beginning of contraction) is very long—200-250 msec. Nearly half the swimming beats in *Cyanea* are associated not with single impulses but with a pair of impulses arising spontaneously from marginal ganglia; the interval between the two impulses does not fluctuate but is constant at about 200 msec. This is not the same as the interval, also uniform, between arrival of an exogenous fast spike at a ganglion and the fairly common initiation of an answering fast spike ("echo"); this interval is 225 msec. The mechanisms of both are obscure but Passano notes that the conduction velocity of the second spike is slowed and the pair arrive farther apart at the distant muscles than at near ones, giving an asymmetrical contraction. The fast net is capable of carrying two impulses as close together as 10-15 msec.

Successive responses are larger, at moderate frequencies, up to a plateau height; in other words, we have **facilitation.** The first stimulus already produces a small response. Facilitation decays slowly over many seconds (Fig. 8.28), the time course being comparable to that in

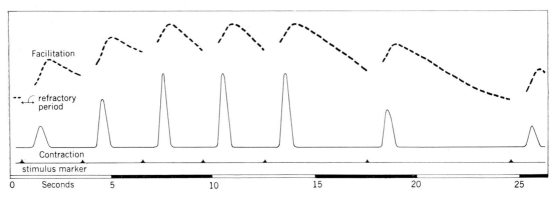

Figure 8.28. Hypothetical curve of the development and decay of facilitation in scyphomedusae. The curve is not continuous because we have no evidence of what is happening during the long absolute refractory period. [Bullock, 1943.]

anemones. Conduction velocity in this system is in the region of 0.5 m/sec, a reasonable value for a through-conducting system if synapses do not have any considerable delay (p. 483); it is not strikingly variable, as is that of the slow net.

The nervous control of muscle is clearly adapted for **short and intermittent contractions**. The refractory period of the muscle is too long to permit smooth tetanus. Swimming muscles of jellyfish are striated, whereas most coelenterate muscle is smooth. Relaxation becomes faster with successive responses. The swimming muscle is not purely phasic, for a tonic component is under local nervous control via a separate conducting system; this is important, for example, in compensatory position reflexes.

With successive stimuli, not only does contraction height increase and duration decrease but latency decreases, velocity of conduction increases, and threshold decreases—as Romanes had already noted in 1877 (see also Bethe, 1935). In a strip preparation additional effects can be seen; for example, **blocks** develop and may be overcome by successive impulses, or velocity may decrease in a narrow strip.

Incisions in the subumbrellar sheet heal rapidly. A medusa thus divided into two independently beating halves can restore nervous coordination in 4–8 hours. Grafts also take rapidly.

In a considerable sample of normal activity recorded from an intact *Aurelia*, the large fluctuations in amplitude of pulsation were nearly completely accounted for by facilitation, or as a function of the interval since the previous contraction and its height (Bullock, 1943).

The properties of this conducting system are consistent with the anatomy of the large, bipolar cell net.

2. Slow diffused nerve nets

But while the swimming beat, as the most conspicuous activity of jellyfish, has attracted the attention of most workers, there has been evidence from the earliest studies of Eimer and Romanes that **other kinds of activity** occur which do not share the all-or-none single motor unit character of the swimming beat. Some of these can influence the swimming beat, bespeaking a one-way interaction between two conduction systems, and some can cross the subumbrellar neuromuscular sheet without producing a contraction wave, suggesting overlying but distinct conduction pathways. Earlier authors, especially Bozler, explicitly suggested the existence of several separate nerve nets which communicate with each other in limited ways and places, but it remained for Horridge to demonstrate in modern times the common occurrence of at least two nets in the bell of both hydro- and scyphomedusae and of additional more local systems or specialized pathways.

The first well-described physiological evidence of **separate conducting systems** was that of Romanes (1877) in *Aurelia;* in this animal gentle stimulation with a camel's-hair brush can cause a slow through-conduction wave (without a contraction wave) to cross the subumbrella and to pass any given complex pattern of cuts. Its passage is manifested by its production of a beat from the first marginal body which it encounters. This is best shown in a strip preparation such as that of Fig. 8.29, with a single marginal body remaining at B and stimulation applied at A. The excitation, which in many cases causes a wave of tentacle retraction along the margin, propagates at about half the speed of that which causes contraction of the swimming muscles. At this point we may quote a passage from Romanes' classical volume, *Jellyfish, Starfish, and Sea Urchins* (1885), which, with admirable scientific caution, imagination, and lucidity, summarizes the almost prescient state of his understanding in those days, before the neuron doctrine and the concept of the synapse (p. 76 and following).

Now this tentacular wave, being an optical expression of a passage of a wave of stimulation, is a sight as beautiful as it is unique; and it affords a first-rate opportunity of settling this all-important question, namely, Will this conductile or nervous function prove itself as tolerant towards a section of the tissue as the contractile or muscular function has already proved itself to be? For, if so, we shall gain nothing on the side of simplicity by assuming that the *contraction*-waves are merely muscle-waves, so long as the *conduction* or undoubtedly *nervous* waves are equally able to pass around sections interposed in their path. Briefly, then, I find that the nervous

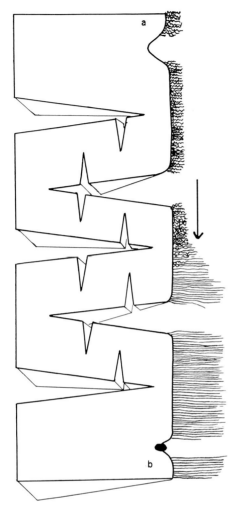

Figure 8.29. *Aurelia.* Preparation used to demonstrate slow conduction of excitation in a separate pathway from that for bell contraction. Light brushing applied at *a* causes a slow wave of excitation, accompanied by tentacle contraction but not bell contraction, to conduct around the margin to *b*. There a marginal body reflexly discharges into the pathway causing bell contraction. [Based on Romanes, 1885.]

waves of stimulation are quite as able to pass around these interposed sections as are the waves of contraction. Thus, for instance, in this specimen [Fig. 8.29], the tentacular wave of stimulation continued to pass as before, even after I had submitted the parallelogram of tissue to the tremendously severe form of section which is represented in the illustration; and this fact, in my opinion, is one of the most important that has been brought to light in the whole range of invertebrate physiology. For what does it prove? It proves that the distinguishing function of nerve, where it first appears upon the scene of life, admits of being performed vicariously to almost any extent by all parts of the same tissue-mass. If we revert to our old illustration of the muslin as representing the nerve-plexus, it is clear that however much we choose to cut the sheet of muslin with such radial or spiral sections as are represented in the illustrations, one could always trace the threads of the muslin with a needle......

This is, no doubt, as I have already observed, a very remarkable fact; but it becomes still more so when we have regard to the histological researches of Professor Schäfer on the structural character of the nerve-plexus. For these researches have shown that the nerve-fibres which so thickly overspread the muscular sheet of Aurelia do not constitute a true plexus, but that each fibre is comparatively short and nowhere joins with any of the other fibres; that is to say, although the constituent fibres of the network cross and recross one another in all directions—sometimes, indeed, twisting round one another like the strands of a rope—they can never be actually seen to join, but remain anatomically insulated throughout their length so that the simile by which I have represented this nervous network—the simile, namely, of a sheet of muslin overspreading the whole of the muscular sheet—is, as a simile, even more accurate than has hitherto appeared; for just as in a piece of muslin, the constituent threads, although frequently meeting one another, never actually coalesce, so in the nervous element of Aurelia, the constituent fibres, although frequently in contact, never actually unite.

Now, if it is a remarkable fact that in a fully differentiated nervous network, the constituent fibres are not improbably capable of vicarious action to almost any extent, much more remarkable does this fact become when we find that no two of these constituent nerve-fibres are histologically continuous with one another. Indeed, it seems to me we have here a fact as startling as it is novel. There can scarcely be any doubt that *some* influence is communicated from a stimulated fibre *A* to the adjacent fibre *B* at the point where these fibres come into close apposition, but what the nature of the process may be whereby a disturbance in the excitable protoplasm of *A* sets up a sympathetic disturbance in the anatomically separate protoplasm of *B*, supposing it to be really such—this is a question concerning which it would as yet be premature to speculate, but I think it may be well for a physiologist to keep awake to the fact that a process of this kind probably takes place in the case of these nerve-fibres, for it thus becomes a possibility which ought not to be overlooked, that in the fibres of the spinal cord, and in ganglia generally, where histologists have hitherto been unable to trace any anatomical or structural continuity between cells and fibres, which must nevertheless be supposed to possess physiological or functional continuity—it thus becomes a possibility that in these cases no such anatomical continuity exists, but that physiological continuity is maintained by some such process of physiological induction as probably takes place among the nerve-fibres of Aurelia.

This passage describes a case of **reflex modulation of the frequency** of swimming beats, via receptors at some distance from the tentaculocysts. In the latter organs the rhythm is determined and access is obtained to the conducting system controlling those muscles. There would appear to be a slow but also through-conducting system, mainly afferent, but efferent to tentacles.

By means of **electrical recording from the slow system** in *Cyanea* and *Cassiopea*, Passano (1958) and Passano and McCullough (1960, 1961) studied its properties. They find its velocity low, variable, and subject to facilitation. It has a very low-frequency background level of activity and fires once or twice in response to a mechanical prod to some parts of the bell. The action potentials are compound and so smoothly graded that there must be many units within reach of the electrodes. They are readily distinguished from the spikes of the fast net. An important observation is that the form of the complex spike is inconstant; indeed it intermittently fails to appear in a given electrode that sees only a small area, when other evidence shows than an impulse has passed. This is compatible with the models of Horridge and Josephson (p. 488) but dictates limits to some of the parameters. The slow net spikes never appear to arise in the statocyst or marginal ganglia but enter the latter to interact with the fast system, as detailed below. Spikes in the slow system cause little if any visible effect on the subumbrellar muscle by themselves in *Cassiopea*, although they modulate fast contractions; but they do cause contraction of the manubrium. The slow system is more sensitive to Mg^{++} narcosis than the fast system.

A second type of reflex pointing to the existence of multiple conducting systems is the **compensatory position reflex** (Fränkel, 1925; Bozler, 1926a, 1926b; Bauer, 1927; Horstmann, 1934a, 1934b). In *Cotylorhiza*, *Aurelia* (Fig. 8.30), *Pelagia*, *Cyanea*, and even in *Cassiopea*—whose normal habit is to lie oral side up on the bottom—tilting brings on an asymmetrical component in the beat and in relaxation in such a sense as to restore the normal position. The uppermost tentaculocyst is necessary and sufficient to cause this local change in the activity of the muscle. A reflex increase in

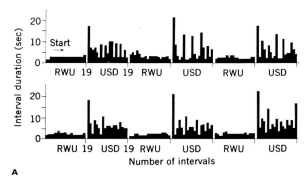
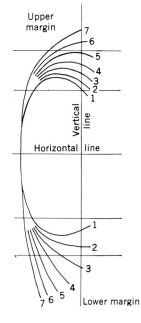

Figure 8.30. A. The effect of inversion on the intervals between swimming beats in *Aurelia*. The ganglion was alternately placed upside down *(USD)* and right way up *(RWU)*, for 20 beats in each position. The durations of intervals between beats are plotted for each successive interval. Beating is much faster right way up. **B.** Diagram of the successive stages, 1–7, in the relaxation of a bell of *Cyanea* mounted on its side. The upper margin contracts to a greater extent and relaxes more slowly and less completely than the lower margin. [Horridge, 1959.]

amplitude of contractions in the trapped circuit wave preparation in *Cassiopea* (see Fig. 8.35) is probably the same mechanism, involving mechanical stimuli, a special conducting system, and possibly double innervation. Bozler has succeeded in removing rhopalia without abolishing the regular beat; compensatory position reflexes are now lost. Thus the beat does not arise in the position sensitive receptor but in a more proximal structure, probably the marginal ganglion. Passano and McCullough (1960) state that the slow, through-conducting nerve net in *Cyanea* and *Cassiopea* does not show electrical spikes arising from the statocysts and therefore is not the mediator of the position reflex. A special local mechanism is required in addition to the two through-conducting systems. The compensatory position reflex is reversed or overruled so that medusae swim bell-downward under some conditions; thus *Pelagia* (Bozler, 1926b; Horridge, 1959) and other forms turn over and descend when disturbed. This is a good example of a higher level of integration in these animals with the simplest nervous systems and probably represents about the level of achievement of their ganglia. It has been reported (Murbach, 1903) that the hydromedusan *Gonionemus* is not dependent on "otocysts" for the maintenance of equilibrium and the suggestion has been made that there is muscle sense in the velum.

Slow local contractions, often spontaneous, are seen in the subumbrella of scyphomedusae. They involve a variable area, often fairly large, suggesting nervous spread. The possibility of intermuscular spread has not been ruled out.

Many authors have seen the **local and graded contractions** of tentacles and mouth parts (manubrium, pseudomanubrium, oral arms, and so on) to local stimulation and to distant stimulation. Such stimuli do not necessarily elicit a swimming beat. Horridge (1955a) has given a careful analysis in the hydromedusan *Geryonia*, which several earlier authors (especially Nagel, 1894; Veress, 1911; Bauer, 1927) also studied, under the name *Carmarina*. Weak stimuli applied to a tentacle cause a very local contraction; stronger and stronger stimuli cause more general contraction of that tentacle, then vigorous shortening of that tentacle and of the other tentacles, and finally a contraction of radial muscles on that side of the manubrium, which bring about an accurate placing of the mouth to the base of the stimulated tentacle (Fig. 8.31). If two points in the margin are stimulated at once the manubrium points somewhere between them. Conduction is occurring across the subumbrella but without directly producing a pulsation. Moreover, cuts show this conduction to be **strictly radial**, that is, not in a nerve net. The pointing reflex can even occur during a continuous series of waves of excitation in the swimming beat conduction system. It is justified to conclude that it travels in a separate overlying system.

Indeed, we cannot simply speak of **two conducting systems** in *Geryonia*. (a) Stimuli to the underside of the bell elicit excitation which propagates outward to the margin causing tentacle shortening and acceleration of the rhythmical

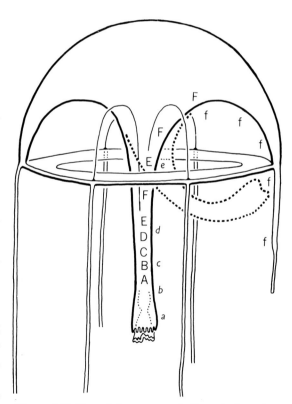

Figure 8.31. The pointing response in *Geryonia*. A mechanical stimulation at the points on the bell indicated by *small letters* leads to a contraction in the region indicated by the corresponding *capital letters*. [Horridge, 1955.]

beat. This centrifugal conduction is strictly radial, like the centripetal, but its velocity is much slower and we may not without further evidence assume that it travels in the same anatomical pathways. (b) When tentacle excitation is strong enough to spill over into other tentacles, all the tentacles around the nearby margin simultaneously contract. This high-speed spread is easily shown to occur in the nerve rings. (c) The velum, while active as a whole in each beat, can contract locally and in the compensatory position reflex it does both at once. Its local conduction system can be activated from the gravity receptors but excitation cannot spread from the velum into the subumbrella.

Similarly, in the scyphozoan *Pelagia*, Bozler (1926b) demonstrated a purely radial centrifugal pathway across the subumbrella, conducting **unidirectionally and without lateral spread** at the low velocity of 0.01 m/sec.

3. Interaction of nerve nets

Clearly we cannot speak of the nervous system as being *a* nerve net. Nor is it readily apparent what we mean by speaking of nerve nets A, B, C, and so on. The tentacle-radial-manubrial system of *Geryonia*, for example, consists of (a) extremely local, presumably interneurally facilitating nerve nets in the tentacles which are partially polarized and strongly oriented longitudinally; (b) a similar net in the manubrium, which has some, weak transverse conduction and its own pattern of polarized connections with sense cells; and (c) two polarized through-conduction, radial, non-netlike pathways across the bell (Fig. 8.32). One of the (c) pathways—the centrifugal—is in functional connection with the circular through-conduction tentacle pathway and thence, irreciprocally, with the swimming beat system. The other, centripetal pathway must not arise directly from or connect with the circular through-conduction tentacle pathway, for if it did, accurate pointing would be spoiled. It appears that the complexity now known forces us to use language like that necessary in higher animals, to speak of pathways, systems, centers, and the like, the antecedents not being always equivalent, sometimes overlapping, and now functionally, now anatomically defined. The parallel is emphasized when we recall that the **various systems are in communication** (a) only at the ring nerves or marginal ganglia, which also receive special sensory input and contain centers of spontaneous discharge and (b) generally in a unidirectional or irreciprocal way. Thus, the tentacle-radial system can influence the pulsation system but the reverse action does not occur. Just as in higher animals (and on the same grounds as those for which Sherrington invented the term), we may recognize the existence of **polarized synapses** in the nerve rings.

The **nerve rings** in this hydromedusan are not simply condensations of a general plexus. They function as integrative centers where various inputs converge and determine output. The two anatomically distinct rings are differentiated: the upper receives afferents from tentacles, nearby sensory cells, ocelli, and statocysts (depending on the species) and probably contains the circular high-speed pathway between tentacles; the lower ring is largely motor to subumbrella and velum. But these are two distinct systems with different —and in each case unidirectional—connections with other systems. We do not know where the through-conduction radial pathways across the subumbrella arise; it seems likely the centrifugal one arises in the upper ring while the centripetal must not make direct connection with it. Nor do we know where the pacemakers for the rhythmic beat lie, except they are highly concentrated in the margin (presumably in the rings) and are distributed all around the margin, not confined just to the radii of the sense organs.

Some further examples of interaction between systems are necessary to illustrate level of organization and the possible variety in the medusoid coelenterates (Horridge, 1955b, 1955c, 1956a, 1956b). In *Aequorea* and many other hydromedusae, but not in *Geryonia*, Horridge showed that when the radial or feeding response system is excited there is inhibition of the rhythmic pulsation system, in two forms. The rapid marginal through-conduction becomes slow and local and then the rhythmic beat fails altogether (Fig. 8.33). This is the **first clear case of nervous inhibition** in the sequence of animal phyla used here. Reflex

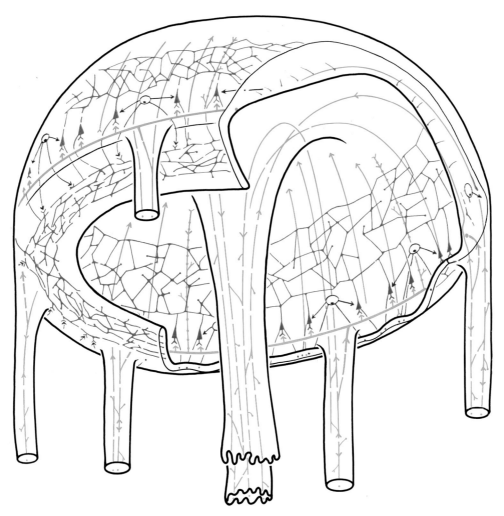

Figure 8.32. Diagram of the nervous system of *Geryonia*, based on physiological experiments by Horridge (1955). The networks and density of fibers are drawn quite arbitrarily. *Blue* = swimming beat system and associated local system in the velum. *Green* = tentacle-manubrium system. *Red* = statocysts. The *blue triangles* are pacemakers of the rhythmic swimming beat. Their frequency can be increased by tilted statocysts and by tentacle stimulation or subumbrella stimulation (via *beaded green pathways* and *solid green ring*). This is an irreciprocal influence of one system on the other. The pacemakers and statocysts also excite the velum to symmetrical and local contraction, respectively. *Dotted lines* are local pathways, conducting decrementally; *solid lines* are through-conducting. Note that the green radial lines conduct in only one direction.

arrest can be obtained in various species by light, (Irisawa et al., 1956), dark, jarring, food, and other stimuli. Again the connection between the systems is one-way. Unlike *Geryonia*, the radial conduction system in the subumbrella is a net with some limited ability to spread in a circular direction around cuts (at 0.002–0.02 m/sec); this is distinct from the circular communication between tentacles, which is faster and confined to the nerve rings. The swimming beat system is also different from *Geryonia;* conduction in *Geryonia* is as fast (up to 0.9 m/sec at 20°C) in the absence as in the presence of the ring nerves, whereas in *Aequorea* conduction is slow after cutting the ring, and circular conduction requires several stimuli to travel as far as a single impulse radially.

Scyphomedusae prove to conform to the same pattern. We have seen above the **anatomically**

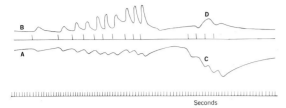

Figure 8.33. Inhibition, facilitation, and temporal summation in a hydromedusan. Responses of the bell of *Aequorea* to electrical stimuli; simultaneous records from two adjacent parts of the bell, A and B. Contraction is registered upward in B and downward in A. Electrical stimuli at A produce a response propagated to B. But the radial muscle at A then contracts, as shown by the maintained deflection C, and now the propagation to B is inhibited and the responses fail (D). B shows facilitation; both A and B show summation. [Horridge, 1955.]

distinct nerve nets in the ephyra larva of *Aurelia;* one mediates the beat and the other feeding and —with strong stimuli—a general, maintained "spasm." Each response can occur without the other and, due to the difference in topography of the nets, cuts can desynchronize the beat of different parts without disturbing the coordination of the spasm. Both systems act on both radial and circular subumbrellar muscle, the one causing a quick twitch, the other a slow, maintained contraction. Double innervation of some muscle fibers is strongly suggested, though inconclusively. The neuronal organization in the marginal ganglia permitting interaction of the two systems has been described above, p. 477. Curiously, at metamorphosis into the perfect medusa, the conspicuous action of the local system controlling feeding of the ephyra is lost in this species; there is no asymmetrical component with tilting and no very slow contraction waves cross the bell, though a slow through-conduction system is still present (see the Romanes quotation, above). But in *Cyanea* and *Cassiopea* there are both slow and very slow systems. In the latter genus the slow system increases the amplitude of the beat, perhaps influencing facilitation. Passano claims there is no visible effect of the slow through-conduction system alone, in muscle response; but it modulates the fast response.

This effect, probably exerted at the muscle and compatible with the above postulate of double innervation, is distinct from the **effect of the slow system on the fast**, through the ganglia. Here is the only point of direct interaction of the two and it is irreciprocal. Passano and McCullough (1961) measured the delay between arrival of a slow impulse at a ganglion and the appearance of a fast impulse, a common but not invariable response. Recording many widely varying cases, they reported that values in a given ganglion of one preparation clustered about a few modes between 0.3 and 5 sec (Fig. 8.34). A possible explanation of these classes of delays is that they represent the activity of different pacemaker units within the same ganglion. Delay is prolonged if there has been a previous fast net impulse too recently (1–1.5 sec). If none has occurred for some time (10 sec), the pacemakers are ipso facto rather unresponsive. Minimal delays between slow spike input and fast spike output occurs about 2 or 3 sec after a previous fast spike—recalling the facilitation of other temporal aspects of response noted by Romanes, Bethe, and Horridge. Two slow net impulses are much more

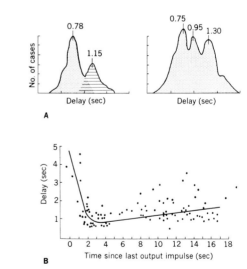

Figure 8.34. The reflex time or ganglionic delay in a scyphomedusan, *Cassiopea*. Delay is the time between arrival at a marginal ganglion of an afferent impulse in the slow nerve net and departure from the ganglion of an impulse in the fast nerve net (see Fig. 8.29). **A.** Two typical preparations that exhibit distinct classes of delays, possibly due to distinct units in the ganglion. **B.** Delay is long if there has been activity in the ganglion within 1 sec or if there has been no activity for 12 or more sec. [Courtesy L. M. Passano.]

effective than one in calling up a fast spike in a ganglion, judging by delay and probability of a response; this priming of the effect of a second by a first slow impulse, which itself was unsuccessful, lasts at least 5 sec. It represents a special form of interneural facilitation, of interest because it occurs in the first ganglion encountered in the present sequence of treatment and in the first one-way junction.

The slow and fast through-conducting nets **do not exhaust the probable conducting systems,** even in the subumbrellar sheet. The compensatory position reflex is still not explained. It is not clear whether the very slow waves (0.02 m/sec in *Cyanea;* not measured in *Cassiopea*) that spontaneously cross the circular and radial muscle and the slow waves that follow stimulation of marginal lappets (0.08 m/sec in *Cassiopea;* not measured in *Cyanea*) are both conducted in nervous pathways or mechanically. Possibly they are in separate nerve net systems. Careful comparison of their relative effects and properties may permit a decision. It has not been excluded that cell-to-cell transmission in muscle is involved in certain of these effects (Veress, 1938).

Besides effects of nervous excitation upon amplitude of contraction, rate of contraction, and rate of relaxation, there is a clear **shortening of the refractory period** of muscle with repeated indirect stimulation (Horridge, 1955b).

The **entrapped circuit wave,** mentioned earlier, has been used not only by Mayer and Harvey in the classical studies but more recently by Kinosita and by Horridge. Kinosita corrected Mayer's conclusion—which amounted to a denial of the all-or-none law in this case—by pointing to the absolute refractory period, which theoretically implies all-or-none response. He then proved the neural character of the essential circulating wave by raising the temperature until the muscle was unable to recover from its shortened refractoriness before the now faster wave had completed a circuit. At this point muscular waves suddenly became half as frequent and he was able on occasion to get one contraction wave every third circuit. Horridge found this preparation suitable for demonstrating the **modulation of amplitude by a slow conduction system** overlying the beat system, and took this to mean double innervation (Fig. 8.35).

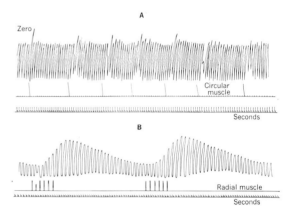

Figure 8.35. Modulation of a regular contraction indicative of a second conduction system. Mechanogram of a ring preparation of *Cassiopea* carrying a continuously circulating contraction wave. **A.** Single shocks applied 18 cm from the recording point produce both a momentary dislocation of the circulating wave and also a shortening of the muscle for several cycles. **B.** Five shocks at 1 per second produce a long-lasting effect on the height of the contraction but here do not disturb the circulation of the wave. Both the radial and the circular muscle give the above reactions. [Horridge, 1956b.]

IV. SPONTANEITY, INTERNAL STATE, AND PHASES OF ACTIVITY

Historically, the interest in jellyfish pulsation arose early, but definite knowledge of it awaited the conclusion, first, of the debate on whether coelenterates have a nervous system and later, on whether the regular beat is reflex or automatic. Haeckel (1866), Eimer (1878), Romanes (1877, 1878), the Hertwigs (1878, 1879), and Schäfer (1879) had all satisfactorily answered the first question, but lingering doubts as to the second persisted as late as Bauer (1927), who concluded that von Uexküll's reflex theory was wrong and that the beat is the intrinsic result of steady-state stimulation. This conclusion meant that the jellyfish was the **simplest animal with a**

IV. SPONTANEITY

true nervous pacemaker; in consequence, some of the earliest general physiology was devoted to analyzing this material. We cannot here detail the results and theories but many experiments, especially by Mayer and Bethe, led to a theory of continually renewed local excess of sodium, producing a relaxation oscillation.

The notion that **a refractory period** of the nerve net determines the frequency of intermittent beating by determining the interval has been widely held but can be dismissed on present views. Bethe and Bozler have both placed one limb of a Y-preparation in an experimental bath which alters the refractory period (for example by temperature or Na^+ or Mg^{++} content) and the other limb in normal sea water, and have observed that coordination is still present between them. The experiment is the same as that which Gaskell used, with the same result, on the vertebrate heart. The conclusion is also the same, that the rhythm comes from a discontinuous pacemaker which itself controls the generation of discrete impulses.

Modern **analyses of the rhythm** include those of Bullock (1943), Pantin and Vianna Dias (1952b), and Horridge (1959). There is a high variability of the intervals between beats in *Aurelia* and a negative first-order serial correlation coefficient. The presence of several equivalent pacemakers, in the eight marginal bodies of scyphomedusae and all around the marginal ring of hydromedusae, not only makes for a higher over-all frequency but more regularity. A single artificially induced beat injected in a rapid uniform rhythm usually resets at the same frequency and usually causes a slight pause, accentuated if several beats are induced artificially. These properties speak for a **self-regulation process** in the pacemaker structures. Variability in intervals is correlated with variability in strength of beat and nearly all the latter could be accounted for as neuromuscular facilitation by considering both preceding interval and the height of the last contraction.

The **pacemakers are accessible to sensory input** which can accelerate or inhibit the rhythm. Larger specimens beat less frequently. A single marginal body is said to slow its pace progressively with decrease of the size of the preparation it is allowed to drive (to $1/2$ when $1/16$ the area remains) whether gradually reduced by cutting or by block (Eimer, 1878; Cary, 1917). Of the 4, 8, or many pacemakers available, the one which is fastest at the moment controls the frequency. This is accordingly lowered by cutting out some of them and raised by grafting on another individual even a larger, slower one.

The variety of **influences that play upon the pacemakers,** even excluding "nonphysiological" experimental conditions is considerable: light, shade, mechanical contact, food juices (including proteins, some amino acids and fatty acids, but no carbohydrates), vibration, agitation of the water, even small changes in hydrostatic pressure (0.3–0.7 atmosphere, Baylor and Smith, 1957). Acetylcholine, adrenaline, curare, ephedrine, histamine, 5-hydroxytryptamine, and physostigmine have no effect; tryptamine accelerates (Horridge, 1959).

There are a number of studies on the functions of the sense organs and the **sensory capacities** of coelenterates (see Verworn, 1891; Nagel, 1892, 1893, 1894; Loeb, 1895; Yerkes, 1902a; Yerkes and Ayers, 1903; Murbach, 1903; Lehmann, 1923; Fränkel, 1925; Bozler, 1926a, 1926b; Henschel, 1935; Warden, Jenkins and Warner, 1940; Edney, 1939; Hyman, 1940; Skramlik, 1945; Zubkov and Polikarpov, 1951; Zhirmunsky, 1958; von Buddenbrock, 1952).

But an important question is still open whether we can account, with these sensibilities to environmental change, for the **phases of quiescence** exhibited by many species of medusae (but not others) in the absence of apparent stimulation. Indeed the more elementary question is also open, as to whether during quiescence the normal intrinsic pacemaker is nonfunctional or is countered, while it continues to act, by a superimposed, opposite "command," as is common in higher invertebrates and vertebrates. The same issue can be formulated for the "spontaneous" deviations from the "normal" position: is there a shutting off of the compensatory position reflex or is there a superimposed counter influence?

There are other cases of short-term rhythmicity in nervously controlled behavior of coelenterates. The fishing tentacles of *Physalia* shorten and then lengthen, about once a minute (Parker, 1932). The nectocalyces of other siphonophores propel the colony by rhythmic contraction. The tentacles of each zooid of the alcyonarian coral *Heteroxenia* rhythmically beat in unison (see p. 519).

As for sea anemones, those simple and slow-moving flowers of the sea, we have seen in the decade just past a sweeping reversal of views. In his classic on *The Elementary Nervous System*, Parker (1919b) considered and rejected the view, largely due to Jennings (1905), that **fluctuations of internal state** are an important constituent of the determination of activity in these animals. He concluded that "an actinian is much more nearly an organism whose internal state is one of general uniformity" and that on this uniform background the changing environment calls forth relatively consistent responses, "in strong contrast with what is found in higher animals." Spontaneity is not a characteristic of actinian behavior, according to Parker. This picture was indeed justified by the evidence at hand, discounting Jennings' and a few other authors' arguments; and it seemed strengthened by the great advances of the thirties in the physiological analysis of actinian nerve nets.

But is is now clear—and this is one of the main advances in comparative neurology—that, on the contrary, the internal state is subject to great and apparently **spontaneous variations,** as well as environmentally induced variations, in these nervously lowly animals. It is not always changing but the state maintained for many hours is only a phase which gives way to others, each accompanied by quite different reactivity to given external stimuli (Fig. 8.36). Only with methods that compress the time axis—60 times speeded up time-lapse cinematography and slow kymograph recording—has this been appreciated (Batham and Pantin, 1950a, 1950b, 1954; Pantin, 1950, 1952, 1955; Needler and Ross, 1958; Hoyle, 1960). Phases of activity often alternate in a rhythm but only for a few cycles in constant conditions. Even then nonrhythmic changes occur, rather than a steady state. Often the spontaneous activity consists of seemingly

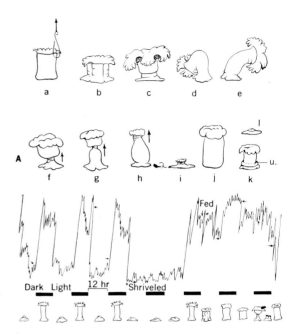

Figure 8.36. The appearance of *Metridium* in various phases during and after feeding. In **B** the marked light-controlled rhythm prior to feeding is largely abolished by a meal; the upper trace is a kymograph record made by means of a hook attached as shown in **A**a. The stages in **A** are: b, ingesting food; c, $\frac{1}{4}$ to $\frac{1}{2}$ hour after ingestion; d, e, swaying movements 6 hours after food-extract stimulus; f, g, h, successive stages in antiperistaltic constriction, leading to defecation and (i) shriveling; j, subsequent refilling, showing extreme expansion 4 hours after shriveling; k, specimen showing excretory ring of mucus and uric acid (u); l, state of extreme contraction, typical of unfed specimens exposed to sunlight. [Batham and Pantin, 1950.]

uncoordinated contractions of different muscle groups but again the coordination is so striking as to have a "decidedly purposive character." Peristaltic contortions, defecation, swayings and sweepings and the complete feeding sequence, locomotion, and extremely dilated or shriveled states **follow each other without external cause** being apparent. The view that each movement is stimulated by the preceding as in a chain reflex is rejected. It is proposed that slow, phasic activity is the main form of behavior in actinians, the relatively quick responses to touch and the like being for the most part rare and specialized forms. Piéron (1906, 1908, 1909) and Bohn (1906) discussed possible tidal rhythms in behavior.

Some features of the basic patterns are worth special note. When coordinated movements are obvious, one part of the body may act as a **leader** time and again for long periods. The parts which follow may do so with long delays—many minutes (Fig. 8.37). External stimuli have differ-

Figure 8.37. Two responses to series of 10 shocks at 10 sec intervals and two spontaneous contractions. Ring of circular muscle of *Metridium* from half way up the column. Electrodes on a tongue of body wall, 1.5 cm from the muscle ring. Note the regular latent period of about 110 sec. [Batham and Pantin, 1954.]

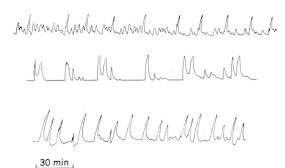

Figure 8.38. Examples of spontaneous activity recorded from different specimens of *Stomphia coccinea*. [Hoyle, 1960.]

ent effects according to the phase and in turn may initiate a change of phase, as though merely **releasing a complex pattern** that can as well be spontaneous. A small trace of food juice or 6 electrical shocks per minute for 15 min may cause a change of pattern lasting for hours. Different phases may reinforce each other or may conflict (Fig. 8.38). To external stimuli there is often an enormous latent period—minutes, possibly hours. Excessive stimulation during the day may be followed by locomotion during the ensuing night. The thresholds for such phase-changing stimuli vary widely. Koshtoyants (1959) and Koshtoyants and Smirnova (1955) studied the complex rhythm of actinians and proposed that it depends on the availability of -SH groups. Needler and Ross (1958) could not confirm the effects of cadmium chloride or cysteine on *Calliactis*.

Apparent **habituation** occurs in hydroids and anemones (Kinoshita, 1911a). Mechanical and chemical stimuli, each repeated again and again as soon as the response to the last had passed off (usually some minutes), resulted in a succession of responses whose duration systematically fell. Sensory adaptation probably plays a large part (Kinoshita, 1911b), but since the stimuli are physiological and the intervals relatively long, or more than response duration, this property is operationally comparable to habituation in higher groups.

The question suggests itself, if we have complex patterned activity requiring only to be released by an adequate stimulus and occasionally even self-triggering as spontaneous phasic activity, is there not some **region of the nervous system more crucially involved** than others, apart from the efferent and afferent pathways? What effect, for example, would massive damage to the oral disc or stomodeal nerve fiber layer have upon such activity? We lack a systematic series of such experiments. Peristalsis in *Aiptasia* continues to be spontaneous, unidirectional, and symmetrical after removal of the whole oral end (Portmann, 1926). Pedal discs cut away from oral discs can still creep and orient to light (*Metridium*, Parker, 1919b). Simple spontaneity is inherent in isolated rings of the column. But the more complex activities remain to be so studied. The tentacles and mouth region are likely to be of special significance; they are more sensitive and also dominant in some sense; they initiate much of the activity of actinians and others.

What we have been emphasizing under terms such as phases of internal state corresponds to the **"Stimmung"** of German writers and in essence to "set" or "affect," if these can be applied to infrahuman species. It is interesting that the early naturalist, Trembley, in 1744 had already clearly recognized, on grounds of the most careful observations, that the simple *Hydra* **reacts quite differently according to its state** or "Stimmung," for example when hungry or sated (compare also Haug, 1933).

It appears to be true that coordinated spontaneous activity, including that inherent in effectors themselves, is a **fundamental feature of** the most primitive nervous systems. Pantin is led to consider direct reflex responses as secondary simplifications to meet specific demands, rather than as examples of an elementary unit of all behavior patterns (see also Wells, 1955, p. 351).

The **differences** between the nervous system of coelenterates and of higher groups have become less qualitative and less identifiable with any recognized structure or elementary property. We are probably more correct in expressing the differences by a relatively smaller number of sensory discriminations (including analyses of patterned stimuli) and of distinguishable forms of efferent response. But at the time of writing there is impressive evidence of how complex can be the achievements of such a decentralized and sheetlike system. Perhaps this only expresses our ignorance of the complexity of the behavior of flatworms and other lower phyla with centralized nervous concentrations.

V. DETAILS OF THE GROUPS OF COELENTERATES

A. Hydrozoa

1. *Hydra*

Since 1860 upward of twenty papers have specifically addressed themselves to the question of the organization of the nervous system of *Hydra*. Several workers have successfully applied methylene blue or its leuco-derivative Rongalit white, and maceration methods have proven useful. In spite of all this, the anatomy of the nerve net in *Hydra* is commonly spoken of as in an unsatisfactory state of knowledge, perhaps mainly for two reasons; its extreme simplicity and the report of continuity between nerve cells. As we approach the simplest nervous systems it becomes increasingly difficult to be convincing about the **identification of nerve cells**—as we have remarked earlier. Particularly in sections, authors have questioned each other's identification since the mere possession of processes is not specific to nerve cells and cell boundaries are indefinite. But the **problem of anastomoses** and of continuity throughout the nerve net has probably roused the principal question, especially in the last quarter century, for in other coelenterates so few authors have claimed this feature and so many, both old and new, have specifically denied it. Although several early authors (Schneider, 1890; Zoja, 1892; Wolff, 1904; Hadzi, 1909; Marshall, 1923) described the net as continuous, recent authors find it difficult to see continuity or discontinuity in other coelenterates when dealing with fine processes. But our confidence in the fusion of processes in *Hydra* is increased not only by the agreement among these reports but by the more recent ones of McConnell (1932), of Semal-Van Gansen (1952), and of Spangenberg and Ham (1960), who confirm them in full knowledge of the contrary findings with medusae (Fig. 8.39). Electron microscopy is urgently needed but faces a formidable task in following a nerve fiber from one cell to another; up to the present nerve fibers have not been identified at all with this tool, in *Hydra* (but for results on medusae, see Figs. 2.26, 2.45). The evidence available by light microscopy is strongly in favor of anastomosis between long processes of widely separated cells. This, together with the reasonable agreement between the disposition of the net and the physiological evidence of conduction gives us some reason to accept the identification of nerve cells and hence the following picture of the arrangement of the nervous system.

The **nervous system is chiefly epidermal.** In the gastrodermis a few sensory cells and ganglion cells have been seen, but no connections among them or between them and the epidermal net. It is usually stated there are no nerve cells in the mesoglea. They ordinarily lie just outside the longitudinal muscle; in the pedal disk they are often more superficial. Kepner and Hopkins (1924), on the basis of experiments with chloretone, concluded that activities mediated by ectodermal muscles are nervously controlled, whereas these mediated by endodermal muscles are not nervous but neuroid. The type of evidence and the argument does not deserve detailing; hopefully it will attract restudy (see Semal-

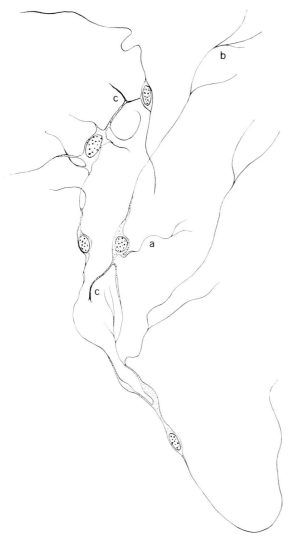

Figure 8.39. *Hydra*. Cells of the ectodermal nerve net, showing (a) very delicate processes, (b) long processes, and (c) short thick processes. [Semal-Van Gansen, 1952.]

Van Gansen, 1952). The epidermal plexus is in the form of a net which is simple and single; there is no evidence of separate or overlying nets.

There are two kinds of cells, sensory cells and ganglion cells, and transitions between them. The **sense cells** are narrow and superficial, usually with two proximal processes. They are chromatin-rich and plasma-poor and usually have no nucleolus. Some have one to several sensory hairs; others terminate distally in a small sphere or bluntly, and these forms may be selectively distributed over the body. This is possibly of interest in view of the report that while the foot is one of the most sensitive regions to mechanical stimuli it is perhaps the least sensitive to certain chemical stimuli (Marshall, 1923). The proximal processes join the net of processes of the **ganglion cells,** which give off from two to seven nerve fibers, mostly two or three; they are small cells with a dense nucleus, usually no nucleolus, and with discernible neurofibrils. Three types of processes are recognized: (a) short, fine processes seeming to end on epithelial and sense cells and presumably afferent, (b) long, fine, dichotomizing processes appearing to join with those of other cells to make the net, and (c) short, thick processes seeming to terminate upon muscle fibers and presumed to be motor. Although these last are simple and undifferentiated it is said that their relation to the muscle is so intimate that it cannot be broken in some cases by a tap on the cover glass of a macerated preparation, which commonly separates cells even of the nerve net. Some ganglion cells have both sensory and motor processes; some have neither and are therefore pure interneurons; many have one or the other. Ganglion cells on the body are said to be simpler and less branched than those on the peristome, tentacles, and foot. The sensory and motor processes may arise from the main processes instead of from the cell body.

McConnell (1932) describes the **development of the net in buds** (Fig. 8.40). The bud does not inherit its nerve cells from those of the parent, but forms new ones from undifferentiated interstitial cells in a wave of development which proceeds from the already forming tentacles and peristome downward. The coalescence of processes is secondary (contra Parker) and the motor processes seem to develop late, after the net is formed.

Hydra has been regarded as a clear confirmation of the Hertwig hypothesis of the phylogeny of ganglion cells, since there is a clear series of **transitions** between sensory neurons and typical ganglion cells.

The nerve net is distributed throughout the epidermis and a relative concentration claimed for the hypostome is doubtful. Cells are more numerous in the peristome. In the tentacles there is a preferential orientation of the net in the lon-

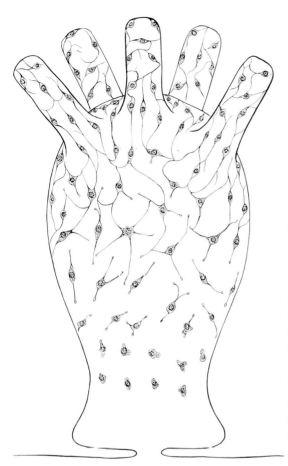

Figure 8.40. A bud of *Hydra*, showing stages in the maturation of the nerve net and the secondary fusion of processes. Methylene blue by Rongalit white method. [McConnell, 1932.]

gitudinal axis and in the foot and to a lesser degree in the oral disc in the circular direction (Hadzi, 1909). Semal-Van Gansen (1952) has longitudinally split *Hydra*, leaving the two moieties connected by a piece of the base or of the hypostome or of the column between; in every case, stimulation of one side can cause contraction of both—confirming the **diffuseness of conduction.** Satisfactory data are lacking on the abundance of cells, their spacing, the diameter of the nerve fibers, the difference in latency of response between near and distant stimulation, and the like.

Responses can be local or general. Apparently only strong stimulation elicits general responses, while weaker stimuli elicit progressively more limited areas of response; those areas which drop out are the ones farthest from the site of stimulation. This fact has given credence to the concept of decremental conduction, which—though it has been rather generally abandoned for the other coelenterates—deserves to be separately examined in this animal. There is some reason to doubt that all responses are simply a function of distance from the site of stimulation. For example, a gentle touch repeated every 2 to 5 sec causes a bending away from the side of stimulation by local contraction of muscles on another side.

It is possible that the nerve net is in fact synaptic and that the principles worked out for higher coelenterates apply also here—local response depending on interneural facilitation and thresholds of frequency of junctions—without invoking distance or decrement of conduction. But on the present evidence we must be prepared to consider the nerve net of *Hydra* as syncytial and hence to entertain the possibility that **all-or-none propagated impulses** are not involved in subtotal response, but that graded, decrementally spread excitation occurs in the nerve net. This does not necessarily mean purely passive electrotonic spread, for we can imagine an active but decremental propagation as in the abortive spikes or local potentials in higher animals (Pantin, 1950; Grundfest, 1956; Bishop, p. 130). Pantin suggested this possibility on the basis of the small size of *Hydra*, but it is more difficult to imagine decrementing excitation spreading over many centimeters as is true in the case of *Corymorpha*, a large solitary marine hydroid. Josephson (1961c) clearly showed all-or-none propagated spikes in colonial hydroids.

There is a large literature on the **behavior and activities** of *Hydra*, going back to Trembley (1744), and many of the observations are of interest in attempting to picture its nervous organization. Rhythmic behavior is said to occur as a preferential sequence of phases (Reis, 1953). The peristome and tentacles become hyperactive under conditions of anesthesia, which are believed to affect only the epidermal nerve net of the body. Sphincters at the bases of the tentacles can be seen to close and open and will allow material into the body from the cavity of a tentacle but prevent the reverse movement (Kepner and

Hopkins, 1924); most of the observations, however, are susceptible to different interpretations or do not immediately illuminate the physiology of the nervous system, and space does not permit detailing them (see for example Schlünsen, 1935; Haase-Eichler, 1931).

2. Corymorpha

The work of Parker (1917) on this large solitary hydroid shows the value of examining related species. The **nerve net is probably chiefly epidermal.** A light touch upon one side of the long stalk results in the hydranth curling down to apply the mouth accurately to that spot (Fig. 8.41). A head-

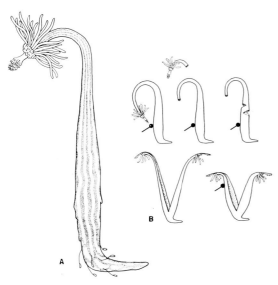

Figure 8.41. A. *Corymorpha*, intact. **B.** Experiments with tactile stimulation. *Above,* the placing response can still occur after decapitation and overlapping notches in the stalk. *Below,* the shortening response occurs on both sides of a split preparation, showing transverse conduction at the base. [Based on Parker, 1919, The Elementary Nervous System, J. B. Lippincott Co., Philadelphia.]

less stalk will do the same. A Y-preparation made by splitting an animal from the hydranth end down almost to the base, when stimulated in one arm of the Y, shows that conduction of excitation can occur both longitudinally and then transversely across the base of the stalk. The same is true if the split is made from the base almost to the hydranth, but if the stem of the Y includes neither base nor hydranth no transverse conduction occurs. The nerve net in the stalk is strongly **preferential for longitudinal conduction.** Transverse conduction is not absent but can be made to occur by notching the stalk halfway through and stimulating on the same side, whereupon bending toward this side on the part of muscles of the whole stalk can still occur. Even after a series of several partial notches from different sides there is some conduction from one end of the stalk to the other.

Positive geotaxis is strongly developed. A stalk with or without hydranth or a short piece of stalk lying horizontally in the bottom of a dish will, as soon as the base has attached itself, slowly stand up over the course of an hour or so; the longer the piece, the sooner will it stand. Suspended from the middle or even from the basal end and hanging straight down, working against specific gravity, the animal will nevertheless achieve the upright posture. Torrey (1904b) believed this to be a growth response of the turgid endodermal cells, but Parker found it is reversibly abolished by anesthesia and occurs after destroying much of the endodermal pith with a needle. He concluded it is neuromuscular even though it is still possible after a series of staggered partial transections of the stalk. What receptors may be involved has not yet been suggested.

Autonomy is highly developed, as in other coelenterates. Not only the stalk but the isolated proboscis and the proximal and distal tentacles each reacts in its characteristic fashion as well when isolated as when attached. **Spontaneous activity** is beautifully demonstrated. When the current of water is turned off in an aquarium a specific feeding response begins in a few moments, involving bending down, pressing the mouth into the mud, and straightening up again within about a minute. This recurs at fairly regular intervals and the isolated stalk and the isolated head both continue rhythmically to perform their normal movements, although at a lower frequency than before they were separated. The intact animal—but not the beheaded stalk—ceases this form of activity as soon as the current of water is restored. Further work on this form would seem to be particularly fruitful.

3. Colonial hydroids

The limited older work on colonial hydroids (Jickeli, 1882, 1883; von Lendenfeld, 1883; Zoja, 1891; and the single recent account, Leghissa, 1950) suggested that their nervous system is similarly organized. Recently Josephson (1961b, 1961c, 1961d) has examined several species physiologically and found wide differences in the development of the **colonial nervous system.** The stems and stolons of all species have conducting systems with an over-all velocity of 0.01–0.16 m/sec. They fall into two physiological classes: through-conducting and local, graded conduction systems. Some forms, like *Tubularia* and *Syncoryne*, rarely have any colonial coordination; stems are not nervously connected. In others, like *Hydractinia echinata*, there are two distinct colonial nerve nets, a through-conducting and a local system (Fig. 8.42), besides the system or systems within each zooid. The through-conducting colonial network has a sharp threshold, responds to a single electrical shock with no effect of strength, above threshold, conducts faster, and mediates the synchronized lashing stroke of all the spiral zooids. This is a startling sight in so passive and naked an animal colony; on snail shells occupied by hermit crabs the spiral zooids form a single line around the aperture, all oriented to lash in toward the aperture. This response requires electrical or injurious mechanical stimuli. It is accompanied by a slight twitchlike retraction movement of the other zooids.

The local system is brought in by gentle mechanical stimuli, repeated electrical stimuli, and even spontaneously; it mediates a slower conducting and different type of response. This may be spread over the whole colony or limit itself to a small region according to the stimulus strength.

Podocoryne has only a through-conducting system, apparently, confirming Zoja (1891), as also have *Tubularia* and *Syncoryne* in the stems of individuals. *Cordylophora*, *Pennaria*, and *Hydractinia aggregata* have only local, graded con-

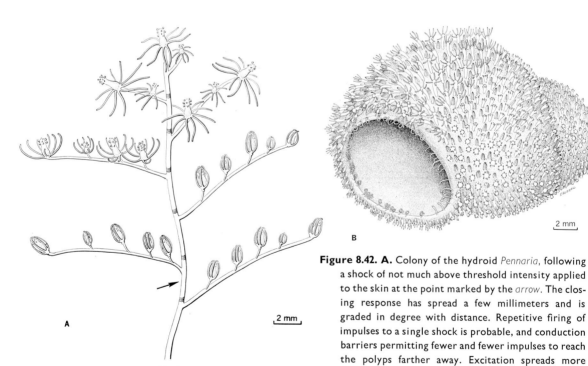

Figure 8.42. A. Colony of the hydroid *Pennaria*, following a shock of not much above threshold intensity applied to the skin at the point marked by the *arrow*. The closing response has spread a few millimeters and is graded in degree with distance. Repetitive firing of impulses to a single shock is probable, and conduction barriers permitting fewer and fewer impulses to reach the polyps farther away. Excitation spreads more readily distally than toward the base. **B.** Colony of *Hydractinia* living on a snail shell, as it would appear shortly following stimulation at the *arrow*. A wave of polyp contraction that will affect all the polyps is shown sweeping across the colony. The dactylozoids (around aperture of shell) are lashing out from their coiled resting position. These two kinds of response travel in separate nerve nets. [Josephson, 1961a.]

duction, but may spread excitation over a good many hydranths (Fig. 8.42). Zoja reported the same result in *Eudendrium, Corydendrium, Campanularia*, and, less distinctly, in *Tubularia*.

With the exception of polyps of *Syncoryne* and of the spiral zooids of *Hydractinia* and *Podocoryne*, the polyp responses of all these species are symmetrical, implying a through-conducting system **in the individual hydroid polyps,** as in the case of coral polyps. Such systems typically command all-or-none responses, but the strength of the polyp retraction is graded with distance from the point of stimulation in those cases involving the local, graded system. Josephson gave reasons for believing that the explanation is repetitive firing (Fig. 8.43); this grades the distance of

Stimulus intensity	Polyp response	Time to polyp relaxation (seconds)	Electrical response recorded near base of polyp (2 sec)
<7		No response	
7		20	
8		35	
10		45	
15		55	

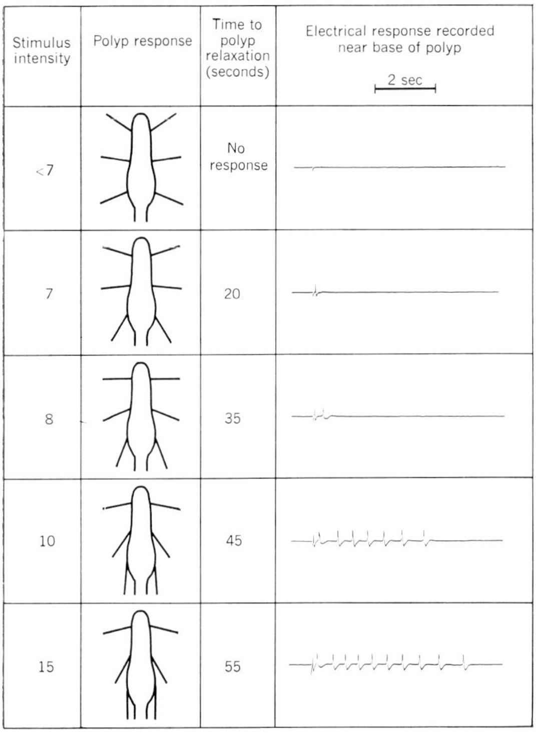

Figure 8.43. Comparison of the response of a polyp of the colonial hydroid *Cordylophora*, and the electrical potentials recorded near the polyp stalk. Only tentacle depression is shown in the diagrams of the responding polyp; hydranth shortening, which also occurs, is not shown. The times to relaxation are approximate. The deflection in the first electrical record is the stimulus artifact. Note repetitive firing of impulses following a single shock slightly above threshold; this provides a gradation of response with intensity and with distance. [Josephson, 1961b.]

spread by interneural facilitation in the colonial net and grades the strength of contraction by neuromuscular facilitation in individual polyps. *Tubularia* has two conducting systems in each individual, one passing through stem and hydranth to its distal circlet of tentacles, the other controlling synchronized responses of its proximal tentacles. There is little evidence of interaction between them.

A model of a nerve net based on non-through-conducting systems in colonial forms is discussed on pp. 488-490.

Surprisingly, hydroids have proved favorable for recording **action potentials** (Josephson, 1961c, 1962). Semimicroelectrodes of metal inserted in the stolon of *Cordylophora* or in the neck region of *Tubularia* pick up all-or-none spikes of 0.05–15 mv height and of 20–120 msec duration. Those of *Cordylophora* occur only after stimulation; frequently they appear as a prolonged burst, even following a brief electric shock of less than twice threshold intensity. Each spike is of compound origin even though all-or-none; the shape systematically changes between the earlier and later members of a burst, and occasionally they are jagged. The reasonable interpretation is that many parallel nerve fibers in the tubelike stolon are conducting and summing their potentials; they are kept in (more or less) synchrony by the cross-connections of the nerve net.

Of interest for general neurology is the observation that there may appear to be no refractory period. In one preparation Josephson could apply two, barely suprathreshold shocks at any interval down to the shortest available (0.2 msec) and still evoke two spikes; the interval between the two responses was never less than about 200 msec. Josephson argues that some element was **summing arriving stimuli** in a lasting state which then determined the number of new spikes to be initiated. This state could be a graded depolarization influencing a spike initiation locus which is not fully invaded by spikes.

There is a strong **tendency to repetitive firing** (Fig. 8.43). Single spikes can be obtained just at threshold with both electrical and mechanical stimuli. But the number of spikes in a burst is an important variable under the control of stimulus intensity. The more spikes there are, the quicker and stronger is the polyp contraction and the farther is the spread through the colony. Recording at increasing distances from the stimulated site shows that the spikes do not decrement in size but the burst decrements in number, and therefore by discrete steps. It is reasonable to suppose these steps correspond to synapses in the nerve net which require interneural facilitation. This analysis of the effect of strength of stimulus and the explanation of graded contraction with stimulus and distance, in terms of all-or-none impulses normally fired repetitively, is a powerful and illuminating principle and follows logically from the results of Pantin in the thirties.

Recording from *Tubularia*, Josephson (1962) found **spontaneously recurring patterns** of spikes. Sometimes two rhythms were evident, a burst of the order of 10 spikes up to 2 per second, recurring about once a minute, and single spikes occurring between bursts about once every 5 sec. The correlation with motor behavior, the location of the pacemakers, and the properties of the conducting system remain to be worked out.

4. Siphonophora and Chondrophora

The anatomy has been studied by Chun (1881), Conn and Beyer (1883), Schneider (1892), Schaeppi (1898), and Mackie (1959, 1960). The greatest detail is available for the chondrophoran *Velella*; *Porpita* is probably very similar. Using silver impregnation on *Velella*, Mackie beautifully stained **two distinct networks** of neurons; one consists of larger cells and fibers (1–5 μ) that meet at junctions, a high proportion of which are believed to be protoplasmic anastomoses (Fig. 8.45); the other network is made up of smaller elements (0.25–0.5 μ fibers) with discontinuous junctions en passant (Fig. 8.6). The two networks may be called the large or closed net and the small or open net, respectively. Fibers of the latter often twine around those of the former. Sensory neurons make contact with both nets. No nerve rings or concentrations have been found. The nerve fibers are chiefly in the ectoderm, sparsely in the endoderm; they freely cross mesoglea to connect separated parts of the ectoderm.

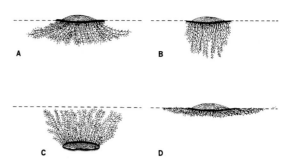

Figure 8.44. Behavior of *Porpita*. **A.** Resting position, from the side. **B.** Food-collecting behavior, tentacles fully lowered. Above a threshold number of tentacles, they all act in concert, lowering by a series of jerks with characteristic pattern; the whole cycle may recur spontaneously every half minute. **C.** Righting activity, from the side. **D.** Aboral or protective response, a single, sustained flexion without gradation. *Dashed line* indicates water surface. (Mackie, 1959.)

Functional studies of *Porpita* point to distinct through-conduction and local systems (Mackie, 1959), and these pathways may tentatively be identified with the large-celled and small-celled nets, respectively. They mediate rather **complex behavior** in this curious floating hydranth (Fig. 8.44). Righting is effected by a prolonged series of jerks and graded contraction of the mantle flap. A "protective response" to strong mechanical stimuli is a single all-or-none aboral bending of all tentacles. Food collecting also involves synchronous jerks of few or many tentacles but orally. If a tentacle is touched, after 1 or 2 sec it will flex downward in four or five summating twitches. If two or more tentacles are touched and respond, their twitches are synchronous, whether they are near together or not. If a larger group of tentacles is stimulated, the response takes on a new character; it becomes sustained, giving a patterned sequence of 40–50 jerks at about 3 per second, in which all tentacles take part, whether stimulated or not (**"concerted activity"**). If a critical number of tentacles is not involved, the jerks die out after four or five, that is, there is a threshold number. Spontaneous repetition of **food-collecting activity cycles** occurs in some specimens at intervals of about 30 sec. Isolated segments down to 45° behave like intact animals; segments attached by thin strips are coordinated. In addition to these movements,

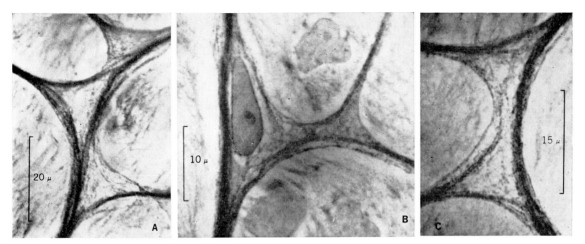

Figure 8.45. *Velella* (Chondrophora); junctions in the "closed" system stained with Holmes' silver method. These images (see also Fig. 8.6) are the basis for calling the large-fiber nerve net a syncytium, in contrast to the small-fiber, "open" system. [Mackie, 1960.]

local responses occur in the organs of ingestion (Mackie, 1961). Electrical stimulation of *Velella* showed that feeding zooids are excited at a different, lower level of stimulation from tentacles. Tentacle twitches follow single shocks, suggesting that a single impulse in a through-conducting system causes each jerk.

In the siphonophore *Nanomia*, behavioral analysis indicates **three distinct types of conduction** of excitation in the coordination of colonial responses (Mackie, 1961). There is through-conduction in the forward swimming by jet propulsion due to simultaneous contraction of all nectophores. Backward swimming is another simultaneous, through-conducted activation, in which two muscle bands deform the velum. Local responses occur to stimulation of nematocyst batteries or tentillae, and these may spread across several cormidia without response from the intervening stem and appendages. Local contraction of tentillae and of tentacles, searching movements of gastrozooids, and wide opening of mouths are involved. The forward swimming is evoked by certain stimuli; also, from time to time *Nanomia* colonies burst into life spontaneously and swim violently around forward. The synchronous jets from all nectophores give way to asynchronous beating and dropping out one by one; therefore, the colonial conduction system is distinct from that in each nectophore. Reverse swimming is explosively precipitated by hitting the surface of some object and, like the foregoing, appears to involve single impulses in a fast colonial conduction system for each beat. The two systems are polarized by their sensory connections; reverse swimming is not initiated by stimulation of hinder regions, nor forward swimming by stimulation of the float. Strong stimulation leads to autotomy of nectophores and bracts. Some nerve fiber pathways were found but not enough to account for these conduction systems. Mackie entertains the possibility of muscle cell-to-cell or epithelial cell-to-cell conduction for this reason. Certainly this possibility ought to be tested, for example by the methods outlined on p. 463; it can hardly be accepted otherwise unless we are confident the anatomy is completely known— in this case, of the colonial tissues.

Physalia reveals local activities of tentacles, gastrozooids, and float and a rapid generalized protective withdrawal of all mobile appendages following abrupt stimulation. Two conduction systems are suggested but only one has been found histologically: a diffuse nerve net in the ectoderm only, with many sensory nerve cells (Totton and Mackie, 1960). The tentacles show an inherent rhythmic shortening and lengthening and evidence of a through-conducting and a local conduction. Ebbecke (1957) has contributed some details on reflex responses of siphonophores.

B. Anthozoa

1. Actiniaria

Species differences are considerable and instructive in the forms examined. *Anemonia* apparently **lacks a through-conduction system** and cannot be made to exhibit a symmetrical response. All its responses, even to strong or prolonged stimulation, are more or less local and none has the character of a protective response. There is evidence of repetitive discharge to single electric shocks. *Calliactis*, on the other hand, has a **well-developed symmetrical response** to column stimulation via a through-conduction system which, in the first instance, activates the sphincter muscle, closing over the oral disc in what is apparently a

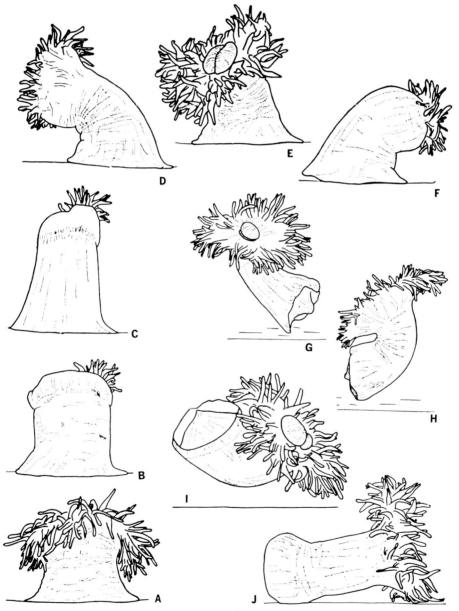

Figure 8.46. The swimming sequence in *Stomphia coccinea*. **A.** Normal position, attached. **B.** Response to stimulation by contact with starfish. **C.** Extension of column by circular muscles after contraction of longitudinals. **D, E, F.** Lateral bending by parietobasilar muscles. **G, H, I.** Lateral bending. **J.** Inactivity after swimming. [Sund, 1958.]

protective response; it lacks fast column responses and on the mesenteries only a sparse, fine-fiber net is present. *Metridium* and *Sagartia* have a fast, symmetrical response mechanism, but the response in the first stages is retraction of the oral disc by longitudinal musculature in the mesenteries. Dense networks of heavy fibers are present on the mesenteries, especially in *M. canum* (identification doubtful) which has 13 μ axons and a conspicuously fast longitudinal mesenteric contraction. *Peachia* also has well-developed mesenteric nets of stout fibers. *Actinia* is **intermediate** between these and *Anemonia*.

Stomphia is an anemone that can be aroused by simple electrical stimulation or contact with certain starfish to exhibit a complex sequence of actions (Fig. 8.46) including elongation, release of hold on substratum, and vigorous side-to-side bendings that bring about actual swimming (Sund, 1958). Wilson (1959) claimed it can under certain conditions show facilitation lasting several days; Hoyle (1960) could not repeat this aspect. Great variability in amplitude of response of certain muscles to constant electrical stimuli suggests repetitive firing and an excitable system separate from the through-conduction one.

The presence of a distinct nerve fiber layer in the epidermis of the column wall was at one time proposed (Carlgren, 1893) as the basis of a classification of the Actiniaria into the more primitive Protantheae, which retain this feature and the correlated ectodermal longitudinal muscles, and the Nynatheae, which are derived by reduction of both. Most anemones which have been studied are of the latter general type. Leghissa (1949a) reported a considerable difference in the nervous development in the tentacle among several actinian species.

2. Ceriantharia

The Ceriantharia with their very differently arranged and epidermal muscles, have an epidermal nerve fiber layer.

The peculiar burrowing anemone *Cerianthus* (Horridge, 1958) shows a rather typical brief facilitation (less than 1 sec), but the first stimulus evokes appreciable contraction. The chronaxie of this system is about 1/20 of that in *Calliactis*. Horridge gives curves of strength-duration relation and of the time course of refractoriness. The neuromuscular system appears to be highly specialized, since almost the only response that can be obtained is a **symmetrical longitudinal contraction** of the column. This is rapid and the excitation is conducted longitudinally through the nerve net with the highest velocity known in the phylum (1.3 m/sec); it appears to be a defensive withdrawal quite similar to that in serpuliform polychaetes. There is in addition a slow, local path but not spreading transversely; spread from tentacle to tentacle around the oral disc is so slow— 4 cm in 4 sec—that nonnervous means are suspected.

3. Madreporaria

Madreporarians have been seldom studied with respect to the organization of responses, but Horridge has added greatly to our knowledge in a recent memoir on coordination of the protective retraction (1957). The anatomy of the nervous system appears to be similar to that in anemones, based on examination of the thickness of the nerve fiber layer in sections. Functionally there is probably **a through-conducting system and a non-through-conducting system,** the former manifested by symmetrical, jerky shortening of the column via endodermal muscles, the latter by local and graded movements of disc, tentacles, and column in feeding responses. These nonsymmetrical and slower movements involve, at most, several neighboring polyps. Feeding movements differ among genera, for example, in the lack of participation of tentacles (Abe, 1938), the absence of ciliary reversal, different degrees of eversion of pharynx and mesenteric filaments and of secretion of mucus (Matthai, 1918; Yonge, 1930; Abe, 1938); but basically feeding movements are much like those of actinians.

Most corals **expand at night** and withdraw their polyps during the day. *Caulastrea* will promptly expand if placed in the dark during daytime and withdraw if illuminated at night. Abe (1939) has studied the well-known cleaning and righting reactions of the unattached solitary fungid corals. Cleaning is not only by ciliary action; heavy objects can be thrown off by a great expansion of the body, which requires about one minute and which is repeated if unsuccessful at fairly regular intervals of 15–20 min. **Righting** is performed by a similar series of efforts and is believed to be a cleaning response rather than a geotaxis.

The **retraction response** does not occur to food or ordinary mechanical stimulation, but only to electrical shocks and injurious mechanical stimuli. We may conclude that it does not depend on a protracted burst of impulses, and can not use the same net which mediates the feeding responses, for a single shock activates the whole through-conducting system in species possessing one, as shown by delivering a second shock to some other point, causing a symmetrical shortening. A number of corals require the facilitation of a prior impulse but respond on the second. (See further, p. 488). Two conducting systems, then, coexist in individual polyps, not only in colonial corals but in the large solitary *Fungia* as well.

Horridge's work shows the interesting fact that the **retraction system between polyps** is in most corals examined only locally conducting. Exceptions are the astreids *(Favia* and *Coelaria)* and the perforate corals *(Acropora)* where the first shock spreads at least to a large area; even here further shocks can spread to still larger areas. In *Acropora* a single shock is sufficient to cause a widespread jerk of the polyps, and in the astreids the first shock causes a slight waving of tentacles over a wide area; however, retraction occurs only on the second shock. A number of alcyonarians (see below) exhibit colonial through-conduction. At the other extreme the large polyps of *Lobophyllia* are physiologically and histologically quite separate, with little or no coordination to any form of stimulation. In between, the non-through-conducting types were of three general kinds, as already analyzed with theoretical models (see p. 488).

(a) *Galaxea* as well as the zoanthid *Palythoa* exhibits increments of area with **successive stimuli,** about one polyp in width; they behave like the classical cases of interneural facilitation of Pantin, in which all junctions may be supposed to require priming by a single impulse before they will transmit. However, there is some form of saturation, in that spread stops short of the whole colony and in fact is limited to a characteristic distance which is smaller for lower frequencies.

(b) Other species like *Porites* respond to the **first shock** over a considerable area (about 2×2 cm = 200–300 polyps) and to successive shocks with only a slight increment in area. Furthermore, the retraction close to the electrode is stronger than that farther away, even to a single shock; whether this is due to repetitive discharge or to a density or other effect is not known.

(c) The **third type of spread** was only found in alcyonarians like *Sarcophyton* and *Heteroxenia* but may be mentioned here for convenience. In *Sarcophyton* the spread resembles interneural facilitation for the first few shocks but then the increments of area become much greater and a colony a meter across may be totally involved—if the frequency is high enough, say 1 or 2 per second; the possibility can not yet be eliminated that a separate through-conduction system is brought in. At low frequencies spread is like *Porites*. Besides polyp retraction, low frequencies and even single shocks cause slow and maintained contractions of the colony which are local and graded.

These **several types of colonial nervous systems** are compared in graphic form in Fig. 8.47.

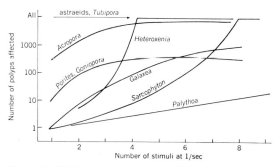

Figure 8.47. The relations between the number of stimuli at 1 per second and the number of polyps affected, for a variety of corals. There are three kinds of response: the ones in which the whole colony is active from the first stimulus, shown by a *horizontal line* at the top of the picture; the ones with an increasing slope such as *Heteroxenia* and *Sarcophyton* in which ultimately a wave spreads over the whole colony; and those with a decreasing slope in which the response spreads to a limited number of polyps no matter how many stimuli are given. [Horridge, 1957.]

Values for the **maximum interval** between shocks still permitting facilitation to be expressed are in the vicinity of 5 sec (about 26°C), in agreement with sea anemones from both tropical and temperate waters. Absolute **refractory periods** for through-conduction systems may be as short as 20 msec or as long as 80 msec in different species. **Conduction velocity** in different species varies between about 0.02 and 0.5 m/sec.

4. Alcyonaria

The Alcyonaria have been more frequently studied. Kassianow (1908a, 1908b) and earlier workers showed that the distribution of nerve cells is similar to that in anemones. There is a conspicuous discrepancy between the two sides of the pinnate tentacles, the nervous system being highly developed on the oral side and much reduced on the aboral. There are radial tracts of nerve fibers in the oral disc over the septal attachments. There may be large bipolar and multipolar cells in the column epidermis in addition to small nerve cells. The principal concentration of nervous tissue is in the oral disc and upper end of the pharynx. A nerve fiber layer is present on the mesenteries, but is much reduced on the lower pharynx and gastral filaments.

The **colonial nervous system** appears to be markedly different in different alcyonarians. Parker (1925) described responses in three species of Panamanian gorgonians as being "as local as one could well imagine," only a few zooids in the immediate proximity of the stimulus responding, and offering no evidence of conduction. The gorgonian *Acabaria* can be excited to a distance of a centimeter by apparently maximal electrical stimulation (Horridge, 1957). A plexaurid gorgonian in the Marshall Islands which J. Thorson and I examined showed conduction at relatively high velocity (0.4 m/sec, 29°C) throughout the colony —involving distances of more than 60 cm; single electric shocks elicited graded response out to a few centimeters, the third or fourth at 1 per second became through-conducting. Besides electric shocks, merely crushing or cutting the coenenchyme also caused this wide-spreading retraction. The alcyonacean *Veretillium* is described as possessing nervous continuity between individuals, on anatomical evidence (Niedermayer, 1914). Documentation is given above for *Sarcophyton*. The stoloniferan *Tubipora* studied by Horridge (1957) in the Red Sea, exhibits a **through-conducting system** for the whole colony which can be completely excited by a single shock as shown by the complete wave of retraction to a second shock applied anywhere. This system is difficult to excite nonelectrically—only injurious stimuli such as severe pinching or beheading a polyp with scissors is adequate to release through-conduction over the colony, all gentler mechanical stimuli causing only local spread or response of the individual polyp; we saw the same feature in some hydroids, above. Similar behavior has been found in *Alcyonium, Solenopodium,* and some colonies of *Heteroxenia*. In contrast to *Sarcophyton*, where summation of successive stimuli applied at one locus is needed to bring about through-conduction, in *Alcyonium* there is no difference in response between eight shocks at 1 per second applied to one point and eight shocks at 1 per second applied to widely separated points.

We have several studies of the nervous physiology of the California sea pansy, *Renilla* (Parker, 1920; Nicol, 1955, 1960). There is a **through-conducting nerve net** over the entire animal and it seems probable that the same net mediates luminescence, retraction of the autozooids, and contraction of the whole colony. These responses have different thresholds of intensity of mechanical stimulation, ascending in the order given and we may suppose that this represents a hierarchy of sensitivity to frequency and number of impulses at the respective neuroeffector junctions. Conduction in this net occurs at 0.04 to 0.08 m/sec (21°C) and does not normally exhibit interneural facilitation but, as in medusae, this may be brought out by magnesium or repetitive conduction across a narrow bridge of tissue (Fig. 8.48).

Facilitation is conspicuous in *Renilla* and is at the neuroeffector junction. At frequencies of one per second, two or three stimuli are necessary to produce the first response. Three shocks per second are already too high to follow, but it is not clear whether the **refractoriness** is in the conducting or the responding system. Following

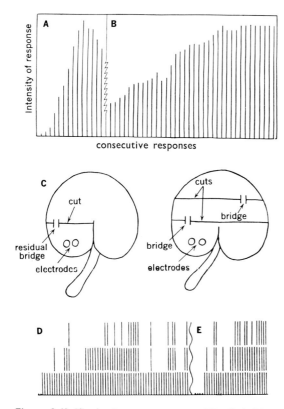

Figure 8.48. The luminescent response of *Renilla köllikeri* (Pennatulacea) to electrical stimulation. **A.** Response to a burst of shocks at a frequency of 1 per second, recording from the entire animal; the first response occurs on the second stimulus. **B.** Responses to a burst of 42 shocks at a frequency of 42 per minute; slit recording. Measurable responses above background were evident only on the sixth shock. **C.** Diagram of patterns of incisions made in the rachis of *Renilla* to investigate facilitation. **D** and **E.** Responses of an animal prepared as in the right-hand diagram in **C**; the bottom row represents responses in the proximal region (under the electrodes) to repetitive stimulation, the middle row represents the central piece, and the upper row, the distal region. The small *pips* at the bottom are initial stimuli before the onset of responses. In **C** stimulation occurs at 42 per minute; in **D**, at 1 per second. [Nicol, 1955.]

maximal stimulation for some seconds, there is commonly an **afterdischarge** and there may be a long continued "frenzy" lasting up to 30 min or more during which waves of luminescence cross the upper surface of the colony; these waves appear to reflect from the opposite border, arise from new loci of origin, meet and cancel each other, to be succeeded by new waves arising from multiple loci. The loci of origin are either in the nerve net or accessible to it since they become active as a result of activity in the net. When the "frenzy" is well developed the loci are perhaps autorhythmic, but during the crucial stage at the beginning of the episode, when the new pacemaker loci are developing, the more activity there is in the nerve net, the more likelihood there is that new pacemakers will arise. **Retraction** of the autozoids is readily elicited by stimulation of the rachis mechanically or by electrical stimulation at any point, but it is difficult to elicit by mechanical stimulation of the autozooid itself, and it then does not spread. It may be that there is only a local nerve net in the autozooid itself, which does not make contact functionally with the through-conducting net, although both control the muscles of retraction. The response to food is also confined to the autozooids stimulated. **Peristalsis** of two kinds occurs and was believed by Parker to be myogenic; in the rachis it can pass around any number and pattern of cuts. *Cavernularia* has several more types of peristalsis and exhibits complex combinations of them (Honjo, 1940).

Luminescent waves can pass across a cut through the superficial tissue of *Cavernularia* (Harvey, 1917); therefore the **deep plexus in the gastrodermis** must be capable of conducting this form of excitation. Hasama (1943) recorded action potentials associated with luminescence in this form (see further, Ctenophora, p. 525). Honjo (1944) found two kinds of conducted luminescent waves: a faster (0.10 m/sec) and a slower (0.075 m/sec), producing a weaker and a stronger glow and elicited from different parts of the colony, general and peduncular, respectively. He concluded that there must be **separate conducting systems**. The faster one, but not the slower, will cross a superficial incision. In addition there is local luminescence to weak stimuli. The sea pens *Leioptilus* (Davenport and Nicol, 1956) and *Pennatula* (Moore, 1926a; Nicol, 1958) respond with luminescence much like *Renilla*. A curious refractoriness is described as a result of strong stimulation, but subsequent repeated stimuli gradually overcome this by progressively spreading in the manner of interneural facilitation.

The xenid corals include species with the startling habit of **rhythmically pulsating polyps** (Fig.

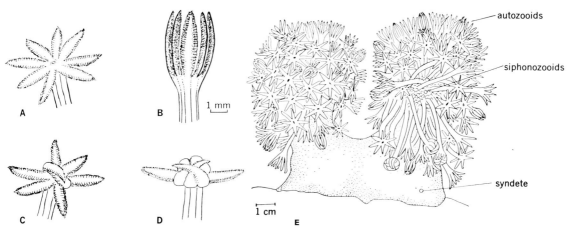

Figure 8.49. Responses of *Heteroxenia* (Alcyonaria). **A.** Relaxed autozooid. **B.** Autozooid contracted in a beat; this symmetrical movement recurs day and night at 30–45 per minute, uncoordinated between zooids. **C, D.** In a partial spasm; either in response to touch or spontaneously, an asymmetrical, local, maintained contraction can affect one or a few tentacles without interfering with the beat of the others, or all tentacles can contract, in a spasm, usually spreading to other zooids. The beat and the spasm involve two different nerve nets. **E.** Entire colony; individuals here and there are beating. [Horridge, 1956d.]

8.49; Horridge, 1956d; Krukenberg, 1887). Each individual has its own pacemaker at one per 3 seconds. While there must be a through-conducting system within the polyp to assure the rhythmic beat of all eight tentacles, it is strictly limited: the rhythm becomes independent in two halves of a Y-split zooid. Conversely, a non-spontaneous non-through-conducting system mediates a spasm of fewer or more tentacles, according to stimulus strength, and this net can spread the excitation to the other limb of a Y-preparation, to other polyps, and, with increasing increments, to the whole colony, via a deep tissue layer. The stronger stimuli which spread excitation farther also cause a longer-lasting spasm of tentacles and zooid retraction, apparently by an increased number of impulses in the net. This presumably acts both to penetrate farther into unexcited elements of the net, by interneural facilitation, and also to excite muscles progressively, manifesting their graded thresholds. The non-through-conducting net which becomes a through-conducting one shows its character well in the simple experiment of delivering two electrical shocks in succession to different regions: if these are more than 2 cm apart (sometimes much less) there will be no spread, whereas if they are given to the same region within this radius, extensive spread can result. **Interaction between the two** systems is not conspicuous, but the spreading system can influence the rhythmic system, for example it depresses its frequency.

C. Scyphozoa

These animals exhibit variety in the development of radial and circular musculature of the bell, in the activities of tentacles, and in the movements of feeding, but these properties have been little studied. Luminescence appears to be nervously controlled in *Pelagia* (Heymans and Moore, 1924; Moore, 1926b).

The most extreme departure from the typical plan is seen in the order, Stauromedusae, which includes the sessile form *Haliclystus* with an aboral stalk. **Rhythmic pulsations are lacking** and Gwilliam (1960) has provided the basis for an explanation in the absence of sufficient specialization for through-conduction to single shocks. The calyx as a whole is incapable of responding in a repetitive manner to repeated shocks at regular intervals. Through-conduction is achieved but only in the subumbrella and only at relatively high frequencies of stimulation—much too high to follow regularly. At low frequencies the radial parts of the calyx respond only locally. Apparently, with interneural facilitation a **non-through-conducting system is converted into a through-conducting one.** Velocity is low—0.07–0.15 m/sec at 12°C. In addition to this system, there appears to be a more **locally conducting subumbrellar nerve net** which coordinates the manubrium with the arms and tentacles. Graded and asymmetrical spontaneous activity arises from many loci and is probably inherent in many, widely scattered ganglion cells. It elicits only local movements, varying in extent with the intervals between impulses in the net. The exumbrellar surface is provided with nervous elements but is relatively insensitive. Evidence indicates that nerve fibers pass through the mesoglea; the nerve net is ectodermal. The properties of the nervous system of *Haliclystus*, though so different from both free-swimming jellyfish and sessile anemones, are adaptive in terms of the habit of life: in particular, the emphasis is on local, nonrhythmic spontaneity and the sequential calling up of more and more tentacle movement with mechanical stimulation of progressively greater intensity, followed by local and more general marginal contraction, involving eventually the whole calyx, and leading finally to symmetrical shortening of the stalk.

The **adaptiveness to the habit of life** is clear in many of the features of nervous organization. Unfortunately information is often quite incomplete at critical points; for example, what are the differences in habit among species of anemones whose neurophysiological differences are known?

We will not attempt in this work to develop arguments from the nervous system as to **general phylogenetic questions.** Some workers view the coelenterates as secondary to flatworms, having lost their heads and become sessile. While involving questions about the nervous system, such problems cannot be considered apart from many nonneurological aspects and we can only hope that the present work will assist those who assemble facts from all relevant fields.

CTENOPHORA

In spite of a rather large bibliography the basic evaluation of the nervous organization of this group is still not possible in the terms that can be used for the preceding and following groups. The evidence available has been taken to mean a nervous system essentially like that of coelenterates. The elaboration of those nervous parts concerned with the dominant form of activity, ciliary locomotion, is usually regarded as but a condensation of the general nerve net without particular significance. No other concentrations or ganglia are currently admitted. But ganglia have been claimed, long ago, in not unreasonable positions. No good information exists on the finer anatomy of the nervous condensations under the comb plates. The specialized coordination of cilia and luminescence as well as of muscular movements suggests a different line or degree of evolutionary development from coelenterates in the meridional strands. Centralization is certainly present to a degree in the apical region but is not yet satisfactorily understood.

The most important anatomical account is that of Hertwig (1880); others are listed in the bibliography. The general correctness of Hertwig's description is supported by the chief modern students, Heider (1927a, 1927b) and Korn (1959), who successfully applied methylene blue and silver methods. A long list of workers have examined the nervous control of the beat of comb plates and at least half a dozen workers the production of luminescence (Chang, 1954). Other aspects of physiology, including the nerve net, have hardly been studied.

I. ANATOMY

A **general subepidermal plexus** is distributed throughout the surface. Isopolar ganglion cells, mainly multipolar, send their prolongations to make contact with each other forming 3-, 4-, or

I. ANATOMY

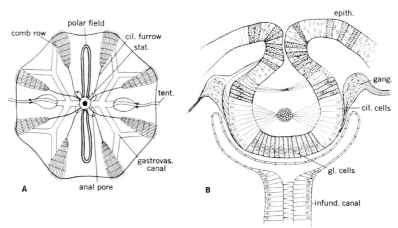

Figure 8.50. **A.** Aboral view of the cydippid ctenophore *Pleurobrachia*, showing aboral sense organ and polar fields. *cil. furrow*, ciliated furrow; *gastrovas. canal*, gastrovascular canal; *stat.*, statocyst; *tent.*, tentacle. [Hyman, 1940.] **B.** Vertical section through statolith of *Coeloplana*. *cil. cells*, ciliated cells whose cilia here support the statolith, and above form a covering membrane; *epith.*, epithelium of aboral surface; *gang.*, ganglion of sensory nerve cells, presumably innervating the statocyst (Horridge doubts there is a ganglion here, based on unpublished electron micrographs); *gl. cells*, gland cells; *infund. canal*, infundibular canal of gastrovascular system. [Abbott, 1907.]

5-cornered, nearly isodiametric meshes. The system is probably (Hertwig, Korn, contra Bethe) synaptic and not syncytial (Fig. 8.50). Only one nerve net is indicated. Samassa's (1892) arguments against a nervous system in ctenophores have all been adequately disposed of by later workers (Bethe, 1895; Child, 1933; Heider, 1927b; Korn, 1959). Specialization takes place **under the comb plate rows** (Fig. 8.51). The meshes of the network elongate meridionally and become con-

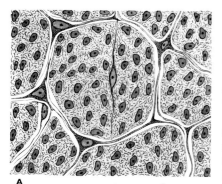

Figure 8.51.

A. The surface of *Beroë*, showing the nerve net; the processes of nerve cells do not appear to fuse. [Hertwig, 1880.] **B.** The basal cushion of a ciliary comb plate of *Pleurobrachia* with the strong strand of fibrous nerve tissue beneath it. [Korn, 1959.] **C.** The strand of fibrous nerve tissue under the ciliated groove and the surrounding general plexus of the body surface. [Heider, 1927.] In electron micrographs Horridge finds nerve fibers making synapses with the bases of ciliated cells in comb plates; vesicles are on the nerve side only. These are presumed to be inhibitory synapses from the ectodermal net which stops the beat. Apical cilia are of the 9+0 pattern and some are specialized as photoreceptors, with lamellate whorled bodies.

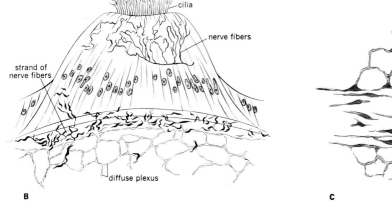
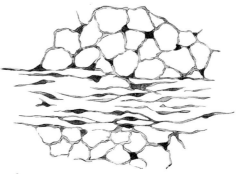

densed into a strong strand resembling a nerve or simple cord. This appears different immediately under the basal cushion of each comb plate and in the intervals between these. Under the cushion, unipolar cells are seen—a notable sign of specialization—and the density of nerve cells is high (Heider). Numerous finely branched bunches of nerve fibers penetrate between the epithelial cells of the cushion and end; in the other direction they join the dense strong strand underneath.

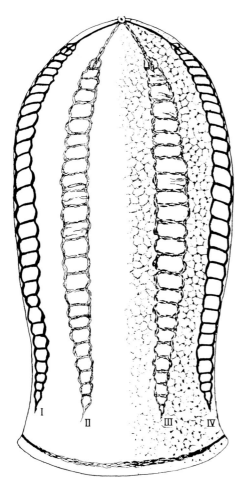

Figure 8.52. Semidiagrammatic representation of the nervous strands and general plexus of *Beroë*. Only four of the eight series of strands are shown. *I* and *IV* are schematically shown as solid black. *II* shows the directions of nerve fibers and *III* the nerve cells. Here and there are indicated nerve twigs that run up into the basal cushions. The general plexus of the body surface is only shown on the right; nuclei of nerve cells are shown between *III* and *IV*. The strand surrounding the mouth is shown with the same connections as the meridional systems just above it. [Heider, 1927.]

We shall see that the physiology points to some degree of complexity here and further close anatomy is needed. Child (1933) emphasizes differences between species in these structures although he could not get good stains showing nerve fibers.

At the aboral pole there is a **complex of apical structures** (Fig. 8.52), notably including a single statocyst in the center. This is a cavity lined by ciliated epidermal cells. From four symmetrical parts of the floor, in the interradii, four very long tufts of cilia ("balancers") project upward and meet in the center, there to support a mass of spherules, the statolith. Presumably this is heavy and places a differential stress on the cilia when the animal is tilted—just as happens in the organs of equilibrium of many higher groups. Komai mentions a constant vibrating motion of the statolith. The whole is enclosed in a transparent dome, called a bell or cupula. Abbott (1907) described **four ganglia** of large nerve cells at the angles of intersection of the tentacular and sagittal planes, just opposite the points of insertion of the long ciliary tufts but outside the statocyst capsule. Komai (1936) regarded these masses as muscle fibers, but his report cannot be decisive without further facts; such a gross disagreement as they have occasioned should be resolved by careful observation. It is to be expected that sensory cell bodies will lie somewhere near the statocyst and quite likely that they will have the quadriradiate symmetry of this region stamped upon them. Korn did not see this region clearly but believed that a well-formed layer of nerve fibers occurs under the sensory epithelium. Two **ciliary furrows** run out in the interradii from each balancer to connect to the aboral ends of the comb plate rows. A continuity of the nerve strand of the latter (via the furrows) with the apical nervous structures, whatever they include, is indicated but not described in detail.

The sensory epithelium of the statocyst is continuous on each side in the sagittal plane with an elongated ciliated depression termed the **polar field** or polar plate. This is presumed to be sensory also but only Hertwig saw a nervous strand associated with it, running along its edge.

A central nerve strand is found in the two long retractile **tentacles** that certain species possess.

The strand has nerve cells and sends fine fiber bundles to a plexus under the adhesive cells at the surface. Probably there are sensory nerve cells and local reflexes since the tentacles are muscular.

Around the mouth of some species (*Beroë*) is a ring of condensed nerve plexus and sensory cells but other species (*Pleurobrachia*) lack this. A plexus is said to lie beneath the stomach epithelium (Korn). No nerve fibers have been unequivocally seen in the jelly but they are believed to be there in order to reach the muscle fibers. Innervation of muscle has not been certainly seen; it has not even been determined whether the main pathways of distribution coincide with comb row nerve strands or not.

Two kinds of **sensory nerve cells** are mentioned: those with several stiff bristles and those with a single heavier projection (Fig. 8.53). Some members of the Lobata and Cestida have contractile sensory papillae at various places; the tips are loaded with sensory and gland cells and appear sensitive to touch. Schmidt (1954) called attention to four groups of large lipid-containing refractile cells in the aboral pole epithelium that he believed to be sensory; Korn could not support this interpretation. Komai (1936) referred

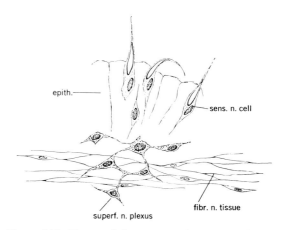

Figure 8.53. Margin of the sensory ridge in vertical section, showing sensory nerve cells *(sens. n. cell)* in the epithelium *(epith.);* beneath, the fibrous nerve tissue *(fibr. n. tissue)* of the nerve strand, and the superficial nerve plexus *(superf. n. plexus).* [Heider, 1927.]

to a rich array of apparently sensory vertical spindle-shaped cells solely on the ventral side and margins of the dorsal side of the peculiar form *Coeloplana*. His description is unconvincing; the fibers do not form a plexus and deeper lying horizontal fibers called "motor nerves" look like muscle cell outlines.

II. PHYSIOLOGY

That the **ciliary comb plates** are under nervous control has been shown in a number of ways by many workers (for example Bauer, 1910; Göthlin, 1920; Fedele, 1926; Coonfield, 1934). Earlier authors (Chun, 1880; Eimer, 1880; Verworn, 1889, 1891; Parker, 1905a) had already shown that the plates of a row normally beat in metachronal waves synchronous with the other rows as long as there is no interruption of the meridional row or ciliary furrow or ablation of the apical organ. Hence an impulse was presumed to spread from the latter to all eight rows. Nevertheless, after isolation a row can beat apparently normally but out of synchrony with others; indeed, any part of a row down to a single plate can beat. This agrees with the general experience from all groups that cilia have an autogenic tendency to beat and superimposed control only stops, reverses, or times the beat. Transmission is usually from aboral to oral but is possible in the reverse direction—in some species more than others (least in *Mnemiopsis*); reversed metachronal waves travel more slowly than normal oralward waves. Preventing a plate from beating, or removing one plate, stops transmission in *Pleurobrachia* and *Beroë*; in the lobates like *Mnemiopsis* holding, cooling, stretching, or removing two or three plates still permits transmission; and in *Cestus* (Verworn) even the removal of several plates is said not to block. The hypothesis of purely mechanical transmission, held by some earlier writers, is abandoned today.

The **hierarchy of synchronism** may represent a gradient of frequency of rhythmic autoexcitation. In *Mnemiopsis* Coonfield noted that when the rate of beating is low, each plate row becomes independent. The apical region presumably contains a pacemaker (or many equivalent pacemakers, the fastest of which controls) and its frequency may be generally higher than that of potential pacemakers resident in the structures common to pairs of rows—which often coordinate when total coordination is lost. In turn these pacemakers dominate lower ones that exist at all levels in each row, and are slower toward the oral end. The lowest level is the individual plate and it can beat without external impulse. The gradient with its high end aborally corresponds to other signs of a gradient shown by Child (1933). Closely adjacent levels may have similar natural periods according to this view. Therefore the result of Coonfield—that soon after cutting across a plate row (in one case in five minutes), synchrony of metachronal waves on the two sides is re-established—may mean that a slight mechanical factor and a similarity of frequency can lead to beating in phase without nervous continuity. In addition a rapid regeneration and reconstitution of connections is probable, but how rapid is not known. An oral half of *Mnemiopsis* can regenerate a new apical body to a functional degree so that the eight rows are coordinated within 48 hours. In this lobate form the plates on the auricles may even normally be unsynchronized with the rest.

The **normal polarity** of the gradient and the usual metachronal wave has been mentioned. In *Mnemiopsis* this is strong and although the comb plate beating can be stopped by a touch and reversed in effective stroke so that locomotion is reversed, the direction of the metachronal wave is never reversed. In other forms (*Pleurobrachia*) it can be, but it returns shortly to the aboral-oral direction. The ability to reverse is regarded by Child as a sign of a lower level of differentiation of the nervous transmitting mechanism.

It is noteworthy that we have here an apparently **true nervous inhibition,** like that already seen in certain coelenterate reactions. The ability to stop ciliary beat disappears in ctenophores with light narcosis.

Besides stopping and reversing effective stroke, the nervous system can control the velocity of the metachronal wave. In *Pleurobrachia* this is variable by a factor of two and is correlated with the frequency of the beating (Child) and the direction of the wave, being faster oralward. Lobate forms are not so variable but show higher velocities.

The fact that reverse metachronal waves can be elicited in some species by stimulation raises the question of **whether the two pathways are different.** Certainly they have different properties. Besides velocity and stability (the reverse waves are soon replaced by normal ones), it is significant that the aboral-oral waves are normally through-conducting whereas reverse waves are easily elicited which run for a short distance. In the latter case, repeated stimulation can build up the distance so that interneural facilitation is evident, overcoming a decremental spread and signifying the probability of synapses.

Particularly interesting as a sign that the pathways are not the same is the **result of clash of waves** from two directions. Commonly both are abolished but it is reported (Child) that one may survive!

Stimulation of the body surface adjoining a plate row leads to one or more waves in the row, traveling only orally. This is probably a physiological sign of the continuity of general nerve net and specialized meridional nerve strands. The farther from a plate row it is applied, the more stimulation is needed. Some **reflexes** have been described. The most prominent is a simple cessation of comb plate beating following weak stimuli; following stronger stimuli there is a muscular retraction of comb plate rows until they are roofed over, mechanically impeding the ciliary beat. The pathways may be different since the second goes to muscle fibers embedded in the jelly and is more likely to be local, whereas simple cessation is usually global (Göthlin, 1920). Kinoshita (1910) found in *Beroë* that mechanical stimulation of the oral end caused cessation of ciliary beat but stimulation of the aboral end caused acceleration; both effects were confined to the stimulated side of a cut across a comb plate row. Inhibition thus travels aborally after such stimuli. Directed locomotion involves differential inhibition of rows on one side to turn the animal and probably arises in the apical organ, perhaps with

II. PHYSIOLOGY

gravity acting asymmetrically on the statolith. The wave of inhibition conducts orally in this case.

The brilliant **luminescence is nervously controlled**, as already indicated by Peters (1905) and early authors, who noted that it occurs not with movement of the animal but upon stimulation and spreads along defined lines under the comb rows. Moore (1924) placed the path of conducted excitation in these meridians, and stated that the general nerve net is not involved in luminescence. *Mnemiopsis* has been most carefully studied from this point of view (Chang, 1954). The response to single brief electric shocks is a flash of a few hundred milliseconds that actually comprises a complex series of peaks of light output, even when recorded through a very small window. These are doubtless from many unsynchronized photogenic effectors (Fig. 8.54). Simple, consistent wave forms of light output can only be obtained by what is probably direct electrical stimulation of the effectors; such responses show gradation, summation, smooth tetany, and short

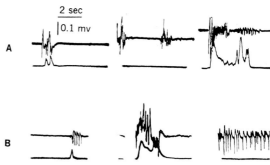

Figure 8.55. Simultaneous recordings of light output and electrical potentials in *Mnemiopsis*. *Upper trace*, potential change, surface macroelectrodes. *Lower trace*, luminescence recorded by photometer, which covers a larger area than the potential electrodes. **A.** Responses resulting from stimulation. **B.** Responses to agitation or spontaneous activity. There can be action potentials without light production and light out of proportion to the action potentials. [Chang, 1954.]

refractoriness. Apparent facilitation may in reality be treppe or increased effector action under this form of stimulation. Direct stimuli can be followed up to 30 per second or more; indirect response via the nerve pathways fails to follow more than once in 15 sec! Velocity of spread along the canals varies between 4.7 and 26.6 cm/sec (24°C); it is the same in both directions. This is similar to velocities in many coelenterate nerve nets. Action potentials were recorded by Chang synchronously with the flashes but of even more complex form (Fig. 8.55). A burst of deflections up to 100 per second accompanies much smoother peaks of light output. He concludes that these are nerve impulses in the meridional strands. (a) They do not correspond in form to the luminous effector response. (b) They are not likely to be ciliary since they do not occur regularly while the plates are active. (c) They occasionally occur without luminescence as though cilioregulatory impulses were going by.

Muscular responses have not been systematized. Some are mentioned above. *Coeloplana* is a flat creeping form whose first response to touch of the margin is retraction of the two tentacles, evidencing a pathway, probably nervous. It also rights itself by muscular action. Lobate species actively move the oral lobes and many forms strongly contract the body and the tentacles.

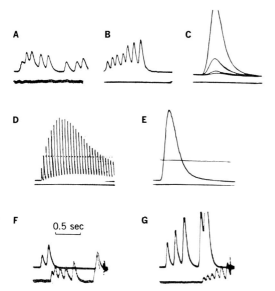

Figure 8.54. Photometer recordings of the light output of luminescence responses of *Mnemiopsis*. **A.** Multiple flashes to a single shock, from part of an intact meridional canal. **B.** The same. **C.** Responses from a small section of canal to stimulation at 128 v with durations of 1, 2, 5, and 10 msec. **D.** Responses to stimuli at 5 per second. **E.** Response to stimuli at 30 per second. **F.** Measurement of velocity of conduction of the luminous response along a canal; fast conduction. **G.** Slow conduction. [Chang, 1954.]

The surface of *Mnemiopsis* is **sensitive to chemicals,** such as food juice, in definite places (Coonfield, 1936a) and also to temperature and contact.

Moore (1933) ascribes significance—as a sign of phyletic level—to the observation that *Coeloplana* reverses its normal reaction to touch under strychnine, releasing its hold of the bottom, flexing dorsally, knotting up, and twitching in unmistakable strychnine spasms. This and other drug reactions are said to be advanced over those of coelenterates and comparable to those of echinoderms and flatworms. The structural level is closer to the former and seems not to go hand in hand with differentiation in reaction to drugs.

The **statocyst influences regeneration.** Pieces with a statocyst regenerate faster. Pieces without one must first regenerate that organ; then the new plate rows and other parts arise in relation to it. A grafted statocyst can inhibit regeneration of the host's statocyst. Pieces of plate rows retain their original polarity with respect to metachronal waves even if grafted in reversed orientation (Coonfield, 1936).

No physiological evidence of diffuse conduction and hence of a true nerve net appears to have been reported. The properties of the meridional nerve pathways—exhibited, for example, by allowing one of two clashing waves to survive and pass on—show that this is something more than a concentrated nerve net of a coelenterate type. The few comments on stimulation between meridians indicate a highly decremental spread; here there may be a typical local net.

Classified References

COELENTERATA

General

Bacq, 1941, 1947; Buddenbrock, 1952; Bullock, 1940, 1951; Fedele, 1924, 1926; Fortuyn, 1920; Grundfest, 1956; Hempelmann, 1926; Hilton, 1921; Jennings, 1906; Loeb, 1902, 1905; Meyer, 1955; Nagel, 1896; Pantin, 1950, 1952, 1955, 1956; Parker, 1919b; Schneider, 1902; Uexküll, 1921; Verworn, 1891; Warden et al., 1940.

Anatomy of Polyps

Allman, 1871–1872; Ashworth, 1899; Batham, 1956; Batham et al., 1960, 1961; Carlgren, 1893; Chun, 1881; Citron, 1902; Claus, 1878; Conn and Beyer, 1883; Duerden and Ayers, 1905; Föyn, 1927; Gelei, 1925; Grimstone et al., 1958; Grošelj, 1909; Hadzi, 1909; Haime, 1854; Hardy, 1891; Harm, 1903; Havet, 1901, 1922; Heider, 1877, 1879, 1881, 1886, 1895, 1899; Hertwig and Hertwig, 1879, 1880; Hess et al., 1957; Hickson, 1895; Hilton, 1927; Jickeli, 1882, 1883; Jourdan, 1879; Kassianow, 1908a, 1908b; Kleinenberg, 1872; Komai, 1942; Korotneff, 1876, 1884, 1887; Kukenthal and Broch, 1911; Leghissa, 1949a, 1949b, 1950; Lendenfeld, 1882, 1883; Lipin, 1909, 1911; Mackie, 1960; Marshall, 1923; May, 1903; McConnell, 1931, 1932; Miyoshima, 1898; Niedermeyer, 1914; Parker and Titus, 1916; Pax, 1914; Robson, 1961a; Schaeppi, 1898; Schneider, 1890, 1892; Semal-Van Gansen, 1952; Spangenberg, 1960; Torelli, 1952; Torrey, 1902; Totton and Mackie, 1960; Wolff, 1904; Woollard and Harpman, 1939; Wulfert, 1902; Zoja, 1890, 1892.

Anatomy of Medusae

Berger, 1900; Bohm, 1878; Bozler, 1927a, 1927b; Claus, 1864, 1878a, 1878b, 1883; Dawydoff, 1953; Eakin and Westfall, 1962; Haeckel, 1866; Hertwig and Hertwig, 1878; Hesse, 1895; Horridge, 1954a, 1955a, 1956a; Hyde, 1902; Kassianow, 1901; Komai, 1942; Korotneff, 1876; Krasinska, 1914; Lehmann, 1923; Linko, 1900; Maaden, 1939; Schäfer, 1879; Thill, 1937; Wolff, 1904; Wu, 1927; Yamashita, 1957a.

Physiology of Polyps

Abe, 1938, 1939a, 1939b; Allabach, 1905; Baker, 1952; Batham and Pantin, 1950a, 1950b, 1950c, 1954; Bohn, 1906, 1907a, 1907b, 1907c, 1908a, 1908b, 1908c, 1909, 1910a, 1910b; Bott, 1942; Broch and Horridge, 1957; Brock, 1927; Buck, 1955; Buck and Coyle, 1955; Bullock and Nachmansohn, 1942; Cardot, 1927; Chapman, 1949; Davenport and Nicol, 1956; Delphy, 1939; Ewer, D. W., 1960; Ewer, R. F., 1947; Feldman and Lenhoff, 1960; Fleure and Walton, 1907; Florey, 1951; Gee, 1913; Glumac, 1953; Haase-Eichler, 1931; Hall and Pantin,

1937; Hargitt, 1907; Harvey, 1917a, 1917b, 1921; Hasama, 1943; Haug, 1933; Honjô, 1940, 1944; Horridge, 1956c, 1956d, 1957b, 1958; Hoyle, 1960; Jennings, 1905; Jordan, 1908, 1934; Josephson, 1961a, 1961b, 1961c, 1961d, 1962; Josephson et al., 1961; Kepner and Hopkins, 1924; Koshtoyants, 1959; Koshtoyants and Smirnova, 1955; Krukenberg, 1887; Leghissa, 1949b; Lenhoff and Schneiderman, 1959; Loeb, 1895, 1900; Loomis, 1955; Mackie, 1959; Marshall, 1923; Mathias et al., 1957; Matthai, 1918; Mitra and Mukerjee, 1924; Mitropolitanskaya, 1941; Moore, 1926a, 1927; Nagel, 1892; Needler and Ross, 1958; Nicol, 1955a, 1955b, 1955c, 1958a, 1960; North, 1957; Norton and Beer, 1947; Nussbaum, 1887; Palombi, 1940; Panceri, 1872a, 1872b; Pantin, 1935a, 1935b, 1935c, 1935d, 1942; Pantin and Pantin, 1943; Pantin and Vianna Dias, 1952a; Parker, 1896, 1905, 1912, 1916, 1917a, 1917b, 1917c, 1917d, 1917e, 1917f, 1918, 1919a, 1920a, 1920b, 1925, 1932; Parker and Marks, 1928; Passano and Pantin, 1955, 1956; Pearse, 1906; Piéron, 1906a, 1906b, 1906c, 1908e, 1909; Pollock, 1883; Portmann, 1926; Pütter, 1903; Rand, 1909; Regnart, 1927; Reis, 1943, 1953; Retterer, 1907; Riddle, 1911; Robson, 1961a, 1961b; Ross, 1945a, 1945b, 1952, 1955, 1956, 1957a, 1957b, 1957d, 1960a, 1960b; Ross and Pantin, 1940; Schlünsen, 1935; Schmid, 1911; Siedentop, 1927; Sund, 1958; Torrey, 1904a, 1904b, 1907; Totton and Mackie, 1960; Trembley, 1744; Willem, 1927, 1928; Wilson, 1959; Yanagita, 1959; Yonge, 1930; Zhirmunsky, 1958; Zoja, 1890, 1891; Zubkov and Polikarpov, 1951.

Physiology of Medusae

Bauer, 1927; Baylor and Smith, 1957; Berger, 1900; Bethe, 1908, 1909a, 1909b, 1935, 1937, 1940; Bozler, 1926a, 1926b; Bullock, 1943; Bullock and Nachmansohn, 1942; Cary, 1914, 1915, 1917; Chapeaux, 1892; Ebbecke, 1957; Edney, 1939; Eimer, 1878; Fränkel, 1925; Gwilliam, 1960; Harvey, 1911, 1912, 1914, 1922; Henschel, 1935; Heymans and Moore, 1924; Horridge, 1953, 1954b, 1955a, 1955b, 1955c, 1956b, 1959; Horstmann, 1934a, 1934b; Hyman, 1940; Irisawa et al., 1956; Jordan, 1912; Kinoshita, 1911a, 1911b, 1937, 1941; Mayer, 1906, 1908b, 1911, 1912, 1914, 1915, 1916a, 1916b, 1916c, 1917a, 1917b; Milne, 1938; Moore, 1926b; Murbach, 1903; Nagel, 1893, 1894, 1898; Pantin and Vianna Dias, 1952b; Passano, 1958; Passano and McCullough, 1960, 1961; Romanes, 1877, 1878, 1885; Sanzo, 1903a, 1903b; Siedentop, 1927; Skramlik, 1945; Steiner, 1934; Terry, 1909; Uexküll, 1901; Veress, 1911, 1938; Vetochin, 1926; Wolf, 1928; Yamashita, 1957b; Yerkes, 1902a, 1902b, 1902c; Yerkes and Ayer, 1903; Zhukov, 1939.

CTENOPHORA

Abbott, 1907; Arsharskii and Rozanova, 1940; Bauer, 1910; Bethe, 1895; Bradfield, 1955; Chang, 1954; Child, 1933; Chun, 1871, 1880, 1882; Claus, 1864; Coonfield, 1934, 1936a, 1936b; Dawydoff, 1933; Eimer, 1873, 1880; Fedele, 1924, 1926; Göthlin, 1920; Harvey, 1917a, 1921; Heider, 1927a, 1927b; Hertwig, 1880; Heymans and Moore, 1925; Hykes, 1927; Kinoshita, 1910; Komai, 1922, 1936; Korn, 1959; Moore, 1924, 1933b, 1936; Mortensen, 1912; Nagel, 1893; Parker, 1905; Peters, 1905; Samassa, 1892; Schmidt, 1954; Verworn, 1889, 1891; Wolf, 1937.

Bibliography

ABBOTT, J. F. 1907. The morphology of *Coeloplana. Zool. Jb. (Anat.)*, 24: 41-70.

ABE, N. 1938. Feeding behavior and the nematocyst of *Fungia* and 15 other species of corals. *Palao trop. biol. Stud.* 1:469-521.

ABE, N. 1939a. On the expansion and contraction of the polyps of a reef coral. *Caulastraea furcata* Dana. *Palao trop. biol. Stud.*, 4:651-669.

ABE, N. 1939b. Migration and righting reaction of the coral, *Fungia actiniformis* var. *palawensis* Döderlein. *Palao trop. biol. Stud.*, 4:671-694.

ALLABACH, L. F. 1905. Some points regarding the behavior of *Metridium. Biol. Bull., Woods Hole*, 10:35-43.

ALLMAN, G. J. 1871-1872. *A Monograph of the Gymnoblastic or Tubularian Hydroids*. Ray Society, London.

ARSHARSKII, I. A. and ROZANOVA, V. D. 1940. Chronaxie and rhythm of the medusa (*Pileme pulmo*) in ontogenesis. *Bull. Biol. Med. exp. URSS*, 9:412.

ASHWORTH, J. H. 1899. The structure of *Xenia hicksoni*, nov. sp., with some observations on *Heteroxenia elizabethae. Quart. J. micr. Sci.*, 42: 245-304.

BACQ, Z. M. 1941. Physiologie comparée de la transmission chimique des excitations nerveuses. *Ann. Soc. zool. Belg.*, 72:181-203.

BACQ, Z. M. 1947. L'acetylcholine et l'adrenaline chez les Invertébrés. *Biol. Rev.*, 22:73-91.

BAKER, J. R. 1952. *Abraham Trembley of Geneva—Scientist and Philosopher 1710-1784*. Arnold, London.

BATHAM, E. J. 1956. Note on the mesenteric nerve net of the anemone *Metridium canum* (Stuckey). *Trans. roy. Soc., N. Z.*, 84:91-92.

BATHAM, E. J. and PANTIN, C. F. A. 1950a. Muscular and hydrostatic action in the sea-anemone *Metridium senile* (L.). *J. exp. Biol.*, 27:264-289.

BATHAM, E. J. and PANTIN, C. F. A. 1950b. Inherent activity in the sea anemone, *Metridium senile* (L.). *J. exp. Biol.*, 27:290-301.

BATHAM, E. J. and PANTIN, C. F. A. 1950c. Phases of activity in the sea anemone, *Metridium senile* (L.), and their relation to external stimuli. *J. exp. Biol.*, 27:377-399.

BATHAM, E. J. and PANTIN, C. F. A. 1954. Slow contraction and its relation to spontaneous activity in the sea anemone *Metridium senile* (L.). *J. exp. Biol.*, 31:84-103.

BATHAM, E. J., PANTIN, C. F. A., and ROBSON, E. A. 1960. The nerve-net of the sea anemone *Metridium senile*; the mesenteries and the column. *Quart. J. micr. Sci.*, 101:487-510.

BATHAM, E. J., PANTIN, C. F. A., and ROBSON, E. A. 1961. The nerve-net of *Metridium senile*: artifacts and the nerve-net. *Quart. J. micr. Sci.*, 102:143-156.

BAUER, V. 1910. Über die anscheinend nervöse Regulierung der Flimmerbewegung bei den Rippenquallen. *Z. allg. Physiol.*, 10:230-248.

BAUER, V. 1927. Die Schwimmbewegungen der Quallen und ihre reflektorische Regulierung. *Z. vergl. Physiol.*, 5:37-69.

BAYLOR, E. R. and SMITH, F. E. 1957. Diurnal migration of plankton crustaceans. In *"Recent Advances in Invertebrate Physiology"*, B. T. Scheer (ed.). Univ. of Oregon Press, Eugene.

BERGER, E. W. 1900. Physiology and histology of the Cubomedusae including Dr. F. S. Conant's notes on the physiology. *Mem. Biol. Lab., Johns Hopk. Univ.*, 4(4):1-84.

BETHE, A. 1895. Der subepitheliale Nervenplexus der Ctenophoren. *Biol. Zbl.*, 15:140-145.

BETHE, A. 1903. *Allgemeine Anatomie und Physiologie des Nervensystems*. G. Thieme, Leipzig.

BETHE, A. 1908. Die Bedeutung der Elektrolyten für die rhythmischen Bewegungen der Medusen I. Die Wirkung der im Seewasser enthaltenen Salze auf die normale Meduse *Pflüg. Arch. ges. Physiol.*, 124:541-577.

BETHE, A. 1909a. Die Bedeutung der Elektrolyten für die rhythmischen Bewegungen der Medusen II. Angriffspunkt der Salze, Einfluss der Anionen und Wirkung der OH- und H-Ionen. *Pflüg. Arch. ges. Physiol.*, 127:219-273.

BETHE, A. 1909b. Abweichungen vom gewöhnlichen Verlauf der Extrasystole beim Herzen und bei der Meduse. *Arch. Anat. Physiol., Physiol. Abt.*, 1909:385-401.

BETHE, A. 1935. Versuche an Medusen als Beispiel eines primitiven neuromuskulären Reaktionssystems *Pflüg. Arch. ges. Physiol.*, 235:288-315.

BETHE, A. 1937. Experimentelle Erzeugung von Störungen der Erregungsleitung und von Alternans- und Periodenbildungen bei Medusen im Vergleich zu ähnlichen Erscheinungen am Wirbeltierherzen *Z. vergl. Physiol.*, 24:613-637.

BETHE, A. 1940. Die biologischen Rhythmus-Phänomene als selbständige bzw. erzwungene Kippvorgänge betrachtet. *Pflüg. Arch. ges. Physiol.*, 244:1-42.

BOHM, G. 1878. Helgolander Leptomedusen. *Jena. Z. Naturwiss.*, N. F., 5:68-203.

BOHN, G. 1906. La persistance du rythme des marées chez l'*Actinia equina*. *C. R. Soc. Biol., Paris*, 61:661-663.

BOHN, G. 1907a. L'influence de l'agitation de l'eau sur les actinies. *C. R. Soc. Biol., Paris*, 62:395-398.

BOHN, G. 1907b. Le rythme nycthéméral chez les actinies. *C. R. Soc. Biol., Paris*, 62:473-476.

BOHN, G. 1907c. Introduction à la psychologie des animaux a symétrie rayonnée. I. Mémoire. Les états physiologiques des actinies. *Bull. Inst. gén. psychol.*, 7:81-129.

BOHN, G. 1908a. De l'influence de l'oxigène dissous sur les réactions des actinies. *C. R. Soc. Biol., Paris*, 64:1087-1089.

BOHN, G. 1908b. Les facteurs de la rétraction et de l'épanouissement des actinies. *C. R. Soc. Biol., Paris*, 64:1163-1166.

BOHN, G. 1908c. L'épanouissement des actinies dans les milieux asphyxiques. *C. R. Soc. Biol., Paris*, 65:317-320.

BOHN, G. 1909. Les rythmes vitaux chez les actinies. *C. R. Ass. Franç. Av. Sci.*, 37:613-619.

BOHN, G. 1910a. Comparaison entre les réactions des actinies de la Méditerranée et celles de la Manche. *C. R. Soc. Biol., Paris*, 68:253-255.

BOHN, G. 1910b. Les réactions des actinies aux basses températures. *C. R. Soc. Biol., Paris*, 68:964-966.

BOTT, R. 1942. Über die Bewegungsarten der Pferde Aktinie (*Actinia equina*). *Natur u. Volk*, 72:192-199.

BOZLER, E. 1926a. Sinnes- und nervenphysiologische Untersuchungen an Scyphomedusen. *Z. vergl. Physiol.*, 4:37-80.

BOZLER, E. 1926b. Weitere Untersuchungen zur Sinnes- und Nervenphysiologie der Medusen: Erregungsleitung, Funktion der Randkörper, Nahrungsaufnahme. *Z. vergl. Physiol.*, 4:797-817.

BOZLER, E. 1927a. Untersuchungen über das Nervensystem der Coelenteraten. I. Teil: Kontinuität oder Kontakt zwischen den Nervenzellen? *Z. Zellforsch.*, 5:244-262.

BOZLER, E. 1927b. Untersuchungen über das Nervensystem der Coelenteraten. II. Teil: Über die Struktur der Ganglienzellen und die Funktion der Neurofibrillen nach Lebenduntersuchungen. *Z. vergl. Physiol.*, 6:255-263.

BRADFIELD, J. R. G. 1955. Fibre patterns in animal flagella and cilia. In: Fibrous Proteins and their Biological Significance, *Symp. Soc. exp. Biol.*, 9:306-334.

BROCH, H. and HORRIDGE, A. 1957. A new species of *Solenopodium* (Stolonifera: Octocorallia) from the Red Sea. *Proc. zool. Soc. Lond.*, 128:149-160.

BROCK, F. 1927. Das Verhalten des Einsiedler-Krebses *Pagurus arrosor* Herbst während des Aufsuchens, Ablösens und Aufpflanzens seiner Seerosen *Sagartia parasitica* Gosse. *Roux Arch. Entwicklungs Org.*, 112:204-238.

BUDDENBROCK, W. VON. 1952. *Vergleichende Physiologie. I. Sinnesphysiologie*. Birkhäuser, Basel.

BULLOCK, T. H. 1940. The existence of unpolarized synapses. *Anat. Rec.*, 78(Suppl.):67.

BULLOCK, T. H. 1943. Neuromuscular facilitation in scyphomedusae. *J. cell. comp. Physiol.*, 22:251-272.

BULLOCK, T. H. 1951. Facilitation of conduction rate in nerve fibers. *J. Physiol.*, 114:89-97.

BULLOCK, T. H. and NACHMANSOHN, D. 1942. Choline esterase in primitive nervous systems. *J. cell. comp. Physiol.*, 20:239-242.

CARDOT, H. 1927. De la spécificité dans les phénomènes de capture chez les actinies. *C. R. Soc. Biol., Paris*, 97:1224-1225.

CARLGREN, O. 1893. Studien über nordische Actinien. *K. svenska Vetensk-Akad. Handl.*, 25(10):1-148.

CARY, L. R. 1914. Studies on regeneration: the influence of the sense-organs on the rate of regeneration in *Cassiopea xamachana*. *Yearb. Carneg. Instn.*, 13:199-200.

CARY, L. R. 1915. Studies on the physiology of the nervous system of *Cassiopea*. *Yearb. Carneg. Instn.*, 14:202-204.

CARY, L. R. 1917. Studies on the physiology of the nervous system of *Cassiopea xamachana*. *Publ. Carneg. Instn.*, 251:121-170.

CHANG, J. J. 1954. Analysis of the luminescent response of the ctenophore, *Mnemiopsis leidyi*, to stimulation. *J. cell. comp. Physiol.*, 44:365-394.

CHAPEAUX, M. 1892. Contribution à l'étude de l'appareil de relation des hydromeduses. *Arch. Biol., Paris*, 12:647-682.

CHAPMAN, G. 1949. The mechanism of opening and closing of *Calliactis parasitica*. *J. Mar. biol. Ass. U.K.*, 28:641-649.

CHESTER, W. M. 1912. Wound closure and polarity in the tentacles of *Metridium marginatum*. *J. exp. Zool.*, 13:451-470.

CHILD, C. M. 1933. The swimming plate rows of the ctenophore, *Pleurobrachia*, as gradients: with comparative data on other forms. *J. comp. Neurol.*, 57:199-252.

CHUN, C. 1871. Das Nervensystem und die Muskulatur der Rippenquallen. *Abh. senckenb. naturf. Ges.*, 11:1-50.

CHUN, C. 1880. *Die Ctenophoren des Golfes von Neapel und der angrenzenden Meeres-Abschnitte*. Engelmann, Leipzig.

CHUN, C. 1881. Das Nervensystem der Siphonophoren *Zool. Anz.*, 4:107-111.

CHUN, C. 1882. Die Verwandtschaftsbeziehungen zwischen Würmern und Coelenteraten. *Biol. Zbl.*, 2:5-16.

CITRON, E. 1902. Beiträge zur Kenntnis des feineren Baues von *Syncoryne*

Bibliography

sarsii. *Arch. Naturgesch.*, 1:1-26.
CLAUS, C. 1864. Bemerkungen über Ctenophoren und Medusen. *Z. wiss. Zool.*, 14:384-393.
CLAUS, C. 1878a. Studien über Polypen und Quallen der Adria. *Denkschr. Akad. wiss. Wien*, 38:1-64.
CLAUS, C. 1878b. Über *Charybdea marsupialis*. *Arb. Zool. Inst. Univ. Wien*, 1:221-276.
CLAUS, C. 1883. *Untersuchungen über die Organisation und Entwicklung der Medusen*. Tempsky, Prague.
CONN, H. W. and BEYER, H. G. 1883. The nervous system of *Porpita*. *Johns Hopk. Univ. Stud. Biol.*, 2:433-445.
COONFIELD, B. R. 1934. Coördination and movements of the swimming plates of *Mnemiopsis leidyi* Agassiz. *Biol. Bull., Woods Hole*, 66:10-21.
COONFIELD, B. R. 1936a. Regeneration in *Mnemiopsis leidyi* Agassiz. *Biol. Bull., Woods Hole*, 71:421-428.
COONFIELD, B. R. 1936b. Apical dominance and polarity in *Mnemiopsis leidyi*, Agassiz. *Biol. Bull., Woods Hole*, 70:460-471.
DAVENPORT, D., and NICOL, J. A. C. 1956. Observations on luminescence in sea pens *(Pennatulacea)*. *Proc. roy. Soc. (B)*, 144:480-496.
DAWYDOFF, C. 1933. Quelques observations sur la morphologie externe et la biologie des *Ctenoplana*. *Arch. Zool. exp. gén.*, 75:103-128.
DAWYDOFF, C. 1953. Contribution à nos connaissances de l'*Hydroctena*. *C. R. Acad. Sci., Paris*, 237:1301-1302.
DELPHY, J. 1939. Sur quelques problèmes d'actinologie. I. Anesthésie (physiologie et technique). *Bull. Mus. Hist. nat., Paris*, 11:479-483.
DUERDEN, J. E. and AYERS, S. A. 1905. The nerve-layer in the coral *Coenopsammia*. *Rep. Mich. Acad. Sci.*, 7:75-77.
EAKIN, R. M. and WESTFALL, J. A. 1962. Fine structure of photoreceptors in the hydromedusan, *Polyorchis penicillatus*. *Proc. nat. Acad. Sci.*, 48:826-833.
EBBECKE, U. 1957. Reflexuntersuchungen an Coelenteraten. *Pubbl. Staz. zool. Napoli*, 30:149-161.
EDNEY, E. B., 1939. Notes on the behavior and reactions to certain stimuli of the fresh-water jellyfish, *Limnocnida rhodesia* Boulenger. *Occ. Pap. nat. Mus. S. Rhod.*, 8:1-11.
EIMER, T. 1873. Zoologische Studien auf Capri. I. Ueber *Beroë ovatus*, ein Beitrag zur Anatomie der Rippenquallen. Engelmann, Leipzig.
EIMER, T. 1878. *Die Medusen Physiologisch und Morphologisch auf Ihr Nervensystem Untersucht*. Tübingen.
EIMER, T. 1880. Versuche über kunstliche Teilbarkeit von *Beroë ovatus*. Angestellt zum Zweck der Controle seiner morphologischen Befunde über das Nervensystem dieses Thieres. *Arch. mikr. Anat.*, 17:213-240.
EWER, D. W. 1960. Inhibition and rhythmic activity of the circular muscles of *Calliactis parasitica* (Couch). *J. exp. Biol.*, 37:812-831.
EWER, R. F. 1947. The behavior of *Hydra* in response to gravity. *Proc. zool. Soc., Lond.*, 117:365-376.
FEDELE, M. 1924. Le prove sperimentali di una regolazione nervosa del movimento ciliare. *Pubbl. Staz. zool. Napoli*, 5:275-291.
FEDELE, M. 1926. Il problema della regolazione dell'attività vibratile nei metazoi. *Riv. Biol.*, 8:360-375.
FELDMAN, M. and LENHOFF, H. M. 1960. Phototaxis in *Hydra littoralis*: Rate studies and localization of the "photoreceptor". *Anat. Rec.*, 137:354.
FLEURE, H. S. and WALTON, C. L. 1907. Notes on the habits of some sea anemones. *Zool. Anz.*, 31:212-220.
FLOREY, E. 1951. Vorkommen und Funktion sensibler Erregungssubstanzen und sie abbauender Fermente im Tierreich. *Z. vergl. Physiol.*, 33:327-377.
FORTUYN, A. E. B. DROOGLEEVER. 1920. *Vergleichende Anatomie des Nervensystems. I. Teil. Die Leitungsbahnen im Nervensystem der wirbellosen Tiere*. Haarlem.
FÖYN, B. 1927. Studien über Geschlecht und Geschlechtszellen bei Hydroiden. *Arch. EntwMech. Org.*, 109:513-534.
FRÄNKEL, G. 1925. Der statische Sinn der Medusen. *Z. vergl. Physiol.*, 2:658-690.
GEE, W. 1913. Modifiability in the behavior of the California shore-anemone *Cibrina xanthogrammica* Brandt *J. Anim. Behav.*, 3:305-328.
GELEI, J. VON. 1925. Bemerkungen zu der morphologischen und physiologischen Gliederung des Körpers unserer Süsswasserpolypen. *Zool. Anz.*, 64:117-125.
GLUMAC, S. 1953. Contribution à la connaissance du fonctionnement des cellules à nématocystes chez l'Hydre d'eau douce. *Glasn. Mus. zrpsk. Zeml. (B)*, 5-6:503-511. (In Russian, with French summary.)
GÖTHLIN, G. F. 1920. Experimental studies on primary inhibition of the ciliary movement in *Beroë cucumis*. *J. exp. Zool.*, 31:403-441.
GRIMSTONE, A. V., HORNE, R. W., PANTIN, C. F. A., and ROBSON, E. A. 1958. The fine structure of the mesenteries of the sea-anemone *Metridium senile*. *Quart. J. micr. Sci.*, 99:523-540.
GROŠELJ, P. 1909. Untersuchungen über das Nervensystem der Aktinien *Arb. zool. Inst. Univ. Wien.*, 17:269-308.
GRUNDFEST, H. 1956. Some properties of excitable tissue. In: *Nerve Impulse*. D. Nachmansohn and H. H. Merritt (eds.). Josiah Macy, Jr. Foundation, New York.
GWILLIAM, G. F. 1960. Neuromuscular physiology of a sessile scyphozoan. *Biol. Bull., Woods Hole*, 119:454-473.
HAASE-EICHLER, R. 1931. Beiträge zur Reizphysiologie von *Hydra*. *Zool. Jb. (allg. Zool.)*, 50:265-312.
HADZI, J. 1909. Über das Nervensystem von *Hydra*. *Arb. zool. Inst. Univ. Wien.*, 17:225-268.
HAECKEL, E. 1866. Die Familie der Rüsselquallen *(Medusae Geryonidae) Jena. Z. Naturwiss.*, 2:93-202.
HAIME, J. 1854. Mémoire sur le cérianthe *(Cerianthus membranaceus) Ann. Sci. nat. (Zool.)*, (4)1:341-389.
HALL, D. M. and PANTIN, C. F. A. 1937. Nerve net in actinozoa V, Temperature and facilitation in *Metridium senile*. *J. exp. Biol.*, 14:71-78.
HARDY, W. B. 1891. On some points in the histology and development of *Myriothela phrygia*. *Quart. J. micr. Sci.*, 32:505-537.
HARGITT, C. W. 1907. Notes on the behavior of sea-anemones. *Biol. Bull., Woods Hole*, 12:274-284.
HARM, K. 1903. Die Entwicklungsgeschichte von *Clava squamata*. *Z. wiss. Zool.*, 73:115-165.
HARVEY, E. N. 1911. Effect of different temperatures on the medusa *Cassiopea* with special reference to the rate of conduction of the nerve impulse. *Pap. Tortugas Lab.*, 3:27-39.
HARVEY, E. N. 1912. The question of nerve fatigue. *Yearb. Carneg. Instn.*, 10:130-131.
HARVEY, E. N. 1914. The relation between the rate of penetration of marine tissues by alkali and the change in functional activity induced by the alkali. *Pap. Tortugas Lab.*, 6:131-157.
HARVEY, E. N. 1917a. The chemistry of light-production in luminous organisms. *Publ. Carneg. Instn.*, 251:171-234.
HARVEY, E. N. 1917b. Studies on bioluminescence. VI. Light production by a Japanese pennatulid, *Cavernularia haberi*. *Amer. J. Physiol.*, 42:349-358.
HARVEY, E. N. 1921. Studies on bioluminescence. XIII. Luminescence in the coelenterates. *Biol. Bull., Woods Hole*, 41:280-287.
HARVEY, E. N. 1922. Some recent experiments on the nature of the nerve impulse. *Arch. Neurol. Psychiat. Chicago*, 7:778-779.
HASAMA, B. 1943. Über die Bioluminesz des *Plocamophorus tilesii* Bergh sowie der *Cavernularia habereri* Moroff im Aktionsstrombild sowie im histologischen Bild. *Cytologia*, 13:146-154.

Haug, G. 1933. Die Lichtreaktionen der Hydren (*Chlorohydra viridissima* und *Pelmatohydra oligactis* (P.) Typica). *Z. vergl. Physiol.*, 19:246-303.

Havet, J. 1901. Contribution à l'étude du système nerveux des actinies. *Cellule*, 18:385-419.

Havet, J. 1922. La structure du système nerveux des actinies. Leur mécanisme neuro-musculaire. In: *Libro en honor de D. S. Ramón y Cajal*, Vol. I. Jiménez y Molina, Madrid.

Heider, A. R. von. 1877. *Sagartia troglodytes* Gosse, ein Beitrag zur Anatomie der Actinien. *S.B. Akad. Wiss. Wien, Math.-Nat. Kl.*, Abt. 1, 75:367-418.

Heider, A. R. von. 1879. *Cerianthus membranaceus* Haime. Ein Beitrag zur Anatomie der Actinien. *S.B. Akad. Wiss. Wien, math.-nat. Kl.*, Abt. 1, 79-204-254.

Heider, A. R. von. 1881. Die Gattung Cladoceora. *S.B. Akad. Wiss. Wien. math.-nat. Kl.*, Abt. 1, 84:634-667.

Heider, A. R. von. 1886. Korallenstudien. *Z. wiss. Zool.*, 44:507-535.

Heider, A. R. von. 1895. *Zoanthus chierchiae* n. sp. *Z. wiss. Zool.*, 59:1-28.

Heider, A. R. von. 1899. Über zwei Zoantheen. *Z. wiss. Zool.*, 66:269-288.

Heider, K. 1927a. Über das Nervensystem von *Beroë ovata*. *Nachr. Ges. Wiss. Göttingen*, 1927:144-157.

Heider, K. 1927b. Vom Nervensystem der Ctenophoren. *Z. Morph. Ökol. Tiere*, 9:638-678.

Hempelmann, F., 1926. *Tierpsychologie vom Standpunkte des Biologen*. Akad. Verlags., Leipzig.

Henschel, J. 1935. Untersuchungen über den chemischen Sinn der Scyphomedusen *Aurelia aurita* und *Cyanea capillata* und der Hydromeduse *Sarsia tubulosa*. *Wiss. Meeresuntersuch, Abt. Kiel*, N.F. 22:21-42.

Hertwig, O. and Hertwig, R. 1878. *Das Nervensystem und die Sinnesorgane der Medusen*. Vogel, Leipzig.

Hertwig, O. and Hertwig, R. 1879 and 1880. Die Actinien anatomisch und histologisch mit besonderer Berücksichtigung des Nervenmuskelsystems untersucht. *Jena. Z. Naturw.*, N.F., 6:457-586 and 7:39-89.

Hertwig, R. 1880. Ueber den Bau der Ctenophoren. *Jena. Z. Naturw.*, 14 (NF 7):313-457.

Hess, A., Cohen, A. I., and Robson, E. A. 1957. Observations on the structure of *Hydra* as seen with the electron and light microscopes. *Quart. J. micr. Sci.*, 98:315-326.

Hesse, R. 1895. Uber das Nervensystem und die Sinnesorgane von *Rhizostoma cuvieri*. *Z. wiss. Zool.*, 60:411-457.

Heymans, C. and Moore, A. R. 1924. Luminescence in *Pelagia noctiluca J. gen. Physiol.*, 6:273-280.

Heymans, C. and Moore, A. R. 1925. Note on the excitation and inhibition of luminescence in *Beroë*. *J. gen. Physiol.*, 7:345-348.

Hickson, S. J. 1895. The anatomy of *Alcyonium digitatum* (Alyonaria). *Quart. J. micr. Sci.*, 37:343-388.

Hilton, W. A. 1921. Nervous system and sense organs. IV. Coelenterata. *J. Ent. Zool.*, 13(suppl.):15-33.

Hilton, W. A. 1927. The muscular sense of invertebrates. *J. Ent. Zool.*, 19:75-76.

Honjo, I. 1940. Beiträge zur Nervenmuskelphysiologie der koloniebildenden Tiere I. Die Peristaltik von *Cavernularia*. *Annot. zool. jap.*, 19: 301-308.

Honjo, I. 1944. Supplementary knowledge of the neural physiology of *Cavernularia obesa* Valenciennes. *Seiro-Seitai*, 9:1-13. (In Japanese.) *Biol. Abstr.* 1951, 25, no. 9895.

Horridge, G. A. 1953. An action potential from the motor nerves of the jellyfish, *Aurellia aurita* Lamarck. *Nature, Lond.*, 171:400.

Horridge, G. A. 1954a. Observations on the nerve fibers of *Aurellia aurita*. *Quart. J. micr. Sci.*, 95:85-92.

Horridge, G. A. 1954b. The nerves and muscles of medusae. I. Conduction in the nervous system of *Aurellia aurita* Lamarck. *J. exp. Biol.*, 31:594-600.

Horridge, G. A. 1955a. The nerves and muscles of medusae. II. *Geryonia proboscidalis* Eschscholtz. *J. exp. Biol.*, 32:555-568.

Horridge, G. A. 1955b. The nerves and muscles of medusae. III. A decrease in the refractory period following repeated stimulation of the muscle of *Rhizostoma pulmo*. *J. exp. Biol.*, 32:636-641.

Horridge, G. A. 1955c. The nerves and muscles of medusae. IV. Inhibition in *Aequorea forskalea*. *J. exp. Biol.*, 32:642-648.

Horridge, G. A. 1956a. The nervous system of the ephyra larva of *Aurelia aurita*. *Quart. J. micr.Sci.*, 97:59-74.

Horridge, G. A. 1956b. The nerves and muscles of medusae. V. Double innervation in Scyphozoa. *J. exp. Biol.*, 33:366-383.

Horridge, G. A. 1956c. A through-conducting system coordinating the protective retract on of *Alcyonium* (Coelenterata). *Nature, Lond.*, 178:1476-1477.

Horridge, G. A. 1956d. The responses of *Heteroxenia* (Alcyonaria) to stimulation and to some inorganic ions. *J. exp. Biol.*, 33:604-614.

Horridge, G. A. 1957. The co-ordination of the protective retraction of coral polyps. *Philos. Trans. (B)*, 240:495-529.

Horridge, G. A. 1958. The co-ordination of the responses of *Cerianthus* (Coelenterata). *J. exp. Biol.*, 35:369-382.

Horridge, G. A. 1959. The nerves and muscles of medusae. VI. The rhythm. *J. exp. Biol.*, 36:72-91.

Horstmann, E. 1934a. Untersuchungen zur Physiologie der Schwimmbewegungen der Scyphomedusen. *Pflüg. Arch. ges. Physiol.*, 234:406-420.

Horstmann, E. 1934b. Nerven- und muskelphysiologische Studien zur Schwimmbewegungen der Scyphomedusen. *Pflüg. Arch. ges. Physiol.*, 234:421-431.

Hoyle, G. 1960. Neuromuscular activity in the swimming sea anemone, *Stomphia coccinea* (Müller). *J. exp. Biol.*, 37:671-688.

Hyde, I. H. 1902. The nervous system in *Gonionema murbachii Biol. Bull., Woods Hole*, 4:40-45.

Hykes, O. V. 1927. K fysiologii nervstva zebernatek (Ctenophora). *Biol. Listy, Prague*, 13:48-52. (*Čas. Lék. čes.* Suppl.)

Hyman, L. H. 1940a. *The Invertebrates: Protozoa through Ctenophora*. McGraw-Hill, New York.

Hyman, L. H. 1940b. Observations and experiments on the physiology of medusae. *Biol. Bull., Woods Hole*, 79:282-296.

Irisawa, H., Irisawa, A. F., and Nishita, Y. 1956. The responses of sensory body of medusae to light stimulation. *Kagaku (Science)*, 26:312-313. (In Japanese.)

Jennings, H. S. 1905. Behavior of sea anemones. *J. exp. Zool.*, 2:447-472.

Jennings, H. S. 1906. *Behavior of the Lower Organisms*. Columbia Univ. Press, New York.

Jickeli, C. F. 1882. Vorläufige Mittheilung über das Nervensystem der Hydroidpolypen. *Zool. Anz.*, 5:43-44.

Jickeli, C. F. 1883. Der Bau der Hydroidpolypen. *Morph. Jb.*, 8:373-416.

Jordan, H. J. 1908. Uber reflexarme Tiere. II. Die Physiologie des Nervenmuskelsystems von *Actinoloba dianthus*. *Z. allg. Physiol.*, 8:222-266.

Jordan, H. J. 1912. Über reflexarme Tiere (Tiere mit peripheren Nervennetzen). III. Die acraspeden Medusen. *Z. wiss. Zool.*, 101:116-138.

Jordan, H. J. 1934. Die Muskulatur der Aktinie *Metridium dianthus*, ihr Tonus und ihre Kontraktion. *Arch. néerl. Zool.*, 1:1-34.

Josephson, R. K. 1961a. The response of a hydroid to weak water-borne disturbances. *J. exp. Biol.*, 38:17-27.

Josephson, R. K. 1961b. Colonial responses of hydroid polyps. *J. exp. Biol.*, 38:559-577.

Josephson, R. K. 1961c. Repetitive potentials following brief electric stimuli in a hydroid. *J. exp. Biol.*, 38:579-593.

Josephson, R. K. 1962. Spontaneous electrical activity in a hydroid polyp. *Comp. Biochem. Physiol.*, 5:45-58.

Bibliography

JOSEPHSON, R. K., REISS, R. F., and WORTHY, R. M. 1961. A stimulation study of a diffuse conducting system based on coelenterate nerve nets. *J. theoret. Biol.*, 1:460-487.

JOURDAN, E. 1879. Recherches zoologiques et histologiques sur les Zoanthaires du Golfe de Marseille. *An. Sci. nat. (Zool.)*, (6), 10:1-154.

KASSIANOW, N. 1901. Studien über das Nervensystem der Lucernariden. *Z. wiss. Zool.*, 69:287-377.

KASSIANOW, N. 1908a. Untersuchungen über das Nervensystem der Alcyonaria. *Z. wiss. Zool.*, 90:478-535.

KASSIANOW, N. 1908b. Vergleich des Nervensystems der Octocorallia mit dem der Hexacorallia. *Z. wiss. Zool.*, 90:670-677.

KEPNER, W. A. and HOPKINS, D. L. 1924. Reactions of *Hydra* to chloretone. *J. exp. Zool.*, 38:437-448.

KINOSHITA, T. 1910. Über den Einfluss mechanischer und elektrischer Reize auf die Flimmerbewegung von *Beroë forscalii*. *Zbl. Physiol.*, 24:726-728.

KINOSHITA, T. 1911a. Über den Einfluss mehrerer aufeinanderfolgender wirksamer Reize auf den Ablauf der Reaktionsbewegungen bei Wirbellosen II. *Pflüg. Arch. ges. Physiol.*, 140:167-197.

KINOSHITA, T. 1911b. Über den Einfluss mehrerer aufeinanderfolgender wirksamer Reize auf den Ablauf der Reaktionsbewegungen bei Wirbellosen III. *Pflüg. Arch. ges. Physiol.*, 140:198-208.

KINOSITA, H. 1937. Entrapped circuit wave. *Zool. Mag., Tokyo*, 49:437-455.

KINOSITA, H. 1941. Initiation of entrapped circuit wave in a scyphomedusan *Mastigias papua*. *Jap. J. Zool.*, 9:209-220.

KLEINENBERG, N. 1872. *Hydra, eine anatomisch-entwickelungsgeschichtliche Untersuchung*. Engelmann, Leipzig.

KOMAI, T. 1922. *Studies on Two Aberrant Ctenophores, Coeloplana and Gastrodes*. Kyoto, Japan.

KOMAI, T. 1936. The nervous system in some coelenterate types. I. *Coeloplana*. *Mem. Coll. Sci. Kyoto* (B), 11:185-191.

KOMAI, T. 1942. The nervous system of some coelenterate types. 2. Ephyra and scyphula. *Annot. zool. Jap.*, 21:25-29.

KORN, H. 1959. Zum Nervensystem der Ctenophore *Pleurobrachia pileus* O. Müller. *Zool. Anz.*, 163:351-359.

KOROTNEFF, A. 1876. Histologie de l'hydrae et de la lucernaire. *Arch. Zool. exp. gen.*, 5:369-400.

KOROTNEFF, A. 1884. Zur Histologie der Siphonophoren. *Mitt. Zool. Stat. Neapel*, 5:229-288.

KOROTNEFF, A. 1887. Zur Anatomie und Histologie des *Veretillum*. *Zool. Anz.*, 10:387-390.

KOSHTOYANTS, KH. S. 1959. A comparative-physiological analysis of the periodical activity of certain invertebrates. *Internat. Cong. Zool. 15th, London*, 1958:841-844.

KOSHTOYANTS, KH. S. and SMIRNOVA, N. A. 1955. On periodic activity of sea anemones. *C. R. Acad. Sci. URSS (Dokl. Akad. Nauk)*, 104:662-665. (In Russian.)

KRASINSKA, S. 1914. Beiträge zur Histologie der Medusen *Z. wiss. Zool.*, 109:256-348.

KRUKENBERG, C. F. W. 1887. Die nervösen Leitungsbahnen in dem Polypar der Alcyoniden. In: *Vergleichend-Physiologische Studien*, Experimentelle Untersuchungen von C. F. W. Krukenberg, Reihe 2, Abt. 4. Carl Winter's Universitätsbuchhandlung, Heidelberg.

KUKENTHAL, W. and BROCH, H. 1911. Pennatulacea. *Wiss. Ergebn. "Valdivia"*, 13(2):113-576.

LEGHISSA, S. 1949a. Contributo allo studio del tessuto nervoso e del sistema nervoso dei celenterati. I. Il tessuto nervoso delle attinie *Pubbl. Staz. zool. Napoli*, 21:272-308.

LEGHISSA, S. 1949b. Considerazioni morfofisiologiche sul tessuto e sistema nervoso delle attinie. *Riv. Biol.*, 41:317-330.

LEGHISSA, S. 1950. L'evoluzione morfologica del tessuto nervoso nei celenterati fissi. *Boll. Zool.*, 17 (Suppl.):213-253.

LEHMANN, C. 1923. Untersuchung über die Sinnesorgane der Medusen. *Zool. Jb. (allg. Zool.)*, 39:321-391.

LENDENFELD, R. VON. 1882. Über Coelenteraten der Südsee. *Z. wiss. Zool.*, 37:465-552.

LENDENFELD, R. VON. 1883. Über das Nervensystem der Hydroidpolypen *Zool. Anz.*, 6:69-71.

LENHOFF, H. M. and SCHNEIDERMAN, H. A. 1959. The chemical control of feeding in the Portuguese man-of-war, *Physalia physalis* L. and its bearing on the evolution of the Cnidaria. *Biol. Bull., Woods Hole*, 116:452-460.

LINKO, A. 1900. Über den Bau der Augen bei den Hydromedusen. *Mem. Acad. Sci. St.-Pétersb. (Sci. math., phys., nat.)*, (8) 10:1-20.

LIPIN, A. 1909. Über den Bau des Süsswasser—Coelenteraten *Polypodium hydriforme* Uss. *Zool. Anz.*, 34:346-356.

LIPIN, A. 1911. Die Morphologie und Biologie von *Polypodium hydriforme* Uss. *Zool. Jb. (Anat.)*, 31:317-426.

LOEB, J. 1895. Zur Physiologie und Psychologie der Actinien. *Pflüg. Arch. ges. Physiol.*, 59:415-420.

LOEB, J. 1900. On the different effect of ions upon myogenic and neurogenic rhythmical contractions and upon embryonic and muscular tissue. *Amer. J. Physiol.*, 3:383-396.

LOEB, J. 1902. *Comparative Physiology of the Brain and Comparative Psychology*. Putnam's, New York.

LOEB, J. 1905. *Studies in General Physiology*. Univ. of Chicago Press, Chicago.

LOOMIS, W. F. 1955. Glutathione control of the specific feeding reactions of *Hydra*. *Ann. N.Y. Acad. Sci.*, 62:209-228.

MAADEN, H. VON DER. 1939. Über das Sinnesgrübchen von *Aurelia aurita*. *Zool. Anz.*, 125:29-35.

MACKIE, G. O. 1959. The evolution of the Chondrophora (Siphonophora-Disconanthae): New evidence from behavioural studies. *Trans. roy. Soc. Canada (V)*, (3), 53:7-20.

MACKIE, G. O. 1960. The structure of the nervous system in *Velella*. *Quart. J. micr. Sci.*, 101:119-131.

MARSHALL, S. 1923. Observations upon the behavior and structure of *Hydra*. *Quart. J. micr. Sci.*, 67:593-616.

MATHIAS, A. P., ROSS, D. M., and SCHACHTER, M. 1957. Identification and distribution of 5-hydroxytryptamine in a sea anemone. *Nature, Lond.*, 180:658-659.

MATTHAI, G. 1918. On reactions to stimuli in corals. *Proc. Camb. phil. Soc. biol. Sci.*, 19:164-166.

MAY, A. J. 1903. A contribution to the morphology and development of *Corymorpha pendula* Ag. *Amer. Nat.*, 37:579-599.

MAYER, A. G. 1906. Rhythmical pulsation in scyphomedusae I. *Publ. Carneg. Instn.*, 47:1-62.

MAYER, A. G. 1908. Rhythmical pulsations in scyphomedusae. *Pap. Tortugas Lab.*, 1:113-131. (Publ. Carneg. Instn. 102.)

MAYER, A. G. 1911. The converse relation between ciliary and neuromuscular movements. *Pap. Tortugas Lab.*, 3:1-25. (Publ. Carneg. Instn. 132.)

MAYER, A. G. 1912. The cause of rhythmical pulsation in scyphomedusae. *Int. Congr. Zool.*, 7:278-281.

MAYER, A. G. 1914. The relation between degree of concentration of the electrolytes of sea water and rate of nerve-conduction in *Cassiopea*. *Pap. Tortugas Lab,.* 6:25-54. (Publ. Carneg. Instn. 183.)

MAYER, A. G. 1915. The nature of nerve conduction in *Cassiopea*. *Proc. nat. Acad. Sci., Wash.*, 1:270-274.

MAYER, A. G. 1916a. A theory of nerve-conduction. *Proc. nat. Acad. Sci., Wash.*, 2:37-42.

MAYER, A. G. 1916b. Further studies of nerve conduction in *Cassiopea*. *Proc. nat. Acad. Sci., Wash.*, 2:721-726.

MAYER, A. G. 1916c. Nerve conduction, and other reactions in *Cassiopea*. *Amer. J. Physiol.*, 39:375-393.

MAYER, A. G. 1917a. Nerve-conduction in *Cassiopea xamachana*. *Pap. Tortugas Lab.*, 11:1-20. (Publ. Carneg. Instn. 251.)

MAYER, A. G. 1917b. Further studies of nerve conduction in *Cassiopea*. *Amer. J. Physiol.*, 42:469-475.

MCCONNELL, C. H. 1931. The successful application of Rongalit white for the study of the development of the nerve net of *Hydra*. *Zool. Anz.*, 93:279-281.

MCCONNELL, C. H. 1932. Development of the ectodermal nerve net in the buds of *Hydra*. *Quart. J. micr. Sci.*, 75:495-509.

MEYER, G. F. 1955. Vergleichende Untersuchungen mit der supravitalen Methylenblaufärbung am Nervensystem wirbelloser Tiere. *Zool. Jb. (Anat.)*, 74:339-400.

MILNE, L. J. 1938. Some aspects of the behavior of the freshwater jellyfish *Craspedacusta* sp. *Amer. Nat.*, 72:464-472.

MITRA, K. and MUKERJEE, H. K. 1924. Reversal of thigmotropism in *Hydra*. *J. Dep. Sci. Calcutta Univ.*, 6:11-62.

MITROPOLITANSKAYA, R. L. 1941. On the presence of acetylcholine and cholinesterase in the Protozoa, Spongia and Coelenterata. *C. R. Acad. Sci. URSS (Dokl. Acad. Nauk)*, 31:717-718. (In Russian.)

MIYOSHIMA, M. 1898. Über das Nervensystem von *Hydra*. *Zool. mag., Tokyo*, 10:141-145.

MOORE, A. R. 1924. Luminescence in *Mnemiopsis*. *J. gen. Physiol.*, 6:403-412.

MOORE, A. R. 1926a. On the nature of inhibition in *Pennatula*. *Amer. J. Physiol.*, 76:112-115.

MOORE, A. R. 1926b. Galvanic stimulation of luminescence in *Pelagia noctiluca*. *J. gen. Physiol.*, 9:375-379.

MOORE, A. R. 1933. On function and chemical differentiation in the nervous system of *Coeloplana bockii*. *Sci. Rep. Tohoku Univ.* (4), 8:201-204

MOORE, A. R. 1936. Reciprocal inhibition and its reversal by strychnine in the modified ctenophore, *Coeloplana bockii*. *Physiol. Zool.*, 9:240-245.

MOORE, M. M. 1927. The reactions of *Cerianthus* to light. *J. gen. Physiol.*, 8:509-518.

MORTENSEN, T. 1912. Ctenophora. In: *Danish Ingolf-Expedition*, Vol. 5 (2) B. Luno, Copenhagen.

MURBACH, L. 1903. The static function in *Gonionemus*. *Amer. J. Physiol.*, 10:201-209.

NAGEL, W. A. 1892. Der Geschmackssinn der Actinien. *Zool. Anz.*, 15: 334-338.

NAGEL, W. A. 1893. Versuche zur Sinnesphysiologie von *Beroë ovata* und *Carmarina hastata*. *Pflüg. Arch. ges. Physiol.*, 54:165-188.

NAGEL, W. A. 1894. Experimentelle sinnes-physiologische Untersuchungen an Coelenteraten. *Pflüg. Arch. ges. Physiol.*, 57:494-552.

NAGEL, W. A. 1896. *Der Lichtsinn augenloser Tiere*. Fischer, Jena.

NAGEL, W. A. 1898. Notiz betreffend den Lichtsinn augenloser Thiere. *Pflüg. Arch. ges. Physiol.*, 69:137-140.

NEEDLER, M. and ROSS, D. M. 1958. Neuromuscular activity in sea anemone *Calliactis parasitica* (Couch). *J. Mar. biol. Ass. U. K.*, 37:789-806.

NICOL, J. A. C., 1955a. Observations on luminescence in *Renilla* (Pennatulacea). *J. exp. Biol.*, 32:299-320.

NICOL, J. A. C. 1955b. Nervous regulation of luminescence in the sea pansy *Renilla köllikeri*. *J. exp. Biol.*, 32:619-635.

NICOL, J. A. C. 1955c. Physiological control of luminescence in animals. In: *The Luminescence of Biological Systems*. F. H. Johnson (ed.). Amer. Ass. Adv. Sci., Washington.

NICOL, J. A. C. 1958. Observations on the luminescence of *Pennatula phosphor*, with a note on the luminescence of *Virgularia mirabilis*. *J. Mar. biol. Ass. U.K.*, 37:551-563.

NICOL, J. A. C. 1960. The regulation of light emission in animals. *Biol. Rev.*, 35:1-42.

NIEDERMEYER, A. 1914. Beiträge zur Kenntnis des histologischen Baues von *Veretillum cynomorium* (Pall.). *Z. wiss. Zool.*, 109:531-590.

NORTH, W. J. 1957. Sensitivity to light in the sea anemone *Metridium senile* (L.). II. Studies of reaction time variability and the effects of changes in light intensity and temperature. *J. gen. Physiol.*, 40:715-733.

NORTON, S. and BEER, E. J. DE. 1947. Use of the *Hydra* for pharmacological study. *Science*, 106:328.

NUSSBAUM, M. 1887. Über die Theilbarkeit der lebendigen Materie. II. Beiträge zur Naturgeschichte des Genus *Hydra*. *Arch. mikr. Anat.*, 29:265-366.

PALOMBI, A. 1940. Studii sugli idroid. I. L'azione delle radiazioni luminose. *Boll. Soc. Nat. Napoli*, 50:149-182.

PANCERI, M. 1872a. The luminous organs and light of the Pennatulae. *Quart. J. micr. Sci.*, N.S., 12:248-260.

PANCERI, M. 1872b. Études sur la phosphorescence des animaux marins. *Ann. Sci. nat. (Zool.)*, (5), 16:1-67.

PANTIN, C. F. A. 1935a. The nerve net of the Actinozoa. I. Facilitation. *J. exp. Biol.*, 12:119-138.

PANTIN, C. F. A. 1935b. Nerve net of Actinozoa. II. Plan of the nerve net. *J. exp. Biol.*, 12:139-155.

PANTIN, C. F. A. 1935c. Nerve net of the Actinozoa. III. Polarity and afterdischarge. *J. exp. Biol.*, 12:156-164.

PANTIN, C. F. A. 1935d. The nerve net of the Actinozoa. IV. Facilitation and the "staircase". *J. exp. Biol.*, 12:389-396.

PANTIN, C. F. A. 1942. The excitation of nematocysts. *J. exp. Biol.*, 19:294-310.

PANTIN, C. F. A. 1950. Behaviour patterns in lower invertebrates. *Symp. Soc. exp. Biol.*, 4:175-195.

PANTIN, C. F. A. 1952. The elementary nervous system. *Proc. roy. Soc. (B)*, 140:147-168.

PANTIN, C. F. A. 1955. The primitive nervous system. *Proc. roy. Instn. G.B.* 36:1-7.

PANTIN, C. F. A. 1956. The origin of the nervous system. *Pubbl. Staz. zool. Napoli*, 28:171-181.

PANTIN, C. F. A. and PANTIN, A. M. P. 1943. The stimulus to feeding in *Anemonia sulcata*. *J. exp. Biol.*, 20:6-13.

PANTIN, C. F. A. and VIANNA DIAS, M. 1952a. Excitation phenomena in an actinian (*Bunodactis* sp?) from Guanabara Bay. *Ann. Acad. bras. Sci.*, 24:335-349.

PANTIN, C. F. A. and VIANNA DIAS, M. 1952b. Rhythm and afterdischarge in medusae. *Ann. Acad. bras. Sci.*, 24:351-364.

PARKER, G. H. 1896. The reactions of *Metridium* to food and other substances. *Bull. Mus. comp. Zool. Harv.*, 29:107-119.

PARKER, G. H. 1905a. The movements of the swimming plates in ctenophores, with reference to the theories of ciliary metachronism. *J. exp. Zool.*, 2:407-423.

PARKER, G. H. 1905b. The reversal of the effective stroke of the labial cilia of sea-anemones by organic substances. *Amer. J. Physiol.*, 14:1-6.

PARKER, G. H. 1912. Nervous and non-nervous responses of actinians. *Science*, 35:461-462.

PARKER, G. H. 1916. The effector systems of actinians. *J. exp. Zool.*, 21:461-484.

PARKER, G. H. 1917a. Nervous transmission in actinians. *J. exp. Zool.*, 22:87-94.

PARKER, G. H. 1917b. The movements of the tentacles in actinians. *J. exp. Zool.*, 22:95-110.

PARKER, G. H. 1917c. Pedal locomotion in actinians. *J. exp. Zool.*, 22:111-124.

PARKER, G. H. 1917d. Actinian behavior. *J. exp. Zool.*, 22:193-229.

PARKER, G. H. 1917e. The activities of *Corymorpha*. *J. exp. Zool.*, 24:303-331.

PARKER, G. H. 1917f. The responses of hydroids to gravity. *Proc. nat. Acad. Sci., Wash.*, 3:72-73.

PARKER, G. H. 1918. The rate of transmission in the nerve-net of the coelenterates. *J. gen. Physiol.*, 1:231-236.

PARKER, G. H. 1919a. The organization of *Renilla*. *J. exp. Zool.*, 27:499-505.

PARKER, G. H. 1919b. *The Elementary Nervous System*. Lippincott, Philadelphia.

PARKER, G. H. 1920a. The phosphorescence of *Renilla*. *Proc. Amer. phil. Soc.*, 59:171-175.

Bibliography

PARKER, G. H. 1920b. Activities of colonial animals. II. Neuromuscular movements and phosphorescence in *Renilla*. *J. exp. Zool.*, 31:475-513.

PARKER, G. H. 1925. Activities of colonial animals. III. The interrelation of zoöids in soft corals. *Proc. nat. Acad. Sci., Wash.*, 11:346-347.

PARKER, G. H. 1932. Neuromuscular activities of the fishing filaments of *Physalia*. *J. cell. comp. Physiol.*, 1: 53-63.

PARKER, G. H. and MARKS, A. P. 1928. Ciliary reversal in *Metridium*. *J. exp. Zool.*, 52:1-6.

PARKER, G. H. and TITUS, E. G. 1916. The structure of *Metridium (Actinoloba) marginatus* Milne-Edwards with special reference to its neuromuscular mechanism. *J. exp. Zool.*, 21:433-458.

PASSANO, L. M. 1958. Intermittent conduction in scyphozoan nerve nets. *Anat. Rec.*, 132:486.

PASSANO, L. M. and McCULLOUGH, C. B. 1960. Nervous activity and spontaneous beating in scyphomedusae. *Anat. Rec.*, 137:387.

PASSANO, L. M. and McCULLOUGH, C. B. 1961. Pacemaker activity in jellyfish ganglia. *Fed. Proc.*, 20:338.

PASSANO, L. M. and PANTIN, C. F. A. 1955. Mechanical stimulation in the sea-anemone *Calliactis parasitica*. *Proc. roy. Soc. (B)*, 143:226-238.

PASSANO, L. M. and PANTIN, C. F. A. 1956. Sensory stimulation in actinians. *Int. Congr. Zool.*, 14:308.

PAX, F. 1914. Die Actinien. *Ergebn. Zool.*, 4:339-640.

PEARSE, A. S. 1906. Reactions of *Tubularia crocea* (Ag.). *Amer. Nat.*, 40:401-407.

PETERS, A. W. 1905. Phosphorescence in ctenophores. *J. exp. Zool.*, 2:103-116.

PIÉRON, H. 1906a. Contribution à la psychologie des actinies. *Bull. Inst. gén. psychol.*, 6:40-59.

PIÉRON, H. 1906b. La réaction aux marées par anticipation réflexe. *R. Soc. Biol., Paris*, 2:658-660.

PIÉRON, H. 1906c. Contribution à la psychophysiologie des actinies. Les réactions de l'*Actinia equina*. *Bull. Inst. gén. psychol., Paris*, 6:146-169.

PIÉRON, H. 1908. La rythmicité chez *Actinia equina* L. *C.R. Soc. Biol., Paris*, 2:726-728.

PIÉRON, H. 1909. Des réactions de l'*Actinia equina* à la déoxygénation progressive du milieu. *C. R. Soc. Biol., Paris*, 1:626-628.

POLLOCK, W. H. 1883. On indications of a sense of smell in Actiniae. With an addendum by G.H. Romanes. *J. Linn. Soc. (Zool.)*, 16:474-476.

PORTMANN, A. 1926. Die Kriechbewegung von *Aiptasia cornea*. Ein Beitrag zur Kenntnis der neuromuskulären Organisation der Actinien. *Z. vergl. Physiol.*, 4:659-667.

PÜTTER, A. 1903. Die Flimmerbewegung. *Ergebn. Physiol.*, 2:1-102.

RAND, H. W. 1909. Wound reparation and polarity in tentacles of actinians. *J. exp. Zool.*, 7:189-238.

REGNART, H. C. 1927. The projected chemical sense in the coelenterata and echinodermata. *Proc. Univ. Durham Phil. Soc.*, 8:61-65.

REIS, R. H. 1943. Interruption of the nervous system in *Pelmatohydra oligactis*. *Trans. Amer. micr. Soc.*, 62:122-126.

REIS, R. H. 1953. Rhythmic behavior patterns in *Pelmatohydra oligactis*. *Trans. Amer. micr. Soc.*, 72:1-9.

RETTERER, E. 1907. A propos du rythme des marées et de la matière vivante. *C. R. Soc. Biol., Paris*, 1:186.

RIDDLE, O. 1911. On the cause of autotomy in *Tubularia*. *Biol. Bull., Woods Hole*, 21:389-395.

ROBSON, E. A. 1961a. Some observations on the swimming behavior of the anemone *Stomphia coccinea*. *J. exp. Biol.*, 38:343-363.

ROBSON, E. A. 1961b. The swimming response and its pacemaker system in the anemone *Stomphia coccinea*. *J. exp. Biol.*, 38:685-694.

ROBSON, E. A. 1961c. A comparison of the nervous systems of two sea-anemones, *Calliactis parasitica* and *Metridium senile*. *Quart. J. micr. Sci.*, 102:319-326.

ROMANES, G. J. 1877. Preliminary observations on the locomotor system of medusae. *Philos. Trans.*, 166:269-313.

ROMANES, G. J. 1878. Further observations on the locomotor system of medusae. *Philos. Trans. (B)*, 167:659-752.

ROMANES, G. J. 1885. *Jellyfish, Starfish, and Sea Urchins, being a Research on the Primitive Nervous Systems.* Internat. Sci. Ser., Appleton, New York.

ROSS, D. M. 1945a. Facilitation in sea anemones. I. The action of drugs. *J. exp. Biol.*, 22:21-31.

ROSS, D. M. 1945b. Facilitation in sea anemones. II. Tests on extracts. *J. exp. Biol.*, 22:32-36.

ROSS, D. M. 1952. Facilitation in sea anemones. III. Quick responses to single stimuli in *Metridium senile*. *J. exp. Biol.*, 29:235-254.

ROSS, D. M. 1955. Facilitation in sea anemones. IV. The quick response of *Calliactis parasitica* at high temperatures. *J. exp. Biol.*, 32:815-821.

ROSS, D. M. 1956. Neuromuscular transmission in sea animals. *Int. Physiol. Congr., Abstr.* XX. 1956:780.

ROSS, D. M. 1957a. The action of tryptamine and 5-hydroxytryptamine on muscles of sea anemones. *Experientia*, 13:192-194.

ROSS, D. M. 1957b. Quick and slow contractions in the isolated sphincter of the sea anemone, *Calliactis parasitica*. *J. exp. Biol.*, 34:11-28.

ROSS, D. M. 1957c. Responses of *Cerianthus* (Coelenterata). *Nature, Lond.*, 180:1368-1370.

ROSS, D. M. 1960a. The effects of ions and drugs on neuromuscular preparations of sea anemones. I. On preparations of the column of *Calliactis* and *Metridium*. *J. exp. Biol.*, 37:732-752.

ROSS, D. M. 1960b. The effects of ions and drugs on neuromuscular preparations of sea anemones. II. On sphincter preparations of *Calliactis* and *Metridium*. *J. exp. Biol.*, 37:753-774.

ROSS, D. M. and PANTIN, C. F. A. 1940. Factors influencing facilitation in Actinozoa. The action of certain ions. *J. exp. Biol.*, 17:61-73.

SAMASSA, P. 1892. Zur Histologie der Ctenophoren. *Arch. mikr. Anat.*, 40:157-243.

SANZO, L. 1903a. Sur un processus d'inhibition dans les mouvements rythmiques des méduses. *Arch. ital. Biol.*, 39:319-324.

SANZO, L. 1903b. Su di un processo d'inibizione nei movimenti ritmici delle meduse. *Riv. Biol. gen.*, 3:592-597.

SCHAEPPI, I. 1898. Untersuchungen über das Nervensystem der Siphonophoren. *Jena. Z. Naturw., N.F.*, 25:483-550.

SCHÄFER, E. A. 1879. Observations on the nervous system of *Aurelia aurita*. *Philos. Trans.*, 169:563-575.

SCHLÜNSEN, A. 1935. Lokomotionen und Orientierungsbewegungen von Hydren unter Lichteinfluss. *Zool. Jb. (allg. Zool.)*, 54:423-458.

SCHMID, B. 1911. Ueber den Heliotropismus von *Cereactis aurantiaca*. *Biol. Zbl.*, 31:538-539.

SCHMIDT, W. J. 1954. Sinneszellen mit "Lipoidstiften" im Epithelpolster der Ctenophorenstatocyste. *Zool. Anz.*, 152:99-105.

SCHNEIDER, K. C. 1890. Histologie von *Hydra fusca*, mit besonderer Berücksichtigung des Nervensystems der Hydropolypen. *Arch. mikr. Anat.*, 35:321-379.

SCHNEIDER, K. C. 1892. Einige histologische Befunde an Cölenteraten. *Jena. Z. Naturw.*, 27(N.F. 20):379-462.

SCHNEIDER, K. C. 1902. *Lehrbuch der vergleichenden Histologie der Tiere.* Fischer, Jena.

SEMAL-VAN GANSEN, P. 1952. Note sur le système nerveux de l'hydre. *Bull. Acad. Belg. Cl. Sci.*, (5), 38:718-735.

SIEDENTOP, W. 1927. Die Kriechbewegung der Actinien und Lucernariden. *Zool. Jb. (Allg. Zool.)*, 44:149-210.

SKRAMLIK, E. VON. 1945. Beobachtungen an Medusen. *Zool. Jb. (allg. Zool.)*, 61:296-336.

SPANGENBERG, D. B. and HAM, R. G. 1960. The epidermal nerve net of *Hydra*. *J. exp. Zool.*, 143:195-201.

STEINER, G. 1934. Der Verlust der Glockenautomatie bei randorganlos aufgezogenen Ohrenquallen *(Aurelia aurita)*. *Biol. Zbl.*, 54:102-105.

SUND, P. N. 1958. A study of the muscular anatomy and swimming behavior of the sea anemone, *Stomphia coccinea*. *Quart. J. micr. Sci.*, 99:401-420.

TERRY, O. P. 1909. The production by hydrogen peroxide of rhythmical contractions in the marginless bell of *Gonionemus*. *Amer. J. Physiol.*, 24:117-123.

THILL, H. 1937. Beiträge zur Kenntnis der *Aurelia aurita* (L.). *Z. wiss. Zool.*, 150:51-96.

TORELLI, B. 1952. Su alcuni particolari aspetti della istologia di *Cerianthus*. *Pubbl. Staz. zool. Napoli*, 23:141-162.

TORREY, H. B. 1902. The Hydroida of the Pacific Coast of North America *Univ. Calif. Publ. Zool.*, 1:1-104.

TORREY, H. B., 1904a. On the habits and reactions of *Sagartia davisi*. *Biol. Bull. Woods Hole*, 6:203-215.

TORREY, H. B. 1904b. Biological studies on *Corymorpha*. I. *C. palma* and environment. *J. exp. Zool.*, 1:395-422.

TORREY, H. B. 1907. Biological studies on *Corymorpha* II. The development of *C. palma* from the egg. *Univ. Calif. Publ. Zool.*, 3:253-298.

TOTTON, A. K. and MACKIE, G. O. 1960. Studies on *Physalia physalis* (L.). *"Discovery" Rep.*, 30:301-408.

TREMBLEY, A. 1744. *Mémoires pour servir à l'histoire d'un genre de polypes d'eau douce, à bras en forme de carnes*. Durand, Paris.

UEXKÜLL, J. VON. 1901. Die Schwimmbewegungen der *Rhizostoma pulmo*. *Mitt. zool. Sta. Neapel*, 14:620-626.

UEXKÜLL, J. VON. 1921. *Umwelt und Innenwelt der Tiere*. 2 Aufl. Springer, Berlin.

VERESS, E. 1911. Sur les mouvements des méduses. *Arch. int. Physiol.*, 10:253-289.

VERESS, E. 1938. Studien über die rhythmischen Bewegungen der Medusen. *Állatorv. Közl.*, 35:153-170. (Hungarian, with German summary.)

VERWORN, M. 1889. *Psycho-Physiologische Protisten-Studien. Experimentelle Untersuchungen*. Fischer, Jena.

VERWORN, M. 1891. Studien zur Physiologie der Flimmerbewegung. *Pflüg. Arch. ges. Physiol.*, 48:149-180.

VETOCHIN, J. A. 1926. Über die Erregungsprocesse im Schirm der Qualle *Aurelia aurita* und über die Regulation der Bewegung dieses Tieres im Meerwasser. *Russk. fiz. Zh.*, 9:517-536. (In Russian, with German summary.)

WARDEN, C. J., JENKINS, T. N., and WARNER, L. H. 1940. *Comparative Psychology*. Vol. 2. Ronald Press, New York.

WILLEM, V. 1927. Observations sur la locomotion des actinies. *Bull. Acad. Belg. Cl. Sci.*, 13:630-650.

WILLEM, V. 1928. Note sur la polarité de l'appareil locomoteur des actinies. *Bull. Acad. Belg. Cl. Sci.*, 14:296-306.

WILSON, D. M. 1959. Long-term facilitation in a swimming sea anemone. *J. exp. Biol.*, 36:526-532.

WOLF, E. 1928. Temperature characteristics for pulsation frequency in *Gonionemus*. *J. gen. Physiol.*, 11:547-562.

WOLF, E. 1937. Nerven- und Sinnesphysiologie. *Fortschr. Zool.*, NF 2:421-444.

WOLFF, M. 1904. Das Nervensystem der polypoiden Hydrozoa und Scyphozoa. *Z. allg. Physiol.*, 3:191-281.

WOOLLARD, H. H. and HARPMAN, J. A. 1939. Discontinuity in the nervous system of coelenterates. *J. Anat., Lond.*, 73:559-562.

WU, H. W. 1927. Preliminary observations on the sense organs and the adjacent structures of two scyphomedusae at a young stage. *Contr. biol. Lab. Sci. Soc. China*, 3:1-5.

WULFERT, J. 1902. Die Embryonalentwicklung von *Gonothyrea loveni* Allm. *Z. wiss. Zool.*, 71:296-327.

YAMASHITA, T. 1957a. Über den Statolithen in den Sinneskörpern der Meduse *Aurelia aurita*. *Z. Biol.*, 109:111-115.

YAMASHITA, T. 1957b. Das Aktionspotential der Sinneskörper (Randkörper) der Meduse *Aurelia aurita*. *Z. Biol.*, 109:116-122.

YANAGITA, T. M. 1959. Physiological mechanism of nematocyst responses in sea-anemone. VII. Extrusion of resting cnidae—its nature and its possible bearing on the normal nettling response. *J. exp. Biol.*, 36:478-494.

YERKES, R. M. 1902a. A contribution to the physiology of the nervous system of the medusa *Gonionemus murbachii*. Part 1—The sensory reactions of *Gonionemus*. *Amer. J. Physiol.*, 6:434-449.

YERKES, R. M. 1902b. A contribution to the physiology of the nervous system of the medusa *Gonionema murbachii*. Part 11—The physiology of the central nervous system. *Amer. J. Physiol.*, 7:181-198.

YERKES, R. M. and AYER, J. B. 1903. A study of the reactions and reaction time of the medusa *Gonionema murbachii* to photic stimuli. *Amer. J. Physiol.*, 9:279-307.

YONGE, C. M. 1930. Studies on the physiology of corals. I. Feeding mechanisms and food. *Sci. Rep. Gr. Barrier Reef Exped.*, 1:13-59.

ZHIRMUNSKY, A. W. 1958. Reactions encountered in natural and in artificial irritation of Actinia. In: *Evolution of the Functions of the Nervous System*. D. A. Biryukov (ed.). Medgiz, Leningrad.

ZHUKOV, E. K. 1939. On the problem of the architecture of physiological functions of the locomotor apparatus of the medusa. *Ann. Leningr. Univ. (Uchen. zap. leningr. Univ.)*. 10:111.

ZOJA, R. 1890. Alcune ricerche morfologiche e fisiologiche sull'*Hydra*. *Boll. Sci.*, 12:65-92, 97-134; 13:1-20.

ZOJA, R. 1891. Sulla trasmissibilità degli stimoli nelle colonie di Idroidi. *R. C. Ist. lombardo*, (2)24:1225-1234.

ZOJA, R. 1892. Intorno ad alcune particolarità di struttura dell'*Hydra*. *R.C. Inst. Lombardo*, 25:700-712.

ZUBKOV, A. A. and POLIKARPOV, G. G. 1951. Conditional reflex in coelenterates. *Adv. mod. Biol., Moscow (Up. sovr. Biol.)*, 32:301-302. (In Russian.)

CHAPTER 9

Platyhelminthes

Summary	536	IV. Trematoda	**561**
I. Introduction	**537**	A. Central Nervous System	561
II. The Anatomy of Turbellaria	**537**	B. Peripheral Nervous System	563
A. Central Nervous System	537		
1. The longitudinal cords and commissures	537	**V. Cestoda**	**566**
2. The brain	542	A. Central Nervous System	566
B. Peripheral Nervous System	550	B. Peripheral Nervous System	571
1. The plexuses	550		
2. The receptors	552		
III. The Physiology of Turbellaria	**556**	Classified References	573
A. The Role of the Brain	556		
B. The Role of the Cords and Plexuses	559	Bibliography	573

Summary

The lowest group of bilaterally symmetrical metazoans already has a well-developed central nervous system consisting of a distinct brain and a set of longitudinal medullary cords connected by commissures, forming a typical **orthogon,** though there is much variety in detail. Notable developments are the outer rind of nerve cell bodies and inner core of fiber matter in the ganglia, unipolar neurons, globuli cell masses in the highest forms (polyclads), neuropile, neuroglia, a brain capsule, deep-lying sensory cells with long, branched, free nerve endings, and at least six differentiated types of sensory neurons.

The **basic turbellarian plan** seems to be three to five pairs of cords with fairly regular commissures, plus a brain. If this is primitive, then derived from it are the forms with no obvious cords but a coarse meshed plexus, as well as those forms with only a single pair of ventral cords. Species with the primitive plan occur among acoels, rhabdocoels, and alloeocoels, along with much derived ones; triclads are the most specialized (but not the highest), especially land planarians with a novel structure, the ventral nerve plate. Polyclads are the most complex, mainly in the brain, which has cell masses, neuropile areas, and pathways on a par with some annelids. The brain in planarians is relatively simple; certain other forms almost or quite lack a brain, perhaps by secondary loss. Interpretations of the origin and **early history of the brain** are still fluid. Seven types of nerve cell bodies are distinguished and two types of glial cells. Golgi preparations show bipolars, multipolars, and unipolars, long and short interneurons, decussations, commissural neurons, motoneurons and a variety of sensory neurons, localized tufted terminals, and (with the exception of a report by Gelei) evidence of a discontinuous synaptic system.

Extirpation of the brain is more severe in some polyclads and temnocephalans than in triclads. Locomotion is slowed or stopped in the absence of strong stimulation; more complex types, such as the swimming and leechlike looping, are particularly vulnerable; food finding or recognition is often impossible. Nevertheless the cords in the absence of the brain can mediate some locomotion, righting, and avoidance reactions. Movement toward dark places, though less precise, occurs without eyes or brain. Conditioned planarians, cut in two transversely and allowed to regenerate, show retention by relearning in as few trials as uncut controls; tail and head regenerates are equal in relearning speed.

There is a very tenuous nervous layer at the base of the epidermis, thickened rostrally in a few forms with minute or no brains. The main **peripheral nervous system** is the submuscular plexus, which takes the place of nerves, and is continuous with the main cords and commissures. In the Terricola only, there is an additional nerve layer, the ventral nerve plate, possibly to be regarded as equivalent to the cords elsewhere, sunk deeper and only continuous with the submuscular plexus at the sides of the body. The peripheral nervous system does not mediate responses or conduct around obstacles or past transections of the cords; hence there is **no general nerve net.** The pharynx is an exception; its plexus is well developed, capable of coordinating feeding movements after isolation of the organ, and is normally inhibited from its spontaneity by a connection from the ventral cord. Tactile, chemo-, and rheoreceptors are both histologically and physiologically demonstrated—the first clear **modality distinction** in general epidermal receptors. Eyes as well as scattered photoreceptors are general and statocysts are present in more primitive groups.

Trematodes are similar to rather poorly developed free-living turbellarians, better developed than cestodes and some turbellarians. Without much variation, they exhibit three pairs of cords (with emphasis on the ventral), plus fairly regular commissures and a brain with a dorsal commissure. There is no circumesophageal ring including the brain but the beginnings of one are present in the commissures between cords. The histological differentiation in the central nervous system is low; the nerve cells do not form a clear rind. There are few or no multicellular sense organs. Single sensory neurons are abundant but few differentiated types; none has long hairs projecting beyond the surface and no clusters suggesting chemoreceptors are known. No obvious influence of the parasitic habit can be recognized aside from this relatively low level of development and the rich nerve supply of the suckers. The deep and superficial plexuses are here extensive and may be regular in pattern. An enteric nervous system, with accumulations of nerve cells forming two rings around the esophagus and plexuses in the walls of pharynx, esophagus, and intestine, is connected directly to the brain.

In the highly specialized parasitic class **Cestoda** the nervous system is somewhat but not extremely simplified in comparison with the better free-living groups. This is expressed in the lack of multicellular sense organs and the low degree of histological differentiation of the nerve cells, neuropile, cords, and ganglia. In general a capsule around the ganglia is missing and the cells of the central nervous tissue are not easily distinguished from mesenchyme. Grossly, there is a trend toward elaboration by the addition of commissures and nerves in a complex, four-rayed, radially symmetrical pattern related to the brain. This goes with the development of muscular attachment organs on the scolex. Basically there is a simple brain in the scolex and a number of longitudinal cords that run without interruption through all the proglottids of the strobila. The main cords and minimum complement are a pair of laterals—often five pairs or more, up to several dozen, occur. Ring commissures are found in each proglottid, one or several or merely a circularly oriented plexus. Nerve cells appear to be few in the cords; tiny ganglionic accumulations sometimes appear at the intersections with commissures. Nerve cells are not numerous in the brain and are chiefly collected medially on the main transverse commissure. The basic invertebrate pattern of neuropile core and ganglion cell rind is obscure. Giant cells and fibers are rather common—at least in tetrarynchs. There may be two plexuses: a tenuous subcuticular one, chiefly of sensory neurons, and a deep one of thin, widely spaced strands, perhaps chiefly for distributing efferent fibers. Superficial bipolar sensory neurons with unbranched distal processes are abundant. Deeper cells with branching free endings are doubtful.

I. INTRODUCTION

The flatworms have attracted a good deal of attention, perhaps in part because, in most arrangements of the phyla, they stand in a strategic position at the bottom of the animals of organ-grade construction, the Bilateria. Furthermore, here are found the first centralized nervous systems—including, indeed, distinct brains. Fortunately, too, they have repeatedly yielded to silver stains, particularly the incomparable Golgi impregnation. Added to this, the nervous system is quite heterogeneous among the orders and families, varying from a mere plexiform diffuseness without ganglia or cords to a brain of very respectable complexity and as many as ten regular cords to as few as two. Thus it is not astonishing that a sizable literature has been accumulated: upward of 170 publications contain contributions worthy of our attention. The rate of addition has slowed down, and yet in many areas our information is still extremely primitive; in physiology the simplest techniques have not been exhausted or comparisons among types carried to any considerable degree; finer anatomy is needed, especially on the lowest and the highest levels—the connections in the peripheral plexus and in the higher brain centers.

The divergence in body plan between the three classes of flatworms is great enough to require separate treatment of turbellarians, trematodes, and cestodes.

II. THE ANATOMY OF TURBELLARIA

A. Central Nervous System

1. The longitudinal cords and commissures

Turbellarians are not uniform in the number or position of the main nerve cords. Without implying a series or evolutionary sequence, the following are the principal types in this respect (Figs. 9.1, 9.2). (The discussion employs the older terminology for the higher categories—for example Alloeocoela—but without asserting its superiority. The newer terminology, as modified by Ax (1961), is perhaps an improvement, though it has not had time to be tested by usage.) (a) *Bothrioplana* (Alloeocoela) has four pairs of well-defined cords: a pair each of ventral, ventrolateral, lateral, and dorsal cords (Reisinger). These are connected by circular commissures characterized by being regular, evenly spaced, unbranched, and at right angles to the cords. This pattern is termed **orthogonal** because of its rec-

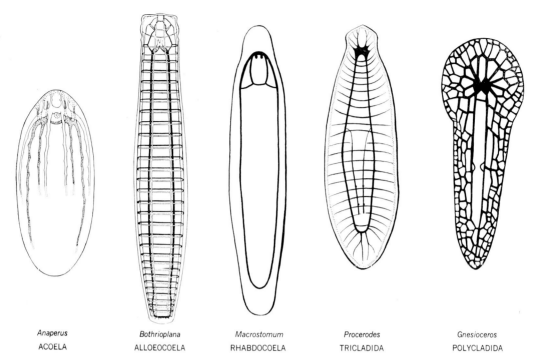

| *Anaperus* | *Bothrioplana* | *Macrostomum* | *Procerodes* | *Gnesioceros* |
| ACOELA | ALLOEOCOELA | RHABDOCOELA | TRICLADIDA | POLYCLADIDA |

Figure 9.1. Examples of the plan of the nervous system in representative turbellarians. [Left to right: from Westblad, 1949; Reisinger, 1926; Hyman, 1943; Lang, 1881; and Hyman, 1951.]

tilinearity, nearly radial symmetry, and evenness. (b) *Convoluta* (Acoela) has five pairs of cords, three of which are on the dorsal side and somewhat better developed than the two on the ventral side. The commissures are irregular. (c) *Hofstenia* (Alloeocoela) cannot be said to have distinct longitudinal cords; its submuscular plexus is well developed all over the body and more so anteriorly, not as a sheet but as an open meshwork. Moreover its subepithelial plexus is also thick, anteriorly and dorsally; it has a neuropile layer surrounded by ganglion cells and has been regarded as the central nervous system of this form. (d) *Planocera* (Polycladida) has a large meshed submuscular plexus in which two pairs (subventral and subdorsal) of longitudinal cords can be discerned. (e) *Gyratrix* (Rhabdocoela) has three pairs of distinct cords and rectilinear commissures, making a good orthogon. (f) *Crenobia* (formerly *Planaria*) *alpina* (Tricladida) has three pairs: small dorsal and large ventral cords, and lateral or marginal strands—not regarded as cords by some authors because they are merely sensory concentrations without ganglion cells. (g) *Mesostoma* (Rhabdocoela) has only dorsal and ventral pairs of cords but is remarkable in having only one commissure between the ventrals. (h) *Planaria polychroa* (Tricladida) has but one pair, the ventral cords; its commissures are numerous and fairly regular. Finally, (i) the type represented by *Geoplana* (Tricladida) has a thick internal plexus deep to the submuscular plexus, called the ventral nerve plate, and in this there are indications of a heavier medial and several indistinct, lateral longitudinal bulges.

These variations cannot tenably be arranged in a **probable evolutionary order** without supposing more than one direction of change from a common type. The history of interpretation of the most primitive form of nervous system among the bilateral animals is complex and present views are undoubtedly not secure. The prevailing opinion is against either the undifferentiated plexus (*Hofstenia*) or the smallest number of longitudinal cords as being ancestral among the platyhelminths—and in favor of the *Convoluta* type, with a considerable number of cords.

The distribution of types among the orders of

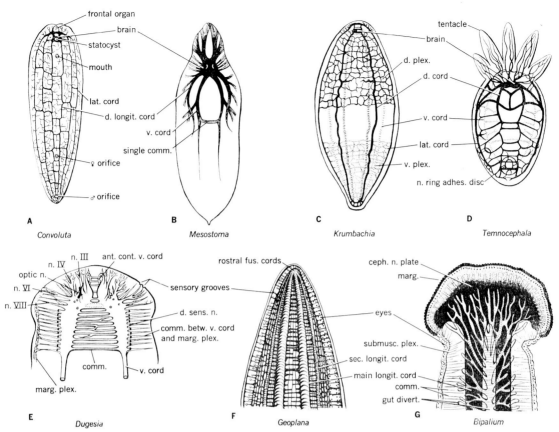

Figure 9.2. The nervous systems of some turbellarians, especially of the anterior end. **A.** Acoela. **B.** to **D.** Rhabdocoela. **E.** Tricladida Paludicola. **F.** and **G.** Tricladida Terricola. [A, Delage, 1886; B, Bresslau and von Voss, 1913; C, Reisinger, 1933; D, Merton, 1914; E, Micoletzky, 1907; F and G, von Graff, 1899.] *ant. cont. v. cord,* anterior continuation of ventral cord; *ceph. n. plate,* cephalic nerve plate; *comm.,* commissure; *comm. betw. v. cord and marg. plex.,* commissure between ventral cord and marginal plexus; *d. longit. cord,* dorsal longitudinal cord; *d. plex.,* dorsal plexus; *d. sens. n.,* dorsal sensory nerves; *gut divert.,* gut diverticulum; *lat. cord,* lateral cord; *marg. plex.,* marginal plexus; *n.,* nerve; *n. ring adhes. disc,* nerve ring in adhesive disc; *rostral fus. cords,* rostral fusion of cords; *sec. longit. cord,* secondary longitudinal cord; *single comm.,* single commissure; *submusc. plex.,* submuscular plexus; *v. cord,* ventral cord; *v. plex.,* ventral plexus.

turbellarians is not simple. The order **Acoela,** usually regarded as the **most primitive,** have three to six pairs of longitudinal strands (usually five) and irregular commissures—represented by *Convoluta.* The longitudinal strands may be indistinct parts of a general superficial plexus as in *Tetraposthia* and *Nemertoderma* (Steinböck, 1931; Lan, 1936; Westblad, 1937). The order **Rhabdocoela** are heterogeneous but include a few forms with a typical orthogon; most members have considerably or drastically reduced the number of cords and sometimes the number of commissures, too. The suborder Notandropora, including *Stenostomum* and *Rhynchoscolex,* have four pairs of cords and are a relatively primitive suborder. The Opisthandropora have lost all but one pair of cords (*Macrostomum, Microstomum*), which may be rather far apart, with commissures nearly or quite absent. The Lecithophora include *Mesostoma* and *Gyratrix,* with two or three pairs of cords and good commissures or almost none; in the same suborder is *Krumbachia,* with primitive-looking, radially symmetrical, four pairs of cords and a coarse-meshed plexus. Temnocephala have three pairs of cords and somewhat irregular commissures.

The order **Alloeocoela** embraces forms as diverse as *Bothrioplana* and *Hofstenia,* described above. The former is more typical and three or four pairs of cords with numerous commissures can be said to be primitive for the order. Even in the same suborder with *Hofstenia* are forms like *Geocentrophora* and *Prorhynchus*, with well-formed orthogonal systems. The order **Tricladida** includes the larger fresh-water and terrestrial planarians and it is not surprising to find specialization of the nervous system. There may be one, two, or three pairs of cords with uniform emphasis on the ventral and reduction or loss of the lateral and (less often) of the dorsal. The suborder **Maricola**, the marine planarians, have three: dorsal, ventral, and lateral pairs of cords, although Hanström denies cord status to the lateral or marginal nerve, finding with his successful Golgi stains that it consists solely of sensory neurons and their axons, with no ganglion cells of internuncial or motor nature. It is certainly important to know the difference in functional significance of this strand of nervous tissue from others but, as we shall see in the nematodes, a purely sensory nerve can with some reason be regarded as the descendant of a member of the orthogonal series of longitudinal strands; there is little reason other than this origin for a sensory nerve to retain its continuity over a distance longitudinally because the fibers do not in the main run for considerable distances in that direction but run into the lateral commissures and the ventral cord. The suborder **Paludicola** or fresh water triclads are not uniform: some have dorsal cords and the same kind of lateral or marginal cord as in the maricolans, as well as the chief pair, the ventrals; others lack the lateral or dorsal or both. Some observers see the cord as a series of ganglia or intermittently more abundant cells.

The suborder **Terricola** (land planarians) are quite specialized and have been called the Diploneura—in contrast to the Haploneura, which embrace marine and fresh-water triclads (Steinböck, 1925)—because they all have a second strong layer of nervous tissue, the ventral nerve plate, in addition to the submuscular plexus (this formulation does not count the delicate subepidermal plexus). The Terricola have no cords in the submuscular layer and no dorsal or marginal cords, but some have several longitudinal thickenings in the ventral nerve plate (*Geoplana pulla*, Figs. 9.2, 9.3), some have but one pair of huge longitudinal thickenings (*Bipalium kewense*), and some have no sign of cordlike structures (*G. rufiventris, G. atra*). Because the cords are developed in different plexuses, Steinböck insisted that they are not homologous in diploneurans and haploneurans. While the nerve plate is a remarkable new structure, it doubtless has an origin in common with something in other turbellarians, perhaps a further sinking in of part of the submuscular layer, so that homology of all turbellarian cords is still possible.

The order **Polycladida** has achieved the most

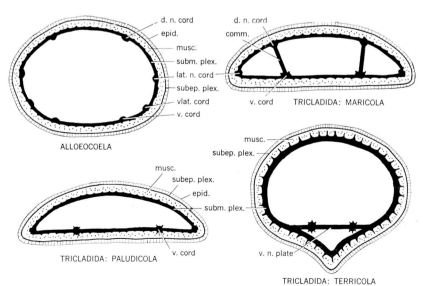

Figure 9.3.

Diagrams comparing cross sections of different Turbellaria. [Hyman, 1951.] *comm.*, commissure between dorsal and ventral nerve cords; *d. n. cord*, dorsal nerve cord; *epid.*, epidermis; *lat. n. cord*, lateral nerve cord; *musc.*, musculature; *subm. plex.*, submuscular nerve plexus; *subep. plex.*, subepidermal nerve plexus; *vlat. cord*, ventrolateral nerve cord; *v. cord*, ventral nerve cord; *v. n. plate*, ventral nerve plate, with thickenings simulating ventral cords.

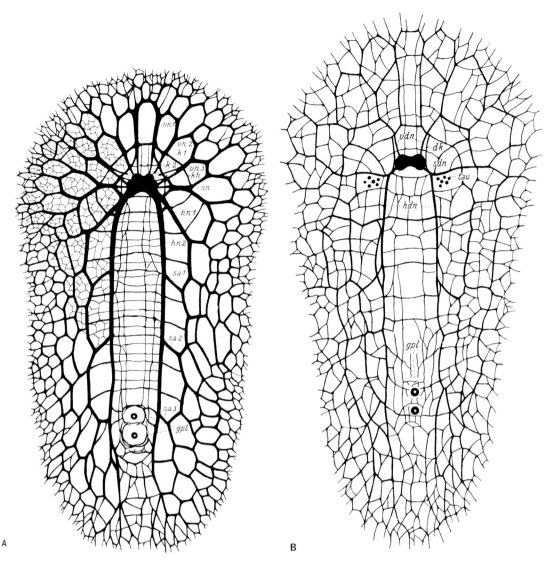

Figure 9.4. *Notoplana atomata* (Polycladida). **A** and **B.** Ventral and dorsal nerve plexuses. [Hadenfeldt, 1929.] *dk.,* dorsal commissure; *gpl,* genital plexus; *hdn,* posterior dorsal nerve; *hn 1* and *hn 2,* first and second longitudinal nerve cords; *k 1* and *k 2,* outer and inner ring commissures; *kh,* globuli cell lobe of the brain; *sa 1–3,* lateral branches of the longitudinal cords; *sdn,* lateral dorsal nerves; *sn,* lateral nerves; *tau,* tentacle eyes; *vdn,* anterior dorsal nerves; *vn 1–3,* anterior nerves 1 to 3.

advanced and complex brains of the phylum, though its pattern of medullary cords is less specialized than some rhabdocoels and terricolans. *Stylochoplana* and *Notoplana* (Hadenfeldt, 1929), *Planocera* (Lang, 1879), and *Gnesioceros* (Hyman, 1951) agree in exhibiting a number of cords radiating from the brain; all but the most medial soon branch and lose identity in the coarse-meshed plexus; hence but two pairs of longitudinal cords can be picked out—a clear ventral and a less distinct dorsal (Fig. 9.4). The larger strands of the plexus may be regarded as homologous to the cords and commissures of the orthogon of other forms. The pattern of the plexus is fairly characteristic; for example in the four genera above, branching is dichotomous such that meshes get smaller laterally and posteriorly, and the meshes in the radial-concentric direction predominate anteriorly but are lost posteriorly. Transverse and longitudinal directions dominate medially, between the ventral cords, like proper commissures, but the last are very fine if not absent in some species. The strands on the dorsal side are more delicate and irregular than on the ventral. At the margin where they meet there may be a distinct marginal strand, possibly deserving recognition as equivalent to a longitudinal cord, like that in planarians. There are vertical strands here and there, passing through the body and connecting dorsal and ventral plexuses. The plexuses just described are the heavier strands of the submuscular nervous layer.

The condition exhibiting three or **four pairs of longitudinal strands** emphasized as cords is probably **best regarded as primitive.** Each of the others can be assumed to be derived, at one or more removes. The *Hofstenia* condition, primitive as it seems, is probably a case of secondary simplification since it is surrounded by relatives of higher attainment. But this cannot readily be said for Acoela, and it is somewhat arbitrary to point, as we are here doing, to the clear indications of longitudinal cords in some species (*Convoluta*) rather than to their faintness in others (*Nemertoderma*). This bias unquestionably owes something to the idea that radially symmetrical ancestors were probably not completely undifferentiated but had long condensations of the general plexus, as in ctenophores, or at least local condensations, like the marginal bodies of medusae.

Dorsoventral or vertical commissures, directly connecting dorsal and ventral cords and traversing the body parenchyma, are found in some marine triclads. They are interesting as examples of the several but not commonplace secondary departures from the morphological plan of a group, like zygoses in gastropods and preoral tritocerebral commissures in crustaceans.

Apart from the commissures and longitudinal cords, sometimes called connectives, and the open-meshed plexuses, **nerves are scarce** in platyhelminths. The structures most suggestive of nerves are those extending laterally from the ventral cords in planarians that lack lateral cords; these "lateral nerves" are medullary and, as Hanström pointed out, should be considered part of the central nervous system. They surely represent the lateral commissures that connected the ventral and erstwhile lateral cords. In lieu of nerves, connections between central structures and the periphery are established by many fine branches, chiefly via the submuscular plexus.

2. The brain

An internal ganglionic mass is **not seen in the simplest** nervous systems, for example in such acoels as *Tetraposthia colymbetes* (An der Lan, 1936) and *Nemertoderma* (Westblad, 1937), (Fig. 9.5). Most acoels (and all other platyhelminths) have an internal brain, though it may be minute and simple, as in *Hofstenia, Archiproporus, Haploposthia,* or *Proaphanostoma* (An der Lan); it may be little more than a node of nerve fibers in front of or surrounding the statocyst. In *Tetraposthia, Nemertoderma,* and *Hofstenia*—but not in the other three genera, nor in most forms— there is a prominent **thickening of the subepidermal** nervous layer subterminally, under the dorsal epidermis. Acoels such as *Convoluta* have a **good internal brain;** in this genus it is bored through by the median frontal gland. Another acoel, *Poly-*

Figure 9.5. Longitudinal sections of acoels. **A.** Young *Nemertoderma.* [Westblad, 1937.] **B.** *Haploposthia.* [An der Lan, 1936.] *cerebr. n.,* cerebral nerves; *epid.,* epidermis; *fr. gl.,* frontal gland; *gast.,* gastrovascular cavity; *sem. ves.,* seminal vesicle; *stat.,* statocyst; *subep. plex.,* subepidermal plexus; *tes.,* testes; *th.,* thickening in the subepidermal plexus.

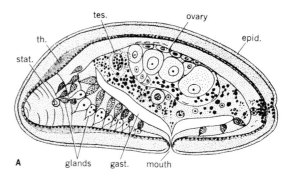

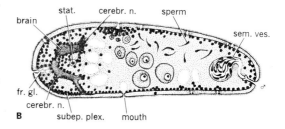

choerus, has a medial pair of ganglia usually treated as the brain, and in addition has a pair of lateral ganglia from which the dorsal cords arise.

The **question of the original or most primitive condition** is not settled. Bock (1923) and Westblad (1937) believe the simplest internal brains (*Haploposthia, Hofstenia*) are secondarily reduced, contrary to Steinböck (1924, 1931) and An der Lan (1936), who regard them as primitive. All agree that a subepithelial layer is primitive but Westblad and Bock take its thickening and becoming the main nervous concentration to be secondary; the feature is scattered unsystematically among genera and families of acoels. Westblad, Bock, and Hanström (1928) are dissatisfied with the idea that the main factor in bringing together the early internal brain was the statocyst; they argue instead that an anterior accumulation of various receptors must have been necessary. The very simple brains occur in species lacking eyes and extensive epidermal sense organs, but this negative feature also may be secondary or primary—coelenterates, which are not necessarily ancestral but are simpler, have eyes, statocysts, and epidermal sensory concentrations. Although the question of the origin of the brain in the most primitive group possessing one is a fundamentally interesting problem, we can do no more at this time than set forth the main facts, as above, and point up the issue, which has gone without new facts for twenty years (see Ax, 1964).

In all the better-developed nervous systems the brain or cerebral ganglion is a sizable internal mass and is the **only ganglion** (Fig. 9.6). It is often at or close to the anterior tip but is often considerably subterminal, even one-third of the way back. The brain is generally at the end of or along the course of the ventral cords; the dorsal and lateral cords may curve around to run into its anterior aspect or, if the brain is subterminal, may run right past it and receive connections from it as side branches. Reisinger (1926) applies the term endon to the brain, to set it off from the orthogon of cords and commissures. He and Hanström (1928) discuss whether the brain developed from a ring of commissures—and is thus an appendage of the orthogon—or developed as a terminal swelling of the longitudinal cords or as an independent structure secondarily connecting with the orthogon. The arguments are not unambiguous and must be left without review, pending new evidence. In the embryology of some forms the cords are outgrowths of the brain and thus achieve histological differentiation long after the brain. The latter is nearly always bilobed, attesting to a paired origin, but the lobes are never far apart. When they are broadly attached there may be no externally visible **commissure,** but in some forms like *Planaria polychroa* there are up to three separate short commissures, large and small, and the same number is recognized in sections of other externally unified triclads.

Delimiting the brain is difficult in some species where the connection with the ventral cord is a gradual transition; some authors have found that the limit of origin of sensory nerves is of some aid. The brain is clearly set off by a sheath only in some forms (and even in these not without complication). In fresh-water planarians, among others, there is generally no clear demarcation from the mesenchyme. Some alloeocoels (*Monoophorum, Pseudostomum*, otoplanids), polyclads (*Notoplana, Stylochoplana*), and others have a thin, but distinct and continuous, structureless **capsule** investing the brain, interrupted by the nerves and not continuing onto them. But in the last two genera named there is a pair of globuli cell masses situated outside the capsule; in *Planocera grafii* many tracts are outside (Hadenfeldt, 1929). In many forms there are sizable islands of mesenchyme and muscle cells inside the brain ("Substanzinseln").

The shape of the brain is too diverse among turbellarians to deserve any special comment. From the strategic position of this class, the volume in relation to body weight should be of some interest, but it has not been measured. Roughly speaking, comparison with animals of a similar size among higher phyla would not be grossly unfavorable to the turbellarian in this respect. To give some idea of **linear dimensions,** published illustrations lead one to expect that a 20 mm *Notoplana* (Polycladida) will have a brain about 700×1300 μ; a 10 mm *Crenobia alpina* will have a brain about 416×680 μ.

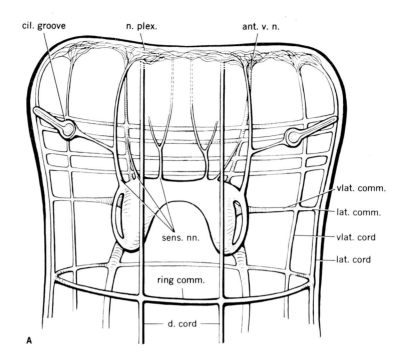

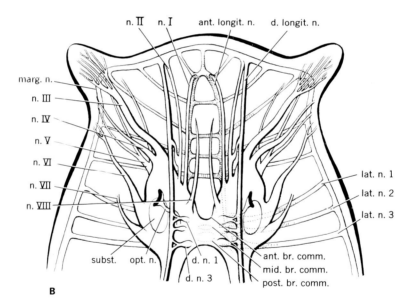

Figure 9.6. Diagrams of the brain and anterior nerves. **A.** *Geocentrophora baltica*. [Steinböck, 1927.] **B.** *Planaria alpina*. [Micoletzky, 1907.] *ant. br. comm.*, anterior brain commissure; *ant. longit. n.*, anterior longitudinal nerve; *ant. v. n.*, anterior ventral nerve; *cil. groove*, ciliated groove; *d. cord*, dorsal cord; *d. conn.*, dorsal connective; *d. longit. n.*, dorsal longitudinal nerve; *d. n. 1* and *d. n. 3*, dorsal nerve 1 and 3; *lat. comm.*, lateral commissure; *lat. cord*, lateral longitudinal cord; *lat. n. 1–3*, lateral nerves 1–3; *marg. n.*, marginal nerve; *mid. br. comm.*, middle brain commissure; *n. plex.*, peripheral nerve plexus; *n. I–VIII*, cerebral nerves 1–8; *opt. n.*, optic nerve; *post. br. comm.*, posterior brain commissure; *ring comm.*, ring commissure; *sens. nn.*, sensory nerves; *subst.*, "Substanzinsel"; *v. comm.*, ventral commissures; *vlat. comm.*, ventrolateral commissures; *vlat. cord*, ventrolateral cord.

An **exception to the distinctness of the brain** in more advanced flatworms is the group Terricola, the land planarians (= Diploneura). The anterior end of the ventral nerve plate would be expected to represent the brain—from its position and the presence of cords in some species. In *Geoplana atra* there is a swelling here and in *Bipalium kewense* the paired cordlike thickenings widen out into a single thick plate coextensive with the broadened head (Fig. 9.3). *Rhynchodemus* has a good, rounded bilobed brain with fine-grained central neuropile and small cells both around it and invading it in nests (Kennel, 1882). These cells lack visible cytoplasm and resemble those in polyclads and nemertineans which are called globuli cells—almost a diagnostic mark of a brain. But *G. pulla* represents many forms in which the ventral plate tapers to a pointed head end and in which no sign of a specialized brain-like region has been detected. Moreover it is said that the sense organs are innervated from the submuscular plexus, not the nerve plate. Since these plexuses run together at the margin around the head, this may not be a real difficulty. But the plate's exact relation to head sense organs—and the possibility of special nerve cell groups—calls for detailed histology. This, together with experimental work on the difference in function of the head nervous system between land and familiar fresh-water planarians, would be a worthwhile study.

Nerves from the brain naturally vary widely among species. To take but one example, we will extract from Micoletzky's (1907) description of *Crenobia alpina* (Fig. 9.6). Since the brain is subterminal in this species, the ventral cord continues forward under it and nerve I, arising anteromedially, passes forward to join this cord just under the anterior marginal nerve. Nerve II arises beside the first nerve and immediately bifurcates to send communicating branches to the dorsal longitudinal cord and to the anterior marginal nerve, which is a continuation of the lateral (marginal) cord. Nerves III, IV, and V are less connective in nature and branch repeatedly to supply the body surface, especially that of the tentacle and surrounding areas; III is the main tentacular nerve. This species has three further

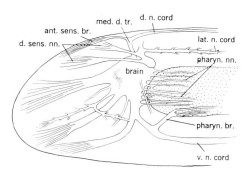

Figure 9.7. Side view of the anterior end of the alloeocoel *Hydrolimax*, showing the main brain nerves and the manner of origin of the three main nerve cords. [Hyman, 1951.] *ant. sens. br.*, anterior sensory branches; *d. n. cord*, dorsal nerve cord; *d. sens. nn.*, dorsal sensory nerves; *lat. n. cord*, lateral nerve cord; *med. d. tr.*, medial dorsal trunk; *pharyn. br.*, pharyngeal branch of ventral nerve cord; *pharyn. nn.*, pharyngeal nerves of the brain; *v. n. cord*, ventral nerve cord.

nerves of this series, VI, VII, and VIII, the first two supplying the area behind the tentacle and the last, the medial dorsal body surface; corresponding nerves are lacking in another species, *Planaria polychroa*. The latter form has no tentacles but a series of special sensory grooves on the side of the head, which are innervated by three pairs of lateral nerves (counterparts are found in *C. alpina*) going to the marginal nerve. From the dorsal surface of the brain arise the optic nerve and three short connectives to the dorsal cord. The nerves are undoubtedly mainly sensory and Micoletzky found support for this in the presence of nerve cell bodies along many of them. Some he described as lacking cells and these he believed to be motor, but this cannot be accepted without more evidence. Statocysts are lacking in triclads but in the groups possessing them they are generally in or attached to the brain and presumably supplied directly therefrom.

The nerves are described for several other turbellarians: see Hadenfeldt (1929) for references to a number of earlier accounts of polyclads (Fig. 9.4); Wilhelmi (1909) for triclads; Steinböck (1927) and Reisinger (1925) for alloeocoels (Fig. 9.7). Hadenfeldt tried to homologize nerves in nine genera of polyclads, which have from six to eleven pairs, but concluded that there is real variation between them, not just fusion or splitting of a moderate degree.

Turning from gross anatomy to the **histological composition of the brain** and cords, we are still heavily dependent on very old accounts. Except for the important works of Micoletzky (1907) on *Planaria*, Gelei (1912, in Hungarian) on *Dendrocoelum*, Hanström (1926) on *Bdelloura*, and Hadenfeldt (1929) on *Notoplana*, we must still rely upon the descriptions of Graff (1882, 1899, 1904, 1914) on rhabdocoels and triclads, Iijima (1884) on *Planaria*, and Lang (1879, 1881, 1884) on *Thysanozoon* and *Procerodes*. The only investigations employing special nerve stains on central nervous tissue with particular success, apart from the alizarin method of Reisinger (1926) for visualizing the cords and nerves in whole mounts, are the Golgi studies of Monti (1900) and Hanström (1926a). Turner (1946) used reduced silver to observe some aspects of the cell bodies.

The **nerve cell inclusions** are similar to those elsewhere. Steopoe (1934) describes the chondriome, vacuome, Golgi apparatus, a centrosome, and a trophospongium of sinuous, branching glia processes reaching almost to the nucleus, as in molluscs and arthropods. Cells up to 40 µ occur.

Nerve fibers have been figured up to 20 µ in diameter (Haswell) but such a value is questionable. In better-known forms the largest fibers are about 5 µ. Cross sections of nerves look much like those in annelids and arthropods.

The **types of nerve cells** show advance over coelenterates in the presence of abundant unipolar neurons in addition to bipolars and multipolars. The unipolars, moreover, are quite like those in higher invertebrates (Fig. 9.8) in having ramifications of differentiated kinds localized in the fibrous core of the central nervous system and sometimes also in the periphery, presumably as effector endings. Some authors assert that unipolars are only to be found in the brain, for example Monti in *Planaria*, Haswell in *Temnocephala*; on the evidence we should only conclude that there is at least a preponderance of them there. Even so, it makes an interesting chapter in histological evolution.

With the aid of successful Golgi impregnations, Hanström and Monti showed a number of cell types significant as the first of their kinds in the scale of complexity of nervous systems. Speaking here only of the central neurons, there are unipolar motoneurons with cell bodies in the ventral cords, several central branches sometimes quite long, and one peripheral process. There are unipolar interneurons with ascending or descending processes or both and sometimes commissural processes as well. The highest type reported by Hanström is a decussating, descending unipolar interneuron with branching collaterals on both ipsi- and contralateral sides, in the brain and extending three or more commissures back. Hanström saw an analogy between these long, crossed neurons and the giant neurons of nemertineans and annelids, and proposed that they function in movement coordination. Multipolars and bipolars were also found, the latter especially in the commissures, where they appear to be isopolar—that is, unpolarized structurally. Monti several times mentions central cells with sensory-type branched endings in the epidermis, but Hanström did not see these.

With ordinary cytological methods Micoletzky distinguished four types of nerve cell bodies in *Planaria*, Hadenfeldt added one more from the complex brain of polyclads, and Turner added two more, thus totaling **seven cell types** besides glia cells. Following Turner, who had the use of reduced silver preparations of the polyclad *Leptoplana* as well as the benefit of the earlier reports, we may call **type A** the large cells (12–16 µ), plasma-rich, chromatin-poor, with an even, fairly deeply staining cytoplasm, round nuclei, central nucleolus, and the peripheral parts of the nucleus staining moderately densely with aniline dyes; the majority of these cells seems to be unipolar. **Type B** cells (also generally 12–16 µ) include the largest cells—about 23 × 29 µ; most often with a single large neurite, many are found with two or more processes, one of which can often be observed passing out through the brain capsule. The cytoplasm is almost without exception vacuolated at the apical end. The nucleus is even more chromatin-poor and has a single, eccentric nucleolus. **Type C** cells are considerably smaller and apparently all unipolar. Their nuclei have scattered chromatin granules and no definite nucleolus. **Type D** is slightly larger than C and mostly unipolar; their nuclei are relatively denser

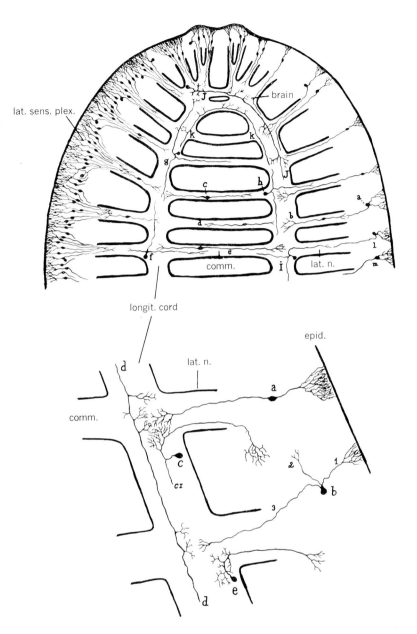

Figure 9.8.

Drawings from Golgi preparations of *Bdelloura candida*. **Above,** anterior end; **below,** conduction paths in the right half of the body. [Hanström, 1926.] *comm.*, commissure; *epid.*, epidermis; *lat. n.*, lateral nerve; *lat. sens. plex.*, lateral sensory plexus; *longit. cord*, longitudinal cord; *a–m* and *1–3*, individual neurons and branches referred to in original description.

and almost invariably contain conspicuous nucleoli. **Type E** is applied to nuclei of what must be supporting cells; the nuclei are irregular in shape and are located between fiber bundles and the brain capsule; the cytoplasm has not been recognized.

Type F are the most interesting and their full significance has not apparently impressed itself on previous authors. These are the cells of the inner and outer "granular masses," or "Körnerhaufen"(Fig. 9.10). "They have but a thin rim of cytoplasm surrounding a dense chromatin-rich nucleus without nucleolus" (Turner). They are 3–4.5 μ across and, apart from some scattered in small groups in the lateral and posterior rind of the brain, are only found in the relatively packed masses mentioned. They are unquestionably neurons and send processes in a large tract into the central neuropile, even though the external mass has to send its tract through an interruption in the brain capsule. These characters are quite sufficient to identify the cells as **globuli cells**, known heretofore only in the highest centers of the brain in the more complex nemertineans, poly-

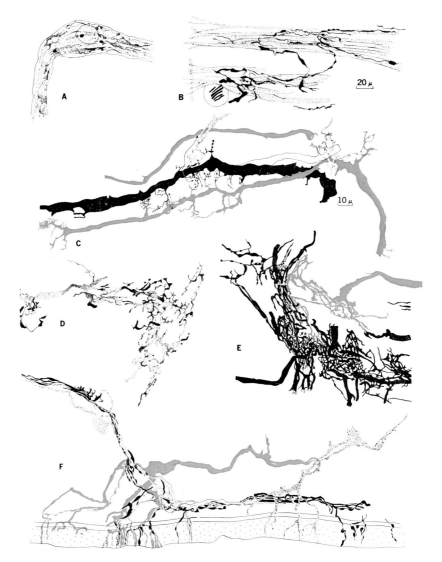

Figure 9.9. Nerve cells and endings in *Dendrocoelum*. Methylene blue stain and Golgi impregnation. **A.** Large bipolar nerve cell from cerebral commissure, showing trophospongium. Methylene blue. **B.** Innervation of dorsoventral muscle fibers from a branch of a transverse nerve. Golgi. **C.** Extensive connections between neurons by means of fine lateral processes. The lower nerve fiber is an intermuscular one with mosslike processes on its surface. Golgi. **D.** Part of a nerve cell from the main longitudinal cord near the brain to show the richly branched telodendria. Golgi. **E.** The tangle of endings of four nerve fibers in the nerve cord. Golgi. **F.** Sensory cells and their processes in the skin; distal processes downward, central processes upward. Golgi. The scale in **C** applies to **C–F**; no scale for **A**. [Gelei, 1909.]

chaetes, molluscs, and arthropods. In all groups they are nearly or quite confined to the brain and to distinct masses therein; their processes where known do not extend beyond the brain but end in the denser parts of the neuropile, typically in association with pathways from the higher sense organs. As in the nemertineans and annelids, so also in Turbellaria, the globuli cells are present only in some members of the group—namely, those with the most complex brains, judged on other criteria. Only polyclads so far have shown them and among this order only *Notoplana* and *Stylochoplana* are known to have two pairs of globuli masses—an internal and an external. *Meixneria*

(Bock), *Planocera* (Lang), *Thysanozöon* (Lang), and *Leptoplana* (Kennel, Turner) have only a single pair, probably comparable to the internal. Nothing of the kind has been found in planarians. There is no reason to consider these cells sensory as some authors have done; they are probably interneurons.

Type G cells are doubtless **glial**, occurring in rows in fiber tracts and applied to the inside of the capsule as well as along the peripheral trunks. They are thin and spindle-shaped with fine processes. **Type H** designates a single pair of large spindle-shaped neurons whose processes decussate; they are known only from *Leptoplana*, near the posterior end of the brain and on the ventral aspect, just inside the capsule (Turner). **Type J** is also known only from Turner's account and stands for occasional cells in the brain loaded with uniform round granules, possibly **neurosecretory**. Micoletzky's enumeration of four cell types in *Planaria* can probably be roughly provided for between types A and D, except for his small bipolars.

In an important monograph in Hungarian, ignored by almost all later authors, Gelei (1909) extensively illustrated details of the finer ramifications revealed by Golgi, methylene blue, and hematin methods in *Dendrocoelum* (Fig. 9.9). Among other points, the most important is Gelei's assertion that here and there, but rather commonly, he saw protoplasmic continuity between smaller branches of nerve fibers. The Golgi impregnation is not without pitfalls in just this matter, and the influence of his mentor, Apáthy, who had in Gelei's view already proved the reality of continuity, must be considered. But these are not sufficient grounds to dismiss the report, although Hanström did not see any anastomoses. On present evidence it appears that just possibly there are sometimes protoplasmic bridges of fine caliber, a conclusion to which we come several times in other groups on different technical bases.

The outstanding advance in **arrangement of nervous tissue** in flatworms over less complex animals is the separation of nerve cell bodies from the main mass of nerve fibers in the large cords and cerebral ganglia. Therewith the typical feature of higher invertebrates is established—an **outer rind of cells and an inner core of fibers** of passage and terminal branchings. Accompanying this separation is the development of glial cells, already noted, along tracts and between cells (Joseph, 1902; Sabussow, 1905; Clayton, 1932). There is a great range among turbellarians in brain development; the simpler species in this respect give no obvious sign of differentiation of cell rind or fiber core into special regions. Hanström (1926b) found the terminal arbors of optic fibers in the upper lateral part of the brain of *Procerodes* and speaks loosely of visual centers, but the only connection of this region he found is

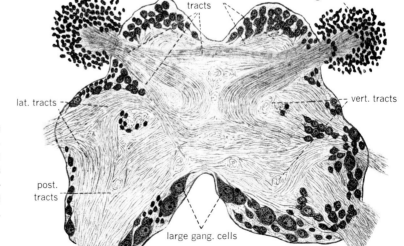

Figure 9.10.

Transverse section through the brain of the polyclad *Notoplana*. Note the grouping of fibers in the core to form distinct tracts and the mass of globuli cells contrasting with medium-sized and large ganglion cells in the rind. [Hadenfeldt, 1929.]

a commissural one to the same region of the opposite side.

Notoplana and *Stylochoplana* have the **most complex brains in the phylum.** Comparison of Fig. 9.10 with similar sections of nemertinean, annelid, mollusc, and some arthropod brains shows this polyclad to be in the same general class of histological differentiation, with some members (not the highest but not the lowest, either) in each of these phyla. Five distinct cell masses can be made out: the two globuli groups, internal and external, the dorsal rind of type C cells, the anterior and lateral rind of types C and D cells, and the ventral and posterior rind of A and B cells. The fiber core has at least ten named tracts distinguishable (Fig. 9.11); their connections are not known. The neuropile is nowhere extremely fine-textured or dense and there are no glomeruli. The internal structure of the brain in fresh-water planarians, of such interest in experimental studies on behavior, is very little known. It is clearly not as complex as the polyclads just described. Three tracts have been distinguished; three commissures and a broad fiber bridge between left and right halves are described; and decussating bundles from the dorsal aspect of one side to the ventral aspect of the other occur. No distinct cell masses are recognized.

B. Peripheral Nervous System

1. The plexuses

Here for convenience we will treat the plexuses, although there is no line of division naturally between central and peripheral cords of nerve cells and fibers. Not only are many of the structures treated below quite the same in structure as those treated above, but there is probably a direct homology in the sense of a phylogenetic continuum between the coarse-meshed plexuses in or deep to the body wall and the cords and commissures (Fig. 9.12).

It has been traditional to recognize in most turbellarians **two general plexuses** or nervous layers, the subepidermal ("äusseres Hautnervengeflecht") and the submuscular ("Hautnervenplexus"); however, we still await application of selective nerve stains to these structures. Several authors have seen many peripheral nerve cells in Golgi preparations (Monti, Sabussow, Botezat and Bendl, Gelei, Hanström) but none has seen more than fragments of the plexus, and then only the submuscular.

The **subepidermal layer** is not indicated on most

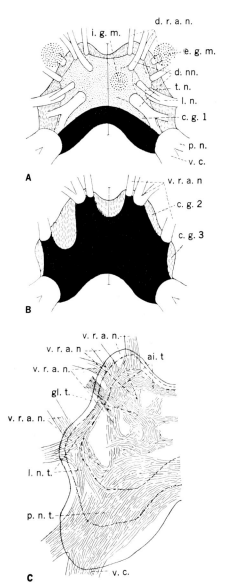

Figure 9.11. The arrangement of cell groups [**A,** dorsal view; **B,** ventral view] and tracts (**C**) in the brain of *Notoplana*. [Hadenfeldt, 1929.] *ai.t.,* anterior-interior tract; *c.g. 1–3,* cell groups 1 to 3; *d.nn.,* dorsal nerves; *d.r.a.n.,* dorsal root of antennal nerves; *e.g.m.,* external globuli mass; *gl.t.,* globuli tract; *i.g.m.,* internal globuli mass; *l.n.,* lateral nerve; *l.n.t.,* lateral nerve tract; *p.n.,* posterior nerve; *p.n.t.,* posterior nerve tract; *t.n.,* tentacular nerve; *v.c.,* ventral cord; *v.r.a.n.,* ventral root of antennal nerve.

Figure 9.12. Histological appearance of ordinary sections of turbellarians. **A.** *Rhynchodemus*, ventral side showing the nerve plexus in the mesenchyme. **B.** *Procerodes* (= *Gunda*), cerebral ganglia in transverse section. **C.** The same, ventral cord. **D.** *Planaria*, longitudinal section of the ventral cord. *gang. cells*, ganglion cells; *d-v. musc.*, dorsoventral musculature; *longit. m.*, longitudinal muscle; *longit. tract*, longitudinal tract; *strands of n. plex.*, strands of nerve plexus; *v. cord*, ventral cord. [Lang, 1881.] **E.** Frontal section of the brain, the origin of the longitudinal nerve cords and of some of the main nerves entering the peripheral plexus, in *Tristomum molae* (Trematoda). [Lang, 1881.] *gang. cell*, ganglion cell in brain; *gang. c. p.*, ganglion cell in the muscle of the parenchyma; *m.*, muscle strands running through the brain; *n. f. tracts*, nerve fiber tracts; *n. I*, first cerebral nerve, to region between oral suckers; *n. II*, second cerebral nerve, to the oral sucker; *n. III*, third cerebral nerve, which joins the preceding and forms a cross connection between *I* and *II*; *plex.*, nerves forming the main plexus; *v. cord*, beginning of the ventral cord.

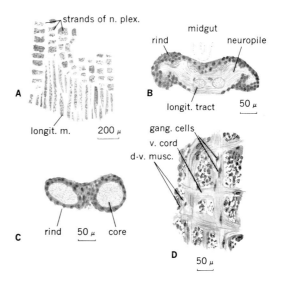

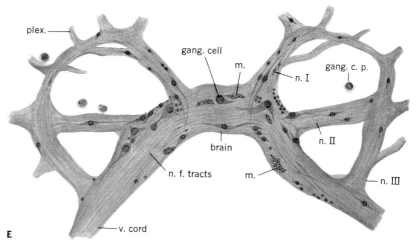

figures of the body wall or integument and must be extremely tenuous if indeed it is commonly present. Such a layer, outside the muscles and basement membrane, at the base of the epithelium, is prominent in the anterior end of many small acoels, resembling the proboscis of enteropneusts. It is locally quite thick dorsally and rostrally in some acoels (*Nemertoderma* and *Hofstenia*), all of which lack or almost lack an internal brain; the increment is sometimes treated as an additional layer outside the general subepidermal plexus (Bock, 1923; Steinböck, 1931; An der Lan, 1936; Westblad, 1937).

The **submuscular plexus** is heavy but it is usually not a continuous layer. In it generally are the longitudinal thickenings forming the cords and the transverse ones forming the commissures or the irregular meshes of the first order in polyclads. Between these thick strands, delicate strands form a small-meshed plexus. But what the arrangement is at a still smaller level, and the course and destination of the constituent fibers, is not known. It has been more than once remarked how few nerve cell bodies are to be found along even the larger strands. Cells believed to be glial are distributed along them but there is no clear sheath. Comparing the acoels with higher groups, a tendency is noted for the submuscular

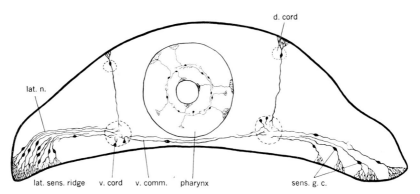

Figure 9.13. Diagrammatic cross section through the entire body in the region of the pharynx, combined from Golgi preparations of *Bdelloura candida*. [Hanström, 1926a.] *d. cord*, dorsal cord; *lat. sens. ridge*, lateral sensory ridge; *lat. n.*, lateral nerve; *sens. g. c.*, sensory ganglion cells; *v. comm.*, ventral commissure; *v. cord*, ventral longitudinal cord. In the pharynx note the sensory ganglion cells with free nerve endings in the outer and in the inner epithelium, internuncial neurons of the pharyngeal plexus, and motoneurons of the pharyngeal plexus sending axons out to muscles of the pharynx.

plexus and cords to sink deeper into the body. The limited physiological evidence (p. 560) indicates almost no ability of the plexus to conduct around obstacles and in the present state of knowledge we should not speak of a nerve net.

Still deeper and found only in the terricolous triclads is the strong plexus called the **ventral nerve plate** (Fig. 9.3); it is considered to be the central nervous system in the land planarians and has been discussed above (p. 539).

The **nervous system of the pharynx** may be fairly elaborate (Fig. 9.13). It is mainly a plexus beneath the muscle layers but there are commonly one or several rings of thickenings, especially near the distal end. There may be also longitudinal strands so that a regular lattice is formed, as in *Phaenocora* (Rhabdocoela) (Fig. 9.14). Hanström obtained Golgi impregnations of the pharynx of *Bdelloura* and saw sensory, motor, and associative cells in the ring; therefore the apparatus for autonomous behavior is present. The isolated pharynx has been observed to carry out all normal activities, including ingestion. The connection of the pharyngeal with the central nervous system is sometimes with the brain, more often with the ventral cords, or both (*Hydrolimax*, Alloeocoela; Fig. 9.7).

A **genital plexus** provides a rich supply to the copulatory organs. Adhesive organs, when developed, may have their own specialized plexus.

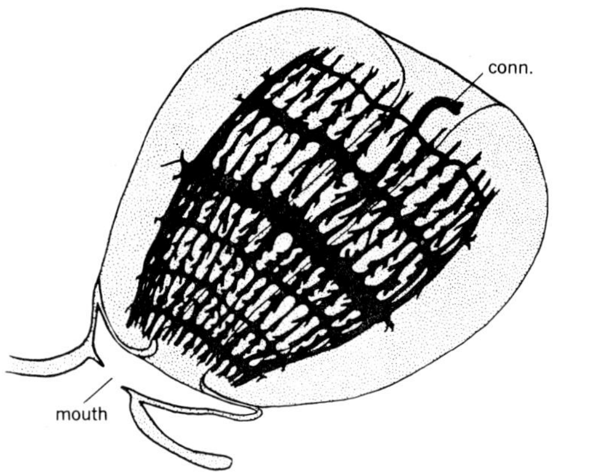

2. The receptors

Sense cells are not only extremely numerous in the turbellarians but a given species has at least six differentiated types, in the better-developed forms. Least specialized are the widely scattered single sense cells (Fig. 9.15). These may be divided into forms with a single distal process and those with a branching distal process. Both have their cell bodies well below the surface, in or beneath the body wall—probably a derived rather than the primitive position. The cells with extensively branching free nerve endings are believed to be even more derived or advanced. The terminations vary, partly as a function of the method by which they are visualized, but at least some seem

Figure 9.14. The pharyngeal nervous system of *Phaenocora* (Rhabdocoela). [Luther, 1921.] *conn.*, connective to the central nervous system.

II. ANATOMY OF TURBELLARIA B. Peripheral Nervous System 2. Receptors

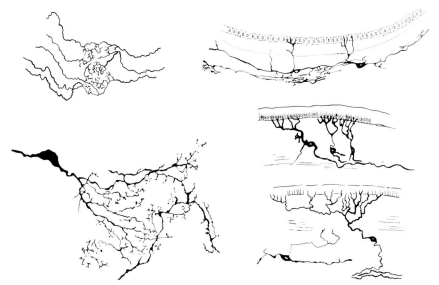

Figure 9.15. Primary sensory neurons in the body surface revealed by Golgi impregnation. Note the freely branching distal processes which can penetrate the superficial epithelium; the cell bodies may lie deep to the epidermis. The central process is usually unbranched and turns to run tangentially in the subepidermal plexus. Some of the neurons are multipolar. [Monti, 1897, 1900.]

to end below the surface of the epidermis while others have one or several hairs or bristles projecting through the epithelium. Some of the long hairs are supposed to move slowly but some are immobile. They often extend far beyond the cilia of the body surface. There are very likely several functional categories but nothing can be said except that tactile sensitivity is reasonably supposed to be one. The axons in a few cases have been traced into the submuscular plexus. Such sense cells are distributed widely with special concentrations on the various head lobes, tentacles, and "ears," and along anterolateral margins where a definite sensory strip ("Sinneskante") is sometimes developed, especially in some acoels and many terricolans.

Three types of single primary sense cells have been described by Gelei (1930), based on study of *Mesostoma* by an osmium-toluidin method good for cell outlines and certain organelles like sensilla, in whole mounts (Fig. 9.16). Inferring the functions purely from the anatomy, he calls (a) **tactile receptors** those with a few widely spreading terminal branches bearing bristles of medium length; such cells are scattered over the whole body. (b) **Rheoreceptors,** supposedly sensitive to water currents, are four pairs of large cells under the epidermis, each with 20–100 very long bristles projecting beyond the cilia and closely clumped in four constant positions on the body: one dorsal and one ventrolateral to the eye, one lateral to the mouth and one dorsolaterally, at the end of the third quarter of the body. The terminal processes, one for each bristle, have an intraepithelial collar, a cone, and an axial thread. The central process or axon of the cell goes to the ventral cord. (c) **Chemoreceptors** appear only on the anterior end of the body and each has a single dendrite and short projecting

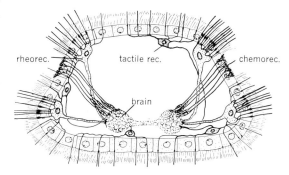

Figure 9.16. Diagrammatic transverse section through the head of *Mesostoma* (Rhabdocoela) to show four of the eight rheoreceptors *(rheorec.)*, the two groups of chemoreceptors *(chemorec.)*, and four tactile receptors. [Gelei, 1930.]

peg. They occur in a concentrated group on the auricle and send axons into the brain. Gelei gives good evidence that the bristle- or peg-bearing terminal processes pass through the epithelium to the surface—not simply between ordinary epithelial cells but through them; a minimum of 3-5 and a maximum of several hundred such sensory processes penetrate every single epithelial cell (Fig. 9.17)! Taking the evidence at face value, and there is little reason to doubt it, these are extraordinary sense cells and certainly deserve further examination. Again, as in coelenterates, we are faced with an enormous abundance of sensory elements but in turbellarians there are fewer cells, generally lying below the epithelium, and most have extensively branched distal processes. The several accounts based on the Golgi impregnation (Monti, 1897, 1900; Sabussow, 1905; Botezat and Bendl, 1909; Gelei, 1909; Hanström, 1926a) agree on the abundance of richly branched distal sensory processes. The spacing of the final twigs makes Gelei's figure reasonable and agrees with Retzius's (1902) illustration of the epidermis in surface view in which he identified, as sensory endings, certain dots which were several times more numerous than epithelial cells, especially anteriorly and dorsally.

Extirpation of the eight long-bristled organs and of the two short-pegged clusters has been successfully accomplished by Müller (1936). The deficits that result confirm their **functions in water current and chemical detection,** respectively. "Inadequate" stimulation—of either organ by the stimuli effective on the other—is impossible. Similar results, on chemoreception only, are reported by Koehler (1926) from *Crenobia alpina*; this animal will not feed if the skin of the rostral tip, between auricles, is removed, and it recovers the behavior only when this region is regenerated. In contrast, sensibility to currents is widely distributed. Some species can recognize food with the tip of the proboscis but others can not. These are nice examples of clear **modality distinction** in skin receptors in lower forms.

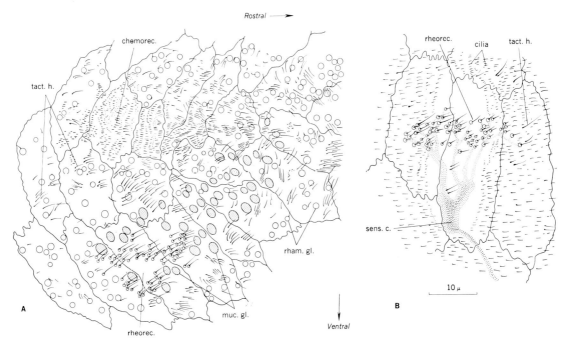

Figure 9.17. Surface views of *Mesostoma lingua* (Rhabdocoela) to show differentiated types of unicellular receptors. **A.** Anterolateral area of the body, showing the localized distribution of three types of receptors. **B.** The rheoreceptor together with adjacent cilia lateral to the mouth and tactile hairs. Toluidin blue, formol osmium. [Gelei, 1930.] *chemorec.*, chemoreceptor hairs; *muc. gl.*, mucus gland openings; *rham. gl.*, rhammite glands; *rheorec.*, rheoreceptor hairs; *sens. c.*, sense cells; *tact. h.*, tactile hairs.

Sense organs composed of multiple sense cells are common; they include various specialized patches of epithelium, depressions, and papillae, as well as statocysts and eyes.

One class of epidermal, grouped endings have traditionally been considered **chemoreceptors;** this class would include Gelei's type (c), above. A few single-celled structures in polyclads are, from their resemblance to the others, also so identified. Generally limited to the head region, this class comprises various grooves, pits, and similar indentations with associated specialized epithelia generally devoid of rhabdoids, either ciliated differently from the general surface or without gland openings—though they may have associated gland cells. The sensory margin of land planarians includes various of these organs. The frontal gland has numerous sensory nerve fibers and together they form a frontal organ in acoels, some alloeocoels, and larval polyclads—one of those curious glandulosensory complexes seen in nemertineans and other phyla. Bipolar cells with a branching distal process ending in a number of short, stiff hairs have been seen in chemoreceptor grooves of *Mesostoma* and *Bothromesostoma* (Gelei, Müller). Müller extirpated ciliated pits in *Stenostomum*, after which it could not detect the presence of food juices or orient toward food; similar results follow removal of a pair of grooves on the head of *Bothromesostoma*. Ciliary currents are supposed to be important in permitting these organs to give directional information on food sources (Steinmann, 1929). A distinction is possible between **smell,** on the one hand—as a distance-sense, for alarm, seeking, and detecting—and **taste,** on the other—as a contact-sense, for testing and for controlling eating movements (Müller, 1936).

A median unpaired **statocyst** is common in marine turbellarians; that is, among Acoela and Alloeocoela and in a few others, and therefore in general among the more primitive orders (Ax, 1961). The innervation appears to be from the brain (Westblad, 1937); the organ is either within the brain or just behind it.

Photoreceptors of unknown sort must be distributed over the body, on physiological evidence. Reaction to light does not require the cephalic eyes in planarians (Beuther, 1927). Rarely, epidermal pigment spot eyes are present. Otherwise the eyes are pigment-cup ocelli lying deep in the mesenchyme or in the brain (Fig. 9.18). Most often there is a single pair, but sometimes two or three pairs are found. But in polyclads there are almost always more than six pairs and up to several hundreds; in land planarians there may be more than 1000 ocelli scattered on the head and far behind it. Primitive ocelli have one to four sensory neurons ending in an expanded knob with a striated or rod border, all inside a cup of one or more pigment cells conferring a directional sensitivity. The nerve cell body is a short distance away and its distal process enters the cup through its opening, making the eye an inverse type. Regarded as derived are eyes with many (up to 150)

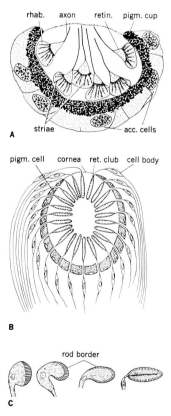

Figure 9.18. Eyes of Turbellaria. **A.** *Dugesia tigrina.* [Taliaferro, 1920.] **B.** Advanced type of eye of a land planarian with retinal clubs. [Hesse, 1902.] **C.** Stages in the evolution of the retinal clubs. [Hofsten, 1921.] *acc. cells,* accessory cells; *pigm.,* pigment; *ret. club,* retinal club; *retin.,* body of retinula cell; *rhab.,* rhabdom; *striae,* striae in rhabdom.

receptor cells, more complex terminals called retinal clubs (Fig. 9.18), and some or many of them entering the cup between pigment cells, making the eye converge in type. The land planarians not only have the largest numbers, but the most advanced eyes.

The **retinula cell** of *Prorhynchus* (Alloeocoela) is described as having three parts, a distal "rhabdom" next to the pigment cell, a middle "ellipsoid," and a proximal "myoid" with the nucleus. After exposure to light the "ellipsoid" disappears and the "rhabdom" rounds up from the elongate shape it adopts in the dark; at the same time the pigment cell contracts and its cytoplasmic lamellae move closer together—whereas they open up in the dark (Kepner and Taliaferro, 1912a; Barrett, 1929). Taliaferro (1920) argues for the notion that light must strike a rhabdom parallel to its long axis to be effective, and therefore that the position of the stimulated rhabdom gives the animal the information about the source of light. He recognizes two main groups of retinulae in *Dugesia*: one lies around the posterior and ventral edge of the cup; the other group consists of all the rest. Illumination of the first group elicits turning toward that side, and illumination of the remainder, turning contralaterally; removal of the posterior portion of the eye was accomplished without disturbing the remainder. The orientation toward light depends on these structures and the central reactions they elicit, not on the symmetry of the eyes. Once oriented in a horizontal beam, the animal receives no orienting stimuli unless it leaves that path. Other details of structure, including retinula cytology and the nonnervous components, cannot be given here (see Press, 1959; Hyman, 1951; Hesse, 1897; Bresslau, 1928; Hanström, 1926b; Jänichen, 1897; Graff, 1882, 1899; Böhmig, 1887; Hertwig, 1881).

Nerve fibers branching upon and **terminating among muscle fibers** have been reported by Monti, Gelei, Hanström, in Golgi preparations (Fig. 9.9B). The endings seem to be profuse, covering a wide area. The cells of origin of these fibers include central, unipolar, and multipolar ones. In addition, Monti claimed to find peripheral cells with some processes ending on muscle, other processes as free nerve endings branching in the epidermis, and one process to the central nervous system. Hanström saw similar cells but could not say whether they ended on muscle and he hesitated to believe that one branch was sensory and another motor, an arrangement which would constitute a "one-neuron arc." It is quite possible that the internal branches are sensory for mechanical deformations.

III. THE PHYSIOLOGY OF TURBELLARIA

A. The Role of the Brain

Over the years a good many workers have done simple experiments on turbellarians; to mention some, Loeb (1893, 1894, 1900), Parker and Burnett (1900), Kepner and Rich (1918), Alverdes (1922), Olmsted (1922), Moore (1923, 1933), Levetzow (1936), and Honjô (1937) have made incisions and transections. In addition, Pearl (1903), Taliaferro (1920), Kawaguti (1932), Ullyott (1936), and Pantin (1950) have contributed partial analyses, especially of locomotion and the reactions to light. Many other authors listed in the bibliography are responsible for relevant findings. Actually our knowledge is still very primitive and the detailed, basic studies of functional capacities of peripheral plexuses, cords, and brain are yet to be done. The great interest in the role of the brain in turbellarians is due largely to its strategic position as the first brain in the usual sequence of animals. The brain perhaps arose first as a sensory recognition device. Perhaps at the same time it became the main seat of selective suppression of spontaneous tendencies of lower neurons. We will see that the present evidence is insufficient for a decision on these suggestions.

An **inventory of actions** presumably nervously coordinated would enable each worker to get more from his specimens but such inventory has not been made complete for any species. Some of those discussed in the literature are: locomotion by ciliary action, by muscular waves passing backward over the ventral surface, forward and backward looping (*Polycelis*, Robertson, 1928; *Rhynchodemus*, Pantin, 1950), leechlike progression (temnocephalans, Honjô, 1937), approach movements and avoidance movements, proboscis protrusion, gradual or very localized lateral bending at different levels of the body, head swinging back and forth, twisting, righting, feeding, swimming by symmetrical or by alternating marginal waves, curling around a transverse axis. The muscular waves in locomotion are mostly monotaxic (across the whole width of the ventral surface) in triclads, ditaxic (across half the width and alternating on the two sides) in polyclads. They may be exaggerated in land planarians like *Rhynchodemus bilineatus* (Pantin, 1950) into "myopodia," which only touch at their lower surface and leave isolated mucous footmarks.

Higher forms of activity have been less studied except orienting movements in directional light and food seeking, including subduing small arthropods, annelids, and the like. Planarians are favorite subjects for demonstration of taxes (Viaud, 1950) and the reversal of sign thereof by other stimuli; this kind of higher integration could be further studied in relation to the role of the brain. In hungry planarians in the dark, geotaxis is negative under low oxygen, positive with adequate oxygen. In *Convoluta* geotaxis is negative when undisturbed, positive when disturbed; in addition a tidal rhythm of geotaxis persists for some days in the laboratory. Upstream rheotaxis becomes strong in streamdwellers, in the presence of food juice (Doflein, 1925). Faced with two lights or sources of chemical, planarians often take a path which is the resultant of the two, but sometimes (Bock, 1936) they pay attention only to one, and they may vacillate. After unilateral blinding some do not circle. It is noteworthy that already here, as we shall see in insects with legs amputated, there are modes of response never normally expressed but ready for immediate expression when some drastic change is forced on the animal. Pearl refers to a method of righting in planarians that is only exhibited by cut pieces.

Removal of the brain in *Planaria torva*, *Crenobia alpina*, and *Polycelis* is said to have little effect on locomotion except to reduce its rate. The brainless pieces are still spontaneously active and reactive to stimuli but with a lengthened reaction time. They are handicapped in recognizing food because of the loss of the chemosensory epithelium of the midrostral tip. The ability to find food after brain removal, as after removal of the rostral epithelium, varies among species of triclads. *Dugesia tigrina* (Bardeen, 1901) and *Crenobia alpina* (Koehler, 1932) are said to be unable to recognize food but some other species are not so seriously affected. Swallowing or ingesting is, however, quite normal in decapitates, when food is accidentally encountered by the pharynx. But even the loss of the eyes does not prevent their tendency to be found in the darkest parts of their container during the day. There is less precision in finding the direction away from horizontal light, but even the eyeless piece has an apparent taxis component—though whether different from that in a gradient of intensity without horizontal beams is not known. The dorsal light reflex can occur without the eyes but not in decapitates (Bock, 1936). Only negative responses to contact occur; positive or approach responses require the brain. The land planarian *Bipalium* (Kawaguti, 1932) is similar in retaining the ability to locomote without being specially stimulated, following decapitation, but it moves slowly and the amplitude of its waving movements is greatly diminished.

The concept has grown up (Loeb, Pearl, Robertson) that the brain in planarians is not so much required for any specific actions, apart from food recognition, as it is for **maintaining a nervous "tone" or level of excitability** and spontaneity. If the level of central excitatory state can be measured indirectly by the speed of locomotion under standard conditions, the level is higher with the brain intact than it is if the brain is removed, but even then the level is not maximal. There is a restraint such that stimuli—for example crushed amphipod juice—induce faster movement. At any temperature between 0°C and its normal maximum of 12°C, *Crenobia alpina* crawls —under Beauchamp's (1935) standard conditions —at about 55–65% of its maximal rate for that temperature.

An interesting question arises when the animal cannot see any horizontal direction in the prevailing stimulus but **finds itself in a gradient**—of light from above, or of temperature, or chemical concentration. The observed fact that planarians gradually collect at one end may be ascribed to a lesser speed of locomotion between turns or to a greater number of turns per unit time, in higher concentrations of the stimulus factor. The latter alternative is supposed automatically to carry worms toward the high end where turns are tighter or, if sensory adaptation to the stimulus is sufficiently rapid, toward the low end where turns are looser (Ullyott, 1936). These are attractive explanations of kineses, dependent only on the level of output (in speed or in turns), which would be set by some function of the mere stimulus quantity. But it has not been investigated quantitatively whether the functions adequate to employ this scheme actually obtain.

The brain is believed to be necessary for the acquisition of **conditioned responses,** following experiments of Hovey (1929). Using a leptoplanid polyclad, he exposed worms for hours to a regime of 5 minutes of light and 30 minutes of darkness. Whenever a worm started to move during the light period it was touched on the rostral margin—the reaction to which is to stop. As the training period went on, the number of touches required to keep the animal from moving during the 5 minutes of light decreased markedly (controls ruled out fatigue, adaptation, or injury of the rostral margin). Hovey called the alteration in behavior "associative hysteresis," and others have called it a kind of "negative adaptation." It can not be brought about after brain removal.

But in a planarian the brain is not necessary for retention of a conditioned response (McConnell et al., 1959). *Dugesia* which had been conditioned to paired light and electric shock were cut in half transversely; after regeneration was long since apparently complete (4 weeks), the regenerated worms were retested and retrained to the same criterion. Not only did anterior pieces do as well as uncut controls but posterior pieces also relearned as fast as the controls. Original training required a mean of 134 trials to satisfy the criterion of learning; head pieces relearned in 40 trials and tail pieces in 43, whereas uncut animals—given the same four weeks to lose their training—needed 40 trials to relearn. These remarkable findings have been confirmed by others, and newer, unpublished results indicate correlations of relearning time with chemical changes in the tissues. Rather than adopting the distinction suggested by McConnell between the role of the brain in acquisition and retention, two alternative explanations should be considered. The deficit after brain removal in Hovey's case might be due simply to the interruption of the only pathway between touch of anterior margin and lower motor centers, and would therefore only tell us that the plexus is not adequate to conduct such excitation around an obstacle; touching in regions where paths are intact may not have equivalent stopping value. Another basis for the different results on leptoplanid and planarian is suggested by the anatomy and the behavior of decapitates: the polyclads have a more elaborate brain and are perhaps more centralized.

The role of the brain in **regeneration** has been studied in several species (Olmsted, 1922a; Lender, 1952–1956). The subject of regeneration is complex and reviews should be consulted (Hyman, 1951; Brøndsted, 1955). Some forms, especially among polyclads, can regenerate only if the brain is present whereas others, such as fresh-water planarians, can regenerate new heads on posterior fragments.

A number of **polyclads** are quite different from the triclads in the effect of brain removal. *Thysanozoon, Planocera, Leptoplana, Phylloplana,* and *Yungia* lose spontaneity more seriously than planarians (Loeb, Olmsted, Moore, Levetzow); they will remain quiet for weeks (since the brain does not regenerate), without crawling by muscular waves and especially without swimming in response to ordinary stimuli. However, in several species it has been shown that strong stimulation —for example cutting, or ordinary stimulation after drugs which lower threshold (phenol or reduced calcium)—will precipitate normal creeping in decapitates, and even swimming in *Yungia* and *Planocera* (Moore). Because this possibility was not specifically tested by Loeb and Olmsted, it is impossible to accept categorically their dictum that the brain is necessary for locomotion. The essentiality of the brain must be doubted although it is relatively **more important here than in triclads for maintaining spontaneity.** This function may be effected by a tonic discharge from the brain—acting like the generalized stimuli that can substitute for it—to maintain a level of permissive central excitatory state in the lower levels of the nervous system, which then introduce the wave pattern, as they clearly can do in strongly stimulated decerebrates. Or it may be that normally the brain paces the wave pattern and the ability of the lower levels to do so is not ordinarily relevant.

Levetzow points out that while decerebrate polyclads exhibit progression, if strongly stimulated, upon encountering an obstacle, they do not turn and proceed but perseverate. *Cryptocelis* cannot dig in the sand. The ditaxic method of crawling characteristic of *Leptoplana* appears to require the brain. If one half of the latter is injured, waves occur on the intact side only. This might mean that the wave pattern cannot cross the midline, because even the requisite level of nervous tone cannot do so; the experiment should be repeated with increased generalized stimulation. Similarly, longitudinally halved animals after regenerating all parts except the missing brain half, which they never regenerate, show mainly waves confined to the old half; occasionally waves spread from the old into the new half, though this might be a function of general excitation.

Besides spontaneity then, the brain may be important in special movements (ditaxic walking, digging), in activity involving a succession of changing directions or phases, and in orientation relative to external stimuli of certain kinds.

The small **temnocephalan** *Caridinicola* has been studied by Honjô (1937), who finds a strong resemblance in behavior to leeches. The only means of locomotion is a leechlike looping by alternate anterior and posterior sucker attachment with extension-contraction movements at the proper phases. The **brain is necessary** for this as well as for normal righting, for approach behavior, and for any movement involving twisting as in seeking or groping. Without the brain simple extension-contraction and alternate lateral bending movements can be executed and the suckers can maintain adhesion; muscle tonus is nearly normal, but the excitability generally is greatly reduced; movements are sluggish and do not include twisting or sucker release. Whereas the ventral cords have some capacity for motor command, the lateral cords appear to be unimportant in affecting reflexes and are believed to be mainly sensory.

Moore (1918) called attention to a common effect of strychnine in reversing certain actions and used this as evidence that there is **active inhibition in this simple nervous system.** Touching *Bdelloura* in locomotion causes it to stop, shorten, and thicken, but under strychnine (1:10,000) a touch causes it to extend itself and to be still more active. Moore describes the common denominator of similar effects of this drug in many animals as reversal of inhibition into excitation—thus the inactive transverse and circular muscles in *Bdelloura* become active. If the drug brings everything into activity indiscriminately, this would mean that extension is more powerful than shortening. While very suggestive, it would be desirable to have further evidence of reciprocal inhibition as opposed to mere alternate excitation.

B. The Role of the Cords and Plexuses

The **nerve cords presumably mediate** the whole coordinated pattern of those movements performed by decerebrate preparations. These movements include locomotion by muscular waves as well as ataxic creeping—by local irregular ripples of the ventral surfaces, swimming in some species, and righting. Cutting the ventral cords in polyclads, according to Olmsted (1922b), destroys muscular wave locomotion behind that level, but Levetzow (1936) disagrees; possibly the issue is the same as that of the supposed immobility of decerebrates—that is, the posterior level can initiate wave movements if its general excitability is high enough. Progression by wave movements does not require the largest trunks, for pieces of the polyclad *Thysanozoon* less than 1/7 of the animal, even from the posterior end, still crawl if excited (Levetzow). Furthermore they crawl anteriorly. *Leptoplana* retains its normal positive photokinesis after decapitation.

The influence mediated by the cords is **primarily ipsilateral.** In ditaxic walking the waves of each side depend on the cord of the same side. The waves are conducted in the cords, not in the muscles or by a chain of reflexes, for in *Leptoplana* if a large piece of the side of the body is cut out so that the muscular wave disappears on that side, it reappears at the expected time at the posterior edge of the wound and continues caudally. Cutting one or two nerve cords close to the brain does not stop swimming movements in the regions of the periphery supplied by more posterior nerves of the series that radiate from the brain in *Planocera*.

Moore (1933) numbered the nerves I-VII, VII being the main or medial ventral cord. When VII is cut, a corresponding sector of the periphery is immobilized, showing that there is **no nerve net** adequate to bring excitation into this sector from the neighboring regions. Curiously, the paralysis is spastic, not flaccid, and there is a ventral flexure; some local source of activation of the muscles can therefore be assumed, whether nervous or in the muscle fibers themselves. The rest of the animal, the anterior and lateral margins, still executes swimming.

If nerves I through VI are cut, leaving VII intact, swimming movements occur in the posterior sector it supplies, which moreover is relaxed and not spastic or curled. Since there is evidently little communication between sectors in spite of the plexiform appearance of the nervous system, the pattern of **initiation of swimming movements** must inhere at the same time in the cords (VII) and in more anterior sets of structures. We are not told whether brainless worms with VII cut, so that its sector is quiet, still swim, and therefore we don't know just how widely distributed the adequate pattern of command is, between brain and lateral nerves. The "nerves" grade from the equivalent of cords to minor peripheral nerves and careful experiments shortening them from the center outward might tell us something of the distribution of central functions.

Animals swimming by means of the posterior sector (VII intact), but with cut lateral nerves, IV-VI, or cut anterior nerves, I-III, have difficulty righting and roll unsteadily while swimming. Righting is conspicuously deficient because two parts at opposite sides of the animal will take hold of the substratum and pull against each other.

Another function apparently resident in the ventral cords in planarians is **release of proboscis activity.** This structure lies quietly in its sheath of body wall—normally and even after brain removal or cord section posterior to its level. But section of the ventral cords anterior to and near the proboscis base causes active movements—often autoamputation and escape from the sheath, followed by wriggling and swimming of the isolated organ. The separated proboscis can carry out all the steps it normally does in food ingestion but in *Planaria albissima* and some others it cannot tell food from nonfood objects (Kepner and Rich, 1918).

The peripheral plexus in planarians, to judge from pieces having no central nervous tissue, has been said to be incapable of nervously mediated responses (Bardeen, 1901), but this appears to be improbable at least as an absolute statement. It should be re-examined with the use of agents heightening the general excitability. Quite possibly the plexuses by themselves are normally not important as sensory-motor coordinating mechanisms.

The **peripheral plexus in polyclads** can mediate simple reflexes locally. Isolated pieces without the brain or large cords respond to mechanical stimulation of the dorsal surface by a dorsal flexure of the natural edge. They therefore right themselves, for the stimulus of contact of dorsal side induces curling and thus brings the ventral side into contact with the bottom, whereupon crawling takes over until the whole piece is dorsal side up. It is difficult to say that these pieces have no central nervous tissue, however, for in polyclads we have seen that the coarse-meshed plexus is equivalent to the orthogon of medullary cords elsewhere.

The probability of **nervous control of cilia** is high but has not been unequivocally shown. Ciliary beat in *Stenostomum* stops and starts and reverses, usually in coordination with muscular movements (Alverdes, 1922; Rampitsch, 1941).

If this animal is cut in two, only the posterior piece swims backwards; Rampitsch believes that the brain inhibits backward swimming, and the more strongly, the more anterior the level of the cilia concerned. Strong stimuli can still cause ciliary reversal after brain removal. The role of cilia and the regulation of cilia may differ among turbellarians; Olmsted and others have observed a very minor role of cilia in polyclad locomotion, and in planarians the lack of change of direction and vigor of beat has led to the supposition (Alverdes) that nervous control is not developed.

IV. TREMATODA

The flukes strongly resemble the more primitive turbellarians in the gross morphology of the nervous system. The group is relatively homogeneous and does not exhibit the variety of the Turbellaria or even the Cestoda. The level of development, judging from the distinctness of the brain and its histological differentiation, is better than the simplest free-living flatworms and the tapeworms, though far below typical triclads and polyclads. This goes with a paucity of differentiated types of sensory cells and especially of multicellular sense organs; the same can be said of the lower turbellarians. There is no pronounced influence of the parasitic habit except in the emphasis on development of suckers, which are richly innervated both on the sensory and motor side. The literature is not extensive and is added to relatively infrequently, doubtless due to the thoroughness of the older monographs both on gross anatomy and with methylene blue and Golgi stains. An unusually high percentage of the authors have successfully employed specific staining methods: Bettendorf, Havet, Zailer, Westblad (alizarin), Ábrahám. The basic works are those of Lang (1881), Gaffron, (1885), Looss (1895), Bettendorf (1897), Brandes (1898), Havet (1900), and Zailer (1914). There is no general difference between the orders Mongenea and Digenea; details are insufficient for the minor order Aspidobothria.

A. Central Nervous System

As for the other classes, the plan of the central nervous system is a brain and set of longitudinal cords and commissures in a simple orthogon (Fig. 9.19). The **number of cords** is generally three pairs: dorsal, lateral, and ventral, the ventral being the largest and least variant. In certain cases one or both of the others is reduced. Thus *Postharmostomum* has only short, weak dorsal cords (Ulmer, 1953), *Köllikeria* has apparently no dorsal cords and but short laterals (Williams, 1959a), *Sanguinicola, Sphyranura, Bothriogaster, Hemiurus appendiculatus (=Distomum appendiculatum)*, and *Lecithodesmus palliatus (=D. palliatum)* usually show no lateral; the cords not named in each case are normal, running the full length of the animal. Not infrequently the ventral cords from the two sides fuse at the caudal end; the dorsal cords may do the same. Occasionally the cords split and rejoin, either as long and regular features (*Gastrothylax, Sanguinicola*), or because short, minor trabeculae of muscle or parenchyma pierce them. Small supernumerary ganglia sometimes develop at intersections of commissures and cords or at forks (*Gastrothylax*).

Commissures are in some species numerous and more or less regular: *Azygia tereticollis (=Distomum tereticolle)* and *Galactosomum lacteum* have several dozen, *Tricotyla molae (=Tristomum molae), Distomum caudatum (Brachylaema caudata* or *Encylometra caudata?), D. sanguineum, Postharmostomum* have from 8 to 15. *Mammorchipedium isostoma (=Distomum isostomum)* has but 3 commissures in front and 3 behind the acetabulum (ventral sucker). Frequently there are only 1 or 2 commissures in front of the acetabulum, more posteriorly. They form complete rings sometimes but may be interrupted either dorsally or ventrally. The most conservative are the commissures between ventral and lateral cords.

In the Trematoda the mouth is anterior, unlike the turbellarians and the mouthless cestodes. For the first time we look for a **circumesophageal nerve ring** such as is nearly universal in higher

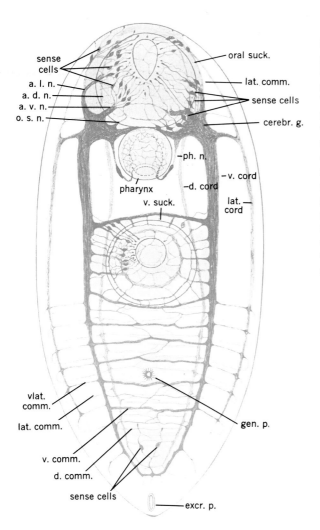

Figure 9.19.

Nervous system of *Cercariaeum*, showing the motor plexuses of the two suckers and the pharynx. In the left half of the oral sucker the deep sensory cells are shown; in the right half, the more superficial ones. The commissural system is shown only in the posterior half of the animal. Methylene blue. [Bettendorf, 1897.] *a. d. n.*, anterior dorsal nerve; *a. l. n.*, anterior lateral nerve; *a. v. n.*, anterior ventral nerve; *cerebr. g.*, cerebral ganglion; *d. comm.*, dorsal commissure; *d. cord*, dorsal cord; *dlat. comm.*, dorsolateral commissure; *excr. p.*, excretory pore; *gen. p.*, genital pore; *lat. comm.*, lateral commissure; *lat. cord*, lateral cord; *ph. n.*, pharyngeal nerve; *o. s. n.*, nerve to oral sucker; *oral suck.*, oral sucker; *v. comm.*, ventral commissure; *v. cord*, ventral cord; *v. suck.*, ventral sucker; *vlat. comm.*, ventrolateral commissure.

groups. The brain and its commissure are dorsal to the pharynx. There is no distinctive ring at the level of the brain, but just behind it, where the normal commissures between the cords begin, a series of complete rings carries out the orthogonal plan; these might be considered available for shifting forward to include the brain. Discussion in early papers concerning a "subesophageal ganglion" relates to loose accumulations of cells on the wall of the gut that is part of the intrinsic enteric nervous system (Marcinowski, 1903).

The **brain** consists of a pair of cerebral ganglia and a commissure, often rather long. Unlike that in cestodes, the commissure does not have a concentration of nerve cells. Although the central nervous masses stand out clearly, this is chiefly due to their fibrous component and it is not clear how well delimited the cellular component is from the parenchyma. A simple (Brandes) or multilayered sheath (Poirier) has been reported around the brain but without sufficient basis for evaluation. It is said—and this is telling for the level of development achieved by the group— that the distinction between outer layer of nerve cell bodies and core of fibrous matter is not clear in trematodes (Poirier and Havet, partly counterbalanced by Ábrahám). Instead, the cells are scattered through the brain, though the fibrous tissue is not likewise diffuse. There is indeed a neuropile made of the processes, collaterals, and ramifying terminals of the nerve cells. Golgi impregnations reveal a few unipolar, mainly bipolar, and multipolar cells, the latter advanced over coelenterates in being heteropolar, with a main neurite and several shorter, more branched processes. Cells measuring 15 μ in diameter are regarded as large (*Galactostomum*, Westblad). There are apparently no giant fibers. The special stains have yielded proportionately little inside the central nervous system compared to their revelations in the periphery, hence little more can be said. Havet and Ábrahám provide additional figures beyond those reproduced here. Brandes is almost alone in speaking of nonnervous cells inside the ganglia and cords; he regards a common type of nucleus as representing glial cells in the spaces between nerve fibers.

The **nerves of the brain** are typically four anteriorly projecting pairs, aside from the cords (often called posterior nerves), which have been discussed already. These are the anterior dorsal, anterior lateral, anterior ventral, and pharyngeal nerves. The first three mainly supply the oral sucker but also the body wall and the pharynx. There is often a connective between the lateral cord behind the brain and the anterior lateral nerve, and a similar connective may join the dorsal cord and anterior dorsal nerve. Occasionally commissures join one anterior nerve with another. The pharyngeal nerve may emerge from the cerebral commissure and proceed anteriorly or posteriorly to join the intrinsic system on the pharynx.

The **miracidium** larva already has a two-lobed brain and pair of nerve cords. **Rediae** have two pairs of anterior nerves and dorsal and ventral pairs of cords. **Cercariae** are not appreciably different from adults in neuroanatomy and many of the details in the preceding account have been worked out on this stage.

Functional studies on the nervous system are still lacking but some **reactions of cercaria** larvae have been studied (Miller and Mahaffy, 1930; Rankin, 1939; Wheeler, 1939). *Cercaria hamata* larvae intermittently swim and rest; shadowing them during rest brings on a bout of swimming, but too frequent repetition of this form of stimulation results in a failure of the response. On the other hand mechanical contact, as by collisions of swimmers with resting individuals, which also stimulates swimming, may be repeated so often without failure that the larvae are kept in continuous locomotion. The authors believe that this points to separate mechanisms but it may only mean that photoreceptors fatigue readily. Yasuraoka (1954) found an interesting reversal of sign of a reaction regarded as a phototropotaxis, whenever miracidia of *Fasciola* are exposed to sudden stimulation by either light or shadow. Cercariae are said to show no sign of chemotactic attraction to hosts, but are extremely sensitive to contact (Ferguson, 1943). They probably find their sites of choice within host tissues by chemical sensitivity of some sort, but the evidence does not necessarily implicate differentiated or extremely sensitive nervous chemoreceptors.

B. Peripheral Nervous System

Plexuses have been a principal preoccupation of investigators undoubtedly because they respond to the elective stains especially well. Still it is not clear just how many plexuses there are or what the connections are within them, and in particular whether conduction over any distance can take place independently of the central nervous system. The most conspicuous latticework of nerve strands is at the same depth as, and stretches between, the cords and their commissures (Fig. 9.20). This is wide-meshed and bears nerve cells,

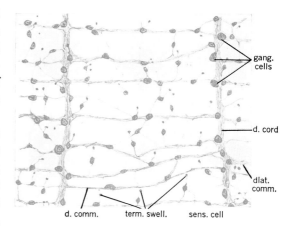

Figure 9.20. Part of the nervous system of *Cercariaeum*, dorsal to the ventral sucker; the sensory cells stand upright. The top of the picture is anterior. Methylene blue. [Bettendorf, 1897.] *d. comm.*, dorsal commissure; *d. cord*, dorsal cord; *dlat. comm.*, dorsolateral commissure; *gang. cells*, ganglion cells; *sens. cell*, sensory cell; *term. swell.*, terminal swellings of the sensory cells.

largely bipolar. Frequently described is an intra- or submuscular plexus beneath the most superficial muscle layer but whether this is a separate nervous layer from the preceding is not ascertained. A delicate subcuticular plexus is also figured (Fig. 9.21).

The **innervation of the suckers,** anterior and posterior (ventral) or oral and acetabular, are the chief feature of the peripheral nervous system (Figs. 9.19, 9.21). Each is supplied by several large nerves, the oral sucker from the brain, the acetabulum from the ventral and sometimes also the dorsal cord. The deep and superficial plexuses are well developed, the former regarded as mainly motor, the latter mainly sensory. Just what these terms mean cannot be stated, since even elementary experiments on the reactivity of isolated suckers are lacking. The nerves sometimes follow a regular pattern, as in the acetabulum of *Gorgoderina* and *Haematoloechus* (Zailer) where a deep and superficial nerve ring each receives branches from three pairs of nerves, sends connectives to the other ring at regular intervals, has six special ganglion cells, and exhibits a characteristic pattern of the main strands of its plexus. Even the sense cells in the oral sucker are constant in number and position.

Other plexuses occur in the walls of the pharynx, esophagus, intestine, and genital ducts. The pharynx and esophagus are the best developed. On the esophageal wall of *Fasciola* two rings of loose accumulations of nerve cells (anterior and posterior) with a connecting strand between them, are reported (Marcinowski, 1903).

Receptors are all or nearly all unicellular (Fig. 9.22), either in the cuticle or in the deep tissues, and include photo- and presumed mechanoreceptors. There are apparently no clusters of

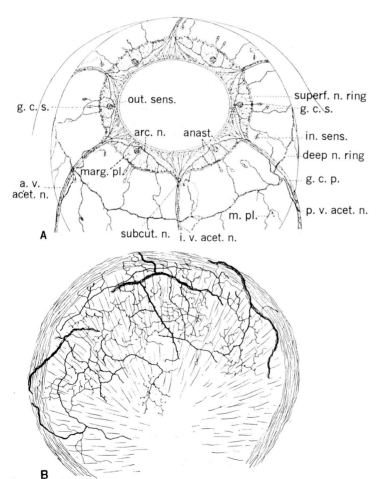

Figure 9.21.

A. Ventral sucker of *Gorgoderina vitelliloba*, ventral side turned up, seen from the side. [Zailer, 1914.] *anast.,* 5 anastomoses in each sextant between arcuate nerve and superficial nerve ring; *arc. n.,* arcuate nerve; *a. v. acet. n.,* anterior ventral acetabular (sucker) nerve; *deep n. ring,* deep nerve ring embedded in musculature; *g. c. p., g. c. s.,* ganglion cells of the deep (p) or the superficial (s) nerve rings, the former at the intersections of the six main nerves with the deep ring, the latter only in the middle of the sextants; *in. sens.,* inner sensilla; *i. v. acet. n.,* intermediate ventral acetabular nerve; *marg. pl.,* marginal nerve plexus of fibers from superficial nerve ring lying between circular muscle layers; *m. pl.,* plexus of ganglion cells in muscle layers; *out. sens.,* outer sensilla making an elevation of the cuticle in each sextant; *p. v. acet. n.,* posterior ventral acetabular nerve; *subcut. n.,* subcuticular nerve branching into free endings under the cuticle in the border of the sucker; *superf. n. ring,* superficial nerve ring. **B.** Cross section of the ventral sucker of *Distomum hepaticum,* showing the plexus of fine nerve fibers and endings in the musculature. [Havet, 1900.]

IV. TREMATODA B. Peripheral Nervous System

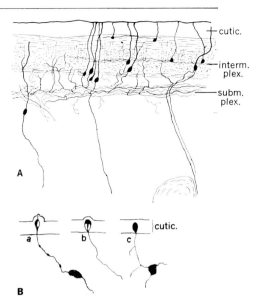

Figure 9.22. A. Transverse section of the skin of *Distomum*. [Havet, 1900.] *cutic.*, cuticular layer; *interm. plex.*, intermuscular plexus; *subm. plex.*, submuscular plexus. **B.** Three types of sense cell in the oral sucker of *Cercariaeum*. [Bettendorf, 1897.] *a*. A tactile papilla is formed by the cuticle bulging over the terminal vesicle and a spike of the distal sensory process. *b*. Nail head type of fiber ending in a vesicle. *c*. Plain end vesicle, staining solidly.

sensory neurons such as those considered to be chemoreceptors in turbellarians and in other phyla. This does not mean there are no single-celled chemoreceptors but it has been so thought and it is interesting that efforts to show chemotactic attraction to hosts have yielded negative results (Ferguson, 1943); chemical determination of the site of settling in a host is of course probable and may or may not involve neural receptors.

Bipolar sensory neurons with unbranched or little-branched processes are abundant under the cuticle, especially in the suckers. The cells are often deep in the subcuticle or even muscular layer, upright or parallel to the surface, and send a long distal process to end in a terminal vesicle in the base of the cuticle. Some endings penetrate the cuticle and may even project beyond it as a minute spike. But there are no long hairs like the rheoreceptors or tangoreceptors of *Mesostoma* (p. 553). Papillae of relatively simple construction are scattered around the mouth and in definite places on the oral sucker (Zailer, Brandes); more complicated tactile organs are described in Aspidobothria. Transitional forms occur, even toward cells with branching free nerve endings, but this tendency is not well developed. Bettendorf and Havet suggest the occurrence of protoplasmic fusion between collaterals of sensory neurons but the evidence is not compelling.

Nerve cells in the muscular layers have been a special problem. Some are believed to be typical bipolar sensory neurons with distal process ending in the deep tissues, presumably sensitive to deformation. Small multipolar cells are not uncommon, particularly in the suckers, and have been thought to be motor. Large multipolar nerve cells, easily confused with the extraordinary myoblasts (Fig. 9.23), have been confirmed by

Figure 9.23.

Left, cross section of the intestine of *Distomum*, showing a highly ramified multipolar nerve cell (*a*) which supplies the intestinal wall (*b*). [Havet, 1900.] **Right,** circular muscles of *Cercariaeum*, showing the myoblasts—these are not nerve cells. [Bettendorf, 1897.]

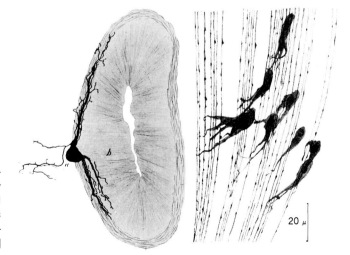

Brandes and Havet and are perhaps motor; they are found here and there in the musculature of suckers, pharynx, and other organs or in the parenchyma.

Unicellular photoreceptors or eye spots are frequent in trematode larvae and occasional in the adult (Lang, André, Faust, and references therein). The nucleus of the sensory neuron is in or near the swollen receptive end which pushes into the mouth of a pigment cup; the axon emerges from the other end, making the eye inverse as in many turbellarians. "The preservation of the pigmentation and of the 'lens' in this or that group [of Digenea] has been decidedly erratic and apparently unrelated to the systematology of the group. The great trend has been one of degeneration" (Faust, 1918). The eye spots generally number two or three and are situated close to the brain, innervated by posterior dorsal cords or anterior dorsal nerves.

Nerve endings on muscle are pictured by Bettendorf, Havet, and Ábrahám in metallic impregnations, and are believed to be motor. They are ramifying terminals and may be slightly swollen at the end. Bettendorf recognizes two kinds of muscle innervation: nerve endings on the plasmatic portion or myoblast cell body (Fig. 9.23) and those on the contractile fibers.

V. CESTODA

As one of the most extremely modified parasitic groups, tapeworms have been examined repeatedly with special reference to the nervous system and found to be considerably simpler than the highest free-living members of the phylum but not as simple as the lowest. The principal authors are Pintner (1880, 1925, 1934), Lang (1881), Niemiec (1888), Zernecke (1895), Cohn (1898), Tower (1900), Johnstone (1912), Becker (1922), Subramaniam (1940, 1941), and Rees (1941, 1951, 1956, 1959). Golgi impregnations and some methylene blue stains are reported by Blochmann, Zernecke, and Tower; Subramaniam and Lacey used reduced silver and gold chloride but with little novel result. Many writers comment on the difficulty of distinguishing nervous tissue, even in the main cords, in microscopic sections. One reason is the prevailing absence of a capsule delimiting the ganglion or cord; in addition, its texture is not sharply different from that of the parenchyma. It is not surprising that the information available is almost confined to gross anatomy; histology and physiology have hardly been touched. Most work has been done on members of the orders Tetraphyllidea, Tetrarhynchoidea, Tetrabothridea, Pseudophyllidea, and Taenioidea of the subclass Eucestoda. Dollfus (1942) provides a convenient summary of the arrangement of the nervous system (including the giant fibers) and receptors in members of the order Tetrarhynchoidea. The several smaller orders are little or not at all known and the subclass Cestodaria is only slightly known in respect to the nervous system.

A. Central Nervous System

The main pattern of the gross anatomy is biradially symmetrical: a series of longitudinal cords of varying number, with commissures connecting them in each proglottid, and a brain in the scolex with a sometimes elaborate system of transverse and ring commissures (Fig. 9.24). We will consider first the **longitudinal cords of the strobila** or chain of proglottids. There does not stand out a basic number. One pair of lateral cords, as the minimum, is typical of several groups: Cestodaria, Trypanorhyncha, Proteocephaloidea, Diphyllidea (Fig. 9.25). In some of the tetraphyllideans, the most primitive order of Eucestoda (for example *Acanthobothrium*), as well as in some of the taenioids, the most specialized order (for example *Taenia*, *Anoplocephala*, and *Thysanosoma*), there are 5 pairs (Figs. 9.24, 9.26): a lateral; two accessory laterals near it, far to the side, by the excretory canals; and a dorsal and a ventral medial cord (dorsal and ventral according to the convention based on the position of reproductive organs). Some taenioids have fewer cords; *Moniezia* has 3 pairs; *Avitel-*

V. CESTODA A. Central Nervous System

Figure 9.24.

The nervous system of *Tetrabothrius* (Tetrabothroidea). **A.** The scolex of *T. macrocephalus*. **B.** The proglottids of *T. laccocephalus*. Reconstructed from sections. [Spätlich, 1909.]

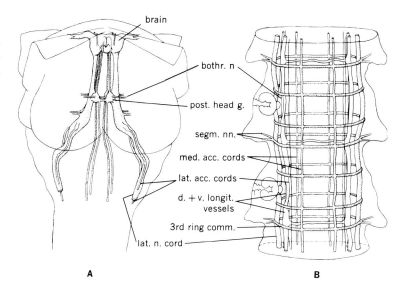

Figure 9.25.

Diagrams of the central nervous system of the anterior ends of three cestodes, which are relatively simple in this respect. Watson (1911) regarded the rosette of *Gyrocotyle* as homologous to the scolex but later authors do not. [**A**, Watson, 1911; **B**, Williams, 1959; **C**, Rees, 1959.] *acetab.*, acetabulum; *ant. cerebr. g.*, anterior portion of cerebral ganglion; *apical org.*, apical organ; *bothrid. n.*, bothridial nerve; *cerebr. g.*, cerebral ganglion; *d. acc. cord*, dorsal accessory cord; *d. comm.*, dorsal commissure; *d. n.*, dorsal nerve; *lat. cord*, lateral cord; *longit. conn.*, longitudinal connective; *longit. n. cord*, longitudinal nerve cord; *n. cord*, nerve cord; *post. cerebr. g.*, posterior portion of cerebral ganglion; *post. g.*, posterior ganglion; *post. rosette*, posterior rosette; *v. acc. cord*, ventral accessory cord; *v. comm.*, ventral commissure; *v. n.*, ventral nerve.

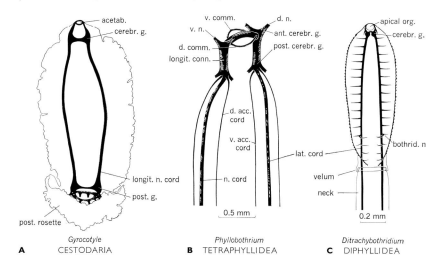

lina, only the lateral pair. In several groups, however, there are many more. In *Tylocephalum* (Lecanicephaloidea) there are a large number, usually 38, which are constant from the neck back; in *Tentacularia* (Tetrarhynchoidea) there are 60 longitudinal cords, all about equal. Subramaniam proposed that this condition is primitive but the evidence for the phylum as a whole is against this. In *Ligula* and *Schistocephalus* (Pseudophyllidea) there is a main, lateral pair plus 7 or 8 smaller pairs of cords in the anterior proglottids, dividing to increase that number about threefold in more mature segments. It is not clear whether the same is true of *Bothriocephalus* and

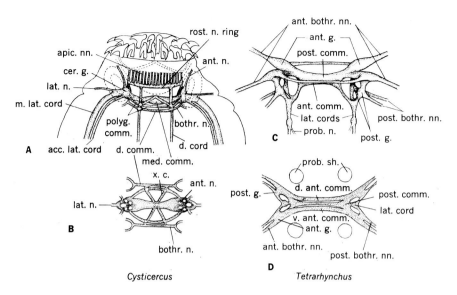

Figure 9.26. The nervous systems of two cestodes. **A.** *Cysticercus*, nerves of the scolex from the dorsal side. **B.** Diagram of an anterior view of the cerebral ganglia and associated commissures of *Cysticercus*, with dorsal and ventral commissures displaced outward. [Rees, 1951.] **C** and **D.** Central nervous system of *Tetrarhynchus erinaceus*. [Johnstone, 1912.] *acc. lat. cord*, accessory lateral cord; *ant. bothr. nn.*, anterior bothridial nerves; *ant. comm.*, anterior commissure; *ant. g.*, anterior ganglia; *ant. n.*, anterior nerve; *apic. nn.*, apical nerves; *bothr. n.*, bothridial nerve; *cer. g.*, cerebral ganglion; *d. comm.*, dorsal commissure; *d. cord*, dorsal cord; *lat. cord*, lateral cord; *lat. n.*, lateral nerve; *m. lat. cord*, main lateral cord; *polyg. comm.*, polygonal commissure; *post. bothr. nn.*, posterior bothridial nerves; *post. comm.*, posterior commissure (= median commissure); *post. g.*, posterior ganglia; *prob. sh.*, proboscis sheath; *prob. n.*, proboscis nerve; *rost. n. ring*, rostellar nerve ring; *x. c.*, x-shaped dorsoventral commissure.

others of this order. The diversity in respect to number of cords extends also to the other features—commissures, scolex nerves, and ganglia; the distribution of variation by orders, or even the range of variation, can not be given in the space available.

Commissures in the proglottids are sometimes reported to be absent (*Schistocephalus*, *Bothriocephalus*); this presumably means they are fine and not visible by the methods available but it may be that the cords are only connected by a diffuse plexus. A single commissure encircling the proglottid is described in some forms and in this event it lies posteriorly near the transverse excretory canal. In others, two, three or more ring commissures are noted. Younger segments may show no clear commissural rings while more posterior segments of the same worm have a characteristic number. In *Tylocephalum* the last commissure in a proglottid is represented by a platelike meshwork.

Ganglionic swellings of minor proportions may occur at intersections of rings and cords but are often not noted. The cords are supposedly medullary strands but in fact authors have several times commented on the scarcity of nerve cells anywhere in or on the cords. Of course, without selective stains this means little except that they are not conspicuous. The cords are usually situated internal to all the main musculature, and therefore in the central mesenchyme. A few segmental nerves are given off, in some species especially toward the margin but in some descriptions all traces of recognizable bundles are denied; in general, most outflow and inflow is presumably by single nerve fibers or fine connections with a plexus.

The arrangement of **nervous structures in the scolex** ranges from simple to intricate within the class. Most of the orders have simply a unified mass, the brain, into which all the longitudinal cords run, whether there are but one pair (*Gyro-*

cotyle, Amphilina: Cestodaria; *Ditrachybothridium* whose affinity is uncertain but related to the next following; Diphyllidea) or several dozen (*Tylocephalum:* Lecanicephaloidea). The brain may be more or less clearly paired with a broad commissure between the two sides. Ring commissures comparable to those in the proglottids may occur close to the brain. This is the condition typical of Tetraphyllidea (*Acanthobothrium, Phyllobothrium*) and Pseudophyllidea (*Ligula, Schistocephalus, Bothriocephalus*). Among the Trypanorhyncha the nearest ring commissure becomes intimately part of the brain complex and ganglionic swellings just behind the brain may appear on four of the cords, associated with the development of a nerve supply to the proboscises, their sheaths, bulbs, and retractors. With the progressive muscularization of sucking organs, more numerous and heavier nerves go to structures in the scolex and concomittantly additional commissures are developed, including dorsoventral and radial or oblique bands. Thus the Taenioidea present apparently complex arrangements, but the main feature is still the termination of the longitudinal cords of the strobila in a pair of cerebral ganglia or a ring commissure connected to it and the concentration of nervous tissue in a heavy transverse commissure.

Characteristic of Taenioidea and Tetrabothridea is a superstructure effect, due to the possession of **two levels of ring or polygonal commissures**—a more rostral, often tighter ring and a more caudal, usually larger ring (Figs. 9.24, 9.26). Each ring may have ganglionic swellings; sometimes one, sometimes the other is the chief head ganglion, and so is called the brain. The caudal ring typically has a complex commissural pattern including diagonal as well as transverse, dorsoventral, and ring members.

Numerous stout and fine **nerves are given off** at both levels, chiefly to supply whatever organs of attachment—sucking depressions and hooks—the species possesses. The sucking depressions, such as bothria, bothridia (phyllidea), or acetabula (suckers), may be very muscular and each of the set of four organs may receive two or three separate nerves. In addition, the species of the order Trypanorhyncha have four characteristic spiny eversible proboscises whose sheath, bulb, and retractor muscle are innervated from the brain. Apical nerves supply the tip of the scolex and sometimes a small circular apical organ of uncertain function, possibly sensory. Some taenioids have an apical protrusible structure or a mobile cone-shaped rostellum and with it a set of nerves and another nerve ring.

The nerves given off by the central nervous complex in the scolex vary widely, just as the organs of attachment and their muscles vary; the pattern is based on a **four-rayed radially symmetrical plan** in contrast to the flattened biradial strobila. Whereas the nervous system is degraded elsewhere, it appears to have become augmented in association with the adhesive organs. Writers here and there have been led to remark on the relative complexity of the anterior nervous system in many Cestodes, although it is clear that this represents a comment on the elaboration of the gross anatomy of nerves and commissures. The histological level of differentiation remains low, compared to the better polyclad turbellarians, for example, but above that in some acoel turbellarians.

Histologically the cestode central nervous system is mainly known for the following features. It is not compact but loose in texture and for the most part is not set off from the mesenchyme by a sheath or capsule. A delicate sheath over a limited part of the brain complex has been noted several times. The nerve cells have so far been identified chiefly from nonselective stains; but this identification cannot be considered reliable since myoblasts and mesenchyme cells, among others, look much like nerve cells. In addition, the form of nerve cells and their processes are certainly unreliably given by ordinary histological methods. Nevertheless, bipolars and multipolars have been reported to be most abundant; some unipolars also occur, perhaps more commonly in the head ganglia. Remarkably, it has been said several times that nerve cells are absent or scarce along the main longitudinal cords, and hence throughout the strobila, except for peripheral cells like sense cells. Some authors, as Tower, have seen small collections of nerve cells at the intersections of cords and commissures.

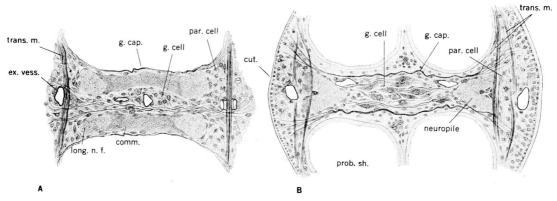

Figure 9.27. Transverse section of the scolex of *Tetrarhynchus*. **A** is slightly anterior to **B** and shows the dorsal and ventral brain commissures; **B** shows the cerebral ganglia and accumulation of large ganglion cells along the median commissure. [Pinter, 1880.] *comm.*, commissure of brain; *cut.*, cuticle; *ex. vess.*, excretory vessel; *g. cap.*, ganglion capsule, which includes muscle fibers; *g. cell*, ganglion cell; *long. n. f.*, longitudinal nerve fibers continuous with cords; *par. cell*, parenchyma cell; *prob. sh.*, proboscis sheath; *trans. m.*, transverse muscle of scolex.

In the scolex, nerve cells and neuropile are said to be primarily concentrated in the main ("median" or "posterior") transverse commissure rather than in the ganglia at each end of it (Fig. 9.27). However, the categorical denials of nerve cells in the ganglia, indulged in by some authors, need carry little weight in view of the technics used and the difficulties of recognizing such cells if they are mingled with mesenchyme: "The so-called ganglia consist of a tissue resembling parenchyma and which seems to be composed of a network of fine fibres enclosing small spaces containing a few nuclei" (Rees, 1941). Tower reports, astonishingly, that the cephalic ganglia consist of a core of cells and a cortex of fibers, in contrast to invertebrate ganglia generally. Johnstone agrees but emphasizes the obstacles to discernment of nervous tissue in this group. It would appear that the loose, spongy texture bespeaks much nonnervous glial tissue, that the neuropile is poorly developed and not compact, that nerve cells are not numerous and are not at least different in type as between the main transverse commissural mass and the "ganglia" at its ends, if not absent from them. However, until suitable methods have been used to reveal the composition of the cestode central nervous tissue, no definite statements should be made.

Nerve cells, as identified and figured in favorable cases, appear quite standard in sizes, forms, and inclusions. There are **giant fibers** or neurocords in a number of genera; our information on these structures is mainly due to Pintner (1925, 1934). He insists they have sometimes been mistaken for excretory vessels. The fibers, which like typical giants are conspicuously empty-looking in sections, are commonly about 12 μ in diameter; the cell bodies, 18 μ; besides slowly tapering bipolars there are tri- and quadripolar cells. They occur in the brain, main cords, and some nerves, such as the proboscidial and bothridial nerves (Fig. 9.28). Pintner believed some of them are centripetal, coming from superficial receptors. He also thought they often fuse or that several cells send processes to join a single fiber. They may leave one nerve and join another, which implies extensive interconnection by protoplasmic continuity. In the middle of the apical nervous mass of *Stenobothrium* (Trypanorhyncha) is described a series of five axial structures, one behind the other, each a quadriradiate X made of giant cells and their processes going out perfectly symmetrically into the nerves. Two of these neurocord crosses consist of four cells, one of eight, one of two, and one is a single huge, central quadripolar cell. Our curiosity as to the responses that these giant fibers might mediate is not satisfied by vague references to the harmony of movements and the synchrony of parts among the proboscises.

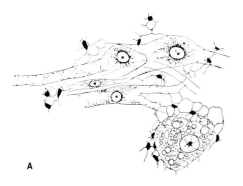

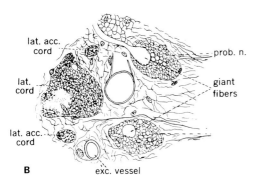

Figure 9.28. Giant cells and fibers of Tetrarhynchoidea in longitudinal and cross section. **A.** *Stenobothrium macrobothrium*. **B.** *Heterotetrarhynchus institatum*, showing lateral and lateral accessory cords and proboscis nerves. [Pintner, 1925, 1934.] *exc. vessel,* excretory vessel.

B. Peripheral Nervous System

At the same level as the cords and ring commissures below the principal muscles in the strobila is a **deep plexus** of thin strands. Its prevalence, density, and variations are not known, for but few workers have made it out, so difficult is the recognition of small islands of poorly defined nervous tissue in sections by the usual methods. There appears to be a continuum between constant, rectilinear commissures and a diffuse plexus connecting the longitudinal cords; in a form like *Tylocephalum*, according to Subramaniam (1941a), younger segments have virtually no commissures but only a meshwork, older segments have a constant commissure pattern and even within one proglottid the commissures are better developed posteriorly, the plexus anteriorly. In some species without distinct commissures the deep plexus is said to be oriented chiefly transversely, hence parallel to the would-be ring commissures. Cells regarded as nerve cells have been encountered along the strands of the plexus (Subramaniam, 1941b), but only rarely.

The existence of a **second, superficial plexus** just beneath the subcuticular cell layer is disputed. At best it is probably delicate, consisting of sensory cells and processes in low density (Zernecke). Negative reports, in the nature of the case, cannot be given much weight (Lacey). Multipolar ganglion cells are reported to occur peripherally in the suckers.

Sensory neurons are abundant, according to Golgi and methylene blue results of Blochmann and Zernecke. Singly or loosely grouped bipolar cells in the subcuticular zone, oriented radially to the surface, send a simple unbranched, thicker process to end in the base of the cuticle and a thinner one which passes centrally. At least some of the latter enter the longitudinal cords. The distal process often ends in a vesicular swelling and may have a small spike; this may be under a funnel-shaped depression in the outer surface of the cuticle. Cells lying in the superficial plexus and also deeper, and which have long, branching processes, some of which end near the cuticle, are seen in Golgi impregnations; they were at first taken for sensory nerve cells but later (Blochmann 1911) they were considered to be mesenchyme cells. Not only these cells but also the so-called Sommer-Landois myoblast cells can be confused with nerve cells. (See Fig. 9.29.)

It is generally believed that cestodes lack sense organs and that their sense cells are only tactile. The latter is simply a presumption, on present evidence.

The **outflow to muscle and organs** is mainly diffuse, except in the scolex as already described. There are some segmental nerves in the strobila in certain species; their destination is mainly unknown but some pass in the direction of the margin, laterally, and a few medially are sometimes called genital. The nerves have been observed to be defined by a thin sheath on occasion and branching, spidery covering cells may turn up in metallic impregnations. But still the principal manner in which efferent nerve fibers leave the cords is probably individually, perhaps com-

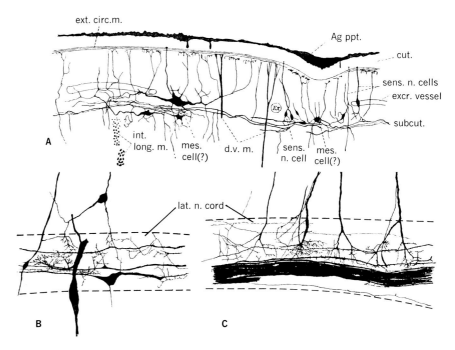

Figure 9.29. **A.** Transverse section of the ventral surface of *Ligula*, impregnated by the Golgi method. [Blochmann, 1895.] *Ag ppt.*, silver precipitate; *cut.*, cuticle; *d.v. m.*, dorsoventral muscle fiber; *ext. circ. m.*, external circular muscle; *excr. vessel*, excretory vessel; *int. long. m.*, internal longitudinal muscle; *mes. cell*, cell presumed to be mesenchyne according to Blochmann (1911), although originally thought to be nerve cells of the plexus with free nerve endings—a good example of the pitfalls of interpreting the most beautiful Golgi images; *sens. n. cell*, sensory nerve cell; *subcut.* subcutaneous plexus. **B.** Portion of a nerve cord in *Ligula* in which a bipolar nerve cell with ascending and descending branches is impregnated and sends a process into the periphery; two transverse bipolars are shown, with both central and peripheral processes, and a multipolar cell of the deep plexus sends branches into the cord as well as into the periphery. **C.** In another portion of the nerve cord, fibers are impregnated; suggestive of centrally terminating afferents. Golgi method. [Zernecke, 1896.] *lat. n. cord*, lateral nerve cord.

monly after passing into a strand of the deep plexus. Only rather unsatisfactory evidence exists for this, however. Likewise in a primitive state is the information on terminations on muscle. Sketchy descriptions propose two modes. Some have claimed to see nerve fibers end on muscle cells. In addition, the plasmatic portions of muscle cells are believed to connect by means of their own branching processes with nerve cells, reminiscent of the curious situation in nematodes.

Rietschel (1935) investigated the role of the nervous system in propagating the slow waves of longitudinal muscle contraction that spread from the scolex, or any anterior cut end backward, in *Catenotaenia*. A series of overlapping partial transections of a proglottid did not interrupt the wave but pressing on a segment so as to immobilize it did block propagation. He concluded in favor of a mechanical chain reflex theory, as in Friedländer's earthworm experiments. Other movements and responses of tapeworms have apparently not been studied with respect to nervous physiology.

Classified References

TURBELLARIA

Anatomy

Ax, 1961, 1964; Baer, 1931; Barrett, 1929; Beklemischev, 1927; Bock, 1913, 1923; Böhmig, 1887, 1895, 1906; Borrini, 1948; Botezat and Bendl, 1909; Brauner, 1926; Bresslau, 1928; Bresslau and Voss, 1913; Bütschli, 1912; Child and McKie, 1911; Chun, 1882; Clayton, 1932; Delage, 1886; Gelei, 1912, 1930; Graff, 1879, 1882, 1899, 1904–1914; Haase, 1927; Hadenfeldt, 1929; Hanström, 1926a, 1926b, 1928; Haswell, 1893; Heath, 1907; Heider, 1914; Hertwig, 1881; Hesse, 1897, 1902; Hofsten, 1907a, 1907b, 1921; Hranova, 1937; Hyman, 1938; Iijima, 1884; Jänichen, 1896; Joseph, 1902; Kennel, 1882; Kepner and Foshee, 1917; Kepner and Taliaferro, 1912a, 1912b; Koehler, 1926; Lan, 1936; Lang, 1879, 1881a, 1881b, 1881c, 1884; Linton, 1910; Löhner, 1910; Luther, 1904, 1905–1907, 1912, 1921; Mark, 1892; Meixner, 1925; Micoletzky, 1907; Monti, 1897, 1900; Palombi, 1926; Pertusa, 1942; Press, 1959; Reisinger, 1926, 1933; Retzius, 1902; Ritter-Zahony, 1907; Sabussow, 1905; Steinböck, 1924a, 1924b, 1925, 1927, 1931; Steinmann, 1929; Steopoe, 1934; Turner, 1946; Ude, 1908; Vejdovský, 1895; Wacke, 1903; Weiss, 1910; Westblad, 1937; Wheeler, 1894; Wilhelmi, 1909, 1913; Woodworth, 1891.

Physiology of the Nervous System

Alverdes, 1922; Bardeen, 1901; Beauchamp, 1935; Behrens, 1961; Benazzi, 1936; Beuther, 1927; Bock, 1936; Doflein, 1925; Honjô, 1937; Kawaguti, 1932; Kepner and Rich, 1918; Koehler, 1932; Lemke, 1935; Lender, 1952, 1954, 1956a, 1956b; Levetzow, 1936; Loeb, 1893, 1894, 1900; Moore, 1918, 1923, 1933; Müller, 1936; Olmsted, 1922a, 1922b; Pantin, 1950; Parker and Burnett, 1900; Pearl, 1903; Rampitsch, 1941; Robertson, 1928; Steiner, 1898; Taliaferro, 1920; Ullyott, 1936; Viaud, 1950; Westblad, 1923.

Conditioned Responses

Angyan and Nemeth, 1957; Dilk, 1937; Hovey, 1929; McConnell et al., 1959; Sgonina, 1939; Soest, 1937; Thompson and McConnell, 1955.

TREMATODA

Anatomy

Ábrahám, 1929; André, 1910a, 1910b; Bettendorf, 1897; Blochmann and Bettendorf, 1895; Ejsmont, 1925; Faust, 1918; Havet, 1900; Heath, 1902; Lang, 1881; Looss, 1894, 1895; Marcinowski, 1903; Poirier, 1885; Ulmer, 1953; Westblad, 1924; Williams, 1959; Zailer, 1914.

Physiology

Chu, 1938; Ferguson, 1943; Miller and Mahaffy, 1930; Rankin, 1939; Wheeler, 1939; Yasuraoka, 1954.

CESTODA

Anatomy and Physiology

Bartels, 1902; Becker, 1922; Blochmann, 1895, 1911; Cohn, 1898, 1907; Dollfus, 1942; Gough, 1911; Johnstone, 1912; Kahane, 1880; Lacey, 1955; Laczkó, 1880; Lang, 1881; Lönnberg, 1891; Lühe, 1896; Lynch, 1945; Niemiec, 1888; Pintner, 1880, 1925, 1927, 1934; Rees, 1941a, 1941b, 1951, 1956, 1959; Rietschel, 1935; Riser, 1949; Roboz, 1882; Spätlich, 1909; Subramaniam, 1940, 1941a, 1941b; Tower, 1900; Watson, 1911; Williams, 1959; Zernecke, 1895.

Bibliography

Ábrahám, A. A. 1929. Das Nervensystem von *Opisthodiscus diplodiscoides nigrivasis* Méhely. *Studia zool.*, 1:147-157.

Alverdes, F. 1922. Untersuchungen über begeisselte und beflimmerte Organismen. *Arch. EntwMech. Org.* 52:281-312.

André, J. 1910a. Zur Kenntnis des Nervensystems von *Polystomum integerrimum* Froel. *Z. wiss. Zool.*, 95:191-203.

André, J. 1910b. Die Augen von *Polystomum integerrimum* Froel. *Z. wiss. Zool.*, 95:203-221.

Ángyán, A. and Németh, G. 1957. Untersuchung der unbedingten und bedingten Reflexe von Planarien vor, während und nach der Regeneration. *Acta physiol. hung.*, 12 (suppl.): 28-29.

Ax, P. 1961. Verwandtschaftsbeziehungen und Phylogenie der Turbellarien. *Ergebn. Biol.* 24:1-68.

Ax, P. 1964. Relationships and phylogeny of the Turbellaria. *The Lower Metazoa: Comparative Biology and Phylogeny.* E. C. Dougherty et al. (eds.). Univ. California Press, Berkeley.

Baer, J. G. 1931. Étude monographique du groupe des Temnocéphales. *Bull. biol.*, 65:1-57.

Bardeen, C. R. 1901. The function of the brain in *Planaria maculata*. *Amer. J. Physiol.*, 5:175-179.

Barrett, W. C. 1929. The retinula cell of the turbellarian *Prorynchus applanatus* Kennel. *Trans. Amer. micr. Soc.*, 48:66-69.

Bartels, E. 1902. *Cysticercus fasciolaris*. Anatomie, Beiträge zur Entwicklung und Umwandlung in *Taenia crassicollis*. *Zool. Jb. (Anat.)*, 16: 511-570.

Beauchamp, R. S. A. 1935. The rate of

movement of *Planaria alpina. J. exp. Biol.*, 12:271-285.

BECKER, R. 1922. Beiträge zur Kenntnis des Nervensystems der Pferdebandwürmer unter besonderer Berücksichtigung von *Anoplocephala magna* (Abildgaard). *Zool. Jb. (Anat.)*, 43:171-218.

BEHRENS, M. E. 1961. The electrical response of the planarian photoreceptor. *Comp. Biochem. Physiol.*, 5:129-138.

BEKLEMISCHEV, W. N. 1927. Über die Turbellarienfauna des Aralsees. Zugleich ein Beitrag zur Morphologie und zum System der *Dalyelliida*. *Zool. Jb. (Syst.)*, 54:87-139.

BENAZZI, M. 1936. Influenza della regione cefalica sul movimento e sulla sensibilità della planarie tricladi. *R. C. Accad. Lincei*, (6), 23:152-155.

BETTENDORF, H. 1897. Über Muskulatur und Sinneszellen der Trematoden. *Zool. Jb. (Anat.)*, 10:307-358.

BEUTHER, E. 1927. Ueber die Einwirkung verschiedenfarbigen Lichtes auf Planarien. *S. B. naturf. Ges. Rostock*, (3) 1:17-57.

BLOCHMANN, F. 1895. Über freie Nervenendigungen und Sinneszellen bei Bandwürmern. *Biol. Zbl.*, 15:14-25.

BLOCHMANN, F. 1911. Die sogenannten freien Nervenendigungen bei Cestoden. *Zool. Anz.*, 38:87-88.

BLOCHMANN, F. and BETTENDORF, H. 1895. Ueber Muskulatur und Sinneszellen der Trematoden. *Biol. Zbl.*, 15:216-220.

BOCK, H. 1936. Lichtrückeneinstellung und andere lokomotorische Lichtreaktionen bei *Planaria gonocephala*. *Zool. Jb. (Allg. Zool.)*, 56:501-530.

BOCK, S. 1913. Studien über Polycladen. *Zool. Bidr. Uppsala.*, 2:31-343.

BOCK, S. 1923. Eine neue marine Turbellariengattung aus Japan. *Uppsala Univ. Årsskr. (Mat. Naturv.)*, 1923(1):1-55.

BÖHMIG, L. 1887. Zur Kenntnis der Sinnesorgane der Turbellarien. *Zool. Anz.*, 10:484-488.

BÖHMIG, L. 1891. Untersuchungen über rhabdocöle Turbellarien. II. *Plagiostomina* und *Cylindrostomina* Graff. *Z. wiss. Zool.*, 51:167-479.

BÖHMIG, L. 1895. Die Turbellaria acoela der Plankton-Expedition. *Ergebn. Plankton-Exped. Humboldt-St*, 2(H. g.):45pp.

BÖHMIG, L. 1906. Tricladenstudien. I. Tricladida maricola. *Z. wiss. Zool.*, 81:344-504.

BORRINI, S. 1948. Sugli occhi della planaria *Polycelis nigra* Ehrenberg. *Mem. Soc. tosc. Sci. Nat. (B)*, 55:123-133.

BOTEZAT, E. and BENDL, W. 1909. Über Nervenendigungen in der Haut von Süsswasser-Tricladen. *Zool. Anz.*, 34:59-64.

BRANDES, G. 1898. Die Gattung *Gastrothylax*. *Abh. naturf. Ges. Halle*, 21:193-225.

BRAUNER, K. 1926. Die Turbellaria acoela der deutschen Tiefsee-Expedition. *Wiss. Ergebn. Valdivia*, 22:27-56.

BRESSLAU, E. 1928-1933. Turbellaria. *Kükenthal's Handb. Zool.*, 2:1:52-320.

BRESSLAU, E. and VOSS, H. v. 1913. Das Nervensystem von *Mesostoma ehrenbergi* (Focke). *Zool. Anz.*, 43:260-263.

BRØNDSTED, H. V. 1955. Planarian regeneration. *Biol. Rev.*, 30:65-126.

BULLOCK, T. H. and NACHMANSOHN, D. 1942. Choline esterase in primitive nervous systems. *J. cell. comp. Physiol.*, 20:239-242.

BÜTSCHLI, O. 1912. *Vorlesungen über vergleichende Anatomie.* Vol. 2. Engelmann, Leipzig.

CHILD, C. 1904a. Studies on regulation. V. The relation between the central nervous system and regeneration in *Leptoplana*: posterior regeneration. *J. exp. Zool.*, 1:463-512.

CHILD, C. 1904b. Studies on regulation. VI. The relation between the central nervous system and regulation in *Leptoplana*: anterior and lateral regeneration. *J. exp. Zool.*, 1:513-557.

CHILD, C. M. and MCKIE, E. V. M. 1911. The central nervous system in teratophthalmic and teratomorphic forms of *Planaria dorotocephala*. *Biol. Bull., Woods Hole*, 22:39-59.

CHU, H. J. 1938. Certain behavior reactions of *Schistosoma japonicum* and *Clonorchis sinensis* in vitro. *Chin. med. J. Suppl* ii:411-417.

CHUN, C. 1882. Die Verwandtschaftsbeziehungen zwischen Würmern und Coelenteraten. *Biol. Zbl.*, 2:5-16.

CLAYTON, D. E. 1932. A comparative study of the non-nervous elements in the nervous system of invertebrates. *J. Ent. Zool.*, 24:3-22.

COHN, L. 1898. Untersuchungen über das centrale Nervensystem der Cestoden. *Zool. Jb. (Anat.)*, 12:89-160.

CORNING, W. C. and JOHN, E. R. 1961. Effect of ribonuclease on retention of conditioned response in regenerated planarians. *Science*, 134:1363-1365.

DELAGE, Y., 1886. Études histologiques sur les planaires rhabdocoeles acoeles. *Arch. Zool. exp. gén.*, (2), 4:109-160.

DILK, F. 1937. Ausbildung von Assoziationen bei *Planaria gonocephala* (Dugès). *Z. vergl. Physiol.*, 25:47-82.

DOFLEIN, I. 1925. Chemotaxis und Rheotaxis bei den Planarien. Ein Beitrag zur Reizphysiologie und Biologie der Süsswassertricladen. *Z. vergl. Physiol.*, 3:62-112.

DOLLFUS, R. P. 1942. Études critiques sur les tétrarhynques du Muséum de Paris. *Arch. Mus. Hist. nat., Paris*, (6), 19:1-466.

EJSMONT, L. 1925. Morphologische, systematische und entwickelungsgeschichtliche Untersuchungen an Arten des genus *Sanguinicola* Plehn. *Bull. int. Acad. Cracovie (Acad. pol. Sci.)*, 1925B:877-966.

FAUST, E. C. 1918. Eye-spots in Digenea. *Biol. Bull., Woods Hole*, 35:117-127.

FERGUSON, M. S. 1943. Migration and localization of an animal parasite within the host. *J. exp. Zool.*, 93:375-400.

FUHRMANN, O. 1931. Vermes Amera. Cestoda. *Kükenthal's Handb. Zool.*, 2:2(2):257-416.

GELEI, J. VON. 1909-1912. Tanulmányok a *Dendrocoelum lacteum* Oerstd. szövettanárol. A Magyar Tudomanyos Akademia, Budapest. (In Hungarian.)

GELEI, J. VON. 1930. "Echte" freie Nervenendigungen (Bemerkungen zu den Receptoren der Turbellarien). *Z. Morph. Ökol. Tiere*, 18:786-798.

GOUGH, L. H. 1911. A monograph of the tape-worms of the subfamily Avitellininae, being a revision of the genus *Stilesia*, and an account of the histology of *Avitellina centripunctata* (Riv.). *Quart. J. micr. Sci.*, 56:317-385.

GRAFF, L. VON. 1879. Kurze Mitteilungen über fortgesetzte Turbellarienstudien. II. Über *Planaria Limuli*. *Zool. Anz.* 2:202-205.

GRAFF, L. VON. 1882. *Monographie der Turbellarien. I. Rhabdocoelida.* Engelmann, Leipzig.

GRAFF, L. VON. 1899. *Monographie der Turbellarien. II. Tricladida terricola (Landplanarien).* Leipzig.

GRAFF, L. VON. 1904-1914. Turbellaria. *Bronn's Klassen*, 4:1 (1):1934-1949; 2164-2216; 4:1 (2):2855-2946.

HAASE, M. R. 1927. Ciliated pits of *Porhynchus stagnalis*. *Biol. Bull., Woods Hole*, 52:185-196.

HADENFELDT, D. 1929. Das Nervensystem von *Stylochoplana maculata* und *Notoplana atomata*. *Z. wiss. Zool.*, 133:586-638.

HANSTRÖM, B. 1926a. Über den feineren Bau des Nervensystems der Tricladen Turbellarien auf Grund von Untersuchungen an *Bdelloura candida*. *Acta Zool. Stockh.*, 7:101-115.

HANSTRÖM, B. 1926b. Eine genetische Studie über die Augen und Sehzentren von Turbellarien, Anneliden und Arthropoden (Trilobiten, Xiphosuren, Eurypteriden, Arachnoiden, Myriapoden, Crustaceen und Insekten.) *K. svenska VetenskAkad. Handl.*, (3), 4 (1):1-176.

HANSTRÖM, B. 1928. Some points on the phylogeny of nerve cells and of the central nervous system of invertebrates. *J. comp. Neurol.*, 46:475-493.

HASWELL, W. A. 1893. A monograph of the Temnocephaleae. *Proc. Linn. Soc. N. S. Wales (Macleay Mem. Vol.).* 1893:93-152.

HAVET, J. 1900. Contribution à l'étude du système nerveux des trématodes

Bibliography

(*Distomum hepaticum*). *Cellule*, 17: 351-380.

HEATH, H. 1902. The anatomy of *Epibdella squamula*, sp. nov. *Proc. Calif. Acad. Sci.* (3) (*Zool*), 3: 109-136.

HEATH, H. 1907. A new turbellarian from Hawaii. *Proc. Acad. nat. Sci. Philad.*, 59:145-148.

HEIDER, K. 1914. Phylogenie der Wirbellosen. In: *Die Kultur der Gegenwart*, Teil 3, Abt. 4, Vol. 4. Teubner, Leipzig.

HERTWIG, R. 1881. Das Auge der Planarien. *Jenaische S. B.*, 1880: 55-56.

HESSE, R. 1897. Untersuchungen über die Organe der Lichtempfindung bei niederen Thieren. II. Die Augen der Plathelminthen, insonderheit der tricladen Turbellarien. *Z. wiss. Zool.*, 62:527-582.

HESSE, R. 1902. Untersuchungen über die Organe der Lichtempfindung bei niederen Thieren. *Z. wiss. Zool.*, 72:565-656.

HILTON, W. A. 1921. Nervous system and sense organs. V. Flatworms. *J. Ent. Zool.*, 13 (suppl.): 34-48.

HOFSTEN, N. VON. 1907a. Zur Kenntnis des *Plagiostomum lemani* (Forel & du Plessis). In: *Zool. Studier tillägnade, Prof. T. Tullberg*. Uppsala.

HOFSTEN, N. VON. 1907b. Studien über Turbellarien aus dem Berner Oberland. *Z. wiss. Zool.*, 85:391-654.

HOFSTEN, N. VON. 1921. Anatomie, Histologie und systematische Stellung von *Otoplana intermedia* Du Plessis. *Zool. Bidr. Uppsala*, 7:1-72.

HONJÔ, I. 1937. Physiological studies on the neuromuscular systems of lower worms. I. *Caridinicola indica*. *Mem. Coll. Sci. Kyoto (B)*, 12:187-210.

HOVEY, H. B. 1929. Associative hysteresis in marine flatworms. *Physiol. Zoöl.*, 2:322-333.

HRANOVA, A. 1937. Sinnesorgane bei *Planaria chichkovi* (Hranova) und *Planaria gonocephala* (Duges). *Annu. Univ. Sofia (phys.-math.)*, (3) (Sci. Nat.) 33:219-226. (In Russian, with German summary.)

HYMAN, L. H. 1938. North American Rhabdocoela and Alloeocoela. II. Rediscovery of *Hydrolimax grisea* Haldeman. *Amer. Mus. Novit.*, 1004: 1-19.

HYMAN, L. H. 1951. *The Invertebrates.* Vol. II. Mc Graw-Hill, New York.

IIJIMA, I. 1884. Untersuchungen über den Bau und die Entwicklungsgeschichte der Süsswasser-Dendrocoelen (Tricladen). *Z. wiss. Zool.*, 40:359-464.

JÄNICHEN, E. 1896. Beiträge zur Kenntnis des Turbellarienauges. *Z. wiss. Zool.*, 62:250-288.

JOHNSTONE, J. 1912. *Tetrarhynchus erinaceus* van Beneden. I. Structure of larva and adult worm. *Parasitology*, 4:364-415.

JOSEPH, H. 1902. Untersuchungen über Stützsubstanzen des Nervensystems nebst Erörterungen über deren histogenetische und phylogenetische Deutung. *Arb. zool. Inst. Univ. Wien*, 13:335-400.

KAHANE, Z. 1880. Anatomie von *Taenia perfoliata* Goze als Beitrag zur Kenntnis der Cestoden. *Z. wiss. Zool.*, 34:173-254.

KARLING, T. G. 1940. Zur Morphologie und Systematik der Alloeocoela cumulata und Rhabdocoela lecithophora (Turbellaria). *Acta. zool. fenn.*, 26:1-260.

KAWAGUTI, S. 1932. On the physiology of land planarians. I. Phototaxis, with a note on the significance of the eye spots. *Mem. Fac. Agric. Taihoku*, 7:15-27.

KENNEL, J. VON. 1882. Die in Deutschland gefundenen Landplanarien *Rhynchodemus terrestris* O. F. Müller und *Geodesmus bilineatus* Mecznikoff. *Arb. zool.-zoot. Inst. Würzburg*, 5:120-160.

KEPNER, W. A. and FOSHEE, A. M. 1917. Effects of light and darkness on the eye of *Prorhynchus applanatus* Kennel. *J. exp. Zool.*, 23:519-531.

KEPNER, W. A. and RICH, A. 1918. Reactions of the proboscis of *Planaria albissima* Vejdovsky. *J. exp. Zool.*, 26:83-100.

KEPNER, W. A. and TALIAFERRO, W. H. 1912a. Organs of special sense of *Prorhynchus applanatus* Kennel. *J. Morph.*, 27:163-177.

KEPNER, W. A. and TALIAFERRO, W. H. 1912b. Sensory epithelium of pharynx and ciliated pits of *Microstoma caudatum*. *Biol. Bull., Woods Hole*, 23:42-58.

KOEHLER, O. 1926. Beiträge zur Sinnesphysiologie der *Planaria alpina*. *Verh. dtsch. zool. Ges.*, 31:182-187.

KOEHLER, O. 1932. Sinnesphysiologie der Süsswasserplanarien. *Z. vergl. Physiol.*, 16:606-756.

LACEY, R. J. 1955. A comparative morphological study on the nervous system of three orders of cestodes. PhD. Thesis, Dept. of Zoology, Univ. Illinois, Urbana. *Dissertation Absts.* 16:410.

LACZKO, K. 1880. Beiträge zur Kenntnis der Histologie der Tetrarhynchen hauptsächlich des Nervensystems. *Zool. Anz.*, 3:427-429.

LAN, H. AN DER. 1936. 7. Acoela I. *Vidensk. Medd. dansk Foren.* Kbh. 99:289-330.

LANG, A. 1879. Untersuchungen zur vergleichenden Anatomie und Histologie des Nervensystems der Plathelminthen. I. Das Nervensystem der marinen Dendrocoelen. *Mitt. zool. Sta. Neapel*, 1:459-489.

LANG, A. 1881a. Untersuchungen zur vergleichenden Anatomie und Histologie des Nervensystems der Plathelminthen. II. Über das Nervensystem der Trematoden. *Mitt. zool. Sta. Neapel*, 2:28-52.

LANG, A. 1881b. Untersuchungen zur vergleichenden Anatomie und Histologie des Nervensystems der Plathelminthen. III. Das Nervensystem der Cestoden im Allgemeinen und dasjenige der Tetrarhynchen im Besondern. *Mitt. zool. Sta. Neapel*, 2:372-400.

LANG, A. 1881c. Untersuchungen zur vergleichenden Anatomie und Histologie des Nervensystems der Plathelminthen. IV. Das Nervensystem der Tricladen. *Mitt. zool. Sta. Neapel*, 3:53-76.

LANG, A. 1881d. Untersuchungen zur vergleichenden Anatomie und Histologie des Nervensystems der Plathelminthen. V. Vergleichende Anatomie des Nervensystems der Plathelminthen. *Mitt. zool. Sta. Neapel*, 3:76-95.

LANG, A. 1881e. Der Bau von *Gunda segmentata* und die Verwandtschaft der Plathelminthen mit Coelenteraten und Hirudineen. *Mitt. zool. Sta. Neapel*, 3:187-251.

LANG, A. 1884. Die Polycladen des Golfes von Neapel und der angrenzenden Meeresabschnitte. *Fauna Flora Neapel*, 11:1-688.

LEMKE, G. 1935. Beiträge zur Lichtorientierung und zur Frage des Farbensehens der Planarien. *Z. vergl. Physiol.*, 22:298-345.

LENDER, T. 1952. Le rôle inducteur du cerveau dans la régénération des yeux d'une planaire d'eau douce. *Année biol.*, 56:191-193.

LENDER, T. 1954. Sur l'activité inductrice de la région antérieure du corps dans la régénération des yeux de la planaire *Polycelis nigra*: Activité de broyats frais ou traités à la chaleur. *C. R. Soc. Biol., Paris*, 148:1859-1861.

LENDER, T. 1956a. Analyse des phénomènes d'induction et d'inhibition dans la régénération des planaires. *Année biol.*, 60:457-469.

LENDER, T. 1956b. L'inhibition de la régénération du cerveau des planaires *Polycelis nigra* (Ehrb.) et *Dugesia lugubris* (O. Schm.) en présence de broyats de têtes ou de queues. *Bull. Soc. zool. Fr.*, 81:192-199.

LEVETZOW, K. G. VON. 1936. Beiträge zur Reizphysiologie der Polycladen Strudelwürmer. *Z. vergl. Physiol.*, 23:721-726.

LINTON, E. 1910. On a new rhabdocoele commensal with *Modiolus plicatulus*. *J. exp. Zool.*, 9:371-384.

LOEB, J. 1893. Ueber künstliche Umwandlung positiv heliotropischer Thiere in negativ heliotropische und umgekehrt. *Pflüg. Arch. ges. Physiol.*, 54:81-107.

LOEB, J. 1894. Beiträge zur Gehirnphysiologie der Würmer. *Pflüg. Arch. ges. Physiol.*, 56:247-269.

LOEB, J. 1900. *Comparative physiology of the brain and comparative psychology*. Putnam, New York.

LÖHNER, L. 1910. Untersuchungen über *Polychoerus caudatus* Mark. *Z. wiss. Zool.*, 95:451-506.

LÖNNBERG, E. 1891. Anatomische Studien über skandinavische Cestoden. *K. svenska VetenskAkad. Handl.*, 24 (6):1-109.

LOOSS, A. 1895. Zur Anatomie und Histologie der *Bilharzia haematobia* (Cobbold). *Arch. mikr. Anat.*, 46:1-108.

LÜHE, M. 1896. Das Nervensystem von *Ligula* in seinen Beziehungen zur Anordnung der Musculatur. *Zool. Anz.*, 19:383-384.

LUTHER, A. 1904. Die Eumesostominen. *Z. wiss. Zool.*, 77:1-273.

LUTHER, A. 1905-1907. Zur Kenntnis der Gattung *Macrostoma*. In: *Festschrift Prof. Palmén gewidmet*. Helsingfors Akt. handl, 1(5):1-61.

LUTHER, A. 1912. Studien über acöle Turbellarien aus dem Finnischen Meerbusen. *Acta Soc. Fauna Flora fenn.*, 36(5):1-59.

LUTHER, A. 1921. Untersuchungen an rhabdocölen Turbellarien. *Acta Soc. Fauna Flora fenn.*, 48(1):1-59.

LYNCH, J. E. 1945. Redescription of the species of *Gyrocotyle* from the ratfish, *Hydrolagus colliei* (Lay and Bennett), with notes on the morphology and taxonomy of the genus. *J. Parasit.*, 31:418-446.

MARCINOWSKI, K. 1903. Das untere Schlundganglion von *Distoma hepaticum*. *Jena. Z. Naturw.*, 37:544-550.

MARK, E. L. 1892. *Polychoerus caudatus*, nov. gen. et nov. spec. In: *Festschrift R. Leuckarts*. Engelmann, Leipzig.

MCCONNELL, J. V., JACOBSON, A. L., and KIMBLE, D. P. 1959. The effects of regeneration upon retention of a conditioned response in the planarian. *J. comp. physiol. Psych.*, 52:1-5.

MEIXNER, J. 1925. Beitrag zur Morphologie und zum System der Turbellaria-Rhabdocoela. I. Die Kalytorhynchia. *Z. Morph. Ökol. Tiere*, 3:255-343.

MERTON, H. 1914. Beiträge zur Anatomie und Histologie von Temnocephala. *Abh. senckenb. naturf. Ges.*, 35:1-58.

MICOLETZKY, H. 1907. Zur Kenntnis des Nerven- und Excretionssystems einiger Süsswassertricladen nebst andern Beiträgen zur Anatomie von *Planaria alpina*. *Z. wiss. Zool.*, 87:382-434.

MILLER, H. M. and MAHAFFY, E. E. 1930. Reactions of *Cercaria hamata* to light and to mechanical stimuli. *Biol. Bull.*, 59:95-103.

MONTI, R. 1897. Sur le système nerveux des dendrocèles d'eau douce. *Arch. ital. Biol.*, 27:15-26.

MONTI, R. 1900. Nuove ricerche sul sistema nervoso delle planarie. *Monit. zool. ital.*, 11:336-342.

MOORE, A. R. 1918. Reversal of reaction by means of strychnine in planarians and starfish. *J. gen. Physiol.*, 1:97-100.

MOORE, A. R. 1923. The function of the brain in locomotion of the polyclad worm, *Yungia aurantiaca*. *J. gen. Physiol.*, 6:73-76.

MOORE, A. R. 1933. On the rôle of the brain and cephalic nerves in the swimming and righting movements of the polyclad worm, *Planocera reticulata*. *Sci. Rep. Tôhoku Univ.*, (D.) 8:193-200.

MÜLLER, H. G. 1936. Untersuchungen über spezifische Organe niederer Sinne bei rhabdocoelen Turbellarien. *Z. vergl. Physiol.*, 23:253-292.

NIEMIEC, T. 1888. Untersuchungen über das Nervensystem der Cestoden. *Arb. zool. Inst. Univ. Wien.*, 7:1-58.

OLMSTED, J. M. D. 1922a. The rôle of the nervous system in the regeneration of polyclad Turbellaria. *J. exp. Zool.*, 36:48-56.

OLMSTED, J. M. D. 1922b. The rôle of the nervous system in the locomotion of certain marine polyclads. *J. exp. Zool.*, 36:57-66.

PALOMBI, A. 1926. *Digenobothrium inerme*, nov. gen. nov. sp. *(Crossocoela).* *Arch. zool. (ital.), Napoli*, 11:143-177.

PANTIN, C. F. A. 1950. Locomotion in British terrestrial nemertines and planarians: with a discussion on the identity of *Rhynchodemus bilineatus* (Mecznikow) in Britain, and on the name *Fasciola terrestris* O. F. Muller. *Proc. Linn. Soc. Lond.*, 162:23-37.

PARKER, G. H. and BURNETT, F. L. 1900. The reactions of planarians, with and without eyes, to light. *Amer J. Physiol.*, 4:373-385.

PEARL, R. 1903. The movements and reactions of fresh-water planarians: a study in animal behaviour. *Quart. J. micr. Sci.*, 46:509-714.

PERTUSA, P. J. 1942. Nota previa sobre el resultado de la investigación de los ojos de los *Planarias*, An. Asoc. esp. Progr. Cienc., 7:351-359.

PINTNER, T. 1880. Untersuchungen über den Bau des Bandwurmkörpers mit besonderer Berücksichtigung der Tetrobothrien und Tetrarhynchen. *Arb. zool. Inst. Univ. Wien*, 3:53-80.

PINTNER, T. 1925. Bemerkenswerte Strukturen im Kopfe von Tetrarhynchoideen. *Z. wiss. Zool.*, 125:1-34.

PINTNER, T. 1927. Kritische Beiträge zum System der Tetrarhynchen. *Zool. Jb. (Syst.)*, 53:559-590.

PINTNER, T. 1934. Bruchstücke zur Kenntnis der Rüsselbandwürmer. *Zool. Jb. (Anat.)*, 58:1-20.

POIRIER, J. 1885. Contribution à l'histoire des trématodes. *Arch. Zool. exp. gén.*, (2) 3:465-624.

PRESS, N. 1959. Electron microscope study of the distal portion of a planarian retinular cell. *Biol. Bull., Woods Hole*, 117:511-517.

PROSSER, C. L. 1960. Mechanical responses of sponges. *Anat. Rec.*, 138:377.

RAMPITSCH, J. 1941. Versuche über die cilioregulatorische Fortbewegung des Turbellars *Stenostomum leucops*. *Zool. Anz.*, 133:253-258.

RANKIN, J. S., JR. 1939. Ecological studies on larval trematodes from western Massachusetts. *J. Parasit.*, 25:309-328.

REES, G. 1941. The musculature and nervous system of the plerocercoid larva of *Dibothriorhynchus grossum* (Rud.). *Parasitology*, 33:373-389.

REES, G. 1950. The plerocercoid larva of *Grillotia heptanchi* (Vaullegeard). *Parasitology*, 40:265-272.

REES, G. 1951. The anatomy of *Cysticercus taeniae-taeniaeformis* (Batsch 1786) (*Cysticercus fasciolaris* Rud. 1808), from the liver of *Rattus norvegicus* (Erx.), including an account of spiral torsion in the species and some minor abnormalities in structure. *Parasitology*, 41:46-59.

REES, G. 1956. The scolex of *Tetrabothrius affinis* (Lönnberg), a cestode from *Balaenoptera musculus* L., the blue whale. *Parasitology*, 46:425-442.

REES, G. 1958. A comparison of the structure of the scolex of *Bothriocephalus scorpii* (Müller 1766) and *Clestobothrium crassiceps* (Rud. 1819) and the mode of attachment of the scolex to the intestine of the host. *Parasitology*, 48:468-492.

REES, G. 1959. *Ditrachybothridium macrocephalum* gen. nov., sp. nov., a cestode from some elasmobranch fishes. *Parasitology*, 49:191-209.

REISINGER, E. 1926. Untersuchungen am Nervensystem der *Bothrioplana semperi* Braun. *Z. Morph. Ökol. Tiere*, 5:119-149.

REISINGER, E. 1933. Neues zur vitalen Nervenfärbung. (Gleichzeitig ein Beitrag zur Kenntnis des Protoplanelliden-Nervensystems.) *Verh. dtsch. zool. Ges.*, 35:155-160.

RETZIUS, G. 1902. Weiteres zur Kenntniss der Sinneszellen der Evertebraten. 2. Die Sinneszellen der Turbellarien. *Biol. Untersuch.*, N.F. 10:31.

RIETSCHEL, P. E. 1935. Zur Bewegungsphysiologie der Cestoden. *Zool. Anz.*, 111:109-111.

RISER, N. W. 1949. Observations on the nervous system of the cestodes. *J. Parasit.*, 35:27.

RITTER-ZÁHONY, R. VON. 1907. Turbellarien: Polycladiden *Ergebn. Hamburger Magalh. Sammelreise*, 3(10):1-19.

ROBERTSON, J. A. 1928. Reaction of *Polycelis* in relation to physiological polarity. *Biol. Zbl.*, 48:427-430.

ROBOZ, Z. VON. 1882. Beiträge zur Kenntnis der Cestoden. *Z. wiss. Zool.*, 37:264-285.

SABUSSOW, H. 1905. Über den Bau des

Bibliography

Nervensystems von Tricladiden aus dem Baikalsee. *Zool. Anz.*, 28:20-32.

SCHAAF, H. 1906. Zur Kenntnis der Kopfanlage der Cysticerken, insbesondere des Cysticercus Taeniae solii. *Zool. Jb. (Anat.)*, 22:435-476.

SGONINA, K. 1939. Vergleichende Untersuchungen über die Sensibilisierung und den bedingten Reflex. *Z. Tierpsychol.*, 3:224-247.

SOEST, H. 1937. Dressurversuche mit ciliaten und rhabdocoelen Turbellarien. *Z. vergl. Physiol.*, 24:720-748.

SPÄTLICH, W. 1909. Untersuchungen über Tetrabothrien. *Zool. Jb. (Anat.)*, 28:539-594.

STEINBÖCK, O. 1924a. Eine neue Gruppe allöocöler Turbellarien: Alloeocoela Typhlocoela (Familie Prorhynchidae). *Zool. Anz.*, 58:233-242.

STEINBÖCK, O. 1924b. Die Bedeutung der *Hofstenia atroviridis* Bock für die Stellung der Alloeocoela im System der Turbellarien. *Zool. Anz.*, 59:156-166.

STEINBÖCK, O. 1925. Zur Systematik der Turbellaria metamerata, zugleich ein Beitrag zur Morphologie des Tricladen-Nervensystems. *Zool. Anz.*, 64:165-192.

STEINBÖCK, O. 1927. Monographie der Prorhynchidae (Turbellaria). *Z. Morph. Ökol. Tiere*, 8:538-662.

STEINBÖCK, O. 1931. 2. *Nemertoderma bathycola* nov. gen. nov. spec., eine eigenartige Turbellarie... *Vidensk. Medd. dansk naturh. Foren.*, 90:47-84.

STEINER, I. 1898. *Die Functionen des Centralnervensystems und ihre Phylogenese: Dritte Abth. Die wirbellosen Thiere*. Vieweg, Braunschweig.

STEINMANN, P. 1929. Vom Orientierungssinn der Tricladen. (Versuch einer Analyse mit Hilfe der vitalen Färbung.) *Z. vergl. Physiol.*, 11:160-172.

STEOPOE, I. 1934. Observations cytologiques sur les cellules nerveuses de *Leptoplana tremellaris* et *Prosthiostomum siphunculus*. *C. R. Soc. Biol., Paris*, 115:1315-1317.

STÉPHAN-DUBOIS, F. and LENDER, T. 1956. Corrélations humorales dans la régénération des planaires paludicoles. *Ann. Sci. nat. (Zool.)*, 18:223-230.

SUBRAMANIAM, M. K. 1940. The nervous system of a proglottid of *Tentacularia macropora*. *Curr. Sci.*, 9:500-501.

SUBRAMANIAM, M. K. 1941a. Studies on cestode parasites of fishes. II. The nervous system of *Tylocephalum dierama* Shipley and Hornell. *Rec. Indian Mus.*, 43:269-280.

SUBRAMANIAM, M. K. 1941b. Sympathetic innervation of proglottides in *Auitellina lahorea* Woodland. *Curr. Sci.*, 10:441-443.

TALIAFERRO, W. H. 1920. Reactions to light in *Planaria maculata*, with special reference to the function and structure of the eyes. *J. exp. Zool.*, 31:59-116.

THOMPSON, R. and MCCONNELL, J. 1955. Classical conditioning in the planarian, *Dugesia dorotocephala*. *J. comp. physiol. Psychol.*, 48:65-68.

TOWER, W. L. 1900. The nervous system in the cestode *Moniezia expansa*. *Zool. Jb., (Anat.)*, 13:359-384.

TURNER, R. S. 1946. Observations on the central nervous system of *Leptoplana acticola*. *J. comp. Neurol.*, 85:53-65.

UDE, J. 1908. Beiträge zur Anatomie und Histologie der Süsswassertricladen. *Z. wiss. Zool.*, 89:308-370.

ULLYOTT P. 1936. The behaviour of *Dendrocoelum lacteum*. I and II. *J. exp. Biol.*, 13:253-278.

ULMER, M. J. 1953. Studies on the nervous system of *Postharmostomum helicis* (Leidy, 1847) Robinson 1949 (Trematoda: Brachylaimatidae). *Trans. Amer. micr. Soc.*, 72:370-374.

VEJDOVSKÝ, F. 1895. Zur vergleichenden Anatomie der Turbellarien. *Z. wiss. Zool.*, 60:90-214.

VIAUD, G. 1950. Recherches expérimentales sur le phototropisme des planaires. *Behaviour*, 2:163-216.

WACKE, R. 1903. Beiträge zur Kenntnis der Themnocephalen. *Zool. Jb. (Syst.)*, 6 (Suppl.) (1):1-117.

WATSON, E. E. 1911. The genus *Gyrocotyle*, and its significance for problems of cestode structure and phylogeny. *Univ. Calif. Publ. Zool.*, 6:353-468.

WEISS, A. 1910. Beiträge zur Kenntnis der australischen Turbellarien. I. Tricladen. *Z. wiss. Zool.*, 94:541-604.

WESTBLAD, E. 1924. Zur Kenntnis der vitalen Alizarinfärbung. *Zool. Anz.*, 61:86-98.

WESTBLAD, E. 1937. Die Turbellarien-Gattung *Nemertoderma* Steinböck. *Acta Soc. Fauna Flora fenn.*, 60:45-89.

WESTBLAD, E. 1949. Studien über Skandinavische Turbellaria Acoela. V. *Ark. Zool.*, 41A(7):1-82.

WHEELER, N. C. 1939. A comparative study on the behavior of four species of pleurolophocercous cercariae. *J. Parasit.*, 25:343-353.

WHEELER, W. M. 1894. *Syncoelidium pellucidum*, a new marine triclad. *J. Morph.*, 9:167-194.

WILHELMI, J. 1909. Tricladen. *Fauna u. Flora Neapel*, 32:1-405.

WILHELMI, J. F. 1913. Platodaria, Plattiere. In: *Handbuch der Morphol. Wirbellosen Tiere*. A. Lang (ed.). Fischer, Jena.

WILLIAMS, H. H. 1959a. The anatomy of *Köllikeria filicollis* (Rudolphi, 1819), Cobhold, 1860 (Trematoda: Digenea). *Parasitology*, 49:39-53.

WILLIAMS, H. H. 1959b. The anatomy of *Phyllobothrium sinuosiceps* sp. nov. (Cestoda: Tetraphyllidae) from *Hexanchus griseus* (Gmelin) the six gilled shark. *Parasitology*, 49:54-69.

WOODWORTH, W. M. 1891. Contributions to the morphology of the Turbellaria. I. On the structure of *Phagocata gracilis* Leidy. *Bull. Mus. comp. Zool., Harv.*, 21:1-44.

YASURAOKA, K. 1954. Ecology of the miracidium. II. On the behavior to light of the miracidium of *Fasciola hepatica*. *Jap. J. med. Sci.*, 7:181-192.

ZAILER, O. 1914. Zur Kenntnis der Anatomie der Muskulatur und des Nervensystems der Trematoden. *Zool. Anz.*, 44:385-396.

ZERNECKE, E. 1895. Untersuchungen über den feinern Bau der Cestoden. *Zool. Jb. (Anat.)*, 9:92-161.

CHAPTER 10

Nemertinea

Summary	580	IV. Physiological Studies	589
I. The Brain	581	V. Phylogenetic Comparisons	593
II. The Lateral Cords	584	Bibliography	594
III. Peripheral Nervous System	586		

Summary

A distinct brain, with differentiation of lobes and of cell types and large numbers of globuli cells, marks the high level of nervous achievement of the ribbon worms. The brain and a pair of lateral cords make up the central nervous system. Commissures of the brain form a ring around the proboscis cavity, instead of the pharyngeal. The brain supplies the proboscis plexus as well as the cephalic end and sends connections to a foregut plexus. Minor longitudinal cords are a pair of dorsolateral nerves, and unpaired median-dorsal and inner dorsal nerves, all communicating with the brain. Some species have a small midventral nerve. There is thus a strong suggestion of an orthogonal plan with emphasis on the laterals. A progressive tendency toward sinking in of the nerve cords is manifest within the orders. There are ventral commissures between the lateral cords and branching nerves, dorsally and laterally. All these nerves make connections with a plexus in the body wall. A few forms of Nemertinea have a series of ganglia on the median dorsal cord; these and the nerves suggest a metamerism. Histological differentiation includes a strong sheath around the central nervous masses and another bounding the fibrous core from the cellular rind of purely unipolar neurons. Besides the small globuli cells in the brain and medium and large cells there and elsewhere, giant neurons are found in many nemertineans. Their axons proceed backward in the lateral cords, from cells both in the brain and in the cords; the destination or function is unknown. Statocysts, eyes, various slits and grooves including frontal and lateral organs, the often elaborate cerebral organs, and clusters of papillae or scattered sense cells are described. The cerebral organs are invaginations involving a mass of gland cells and a mass of nerve cells; a marked sequence of increasing complexity and intimacy of association with the brain can be put together from the diversity of species.

There is fair evidence of nervous control of cilia and suggestive evidence for nervous control of luminescence and slime production. Decapitation in some species causes a lasting cessation of ciliary beating and hence of gliding locomotion. Assuming ciliary automaticity, this is believed to mean that lower nervous levels maintain an inhibitory outflow to the cilia and that the brain normally suppresses this inhibition. Stimulation of the decapitate can restore ciliary activity, and in the correct direction—to carry the worm away from the end stimulated. The lateral cords lack the capacity to initiate or carry out antiperistalsis in the posterior two thirds, though this ability is present in the anterior third. Unilateral transection of a cord gives no asymmetry of movement in *Lineus* but does in *Oerstedia*, pointing to a difference in the role of cross connections. Bilateral cord section reveals properties of the plexus: the descending brain influence, without which locomotion stops, can spread around such transections, as can the order to initiate antiperistalsis after touching the head. But tail stimulation shows an inability of the intact connections at the level of the cuts to transmit the order for peristalsis. A similar but reciprocal polarized ability to spread excitation is exhibited by posterior levels: posteriad spread around cuts in the caudal two thirds is difficult; anteriad spread is easier. The plexus conducts with a rapid decrement and unequally in the two directions; it is not considered to be a typical nerve net.

The nemertinean worms are nervously, as in other ways, a distinct cut above the flatworms in complexity of structure, although quite reminiscent of them in plan. One suspects that nemertinean behavior must be palpably more complicated since their sense organs, central nervous tissue, and muscles are all more differentiated. The group has received little attention, especially in recent years, and is in addition relatively homogeneous; otherwise they would deserve considerably more space. The ribbon worms or rubber-band worms are typically errant predators with a variety of sense organs and a new plane of complexity in nervous histology. Points of interest are the common development of giant cells and fibers, the possession of globuli-type

I. THE BRAIN

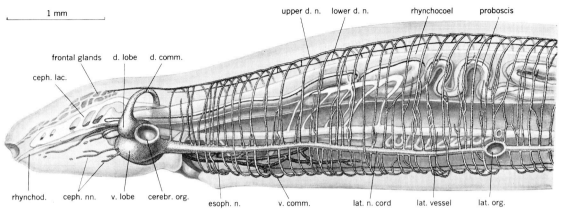

Figure 10.1. *Tubulanus* (= *Carinella*) *annulata*, anterior end, semischematic. [Redrawn from Bürger, 1895.] *ceph. lac.*, **cephalic lacunae**; *ceph. nn.*, **cephalic nerves**; *cerebr. org.*, **cerebral organ**; *d. comm.*, **dorsal commissure of brain**; *d. lobe*, **dorsal lobe of brain**; *esoph. n.*, **esophageal nerve**; *lat. n. cord*, **lateral nerve cord**; *lat. org.*, **lateral organ**; *lat. vessel*, **lateral vessel**; *lower d. n.*, **lower dorsal nerve**; *rhynchod.*, **rhynchodaeum**; *upper d. n.*, **upper dorsal nerve**; *v. comm.*, **ventral commissure**; *v. lobe*, **ventral lobe of brain**.

cells confined to parts of a subdivided brain, the reduced remnants of an orthogon of multiple cords emphasizing in this phylum the laterals, and the presence (severely limited and polarized) of ability to spread excitation around cord transections via a general plexus.

Of the scarcely more than forty references selected as making significant contributions in the last eighty years, the most important are those of Bürger, Montgomery, Böhmig, Coe, Hubrecht, Brinkmann, and Riepen. The only one with histology based on special nerve stains (methylene blue; no silver stains have been successfully applied) is that of Bürger (1890); Riepen used the same stain for gross anatomy. Physiological studies of nervous function are those of Eggers, Friedrich, and Corrêa on movements and transections.

The central nervous system of nemertineans comprises a well-formed brain and one pair of main longitudinal medullated strands, the lateral cords (Fig. 10.1). The distinction between central and peripheral is vague since there are also some lesser medullated strands and some outlying ganglia; these will be treated below under the peripheral nervous system.

I. THE BRAIN

The brain and the lateral cords are epidermal in certain Paleonemertini such as *Carinina* and *Procarinina*; in most other members of this, the lowest order of the four, they are situated in the dermis. In some animals of this order and in the Heteronemertini they are located within the musculature of the body wall. These two lower orders together form the group called the Anopla. In the Enopla, comprising the two highest orders, the Hoplonemertini and Bdellonemertini the brain and cords are in the mesenchyme, deep to the muscle layers. There is evidence therefore, of a **progressive sinking in of the central nervous system,** with both extremes represented. In the lower two orders the brain is anterior to the mouth; in the higher two it is posterior to the mouth. The position of the central nervous system is thus of taxonomic interest at the level of higher categories.

The brain has **two lobes on each side,** except in the Bdellonemertini (Fig. 10.2). The dorsal lobes of the two sides are connected by a **dorsal commissure** which passes above the rhynchodaeum—the cavity dorsal to the mouth—opening anteriorly, and from which the proboscis, characteristic of the phylum, is everted. The two ventral lobes are connected by a **ventral commissure,** so

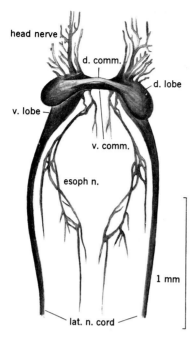

Figure 10.2. The brain of *Tubulanus* seen in dorsal view, reconstructed from serial sections; interior end upwards [Bürger, 1897.] *d. comm.*, dorsal commissure of brain; *d. lobe*, dorsal lobe of brain; *esoph. n.*, esophageal nerve; *lat. n. cord*, lateral nerve cord; *v. comm.*, ventral commissure; *v. lobe*, ventral lobe of brain.

that a ring is formed around the proboscis cavity and not around the buccal or pharyngeal cavity. Additional, accessory commissures sometimes occur ventrally. The ventral lobes are always continuous with the lateral cords; the dorsal lobes send a fascicle into them in some species of hoplonemertines. There appears to be a trend toward increased relative size of the dorsal lobes: they are said to be smaller than the ventral in most paleonemertines and larger in the hetero- and hoplonemertines. Variations in size of the dorsal lobe occur apparently in parallel with the development or absence of the so-called cerebral organ (see below).

Nerves of the brain include some that are in fact medullated strands. A number of **cephalic nerves** pass forward to the epidermis, muscle fibers, and cephalic glands of the anterior end to the frontal organ and to the eyes and cerebral organ, when present. The statocysts, rarely present, are innervated by the brain but are embedded in it. **Proboscis nerves** arise as a pair from the ventral commissure in the paleo- and heteronemertines (Anopla), whose proboscis is unarmed. They enter the proboscis where it attaches to the wall of the rhynchodaeum or proboscis cavity and travel the length of that enormously long organ either between the muscle layers or just below the epithelium that becomes the outside of the proboscis when everted, according to the species. Bdellonemertines likewise have only one pair of proboscis nerves. In the hoplonemertines, many of which have an armed proboscis, from 7 to 50 nerves arise—the number being characteristic, with some variation, for the species— usually directly from the anterior side of the brain and, remaining constant in number, run into the proboscis; in the longitudinal muscle they go as far back as the level of the stylet apparatus. They are evenly distributed around a cross section and are interconnected by a plexus that is sometimes more and sometimes less regular. Behind the stylet region the nerves continue, now much smaller, just below the epithelium.

A pair of **foregut nerves** arises from the ventral lobe and supplies the esophagus and stomach, breaking up into plexuses in the lower orders but not in hoplonemertines, in which group they are also less medullary. The nerve supply to the gut is really twofold: besides this cephalic contribution there are nerves from the body wall plexus, according to Hubrecht. A pair of **dorsolateral nerves** (subdorsal nerves) arises from the caudal surface of the dorsal lobe to run longitudinally between the proboscis sheath and body wall, gradually moving from a mesenchymal to an intramuscular position posteriorly but disappearing by about midbody; these nerves or medullated cords are known in several genera of hoplonemertines, especially the pelagic forms, and in bdellonemertines. An unpaired **median dorsal nerve** is present in most nemertineans except bdellonemertines and usually arises from the dorsal brain commissure, although in the polystyliferous hoplonemertines it has only an indirect connection with the brain. It travels in the same layer of the body wall as the lateral cords except in the cephalotrichids and hoplonemertines, where it lies outside the muscle.

I. THE BRAIN

From this nerve there is sometimes given off an **inner dorsal nerve,** likewise unpaired and apparently supplying the dorsal side of the proboscis sheath. Many paleonemertines possess a **midventral nerve** extending into the muscle layers of the body wall. These several longitudinal strands strongly suggest homology with the several longitudinal strands of platyhelminths and thus support the orthogon theory (p. 538).

The **histological differentiation of nervous tissue** shows advance in several ways over that of all the preceding groups (Fig. 10.3). The brain and the lateral cords have a capsule or ganglionic neurilemma of connective tissue (except in some lower genera). Distinct from the neurilemma are prominent glia cells, especially in the fibrous core.

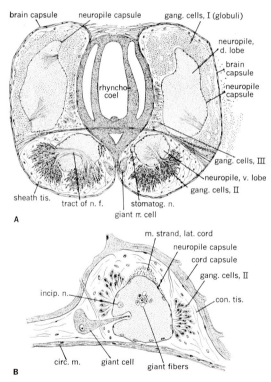

Figure 10.3. Histology of the brain and cord of a nemertinean, *Langia formosa.* **A.** Transverse section of the brain showing the two lobes. **B.** Transverse section of lateral cord. [Modified from Bürger, 1890.] *circ. m.,* circular muscle of body wall; *con. tis.,* connective tissue enveloping rind of nerve cells; *d. lobe,* dorsal lobe of brain; *gang. cells I, II, III,* ganglion cells of type I, II, or III; *incip. n.,* bundle about to give rise to a nerve; *m. strand, lat. cord,* muscle strand of lateral cord; *n.,* nerve; *n. f.,* nerve fibers; *stomatog. n.,* stomatogastric nerve; *v. lobe,* ventral lobe of brain.

There is in many places a kind of limiting layer of glia separating the rind of nerve cell bodies and the fibrous core. The glia cells appear to lay down supporting fibers. Pigment-containing glia cells are common and contribute to the bright red color of the ganglia and cords of many species, especially the pelagic ones but also in the genera *Lineus* and *Cerebratulus.* The nerve cell bodies are sharply separated from the neuropile and fiber tracts, a feature which goes with the seemingly exclusive development of unipolar neurons. In other words, the characteristic condition of higher invertebrates is here exhibited. Moreover, there are signs of differentiation usually associated with relatively higher central structures in the several classes of neurons. Besides (a) fairly large, more superficial cells and (b) somewhat smaller, more deeply lying plasma-rich and chromatin-poor cells, there are both larger and smaller ones. The larger (c) have been called neurocord or giant cells and the smaller (d) are reminiscent of the small, chromatin-rich globuli cells of higher invertebrates. These types are not indiscriminately intermingled but are distributed in definite regions of the brain and lateral cord.

Giant cells are described from a number of hetero- and hoplonemertines (Figs. 10.3, 10.5). They are sometimes as large as 75×117 μ in diameter (*Uniporus*), the axon up to 12 μ, possibly more. Most such cells illustrated measure 20–50 μ. In the hoplonemertines (for example *Prosadenoporus, Drepanophorus, Amphiporus,* and possibly *Uniporus*) and in some heteronemertines (*Zygeupolia* and *Micrura*) there is a single pair located on the medial side of the ventral lobe. Most heteronemertines have several giant cells in the brain and in addition a number in the lateral cords. *Cerebratulus lacteus,* according to Thompson, has 6 pairs and 1 unpaired cell in the brain; Montgomery counted only 3 pairs here but about 80 in each lateral cord; these cells may have a kind of metameric regularity but they are confined to the middle levels of the body. The processes of the giant cells of the brain decussate in the ventral commissure and pass back into the lateral cords where, together with processes arising from ipsilateral cells in the cords, they form a definite bundle of giant fibers. Giant cells are lacking in all pelagic hoplonemertines and a few nonpelagic (*Gononemertes*) and in some heteronemertines like *Eupolia* and *Lineus*, as well as in paleonemertines. As yet no behavioral trait has been correlated with the presence or absence of giant cells, and their function is quite unknown.

The **small chromatin-rich cells** form a special cluster laterally in the dorsal lobe of the brain. The associated neuropile is neither loose-textured nor smoothly dense, two successive grades of elaboration, but is lumpy or patchily more dense. Hanström called attention to the similarity with the specialized masses of globuli cells and their glomerular neuropiles in the cerebral ganglia of higher annelids, arthropods, and molluscs. As in those groups, these structures in nemertineans are confined to the brain. Furthermore, they are not consistently correlated with any particular sense organ. On anatomical grounds they would seem to be regions of **higher associative function.**

II. THE LATERAL CORDS

The chief longitudinal cords of nemertineans are situated at the depth already described, and far laterally or, in some much flattened species, ventrolaterally; they have been called ventral connectives (Figs. 10.1, 10.3, 10.4, 10.8). Although the most obvious **connections to the brain** are

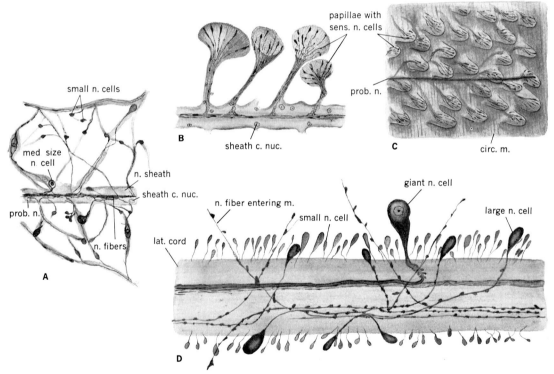

Figure 10.4. Details of the lateral cord, proboscis plexus, and sensory papillae with intra vitam methylene blue staining. **A.** *Amphiporus marmoratus*, proboscis plexus with one of the proboscis nerves. Note scattered individual nerve cells of small and medium size and faint sheath cell nuclei. **B.** *Drepanophorus* sp., anterior proboscis showing sensory papillae, each sending a bundle of axons to a proboscis nerve. **C.** The same at low magnification. **D.** *Cerebratulus marginatus*, lateral cord with several types of nerve cells (in life these are embedded in a yellowish red sheath tissue not shown here), ascending and descending nerve fibers, and efferent fibers to musculature. The large cell and fiber is one of the giant neurons; its cell body is about 25 μ in diameter. Anterior is to the right. [Bürger, 1891.] *c.*, cell; *circ.*, circular; *lat.*, lateral; *m.*, muscle; *med.*, medium; *n.*, nerve; *nuc.*, nucleus; *prob.*, proboscis; *sens.*, sensory.

II. THE LATERAL CORDS

with the ventral lobe, there are—at least in a number of hoplonemertines, especially the pelagic forms—direct connections with the dorsal lobe of the brain as well; the fiber bundle from the latter remains distinct and on the dorsal side of the cord, often separated from the ventral funiculus of the cord by a stratum of nerve cells. There are thus **two funiculi** comprising the fiber matter of the cord, except in a few species that have another, laterally, from the dorsal lobe (*Bürgeriella*).

The lateral cord enlarges into an anal swelling posteriorly, in the Bdellonemertini, which have a sucker well supplied with nerves from this swelling.

Nerve cell bodies (Fig. 10.4) are grouped into two rows of somewhat metameric clumps, one dorsally and one ventrally; these clumps lie not exactly one behind the other but are staggered—alternately a little to the right and to the left. The cells are chiefly of the middle sizes. A number of giant cells (up to several score) sometimes occur in the cords, and in at least some species they are absent from the most rostral and caudal levels, being concentrated in the middle levels.

Nerves of the lateral cord (Fig. 10.5) generally, and sometimes quite regularly, number three on each side between each two gut diverticula; an apparent or incipient metamerism has often been discussed. The dorsal nerves of the cord proceed dorsally and connect with the dorsolateral nerve. The lateral nerves supply the lateral body wall.

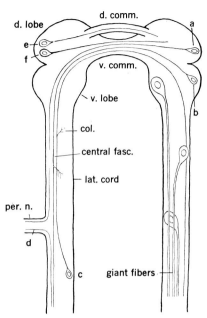

Figure 10.6. *Cerebratulus*, neurons identified in the anterior end of the central nervous system. [Bürger, 1895.] *a–f*, nerve cells representing types whose axons could be followed, besides the giant cells. *central fasc.*, central fasciculus of fibers in the cord; *col.*, axon collateral or dendritic process; *d. comm.*, dorsal commissure; *d. lobe*, dorsal lobe of the brain; *lat. cord*, lateral cord; *per. n.*, peripheral nerve; *v. comm.*, ventral commissure; *v. lobe*, ventral lobe of the brain.

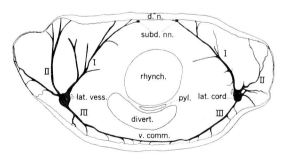

Figure 10.5. *Phallonemertes* (Hoplonemertini). Cross section showing branches from the lateral cords. [Brinkmann, 1917.] *divert.*, blind diverticulum of gut; *d. n.*, dorsal nerve; *lat. cord*, lateral cord with its dorsal, lateral, and medial branches (I, II, and III); *lat. vess.*, lateral vessel; *pyl.*, pyloric tube; *rhynch.*, rhynchocoel; *subd. nn.*, subdorsal nerves; *v. comm.*, ventral commissure.

The ventral nerves proceed medially to join with those of the other side, forming more or less irregular commissures between the cords; they commonly branch and anastomose and contribute to a plexus which runs between the muscle layers completely around to the median dorsal nerve and the dorsolateral nerve. The lower orders show less regularity in the arrangement of nerves than the polystyliferous hoplonemertines.

In most forms an **anal commissure** has been described between the lateral cords; it may be in front of or behind the anus and either dorsal or ventral to the gut.

A longitudinal **muscle of the lateral cord** is found in many genera, intimately associated with the external neurilemma or even internal to it. Nemertineans are notorious for their great extensibility but nothing is known of the histology or physiology of the nerve cords under stretch.

Pathways revealed by methylene blue (Fig. 10.6) include: (a) Giant fibers, arising as heretofore described, form a fascicle containing contralateral brain-originating and ipsilateral cord-originating fibers, all passing posteriorly. Some have been seen to divide dichotomously and at wide intervals fine twigs are given off. (b) Contralateral, nongiant fibers enter the cord from cells in both dorsal and ventral lobes of the brain. (c) Cells in the cord send processes forward to cross in the ventral brain commissure and descend in the other lateral cord. (d) Fibers pass between the cord and the peripheral nerves; they extend both rostrally and caudally but a clear distinction between afferents and efferents has not been made. Ramifying collaterals end in the neuropile.

A good deal of cytological description of the nerve cells and nonnervous tissue is given by Böhmig (1898), Brinkmann (1914), Bürger (1890, 1891, 1895), Montgomery (1897). The embryology of the central nervous system has been worked on by Hammersten (1918), Nusbaum and Oxner (1913), Salensky (1912), and Stiasny-Wynhoff (1923).

III. PERIPHERAL NERVOUS SYSTEM

The main nerves and their distribution have already been given. A special feature which has attracted attention is a series of **ganglia on the median dorsal nerve** in the pelagic *Neuronemertes* (Figs. 10.7, 10.8). This is a genus named for the apparently high degree of specialization of nervous anatomy generally, though it may be questioned whether this is a real elaboration in view of the absence of sense organs and the habit of life. The dorsal ganglia are fairly regular and correspond with the metamerism of the lateral cord; they are well-defined clusters of ganglion cells, like those in the cords. Nothing is known of their function or connections.

Plexuses are formed by branching and anastomosis of all the nerves: cephalic, foregut, proboscis, and body wall. There is a greater tendency to form plexuses in the lower genera; in some of the higher forms, nerves (even the gastric) are said to lack nerve cell bodies and not to anastomose. The body wall plexus (Fig. 10.9), sometimes called a nerve layer, may lie in the epithelium in some paleonemertines (*Carinina*, *Procarinina*) or in the dermis just outside the muscles; it is said to be prominent in *Hubrechtia* and sparse or absent in *Procephalothrix*, both belonging to this order. Many heteronemertines are described as having only one (others two) stratum of plexuses; the single one—or the outer if there are two—lies between outer longitudinal and circular muscles, and the inner plexus lies between the circular and inner longitudinal muscles. Radial branches connect the two. The hoplonemertines have a single plexus corresponding to the inner layer. Interconnections through this intermuscular plexus between lateral cords and dorsolateral, median dorsal, and midventral nerves are in some forms fairly regular, suggesting a progressive development toward a ladder or orthogonal lattice arrangement. We will return to the question of the

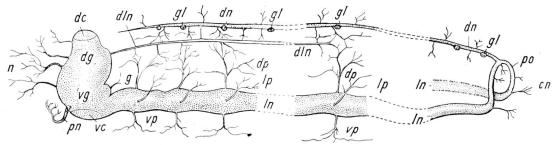

Figure 10.7. Nervous system of *Neuronemertes*. [Coe, 1933.] *cn*, caudal nerves; *dc* and *vc*, dorsal and ventral commissures; *dg* and *vg*, dorsal and ventral ganglia of the brain; *dln*, dorsolateral nerve; *dn*, dorsal nerve, with metameric ganglia *(gl); dp, lp*, and *vp*, dorsal, lateral, and ventral peripheral nerves; *g*, gastric nerve; *ln*, lateral nerves; *n*, cephalic nerves; *po*, posterior commissure; *pn*, proboscidial nerve.

III. PERIPHERAL NERVOUS SYSTEM

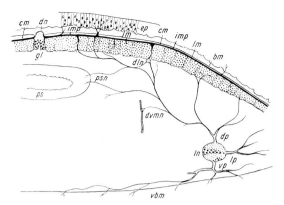

Figure 10.8. *Neuronemertes.* Diagrammatic cross section showing the relation of the dorsal nerve to some other elements of the nervous system. [Coe, 1933.] *bm*, basement membrane; *cm* and *lm*, circular and longitudinal muscle layers of body wall; *dln*, dorsolateral nerve; *dn*, dorsal nerve; *dp*, dorsal peripheral nerve; *dvmn*, nerve to dorsoventral muscles; *ep*, surface epithelium; *gl*, ganglion of the dorsal nerve; *imp*, intermuscular plexus; *ln*, lateral nerve cord; *lp* and *vp*, lateral and ventral peripheral nerves; *ps* and *psn*, proboscis sheath and one of its nerves; *vbm*, ventral body wall.

Figure 10.9. Plexus formed by branches of ventral cords in *Dionemertes*. [Brinkmann, 1917.]

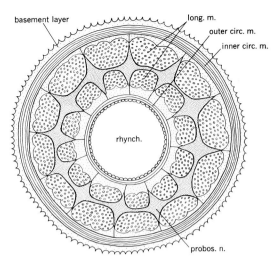

Figure 10.10. Transverse section of everted proboscis of *Cuneonemertes gracilis* Coe, with 12 proboscideal nerves *(probos. n.)* connected by a plexus in the midst of the longitudinal muscle layer *(long. m.)*. The basement layer underlies the inner, glandular epithelium. [Coe, 1927.] *inner circ. m.* and *outer circ. m.*, inner and outer circular muscle layers; *rhynch.*, extension of rhynchocoel lined with endothelium.

function of these plexuses below (p. 592). There is as yet insufficient evidence to apply the term nerve net in the sense of a diffusely conducting system. A profusion of radial nerves springing from the plexus and apparently supplying the musculature is more evident in hetero- than in hoplonemertines. The actual form of the connection of nerve to muscle is not known.

The proboscis nerves, already mentioned, not only form a **proboscis plexus** of circular and commissural connections between themselves (Fig. 10.10) but give off radial branches that contribute to plexuses both superficially, under the lining epithelium, and deeply, inside the outer circular muscle layer. One or more relatively strong ring commissures connect the proboscis nerves in the region of the diaphragm and bulb of the proboscis. This seems to be quite an extensive innervation; it raises the question whether we really appreciate all the functions and activities of the proboscis. In view of the rich supply of primary afferent neurons, described below, we may expect to learn of some moderately involved behavior, involving sensibility or discrimination, after the proboscis has been shot out of its rhynchocoel. C. B. Wilson is one of few who have watched the activities of the proboscis. In *Cerebratulus* it is used in burrowing, being forced into the substrate ahead of the body, which then pulls up to it. It is reported to be quite sensitive to touch. Pantin (1950) observed similar employment of the proboscis in the land species *Geonemertes dendyi*; in a dramatic escape reaction the huge proboscis of this form is everted, adheres to the ground and the animal "flies forward" by a wave of shortening.

Receptors consist of statocysts, eyes, various slits and grooves including frontal organs and lateral organs, the often elaborate cerebral organs, and epithelial sense cells, clustered (papillae) or scattered.

Eyes are present in a few paleo- and most heteronemertines and in most hoplonemertines except the pelagic species, which are deep, mid-water zooplankters. They are lacking in bdellonemertines. Usually confined to the head, anterior to the brain, they sometimes occur also a short distance behind the head; the number varies from two to several hundred. They may be in the dermis, in the musculature, or applied to the brain. The structure is that of an inverted, pigment cup ocellus, like the flatworm eye; in a few species the most complex ocelli are provided with a lens. The nerve always goes to the dorsal lobe of the brain.

Statocysts are found in a single family, the Ototyphlonemertidae; they lie—one on each side (occasionally two)—in the dorsal rind of the ventral lobe of the brain.

A variety of supposedly chemoreceptive or tactile (or combined) organs are found in the form of **cephalic grooves and slits.** Patches of epithelium lacking gland and pigment cells and underlain by a cluster of nerve cell bodies, they are distributed in species, and in group-specific ways. Their often considerable extent, depth, and fluting suggest a high importance. Of this nature apparently is the **frontal organ** of some hoplonemertines, a protrusible flask-shaped pit at the anterior tip, with a specialized ciliated epithelium. Some heteronemertines have three similar organs. They are all well supplied with nerve cells and send their nerves to the brain. In some of the lower genera a pair of depressions near the nephridial openings on the lateral margins are called the **lateral organs** (Fig. 10.11); their nerves enter the adjacent lateral cords.

Scattered **unicellular receptors** occur; especially often mentioned are long slender cells in the epidermis of the head and the anal region, bearing distally so-called tactile bristles projecting from the surface. On the proboscis there are described papillae in large numbers, each with a group of epithelial sensory neurons whose axons run into the subepithelial plexus (Fig. 10.4). Such an extensive sensory equipment points to a role of the proboscis not merely as an effector.

The **cerebral organs** are a pair of invaginations near the brain with the inner end embraced by a mass of gland cells and nerve cells (Fig. 10.12). In its simplest form, as in *Tubulanus annulatus*, the invagination is short, confined to the epidermis, and wide open to the outside; the epithelium in it is modified like other sensory pits but a special group of gland cells opens around the periphery of the central mass of sensory nerve cells. A nerve passes to the dorsal brain lobe. In other forms the canal lengthens and the mass of gland and nerve cells at its end moves through the body wall, coming to lie at various depths, finally in contact with the brain. Many intermediate stages are known in different species, where there are still nerves—indeed, up to four of them—and the canal is long and even forked. The gland cells become very elongate in order to open into the canal and in the more complex forms they are arranged in clusters or tracts but are always segregated rather than intermingled with the nerve cells. In all the deeper penetrating cerebral organs a strong connective tissue capsule surrounds them, often continuous with that around the brain. In all the deeper lying organs there is a close topographic relation with the lateral blood vessel; the organ projects into it or is virtually surrounded and bathed in blood. Although the vascular endothelium is said to be thinned, the capsule of the organ is still present. In the most

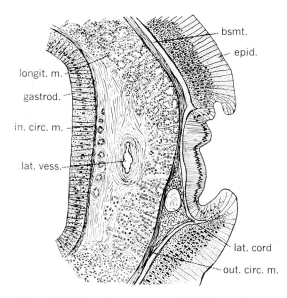

Figure 10.11. Lateral sense organ of *Tubulanus frenatus*. [Coe, 1943.] *bsmt.,* basement layer; *epid.,* epidermis; *gastrod.,* epithelial lining of stomach; *in. circ. m.* and *out. circ. m.,* inner and outer circular muscle; *lat. cord,* lateral nerve cord; *lat. vess.,* lateral blood vessel; *longit. m.,* longitudinal muscle.

IV. PHYSIOLOGICAL STUDIES

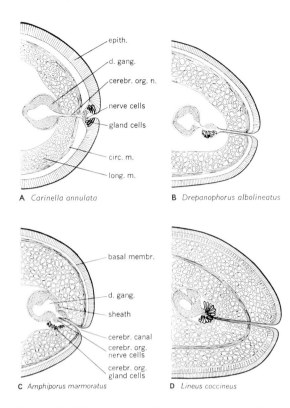

Figure 10.12. Sections through the head regions of four nemertineans, showing progressively greater development of the cerebral organ. In **A** the organ is purely epithelial; in **D** it has become closely associated with the dorsal ganglion. [Scharrer, 1941.] *basal membr.*, basal membrane; *cerebr. canal*, cerebral canal; *cerebr. org.*, cerebral organ; *circ. m.*, circular muscle layer; *d. gang.*, dorsal lobe of cerebral ganglion; *epith.*, epithelium of epidermis; *long. m.*, longitudinal muscle layer; *sheath*, capsule of cerebral ganglion.

advanced stage, the organ is another lobe of the brain, the nerve cell mass being completely continuous with the rind of the dorsal lobe. Now the nerve cells do not look like epithelial sensory cells but are unipolar and indistinguishable from brain cells. The destination and connections of the processes have not been followed in either primitive or specialized forms.

The simpler forms of cerebral organs occur in the lowest paleonemertines (Tubulanidae), the most complex forms in some of the higher heteronemertines (Lineidae), both in the same subclass (Anopla). In the Enopla fairly simple organs are found in the more primitive suborder Polystilifera (Drepanophoridae) and fairly complex ones in the more advanced suborder Monostilifera (Amphiporidae). Cerebral organs are missing in adult Bdellonemertini and in the pelagic Hoplonemertini, as well as in some anoplans.

This **intimate association of nerve cells and glands**—neither has been spoken of as neurosecretory—has aroused interest because of the impressive elaboration and hence its assumed importance in some species. One can only suppose that its functional meaning has changed greatly within the nemertineans. Reisinger (1926) assumed on meager evidence that one function is chemoreceptive for sources of stimuli not in contact; he observed a current of water in the canal and its intensification in the presence of food.

IV. PHYSIOLOGICAL STUDIES

Although very few authors have turned to the questions of functional organization of the nemertinean nervous system, some remarkable facts are at hand concerning control of cilia, differentiation of capacities of the plexuses and cords to spread coordinated excitation, and the role of the brain. These we owe to Eggers, Friedrich, and Corrêa.

In *Lineus ruber* and *Amphiporus lactifloreus* at least, ordinary gliding locomotion when undisturbed is dependent on ciliary activity. A few minutes in 0.5% LiCl in sea water abolishes gliding and ciliary beating while excitability of the animal and normal wavelike muscle movements are unaffected; 0.5% $MgCl_2$ does not affect gliding or ciliary motion but abolishes responses to moderate stimuli—effective stimuli cause only strong contraction, not wavelike movements. The interest of this ciliary locomotion is that a normally gliding animal responds to touch by a brief cessation of ciliary beat followed by reversal of beat and of gliding. The **cilia are under nervous control,** presumably having an intrinsic tendency to beat in one direction and being subject to at least two grades of superimposed nervous influence, stopping the beat and reversing it. In these forms the gliding appears to depend on a slime trail secreted by epidermal glands; the production of slime is greatly increased under certain conditions and it is an open question whether these glands are under nervous control. One species, *Emplectonema kandai* is **luminescent,** giving brilliant flashes along the entire body when stimulated thermally, chemically, or electrically but only locally to touch; epidermal gland cells seem to be responsible and may be under nervous control.

Analysis of functional organization depends on detectable and characteristic responses and nemertineans are notoriously sluggish and limited in repertoire. Some few species, aside from the deep-sea forms, swim; creeping, burrowing, and avoidance movements are the principal locomotor activities. Righting is a useful reaction. The following applies particularly to *Lineus ruber* but largely also to *Emplectonema* and *Amphiporus:* species differences have been found and must be expected even in apparently basic features. The reaction of a worm quietly gliding when touched on the head is to cease ciliary beating and gliding, to flatten the posterior half abruptly (in *Emplectonema* and *Amphiporus* merely to twitch the tail end), and to initiate antiperistalsis. This last is more complex than in earthworms: forward-moving waves begin a short distance behind the head and progressively the origin shifts back until it is at the tail. The net movement is backward, away from the stimulus. Touching the tail induces an abrupt contraction of longitudinal muscles, initiation of peristalsis—backward-moving waves that carry the worm forward—and often there is a visible acceleration of the ciliary gliding forward. These are the main reactions which play a part in the available analysis of brain and cord function. Strong stimuli cause violent coiling—and in many worms proboscis eversion—and eventually autofragmentation.

The **role of the brain** has been inferred from the effects of decapitation, which are quite different according to the species. In one group, including *Lineus ruber, Amphiporus lactifloreus,* and *Oerstedia dorsalis,* under the conditions used by Friedrich (1933), this operation causes a lasting cessation of ciliary beating and hence of gliding; the worms remain motionless except for a small anterior rhythmic muscle contraction, which represents another effect of the removal of the brain. On the assumption of ciliary automaticity, it is supposed that the remaining nervous system —lateral cords or peripheral plexus—now exerts a nervous inhibition on the cilia and hence that the brain was exerting an influence to suppress this tendency of the lower levels! Whether the effect of the brain is to inhibit an inhibition or otherwise, its effect can be imitated in the decapitated worm by stimulation. If this is not too weak it causes the rhythmic contractions to stop, the cilia start up again, and gliding locomotion resumes; moreover this is forward if the stimulation was posterior and backward if anterior, so there is nervous control of direction in the absence of the brain. The ciliary gliding soon ceases if stimulation is not maintained. Besides losing spontaneous gliding, directedness is also lost with the removal of the head sense organs. A sagittal cut separating the right and left halves of the brain had the same effect as decapitation and so did attempts to injure dorsal or ventral lobes or to separate them; this makes one wonder how much the signs of decapitation are due to the trauma of the operation, possibly through blood loss.

Control operations which did not injure nervous tissue were not described; the observation that the same effects follow the slightly less serious manipulation of cutting the two lateral cords just behind the head is not an adequate control. Repetition of the experiment is desirable because of the lesser effect reported for similar bilateral transection a little farther back. In any event, adequate stimulation overcomes the deficit completely, barring directedness. Corrêa found a similar quiescence after decapitation in *Emplectonema gracile*, showing that gliding locomotion depends on the intact head, but the ease of substituting for it by mild stimulation made her put it in another class. Still a third class, characterized by locomotion that is independent of the presence of the brain, is illustrated by *Lineus lacteus*, two species of *Ototyphlonemertes*, and one of *Prostomatella*. Here the cilia continue to beat after decapitation. Quite possibly each species

IV. PHYSIOLOGICAL STUDIES

can under proper conditions act in any of these three ways. Nevertheless it is significant in view of the functions of the brain in higher groups that there is reasonably clear evidence that in some species a tonic influence descending from the brain in these lower worms makes for spontaneity and probably does so by an inhibitory effect on lower neurons.

Turning to **experiments on the cords,** Friedrich has opened the subject with simple transections that point to some notable properties of the body wall plexus as well as of the lateral cords (Fig. 10.13). Referring to *Lineus ruber* and *Amphiporus lactifloreus*, if the anterior third or more of a worm is cut off, the remaining piece is capable of ciliary and peristaltic activity upon stimulation, but not of antiperistalsis; contrary to the response of a piece cut off within the rostral third, stimulating the anterior end elicits peristalsis, carrying it toward the stimulus. The author concluded that a difference between levels of the animal lies in an absence of antiperistalsis-initiating mechanisms caudally.

Unilateral transection of the lateral cord has no effect in the last-named species; there is no circling or other evidence of asymmetry of tonus. Either there are efficient cross connections between the cords or each cord can control the musculature of both sides. By way of contrast, *Oerstedia dorsalis* unilaterally transected just behind the head circles for 15 min and then crawls roughly in one direction, but this movement occurs because the circling is interrupted and corrected at frequent intervals—the head tends to maintain a fixed orientation and corrects the asymmetry due to tonus. This can be taken to mean less efficient cross connections. If two transections are made on the same side in *Lineus*, 1–2 cm apart, the region between still has normal symmetrical waves—again pointing to connections between cords (or possibly, bilateral muscle control via the plexus).

If both lateral cords are cut at the same level and within the rostral third, in *Lineus*, the animal still glides forward when undisturbed. In contrast to the decapitate or cord-transected preparation farther forward, there is evidently sufficient descending influence to make extrinsic stimulation unnecessary and this influence is able to spread around the transections—through the general plexus or one of the minor medullary nerves from the brain (median dorsal in this genus). Stimulation of the head causes a normal abrupt flattening behind the cuts, backward gliding, and then antiperistalsis—as in an intact animal; even the more complex and coordinated excitation required by these actions can spread around cord interruptions in the antero-posterior direction. But tail stimulation shows a deficit in response: a strong contraction behind the level of the cuts may be accompanied by no reaction in front, normal forward gliding continuing, and when peristalsis begins, it originates at the level of the cuts. Very strong stimulation can bring about a weak peristalsis in front of this. There is then a **polarized ability to spread excitation** around obstacles in the cords. Anteriad spread is possible but difficult.

If the bilateral transections opposite each other are behind the rostral third, head stimulation elicits the abrupt flattening, contraction, and antiperistalsis only in front; there is no reaction behind. To very strong stimulation antiperistalsis will occur in the caudal segment but with a different rhythm than the rostral, and there can be ciliary reversal. In this region even **posteriad spread around obstacles** is difficult. Tail stimulation, however, now causes reaction on both sides of the cut even with moderate intensities—peristalsis occurs in front as well as behind and the forward gliding may be quickened. These results are recapitulated in Table 10.1.

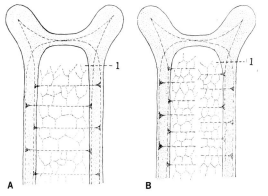

Figure 10.13. The connections between the lateral cords of the nervous system in **(A)** *Lineus* and *Amphiporus* and **(B)** *Oerstedia*. *1* indicates the point of transection (see text). [Friedrich, 1933.]

Table 10.1. POLARIZED SPREAD OF EXCITATION AROUND TRANSECTIONS OF BOTH LATERAL CORDS IN *LINEUS RUBER*. (Friedrich, 1933.)

Region of transection (of both lateral cords at same level)	Moderate Stimulation of Head		Moderate Stimulation of Tail	
	Response anterior to level of cuts	Response posterior to level of cuts	Response anterior to level of cuts	Response posterior to level of cuts
In anterior third of body	Twitch reflex, antiperistalsis, and ciliary reversal	Twitch reflex, antiperistalsis, and ciliary reversal	None	Contraction and peristalsis
Behind anterior third of body	Twitch reflex, antiperistalsis, and ciliary reversal	None	Peristalsis	Contraction and peristalsis

The same requirement for strong stimulation to cross the region of interrupted cords obtains even when the stimuli are applied adjacent to the cuts. If the transections are staggered, one being a centimeter or two behind the other, the results are the same as though they were at the level of the farther one. Clearly the lateral cords do not directly or efficiently communicate with each other. Each can excite all the musculature of its level on both sides, presumably through the body wall plexus. That plexus spreads excitation longitudinally with a rapid decrement. The median dorsal nerve does not provide a direct or efficient communication between brain and regions behind caudally interrupted cords or between tail stimulation and levels in front of more rostrally cut cords.

Perhaps most interesting is the conclusion that the **peripheral nervous system** can spread excitation longitudinally with a rapid decrement and in a polarized fashion that is opposite at the two ends. The more caudal levels of the peripheral nervous system—we can only suppose it is the general plexus and not the dorsal or ventral nerves—can carry impulses adequate to cause peristalsis, and the more rostral levels can carry impulses for antiperistalsis, but the reverse is not possible. Is there a **nerve net?** The definition and criteria have been discussed on p. 462; it is not sufficient to find a plexus of fibers, but neither is it necessary to show through-conduction in all directions. As far as it has been analyzed, the nemertinean body wall system has extremely limited ability to spread excitation longitudinally and at each level can conduct around a cut only for one direction if at all. Nevertheless it can circumvent a cut and transmit a pattern, as for peristalsis. At the most, this system is a borderline case; it seems best to emphasize that its normal function is not to spread excitation diffusely but to conduct from the cord to a limited musculature at that level, like a set of motor nerves.

One experiment was intended to eliminate the possibility of mechanical re-excitation across denervated regions. Threading a worm through a narrow ether chamber and narcotizing only a short segment completely prevents responses of one end from spreading to the other. If the region affected is too long, it does not quite answer the requirement of the experiment; but the better control gained over such mechanical reflex spread is apparent in the results, with their limitation of spread across cuts to certain levels and directions.

Righting is said to occur normally in isolated pieces of the body.

The bdellonemertine genus *Malacobdella* is a **special case** (Eggers, 1935). A commensal in clams, living a quasi-parasitic life and having a posterior sucker (Fig. 10.14), it is interesting that its physiology with respect to locomotion is so reminiscent of *Hirudo*. Decapitation leads to complete quiescence. The sucker holds on even when the cords are cut back to the sucker itself; that is,

V. PHYLOGENETIC COMPARISONS

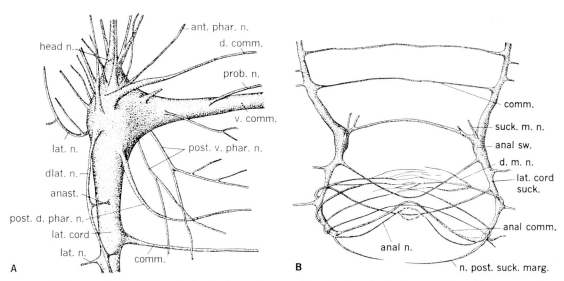

Figure 10.14. *Malacobdella*. **A.** Cerebral ganglion seen from the dorsal side in a methylene blue preparation. **B.** Nerves in the region of the sucker. [Riepen, 1933.] *anal n.*, anal nerve; *anal comm.*, anal commissure; *anal sw.*, anal swelling of the lateral cord; *anast.*, anastomosis between dorsal lateral nerve and lateral cord; *ant. phar. n.*, anterior pharyngeal nerve; *comm.*, commissure between lateral cords; *d. comm.*, dorsal brain commissure; *dlat. n.*, dorsolateral nerve; *d. m. n.*, nerve to dorsal muscle plate; *head n.*, head nerve; *lat. cord*, lateral cord; *lat. n.*, lateral nerve; *n. post. suck. marg.*, nerve of the posterior sucker margin; *post. d.* and *v. phar. n.*, posterior dorsal and ventral pharyngeal nerve; *prob. n.*, proboscis nerve; *suck. m. n.*, nerve to sucker muscle; *v. comm.*, ventral brain commissure.

the entire mechanism needed to maintain that muscular state is in the sucker. Once released, it cannot reattach when isolated. Normally the release of the sucker occurs only when a certain state of contact with substrate of the anterior end has been achieved—indicating a chain-reflex situation as in leeches. The cilia are independent of the brain and do not reverse their beat.

The lateral cord or something associated with it is necessary for **regeneration** in *Lineus socialis*. Most species can only regenerate missing posterior ends and hence can be said to require the brain. This remarkable species can and normally does reproduce by fragmentation. But each fragment must contain a piece of lateral cord; evidently the ubiquitous ganglion cells of the peripheral plexuses are not adequate.

Jenkins and Carlson (1903) estimated the **velocity of propagation** in *Cerebratulus* by the difference in latency of mechanograms to near and far stimulation, as 5–9 cm/sec. Presumably this applies to nongiant and possibly to a multiple relay pathway since it is so slow. The procedure of these workers leaves much to be desired, and the phylum is ripe for modern physiological attack. In view of the great length of many species (one or more meters) a low velocity means a particularly long reaction time.

The literature on taxes and sensory abilities cannot be reviewed here. It will only be noted that photosensitivity is not confined to the anterior end where all the eyes are; according to one report, *Oerstedia dorsalis* in the presence of two equal lights seldom takes a path representing the resultant between them, but weights one over the other (Buddenbrock, 1923).

V. PHYLOGENETIC COMPARISONS

If it were not for the proboscis and the anus, and comparison rested on the neuroanatomy, the present phylum would undoubtedly be united with the Platyhelminthes. The **resemblances to the nervous system of Turbellaria** are numerous and probably not fortuitous. In both, the central nervous system consists of a brain and longitudinal cords which are medullary and do not form ganglia. There are signs in each group of a series of longitudinal cords at several points around the body, communicating by an intervening plexus, with some cords being emphasized and others reduced—the orthogon plan. A series of brain-innervated sense organs is quite similar. Histologically the tendency begun in the flatworms to separate the nerve cell bodies from the neuropile and fiber tracts, placing the former outside, has been continued and extended with more differentiation in the ribbon worms.

Elsewhere the similarities have been noted between the plan of the nervous system of turbellarians and molluscs, especially amphineurans. The same similarities apply to the nemertineans. However, this is a rather general resemblance shared in some degree by all relatively elongated bilateral metazoans, since it is plausible to see common denominators, possibly serial homology, between all main longitudinal nerve cords.

Bibliography

BÖHMIG, L. 1898. Beiträge zur Anatomie und Histologie der Nemertinen. [*Stichostemma graecense* (Böhmig), *Geonemertes chalicophora* (Graff).] *Z. wiss. Zool.*, 64:478-564.

BÖHMIG, L. 1929. Nemertini—Schnurwürmer. *Kükenthal's Handb. Zool.*, 2(3):1-110.

BRINKMANN, A. 1914. *Uniporus*, ein neues Genus der Familie Drepanophoridae Verrill. *Bergens Mus. Aarb.*, 1914(6):1-29.

BRINKMANN, A. 1917. Die pelagischen Nemertinen. *Bergens Mus. Skr.*, NR 3(1):i-vii; 1-194.

BRINKMANN, A. 1927. *Gononemertes parasita* und ihre Stellung im System. *Nyt. Mag. Naturv.*, 65:57-81.

BUDDENBROCK, W. VON. 1923. Untersuchungen über den Mechanismus der phototropen Bewegungen. *Wiss. Meeresuntersuch.*, NF 15(5):1-19.

BÜRGER, O. 1890a. *Beiträge zur Kenntnis des Nervensystems der Nemertinen*. Engelmann, Leipzig.

BÜRGER, O. 1890b. Untersuchungen über die Anatomie und Histologie der Nemertinen. *Z. wiss. Zool.*, 50:1-277.

BÜRGER, O. 1891. Beiträge zur Kenntnis des Nervensystems der Wirbellosen. Neue Untersuchungen über das Nervensystem der Nemertinen. *Mitt. zool. Sta. Neapel*, 10:206-254.

BÜRGER, O. 1895. Die Nemertinen des Golfes von Neapel und der Angrenzenden Meeres-Abschnitte. *Fauna Flora Neapel*, 22:i-xvi; 1-743.

BÜRGER, O. 1897-1907. Nemertini (Schnurwürmer). *Bronn's Klassen*, 4 (suppl.):1-542.

CLAYTON, D. E. 1932. A comparative study of the non-nervous elements in the nervous systems of invertebrates. *J. Ent. Zool.*, 24:3-22.

COE, W. R. 1892-1895. On the anatomy of a species of nemertean (*Cerebratulus lacteus* Verrill), with remarks on certain other species. *Trans. Conn. Acad. Arts Sci.*, 9:479-514.

COE, W. R. 1904. Nemerteans. *Harriman Alaska Exped., Wash. Acad. Sci.*, 11:1-220.

COE, W. R. 1905. Nemerteans of the west and northwest coasts of America. *Bull. Mus. comp. Zool. Harv.*, 47:1-319.

COE, W. R. 1927. The nervous system of pelagic nemerteans. *Biol. Bull., Woods Hole*, 53:123-138.

COE, W. R. 1933. Metameric ganglia connected with the dorsal nerve in a nemertean. *Zool. Anz.*, 102:237-240.

COE, W. R. 1943. Biology of the nemerteans of the Atlantic coast of North America. *Trans. Conn. Acad. Arts Sci.*, 35:129-328.

CORRÊA, D. D. 1953a. Sôbre a locomoção e a neurofisiologia de nemertinos. *Bol. Fac. Filos. Ciên. S. Paulo (Zool.)*, 18:129-147.

CORRÊA, D. D. 1953b. Sôbre a neurofisiologia locomotora de hoplonemertinos e a taxonomia de *Ototyphlonemertes*. *Ann. Acad. bras. Sci.*, 25:545-555.

CRAVENS, M. R. and HEATH, H. 1907. The anatomy of a new species of *Nectonemertes*. *Zool. Jb. (Anat.)*, 23:337-356.

DEWOLETZKY, R. 1887. Das Seitenorgan der Nemertinen. *Arb. zool. Inst. Univ. Wien*, 7:233-280.

EGGERS, F. 1924. Zur Bewegungsphysiologie der Nemertinen. I. *Emplectonema*. *Z. vergl. Physiol.*, 1:579-589.

EGGERS, F. 1935. Zur Bewegungsphysiologie von *Malacobdella grossa* Müll. *Z. wiss. Zool.*, 147:101-131.

FORTUYN, Æ. B. D. 1920. Die Leitungsbahnen im Nervensystem der wirbellosen Tiere. Part I of *Vergleichende Anatomie des Nervensystems*. C. U. Ariëns Kappers and Æ. B. D. Droogleever Fortuyn. Bohn, Haarlem.

FRIEDRICH, H. 1933. Vergleichende Studien zur Bewegungs- und Nervenphysiologie bei Nemertinen. *Zool. Jb. (Allg. Zool.)*, 52:537-560.

HAMMARSTEN, O. D. 1918. Beitrag zur Embryonal-Entwicklung der *Malacobdella grossa*. Inaug. Diss. Uppsala.

HASWELL, W. A. 1893. A monograph of the Themnocephalae. *Proc. Linn. Soc. N. S. Wales (Macleay Mem. Vol.)*, 1893:93-152.

HILTON, W. A. 1921. Nervous system and sense organs. VI. Nemertinea. *J. Ent. Zool.*, 13 (suppl.):49-54.

HUBRECHT, A. A. W. 1875. Some remarks about the minute anatomy of Mediterranean nemerteans. *Quart. J. micr. Sci.*, 15:249-256.

HUBRECHT, A. A. W. 1880a. Zur Anatomie und Physiologie des Nervensystems der Nemertinen. *Natuurk. Afd., Verh. K. Akad. Wet.*, 20(3):1-47.

(HUBRECHT, A. A. W.) 1880b. Hubrecht's researches on the nervous system of nemerteans. *Quart. J. micr. Sci.*, 20:274-282. (Review of Hubrecht, 1880a, anonymous.)

HUBRECHT, A. A. W. 1880c. The peripheral nervous system in Palaeo- and Schizonemertini, one of the layers of the body-wall. *Quart. J. micr. Sci.*, 20:431-442.

HUBRECHT, A. A. W. 1883. Studien zur Phylogenie des Nervensystems. II. Das Nervensystem von *Pseudonematon nervosum*, g. & sp. nn. *Natuurk. Afd., Verh. K. Akad. Wet.*, 22(3):1-19.

Bibliography

HUBRECHT, A. A. W. 1887a. Report on the Nemertea collected by H. M. S. Challenger during the years 1873-1876. *Rep. sci. res. Voyage "Challenger"*, 19(54):1-151.

HUBRECHT, A. A. W. 1887b. The relation of the Nemertea to the Vertebrata. *Quart. J. micr. Sci.*, 27:605-644.

JACKSON, L. W. 1935. Sense organs in *Malacobdella. Nature, Lond.*, 135:792.

JENKINS, O. P. and CARLSON, A. J. 1903. The rate of the nervous impulse in the ventral nerve-cord of certain worms. *J. comp. Neurol.*, 13:259-289.

MCINTOSH, W. C. 1876. On the central nervous system, the cephalic sacs, and other points in the anatomy of Lineidae. *J. Anat., Lond.*, 10:231-252.

MONTGOMERY, T. H. 1897. Studies on the elements of the central nervous system of the Heteronemertini. *J. Morph.*, 13:381-444.

NUSBAUM, J. and OXNER, M. 1913. Die Embryonalentwicklung des *Lineus ruber* Müll. *Z. wiss. Zool.*, 107:78-197.

PANTIN, C. F. A. 1950. Locomotion in British terrestrial nemertines and planarians... *Proc. Linn. Soc. Lond.*, 162:23-37.

PLATE, L. 1922. *Allgemeine Zoologie und Abstammungslehre*. G. Fischer, Jena.

PUNNETT, R. C. 1901. *Lineus. Liverpool mar. Biol. Comm. Mem.*, 7:1-37.

REISINGER, E. 1926. Nemertini. *Biol. Tiere Dtschl.*, 1(7):1-24.

RETZIUS, G. 1902. Weiteres zur Kenntnis der Sinneszellen der Evertebraten. 3. Die Anordnung der Sinneszellen bei den Nemertinen. *Biol. Untersuch.*, N.F. 10:31.

RIEPEN, O. 1933. Anatomie und Histologie von *Malacobdella grossa* (Müll.) I. *Z. wiss. Zool.*, 143:323-424.

SALENSKY, W. 1912. Morphogenetische Studien an Würmern. Zweiter Band. Über die Morphogenese der Nemertinen. I. Entwicklungsgeschichte der Nemertine im Inneren des Pilidiums. *Mem. Acad. Sci. St.-Petersb. (Phys.-Math.), (Zap. Imp. Akad. Nauk)*, (8) 30(10):1-74.

SCHARRER, B. 1941. Neurosecretion. III. The cerebral organ of the nemerteans. *J. comp. Neurol.*, 74:109-130.

STIASNY-WYNHOFF, G. 1923. Die Entstehung des Kopfes bei den Nemertinen. *Acta Zool., Stockh.*, 4:223-240.

STIASNY-WYNHOFF, G. 1926. The Nemertea polystilifera of Naples. *Pubbl. Staz. zool. Napoli*, 7:119-168.

THOMPSON, C. B. 1901. *Zygeupolia litoralis*, a new heteronemertean. *Proc. Acad. nat. Sci. Philad.*, 53:657-739.

THOMPSON, C. B. 1908. The commissures and the neurocord cells of the brain of *Cerebratulus lacteus*. *J. comp. Neurol.*, 18:641-661.

CHAPTER II

Pseudocoelomate Phyla: Acanthocephala, Rotifera, Gastrotricha, Kinorhyncha, Nematoda, Nematomorpha, and Entoprocta

Summary	598	C. The Peripheral Nervous System	616
I. Acanthocephala	**599**	1. Receptors	616
		2. Neuromuscular relations	620
II. Rotifera	**601**	3. The enteric nervous system	622
III. Gastrotricha	**604**	**VI. Nematomorpha**	**622**
IV. Kinorhyncha	**604**	**VII. Entoprocta**	**624**
V. Nematoda	**605**	Classified References	626
A. Introduction	605	Bibliography	627
B. The Central Nervous System	605		
1. Gross anatomy	605		
2. Finer anatomy	609		

Summary

Acanthocephalans are endoparasites and have a poorly developed nervous system and sense organs, notable for eutely and sexual dimorphism. There is a cerebral ganglion and a genital ganglion in the male, each giving off nerves but without a proper longitudinal cord or anterior nerve ring. In a representative genus, *Hamanniella*, 80 cells comprise the cerebral ganglion and 30 the genital. The receptors are based on a plan of one or two nerve fibers ending in a papilla with a central pit.

Rotiferans are minute and commonly active free swimmers. The nervous and sensory apparatus is rather good for the grade of construction of the phylum. There is a brain and several outlying ganglia with connectives between them and the brain, but no real longitudinal cord. *Epiphanes* as an example, has 183 nerve cell nuclei in the brain, 34 and 23 in the next two largest ganglia. There is a clear separation of rind and core but no neuroglia cells can be recognized. Several cell masses and cell types manifest a certain degree of histological advancement. Ciliomotor and muscular innervation are probable. Receptors of several kinds are abundant; the most complex sense organs are hollow, retractable innervated tentacles. Free nerve endings with deep cell bodies are indicated.

Gastrotrichs are quite similar to rotifers in respect to the nervous and sensory equipment, although a pair of lateral nerve cords are distinct and bear nerve cells along their course.

Kinorhynchs are quite different from the two immediately preceding groups. Segmentation is evident in the ganglia. The main nervous masses are in close contact with the epidermis. There is a brain, an anterior circumoral nerve ring, a median ventral, a median dorsal and a pair of lateral nerve cords with metameric ganglia. Receptors are moderately well developed but not as well as in the preceding groups.

Nematodes have many points of special interest, nervously as otherwise. With a high degree of apparent centralization, three-fourths of all nerve cells being concentrated in an anterior circumenteric ring of ganglia, they have at the same time the most completely orthogonal system among the Bilateria. The peripheral nervous system consists of (a) a number of longitudinal cords, called nerves for their poverty of nerve cells, probably basically eight—dorsal, ventral, and a pair each of subdorsal, subventral and lateral nerves, (b) irregular dorsoventral commissures between these, (c) six cephalic nerves, (d) a few special ganglia and nerves in the tail, and (e) two enteric or sympathetic systems, one anterior and one posterior. There are no peripheral plexuses or nets, barring a doubtful description for the gut. There is no nerve tree and but little branching of the main nerves, many of which are perhaps purely sensory or motor. The whole nervous system consists of about 200 neurons in several parasitic and free living forms, but there is a suggestion of a good many more in some marine species. In *Ascaris* and some others, a fixed and constant number of neurons—162—comprise the central nervous system, each of definite form and position. Only a few of these are internuncial, the majority being sensory or motor; most are unipolar and several peculiarities of cytology and sheaths are notable. Anastomosis of neurons is highly developed, many large connections having constant positions. The nerve cells, the course of their fibers and this type of connection being mapped nearly exhaustively, *Ascaris* is the most completely known animal, neuroanatomically. There is little neuropile. Half a dozen types of special sense organs each have a single nerve cell, which is bipolar and deep, with a long distal process, usually unbranched. Muscle innervation is by (a) long fleshy processes of the muscle cells which come right to the nerves and (b) short collateral branches of the motor axons reaching a short distance from the nerves. Widely separate muscle cells of the dorsal "field" may show rhythmic action potentials in phase, as though driven from a common source, while at the same time those of a different "field," the ventral, are showing another rhythm, and are therefore supplied from a different source. Some of the cells and fibers are large and opportunities for central and peripheral physiologic problems appear favorable.

Nematomorphs exhibit some adaptations for their parasitic juvenile stages and some for their free living adult life. Generally there is an anterior circumenteric swelling, the cephalic ganglion, and a single midventral medullary cord. The histology is quite different from nematodes and more conventional; a rind and core can be discerned. There are numerous large, so-called giant cells. Neuroglia cells are conspicuous. A peripheral plexus is dubious, recalling nematodes and kinorhynchs and possibly others. Receptors are identified with great uncertainty and therefore it is impossible to say whether the sensory complement is rich or poor.

Entoprocts are sessile, stalked and very simple in histological grade. There is a single clear ganglionic mass, regarded as ventral on the present interpretation of the body plan; hence there is said to be no supraesophageal or cerebral ganglion. There is no circumenteric ring and no definite nerve cord. A peripheral plexus is described and a continuity between polyps in a colony. Sense cells with bristles, sensory pits and lateral papillae, supplied by several deep-lying sense cells, provide an extremely modest receptive apparatus.

I. ACANTHOCEPHALA

These lowly worms, with their diagnostic armed proboscis, are exclusively endoparasitic. It is not surprising that the sense organs are poorly developed, especially in variety, and with them the nervous system. Among the features of special interest are eutely (constant nuclear number) and a marked sexual dimorphism extending to the nervous system.

Harada (1931) made a special study of the nervous system and obtained some successful methylene blue stains; nevertheless the older general monographic descriptions were not replaced, the chief ones being Greeff (1864), Hamann (1891), Kaiser (1893), Brandes (1898), and Kilian (1932). The nervous system (Fig. 11.1) consists of a single cerebral ganglion with nerves from it, and in the male a genital ganglion with nerves from it. Since a digestive tract is lacking, no circumenteric ring can be recognized. The ganglia and nerves are internal and thinly sheathed, sometimes with a muscle layer called a retinaculum.

The **cerebral ganglion** has a small fibrous core surrounded by large polygonal cells, packed closely (Fig. 11.2). In a specimen of *Hamanniella microcephala* of about 12 cm, the ganglion measured $90 \times 160 \times 220$ μ. The cells number 80 in this species; the fibers leaving the ganglion, 60. In *Macracanthorhynchus hirudinaceus* these figures are 86 and 56; in *Bolbosoma turbinella* the cells number 73. In the first two genera and in *Gigantorhynchus echinodiscus*, the number of nerves given off is 8; in the last named genus, 11. The largest cells are in the two ends of the ganglion and reach 25×50 μ with 5–6 μ fibers; most are bipolar or unipolar. A few small cells lie in the fibrous core. Two binucleate nerve cells are said to be always present.

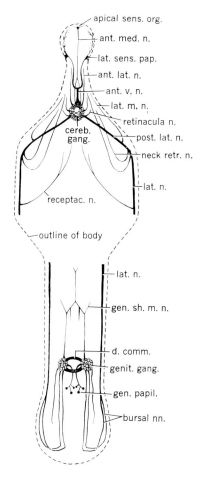

Figure 11.1. Acanthocephala. *Echinorhynchus gigas*, plan of the nervous system. [Brandes, 1898.] *apical sens. org.*, apical sense organ; *ant. lat. n.*, anterior lateral nerve; *ant. med. n.*, anterior median nerve; *ant. v. n.*, anterior ventral nerve; *bursal nn.*, bursal nerves; *cereb. gang.*, cerebral ganglion; *d. comm.*, dorsal commissure of genital ganglion; *genit. gang.*, genital ganglion; *gen. papil*, genital sensory papillae; *gen. sh. m. n.*, nerve to muscles of genital sheath; *lat. m. n.*, nerve to lateral muscle of body wall; *lat. n.*, lateral nerve; *lat. sens. pap.*, lateral sensory papillae; *neck retr. n.*, nerve to neck retractor muscle; *post. lat. n.*, posterior lateral nerve; *receptac. n.*, nerve to receptaculum; *retinacula n.*, nerves to retinacula muscle sheath.

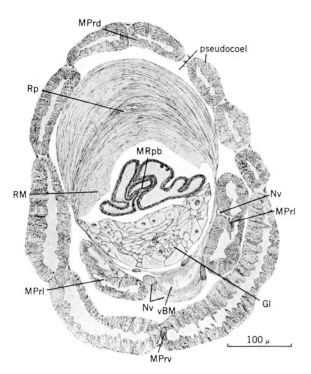

Figure 11.2.

Acanthocephala. *Hamanniella*, transverse section through the cerebral ganglion (*Gl*), in which nerve cells are seen around the outside, nerve fibers in the center. Most of the rest of the tissue seen is muscle; the body wall surrounding the section is not shown. [Kilian, 1932.] *MPrd*, *MPrl*, and *MPrv*, dorsal, lateral, and ventral protrusor muscles of the receptacle into which the proboscis is withdrawn; *MRpb*, retractor muscle of the proboscis; *Nv*, nerve fibers in peripheral nerves; *RM*, wall of receptacle sack; *Rp*, dorsal wall of receptacle of the proboscis; *vBM*, midventral receptacle muscle.

The **genital ganglion** is paired, present in males only, and situated in the penis base. Its structure resembles the cerebral and it may even be larger, though comprising fewer cells—30 in *Hamanniella*, 14 or 15 in *Bolbosoma*. In the former, 7 large nerves are given off from each ganglion; in the latter genus, only 3. In addition, a dorsal and ventral commissure (8 fibers and 15 fibers, respectively, in *Hamanniella*) connect the two ganglia, completing a nerve ring.

The **nerves** have been followed to certain muscles and body wall regions. For the details, the original descriptions of Rauther (1930c), in Kükenthal, or Meyer (1933), in Bronn, may be consulted. The principal nerve pair are the posterior lateral; neither these nor the others give substantial ground for speaking of nerve cords or even of connectives. Some authors saw a few fibers of this nerve pass to the genital ganglion but Harada believed there is no connection between the cerebral and genital ganglia. The remaining nerves in *Hamanniella* and the genera similar to it are: a median anterior nerve and a ventral anterior nerve to the muscles and papillae of the proboscis, a pair of anterior laterals to the lateral protrusor muscles, and a pair of middle laterals to the muscle of the receptaculum. The posterior laterals already mentioned send branches to the dorsal and ventral protrusors and the anterior body wall muscles and continue to the hind end in the longitudinal muscle layer. In the female these nerves split into two on each side and give off genital branches; there is no ganglion, though a single constant cell on one branch has been called a single-celled ganglion. In the male also there are four nerves in the hind end—dorsal and ventral, right and left—derived from the posterior lateral nerves. Besides these four, the genital ganglion has a series of nerves passing forward along the genital tract and posteriorly into the bursa, chiefly to sensory end bulbs in the papillae.

Apparently the only **receptors** are variations on a common plan of one of two nerve fibers ending in a papilla with a central pit. These are usually regarded as tactile but Kilian supports a chemical function; there is no determinant physiological evidence as yet. Typically there is an anterior terminal and two lateral receptors on the proboscis, eight papillae on the bursa copulatrix in the wall in two symmetrical rows, and a dorsal and a ventral organ of similar form on the penis. All sensory endings are supposed to go to these organs. Motor nerve fibers are said to break up into ramifications on the muscle and often push into the muscle in its nuclear region.

II. ROTIFERA

Minute size and frequently great activity are among the points of interest of these animals in the present study. They are, for animals of their grade of construction, quite well provided with nervous apparatus and often extremely well provided with sensory apparatus. A correlation between the development of these structures and the habits of diverse species of rotifers would certainly be instructive. Constancy of number, position, and form of the cells is a feature of the group; *Epiphanes* (= *Hydatina*) *senta* has about 280 nerve cells—or, better, nuclei—out of an approximate nuclear total of 959 (Martini, 1912). With 183 nerve cells in the brain, 34 and 23 in the next two largest ganglia (those of the mastax and foot), and 40 in the peripheral nervous system, we see that there is a pronounced emphasis on the cerebral ganglion. The peripheral nervous system may include several small outlying ganglia. All ganglia are well separated from the epidermis; a series of nerves run freely across the body cavity.

The first detailed accounts of the nervous system were given by Zelinka (1888, 1891) and Hlava (1905). Hirschfelder (1910) carefully observed the histology of the brain and its cell types, Martini (1912) gave a thorough account of *Epiphanes* and Nachtwey (1925) of *Asplanchna;* Nachtwey had some success with methylene blue. Among other workers, Peters (1931) and Stossberg (1932) deserve special mention in the present connection. The best review is in Bronn (Remane, 1929–33).

The **brain** is relatively voluminous and is dorsal to the mastax or the esophagus (Fig. 11.3). It is said by some authors to be clearly sheathed, and this is denied by others, for different species. Neuroglia cells have not been distinguished. In some forms a muscle or the duct of the retrocerebral organ penetrates the brain. There is a distinct separation of cell rind and fiber core (Fig. 11.4). The rind is unequally thick and dense, being sometimes thin or lacking ventrally, but this differs among species. **Four cell types** are recognized, ranging from 3–9 μ (see Nachtwey or Remane); they are distributed unevenly, and Hirschfelder defined several cell masses. These cell groups are not striking or general but the level of development is indicated by *Asplanchna*. In this form there is a dorsolateral bulge, with most of the type *a* or largest cells (mainly superficial and not next to the core); type *b* are sparse

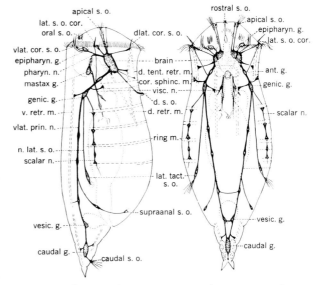

Figure 11.3.

Rotifera. Scheme of the nervous system of the Monogononta, in lateral and ventral views. The mastax covers part of the brain in the ventral view. [Remane, 1929–1933.] *ant. g.*, anterior ganglion of the main ventral connective; *apical s. o.*, apical sense organ; *caudal g.*, caudal ganglion; *caudal s. o.*, caudal sense organ ("antenna"); *cor. sphinc. m.*, coronal sphincter muscle; *dlat. cor. s. o.*, dorsolateral coronal sense organ; *d. retr. m.*, dorsal retractor muscle; *d. s. o.*, dorsal sense organ ("antenna"); *d. tent. retr. m.*, dorsal tentacle retractor muscle; *epipharyn. g.*, epipharyngeal ganglion; *genic. g.*, geniculate (genu) ganglion; *lat. s. o. cor.*, lateral sense organ of the corona; *lat. tact. s. o.*, lateral tactile sense organ (lateral "antenna"); *mastax g.*, mastax ganglion; *n. lat. s. o.*, nerve to lateral sense organ; *oral s. o.*, oral sense organ; *pharyn. n.*, pharyngeal nerve (=mastax connective); *ring m.*, ring muscle; *rostral s. o.*, rostral sense organ; *scalar n.*, scalar nerve; *supraanal s. o.*, supraanal sense organ; *vesic. g.*, vesicle ganglion; *visc. n.*, visceral nerve; *vlat. cor. s. o.*, ventrolateral coronal sense organ; *vlat. prin. n.*, ventrolateral principal nerve (= main ventral connective); *v. retr. m.*, ventral retractor muscle.

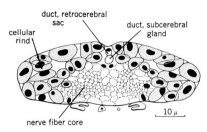

Figure 11.4. Rotifera. *Eosphora digitata*, cross section of brain. [Redrawn from Hirschfelder, 1910.]

and mostly in the ventral rind; type *c* are small and densely chromatic in basic dyes and are commonest next to the fiber core; type *d* are also small and chromatin-rich and are chiefly posterior, though found also in mid and anterior regions of the brain in small numbers. Among each type there are thought to be both unipolar and bipolar cells but the lack of specific nerve stains hampers precise specification of numbers and connections of processes. The number of cells in the brain and other regions is given in Table 11.1.

Scattered ganglion cells occur just outside the brain and send processes to it and to muscle or skin.

The **fiber core of the brain** consists of a tangle of fairly coarse nerve fibers, not yet followed into the cells or to their destinations.

The **nerves of the brain** are numerous but mostly small. The chief ones are the pharyngeal nerve or mastax connective, the ventrolateral principal nerve or, sometimes called the main ventral connective, in some species a dorsolateral principal nerve, a salivary gland nerve, and nerves to the corona, apical field, and nuchal tactile organs.

The **mastax ganglion** is generally the next largest (Fig. 11.3). It is unpaired and is ventral to the complex feeding apparatus. Though it is sometimes called the subesophageal ganglion, and completes a circumenteric ring with its connectives to the brain, it should not be regarded as equivalent to the ganglion of that name in articulates and others but as a local system of cells for the mastax. It is missing—or perhaps is part of the brain—in seisonids. Male rotifers, with a reduced mastax, generally lack this ganglion. As a rule it is a loose swarm of cells within the wall of the mastax, with no distinct core. In the complex malleate mastax of Brachionides, Epiphanidae, and Euchlanidae there is enough differentiation in the ganglion to designate a fibrous core. Details of the topographic variations among species may be found in Remane.

The **nerves of the mastax ganglion** are chiefly two. A so-called pharyngeal nerve is actually a connective to the brain. The visceral nerve on each side goes to a variety of viscera and probably supplies a plexus in the wall of the mastax, where scattered ganglion cells are found. Sometimes other, small nerves go to nearby muscles.

The **caudal ganglion** is an unpaired structure in the foot or hind end of footless species. It is lacking in Seisonidae and some footless Monogononta. Even less well demarcated than the mastax ganglion, the caudal ganglion is a loose collection of cells, often tending to be dispersed in small clusters. It supplies glands, caudal sense organs, and some muscles.

There are many **smaller inconstant ganglia** in some forms. A relatively large one is the epipharyngeal ganglion of *Synchaeta* with eleven nuclei. Along the ventrolateral principal nerve in particular, knots of nerve cells are prone to occur, usually at a branching. Thus in *Epiphanes* there is an anterior ganglion, a geniculate ganglion, an

Table 11.1. NUMBERS OF NERVE CELL NUCLEI IN MAIN GANGLIA OF THREE SPECIES OF ROTIFERS.

Species	Brain	Mastax ganglion	Caudal ganglion	Author
Epiphanes senta	183	34	23	Martini (1912)
Asplanchna priodonta	about 225	about 50	48	Nachtwey (1925)
Synchaeta triophthalma	about 223	38	20	Peters (1931)

Plus about 200 in other ganglia, scattered cells, and sense organs; total for whole body, about 850.

II. ROTIFERA

anterior ovarial, a posterior ovarial, a lateral vesical, and a medial vesical ganglion. Scattered single cells lie on the stomach, intestine, protonephridia, and bladder.

The nerves can only be summarized here (Fig. 11.3). A wealth of detail on topographic anatomy in many species is known; Remane provides a summary and a new, synthetic nomenclature. The largest is the ventrolateral principal nerve, a mixed sensory and motor bundle with nerve cells along its course and a tubular, sheathed structure unlike all other nerves. It gives off many branches and has a ventral commissure. Other nerves from the brain and mastax ganglia have been mentioned.

Motor innervation of two kinds has been discussed: ciliomotor and muscular. The former function is suggested for coronal nerve fibers ending near ciliated cells; no physiologic proof is at hand but the suggestion seems very probable, since cilia stop and start and alter direction. Martini thought that broad, plasmatic innervation processes of the muscle cells pass toward the nervous system as in nematodes, but Zelinka (1888), Nachtwey (1925), and others consider these processes to belong to nerve cells. Each transverse muscle band of the body is said to be supplied by one ganglion cell.

Receptors are numerous and diverse (Fig. 11.5). Of simple construction, they are commonly unicellular and never of many cells. Often the specialization appears to be merely an organelle within a cell, for example pigment spots and membranelles. The most complex sense organs are certain dorsal and lateral antennae with one or several sense cells in a hollow tentaclelike extension of the body surface, retractable by muscles. **Photoreceptors** are indicated sometimes by pigment flecks in a sense cell, sometimes by refractile bodies. Doubtful cases are common where the pigment is believed to be a metabolic product without sensory significance. Although Plate believed the photoreceptors are inverted pigment-cup ocelli with separate pigment cell and sense cell, it is more likely that both are the same cell or syncytium. Such eyes occur in three places: in or on the brain, in or near the lateral coronal sensory field, and in the apical or frontal field. Usu-

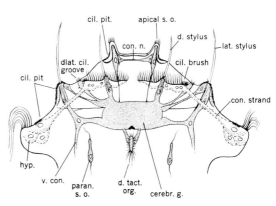

Figure 11.5. Rotifera. *Synchaeta pectinata*, dorsal view of anterior end showing nervous apparatus and sense organs. [Remane, 1929–1933.] *apical s. o.*, apical sense organ; *cerebr. g.*, cerebral ganglion; *cil. brush*, ciliated brush; *cil. pit*, ciliated pit; *con. n.*, nerve connecting apical sense organs; *con. strand*, strand connecting hypodermal cushions; *dlat. cil. groove*, dorsolateral ciliated groove; *d. stylus*, dorsal stylus; *d. tact. org.*, dorsal tactile organ (dorsal "antenna"); *hyp.*, hypodermal cushion; *lat. stylus*, lateral stylus; *paran. s. o.*, paranotal sense organ; *v. con.*, ventral connective or principal nerve.

ally but one or two of these are found in a given species. A general skin sense to light is indicated by irregular and imprecise locomotion in response to light (Viaud, 1940, 1943). Species with good eyes are supposed to move along the resultant between two lights.

Remane classes a diverse assortment of presumed receptors as **ciliary sense organs.** Five general types with intermediates may be distinguished: membranelles of long fused cilia, movable or not; sensory hairs—long, stiff, and single; sensory bristles; rods in patches; ciliary grooves with short, beating cilia. The last is thought to be chemosensory; the others, perhaps mostly tactile—or rheoreceptive. Combinations of these types are variously distributed in dorsal, lateral, caudal, and paranotal sensory fields, and in supra-anal, coronal, and pharyngeal sense organs. Remane summarizes a wealth of detail for many pecies.

So-called **palpal organs** are fingerlike, sometimes movable and retractile structures, without sensory cilia. An innervation is rarely described and their sensory nature is uncertain. **Statocysts** have been several times reported but Remane shows serious reasons for doubting this identification in each case.

Free nerve endings are reported in the epidermis in regions without specialized sense cells (Stossberg, 1932). This is remarkable for such simple animals, for it means the nerve cell body is far removed from its presumed primitive position near the sensory end.

III. GASTROTRICHA

These free-living microscopic forms strongly resemble rotifers in the nervous system and sense organs, as in other ways. We owe nearly all our information to the pioneer observations of Zelinka (1889) and the detailed studies of Remane (1929, 1935), which he has reviewed in Kükenthal and Bronn.

There is a large **brain** near the anterior end, consisting of paired masses on the sides of the pharynx (Fig. 11.6). A dorsal commissure connects the two sides; there is no ventral connection and therefore no complete circumenteric ring. A pair of **lateral nerve cords** extend the length of the body, with ganglion cells along their course. There is said to be no trace of a peripheral plexus —recalling nematodes.

Receptors are of many kinds. Tufts of long cilia and single hairs or bristles which may be motile or stiff are presumably tactile or rheoreceptive. They may be supplied by a nerve fiber coming from a ganglion cell in the brain—an astonishing centralization that is seen only now and then in the highest invertebrates. The lateral

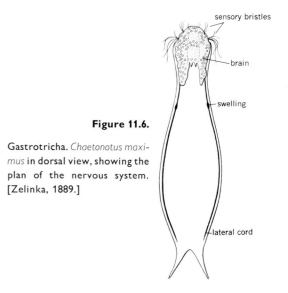

Figure 11.6.

Gastrotricha. *Chaetonotus maximus* in dorsal view, showing the plan of the nervous system. [Zelinka, 1889.]

sense organs consist of a pair of ciliated pits, variously modified and specialized and perhaps chemoreceptive. A few species possess pigment ocelli in the form of brain cells containing red pigment granules. Unicellular statocysts are suggested, but on equivocal grounds.

IV. KINORHYNCHA

Another group of microscopic forms more or less closely related to the two preceding, they differ markedly with respect to nervous anatomy. The segmentation of the body is reflected in the ganglia and the principal nervous structures are closely in contact with the epidermis, quite unlike rotifers. Our knowledge is chiefly due to the work of Zelinka (1908) and is reviewed by Remane (1929, 1935) in Kükenthal and Bronn.

The **brain** is a broad, ribbonlike nervous ring encircling the anterior pharynx or the base of the mouth cone. The fibers are mainly circular; ganglion cells are few or absent ventrally. They are concentrated mainly as two rings on the anterior and posterior edges of the broad ring of nerve fibers and a lesser nerve cell ring in the middle of the latter and on its inner side. The posterior cell ring is produced into 9 lobes.

Medullary **nerves** are given off anteriorly into the epidermal cushions at the bases of the scalids. Posteriorly the nerve ring gives off a median ventral, a median dorsal, and paired lateral **nerve cords** with segmental ganglia. The ventral cord is the largest, and runs in contact with the midventral epidermis. In each segment or zonite a ganglion is found, the cord being cell-free for a short

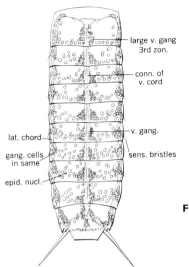

stretch between successive zonites (Fig. 11.7). The same arrangement, on a smaller scale, appears to be repeated middorsally and laterally. Scattered ganglion cells are reported in the epidermis between the nerve cords.

Sense organs may not be as well developed as in the two preceding groups. Eyes are simple cups with aggregations of red pigment granules and a refractile body. Sensory bristles occur in longitudinal rows on the trunk. Each is innervated by a sensory nerve cell in the epidermis.

Figure 11.7. Kinorhyncha. *Pycnophyes communis*, ventral view of Hyalophis stage, with the anterior end withdrawn. [Zelinka, 1928.] *conn. of v. cord*, connective of ventral nerve cord; *epid. nucl.*, epidermal cell nuclei; *gang. cells in lat. chord*, ganglion cells in lateral chord; *large v. gang., 3rd zon.*, large ganglionic swelling on the ventral cord in the third zonite; *lat. chord*, lateral chord of thickened, syncytial epidermis; *sens. bristles*, sensory bristles; *v. gang.*, ganglionic swelling on ventral cord in ordinary zonite.

V. NEMATODA

A. Introduction

The main ganglia and nerves, despite the few cells and fibers, were known to Meissner in 1853. The most important workers on general anatomy have been Schneider (1863), Bütschli (1874), Hesse (1892), Voltzenlogel (1902), Goldschmidt (1908, 1909), Martini (1916), Chitwood and Chitwood (1933, 1940), and Wessing (1953). Useful reviews are provided by Rauther (1930a) and the Chitwoods (1940). The finer anatomy has attracted very few authors. However, Goldschmidt devoted years to a minute examination of *Ascaris* and has left a series of papers (1903, 1908, 1909, 1910, and others on nonnervous structure) whose detail is surpassing and, carefully judged on all available evidence, is probably correct in most points. This is all the more remarkable because essentially he did not employ special methods but laboriously correlated serial sections and whole mounts. He could not see the very fine processes—the connections in the sparse neuropile are still unknown—but the strange anastomotic connections which present such a puzzle for physiology were mapped by Goldschmidt in detail. Excessively polemical, he rejected the results of others, sometimes with good reason but sometimes, as far as we can see today, incorrectly. The only studies based on selective nerve stains are those of Apáthy (gold chloride) and Deineka and Filipjev (methylene blue); we accept the Chitwood statement (1940) that his own description based on this stain in 1930 should be ignored. There is very little physiological information as yet but the nematodes suggest themselves as highly interesting material for microelectrode stimulation and recording.

B. The Central Nervous System

1. Gross anatomy

The basic plan of the nematode nervous system can be seen in Fig. 11.8. A circumenteric nerve ring, some distance behind the anterior end, forms a decided concentration. From it, a series of longitudinal nerves pass backward; possibly their basic number is 8, but species with 4, 6, 10, and 12 are known. The 4 universal nerves are: a dorsal, a ventral, and a pair of laterals; most species have also a pair of subdorsal and a pair of subventral nerves (the latter 2 pairs are spoken of collectively as submedian nerves and the terms laterodorsal and lateroventral are also used). The few species in which these are not reported need

total of 12. (The last case is based only on *Oxyuris* according to Martini, 1916, and is not clearly established.) Six nerves (*Ancylostoma*) are formed where there are 2 pairs of laterals and no submedians. Irregular commissures connect these nerves. To this basic outline is added a set of cephalic sensory nerves, local ganglia and nerves in the anal region, and an anterior and a posterior enteric system.

The circumenteric nerve ring and the ganglia close to it can be **considered a central nervous system.** The ring is more than a large commissure since it includes connections between neurons, whereas most of the ganglia are only clusters of 2–12 cells without neuropile. The whole is a natural unit, with the ganglia being collections of cell bodies localized near the origin of the nerves. Apart from the greater size of the ganglia laterally and ventrally the central nervous system is nearly radially symmetrical. The **ganglia** are listed in Table 11.2 with details applicable to *Ascaris*. Besides the nerve ring, there are 3 symmetrical and 2 asymmetrical **commissures** in this genus (Fig. 11.9). The largest of the symmetrical ones is best called the anterior ventral-to-lateral commissure (although usually and confusingly called ventrolateral or lateroventral); shortly behind it is the posterior ventral-to-lateral commissure. Both connect the ventral nerve with the lateral nerve behind the ring. The symmetrical anterior dorsoventral (= ventrodorsal) commissure connects the dorsal and ventral nerves, passing for much of its course in front of the ring and joining the anterior ventral-to-lateral commissure near its ventral end. On the right side only, a dorsoventral commissure I passes in front of the ring and a dorsoventral commissure II (= oblique dorsoventral) passes diagonally backward from the rostral end of the ventral nerve to the dorsal nerve some distance behind the ring.

It should be noted that the term commissure is used although connecting nonidentical structures and even on the same side, as an extension of its application to the circularly running cross connections between longitudinal cords in these and other animals. The theoretical justification is the proposition that all points at the same level along such longitudinal cords have a kind of equiva-

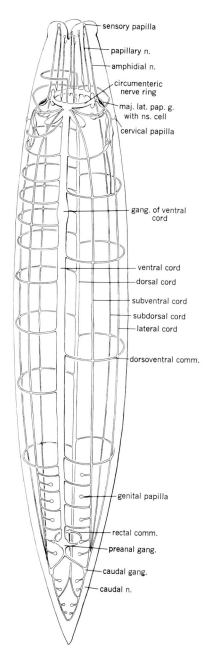

Figure 11.8. Nematoda. *Ascaris*, plan of the nervous system, from the ventral side. [Based on Goldschmidt, 1910 and Voltzenlogel, 1902.] *comm.*, commissure; *gang.*, ganglion; *maj. lat. pap. g. with ns. cell*, ganglion of major lateral papillary nerve, with neurosecretory cell; *n.*, nerve.

re-examination. The lateral nerves may be represented by a single pair or by 2 pairs, dorsolateral and ventrolateral, in which case the total is 10; or they may be represented by 3 pairs, dorsolateral, ventrolateral, and lateral, making a

Table 11.2. ENUMERATION OF THE NERVE CELLS OF *ASCARIS LUMBRICOIDES*. Ostensibly all nerve cells (central, peripheral, and visceral) are included; where a careful count is not available the estimated value is indicated by "appr." From Goldschmidt (1908b) and Chitwood and Chitwood (1940).

	Ganglia, or Peripheral Structure	Nerve Cell Count	Cell Numbers Assigned by Goldschmidt	Number of Sensory Neurons
	A	B	C	D
	Central Nervous System			
1	Subdorsal papillary ganglia	7 × 2 = 14	57-63 rl	all
2	Subventral papillary ganglia	7 × 2 = 14	50-56 rl	all
3	Lateral papillary ganglia	4 × 2 = 8	64-67 rl	all
4	Amphidial ganglia	11 × 2 = 22	68-78 rl	all
5	Internal lateral ganglia	11 × 2 = 22	23-29, 30-32, 49	none
6	External lateral ganglia	13 × 2 = 26	33-45 rl	one
7	Circumenteric nerve ring	4	46 rl, 47, 48	none
8	Dorsal ganglion	2	19, 20	none
9	Subdorsal ganglia	2 × 2 = 4	21, 22 rl	none
10	Ventral ganglion	33	1-15, 17, 18 rl, 16	none
11	Retrovesicular ganglion	13	79-91	none
		Total 162		
	Tail Structures			
12	Preanal ganglion	11		four
13	Preanal sensory cells (Bursal papillae) ♂	appr 15 × 2 = 30		all
14	Lumbar ganglia	appr. 6		all (?)
15	Postanal sensory cells (incl. phasmidial ganglion) ♂	7 × 2 = 14		all
16	Latero-caudal nerves	3 × 2 = 6		all (?)
	Enteric Systems			
17	Esophageal "sympathetic"	appr. 17		(?)
18	Rectal "sympathetic"	2 × 2 + 4 = 8		(?)
19	Grand total	appr. 254		appr. 119

rl: right and left

lence, at least in the unmodified orthogonal pattern. The term nerve is used in the present group for some structures which might merit the term cord. Although cell bodies occur along the ventral nerve and in some species in the dorsal nerve, they are few and far between and in addition tend to clump, leaving long cell-free stretches; the other nerves are generally cell-free. Another reason for using "nerve" is that the special term "chord" applies in nematodes to thickenings of the epidermis in the dorsal, ventral, and lateral lines and between these in four subsidiary lines in some species. Nevertheless, the belief is tenable that the main longitudinal "nerves" represent morphological features as fundamental as nerve cords in other animal groups.

Nematodes other than ascarids have been much less studied. The gross anatomy is fairly well known for *Siphonolaimus* (Strassen, 1904), *Ancylostoma* (Looss, 1905), *Hexamermis* (Meissner, 1853; Rauther, 1906), *Oxyuris* (Martini, 1916), *Camallanus* and *Cucullanus* (Törnquist, 1931), *Cephalobellus* (Chitwood and Chitwood, 1933), *Rhabditis* (Chitwood and Wehr, 1934; Chitwood and Chitwood, 1940), and *Spironoura* and *Oesophagostomum* (Chitwood and Chitwood, 1940). In addition Filipjev (1912) provided a number of details on *Metoncholaimus* and *Paroncholaimus*.

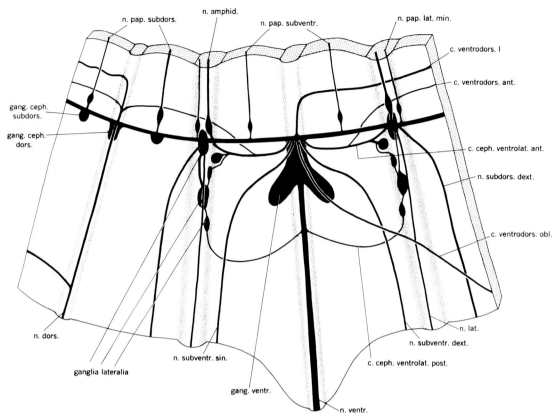

Figure 11.9. Nematoda. *Ascaris*, region of the circumenteric ring viewed from inside, opened by an incision to the right of the middorsal line and laid flat. Anterior is above. [Modified from Goldschmidt, 1909.] *c. ceph. ventrolat. ant.*, anterior ventrolateral cephalic commissure; *c. ceph. ventrolat. post.*, posterior ventrolateral cephalic commissure; *c. ventrodors. ant.*, anterior ventrodorsal commissure; *c. ventrodors. I*, first (asymmetrical) ventrodorsal commissure; *c. ventrodors. obl.*, oblique ventrodorsal commissure; *gang. ceph. dors.*, dorsal cephalic ganglion; *gang. ceph. subdors.*, subdorsal cephalic ganglion; *gang. ventr.*, ventral ganglion; *ganglia lateralia*, several parts of the lateral ganglion; *n. amphid.*, amphidial nerve, with ganglion at the base; *n. dors.*, dorsal nerve or cord; *n.lat.*, lateral nerve or cord; *n. pap. lat. min.*, minor lateral papillary nerve, and ganglion; *n. pap. subdors.*, subdorsal papillary nerve, and ganglion; *n. pap. subventr.*, subventral papillary nerve, and ganglion; *n. subdors. dext.*, right subdorsal nerve; *n. subventr. dext.*, right subventral nerve; *n. subventr. sin.*, left subventral nerve; *n. ventr.*, ventral nerve or cord.

"The general features of the nervous system in all the forms studied [are] very similar even though they represent considerably diverse groups" (Chitwood and Chitwood, 1940). Differences in the central nervous system lie chiefly in the lesser subdivision of the lateral and ventral ganglia in other species than in *Ascaris*, in the frequent paired condition of the ventral nerve anteriorly, and in the absence of the subdorsal ganglia possessed by *Ascaris*. In the peripheral nervous system, besides details of the anal region, there is divergence in the number of longitudinal nerves, as already mentioned, and in most species there have not been seen any dorsoventral commissures in the midbody, whereas they are numerous in *Ascaris*. Not only are the differences between various nematodes, parasitic and free-living, rather trivial in gross anatomy, but as we shall see there is a remarkable agreement in the number and distribution of the neurons. However, Dr. Chitwood permits me to say that marine free-living species, which are the most neglected, have many more nerve cells than the better-known species and may have better-developed nervous systems.

The evidence from the finer anatomy makes it

likely that the papillary nerves are purely sensory and very likely the lateral nerves are also, including the ventro- and dorsolateral. The motor innervation comes from the nerve ring and the median and submedian nerves, and these nerves may be purely motor except perhaps in species with median and submedian bristles or scattered papillae on the body.

2. Finer anatomy

The neuronal composition and histological and cytological details are known almost entirely from *Ascaris*. Cell for cell, *Parascaris equorum* (= *Ascaris megalocephala*) corresponds with *A. lumbricoides*, with trivial exceptions in the form characteristic of certain cells. A number of other parasitic as well as free-living species which have been studied in this respect agree fairly closely (including *Oxyuris*, *Rhabditis*, *Cephalobellus*, *Ascaridia*, *Oesophagostomum*); cell constancy has been repeatedly confirmed for various limited parts of the nervous system (Pai, Schönberg, Frenzen). *Anguilla aceti* has 279 nerve cells; *Rhabditis longicauda* close to 200. *R. anomala*, according to Wessing (1953), has 162 nerve cells in the anterior central nervous system, exactly as in *Ascaris*, during its early postembryonic life, but later amitotic divisions produce an incon-

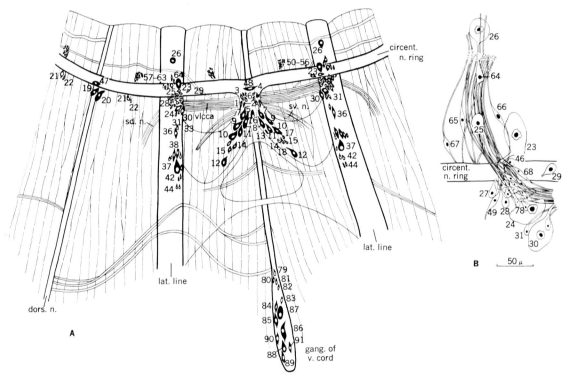

Figure 11.10. Nematoda. Whole mount showing all nerve cell bodies of the central nervous system of *Ascaris lumbricoides*, including sensory cells of the anterior body region (in the ganglia of the papillary and amphidial nerves). **A.** The cylindrical body wall of the pertinent level of the body has been opened by a subdorsal incision and unrolled after removing the esophagus and lips. The figure is only schematized by the approximation of the course of the muscle fibers, the spreading out of nerve cell bodies that lie over each other, the omission of differences in staining of the cells, and an enlargement of the cells relative to the scale of the whole. Most of the cells are marked with their individually identifying numerical designations; the others can be found marked in the original monograph. **B.** The internal and posterior internal lateral cephalic ganglia and the major and minor lateral papillary nerves of the left side are shown enlarged; a section of the fibers has been omitted, as indicated by the dotted lines, to bring the uppermost cell into the picture. *circent. n. ring*, circumenteric nerve ring; *dors. n.*, dorsal nerve; *lat. line*, lateral line; *sd. n.* and *sv. n.*, subdorsal and subventral nerves; *vlcca.*, ventrolateral cephalic commissure. [Redrawn from Goldschmidt, 1908.]

stant number of nuclei; Martini (1916) could directly identify many of the cells of *Oxyuris curvula* with those of *Ascaris*.

(a) Nerve cell bodies. The ganglia consist of from 2 to 33 nerve cells (Fig. 11.10); some are collections purely of sensory cell bodies. Goldschmidt's enumeration of the cells of the anterior region of *Ascaris* (and therefore of the whole central nervous system) assigns a sensory function to 62 cells, motor to 77, and internuncial to 23—altogether 162 cells. The interneuron figure in particular is not entirely certain because his main criterion for interneurons was simply that, with his methods, Goldschmidt could not see direct connections with muscle cells. It can be accepted, however, that in these animals there are **far fewer internuncial than either motor or sensory neurons.**

Motoneurons and interneurons are mostly unipolar, some are multipolar and a few bipolar. Large cells reach 70 μ in diameter (according to Deineka, 150–200 μ, but flattened), medium sized cells are 30–50 μ, and small cells are below 15 μ. A variety of distinctive shapes occurs: funnel-, club-, spherical-, and pear-shaped unipolars, amphora-shaped bipolars, and aranoid multipolars. **Every cell has been numbered and can be recognized** by the combination of these characters, its position, and the destination and connections of the axons. There is a nearly perfect bilateral symmetry; two cells are only on the right and three are median. Goldschmidt provides a complete catalogue and drawings of each ganglion and every cell in it (Fig. 11.10), based on a large number of specimens and with but rare anomalies.

The ganglion cells exhibit some remarkable **cytological features** (Fig. 11.11). In large cells three zones are distinct in the cytoplasm, the middle one being continuous with the axoplasm of the neurite; sometimes but not generally there is a change in consistency in the axoplasm and hence a kind of axon hillock. Most peculiar, however, is the conspicuous radial striation of many cells, especially the large ones but also some small ones, as in the bursal nerve. This has presented a classical puzzle since it was noted by Leuckart in 1876. It seems now to be a variety of the neuroglial invasion seen in many large nerve cells; but this explanation does not clear up the question of the significance of invasion or of its regular radial pattern in this case. Details of the distribution among cells, of the varieties and degrees of its expression, and of the distribution of the radial striae in the cell can be found best in Goldschmidt's memoir of 1910. He and others were convinced that the glial fibers enter the nerve cell and become continuous with the neurofibrils by a T-shaped junction. Some cells have a curious cytoplasmic zone which looks like a capsule a little distance from the nucleus, bounding the inner coarse-foamed cytoplasm; this capsule is also continuous with the neurofibrils and hence appears to be a boundary against which the radial striae abut in those cells having both. Neurofibrils are conspicuous but have quite different aspects in different cells. Rather than repeat the descriptions—which vary with the method—it seems best to refer to them and await newer electron microscope work, contenting ourselves with the hope that such work will not, as so often, overlook the wealth of characteristic variation among cells established by the older workers and will identify the cells examined. The same can be said of the other inclusions, especially the tigroid or basophilic (Nissl) substance; this is supposed to be present in many of the cells in *Ascaris*, in some sparse and diffuse, in others coarse and zoned. The corresponding cells left and right are always in the same condition with respect to Nissl granulation, though adjacent cells commonly are not. This is a nice example of the value of identified cells for it tells us that the variations observed are not meaningless.

(b) Nerve fibers and connections. Each nerve contains a constant number of fibers from one to a few dozen. Differences in diameter are fixed, but most are thick enough to permit tracing in serial sections with ordinary cytological stains. With a few exceptions, sensory fibers are thin; motor and commissural fibers, thick. Actual values are difficult to find but it appears that the thicker neurites are about 8 μ in diameter. Abrupt changes in diameter are not infrequent. Neurofibrils are described in conflicting ways but rarely as absent; most intriguing are the differences sup-

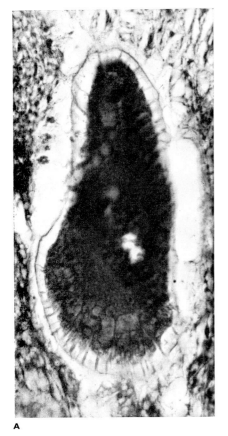

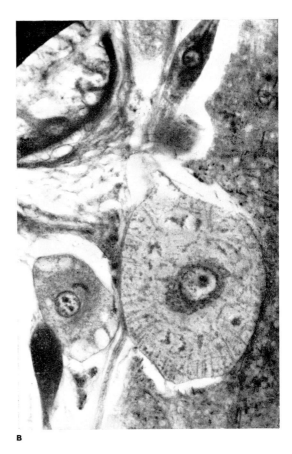

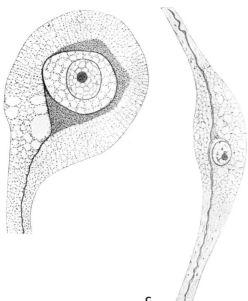

Figure 11.11. Nematoda. Nerve cells in *Ascaris*. Note radial striae, dense zone around nucleus, vacuolar aspect, coarse inclusions, and fibrils. **A.** Cell no. 23, chrome-hematoxylin. The space outside cell is attributed to shrinkage. **B.** Large cell is no. 23, small cell is no. 25, Apáthy's gold chloride. **C.** Cell no. 25 on left, cell no. 51 on right, Apáthy's gold chloride. [Goldschmidt, 1910.]

posed by Goldschmidt to be characteristic of certain fibers: some have fibrils concentrated into an axial thread, others as a loose bundle, and others as a hollow cylinder. Most fibers have no individual glial covering. They are often packed tightly in a cross section (Fig. 11.12), forming polygons and without apparent sheath tissue except around the whole nerve. Here and there

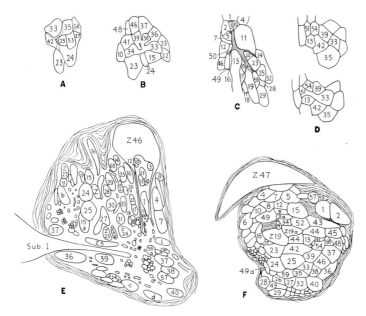

Figure 11.12.

Nematoda. Parts of the nerve ring of *Ascaris* in section, showing individual nerve fibers and their anastomoses. The numbers are arbitrary but designate distinct fibers followed for some distance in reconstructions (see Fig. 11.14). In the plane of section **A**, fiber 24 has short transverse bridges connecting it with both 27 and 42. (The corresponding place is seen in Fig. 11.14, **A** in quadrat 30:25, where the thick fiber numbered 25 at the left is the same fiber as that numbered 24 here; the fiber numbered 32 there is here called 42 and that numbered 34 is here called 27.) In the same section fiber 35 communicates with 42, and 23 with 33. **B** is also found in Fig. 11.14, **A** in quadrat 50:49, where fiber 33 is labeled 43 in the right margin, 34 is 23, 38 is 32, and 39 of the section is the same as 16 in the right margin. **C** shows a great many fibers connected in one plane and in addition a change in stainability of the axoplasm; the stippled, branching fiber was in the adjacent sections fiber 14; the same locus is in the upper left corner of Fig. 11.14, **B**, where the fiber is numbered 1. **D** shows adjacent sections to indicate the abruptness of the anastomosis of 33 and 51. **E** is chosen from a region of the ring near a point of nerve entry; the fibers are loosely arranged with fine branches, neuropile, and glia (not shown) between them. This appears to be the maximum extent of neuropile in the central nervous system of *Ascaris*. **F** is from a region of the ring far from a nerve entry; the fibers are packed in a mosaic with little glia or neuropile; there is some segregation of small fiber bundles peripherally. Z = cell; *Sub. I* — nerve entering. [Goldschmidt, 1909.]

they are loosely packed, chiefly in the limited regions of neuropile development. Exceptional fibers, like the thick process of cell 47 in the dorsal nerve, have a considerable sheath. Groups of like-sized fibers tend to occur, smaller fibers generally on the outside.

The fate of the nerve fibers centrally—that is, apart from muscle innervation—seems to be twofold: anastomotic connection of large branches and neuropile termination of small. **Anastomosis** is said to be common and occurs in specific and constant places according to Goldschmidt. No one had disagreed with him and Deineka, his bitter antagonist on many other points, agrees that fusion occurs widely, as does Chitwood. It is extremely difficult to accept Goldschmidt's conclusion that in the central nervous system "everything is continuously connected with everything else" (1909, p. 324), directly or indirectly, and that "complete plasmatic continuity prevails in the central organ" (1909, p. 341). These statements are based, however, on broad anastomoses, illustrated repeatedly (Figs. 11.12, 11.14), from methylene blue stained whole mounts (Deineka) as well as serial sections, and insisted upon emphatically in a series of papers. They cannot be dismissed or ignored, as has been done for half a century.

Goldschmidt argued in favor of the **neuron doctrine**, at a time when it was being bitterly debated, on the ground—quite correct as we see it now—that the continuity-contiguity question is not central but rather the question whether the nervous system is made of cells and only cells. Since his time many more cases have come to light where neurons fuse, but these are regarded as secondary departures from a typical condition and not as disproof of the reality of the neuron. Assuming that the extensive anastomosis in the nematode nervous system is correct, they acquire a special interest physiologically but no cherished dogma need thereby be upset.

Anastomoses are of different kinds. Parallel fibers may simply flow together—two or even four forming one. A fiber may split and one branch turn across many nearby fibers to join a distant one. Several fibers may fuse into a plasmatic mass and this then give off fibers in all directions. Repeated bridges between the same two fibers are found. A given side branch may go to two others or even five others. Only a few fibers run far in the nerve ring without receiving a bridge. The bridges tend to be concentrated in certain regions. The nerve ring is like a continuous net but not diffuse or without rules; rather it has definite pathways. Many of these points are shown by Figs. 11.14 and 11.15 and detailed exemplification, with cell numbers, may be found in Goldschmidt (1909).

The **neuropile** and the connections it provides are known, in contrast, only slightly. Goldschmidt did not use selective nerve stains and could not visualize the finer branches; Deineka has reported briefly on the terminal ramifications of sensory central processes and dendrites of motor cells which interweave—apparently by contact only—in the nerve ring (Fig. 11.13 A, B). There is clearly a limited amount of neuropile and it probably has no specialized regions or texture but is extremely simple.

Dendritic processes of central neurons present a special problem. Deineka (1908) is the only worker to describe them. He sees heavy, capitate processes on some and much-branched fine twigs on others (Figs. 11.13, 11.17). They may be at the end of a short or a long stem process; when long, the stem process may bear short, fine collaterals or be smooth. Goldschmidt denied all Deineka's cell types, allowing that only one did correspond to a single cell, number 48. In this and every other connection Goldschmidt unmercifully attacked Deineka who, together with his teacher, Dogiel, replied in kind. In the present case there does not seem sufficient ground to reject Deineka's description of structures which Goldschmidt could not be expected to have seen by his methods; his main ground for rejection—that in the supposed location of bipolars there are only unipolars—is not decisive, for Deineka did not carefully assign locations; the stem pro-

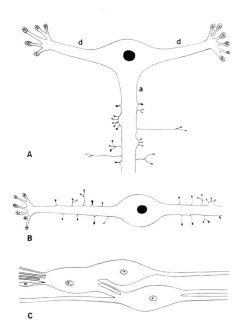

Figure 11.13. Nematoda. Types of motor cells of *Ascaris*. **A.** Cell with two large dendrites (*d*) with knobby endings, and an axon (*a*) with side branches that lead to motor endplates. **B.** Cell with one knob-bearing dendrite; both dendrite and axon have branches with motor end plates. [Deineka, 1908.] **C.** The connection between three nerve cells in the ventral cord. [Goldschmidt, 1908.]

cess of the dendrites might have been overlooked or regarded as part of the soma by Goldschmidt. It would be simpler if Deineka's work were not in other respects much more vulnerable. But the only remarkable feature in respect to the dendrites is the short, little-branched knobbed type and this, on the evidence, is at least as likely to be correct as not.

A few **general statements about connections** can be made, abstracting from the charts. Based on the anastomotic connections, 2-neuron arcs exist, as from L10 to B7, L13 to B4, or L14 to B8 in Fig. 11.14. Interneurons are not primarily intermediaries between sensory and motor neurons; rarely do sensory fibers connect directly, by this means anyway, with cells believed to be interneurons. The latter are then better thought of as elaborating motor accompaniments of the most direct response. Sensory input in these animals must be of an undifferentiated sort since there is much fusion of sensory fibers. There is a great difference between cells in the complexity of their

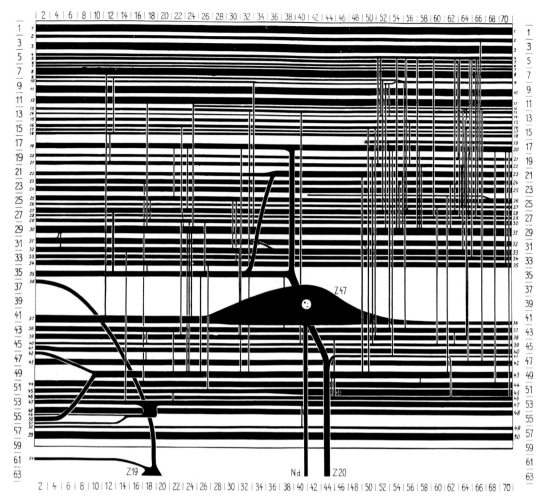

Figure 11.14. Nematoda. Diagrammatic plans of the fiber composition and connections in two parts of the circumenteric nerve ring of *Ascaris megalocephala*. **A.** The region of origin of the dorsal cord. **B.** The region of origin of the ventral cord. In each case the left-right axis of the plan is the flattened circumferential direction of a piece of the ring; rostral is up. Reconstructed from tracings from serial sections, simplified by untangling overlying or twisting fibers; it is therefore without significance whether a fiber is shown above or below another. In **A** every fiber entering or leaving the ring is shown and the number corresponding to the cell of origin is given. In **B** 70 fibers—of the 134 that actually enter or leave—are shown. Of the fibers in the ring, all the larger, most of the medium sized, and but few of the small ones have been traced and shown.

connections; cell 46 is fantastically involved whereas others have characteristically minimal connections. Motor cells, so called because they have muscle connections, are connected in a way that makes it likely that they act also in the capacity of interneurons, exciting other neurons and not only muscle.

Figure 11.15 summarizes Goldschmidt's view of the contrast between the central organization of typical invertebrates and that of nematodes. Clearly, if this view is even approximately correct, the question of the **role of nerve impulses** versus graded, decremental spread of excitation has special force in this group. Before the recognition of the decremental form of activity as a normally useful one, it would have been expected that any impulse arising in this kind of system would spread everywhere; perhaps this is one reason that these relatively very carefully documented histological claims have been glossed

V. NEMATODA   B. Central Nervous System   2. Finer anatomy

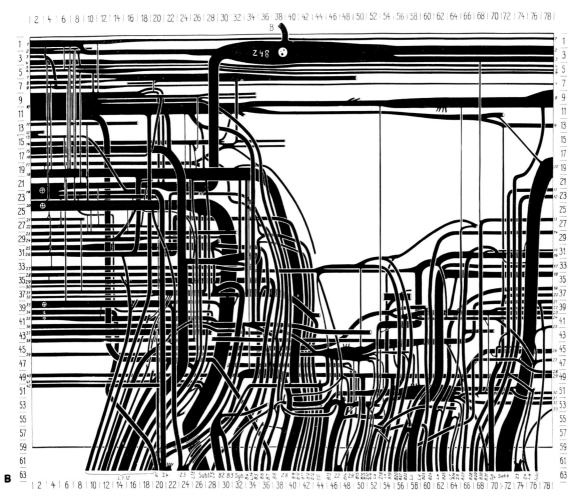

The numbers on the abscissa represent the serial numbers of the 5 μ sections. The odd numbers on the ordinate are for convenience in locating areas. The small numbers at right and left are merely serial numbers of the fibers as they occur on the plan, from above downward. Fibers identified with their cells are given the cell number Z; fibers known simply to come from the anterior lateroventral commissure are designated L; those of the ventral nerve are B; those of the subdorsal and subventral nerves are Sub. The ⊕ represent sites of muscle innervation by the approximation of processes of nearby muscle cells to a thin place in the sheath of the nerve ring. [Goldschmidt, 1909.]

over or forgotten in the interim. Just how far conduction blocks, low safety factors, local responses, facilitation, and the like may help to account for the functioning of this bizarre system—or, reciprocally, how much functional anatomy can be inferred by tracing excitation initiated in controlled ways—would seem to be experimentally approachable questions with modern technics.

(c) Glia and coverings of the central nervous system. Surrounding the circumenteric nerve ring is a mass $2^1/_2$ times as wide as the ring itself, made of lamellae of specialized connective tissue. Little cytoplasm and few nuclei are involved. This tissue is continuous with a similar covering of the lateral ganglia, extending along the lateral line. A somewhat similar lamellated tissue encases the main nerves. Underneath it the circumenteric ring has a dense capsule; this is the product of four fused cells which also send processes into

and in some species, eyes. Statocysts are not known.

Sensory **bristles** are possessed by some groups usually in two submedian rows, and may be scattered widely on the body. They are probably distinct from the adhesive and the stilt bristles, which are regarded as being not sensory. They are cuticular projections jointed at the base and each contains one nerve fiber which is the distal process of a deep lying sensory neuron.

More important are the papillae. The **labial papillae** vary in distribution from a basically hexaradiate pattern of 18. The only satisfactory account is that of Goldschmidt (1903) on the anterior papillae of *Ascaris*; Rauther (1930a) and Chitwood and Chitwood (1940) have useful reviews. The most important features for the present purposes are the following: each sense organ is supplied by a single nerve fiber whose bipolar cell body lies far away, in the papillary ganglia (Table 11.2). This is one of the notable characters of nematodes. The terminal is unbranched and may have swellings and thinnings and dark staining segments; it may end in a bulb underneath the cuticle or in a sharp point projecting through a canal in the cuticle—each is characteristic of certain papillae (Fig. 11.16). The ascending fiber is accompanied by long narrow processes of two nonnervous cells: an inner, so-called supporting cell which surrounds the fiber at the end, and an outer, so-called escort cell. The former comes from a deeply lying cell body and its long thin process running alongside the sensory fiber was mistaken by Deineka for a nerve fiber, although he was puzzled by the lack of connections of its cell except with sensory cells and others like itself. The six papillary nerves consist of the long distal processes of the sensory neurons together with the supporting-cell processes; they differ from other nerves, which are immediately beneath the epidermis, in traversing the pseudocoelom. The papillae account for 16 of the 36 neurons of the papillary ganglia; the others have not been traced to special sense organs and perhaps end freely in the lips; the possibility that some fibers are motor to muscles of the bristles has been proposed.

Genital papillae in the pre- and postanal region

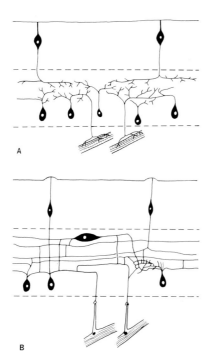

Figure 11.15. Contrast in organization of conventional nervous system, represented by the upper diagram, and nematode system, represented by the lower, according to the view of Goldschmidt (1909). The main feature is the apparent abundance of anastomoses between nerve fibers in the nematode and their supposed rarity in others. There is much less ground for the diagrammatic view of the neuropile shown in a small region of the nematode figure, since special nerve stains have been of little help in the central neuropile. A system like the lower figure might be conceived to work if some of the anastomoses are thin or unable to propagate a spike (see Nonsynaptic Interaction, Chapter 4).

the sensory nerves. The majority of ganglion cells have individual capsules of concentric glial processes sharply set off from the surroundings. A minority, the larger cells, have in addition the radially penetrating glial processes already referred to (Fig. 11.11). Supporting and escort cells of the sensory neurons are mentioned below.

C. The Peripheral Nervous System

1. Receptors

Nematodes are rather well supplied with specialized sense organs as well as free nerve endings. The former include sensory bristles, papillae (labial, cervical, and anal), amphids, phasmids,

V. NEMATODA   C. Peripheral Nervous System   1. Receptors

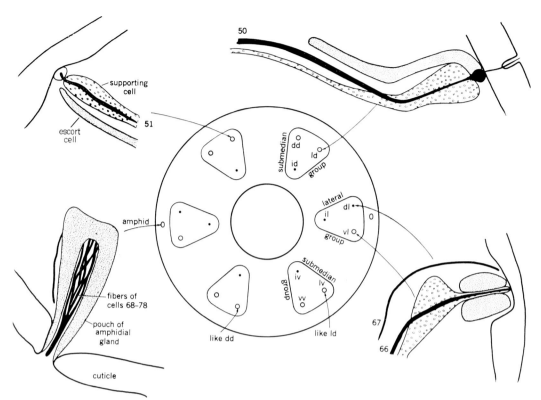

Figure 11.16. Nematoda. Receptors of the anterior end of *Ascaris*. The central figure is a plan of the rostral tip viewed head-on, with the groups of receptors and the names of the three individual organs of each group. The four drawings show the forms of the outer ring of receptors and the two amphids and give the numerical designations of the nerve cells whose distal sensory processes innervate them. These cells are seen also in Fig. 11.10B. [Based on Goldschmidt, 1903.] *dd*, dorsodorsal; *id*, innerdorsal; *ld*, laterodorsal; *dl*, dorsolateral; *il*, innerlateral; *vl*, ventrolateral; *iv*, innerventral; *lv*, lateroventral; *vv*, ventroventral.

(Fig. 11.17) are quite similar in composition but not known as well. Rarely, there may be one or two or three sensory nerve fibers in one papilla, and occasionally a single nerve fiber sends branches to 2 or, rarely, to 3 papillae. Deineka figured branches from many of the distal axons of papillary sense cells, that go down into the muscle and end in terminal enlargements and other, shorter branches that end in the subcuticle; in addition to the two main axons, proximal and distal, he often saw broad, short anastomotic processes from one sensory cell body run into a similar process of another cell (Fig. 11.18). Although these curious findings need confirmation, there is little reason to dismiss them as artifacts. Genital papillae are few in the female but numerous in the male, in which the deeply lying cell bodies and initial processes make up the bursal nerve. The central processes pass from this through the transverse lateroventral commissures to the ventral nerve; it is not known whether they must reach the circumenteric nerve ring or can find functional contact with motor neurons in the ventral nerve, but the latter seems more likely. In the better-known forms there are almost no sense organs between the cephalic end and the anal region. Only one or two pairs of **cervical papillae** or deirids occur (Fig. 11.8), each supplied by a single sensory cell that lies in the lateral ganglia; in a few species numerous cervical papillae are scattered along this region and in others there are bristles, as already mentioned. The papillary sense organs have traditionally been regarded as tactile receptors but this appears highly questionable.

Amphids, which earlier authors called lateral

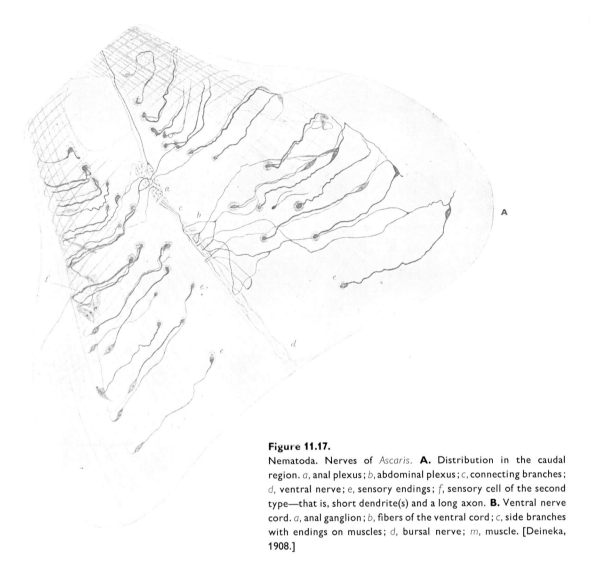

Figure 11.17.

Nematoda. Nerves of *Ascaris*. **A.** Distribution in the caudal region. *a*, anal plexus; *b*, abdominal plexus; *c*, connecting branches; *d*, ventral nerve; *e*, sensory endings; *f*, sensory cell of the second type—that is, short dendrite(s) and a long axon. **B.** Ventral nerve cord. *a*, anal ganglion; *b*, fibers of the ventral cord; *c*, side branches with endings on muscles; *d*, bursal nerve; *m*, muscle. [Deineka, 1908.]

organs or lateral papillae, are glandulosensory organs represented by a single pair on the anterior end (Fig. 11.16). Each consists of a pouch opening to the outside, being the dilated end of the long duct of the amphidial gland; the sensory axons of the eleven cells of the amphidial ganglion are said to penetrate the wall of this pouch, traverse its cuticularly lined cavity as a bundle, and end by fusing into a single fiber lying in the mouth of the cavity. Amphids are supposed to be chemoreceptors but there is no theory of the **intimate relation of receptor and gland,** a problem already encountered in the Nemertinea and again in higher groups. These organs are reduced greatly in parasitic as compared with free-living species. Not only is the composition of the organ different from that of the papillae but the ganglion composed of its bipolar sensory neurons is not directly connected with the nerve ring as are the papillary ganglia; its central processes pass by way of the anterior lateroventral commissure into the ventral ganglion and thence to the nerve ring (Figs. 11.8, 11.9). Perhaps this bespeaks a phylogenetic origin behind the ring—where the amphidial gland itself still is.

Phasmids (= ghost thing, one of the many terms introduced by the crusading nematologist Cobb) are similar to amphids but are located on

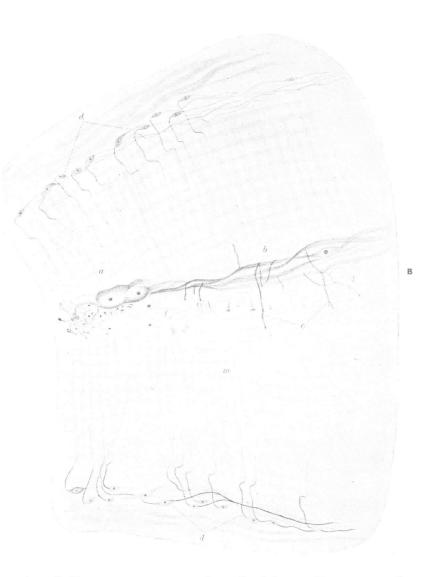

the tail. Fewer axons penetrate the wall of the duct of the phasmidial gland—two in female *Spironoura*, probably a few more in the male. Three sensory cells in the female and five in the male comprise the phasmidial ganglion in this genus and send central axons forward through the lateral caudal nerve to the lumbar ganglion and from there through the anolumbar commissure to the ventral nerve. As with amphids, phasmids are presumed to be chemoreceptive, on no particular grounds. They vary not only between the sexes but among the higher taxonomic categories.

Eyes are present as a single pair lateral to the esophagus, in a number of free-living nematodes, especially marine species. The eyes are remarkable in structure and a resemblance to the cerebral eyes of rotifers has been noted. No nerve supply has been described.

Free nerve endings occur particularly on the anterior end.

2. Neuromuscular relations

The mode of innervation of muscle, known since Schneider (1866), is another of the classical peculiarities of this group. The **muscle is said to go to the nerve** instead of the nerve going to the muscle. Long processes of the muscle cells seeking out the main nerves have been abundantly confirmed by many workers; the most detailed study is that of Martini (1916). In addition, the motor nerve fibers have short side branches (Fig. 11.18), according to the only methylene blue studies, those of Deineka (1908) and Filipjev (1912).

The **muscle fibers** of the body wall, which are exclusively longitudinal, are arranged in a single layer attached throughout their length to the epi-

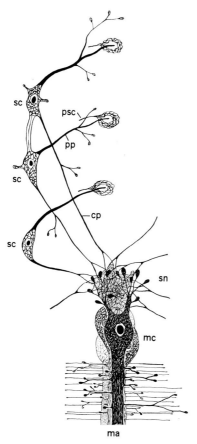

Figure 11.18.
Nematoda. Diagram of the relationship between sensory and motor nerve cells in *Ascaris*. [Modified from Deineka, 1908.] *cp*, central process of sensory cell ending in a network of sensory fibers, *sn*; *ma*, axon of motor nerve cell with collaterals which bear motor end plates; *mc*, motor cell; *pp*, peripheral process of sensory cell, with side branches; *psc*, process of another sensory cell, cut off; *sc*, sensory cell.

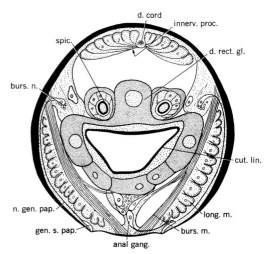

Figure 11.19. *Ascaris*, male. Cross section in the cloacal region. [Rauther, 1930.] *anal gang.*, anal ganglion; *burs. m.*, bursal muscles; *burs. n.*, bursal nerve; *cut. lin.*, cuticular lining of the cloaca; *d. cord*, dorsal cord; *d. rect. gl.*, dorsal rectal gland; *gen. s. pap.*, genital sensory papilla; *innerv. proc.*, innervation processes of muscle cells reaching toward dorsal nerve; *long. m.*, longitudinal muscles; *n. gen. pap.*, nerve to genital papilla; *spic.*, spiculum.

dermis; they are spindles lying side by side and fitted together in constant number and position. The contractile myofibrils are confined to the outer part of the muscle cell and the inner part is a considerable mass of cytoplasm. From this there is given off approximately at the level of the nucleus, one or several broad processes that reach across the pseudocoel to the middorsal or midventral nerve, the submedian nerves, or the circumenteric nerve ring. Here they come into intimate contact with nerve fibers (Figs. 11.19, 11.20). Martini describes a considerable degree of regularity about the number of such innervation processes, their anteroposterior position and their relations to each other, as between more medial and more lateral muscle cells. Some muscle cells have 3 or 4 innervation processes but whether they go to the same or different nerve fibers is not known. They tend to be concentrated at certain levels of the body. Sections through the nerve ring or dorsal or ventral nerve show—here and there, in constant positions—pronounced thinnings of the sheaths and close apposition of muscle process and nerve fiber (indicated in Fig. 11.14).

Deineka and Filipjev saw many short simple collateral **branches projecting laterally from the nerve** fibers and ending in club-shaped terminals (Fig. 11.20). Deineka found that these collaterals are not confined to the loci of apposition of muscle innervation processes; rather, some ended on these muscle processes and others left the sheath of the nerve and ended at a little distance, perhaps on adjacent muscle fibers. Virtually all the authors have been convinced that there was an exchange of fibrils between nerve and muscle—neurofibrils being continuous with an elaborate system of fine fibrils traversing the sarcoplasmic part of the muscle cell. The illustrations offered

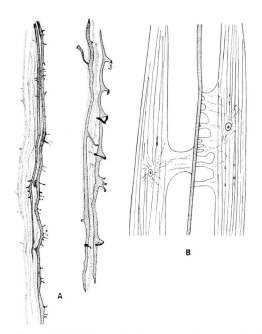

Figure 11.20. Nematoda. Innervation of muscle. **A.** Ventral cord of *Ascaris*, showing the very short axonal collaterals to muscles. [Deineka, 1908.] **B.** *Oxyuris*. The muscle cells send innervation processes to the nerve; this is the particularly broad innervation of the seventh ventral exterior muscle cell. [Martini, 1916.]

on this point are not unequivocal or convincing and the exchange cannot be accepted as probable. Due to the confinement of neuromuscular junctions to the immediate vicinity of the main nerves, nematodes should be favorable material for electron microscope study of these relations.

The larger species also offer **physiological opportunities.** Besides the central problems raised by anastomosis, already referred to, the nervous control of muscle is interesting in an animal with a single layer of solely longitudinal muscle and very few motor neurons. Each such neuron appears to make contact with at least several muscle cells; whether the reverse is true is not known. It is said that all movements are exclusively dorsoventral. Goldschmidt recognized undulatory, pendular, boring, and sexual movements in *Ascaris*. Stauffer, considering free-living species as well, recognizes six types of locomotion: (a) swimming by serpentine undulations, (b) gliding by similar movements but with the aid of some purchase on a substratum, (c) crawling by alternate lengthening and shortening with sinuous curving during the latter, (d) crawling but holding the body straight throughout the cycle, gripping the substratum by annulae of the body wall, (e) walking on stiff bristles like a caterpillar, with waves progressing forward, and (f) looping like a leech using adhesive bristles. End-to-end thrashing is common in small species but is at best inefficient, if it is locomotor.

Baldwin and Moyle (1947) introduced an isolated nerve-muscle preparation in *Ascaris* and reported spontaneous alternating contractions and relaxations as well as an effect of loading on the strength of contractions. Because of the location of neuromuscular junctions, the preparation is bound to include considerable portions of a main nerve and therefore the possibility of reflex connections. Joseph in 1882 said that by local electrical stimulation over the submedian or lateral nerves he could obtain retraction or protrusion of the lips in *Ascaris*. While his report gives little basis for confidence, let alone any distinction between reflex, motor nerve, and direct muscle stimulation, it points to a possibly useful approach.

Jarman (1959) has initiated the exploitation of this group with the intracellular recording electrode by penetrating the sarcoplasmic bag of muscle cells. He finds a resting potential and superimposed depolarizing spikes occurring fairly rhythmically at 2–5 per second, under his conditions; by penetrating two muscle cells, comparisons were made of the synchrony of their action potentials. Cells of the dorsal and ventral muscle fields are not in synchrony while cells of the same field are. We may infer separate motor innervation for the two fields, and hence that an impulse cannot spread throughout the nervous system even if there are anastomoses. There must be at least a common pacemaker, if not a common motoneuron, for the muscle cells of one field. Cutting experiments showed that the nerve cords are necessary for the longitudinal spread of the excitation. Norton and De Beer (1957) tested some drugs on preparations of *Ascaris*: atropine and tubocurarine have no direct effect but block the response to acetylcholine; piperazine not only antagonizes the acetylcholine response but paralyzes the worm, although it is still responsive to direct electrical stimulation; several substances including succinylcholine elicit contractions.

3. The enteric nervous system

Two enteric systems are known, one in the esophageal and one in the rectal region. The former is more complex (Fig. 11.21). There are three longitudinal nerves, a dorsal and two sublateral, extending almost the length of the esophagus. Two or three commissures join them and a plexus is said (Goldschmidt) to occupy the interstices, but this seems somewhat dubious considering the methods used and the small number of total cells—17 or 18 in *Ascaris*. The cells are enumerated for several species: 29 in *Spironoura affine*, 27 in *Angusticacum holopterum*, 20 in *Oxyuris equi* (see Chitwood and Chitwood, 1938). The course or destination of the nerve fibers is unknown but the connection with the central nervous system in *Ascaris* is by a single fiber from the subventral nerve.

The rectal sympathetic system is a term applied to the anorectal commissures and their associated ganglia and nerves. From the anal ganglion at the end of the ventral nerve a pair of commissures passes dorsally around the rectum; a dorsal rectal ganglion where they meet gives rise to an unpaired median caudal nerve extending posteriorly into the tail. This is the basic plan and is all that is present in females (*Spironoura*, Chitwood and Chitwood, 1940). In males of this genus there are additional clumps of nerve cells along the commissure and a nerve is given off laterally to a ganglion for the spicule. Each of these ganglia consists of two to five cells. Further details may be found in the Chitwoods' (1940) work.

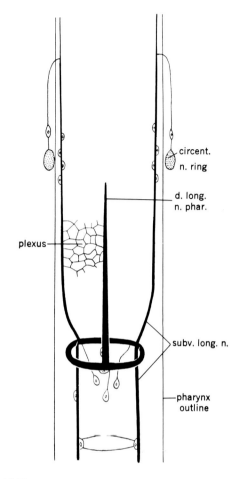

Figure 11.21. Nematoda. Schematic representation of the visceral nervous system. [Goldschmidt, 1910.] *circent. n. ring*, circumenteric nerve ring (main central nervous mass); *d. long. n. phar.*, dorsal longitudinal nerve of the pharynx; *plexus*, nerve plexus on wall of pharynx between dorsal and subventral nerves—shown only in one area; *subv. long. n.*, subventral longitudinal nerve of the pharynx.

VI. NEMATOMORPHA

These extremely attenuated worms are remarkable from the present point of view in being free-living as adults, though parasitic in juvenile stages, and in showing some adaptations associated with each habit; many receptors and a digestive tract degenerate in various degrees. Information on the nervous system is only that contained in general anatomical accounts employing ordinary histological technics, and is basically that known to Villot (1874, 1887, 1889) and Vejdovský (1886, 1888, 1894). Montgomery (1903) and Feyel (1936) are important respectively in consolidating knowledge of the order Gordioidea and in extending it to the other order, the Nectonematoidea. Rauther (1905, 1930b) provides useful reviews. The central nervous system usually comprises an anterior circumenteric swelling (the cephalic ganglion) and a single midventral medullary cord (Fig. 11.22); either may be internal or may be broadly confluent with the epidermis.

VI. NEMATOMORPHA

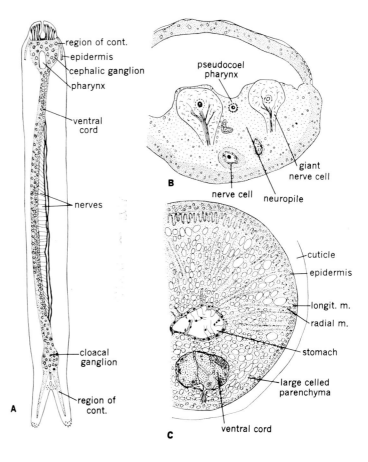

Figure 11.22. Nematomorpha. **A.** *Gordius*, plan of the nervous system. *region of cont.*, the regions of continuity of the nerve fiber matter of the ganglia and that of the subcuticular epidermis. [Brandes, 1898.] **B.** *Nectonema*, transverse section of cephalic ganglion. [Feyel, 1936.] **C.** *Paragordius*, transverse section of the body showing the ventral cord. [Combined from Montgomery, 1903.]

The **cephalic circumenteric ganglion** is a mass surrounding the minute esophagus and cannot naturally be divided into cerebral or supraesophageal and subesophageal (Fig. 11.23). In some gordiids there is said to be no supraesophageal cellular portion and in *Paragordius* not even a dorsal commissure. In *Gordius aquaticus*, however, Rauther figures a prominent supraesophageal cell mass and *Nectonema* exhibits the same (Feyel). In any event the cephalic ganglion is an anterior enlargement of, and not sharply demarcated from, the ventral cord. Most of its cells (the nerve cells correspond to Montgomery's chromophobic cells) are ventral and medial, a few dorsal, especially giant cells. *Nectonema* has four giant cells in the so-called anterior chamber which may represent the supraesophageal portion of the ganglion. A female *Paragordius* had 28–40 giant nerve cells in its subesophageal mass. The cell types and histological differentiation are the same in cephalic ganglion and ventral cord. The central mass is fibrous and continuous right and left, with the lateral funiculi of the ventral cord behind and with the paired cephalic nerves in front. Ventrally and

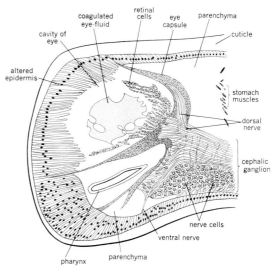

Figure 11.23. Nematomorpha. *Paragordius*, parasagittal section of the head, showing the eye and cephalic ganglion. [Montgomery, 1903.]

dorsally more or less clear commissures may be defined.

The single, median **ventral cord,** well defined and sheathed, usually lies buried in mesenchyme of the pseudocoel connected with the midventral epidermis by a kind of mesentery often called the neural lamella (Fig. 11.22). Müller's (1926) denial that the ventral cord is nervous can be dismissed, so poor are his histological details and figures. Characteristically, the cord is subdivided longitudinally into three, a median and two lateral funiculi, defined by inward projections of the sheath. Nerve cells are on the ventral and lateral surfaces, fiber matter is central and dorsal. Nerve cells of very different size occur: small ones (about 6 µ), unipolar or bipolar, are much the most numerous; the largest (10–14 µ in diameter) are medial and perhaps bipolar and can be called giant. These giants may be quite numerous—for example one or several in most cross sections; they are irregularly scattered along the cord, at least in *Paragordius*. Different authors have reported, from preparations with ordinary stains, that the nerve cells are unipolar, bipolar, or multipolar. Glial cells and particularly fibers are conspicuous, ramifying among the cells, throughout the fiber core and continuous with the cord sheath, the neural lamella, and the epidermis. The fiber core consists mainly of tracts of longitudinal fibers; at frequent, irregular intervals a commissure crosses the midline. No segmental grouping of cells or commissures occurs.

A **cloacal ganglion** forms a swelling at the posterior end of the ventral cord. In *Paragordius* nerve cells per cross section increase and occur even dorsally, but no giant cells are found in the ganglion. In the male the ganglion is less developed and resembles the ventral cord in possessing giant cells and confining nerve cells to the ventral side.

The **peripheral nervous system,** aside from several nerves, is hardly known. A real peripheral plexus is not in evidence by ordinary methods. From the cephalic circumenteric ganglion paired cephalic nerves supply the ventral and lateral epidermis and in *Paragordius* dorsal nerves pass into the eye capsule. From the ventral cord no well-formed nerves emerge but it is believed that nerve fibers pass sparsely through the ventral mesentery (neural lamella) to the epidermis; this is indeed the only route in or out. A longitudinal strand at the place of meeting of the lamella and the epidermis has been called a hypodermis nerve, but proof of its nervous nature is not yet adequate. From the cloacal ganglion in the female a pair of anterior and a pair of posterior cloacal nerves course backward in the cloacal lining; in the male a pair of genital and a pair of caudal nerves are designated.

Sensory nerve cells in the epidermis of several kinds are described (Rauther, 1905). (Montgomery's types are doubtful; this author rather too easily identifies cells as nervous.) Bristles, spines, and papillae on the skin may be sensory but this has not been shown; if they are, these animals are richly provided with diverse receptors.

A large cephalic vesicle in *Paragordius* has been called an **eye** (Montgomery), but the adequacy of the demonstration that the "retinal" elements are receptors and connected to the dorsal nerves leaves much to be desired. No such organ is known in other forms.

How innervation reaches the musculature is unknown. Earlier authors believed it was by way of the neural lamella, hypodermal nerve, and lateral branches therefrom; but each of these stages is questionable.

VII. ENTOPROCTA

Information on the nervous system of these stalked, sessile, tentaculate, and often colonial forms comes from a number of the general anatomical monographs, on the two main genera *Loxosoma* and *Pedicellina*. Nitsche (1869, 1870, 1875), Salensky (1877), Harmer (1886), Foettinger (1887), Ehlers (1890), Nickerson (1901), Retzius (1905), Assheton (1912), and Hilton (1922) are the principal authors. Harmer, Retzius, Assheton, and Hilton have used special nerve stains, obtaining some success on peripheral elements with silver nitrate and methylene blue. Useful general

VII. ENTOPROCTA

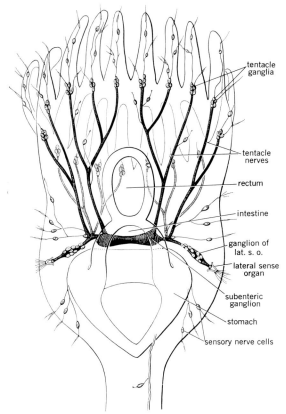

Figure 11.24. Entoprocta. *Loxosoma*, plan of the nervous system of the individual polyp. [Harmer, 1885.]

reviews are provided by Cori (1936), Hyman (1951), and Brien (1959). Since the phylum comprises only about a dozen genera, is generally sessile, extremely simple, and little studied with revealing methods, there is only a limited amount of detail known.

A single clear ganglionic mass and no definite cords are present (Fig. 11.24). The interpretation of this ganglion rests on that of the general body plan. The gut is U-shaped, with mouth and anus opening on the calyx surrounded by the tentacle crown. Unlike the Ectoprocta however, this side of the animal is regarded as ventral and the opposite side, from which the stalk extends, is dorsal. Hence the ganglion, lying under the body wall of the calyx, is ventral, although as in the lophophorate phyla it is between esophagus and rectum, in the loop of the gut. It is sometimes called the subenteric ganglion, or an equivalent name, and the apical ganglion of the larva is lost at metamorphosis. Entoprocta are therefore said to lack a supraesophageal or cerebral ganglion. Furthermore, there appears to be no trace of a dorsal loop over the gut and therefore no circumenteric ring—a rare finding in groups above the flatworms.

The **subenteric ganglion** lies internally (Fig. 11.25, A), free in the pseudocoel, behind the esophagus, and between the protonephridia. Sometimes bilobed, it may be lenticular or rectangular and sometimes shows no clear sign of a paired origin. Cells are concentrated around the outside and the core is fibrous punctate substance receiving the roots of the nerves. Hilton and Retzius proposed that the ganglion acts as a motor center for the tentacles in *Pedicellina*, doubtless integrating sensory input, too. But the solitary species of *Loxosoma* are more active and move by a vigorous kind of leaping, for which a motor innervation is not well worked out.

The **nerves** arise in the ganglion as five main pairs (*Pedicellina*), numbered from oral to anal side. The first, second, and fourth proceed to the

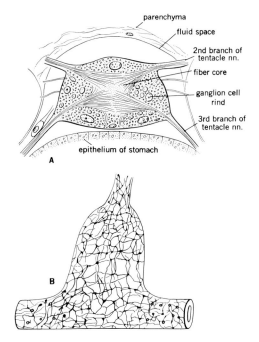

Figure 11.25. Entoprocta. **A.** *Pedicellina*, section of the subesophageal ganglion. [Cori, 1936.] **B.** *Barentsia*, longitudinal section through the muscular socket stained with methylene blue, showing the nerve net. [Hilton, 1922.]

tentacle crown. Each tentacle receives a branch of one of these that forms a small tentacle ganglion. From this one or two nerves enter the tentacle. The sense cells of the tentacle epithelia send axons into these nerves. Nerve three supplies the body wall and nerve five, the stalk. A fine sixth nerve seems to be related to the reproductive organs.

A peripheral plexus, believed to be a nerve net, was described by Hilton in the stalk, stolons, and muscular swellings of *Barentsia* (Fig. 11.25, B). He considered this to be of direct motor influence, not only sensory, but motor nerve fibers have not been definitely seen. The plexus on the stolon provides a **colonial pathway** between individuals and Hilton said conduction between individuals is more evident than conduction from the stem to the tentacles of a given individual.

Sense cells with one or a few long stiff bristles are found on the outer side of the tentacles, around the sides of the calyx, and at the junction of calyx and stalk. The cell body lies below the epithelium. Hilton described sensory pits on the stolons and stalks and found these regions more sensitive than tentacles and body. A weak stimulus causes rotation of the stalk on the base; stronger stimuli cause movements of the body on the stem or tentacle contraction.

A pair of **lateral sense organs** occurs in adults of some species and in larvae of others (loxosomatids). On the sides of the calyx, orally, they are papillae with a simple tuft of bristles supplied by a nerve having a group of bipolar cells in a swelling on its course. Harmer thought that there is always an intermediate ganglion cell between the sense cells and the subenteric ganglion but the evidence suggests that his ganglion cell is in fact the bipolar sense cell.

Classified References

Acanthocephala

Brandes, 1898; Greef, 1864; Hamann, 1891; Harada, 1931; Hilton, 1921; Kaiser, 1893; Kilian, 1932; Meyer, 1933; Rauther, 1930c; Schneider, 1868.

Rotifera

Dehl, 1934; Gast, 1900; Hirschfelder, 1910; Hlava, 1905; Martini, 1912; Nachtwey, 1925; Peters, 1931; Remane, 1929, 1933; Seehaus, 1930; Stossberg, 1932; Viaud, 1940, 1943a, 1943b; Waniczek, 1930; Wesenberg-Lund, 1929; Wierzejski, 1893; Zelinka, 1888, 1891; Zenkevich and Konstantinova, 1956.

Gastrotricha

Remane, 1929, 1935; Zelinka, 1889.

Kinorhyncha

Reinhard, 1887; Remane, 1929, 1930, 1935; Sarasin and Sarasin, 1888; Schepotieff, 1907; Stokes, 1887; Zelinka, 1894, 1908, 1928.

Nematoda

Apáthy, 1893, 1894, 1897, 1898; Baldwin and Moyle, 1947; Brandes, 1892, 1899; Bütschli, 1874, 1885; Chitwood, 1930, 1931; Chitwood and Chitwood, 1933, 1938, 1940; Chitwood and Wehr, 1934; Clayton, 1932; Deineka, 1908; Dosse, 1942; Elsea, 1951; Filipjev, 1912; Frenzen, 1955; Garbowsky, 1898; Goldschmidt, 1903, 1904, 1907, 1908a, 1908b, 1909, 1910; Hesse, 1892; Hilton, 1920, 1921, 1927; Höppli, 1924, 1925; Jarman, 1959; Joseph, 1882, 1884; Kreis, 1934; Looss, 1896, 1905; Magath, 1919; Martini, 1916; Meissner, 1853; Mirza, 1929; Norton and De Beer, 1957; Pai, 1927, 1928; Rauther, 1906a, 1930; Rees, 1941; Rodhe, 1885, 1892, 1894; Salensky, 1909; Schneider, 1860, 1863, 1866; Schönberg, 1944; Schulz, 1931a, 1931b; Stekhoven, 1959; Strassen, 1904; Törnquist, 1931; Voltzenlogel, 1902; Ward, 1892, 1893; Wessing, 1953.

Nematomorpha

Brandes, 1898; Bürger, 1891; Camerano, 1888; Feyel, 1936; Linstow, 1889; Montgomery, 1903; Müller, 1926; Rauther, 1904, 1905, 1930; Rohde, 1892; Schepotieff, 1908; Vejdovský, 1886, 1888, 1894; Villot, 1874, 1887, 1889.

Entoprocta

Assheton, 1912; Atkins, 1932; Brien, 1959; Cori, 1936; Davenport, 1893; Ehlers, 1890; Foettinger, 1887; Harmer, 1885, 1886; Hatschek, 1877; Hilton, 1922; Hyman, 1951; Nickerson, 1901; Nitsche, 1869, 1870, 1875; Salensky, 1877; Zirpolo, 1933.

Bibliography

APÁTHY, S. 1893. Ueber die Muskelfasern von *Ascaris*, nebst Bemerkungen über die von *Lumbricus* und *Hirudo*. *Z. wiss. Mikr.*, 10:36-73.

APÁTHY, S. 1894. Das leitende Element in den Muskelfasern von *Ascaris*. *Arch. mikr. Anat.*, 43:886-911.

APÁTHY, S. 1897. Das leitende Element des Nervensystems und seine topographischen Beziehungen zu den Zellen. *Mitt. zool. Sta. Neapel*, 12:495-748.

APÁTHY, S. 1898. Bemerkungen zu Garbowskys Darstellung meiner Lehre von den leitenden Nervenelementen. *Biol. Zbl.*, 18:704.

ASSHETON, R. 1912. *Loxosoma loxalina* and *Loxosoma saltans*—two new species. *Quart. J. micr. Sci.*, 58: 117-143.

ATKINS, D. 1932. The ciliary feeding mechanism of the entoproct Polyzoa, and a comparison with that of the ectoproct Polyzoa. *Quart. J. micr. Sci.*, 75:393-423.

BALDWIN, E. and MOYLE, V. 1947. An isolated nerve-muscle preparation from *Ascaris lumbricoides*. *J. exp. Biol.*, 23:277-291.

BRANDES, G. 1892. Über das Nervensystem von *Ascaris megalocephala*. *Ber. naturf. Ges. Halle.* 1892:106-107.

BRANDES, G. 1898. Das Nervensystem der als Nemathelminthen zusammengefassten Wurmtypen. *Abh. Naturf. Ges. Halle*, 21:271-299.

BRIEN, P. 1959. Classe des Endoproctes ou Kamptozoaires. *Grassé's Traité de Zoologie*, 5:1:927-1007

BÜRGER, O. 1891. Zur Kenntnis von *Nectonema agile* Verr. *Zool. Jb. (Anat.)*, 4:631-652.

BÜTSCHLI, O. 1874. Beiträge zur Kenntnis des Nervensystems der Nematoden. *Arch. mikr. Anat.*, 10:74-100.

BÜTSCHLI, O. 1885. Zur Herleitung des Nervensystems der Nematoden. *Morph. Jb.*, 10:486-493.

CAMERANO, L. 1888. Recherches sur l'anatomie et l'histologie des gordiens. *Arch. ital. Biol.*, 9:243-248.

CHITWOOD, B. G. 1930. Studies on some physiological functions and morphological characters of *Rhabditis* (Rhabditidae, nematodes). *J. Morph.*, 49:251-275.

CHITWOOD, B. G. 1931. A comparative histological study of certain nematodes. *Z. Morph. Ökol. Tiere*, 23:237-284.

CHITWOOD, B. G. and CHITWOOD, M. B. 1933. The histological anatomy of *Cephalobellus papilliger* Cobb, 1920. *Z. Zellforsch.*, 19:309-355.

CHITWOOD, B. G. and CHITWOOD, M. B. 1938. Esophago-sympathetic nervous system. In: *An Introduction to Nematology*, Sec. 1(2):95. M. B. Chitwood, Babylon, N.Y.

CHITWOOD, B. G. and CHITWOOD, M. B. 1940. Nervous system. In: *An Introduction to Nematology*, Sec. 1(3):159-174. B. G. Chitwood Babylon, N.Y.

CHITWOOD, B. G. and WEHR, E. E. 1934. The value of cephalic structures as characters in nematode classification, with special reference to the superfamily Spiruroidea. *Z. Parasitenk.*, 7:273-335.

CLAYTON, D. E. 1932. A comparative study of the non-nervous elements in the nervous systems of invertebrates. *J. Ent. Zool.*, 24:3-22.

CORI, C. J. 1936. Kamptozoa. *Bronn's Klassen*, 4:2;4:1-119.

DAVENPORT, C. B. 1893. On *Urnatella gracilis*. *Bull. Mus. comp. Zool. Harv.*, 24:1-44.

DEHL, E. 1934. Morphologie von *Lindia tecusa*. *Z. wiss. Zool.*, 145: 169-219.

DEINEKA, D. 1908. Das Nervensystem von *Ascaris*. *Z. wiss. Zool.*, 89:242-307.

DOSSE, G. 1942. Beiträge zur morphologischen und histologischen Untersuchung parasitischer Nematoden. *Z. Parasitenk.*, 12:451-478.

EHLERS, E. 1890. Zur Kenntnis der Pedicellinen. *Abh. Ges. Wiss. Göttingen*, 36:1-200.

ELSEA, J. R. 1951. The histological anatomy of the nematode *Meloidogyne hapla* (Heteroderidae). *Proc. helm. Soc. Wash.*, 18:53-63.

FEYEL, T. 1936. Recherches histologiques sur *Nectonema agile* Verr. Étude de la forme parasite. *Arch. Anat. micr.*, 32:197-234.

FILIPJEV, I. N. 1912. Zur Kenntnis des Nervensystems bei den freilebenden Nematoden. *C. R. Trav. Soc. Imp. Nat. So.-St. Petersb.*, 43(1):205-215; 220-222. (In Russian, with German summary).

FOETTINGER, A. 1887. Sur l'anatomie des pédicellines de la côte d'Ostende. *Arch. Biol., Paris*, 7:299-329.

FRENZEN, K. 1955. Studien zu den Problemen der Zellkonstanz: Untersuchungen an *Ascaridia galli* Schrank 1788. *Z. wiss. Zool.*, 158:304-340.

GARBOWSKY, T. 1898. Apáthys Lehre von den leitenden Nervenelementen. *Biol. Zbl.*, 18:488-509; 536-544.

GAST, R. 1900. Beiträge zur Kenntnis von *Apsilus vorax* (Leidy). *Z. wiss. Zool.*, 67:167-214.

GOLDSCHMIDT, R. 1903. Histologische Untersuchungen an Nematoden. I. Die Sinnesorgane von *Ascaris lumbricoides* L. und *Ascaris megalocephala* (Clequ.) *Zool. Jb. (Anat.)*, 18:1-57.

GOLDSCHMIDT, R. 1904. Ueber die sogen. radiärstreiften Ganlienzellen von *Ascaris*. *Biol. Zbl.*, 24:173-182.

GOLDSCHMIDT, R. 1907. Einiges vom feineren Bau des Nervensystems. *Verh. Dtsch. Zool. Ges.*, 1907:130-131.

GOLDSCHMIDT, R. 1908a. Die Neurofibrillen im Nervensystem von *Ascaris*. *Zool. Anz.*, 32:562-563.

GOLDSCHMIDT, R. 1908b. Das Nervensystem von *Ascaris lumbricoides* und *megalocephala*. I. *Z. wiss. Zool.*, 90: 73-136.

GOLDSCHMIDT, R. 1909. Das Nervensystem von *Ascaris lumbricoides* und *megalocephala*. II. *Z. wiss. Zool.*, 92: 306-357.

GOLDSCHMIDT, R. 1910. III. Das Nervensystem von *Ascaris lumbricoides* und *megalocephala*. *Festschr. für R. Hertwig* Vol. II. Fischer, Jena.

GREEF, R. 1864. Untersuchungen über den Bau und die Naturgeschichte von *Echinorhynchus miliarius* Zenker. (*E. polymorphus.*) *Arch. Naturgesch.*, 30:98-140.

HAMANN, O. 1891. Die Nemathelminthen. I. Monographie der Acanthocephalen. *Jena. Z. Naturw.*, 25:1-119.

HARADA, I. 1931. Das Nervensystem von *Bolbosoma turbinella* (Dies.) *Jap. J. Zool.*, 3:161-199.

HARMER, S. F. 1885. On the structure and development of *Loxosoma*. *Quart. J. micr. Sci.*, 25:261-337.

HARMER, S. F. 1886. On the life-history of *Pedicellina*. *Quart. J. micr. Sci.*, 27:239-263.

HATSCHEK, B. 1877. Embryonalentwicklung und Knospung der *Pedicellina echinata*. *Z. wiss. Zool.*, 29:502-549.

HESSE, R. 1892. Über das Nervensystem von *Ascaris megalocephala*. *Z. wiss. Zool.*, 54:548-568.

HILTON, W. A. 1920. Notes on the central nervous system of a free-living marine nematode. *J. Ent. Zool.*, 12:82-84.

HILTON, W. A. 1921. Nervous system and sense organs. VII. Round worms. *J. Ent. Zool.*, 13(suppl.):55-65.

HILTON, W. A. 1922. The nervous system and sense organs. Endoprocta. *J. Ent. Zool.*, 14(suppl.):76-79.

HILTON, W. A. 1927. The muscular sense of invertebrates. *J. Ent. Zool.*, 19:75-76.

HIRSCHFELDER, G. 1910. Beiträge zur Histologie der Rädertiere. (*Eosphora, Hydatina, Euchlanis, Notommata.*) *Z. wiss. Zool.*, 96:209-335.

HLAVA, S. 1905. Beiträge zur Kenntnis der Rädertiere. I. Über die Anatomie von *Conochiloides natans* (Seligo). *Z. wiss. Zool.*, 80:282-326.

HÖPPLI, R. 1924. Über sechs bisher noch nicht bekannte Sinnesorgane an den Lippen der Ascariden. *Zool. Anz.*, 61:39-42.

HÖPPLI, R. 1925. Über das Vorderende der Ascariden. Vergleichende histologische Untersuchungen unter besonderer Berücksichtigung der Zellkonstanzfrage. *Z. Zellforsch.*, 2:1-68.

HYMAN, L. H. 1951. *The Invertebrates*. Vol. III. McGraw-Hill, New York.

JARMAN, M. 1959. Electrical activity in the muscle cells of *Ascaris lumbricoides*. *Nature, Lond.*, 184:1244.

JOSEPH, G. 1882. Vorläufige Bemerkungen über Muskulatur, Excretionsorgane und peripherisches Nervensystem von *Ascaris megalocephala* und *lumbricoides*. *Zool. Anz.*, 5:603-609.

JOSEPH, G. 1884. Beiträge zur Kenntnis der Nervensystems der Nematoden. *Zool. Anz.*, 7:264-266.

KAISER, J. E. 1893. Die Acanthocephalen und ihre Entwickelung. *Zoologica, Stuttgart*, 2; 7:1-136; 1-148; i-xix.

KILIAN, R. 1932. Zur Morphologie und Systematik der Gigantorhynchidae (Acanthoceph.). *Z. wiss. Zool.*, 141:246-345.

KREIS, H. A. 1934. Oncholaiminae Filipjev 1916. Eine monographische Studie. *Capita zool.*, 4:1-271.

LINSTOW, VON. 1889. Ueber die Entwicklungsgeschichte und die Anatomie von *Gordius tolosanus* Duj. = *G. subbifurcus* v. Siebold. *Arch. mikr. Anat.*, 34:248-268.

LOOSS, A. 1896. Ueber den Bau des Oesophagus bei einigen Ascariden. *Zbl. Bakt.*, (1.) 19:5-13.

LOOSS, A. 1905 The anatomy and life history of *Agchylostoma duodenale* Dub. *Rec. Egypt. Govt. Sch. Med.*, 3:6-159.

MAGATH, T. B. 1919. *Camallanus americanus*. nov. spec., a monograph on a nematode species. *Trans. Amer. micr. Soc.*, 38:49-170.

MARTINI, E. 1912. Studien über die Konstanz histologischer Elemente. III. *Hydatina senta*. *Z. wiss. Zool.*, 102:425-645.

MARTINI, E. 1916. Die Anatomie der *Oxyuris curvula*. *Z. wiss. Zool.*, 116:137-534.

MEISSNER, G. 1853. Beiträge zur Anatomie und Physiologie von *Mermis albicans*. *Z. wiss. Zool.*, 5:207-284.

MEYER, A. 1933. Acanthocephala. *Bronn's Klassen*, 4:2:2:1-582.

MIRZA, M. B. 1929. Beiträge zur Kenntnis des Baues von *Dracunculus medinensis* Velsch. *Z. Parasitenk.*, 2:129-156.

MONTGOMERY, T. H., Jr. 1903. The adult organisation of *Paragordius varius* Leidy. *Zool. Jb. (Anat.)*, 18:387-474.

MÜLLER, G. W. 1926. Über Gordiaceen. *Z. Morph. Ökol. Tiere*, 7:134-219.

NACHTWEY, R. 1925. Untersuchungen über die Keimbahn, Organogenese und Anatomie von *Asplanchna priodonta* Gosse. *Z. wiss. Zool.*, 126:239-492.

NICKERSON, W. S. 1901. On *Loxosoma davenporti* sp. nov. An endoproct from the New England coast. *J. Morph.*, 17:351-398.

NITSCHE, H. 1869. Beiträge zur Kenntniss der Bryozoen. II. Ueber Anatomie von *Pedicellina echinata* Sars. *Z. wiss. Zool.*, 20:13-36.

NITSCHE, H. 1870. Beiträge zur Kenntniss der Bryozoen. *Z. wiss. Zool.*, 20:1-36.

NITSCHE, H. 1875. Beiträge zur Kenntniss der Bryozoen. V. Ueber die Knospung der Bryozoen. *Z. wiss. Zool.*, 25(suppl):343-402.

NORTON, S. and DE BEER, E. J. 1957. Investigations on the action of piperazine on *Ascaris lumbricoides*. *Amer. J. trop. Med.*, 6(5):898-905.

PAI, S. 1927. Lebenszyklus der *Anguillula aceti* Ehrbg. *Zool. Anz.*, 74:257-270.

PAI, S. 1928. Die Phasen des Lebenszyklus und ihre experimentell-morphologische Beeinflussung. *Z. wiss. Zool.*, 131:293-344.

PETERS, F. 1931. Untersuchungen über Anatomie und Zellkonstanz von *Synchaeta* (*S. grimpei* Remane, *S. baltica* Ehrenb. *S. tavina* Hood und *S. triophthalma* Lauterborn). Ein Beitrag zur Frage der Artunterschiede bei Konstantzelligen Tieren. *Z. wiss. Zool.*, 139:1-169.

RAUTHER, M. 1904. Das Cerebralganglion und die Leibeshöhle der Gordiiden. *Zool. Anz.*, 27:606-616.

RAUTHER, M. 1905. Beiträge zur Kenntnis der Morphologie und der phylogenetischen Beziehungen der Gordiiden. *Jena. Z. Naturw.*, 40:1-94.

RAUTHER, M. 1906. Beiträge zur Kenntnis von *Mermis albicans* v. Sieb. mit besonderer Berücksichtigung des Haut-Nerven-Muskelsystems. Studien über die Organisation der Nematoden I. *Zool. Jb. (Anat.)*, 23:1-76.

RAUTHER, M. 1930a. Nematodes. *Kükenthal's Handb. Zool.*, 2(4):249-402.

RAUTHER, M. 1930b. Nematomorpha, Saitenwürmer. *Kükenthal's Handb. Zool.*, 2(4):403-448.

RAUTHER, M. 1930c. Acanthocephala Kratzwürmer. *Kükenthal's Handb. Zool.*, 2(4):449-482.

REES, G. 1941. The scolex of *Aporhynchus norvegicus* (Olss.). *Parasitology*, 33:433-438.

REINHARD, W. 1887. Kinorhyncha (Echinoderes), ihr anatomischer Bau und ihre Stellung im System. *Z. wiss. Zool.*, 45:401-467.

REMANE, A. 1929. Gastrotricha. *Kükenthal's Handb. Zool.*, 2(4):121-186.

REMANE, A. 1929-1930. Kinorhyncha = Echinodera. *Kükenthal's Handb. Zool.*, 2(4):187-248.

REMANE, A. 1929-1933. Aschelminthen. Rotatoria. *Bronn's Klassen*, 4:2:1:1-4:1-516.

REMANE, A. 1935. Gastrotricha und Kinorhyncha. *Bronn's Klassen*, 4:2:1:2:1-160.

RETZIUS, G. 1905. Das sensible Nervensystem der Bryozoen. *Biol. Untersuch.*, N.F. 12:49-54.

RODHE, E. 1885. Beiträge zur Kenntnis der Anatomie der Nematoden. *Zool. Beitr., Berl.*, 1:11-32.

RODHE, E. 1892. Muskel und Nerv. I. *Ascaris*. II. *Mermis* und *Amphioxus*. III. *Gordius*. *Zool. Beitr., Berl.*, 3:69-106; 161-192.

RODHE, E. 1894. Apáthy als Reformator der Muskel- und Nervenlehre. *Zool. Anz.*, 17:38-47.

SALENSKY, M. 1877. Études sur les Bryozoaires Entoproctes. *Ann. Sci. nat. (Zool.)*, (6)5(3):1-60.

SALENSKY, W. 1909. Über die embryonale Entwicklung des *Prosorochmus viviparus* Uljanin (*Monopara vivipara*). *Bull. Acad. Sci. St.-Petersb. (Izv. Imp. Akad. Nauk.)*, 3(6):325-340.

SARASIN, P. and SARASIN, F. 1888. Ueber die Anatomie der Echinothuriden und die Phylogenie der Echinodermen. *Ergebn. Natur. Forsch. auf Ceylon*, 1(3):83-154.

SCHEPOTIEFF, A. 1907. Die Echinoderiden. *Z. wiss. Zool.*, 88:291-326.

SCHEPOTIEFF, A. 1908. Über den feineren Bau der *Gordius*larven. *Z. wiss. Zool.*, 89:230-241.

SCHILLER, J. 1937. Sur le déterminisme des réflexes chez quelques Monascidies et sur leur utilité. *Bull. Inst. océanogr. Monaco*, 721:1-8.

SCHNEIDER, A. 1860. Ueber die Muskeln und Nerven der Nematoden. *Arch. Anat. Physiol. Lpz.*, 1860:224-242.

SCHNEIDER, A. 1863. Neue Beiträge zur Anatomie und Morphologie der Nematoden. IV. Das Nervensystem. *Arch. Anat. Physiol., Lpz.*, 1863:1-25.

SCHNEIDER, A. 1866. *Monographie der Nematoden*. Reimer, Berlin.

SCHNEIDER, A. 1868. Ueber den Bau der Acanthocephalen. *Arch. Anat. Physiol., Lpz.*, 1868:584-597.

SCHÖNBERG, M. 1944. Histologische Studien zu den Problemen der Zellkonstanz an *Rhabditis longicauda fertilior*. *Biol. gen.*, 17:338-366.

SCHULZ, E. 1931a. Betrachtungen über die Augen freilebender Nematoden. *Zool. Anz.*, 95:241-244.

SCHULZ, E. 1931b. Nachtrag zu der Arbeit Betrachtungen über die Augen freilebender Nematoden. *Zool. Anz.*, 96:159-160.

SEEHAUS, W. 1930. Zur Morphologie der Rädertiergattung *Testudinella* Bory de St. Vincent (= *Pterodina* Ehrenberg). *Z. wiss. Zool.*, 137:175-273.

STEKHOVEN, J. H. S. 1959. Nematodes. *Bronn's Klassen*, 4:2:3:661-879.

STOKES, A. C. 1887. Observations sur

les *Chaetonotus*. *J. Microgr.*, 11:77-85.

STOSSBERG, K. 1932. Zur Morphologie der Rädertiergattungen *Euchlanis*, *Brachionus* und *Rhinoglena*. *Z. wiss. Zool.*, 142:313-424.

STRASSEN, O. L. ZUR. 1904. *Anthraconema*, eine neue Gattung freilebender Nematoden. *Zool. Jb.*, *(suppl.)*, 7:301-346.

TÖRNQUIST, N. 1931. Die Nematodenfamilien Cucullanidae und Camallanidae, nebst weiteren Beiträgen zur Kenntnis der Anatomie und Histologie der Nematoden. *Göteborgs VetenskSamh. Handl. (B)*, (5)2(3):i-xi; 1-441.

VEJDOVSKÝ, F. 1886. Zur Morphologie der Gordiiden. *Z. wiss. Zool.*, 43:369-433.

VEJDOVSKÝ, F. 1888. Studien über Gordiiden. II. *Z. wiss. Zool.*, 46:188-216.

VEJDOVSKÝ, F. 1894. Organogenie der Gordiiden. (Zugleich ein Beitrag zur Kenntniss der Metamorphose und Biologie der Zelle). *Z. wiss. Zool.*, 57:642-703.

VIAUD, G. 1940. Recherches expérimentales sur le phototropisme des rotifères. *Bull. biol.*, 74:249-308.

VIAUD, G. 1943a. Recherches expérimentales sur le phototropisme de les rotifères. II. *Bull. biol.*, 77:68-93.

VIAUD, G. 1943b. Recherches expérimentales sur le phototropisme des rotifères. III. *Bull. biol.*, 77:224-242.

VILLOT, A. 1874. Monographie des Dragonneaux (genre *Gordius*, Dujardin). *Arch. Zool. exp. gén.*, 3:181-238.

VILLOT, A. 1887. Sur l'anatomie des gordiens. *Ann. Sci. nat. (Zool.)*, (7) 2:189-212.

VILLOT, A. 1889. Sur l'hypoderme et le système nerveux périphérique des gordiens. *C. R. Acad. Sci., Paris*, 108:304-306.

VOLTZENLOGEL, E. 1902. Untersuchungen über den anatomischen und histologischen Bau des Hinterendes von *Ascaris megalocephala* und *Ascaris lumbricoides*. *Zool. Jb. (Anat.)*, 16:481-510.

WANICZEK, H. 1930. Untersuchungen über einige Arten der Gattung *Asplanchna* Gosse (*A. girodi* de Guerne, *A. brightwellii* Gosse, *A. priodonta* Gosse). *Ann. Mus zool. polon.*, 8:109-322.

WARD, H. B. 1892-1893. On *Nectonema agile*, Verrill. *Bull. Mus. comp. Zool. Harv.*, 23:135-188.

WESENBERG-LUND, C. 1929. Rotatoria, Rotifera-Rädertierchen, in *Kükenthal's Handb. Zool.*, 2:1:4:8-120.

WESSING, A. 1953. Histologische Studien zu den Problemen der Zellkonstanz: Untersuchungen an *Rhabditis anomala*. P. Hertwig. *Zool. Jb. (Anat.)*, 73:69-102.

WIERZEJSKI, A. 1893. *Atrochus tentaculatus* nov. gen. et sp. Ein Räderthier ohne Räderorgan. *Z. wiss. Zool.*, 55:696-712.

ZELINKA, C. 1888. Studien über Räderthiere. II. Der Raumparasitismus und die Anatomie von *Discopus synaptae* n.g. nov. sp. *Z. wiss. Zool.*, 47:353-458.

ZELINKA, C. 1889. Die Gastrotrichen. *Z. wiss. Zool.*, 49:209-384.

ZELINKA, C. 1891. Studien über Räderthiere. III. Zur Entwicklungsgeschichte der Räderthiere nebst Bemerkungen über ihre Anatomie und Biologie. *Z. wiss. Zool.*, 53:1-159.

ZELINKA, C. 1908. Zur Anatomie der Echinoderen. *Zool. Anz.*, 33:629-647.

ZELINKA, C. 1928. Monographie der Echinodera. Leipzig.

ZELINKA, K. 1894. Uber die Organisation von Echinoderes. *Verh. dtsch. zool. Ges.*, 4:46-49.

ZENKEVICH, L. A. and KONSTANTINOVA, M. I. 1956. Locomotion and the motor system of Rotifera. *Zool. Zh.*, 35:345-364. (In Russian.)

CHAPTER 12

Lophophorate Phyla: Ectoprocta, Brachiopoda, and Phoronida

Summary	632	III. Phoronida	640
I. Ectoprocta	**632**	1. Nervous ring and ganglion	641
1. The cerebral ganglion	633	2. The lophophore innervation	642
2. The lophophore innervation	634	3. Receptors	642
3. Nerve supply to the body wall	636	4. The giant fiber system	643
4. Nerve supply to the gut	636	5. Nervous system of the body wall	645
5. The colonial nervous system	636	6. Innervation of the gut and other internal organs	646
II. Brachiopoda	**637**	Classified References	646
		Bibliography	646

Summary

Ectoprocts are marked by extremely simple nervous systems and no sense organs but only simple, unicellular receptors, in association with the sessile, colonial habit and minute size. There is a distinct brain but no main nerve cords; a peripheral plexus is well developed on the lophophore and forms a continuous colonial connection on the body wall. The brain is often hollow. Its histology is typical of small, lower invertebrates. The tentacles have 4 or 5 nerves, of which the inner and outer are perhaps sensory and those on the two sides motor. The gut has an innervation, both from the brain and from the tentacle sheath nerves. There is some evidence of colonial coordination. Histological signs of nerve nets are reported between individuals, within the individual, and in the larva.

Brachiopods also have a simple nervous system and no sense organs, except for a statocyst in some species. The complexity of the nervous system therefore seems to be better predicted by the sedentary life and sensory development than by the rather involved musculature and movements of arms, shell valves, pedicle, and setae. There is a small dorsal supraesophageal ganglion, a nervous ring and a larger subesophageal ganglion. It is not clear whether unipolar cells predominate, but a distinct rind and core are present, although the ganglia are in or just under the outer epidermis and the nerves are strewn with nerve cells. Nerves and plexuses in the body wall, lophophore, and gut connect mainly with the subesophageal ganglion. Specific nerve stains have not been used, and any new information would make a proportionately large contribution.

Phoronids are better known but also offer some of the best opportunities for new work. Seemingly primitive in nervous histology, the system is entirely intraepithelial so far as is known. A general plexus is apparently organized to function as a nerve net. A local thickening around the oral end of the gut represents a nerve ring which is heaviest dorsally; this is the nearest structure to a ganglion. There are no nerve cords, but one or a pair of thick-sheathed giant fibers runs from the ganglion to the main longitudinal muscles and mediates a quick withdrawal response. There are no sense organs, but unicellular sense cells occur commonly in clusters. The motor supply is by axons locally and diffusely perforating the basement membrane to reach the muscles. Action potentials have been recorded from the giant fiber and suggest no intermediate neuron between it and the muscle.

I. ECTOPROCTA

These microscopic, colonial, sessile forms, formerly called Bryozoa or Polyzoa, have extremely simple nervous systems and no sense organs. But the distinctness of a brain bears the stamp of simplification from more complex ancestors, rather than primitiveness. Although there are serious gaps in our knowledge, we are fortunate in having the successful methylene blue stains of Gerwerzhagen (1913), Marcus (1926, 1934), Graupner (1930), and Bronstein (1937), who are at the same time the principal authors. A number of earlier workers saw and described the main nervous concentrations in histological preparations (Nitsche, 1868). The two classes, Gymnolaemata (marine) and Phylactolaemata (fresh water), are not much different in nervous anatomy. The group has been conveniently treated in recent times in German, English, and French by Cori (1941) (in Kükenthal's *Handbuch der Zoologie*), Hyman (1959), and Brien (1960) (in Grassé's *Traité de Zoologie*). Virtually no physiological information is yet at hand. Suggestive are some of the considerations given below under "Colonial nervous system" and the report of

I. ECTOPROCTA

Kato (1950) that *Acanthodesia serrata*, upon stimulation, luminesces from two specific patches. The nervous system can be considered under five rubrics: (a) the cerebral ganglion, (b) the lophophore innervation, (c) the supply to the body wall, (d) the nerves to the gut, and (e) the colonial nervous system.

1. The cerebral ganglion

This organ is hollow in the class Phylactolaemata and is formed from an invagination of the ectoderm in the embryo; in Gymnolaemata it seems to be solid. The ganglion lies on the dorsal side—considered to be that on which the anus opens in the ectoprocts—against the roof of the buccal cavity and at the base of the epistome. It is held in place by peritoneal strands and lies between epidermis and somatopleure, not free in the coelom. The hollow structure (or cerebral vesicle) has thick ganglionic walls dorsally, posteroventrally, and posteriorly (*Cristatella*, *Plumatella*) and an extremely thin wall anteroventrally (Fig. 12.1). The ganglionic walls are formed by three thickenings, the dorsal, ventral, and posterior pads. Each has a central core of fibrous matter and a few dozen nerve cells around the periphery, with their processes directed centrally. The histology appears to be typical for small invertebrate ganglia (Fig. 12.2). There is little or no glial tissue apart from the vesicle wall or ganglionic sheath of a single layer of flat cells. In the few freshwater forms studied, the nerve cells are all much alike and there is no differentiation in the neuropile. The marine forms examined have a small spheroid, depressed ganglion at the extreme anterior end, there being no epistome. The nerve cells, according to Graupner (in *Flustrellidra*), are mostly

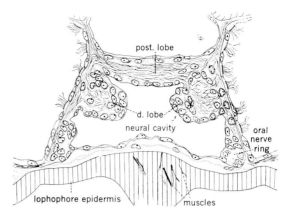

Figure 12.2. Ectoprocta. *Plumatella fungosa*, transverse section of the cerebral ganglion to show its hollow nature and thickened dorsal and posterior walls; the section is in front of the ventral ganglion. *d. lobe*, dorsal thickening of the wall of the cerebral vesicle; *post. lobe*, posterior thickening. [Brien, 1960, in Grassé's *Traité de Zoologie*, Masson et Cie., Paris.]

bipolar, about 25 in number and of three size classes. There appear to be several nerve cell masses here—six are described—but it is difficult to tell whether some are single nerve cells or clusters. Bronstein saw eight similar masses in *Alcyonidium* but he did not add much new understanding as to its composition. A number of nerves arise from the cerebral ganglion, as enumerated below.

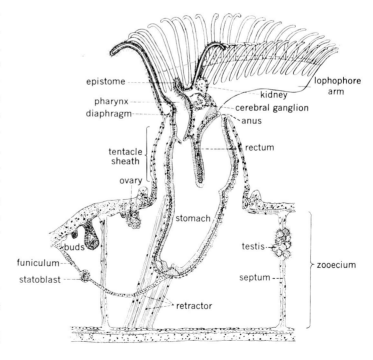

Figure 12.1. Ectoprocta. *Cristatella*, sagittal section of a zooid with the tentacle crown everted and expanded; schematic. [Cori, 1941.]

2. The lophophore innervation

The lophophore is horseshoe-shaped in phylactolaemates, open ventrally, but is a closed circle in gymnolaemates. Accordingly, the **main nerve strand of the base of the lophophore is a** nerve ring in the marine forms and two arms in the fresh-water species (Fig. 12.3). In either case

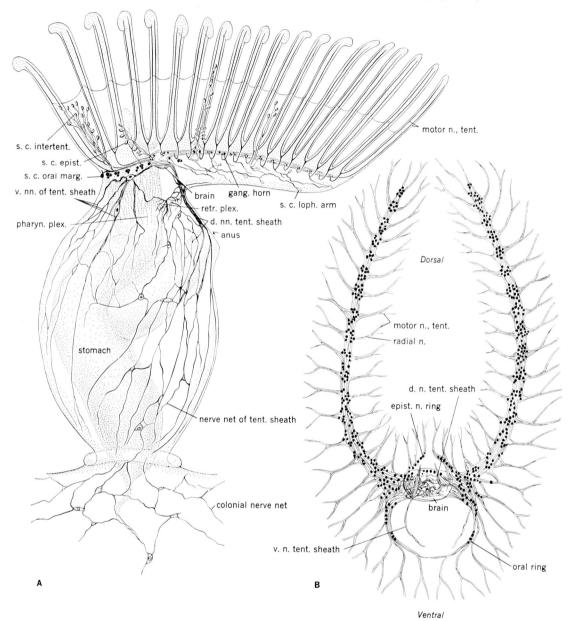

Figure 12.3. Ectoprocta. *Cristatella mucedo.* **A.** Single zooid with lophophore everted, seen from the side, so only one lophophore arm is shown. **B.** View from oral aspect; sensory nerves of the tentacles are omitted (see Fig. 12.4). Methylene blue. [Gerwerzhagen, 1913.] *d. n. tent. sheath*, dorsal nerve of the tentacle sheath (the body wall of the eversible part of the zooid); *epist. n. ring*, nerve ring of the epistome; *gang. horn*, ganglionic horn (continuation of cerebral ganglion, with its cavity, into base of lophophore); *motor n. tent.*, motor nerve of tentacle; *pharyn. plex.*, pharyngeal plexus; *radial n.*, radial nerve; *retr. plex.*, plexus on retractor muscle; *s. c. epist.*, sensory nerve cell on epistome; *s. c. intertent.*, sense cell of intertentacular membrane; *s. c. loph. arm*, sense cell of lophophore arm; *s. c. oral marg.*, sense cell on oral margin; *v. nn. of tent. sheath*, ventral nerves of tentacle sheath.

I. ECTOPROCTA

these are the largest structures of the nervous system apart from the cerebral ganglia. In phylactolaemates they are called ganglionic horns and are lateral extensions from the cerebral ganglion, with a cavity continuous with that of the ganglion. Each proceeds into a lophophore arm just under the epidermis in the groove between tentacle bases, giving off nerves to the tentacles. For some distance they are medullary, then they taper toward the free ends or ventral side of the lophophore arms. The outer wall of the lumen is thin; the ganglionic wall is the inner one, which has peripheral cells and a central fibrous strand. The horns do not supply the medial tentacles; instead there are two pairs of nerves from the cerebral ganglion which, because they meet at their ends in the midline, are called nerve rings. The **oral ring** sends nerves to the outer or ventral, and the **epistomial ring** to the inner or dorsal medial tentacles.

In gymnolaemates there is described only one **lophophore ring nerve**, tapering from the cerebral to the ventral side; it gives off two lateral branches embracing the buccal cavity. The nerve cells are said to be distinguishable from the cerebral cells by a greater basophilia.

Each tentacle has four or five nerves. From the above described rings or ganglionic horns so-called radial nerves arise in between the tentacles and bifurcate to send branches to facing sides of adjacent tentacles; thus each tentacle has two nerves on opposite sides, from adjacent radial nerves (Fig. 12.4). In addition, two other nerves run up the tentacle—inner and outer; they also come from the radial nerves, by way of irregular side branches that coalesce or by a second bifurcation. In some species a fifth central strand is identified. Most interesting is the conjecture that the central and the inner and outer nerves are sensory and the nerves on the sides motor. There is certainly some basis for this in the collecting of axons of sense cells in the tentacle, but it is difficult to be sure of a motor connection or one of an unmixed nature. The literature often speaks of each nerve as a fiber but it is highly unlikely

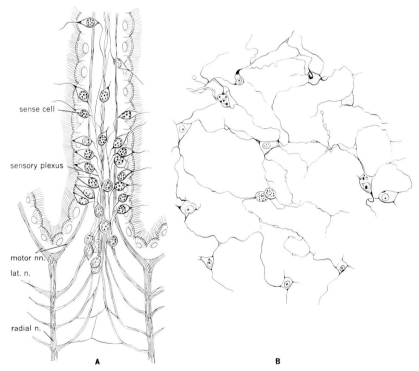

Figure 12.4. Ectoprocta. *Cristatella mucedo*. **A.** The tentacle nerves, two motor and a central sensory, all connected by a radial nerve to the ganglionic horn in the base of the lophophore arm. n., nerve. **B.** Ganglion cell plexus of the tentacle sheath; note the evidence of anastomosis. Methylene blue. [Gerwerzhagen, 1913.]

that the nerves do in fact consist of single neuronal processes.

Sensory nerve cells are abundant on the tentacles, including their outer face, where cilia are lacking. On the inner face they are said to occur in two regular rows, each consisting of two files of cells side by side. At least some bear long stiff hairs and therefore suggest a tactile function. Their axons pass, via the tentacle nerves described, to the ganglionic strand of the lophophore, to the oral or epistomial ring, or to the cerebral ganglion. Occasional ganglion cells occur along the supposedly sensory nerves.

The epistome, found in the fresh-water forms, sends a bundle of sensory axons to the apical side of the cerebral ganglion.

3. Nerve supply to the body wall

The free part of the body wall covering the neck of the protruded polyp is called the tentacle sheath because, by inversion, it becomes just that when the lophophore is retracted (Fig. 12.1). The fixed part of the body wall is the boxlike zooecium and the living layers adherent to it. To these areas go long nerves that arise from the cerebral ganglion; they end in a general epidermal plexus. Typically, two dorsal and two ventral (*Cristatella*) or two lateral nerves (*Flustrellidra*) branch over the tentacle sheath and incidentally form embracing nerve rings near the orifice and about half-way along its length. Some branches are said to be motor and some sensory. The main motor structure to be innervated is the retractor muscle of the tentacle sheath, but in several species the supply to this has not been seen. Circular and longitudinal parietal muscle is said to be innervated from the sheath plexus. Sensory cells are concentrated around the orifice. A distinct plexus has been well seen in methylene blue preparations by several authors. It extends over the tentacle sheath and is continuous with a similar structure on the zoecial wall. Multipolar ganglion cells make contact with their processes and these appear to be protoplasmically continuous; the aspect is that of a nerve net similar to the descriptions for *Hydra*.

A **peripharyngeal nerve collar** is reported in *Farrella* by Marcus, in addition to and beneath the lophophore nerve ring. It is directly connected to the cerebral ganglion. Missing in *Flustrellidra* and the phylactolaemates, it is present in *Membranacea* and *Alcyonidium*.

A nerve net in the larva is illustrated by Marcus (1926c).

4. Nerve supply to the gut

The anterior portion of the digestive tract receives nerves from the cerebral ganglion directly and probably from the tentacle sheath nerves as well. The main distribution is between the gut epithelium and the circular muscle layer.

5. The colonial nervous system

Several older authors have recognized the problem of a possible nervous connection between individuals of the same colony, and refer to Darwin as having noted that vibracula are capable of synchronous movement. But there do not appear to be any careful observations of living animals to see whether excitation spreads without being attributable to a common stimulus or to mechanical re-excitation of successive zooids.

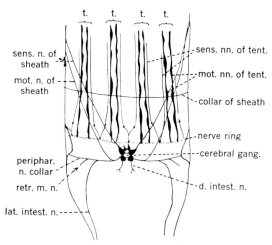

Figure 12.5. Ectoprocta. *Farrella repens*. Plan of the nervous system. Methylene blue. [Marcus, after Brien, 1960, in *Grassé's Traité de Zoologie*, Masson et Cie., Paris.] *cerebral gang.*, cerebral ganglion; *d. intest. n.* and *lat. intest. n.*, dorsal and lateral intestinal nerve; *mot. n. of sheath*, motor nerve of sheath (see explanation in Fig. 12.3); *mot. nn. of tent.*, motor nerves of tentacle; *periphar. n. collar*, peripharyngeal nerve collar; *retr. m. n.*, nerve of retractor muscle; *sens. n. of sheath*, sensory nerve of sheath; *sens. nn. of tent.*, sensory nerves of tentacle; *t.*, tentacle.

II. BRACHIOPODA

Local stimulation certainly causes a spreading withdrawal response through many individuals and for some distance, in some species of phylactolaemates (unnamed), but some writers think that in gymnolaemates there is no nervous connection since usually there is hardly any effect on a neighbor by excitation of one individual (Marcus, Brien, Hiller). Nitsche (1871) admitted spread in the latter group but ascribed it to the mechanical disturbance of a responding zooid transmitted through soft parts to neighbors—therefore to a chain reflex; Bronstein (1937) believed this could be ruled out in some cases.

I have found spread in *Bugula* just as good through regions of already withdrawn zooids or missing or otherwise unresponding zooids—the spread may be regarded as truly nervous. To judge from the corals and hydroids, one may expect among different species diversity of patterns with respect to spread and its basis.

Anatomically, there are grounds for admitting a colonial nervous system. Krukenberg (1887) lists earlier observations of nerve strands between zooids. Brien and Cori also cite observations in both fresh-water and marine species of a plexus radiating beyond the individual and of sense cells in the colony wall. Gerwerzhagen's account is the most convincing; he saw ganglion cells connected to each other in all parts of the colony wall of *Cristatella*, and concluded that it is a motor net. This genus is unique in being motile, creeping over the substratum; it therefore might require nervous coordination of the muscle cells in the colonial tissue. Bronstein (1937) states that fibers occur in the stolon of *Bowerbankia* that are probably colonial. Hiller (1939) saw a plexus of small nerve cells and long fibers which pass between individuals, in *Membranipora*, going through the pore plates and connecting with the nerves of the tentacles (Fig. 12.6). One may suspect a

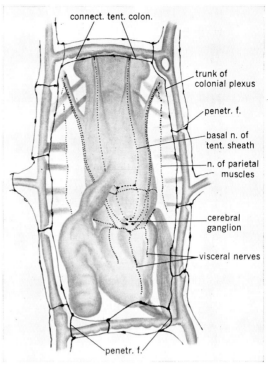

Figure 12.6. Ectoprocta. *Membranipora pilosum,* one zooecium showing the connections forming a colonial nervous system. [Hiller, 1939.] *connect. tent. colon.,* connections of nerves of the tentacle sheath with the colonial nerve plexus; *n.,* nerve; *penetr. f.,* nerve fiber penetrating through pore in communicating plate.

priori that there are connections to the vibracula from neighboring zooids. Bronstein refers to species of *Caberea* where vibracula beat in unison and propel the colony. Polymorphism in the nervous system has not been studied in the groups exhibiting specialized types of individuals, but Marcus (1939) mentions a ganglionlike mass in avicularia. Forbes (1933) found that avicularia of *Bugula* would not respond when electrical stimuli caused all polypides to retract but promptly closed to filtered food juice or inert solid particles; this suggests an independent effector rather than a nervous connection to other individuals.

II. BRACHIOPODA

The remnant phylum of brachiopods—260 living species and 30,000 fossil species are known—has points of considerable potential interest in nervous anatomy and physiology. Sedentary and usually sessile (though lingulids can change location), these are benthic filter feeders innocent of sense organs except for a statocyst in some forms. The nervous system is, as expected therefore, extremely simple. It is significant as an expression of what is conservative and common in nervous organization because the brachiopods have been evolving since pre-Cambrian times. Another aspect of interest is the relatively complex development of musculature and presumably of movements in an animal with so feeble a sensory apparatus. The lophophore arms, the shell valves, the pedicle, and setae are all manipulated by involved movements. *Glottidia* actively burrows and shows a shadow response of retreat. Many species are of considerable size but the largest nervous masses are nevertheless minute and difficult to find.

The literature on the nervous system practically boils down to van Bemmelen (1883) and Blochmann (1892, 1900); van Bemmelen worked on one of the two main subdivisions of the phylum, the articulates (mainly on *Gryphus*), and Blochmann on inarticulates (*Crania, Discinisca,* and *Lingula*). There has been no application of specific nerve stains and very little histological description. Useful recent reviews are provided by Helmcke (1939), Hyman (1959), and Beauchamp (1960).

The **main features** that are recurrent in this and other groups, supporting the probability of a general pattern among bilateral metazoans, are the following (Figs. 12.7, 12.9). A supraesophageal ganglion lies dorsal to the first part of the enteric tract—but note that this is not close to one end of the animal, especially a front end in locomotion of the adult. There is a ring of nervous tissue around the esophagus and a subesophageal ganglion. Each of the two ganglia gives off nerves, probably mainly mixed sensory and motor. The ganglia are composed of a fibrous core and the nerve cell bodies lie around the outside. It is not clearly established whether the cells are unipolar or bipolar or both. Ganglion cells in the peripheral nerves and plexuses are usually multipolar and isopolar. The position of the ganglia and most nerves is superficial—that is, in the base of the outer body epithelium (Blochmann, mainly inarticulates) or, interpreting van Bemmelen (based on an articulate), in connective tissue just under the epithelium. An epithelial position is probably the persistence of or return to a primitive feature. Ganglia and nerves are perhaps misnomers in our present state of knowledge, since these terms usually mean well-delimited internal structures.

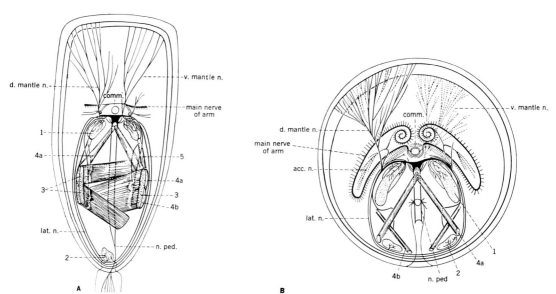

Figure 12.7. Brachiopoda. **A.** *Lingula*, dorsal view of nervous system and muscles. **B.** *Discinisca*, similar view. The numbers identify muscles (see Beauchamp). [Blochmann, after Beauchamp, 1960, in *Grassé's Traité de Zoologie,* Masson et Cie., Paris.] *acc. n.,* accessory nerve of the arms; *d. mantle n.,* dorsal mantle nerve; *lat. n.,* lateral nerve of the body; *n. ped.,* nerve of the peduncle; *v. mantle n.,* ventral mantle nerve.

II. BRACHIOPODA

Little detail is known. The supraesophageal ganglion is smaller than the subesophageal at best (Inarticulata: *Gryphus*) and in some forms does not even make a distinct bulge or thickening but is represented merely by the dorsal side of the circumesophageal ring (all three inarticulates studied by Blochmann). The subesophageal ganglion is a sizable median mass (Fig. 12.8) in *Lingula* and *Discinisca;* in *Lingula* a pair of large lateral ganglia are also described (Beyer, 1866) and in *Crania*, only a pair of lateral masses. Kirtisinghe (1952) noted some giant nerve cells (28×48 μ) in the subesophageal ganglion of *Lingula*. Earlier workers remarked on the occurrence

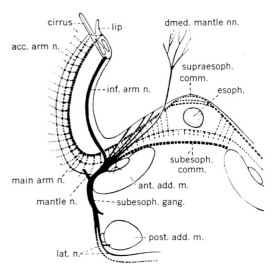

Figure 12.9. Brachiopoda. Diagram of the nervous system of *Crania*. [Blochmann, after Beauchamp, 1960, in Grassé's *Traité de Zoologie*, Masson et Cie., Paris.] *acc. arm n.*, accessory arm nerve; *ant. add. m.*, anterior adductor muscle; *dmed. mantle nn.*, dorsomedial mantle nerves; *esoph.*, esophagus; *inf. arm n.*, inferior arm nerve; *lat. n.*, lateral nerve; *post. add. m.*, posterior adductor muscle; *subesoph. comm.*, subesophageal commissure; *supraesoph. comm.*, supraesophageal commissure.

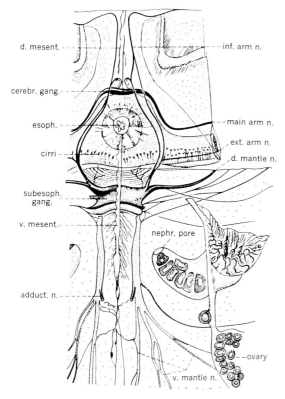

Figure 12.8. Brachiopoda. *Gryphus vitreus*, plan of the nervous system. [Van Bemmelen, after Beauchamp, 1960, in Grassé's *Traité de Zoologie*, Masson et Cie., Paris.] *adduct. n.*, nerve of adductor muscle; *cerebr. gang.*, cerebral ganglion; *d. mantle n.*, dorsal mantle nerve; *d. mesent.*, dorsal mesentery; *esoph.*, esophagus; *ext. arm n.*, external nerve of the lophophore arm; *inf. arm n.*, inferior nerve of the arm; *main arm n.*, main nerve of the arm; *nephr. pore*, nephridial pore, its duct full of eggs; *subesoph. gang.*, subesophageal ganglion; *v. mantle n.*, ventral mantle nerves; *v. mesent.*, ventral mesentery.

of nerve cells along nerves and it is to be expected that in a simple epithelial system like this there will be a gradual transition from the ganglia to medullary nerves; whether any of the nerves is nonmedullary is not clear.

Various **main nerves** are recognized (Fig. 12.9). From the supraesophageal ganglion the principal arm nerves or medullary strands of the lophophore run in the base of the brachial fold. From the single (or multiple) circumesophageal connective accessory and lower area nerves arise, the latter sending a branch to each tentacle (Fig. 12.9). The subesophageal ganglion sends out a main lateral nerve and several minor laterals into the dorsal mantle lobe on each side. Fine nerves also go to the dorsal mesentery. A pair of heavy strands proceed posteriorly and give branches to the ventral mantle lobes and to the adductor muscles, and finally into the pedicle. These nerves agree rather well among the genera described. Apparently most of the muscles are supplied by the subesophageal ganglion. In *Lingula* and *Crania* there is said to be a confluence of the ends of the

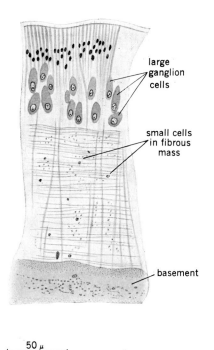

Figure 12.10. Brachiopoda. *Lingula*, part of a sagittal section of the subesophageal ganglion. [Blochmann, 1900.]

much-branched nerves in the margin of the mantle lobe to form a ring nerve there. Nervous plexuses in the wall of the digestive tract, in the body wall, and in the arms are supposed to be present; some figures are given by Joubin (1886). The same author (1892) thought he saw nervous elements ending in striated muscle.

Nerves to the stomach are seen coming from both supra- and subesophageal ganglia (Joubin, 1886).

Receptors are little known. Sensory cells have not been definitely recognized, as may be expected in the absence of specific nerve stains. They are thought to be scattered but most abundant on the tentacles with their supraesophageal innervation. There may be a concentration under an epithelial thickening on the lips (Joubin, 1886) and on the cirri. Statocysts are found in larval and juvenile articulates and, according to Yatsu (1902), in *Lingula* adults, though this was previously denied by Blochmann in the same genus; Morse (1902) said the liths inside are in constant vibratory motion. The existence of a closure response to shadow points to photoreceptors, probably unicellular.

Ciliary control by the nervous system seems likely, on the basis of reports that cilia change speed and amplitude of beat and, in combination with movements of tentacles, seem to function in acceptance and rejection of food particles.

Suggestions that stomach papillae are gustatory and that terminal papillae of the mantle are tactile have been questioned (Delage and Hérouard, 1897).

III. PHORONIDA

The best understood lophophorates, nervously—the phoronids—offer at the same time some of the best opportunities for new work. This small phylum of wormlike tubicolous filter feeders consists of but two genera and hardly a score of species, relatively highly homogeneous in organization. The nervous system is the most completely superficial one among the lophophorate phyla; all known nerve cells are intraepithelial. Over the whole surface an epidermal plexus is either visible or functionally indicated (Fig. 12.11)—and the functional evidence points to a **true nerve net**. This epidermal stratum is thickened around the anterior end of the enteric tract as a nervous ring which is heaviest dorsally and, forming the principal nervous concentration of the body, is there called a ganglion. The **nervous ring** receives the numerous tentacle nerves and sends motor innervation to the main longitudinal muscles, in part via one or a pair of giant fibers. There are no longitudinal nerve cords, according to current interpretation. The extreme simplicity is correlated with an absence of sense organs and of coelomic and intestinal systems of neurons; it does not appear to be a secondarily derived simplicity. All the main parts of the adult nervous system are new formations appearing only during or after metamorphosis of the larva and disappearance of the larval ganglion and tentacles. There is a striking resemblance between phoronids and enteropneusts in the general histological character of the nervous system, but it is doubtless superficial

III. PHORONIDA

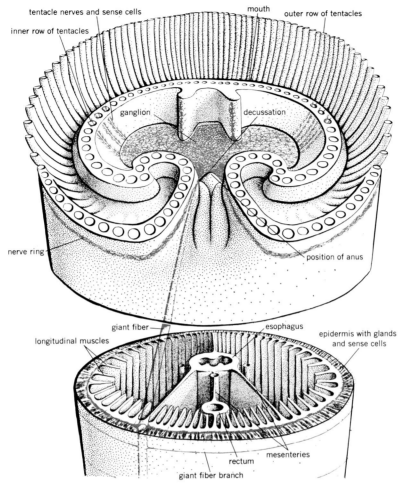

Figure 12.11. Phoronida. *Phoronis,* anterior end viewed from the dorsal side; the tentacles cut away. Nervous tissue in blue. [Based on Silén, 1954.]

and parallel rather than indicative of common origin. Silén should be consulted for a discussion of similarities and differences with several other phyla.

Knowledge of the gross features of phoronid nervous organization has been available since Caldwell (1882), Benham (1889), and Cori (1890); Selys-Longchamps (1907) provided the classical summary, beautifully illustrated. Additions by Hilton (who had some slight success with methylene blue), Marcus, Ono, and others appeared in the next fifty years. Then Silén in 1954 gave a modern, detailed account with the first critical histology, based on silver stains; together with his comparative information, this paper adequately replaces all earlier literature. Nevertheless, special stains did not succeed as well as may be hoped and the connections among the neurons remain unknown.

1. Nervous ring and ganglion

The ring is circumoral and the ganglion is the more or less bulging dorsal side of it, lying between mouth and anus (Fig. 12.11). The ganglion passes, without sharp limits, into the neighboring general epidermal plexus as well as into the nervous ring. The topographic level of the ring and ganglion is that of the diaphragm, and so it slopes obliquely posteroventrally. The region of body epithelium in which the ganglion lies Silén has called the preoral field, delimited dorsally by the anus and nephridial openings, laterally and ventrally by the inner row of tentacles. Besides continuity with the general plexus, the main path-

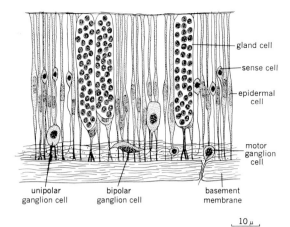

Figure 12.12. Phoronida. *Phoronis hippocrepia*, longitudinal section of the muscle region to show basiepithelial nerve plexus and the types and positions of neurons; diagrammatic. (Silén, 1954.)

ways leading into these nervous concentrations are the nerves—or better, the epidermal tracts of nerve fibers from the tentacles of the lophophore. The main efferent pathways are direct fibers into the anterior ends of the longitudinal muscle bands and the single or paired lateral giant fibers.

The **histological structures** of the ring and ganglion are similar. The nerve fiber stratum is distinct from the nerve cell bodies, the former lying beneath the latter. The cells form a mass, like a primitive rind, three of four deep, and are overlain by ordinary ciliated epithelial cells. Locally, gland cells and supposedly even sense cells are sparse.

As in the general plexus, **three types of nerve cells,** other than sense cells, are distinguished. Large (6–7 μ in *Phoronis hippocrepia*, 10–11 μ in the larger *P. harmeri*), unipolar, cytoplasm-rich cells are perhaps the most abundant but are not peculiar to the ganglion and ring as formerly thought. Their axons go inward, and then turn tangentially to run, little branched, for long distances; they are believed to compose most of the fiber stratum. Sometimes a few cells of this type are notable for their larger size but otherwise no indications of different categories are seen. Bipolar cells with horizontal (that is, tangential) processes are numerous, commonly lying in the fiber stratum. As the only trace of a cluster of cells into a somewhat delimited nucleus, Silén found a distinct group of such cells in the ganglion, on each side of the middorsal line, in the fiber layer. The third cell type resembles the second but lies perpendicularly and sends processes through the basement membrane into the muscles. This composes a special motor supply, chiefly from the ring and going to the most anterior ends of the longitudinal retractor muscles below the diaphragm. A dense group of perforations of the basement membrane is seen, usually corresponding to each muscle.

The **fiber layer** appears to be a very simple neuropile, of loose texture and without special differentiation, at least as far as the methods so far applied can go.

2. The lophophore innervation

Sense cells are abundant in the tentacle epithelium and particularly dense in the interior margins of the lophophore organs, a part which Silén calls the lophophore sense organ. Connecting each tentacle with the circumoral ring are two tracts or thickenings of the general epidermal plexus, one extending along the inner and one along the outer side of the tentacle. In species with spiral whorls of many tentacles, the more distant individual tentacle tracts come from a medullary lophophore tract that connects to the nervous ring. There may be a limited amount of motor innervation to the lophophore but the tentacles are not very mobile individually, and it is probable that most of this innervation is sensory.

3. Receptors

Properly speaking, there are no sense organs. The inner margins of the paired lophophore organs have a dense accumulation of sense cells and —since the rest of the organ is highly glandular— we have another of the curious, intimate associations of sensory and glandular epithelia encountered in many animal groups. There are no suggestions as to the meaning of this association or the role of the sensory part of the lophophore organ. The glandular part, at least, regresses between periods of reproduction. A short thick nerve tract connects the sensory epithelium with the ganglion.

Over the general body surface sense cells occur

commonly in clusters. Each cell is a typical upright bipolar with a distal process extending as a short stiff bristle beyond the surface of the epithelium.

4. The giant fiber system

Silén has presented a strong case for the proposition that the classical "lateral nerves" are actually single giant nerve fibers, one on each side (Fig. 12.13). In several species (and presumably primitively) there is a pair, but in most species (see Silén for a list) there is found only a single

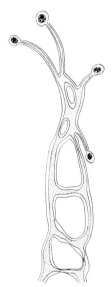

Figure 12.13.

Phoronida. *Phoronis hippocrepia*, reconstruction from horizontal sections, of anterior end of left giant nerve fiber with initial cells and sheath. [Silén, 1954.]

20 μ

fiber, always the left. In one species there is positive evidence that giant fibers are lacking. The giant fibers run basiepithelially through the length of the muscular region of the body, just ventral to the line of attachment of the lateral mesenteries.

The **origin of the giant fibers** is found in the ganglion, on the side opposite to that of its longitudinal course. In *Phoronis hippocrepia* several short thin fibers coalesce to form the giant and each of these initial tributaries is the axon of a cell in the nerve fiber stratum no larger than a common motor neuron. In *P. harmeri* the giant fiber seems to issue from a single cell. No tributary processes from cells during the further course of the giant fiber have been seen. However, in these points, as also in connection with the distributing branches,

uncertainty must be emphasized; these elements do not stain positively with silver but must be followed as negative pale images, a difficult procedure wherever the fiber is less than colossal. The conclusion that cells are only appended to the giant in its anterior end is supported by the report of Schultz (1903) that, following transection, the fibers degenerate and then regenerate from the anterior stump backward. Each fiber decussates middorsally even when there is but one. The route taken from the ganglion to the lateral epithelial position varies among species. In some it leaves the epidermis just ventral to the nephridial opening and passes through the thick basement membrane from the medial to the lateral side of the nephridial duct; here it re-enters the epidermis where a further short obliquely ventral course brings it to the line in which it runs longitudinally, as already mentioned. In other species the fibers do not leave the epidermis in the nephroduct region. After a long course through the muscular region the giant dwindles rapidly and ends as that region gives way to the swollen end bulb. A complication that has confused the literature is the repeated subdivision and reuniting of the giants in species which have a pair of them; these loops are irregular in size and spacing and may involve cross connections, all of which are followed with difficulty. This irregular behavior, reminiscent of conditions in some polychaete giants, is absent in the species with a single giant, in correlation with the larger diameter of the fiber.

The **diameter of the giant** in the trunk in large species (*Phoronopsis harmeri*, 0.8 mm trunk diameter) reaches 27 μ. In a small species (*Phoronis mülleri*, 0.3 mm trunk diameter) it reaches 9 μ. These are examples with a left giant only. Where fibers are present on both sides, they are smaller, and the right is smaller than the left. Thus in *P. australis* (1.2 mm trunk diameter) the left is 5 μ and the right 3 μ. In *P. hippocrepia* (0.4 mm trunk diameter) the left is 4 μ and the right 1.5 μ. The fiber is generally circular and the diameter fairly uniform except near the origin, the termination, and regions of subdivision.

A **thick osmiophilic sheath of concentric laminae**, with circular fibrillae and flattened nuclei, sur-

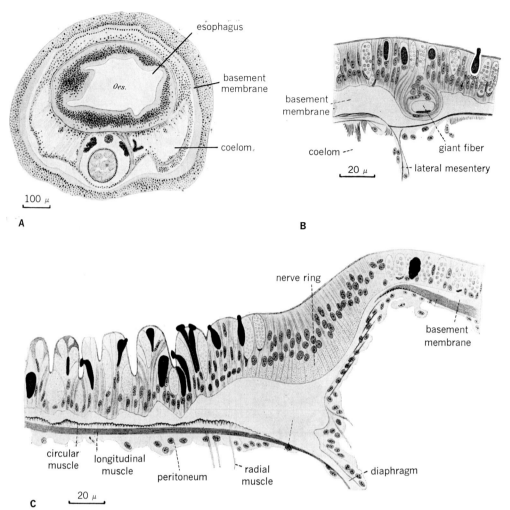

Figure 12.14. Phoronida. *Phoronis.* **A.** Transverse section near anterior end, through the level of the oral nerve ring and diaphragm. **B.** Transverse section of body wall below the left nephridium, showing the "lateral nerve," now regarded as simply the giant fiber. **C.** Longitudinal section of the body wall at the level of insertion of the diaphragm, showing the nerve ring cut across, the glandular character of the ordinary epidermis, the modification over the nerve ring, the close relation of muscles to epidermis, and the continuous sheet of basement membrane intervening. [Selys-Longchamp, 1907].

rounds the giant fiber. Its thickness is from $1/4$ to $3/7$ of the fiber diameter.

Distributing branches are believed to come off the giants at intervals and to reach the muscles. Branches have been seen but they are small (1–2 µ) and are rare and difficult to find; they always proceed at right angles circumferentially in the epidermis. It is presumed that terminal processes enter the muscles.

Functionally the giant fibers are typical, mediating a rapid, synchronous withdrawal response. Stimulation of the tentacles in *P. hippocrepia*, *mülleri*, or *pallida* can cause three kinds of retraction of the lophophore end into the tube. In order, from lowest to highest threshold for mechanical stimulation, they are: first, a quick jerk or a series of them involving only the anterior millimeter or two and causing a partial retraction; second, a slow smooth retraction of eventually large amplitude, and third, a quick jerk of large amplitude. The last is abolished by cutting the giant fiber or fibers high in the nephridial region; or it only occurs down to but not beyond a cut in the trunk region. The first two types of retraction are not affected and may be identified respectively with the motor outflow from the circumoral ring to the anteriormost muscles and the motor outflow from the general body plexus.

Electrical recording from intact animals yields an all-or-none, fast, through-conducting spike assignable to the giant fiber (Wilson and Bullock, 1958). The velocity was found to be 5–6 m/sec (25°C) in relaxed, unstretched specimens of *Phoronopsis harmeri* and *P. viridis* whose fibers had a diameter (not including the thick sheath) of about 25–35 μ. A sharp 5 msec muscle action potential followed the spike after 1.5 msec and declined in amplitude with repetition at one per second. These features resemble polychaetes but suggest no intermediate neurons between giant and muscle.

Phylogenetically, the giant fibers might be imagined to represent a specialization of an original pair of lateral nerve cords. This would be consistent with other phyla that have retained now one, now another pair of a hypothetical ancestral orthogon. In the present phylum this seems unlikely because the giant fibers show no signs of a medullary cord—neither cells nor commissures nor residual nongiant fibers. The nearest relatives, presumed to be ectoprocts and brachiopods, lack any longitudinal cords and the phoronids may represent a secondarily elongated body plan. The giant fibers then would be wholly new adaptations, growing back into the general plexus with the development of the tubicolous habit and the separation of the exposed sensory region from the muscular effector region.

5. Nervous system of the body wall

Aside from the briefest of hints by early observers (Hilton, 1922), we owe to Silén virtually all that is known of this system. Throughout the epidermis is a more or less visible stratum of nerve fibers, just outside the basement membrane and therefore at the base of the epithelium. This stratum is vanishingly thin in the swollen, thin-walled aboral bulb but is distinct and 3–5 nerve fibers thick in the long muscular region between the bulb and the lophophore. We may call this stratum the **epidermal or body wall plexus**. The nerve fibers are said to be thin—hardly reaching 0.2 μ in diameter—and to run mainly longitudinally and circularly. The plexus receives fibers in large numbers from the nucleated strata of the epidermis above it and sends fibers in smaller numbers and singly, down through the basement membrane into the muscle layers. Four types of nerve cells are recognizable. The most superficial are the vertically bipolar sense cells; below them are large (5.5 × 7 μ) unipolars with short axons, which are a surprising component of a simple epidermal plexus and are no doubt internuncial. The third category have their nuclei right in the nerve fiber stratum; they are large horizontal bipolars with long straight processes, and though the cells are uncommon the fibers are thought to make up the bulk of the plexus and to be internuncial. Finally there are, here and there, similar cells that send axons down into the muscle layer—the motoneurons.

Physiological evidence suggests a nerve net in this plexus. Silén cut off the lophophore and anterior nervous concentrations, leaving a long tubular "decapitate" animal with no nervous concentration and, besides the plexus, only the giant fiber in the body wall. Local mechanical stimulation causes local contraction of a ring of circular muscle and this is believed, though without special experimental evidence, to be mediated nervously, and hence to imply adequate abundance of sense cells and of local motoneurons. Stronger stimulation causes slowly spreading response that can occupy the whole length but with decrements in intensity; this spread is also believed by Silén to be nervous and to imply internuncial neurons but muscular spread has not been ruled out. If an oblique cut is made—resulting in a tongue of body wall attached at one end to the rest of the body (see Fig. 27.15 under Enteropneusta)—stimulation can be shown to spread from the tongue to the body or vice versa, and no difference is observed whether the giant fiber is included in the tongue or not. That fiber apparently plays no part in these decrementally spreading responses that arise from sense cells in the general plexus. Similar results were obtained in the aboral bulb (ampulla), where the giant fiber cannot be seen.

This is the best evidence available in favor of a nerve net in Phoronida; it is much more than suggestive, but we ought to have further corroborative details on the directions and extent of spread around obstacles, and the effect of repetition and of controlled electrical stimulation, as well as more evidence on the possibility of muscular spread.

6. Innervation of the gut and other internal organs

The motor supply to the body muscles has already been treated. A short thick nerve tract arises in the ganglion, decussates, and passes into the coelom of the anal region—but to an unknown destination. A basiepithelial nervous layer is visible in the wall of the esophagus, continuous through the mouth with the general body wall plexus. No further internal nervous system is known definitely as yet.

Classified References

Ectoprocta (Bryozoa)

Allman, 1856; Atkins, 1955; Borg, 1926; Braem, 1911; Brien, 1960; Bronstein, 1937; Claparède, 1871; Clayton, 1932; Cori, 1941; Davenport, 1893; Forbes, 1933; Gerwerzhagen, 1913; Graupner, 1930; Harmer, 1887; Hiller, 1939; Hilton, 1922; Hyman, 1959; Kato, 1950; Krukenberg, 1887; Lutaud, 1955; Marcus, 1926a, 1926b, 1926c, 1934, 1939; Nitsche, 1868, 1869, 1871a, 1871b, 1875; Retzius, 1905; Rogick, 1937, 1948, 1949; Saefftigen, 1888; Salensky, 1877; Seeliger, 1906; Silbermann, 1906; Verworn, 1887.

Brachiopoda

Beauchamp, 1960; Bemmelen, 1883; Beyer, 1866; Blochmann, 1892a, 1892b, 1900; Booker, 1933; Delage and Hérouard, 1897; Heath, 1888; Helmcke, 1939; Hilton, 1922; Hyman, 1959; Joubin, 1886, 1892; Kirtisinghe, 1952; Morse, 1902; Schulgin, 1885; Shipley, 1883; Yatsu, 1902.

Phoronida

Andrews, 1890; Benham, 1889; Caldwell, 1882; Clayton, 1932; Cori, 1890, 1939; Harmer, 1917; Hilton, 1922b, 1922c; Koshtoyants, 1956; Lönöy, 1953; Marcus, 1949; Masterman, 1900; M'Intosh, 1888; Ono, 1944; Pixel, 1912; Schultz, 1903; Selys-Longchamps, 1907; Silén, 1954; Torrey, 1901; Wilson and Bullock, 1958.

Bibliography

ALLMAN, G. J. 1856. *A Monograph of the Fresh-water Polyzoa, Including All the Known Species, both British and Foreign.* Ray Society, London.

ATKINS, D. 1955. The ciliary feeding mechanism of the cyphonautes larva (Polyzoa Ectoprocta). *J. mar. biol. Ass. U.K.* 34:451-466.

BEAUCHAMP, P. DE. 1960. Classe des Brachiopodes. *Grassé's Traité de Zoologie,* 5:2:1380-1499.

BEMMELEN, J. F. VAN. 1883. Untersuchungen über den anatomischen und histologischen Bau der Brachiopoda Testicardinia. *Jena. Z. Naturw.,* 16:88-161.

BENHAM, W. B. 1889. The anatomy of *Phoronis australis. Quart. J. micr. Sci.,* 30:125-158.

BEYER, H. G. 1866. A study of the structure of *Lingula (Glottidia) pyramidata* Stein (Dall). *Johns Hopk. Univ., Stud. Biol.,* 3:227-265.

BLOCHMANN, F. 1892a. Ueber die Anatomie und die verwandtschaftlichen Beziehungen der Brachiopoden. *Arch. Ver. Naturg. Mecklenb.,* 1892:37-50.

BLOCHMANN, F. 1892b. *Untersuchungen über den Bau der Brachiopoden. I.* Fischer, Jena.

BLOCHMANN, F. 1900. *Untersuchungen über den Bau der Brachiopoden. II.* Fischer, Jena.

BOOKER, F. W. 1933. Note on the internal structures of *Barrandella* and *Sieberella. J. roy. Soc. N.S.W.,* 66:339-343.

BORG, F. 1926. Studies on recent cyclostomatous Bryozoa. *Zool. Bidr. Uppsala,* 10:181-504.

BRIEN, P. 1960. Classe des bryozoaires. *Grassé's Traité de Zoologie,* 5:2:1053-1335.

BRONSTEIN, G. 1937. Étude du système nerveux de quelques bryozoaires gymnolémides. *Trav. Sta. biol. Roscoff,* 15:155-174.

CALDWELL, W. H. 1882. Preliminary note on the structure, development and affinities of *Phoronis. Proc. Roy. Soc.,* 34:371-383.

CLAPARÈDE, E. 1871. Beiträge zur Anatomie und Entwicklungsgeschichte der Seebryozoen. *Z. wiss. Zool.,* 21:137-174.

CLAYTON, D. E. 1932. A comparative study of the non-nervous elements in the nervous systems of invertebrates. *J. Ent. Zool.,* 24:3-22.

CORI, C. J. 1890. Untersuchungen über die Anatomie und Histologie

Bibliography

der Gattung *Phoronis*. *Z. wiss. Zool.*, 51:480-568.

CORI, C. J. 1939. Phoronidea. *Bronn's Klassen*, 4:4:1:1:1-183.

CORI, C. J. 1941. Bryozoa. *Kükenthal's Handb. Zool.*, 3:2:15:57:263-502.

DAVENPORT, C. B. 1893. On *Urnatella gracilis*. *Bull. Mus. comp. Zool., Harv.*, 24:1-44.

DELAGE, Y. and HÉROUARD, E. 1897. Brachiopodes. In: *Traité de Zoologie Concrète*. Vol. V. Paris.

FORBES, A. 1933. Conditions affecting the response of the avicularia of *Bugula*. *Biol. Bull., Woods Hole*, 65:469-479.

GERWERZHAGEN, A. 1913. Beiträge zur Kenntnis der Bryozoen. I. Das Nervensystem von *Cristatella mucedo* Cuv. *Z. wiss. Zool.*, 107:309-345.

GRAUPNER, H. 1930. Zur Kenntnis der feineren Anatomie der Bryozoen (Nervensystem, Muskulatur, Stützmembran). *Z. wiss. Zool.*, 136:38-77.

HARMER, S. F. 1887. Sur l'embryogénie des bryozoaires ectoproctes. *Arch. Zool. exp. gén.*, (2) 5:443-458.

HARMER, S. F. 1917. On *Phoronis ovalis* Strethill Wright. *Quart. J. micr. Sci.*, 62:115-148.

HEATH, A. 1888. Notes on a tract of modified ectoderm in *Crania anomala* and *Lingula anatina*. *Proc. Lpool. biol. Soc.*, 2:95-104.

HELMCKE, J. G. 1939. Brachiopoda. *Kükenthal's Handb. Zool.*, 3:2: 139-262.

HILLER, S. 1939. The so-called 'colonial nervous system' in Bryozoa. *Nature, Lond.*, 143:1069-1070.

HILTON, W. A. 1922a. The nervous system and sense organs. IX. The Bryozoa. *J. Ent. Zool.*, 14:45-53.

HILTON, W. A. 1922b. The nervous system of Phoronida. *J. comp. Neurol.*, 34:381-389.

HILTON, W. A. 1922c. Nervous system and sense organs. X. Phoronida and Actinotrochia. *J. Ent. Zool.*, 14:65-72.

HILTON, W. A. 1922d. The nervous system and sense organs. XI. Brachiopoda. *J. Ent. Zool.*, 14:79-82.

HYMAN, L. H. 1959. *The Invertebrates*, Vol. V. McGraw-Hill, New York.

JOUBIN, L. 1886. Recherches sur l'anatomie des brachiopodes inarticulés. *Arch. Zool. exp. gén.*, (2)4:161-303.

JOUBIN, L. 1892. Recherches sur l'anatomie de *Waldheimia venosa* (Sol). *Mém. Soc. Zool. Fr.*, 5:554-583.

KATO, K. 1950. On a luminous bryozoan. *Zool. Mag. Tokyo*, 59:9-10.

KIRTISINGHE, P. 1952. Giant nerve cells in *Lingula*. *Nature, Lond.*, 170:206-207.

KOSHTOYANTS, KH. S. 1956. Contributions on the evolution of functions of the nervous system (Phoronida, Enteropneusta, Ascidia). *Adv. mod. Biol., Moscow (Usp. sovr. Biol.)*, 41:306-320. (In Russian.)

KRUKENBERG, C. F. W. 1887. Die nervösen Leitungsbahnen in dem Polypar der Alcyoniden. In: *Vergleichend-physiologische Studien, Experimentelle Untersuchungen von C. F. W. Krukenberg*, Ser. 2, Sect. 4. Winter, Heidelberg.

LÖNÖY, N. 1953. A comparative anatomical study on *Phoronis ovalis* Wright from Norwegian, Swedish, and Brazilian waters. *Univ. Bergen Årb. Naturv. R.*, 1953:1-29.

LUTAUD, G. 1955. Sur la ciliature du tentacule chez les bryozoaires chilostomes. *Arch. Zool. exp. gén.*, 92:13-19.

MARCUS, E. 1926a. Beobachtungen und Versuche an lebenden Meeresbryozoen. *Zool. Jb. (Syst.)*, 52:1-102.

MARCUS, E. 1926b. Beobachtungen und Versuche an lebenden Süsswasserbryozoen. *Zool. Jb. (Syst.)*, 52: 279-350.

MARCUS, E. 1926c. Sinnesphysiologie und Nervensystem der Larve von *Plumatella fungosa*. (Pall.). *Verh. dtsch. Zool. Ges.*, 31:86-90.

MARCUS, E. 1934. Über *Lophopus crystallinus* (Pall.). *Zool. Jb. (Anat.)*, 58:501-606.

MARCUS, E. 1939. Bryozoarios marinhos brasileiros. III. *Bol. Fac. Filos. Ciênc. S. Paulo (Zool.)*, 3:11-353. (In Portuguese, with English summary.)

MARCUS, E. DU B.-R. 1949. *Phoronis ovalis* from Brazil. *Bol. Fac. Filos. Ciênc. S. Paulo (Zool.)*, 14:157-171.

MASTERMAN, A. T. 1900. On the Diplochorda. III. The early development and anatomy of *Phoronis buskii* McI. *Quart. J. micr. Sci.*, 43:375-418.

M'INTOSH, W. C. 1888. Report on *Phoronis buskii*, n. sp., dredged during the voyage of H. M. S. Challenger, 1873-1876. *Challenger Rep.* 27:1-27.

MORSE, E. S. 1902. Observations on living Brachiopoda. *Mem. Boston Soc. nat. Hist.*, 5:313-374.

NITSCHE, H. 1868. Beiträge zur Anatomie und Entwickelungsgeschichte der phylactolaemen Süsswasserbryozoen, insbesondere von *Alcyonella fungosa* Pall. sp. *Arch. Anat. Physiol., Lpz.* 1868:465-521.

NITSCHE, H. 1869. Beiträge zur Kenntniss der Bryozen. I. Beobachtung über die Entwickelungsgeschichte einiger cheilostomen Bryozoen. *Z. wiss. Zool.*, 20:1-13.

NITSCHE, H. 1871a. Beiträge zur Kenntniss der Bryozoen. III. Ueber die Anatomie und Entwickelungsgeschichte von *Flustra membranacea*. *Z. wiss. Zool.*, 21:416-498.

NITSCHE, H. 1871b. Beiträge zur Kenntniss der Bryozoen. IV. Ueber die Morphologie der Bryozoen. *Z. wiss. Zool.*, 21:92-119.

NITSCHE, H. 1875. Beiträge zur Kenntniss der Bryozoen. V. Ueber die Knospung der Bryozoen. *Z. wiss. Zool.*, 25, (suppl.) 343-402.

ONO, K. 1944. Physiological studies of stimuli in *Phoronis*. *Physiol. Ecol. Contr. Otsu hydrobiol. exp. Sta. Kyoto Univ.*, 10:1-11.

PIXEL, H. L. M. 1912. Two new species of the Phoronidea from Vancouver Island. *Quart. J. micr. Sci.*, (2) 58:257-284.

RETZIUS, G. 1905. Das sensible Nervensystem der Bryozoen. *Biol. Untersuch.*, N.F. 12:49-54.

ROGICK, M. D. 1937. Studies on freshwater Bryozoa. VI. The finer anatomy of *Lophopodella carteri* var *typica*. *Trans. Amer. micr. Soc.*, 56:367-396.

ROGICK, M. D. 1948. Studies on marine Bryozoa, II. *Barentsia laxa* Kirkpatrick 1890. *Biol. Bull., Woods Hole*, 94:128-142.

ROGICK, M. D. 1949. Studies on marine Bryozoa. IV. *Nolella blakei* n. sp. *Biol. Bull., Woods Hole*, 97:158-168.

SAEFFTIGEN, A. 1888. Das Nervensystem der phylactolaemen Süsswasser-Bryozoen. *Zool. Anz.*, 11: 96-99.

SALENSKY, M. 1877. Études sur les bryozoaires entoproctes. *Ann. Sci. nat. (Zool.)*, (6) 5 (3):1-60.

SCHULGIN, M. A. 1885. *Argiope kovalewskii*. (Ein Beitrag zur Kenntniss der Brachiopoden.) *Z. wiss. Zool.*, 41:116-141.

SCHULTZ, E. 1903. Aus dem Gebiete der Regeneration. III. Über Regenerationserscheinungen bei *Phoronis mülleri*. Sel. Long. *Z. wiss. Zool.*, 75:391-420.

SEELIGER, O. 1906. Über die Larven und Verwandtschaftsbeziehungen der Bryozoen. *Z. wiss. Zool.*, 84:1-78.

SELYS-LONGCHAMPS, M. DE. 1907. *Phoronis*. *Fauna Flora Neapel*, 30: 1-280.

SHIPLEY, A. E. 1883. On the structure and development of *Argiope*. *Mitt. zool. Sta. Neapel*, 4:494-520.

SILBERMANN, S. 1906. Untersuchungen über den feineren Bau von *Alcyonidium mytili*. *Arch. Naturgesch.*, 72: 1:265-310.

SILÉN, L. 1954. On the nervous system of *Phoronis*. *Ark. Zool.*, (2) 6:1-40.

TORREY, H. B. 1901. On *Phoronis pacifica*, sp. nov. *Biol. Bull., Woods Hole*, 2:283-288.

VERWORN, M. 1887. Beiträge zur Kenntnis der Süsswasserbryozoen. *Z. wiss. Zool.*, 46:99-130.

WILSON, D. M. and BULLOCK, T. H. 1958. Electrical recording from giant fiber and muscle in phoronids. *Anat. Rec.*, 132:518-519.

YATSU, N. 1902. Notes on the histology of *Lingula anatina* Bruguière. *J. Coll. Sci., Tokyo*, 17(5):1-30.

ZIRPOLO, G. 1933. *Zoobotryon verticillatum* (delle Chiaje.) *Mem. Accad. Lincei*, 17:109-442.

CHAPTER 13

Sipunculoidea, Echiuroidea, and Priapuloidea

Summary	650	**II. Echiuroidea**	**656**
I. Sipunculoidea	**650**	**III. Priapuloidea**	**658**
A. The Ventral Nerve Cord	651	Classified References	659
B. The Brain	653		
C. Nerves and Plexuses	654	Bibliography	659
D. Receptors	654		
E. Physiology	655		

Summary

Sipunculids have been little studied by special nerve techniques or physiologically but offer several attractive features for new work. The nervous system consists of a brain and ventral nerve cord, joined by a circumesophageal connective, and peripheral nerves and plexuses in the body wall and gut; all these structures are much more advanced than in the groups treated and resemble those in annelids. The ventral cord is unpaired, unsegmented, internal, with a rind which is entirely ventral and a fiber core dorsal, without notable differentiation of cell groups or neuropile in the cross section or longitudinally. The brain has masses of globuli cells, neurosecretory cells, multipolar giant cells, and perhaps centralized sense cells; the neuropile shows some differentiation, including glomeruli. The nerves are said to be free of nerve cells. They join a plexus beneath the peritoneum and another beneath the skin which supplies many receptors and glands. Another plexus on the gut wall connects with the brain. Although lacking elaborate sense organs, sipunculids are well provided with unicellular and simple multicellular receptors. There is some information on simple physiological experiments concerning ventral cord conduction, the central origin of muscle tone, inhibitory reflexes, brain lesions and peripheral nerve-muscle relations. Fast and slow, graded neuromuscular junction potentials have been recorded intracellularly in muscle fibers, and also, under some conditions, spikes.

Echiuroids have a similar system except that the brain is merely a dorsal commissural medullary strand and there are traces of segmentation. A rich supply of anastomosing nerves enters the proboscis margin. Receptors are known only in the form of primary sense cells, singly or grouped in papillae.

Priapuloids have a still less well developed nervous system and receptive apparatus. A circumenteric nerve ring and a quasisegmental ventral cord are located in the epidermis. Pharynx and proboscis are well innervated. A general superficial plexus is indicated. No sense organs are definitely known.

These three phyla are not regarded as closely related. They are here treated together for convenience as minor coelomate groups. Hyman placed the priapuloids among the pseudocoelomates in the heterogeneous grouping Aschelminthes. But there are reasons to believe that the body cavity is a coelom and certainly the nervous system is more like that of sipunculoids than of nematodes, rotifers, acanthocephalans, and so on.

I. SIPUNCULOIDEA

This group of unsegmented protostomatous coelomates is relatively familiar and accessible, though comprising only about 13 genera. It has been frequently studied for general anatomy but hardly at all with special technics for the histology or physiology of the nervous system. Several features should attract new study: the great extent of apparently undifferentiated, unsegmented neuropile in the long ventral nerve cord, made up of small, perhaps short fibers and with relatively little glia; the annelidlike plan; the presence of good nerves in the form familiar in higher animals; the skin plexus; the association of glands and receptors; and the active introvert and sedentary habit. The main accounts of the nervous system are those of Andreae (1882); Andrews (1890); Ward (1891); Cuenot (1900); Metalnikoff (1900); Mack (1902); Paul (1910); Hérubel (1907); Gerould (1939); and especially Åkesson (1958). Good general summaries are given by Baltzer

(1931b) in Kükenthal's *Handbuch* and by Tétry (1959) in Grassé's *Traité*. Like the articulates, the sipunculids have a supraesophageal ganglion or brain, a circumesophageal connective and an unpaired ventral nerve cord the length of the body, giving off body wall nerves, an enteric system, and a subepidermal plexus.

A. The Ventral Nerve Cord

This is median, unpaired, and internal, connected to the body wall only by its nerves and in some forms by a mesentery anteriorly. It is quite uniform in size and composition throughout the length of the body. There is no sign of segmentation or of a paired origin. In some forms such as *Sipunculus* the cord ends in a slight swelling, sometimes called the caudal ganglion, associated with a protrusible, secretory "terminal organ" (see Åkesson for details). Several scores of **nerves come off the cord** in right and left series; these are sometimes opposite, sometimes alternating, sometimes irregular—all three conditions may succeed each other along the length of one individual (*Phascolosoma*, Hilton, 1917). Each nerve emerges as a number of rootlets which then fuse. The anterior end of the cord is elevated from the body wall and in *Sipunculus* the first 6 or 7 nerves emerge as long unpaired roots which fork right and left to supply the introvert muscles. This part of the cord is accompanied right and left by a pair of longitudinal **paraneural muscles,** tapering posteriorly to different degrees in different species. It can be expected that the fibers in the cord will be zigzagged —even though the cord may not show this in outline—when this muscle is contracted. There is a thin **connective tissue coat** around the cord and nerves; von Mack (1902) gives a detailed account.

Histologically the cord appears as in Fig. 13.2. The cellular rind is all ventral, the fiber core dorsal, and both are unpaired; the boundary between them is sharp, being almost a membrane. The differentiation of cell groups and of neuropile is very slight. Åkesson, Ward, and von Mack note small, medium, and large cells. The first and third are most common in the caudal enlargement of *Sipunculus*. The large cells are like the giant cells of the brain in form and size and in appear-

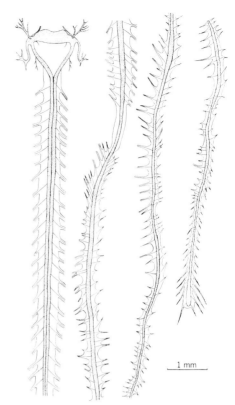

Figure 13.1. Sipunculoidea. *Phascolosoma,* the central nervous system dissected in toto. It has been cut into four pieces, which should be imagined as continuous, the lower end of each attached to the upper end of the next one to the right. The brain is at upper left and shows pigment spots, cerebral tubes, and cerebral nerves. It joins the ventral cord via circumesophageal connectives. The ventral cord is the central, stippled band; it is accompanied by muscle bands, which also send slips to the body wall accompanying each nerve. The nerves are interrupted (upper, second from left) where the introvert ends. [Hilton, 1917.]

ing to be neurosecretory. The cord is a good example of a long stretch of nervous tissue of uniform structure, probably chiefly of relays of short neurons. The inside of the cord sheath is continuous with a network of glia fibers in the cord, but this tissue is less profuse than in annelids and arthropods. Hilton recognizes four classes of nerve fibers: (a) short ascending or descending interneurons, (b) local interneurons, (c) efferent fibers, and (d) afferent fibers from the nerves. No long distance pathways have been reported. Some histological details of regenerating cords are given by Schleip (1934a, 1934b, 1935), Wegener (1938), and Åkesson (1958).

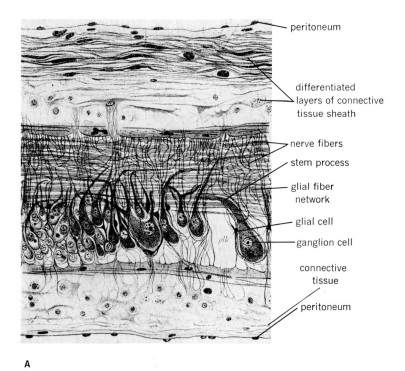
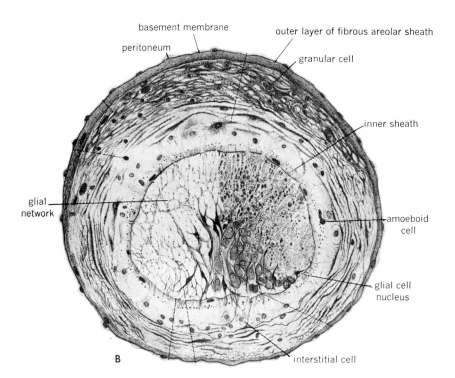

Figure 13.2. Sipunculoidea. *Sipunculus nudus*, sections through part of the anterior third of the ventral cord. **A.** Longitudinal section near the median plane. **B.** Transverse section. [Mack, 1902.]

B. The Brain

The brain is a transverse oval mass sometimes with right and left lobes, just dorsal to the pharynx. It is internal, embedded in connective tissue at the base of the tentacles; close association with the latter seems to be primitive whereas a more posterior position as in *Sipunculus*, is secondary. The brain bears 5 or 6 pairs of nerves

Microscopically the brain is sheathed in a connective tissue tunic described in different books as thin and thick. It differs from the ventral cord in having a cell rind on all sides of the fiber core and in possessing several sizes of cells. Small chromatic cells without apparent cytoplasm occur in dense masses dorsolaterally—justifying the term globuli cell mass; their axons may form distinct bundles. Medium-sized cells, usually uni-

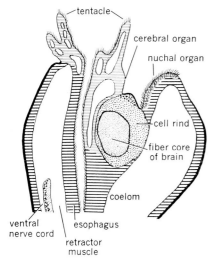

Figure 13.3. Sipunculoidea. Scheme of the head in sagittal section. [Åkesson, 1958.]

to the tentacles, mouth, pharynx, muscular ring at the introvert base, and to the region of the oral disc including the nuchal organ, when present. Nerves to the nearby muscles and to the retractor of the introvert often take exit from the circumesophageal connective. There is one main nerve to each tentacle lobe, running on the inner side of the middle tentacular canal (*Dendrostomum*). Laterally the circumesophageal connective emerges from each side of the brain, encircles the gut, and joins the anterior end of the ventral cord. The anterior surface of the brain, especially in *Sipunculus*, may exhibit a series of papilliform processes or leaflike projections which are thought to represent a sense organ. In some species a median unpaired cephalic tube, or a pair of them, opens onto the dorsal surface and internally ends blindly on the brain in an expansion called the cerebral organ. This structure and the pigment spots in the brain are treated below under receptors.

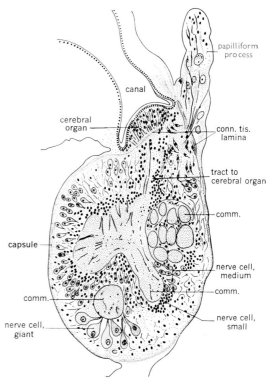

Figure 13.4. Sipunculoidea. *Sipunculus nudus*, longitudinal section of the brain, *comm.*, commissure; *conn. tis. lamina*, connective tissue lamina separating brain from cerebral organ and from papilliform process. [Metalnikoff, after Tétry, 1959, in *Grassé's Traité de Zoologie*, Masson et Cie., Paris.]

polar, are given as 20 μ in diameter, and predominate below the fiber core. Giant cells, 50 μ or more and typically multipolar, lie in definite places in the posteromedian region of the brain and send their axons across the midline. Small multipolars are said to be closely associated with the giant cells. Fusiform bipolars, possibly centralized primary sense cells, occur at the base of the digitiform organ. Metalnikoff (1900) attempted to distinguish centers and pathways (see also

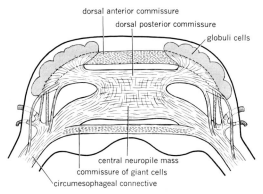

Figure 13.5. Sipunculoidea. *Sipunculus nudus*, scheme of the main fiber bundles of the brain. The nerves shown pass to the walls of the cerebral tube and the dorsal tentacles. [Åkesson, 1958.]

Ward and Åkesson). The distribution of cell types just given represents the centers. Besides the central neuropile and the central commissure within it, there are in the highest forms like *Sipunculus* a posterior dorsal commissure, an anterior dorsal commissure, and a posterior giant cell commissure; the last two contain glomerular masses of fibers. The first two are really common neuropile; the so-called central commissure is the root of the circumesophageal connective. Each of the nerves has a central root or tract. Neurosecretory cells of several types are known, discharging into body fluids and into the neuropile; a storage site is thought to be identified in the digitate processes of *Sipunculus* (see Chapter 6). *S. nudus* is the most differentiated in brain structure; others, especially smaller species like *Golfingia minuta* and *Onchnesoma steenstrupi*, are relatively simple, lacking for example the globuli masses and glomeruli. Details for 14 species are given by Åkesson (1958).

C. Nerves and Plexuses

The nerves from the ventral cord generally lie on the inside of the circular muscle. Each main lateral nerve branches several times, forming rings that are incomplete dorsally, as confirmed physiologically by von Uexküll. The nerves are said to be free of nuclei except for those of the sheath, implying that ganglion cells are not indiscriminately scattered along nerves as in lower forms. Radial nerves are given off inwardly from these circular nerves to form a plexus just beneath the peritoneum and also outwardly to join the subepidermal plexus. The latter is said to be composed mainly of longitudinal fibers and supplies many skin receptors and glands. Åkesson traces a continuity between the subperitoneal plexus of the body wall and that of the retractors and gut. A plexus on the gut wall, in the connective tissue layer and in the mesentery supporting the spindle muscle, is continuous with the visceral nerves from the brain (Andrews, Metalnikoff) and contains large multipolar cells. Posterior regions of the intestine and also the nephridia receive branches from the ring nerves (Stehle, 1953). The spindle muscle is accompanied by a strong nerve and several smaller nerves running parallel to its fibers. The retractors likewise have several longitudinal nerves, derived from the circumesophageal connectives. These nerves, plexuses, and even motor terminations are described from methylene blue and silver preparations by Åkesson (1958).

D. Receptors

For sedentary animals feeding on detritus and the like, sipunculids are rather well provided with receptors. However, they lack elaborate organs of special sense.

Unicellular receptors as bipolar sense cells are probably of general occurrence; among others, Ward (1891) and Shitamori (1936) have seen them, especially about the tentacles.

Multicellular sense organs of several sorts are described but their relations are poorly understood; often the sensory identification is not certain. One **simple type is a shallow pit** in the ciliated epithelium, seen on the tentacular fold and introvert of *Sipunculus* (Ward, Metalnikoff). A common type is a **fusiform bud** of several spindle-shaped cells, probably sensory nerve cells, distally penetrating up through the cuticle to the surface and proximally continuous with a nerve bundle. The distal protrusion may be specialized as a papilla and set off by a groove (*Golfingia*, Cuenot, Nickerson).

Reminiscent of earlier phyla, there are curious

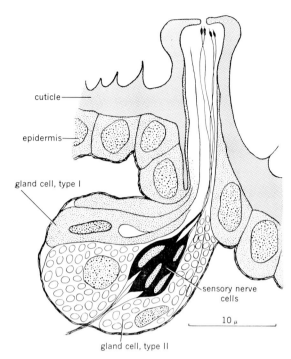

Figure 13.6. Sipunculoidea. *Golfingia procera*, section through an epidermal organ from the trunk region. Normally there is a layer of mud up to the level of the tip of the papilla. [Åkesson, 1958.]

epidermal organs marked by the **association of sense cells and glands,** which may be multicellular. Several types of such organs are described and illustrated in detail by Åkesson (Fig. 13.6).

Associated with the brain are several apparently sensory structures. The **nuchal organ** is a 2- or 4-lobed pad of thickened epithelium lying dorsally at the edge of the oral disc. It is innervated from the nearby anterior surface of the brain, usually by a pair of nerves branching repeatedly. There may be clusters of cells along the branches forming small ganglia; these are probably the bipolar sense cells, sunk below the surface and elongating their distal processes. The tendency extends to incorporation of some of the sense cells into the brain in several species. It is remarkable that exactly the same tendency and series of stages is found in several families of polychaete annelids with respect to a closely analogous sense organ, also called the nuchal organ. Peebles and Fox (1933) report that the most sensitive region for chemical, photic, and mechanical stimuli includes the tentacles and collar and, especially, the nuchal organ. The best guess seems to be that the latter is a chemoreceptor as in polychaetes.

The **cerebral organ** (Gerould, 1939) is the expanded inner end of the cephalic tube or united tubes already mentioned. The tube may be absent and the cerebral organ may be at the surface; in most species the organ is reduced or absent. In *Sipunculus* it is well developed and is richly provided with sensory neurons. More than one author has pointed out that there is a connective tissue separation between the organ and the brain, so that they are not perfectly continuous. A fairly heavy tract of nerve fibers passes from the cells of the cerebral organ to the brain, and it is believed that they are sensory. Åkesson expresses some doubts, based largely on the mainly secretory function in the larva.

A pair of darkly pigmented spots is found in the brain of many species. These are thought to be **pigment-cup ocelli;** each lies at the end of an ocular tube that descends into the brain from the inner end of the cephalic tube (or tubes) or from the dorsal surface when the tubes are missing. The meaning and nature of these epithelial tubes —and of material in the lumen believed to be retractile—is not clear. Ocular tubes are rudimentary in *Sipunculus* but tentacular eyes are developed, as it were, to replace them.

Possibly sensory are a set of **outgrowths from the anterior brain** surface into the coelom in *Sipunculus* and some others (Metalnikoff, 1900). They may be filaments, papillae, or leaflike processes, and are invested with peritoneum and demarcated from the brain by a connective tissue barrier. Through this pass bundles of nerve fibers to the outgrowths, which therefore are taken to be receptive with respect to the coelomic contents. A pair of leaflike outgrowths also projects into the cephalic tube in *S. nudus* (Metalnikoff).

E. Physiology

A few experiments shed light on the functioning of the nervous system in sipunculids. Von Uexküll (1896) made a preparation of part of the cord dissected free with a piece of body wall still attached at both ends. Stimulating in the middle part he showed that conduction occurs both forward and backward and at the same velocity (0.1–0.2 m/sec). The **velocity** is uniform all along the cord but is everywhere strongly decremental and presumably is relayed through many interneurons. In contrast to the ventral cord, the brain is very easily excited and causes prompt retractor response followed by a refractory period. In 1903 the same author studied the problem of **muscle tone** in *Sipunculus*. Whereas a muscle relaxes after its nerve is cut, tone is maintained even after cutting the ventral cord into small sections, as long as one section is attached to the muscle nerve. Drugs on these sections alter the tone, which was therefore conceived of as a product of central ganglion cells locally and throughout the cord. Reflex connections exist such that mechanical **stimulation elicits inhibition** locally and hence thinning; if however there is a transection nearby, the thinning is not seen. There seems to be some antagonistic effect of a rise in tonus on the tone of adjacent centers. Electrical stimulation of the ventral cord can elicit relaxation of retractor muscles as well as body wall. Magnus (1903) gave some effects of drugs on the body wall of *Sipunculus*.

Peebles and Fox (1933) attempted the difficult operation of destroying the brain. In a few successful cases they say that burrowing and retraction of the introvert are not prevented by this maneuver, though the movements are slow and abnormal and tone is low. Since the retractors are innervated from the brain it would be desirable to have further evidence of the completeness of the destruction or of the definite activity of these muscles before accepting this finding. Cutting the ventral nerve cord is accompanied by severe loss of internal pressure and the inability to extend the introvert beyond the level of the cut may be due in part to this.

Detailed studies of the **mechanism of control of muscle,** using the proboscis retractor and the intestinal retractor, have been accomplished by Prosser et al. (1954, 1959). The spread of activation through the proboscis muscle is clearly by long parallel nerves; through the intestinal muscles it is by mechanical pull on successive contractile elements. The proboscis retractor receives two classes of nerve fibers, thick and thin—about 2 μ and less than 1 μ. Focal stimulation can elicit either fast muscle action potentials alone or slow ones alone, each preceded by a nerve impulse. The muscle response is blocked by tubocurarine as well as by procaine, is enhanced by physostigmine, and is made repetitive by veratrine. Intracellular recordings have been achieved in the muscle fibers and show fast and slow graded junctional potentials (e.p.s.p.'s) and, under certain conditions, spikes. At least many of the muscle fibers must receive a dual innervation.

II. ECHIUROIDEA

Here also a single median ventral nerve cord sends a circumesophageal connective around each side of the proboscis, often much elongated with that structure, to a dorsal commissural medullary strand representing the cerebral ganglion. The general pattern resembles the last group and the annelids. In the adult there is usually said to be no sign of segmentation, but Baltzer (1931c) mentions quasi-metameric accumulations of ganglion cells in *Echiurus*. In the larva there is a clearly metameric, paired row of ganglionic rudiments, like two straight strings of beads; the group clearly has traces of segmentation. Among the most important monographs treating the nervous system are Spengel (1880), Rietsch (1886), Skorikow (1909), and Baltzer (1917, 1931c); Dawydoff in Grassé (1959a) gives a useful summary. (See Figs. 13.7, 13.8.)

The ventral nerve cord runs the length of the body internal to the epidermis, with a short ventral mesentery. It sends branches to surround the anus. A generalized cell rind lies dorsolaterally on each side, thus encouraging the idea of a paired origin. A tiny "neural canal" occurs in the midline of the dorsal region of the cord and may be followed into the circumesophageal connectives. Spengel thought it might be a giant nerve fiber. A multilayered sheath surrounds the cord:

II. ECHIUROIDEA

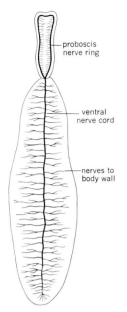

Figure 13.7. Echiuroidea. *Echiurus pallasii*, plan of the nervous system. [Greeff, 1879.]

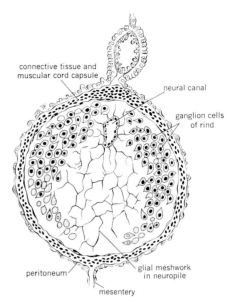

Figure 13.8. Echiuroidea. *Bonellia*, transverse section of ventral nerve cord. [Modified from Baltzer, 1931c, in Grassé's *Traité de Zoologie*, Masson et Cie., Paris.]

a double or triple connective tissue investment, a muscle layer, and externally the peritoneum.

The cerebral ganglia do not show as ganglionic swellings—nor are there such swellings anywhere in the nervous system. But the circumesophageal connectives are extraordinarily long and meet dorsally at the front of the proboscis. They are not cell-free but medullary, each containing dorsally one of the lateral cell columns of the cord. There may be several commissures during the course of the connectives through the proboscis (*Echiurus*), and a rich supply of small anastomosing nerves is given off into the proboscis margin (Baltzer, 1931c), best developed in the glandular

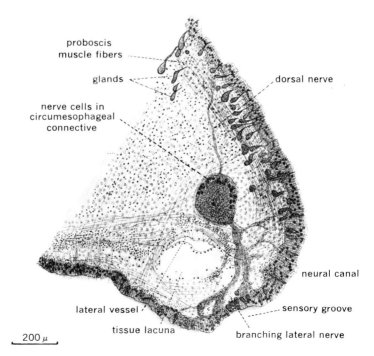

Figure 13.9. Echiuroidea. *Echiurus pallasii*, transverse section of the lateral border of the proboscis. The greatly elongate circumesophageal connectives are medullary and run in this border and give off nerves, apparently chiefly sensory. [Rietsch, 1886.]

forward zone of *Bonellia*. Nothing is known histologically about the anterior nervous system (Fig. 13.9).

Peripheral nerves emerge from the cord at short intervals and, with mates on the other side, form complete nerve rings in the body wall, that is, fuse middorsally (*Echiurus*). But the nerves are not regular nor exactly paired. Branches have been seen to pass into papillae in the epidermis. There is said to be a plexus of nerve fibers with ganglion cells in the skin, especially in the proboscis of *Bonellia*.

Receptors are known only as bipolar sensory neurons isolated or grouped in papillae; they are especially numerous along the edge of the proboscis.

III. PRIAPULOIDEA

In spite of their muscular, sipunculid appearance, heavily armed introversible proboscis, and predaceous habit, the two genera of this phylum have a meager nervous system with poor sensory elaborations. It is reminiscent grossly of both annelids and kinorhynchs, and is located in the epidermis. A **circumenteric nerve ring** encircles the pharynx and consists of ganglion cells and fibers aggregating in thickness considerably more than the ventral cord. It gives off four nerves to the pharynx which run posteriorly in the epidermis of that organ and connect by several nerve rings. Thirteen pairs of proboscis nerves also arise from the main nerve ring and run backward in the epidermis of the longitudinal ribs. The **ventral nerve cord** is the chief nervous structure of the trunk and runs in the midventral line, as a thickening of the epidermis. It is a medullary cord, slightly wider in each of the superficial annuli, and therefore quasi-segmental. Its cells lie chiefly in a ventral column on each side. Caudally, it ends in a widening, an accumulation of cells sometimes called the anal ganglion. A pair of main lateral nerves leaves each annular thickening and proceeds circumferentially, making intracutaneous nerve rings incomplete dorsally. The last pair is heavier than the others. Other nerves have been seen only fragmentarily; their delicacy and superficial position make them difficult to work out. A general superficial plexus is reported basiepidermally. The main authors are Apel (1885), Scharff (1885), and Théel (1906, 1911); see also Baltzer (1931a) in Kükenthal's *Handbuch* and Dawydoff in Grassé (1959b) (Fig. 13.10).

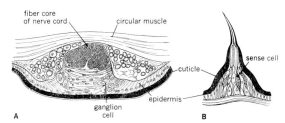

Figure 13.10. Priapuloidea. **A.** *Halicryptus spinulosus*, transverse section of midventral epidermis of body at the beginning of the posterior third, with the ventral nerve cord. [Apel, 1885.] **B.** *Priapuloides*, cutaneous papilla with a tactile hair. [Scharff, 1885.]

Sense organs are not clearly identified. Abundant spines, papillae, and warts beset the surface and possibly some of these are sensory (see especially Scharff, 1885). Some simple reactions to light, touch, and substratum were described by Langeloh (1936); priapuloids seem remarkably indifferent to chemical stimuli.

Classified References

Sipunculoidea

Åkesson, 1958; Andreae, 1882; Andrews, 1890; 1887; Awati, 1936; Baltzer, 1931b; Brandt, 1870; Clayton, 1932; Cuenot, 1900; Fischer, 1893; Gerould, 1939; Haller, 1889; Hérubel, 1907; Hilton, 1917; Mack, 1902; Magnus, 1903; Metalnikoff, 1900; Nickerson, 1901; Paul, 1910; Peebles and Fox, 1933; Prosser, 1960; Prosser and Melton, 1954; Prosser and Sperelakis, 1959; Prosser et al., 1959; Schleip, 1934, 1935; Sluiter, 1902; Shipley, 1890; Shitamori, 1936; Stehle, 1953; Tétry, 1959; Théel, 1911; Uexküll, 1896, 1903; Ward, 1891; Wegener, 1938.

Echiuroidea

Baltzer, 1917, 1931c; Dahl, 1958; Dawydoff, 1959a; Greeff, 1879; Jameson, 1899; Jourdan, 1891; Quatrefages, 1847a; Rietsch, 1886; Skorikow, 1909; Sluiter, 1902; Spengel, 1880, 1912; Théel 1906.

Priapuloidea

Apel, 1885; Baltzer, 1931a; Dawydoff, 1959b; Hammersten, 1915; Horst, 1882; Langeloh, 1936; Scharff, 1885; Schauinsland, 1887; Théel, 1906, 1911.

Bibliography

ÅKESSON, B. 1958. A study of the nervous system of the Sipunculoideae. *Unders. över Öresund*, 38:1-249.

ANDREAE, J. 1882. Beiträge zur Anatomie und Histologie des *Sipunculus nudus* L. *Z. wiss. Zool.*, 36:202-256.

ANDREWS, E. A. 1887-1890. Notes on the anatomy of *Sipunculus gouldii* Pourtalès. *Johns Hopk. Univ. Stud. Biol.*, 4:389-430.

APEL, W. 1885. Beitrag zur Anatomie und Histologie des *Priapulus caudatus* (Lam.) und des *Halicryptus spinulosus* (v. Sieb.). *Z. wiss. Zool.*, 42:459-529.

AWATI, P. R. and PRADHAN, I. B. 1936. The anatomy of *Dendrostoma seignifer* Selenka et de Man. *J. Univ. Bombay* 4:114-131.

BALTZER, F. 1917. Monographie der Echiuriden des Golfes von Neapel. I. *Echiurus abyssalis* Skor. *Fauna Flora Neapel*, 34:1-234.

BALTZER, F. 1931a. Priapulida. *Kükenthal's Handb. Zool.*, 2:2(9):1-14.

BALTZER, F. 1931b. Sipunculida. *Kükenthal's Handb. Zool.*, 2:2(9): 15-61.

BALTZER, V. F. 1931c. Echiurida. *Kükenthal's Handb. Zool.*, 2:2(9): 62-168.

BEAUCHAMP, P. DE. 1960. Classe des Brachiopodes. *Grassé's Traité de Zoologie*, 5:2:1380-1499.

BRANDT, A. 1870. Anatomisch-histologische Untersuchungen über den *Sipunculus nudus* L. *Mém. Acad. Sci. St.-Pétersb. (sci. math., phys., nat.) (Zap. Imp. Akad. Nauk.)*, (7) 16(8): 1-46.

CLAYTON, D. E. 1932. A comparative study of the non-nervous elements in the nervous systems of invertebrates. *J. Ent. Zool.*, 24:3-22.

CUENOT, L. 1900. Le phascolosome. In: *Zoologie Descriptive des Invertébrés*, Vol. I. L. Boutan (ed.). 1:386-422.

DAHL, E. 1958. The integument of *Echiurus echiurus*. *Föhr. K. fysiogr. Sällsk. Lund*, 28:33-44.

DAWYDOFF, C. 1959a. Classe des Echiuriens. *Grassé's Traité de Zoologie*, 5:1:855-907.

DAWYDOFF, C. 1959b. Classe des priapuliens. *Grassé's Traité de Zoologie*, 5:1:908-926.

FISCHER, W. 1893. Weitere Beiträge zur Anatomie und Histologie des *Sipunculus indicus*, Peters. *Jb. hamburg. wiss. Anst.*, 10:1-12.

GEROULD, J. H. 1939. The eyes and nervous system of *Phascolosoma verrillii* and other sipunculids. *Trav. Sta. zool. Wimereux*, 13:313-325.

GREEFF, R. 1879. Die Echiuren (Gephyrea Armata). *Nova Acta Leop. Carol.*, 41(2):1-172.

HALLER, B. 1889. Beiträge zur Kenntnis der Textur des Central-Nervensystems höherer Würmer. *Arb. zool. Inst. Univ. Wien*, 8:175-312.

HAMMERSTEN, O. D. 1915. Zur Entwicklungsgeschichte von *Halicryptus spinulosus* (von Siebold). *Z. wiss. Zool.*, 112:527-570.

HÉRUBEL, M. A. 1907. Recherches sur les sipunculides. *Mém. Soc. Zool. Fr.* 20:107-418.

HILTON, W. A. 1917. The central nervous system of a sipunculid. *J. Ent. Zool.*, 9:30-35.

HORST, R. 1882. II Priapulidae. *Nederland Arch. Zool.*, 1 (suppl.) (4):13-42.

JAMESON, H. L. 1899. Contributions to the anatomy and histology of *Thalassema neptuni* Gaertner. *Zool. Jb. (Anat.)*, 12:535-566.

JOURDAN, É. 1891. Les corpuscules sensitifs et les glands cutanées des géphyriens inermes. *Ann. sci. nat. (Zool.)*, (7) 12:1-14.

LANGELOH, H. P. 1936. Sinnesphysiologische Untersuchungen an Priapuliden *(Priapulus caudatus* und *Halicryptus spinulosus)*. *Biol. Zbl.*, 56: 260-268.

MACK, H. VON. 1902. Das Centralnervensystem von *Sipunculus nudus* L. (Bauchstrang). Mit besonderer Berücksichtigung des Stützgewebes. Eine histologische Untersuchung. *Arb. zool. Inst. Univ. Wien*, 13: 237-334.

MAGNUS, R. 1903. Pharmakologische Untersuchungen an *Sipunculus nudus*. *Arch. exp. Path. Pharmak.*, 50:86-122.

METALNIKOFF, S. 1900. *Sipunculus nudus*. *Z. wiss. Zool.*, 68:261-322.

NICKERSON, M. L. 1901. Sensory and glandular epidermal organs in *Phascolosoma gouldii*. *J. Morph.*, 17: 381-398.

PAUL, G. 1910. Über *Petalostoma minutum* Keferstein und verwandte Arten nebst einigen Bermerkungen zur Anatomie von *Onchnesoma steenstrupii*. *Zool. Jb. (Anat.)*, 29:1-50.

PEEBLES, F. and FOX, D. L. 1933. The structure, functions, and general reactions of the marine sipunculid worm *Dendrostoma zostericola*. *Bull. Scripps Instn. Oceanogr. tech.*, 3: 201-224.

PROSSER, C. L. 1960. Comparative physiology of activation of muscles, with particular attention to smooth muscles. In *"The Structure and Function of Muscle"*, G. H. Bourne, ed. Academic Press, New York.

PROSSER, C. L. and MELTON, C. E. 1954. Nervous conduction in smooth muscle of *Phascolosoma* proboscis retractors. *J. cell. comp. Physiol.*, 44:255-275.

PROSSER, C. L. and SPERELAKIS, N. 1959. Electrical evidence for dual innervation of muscle fibers in the sipunculid *Golfingia* (= *Phascolosoma*). *J. cell. comp. Physiol.*, 54: 129-133.

PROSSER, C. L., RALPH, C. L., and STEINBERGER, W. W. 1959. Responses to stretch and the effect of pull on propagation in non-striated muscles of *Golfingia* (= *Phascolosoma*) and *Mustelus*. *J. cell. comp. Physiol.*, 54:135-146.

QUATREFAGES, A. DE. 1847. Études sur les types inférieurs de l'embranchement des annelés; mémoire sur l'échiure de Gaertner *(Echiurus gaertnerii* Nob.). *Ann. sci. nat. (Zool.)*, (3) 7:307-343.

RIETSCH, M. 1886. Étude sur les géphyriens armés ou échiuriens. *Rec. zool. Suisse*, 3:313-515.

SCHARFF, R. 1885. On the skin and nervous system of *Priapulus* and *Halicryptus*. *Quart. J. micr. Sci.*, 25:193-213.

SCHAUINSLAND, H. 1887. Zur Anatomie der Priapuliden. *Zool. Anz.*, 10:171-173.

SCHLEIP, W. 1934a. Die Regeneration des Rüssels von *Phascolion strombi* Mont. (Sipunculidae). *Z. wiss. Zool.*, 145:462-496.

SCHLEIP, W. 1934b. Der Regenerationsstrang bei *Phascolosoma minutum* Kef. (Sipunculidae). *Z. wiss. Zool.*, 146:104-122.

SCHLEIP, W. 1935. Die Reparationsvorgänge nach Amputation des Hinterendes von *Phascolosoma minutum* Kef. (Sipunculidae). *Z. wiss. Zool.*, 147:59-76.

SHIPLEY, A. E. 1890. On *Phymosoma varians*. *Quart. J. micr. Sci.*, 31:1-27.

SHITAMORI, K. 1936. Histology of the integument of *Siphonosoma cumanense* (Keferstein). *J. Sci., Hiroshima Univ.*, (B), Div. 1, Zoology, 4:155-175.

SKORIKOW, A. S. 1909. Echiurini, sousfamille nouv. des *Gephyrea armata*. Aperçu systématique et monographique. *St. Petersburg Ann. mus. zool.*, 14:77-102.

SLUITER, C. P. 1902. Die Sipunculiden und Echiuriden der Siboga-Expedition. *Siboga Exped.*, 11:1-53.

SPENGEL, J. W. 1880. Beiträge zur Kenntnis der Gephyreen. II. Die Organisation des *Echiurus pallasii*. *Z. wiss. Zool.*, 34:460-538.

SPENGEL, J. W. 1912. Beiträge zur Kenntnis der Gephyreen. III. Zum Bau des Kopflappens der armaten Gephyreen. *Z. wiss. Zool.*, 101: 342-385.

STEHLE, G. 1953. Anatomie und Histologie von *Phascolosoma elongatum* Keferstein. *Saar. Univ. Ann. Naturwiss.*, 2:204-256.

TÉTRY, A. 1959. Classe des Sipuculiens. *Grassé's Traité de Zoologie*, 5:1:785-854.

THÉEL, H. 1906. Northern and Arctic invertebrates in the collection of Swedish State Museum (Riksmuseum). II. Priapulids, Echiurids, etc. *K. svenska VetenskAkad. Handl.*, 40(4):1-26.

THÉEL, H. 1911. Priapulids and Sipunculids dredged by the Swedish Antarctic Expedition 1901-1903 and the phenomenon of bipolarity. *K. svenska VetenskAkad. Handl.*, 47(1):1-36.

UEXKÜLL, J. VON. 1896. Zur Muskel- und Nervenphysiologie von *Sipunculus nudus*. *Z. Biol.*, 33:1-27.

UEXKÜLL, J. VON. 1903. Studien über den Tonus. I. Der biologische Bauplan von *Sipunculus nudus*. *Z. Biol.*, 44:269-344.

WARD, H. B. 1891. On some points in the anatomy and histology of *Sipunculus nudus*. *Bull. Mus. comp. Zool. Harv.*, 21:143-184.

WEGENER, F. 1938. Beitrag zur Kenntnis der Rüsselregeneration der Sipunculiden. *Z. wiss. Zool.*, 150:527-565.

CHAPTER 14

Annelida

Summary	662	(h) Other midbrain structures	718
I. Introduction	**666**	(i) Nuchal ganglia and centers	719
II. The Ventral Cord	**669**	(j) Roots and "commissures" of the circumesophageal connective	720
A. Position, Form, and Number	669	(k) Representation of giant fiber systems in the brain	720
B. Ganglia	671	2. Oligochaeta	722
C. Histologic Structure	673	3. Hirudinea	723
D. Neuronal Composition, Cell Groups, and Pathways	676	E. Nerves of the Brain	725
1. Polychaeta	676	F. Comparison of the Brain in Polychaete Families	727
2. Oligochaeta	680	1. Amphinomidae	727
3. Hirudinea	687	2. Euphrosynidae	727
E. Giant Fibers: Anatomy and Physiology	689	3. Eunicidae	732
1. Polychaeta	689	4. Lumbrineridae	732
2. Oligochaeta	701	5. Nephtyidae	732
		6. Glyceridae	733
III. The Brain	**708**	7. Nereidae	735
A. Generalities	708	8. Aphroditidae and allies	735
B. Histology	710	9. Sedentary families	736
C. Major Divisions	711	10. Other groups	737
D. Internal Structures and Connections	712	IV. The Peripheral Nervous System	**737**
1. Polychaeta and Archiannelida	712	A. Nerves of the Ventral Cord	737
(a) Stomodeal center	712	B. Peripheral Ganglia	741
(b) Palpal center	713	C. Lateral Longitudinal Nerves	742
(c) Other forebrain structures	714	D. Receptors	743
(d) Antennal centers	715	1. Generalized epidermal sensory nerve cells and sense organs	743
(e) Optic centers	715	2. Mechanoreceptors	746
(f) Corpora pedunculata and related structures	716	3. Photoreceptor cells	747
(g) Median mass	718		

4. Eyes	748	3. Walking	764
5. Statocysts	749	4. Swimming	765
6. Nuchal organs	750	5. Writhing	766
E. Nervous Supply and Control of Effectors	750	6. The twitch reflex	766
F. The Peripheral Plexuses	755	7. Reflex arrest of creeping	766
G. The Stomodeal System	756	8. Local events	766
V. Physiological Studies on Central Organization	**760**	B. Roles of Brain and Subesophageal Ganglion	767
A. Reflexes and Cord Activity	761	**VI. Phylogenetic Comparisons**	**768**
1. Peristaltic creeping	762	Classified References	774
2. Parapodial creeping	763	Bibliography	776

Summary

The basic morphologic plan of the nervous system is common to annelids and arthropods and comprises a ventral rope-ladderlike chain of paired segmental ganglia, paired supraesophageal ganglia, and circumesophageal connectives, plus a stomodeal supply from the brain and first ventral ganglia. The ventral cords are derivable from a presumed primitive orthogon plan. The trochophore larva of certain species exhibits something like an orthogon of connectives and commissures, but this becomes chiefly the brain. The ontogeny of the brain and cords is not uniform throughout the phylum.

Values are available for the numbers of nerve cells (and other quantities) in one species of earthworm at various ages; the most notable feature between hatching and maturity is the 50-fold increase in a small characteristic brain cell.

The **ventral cord** is usually internal but in some species basiepithelial. There are two separate cords in a number of genera but in most only one, fused, midventral cord. The primitive condition was probably superficial and paired, but it is not necessary that forms now showing these features are primitive. A third longitudinal connective, the median connective, occurs in many polychaetes and leeches; it may be related to the stomodeal system. Hirudineans are distinctive in having long, cell-free connectives separating beadlike, sharply demarcated ganglia. Oligochaetes and many polychaetes, in contrast, have ill-defined segmental ganglia that virtually run into each other; the cord is called medullary because nerve cells are scattered all along its length. Leeches are further peculiar in the marked compartmentation of their ganglia. Concentration of ganglia is found in some annelids but not nearly so highly developed as in many arthropods. Often the subesophageal ganglion is compound. In a few cases, the first ventral ganglia have pushed up the circumesophageal connectives and united with the brain, an event of great importance in some views of the origin of the arthropod brain. In branchiobdellids and hirudineans, the caudal ganglia are also compound.

The circumesophageal connectives each have two roots in the brain and often bear nerves and contain nerve cells, as if containing the first ventral ganglia. The commissures that would answer this question have not been studied.

The **histologic structure of the ventral cord** is basically similar to that of most invertebrates and unlike anything in the vertebrates. Neuropile typically occupies the central and greater part of the ganglia; it is not clearly demarcated from tracts. The rind of cell bodies is generally loosely packed and confined to ventral and lateral aspects of the ganglia; very few, if any, synapses occur in the rind. Among other features, a layer of longitudinal and circular muscle is commonly found in the sheath of the cord. Leech and earthworm nerve cells and large axons are classical objects to show neurofibrils in reduced silver preparations; polychaetes are probably the same.

A property prominent in this phylum is **tolerance of stretch.** From the most contracted state, extension first involves straightening the zigzagged nerve cord, then straightening zigzagged nerve fibers and finally neurofibrils. Considerable stretch can occur beyond this stage, and nerve fibers are measurably reduced in diameter. Nevertheless, the velocity of nerve impulse conduction is not reduced but remains constant, from some length presumed to represent the complete straightening, up to a maximum tolerance that varies with the species and the nerve fiber.

Neuroglia is not clearly distinct from ordinary connective tissue. There is a nerve cord sheath, well marked in some forms like earthworms, and a framework of glial cells and fibers forming more or less distinct partitions vertically, horizontally, between rind and core, and between smaller masses of cells and of fibers. Larger cells and fibers have individual sheaths. The glial cells were thought by most writers to be ectodermal and a discrete cell type or more than one type; close relations with blood vessels have been described.

Neuronal composition, cell groups, and pathways of the ventral cord are known in some detail for a few forms, based mainly on successful use of methylene blue as an intravitam stain. Afferent fibers are generally small and enter the cord in all the nerves after great reduction in number from the number of sensory neurons in the periphery (according to some authors but denied by others). They usually dichotomize and send branches up and down the cord for short distances ipsilaterally in one of three longitudinal neuropile strips. There is perhaps some segregation by function. Internuncial cells lie in the rind ventrally and laterally and are chiefly unipolar, but a few are special multipolar cells; there are ipsilateral and contralateral ones, short intrasegmental and long intersegmental units, large and small cells. Motoneurons are very few, at least in *Nereis;* they send large axons into all the nerves, mainly of the side opposite the cell body. Certain cell groups can be defined topographically and some individual cells can be consistently recognized, as in arthropods. Endings of nerve fibers among the cells in the rind are occasionally reported but are at best rare and may be neuroglia. Afferent cell bodies inside the central nervous system are not definitely known for the cord. A typical midbody segmental ganglion of the earthworm, *Pheretima*, contains about 1000 nerve cells. No clue is at hand as to the meaning of the cell packets in leech ganglia. The ratio of presumed motor axons in the nerve roots to total nerve cells in the cord is about 1:3.4 in an earthworm and in a leech.

Another constituent of the ventral cord in many polychaetes and nearly all oligochaetes (but not in hirudineans) is the **system of giant fibers.** A few giant fibers are final motor axons, but most are purely central and internuncial. Some are multicellular, others unicellular. A few have segmental septa. When there are several giant fibers, they have different afferent connections and probably motor connections in part. The limited evidence supports a generalization that giant systems mediate an abrupt, symmetrical, over-all withdrawal response to startle stimuli. It is not certain that high conduction velocity is the only significance of the large axon diameter; possibly spike amplitude is also important. Velocity appears to vary as some function of diameter closer to the square root than to simple proportionality. Connections between paired fibers of typical, delaying, but unpolarized synaptic type are known and also low-resistance electrotonic pathways and others of more puzzling nature.

The **brain** in certain polychaetes attains a remarkably high level of complexity, exceeding some lower arthropods and molluscs. Differentiation of parts means that we face a question of independent evolution or homology of brain divisions; little can be said confidently about this. An example is the suggestion from the anatomy of each group that the first ventral ganglia can be seen in various stages of migration up the circumesophageal connections and approximation to the brain, as in arthropods. Even the correspondence of sense organs of the head is still controversial. Both oligochaetes and hirudineans as well as many polychaetes have, by contrast, exceedingly simple brains, often superficial, in correlation with the virtual absence of organs of special sense.

Histologically the brain shows more advanced features than the cord in both the rind and core. Notably this includes clusters of small, plasma-poor, chromatin-rich cells, called globuli cells, and dense, fine-textured neuropile masses; these are present only in some polychaetes having better developed sense organs. Oligochaetes have small, characteristic cells in the brain but not masses of globuli cells or dense neuropiles. Leeches are still simpler.

In the more advanced polychaetes, there is commonly recognized a forebrain, midbrain, and hindbrain; these are not necessarily homologous or even morphologically significant divisions, but they may be, since, in several better known cases, they can be defined by their major constituents. The forebrain includes palpal and stomodeal centers and anterior roots of the circumesophageal connectives. The midbrain includes antennal and optic centers and the posterior roots of the same. The hindbrain includes centers for the nuchal sense organs. Among the polychaetes are found the first optic ganglia—distinct masses between the eye and brain, preshadowing a typical arthropod feature. More widespread among advanced polychaetes are corpora pedunculata, representing the highest structures in the nervous system and perhaps homologous to structures of the same name in arthropods. Somewhat less specialized are the glomerular neuropiles, found in better developed palpal and nuchal centers (but not in antennal or optic), and the median mass of the midbrain. Various commissures are identified; they appear to be synaptic neuropile bands rather than simply tracts. The giant system is sometimes represented in the brain by cells of origin, sensory connections, and a decussation with a two-way synapse. Motoneurons are present but few in number; the brain is mainly a receiving area for sensory nerve fibers and a mass of interneurons, presumably integrating influx and formulating descending commands. Brain nerves are variable in number and not named or homologized.

Because of their diversity, comparison of polychaete families is instructive. Table 14.3 summarizes such a comparison for giant fiber systems and Table 14.5 for the brain. Some of the highlights follow. Amphinomids have a fantastic elaboration of the nuchal organ and its brain lobe; some primary sensory cells have moved into the brain. Eunicids, alciopids, phyllodocids, and onuphids have optic ganglia but, at most, only the beginnings of corpora pedunculata. Lumbrinerids have lost eyes, palps and antennae, but there is a surprisingly small effect in brain structure. Nephtyids show various degrees of elaboration of curious posterior lobes. Glycerids have lost most head sense organs and have a very simple brain but an elaborate peripheral nervous system of the prostomium. Nereids are excellent types of higher polychaetes, though other families have carried each feature farther. Enough detail is known to invite new anatomical and physiological work. Aphroditids and the related families of scale worms agree in displaying the highest annelid brains. Sedentary families generally have poor brains, but serpulids are exceptional.

The **peripheral nervous system** in annelids is rather distinctly defined from the central nervous system. Nerves of the ventral cord are rather constant for each species and, throughout the phylum, are either three or four on each side of each segment; all are mixed motor and sensory, but some have dorsal and ventral roots, thought to be mainly motor and mainly sensory respectively. Each nerve supplies a definite area. Nerve cell bodies lie along many nerves. The exact means by which the number of afferent axons is reduced centrally and the number of motor axons is multiplied peripherally are not agreed upon, but perhaps involve peripheral synapses in some cases. Peripheral ganglia in the form of small nodes of cells on the nerves are rather common. The podial ganglia are the most consistent and seem to be capable of mediating certain reflexes. The elytral ganglion of scale worms is probably a collection of peripheral motoneurons.

A lateral longitudinal nerve occurs in a few polychaete families and in acanthobdellids; by some it is thought to be primitive and homologous to members of the orthogonal series of cords of polyneurous flatworms.

Receptors include prominently the generalized epidermal sense cells, occurring singly and in smaller and larger sense organs. Deep sense cells with long, branched free nerve endings are common and are thought to be mechanoreceptors. Numbers of integumental sense organs are available for some earthworms. Photoreceptors occur as widely distributed isolated cells, organized clusters, and moderately good eyes in some forms. Experiments have suggested separate systems of receptors for shadow and for light detection. Shadow reception is prominent in some tube dwellers, and the possible anatomical basis is known in *Protula*. Statocysts are known in a number of polychaetes, mainly sedentary forms; they may be numerous. Their axons go to the subesophageal ganglion, not the brain; they do not signal vibration, but ablation of both sides causes deficiency symptoms in movement and orientation. Nuchal organs, innervated from the brain, are dorsal pits, grooves, or folds at the posterior edge of the prostomium or elongated and extending back over many segments; they are reduced or absent in some groups (terebellids, sabellids, serpulids, and others) and elaborated and even erectile in others (ophelliids, ariciids, spionids, capitellids, and others). A function as chemoreceptors is merely inferred from the histology.

The **nervous control of muscle** has been examined in a few species. Two kinds of responses can be distinguished, differing in threshold of nerve stimulation, latency, and form of contraction: one is fast, initially large, and rapidly declines; the other is slow, initially small, and facilitates. There is probably peripheral inhibition, at least in earthworms and leeches. *Nereis* has about seven motoneurons in the cord on each side, per segment; extensive multiplication of axons is said to involve relaying via numerous peripheral motoneurons in this form. Another report doubts this and emphasizes rich branching of the former neurons. Connections between sensory and motor fibers in the nerves are claimed, permitting the possibility of peripherally mediated reflexes. The earthworm, *Pheretima*, has about 474 motor axons in a midbody segment. Luminescence is nervously controlled in several polychaetes; in the elytra of polynoids it involves central reflexes and a peripheral relay ganglion, which tends to fire repetitively to single electric shocks or after isolation from the central nervous system. Facilitation is evident.

Peripheral plexuses are not well understood. Under the epidermis, and perhaps elsewhere, there is a plexus of nerve fibers, probably not anatomically continuous and probably not conducting as a nerve net for any considerable distance. It has been claimed—and more recently denied—that convergence of many sense cell axons upon a few second-order afferent neurons occurs. *Nereis* is said (but without EM) to have 36–40 cord afferents per segment but many

Summary

thousands of sense cells. Also it seems indicated that some connections between sensory and local motoneurons may occur in these plexuses, at least in some polychaetes and leeches.

The **stomodeal system** is well developed in annelids as a set of nerve cells and fibers, forming a plexus on the wall of the anterior parts of the alimentary canal and connecting primarily with the brain and first ventral ganglion. Possibly it is continuous with a plexus in the rest of the gut and other viscera and through them with nerves from the ventral cord. It becomes a specialized development in the protrusible proboscis of some polychaetes. It is doubtless motor, sensory, and integrative as well, and largely autonomous. The only functional studies concern the connections of the segmental ganglia with the gut; two sets of fibers exert antagonistic effects on muscle tone and one set causes secretion of digestive enzymes.

Physiological studies on central organization in annelids raise many interesting questions. The physiology of nerve-muscle relations, giant systems, and neuronal elements is treated above. The ventral cord is continually active electrically, chiefly as spikes, and largely independent of removal of the adjacent segments and of receptors. Some properties of the leech ventral cord to electric shocks and to stretch are known. The simpler coordinated movements that compose complex behavior have received some attention; they are the following. (a) Peristaltic and antiperistaltic creeping is normally coordinated and conducted through the nerve cord; in its presence, muscle-to-muscle or other peripheral conduction probably plays no significant role. The sequence of reciprocal excitation and inhibition seems to be inherent in the cord and requires only tonic input from receptors (contact or stretch). Reflexes as a result of the movements enhance and facilitate the movements. (b) Parapodial creeping is even more dependent on central automaticity and less dependent on reflex enhancement. (c) Walking in leeches, by alternate use of the suckers, is, on the contrary, dependent on a chain of reflexes. Both central and peripheral stretch reflexes are known. Leeches show a "catch action" or apparently frozen state of muscle contraction, but this is under constant reflex maintenance. (d) Swimming is probably initiated by an intrinsic central rhythm requiring a certain general level of central excitatory state. (e) Writhing may actually include several distinct types of movement; in the present context, they are interesting because they show the presence of fast intersegmental paths other than giant fibers. (f) The twitch reflex is the response to giant fiber activation. As a result of highly labile and integrative afferent activity, a specialized intersegmental response briefly overrides segmental individuality. (g) Reflex arrest of peristalsis and of antiperistalsis have separate fast pathways. These and other nonlocomotor activities manifest in the cord phenomena of reciprocal excitation and inhibition: intrinsic spatial and temporal patterns, local, intersegmental, and chain reflexes, occlusion, facilitation, central and peripheral inhibition, fast and slow pathways, afterdischarge, and other features familiar in higher forms.

The **role of the brain** has been studied mainly by total ablation. Many activities survive this loss, such as locomotion, coitus, righting, and maze learning; feeding and burrowing are possible but awkward. Notable are heightened excitability, restlessness, and sensory deficits attributable to cephalic receptors. Some restricted brain lesions have been reported; one conclusion for *Hirudo* is that crossed and uncrossed pathways descending from the brain must be of nearly equal value. The brain plays a role for a critical initial period in permitting regeneration of lost posterior segments; this and other signs suggest neurosecretion.

Removal of the **subesophageal ganglion** reduces spontaneity, muscle tone, search movements, and backward locomotion.

Some **phylogenetic comparisons** are possible but very tentative. Like platyhelminths, the main pattern of cords can be considered as derived from an orthogon and possibly homologous. Metameric repetition in the nervous system is thought by one theory to have developed from the numerous commissures between cords in unsegmented ancestors. According to another theory, the three ring nerves of the prototroch of trochophore-like ancestors became serially repeated. Although there are many striking analogies, no detailed homologies should at present be claimed between parts of the brain or stomodeal nervous system of annelids and lower phyla. There are various conflicting views on the morphological composition of the brain in annelids and its relation to that of arthropods. Two of the main contenders are the theory of an essentially unitary origin (Hanström) and that of stepwise increments from segmental sources (Raw). Even Hanström believes in one increment from the first ventral ganglia, comparable to the arthropod tritocerebrum, but newer evidence makes it doubtful that any such move has been completed in annelids. The annelid brain is reasonably supposed to be equivalent to the combined protocerebrum and deutocerebrum of arthropods. Within the brain, both extreme views agree on the homology of such constituent structures as stomodeal, palpal, antennal, optic, and nuchal centers and possibly corpora pedunculata. The last named are regarded as independently evolved in polychaetes and only convergent, not homologous, with globuli masses in molluscs, nemertineans, and others but probably homologous with arthropod corpora pedunculata.

I. INTRODUCTION

The impressive similarity in plan among the nervous systems of the whole annelid-arthropod line (the Articulata) is certainly one of the most widely ranging common morphological schemes among organ systems in the animal kingdom, though it is itself merely a special case of the underlying similarity noted earlier (p. 11) of all nervous systems. The plan we deduce to be basic is the ventral, rope-ladderlike system resulting from paired longitudinal cords and metameric pairs of ganglia, each with a commissure, together with a dorsal anterior ganglionic mass (the brain) and circumesophageal connectives joining the two. The similarities go deeper, including special sensory portions of the brain, three pairs of nerves per segment, the second pair being chiefly sensory, and finally a typical histologic composition of central neuropile and peripheral, asynaptic, unipolar cell bodies.

In view of the conservatism of neural morphology, one might hope for some evidence from this system bearing on the question of the origin of the metameric articulates. This will be briefly discussed in the final section (p. 768), and although no entirely satisfactory interpretation is yet agreed upon, the nervous system contributes especially significantly to this general problem.

1. Ontogeny

Unlike many features of free-swimming larvae, the plan of the nervous system of the trochophore seems not to be coenogenetic but probably primitive. It is well exemplified by the archiannelid trochophore (Fig. 14.1) and displays a series of longitudinal cords and circular commissures basically similar to the **orthogonal plan** in many turbellarians. Both sets of cords are more numerous in the apical half; some of the longitudinals are therefore incomplete, but close to eight are complete. Ganglion cells are scattered in irregular clumps along the course of all the cords. Certain nerve rings, such as that under the equatorial ciliary ring (the prototroch), are heavier. A concentration under the apical organ, presumably a sense organ, can be regarded as the simplest known cerebral ganglion.

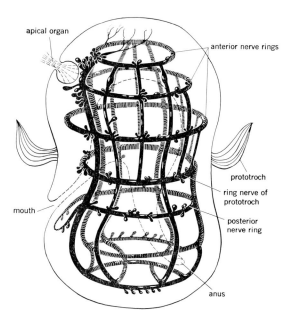

Figure 14.1. Trochophore larva of *Lopadorhynchus*, showing the nervous system with its three nerve rings anterior to and one posterior to the prototroch nerve ring and its ciliated apical sense organ. [Hanström, 1928.]

The trochophoral nervous system becomes a little more than brain and part of the circumesophageal connectives in the adult polychaete. The **ventral cords develop separately** from a pair of ectodermal ridges, which only later unite with the brain through these connectives. This secondarily formed and perhaps coenogenetically delayed set of cords includes most or all of the segmental nervous system, which in turn can be considered not only serially homologous within itself but continuously homologous with the larval cord and hence with turbellarian and ctenophore cords. Aggregations of ganglion cells from a continuous distribution into discrete ganglia may be seen in some forms, suggesting that this was the phylogenetic sequence. Oligochaetes and leeches, which lack a trochophore, seem to have

I. INTRODUCTION

a unified development of brain and ventral cord, both arising from a teloblastic strip of the ectoderm (Penners, 1924; Schmidt, 1926), the "n row" from primordial teloblasts (from cell $2d^{iii}$). The neural strips are paired, though the adult oligochaete gives little evidence of it in most families. The brain of leeches, according to Whitman (1887), arises separately from the cord but some doubt has been thrown upon this by Müller (1932). Mencl (1908) contributes the interesting statement that the interganglionic connectives are at first formed wholly of crossed fiber systems, uncrossed tracts being added later.

Hanström points out an interesting contrast in that the main longitudinal cords of turbellarians, nemertineans, and chitons **develop by outgrowth** from the brain, whereas annelids and arthropods form cords directly from ectoderm. He regards the latter as primitive, the former manifesting an overriding dominance of the brain which, as it were, swallows or attracts into it neurobiotactically the anterior ends of the cords. The articulates then would be atavistic as a result of the development of segmentation.

Later embryology of the polychaete brain and nerves has been studied, among others by Gilpin-Brown (1958) and Korn (1958, 1960), who give the earlier references. Prosser (1934a) correlated the development of behavior in *Eisenia* with neuromuscular differentiation.

Ogawa has laboriously counted and measured cells and fibers during various stages of growth from hatching to maturity in the megascolecine *Pheretima*. The newly hatched worm (*Ph. communissima*) has practically the full number of segments (98; in the adult 109) and has the full number of setae (60 in segment XXX), but is only 28×1.8 mm and 36 mg, while the adult is 171×7.8 mm and 7,510 mg. The **number of ganglion cells** in typical ventral and in subesophageal ganglia increases between these stages by less than a factor of two (see Fig. 14.2, from which

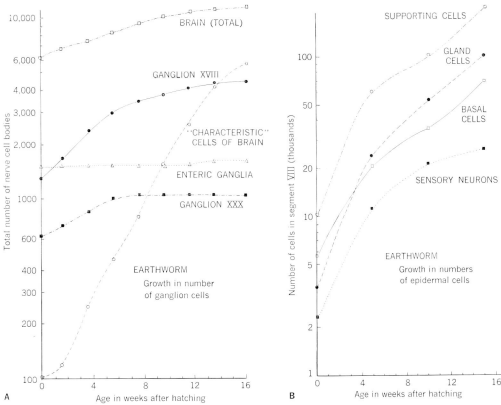

Figure 14.2. Growth curves of numbers of nerve cells and other types of cells in the earthworm, *Pheretima communissima*. **A.** The total number of ganglion cells on both sides in the ganglia indicated. **B.** The numbers of cells of the types named, in segment VIII. [Replotted from Ogawa, 1939.]

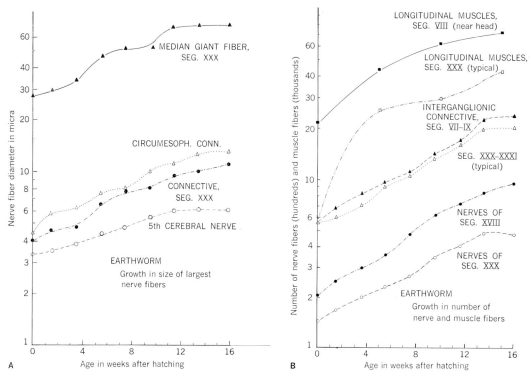

Figure 14.3. Growth curves of size of largest nerve fibers and numbers of nerve and muscle fibers in various segments of the nervous system, in *Pheretima communissima*. [Replotted from Ogawa, 1939.]

absolute values may be read), reaching practically a final value before the clitellum develops or sexual maturity is reached. The number of motor nerve fibers in the segmental nerves increases appreciably more—3 to 4 times. But the cells of certain ganglia in reproductive segments like that of the prostate gland (XVIII) increase about 3.5-fold and the "characteristic" cells in the brain more than 50-fold; furthermore both of these counts continue to increase significantly long after the clitellum has formed and sexual maturity has been attained, rather than plateauing at an early age. In contrast, the cells of the stomodeal ganglion hardly increase 10 per cent. The so-called intermuscular neurons—deep, peripheral, probably motor relay cells—also develop early.

Other cells increase in number more than nervous cells (except for the characteristic cells of the brain): muscle fibers during the same interval increase 5.5 (circular) to 7.2 (longitudinal) times; gland cells in the epidermis, 12.2 times; supporting cells, by a factor of 20; and basal cells, 27.5. Sensory cells in the epidermis, although increasing far more than central nervous cells—by a factor of 11.5—have the lowest ratio of any cell type in the epithelium and are therefore proportionally more abundant in the young animal. All the cell types mentioned increase in greatest dimension about the same amount (2.0–2.5 times) except for longitudinal muscle fibers, which are said to increase 4.4 times in length, and neurons of the stomodeal ganglion, which hardly increase in size at all. Volume may, however, show greater differences, typical ganglion cells increasing considerably more than epidermal cells but not as much as muscle fibers. Figure 14.3 shows the growth in nerve fiber diameter in various places.

Physiologically the annelids are of special interest because this is the lowest group amenable to experimental analysis by standard procedures that depend on well-formed nerves and a body cavity.

2. Polychaeta

The first and largest class, essentially the marine annelids, includes most of the evidences of what is primitive or ancestral to the higher groups, not only of the same phylum but also of the onychophorans and arthropods. It also includes the most advanced, arthropodlike achievements of the annelids, especially in the nervous system. Whereas the other classes are relatively homogeneous and specialized, the polychaetes include endless variation, primitive and advanced features being mixed as though evolving independently. It is therefore more difficult than it is for most groups to give the anatomy or functional organization for the class. We will have to content ourselves with a limited indication of the variety of forms actually found and with an attempt to propose the most common or the least derived conditions, so far as the literature permits.

3. Archiannelida

Whether they are primitive or reduced, this class has a nervous system falling within the range of those in polychaetes and our meager knowledge of it will be treated in general with that group.

4. Oligochaeta

The earthworms and their fresh-water allies are marked by a much greater homogeneity than the polychaetes. Although they range in habitat from glaciers to jungles and are divisible into terricolous and limicolous groups totaling some seven superfamilies (Naidina, Enchytraeina, Tubificina, Lumbriculina, Phreoryctina, Lumbricina, Megascolecina), the diversity in nervous organization is much less. Moreover, the nervous system shows few signs of being farther removed from the stem form of the annelids than the polychaetes, from which they supposedly are derived. In most ways, it appears to be a rather generalized system, although the group in other ways is regarded as specialized and not primitive; it is therefore deserving of the great volume of work that has been devoted to it (upward of 175 papers).

5. Hirudinea

The leeches comprise a rather homogeneous group of specialized structure and habit. The nervous system and sense organs are fairly uniform in correspondence with the unusually uniform food habit, type of locomotion, and constancy of number of segments (at 34). The brain is extremely simple and there has been considerable concentration of ganglia at the two ends of the ventral chain in association with the development of suckers.

Except for the poorly developed brain in the last two classes, the features of the annelid nervous system are extraordinarily similar in the several groups and we will therefore divide this chapter by parts of the system rather than by classes, only recognizing the latter under each region when necessary.

II. THE VENTRAL CORD

A. Position, Form, and Number

It is curious that although some lower forms (platyhelminths, nemertineans) have achieved an internal position of the central nervous system, many archiannelids and polychaetes exhibit the primitive basiepithelial position (some spionids, syllids, opheliids, maldanids, polygordiids, protodrilids, dinophilids). There are various intermediate forms in which the cord lies in or between muscle layers (*Terebella* and *Pista* of the Terebellidae, *Neanthes* of the Nereidae, *Eunice*, *Diopatra* of the Onuphidae, *Glycera*, *Chaetopterus*, *Lepidametria* of the Polynoidae), but in many families it is inside all the muscle layers, free in the coelom. Occasionally the character is inconsistent within a family or even a genus (for example Amphinomidae, Gustafson, 1930).

(In polychaetes, due to the unusual situation of a lack of generally recognized orders, the families assume a special practical importance. The families will therefore ordinarily be given for genera which are not type genera; the family for type genera is indicated by the generic name. In oligochaetes the superfamily will generally be given.)

The ventral cord in oligochaetes (Fig. 14.4) and

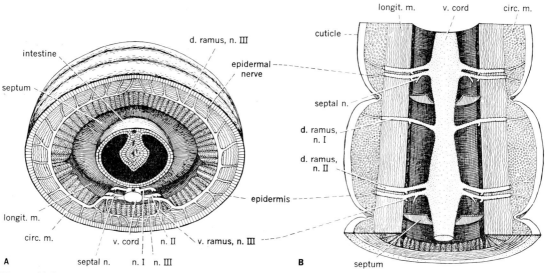

Figure 14.4. Ventral cord and nerves of the earthworm *Lumbricus*, in transverse and frontal views. [Hess, 1925.] *circ. m.*, circular muscle; *d. ramus n. I, n. II, n. III*, dorsal ramus of anterior, middle, or posterior segmental nerve; *longit. m.*, longitudinal muscle layer; *septal n.*, septal nerve; *v. ramus n. III*, ventral ramus of posterior segmental nerve.

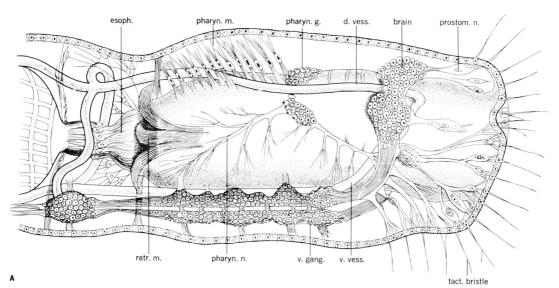

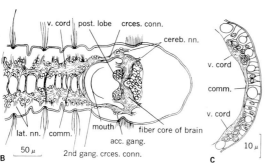

Figure 14.5. The central nervous system of two lower oligochaetes. **A.** *Chaetogaster*, lateral view of anterior end. [Vejdovský, 1884.] **B.** *Aeolosoma*, frontal section of anterior end. **C.** The same, transverse section of midventral epidermis, ventral to the right. [Brace, 1961.] *acc. gang.*, accessory ganglion; *cereb. nn.*, cerebral nerves; *comm.*, commissure; *crces. conn.*, circumesophageal connective; *d. vess.*, dorsal blood vessel; *esoph.*, esophagus; *2nd gang. crces. conn.*, second ganglion on circumesophageal connective; *lat. nn.*, lateral nerves; *pharyn. g., m., n.*, pharyngeal ganglion, muscle, and nerve; *post. lobe*, posterior lobe of brain; *prostom. n.*, prostomial nerve; *retr. m.*, retractor muscle; *tact. bristle*, tactile bristle; *v. cord*, ventral nerve cord; *v. gang.*, ganglion of ventral cord; *v. vess.*, ventral vessel.

hirudineans is generally completely internal, free in the coelom, deep to all the body wall muscles, and supported by no mesentery but by its nerves, blood vessels, and sheath. However, in some forms (*Aeolosoma tenebrarum*, Naidina) it is within the epithelium of the body surface.

There are **two separate cords** in a number of cases (the archiannelids of the Saccocirridae and Dinophilidae, the polychaetes of the Sabellariidae, Sabellidae, Serpulidae) and this is doubtless the primitive condition, though the embryology of some of these families with particularly widely separated cords and consequent elongated commissures suggests that the primitive condition has been secondarily exaggerated. Usually, however, a different secondary process has been expressed, resulting in **fusion** of the paired cords into one midventral cord with only an internal commissural pathway and a slightly paired appearance in the cross section of the interganglionic connective, to suggest the original condition (most polychaetes and some archiannelids). In oligochaetes the cord is generally a median, unpaired structure, all evidence of paired origin being lost externally. But in both the Aeolosomatidae and Naididae of the Naidina the primitive rope-ladder form is visible. The number of commissures here is not perfectly regular and is larger than the number of segments (Fig. 14.5). In leeches, as in most other oligochaetes, the ontogenetically paired cords are united in the adult, but the ganglia, quite unlike oligochaetes, are sharply delimited and far apart, with the cells confined to them ventrally and laterally, as is usual elsewhere.

A third longitudinal connective, the **median connective,** occurs in many polychaetes and leeches. It runs between and dorsal to the paired interganglionic connectives, may contain cell bodies, and at least in leeches apparently communicates with the stomodeal system by some branches to the plexus in the gut wall.

B. Ganglia

Ganglia are generally clearly evident with cell-free connectives between them, but in some groups the whole cord is a medullary cord and the nerve cell bodies are scattered throughout its length (the whole class Oligochaeta, *Polyophthalmus* of the Opheliidae and *Clymenella* of the Maldanidae among polychaetes, *Polygordius*, *Chaetogordius*, *Saccocirrus* and *Protodrilus* of the archiannelids—but not *Dinophilus* and *Trogochaetus*, which have ganglia). In earthworms, which we called above a generalized group, the ganglia are gradual and inconspicuous bulges in the cord, rather than discrete collections of cells with nervelike connectives between them (Fig. 14.4). The bulge may be so small that the number of cell bodies therein is less than the total number between ganglia, as in *Tubifex*. The specialized class of leeches have strikingly discrete, **beadlike ganglia** and long cell-free connectives between them (Fig. 14.6). A peculiarity within the ganglia is the compartmentation of the cell bodies into externally visible packets, usually two median ventral ones, one behind the other,

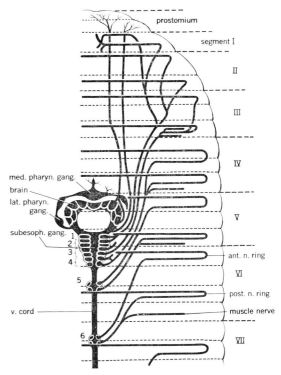

Figure 14.6. Anterior end of the nervous system of *Hirudo*, dorsal view. [Modified from Livanow, 1904.] *ant. n. ring*, anterior nerve ring; *lat., med. pharyn. gang.*, lateral or median pharyngeal ganglion; *post. n. ring*, posterior nerve ring of the segment; *subesoph. gang.*, subesophageal ganglion; *v. cord*, ventral cord; 1–6, presumed segments of the ganglia.

and two lateral groups on each side, one behind the other. The commonest and possibly primitive number of nerves given off by each ganglion is probably three on each side, but the number is often two and occasionally four. Details of the segmental nerves are given below (p. 737).

It would be significant to know whether the **medullary condition** is primitive or derived, for this might help to decide among the theories of origin of metamerism. The distribution among classes does not suggest clearly that it is primitive; on the contrary, a suggestion exists from the embryology of echiuroids and priapuloids and from the phylogeny of Onychophora (none regarded here as annelids but all probably related to them) that it can be derived secondarily.

The ganglia are **generally unitary;** that is, the right and left members of the embryological pair have fused completely. Sometimes two pairs of ganglia occur in a true segment (Sabellariidae, Serpulidae, Sabellidae, Pectinariidae) and occasionally even three (Pectinariidae). The ganglia are typically located in the segment which their nerves supply but exceptions occur (a) in most species in the first few segments, because of a displacement posteriorly of the central nervous system by a segment or two; (b) in a few species in posterior segments where the interganglionic connectives become shorter than the segments and the ganglia in consequence lie anterior to their own segments (Tubificina); (c) in *Nereis* each ganglion extends from a little in front of the intersegmental septum to a point about halfway through the succeeding segment.

Concentration of ganglia into compound ganglionic masses, so common in arthropods, is found already in some annelids. Among polychaetes it is seen in opheliids, eunicids, aphroditids, and polynoids and reaches an extreme in *Myzostomum*. Often the other ganglia are discrete but a subesophageal compound ganglion representing a fusion of the first few ventral ganglia occurs (*Hermione* of the Aphroditidae, *Myxicola* of the Sabellidae, *Petta* of the Pectinariidae, *Nereis*). In a few cases the first pair of ventral ganglia has pushed up the circumesophageal connectives to unite with the brain, an event of great importance in some views of the origin of the arthropod brain (see below,

p. 770). In oligochaetes concentration of ganglia is rare except in the subesophageal ganglion, which in the lumbricids represents the fusion of two or three of the first segmental ganglia. Hess considers the ganglion of the caudal segment to represent a fusion of two, but this region varies considerably. The rare case is the parasitic family Branchiobdellidae, where the anal ganglion is composed of five fused segmental ganglia and lies in the ninth segment, several segments in front of the destination of its nerves. In Hirudinea the last seven ganglia are fused into an anal ganglionic mass under the posterior sucker, and the first four ganglia behind the brain are fused into a subpharyngeal ganglionic mass. In both cases the cell packets or follicles, defined by connective tissue partitions, are preserved and help to identify the component segments. Both ganglionic masses are displaced longitudinally toward the middle but their nerves still supply the segments to which they belong.

The **circumesophageal connectives** have two roots in the brain, basically. The whole connective is commonly medullary in oligochaetes and may be considered to contain the ganglia of the first segment, one on each side, at least in lumbricids, where the nerves for this segment come from the connective. But the commissure for this ganglion has not been identified, nor has that for the two other ganglia presumably comprising the subesophageal ganglion, owing to the difficulty of assigning morphologic identity to the numerous, indistinctly separated, commissural pathways in such a fused compound ganglion. A complication of common occurrence in polychaetes seems to be due to enlargement of the pharynx or to a caudal extension of the split between right and left cords, in effect lengthening the circumesophageal connective by including in it several segments (Fig. 14.7). The result is that ganglia, apparently quite like ordinary segmental ganglia (Eunicidae, Aphroditidae, Amphinomidae) or reduced in association with loss of appendages in those segments (Sabellariidae, Nereidae), lie along the extended connective. Of course, the commissures between the left and right members of each pair of ganglia remain to show their status and are now greatly stretched, running posterior or ventral to the gut.

II. VENTRAL CORD C. Histologic Structure

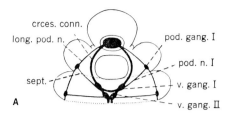
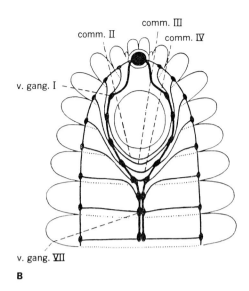

Figure 14.7. A primitive and a more highly derived arrangement of circumesophageal nervous structures. **A.** *Paramphinome*. **B.** *Hermodice*. [Gustafson, 1930.] *crces. conn.*, circumesophageal connective; *comm. II–IV*, commissures of ventral ganglia II–IV; *long. pod. n.*, longitudinal podial nerves (lateral nerve cord); *pod. gang. I*, first podial ganglion; *pod. n. I*, first podial nerve; *sept.*, intersegmental septum; *v. gang. I–II*, first or second ventral ganglion.

C. Histologic Structure

A conspicuous uniformity of texture and composition characterizes the central nervous tissue of annelids, arthropods, molluscs, and other ganglionated invertebrates. The main common denominators are the following. Ganglia are always composed of a fibrous core and a cellular rind. A tissue which in ordinary stains appears nearly homogeneous, almost without nuclei, acidophilic, of unresolvable texture, variously (according to fixation and later treatment) quite smooth, fine-grained, punctate, "soft-focus" punctate, or when poorly preserved frothy, usually loose and suggestive of a section through a soggy haystack, with wisps of one orientation contrasting with others cut in a different plane or with masses having no directionality—this tissue dominates and forms the bulk of the core. It is often called a **neuropile** and this word is useful if we do not forget that it should properly exclude those areas which are purely made of fibers in transit—that is, tracts or fascicles not giving off terminating collaterals and thus making no synaptic connections. Since we can rarely say that a tract is without functional contacts, and the long pathways are only locally and in the best-developed centers really discrete from neuropile, there is little that is strictly comparable to the white matter of vertebrates. The neuropile is central and generally occupies the greater part of the volume of a ganglion (Fig. 14.8). The special case of giant nerve fibers is treated separately below (p. 689).

Superficial to the fiber matter lies the **cell body rind**, almost entirely composed of unipolar, primitively cytoplasm-rich cells with chromatin-poor nuclei; they are usually loosely packed, with considerable connective tissue and intercellular space. Bipolar and multipolar cells occur and are noted below in the section on Neuronal Composition. A minority of small cytoplasm-poor cells, chiefly internuncials, also occurs. In most ganglia the rind is confined to the ventral and lateral aspects (Fig. 14.9). A number of significant comparative facts about the cells are summarized in Chapter 2. Neurofibrils have been seen by silver as well as by methylene blue in unstained living tissue in oligochaetes and leeches but are rare in polychaetes (Michel, 1899). The boundary between cells and neuropile is commonly sharp and emphasized by a layer of neuroglia.

When the ganglia are in the body cavity a thin **peritoneum** covers the cord, and, in annelids, immediately beneath this is typically a layer of **longitudinal muscle**.

A histological property prominent in this phylum is the great **tolerance of stretch.** We have noted this before, especially in the nemertineans. In the earthworm one can ascertain by silver impregnation of individuals fixed in various degrees of stretch that a part of the extension from minimum length consists in straightening out kinked, zigzagged or irregularly coiled struc-

Carlson (1904b) discovered the lack of effect of stretch on velocity and concluded that the conducting substance is fluid. Bethe (1908, 1910) claimed that velocity actually increases with stretch in *Hirudo* and implicated the neurofibrils which he believed to have a constant length, uncoiling with stretch. Bullock, Cohen, and Faulstick (1950) and Goldman (1963, *J. cell. comp. Physiol.* 62:105–112) found conduction velocity to be constant over a wide range of length under stretch even while diameter is markedly reduced in giant fibers of *Lumbricus* (p. 149). (The length-velocity curve has a long plateau; at the shortest lengths a positive slope is presumably due to uncoiling.) We cannot conclude today that either the fibril or the axoplasm is the conducting element, but the finding is unexpected on present theory. Goldman (1964, in MS) has examined axon properties from the point of view of the Hodgkin-Huxley theory. Resting potential, spike amplitude, rates of rise and decline, threshold, specific membrane resistance, and capacitance all remain constant under stretch. Membrane unfolding may account for part of the result; decreased specific axoplasm resistivity and measurable volume loss was found in sufficient degree to account for the rest.

Nonnervous tissue has been an important constituent of all the central nervous organs heretofore treated, but it has been more carefully studied in the present group than in any other so far (Joseph, 1902; Cajal, 1904; Krawany, 1905; Boule, 1908, 1913; Schneider, 1908; Sánchez,

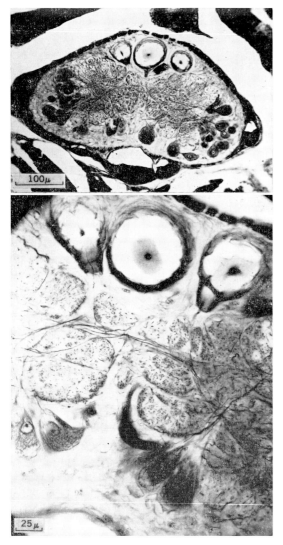

Figure 14.8. Ventral nerve cord of *Lumbricus*, transverse section, Bodian protargol stain. The three giant fibers are seen dorsally, the subneural vessel ventrally, the fiber core and cell rind, muscular sheath, and neurofibrils in the cells and large fibers.

tures—at first the whole cord, then nerve fibers themselves, and finally the neurofibrils within fibers (that is, there is a stage in stretching when the outlines of fibers are microscopically straight but the fibrils within are still zigzagged). This property differs not only among fibers in the same animal—for example, the median giant fiber of *Lumbricus* is more tolerant than the laterals—but between animals—squid giants are much less tolerant and frog sciatic still less so. In experiments on a gastropod, Jenkins and

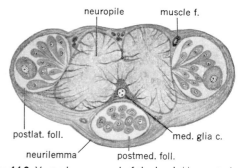

Figure 14.9. Ventral nerve cord of the leech *Haementeria*, transverse section. Note the thick neurilemma capsule, compartmentation, and large median glial cell *(med. glia c.)*. [Scriban and Autrum, 1932.] *muscle f.*, muscle fiber; *postlat. foll.* and *postmed. foll.*, posterolateral and posteromedian follicles or compartments.

II. VENTRAL CORD C. Histologic Structure

1909; Keyl, 1913; von Szüts, 1914; Holmes, 1930; Clayton, 1932). Nevertheless, it remains uncertain which supporting cells within the ganglia are distinct, embryologically or otherwise, from connective tissue generally. As seen in the earthworm, beginning with the surface of the ventral cord there is a sheath covered with peritoneum and under this a thin layer of connective tissue consisting of a few nuclei and a large amount of apparently intercellular substance. Thereunder lies the muscle layer, usually described as of longitudinal but, according to Holmes, including also circular fibers, and responsible for the shortening of the cord when the somite shortens, without a great deal of fluting. There has been described a plexus of nerve fibers supposedly for innervating this muscle layer (Krawany). Beneath this occurs a thicker layer of connective tissue adjoining the giant fiber sheaths on the dorsal side of the cord and the nerve cell bodies laterally and ventrally. From it extend inward numerous trabeculae, putting the whole supporting framework in connection. This **glial framework** includes as its main constituents (a) a horizontal septum dividing the dorsal giant fiber portion of the cord from the central neuropile. (b) A median vertical septum in the interganglionic connectives only, dividing the ventral portion—that is, the neuropile and the few cell bodies—into lateral halves. This septum may be double, the lamellae diverging dorsally, and the space between may be occupied by nerve fibers which have been homologized with the median nerve of other articulates. This median vertical septum, preventing lateral continuity between ganglia but interrupted so as to permit commissures within them, represents a remnant of the primitive rope-ladder arrangement. (c) The giant fibers each have a thick connective tissue sheath with many nuclei. (d) Finally, a diffuse meshwork of neuroglia permeates the whole cell body layer and, much more sparsely, the neuropile. Schneider described some of the axons as having a cuticular (inner) sheath and commonly a cellular external one, too, as in the case of the giant fibers. Glia form the characteristic pockets of ganglion cells in the ganglia of hirudineans.

The glial cells which compose this extensive framework are at least very much like connective tissue cells such as the interstitial cells in the body muscles. But even Holmes, who believed they are really the same, described some general differences in nucleus and cytoplasm, and most earlier writers seemed to regard them as ectodermal and of distinct cytology. Joseph (1902) explicitly compared them with the astrocytes of vertebrates. In both silver and iron hematoxylin stains they present certainly an appearance very much like fibrous astrocytes. Numerous processes of widely differing lengths and sometimes branching are, or contain, heavily chromatic fibers; they can be followed into the cytoplasm (Fig. 14.10),

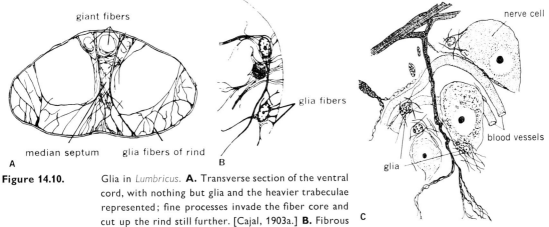

Figure 14.10. Glia in *Lumbricus*. **A.** Transverse section of the ventral cord, with nothing but glia and the heavier trabeculae represented; fine processes invade the fiber core and cut up the rind still further. [Cajal, 1903a.] **B.** Fibrous glial cells with processes that attach to or end upon the bounding membrane of a nerve fiber bundle. [Boule, 1908.] **C.** Higher magnification of the same, showing a glial trabecula, several glial cell satellites of nerve cells, and fine glial fibers. [Havet, 1916.]

which sparsely surrounds an oval nucleus with scattered chromatin granules and inconspicuous or no nucleolus. The process and fibers fit up against nerve cell bodies or between fiber fascicles. They often end in expanded feet applied in rows to a fascicle boundary or a blood vessel. The fibers have been likened to the cytoplasmic fibrils of ciliated epithelial cells, which begin in basal granules close to the distal end and run to a footlike attachment on the basement membrane. Havet distinguished protoplasmic and fibrous types, the former with more granular and alveolar cytoplasm in metallic impregnation, the latter with more fibers and a more indefinite contour. Both have close relations with the blood vessels, though differing in the details.

D. Neuronal Composition, Cell Groups, and Pathways

1. Polychaeta

Reliable knowledge of the types of neurons and the organization and distribution of their processes begins with—though it does not end with—the use of the specific nerve stains. To be sure, there can be extracted without such methods fairly safe conclusions in favorable cases, as in the excellent work of Nansen (1887b), where we will begin the account of the polychaetes. This author recognizes six cell groups at the height of the broader of the two commissures of each segment in *Myzostomum* (an atypical form whose abdominal ganglia are concentrated into one compound mass). Four groups of **motoneurons** are distinguished. (a) There is a median group lying both dorsal and ventral to the commissure, most of which send their axons into peripheral nerves; probably this group is inhomogeneous. (b) A medial dorsal group and (c) a lateral dorsal group send processes into ipsilateral peripheral nerves, and (d) another dorsal group gives origin to crossed fibers entering contralateral peripheral nerves. Note that dorsal cell bodies are unusual among annelids. (e, f) Two ventral groups appear to be mainly **interneurons**; their processes can be followed into the neuropile and are then lost.

The bulk of our information on the neural

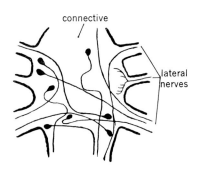

Figure 14.11. Pathways in a ventral ganglion of a polychaete, *Aphrodita*. A selection of representative neurons seen in methylene blue preparations, showing the main types according to direction of their axons. Dendrites not shown. An afferent fiber is shown in the right middle nerve. [Schematized by Fortuyn, 1920, from data of Retzius.]

makeup of polychaete ganglia we owe to the superlative work of the Swedish neurologist Retzius (1891, 1892). In *Aphrodita* he described the pathways summarized in Fig. 14.11. In addition to neurons equivalent to those listed above, there may be noted the following features. (g) Motor fibers enter the first and third segmental nerves not only from local cells but also from some located in other ganglia, both in front of and behind the given segment. No motor outflow was found in the second nerve. This corresponds to the generalization given on p. 737 but overlooks the later discovered motor supply to parapodia, via the podial ganglia, typically superimposed on the primarily sensory second nerve.

(h) Fibers occur which originate in other ganglia, pass through this one, and end elsewhere; that is, there are **intersegmental interneurons** of more than one segment in length. However, in an electrophysiological search for giant fibers (Bullock, 1948) no sign of through-conduction over many segments was found, although the methods used would detect even non-giant activity if a considerable number of fairly synchronous or a very few long fibers exist in this relatively sluggish form. (i) Longitudinal interneurons originate in the ganglion and pass into the connectives to end in other ganglia; how far they can go is not certain. These are usually either ascending or descending neurons but in *Lepidonotus* (Polynoidae) Retzius found (j) some cells whose process divides dichotomously, one branch going forward and the other caudally, both crossing to the side opposite the cell body. (k) In this genus Retzius remarks that, in contrast to the specialized condition in *Aphrodita*, most of the cell groups are paired; this is the general case in annelids and arthropods. (l) In *Nephtys* (Fig. 14.12) he found the unusual condition of terminal ramifications of interneuronal axons and motoneuron **dendrites in the cell body layer,** not confined to the neuropile. The opportunity is thus created for functional contact between telodendria and cell body, a feature almost peculiar to vertebrates.

Retzius described a **peculiar superficial terminal** apparatus in both the subesophageal and the abdominal ganglia in *Nereis*, consisting of a tuft of telodendria resembling the rounded crown of a tree (Fig. 14.12, A). The remarkable aspect is their location—just inside the sheath of the ganglion and external even to the cell body layer. Their form seems to exclude their being motor nerve endings serving the thin muscle layer of the sheath, and no obvious relation to nerve cell bodies is evident. Their cells of origin are not identified but they are at the end of collaterals of fibers coursing longitudinally through the ganglion. No functional meaning can at present be attached to them. Smith (1957) cast doubt on these endings on technical grounds, suggesting that they are artifacts.

From Retzius' study of *Nereis* the following further findings may be noted. (m) The majority of the fibers in the ganglion decussate and three great commissures are formed. This situation seems unlikely to have been true in the primitive

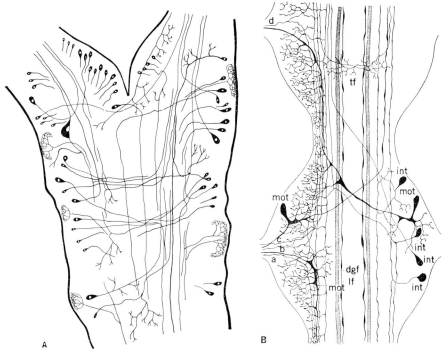

Figure 14.12. Pathways in ventral ganglia of two polychaetes. A selection of representative neurons seen in methylene blue preparations. **A.** *Nereis*, subesophageal ganglion showing four commissure regions and several types of longitudinally directed neurons. The terminals just outside the cell rind are questioned by recent authors. The circumesophageal connectives take origin at the top. None of the fibers is shown that enter this ganglion, whether from peripheral nerves or from the brain or cord. [Schematized by Hanström, 1928, from data of Retzius.] **B.** *Nephtys*, ventral cord ganglion. *a, b, c, d,* sensory fibers; *dgf,* dorsal giant fibers; *int,* interneurons; *lf,* long intersegmental interneurons; *mot,* motoneurons; *tf,* terminal fibers, whether axonal endings or dendrites. [Schematized by Zawarzin, 1925, from data of Retzius.]

ladder type of nervous system or to be true of the few species which have preserved or secondarily reacquired such an arrangement. It can therefore be looked upon as a corollary of the evolution of the nervous system and possibly as one of the causal factors in the fusion of the paired ganglia. The embryology of this feature in species starting with quite separate neural ridges should be interesting. Of course, decussation occurs in much lower animals than annelids—in fact, in the lowest Bilateria (p. 546 and Fig. 9.8).

(n) Most of the sensory fibers entering divide in a T-shaped manner reminiscent of that in the vertebrate cord, which figured in a famous controversy following Cajal's description a few years earlier. (o) Rarely bipolar neurons occur, for example an interneuron type in the subesophageal ganglion, one of whose processes goes forward into the circumesophageal connective, the other passing backward. (p) A contribution to the curious median nerve (p. 740) was found, the cells of origin being a median unipolar group, but whether motor or associational in destination is not clear.

Hamaker (1898) confirmed and extended Retzius' studies, using the closely related genus *Neanthes* (Nereidae). Among his findings was (q) a motor cell type whose axon splits after decussating, the two divisions entering two separate peripheral nerves. The two main processes have dendritic offshoots in the region of their decussation with the result that the dendrites intermingle with each other. This suggests bilateral symmetry of action but leaves a source of activation of both units to be found. Fortunately Hamaker made the further observation that the processes, after splitting, tunnel through the lateral giant fibers, a relation very similar to the simple synapses between giant and motor fibers in crayfish and hence suggesting a role in the startle response (Bullock, 1945b, 1948).

Building on the earlier accounts, with a new wealth of detail, is the most coherent picture so far, that of Smith (1955, 1957) on *Nereis*. His contributions to the composition of the segmental nerves and the character of sensory and motor innervation are given in their places below. The composition of a typical ventral cord ganglion in terms of afferent, internuncial, and motor neurons is summarized in Fig. 14.13. (r) **Afferent fibers** enter after great reduction in number, he claims, largely by relaying via peripheral internuncials, as 3–4 fibers in nerve I, 6–8 in II, 2 in III, and 6 in nerve IV. Only rarely do entering afferents cross the midline; they turn up or down or dichotomize and project both ways, usually giving off slender branching transverse collaterals

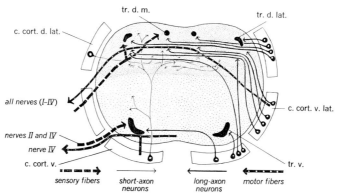

Figure 14.13. Diagrammatic transverse section of a nereid ganglion to show (1) the position of the long-axon longitudinal tract, (2) the directions of the long internuncial axons through the neuropile prior to their entry into the tracts, (3) the distribution of the short-axon internuncials, primarily confined in their collateral branchings to the dorsal neuropile, (4) the course and destination of the sensory fibers in fine-fiber longitudinal tracts, and (5) the course of the motor axons through the dorsal neuropile. The dorsal neuropile containing the greatest concentration of internuncial branchings is the more densely stippled. [Smith, 1957.] *c. cort. d. lat.*, dorsolateral sector of the cell cortex of the ganglion; *c. cort. v.*, ventral sector; *c. cort. v. lat.*, ventrolateral sector; *tr. d. lat.*, dorsolateral fine-fiber longitudinal tract; *tr. d. m.*, dorsomedial tract; *tr. v.*, ventral tract.

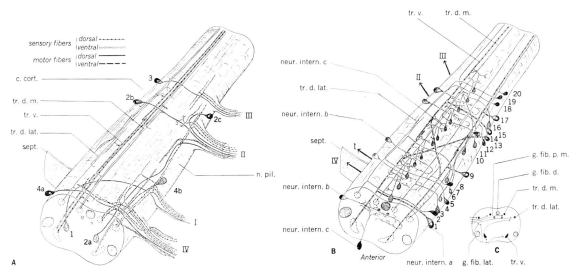

Figure 14.14. Diagrammatic stereograms of a nereid ganglion and connective.

A. Showing the sensory and motor fibers of the segmental nerves, the position in the cell cortex of the principal motor cells, the distribution of their axons in the neuropile, the levels of entry into the cord of the sensory fibers, and the fine-fiber longitudinal internuncial tracts with which the afferent fibers are associated. Sensory fibers of dorsal and ventral entry are distinguished in the manner shown in the "key." Motor fibers, which may traverse the neuropile either dorsally or ventrally in their efferent passage, are also distinguished by different types of lines. The anterior end of the ganglion is at the lower end of the figure and the cut surface shows, in section, the giant fibers *(stippled)* and the internuncial tracts *(broken line)*. I–IV, the segmental nerves; 1–4b, the segmental nerves supplied by the motor cells opposite which the numbers appear; *c. cort.*, cell cortex; *n. pil.*, neuropile; *sept.*, intersegmental septum; *tr. d. lat.*, dorsolateral fine-fiber longitudinal internuncial tract; *tr. d. m.*, dorsomedial tract; *tr. v.*, ventral tract.

B. Showing the principal neurons and fibers of the internuncial system of the nerve cord. The anterior end of the ganglion is at the bottom of the figure and the ganglion extends posteriorly to the narrower connective. The cell bodies of internuncial neurons with long axons supplying the longitudinal fine-fiber tracts are shown (1–20) to the right of the figure. Internuncial neurons with horizontal branching axons *(neur. intern. b)* are on the left of the figure and internuncial neurons with vertical axons extending from ventral cell bodies *(neur. intern. c)* are shown in a longitudinal row along the length of the ganglion. All three systems, shown here on the one or other side of the cord, are duplicated bilaterally. The cut end of the cord shows giant fibers *(stippled)*, the position of the fine-fiber tracts *(broken lines)*, and one each of the two types of branched axon neurons. I–IV, position of the segmental nerve roots; *neur. intern. a*, long-axon internuncial neuron; *neur. intern. b* and *neur. intern. c*, branched axon internuncial neurons; *sept.*, intersegmental septum; *tr. d. lat.*, dorsolateral fine-fiber longitudinal tract; *tr. d. m.*, dorsomedial tract; *tr. v.*, ventral tract.

C. Outline transverse section of the ganglion showing the position of the giant- and fine-fiber longitudinal tracts and the principal directions taken by the long internuncial axons of one side of the cord in passing to one or other of the fine-fiber tracts. *g. fib. d.*, dorsal giant fiber; *g. fib. lat.*, lateral giant fiber; *g. fib. p. m.*, paramedial giant fiber; *tr. d. lat.*, dorsolateral longitudinal fine-fiber tract; *tr. d. m.*, dorsomedial tract; *tr. v.*, ventral tract. [Smith, 1957.]

promptly. They then join one of the three main longitudinal tracts that traverse the cord, and run for short distances (Fig. 14.14, A). The presumed proprioceptor fibers probably enter the dorsomedial tract, exteroceptors dorsolaterally, and others of unknown function enter the ventral tracts.

Internuncial cells (Fig. 14.14, B), mostly ventral or lateral and all unipolar, are in two series. (s) One series have long axons that enter one of the three longitudinal tracts, about equally ipsilaterally and contralaterally. They bear short collaterals (dendritic, presumably) and may run through

1, 2, 3, or more segments. The dorsolateral tract has about 15 or 20 axons, including 9 or 10 sensory fibers from nerves I and II. The dorsomedial tract on each side has only about 4 internuncial axons and 5 or 6 sensory axons. The ventral tracts are larger but an accurate accounting of its axons can not be given. (t) The other series have short processes, much-branched and confined to the ganglion of origin; ipsilateral or contralateral ramifications form a localized plexus in the dorsal neuropile. These are the most abundant cells of the ganglion. (u) The system of giant interneurons is treated in the next section.

Motoneurons are only six or seven in number on each side of each ganglion (Fig. 14.15), one each to nerves I and III, one or sometimes two to nerve IV, and 3 to nerve II. Their pattern is substantially constant. To supply the extensive musculature peripheral multiplication of fibers is believed to take place by interpolation of relay neurons (p. 750). Each nerve includes one axon from a contralateral cell body via a dorsal commissure, where its dendritic branches interlace with the main internuncial and afferent neuropile. Synapses between the dorsally crossing motor axons and giant fibers have not been found, but a motor fiber emerging ventrally in nerve IV is synaptically connected with the lateral giant fiber. The giant fibers are believed not to be excited directly by afferents but indirectly, via internuncials.

2. Oligochaeta

Afferent fibers enter all the segmental nerves, most heavily the second (Fig. 14.16). Their cell bodies are always in the periphery, in or under the epidermis, in the muscles or other tissues along the course of nerves, but not in the central nervous system; exceptions, such as those noted below in the polychaete brain, have apparently no counterpart here. Their form, distribution, and relations are treated below (p. 743). Immediately upon entering the cord almost all the afferent fibers divide in a T-shaped manner, as in vertebrates, and the long ascending and descending branches form **three sensory fascicles**, discovered by Retzius. The lateral sensory fascicle is composed of fibers from all three segmental nerves and is the thickest; the middle fascicle receives fibers only from the first and second nerves; the medial fascicle contains fibers only from the second nerve and is the thinnest bundle. The fibers in these tracts are rather uniformly thin in comparison with the tracts of ascending and descending interneurons. Judging from the drawings from methylene blue preparations given by Retzius (1892) and Ogawa (1939), most of the afferents end within a segment or two of the level of their entry and the great majority remain ipsilateral. There are, however, a substantial number of fibers which cross to the opposite side, and at least some of these without first giving off ipsilateral collaterals. Promptly after bifurcating the incoming fibers begin to bear short terminal processes at angles to the main collateral. These are numerous and rather evenly spaced, mostly borne on only one side of the fiber and generally of a few definite lengths, so they end in simple terminal forks or tufts in a few longitudinal planes as

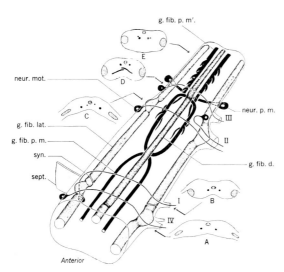

Figure 14.15. Stereogram of a ganglion and connective of *Nereis diversicolor*, showing the form and arrangement of the giant internuncial fibers and the principal motor fibers of the segmental nerves. The anterior (preseptal) end of the ganglion is at the bottom of the figure. **A, B, C, D,** and **E,** outline transverse sections through the cord to show the position of the giant fibers at the levels indicated by the arrow. **A** to **D** are through the ganglion, **E** through the connective. [Smith, 1957.] *I–IV*, the roots of the segmental nerves; *g. fib. d.,* dorsal giant fiber; *g. fib. lat.,* lateral giant fiber; *g. fib. p. m.* and *g. fib. p. m′.,* paramedial giant fiber; *neur. mot.,* motor neuron; *neur. p. m.,* cell body of paramedial fiber; *sept.,* septum; *syn.,* macrosynapse.

II. VENTRAL CORD D. Neuronal Composition 2. Oligochaeta

of the segmental nerves, the third being the most heavily supplied and the first the least. The fibers are for the most part the largest in the nervous system, aside from the giants; though varying in diameter along their course, 4 μ is perhaps a typical figure at the base of the segmental nerve, while the largest are from 6–12 μ in different nerves (smallest in cerebral nerves). According to Ogawa, each has one neurofibril (contra Hirudinea, which see). In some of the published figures the efferent fibers take exit, on the whole, dorsally or deep to the entering afferents, but there does not appear to be good justification for recognizing dorsal motor and ventral sensory portions of the cord.

The cell bodies of the motoneurons lie in the ventral and lateral mantle of the ganglion, not segregated from interneurons or grouped in defined fundamental groups. They are large (20–45

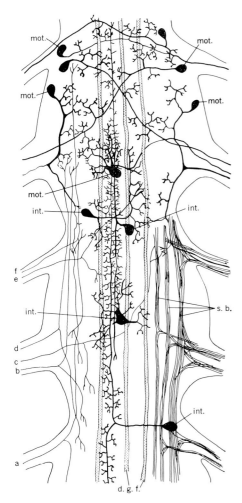

Figure 14.16. Motoneurons and interneurons in the ventral cord of the earthworm. The posterior nerve of each ganglion is actually a double nerve. [Schematized by Zawarzin, 1925, from Retzius, 1892.] *a–f*, sensory fibers; *dgf*, dorsal giant fiber; *int.*, interneurons; *mot.*, motoneurons; *s. b.*, sensory bundles.

though against definite boundaries (Ogawa, Smallwood). The histological evidence that some authors have offered, of endings in connection with lateral and ventral giants in particular, is not yet satisfactory (but see Giant Fibers, below). But the region in which the afferent endings occur is a region also invaded by many processes of both interneurons and motoneurons as well as of the giants. The anatomy is not suggestive of extremely circumscribed reflexes but every sensory fiber connects with many efferent and interneurons and every such neuron connects with many afferents.

Efferent neurons are represented by axons in all

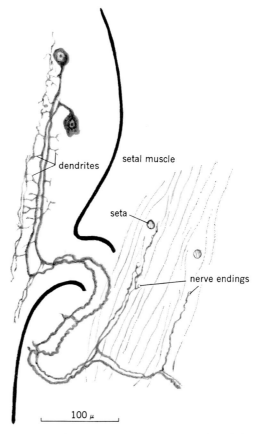

Figure 14.17. Two motoneurons to the muscles of the setae, in *Pheretima*, showing the manner of branching. [Redrawn from Ogawa, 1939.]

μ, rarely up to 60 μ in diameter), pear-shaped or polygonal, mostly unipolar, plasma-rich and chromatin-poor, with one or more conspicuous nucleoli, a neurofibrillar net, Golgi apparatus, and other inclusions as described in Chapter 2. A minority are bipolar or multipolar and all transitions can be found (Smallwood). Typically, a large number of extensively branching **dendritic processes** ramify in the neuropile of the same side as the cell body or of both sides. This arrangement (Figs. 14.16, 14.17) suggests a great role of summation of spatially separate subthreshold responses, discussed more fully below; see also Chapter 5. Many of these dendritic endings lie in the region of termination of incoming afferents and many, belonging to the same motoneurons, are more widely distributed, presumably ending in relation to interneurons. No specific synaptic relations with the dorsal giant fibers have been established but Smallwood (1930) shows figures indicating a close but not exclusive relation between certain motoneurons and the ventral giant fibers (see further under Giant Fibers, p. 708).

The axon often has **central collaterals** that ascend or descend in the cord to neighboring segments. These could be cellulipetal—that is, dendritic or afferent for that neuron—or cellulifugal, acting either as internuncial branches to other segments carrying information of the motor outflow in its segment or as parallel efferents passing into segmental nerves of other segments. The last alternative should have been seen if it occurs, at the least as disembodied motor fibers entering the nerves. Their absence from the published figures may therefore signify that they do not occur, if the authors have not discriminated against such elements in drawing. The question cannot be decided on present evidence and remains an interesting problem for the future, for the other alternatives would have theoretical importance in neuronal excitation. Available evidence indicates that efferent fibers do not cross segmental lines to excite an extensive musculature.

Each efferent fiber in a segmental nerve has a very large number of **peripheral collaterals** (Fig. 14.17) innervating a large number of muscle fibers, in agreement with Ogawa's calculation that there are about 100 times as many muscle fibers per segment as motor nerve fibers. Actually, the discrepancy is greater than this; details of the innervation and manner of ending are given below, p. 750.

We owe our knowledge of the **types of neurons** and their intimate pattern to the methylene blue and Golgi staining methods and to the workers who have applied them—Cerfontaine (1892), Retzius (1892), Krawany (1905), Haller (1889, 1910), Stough (1926), Smallwood and Holmes (1927), Smallwood (1930), and Ogawa (1939). The first authors found a small number of kinds of cells, and different kinds for each of the three nerves. Each author found a few more kinds than the last and failed to confirm certain kinds. The various findings are summarized in diagrams by Fortuyn (1920) and Prosser (1934c). The new work of Ogawa not only confirms most of the old results but adds many new types of cell, so it is no longer feasible to show the specificity of the pattern by a diagram containing one of each type. Motoneurons may be unipolar, bipolar, or multipolar. They may send their axons across the midline into nerves of the opposite side or into ipsilateral nerves. They sometimes have collaterals that may ascend into more anterior segments, descend into posterior levels, or both. These collaterals may have, moreover, various destinations, but these are unknown. They may have extensive dendritic ramifications or restricted ones. The cell body may lie at the level of the nerve of exit or up to half a segment anterior or posterior to it in the lateral aspect of the ganglion or, when in the midline, may send its stem process (if unipolar) or its axonic process (if bipolar or multipolar) directly toward the side of the nerve of exit or in the opposite direction so that a doubling back and double decussation is necessary. All these characters have been used for designating types of neurons. Obviously the total number of permutations is very great and in view of this the observed types represent a small and specific selection of the possibilities.

Using only the first three sets of alternatives given above there are some fifty-four possible combinations, but only twenty (Table 14.1) have been found and many of these are rare. The **major types of motoneurons** are unipolar cells

II. VENTRAL CORD D. Neuronal Composition 2. Oligochaeta

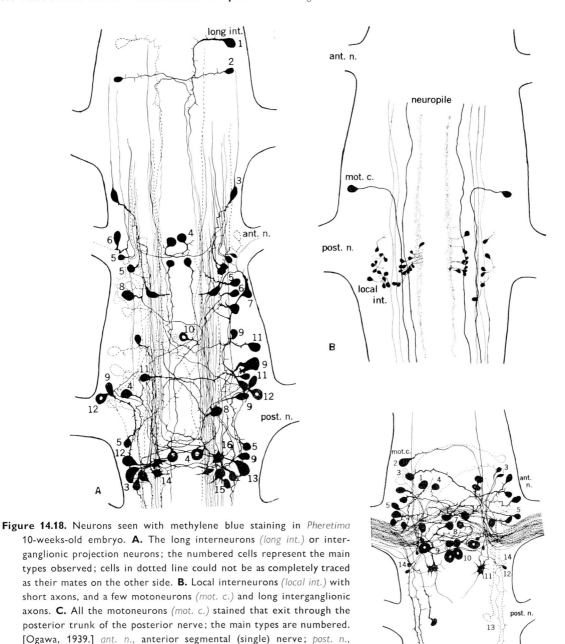

Figure 14.18. Neurons seen with methylene blue staining in *Pheretima* 10-weeks-old embryo. **A.** The long interneurons *(long int.)* or interganglionic projection neurons; the numbered cells represent the main types observed; cells in dotted line could not be as completely traced as their mates on the other side. **B.** Local interneurons *(local int.)* with short axons, and a few motoneurons *(mot. c.)* and long interganglionic axons. **C.** All the motoneurons *(mot. c.)* stained that exit through the posterior trunk of the posterior nerve; the main types are numbered. [Ogawa, 1939.] *ant. n.*, anterior segmental (single) nerve; *post. n.*, posterior segmental (double) nerve.

without ascending or descending collaterals, contralateral neurons being more common than ipsilateral. The smallest variety as well as absolute number belongs to the first nerve, and these cells are nearly all close to but just posterior to the level of the nerve exit and in the lateral cell group (Fig. 14.18). The second nerve motoneurons are more numerous and varied, especially in the increase of ascending collaterals. The cell bodies are almost all anterior to the level of the nerve and widely spread between it and the first nerve. The third nerve, with the greatest number and variety of neurons, has almost all those with descending collaterals, still a small number, and the greatest proportion of medially situated cells. Its cells are mostly opposite the nerve, which is at the posterior edge of the ganglion, and closer to the next following first nerve than to that of its

Table 14.1. TYPES OF NEURONS IN TYPICAL VENTRAL CORD GANGLION OF THE EARTHWORM BY FORM AND PROCESSES.[1]

Position, form, and branches	Nerve in which each leaves	Authors	Comment
Motoneurons			
1. Contra-uni	1, 2, 3	CRKO	dominant; appr 12 subtypes
2. Contra-bi	1, 3	CK	
3. Contra-multi	1, 3	KH	
4. Contra- and ipsi-uni	3	O	cell with symmetrical outflow
5. Ipsi-uni	1, 2, 3	CRKHO	appr 10 subtypes; common
6. Ipsi-bi	1, 2, 3	CHO	may send axons into 2 nerves
7. Ipsi-multi	2	HO	
8. Contra-uni with contra-asc-collat	1	C	
9. Contra-multi with contra-asc-collat	3	C	
10. Contra-uni with ipsi-asc-collat	3	O	
11. Ipsi-uni with ipsi-asc-collat	2, 3	RK	
12. Ipsi-bi with ipsi-asc-collat	2	CRO	
13. Ipsi-multi with ipsi-asc-collat	1, 3	HO	
14. Ipsi-bi with contra-asc-collat	2	O	
15. Contra-bi with ipsi-desc-collat	3	C	
16. Ipsi-uni with ipsi-desc-collat	3	C	
17. Contra-multi-ipsi-desc and asc-collat	1	H	
18. Ipsi-uni-ipsi-desc and asc-collat	3	R	
19. Ipsi-multi-ipsi-desc and asc-collat	2, 3	HO	may send axons into 2 adj nn
20. Uni-to-contra lateral giant fiber	3	SSmO	giant sends eff into contra-n
Interneurons			
Large cells:			
1. Contra-uni-desc		KO	appr 4 subtypes
2. Contra-uni-asc		RK	appr 2 subtypes
3. Contra-uni-desc and asc		RKO	appr 2 subtypes
4. Contra-bi-local		R	
5. Contra-bi-desc		C	
6. Contra-bi-asc		O?	
7. Contra-bi-desc and ipsi-asc		C	
8. Contra-bi-asc and ipsi-desc		O	
9. Contra-multi-asc		O	
10. Contra-multi-asc and desc		HO	
11. Ipsi-uni-local		C	? small
12. Ipsi-uni-desc		KO	appr 2 subtypes
13. Ipsi-uni-asc		KO	appr 3 subtypes
14. Ipsi-uni-desc and asc		CRO	appr 2 subtypes
15. Ipsi-bi-local		K?	
16. Ipsi-bi-desc		K	
17. Ipsi-bi-desc and asc		CO	
18. Ipsi-multi-local		CRO	
19. Ipsi-multi-desc and asc		HO	
20. Median-uni-to-median giant fiber		SSmO	
Small cells:			
21. Ipsi-uni-local		CO	appr 50% of all cells in ganglion

[1] Data pooled from *Lumbricus*, *Eisenia*, and *Pheretima*. C = Cerfontaine, 1892; R = Retzius, 1892a; H = Haller, 1889, 1910; K = Krawany, 1905; S = Stough, 1926; Sm = Smallwood and Holmes, 1927; O = Ogawa, 1939; adj. = adjacent; asc = ascending; bi = bipolar; appr = approximately; contra = contralateral; collat = collateral; desc = descending; eff = efferent; ipsi = ipsilateral; multi = multipolar; nn = nerves; uni = unipolar.

own segment. Besides the general paucity of descending collaterals there may be noted a tendency for all collaterals to be on the side of the cell body. In the latest and most complete work, that of Ogawa, there are fewer collaterals shown than in previous works, only one descending and six ascending among 59 cells drawn on each side in the nearly adult *Pheretima*, for all three nerves.

The best figures available on the **number of cells** in a ganglion are those of Ogawa. According to her, something less than half of the total of 1138 cells in a typical segment (the 30th) in the adult (approximately 17 cm long) *Pheretima communissima* are motor and 474 motor fibers enter the three pairs of nerves and supply 59,856 muscle fibers whose nuclei lie in that segment (see p. 750).

Interneurons have been variously called "Schaltzellen," "Binnenzellen," commissural cells (incorrectly), intercalary, interganglionic, or association elements. The penultimate study covering these matters, up to the present, concluded that these neurons are extremely few and unimportant and did not even list them in a summary of the contributions to the neuropile (Smallwood, 1930). The work of Ogawa (1939), however, would seem to place on a firm footing the abundance and variety of interneurons. Besides confirming what most of the early writers had described, she added several, so a total of some 21 types have been recognized (Table 14.1), based on the same three characters above-mentioned under motoneurons; a number of these include several subtypes and doubtless the possibilities of classification and the actual diversity have not been exhausted. One type alone, **small, unipolar, local neurons,** whose processes are very short, make up about one half of all the nerve cells, motor and internuncial, in typical ganglia. In certain segments, such as that of the prostate (XVIII in *Pheretima*), the proportion must be even larger for there is a much greater increase over typical segments in the total nerve cell count than in the motor nerve fiber count in the segmental nerves (1138 cells: 474 fibers in segment XXX; 4490 cells: 976 fibers in segment XVIII). Also, the relatively complex ganglia of the anterior cord (for example, segment VIII) and the subesophageal are described as noticeably richer in these small interneurons than typical

segments. This impression is supported by further counts of the fibers in the interganglionic connectives. We may perhaps doubt the completeness of fiber counts but they are useful as between levels of the cord. Whereas the fibers in the interganglionic connective total about four times the efferent fibers in the nerves of segment XXX (2090:474) and three times those of the nerves in segment XVIII (3094:976), they are twice as numerous as all nerve cells in segment XXX (2090:1138) but only two thirds of the number of nerve cells in the prostatic segment XVIII. The proportionate increase in cells in the latter must be in elements which do not send processes out into nerves or up and down the cord.

Inspection of the published figures of the less numerous **large interneurons,** which are almost all **intersegmental projection units,** and of Table 14.1 suggests that there is approximately an even distribution as between ascending and descending fibers and between crossed and uncrossed ones. The cells are well scattered through the ganglion longitudinally and mediolaterally and the terminal ramifications occur, apparently, throughout the neuropile. They do not form any highly specialized textural differentiations in the neuropile. Unfortunately the direct evidence at hand does not permit a useful statement of the average or extreme number of segments spanned by longitudinal projection fibers. The figures just given, however, permit a crude estimate. The drawings published support the estimate that the number of ascending and descending fibers from a ganglion is to the number of large interneurons approximately as 1:1. But if the small, local interneurons are nearly half the total of all cells and motoneurons are also nearly half in a typical ganglion, the remainder must be almost solely responsible for the interganglionic fibers, which are nearly twice as numerous as all cells in the ganglion (XXX). The discrepancy must be due to fibers from other ganglia. If we assume that the large internuncials are not more than a quarter of all the cells in the ganglion, the average span of ascending or descending **projection fibers will be four segments** and of an average neuron with both ascending and descending processes, eight segments.

The data from these studies with selective **nerve**

stains must form the basis of our intimate knowledge of the **synaptic relations between neurons,** a topic of fundamental interest in the evolution of nervous structure and function and to which we devote a special chapter (Chapter 4). Details may be found there and it will suffice here to point out three generalities about the oligochaete central nervous system. (a) First, the terminal ramifications of a given neuron, especially the presumed dendritic or afferent branchings, are widespread relative to the size of the ganglion. Without precluding specific connections, this raises the questions: what is the critical area for initiation of a postsynaptic impulse and how independent or interdependent are the several branches? (b) Second, although the character of the endings is intricate, fine, and extensive, they are not highly specialized. The bottle brushes, bushy tufts, and other distinct varieties of endings of higher articulates are not yet developed. (c) Finally, it is of interest whether synapses are confined to the neuropile or can also occur on cell bodies. Unfortunately the evidence on this point is not as direct and unequivocal as would be desired, and especially cannot be taken as exclusive but only permissive, since it is probable that finer endings of many of the neurons that authors have drawn were incompletely stained. At their face value, many of the neurons illustrated would be quite smooth and devoid of dendritic branches or collaterals. Nevertheless, one can only conclude from inspection of the published drawings that a fair number of fibers or branches appear to end in a region that seems to be purely cell body mantle. Of course, these could be making synapses with each other, but at present the possibility

Table 14.2.   CELL TYPES IN SUBESOPHAGEAL GANGLION OF EARTHWORM BY FORM AND PROCESSES.[1]

Types of neurons	Authors	Comment
Motoneurons		
Passing out segmental nerves		
to segments I, II, and prostomium (*Ph*)	O	
types similar to those in typical ganglion	K, O	see Table 14.1
Passing out other nerves		
asc in c-e-c and eff in brain nerve to prostom	O	incl cells in c-e-c
asc in c-e-c, passing into stomodeal gang	O	perhaps actually interneurons
Interneurons		
Ascending to brain, through c-e-c		
contra-uni-smooth (no dendritic branching)	K, O	3 or 4 subtypes
ipsi-uni-smooth	K, O	appr 2 subtypes
ipsi-uni with extensive dendrites	O	
contra-uni-asc and contra-desc	K, O	
ipsi-uni-asc and ipsi-desc, smooth	O	
ipsi-uni-asc and ipsi-desc, dendritic	O	
ipsi-bi	K	
contra-multi	K	
some terminate in central knot in brain	O	
some terminate in widespread branches in brain	O	
Descending into ventral cord		
large-ipsi-uni	O	3 subtypes
large-contra-uni	O	
Local		
small-ipsi-uni	O	more than 50% of all cells

[1] Data pooled from *Eisenia* and *Pheretima* (= *Ph*); K = Krawany, 1905; O = Ogawa, 1939; c-e-c = circumesophageal connective; eff = efferent; gang = ganglion; prostom = prostomium; other abbreviations as in Table 14.1.

of synapses on cell bodies has to be admitted, though they are certainly an extremely minor proportion of the junctions.

The **subesophageal ganglion** presents a special case among the ventral ganglia. It contains about twice as many nerve cells as a nearby segmental ganglion (segment VIII), though hardly any more than the prostatic segment XVIII and four times as many as a typical midbody ganglion like segment XXX. Together with the posterior half of the circumesophageal connective, it sends nerves to two segments (I and II), and in addition some to the prostomium (Ogawa, 1939, *Pheretima*, Megascolecidae) or to three segments (I, II, III, Henry, *Diplocardia*, Megascolecidae and Hesse, Hess, Henry, *Lumbricus*, Lumbricidae) or only to segment I (Isossimow, Henry, *Lumbriculus*, Lumbriculidae). The types of motoneurons according to Ogawa are similar to those of ordinary ventral ganglia. In addition to its segmental functions, the subesophageal ganglion sends nerve fibers (a) into the brain, (b) through the brain and out the cerebral nerves presumably to muscles of the prostomium, (c) through the circumesophageal connective and into the stomodeal ganglion, and (d) into the ventral cord behind it. The cell types involved in these and the intrinsic interneurons are summarized in Table 14.2.

3. Hirudinea

Retzius (1891b), Biedermann (1891), Rohde (1892), Havet (1900), and Sánchez (1909, 1912) have given us beautiful figures (Fig. 14.19) of methylene blue and silver stained neurons of leeches, which are classical objects for neurofibrils and for the neuron doctrine (Apáthy, 1897; Cajal, 1904; Sánchez, 1909, 1912). Ogawa (1930) contrasts the large peripheral motor fibers of

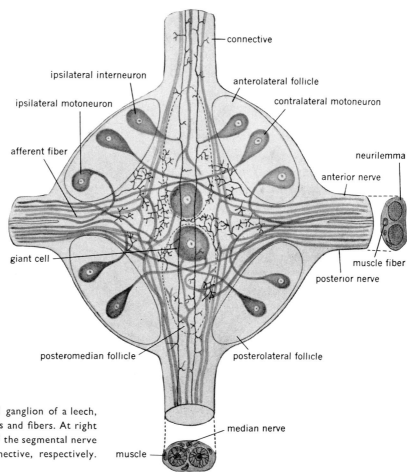

Figure 14.19.

Diagram of a typical ventral ganglion of a leech, showing the main nerve cells and fibers. At right and below, cross sections of the segmental nerve and the intersegmental connective, respectively. [Scriban and Autrum, 1932.]

Hirudo with those of oligochaetes, the former having often **two or three neurofibrils** in a cross section (hematoxylin), the latter always but one. A number of types of motoneurons and of interneurons have been recognized.

It is worth noting that in 1849 Bruch had already published an analysis of the organization and cell types of the hirudinean ganglion that in its main outlines was astonishingly correct and represents one of the first studies of **conduction paths** in invertebrates. He found virtually all the neurons are unipolar and virtually none of the motoneurons have collaterals which ascend or descend in the ventral cord. This contrasts with the oligochaetes and may be of phylogenetic or functional significance; at present it is difficult to evaluate. Most of the authors found only ascending interneurons. (Figures 315 and 316 in Hanström, 1928b, which appear to the contrary, are printed upside down.) Several authors have seen interganglionic fibers passing through ganglia and not ending in a cell body, as far as they could be followed. These suggest long, through-conducting, possibly descending paths and this is in agreement with the conduction rate in the cord, 0.3–0.4 m/sec (Bethe, 1908), which is reasonably high for nongiant fibers without synaptic delays. The findings of Bethe, Schwab, and Dittmar on the effect of stretch are highly questionable (pp. 149, 674).

Rather than having any simple anatomical relationship, the six packets or **follicles of cell bodies** (Fig. 14.9) contain cells sending axons into contralateral as well as ipsilateral nerves, and the three anterior as well as the three posterior packets send axons into both anterior and posterior nerves. Both median and lateral packets include motor as well as internuncial cells. The primitive subdivision of segments is probably into three annuli (Mann, 1953) but possibly two. No clue therefore has been uncovered as to the meaning of the cell packets. From the paths of the neurons, certainly each ganglion is a unit, as proposed in the Castle and Moore interpretation of the segments (Mann, 1953), and not a compound of parts of two segmental units as required by the widely employed Whitman method of defining segments.

Details of the **types of neurons,** which are essentially like a selection of those listed under Oligochaeta, may be found in Fortuyn's review (1920) and in Fig. 14.19. Crossed and uncrossed neurons are common as well as cells sending two axons into separate peripheral nerves—apparently always on the same side. The latter is especially exemplified by the two "Kolossalzellen" in each ganglion, which are large, median cells but with nongiant axons and probably (Biedermann, Havet) no longitudinal process in the cord. The dendritic processes of motoneurons are very widespreading (Fig. 14.20). There are no real giant

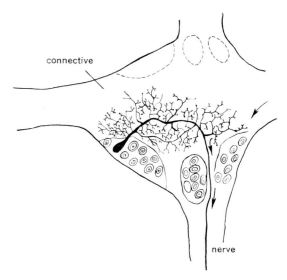

Figure 14.20. Motoneuron from leech, *Nephelis*, in a longitudinal section of a ventral ganglion. Golgi stain. The rich branching of presumed dendrites, bearing pyriform terminals, is seen. Arrows indicate probable direction of conduction of impulses. [Havet, 1900.]

fibers in leeches (but see Rohde's large fiber in the median nerve). Fast through-conduction units recorded physiologically point to long interneurons of good size, however.

Ogawa (1930) found the same ratio of large fibers in the nerves to total nerve cells in the ganglion (1:3.4) as in the earthworm *Allolobophora foetida*. The majority of cells are therefore probably interneurons and from the paucity of interganglionic fibers in the studies employing methylene blue, probably local, though Sánchez could not confirm the existence of such cells.

The afferent fibers, just as in oligochaetes, do not branch in the peripheral nerves but upon entering the ganglion bifurcate into ascending and descending processes which run to other segments or end in the same ganglion, ipsi- or contralaterally or both. They form three longitudinal bundles as in earthworms and like the latter are mostly very fine fibers (contra Scriban and Autrum, 1932, after van Gehuchten). They may be delimited in discrete bundles in the nerves (Ogawa, 1930). A few large afferents are described.

E. Giant Fibers: Anatomy and Physiology

1. Polychaeta

Although at least a good many of the giant fibers originate in the brain, we may treat them at this point since most of the neuron is found in the cord and indeed any exact knowledge of the portion in the brain is confined to three or four species. We remarked earlier upon the great diversity manifest in this class in contrast to the other classes of the phylum; this is well illustrated by the giant systems which vary from complete absence or slight development to elaboration to fantastic proportions, from many units to a single unpaired one. Some are compound units with many cell bodies, others simple; some are purely central, others are final motor neurons. A detailed review has been provided by Nicol (1948).

The following families we may presume on present evidence to **lack any outstanding fibers** in either absolute or relative size: Hesionidae, Phyllodocidae, Syllidae, Amphinomidae, Euphrosynidae, Dorvilleidae, Sphaerodoridae, Chaetopteridae, Scalibregmidae, Pectinariidae, Oweniidae, Acoetidae.

Large fibers, often called giants, but which are **confined to one or a few segments** and not accompanied by long through-conduction giant systems, are found in the Aphroditidae, Arabellidae, and Myzostomida. We may describe these first. In *Hermione* (Aphroditidae) large intrasegmental fibers originate in cell bodies in each segment and enter the peripheral nerves after decussating (Rohde, 1887). *Arabella* appears to have from none to six variable longitudinal giants in sections of the cord (Livanoff, 1924; Fedorow, 1927; Spengel, 1881) and these are believed to be associated with large cells in segments II to XX. There are many more cells than fibers in any one cross section, so Spengel supposed the fibers to be multicellular, as is true for many other families. However, Bullock (1948) concluded that the fibers each extend a short distance and he could not obtain an action potential over any considerable distance, as is easily done in all the families having typical giants. Some kind of rapid conduction (1–1.5 m/sec) was evidenced by motor response, however. In *Myzostomum* there are groups of large longitudinal fibers but the ventral chain is consolidated into a single ganglionic mass so they are not really long through-pathways. There are also half a dozen pairs of large unicellular motor axons which decussate and leave in peripheral nerves.

Similar large, intrasegmental motor fibers occur in the Sigalionidae, Polynoidae, Ichthyotomidae, and Nereidae. In these families, however, there are **intersegmental through-conduction fibers** in addition to the intrasegmental units. In the sigalionids the long giants are perhaps less specialized than in some polychaetes; there is not a fixed, small number but several giant fibers, and most of them are unicellular. Several have their cell bodies anteriorly in the subesophageal region and then run backward after decussating. Several others have cell bodies at various posterior levels, decussate twice, and pass forward. An additional fiber or pair of fibers originates in a pair of cell bodies in the brain, without decussating (p. 720). Polynoids have one or two pairs of giant fibers, the laterals are said to send segmental branches into peripheral nerves. These would be motor neurons additional to the intrasegmental ones. Very likely they both innervate largely the same muscles; that is, neither is a final common path but some motor commands reach the effectors through the intersegmental long neurons which produce a nearly synchronous contraction of a great anteroposterior extent of musculature and allow no fine control or varied response; other commands would reach the effectors through repeated unit efferent neurons whose central synapses may by convergence of various paths integrate and do so differently in successive segments, introducing local variation into the response. In *Lepidametria*, with one pair of giants, Bullock found a single

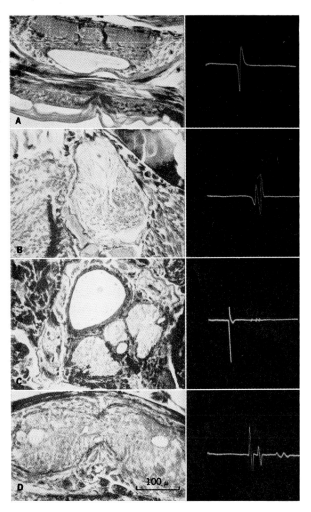

Figure 14.21.

Anatomical and electrical signs of giant nerve fibers in some polychaetes. Cross section of ventral nerve cord (left) and action current pattern on maximal direct stimulation (right) for each species. Dorsal side up; scale applies to all figures. **A.** *Diopatra.* Single giant fiber is the large, flattened space below the punctate tissue of the cord. **B.** *Haploscoloplos.* The two median pale areas, one under the other, are the giant fibers. **C.** *Lumbrineris.* A large middorsal, a small midventral, and two lateral fibers are visible. **D.** *Neanthes.* Two large laterals, conducting at the same velocity, a smaller median, and two small medial fibers are visible. [Bullock, 1948.]

giant spike conducting at 5 m/sec and, relative to other polychaetes, quite a high resistance or threshold to mechanical stimulation. Ichthyotomids have been reported to have, besides the intrasegmentals, several small giants (or large ordinary) fibers which extend for a considerable distance.

In the Nereidae we encounter not only a fixed pattern and a small number of fibers, as in the polynoids, but definite evidence of multicellular or syncytial fibers and segmental septa, as in the earthworm and crayfish (see pp. 702 and 868). The characteristic pattern of five giant fibers (Fig. 14.21) consists of large paired laterals, small paired paramedials, and an intermediate-sized median fiber. The laterals are divided segmentally or at frequent intervals by complete, obliquely transverse septa, so a macrosynapse is formed; the presumption of cell bodies in every segment is introduced. These are the only **complete septa** so far described in polychaetes, but it is probable there are some other giant fibers so divided (for example *Lumbrineris* laterals; Bullock, unpublished). Septa are clearly absent, however, from many of the large polychaete giants carefully studied on this point by the writer (*Protula*, Serpulidae; *Spirographis*, Sabellidae; *Marphysa*, Eunicidae; *Lumbrineris* dorsomedian; *Clymenella*, Maldanidae). The significance of septa is discussed on pp. 227 and 704. The median fiber in nereids may be compound like the laterals but the paramedial pair is described by Hamaker (1898) as a chain of unicellular elements each decussating twice and synapsing by "fibrillation." This description needs confirmation in view of the relatively fast and unpolarized conduction (2.5 m/sec).

The physiologic data from **electrical recording** agree with the anatomical in the number and relative sizes of the units and in addition permit conclusions inaccessible by histologic methods.

(a) Each of the fibers is an **independent unit,** without anastomoses or even physiologic connection with others (with the possible exception of the two paramedials in nereids) and is through-conducting—that is, not a local efferent mechanism for some other pathway between loci of stimulation and recording. The evidence for this may be instructive as an example of "physiological neuronography." (i) No other sign of excitation arrives at any point earlier than the giant spikes; therefore it is probable that they are all independent through-pathways. (ii) Each of the spikes can be elicited alone by suitable mechanical stimulation, presumably through sense cells. (iii) The interval between earlier and later spikes is not independent of conduction distance but directly proportional to it. (iv) The arrival of one of the earlier spikes at any point is not a sufficient condition to cause one of the later spikes, but a stronger electrical stimulus at the distant point of excitation will do so. (v) Later spikes may be obtained without the earlier ones by electrical stimulation with square waves of long duration (2 msec), due to crossing of the strength-duration curves.

(b) It is also found that each of the fibers is **unpolarized,** conducting at the same rate in both directions.

(c) The **connections** of the giant fibers are such as to mediate centrally an abrupt, symmetrical, over-all withdrawal response to certain kinds of commonplace stimuli whose common denominator seems to be a startle quality. This statement implies several things. (i) The giant neurons are probably not sensory. One of the arguments for this is the usually found precise synchrony of the firing of the two independent lateral giants in nereids. (ii) They are apparently not motor in nereids as they are in some polynoids, arenicolids, and sabellids. There is at least one very labile junction between giant spike and muscle contraction that does not show facilitation but rapidly fatigues after following giant spikes at a high frequency for a time. The muscular response to giant firing is only the first part of the response to an adequate stimulus in the unanesthetized or lightly anesthetized animal; many responses which seem prompt to the eye and probably the major part of the response after giant fiber stimulation, are not the immediate result of giant firing but may occur in the absence thereof, or as a relatively indirect aftermath. The intrasegmental large fibers described by Hamaker are doubtless a motor path for giant activity. The pair of neurons in each segment anastomoses in the midline (a decussation with secondary fusion?) and sends axons into two nerves on each side, assuring symmetrical response—which is perhaps only desired on occasions calling for rapid withdrawal.

(iii) The giant neurons are then probably purely central interneurons. Nevertheless, they are high-speed through-conduction pathways, not truly associative neurons of a high order. They do not participate in the spontaneous activity of the intact or isolated nervous system but fire only when a special pattern of afferent neuron activity arrives. They are evidently **non-final motor paths** rather than sensory pathways, for they achieve rapid distribution of excitation to responding centers at all levels rather than delivery of impulses to a sensorium for integration and subsequent initiation of motor command. The giant fiber burst is that discharge to lower motor neurons and it usually operates at a reflex level but occasionally from a sensorium of a cerebral level. Synapses to motor axons are not yet known except one between the lateral giant and a large fiber in nerve IV. (iv) The **startle quality** of the adequate stimuli is an observer's subjective interpretation. Prodding, squeezing, and similar mechanical stimuli, weak or strong, usually fail to elicit giant firing, whereas light mechanical stimuli of sudden onset usually do; adaptation sets in soon in this case. These adequate stimuli may be quite local. A light tap with a thin glass rod or a single drop of water falling from a pipette a few centimeters above a specimen lying partially out of water typically produces a brief burst of from 1–20 spikes at a frequency of 200 per second.

Table 14.3. ANNELID GIANT THROUGH-CONDUCTION SYSTEMS.

Group	Fibers No.	Fibers Detail	Motor branches	Cell bodies Uni- or multi-cellular	Cell bodies Location	Additional Features	Fiber diameter (μ)	Conduction Velocity, 18–24° C (m/sec)
A. Archiannelida								
Polygordiidae	0–1	sev (?)				followed into brain		
B. Polychaeta Errantia								
Nereidae (3)	5					*Neanthes*:		
		2 lg lat	0	mult	each seg (1)	with segmtl septa	30–37	5
		1 median	0	mult	each seg	with segmtl septa (?)	15–18	4.5
		2 medial	0	uni	each seg	synaptic but fast thru-cond	7–9	2.5
Syllidae	0							
Hesionidae	0							
Aphroditidae (3)	0							
Polynoidae (3)	2 or 4	2 lats, lg	+			lg intrasegmtl motor units	20–30	0.5 (ng) (2)
		2 + medials, sm	0			*Lepidametria*, with 2		5
						Lepidasthenia: enter c-e-c, end in subesoph g; (+ segmtl motors)		
Sigalionidae	22+	14 ap, long	0	uni, bi	brain and cord	*Sthenelais*: (+ segmtl motors)		2 (2)
		8 + pa, short sm						
Polyodontidae	2							
Chrysopetalidae	0							
Phyllodocidae	0							
Tomopteridae	1	possibly more						
Nephtyidae (3)	4	2 lg and sev sm				*Nephtys*:	20–25	5
Amphinomidae	0							
Euphrosynidae	0							
Eunicidae (3)	1	very lg			no lg cells	*Marphysa*:	170	10
Onuphidae (3)	1	very lg				*Diopatra*:	110	10
Lumbrineridae (3)	4	1 very lg, 3 sm			ant g, lg	*Lumbrineris*:	130	10
							20–25	4–5
Arabellidae (3)	0	many short giants		mult (?)	segs II-XX, lg	probably short, relaying, 0–6 per xs		1.5 (ng) (2)

Family								
Dorvilleidae	0							
Glyceridae (3)	8–14	8 + lg / 2–6 sm		in brain (?), lg	Glycera:		15–20	3–4
Ichthyotomidae	sev	sm						
Lysaretidae	20–30		uni / uni	2± each ant, lg / post cord, sm	Halla: / / Oligognathus	fibers run posteriorad, maybe fuse / fibers run anteriorad	8–40	
Sphaerodoridae	0							
Paronidae	0							
Typhloscolecidae								

C. Polychaeta Sedentaria

Family								
Orbiniidae (3)	1–2	both median		in brain (?)	Haploscoloplos:		20–35	7
Spionidae	1–2	lg			Polydora:		22–30	
Chaetopteridae (3)	0							
Magelonidae	1	lg	mult	brain (?), lg		double ant to VII, enters c-e-c		
Oweniidae	0							
Terebellidae (3)	0				Amphitrite			
	1	lg		cord paired	Pista:		48	2–4
	1	lg	mult	cord, lg	Lanice			
Ampharetidae	sev	0–3 per xs each 1–2 seg	uni (?)	cord	Ampharete, Amphicteis	branch and anast (?)		
		lg			Melinna			
Pectinariidae	2							
Cirratulidae (3)	0				Cirriformia			
	0	lg	mult	cord, lg	Dodecaceria			0.9 (ng) (2)
Capitellidae	1				Notomastus, Mastobranchus	possible anast and discont fibers, good "myelin" sheath		
	0–4	varies along cord			Capitella sp			
Opheliidae	0							
Maldanidae (3)	sev	sm, irreg unpaired	mult	cord (in epith)	Clymenella:	fibers unlike, tapering	35 max	4 max
	2				Maldane: 1 ant, 1 post			
(3)	2	lg, 1 per xs	+	each seg, 50–80 μ	all fibers anast		65	
Arenicolidae	1–3	varies along cord	mult		Arenicola cristata		15–25	2
	0				Branchiomaldane			

(Table continues on following page)

(Continuation of Table 14.3)

Group	Fibers		Motor Branches	Cell bodies		Additional Features	Fiber diameter (μ)	Condution Velocity, 18-24° C (m/sec)
	No.	Detail		Uni- or Multi-cellular	Location			
C. Polychaeta Sedentaria (cont.)								
Scalibregmidae	0							
Flabelligeridae	2	paired						
Sternaspidae								
Sabellidae	2	very lg	0	mult		*Sabella* anast each seg	100–130	4–6
(3)	2	very lg	0			*Spirographis*: anast repeatedly in thorax		
(3)	2	very lg				*Eudistylia*: 1 anast or none; reciprocal synaptic transm; enters c-e-c	200–250	4.5–7
(4)	1	very lg	+	mult	10 + sm cells per seg plus 1 pr in brain	*Myxicola*: starts in brain, paired, decussates; fuses in cord; sheath = 1% of fiber diameter	100–1000 (max 1700)	3.2–21
Serpulidae (3)	2	very lg	0	uni	in brain, lg	*Protula*: synapse at decussation	250–350	10
Sabellariidae	2	lg						
Incertae sedis								
Goniadidae	2 (?)	sm						
Disomidae	2	very lg				*Poecilochaetus*	83	
Myzostomida	sev 12–14		0	uni	in groups, longit 6–7 prs cord	*Myzostoma* extends thru single lg ganglionic mass segmtl motor fibers		
Acoetidae	0							
D. Oligochaeta (Superfamilies:)								
Lumbricina (3) (5)	3	1 lg median 2 lg lats	0 +	mult mult	sev each seg sev each seg	*Lumbricus*: segmtl septa segmtl septa synaptic connections	60–75 30–50	15–45 7–15

Phreoryctina	2				between rt and left lats. lg ventral fibers, segmtl, not thru-conducting (?)	
Megascolecina (6)	3				Pheretima:	
		1 lg median	0	mult	septa as above	75 40
		2 lg lats	+	mult	septa as above	30 15
		2			lg ventral fibers, segmtl median divides into c-e-c	
Lumbriculina	3			sev each seg		
Branchiobdellidae	0			sev each seg		
Naidina	3				Chaetogaster, Nais, Aeolosoma	
Enchytraeina	1–3					
Tubificina	1–3	very lg			3 posty fuse to 1 anty	
E. **Hirudinea**	0	sev thru-conducting fibers		uni	demonstrated electrophysiologically	1

anast = anastomose; anty = anteriorly; ap = anteroposterior; bi = bicellular; c-e-c = circumesophageal connective; discont = discontinuous; g = ganglion; lat = lateral; lg = large; max = maximum; mult = multicellular, syncytial; ng = nongiant conduction in cord; pa = posteroanterior; posty = posteriorly; pr = pair; seg = segment; sev = several; sm = small; uni = unicellular; xs = cross section.

(1) By supposition from the fact of septa; not demonstrated.
(2) Determined by kymograph by Jenkins and Carlson (1903).
(3) Confirmed in one or more species of the genus by electrical recording in Bullock (1945a, 1945b, 1948, 1953).
(4) Confirmed as above, in Nicol (1948a); Nicol, and Whitteridge (1955).
(5) Confirmed as above, in Eccles, Granit, and Young (1932); Rushton and Barlow (1943); Rushton (1945, 1946).
(6) Confirmed as above, in Adey (1951).

(v) The pattern of **connections to sensory sources** is different for each of the giants. Smith (1957) believes that there are not direct connections with afferent fibers but indirect excitation of giants through the internuncial neuropile. The median fiber spike of nereids (third to direct electrical stimulation) can be elicited by stimuli to approximately any part of the anterior quarter of *Neanthes virens*. The smallest, slowest spike from the paramedial fibers is fired by stimuli anywhere in the posterior three-quarters. A region of overlap of a few segments occurs. This combination is exactly like that in the (doubtless) independently evolved oligochaete, *Lumbricus*. The fast, lateral giants of *Lumbricus* or *Neanthes* can be fired from any level but require stronger stimuli. The threshold varies within each fiber's zone of responsiveness. Once fired, any giant conducts its impulse throughout the cord, both anteriorly and posteriorly, so in normal function a given stretch of fiber may conduct now one way and now the other. The two laterals usually fire together. There are almost certainly **specific connections on the motor side** also, so each fiber produces a different response, presumably of setae, though all result in shortening. Some details of the motor connections of two polychaetes are given by Horridge (1959).

The presence and number of **giant fibers in other polychaetes** of some two dozen families is summarized in Table 14.3. It will be noticed that many of these have been confirmed by recording the pattern of action potentials. Actually the amplifier and oscilloscope offer a more sensitive method of detecting the sometimes small and uncertain or possibly anastomosing giants than histologic means, owing to the facts that the height of the spike increases disproportionately in the large fibers, through-conducting long fibers are automatically selected, and controls against artifacts are so simple—not to mention the great speed and simplicity of the demonstration. Of course the electrical method can not detect, for instance, paired giants with frequent anastomoses and may at first even fail to separate independent units of identical properties.

The tables should be consulted for an impression of the range of development of giant systems in the group. We may here call attention to some features of general interest. Many of the giant fibers are **unicellular** and many are **multicellular**, referring to the number of cell bodies whose processes contribute to the fiber. These cell bodies are sometimes themselves giant (for example 80–150 μ in diameter) but are more often ordinary or only somewhat larger than ordinary in size. They are of the plasma-rich, chromatin-poor type. **Giant cell bodies** may give rise to unicellular giant fibers (*Halla*, Lysaretidae; *Protula*, Serpulidae) but even multicellular fibers are supposed in some instances to require very large cell bodies (for example *Arenicola*, where one or two 50–90 μ cells in each segment contribute to the syncytial fiber). Contrariwise, cell bodies of hardly extraordinary size may not only be found with multicellular giant fibers but also with unicellular giant fibers (*Sigalion*, *Halla*). It should be of broad biological interest to compute the ratio of volume of axon to volume of cell bodies in a number of forms. This has been done by Nicol for *Myxicola* (Sabellidae) and he finds that although the single fiber in this form may have 1300 cells, the volumes are about 4×10^{10} and $8 \times 10^6 \mu^3$, respectively, or 5000:1 (Fig. 14.22). This may be compared with an anterior horn motoneuron in the rhesus monkey, for which Bodian and Mellors gave values of $12.5 \times 10^6 : 5 \times 10^4 \mu^3$ or 250:1. The cells have been described carefully in some forms; they may have well-developed neurofibrils, like ordinary plasmatic annelid nerve cells, and some are regarded as having Nissl substance (*Lumbrineris*, Fedorow, 1927).

Most polychaete giants are purely central interneurons but a few have **peripheral, efferent branches** (*Myxicola*, Sabellidae; claims for *Arenicola* and *Lepidasthenia* are doubtful). These motor branches are probably not final common paths because the same muscles that they innervate are probably the effectors for ordinary locomotion and other responses which must be mediated by nongiant systems. These responses probably have separate peripheral pathways; that is, the muscles probably are doubly or multiply innervated. The same may be said of at least some of the purely central giants which act through motor neurons that anastomose across the midline

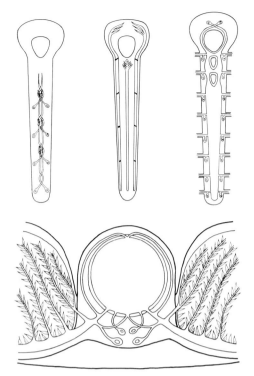

Figure 14.22. Diagrams, not to scale, of the plan of the giant fibers in some polychaetes. **Above, left:** *Neanthes*, the small median giant fibers which extend anteriorly through two segments only. **Above, center:** *Neanthes*, the medial and lateral giant fibers—the medial fiber is connected with cells in the subesophageal ganglion, the laterals pass into the esophageal connectives and possibly into the supraesophageal ganglion. **Above, right:** *Myxicola infundibulum*: the nerve cord in this animal is double in the anterior thoracic region and single posteriorly; the giant fibers conform to this pattern. In the first few setigers two fibers, arising from cells in the supraesophageal ganglion, interconnect by transverse anastomoses, and in posterior thorax and abdomen they fuse to form a single large fiber that is in protoplasmic continuity with many small cells along the cord and gives rise in each segment to several peripheral branches (**below**, in cross section), which directly innervate the longitudinal musculature. [Nicol, 1948.]

(nereids), this dictates symmetrical contraction, whereas much ordinary movement is alternating.

The **number** of giant fibers varies from none to twenty-five or more (*Halla*, Lysaretidae) but when there are a number of fibers they are never very large. There are never more than two of the very large fibers, but there may be one of them and several smaller giant fibers in addition (*Lumbrineris*; Fig. 14.21). The **extent** of the giant fibers anteroposteriorly in the body is almost always the whole length or close to it. The large intrasegmental fibers often called giants are, of course, not included in this statement; they are local motor neurons. A good many of the giant fibers ascend to or originate in the brain. These are treated below (p. 720).

Neurofibrils have not been found in any polychaete giant fiber, quite in contrast to the earthworm, although a good many species have been examined by methylene blue or silver or by other methods which should bring them out, if present. The axoplasm is smooth or textured with fine striations. Segmental constrictions have been described and often the fibers appear flattened but these are probably temporary results of the movements of the animal; while they may be expected to influence the impulse in some way, they probably have no great morphologic or functional significance.

The fibers are typically surrounded by a **sheath** of several layers—a thick outer lamellar connective tissue coat and at least sometimes (perhaps always) a thin myelinlike layer. The fibers are unmyelinated, by old histological standards, and none has been examined with the electron microscope. Sheath cell nuclei are sometimes seen in the innermost layers of the sheath. Neuroglial trabeculae frequently penetrate and sometimes seem to cross or divide a giant fiber. Serial sections—or, in whole preparations, careful illumination and focusing or a change in light angle—generally show these to be incomplete (for example maldanids, Lewis, 1898). The sheath varies a good deal from species to species but no correlation with velocity of conduction is apparent (Nicol, 1948a). *Haploscoloplos* (Orbiniidae), for example, has a very thin sheath but a good velocity for its size (see below).

Branches, collaterals, or processes of the giant axons for bringing impulses in or conducting them out to other neurons are remarkable for their scarcity. It is possible that certain minute thornlike outgrowths (*Protula*, Bullock, 1953b) are functional processes (Fig. 14.23). Perhaps there are ordinary terminal ramifications along the course of the stem process of the cell bodies that contribute to the axon. In the present state of

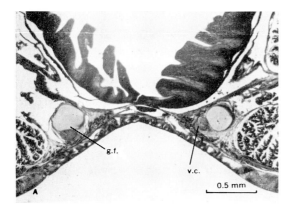

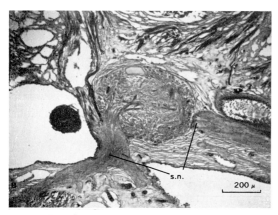

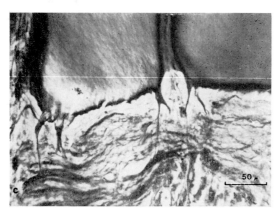

Figure 14.23. The giant nerve fibers in the serpulid *Protula.* **A.** Low magnification of transverse section of the midventral body wall in the trunk region—gut above, outer epithelium below, main longitudinal muscles at the sides. g. f., giant fiber; v. c., ventral cord. **B.** Subesophageal ganglion in transverse section, with enormous sensory nerves (s. n.) from the epithelium of the prostomium; giant fiber dorsally, just beginning to swell posteriorly. Midline to the left. **C.** The giant fiber (above) and some of its rare processes, seen in a frontal section. The collateral processes taper rapidly and lose themselves in the neuropile. Holmes silver. [Bullock, 1953.]

knowledge we are not sure of the anatomical basis for impulses entering or leaving the giant fiber in any single case (excepting those with peripheral branches). Elsewhere we describe in a serpulid, *Protula*, the clearly functional afferent connections of the cell body and its dendrites at the origin of the giant fiber in the brain (p. 720).

Turning to some functional aspects, Table 14.3 summarizes **conduction velocities and axon diameters** in annelid giants. It can be seen that at present no support can be given for any particular value relating the two, for example an exponent of 0.6 or 1.0. It is clear that factors other than diameter are effective. Thus the 20 μ fibers in *Haploscoloplos* conduct at 70% of the velocity of the 100 μ fibers in *Diopatra* (Onuphidae) and *Lumbrineris*, and these two in turn conduct fully as fast as the 200–400 μ fibers in serpulids and sabellids. The enormous axon in *Myxicola*, the largest known nerve fiber, achieves 20 m/sec in a region with a diameter of 1000 μ (Nicol and Whitteridge, 1955). The earthworm, with a much lower salinity and thus conductivity and a much smaller diameter (60–90 μ), achieves speeds at comparable temperature of 25–40 m/sec. Even when we compare fibers within the same individual, diameter may seem to play a minor role. The ratio of diameters of the largest and the next largest giant axons in *Lumbrineris hebes* is about 6:1 and that of their velocities is 2:1; very strikingly, the necessary consequence of this velocity and hence wavelength—the ratio of **spike heights** —is about 30:1. The matter deserves further study in this group, for the tentative conclusion is inescapable that increasing velocity by increasing diameter is inordinately expensive and all the more difficult to accept when we see that other means of increasing speed are already being exploited by the annelids.

The magnitude of conduction velocities seems low in comparison with fish and frog and higher vertebrates (from 40 m/sec up). They are, however, in the same range as the nongiant axons of about 35–75 μ in crustaceans (6–12 m/sec) and within a factor of two of the 300–600 μ axons of cephalopods (10–20 m/sec). The earthworm in fact exceeds the latter in velocity. It may be of significance that recent measurements, presumably on Müller's fibers in the spinal cord of lampreys (*Entosphenus:* Cyclostomata), show a maximum velocity of 5 m/sec for fibers of about 45 μ. Clearly a great deal of **physiological evolution,** over and above the anatomical, has taken place between this point and that of teleosts, which achieve values of 90 m/sec, presumably in Mauthner's fibers of about 60 μ.

In those sabellids having two giant axons and in serpulids, which all have two, a phenomenon of general as well as local interest, and often indeed puzzling, is the **interaction between paired fibers.** This is of several kinds (Bullock, 1953b). Genera which like *Sabella* have anastomoses at frequent intervals along the whole cord have in effect one functional unit. The same is almost true of *Spirographis,* in which anastomoses are present in several of the commissures in the anterior eight segments, and of *Eudistylia,* where about half the individuals have an anastomosis in the subesophageal ganglion, the other half having a synaptic connection. In *Protula* there is no connection except a synaptic one in the brain. In all these cases, when stimulation occurs anteriorly, both giants will be activated as though there were one unit. When adequate stimuli impinge on posterior segments, the sensory connections usually fire both giants but they may fire only one; in this case the other side is activated only after a delay during which conduction to the anterior cross connection and back on the other side transpires. This seems maladaptive and we may suspect it rarely occurs in nature. In *Protula*, in fact, the giant spike cannot be elicited except by stimuli to the anterior end and this may be the condition that permits the lack of anastomoses. Synaptic connection between the two giants would then be normally used in the case of unilateral head stimulation. The properties of the synapse fit this supposition—a 1:1, unpolarized, relay junction. If this line of reasoning has any validity, it tells us that nature does not use an anastomosis unnecessarily or that a synaptic connection is the simpler for evolution to achieve, a conclusion in conformity with expectation from the neuron doctrine.

But another kind of interaction has been observed—cross-over or activation of one giant by the other via **reciprocal, distributed transmission.** This means transmission left to right or right to left can occur at very many points along a great length of the body; the point of cross-over or effective transmission can appear to slide gradually up and then down the cord, several millimeters in some seconds. The junctions so manifested in whole intact regions of many but not all specimens of certain sabellids and serpulids are highly labile and unpolarized with a delay of 1–2 msec. They appeared reminiscent of ephapses, although in undamaged tissue, and were first called quasi-artificial synapses (Bullock, 1953b). In still unpublished work on *Eudistylia* with intracellular electrodes in both fibers, Hagiwara and Morita find typical synaptic potentials at any point penetrated along either fiber, upon arrival of an impulse in the other; transmission is not electrotonic. They conclude that this is a case of very numerous, bilateral (or reciprocal), chemically transmitting synapses. It can not be stated whether each synapse is two-way or there are many junctions polarized in each direction. The anatomical basis has not been found; transverse connections across the one millimeter separating the right and left giants must be very fine, for they have escaped notice in excellent silver preparations.

Interaction which results in keeping the impulses in two fibers exactly in step seems to occur here, too, but is better studied in earthworms (in situ) and in crustaceans (experimentally approximated fibers).

The functional organization reflected in segmental regions or **levels in which sensory connections occur** was broached above (p. 696), in the case of *Neanthes* (Nereidae). Similar information is available for a few other species. The single axons of *Diopatra* (Onuphidae) and of *Myxicola* (Sabellidae) can be fired from any level by mechanical stimuli. It is remarkable that the same is true, apparently, of *Lumbrineris;* at least up to segment LX, all three or four giant spikes are readily elicited. This raises the question: what functional distinction is there between them or what meaning can be attached to having several units unless the qualities of the adequate stimuli in nature are different? *Haploscoloplos* (Orbiniidae) shows a topographic separation: the faster spike is fired from any point in the anterior 25–40 segments (approximately the anterior quarter of the worm), the slower from any level behind this, with a small zone of overlap. *Clymenella* (Maldanidae) also produces one spike from the first quarter or so (which means 5–10 segments in this case) and another from any level farther back. There was no overlap nor any blind zone in most individuals tested. As in the other species (*Lumbricus*, *Neanthes*, *Haploscoloplos*), the fiber is thickest in the zone in which it is readily fired and the thickest fiber is the anteriorly excited one.

Also like the others, any spike is conducted to all levels, both forward and backward.

Species differ greatly in the ease with which the giants can be **fired by mechanical stimuli** and in the vigor of the response. In all cases tapping rather than prodding is necessary; that is, either the sense organs or some central synapse is sensitive to rate of rise of excitation and selects only the abrupt stimuli. *Diopatra*, *Lepidametria*, and *Nephtys* do not respond readily whereas *Neanthes* and especially *Clymenella* and *Haploscoloplos* are prone to high frequency, sometimes long-continued discharge.

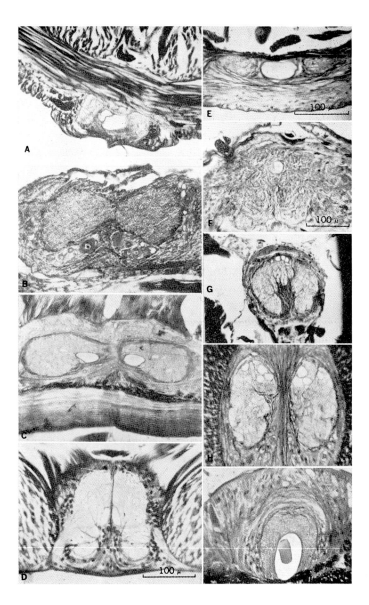

Figure 14.24.

The variety of giant fiber complements in polychaetes. Transverse sections of ventral nerve cords; from top to bottom, left column: **A.** *Clymenella* (Maldanidae, two fibers asymmetrically, in epithelial cord). **B.** *Aphrodita* (Aphroditidae, no giants). **C.** *Lepidametria* (Polynoidae, two fibers); **D.** *Nephtys* (Nephtyidae, two large and two small giants). Top to bottom, right column: **E.** *Pista* (Terebellidae, one fiber). **F.** *Arenicola*, (Arenicolidae, two fibers). **G.** *Arabella* (Arabellidae, several short fibers). **H.** *Glycera* (Glyceridae, 6 or more fibers). **I.** *Maldane* (Maldanidae, one fiber; shown dorsal side down). Scale in **E** applies to **H** and **I**; scale in **D** applies to **A, B, C, G**. [Bullock, 1948.]

Referring to Table 14.3, a word may be said concerning the **distribution among families and genera.** It is obvious that both sedentary and errant forms possess and lack giant systems (Fig. 14.24). Their common functional significance is therefore not directly related to locomotion or its absence. The swimming and pelagic forms (Tomopteridae, Syllidae) lack them. They are not related to evolutionary advancement, for the Aphroditidae lack them and some archiannelids (*Polygordius*) have them, while several families have some species with and others without giants (Terebellidae, Lysaretidae, Cirratulidae). Giant fibers are obviously more affected by body size than ordinary cells at least in some cases. Nevertheless, many small species possess giants while their nongiant fibers are probably not particularly small.

The question becomes insistent—as in the case of giant fibers in small dipterous insects (p. 1234): is the saving of time due to large diameter really significant where the giant is small and the distances are short? The only other suggestion that has been offered (Bullock, 1948, 1950) makes the increased spike height (as seen by external electrodes or neighboring cells) the significant factor, but no evidence for this has been brought forth. The **biological meaning of giant systems** that appears most plausible is the special importance to the species of a fast withdrawal or startle response in which a large mass of musculature must be nearly synchronously and symmetrically excited without integration, facilitation, or local control and with a minimum number of junctions. This idea needs to be tested by a wider correlation than our present knowledge of behavior permits. So far as that goes, it does appear that the species with giant fibers have a startle response and those without do not. The giant response fails after a few repetitions but without impairing use of the same muscles for ordinary movements, as in locomotion.

Further discussion on giant systems in annelids, including data on transection and **regeneration,** may be found in Nicol's valuable review (1948).

2. Oligochaeta

The giant nerve fibers of articulates were doubtless the first nervous units to be seen (Ehrenberg, 1836, on crustaceans) and those of earthworms had already become classical objects of study, with a considerable literature before the neuron doctrine was conceived. These well-named "Kolossalfasern" are so large relative to other fibers in the same animal and to vertebrate fibers that for a long time many refused to believe they were nervous in nature and the noncommittal term Neurochorde became well established. The theories advanced interpreting them as degenerate nerve fibers, supporting structures, canals—even to homology with the notochord—excretory or nutritive tubes are reviewed by Ashworth (1909) and Stough (1926). It is thus remarkable that Leydig (1864, 1865, 1886) had at an early date so nearly our own conception of them. Following his identification of them as nerve fibers, nearly half a century of controversy over this point transpired in which, haltingly and by spurts, the evidence for Leydig's view piled up: neurofibrils, myelin sheaths, neuronlike branches, and origin from nerve cells. Only after some fifty-one years was critical physiologic evidence adduced that the giant fibers actually conduct excitation. Bovard, in 1915, found a coincidence of the delayed regeneration of the fibers and the recovery of end-to-end contractions; Yolton, in 1923, and Stough, in 1930, cut the giants alone and noted that the twitchlike startle reaction involves the moiety stimulated and only up to the level of the cut. Finally Eccles, Granit, and Young (1932) recorded action potentials. These developments are reviewed by Nicol (1948) and Prosser (1934c). The story is a history of zoology in miniature; in particular we may note that much of the controversy would have been shortened if direct observation of the living object, unstained or colored with methylene blue (Fig. 14.25), had been more widely used (compare the beautiful figures of Smallwood and Holmes, 1927, and Ogawa, 1939). The simple experiments of Bovard, Yolton, and Stough could have been done, technically, at any earlier period.

The exact knowledge of the anatomy in oligochaetes waited, even after the excellent and modern work of Boule (1908a, 1908b, 1913), until 1926,

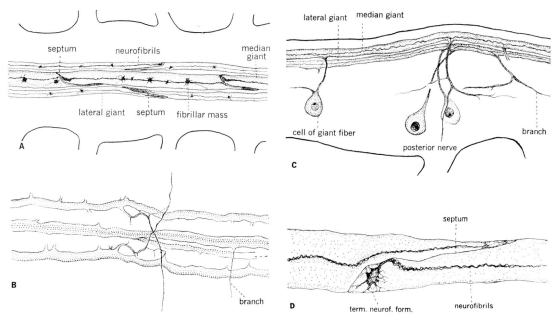

Figure 14.25. Giant fibers in oligochaetes; embryos of *Pheretima*. **A.** A typical ganglion, dorsal view, showing the oblique septa and the neurofibrils. **B.** Eighth segment, dorsal view, showing the manner of branching of the giant fibers. **C.** The connection of giant fibers with cells in a typical segment. **D.** A segmental septum of the median giant fiber. [Ogawa, 1939.] *term. neurof. form.*, terminal neurofibril formation.

when Stough provided a thorough account of the fibers, their cells, and their septa. His description has been confirmed and extended with the electron microscope by Hama (1959; see also de Robertis and Bennett, 1954, 1955; Issidorides, 1956). In almost all oligochaetes there are three dorsal giant fibers (exceptions are noted in Table 14.3), heavily sheathed and separated from the rest of the ventral cord by a transverse septum of connective tissue. In the ventrolateral part of the neuropile there are, at least in the terricolous earthworms, a pair of **ventral giant fibers** of much smaller size than the dorsal.

The **dorsal giants** extend through virtually the whole length of the cord. They are relatively small in the anteriormost and posteriormost segments; the median fiber reaches its maximum diameter (about 75 μ in *Lumbricus*, *Pheretima*, and even in a large Australian species of *Megascolex*) in the first or second quarter of the body; the lateral pair reaches theirs (about 50 μ) in the third quarter, roughly speaking. The statement is often made that the diameters of the laterals add up to that of the median but this can be at best an approximation, true at some one level and increasingly untrue away from it. A quantitative study in relation to fiber diameter and height and form of action potential is needed. The median fiber seems to come to an end in the cephalic portion of the subesophageal ganglion. The laterals are said to decussate here and to enter the circumesophageal connectives, but they cannot be followed any considerable distance toward the brain.

Cell bodies in every segment send their simple processes into the giants and indeed they are syncytial, as Friedlander (1888 and following) and others after him clearly showed by demonstrating that many cells contribute to each giant fiber (Fig. 14.26). Stough described the median as receiving the processes of four large cells lying close to the midventral line of each ganglion. The stem processes of these cells often fuse before joining the giant and there may be dendritic branches ending in the neuropile. The lateral giant fibers each receive the processes of two large cells, one contralateral and one ipsilateral, and of several small nerve cells in each segment.

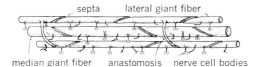

Figure 14.26. Diagram of the dorsal giant fibers of *Lumbricus* in a view from below, showing the septa, anastomoses between laterals, the cell bodies, and the terminating branches. [Bullock, 1945.]

To the lateral giant fiber, Ogawa assigned cells at the caudal end of each ganglion, and to the median giant fiber, cells in the rostral end. Adey (1951) described the several large and many small cells in each ganglion as joining only the median giant anterior to segment 60 (*Megascolex*) and only the lateral giants, chiefly contralateral, behind this; it remains to be learned where the cells are for the rostral lateral and caudal median giant segmental units.

The **giant axons** themselves show no trace of their compound origin. They are composed of an axoplasm that is quite homogeneous by ordinary methods except at the center where, both in fixed preparations (ordinary light microscopy) and in life, one can see a denser core, revealing in methylene blue or reduced silver a small number of heavy **neurofibrils** (Figs. 14.8, 14.25), zigzagged or coiled in any but the most stretched condition, even when the surface of the fibers is quite straight to the microscope. These fibrils can be traced to an origin in the cell bodies and an end in a splayed-out tangle near the septum (Fig. 14.27). No trace of them or adequate explanation of them has yet come from electron microscope studies! The **sheath** (Fig. 2.47) is composed of lamellae of concentric supporting cell membrane, similar to vertebrate myelin even to the dimensions and the spiral arrangement; this is the heaviest sheath known in invertebrates (with the possible exception of prawns, which have not been examined with the electron microscope). The lamellar sheath is about 6% of the fiber diameter (Taylor, 1940; Adey, 1951), increasing linearly with fiber size. Unlike vertebrate myelin sheaths, however, the lamellae are separated in the earthworm giant by interlamellar sheath cell cytoplasm, and the inner surfaces of the lamellar membranes (which represent invaginated sheath cell membrane) do not fuse. In all these respects there is a strong resemblance to embryonic vertebrate myelin. One peculiarity is that so-called attachment areas of the lamellae occur in rows or columns of defi-

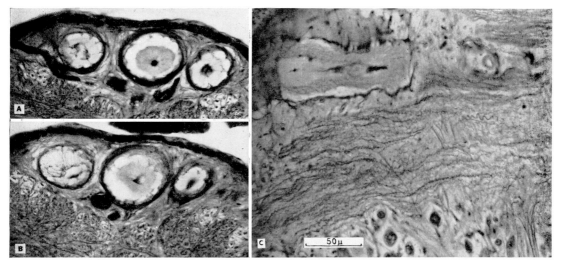

Figure 14.27. The giant fibers of *Lumbricus* in the region of the segmental septum. Bodian protargol stain. **A** and **B**. Transverse sections at the level of the left lateral giant fiber septum, which is unstained but outlined by neurofibrils splaying out on one or both sides. **C**. Frontal section at the level of the median fiber septum, seen as a region of frayed-out fibers between regions of condensed fibrils. Below: the ordinary fibers; this is a moderately stretched worm but the characteristic zigzagged appearance is still visible. Scale applies to all three. [Bullock, 1945.]

nite, slightly oblique arrangement (de Robertis and Bennett, 1954a; Hama, 1959). It would not be strictly correct to call these giant fibers myelinated without qualification but "almost" or "imperfectly myelinated" is accurate (see p. 104).

The **segmental septa** which completely divide each of the fibers at intervals of a small fraction of the wavelength of an impulse (the length of fiber occupied by action potential at any moment) are one of their most remarkable features, not less so because they were overlooked by all the microscopists between Leydig and Stough (with the exception of Schneider in his amazingly original textbook of histology of 1902). Easily visible in methylene blue, intravitam, whole mounts and in iron hematoxylin sections, they are long, oblique partitions, in length about three times (median) to five times (lateral giants) the diameter of the fiber. The median fiber septa occur at a level from the posterior end of each ganglion to the anterior end of the next following one; the laterals occur in the middle of each ganglion. The direction of the obliquity is the same in successive segments of each fiber, and the two laterals are alike but the median and laterals are opposite. As seen in the electron microscope (Fig. 14.28), the septa are simply the apposed axon surface membranes of the adjacent units, 65 Å apart, the whole being 200 Å thick; no sheath lamellae enter (contrary to Taylor's interpretation of polarized light results). Some synaptic vesicles are encountered, often lined up just inside the membrane. The whole structure is quite symmetrical. Scattered along the fibers, as we noted also in many polychaete giants, are numerous trabeculae of the sheath projecting into the axoplasm and partially interrupting the fiber. These can be mistaken for the septa in transverse section. The neurofibrils do not cross the septum but do show a specialization in its vicinity, splaying out into an elaborate tangle of much finer threads. Stough described a consistent difference in staining quality on the two sides but this has not been confirmed (Hama, 1959); the propagation of impulses has been found to be unpolarized and of equal velocity in the two directions (Eccles, Granit, and Young; Bullock; Rushton) and the transmission across the septal synapses is probably electrical (see p. 227).

The **lateral giants** are connected to each other at frequent intervals. Stough and others have traced seemingly anastomotic fibers both dorsal

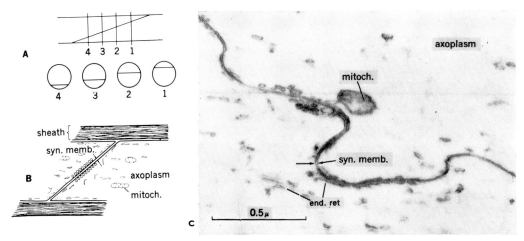

Figure 14.28. Electron microscope findings on the giant fiber septum in the earthworm. **A.** Diagrams of the appearance of the septum in a series of cross sections of the fiber, as indicated in the longitudinal section. **B.** Diagram of longitudinal section as revealed in electron micrographs; the diameter of the fibers is not to scale. The septum consists of two unit plasma membranes, together forming a complete partition and an extensive synapse *(syn. memb.)* [Hama, 1961.] **C.** High-magnification micrograph of a portion of a septum, showing the close apposition of the synaptic membranes, endoplasmic reticulum *(end. ret.)* associated with the membranes, and the symmetrical arrangement. In this example vesicles are at least not abundant (compare Fig. 2.27). [Hama, 1960.]

and ventral to the median giant and each of the investigators recording action potentials has noted that a single spike, slower than the median spike, represents both laterals and obeys the usual criteria for an all-or-none response. However, Rushton, while confirming the reality of a functional connection by cutting alternately the left and right lateral giants at different levels, found that each time the impulse was forced to cross, it was delayed about 0.8 msec. Bullock has reported that fatigue and electrotonus often lead to a dissociation of the two laterals, their all-or-none spike changing into a labile, toothed, or clearly double spike. Wilson (1961) inserted a microelectrode into each lateral giant fiber, impressed a voltage upon one, and observed one-third of that voltage in the other; there must be a very low-resistance electrotonic connection between them—though the method does not disclose whether it is a low-resistance synapse or an anastomosis. Hama (1959) found and illustrated a synaptic membrane between one lateral giant fiber and the commissural process of the other, using the electron microscope. Stimulation must occasionally excite only one of them, so the second lateral must be excited from the first, but no delay is apparent normally. Rushton found that even when out of step by 0.8 msec, the two spikes get back into step if sufficient length of cord is available. This is now understandable because the electrotonic current across the midline is enough to excite directly as long as the safety factor is normal. Fatigue and the like may lower the safety factor and thus introduce delay. There is no electrical interaction between median and lateral giants (Kao, 1956).

The **functional properties of the giants** have been examined in several respects. Strength-duration curves for the two units cross, the median being more excitable with short stimuli, the lateral having the lower rheobase (Niki et al., 1953; Sone, 1953; Katsuki et al., 1954). Maintained electrical polarization causes repetitive firing that lasts for a dozen or two spikes (Katsuki et al., 1954; Amassian and Floyd, 1946); the two giant units differ in this iterative tendency. Turner (1955) reported effects of temperature on the spike parameters. Several workers have penetrated the axons with microelectrodes (Svaetichin, 1950; Kao and Grundfest, 1957; Wilson, 1961); typical overshooting spikes are recorded but, in contrast to squid giants, lacking an undershoot (see Chapter 3).

The **velocity** of conduction is 15–45 m/sec for the median and 7–15 m/sec for the lateral giant of *Lumbricus* at 20–24° C and about the same at 10–15° C in *Megascolex*. These values are among the highest of the invertebrates although the diameters of the fibers are not at all extraordinary among giants (see Table 14.3). The speed has been compared with diameter in different regions of worms of the same species (*Megascolex* sp., Adey, 1951) but without careful control of stretch; the relation found is the first case in any animal in which, as diameter is increased by a given percentage, velocity increases by a greater percentage. In view of the history of this subject, in which vertebrate fibers were at first assigned similar high exponents and earlier squid fiber values were replaced by an exponent of 1.0, we should perhaps reserve judgment on the present figure.

It is now quite clear (Bullock, Cohen, and Faulstick, 1950) that **stretch** produces no change in velocity (m/sec) even when microscopic control shows the diameter to be greatly reduced. This interesting cellular mechanism deserves attention by transmembrane impedance measurements (see further, p. 674). Of course this mechanism only partly compensates for the stretch in terms of elapsed time between given landmarks; it still takes longer for an impulse to go from head to tail when the worm is stretched out—contra Bethe, see Holst, 1932; the results of Dittmar (1954) and Schwab (1949) on leeches require re-examination to assure that the shortened condition did not involve some zigzagging; see also p. 152. The earthworm presents a case of interest in connection with the factors that determine **spike form** as well as conduction rate: the slower lateral spike is broader and taller than the median. The usual explanation, which has not been quantitatively tested, however, is that the height of the laterals is greater because two spikes sum in amplitude, but the speed is slower because the individual diameters of the laterals are smaller than the median. But under current theory the summed spikes should only be greater than the median when that has a diameter less than the sum of the laterals, and this seems unlikely, for the relation mentioned obtains even in the anterior quarter where the median is largest in relation to the laterals.

The **afferent connections** of the giants are known only from physiological data. The visible branches of the giants, which seem to be their only contact with other neurons, are described below under efferent connections, but they or some of them could as well be regarded as affer-

ent on present evidence. Rushton and Bullock independently showed in 1945, in agreement with Stough's (1930) data, that only the median is excited by mechanical stimuli to the anterior 40 segments (*Lumbricus*) and only the laterals to similar stimuli behind this level, though each carries its impulse both ways from the starting point throughout the worm and is electrically excitable at any level (Fig. 14.29). Evidently there are only sensory connections to the median in the anterior and to the laterals in the posterior levels. A small, variable zone of overlap occurs and, more interestingly, the transition level can evi-

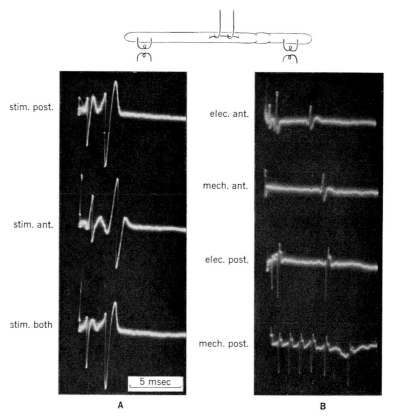

Figure 14.29. Responses of giant fibers of *Lumbricus*; arrangement of stimulating electrodes at the two ends and recording electrodes in the middle, shown at the top. **A.** Evidence that the same fibers conduct both ways; stimulation at the posterior end *(stim. post.)* elicits a smaller median giant spike with a short latency and a larger, slower lateral giant spike. Stimulation at the anterior end also elicits two spikes, but recorded differently since they arrive from the other direction. If these were distinct pathways, stimulating both ends at once should give four spikes but it only gives two, namely those from the physiologically closer, posterior, end. **B.** In another preparation, mechanical and electrical stimulation are compared. Electrical stimulation anteriorly with a strong shock gives the median and lateral spikes and sometimes a late, extra median spike, owing to prolonged synaptic activity. A mechanical stimulus (tap) anteriorly gives only median spikes. Posteriorly, electrical stimulation gives the median and lateral and sometimes a late, extra lateral spike; a mechanical tap gives a burst of lateral spikes only. [Bullock, 1945.]

dently shift some distance, because under natural conditions Rushton found that extended worms, with only the tail remaining in the burrow, give the median fiber response to stimuli applied some distance behind segment XL. Potential sensory connections exist for both, then, in a broad zone behind the clitellum, and central influences from higher levels perhaps determine which shall function. It will be clear that the fibers conduct both ways not only to electrical stimulation but in normal responses at least within their region of sensory connections.

The twitch reflex, presumably due to **excitation of the giants** by any of various forms of mechanical, thermal, or galvanic stimuli, has been studied as a function of repetition of stimuli and state of the earthworm by Kuenzer (1958). Exhaustion of response to one mode of stimulus is specific and does not interfere with excitation by another; the nervous system can distinguish among the modes named. Exhaustion by repeated stimulation of one segment reduces responsiveness of neighboring segments, with less reduction as the separation increases. Extirpation of the brain makes the preparation less rapidly exhaustible, more excitable; cutting the cord in front of the clitellum does the same for segments posterior to the cut but the opposite anterior to the cut; transection behind the clitellum depresses both ends. Roberts (1962) found that the junction between afferent fibers and giant fiber very quickly fatigues, as does the junction between giant and motoneuron. Kao and Grundfest (1957) recorded complex excitatory postsynaptic potentials with internal electrodes in the median giant, manifesting the activity of small presynaptic fibers impinging on the giant near the electrode. The most interesting way in which this small presynaptic activity could be produced was by simple stimulation of the median giant itself, far away from the region of recording. The evidence indicates that efferent connections from the giant arouse certain small neurons which continue to fire for some time and which, more remarkably, connect to afferent (= dendritic) processes of the same giant nearby, so the e.p.s.p.'s are visible to the microelectrode in the giant fiber. The maintained depolarization as a result is often able to initiate a new spike after the relative refractoriness has sufficiently passed off. This is a rare case of a true reverberating circuit or re-entry system and tends to cause repetitive firing, but it is normally heavily damped.

The **efferent connections** must also be different for the median and lateral giants if the presence of two separate units is to have biological meaning. The only evidence of this consists in the report (Rushton, 1946) that on rough substrates median fiber excitation is followed by such rotation and fixation of posterior setae as to permit retraction of the head, lateral fiber excitation by $180°$ contrary rotation, and fixation of anterior setae permitting retraction of the tail. Seemingly the longitudinal muscles are fired extensively, nearly simultaneously and symmetrically by either, and the contraction to both is said to be stronger than to either one alone; but the two giant units cause opposite effects on setae of the same segment. Anatomically the situation is not so clear. Several authors, most recently Ogawa (1939), have described branches of the lateral giants of ordinary size passing into the third nerve of each segment, after decussating (note this makes a double decussation, beginning with the cell body), so that we must consider the laterals as not only a through-conduction pathway for distributing excitation on the efferent side of a reflex arc but also as a final motor pathway. No such evidence exists for the median and we will therefore call it a chain of compound interneurons. Both, however, give off in each segment a number of small processes which soon lose themselves in the neuropile (Fig. 14.25).

As a first approximation, the **function of the giant system** is quite clear—the mediation of withdrawal responses to startle stimuli. On the evidence of ten Cate (1938) and especially of Collier (1938), it seems probable that they do not mediate the abrupt thinning and lengthening of the end opposite the site of stimulation that heralds initiation of a new peristaltic wave, as was proposed by von Buddenbrock and by ten Cate. The significance of the giant size of the fibers for the function mentioned is always given as the shortening of response time consequent upon the relation between velocity and diameter.

This is elaborated a little further under polychaetes, above (pp. 698, 701).

The **ventral giant fibers** are also segmental units; a clear boundary like a septum, not crossed by the neurofibrils, is evident in each ganglion. They are much smaller and lightly sheathed, are best developed in the mid- and hind-portions, but are not sufficiently known to be assigned definitely to the motoneuron or interneuron pool. Nor can anything be said of their function.

III. THE BRAIN

A. Generalities

1. Polychaeta

The brain in the polychaetes attains a remarkably high level of complexity. This statement is based on the degree of differentiation—that is, the number of cell and fiber masses defined by distinguishing texture or natural boundaries. The remarkableness rests on the comparison with lower and higher groups; many of the errant polychaetes approach relatively advanced arthropods and exceed in these respects both simpler arthropods and molluscs; they far surpass the groups considered heretofore. This complexity means that for the first time we encounter seriously the problem of which topographical features, brain divisions, nerves, cell and fiber masses are homologous in various species, and which are the result of the peculiar evolution of the given species or family. These morphological problems have been extensively discussed in a series of older monographs but their present status cannot be regarded as definitive. One factor, by no means the most important, is the difficulty of homologizing the appendages and sense organs of the head, depending as that does largely on prior or simultaneous judgments as to the homology of parts of the brain or of its nerves.

In position, the brain is generally completely internal, in the prostomium and dorsal. It is, however, sometimes quite superficial, lying actually in the epithelium (*Cirratulus*, *Terebella*, certain other sedentary forms), and in these cases is also exceedingly simple.

2. Oligochaeta

In contrast to the polychaetes, the oligochaetes and hirudineans never attain any considerable degree of development of the brain. In general, this organ is very simple in earthworms and leeches—on a par with that of sedentary polychaetes, but far simpler than some, for example sabellids and serpulids. This goes with the virtual absence of specialized sense organs (as opposed to unicellular receptors) and is reflected in the lack of lobes or internal divisions (except in some of the small, aquatic forms comprising the Limicola), the paucity of cell types and specialized neuropile regions and the less profound behavior signs upon removal of the brain. Although many workers have studied the ventral cord, virtually the only microscopic studies of the brain of oligochaetes are those provided by Krawany (1905), Ogawa (1939), and Adey (1951).

The brain is still attached to the epidermis and in its original prostomial position in the Aeolosomatidae (*Naidina*). It is elsewhere internal, and in the Terricola is pushed back into the first, second, or third segment, stretching its nerves, which of course have the same distribution as before. Characteristic lobes—for example a pair laterally, a pair posterolaterally, and three anteriorly in *Tubifex*—are common among the Limicola, where the brain is said to be larger in proportion to the body than in the much larger Terricola.

3. Hirudinea

The supraesophageal ganglion in the leeches is displaced far posteriorly, lying in segment V or VI and sending long nerves forward (Fig. 14.6). It is a small, straplike ganglion, elongate from side to side. Connective tissue partitions divide the cells into compartments as in the ventral ganglia; this is of special interest since the constancy of the number per segment (three pairs) offers a means of estimating the number of segments represented in the brain. Mann (1953) finds twelve

Figure 14.30.

The anterior nervous system of leeches. **A.** Lateral view of the brain, subesophageal ganglion, and nerves of the head of *Haemopis*. Nerve roots are lettered *a–f*, annuli are numbered *1–15*, segments *I–VI*. **B.** Transverse section through the nerve collar, consisting of brain and subesophageal ganglion of *Haemopis*. [Mann, 1953.] **C.** Lateral view of the brain of *Theromyzon*. *n1–n12*, cephalic nerves; cell compartments numbered *L* and *R*. [Hagadorn, 1958.]

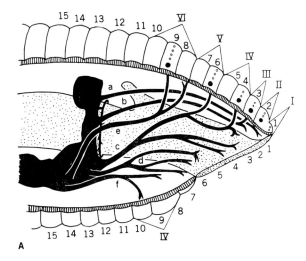

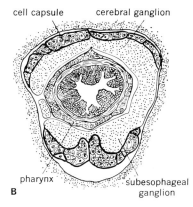

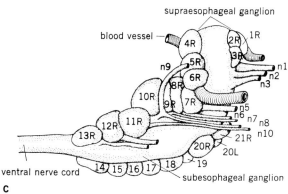

in *Haemopis*, clearly indicating two segments, but Scriban and Autrum (1932) give three pairs and hence one segment as the typical number. Since they show five ganglia in the subesophageal ganglionic mass while Mann assigns four thereto, it may be that various degrees of migration of the second ganglion up the circumesophageal connective are in evidence. Certainly all figures show a short, nearly vertical connective around the pharynx with nerves coming from it and cell packets of the proper number spread along it, so the boundaries of the brain are likely to include some morphologically ventral ganglia. Mann's evidence (Fig. 14.30) clearly indicates two pairs of ganglia in the brain, which has but one pair of nerves; since this was the same genus as Hanke (1948) studied it seems probable that the latter author was incorrect in assigning only one segment to this nerve—and hence only five to the supra- and subesophageal ganglia together—

and that Miller (1945) correctly assigned two segments to the cerebral nerve on the basis of its early division in his species into two trunks. Sánchez (1912) asserts that there are three pairs of nerves.

The situation in *Haemopis*, according to Mann, may be summarized thus (Figs. 14.6, 14.30 A):

Segment	I	II	III	IV	V	VI	VII
Sensilla	+	+	+	+	+	+	+
Eyes	0	+	+	+	+	+	+
Annulus	1	2	3	4,5,d6	v6,7	d8,d9	v8, v9,10
Nerve	1	*	2**	3	4	5	6
Origin	brain			subesophageal ganglion			

* Origin of connective to stomodeal ganglion between nerves 1 and 2.
** Second nerve arises in subesophageal ganglion, runs upward and emerges from circumesophageal connective.
d = dorsal part of-, v = ventral part of-.

Mann and many other writers do not refer to or recognize a prostomium. Indeed, there seems no reason to do so from the anatomy of the group. But neither does the group offer any conclusive evidence against an homology between the first segment and the prostomium of other annelids. It should especially be noted that the connective to the stomodeal ganglion joins the region of transition between brain and circumpharyngeal connective below the cerebral nerve to segments I (= prostomium) and II (= homologue of first ventral ganglion) and is therefore not exclusively related to the prostomial ganglion but also to that of the next segment. Unfortunately no neuronal analysis is available to offer confirmation or otherwise of the direct neural connections, both upstream and down, which are known in earthworms. Nor is it yet known whether postoral commissures exist, as they should, for the ganglia of segments II and III as well as those which are entirely ventral. These simple keys to morphology—and in the present case, in particular, the cell packets—illustrate the false economy of attempting to interpret external anatomy while explicitly omitting as "confusing" or "premature" the use of microscopic anatomy, even that which is already known (Hanke, Henry, Ferris).

B. Histology

The remarks made under the Ventral Cord (p. 673), concerning central histology, apply to the brain as well. The mantle of unipolar nerve cell bodies more nearly surrounds the fiber matter here, though commissures, tracts, and less often neuropile lie exposed in certain areas. Both cell bodies and fiber matter, however, show a greater differentiation, at least among the more advanced brains, than is found in ventral ganglia (Fig. 14.31). In general, **two types of cell bodies** occur: (1) a large plasma-rich, chromatin-poor, loosely packed type, both more primitive and more widespread anatomically and phylogenetically, and (2) a small, plasma-poor, chromatin-rich, densely packed type with fewer neuroglia, frequently occurring in special masses but also found mixed with the first type. The latter or globuli type is, from its distribution, more specialized and associative.

A variety of differently textured **fiber matter** occurs. Not only are some tracts and commissures more clearly defined (though it is still not possible in general to conclude that functional endings do not occur along their course) but neuropile of variously specialized appearance suggests specific types of synaptic apparatus. One type of neuropile has received the name, glomerular (Fig. 14.32) and has a superficial similarity to structures of that name in the vertebrate olfactory bulb. It is regarded as indicating a rather high degree of functional complexity, though not as high as that in certain smooth, fine-grained neuropiles containing the endings of the globuli cells. Further details concerning the types of endings and interrelations in specific neuropiles will be given under the anatomical headings below.

Blood vessels penetrate the brain in some polychaetes (*Notopygos*, Amphinomidae, Gustafson, 1930; Malaquin and Dehorne, 1907). Scharrer

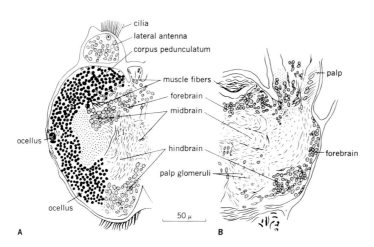

Figure 14.31.

The brain of *Harmothoë*, frontal sections of young animals. **A.** Section at a level where the hindbrain and corpora pedunculata are well formed. **B.** Level of the palp base, showing the glomeruli of neuropile associated with palp afferents. [Korn, 1958.]

III. BRAIN   C. Major Divisions

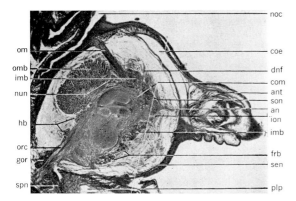

Figure 14.32. Longitudinal section through the brain of the polychaete *Hermione*. [Bernert, 1926.] The oblique line gives the plane of section of another figure in the original work. *an*, antennal nerve; *ant*, antenna; *coe*, coelom; *com*, commissure of lateral midbrain; *dnf*, dorsal nerve fibers; *frb*, forebrain; *gor*, ganglion of the outer root of the circumesophageal connective; *hb*, hindbrain; *imb*, medial midbrain; *ion*, inferior ommatophore nerve; *noc*, nuchal organ cleft; *nun*, nuchal nerve; *om*, ommatophore muscle; *omb*, outer midbrain; *orc*, outer root of circumesophageal connective; *plp*, palp; *sen*, sensory epithelium; *son*, superior ommatophore nerve; *spn*, subpalpal nerve.

(1944) found the earthworm central nervous system has double vessels ending in loops like those in the opossum, lizard, and lamprey, in contrast to the network arrangement in other reptiles, other mammals, and squid.

C. Major Divisions

Three main divisions of the brain have been recognized by topographic features in many of the errant families of polychaetes; oligochaetes, leeches, and the remaining polychaetes present no comparable major divisions. We may use the terms forebrain, midbrain, and hindbrain, emphasizing at the outset that not only are no homologies intended with vertebrate structures for which these names are used but that no fundamental morphological reality can be assured in these polychaetes on present phylogenetic or embryologic evidence. Certainly the terms should not be taken to imply a true segmentation of the brain or that it is compounded of the fused ganglia of three hypothetical preoral segments. Nor even should it be assumed that the original prostomial supraesophageal ganglion is basically divided into three parts which are homologous as such between various families. The claim is, however, that at least some of the specific structures included in these three topographical main divisions can be homologized among the polychaetes, although the nonspecific or unnamed cell and fiber masses and hence the boundaries between fore-, mid-, and hindbrain cannot (Pruvot, 1885; Racovitza, 1896; Holmgren, 1916; Bergquist, 1925; Heider, 1925; Bernert, 1926; Hanström, 1927, 1928a, 1928b, 1929; Johansson, 1927; Söderström, 1927a; Binard and Jeener, 1926; Gustafson, 1930). We may assign arbitrary boundaries of our own, emphasizing that they are not based upon any supposedly fundamental external anatomy. On this basis, the terms **fore-, mid-, and hindbrain** may be defined loosely by their inclusion, respectively, of (a) palpal and stomodeal centers, and anterior roots of the circumesophageal connectives, (b) antennal and optic centers and posterior roots of the circumesophageal connectives, and (c) nuchal centers.

The **external features** of the brain (Fig. 14.33) present almost no common denominators in any wide selection of families. The main nerves tend

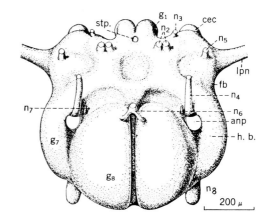

Figure 14.33. Reconstruction of the brain of *Pherecardia striata*, dorsal view. [Gustafson, 1930.] *anp*, antennal perforation; *cec*, circumesophageal connective; *fb*, forebrain; g_1, stomatogastric lobe; g_7, lateral posterior, and g_8, median posterior, lobe of the brain; *hb*, hindbrain; *lpn*, longitudinal podial nerve; n_2, oral lip nerve; n_3, nerve to the sides of the oral lips; n_4, paired antennal nerves; n_5, palpal nerves; n_6, median antennal nerve; n_7, anterior optic nerve; n_8, nuchal nerve; *stp*, stomatogastric perforation.

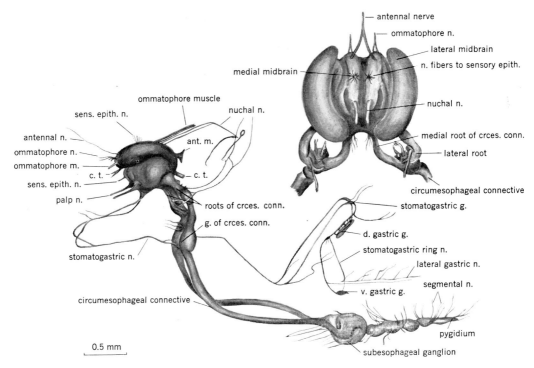

Figure 14.34. Reconstruction of the central nervous system of *Hermione*. Lateral view and dorsal view of brain. [Bernert, 1926.] *ant. m.*, antennal muscle; *crces. conn.*, circumesophageal connective; *c. t.*, connective tissue strand; *d. gastric g.*, dorsal gastric ganglion; *g.*, ganglion; *m.*, muscle; *n.*, nerve; *sens. epith. n.*, nerve fibers from sensory epithelium of head; *v.*, ventral.

to occur in roughly the same order, indicated by the list of centers above. In a number of families (Fig. 14.34) three lobes corresponding to our arbitrary definition are externally apparent (eunicids, polynoids, sigalionids, aphroditids) but in others they are not, or only one of the regions is set off (amphinomids (Fig. 14.33), euphrosynids, nereids). Drawings of the external anatomy of the brain in the archiannelid *Polygordius* show (Fraipont) three divisions but according to Hanström (1929) the anterior of these is a sensory cell accumulation; the other two are not really identified with the lobes of polychaetes. Common to a number of the better-developed families are holes through the substance of the brain, occupied by muscle bundles in these positions: (a) anteriorly in the midline through the stomodeal commissure or a little farther back between fore- and midbrain and (b) laterally between the two roots of the circumesophageal connectives. The brain is sometimes very markedly encapsulated, in other cases loosely bounded or very thinly sheathed.

D. Internal Structures and Connections

1. Polychaeta and Archiannelida

Below are listed the principal cell and fiber masses, commissures, and connections that have been recognized in the most advanced or perfect polychaete brains. Virtually all have been described in each of a series of forms among the higher errant families, and each is probably homologous throughout these groups. They are given in approximately rostro-caudal order. Differences between species will be noted partly in passing and partly in the section comparing families (p. 727).

(a) Stomodeal center. Near the midline in the forebrain where the stomodeal nerves emerge (see below, p. 725) is an area of relatively undifferentiated neuropile and a mantle of cell bodies supposed to be primarily associated with those nerves; these together can be called a stomodeal center.

III. BRAIN D. Internal Structures 1. Polychaeta

The term stomodeal (= stomatogastric, enteric, visceral, sympathetic), encountered here for the first time, arose historically for a part of the nervous system of arthropods and annelids, easily visible in the gross anatomy, which lies along the anterior regions of the digestive tract, extending various distances posteriorly and connected with the brain or/and the subesophageal ganglion anteriorly. It is in general a visceral system of nerves, plexuses, and ganglia. It is doubtless both motor and sensory and independently integrative as well. In the polychaetes it is variously developed (see Peripheral Nervous System, p. 756) but usually associated chiefly with the pharynx, which is often a complex, protrusible organ for offense, defense, burrowing, or feeding.

The stomodeal center exhibits an apparently **slight development** in the pectinariids (Nilsson, 1912), sabellariids, sabellids, serpulids (Johansson, 1927) and nereids (Holmgren, 1916). Holmgren regards the stomatogastric system as entirely originating in the subesophageal ganglion but we may consider, with Gustafson (1930) that the first cerebral nerve which innervates the proboscis (the eversible pharynx), in company with nerves from the subesophageal ganglion, represents a cerebral contribution to the system. It exhibits a fairly **strong development** in eunicids, aphroditids, polynoids, syllids, phyllodocids, nephtyids, and amphinomids, and reaches very large proportions in euphrosynids (Gustafson, 1930). Gustafson suggested a relation between the degree of development of this brain center and that of the oral lips, but no attempt seems to have been made to relate it with the elaboration or use of the proboscis.

In the euphrosynids the cells are of the specialized plasma-poor, chromatin-rich type and are said to be arranged partly in globuli form. But in most families both cells and neuropile are rather unspecialized and not clearly demarcated from adjacent forebrain. Curiously, almost nothing is known of the **connections** of this region. Gustafson (1930) described a long tract to the nuchal center of the hindbrain. A stomodeal commissure (Fig. 14.36) connects the two sides and is divided into anterior and posterior parts by a hole which transverses the brain in the midline. We may hazard a guess that the **functions** of the center and its nerves are only partly motor to the pharynx and largely sensory—based on the principles on which the rest of the brain is constructed—and that it has to do with the proboscis, including in particular exteroception when that organ is everted. It should be noted here that the rest of the stomodeal system described elsewhere (p. 756) includes extensive medullary cords or ganglia, so we must be dealing, as in the cerebral autonomic centers of vertebrates, with **higher-order interneurons**; there may be first-order sensory or final-order motor cells or some of each in the stomatogastric nerves, but they cannot consist entirely of these categories.

(b) Palpal center. The other most characteristic and defining structure of the forebrain is the area of neuropile receiving the central processes of the primary sensory neurons of the palpi (Figs. 14.32, 14.35). These are not clearly marked centers and special cell body masses have not been identified with them. They are best indicated by the entry of the nerves of the palps; these nerves vary in number and are perhaps not to be clearly distinguished morphologically or in brain connections from the commonly found diffuse cluster of fine nerve filaments from the sensory epithelia of the adjacent part of the prostomium which enter the forebrain (including probably n_2 and n_3 of Gustafson, 1930, and some of the first five nerves of Holmgren, 1916). The fibers of these nerves disappear in the neuropile but Gustafson, though without special nerve stains, describes them as partly decussating in a **palpal commissure.** Other connections include a bundle extending toward the nuchal center (amphinomids) and a tract communicating with the stalks of the corpora pedunculata (*Hermione, Sthenelais, Nereis*).

This description is based on aphroditids, hesionids, polynoids, polyodontids, sigalionids, nereids, eunicids, onuphids, and amphinomids. It cannot hold without modification for the families such as the lumbrinerids and nephtyids which lack palps (see Table 14.5). In the glycerids, which have either no palps or minute ones, according to the interpretation of the tiny prostomial appen-

Figure 14.35. Brain and cephalic sensory nerve endings in *Nereis diversicolor*, dorsal view. Methylene blue. [Retzius, 1895b.] *a*, antenna; *an*, antennal nerve; *au*, ocellus; g_1, anterior group of ganglion cells; g_2, lateral group; g_3, posterior group; *k*, clubbed nerve terminals of granular fibers; *m*, dendriform muscle nerve endings; *p*, palp; *pn*, palp nerve; *pr*, globuli cell mass of corpora pedunculata; *s*, sensory patches of the epidermis; *sn*, bipolar neurons whose distal processes end in these sensory patches. The patches just behind the posterior eyes are the nuchal organs.

dages, a very special nervous development in the prostomium has been homologized with the palpal centers but is described below (p. 734).

The **sedentary families** probably have homologues of palps. The innervation from the anterior region of the much simplified brain of certain structures—the tentacular filaments of the head in cirratulids, the oral tentacles of pectinariids and ampharetids, the "Stirnfühler" of sabellariids (Meyer, 1888), the "Fühler" of flabelligerids (Günther, 1912), the crown of feathery tentacles of sabellids and serpulids (Johansson, 1927), the palms of scalibregmids, and small areas of the epithelium of the prostomium in opheliids, maldanids, and spionids—is at least compatible with such an homology. In some cases (sabellariids, serpulids, flabelligerids) the innervation includes a contribution from the circumesophageal connective (Johansson, 1927) and has been likened (Hanström, 1928) to the subesophageal contribution to the stomodeal nerves.

(c) Other forebrain structures. The remaining cell bodies of the forebrain are not distinctly grouped into cell masses; they comprise chiefly the **dorsal mantle,** made up mostly of large plasmatic cells, but are described as quite various in topographic anatomy in different species (g_3 of

III. BRAIN   D. Internal Structures   1. Polychaeta

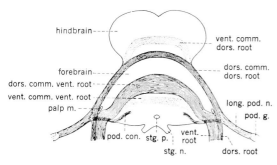

Figure 14.36. Diagram of the commissures of the esophageal connective and the podial longitudinal nerves in Amphinomidae. [Gustafson, 1930.] *dors. comm. dors. root* and *vent. comm. dors. root,* dorsal and ventral commissures of the dorsal roots and the longitudinal podial nerves; *dors. comm. vent. root* and *vent. comm. vent. root,* dorsal and ventral commissures of the ventral root; *dors. root,* dorsal root of circumesophageal connective; *long. pod. n.,* longitudinal podial nerve (lateral nerve cord); *pod. con.,* connective between longitudinal podial nerve and forebrain; *pod. g.,* podial ganglion; *stg. n.* and *p.,* stomatogastric nerve and perforation; *vent. root,* ventral root of circumesophageal connective.

Gustafson). At least one **commissure** besides the two already mentioned and two belonging to the circumesophageal connectives, described thereunder, is present in higher errant forms (Fig. 14.36). Holmgren recognized a well-delineated commissure just behind that of the stomodeal centers in *Nereis.* Gustafson's k_2 is a dense transverse neuropile in amphinomids, suggestive of associative rather than the simplest functions, and in euphrosynids the same author distinguishes what is possibly a separate "commissure," a ventral, stomodeal association mass (stb). This last is related to the glomeruli and nerves supplying the ventral side of the oral lips and connects by a strong paired tract (k_9) with the more anterior of the glomeruli of the hindbrain nuchal center.

(d) Antennal centers. Like the palpal centers, these are simple sensory end stations (Figs. 14.31, 14.32, 14.35). They are included in the neuropile of the region here called the midbrain. Groupings of cell bodies are usually not clearly set apart but Holmgren (Fig. 14.41) and Clark recognize a number in errant forms; none can be called clearly antennal without further ado. No definite connections or commissures of the antennal centers can be recognized.

Antennae vary greatly in number and their nerves accordingly (Table 14.5, p. 728). It need only be added here that although median unpaired antennae are common (amphinomids, onuphids, eunicids, syllids, polynoids, and others), the centers are always paired and that there is probably only one center on each side; at least there cannot be distinguished separate centers in forms having two or more antennal nerves on each side.

(e) Optic centers. When eyes are present, they send paired nerves to the midbrain. These perhaps end in the optic center of the same side but appear more likely to enter immediately the **optic commissure** (Figs. 14.37, 14.40), judging from a variety of descriptions based on nonspecific stains. Presumably the incoming fibers end in part in the contralateral optic center. However, the paired centers are usually not clearly defined and their size is therefore not apparent, whereas the commissure is clearly demarcated, well developed, and strongly suggestive of an associative neuropile with many synaptic endings rather than simply a fiber tract. So it seems likely that many of the afferent fibers end in the so-called commissure. This structure is itself differentiated in some species (*Pherecardia,* Amphinomidae) into two kinds of fibrous tissue of different texture, one largely surrounding the other. In another type there are two commissures, one above the other, separated by a thin layer of cell bodies and loose neuropile (*Hermione*). The cell bodies seemingly associated with these centers are immediately dorsal and lateral to them and sometimes at least partly of the globuli type.

The optic commissure appears to have **connections** with the forebrain and with the hindbrain neuropiles. In addition, Gustafson (1930, Amphinomidae) described a special connection with a tall, narrow neuropile mass close to the midline, posterodorsally between the cell body masses of the midbrain globuli and the hindbrain nuchal center. The most certain connection, however, is with the stalks of the corpora pedunculata, where those exist.

Many polychaetes, especially sedentary fam-

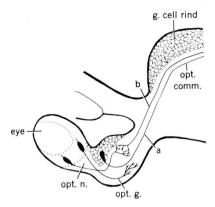

Figure 14.37. *Leodice*, cross section through the optic centers. [Hanström, 1928.] *a*, neuron with an axon running into the optic commissure; *b*, fiber entering the optic glomerulus from the commissure; *g. cell rind*, ganglion cell layer of the brain; *opt. comm.*, optic commissure; *opt. g.*, optic ganglion; *opt. n.*, optic nerve.

ilies but also errant ones (Lumbrineridae, Nephtyidae, Glyceridae, some Onuphidae), lack eyes or possess only pigment cup ocelli and correspondingly lack these central structures. But special mention should be made of the highly developed centers in *Leodice* (Eunicidae, Hanström, 1926). Here the primary sensory neurons in the eye send axons to a special **optic ganglion** between the eye and the brain (Fig. 14.37), where they end in synaptic relation with (i) dendrites of neurons whose cell bodies are in the ganglion and whose axons also end in the neuropile of the ganglion—that is, local neurons, (ii) dendrites of neurons whose cell bodies are in the ganglion but whose axons pass into the optic commissure, some to end in the contralateral ganglion—that is, commissural and projection neurons, and (iii) dendritic endings of fibers of distant origin—that is, foreign projection neurons. These well-defined structures form close analogues and possibly homologues with the optic ganglia of arthropods and are the first such structures in the articulates. Similar and possibly better optic ganglia occur in the pelagic Alciopidae (p. 748) (Demoll, 1909; Rádl, 1912) and likewise in the Phyllodocidae (Pruvot, 1885).

(f) Corpora pedunculata and related structures (Figs. 14.31, 14.38, 14.39, 14.40). In a number of the errant families (Aphroditidae, Polynoidae, Sigalionidae, Polyodontidae, Nereidae, Hesionidae but not Eunicidae, Onuphidae, Lumbrineridae, Glyceridae) and in one sedentary family (Serpulidae) distinct corpora pedunculata with stalks occur. These are structures of characteristic appearance that strongly suggest homology with similar structures in arthropods (see p. 818), regarded as among the highest associative centers in the brain in that phylum. They are formed, mushroomlike, of a compact mass of globuli-type cell bodies and a stalk of finely textured neuropile. Since similar cell bodies are found elsewhere in the brain, often even collected into small dense masses ("globuli"), it is the large compact mass of them and especially the smooth, dense stalks which define corpora pedunculata. They are superimposed on the midbrain (external midbrain of Bernert in *Hermione*). Golgi impregnation shows the typical specialized associative neu-

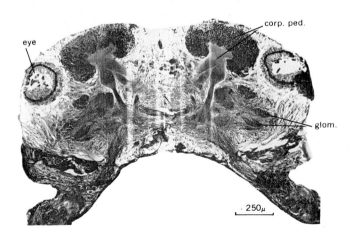

Figure 14.38.

Harmothoë (Polynoidae), transverse section of the brain, showing common neuropile in center, three stalks of the corpora pedunculata *(corp. ped.)* surmounted by globuli cell masses, palpal glomeruli *(glom.)*, and tracts entering the circumesophageal connectives below. [Courtesy B. Hanström.]

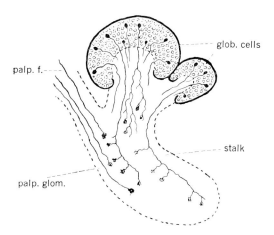

Figure 14.39. *Nereis.* Section of the corpora pedunculata showing the connection of sensory palpal fibers *(palp. f.)* with dendrites of globuli cells in the region of the palpal glomeruli *(palp. glom.)*. *glob. cells*, globuli cells. [Hanström, 1927.]

ron of invertebrates (Fig. 14.39). Small, unipolar cell bodies send out short fine processes with very numerous side branches, ending in terminal tufts, mostly limited to the stalk of the same side as the cell body. A significant number of commissural processes occur, however, connecting the stalks of the two sides; also, projecting processes pass to palpal, optic, antennal and nuchal centers. We may remind ourselves here that, unlike vertebrates, it can not be assumed that impulses normally proceed from the region of the cell body centrifugally along long processes. Thus the projection processes of the corpora pedunculata may be gathering information from afferent centers and conducting it centripetally to the complex associating centers of the stalks. The synapses with palpal and nuchal neurons often occur in characteristic knots of neuropile, the **palpal and nuchal glomeruli,** but these are only found in cases where the respective sense organ and its center is well developed; such glomerular synaptic areas are regarded as a sign of relative advancement. Usually only one of the two is glomerular, and antennal and optic centers in annelids do not seem to employ this type of connection. We may hazard a guess that the antennal functions have not attained a sufficient level of advancement but that the optic connections have bypassed the glomerular device in favor of others in the optic

ganglia and commissure. Crossed and uncrossed connections with the circumesophageal connectives take place by means of fibers from those connectives entering and ending in the stalks.

Among the polychaetes having corpora pedunculata there is some **variation** in the number of globuli masses and stalks (Fig. 14.40). Thus *Nereis diversicolor* has three globuli on each side, *N. pelagica* two larger globuli and *Neanthes virens* several small globuli; all three forms have four stalks. *Harmothoë, Hermione, Polynoë,* and *Laetmonice* have only one globuli mass but have subdivided stalks which approach in various degrees the condition in nereids. *Lepidonotus* has two well-separated globuli; the stalk of the anterior one is partly split, which may mean either incipient subdivision or partial fusion. In all cases the stalks, after a more or less pronounced flexure from the initial dorsoventral direction toward a transverse axis, stop short of the midline and therefore of meeting each other. Commissural fibers, however, were noted above.

Of greater interest are **comparisons** with fam-

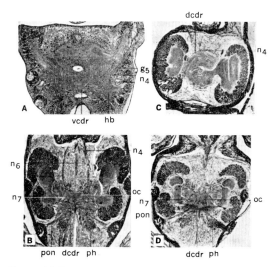

Figure 14.40. Horizontal sections through the brains of some members of the two families Amphinomidae and Polynoidae. **A.** *Leocrates japonicus.* **B.** *Laetmonice benthalina.* **C** and **D.** *Lepidonotus squamatus.* [Gustafson, 1930.] *dcdr*, dorsal commissure of the dorsal roots; g_5, lateral ganglia in the forebrain; *hb*, hindbrain; n_4, paired antennal nerve; n_6, unpaired antennal nerve; n_7, anterior optic nerve; *oc*, optic commissure; *ph*, posterior horns of the brain, *pon*, posterior optic nerve; *vcdr*, ventral commissure of the dorsal root of the circumesophageal connective.

ilies lacking proper corpora pedunculata. Thus the active, predaceous glycerids completely lack such features. But the Eunicidae, Onuphidae, Amphinomidae, and Euphrosynidae, while lacking stalks and clearly defined globuli, have neurons suggestive of rudimentary corpora pedunculata. In the first two families Hanström (1927) found neurons by Golgi staining, situated in the forebrain and connecting with palpal afferents in the same manner as those in the glomeruli of nereids and polynoids, though without formed glomeruli. He regarded these as significant of the probable **origin of corpora pedunculata** (see p. 772). Gustafson (1930) described several "ganglia" of mixed, small, globulilike cells and larger, plasmatic cells in the forebrain (g_2, g_4) and midbrain (g_9, g_6) of higher species of amphinomids, which suggests a tendency to clump into globuli. They have no real stalks, but coarse strands connect with the hindbrain nuchal centers. Though no special stains were used, connections with palpal centers do not seem as important as those with the elaborately specialized nuchal organ afferents. There are even glomeruli in this hindbrain center, presumably for the connections with anterior association neurons. We cannot say without further evidence from selective stains whether nuchal association centers offer another source from which corpora pedunculata may have evolved or whether these nuchal centers in amphinomids are really the same centers as those in eunicids mentioned just above. Possibly only a change in emphasis due to relative enlargement of the nuchal organ is responsible for the apparently new center. Nephtyids are described as having distinct globuli of typical cell type but no stalks.

Among thirteen genera of eight families of sedentary polychaetes (Cirratulidae, Terebellidae, Ampharetidae, Opheliidae, Maldanidae, Scalibregmidae, Sabellidae, and Serpulidae) which Hanström investigated particularly on this point, he found corpora pedunculata in only one, *Serpula vermicularis* (confirmed by Johansson, 1927). Here the stalks come close to glomeruli of the afferents from the tentacular crown (modified palps).

The probable evolution of these structures and sensory factors associated with it are discussed below (p. 772).

(g) Median mass. Of phylogenetic interest, in the neuropile of the midbrain, anterior to the optic commissure, Gustafson (1930) found in amphinomids and euphrosynids a well-marked, unpaired structure which may be called the median mass. It is characterized by a dense neuropile, though not approaching that of the stalks of corpora pedunculata. It has a strong tract which begins as a midline structure (therefore called a commissure by Gustafson—his k_5), then divides into paired ventral tracts proceeding posteriorly to the nuchal glomeruli of the hindbrain; they may also have forebrain connections.

(h) Other midbrain structures. A large, lateral plasmatic midbrain group (Gustafson's g_5, Holmgren's G 14) is the only other defined group which appears to be found in many families (Fig. 14.41). The destinations of its fibers are mainly unknown. With methylene blue Holmgren found unipolar cells in *Nereis* whose short stem process divides promptly into three. One branch, the outer, gives off a number of terminating twigs in the region of the cell and then sends a long fiber into the circumesophageal connective and another into nerve IV, a small motor nerve to the side of the prostomium. A second branch divides into a number of terminating twigs a little anterior to the region of the cell body, and the third, an inner branch, crosses the midline and breaks up into terminals in the corresponding center of the opposite side. We can only presume that all three sets of terminals are dendritic—that is, afferent for that neuron—and initiate impulses independently according to the excitatory state in that part of the neuropile, for the twin final common paths.

Two **commissures** can be recognized besides those already mentioned and those described below belonging to the circumesophageal connective. (i) Behind the median mass and dorsally is one (k_3 of amphinomids and k_{10} of euphrosynids, according to Gustafson) that is presumed to derive fibers from cell groups of globuli character in the forebrain. (ii) The other is the "great commissure" of Gustafson (k_4), a massive bundle ventrally situated just above the undivided portion of the tract (k_5) from the median mass to the hindbrain and providing for most of the crossings

III. BRAIN D. Internal Structures 1. Polychaeta

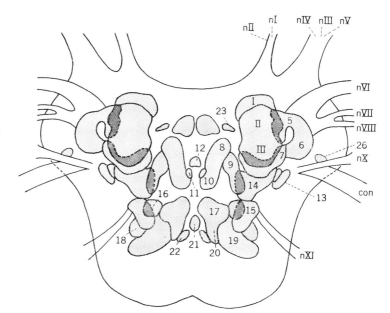

Figure 14.41.

The cell groups in the brain. Reconstruction from *Nereis*, seen from above. [Simplified from Holmgren, after Hanström, 1928.] *con*, circumesophageal connective; *nI–nXI*, cerebral nerves; *I–III*, globuli cell masses of corpora pedunculata; *2–26*, ganglion cell groups recognizable anatomically.

of the midbrain and possibly some of the forebrain (Fig. 14.36).

(i) Nuchal ganglia and centers (Figs. 14.32, 14.35). The principal structures of the hindbrain are related to the afferent nerve supply from the nuchal organ (see also Fig. 14.65).

The proximal processes of primary sensory neurons in or near the receptive epithelia enter the hindbrain as nerves which are sometimes long and discrete, sometimes short and numerous. Still again, the bipolar sensory cell bodies may lie in a large clump, the nuchal ganglion, which is in various degrees incorporated into the brain, reaching a condition of virtually complete amalgamation, lying internal to the connective tissue capsule or at least having no formed barrier of connective tissue or neuroglia between it and the brain. The presence of sensory cell bodies inside the central nervous system, while not unique (they have been noted in *Promonotus* of the Turbellaria, and they occur in man in the proprioceptive nuclei of the brain stem and occasionally in arthropods) is extremely unusual. This condition occurs in *Nereis* and in some amphinomids and euphrosynids, ampharetids, and pectinariids.

The **nuchal centers** have their cell bodies in the mantle of the hindbrain and are large, plasmatic ganglion cells in forms without specially developed nuchal organs, as in *Hermione*; but globuli cells become common and even the principal type in the amphinomids and euphrosynids, where this region is the outstanding feature of the brain (g_7, g_8, and g_{10} of Gustafson). The neuropile of the nuchal centers is commonly glomerular and attains relatively elaborate development in the families just named, with almost stalklike strands connecting them to the clumps of globuli cell bodies. In this neuropile the primary sensory neurons synapse with the second-order neurons of the nuchal centers. The degree of elaboration is said to parallel that of the nuchal organ, within these families.

Among the families the same is true. The hindbrain centers are relatively well developed in the Hesionidae, Syllidae, Phyllodocidae, and Eunicidae, less well in the nereids, spionids, tomopterids, and aphroditiform families, and not distinguishable as discrete parts of the brain in cirratulids, pectinariids, and opheliids, which nevertheless still have a nuchal organ.

Only a **diffuse commissure** seems to occur but several strong **tracts** have already been described connecting this afferent center with association centers of the forebrain and midbrain, especially the median mass, optic commissure, stomodeal association mass (stb), forebrain globuli (g_4) and corpora pedunculata.

(j) Roots and "commissures" of the circumesophageal connective (Figs. 14.33, 14.36). There are typically two roots of the circumesophageal connective on each side, the anterior (or ventral root of some authors) emerging from the forebrain, the posterior (or dorsal) from the midbrain. The two are commonly separated by a bundle of muscle fibers of the palps. Each of the two pairs of roots is directly connected to two "commissures," according to Gustafson (1930). (This word is loosely used even among those who confine it to transverse fascicles and do not call connectives commissures. When the structure in question appears to connect paired fiber tracts or connectives, not synaptic regions, one suspects it is a decussation, its fibers proceeding to destinations not symmetrical with their origins.) Previous authors have recognized various of these commissures of Gustafson but he describes four as being general among the errant families.

The ventral commissure of the anterior roots is the most anterior of the four and forms a constituent of the "great commissure" (k_4), the dorsal commissure of the same roots being next, both traversing the forebrain. The second one is the largest of the four; it is said to be usually divided into two separate bundles but is enlarged and single in amphinomids. The third is the dorsal commissure of the posterior roots, situated just posterior to the optic commissure in the midbrain. Into it also runs the podial longitudinal nerve in the families that have that nerve (Amphinomidae, Euphrosynidae), showing that neither root represents a remnant of it. Below and behind, in the hindbrain, is the ventral commissure of the posterior roots (part of k_8) and this also contains fibers from the podial longitudinal nerve; in fact, that nerve contributes the greater part of the last two commissures. It also sends a bundle into the forebrain which passes from behind the circumesophageal connectives to the anterior and dorsal part of the brain.

(k) Representation of giant fiber systems in the brain. The evidence suggests that these specialized through-conduction systems are generally represented or originate in the brain. However, the evidence is meager in most species and details are available in only one or two. Giant fibers have been followed into the brain from the ventral chain or into the circumesophageal connectives in the Archiannelida, Sigalionidae, Polynoidae, Nereidae, Eunicidae, Spionidae, Magelonidae, Sabellidae, and Serpulidae, or in about one-third of the families possessing giant fibers (see p. 692).

In *Euthalenessa* (Sigalionidae) two cell bodies in the brain send processes, without decussating, into the connectives, and they fuse subesophageally into one giant fiber. In *Sigalion* similar cells and processes do not fuse but continue as paired axons in the cord (Rohde, 1887). Gravier (1898) and Hanström (1926) believed that some large cells in the brain of *Glycera* give rise to the giant fibers in the cord but Nicol (1948a) was unable to follow the one into the other. The giants taper and sometimes divide, so it is difficult to follow them without the use of special nerve stains.

The single giant fiber on each side in the serpulid *Protula* (Bullock, 1953b) can be traced in silver impregnations to a single cell of origin in a special **giant cell lobe** projecting from the anterodorsal aspect of the brain (Fig. 14.42). Here a large multipolar, heteropolar cell body (a rare type in invertebrates, see p. 46) nearly fills the lobe and lies close under a sensory epithelium. The short proximal processes of many primary sensory neurons in the epithelium end around the cell body and the short dendrites. The axon decussates promptly, in the anterior part of the brain. The decussation of the axons of the two sides occurs as nearly as may be in the same plane, so in every specimen examined they come into intimate though simple contact, and to this connection, the only one found between the two giants, was assigned the physiologically demonstrated **two-way synapse.** In this region the fibers are only 8–30 μ in diameter. From the decussation the giants course directly into the circumesophageal connectives, swelling within two millimeters to three or four hundred micra. Scattered in the region anterior to the decussation are six or eight large cells on each side, of multipolar, heteropolar form, whose axons parallel the giants, decussate, and follow them for some distance into the ventral cords. The axons are nearly as large as the giant at the decussation but remain

III. BRAIN   D. Internal Structures   1. Polychaeta   721

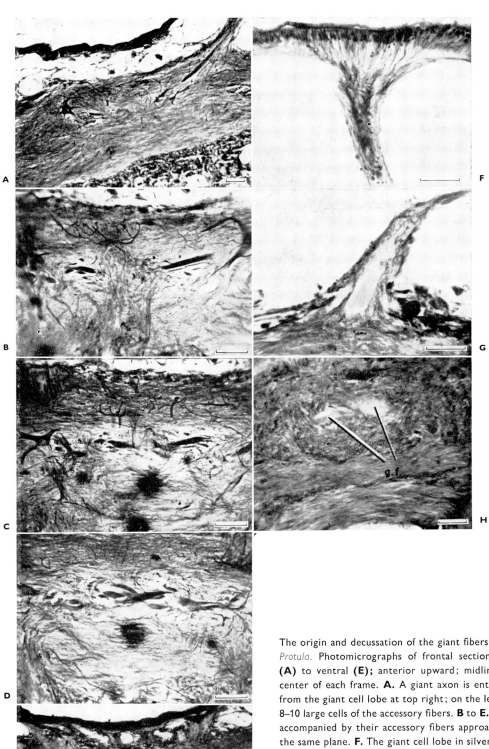

Figure 14.42.

The origin and decussation of the giant fibers in the brain of *Protula*. Photomicrographs of frontal sections, from dorsal (**A**) to ventral (**E**); anterior upward; midline through the center of each frame. **A.** A giant axon is entering the brain from the giant cell lobe at top right; on the left is one of the 8–10 large cells of the accessory fibers. **B to E.** The two giants accompanied by their accessory fibers approach and cross in the same plane. **F.** The giant cell lobe in silver stain, showing the cell, its dendritic processes, and numerous sensory axons converging upon it from the sensory epithelium above. **G.** The same by a cytological method (Flemming's; Masson), showing the soma better. **H.** The decussation and point of contact of the giant fibers (g. f.) by a cytological method. Calibrations, 50 μ. [Bullock, 1953.]

this size; they may be called **accessory neurons to the giants.** Their dendrites are widely spread in the anterior neuropile and could bring other influxes to bear on the system. They could have functional contact with the giants either at or near the decussation by the same simple type of synapse or, later, by collaterals, but the latter have been searched for in good reduced silver preparations and must be extremely minute if present. In any case, the extraordinary arrangements for bringing the giant neuron itself into such immediate functional contact with a sensory epithelium argues for the primacy of that source of input for control of the withdrawal reflex. This reflex is readily and naturally elicited by shadow or decrease in light intensity. But the adequate stimulus for the described patch of sensory epithelium has not been ascertained.

In the sabellid *Myxicola* Nicol was able to trace the origin of the giant fiber (single in most of the body, divided into two for short stretches in the thorax and in the circumesophageal connectives) into a pair of cell bodies in the brain. The processes of these cells decussate, but no special relations with other neurons or with afferent sources were noted. The brain is simple in these forms, as in the serpulids, and we cannot assign the giant cell or the decussation to any of the previously named regions of the brain or "commissures."

In contrast, it has not been possible to follow the giant fibers of most sabellids to their ends in the brain. The single giant axon in *Spirographis* (Bullock, 1953) has been followed in silver impregnations. It is much reduced in size in the circumesophageal connective and, following it forward as this connective gradually gives way to brain, the fiber tapers to 2 μ and is then lost in the neuropile before it has had opportunity to decussate. Running parallel to it are three or four large fibers in a loose group of about the same size; these are lost superiorly in five tapering ends in the brain and end abruptly inferiorly in the connective. They not infrequently anastomose with each and with the giant. Thus it can not be said whether there are cell bodies for the giants in the brain or what opportunities for connections with prostomial receptors may exist.

2. Oligochaeta

The internal structure of the brain in earthworms is exceedingly simple (Fig. 14.43). The most advanced feature is the presence of **"characteristic" cells** of Ogawa (1939), concentrated

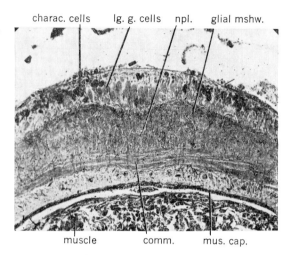

Figure 14.43. The brain of an earthworm. Transverse section of 5-weeks-old *Pheretima*, stained with hematoxylin. [Ogawa, 1939.] *charac. cells*, "characteristic cells" of the brain, which are small, dark ganglion cells (see Fig. 14.2); *comm.*, cerebral commissure; *glial mshw.*, glial meshwork, especially at inner boundary of rind; *lg. g. cells*, large ganglion cells; *mus. cap.*, brain capsule with a muscle sheet; *npl.*, neuropile.

anterodorsally and arranged in crude layers. These cells are almost all multipolar and heteropolar, a fact which sets them apart at once (see p. 46); they send their short, varicose processes into a plexus at the surface of the neuropile. They are small cells but have, even so, a relatively small nucleus. They are thus quite different from the globuli cells that are characteristic of the brain of polychaetes. That they are the "highest" cells in the earthworm nervous system is suggested by the finding that they increase in number during growth far more than any other cell—nervous or nonnervous—which Ogawa counted. At the time of hatching in *Pheretima* there are about 1000 characteristic cells among the 6241 cells in the brain; in the adult the total has doubled, as in most ganglia, but the characteristic cells have increased to 5865, or about 50% of the total.

Besides these cells there are **motoneurons,** men-

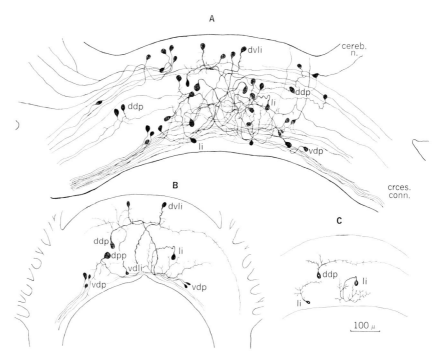

Figure 14.44. Neurons in the brain of earthworms. Transverse sections of *Pheretima* at 8 weeks **(A)** and at hatching **(B** and **C)**, showing some of the neurons seen in the cerebral ganglion. [Ogawa, 1939.] *cereb. n.*, cerebral nerve; *crces. conn.*, circumesophageal connective; *ddp*, dorsal descending projection neuron, sending axon into circumesophageal connective; *dvli*, dorsoventral local interneuron; *li*, local interneuron; *vdli*, ventrodorsal local interneuron, sending axon into dorsal part of the neuropile; *vdp*, small ventral descending projection neurons.

tioned already under nerves (and see Table 14.4), and two groups of interneurons, small and large (Fig. 14.44). The **small interneurons** are apparently devoid of dendrites or collaterals but have long axons without varicosities. Some end in the cerebral neuropile and others are projection fibers to the subesophageal ganglion. Still others pass through the stomodeal connectives to the ganglion or plexus of that system but whether they are preganglionic efferents or final motor fibers we do not know. The **large interneurons** lie chiefly in the ventral cell body mantle of the brain and include several categories of crossed and uncrossed local and projection neurons (see Table 14.4).

The **cell body mantle** can be divided (*Megascolex*) into two regions, dorsal and anteroventral, and the dorsal group of cells falls into two masses, lateral and medial. The lateral-dorsal group embraces most of the "characteristic" cells and is said (Adey, 1951) to receive most of the ascending fibers of the circumesophageal connectives, but this is not based upon selective nerve fiber staining. The medial-dorsal group consists of large cells whose axons decussate and are believed to descend into that connective. The anteroventral group consists of medium-sized, very granular cells and some larger, more medial ones; the latter are believed to send axons to the medial-dorsal group.

There can be recognized **three regions of the neuropile**—anterior, middle, and posterior. The first is of fine texture and is chiefly a network of the characteristic cell processes. The second looks like the neuropile of ordinary ganglia of the ventral cord and includes the terminals of many afferent fibers. The posterior neuropile is coarse and formed largely of the processes of the large interneurons ascending from and descending into the subesophageal ganglion.

3. Hirudinea

The brain is simpler in leeches than in typical

Table 14.4.   TYPES OF NEURONS IN THE BRAIN OF THE EARTHWORM BY FORM AND PROCESSES.[1]

Types of Neurons	Comment
Motoneurons	
large, bushy, dorsal, contra, uni or ipsi, uni	but brain nerves have no large fibers
small, smooth, ipsi, uni into brain nerves	
small, smooth, ipsi, uni into stomodeal ganglion	probably many are interneurons ending in stomodeal ganglion
Interneurons	
large, mostly ventral and bushy:	
contra, uni, ventral, axons to dorsal brain neuropile	
ipsi, uni, ventral, axons to dorsal brain neuropile	
contra, uni, dorsal, axons to ventral brain neuropile	
ipsi, uni, local	
contra, uni, desc into circumesophageal connective	
ipsi, uni, desc into circumesophageal connective	
Small, smooth (no dendrites apparent):	
local	
desc to subesophageal ganglion	
"Characteristic" cells, small, nucleus relatively small:	great increase with age from 10% in just hatched to 50% of all brain cells in adult
ipsi, multi, local with short varicose processes	
ipsi, bi, local with short varicose processes	

[1] Data mostly from *Pheretima* (Ogawa, 1939), some from *Eisenia* (Krawany, 1905). Abbreviations as in Table 14.3.

polychaetes or even earthworms. It does not form a sizable bulge but merely a long transverse band stretched across the dorsal or anterior side of the arching esophagus. The internal composition is known chiefly from the work of Sánchez.

The prodigious two-part monograph (1909, 1912) of this pupil of Cajal contains enormously detailed descriptions of the typical ventral ganglion, the anal ganglia, the subesophageal ganglia, the brain, the stomodeal system, connectives, and nerves in each of several species, stained with reduced silver, methylene blue, and other techniques, together with comments on the works of Retzius, Havet, and other previous authors! Incredibly verbose and innocent of any attempt to distill the important from the less important, the new additions to knowledge from the mere confirmations, this mine of illustrations and details has been understandably neglected by later writers. In an effort to examine it for the principal results, we have had a translation prepared but even so it is certain that some nuggets have escaped notice by protective coloration. The facts concerning ventral ganglia, the stomodeal system, and peripheral anatomy are treated in the respective sections.

The brain is **without marked signs of differentiation**, especially in its fiber core. The cell rind is of course divided by connective tissue partitions, to form compartments, as in other ganglia of these animals. These are treated above with the gross anatomy (p. 671). A thick external capsule and a thinner internal capsule between rind and core meet at places forming the partitions; some partitions are incomplete or secondary. Neuroglia fibers ramify in tapering bundles from these partitions toward the center of the cell packets and fiber core, where they become rather scarce. Large, glial stellate cells are rarer than in ventral chain ganglia.

Most **nerve cells** are unipolar, none are multipolar (though these are abundant in the stomodeal system), and a few bipolar. Each major cell process has one or two conspicuously clear neurofibrils (Cajal reduced silver) from the coarse network of them in the cell body. This is one of the most outstanding classical objects for demonstration of these organelles, which are not found, for example, in polychaete fibers; electron microscopy is needed. The cell types distinguished in the brain are few. Medium-sized somata (30–60 μ) dominate; there are some large but no giant cells, as there are in the ventral ganglia. Small cells are not abundant; there is a conspicuous absence of signs of local, short-axon, globuli-type cells. Sánchez described (a) cells whose axon passes medially to cross the midline and bears few or no branches, (b) cells whose axon passes back into the circumesophageal connective and bears few or no branches, (c) cells with a bifurcating process sending an axon in each of these directions and bearing more or less numerous side branches, and (d) motoneurons sending axons into the nerves. Category (c) is commonest. Either dendrites are quite scarce or the beautiful silver stains are seriously incomplete. There are only a few bundles or formed groups of fibers but most of the core is an undistinguished neuropile. Two commissures, anterior and posterior, can be discerned. Afferent fibers enter the brain from the connectives and nerves.

The **connectives** resemble nerves in general; both are heavily invested by neurilemma and penetrated by trabeculae of glia fibers, especially the proximal regions of the peripheral nerves. Apáthy thought that each large fiber has its own sheath and that small fibers travel in bundles in a common sheath, but Sánchez sees only a continuous glial meshwork without private sheaths. Axons fall into two fairly distinct groups. Thick, undulating, dark staining (reduced silver) ones tend to travel in groups and near the periphery of the nerves; these are at least chiefly motor. Thin, pale, straight fibers are more numerous, not grouped and chiefly afferent; they bifurcate just after entering the central nervous system, like so many sensory fibers of higher animals.

E. Nerves of the Brain

1. Polychaeta

The exact number of nerves varies greatly. It will exceed the following list when more than one pair of nerves goes to the same general structure in the periphery or when such structures are repeated (for example antennae). Often the innervation of certain regions is by many fine filaments not gathered or sheathed into proper nerves. There will be fewer than this list when certain sense organs or appendages are not present. The exact combination of destinations also varies, so this is a generalized summary.

I. *Stomodeal nerves.* Medial, anterior, or anteroventral, from the forebrain to proboscis, proboscis ganglia, oral lips, ventral prostomium.

II. *Sensory nerves to prostomium, oral lips, and special sensory epithelia*, sometimes mixed with motor bundles to muscles of the same region.

III. *Palpal nerves.* From the lateral aspect of the forebrain.

IV. *Antennal nerves.* Especially variable in number and position, generally posterior to the preceding, medial, or far lateral, paired even for unpaired antennae.

V. *Optic nerves.* From midbrain, close to IV, varying with position and character of eyes, often diffuse even when a discrete nerve goes to another pair of eyes (Amphinomidae) or several roots may unite into one nerve on each side which then splits to supply two eyes and also patches of generalized sensory epithelia (*Hermione*).

VI. *Sensory nerves or filaments to lateral aspects of prostomium.*

VII. *Nuchal nerves.* The only hindbrain nerves, sometimes long, sometimes absent as the result of movement of the sensory cell bodies into the brain; in this case the bundles of distal processes of these cells may be termed nuchal nerves (*Nereis*).

VIII. *Podial longitudinal nerve.* Joins posterior root as it enters; only in amphinomids and euphrosynids.

IX. *Circumesophageal connective.* Attached by two roots as described on p. 720 to forebrain and midbrain.

Details of the actual anatomy of some representative species can be seen in the illustrations (Figs. 14.32, 14.34, 14.35).

2. Oligochaeta

The cerebral nerves are of variable number and cannot be named or homologized among the

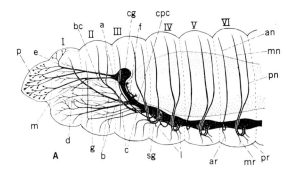
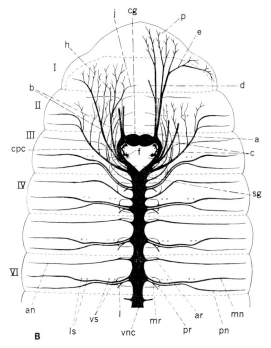

Figure 14.45. Nerves of the anterior end in *Lumbricus*. **A.** Lateral view. **B.** Dorsal view; branches of some of the paired nerves have been omitted from one side or the other of segments I and II. [Hess, 1925.] *a*, nerve from lateral region of cerebral ganglion which passes to prostomium; *an*, dorsal ramus of anterior segment nerve; *ar*, ventral ramus of anterior segmental nerve; *b*, nerve from near middle region of circumpharyngeal connective, which passes to segment 1; *bc*, buccal cavity; *c*, nerves from ventral region of circumpharyngeal connective which pass to segment 2; *cg*, cerebral ganglion; *cpc*, circumpharyngeal connective; *d*, branch of nerve to prostomium that supplies tissues of dorsal region of buccal cavity; *e*, nerve that supplies the portion of the prostomium in the dorsomedian region of segment 1; *f*, ganglionated thickening of enteric nerve plexus; *g*, branch of nerve to segment 1 that supplies tissues of ventral region of buccal cavity; *h*, nerve to ventral region of buccal cavity; *j*, medial ventral nerve; *l*, septal nerve; *ls*, lateral setae; *m*, mouth opening; *mn*, dorsal ramus of median segmental nerve; *mr*, ventral ramus of median segmental nerve; *p*, prostomium; *pn*, dorsal ramus of posterior segmental nerve; *pr*, ventral ramus of posterior segmental nerve; *sg*, subpharyngeal ganglion; *vnc*, ventral nerve cord; *vs*, ventral setae; *I–VI*, segments 1 to 6.

families because of the absence of both external and internal landmarks. From three to eleven pairs have been recorded (Fig. 14.45). They contain sensory and small motor fibers but none of the coarsest motor fibers common in segmental nerves. The fibers include (a) afferents from the epidermis of the prostomium; also the first and second segments according to Ogawa (1939), the first segment according to Hess (1925b), but serving only the prostomium according to Henry (1947), (b) afferents from the wall of the buccal cavity, (c) afferents from a ciliated groove in some aeolosomatids and tubificids (Vejdovský, 1884; Aiyer, 1926; Beddard, 1901), reminiscent of the nuchal organ of polychaetes, (d) efferents from the larger cells in the dorsal layer of the brain, which branch extensively before passing into the nerves, (e) efferents from small cells in the lateral borders of the brain which do not branch before leaving, and (f) efferents originating in the subesophageal ganglion, with dendritic processes there plus extensive branches in the brain (alternative source of excitation, summation, or modification of ascending impulses?) and passing out to muscles of the prostomium. Not only those of (f) but some of the afferent fibers traverse the circumesophageal connective and make direct connection with the subesophageal ganglion. The distribution of the motor fibers cannot be specified more precisely than to state that it includes muscle fibers in the prostomium.

From the ventral surface and laterally, in the angle of the root of the circumesophageal connective, there arise from the brain one or several short connectives to the stomodeal ganglion, just as in polychaetes. The fiber composition of these nerves is treated with that system on p. 758. Also described therewith is a nerve, sometimes called the recurrent nerve, which distributes to the dorsal wall of the enteron and arises in association with the stomodeal connectives—that is, from the brain close to them or from the connectives themselves.

3. Hirudinea

The nerves of the brain of leeches are treated above, on p. 709.

F. Comparison of the Brain in Polychaete Families

The great diversity among polychaetes has already been commented upon. This made it impossible to characterize the nervous system of the group without continual reference to variation or without restricting the description to the most elaborate form of the given structure. Some review seems necessary, arranged by family rather than by anatomical parts. This has zoological as well as neurological interest, especially when correlated with ecology and habit of life, sensory abilities, and complexity of behavior. Space and the rather surprising limitations of our knowledge of natural history dictate an abbreviated form of this review. Most of the differences of any moment are concerned with the brain; a selection of these differences, for which comparable facts for a number of families are available, is summarized in Table 14.5. Others concerned with giant fiber systems are in Table 14.3, p. 692. Only certain of the noteworthy characteristics of selected families will be given in the following.

1. Amphinomidae

Often regarded as a fairly generalized type, this group is exceedingly special in the fantastic elaboration of the nuchal organ (p. 750). The corresponding parts of the brain are not only large but rather well differentiated, including masses of globuli-type cells and fairly dense neuropile areas of associative nature (Figs. 14.33, 14.36, 14.40). The bipolar primary sensory neurons, whose distal processes are receptive in the nuchal epithelium, tend to cluster into a **nuchal ganglion** and thus to crowd against the hindbrain, even to invading its connective tissue sheath; this results in the anatomically remarkable consequence of sensory cells incorporated into the central nervous system (see Table 14.5 for other cases). Since the family has a full complement of prostomial sense organs (see Table 14.5), Hanström in 1926 predicted corpora pedunculata would be found; perhaps the masses of globuli cells, sometimes undiluted by plasma-rich cells, described by Gustafson (1930) deserve this name as much as those of *Nephtys, Podarke,* or *Serpula,* but the latter author denies such an identification and certainly, distinctly delimited globuli and stalks are lacking. One could readily suppose this to be a stage in the development of these great associative structures, as Hanström did for eunicids.

This family, together with the next, is almost unique in possession of **lateral longitudinal nerves** connecting the podial ganglia (Fig. 14.7)—that is, in being tetraneurous (Storch), perhaps a primitive condition. It is surely a secondary condition, however, that the circumesophageal connectives are unusually elongated, as though by the enlargement of the proboscis, so the first two to five ventral ganglia are situated on the connectives, their commissures greatly stretched in passing below the proboscis (Fig. 14.7). The **stomodeal system** is well developed, according to Gustafson, as a result not only of the great proboscis but of the important sensory epithelia of the oral lips. We may venture the suggestion that the latter represent neurologically the sensory field ordinarily included in the palps and therefore some of the central structures identified as stomodeal centers by this author correspond to the palpal centers of others. Actually we may take it as a lesson that there is no sharp line between stomodeal and palpal areas or functions.

2. Euphrosynidae

The last point is all the more cogent in this family, where the outstanding feature is the great size and elaboration of the forebrain centers, called by Gustafson **stomodeal ganglia,** with glomeruli and globuli cell masses quite reminiscent of the palpal centers of *Nereis* and others. Otherwise this family is very similar to the last.

Table 14.5. COMPARISON OF SOME FEATURES OF THE BRAIN IN VARIOUS POLYCHAETES.

Family and Sample genus	Head Sense Organs				Glomeruli	Corpora Pedunculata	Degree of Differentiation of Brain
	Eyes (pairs)	Palps (pairs)	Antennae (total)	Nuchal Organ			
ERRANTIA							
Amphinomidae *Eurythoë*	2	1 tiny	5 tiny	+++	+ nuchal	0 dvlpg (?)	+++
Euphrosynidae *Euphrosyne*	2	1 tiny	5 tiny	++	+ stomod	0 "globuli"	+++
Eunicidae *Eunice*	1	1	5	+	0	0 dvlpg (?)	++
Onuphidae *Hyalinoecia*	1	1	5	+	+	0	++
Lumbrineridae *Lumbrineris*	0	0	0	+	0	0	++
Nephtyidae *Nephtys*	0 or 2 (in brain)	0	4	+	0	+ (no stalk) (some spp 0)	++
Glyceridae *Glycera*	0	0	4 sm	+	0	0	+
Nereidae *Nereis*	2	1 lg, seg'd	2	+	+ palpal	++	++++
Aphroditidae *Hermione*	2	1	1	+	+ palpal	+++	++++
Polynoidae *Lepidonotus*	2	1	3	+	+ palpal	+++	++++
Sigalionidae *Sthenelais*	1	1 long	1	+	+ palpal	+++	++++
Hesionidae *Podarke*	2	1 seg'd	2	0 ?	+ palpal	+	++
SEDENTARIA							
Terebellidae *Terebella*	sev oc subcut	(?) lip tents	0	0 ?	0	0	—
Cirratulidae *Cirratulus*	same	(?) tentac filmts	0	+	0	0	—
Ampharetidae *Amphicteis*	1	sev. oral tents	0	+	0	0	—
Pectinariidae *Pectinaria*	1	(?) oral tents	(?) ant memb	+	0	0	—
Opheliidae *Ammotrypane*	0 (3 in other spp)	plpds	0	+	+ nuchal	0	+

Brain Subdivided	Brain Internal	Origin of Stomodeal Nerves in Brain or Ventral Ganglia	Nuchal Ganglia Primary Sense Cells Join Brain	Special Features
+	+	br and vg	+ partly	elaborate nuchal + optic assoc centers; median body; lat long nerv betw podial g; long c-e-c
+	+	br and vg	+ partly	great dvlpt stomod center, with globuli cells; otherwise as above
+	+	br		first optic gang (separate from brain) in articulates; also in Alciopidae, Phyllodocidae, Onuphidae
+	+	br		as above; lacking in the blind *Diopatra*
+	+	br		lack of sense organs has surprisingly small effect on brain; well devlpd gnl sensory epith
0	0	br and vg	0 ?	post lobes in some spp, of mucous cells with duct
0	+	br	+ ?	elaborate prostomial sensory apparatus with paired medullary strands and annex gang
+	+	br and vg	+	corp ped may have 3 globuli and 4 stalk branches
+	+	br	0	corp ped with 1 globuli mass, 2 stalks; nuchal glomeruli also
+	+	br and vg		corp ped with 2 globuli, 3 stalks
+	+			corp ped with 3 globuli, 3 stalks
+	+			corp ped with 1 globulus, 1 stalk
0	0	v. c-e-c		brain in epith; like a broad commissure between c-e-c's
0	0			same
0	0	vg	+	
0		vg		
0	+	vg (c-e-c in other spp)		segmental eyes, laterally in some spp, innervated from ventral cord

(continued on next page)

(Conclusion of Table 14.5)

Family and Sample genus	Head Sense Organs				Glomeruli	Corpora Pedunculata	Degree of Differentiation of Brain
	Eyes (pairs)	Palps (pairs)	Antennae (total)	Nuchal Organ			
Maldanidae *Maldane*	0	plpds	0	+		0	+
Scalibregmidae *Eumenia*	0	+ sm	0	++	0	0	++
Sabellidae *Sabella*	2 oc	tentac crown	0	0	0	0	++
Serpulidae *Serpula*	sev oc	tentac crown	0	0		+	++
Sabellariidae *Phalacrostemma*				+ (?)		doubtful	++
Spionidae *Scololepis*	2 cup, in br			+++ segmty repeated	0	0	+
Capitellidae *Stygocapitella*	0				0	0	+
Sternaspidae *Sternaspis*					0	0	+
Tomopteridae *Tomopteris*	1 with lens, in br	0	2 lg. 2 sm.	+	0	0	+
ARCHIANNELIDA *Polygordius*	0	sensory tents.	0	0	+	0	+

Abbreviations:

ant – antennular
assoc. – association
br – brain
c-e-c – circumesophageal connective

corp ped – corpora pedunculata
d – dorsal
dvlpg – developing; dvlpt – development
epith – epithelium

filmts – filaments
gnl – general
lat long – lateral longitudinal
lg – large

*Fore- and midbrain only.

giant fibers and several smaller ones, but the latter do not give evidence of being through-conducting while the former do.

6. Glyceridae

These also have lost eyes and palps, the antennae are in miniature, and the entire pointed prostomium is small. The brain is the simplest among all the errant polychaetes studied, undivided and almost entirely composed of large plasma-rich cells. There are few nerves and no glomeruli or corpora pedunculata.

Of the two pairs of nerves, the anterior, supplying the amazing nervous system of the prostomium, is described below. The posterior is supposed to supply, by one branch, the proboscis (representing therefore a **stomodeal nerve)**, and by another branch an organ which Gravier (1898) identified as a nuchal organ but about which Hanström expresses some reservations. In any case the nerve in question is correspondingly of double origin as shown by Golgi stains. There are large unipolar plasmatic cells and small bipolars sending processes into the nerve (Fig. 14.46). Bipolars are rare in the central nervous system of invertebrates and when one sends a process into a peripheral nerve it suggests a centralized primary sense cell. We have already noted the tendency in nereids, amphinomids, and others for the nuchal ganglion to join the hindbrain, and here it looks as though the incorporation into the brain of nuchal sense cells is complete—if the cells described do in fact supply such an organ!

There are three or four pairs of larger **giant fibers** and about three pairs of smaller fibers, the one grading into the other and the smaller grading into ordinary fibers. This is a most unusual situation, anticipating the cockroach and lamprey, where also a not-precisely-fixed number of large fibers occur with others of intermediate size. Most animals with giant fibers have a

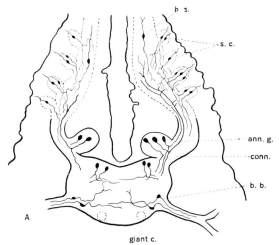
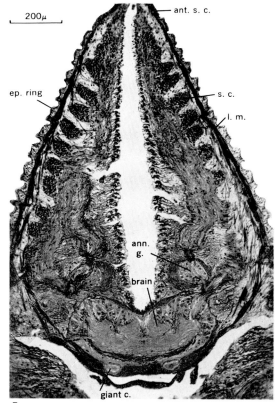

Figure 14.46. The brain of the polychaete *Glycera*. **A.** Diagrammatic frontal section of part of the head. Golgi impregnation. **B.** Photomicrograph of similar section. Ordinary histological stain. [Hanström, 1928.] *ann. g.*, annex ganglion; *ant. s. c.*, antennal sense cells (the minute antennae are not seen in this section); *b. b.*, bipolar ganglion cell of the brain; *b. s.*, bipolar ganglion cell of the medullary strand; *conn.*, connective between brain and medullary strand; *ep. ring*, epithelial ring, one of the grossly visible annuli of the prostomium, which correspond in number to the sense cell groups; *giant c.*, giant cell; *l. m.*, longitudinal muscle; *s. c.*, sense cell of one of the annular concentrations.

sharply discontinuous fiber-size spectrum and a definite number of giants. Perhaps the present condition is transitional, or a stage in the evolution of giant systems. Physiologically, at least a good number of the large fibers are independent and through-conduction units are as fast-conducting as many better-defined giants (Table 14.3, p. 692). The provision of an efficient giant system is, then, not necessarily correlated with a high level of cerebral complexity.

Glycera presents a unique feature and problem, in spite of the lowly condition of the brain, in the great development of the peripheral nervous system of the **prostomium**. This member is conical, superficially annulated, and divided internally, by a dorsoventral coelomic cleft, into lateral halves. Inside the subepithelial longitudinal muscle layer there runs on each side a strong **medullary strand** of neuropile strewn with cell bodies. These strands are beset laterally with a longitudinal row of clusters of sense cells, each of the 14 or so clusters consisting of a closely packed collection of extraordinarily chromatin-rich cell bodies, which Hanström, the discoverer of these details, calls the most remarkable feature of the glycerid head nervous system. The clusters correspond to the superficial annuli and are each connected by two short nerves to its epithelial annulus. The anterior nerve of each annulus carries only distal processes of the sense cells in the clusters and almost all such processes. It supplies only the anterior part—a distinct subannulus and its endings are of the type of the much-branched, free nerve endings often associated with subepithelial cell bodies. But the posterior nerve carries proximal processes of a densely packed girdle of epithelial sense cells, whose short, unbranched, distal processes end in the cuticle of the posterior subannulus (Fig. 14.47). Both types of sensory neurons are common on the parapodia and elsewhere in *Glycera* and in polychaetes generally and the posterior type exclusively supplies the enormous proboscis of *Glycera;* we may suggest with Hanström that it is chemoreceptive, the free-branching type perhaps tactile. What distinguishes the sense cells in this prostomial system is the great concentration, the discrete cluster of cell bodies, the almost mutually exclusive

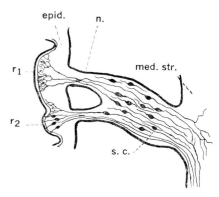

Figure 14.47. Detail of the sense cell concentration *(s. c.)* in the prostomium of *Glycera,* in frontal section. Golgi preparation. Note both free-branching sensory endings from deep-lying cells and simple distal processes of intraepithelial cells. [Hanström, 1927.] *epid.,* epidermis; *med. str.,* medullary strand of prostomium; *n.,* nerve; r_1 and r_2, subannuli of one of the superficial rings of the prostomial epithelium, each subannulus corresponding to one nerve, each ring to one sense cell group.

areas of ending of two types of sense cell, and the peripheral neuropile into which they run.

At the anterior end of the medullary strand are a quite different, scattered group of sense cells belonging to the tiny antennae on the point of the prostomium. In the medial side of the medullary strand, posteriorly before it enters the brain, it has a ganglion consisting of large plasma-rich cells like those of the brain. This is called the **annex ganglion** and is regarded as a detached portion of the brain itself. Its processes go out after a T-shaped bifurcation, one into the brain and one into the medullary strand. The possibility could not be eliminated, even in Hanström's Golgi preparations, that they are motor neurons for the prostomial muscle layer, but their number and the few instances of terminals seen in the medullary strand suggest a role in mediating the influx from the prostomial sensory apparatus. There are also purely internuncial bipolars in the medullary strand, so considerable synaptic complexity is possible in its neuropile.

Based on the supply to sensory epithelia of the prostomium and on the anteroventral origin of the nerve from the brain, Hanström considers the medullary strand equivalent to **palpal centers** that have moved out along the palpal nerves. The terminal appendages can then be regarded as antennae only by supposing their nerves to have fused with the palp nerves. Whatever its morphologic meaning this peripherally extended central (that is, synaptic) apparatus is most remarkable and is not known in any other annelid. It suggests the process that has led to detached optic ganglia in eunicids (p. 732) and later in arthropods. A peculiar saccular apparatus in the brain seems to be secretory and to retain a primitive connection with the epidermis (Simpson, 1959).

The habit of life is a burrowing, wandering one; the food is supposed to be small, living animal prey. No adequate knowledge of the natural history is at hand, therefore, to explain the unusual condition of the sensory and central nervous equipment.

7. Nereidae

From the standpoint of comparative neurology there could hardly have been a better choice of a polychaete as the common classroom type than *Nereis*. Well provided with sense organs, its brain and hence the rest of the nervous sytem is well differentiated and exhibits virtually all the features of advanced polychaetes. No special description here is necessary, therefore, as the general accounts of abdominal cord, giant fiber system, brain, and stomodeal and peripheral systems apply. Nereids are not the most highly advanced polychaetes; at least there are other families that carry each feature farther—corpora pedunculata, median body, proboscis (that is stomodeal) nervous system, eyes, antennae, nuchal system; but it is relatively advanced in all of them and has not become specialized at the expense of any considerable feature.

Holmgren's (1916) description is classical. He distinguishes 23 cell groups in the brain, 6 commissures, and numbered brain nerves up to XVI. Most of these appear in some form in the descriptions and figures given above (p. 712 and Figs. 14.35, 14.41).

Gilpin-Brown (1958) worked out much of the embryology of the brain and its nerves, contradicting Henry and Ferris on the origins of ganglia and peristomium. The circumesophageal connective has two roots in the brain, but a purported third is really part of the subepithelial plexus of the head; details of the associated commissures are given. Five stomodeal nerves join the brain at widely different points.

8. Aphroditidae and allies

The aphroditiform worms, scale worms, sea mice, and others have recently been separated into several families. Those for which some knowledge of neuroanatomy is available include the Aphroditidae, the Polynoidae, Sigalionidae, and Hesionidae. Although these vary in detail, in number of eyes and antennae, in conformation of corpora pedunculata, in origin of stomodeal nerves, and in giant fibers, they are all much alike in the very high advancement of the brain—the highest in the phylum. *Podarke* of the hesionids is an exception in being of very modest advancement. Again the development seems balanced rather than heavily dependent on one specialized feature. Bernert (1926) calculated the brain of *Hermione* to occupy a 9000th part of the volume of the whole body. And Hanström found that the corpora pedunculata occupy 30% of the brain in *Sthenelais* (Sigalionidae). Unfortunately comparative figures for other annelids are not available.

The **eyes** are, for annelids, highly developed, especially in a form like *Hermione*, where a dorsal and a ventral eye occur in ommatophores on each side (Bernert). Specialized patches of sensory epithelium are widely distributed and supplied sometimes from branches of other nerves, such as the ommatophore nerve, and sometimes from diffuse bundles of fibers directly from the brain (all three divisions). It was in this group that Hanström (1926) found **trichogen sense cells** in the elytra, which seem to be forerunners of the characteristic arthropod sensory hair (p. 1005).

The habits of life of these families vary widely: some burrow in mud, others freely wander on the surface, many are crevice dwellers, and a number are commensals living on asteroids and other larger invertebrates. The food habits and behavior seem not sufficiently well known to permit a correlation with or an explanation of the high development of the brain.

9. Sedentary families

These include the following, about which some knowledge of the brain is at hand: Terebellidae, Cirratulidae, Ampharetidae, Pectinariidae, Opheliidae, Maldanidae, Scalibregmidae, Sabellidae, and Serpulidae. They are all equipped with brains of simple form and low degree of differentiation, undivided and in all except the last without glomeruli or corpora pedunculata. In many, the brain is hardly more than a broad commissure between the two circumesophageal connectives, actually in the epithelium. Generally nothing is found but large, plasma-rich cell bodies and loosely organized neuropile without local differentiation. These central characteristics are correlated with the notable **reduction in sense organs**—antennae are uniformly absent, eyes are absent or represented only by subepithelial pigment cup ocelli in the brain (opheliids may have lateral line ocelli; see Table 14.5), and palps are always modified; indeed often the appendages are so bizarre that their homology with palps is quite uncertain. **Giant fiber systems** are quite often well developed, reaching in fact the greatest diameters

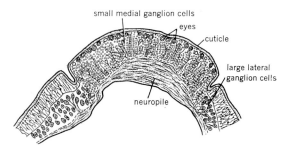

Figure 14.48. A simple polychaete brain. Transverse section of the brain of *Terebella debilis*. [Hanström, 1927.]

in the animal kingdom (*Myxicola*, Sabellidae) as well as the ultimate in concentration, fusion, and directness of path to the musculature. But, just as often, giants are absent (Table 14.3, p. 692).

Among these families the terebellids (Fig. 14.48) and cirratulids are lowest in elaboration of the central nervous system; sabellids and serpulids are the highest. These last have fairly sizable, internal brains which, though simple relative to most errant polychaetes, have some textural differentiation in the neuropile, attaining the peak in *Serpula* where, quite surprisingly, **corpora pedunculata** were described by Hanström

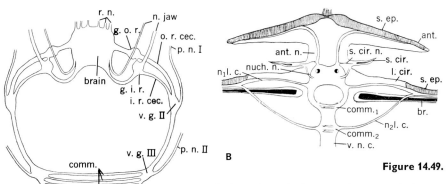

Figure 14.49.
Brain and cerebral nerves in two polychaetes. **A.** Scheme of the anterior nervous system and innervation of the head appendages in *Scolecolepis* (Spionidae). *comm.*, postoral commissures of the first several ventral ganglia; *g. i. r.* and *g. o. r.*, ganglion of inner and outer root of circumesophageal connective; *i. r. cec.* and *o. r. cec.*, inner (posterior) and outer (anterior) root of circumesophageal connective; *n. jaw*, nerves to the jaw apparatus; *p. n. I–III*, nerves to parapodia I–III; *r. n.*, rostral nerves; *v. g. II–IV*, ventral ganglia II to IV or ganglia of the first to third parapodia. **B.** Scheme of the anterior nervous system and innervation of the head appendages in *Tomopteris* (Tomopteridae). *ant.*, antenna or palp; *ant. n.*, antennal nerve; *br.*, bristle; $comm._1$ and $comm._2$, first and second postoral commissure; *l. cir.*, large cirrus or cirral antenna; *nuch. n.*, nuchal nerve; $n_1 l. c.$ and $n_2 l. c.$, first and second nerves to large cirrus; *s. cir.*, small cirrus; *s. ep.*, sensory epithelium; *v. n. c.*, ventral nerve cord. [Hanström, 1928.]

(1926), each with one globulus and one stalk. These bodies are probably associated not alone with the heavy nerves supplying the branchial crown of feathery tentacles (Hanson, 1949) but also with the diffuse supply of the prostomial epithelium, for from here appears to be activated the shadow response so characteristic of these shy tube dwellers (Hargitt, 1912; Nicol, 1950) as well as the giant fiber response to mechanical stimulation (Bullock, 1953b). An indication of the central character of these responses is their notable lability (Nicol), which possibly represents the mediation of such higher cerebral centers.

As one might expect from the lack of development of the protrusible proboscis of the errant families, the **stomodeal system** of the sedentary forms is much less developed, although clearly present and exhibiting a ganglion and plexus on the pharynx, and connects to the central nervous system either via the brain or the first ventral ganglion.

10. Other groups

It may be noted, in hopes of attracting future workers, that nothing in detail can be said about the internal anatomy of the brain in capitelliform, orbiniid, and indeed a number of other polychaete groups. Isolated facts have been noted in the several sections about a few families—for example syllids, phyllodocids, alciopids, spionids, tomopterids (see Fig. 14.49), archiannelids, and others; (see Tables and Index). Hanström (1929) contributed substantially to some morphological questions on the last three groups named.

IV. THE PERIPHERAL NERVOUS SYSTEM

A. Nerves of the Ventral Cord

At this point in the evolutionary scale we are intermediate between the great regularity in the number, pattern, composition, and distribution of the main nerves in higher phyla and the irregularity of the lower. There does appear to be in the annelids a common scheme, but a considerable degree of variation exists. Our knowledge is not complete enough to say for each aspect of any species chosen as a type that this is the basic or the primitive or the common condition. We shall, however, describe a particular form which seems to be as representative as any other.

The amphinomid *Hermodice* (Storch, 1912, 1913; Gustafson, 1930) has **three main nerves** for each segmental ganglion on each side, plus some small nerves to the skin. The first (anterior) and third nerves have double roots, dorsal and ventral, which may be of distinct function, the dorsal mainly motor, the ventral mainly sensory (Nilsson, 1912 on *Petta* of the Pectinariidae). Livanow (1904) calls the first and third nerves in eunicids motor, though we may doubt that they are purely so. These nerves lie in the circular muscle layer and, meeting in the middorsal line, form complete rings, communicating with the continuous subepithelial plexus. The second (middle) nerve (Fig. 14.50) runs to the podial ganglion and is supposedly mixed but chiefly sensory; its motor fibers are said to supply parapodial muscles, and its size, fiber composition, and extent, including the formation of or failure to form a complete ring, vary greatly with the degree of development of sensory structures (Livanow). From the **podial ganglion** it divides

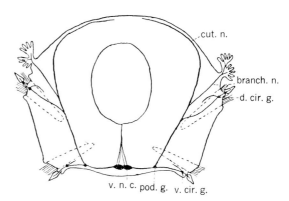

Figure 14.50. Scheme of the segmental nerves seen in a transverse section of *Hermodice* (Amphinomidae). [Storch, 1914.] *branch. n.*, branchial nerve; *cut. n.*, cutaneous nerve; *d. cir. g.* and *v. cir. g.*, dorsal and ventral cirrus ganglion; *pod. g.*, podial ganglion; *v. n. c.*, ventral nerve cord.

and a ventral nerve goes to the small ganglion of the neuropodium, which sends twigs to the ventral cirrus, the setal sack, and the skin adjacent (Storch). The dorsal branch from the podial ganglion sends a division to the small notopodial ganglion, which supplies the muscles of the setal sack and the dorsal cirrus, a division to the gill, and one to the dorsal body wall and subepithelial plexus. This description leaves most of the muscles of the parapodium and setae unsupplied, though doubtless they are all innervated from the middle segmental nerve and podial ganglion, as is specified in part for nereids (Hamaker 1898; Smith, 1957); for a polynoid, see Horridge (1959).

The exact supply to the protractor and retractor **muscles of the setae** would be of special physiological interest, for the functioning of giant fibers depends on anchoring the proper part of the worm by means of setae as much as on contraction of the longitudinal muscles of the body wall. In those species having more than one giant specialized for different startle responses, depending on the origin of the stimulus, it should be possible to show specific functional connections of each to setal muscles in specific segments (see pp. 699, 707).

There are other muscles of the setae, and branches of the parapodial nerves have been identified which, though partly sensory (Retzius, 1891a, 1892b), are motor to two sets of muscles that encircle the setal bundle, spreading and compacting the fascicle (*Petta*, Eisig, 1906; Nilsson, 1912). It should be emphasized that the second segmental nerve, which includes the parapodial supply, is primarily sensory in composition. Storch also claims to have found a small prenephridial ganglion on the lateral longitudinal connective, giving a **nerve supply to the nephridium** in *Hermodice*.

Nereis has **four pairs of segmental nerves** (Smith, 1955, 1957), of which the anteriormost from each ganglion is just preseptal; therefore it is posteriormost of a segment and is called nerve IV (Fig. 14.51). Nerves I, II, and III are postseptal; II is

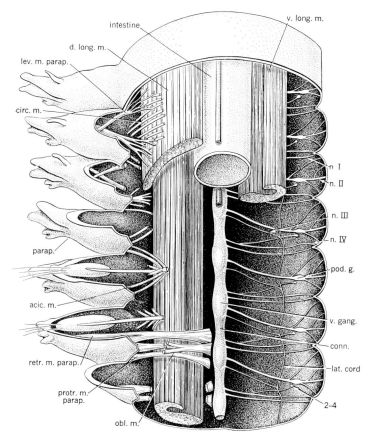

Figure 14.51.

Stereogram of seven body segments of *Nereis virens*, dissected at various levels to show the nerve cord and segmental nerves and the muscles of the body wall and parapodia. The anterior end is at the top of the figure. [Smith, 1957.] *n. I–IV*, the segmental nerves in antero-posterior succession; *2–4*, peripheral connection between nerves II and IV; *acic. m.*, acicular muscle of the parapodium; *circ. m.*, circular muscles of the body wall; *conn.*, connective of the ventral nerve cord; *d. long. m.*, dorsal longitudinal muscle; *lat. cord*, lateral cord; *lev. m. parap.*, levator muscle of the parapodium; *obl. m.*, oblique muscle; *parap.*, parapodium; *protr. m. parap.*, protractor muscle of the parapodium; *pod. g.*, podial ganglion, on nerve II; *retr. m. parap.*, retractor muscle of the parapodium; *v. gang.*, ganglion of the ventral cord; *v. long. m.*, ventral longitudinal muscle.

the largest and I and III are very thin. Nerves I and IV traverse the floor of the body wall transversely and terminate in the dorsal integument; II supplies the parapodium and III links with the lateral longitudinal nerve that connects the homologous nerves of other segments. All four are mixed and contain few fibers. Each **receives afferents** from a limited region of the skin: I and IV between them draw on integumentary receptors over most of the ventral and all of the dorsal surface: II supplies the parapodial epidermis; III is said to be primarily proprioceptive from the dorsal and ventral longitudinal muscles. Reduction in number of afferent axons between primary sensory neurons and segmental nerve fibers is suggested by Smith but is refuted by Horridge (see further, p. 755). **Motor fibers** are fewer than sensory in all nerves. As seen in methylene blue preparation, nerves I and III contain a single motor axon each, IV contains one or sometimes two, and II contains three motor axons (Fig. 14.52). The axon or axons of IV are mainly concerned with the longitudinal muscles which execute locomotor flexures (Fig. 14.53); those of III command

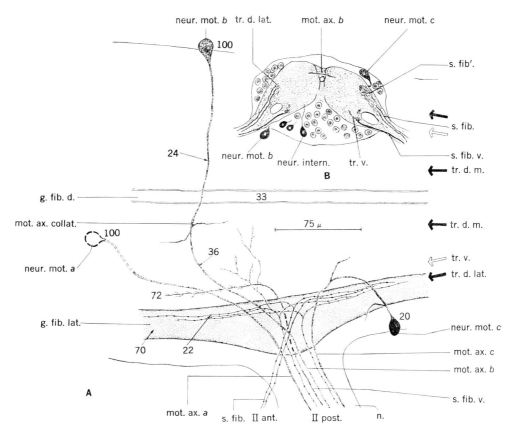

Figure 14.52. Components of a ganglion of the ventral cord: afferent and efferent fibers of nerve II, the segmental parapodial nerve. **A.** The base of nerve II and the ventral cord seen in dorsal view. The positions and orientations of the dorsal longitudinal fine fiber tracts are indicated by the *black arrows*, the ventral tract by *outlined arrows*. The *numbers* adjoining different structures indicate their depth in the cord from the dorsal surface of the neuropile: 0 is dorsal, 100 is ventral. Anterior to the left. **B.** Transverse section of a ganglion of *Platynereis dumerilii* at the level of the second segmental (parapodial) nerves. [Smith, 1957.] *II ant.* and *II post.*, anterior and posterior branches of nerve II; *g. fib. d.*, dorsal giant fiber; *g. fib. lat.*, lateral giant fiber; *mot. ax. (a, b,* and *c)*, motor axons of the correspondingly lettered neurons; *mot. ax. collat.*, motor axon collateral; *n.*, nerve root; *neur. intern.*, internuncial neuron; *neur. mot. (a, b,* and *c)*, motor neurons; *s. fib.* and *s. fib'.*, sensory fibers with a dorsal entry; *s. fib. v.*, sensory fiber with a ventral entry; *tr. d. lat.*, dorsolateral fine-fiber longitudinal tracts; *tr. d. m.*, dorsomedial tract; *tr. v.*, ventral tract.

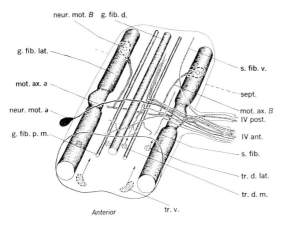

Figure 14.53. Components of a ganglion of the ventral cord: afferent and efferent fibers of nerve IV and their central relationships. Anterior toward the bottom of the figure. [Smith, 1957.] *IV (ant.* and *post.),* the anterior and posterior roots of nerve IV; *g. fib. d.,* dorsal giant fiber; *g. fib. lat.,* lateral giant fiber; *g. fib. p. m.,* paramedial giant fiber; *mot. ax. a,* motor axon with a dorsal commissural fiber; *mot. ax. B,* **motor axon with a ventral commissural axon uniting the motor fibers of the two sides;** *neur. mot. a,* **motor neuron with a dorsolateral cell body;** *neur. mot. B,* **motor neuron with a ventral cell body;** *s. fib.,* **sensory fiber with dorsal entry;** *s. fib. v.,* **sensory fiber with ventral entry;** *sept.,* **intersegmental septum;** *tr. d. lat.,* **dorsolateral fine-fiber longitudinal internuncial tract;** *tr. d. m.,* **dorsomedial tract;** *tr. v.,* **ventral tract.**

parapodial movements. The peripheral connections are discussed further on p. 750.

In many annelids there is a **median nerve** running between the pairs of ventral abdominal ganglia, which may contain nerve cell bodies. Its significance is unknown.

The common pattern of three pairs of nerves per segment obtains in most **oligochaetes,** for example in *Tubifex* (Tubificina, Yamamoto and Okada, 1940), and in lumbricids (*Lumbricus*) and megascolecids (*Pheretima, Diplocardia*), but in the small naids the pattern is irregular (Fig. 14.5) and in the lumbriculids there is an extra nerve which Isossimov regards as secondary to the development of a posterior subannulus in each segment. The typical pattern in most species is not attained in the first five segments, where there is a reduced number of nerves. The terminal ganglion, in the last segment, is an extra large one and six pairs of nerves arise from it.

The first of the **segmental nerves of earthworms** comes off anterior to the ganglion proper although, as we have noted, the whole cord is medullary in this group. The nerves from the two sides meet in the middorsal line, thus forming a nerve ring. They are provided, as are all the nerves, with nerve cells scattered along their course, possibly in a characteristic number. These are described below as intermuscular cells (p. 751). The second nerve does not form a complete ring in *Lumbriculus* where, however, it bears the most conspicuous peripheral ganglia (treated in the following section). In lumbricids and megascolecids all the nerves form complete rings. The second is the largest nerve and it arises from the ganglion itself in the caudal half of the segment. The third nerve arises so close to the second that the two together are often called the double nerve, in contrast to the single nerve or first segmental nerve (Fig. 14.54). The third forms a ring and has two grossly visible peripheral ganglia (*Lumbriculus*). The fourth nerve, peculiar to lumbriculids, lies in a posterior subannulus, has a very small peripheral ganglion, and completes a ring only in midbody segments.

There are only two **segmental nerves in leeches,** on each side in most species, but the posterior one typically (*Hirudo*) splits into two, recalling the "double nerve" of oligochaetes. Mann has presented an argument (1953) leading to the conclusion that the basic pattern is three pairs of

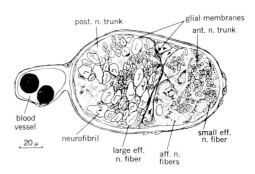

Figure 14.54. The general features of a peripheral nerve. Transverse section of the posterior (double) nerve of segment XXX in the earthworm *Pheretima.* [Ogawa, 1939.] *aff. n. fibers,* **afferent nerve fibers;** *ant. n. trunk,* **anterior nerve trunk;** *large eff. n. fiber,* **large motor axon;** *post. n. trunk,* **posterior nerve trunk;** *small eff. n. fiber,* **small motor axon.**

nerves, as in the other annelids. The middle pair serves the middle annulus, which includes the ring of sensilla corresponding to the parapodia or setal row in other annelids. Not far from its origin each nerve branches into dorsal and ventral divisions. These run deeply, giving off branches which pass through the longitudinal muscle toward the periphery and into an intermuscular nerve ring encircling the body between circular and longitudinal muscles and a subepidermal nerve ring; these rings supply the muscles and receive sensory fibers from the epidermis (Bristol). The histological structure of the nerves and two types of fibers are shown in Fig. 14.54.

Prosser (1935) recorded impulses in the segmental nerves of earthworms. That this can be done is important, for various studies on reflexology and central control are awaiting attention. The relevant information is treated below with effector control (p. 750) and reflex organization (p. 761).

B. Peripheral Ganglia

The families Polynoidae, Sigalionidae, Hesionidae, Nereidae, Eunicidae, Phyllodocidae, Amphinomidae, Euphrosynidae, Syllidae, and in general the errant polychaetes are reported to possess distinct **podial ganglia,** one pair to a segment, located on the course of the second segmental nerves (Figs. 14.50, 14.51, 14.55) and proximal to the branchings which supply the parapodia. The Glyceridae and Alciopidae are uncertain in this respect and the sedentary families lack podial ganglia in some or all of their species (spiomorphs, drilomorphs, terebellomorphs, serpulimorphs). Some terebellids (as we shall see, these are in other ways neurologically heterogeneous forms) and some ampharetids have them. The podial ganglia, being regular segmental features and not subject to the distortions that the mouth may inflict on the anterior abdominal ganglia, have been of importance in establishing that the ganglia on the circumesophageal connectives are indeed members of the abdominal chain, each belonging to a segment (Fig. 14.7).

There are rather commonly **other ganglia** besides the podial ganglia of Storch—for example small nodes of cells along the course of the nerves supplying notopodium and neuropodium, themselves coming from the podial ganglion. Or there may be scattered cells or groups of cells distributed widely along the peripheral nerves (Tomopteridae, Pectinariidae).

The designation ganglion or ganglion cell implies that these peripheral sites contain at least motor and possibly interneurons as well, and this in turn raises the question **whether they mediate reflexes** and perform integrative functions independently of the central nervous system. No answer to this question can be given at present, although it is one of general interest in the evolution of centralization, and of particular interest in view of the facts available on the intermuscular nerve cells of oligochaetes (see p. 751). Hamaker (1898) gave evidence in *Nereis* of (a) motor fibers from cells in the ventral ganglia traversing the podial ganglia without stopping, (b) fibers of central origin which end like preganglionic autonomics in vertebrates, and (c) motor fibers originating in cell bodies in the podial ganglia. (d) Sensory fibers from primary sensory neurons in the parapodial cirri and lobes are said to enter the ganglion and end there, giving the basis for reflexes, but all these results were reported without the use of specific nerve stains and therefore cannot be taken as demonstrated. Maxwell (1897) describes certain movements of parapodia in the absence of the ventral cord. Wilson (1960) found physiological evidence for reflexes mediated by podial ganglia in nereids.

In the scales of many polynoids a tiny **elytral ganglion** has been described (Nicol, 1953, 1954 b), apparently consisting of only a few cells with afferent fibers from the epidermis and efferent fibers

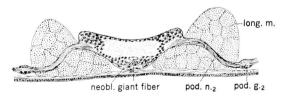

Figure 14.55. Transverse section through the second segmental ganglion of the ventral nerve cord of *Hyalinoecia* (Onuphidae). [Haffner, 1959b.] *long. m.,* longitudinal muscle; *neobl.,* neoblasts; *pod. g. 2* and *pod. n. 2,* second podial ganglion or nerve.

from the central nervous system. This ganglion controls the luminescence of the scale which occurs in series of brief flashes. Mechanical stimulation of intact worms elicits flickering on the elytra, spreading in both directions along the body; transecting the body elicits only posterior luminescence. Besides reflex flashing involving the central nervous system, autotomy or breaking off of a scale, which are common enough occurrences, produce protracted luminescence therein. A single shock to a scale isolated under anesthesia usually elicits a prolonged series of flashes, starting at about 10 per second and slowing to 1 per second. Facilitation at or in the effector is marked and lasts several minutes. Strength of stimulus has no effect above nerve threshold and below effector threshold. There appears to be a rhythmic nervous pacemaker activated by stimulation and capable of sustained oscillation even if not rising to firing level. The evidence does not indicate whether the ganglion mediates reflexes within the scale or contains synapses.

In oligochaetes, at least in some genera, there are miniature ganglia in the form of clusters of nerve cells in definite places on each of the main nerves, essentially two such **ganglia on each nerve.** The more ventral are situated opposite the setal bundles, the more dorsal at the intersection of the lateral longitudinal nerve. Similarly in leeches, two groups have been recognized, the larger with six to eight cells. The neuronal composition is not adequately known and even the presence of the ganglia in most oligochaete families is not certain.

The question of homology of the peripheral ganglia with the podial ganglia of polychaetes is not simple because of the greater number of them, dorsoventrally, in the oligochaetes and leeches, and the absence of setae and their muscles in leeches. However, the definite pattern, and the relation to setae in earthworms and to the lateral longitudinal nerve—although each is true only for some of the ganglia—strongly suggest a common origin with the polychaete ganglia, perhaps by subdivision and moving apart, secondary to the wide separation of setal bundles in lumbriculids.

C. Lateral Longitudinal Nerves

In the families Amphinomidae, Euphrosynidae, Nereidae, and apparently in very few if any other polychaetes, there are longitudinal nerves connecting the podial ganglia (Fig. 14.7, 14.51). They run continuously the length of the body, ending anteriorly by entering the brain just lateral to, but independent of, the dorsal root of the circumesophageal connective. A lateral longitudinal nerve runs just beneath the epidermis, external to the circular muscles and dorsoventrally between the two rows of setae in lumbriculid oligochaetes. En route, it makes connection with all four segmental nerves, and a tiny ganglion occurs at each such intersection. It extends anteriorly to the prostomium and according to Siazov (1909) begins in the brain, as in polychaetes, but Isossimow (1926) could not confirm this. There appears to be a lateral longitudinal nerve in hirudineans at least in the Acanthobdellidae, a very primitive family sometimes listed with oligochaetes.

Storch, and Hanström after him, believed this lateral longitudinal nerve to be homologous to members of the orthogon of polyneurous platyhelminths and therefore a homologue-in-parallel or a reduced lateral counterpart of the main ventral cords; it would thus be a primitive feature. Storch regarded its retention in these families of sufficient importance that he proposed a major division of the polychaetes into those possessing these nerves **(Tetraneura)** and those lacking them **(Dineura);** the latter including podogangliate families, having podial ganglia, and apodogangliate families, lacking them. These groups of course can hardly expect any formal taxonomic recognition, ignoring all other characters as they do, but we may use them as convenient adjectival expressions. Gustafson (1930) argues against the concept that tetraneury is primitive, proposing that it is a secondary result of the great elaboration of the dorsal parapodium for defense in amphinomids and euphrosynids. In particular, he gives evidence against the interpretation of Storch that certain nereids and spionids show a remnant of tetraneury in the anterior end and against the claim that there are any ordinary segmental podial ganglia at the bases of the prostomial palps. The suggestion that one of the two pairs of roots of the circumesophageal connective in the brain, so common in many families, represents a remnant of the lateral longitudinal nerve (Hanström, 1928b) is also ruled out by the finding that both roots, with their characteristic form and connections inside the brain, are present also in the amphinomids, side by side with the lateral longitudinal or podial nerve, whose entry is quite independent although internally most of its fibers immediately enter the commissures of the dorsal root of the circumesophageal connective.

If we accept Gustafson's evidence against intermediate forms and remnants, it is still **possible that tetraneury is primitive.** To be sure, the only arguments for it are (a) the suggestive resemblance to turbellarians, (b) the relatively unspecialized character of the amphinomids and euphrosynids in many respects, (c) the presence of structures strongly resembling homologues in at least some oligochaetes and hirudineans, and (d) the difficulty of accepting a secondary or de novo origin of the lateral longitudinal nerves in view of the usual conservatism of the nervous system in such matters. Similarly, the arguments that have been offered against it amount to the following: (a) these families are not altogether unspecialized, (b) the parapodia are in particular rather well developed, and (c) it is difficult to imagine that so many families have completely lost an original tetraneury and that these few so perfectly retained it. Unfortunately, the pertinent embryology is not yet available or it might shed some light on this phylogenetic question.

D. Receptors

1. Generalized epidermal sensory nerve cells and sense organs

Sensory structures are readily stained with methylene blue, Golgi, and reduced silver. As a result many authors have contributed information on the superficial nervous elements, especially in earthworms (Leydig, 1865; Ude, 1886; Kulasin, 1888; Cerfontaine, 1890; Retzius, 1891a, 1892c; von Lenhossék, 1892; Smirnow, 1894; Langdon, 1895; Hesse, 1896; Kolmer, 1905; Dechant, 1906; Hess, 1922–1925; Smallwood, 1923, 1926; Hanström, 1926; Zyeng, 1930; Ogawa, 1928–1939; Smith, 1957; Hartmann-Schröder, 1958). While there is satisfactory agreement concerning the nerve cells concerned—epidermal, photoreceptive, and intermuscular—the situation is not yet clear as to their relationships and the organization and function of the widespread plexus formed by their processes and multipolar ganglion cells.

(a) Scattered throughout the epidermis are **solitary primary sense cells** (Fig. 14.56), whose axons run singly into the epidermal plexus or run through the circular muscle towards a segmental nerve. Ogawa (1939) estimates that 30% of all

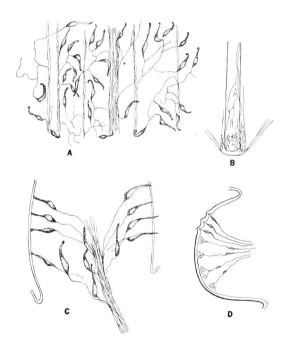

Figure 14.56. Unicellular receptors in polychaetes. **A.** Surface view of epidermis in *Nereis*. Methylene blue. Numerous bipolar sense cells are seen ending near the surface; they are upright and therefore partly shortened in perspective. The axons appear to unite and then run into the nerve fiber bundle, lying deeply. **B.** A seta in *Nereis*, showing free-branching nerve endings with knobbed tips in its sack; the nerve cells are some distance away and are not shown. Methylene blue. **C.** Basal part of a parapodial lobe in *Nereis*, primary sensory neurons ending under or penetrating the cuticle. Methylene blue. [Retzius, 1892.] **D.** Horizontal section of a parapodial lobe of *Glycera* with sense cells and free nerve endings. [Hanström, 1927.]

sensory cells are such solitary ones. They are most numerous in the anterior segments and on the prostomium; the ratio of the number in anterior segments to those in segments behind the tenth in *Aulophorus* is 6:1, according to Brode (1898). The bipolar cell body varies from 5–60 μ in length; the longest ones lie in the deep parapodial epithelium of polychaetes. They are commonly regarded as chemoreceptive or nonspecific.

(b) In addition to these there are a large number of **sensory neurons with branched free nerve endings** in the epidermis (Fig. 14.56), the cell bodies lying deep to the epidermis. These are frequently thought to be tactile or mechanoreceptive. In the oligochaetes the cells are not yet identified but have been thought to be central (Langdon). In the polychaetes they are readily identifiable cells shortly beneath the epithelium and especially numerous on lobes of the parapodia and on elytra in the scale worms. Smith (1957) found no such endings in nereids. It has been proposed that some of the endings are efferent fibers to epithelial glands and evidence exists that these are under nervous control. Waves of secretion are stopped in segments from which the cord has been removed (Coonfield, 1932).

(c) A **special type** of primary sense cell has been recognized by Hanström (1927) in the scale worm *Sthenelais* (Sigalionidae), associated with microscopic, hollow thorns on the dorsal surface of the elytra (Fig. 14.57). The cavity of the thorn is filled by a distal, swollen plasmatic process of the sensory nerve cell body whose nucleus-containing soma is situated directly beneath the cuticle, under the thorn. The two portions of the neuron are connected by a thin filament traversing a fine tube through the cuticle. The sensory nerve cell has itself secreted the distal cuticular structure of the thorn, in Hanström's view, so he proposes the term **trichogen sense cell**. The significance of this element lies in the belief that it is the forerunner of the sensory hairs of arthropods.

(d) The **integumental sense organs** comprise a variety of structures in which usually a number of spindle-shaped primary sense cells are clustered closely in a group in the epidermis. These are described in nereids, maldanids, amphictenids, opheliids, and other polychaetes; they are no more common on appendages than on the body but appear to be most frequent in tube dwellers. Included in this category are papillae and taste organs (on the prostomium), lateral organs (metameric), and anal papillae in opheliids, each with a cluster of sense cells in a patch of epidermis (Hartmann-Schröder, 1958). The greatest variety among oligochaetes exist in the small aquatic families but in the terricolous earthworms it has been well established that "sensory buds," or simply "sense organs" are generally distributed throughout the animal. In these families the organs are simple (Fig. 14.58), without or nearly without specially arranged supporting cells, composed of a few dozen (16–45) sensory nerve cells and disposed on the segments in three rows or girdles. Each cell gives rise to a nerve fiber, so a bundle of axons passes into the subepithelial plexus. They have been repeatedly enumerated, and some of the results, as examples, are: 1009 in a typical body segment of *Lumbricus terrestris* (Langdon); 1200 in the 10th and 1900 in the first, compared to 218, 324 and 530 in the same segments of *Microscolex elegans* (Bovard, 1904); 302 in a typical segment in the half-grown *Pheretima communissima* (Ogawa, 1939). They are more numerous in the anal segments and especially in the prostomium,

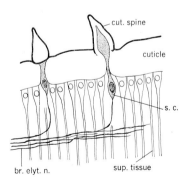

Figure 14.57. A special form of unicellular receptor foreshadowing arthropods: the trichogen sense cell. Dorsal part of the elytra of *Sthenelais picta* (Sigalionidae) at high magnification. [Hanström, 1927.] *br. elyt. n.*, branch of elytral nerve; *cut. spine*, cuticular spine, which is hollow and occupied by a distal process of a sensory cell *(s. c.); sup. tissue,* fibrous supporting tissue of the elytra.

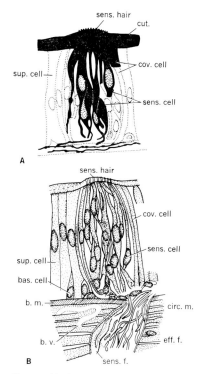

Figure 14.58. Sense cell clusters in the earthworm *Lumbricus*. **A.** Sense bud in the epidermis by the Golgi method. **B.** The same in ordinary histological preparation. [Langdon, 1895.] *bas. cell*, basal cell; *b. m.*, basement membrane; *b. v.*, blood vessel; *circ. m.*, circular muscle; *cov. cell*, cover cell; *cut.*, cuticle; *eff. f.*, efferent nerve fiber; *sens. cell*, sensory cell; *sens. f.*, sensory fiber; *sens. hair*, sensory hair; *sup. cell*, supporting cell.

where they are crowded together densely, reaching the greatest concentration in the upper lip. On the average Hesse found 686 organs per square millimeter in the prostomium, which means something over 20,000 sensory nerve cells per square millimeter—and each of the other receptor elements is also most crowded in the same region. Further details of the concentration in the three girdles of each segment, of the increase with age, of the special varieties in the Limicola, and of sensory abilities, possible division of function, and natural history can be found in the articles referred to and in a number of others which are reviewed by Stephenson (1930) and Stolte (1935). Laverack (1960) showed regional differences in the sensitivity of the body wall to touch and chemicals.

Hirudineans exhibit at least as wide a range of solitary sense cells as do oligochaetes: neurons with free nerve endings and structures composed of grouped sensory nerve cells—sense organs, sense buds, and sensilla, some of them protrusible (Retzius, Havet, Hesse, Ascoli, and others; see Scriban and Autrum). Certain sensilla are distributed in species-specific and phylogenetically significant ways.

Physiological work relevant to dispersed epidermal receptors has been considerable but for the most part concerns the demonstration of various forms of sensibility and is to that extent beyond our present scope. Responses to stimuli of many kinds have been reported (some attributable to receptors in the following sections), including contact, gravity, water, oxygen tension, pH, light, chemicals; a partial list of authors is Darwin (1881); Harper (1905, 1908, 1909); Jennings (1906); Yerkes (1906); Hargitt (1912); Baglioni (1913b); C. Hess (1913); W. N. Hess (1922, 1924a); Alsterberg (1924); Frisch (1926); Herter (1926); Ameln (1930); Nomura et al. (1932); Prosser (1934b); Unteutsch (1937); Howell (1939); Nicol (1950); Clark (1956); Laverack (1961a). Kaiser (1954), as an example, reported evidence of sensibilities in a leech to light, to shadow, to warmth, and to chemicals. *Hirudo* reacts positively to several constituents of human sweat; the anterior sucker is most sensitive. Commensal polychaetes may manifest a high degree of chemical discrimination in their responses to specific host species (Hickok and Davenport, 1957). Earthworm single afferent fibers in segmental nerves have been found which fire for many seconds when the skin is bathed in buffer solutions of pH 4.0 but do not fire at pH 4.2 (Fig. 14.59); other units have slightly different thresholds but likewise very sharp and not sensitive to other chemicals tested. Measured electrophysiologically, differences in pH threshold between three species of earthworms correlate with characteristic differences between them in tolerance of acid soils (Laverack, 1961a).

Horridge (Fig. 14.67, C) has added new findings on the polynoid *Harmothoë*. In the second segmental (parapodial) nerve there are 3 fast motor axons of 6–8 μ, 6 of 4–6 μ, 57 of 2–4 μ, 176 of 1–2 μ, and about 2100 of less than 1 μ. Most axons

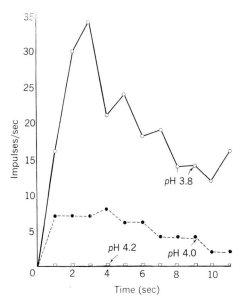

Figure 14.59. Sensory physiology of annelids. The response of a single unit in *Lumbricus* to solutions of different pH. No impulses were caused by pH 4.2 though 4.0, and pH 3.8 caused firing for a number of seconds. [Laverack, 1961.]

over 2 μ have a single-layered sheath. All those over 1 μ and many of the smaller ones have a neurofibril which consists of densely packed neurotubules. The large number of axons is in reasonable agreement with the number of peripheral sensory cells, rather than being fewer as claimed for *Nereis* and earthworm by others. Recordings from axons of bristle receptors show them to be highly sensitive, rapidly adapting mechanoreceptors. Proprioceptors have been identified with sensory cells having elongated dendrites lying across the ventral surface of the parapodium; they respond with adapting trains of impulses to appropriate strains in the cuticle. Touch receptors are scattered over the body surface. The bipolar neurons of the parapodial cirri are inferred to be chemoreceptors.

Impulses emerge from the ventral cord in each segment on both sides when bristle receptors on any part of the worm are touched. These impulses are inferred to be efferent in the sensory axons of the bristle receptors. They emerge from a synaptically connected system which runs throughout the worm, and the excitation within this system can not be labeled as beginning or terminating in any one segment. This nonspecific system synapses with the giant fibers at many rapidly adapting synapses and excitation in it activates and necessarily accompanies the sequence of locomotory movements.

2. Mechanoreceptors

Proprioceptors are indicated, on physiological grounds. Proprioceptive nerve fibers are claimed by Smith (1957) in nereid polychaetes, showing as fine fibers ramifying on the surface of the longitudinal muscles. They do not seem to penetrate the sheath of the muscle fiber and on that account are believed not to be motor. The cell bodies have not been found. Horridge regards this nerve supply as not afferent but efferent, representing the slow motor innervation. Mechanoreceptor fibers have been found electrophysiologically by Hagiwara and Morita (unpublished) in *Hirudo* in all four roots of each ganglion of the ventral chain; to judge from spike amplitude there are two or three large and a number of much smaller mechanoreceptor fibers. A touch to the skin with a fine rod causes a burst discharge with rapid adaptation. A stretch of the whole animal or of the skin of the segment concerned produces a rapidly adapting response also; sometimes a short discharge is seen at the release from the stretch. The mechanoreceptor fibers of the ganglion in each segment innervate their own segment, the posterior half of the preceding segment, and the anterior half of the following segment. Therefore each portion of skin is supplied by overlapping innervation from two successive ganglia. The innervation is strictly ipsilateral. The innervation area of each large mechanoreceptor fiber is surprisingly wide. Sometimes a single fiber can be activated by gentle touch from any point in the entire innervation area of a ganglion—an area two segments in length, on one side; this is probably not due to mechanical spread of the stimulus, for the boundaries are rather sharp. The areas of separate axons therefore overlap greatly.

The mechanoreceptor impulses initiate ascending as well as descending discharges in the ipsilateral and contralateral connectives; from any given segment these can be recorded throughout the length of the ventral cord.

IV. PERIPHERAL NERVOUS SYSTEM D. Receptors 3. Photoreceptors

3. Photoreceptor cells

Sensitivity to light is not confined to the region of eyes but is widely distributed over the body in annelids, especially earthworms and leeches; see for example Hess (1922–1925); Howell (1939); Stephenson (1930). Besides a variety of little understood eyelike organs (especially on the head) in limicolous species (see Stolte and Stephenson) but apparently lacking in Terricola, widely scattered single-celled photoreceptors were discovered by Hesse (1896) and studied in particular by Hess (1922–1925). These peculiar cells, about half the height of the epidermis, have a refractile inclusion body, a "retinella," and a proximal nerve fiber (Fig. 14.60). Like all the sensory nerve cells, they have a well-developed network of neurofibrils which are continued into the axons. They are most numerous on the upper lip, fewer in the body segments. Internal clusters of photoreceptors of the same cell type occur along the course of nerves in the head, in some caudal segmental nerves, and under the brain capsule. Their function has been deduced from the cell structure, the occurrence deep in the epithelium and below it, and the correlation of abundance with demonstrable light sensitivity. Hess counted 440 in the prostomial epidermis and 700 on the cerebral nerves. The axons enter the peripheral

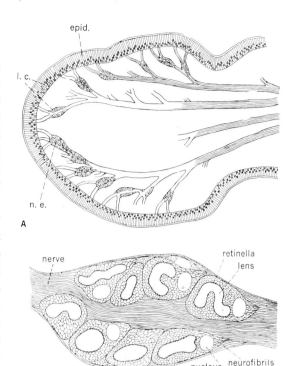

Figure 14.60. Photoreceptor nerve cells in the periphery, *Lumbricus*. **A.** Diagrammatic drawing of a sagittal section of the prostomium, showing the light cells both in the epidermis *(epid.)* and as enlargements along the course of nerves. *l. c.*, light or photoreceptor cell; *n. e.*, enlargement on nerve due to cluster of light cells. **B.** One such swelling on a nerve, in longitudinal section. Reduced silver stain. [Hess, 1925.]

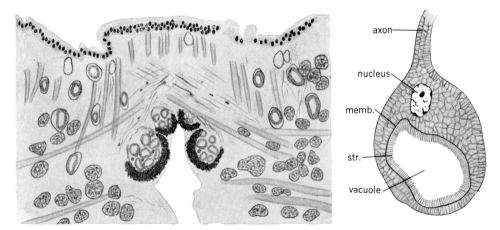

Figure 14.61. Photoreceptor nerve cells in leeches. **Left,** *Haementeria;* four pigmented eye cups and scattered single photoreceptor cells, recognized by the radially striated border around a large vacuole. **Right,** *Herpobdella;* a receptor cell from an eye showing the nucleus, axon, vacuole, and striated border *(str.)* defined by an outer limiting membrane *(memb.)*. [Scriban and Autrum, 1932.]

plexus or nerves but generally cannot be followed very far. Solitary photoreceptors are well developed in leeches; in addition, simple ocellar eyes are found in certain places in all leeches, consisting of clusters of these same large cells with receptive vacuole, striated border (Fig. 14.61), and axon, the whole in a pigment cup. The shadow response does not depend on the presence of the anterior eyes. Scattered unicellular photoreceptors are known also in polychaetes and archiannelids, while the former also possess clustered photoreceptor cells—sabellid tentacles, for example.

Response to light and shadow have been studied most carefully in earthworms; see, for example, Hess (1922–1925); Unteutsch (1937); Howell (1939). A separate receptor system for shadow and for light detection is inferred, with somewhat different distribution over the surface and different action spectra. Systematic variation in latency of reaction to exposure of parts of the body occurs, according to the number of light and of shadow receptors stimulated and remaining unstimulated; no center is postulated but an over-all integration of the number of segments and density of reception in each must be performed. The latencies can be many seconds (Unteutsch). These experiments should be repeated and the interpretation scrutinized. Howell provides evidence of a descending pathway for withdrawal from light that starts in the brain and crosses to the contralateral circumesophageal connective. Shadow responses are conspicuous in certain polychaetes, especially among the sabellids and serpulids; Nicol (1950) has provided a careful study. The connection to the giant fiber response is very direct and in *Protula* this fiber arises in a special brain lobe close to a sensory epithelium and receives numerous short afferent axons—whether photoreceptor or other remains to be learned.

Hagiwara and Morita (unpublished) find small spikes in each nerve of the ventral ganglia of *Hirudo* in response to illumination of the skin. Both small and large fibers in the connectives fire to illumination.

4. Eyes

Simple as well as lenticulate eyes are found in some polychaetes (Andrews, 1891, 1892; Hesse, 1899; Nilsson, 1912); see Fig. 14.62. *Nereis* has a cuticular lens fitting into a cupped retina in each of two pairs of eyes near the brain. The anterior pair are thought to mediate negative responses to light; the posterior pair, positive responses (Langdon, 1900; Mosella, 1927; Herter, 1926; Brand, 1933). *Tomopteris* (Fig. 14.63), *Eupolyodontes*,

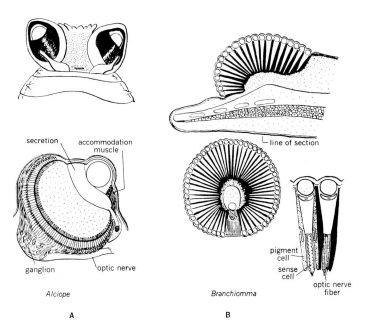

Figure 14.62.

The eyes of some polychaetes. **A.** *Alciope* (Alciopidae), camera type eye; above, ventral view; below, section through optical axis. There is a secretory gland (perhaps creating turgor pressure) and a muscle pulling the lens back. **B.** *Branchiomma* (Sabellidae), compound eye spots arranged quasiradially; above, longitudinal section; below, transverse section and detail. [**A.** Demoll. **B.** Hesse. Both after Milne and Milne, 1959. *Handbook of Physiology*, Sect. 1, Vol. 1.]

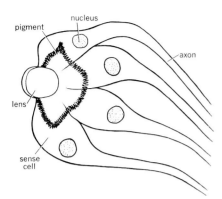

Figure 14.63. The eye of *Tomopteris* (Tomopteridae) in long section. [Hanström, 1929.]

and other planktonic forms have remarkable eyes. Well-developed camera-type eyes are displayed by the pelagic alciopids. The retina is upright; near and far accommodation of the lens is described, the latter by muscles; muscles also move the eyes (Demoll, 1909; Greeff, 1877; Pflugfelder, 1932; Hess, 1918). *Eunice* appears to have a midventral eye spot in each segment, of curious structure (Schroeder, 1905) and doubtful function (Hess, 1913). Multipolar, heteropolar cell bodies in the central nervous system are supposed to be the primary sensory neurons. (The ocelli of leeches are mentioned in the account of photoreceptor cells, above.)

5. Statocysts

Vesicles with concretions that bear upon a sensory epithelium are known in a number of polychaetes, but essentially only among sedentary families living in tubes or burrows (Ariciidae, Arenicolidae, Terebellidae, Orbiniidae, Sabellidae). There may be 5 or 6 pairs (ariciids) up to about 20 (orbiniids), all in anterior segments and dorsally, near the gills or notopodia. The other families named have a single pair situated in the first or second segment, postorally and usually just above the first bundle of setae. The axons run to the circumesophageal connective or an anterior ventral ganglion, never to the brain.

The statocyst may be in the epithelium or deep and may communicate by a duct with the outside or be closed—and these characters may be mixed in various combinations (Fig. 14.64). In many sabellids, such as *Branchiomma*, the statocyst is in the epidermis and has a ciliated duct; in *Myxicola*, of the same family, it is subepidermal and closed. *Arenicola marina* has its statocyst deep in the longitudinal muscles with a long canal, whereas *A. grubii* and *A. ecaudata* have closed vesicles. *Scoloplos* (Ariciidae) has open grooves, called "statocrypts," which sometimes contain sand grains. *Nainereis* (Orbiniidae) may have both open and closed cysts in the same individual. The receptor cells are not described exactly but appear to be numerous, epithelial primary sensory neurons.

Buddenbrock (1913) studied some responses

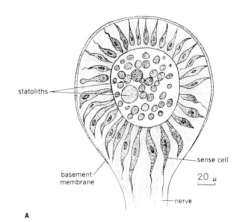

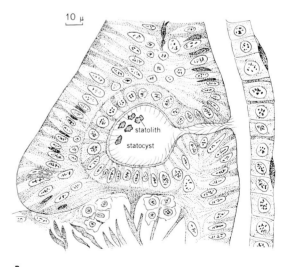

Figure 14.64. Statocysts in polychaetes. **A.** *Arenicola ecaudata* (Arenicolidae), optical section. Methylene blue. **B.** *Branchiomma vesiculosum* (Sabellidae), transverse section of a young specimen. Hematoxylin. The statocyst communicates with the outside by a ciliated canal. [Fauvel, 1907.]

mediated by statocysts in *Branchiomma* and *Myxicola* (Sabellidae) and in *Arenicola*. Both statocysts must be removed to cause deficiency symptoms. These organs do not appear to be responsible for vibration perception but do mediate geotaxis. They work in conjunction with muscle stretch receptors, demonstrable after statocyst ablation, in determining posture and orientation.

6. Nuchal organs

These vary from a simple pit at the posterior edge of the prostomium to an elaborate folded and ciliated affair that extends over many segments (ariciids, spionids, amphinomids, euphrosynids); see Fig. 14.65. It is always dorsal and paired or, when apparently unpaired, supplied by paired nerves. The maximum development occurs in certain sedentary families where the organ may even be erectile (opheliids, capitellids, ariciids, scalibregmids). It is reduced in some sessile forms (terebellids, sabellids, sabellariids) and absent in some others (serpulids, oweniids). Nuchal organs are innervated from the brain, even when segmental; that is, the axons of the primary sensory neurons pass into the hindbrain.

These curious organs have been regarded often as prostomial structures like eyes and palps, although sometimes secondarily extended back over the body. Others argue that they are true segmental organs, secondarily connected to the brain (Söderström, 1927b; Gustafson, 1930; Rullier, 1950). The latter view is based particularly on the Spionidae. A resolution of this problem is not possible here.

From the histology, these organs are supposed to be chemoreceptors, mediating perhaps a control over food selection. An obvious analogy exists in the ciliated grooves innervated by the brain in turbellarians and nemertineans. Similar organs occur in archiannelids (Polygordiidae, Protodrilidae, Saccocirridae, Nerillidae, Dinophilidae) and in some oligochaetes.

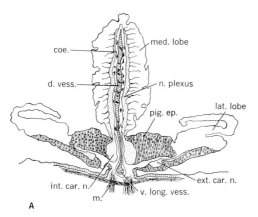

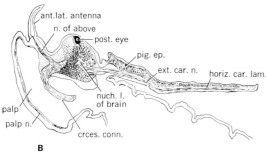

Figure 14.65. Nuchal apparatus in *Notopygos labiatus* (Spionidae). **A.** Transverse section of the nuchal organ or caruncle. **B.** Parasagittal section of the head through the caruncle and nuchal nerve. [Malaquin and Dehorne, 1907.] *ant.lat. antenna*, anterolateral antenna and its nerve; *crces. conn.*, circumesophageal connective; *coe.*, coelomic diverticulum; *d. vess.*, dorsal caruncular vessel; *ext. car. n.*, external caruncular nerve; *horiz. car. lam.*, horizontal caruncular lamella; *int. car. n.*, internal caruncular nerve; *lat. lobe*, lateral lobe of the caruncle; *med. lobe*, median lobe; *m.*, muscle; *n.*, nerve; *nuch. l. of brain*, nuchal lobe of brain; *pig. ep.*, pigmented epithelium; *post. eye*, posterior eye; *v. long. vess.*, ventral longitudinal vessel of the caruncle.

E. Nervous Supply and Control of Effectors

Polychaetes have been little known until the work of Smith (1957) on *Nereis*. Here the number of motoneurons sending axons to the body muscles is small—about 7 on each side per segment (p. 680). Of these, 3 or 4 supply the 500 or more longitudinal muscle fibers on each side. As in the case of sensory axons, there is evidence of multiplication of fibers in the periphery but, according to Smith, not by extensive branching. Rather, relay neurons are said to be interpolated into motor pathways (Fig. 14.66), much as earlier authors suggested for the intermuscular nerve cells. Horridge doubts this relay and believes that there is sufficient branching of the primary motor fibers to supply the muscles; this would be the fast motor system. In the special case of the parapodial

IV. PERIPHERAL NERVOUS SYSTEM E. Control of Effectors

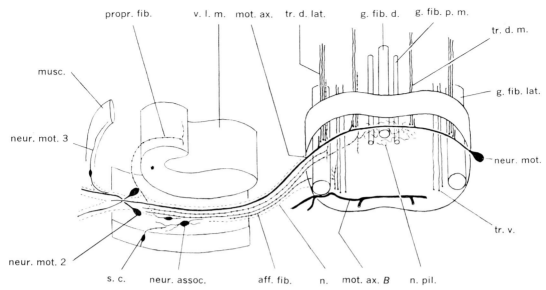

Figure 14.66. Components of a segmental nerve; the central and peripheral connections of a schematic nerve from studies of nereid polychaetes. [Smith, 1957.] *aff. fib.*, afferent fibers; *g. fib. d.*, dorsal giant fiber; *g. fib. lat.*, lateral giant fiber; *g. fib. p. m.*, paramedial giant fiber; *mot. ax.*, motor axon of dorsal emergence; *mot. ax. B*, motor axon of ventral emergence; *musc.*, muscle; *n.*, segmental nerve; *n. pil.*, neuropile; *neur. assoc.*, association neuron of the subepithelial plexus; *neur. mot.* and *neur. mot.* (2 and 3), motor neurons of the first, second, and third order; *propr. fib.*, proprioceptor fiber; *s. c.*, sensory cell; *tr. d. lat.*, dorsolateral fine-fiber longitudinal internuncial tract; *tr. d. m.*, dorsomedial tract; *tr. v.*, ventral tract; *v. l. m.*, ventral longitudinal muscle.

nerve, such second-order motor neurons occur distal to its ganglion and in turn connect with third-order neurons that supply the muscles. Peripheral connections are said to occur between the parapodial ganglion on nerve II, which supplies the parapodium, and nerve IV, which supplies the main longitudinal muscles of the body, and also between the lateral longitudinal nerve and nerve IV, permitting the possibility of coordination not entirely dependent on central nervous linkages. Connections are also claimed between the sensory and motor fibers of a given segmental nerve, directly and via internuncial relay neurons in the periphery. In order to show the character of good methylene blue preparations on which the anatomy is based, Fig. 14.67 is given, showing low-power views of whole dissections in *Harmothoë*.

In oligochaetes, as we have seen (Fig. 14.17), motoneurons give off a large number of collaterals into the muscle, so it is not surprising that, as Ogawa found, some 85,728 longitudinal muscle fibers and 33,984 circular fibers in a typical (XXX) segment are innervated by 474 nerve fibers, a ratio of 250:1 (*Pheretima*). The muscle fiber estimations were made by counting nuclei of the (uninucleate) muscle fibers, so they do not include the fibers in the segment whose nuclei lie in other segments. In this form nearly all the longitudinal fibers extend over 2–4 segments.

The term intermuscular nerve cells has been applied to the cells in oligochaetes seen by a number of earlier authors but particularly studied by Dawson (1920), Zyeng (1930), and Ogawa (1939) and corresponding to the cells in polychaetes studied by Smith (1957). They are of several kinds, not necessarily with the same functional meaning but with the common features that they are found in or between muscle layers or along the peripheral nerves and have relatively long processes from each pole (Fig. 14.68). They are bipolar or multipolar. One group (a) is distributed between circular and longitudinal muscles, not in the nerve trunks but in two rows belonging respectively to the first nerve and the second and third nerves together. Those for the first nerve send an axon thereto, a distal process through

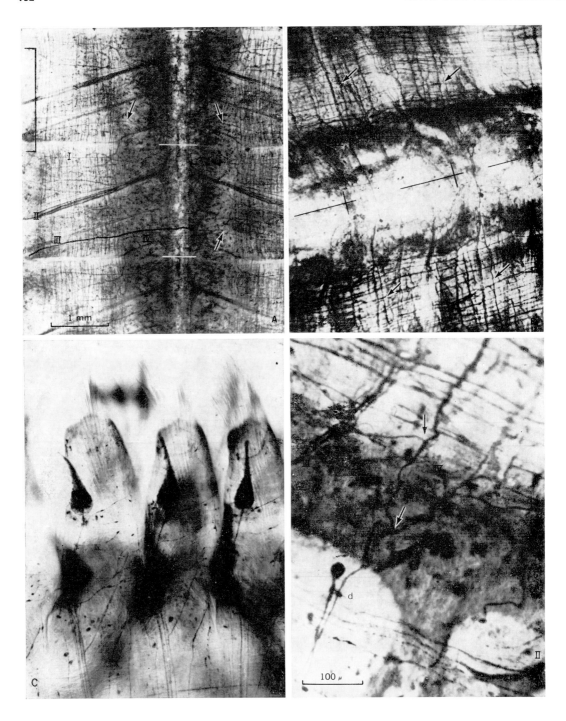

Figure 14.67. Peripheral nerves in Polynoidae; methylene blue preparations of the ventral body wall. **A** and **B.** Segmental nerves are indicated by roman numerals *I–IV;* the motor axon of *IV* is drawn in ink in one segment and indicated by arrows elsewhere. (The lower right arrow in **A** is 1 mm too high.) The transverse lines give the positions of intersegmental septa. Scale in **A** applies as well to **B. C.** Segmental sensory fibers and bipolar sense cells. **D.** The cell body of one motoneuron *IV*₁ and the initial branches of the axon of the corresponding cell body of the other side. The point where the fibers cross on the midline and the tuft of dendrites *d* near the cell body can be seen. [**A, B, D,** *Harmothoë,* Horridge, 1959; **C,** *Lagisca,* Horridge, 1963, *Proc. roy. Soc.* (B) 157:199–222.]

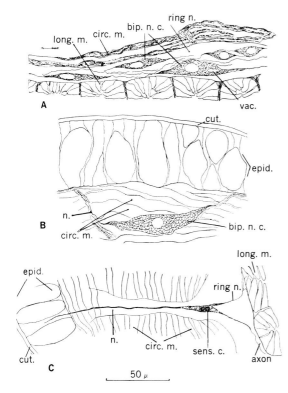

Figure 14.68. Nerve cells in peripheral nerves of *Eisenia* (Oligochaeta). **A.** Group of spindle shaped bipolar cells *(bip. n. c.)* in the intersetal area of a segmental nerve *(ring n.)*. **B.** Bipolar nerve cell in the circular muscle between the ventral setae. **C.** Bipolar cell thought to be sensory, in a superficial nerve penetrating circular muscle; it sends one process to the epidermis and one onto a muscle fiber, besides an axon centrally. [Dawson, 1920.] *bip. n. c.*, bipolar nerve cell; *circ. m.*, circular muscle; *cut.*, cuticle; *epid.*, epidermis; *long. m.*, longitudinal muscle; *n.*, nerve; *ring n.*, ring nerve (ordinary segmental nerve that meets its mate middorsally); *sens. c.*, presumed sensory nerve cell; *vac.*, vacuole.

the circular muscle to the epidermis ending there in a budlike swelling and one or two processes into the longitudinal muscle. Those for the double nerve are said to lack the epidermal ending. This group is supposed to occur in a rather definite number—for example 20 for the first nerve, 18 for the other two in *Pheretima communissima*; 6 for the first, 5 for the second, and 15 for the third in *Eisenia foetida*. (b) A second group occurs along the course of the nerves, is bipolar (or, at bifurcations, tripolar) and sends its distal processes into the epidermis, the proximal ones toward the ventral cord. (c) The third group are crescentic bipolars lying just under the epidermis or in the circular muscle. One of their processes extends outward into the epithelium; the other is lost in the plexus. Contrary to earlier descriptions, Ogawa reports that they do not anastomose. Details of the cytology, distribution, and increase with growth are given by the several authors. Conflicting functional interpretations have been offered and no firm conclusion can be drawn even as to the possibility that they are sensory, as the first group is possibly proprioceptive as well as exteroceptive.

It is claimed that unipolar stomodeal neurons in the pharyngeal plexus send long axons across to the somatic musculature of the body wall (Nevmyvaka, 1956).

The **nerve endings,** if they are the terminal ends of the methylene blue stained efferent fibers (rather than still finer continuations as yet unstained) are simple, free, or slightly knobbed ends (Fig. 14.69); no real evidence of multiple endings on each muscle fiber is at hand (but see Nevmyvaka, 1956). The motor unit is large, especially in *Nereis*, where, as in arthropods, there are very few efferent nerve fibers; perhaps there is less need of fine gradation. Facilitation, as one might expect, is not as well developed.

The **physiology of nerve-muscle relations** has been examined in several annelids by Horridge (1959), Horridge and Roberts (1960), and Wilson (1960). Stimulation of nerve I in *Neanthes* (Nereidae) gives a single, all-or-none response of the parapodium, as though by way of a single efferent fiber and motor unit. Nerve II gives facilitating (slow) responses of the parapodium; the

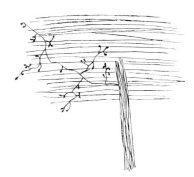

Figure 14.69. Nerve endings in muscle in the leech *Nephelis*. [Havet, 1900.]

impulses seem to go right through the podial ganglion and not to relay there (contra Smith's histological conclusions). No responses were obtained to nerve III. Nerve IV stimulation produces at least two responses in the longitudinal musculature, differing in threshold, latency, and type; one is fast, initially large, and rapidly declines; the other is slow, initially small, and facilitates. Local reflexes are mediated by the podial ganglion, involving the same muscles as those controlled by nerve II. Stimulation of segmental nerves of *Hirudo*, *Lumbricus*, and *Aphrodita* gives slow or facilitating contractions, whereas polynoids and sabellids stimulated through the giant central fibers give fast, declining muscle action potentials. The locus of the failure is probably the giant-to-motoneuron synapse.

Polychaetes like *Nereis* and *Harmothoë* (Polynoidae) (Fig. 14.70) offer special advantages for experimental analysis; the small number of units shown by Smith (1957) is one. Horridge (1959) worked out several preparations for studying reflexes and many details of neuron distribution revealed by methylene blue. Among other features, he uncovered a synergism between lateral and medial giant fibers; the latter can restore failing response to the former, probably by summation at the giant-to-motoneuron synapse.

Muscle innervation in hirudineans has been occasionally studied. Motor nerve endings are like those in earthworms. According to Gaskell (1914) the longitudinal muscles are excited and the circular and radial muscles possibly inhibited by the posterior of the two segmental nerves, whereas the circular and radial muscles are excited, the longitudinal possibly inhibited by the anterior. Schwab (1949) gave some details on muscle properties, conduction, and pharmacology in *Haemopis*.

Many leeches exhibit a high degree of **control of color**. At least some of the chromatophores are under the control of the nervous system (Smith, 1942) and the response to illumination or background is probably mediated by the eyes and brain (Janzen in Herter, 1926).

The **nervous control of luminescence** offers another means of investigating the organization of effector control. Polynoid worms and chaetopterids are the only forms, among the numerous luminescent polychaetes, that have been investigated in this respect; Nicol provides an excellent review (1960) as well as a series of original studies. Upon stimulation polynoids flash from the elytra (scales), each of which contains a small elytral ganglion (see p. 741). It seems probable, although the evidence is scanty, that in the intact animal luminescence is mediated by a reflex

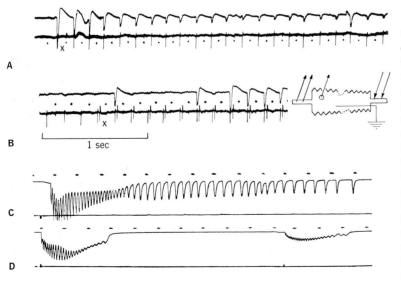

Figure 14.70. Control of effectors in polychaetes. **A.** Muscle action potentials from ventral longitudino muscle of *Nereis* (upper trace), together with impulses of the lateral giant fiber (lower trace), in response to shocks (indicated by *dots*) delivered to the nerve cord. Repeated stimulation causes failure of muscle response. **B.** The introduction of impulses in the paramedial giant fiber (x) restores the response to the lateral giant fiber impulse. Arrangement of electrodes shown in inset. [Horridge, 1959.] **C.** Photoelectrically recorded light output of a single elytrum of *Lagisca*, showing rhythmic flashing induced by a single electric shock. Time in seconds. **D.** The same in a scale of *Harmothoë* [Nicol, 1953.]

through the ventral nerve cord; in the isolated elytrum repetitive flashing is attributed to iterative firing of one or several ganglion cells in the elytral ganglion as a consequence of direct excitation of the cut efferent nerve or of the ganglion. Even a single electric shock to that ganglion can elicit repetitive discharge (Fig. 14.70). The elytral ganglion acts as though it were a multiplying relay station in the efferent pathway, as Smith (1957) proposed for *Nereis* peripheral ganglion cells and podial ganglia. Facilitation is a feature of the repetitive flashing and may occur at more than one site along the pathway. Some curious results of transection of the cord, suggesting a preference for posterior spread of excitation (Kutschera, 1909; Nicol, 1957b), are not yet satisfactorily interpreted. Luminescence in chaetopterids is localized to the region of a tactile stimulus or spread reflexly to more and more distant regions after stronger and stronger electrical stimulation. Pathways in the ventral cord are implicated with interneural facilitation and a preference for posterior propagation.

F. The Peripheral Plexuses

Many descriptions, chiefly from oligochaetes, agree that a plexus of nerve fibers occurs under the epidermis over the whole body and some show it continued into the muscles, over the segmental coelomic septa, gut, blood vessels, and nephridia (Fig. 14.71). Dechant (1906) argued for a subcuticular, supraepithelial nerve net in oligochaetes, but this seems unlikely. There seems to be some reason for doubting that all the argentophile fibers that have been seen are nervous, but the subepidermal plexus is satisfactorily established. It is not at all satisfactorily understood, however. Opinions up to the 1920's held it to be a true anastomosing nerve net, but it seems most probable today, in view of a similar history in the coelenterates and of the experimental evidence, that it is no true anatomically continuous net. The **possibility of fusion of sensory processes** or of their ending in the plexus in synaptic relation to interneurons (perhaps some of the intermuscular cells or of the nerve cells described by various authors in the plexus itself) is another matter; it

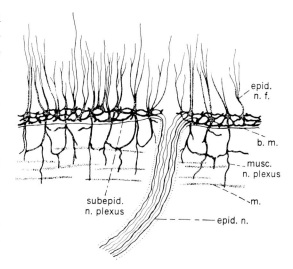

Figure 14.71. A view of the subepidermal nerve plexus *(subepid. n. plexus)* of *Lumbricus*, showing connection with the central nervous system via epidermal branches of segmental nerves *(epid. n.)*, with the epidermal nerve fibers *(epid. n. f.)* from sense cells and with the muscular nerve plexus *(musc. n. plexus)*. *b. m.*, basement membrane; *m.*, muscle. Based on reduced silver preparations. [Hess, 1925.]

is chiefly kept open because of the apparent discrepancy in numbers between the sensory nerve cells and the nerve fibers in the segmental nerves, close to the cord. The sensory nerve cells total at least 10,000–30,000 per segment (*Lumbricus*), due to large numbers of sense organs, each with 16–45 nerve cells. The nerve fibers in the nerves are hard to count, especially to include all the finer fibers, among which the afferents are included, but pending electron microscopy we have only the estimate of Prosser (1935) of less than 3,000 per segment. In *Nereis* Smith (1957) counted only 36–40 afferent fibers per segment on both sides but 1,000 sensory cells per mm². Electron micrographs of *Harmothoë*, however, throw doubt on the whole idea of reduction (p. 745).

The physiological **evidence is against the conception of a net,** even a discontinuous one, in which excitation can propagate for some distance. To be sure, if the nerve cord is cut, or one of two segments of it excised, in an earthworm pinned down in these segments to prevent passive stretch from stimulating across the cut, waves of peristalsis can still pass uninterruptedly over the worm. This, however, received its explanation when Prosser (1935) found that the overlap of innervation was such that he could detect impulses in each segmental nerve from three segments; the different nerves serve fairly discrete but overlapping fields. Similarly, stimulating a segmental nerve causes contraction not only in its own segment but in adjacent segments. This also should signify that no special morphologic meaning need be read into the overlapping supply of the prostomium from the subesophageal ganglion and of the first segment or two from the cerebral nerves (p. 725). If more than three segments are denervated by excising the cord, no contraction wave passes, nor can a contraction wave pass around the end of a T-shaped cut in the body wall in the absence of the ventral cord (Janzen, 1931). As Prosser (1950) points out, there is no reason to implicate the nerve fiber plexus in the phenomenon of spontaneous contraction which can occur in conducted waves in isolated strips of body wall from the extreme posterior end in certain species; until there is reason to doubt it we may suspect intrinsic activity in the muscle and conduction from cell to cell therein. [See also Autrum (1932) on leeches.]

Another question of evolutionary as well as physiological interest is whether there are **direct nervous connections from receptors to muscle**, via the plexus, without requiring an arc through the ventral cord. The answer seems to be in the affirmative, as was long suspected. The experiment of Prosser (1950) on earthworms, in which isolated strips of body wall gave a clear-cut response to bright illumination, appears decisive. He emphasized that these responses were few, occurred only in preparations of the posterior end, and were weak and local. Similar results have been reported for tactile stimulation and the responses may even be conducted some distance if the preparation is prone to the spontaneous activity mentioned above. The conclusion to which we are forced is that such direct connections from receptor to muscle must be insignificant in comparison to those indicated through the central nervous system, at least in oligochaetes. They may be more important in polychaetes (Smith, 1957) and in leeches (Beritov and Gogava, 1945; Beritov, 1945). The leech *Hirudo* shows well-developed responses of the isolated skin and body wall and these are regarded by the latter authors as nervously mediated via peripheral synapses between sensory and motor neurons.

The question whether peripheral ganglia on the segmental nerves are reflex centers is more difficult. In general, the evidence seems to be that podial ganglia of nereids are, but that sensory neurons do not make functional contact with the motoneurons in elytral ganglia of polynoids (see p. 741).

G. The Stomodeal System

Sometimes called the sympathetic, stomatogastric, visceral, or enteric nervous system, this set of nerves and ganglia is a characteristic feature of articulates. It is fully developed in the archiannelids *Protodrilus* and *Saccocirrus*. It forms an almost independent nervous system for the anterior part of the alimentary canal, probably continuous with a plexus in the rest of the gut and some other viscera, and particularly, in polychaetes, the pharynx, becoming highly developed in many of the errant polychaetes with a protrusible proboscis. It is doubtless both motor and sensory and independently integrative as well. This is based upon very meager knowledge of the neuronal composition. Thus it is not strictly analogous to the autonomic system of vertebrates, which is defined as an efferent system and is mostly quite incapable of integrated action independent of the central nervous system.

The presence of two or three rings around the pharynx (Fig. 14.72), with numerous longitudinal nerves beset with ganglion cells strongly suggests the platyhelminth pharyngeal rings.

The system is generally connected to the brain through one or two pairs of nerves from the forebrain (see pp. 712, 725) and to the subesophageal or first ganglion of the ventral cord by a pair of nerves, but occasionally one or the other of these appears to be absent.

After these common features the system varies greatly in detail. In *Eunice* Heider (1923, 1925) described what must be one of the best developed. Two separate nerves, an outer and an inner, spring from the forebrain. The outer pair (pharyngeal cords) is distributed in fine strands to the

IV. PERIPHERAL NERVOUS SYSTEM G. Stomodeal System

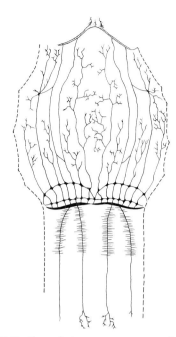

Figure 14.72. Part of the nervous system of the gut in *Nephtys*. [Quatrefages, after Hilton, 1925.]

wall of the pharynx over the jaws, is without its own ganglia, and no doubt represents a motor pathway to the jaw muscles. The inner or esophageal cords pass posteriorly under the brain and then dorsally behind it to a ganglion, the supraesophageal ganglion of the stomodeal system. From this, two strong medullary strands continue backward in the wall of the esophagus. They divide, the dorsal moiety continuing posteriorly as the principal nerve of the esophagus, the ventral running to an infraesophageal stomodeal ganglion, which thus completes a nerve ring. From this ganglion, in turn, there proceeds caudally an unpaired strand ending in a third ganglion whose lateral corners extend dorsally around the roof of the jaw sack. A paired fourth ganglion lies between the second and third on the lateral walls of the jaw sack, each one in the form of an X; it contains neuropile as well as ganglion cells and probably receives a connection from the first mentioned outer or pharyngeal nerve.

In *Hermione* two roots in the forebrain are found, both supposed to correspond to the inner, esophageal cord of eunicids, but there is no outer cord. The main cord nearly touches nerves of the ganglion on the circumesophageal connective and this may represent an original contribution from the first ventral ganglion. This would make more plausible the homology suggested by Hanström between the further course of the main stomodeal cord and the nervus recurrens of arthropods, which comes from the tritocerebrum —a ventral ganglion united to the supraesophageal ganglion. *Hermione* also has a nerve ring surrounding the esophagus and ganglia which may correspond to some of those in *Eunice*. A principal esophageal nerve continues, as in the latter genus, and supplies the glandular epithelium and musculature of the stomach and intestine.

In amphinomids (Fig. 14.73) the **double origin,** from brain and first ventral ganglion, is quite clear. Otherwise the system is similar to what has been described. There is a single large proboscis ganglion on each side, with a nerve loop connecting the pair around the ventral side of the pharynx. A pair of principal esophageal nerve cords strewn with ganglion cells continues back at least as far as the midgut.

In **sedentary forms** the stomodeal system is simpler, corresponding to the lack of development of the pharynx. There is a pharyngeal ganglion and plexus (no proboscis) but a simpler and weaker set of nerves. In contrast to the eunicids, however, the only connection with the central nervous system in some species (*Petta*, Pectinariidae, Nilsson, 1912) is by a nerve from

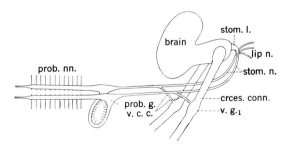

Figure 14.73. Diagram of the stomodeal system of Amphinomidae (Polychaeta). [Gustafson, 1930.] *crces. conn.*, circumesophageal connective; *lip n.*, nerves to oral lips from the anterior part of the stomodeal lobe of the brain; *prob. g.*, proboscis ganglion; *prob. nn.*, posterior proboscis nerves; *stom. l.*, stomodeal lobe of the brain; *stom. n.*, stomodeal nerve; $v.\ g._1$, first ventral ganglion of the cord; *v. c. c.*, ventral cord contribution to the stomodeal system.

the circumesophageal connective close to the subesophageal ganglion. We may perhaps regard both conditions as derived by reduction from a primitive twofold origin. In serpulids (Johansson, 1927) only the brain connection is apparent, but there are three different nerves from the brain going to the stomodeal apparatus.

A **caudal enteric system** has been described (in *Petta*, Nilsson). A pair of medial nerves run from the last ventral chain ganglion back to the anus and there turn forward to run on the ventral side of the gut as nervi gastrici posteriores. They join a pair of small ganglia, ramify, and form an intestinal plexus. This corresponds to the situation in Crustacea and other arthropods and is probably common among polychaetes.

Details of the **peripheral terminations,** sensory or motor, in the viscera were long lacking except for some studies on the epithelium of the proboscis in *Nephtys, Phyllodoce, Glycera,* and *Goniada* (Wallengren, 1902; Retzius, 1902). These organs are virtually somatic rather than visceral in their main role as eversible structures for offense, defense, or feeding. It is therefore not surprising that they are much like terminations in skin elsewhere. The sensory neurons offer an excellent case of the transition from simple primary sense cells with undivided distal process to those with branched free nerve endings. Hanson (1951) cited earlier authors who reported nerve fibers along the blood vessels of polychaetes. Whitear (1953) has now provided a detailed study of the plexus and its sensory and motor elements in *Arenicola,* based on methylene blue intra vitam staining; this is interesting as a possible anatomical basis for the complex intrinsic rhythms of activity analyzed by Wells (1955). Nevmyvaka (1956) reports some unipolar cells of the pharyngeal plexus of earthworms send long axons across to the somatic musculature of the body wall.

The stomodeal system is **well represented in oligochaetes.** It has been described in various members of the Naidina (Brode, Vejdovsky), Enchytreina (Hrabě, 1932), Tubificina, Lumbriculina (Henry, 1947), Lumbricina (Chen, 1944; Henry, 1947), and Megascolecina (Imai, Ogawa, Henry). *Branchiobdella* is said to lack such a system (Schmidt, 1905). The origin of the system is from the central nervous system in connections with the brain and also with the circumesophageal connective or subesophageal ganglion by a variable number of connectives (for example seven in *Lumbricus*, two in *Mesenchytraeus*). The common connection with that region of the circumesophageal connective in which the ganglion of segment I lies seems to have been overlooked in some writings in which it is pertinent, but this connection is compatible with the homology of the first segmental ganglion of annelids with the tritocerebrum of arthropods.

There is considerable variation in the **plan of the nerves and cell groups,** but according to Chen (1944) the pharyngeal plexus, by implication equated with the stomodeal or anterior enteric nerve supply, consists essentially of three parts: a ganglionated or medullary chain directly under and behind the circumesophageal connective, five to seven connectives joining these two, and a large number of nerve trunks that leave the chain and run both anteriorly and posteriorly over the pharynx. These trunks branch, anastomose, and form a plexus. In some forms, such as *Lumbriculus* and *Mesenchytraeus*, the stomodeal nerves complete a ventral commissure in the pharyngeal wall, but in this as in most topographic features great variation occurs.

The stomodeal ganglion has been analyzed with methylene blue in *Pheretima* (Ogawa) and presents some interesting features of **neuronal composition** (Fig. 14.74). Fibers enter it via the connectives coming from (a) cells in the subesophageal ganglion and ending in the stomodeal ganglion, (b) cells in the brain and ending in the ganglion, (c) cells in the brain but continuing through the ganglion to the wall of the enteron. Coming the other way, from the periphery, (d) afferent fibers from sensory nerve cells in the enteron end in the ganglion and (e) similar afferents proceed through it to end in the brain. The unipolar cells of the ganglion are curious in that about as many of them send their axons back into the brain or circumesophageal connective—quite unlike familiar visceral systems—as send their axons into the enteric plexus.

Of considerable interest is the finding that this portion of the whole nervous system is most

IV. PERIPHERAL NERVOUS SYSTEM G. Stomodeal System

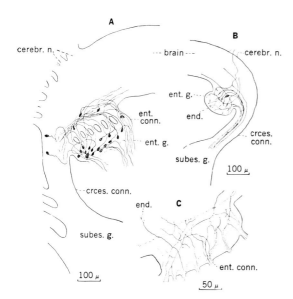

Figure 14.74. Neurons of the stomodeal system of the earthworm *Pheretima*. **A.** Ganglion cells that send axons from the enteric ganglion to the cerebral ganglion and vice versa; newly hatched worm. **B.** Fibers from the subesophageal ganglion, ending in the enteric ganglion; 10 mm embryo. **C.** Afferent fibers coming through the enteric connectives from alimentary organs; newly hatched worm. [Ogawa, 1939.] *cerebr. n.*, cerebral nerves; *crces. conn.*, circumesophageal connective; *end.*, nerve ending; *ent. conn.* or *g.*, enteric connective or ganglion; *subes. g.*, subesophageal ganglion.

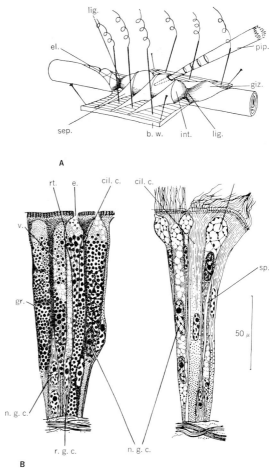

Figure 14.75. Influence of nerve supply upon digestive function in *Lumbricus*. **A.** Diagram illustrating the technique for withdrawing intestinal juice to compare proteolytic power before and after electrical stimulation of the body wall. *b. w.*, body wall of anterior intestinal region, pinned out; *el.*, electrode; *giz.*, gizzard; *int.*, intestine; *lig.*, ligature; *pip.*, tip of graduated pipette inserted into intestine; *sep.*, septum. **B.** Transverse sections of portions of the intestinal epithelium of *Lumbricus*. *Left*, resting epithelium from an unstimulated worm, previously starved for several days. *Right*, the epithelium from the same region of the intestine of a similarly treated worm, fixed after stimulation of the secretory nerves. *cil. c.*, ciliated cell; *e.*, apex of gland cell; *gr.*, zymogen granule; *n. g. c.*, nucleus of gland cell; *rt.*, cytoplasmic reticulum; *r. g. c.*, restitution gland cell; *sp.*, intercellular space; *v.*, vacuole. [Millott, 1944.]

nearly in its perfect form at the time of hatching in *Pheretima*, the number of nerve cells in the ganglion showing virtually no change between that time and the adult (Fig. 14.2).

Although there have been various studies upon the movements of the gut and its pharmacology (Millott, 1943; Wu, 1939a; Ambache et al., 1945), we still await an elucidation of the **functional role** of the enteric nervous system. Millott has shown that, superimposed on the intrinsic nervous system of the gut, two sets of nerve fibers from the central nervous system exert antagonistic effects on the tone of the gut. Traveling by the same route, through segmental nerves and then via septa, fibers giving histologic (Fig. 14.75) and physiologic evidence of causing secretion of digestive enzymes by intestinal glands were also found.

A well-developed stomodeal system is present in **hirudineans,** with several ganglia in the walls of the pharynx, an extensive plexus of ganglion cells and fibers in the wall of the gut (Fig. 14.76), and, in the rhynchobdellids, serving the proboscis. The cells are chiefly unipolar and bipolar and

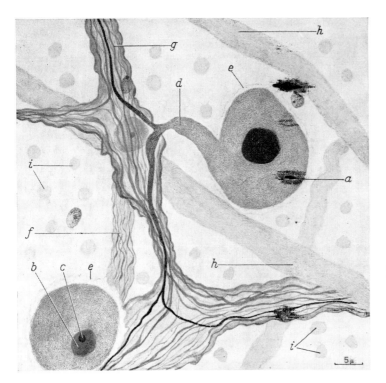

Figure 14.76.

Nerve plexus in the wall of the midgut of *Hirudo medicinalis*, stained by the Bielschowsky-Ábrahám technique. [Ábrahám, 1958.] *a*, nerve cell; *b*, nerve cell nucleus; *c*, small body in nerve cell nucleus; *d*, process of nerve; *e*, capsule; *f*, nerve trunk; *g*, nerve fiber; *h*, muscle fiber; *i*, nucleus of epithelial cell.

do not anastomose; numerous fine terminals lie parallel to the muscle fibers (Ábrahám and Minker, 1958); Sánchez (1909, 1912) gave a wealth of detail from reduced silver preparations. Contact with the central nervous system takes place by means of one pair of short connectives from the circumpharyngeal connectives between the cerebral and the next pair of nerves. Bristol identified three cell follicles on each side associated with this enteric connective and apparently supernumerary to the follicles normal to the segments (in *Herpobdella* = *Nephelis*).

Gaskell (1914) has shown experimentally that the rate of pulsation of the lateral lacunae of the vascular system in *Hirudo* is under nervous control, mediated by anterior accelerator and posterior inhibitor nerves. Drug actions and chromaffin cells have been cited as increasing the analogy with the vertebrate autonomic system (Lancaster, 1939; Vialli, 1934). However, it is not clear that any relation exists between this vascular control and the stomodeal nervous system; by analogy with arthropods, it seems unlikely.

V. PHYSIOLOGICAL STUDIES ON CENTRAL ORGANIZATION

We have seen in the systematic accounts that the peripheral plexus is probably not a true (that is, diffusely conducting) nerve net, and though it may under certain experimental conditions mediate response directly, it probably plays no significant role in this respect normally. Thus, apart from local reflexes mediated by podial ganglia in polychaete parapodia, normal responses are manifestations of central nervous organization. We have included as much physiology as possible in the systematic accounts—for example peripheral nervous function, giant systems, nerve components (and see Conduction and Transmission in Chapters 3 and 4)—but the extensive literature on reflexology, locomotor and other response mechanisms, and the role of higher centers is better treated separately. Three difficult lines have to be drawn, excluding sensory abilities, pure muscle physiology, and complex behavior. Summaries of these matters are to be

V. CENTRAL ORGANIZATION A. Ventral Cord

found in Hempelmann (1934), Warden, Jenkins, and Warner (1940), Stolte (1935), and Herter et al. (1936-1939).

A. Reflexes and Cord Activity

The nerve cord is continually active, as seen by electrodes placed on its surface, whether in situ or isolated from the body (Holst, 1933; Prosser, 1935; Rushton 1945; Moore and Bradway, 1945; Bullock, 1945; Beritov and Gogava, 1949, 1950; Vereshchagin and Sytinsky, 1960). The character of this ongoing or **background activity** is an apparently random sequence of spikes of 1 to 5 msec duration and widely varying amplitudes, with a smaller component (in peak voltage) of slow oscillations, in the range of 40-100 msec duration. A feature pointed out by Beritov and Gogava (1949) is of considerable potential value; activity recorded in different nerves of the same segment occurs at quite different times (*Hirudo*); also, it is rather independent of removal of the rest of the nerve cord if its own ganglion is left intact. Closer study of the output of several nerves at once may teach us more about coordination and central control.

The physiology of the giant fiber systems, relevant to the present heading, is treated on pp. 689-707.

Non-giant fiber pathways have been very little exploited. Beritov and Gogava (1950) distinguish between faster, through fibers in the leech, conducting at 1 m/sec, and slower association fibers, with broader spikes and longer refractory periods. Kao and Grundfest (1957) found evidence of small-fiber activity both consequent to and causing giant spikes (p. 707).

A single shock applied to a nerve of a ventral ganglion of *Hirudo* (Hagiwara and Morita, unpublished) does not usually result in a discharge in the cord; repetitive stimuli are necessary. As the frequency of stimuli is increased, the discharge of the cord becomes more marked, but the impulses in the cord do not show any regular phase relation to the stimuli. Each of the nerve roots shows spontaneous efferent discharges. Electrical stimulation applied to the connective results in an increase of the discharges. A single shock often causes multiple impulses in the same efferent fiber. The cell bodies of these fibers are probably those found in the outer layer of the ganglion. Intracellular recording shows that these cells receive presynaptic impulses from the connective and send impulses out through the roots.

Stimulation of the cord at any level results in a discharge that can be recorded at any other level; the spread over long distances is probably through a chain of synapses. Impulses of one or two fibers always show a short latency, and follow each stimulus, one to one, up to high frequency (300 per second), suggesting the presence of a few through fibers, but as judged from the amplitude of recorded impulse the diameters of these fibers do not seem to be much larger than those of others in the connective, according to Hagiwara and Morita. Electrical stimulation applied to a nerve root increases the spontaneous discharge of the isolated cord and the increase often lasts a long time (10-20 sec) after the end of stimulation. Sometimes rhythmic burst discharge is seen in the isolated cord, but it is not certain whether this rhythm is similar to that of the swimming movement of the animal. A stretch applied to the isolated cord causes a discharge of fibers in the connectives. Adaptation is seen, but it seems to be slower than that found for the mechanoreceptor. The stretch-sensitive part is in the ganglion, not in the connective.

Besides the more complex orientations, selections, and feeding and reproductive activities, the chief **kinds of responses** are (1) peristaltic and antiperistaltic creeping (oligochaetes only), (2) parapodial creeping (polychaetes only), (3) walking (i.e. inchwormlike locomotion; hirudineans only), (4) swimming, (5) lashing and writhing in various forms, (6) the twitch reflex ("Zuckreflex"), (7) reflex arrest of creeping, and (8) various local events such as proboscis protrusion, luminescence (p. 754), opening and closing of dorsal pores and of anus, and movements of suckers, branchial filaments, tentacles, and other appendages. The first four are locomotor, though a given species usually manifests but one or two of them (including burrowing, in in-fauna species). The next three are defensive or withdrawal movements; in a given species they may all be present, or none of them, or in almost any combination.

1. Peristaltic creeping

This has been studied by at least twenty investigators and a variety of ad hoc hypotheses, centers, and tonic influences proposed. The kinetics have been well described by Gray and Lissmann (1938) and may be summarized as follows. From the resting condition peristalsis always begins as a wave of thinning (circular muscle contraction) at the anterior end and proceeds caudally, progression going opposite to the direction of the wave. This thinning wave normally occupies about half the length of the worm (60–70 segments in *Lumbricus*). The thickening phase (longitudinal muscles contracting) is stationary relative to the ground. Each segment moves forward in steps 2–3 cm long, and at an average speed of this length times the frequency of waves over the body, which in their study was 7–10 per minute (no temperature given). There are usually then one or two "points d'appui" or "feet" at the front of the thickening wave, so segments anterior to them are actively thrust forward and posterior segments are pulled up; both sets of muscles then participate in forward propulsion. Each cycle lasts from 4–11 sec and each segment is contracted for about half of this time; the wave is propagated at about 25mm/sec. Setae have been said to play a minor role in earthworms (Bohn, 1901; Yapp and Roots, 1956), but an adequate examination of their movements appears to be lacking. They certainly participate in the withdrawal resulting from giant fiber activation (p. 707) and in polychaetes like *Arenicola* (Just, 1924).

Lengths of body wall **free of nerve cord** do not, in most cases, exhibit spontaneous waves of contraction (Budington, 1902; Collier, 1939a, Beritov and Gogava, 1945; Beritov, 1945), nor, if prepared between intact regions, can they conduct waves if passive stretch is prevented (Garrey and Moore, 1915; Bovard, 1918a; Prosser, 1934c), though the contrary has been claimed (Hess, 1925). Isolated strips at least of certain species or levels (caudal) do exhibit spontaneous, conducted waves (Furst, 1890; Straub, 1900; Hatai, 1922; Hart, 1924; Janzen, 1931; Prosser, 1950). But Janzen (1931) and Prosser (1934b, 1934c, 1950) conclude that any muscle-to-muscle conduction or supposed coordination by the peripheral plexus plays no significant role in the presence of the nerve cord.

The **pattern of reciprocal excitation and inhibition** (there is evidence of active peripheral inhibition—for example Budington, von Holst, Beritov) is probably inherent in the ventral cord and does not require rhythmic stimulation. It does require maintained exteroceptive or proprioceptive stimulation, from either contact with the substratum or stretch. An earthworm suspended in air may exhibit regular ambulatory waves, but these can be stopped by immersing it in water, then started up again by passive stretch or contact with substratum or by certain other forms of mechanical stimulation (Collier, 1939a; Garrey and Moore, 1915). Although tension cannot normally explain the rhythm at the front end or on a slippery surface, it can and doubtless does facilitate the spread of the wave of thinning in segments behind the leading ones. This is dramatically shown in the classical experiments of Friedländer (1894) where peristalsis proceeds over a complete interruption of the cord or even of the whole worm after the two pieces are tied in tandem by threads. Polychaetes like *Arenicola* are more dependent on the cord; passive coordination across a cut cord is absent. Artificial stretch can initiate the contraction of circular muscle in *Lumbricus*, and the force required is within the 2–8 g of pull normally exerted in locomotion, as measured upon an ingenious, recording, traction balance (Gray and Lissmann, 1938b). Therefore, stretch is normally significant but only in facilitating the spread; it cannot be responsible for initiating the wave of contraction. This reflex response of circular muscle contraction to stretch only occurs in segments where both sets of muscles are inactive. If tension is applied when longitudinal muscles are contracting, a different reflex occurs—a reinforcement of that contraction (up to 70 g)—and, if applied during early stages of longitudinal relaxation, the rate of elongation is accelerated. Von Holst shows that the rate of conduction of the wave of peristalsis is increased by stretch and slowed, even to dying out, by conditions preventing elongation. In sum, the several reflexes to proprioceptive stimuli are such as to enhance the intrinsic action at that moment coming from the ganglion to the muscles. This will help to maintain the rhythm against resistance (as over rough ground) and, once initiated, ambulation automatically sets up a pattern of stimulation correct for producing the same events in successive segments.

These reflexes, evidence from differential anesthesia and, in *Harmothoë* spikes recorded from elongate bipolars (Horridge) indicate **stretch receptors** in the body wall, below the epidermal receptors. We shall see below, in connection with the walking of leeches, that direct electrical evidence is available in these animals. The intermuscular nerve cells are suggested, but no direct identification can be made. Some of the properties of the stretch reflex are of general interest. Maintained tension produces peristalsis lasting up to 25 minutes, representing a slow adaptation; further stretching can then renew the response. Tactile and tension effects are additive but after a maximum frequency of peristaltic contractions is reached (about one in 2 sec) further stimulation decreases the frequency. After removal of tension, peristalsis continues, but seldom beyond three minutes, the period depending not on the amplitude of the contractions but on the magnitude of the stimulus. A true afterdischarge seems indicated somewhere in the reflex arc and it may be inhibited or increased in various ways (Fig. 14.77).

Although normally a chain of reflexes accompanies it, the pattern can cross a stretch of immobilized or denervated segments or segments whose body wall has been cut completely away. Von Holst points out that this conduction is of unusually high velocity, like that in artificially stretched preparations. A denervated length of cord, isolated except for its connection to intact segments at one end, can pace those segments and shows rhythmic electrical bursts synchronous with the muscle contractions (phase relation not examined). Completely isolated cords show rhythms but it cannot be established that they are locomotor. Many lines of evidence, then, without unequivocally demonstrating it, do indicate an **intrinsic rhythm of reciprocal excitation,** probably in all ganglia and enhanced by steady stimulation (maintained contact or stretch or direct current, anterior anodal).

Further evidence against chain reflexes for the rhythm itself is that adding weights at different moments never breaks the cycle and good rhythmicity persists after transecting the circular muscle (Collier, 1939b). Horridge (1963, p. 752) injected oil into the coelom of polychaetes without changing creeping sequence, which discounts proprioceptors. The reflexes permit, enhance, and adjust the rhythm of peristalsis to varying resistance; the rhythm seems to be neurogenic.

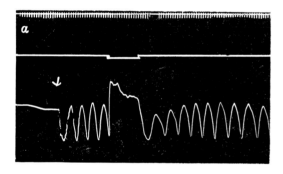

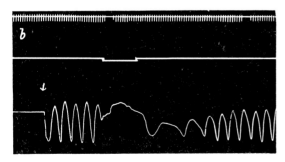

Figure 14.77. The inhibition of peristalsis by tactile stimuli in the earthworm. A one-gram tension is applied at the arrow to a preparation with denervated suspension. The lower signal indicates the duration of touch with a camel's hair brush. *a.* Brush applied to dorsal surface. *b.* Brush applied to ventral surface. [Collier, 1939a.]

2. Parapodial creeping

An excellent account of the locomotion in *Nereis* has been provided by Gray (1939). There are two general kinds: slow creeping over solid substratum, employing parapodia alone, and more rapid ambulation grading into swimming, with longitudinal muscles acting in coordination with parapodia. Unlike the earthworms, the alternating musculature is here right and left and the progression of the wave of contraction is cephalad in forward locomotion—that is, in the same instead of the opposite sense. From a motionless start locomotion begins with movement of the anterior segments (parapodia alone or with longitudinal muscles; the two sides of a segment always in opposite phase). Then occurs a very rapid spread caudally, activating every fourth to eighth segment, establishing the number of waves —typically many instead of one as in earthworms. Now begins the slow, effective wave, successively activating more anterior segments one by one (Fig. 14.78).

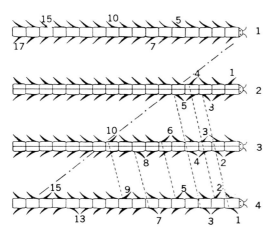

Figure 14.78. Diagram showing the start of slow ambulation in *Nereis*. Note the rapid spread (—·—·—·—) of the ambulatory pattern over the whole body from head to tail, and the movement of this pattern from tail to head at a much slower rate (------------). [Gray, 1939.]

The mechanics of the parapodia (Foxon, 1936) require coordinated action of adductors of the appendage and protractors of the aciculum (setae) on the power stroke, adductors, and retractors on the recovery stroke. In rapid creeping the longitudinal muscles under a given parapodium are relaxed during the effective stroke—that is, while that parapodium is stationary relative to the substratum and the other side of that segment is pivoting around it, with longitudinal muscle contracting and parapodium at rest. In this form of locomotion the parapodial adductors furnish little motive power; the longitudinals do the work. Note the contrast with earthworms, where the fixation relative to the ground occurs in whole segments and where longitudinals are contracting.

The evidence suggests even **greater dependence on central automaticity** and independence of reflexes than in oligochaetes. Cord interruption destroys all coordination between regions, but both regions, (indeed, even short pieces) can exhibit all the phases of normal ambulation and loss of several adjacent parapodia does not interfere with the wave progression.

Another use of parapodia is in the form of burrowing shown, among others, by *Nephtys*. Alternate anchoring with setae while the prostomium is pushed forward into wet sand and anchoring by the protruded proboscis while the body is pulled up characterizes this type. This, of course, is not a segmental pattern but probably chiefly inheres in the subesophageal ganglion.

3. Walking

Unlike the two just described patterns, the ambulation of leeches is probably a closed chain of reflexes. (See Herter, 1929, for species differences in leech movements.) According to Gray, Lissmann, and Pumphrey (1938), *Hirudo* can walk only when a certain sequence of exteroceptive stimulation via suckers and ventral surface is present—and then cannot swim; they may swim when this stimulation is removed, as by suspension in water—but then cannot execute walking movements.

Adhesion of a sucker only occurs if it is in the protruded position. Protrusion of the anterior sucker can only occur when the circular muscle is contracted isotonically (body elongate). This position, with longitudinals relaxed, is adequately stimulated by fixation of the posterior sucker. This fixation depends on the sucker first being protruded and protrusion of the posterior sucker depends on prior isotonic contraction of longitudinals which in turn is stimulated by fixation of the anterior sucker. It has not been clearly established what the simplest condition leading to release of either sucker is, or whether it is reflex rather than central.

If isotonic contraction of longitudinals is impeded by a load, the frequency of walking decreases. If the load is large, and contraction therefore isometric, movement ceases and a tension fifty times the normal is maintained for 60 sec or so. **Stretch reflexes are thus indicated,** and the presence of an afferent discharge of electrical spikes in severed distal stumps of nerves of stretched segments, as well as of efferent discharge in proximal stumps of unstretched segments near a stretched region, has been recorded (Gray, Lissmann, and Pumphrey).

Beritov (1944a, 1944b, 1945) showed that with the ventral cord intact *Hirudo* can maintain a constant length in the face of variations in load; this must be a **central reflex** because without the cord there is no exact adjustment, although responses to skin stimulation and even to increments in load occur, which Beritov interpreted as **peripheral reflexes.** The maintained length under varying load is best if the ventral cord is intact and the animal is attached by a sucker rather than suspended by a thread; middle segments alone are not so good at this adjustment. Presumably stretch receptors in the sucker are involved.

Conduction of waves of contraction can occur through several denervated segments and is independent of segmental reflexes. In the absence of suckers, walking occurs when the ventral surface is in contact with the substratum; it can even occur in the absence of the subesophageal ganglion and brain but then requires stronger stimulation, owing to the severe loss of tone. An intrinsic rhythm is therefore possible, but it seems best on present evidence to attribute ambulation to a **closed chain of reflexes,** usually associated with the suckers.

Beritov and Gogava (1945; see also Beritov, 1945) presented evidence on *Hirudo* which they believed indicated **inhibitory efferent nerve fibers** going out to muscles, as in crustaceans, as well as inhibitory central effects in normal reflexes. Moreover, these forms of inhibition were regarded as occurring automatically at the correct phase of a movement so as to be reciprocal with the excitation of the antagonistic muscle, whether circular versus longitudinal or dorsal longitudinal versus ventral. The stimulation of segmental nerves in isolated body wall preparations sometimes gave relaxation of a contracted muscle; this peripheral inhibition often showed a long after-inhibition and on that account was thought to be humoral. In contrast, the central inhibition was thought to be anelectrotonic as a result of slow waves in the neuropile. Incidental to the experiments the authors describe certain central effects of curare but little or no neuromuscular block.

The so-called **"catch action" of longitudinal muscles** of *Hirudo* may be mentioned at this point. Studies by Gogava (1944) and Beritov (1944a, 1944b) and Kometiani (1945) have revealed that this apparently frozen state of contraction is under constant reflex maintenance and that increased load causes increased oxygen consumption.

4. Swimming

In *Nereis* this form of movement is similar to the rapid, undulatory creeping already described but with increased amplitude, frequency, and wavelength of the undulating waves. Propulsion is due to the parapodium acting as a paddle, partly to its movement relative to its segment, and partly to the movement of the whole segment (longitudinal muscles). In *Nephtys* the parapodia are probably entirely passive. According to Gray (1939), if *Nereis* had no parapodia it would swim in the other direction—that is, opposite to the direction of waves over the body. Although no analysis has been made, we may suppose that the **pattern is intrinsic,** perhaps requiring, or modified by, certain general kinds of excitation. Many errant polychaetes and some sedentary ones have peculiar forms of swimming not yet analyzed.

This conclusion not only agrees with that reached for rapid creeping but with the analysis of **swimming in the leech** *Hirudo*. This is accomplished quite differently from that of *Nereis*. Bilaterally symmetrical contraction of longitudinals, alternately the dorsal and ventral sets, passes backward as a wave, opposite the direction of progression; the circulars remain relaxed and the dorsoventral muscles contracted, flattening the leech. The wave is conducted over the cord after removal of the skin and musculature of at least three segments, with prevention of passive coordination (for example by pinning or by rotation of one-half of the body 180° around the long axis and ligaturing). Pieces of the body can swim. An electrical rhythm can be recorded in a deafferented portion of the cord at least five segments from a still intact portion, the rhythm corresponding to swimming movements of the latter. No direct evidence of a rhythm in completely deafferented cord is available. Exteroceptive stimuli of certain kinds (contact of suckers or ventral surface) usually inhibit swimming completely, but stimuli of other kinds (gentle contact on dorsal surface) enhance it; however, Kaiser (1954) points out that the so-called breathing movements are essentially like swimming and occur even during sucker attachment, so tactile stimuli do not necessarily prevent these movements. Proprioceptive stimuli affect it (for example slowed swimming in viscous media). After cord transection, one half may swim while the other half walks, or both halves may swim independently, so mechanical coordination is not great.

The best hypothesis to cover these facts seems to be that an intrinsic rhythm becomes expressed if the **general level of excitation** of certain kinds is maintained above a certain level. This has also been claimed for the stepping rhythm of mammals (Sherrington, 1931) and for swimming in the dogfish and eel. Although in the eel exteroceptive stimulation (which can be nonrhythmic) is necessary to maintain this level, in the leech and dogfish proprioceptive stimuli normally suffice; these may reflexly enhance the rhythm but are probably not the pacemakers.

Swimming in limicolous oligochaetes has not been analyzed but is probably like that of polychaetes, minus parapodia.

5. Writhing

The forms taken by defensive response to strong stimulation have not been separated clearly. Von Holst recognized lashing and coiling ("Winden" and "Schlängeln"). Though they are not locomotor, they may have something in common with swimming. For purposes of this account they are of interest as showing the presence of fast intersegmental paths (other than the giant fibers) which can nearly simultaneously call up a musculature along the whole length—in this case first on one side, then another—far outlasting the stimulus.

6. The twitch reflex

This is the response to giant fiber activation and has been discussed on pp. 689–707. It is lacking in hirudineans and in polychaetes without giants. Usage of the term has varied owing to lack of recognition of the moment when giant fiber response, strictly speaking, is followed by secondary, more noticeable movements—which yet seem rapid but are probably carried in other pathways. The twitch reflex is a simple but highly specialized and outstanding example of a central, intersegmental pattern, overriding metameric individuality (Holst, 1932; Studnitz, 1937, 1938).

7. Reflex arrest of creeping

Though known to many earlier writers, Collier (1938) has given us the most information on this interesting response. Stimuli of various kinds to the anterior half of a creeping earthworm produce a freezing of the peristaltic pattern, without changing the shape of the wave. From the fact that this response recovers following cord section before giant fibers regenerate but after the return of reflex antiperistalsis (to head stimulation at rest), it must have a pathway separate from these. There are specific inhibitory responses for reflex arrest of peristalsis and of antiperistalsis, each with a separate pathway. The immobilization of antiperistalsis due to a spontaneously developed wave of peristalsis is in still a third pathway. The first two—reflex inhibitions—persist for some time after a stimulus. Among the adequate stimuli for arrest of peristalsis (for example certain forms of touch, chemicals, and vibration), it is interesting to note that loss of contact with the ground by anterior segments is one. Unlike the others, this one requires the presence of the brain; for the others even a number of anterior segments can be missing.

8. Local events

Various nonlocomotor activities present themselves for study. Rhythmic bursts in *Arenicola* of feeding movements, of irrigation of the tube, of defecation, and other events have been found (Wells, 1955, and p. 320) not to be reflexes to accumulating needs but more likely based on spontaneously active "clocks," some of which are located in the organs (for example the esophagus), others in the nerve cord; the brain is unnecessary. Luminescence has been referred to already (p. 754). Influence of the ventral cord upon the tone of the gut and upon secretion of digestive enzymes was mentioned on p. 759. Carlson (1908) found that weak stimulation of the ventral cord of *Arenicola* inhibits the esophageal heart but augments the pulsations of the dorsal vessel in both frequency and amplitude; he saw some nerve cells in the walls of vessels. Opportunities for analysis are in the control of dorsal coelomic pores, of the rectum, of suckers, branchial filaments, tentacles (Welsh, 1934; Dales, 1955), setae, branchial fans (Berrill, 1927), mouth parts (Stolte, 1932; Wells, 1937b), and reproductive movements.

In sum, we find evidence in the cord of reciprocal excitation of antagonistic muscles, intrinsic patterns in space and time, local, intersegmental, and chain reflexes, occlusion, facilitation, central (as well as peripheral) inhibition, rapid pathways, slow pathways, afterdischarge, alternation of sides and symmetrical activity, wave progression with and against the direction of locomotion, the requirement for an adequate state of central excitation, dynamic polarity, and, as we shall soon see, suprasegmental influences.

B. Roles of Brain and Subesophageal Ganglion

The syndrome after **removal of the brain** has been described by many workers in earthworms and by a few in hirudineans and polychaetes. Earthworms feed and burrow somewhat slowly and awkwardly. The head is held high. The animals are restless; they execute motor activity quite normally, enter into coitus, right themselves, and can learn mazes. There is noticeable sensory deficiency: response to moderate light is reversed from negative to positive, and normal arrest of creeping upon loss of contact between ventral surface and substratum is lost, so worms crawl off the edge of a horizontal surface and fall. Leeches are similar, but especially emphasized are the heightened excitability, restlessness, greater amplitude of movements, more rapid righting, swimming (in circles, toward the dorsal side) of more regular rhythm than normally and more prolonged. The last may be incidental to the lessened chance of contact in a circular course.

More **restricted lesions** of the brain have some points of interest. Sagittal section, dividing the brain in lateral halves (Schluter, 1933, *Hirudo*), has little effect. In walking the suckers hold on longer than usual, swimming is more often interrupted, and righting is slowed by searching movements without turning the head. Removal of one lateral half of the brain causes asymmetry at rest and in locomotion which gradually disappears (about 4 weeks). But cutting one circumesophageal connective has less effect—asymmetry is often not noticeable but righting is slowed. This must mean that contralateral and ipsilateral paths are of nearly equal value, and that either one is nearly as good as both. However, splitting of the brain or severing one esophageal connective may reverse the sign of phototaxis.

Removal of the subesophageal ganglion as well as the brain changes the picture drastically. Righting and creeping on stimulation can still occur. Yerkes and Heck found that maze learning is still retained in *Lumbricus*, but spontaneity and muscle tone are greatly reduced in all three classes. Leeches remain unmoving all day but strong stimulation will induce walking, though of course without anterior sucker attachment. Swimming is not hindered and seems to be prolonged, probably on account of the loss of contact inhibition from the anterior sucker; the dorsal light reflex still occurs, the preparations swimming on their backs when illuminated from below. Janzen reported that earthworms are deficient in ability to locomote backward. Orientation, already damaged from brain excision, is further hindered. Earthworms find places to creep into only by accident, though search movements are not altogether lost and burrowing occurs in very loose soil. They aggregate in dark corners nearly as regularly as normal animals; this is probably a kinesis.

In sum the subesophageal ganglion is especially responsible for producing tone and spontaneity, search movements, and backward flight; the brain, for regulating these movements both by inhibiting them and by directing them in accordance with special sensory influx. It also is chiefly responsible for the great variations in inner state emphasized by Jennings (1906) and, we may suspect, for the preferences and choices which determine normal habitat, food, and long-term behavior. But many movements are adequately **represented in ventral cord** ganglia—swimming, walking, creeping, righting, twitch response, and withdrawal.

A **role of the brain in regeneration** of lost posterior segments is indicated in both polychaetes and oligochaetes. For a critical initial period of about 3 days after loss of posterior segments, the brain must be present, but its importance declines with time; implantation of brains from regenerating worms into decerebrated hosts can result in regeneration. Neurosecretory activity is implicated in this function. Clark and Bonney (1960) and Clark and Evans (1961) give evidence and also earlier references. Strelin (1955) investigated the role of the nervous system in asexual reproduction in some oligochaetes, and Bennett and Suttle (1960) reported changes in secondary sex characters after brain ablation.

Studies of **behavior,** habituation, and learning are of great interest in these animals because the nervous basis may thus be approached in some ways more easily than in higher forms. Behavior is now moderately variegated in comparison with lower and higher phyla (for example, Clark, 1960a, 1960b, on nereids; Wells, 1955, on *Arenicola*; Gee, 1913, on leeches; Stephenson, 1930, on oligochaetes). Learning is demonstrable in several ways and habituation as well as other aspects have been studied (Yerkes, 1912; Heck, 1920; Swartz, 1929; Copeland, 1930; Copeland and Brown, 1934; Raabe, 1939; Bharucha-Reid, 1956; Krivanek, 1956; Arbit, 1957; Ratner and Miller, 1959a, 1959b).

VI. PHYLOGENETIC COMPARISONS

The only lower group with which we can make meaningful comparisons is the Platyhelminthes. Other intermediate groups (nemertineans, nematodes, and so on) are doubtless derivative. The same, indeed, may be true of present-day platyhelminths but they are probably the closest available animals to the line of evolution of articulates, in particular the class Turbellaria.

As we have seen earlier, the most characteristic morphologic feature of the nervous system of turbellarians is the orthogonal arrangement of longitudinal connections and transverse commissural ring nerves. Annelids also possess an **orthogonal nervous system**, most clearly manifest in the trochophore larva but also evident in the adult, if somewhat one-sided. Thus the ventral cord is doubtless derived from a pair of longitudinal cords, and in a few oligochaetes, hirudineans, and polychaetes there is an additional pair of lateral longitudinal nerves that could well represent another pair of the original longitudinal medullary connectives. In this way four of the turbellarian six or eight cords can be accounted for. The segmental nerves can be looked upon as old ring nerves, especially as some of them regularly meet in the middorsal line.

Granting a genetic relation between the turbellarian orthogon and the annelid nervous system, there remains the question how the changes —in particular, **metamerism**—took place to result in the nerve cell bodies' being confined to ganglia at intervals so the connectives between them as well as the transverse nerves are no more medullary. This problem is, of course, far larger than a purely neurological one and cannot be fully discussed here. But the nervous system plays a particularly important role in the arguments that have been advanced. On the pseudometameric theory (Meyer, 1890; Racovitza, 1896; Plate, 1922; Lang, 1881, 1904; Sedgwick, 1884; Binard and Jeener, 1930; Raw, 1949), the annelid nervous system would be directly derived from that of the turbellarian, or anthozoanlike ancestors, whose numerous, closely spaced trans-

VI. PHYLOGENETIC COMPARISONS

verse commissures are regarded as incipient segmentation. Many later writers, however, prefer the Korm or orthogon theory (Heider, 1914; Söderström, 1920-1927; Hanström, 1928b), which is closely related to the trochophore theory of Hatschek (1878), Kleinenberg (1886), and Eisig (1899). According to this theory, annelids came from unsegmented trochophorelike ancestors by a kind of tandem repetition, and in one proposal, due to Hanström (1928b) although anticipated in part by Söderström (1920), the ring nerve of the prototroch and the adjacent ones in front and behind it are repeated to form the metameric annelid system. This provides the three rings or pairs of peripheral nerves per segment common in articulates (annelids, tardigrades, crustaceans, insects) and probably fundamental to them. Incidentally it also places the nuchal organ as being probably in the segmental series of organs instead of being a purely preoral head organ, since the nuchal organ is described by Meyer as associated in ontogeny with the nerve ring in front of the prototroch. This is an advantage according to the idea of Söderström that the nuchal organ must be considered a regular appendage, based on its form in the Spionidae, Ariciidae, Amphinomidae, and Euphrosynidae. Why it should continue in these families to be innervated only from the brain is not explained.

This theory, that segmentation of the nervous system took place by repetition of the three ring nerves of trochophore larvae including that of the prototroch and the ones next in front and behind, is remarkable in putting preoral structures into the basis of the somatic segment. To this extent the **brain is serially homologous,** on this theory, with abdominal ganglia (as also proposed by Henry, 1947) leaving the ganglion of the apical organ and the first few nerve rings as uniquely brain primordia. Söderström described the brain as coming from a protosegment, and therefore homologous with the segments. It must be said that no evidence except the nuchal organ (and that of doubtful value) has been offered for the choice of nerve rings which are supposed to be repeated to obtain segmentation. In this theory, although the genetic relation with turbellarians is indirect, the derivation of the nervous system from an orthogon and the general homology of annelid and turbellarian nervous systems is still emphasized. In contrast to the pseudometameric theory, wherein the ganglia and commissures acquire a kind of secondary homology by regularization of the nodes and commissures of the irregular series in unsegmented flatworms, the serial homology is strict in the trochophore theory due to the common origin of the segments. In both, the longitudinal cords are regarded as homologous by symmetry.

While these considerations provide a derivation for the central nervous system as a whole, specific homologies between the brains of annelids and platyhelminths are very difficult aside from their both being anterior enlargements taking in several meshes of the orthogon. This is mainly owing to the simplicity of the brain of the flatworm and its lack of characteristic structural features. To be sure, both turbellarians and polychaetes, and in fact nemertineans and molluscs as well, can be described as having brains which innervate (a) special apical sensory epithelia, sometimes protruded to form sensory tentacles, (b) ciliated chemoreceptive grooves, (c) brain eyes and (d) statocysts. The last, to be sure, are brain-innervated only in forms like *Protodrilus* in the annelids; in the few other families possessing statocysts (Arenicolidae, Terebellidae, Sabellidae) they are innervated, perhaps secondarily, by the circumesophageal connectives or anterior abdominal ganglia. But no detailed homology between these sense organs or their brain nerves and centers need be claimed. Similarly, the pharyngeal nervous system of turbellarians may be homologous with the stomodeal system of articulates but we need not claim it to note the striking resemblances. As in articulates, the turbellarians have the point of contact with the central nervous system in various places in different species, sometimes on the brain, sometimes well back on the longitudinal connectives.

We may now turn our attention to the question of the **composition of the brain** in annelids in order to be able to make comparisons with the arthropods. A great diversity of views has been expressed (Fig. 14.79), from Nielson's idea that the brain represents three fused pedal ganglia, to

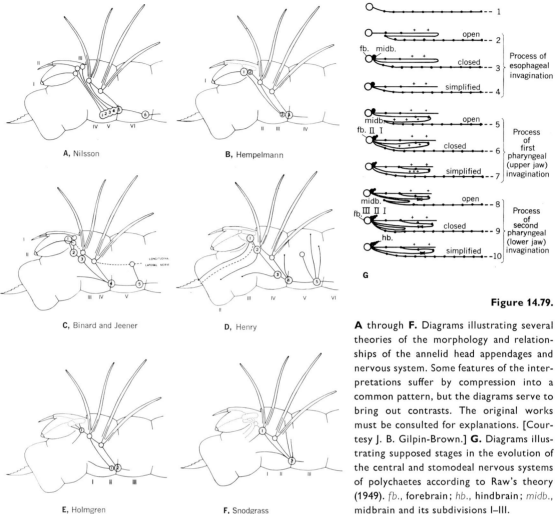

Figure 14.79.

A through **F.** Diagrams illustrating several theories of the morphology and relationships of the annelid head appendages and nervous system. Some features of the interpretations suffer by compression into a common pattern, but the diagrams serve to bring out contrasts. The original works must be consulted for explanations. [Courtesy J. B. Gilpin-Brown.] **G.** Diagrams illustrating supposed stages in the evolution of the central and stomodeal nervous systems of polychaetes according to Raw's theory (1949). *fb.*, forebrain; *hb.*, hindbrain; *midb.*, midbrain and its subdivisions I–III.

various ideas that it is compound in nature, mostly including at least a part supposed to have come from a somatic segment or its equivalent (Hempelmann, Söderström, Binard-Jeener, Meyer, Hanström, Raw). These ideas have been reviewed by Hanström (1928b), Gustafson (1939), briefly by Korn (1958), and each of the earlier writers. Figure 14.79 and Table 14.6 summarize the main views. The question has not been settled and we can neither repeat the lengthy arguments nor adjudicate among them in this case. Hanström's view of an essentially unitary origin and Raw's view of a complex sequence of stepwise increments from segmental sources are the two principal contenders; each is internally consistent and tenable, with certain attractive features.

Hanström disposed of previous **suggestions of cephalized segments** contributing to the brain, but retained one. He believed that the diverse point of origin of the stomodeal nerve among the polychaete families indicated a cephalad shift has taken place in some, so one or even more ventral ganglia have moved up the circumesophageal connective and fused with the brain, like the tritocerebrum of arthropods. The evidence that not only stomodeal centers but whole ganglia have moved consisted solely in reference to those cases where the circumesophageal connectives, pushed apart for an unusually long distance posteriorly, include one or more of the ventral abdominal ganglia. There is no indication that they have actually moved up to the brain in any family. As

VI. PHYLOGENETIC COMPARISONS

for the stomodeal nerves, there are two on each side, arising from the brain and from the first ventral ganglion, in amphinomids, euphrosynids, phyllodocids, polynoids, nereids, nephtyids, and syllids; there is a single origin from the brain or superior part of the circumesophageal connective in sabellariids, flabelligerids, serpulids, eunicids, and aphroditids; there is an origin from the first ventral ganglion only in opheliids, pectinariids, and ampharetids. There is no evidence that the last condition is primitive. Most telling of all, the commissure of the stomodeal centers of the brain, which had not been described in 1928, and which would have to remain postoral, as Hanström recognized, in a cephalized ganglion like the arthropod tritocerebrum, has now been identified clearly in amphinomids, euphrosynids, and others (Gustafson, 1930) and is directly across the anteroventral part of the brain; that is, it is preoral. We may follow Gustafson in assuming that the primitive condition comprises two connections between stomodeal and central nervous systems, one from the brain and one from the first pair of ventral ganglia. One or the other connection has then been reduced in various families. There is then, on this evidence, no cephalized segmental ganglion and the **brain is a unitary derivative** of the preoral, episphaeric nervous system of the trochophore, very much as Holmgren (1916) conceived it to be.

Following Holmgren still further, we may say that the annelid brain is regarded by Hanström as morphologically undivided. The three divisions used in the anatomical description do not have a sufficiently general basis among the polychaetes to be considered as really homologous, as divisions. They are anatomical subdivisions of convenience (compare arthropods. Table 16.1).

Raw (1949) forcefully defends a view, on the contrary, that a series of segmental ganglia have moved up to join the forebrain, which is the only primitively preoral ganglion of the annelid ancestor. The midbrain is supposed to have come from three separate increments, the second being

Table 14.6. SUMMARY OF OPINIONS REGARDING THE MORPHOLOGIC COMPOSITION OF THE ANNELID BRAIN. (From Raw, 1949).

Brain divisions first suggested by Hatschek	Forebrain ("Stomatogastric lobes")	Midbrain or antennal brain			Hindbrain ("Nuchal brain")
		Anterolateral Antennae	Posterolateral Antennae	Median Antenna	
Pruvot (1885)	seg. 1 with mouth lips and palps	seg. 2	seg. 3		
Hatschek (1891)		seg. 1			
Racovitza (1896) Holmgren (1916)		nonsegmental brain			
Nilsson (1912)	seg. 1	seg. 2			seg. 3
Lameere (1925) Binard and Jeener (1928)		seg. 3	seg. 2	seg. 1	
Hanström (1927–1928)	seg. 1, added in some only	nonsegmental brain			
Söderström (1920)	nonsegmental brain				outside the brain
Gustafson (1930)	(?) seg. 1 after	the nonsegmental brain			added to the brain
Raw (1949)	ancient brain A	BIII (Gap)	BII (Gap)	BI (Gap)	C

inserted between the first and the forebrain, the third between the second and the forebrain. The hindbrain was added last. These major events were associated with three successive invaginations of the oral region, supposed to have caused the development of the esophagus, the sacs connected with upper jaw armature, and those connected with lower jaw armature, respectively, separated by gaps of evolutionary time. The invaginations carried loops of ventral cord and ganglia into visceral (stomodeal) service and brought new, nonconsecutive ganglia close to the brain, with which they fused, bringing with them new stomodeal nerve roots. Raw thus regards the evolution of the brain to be intimately related to that of the stomodeum and its innervation. Another important feature of this theory is that the nuchal organ is considered to be segmental and its pair of nerves to be a dorsal pair of nerve cords equivalent to the ventral cords. Together with the lateral or podial nerve cords still visible in several families, polychaetes then were characterized not by tetraneury as per previous writers, but by sexneury, bringing them closer to presumed orthogonal ancestors. The anterior members of the hypothetical segmental series of ganglia of the nuchal system are believed to have been pushed forward and packed together during the invaginations mentioned, finally to join the brain (step 10 of Fig. 14.79, G). The whole theory is an involved tissue of drastic suppositions with little but plausibility to support it.

These extreme views agree on a major point, that whether major divisions of the brain are homologous or not, some of the constituent **centers may well be homologous,** in particular the stomodeal, palpal, antennal, optic, and nuchal centers and possibly the corpora pedunculata, at least in many families. The latter structures present a difficulty, since such widely different and presumably unrelated families as nereids and serpulids exhibit them according to Hanström. The only alternative is to suppose that similar structures have evolved independently in different groups; this is not implausible for the corpora pedunculata. Except for the serpulids, the families having corpora pedunculata are not widely dissimilar and perhaps come from one stem; un-

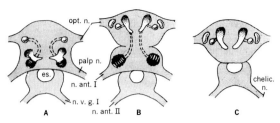

Figure 14.80. Scheme of the relations of parts of the brain in **(A)** polychaete, **(B)** mandibulate arthropod, and **(C)** chelicerate arthropod, according to Hanström (1928). Visual masses *black*, corpora pedunculata *stippled*, palpal and antennal glomeruli *hatched*. *chelic. n.*, nerve to chelicerae; *es.*, esophagus; *n. ant. I* or *II*, nerve for first or second antennae; *n. v. g. I*, nerve of appendage of the first ventral ganglion; *opt. n.*, optic nerve; *palp n.*, nerve for palp.

fortunately there is not a generally accepted phylogeny of polychaetes with which to compare the distribution.

Whether strictly homologous or independently developed, it seems better to believe that the **origin of the corpora pedunculata,** which are structures of such outstanding importance later in the arthropod brain (Fig. 14.80), took place within the polychaetes than to think of them as present in ancestors of annelids and secondarily lost in most polychaetes and all oligochaetes and hirudineans. The globulilike structures of the brain of nemertineans and of molluscs can probably not be genetically related to any part of the annelid brain. There is to be sure something very similar in the globuli cell mass of polyclad turbellarians. The families of polychaetes possessing corpora pedunculata were discussed above (p. 718); they include the hesionids, aphroditids, polynoids, sigalionids, nereids, and serpulids. Not all the remaining families have been examined but the corpora pedunculata are certainly absent in eunicids, lumbrinerids, onuphids, glycerids, amphinomids, euphrosynids, and a sample of seven families of sedentary worms. Secondary loss of elaborate association centers once present does seem quite possible, even likely in the lumbrinerids, glycerids, and Sedentaria, and reduction to the extent of loss of the stalks may account for the condition in nephtyids. In all these groups palps are absent, reduced, or modified into feeding organs; eyes are absent or reduced to pigment cup ocelli; and antennae are absent or tiny. Hanström has accordingly enunciated a causal correlation between the development of the sense organs and the corpora pedunculata. He believes the latter were at first developed from the anterior ventral ganglion cell mantle of the brain, from among those which made connections particularly with palpal nerve fibers. Later other sense organs also became connected to the same associative neuropile and the cell bodies became small at the expense particularly of cytoplasm, nuclei became concentrated with respect to chromatin, and processes formed a sharply delimited region of dense, fine-fibered neuropile. Parallel with these developments the palpal centers became more differentiated, as manifested by the special formations in the neuropile called glomeruli. Nuchal centers do not appear to be as closely correlated with development of corpora pedunculata, but when nuchal organs are especially elaborate, clusters of cells—including a high proportion of globuli cells—develop in intimate relation to these centers. This situation in amphinomids and euphrosynids and the somewhat similar diffuse masses of globuli cells in eunicids and onuphids are suggestive of stages in the development of corpora pedunculata but they might instead be stages in their reduction.

Among the species displaying corpora pedunculata there was noted some diversity, particularly in the number of globuli groups and of stalks. Hanström presented the case for believing that the single globulus of *Hermione* (Aphroditidae), *Harmothoë*, and *Polynoë* (Polynoidae) is primitive, the most primitive or at least simple being *Podarke* (Hesionidae). The subdivision into two (*Lepidonotus*, Polynoidae; *Nereis pelagica*) and into three (*Sthenelais*, Sigalionidae; *Nereis diversicolor*) can be regarded as an indication of advancement independently evolved in the several families. Size is also an indication of the degree of development, and on this count also sigalionids excel, *Sthenelais* having corpora pedunculata which aggregate 30% of the brain volume.

Of the sensory structures innervated by the brain, the **nuchal organ** (or dorsal organ) is supposed by Hanström and Söderström to be oldest but there is no real evidence for this; it is a consequence of the elaborate explanation of the segmentally repeated nuchal organ in certain families, mentioned above. This explanation, may, of course, be so; certainly these organs present a most remarkable problem which deserves a coordinated study based on innervation, comparative morphology, and embryology.

With respect to comparisons and **homologies with arthropods** and onychophorans, it seems justified at the least to admit that the annelid brain is just what it should be to form a primordium or starting point for the higher forms. Its histologic structure, cell types, system of sensory centers, association centers, commissures, glomeruli, and globuli are quite compatible with the more highly differentiated brains of the Onychophora and Arthropoda, which carry the same forms and structures further in the same direction. A consequence of the general homology of articulate central nervous structures and of the conclusion in favor of a unitary annelid brain, both enunciated earlier, is that the annelid brain can be supposed to be homologous with the combined proto- and deutocerebrum of the higher groups. Haffner (1959) seems to be opposed to this correspondence but without compelling grounds. The tritocerebrum is represented by the first ventral ganglion of annelids, in our view.

More detailed homologies may well be present, for example the sensory centers, but there is not enough definite evidence to do more than suggest them. Holmgren, and after him Bergquist and Hanström, suggested a homology **between the nuchal center and central body** of Onychophora and arthropods, an unpaired median associational neuropile. Gustafson (1930) has made a number of factual objections to this and we must let the suggestion remain unsupported unless further evidence is brought for it. The **corpora pedunculata** certainly look like truly corresponding structures in the several phyla, as Hanström proposed. If they are not homologues, the forces making for parallel development of strikingly similar specialized association centers are indeed remarkable. Söderström objected to this on the ground that the part of the brain occupied by corpora pedunculata in polychaetes is not what can reasonably be identified with the protocerebrum of arthropods, where these structures lie in that phylum. But Hanström pointed out in reply that the division into protocerebrum and deutocerebrum could have taken place after present-day polychaetes evolved and so need bear no relation to the observed lobes or topographic subdivisions of these animals, or that the corpora pedunculata could have shifted, for example by neurobiotaxis.

Classified References

General

Apáthy, 1892, 1897; Bethe, 1903; Biakowska and Kulikowska, 1911; Biedermann, 1891; Binard and Jeener, 1926; Blanchard, 1848; Buddenbrock and Studnitz, 1936; Cajal, 1904; Clayton, 1932; Faivre, 1857; Ferris, 1953; Gemelli, 1905; Goodrich, 1897; Häckel, 1866; Haffner, 1959a; Haller, 1887; Hanström, 1928a, 1928b; Havet, 1916a, 1916b; Heider, 1914; Hempelmann and Woltereck, 1912; Henry, 1947; Herrick, 1924; Hilton, 1926a, 1926c; Holmgren, 1916; Iwanoff, 1928; Joseph, 1900, 1902; Kolmer, 1905; Lang, 1904; Lenhossék, 1895; Leydig, 1862, 1864; Meyer, A., 1927; Meyer, E., 1890, 1901; Michaelsen, 1919, 1925; Nansen, 1887a; Nierstrasz, 1922; Pantin, 1937; Pflücke, 1895; Plate, 1922a; Quatrefages, 1850, 1865; Raw, 1949; Roule, 1889; Salensky, 1882, 1883, 1885, 1907; Schneider, 1902, 1908; Semper, 1876; Söderström, 1926, 1927a; Szüts, 1914b; Vejdovský, 1879; Walter, 1863; Wawrzik, 1892; Willis, 1672; Woltereck, 1904, 1905; Zavarzin, 1950; Zawarzin, 1925.

Anatomy of Polychaetes *(see also General; Anatomy of Giant Fiber System of Polychaetes; Anatomy of Receptors)*

Aiyar, 1933; Allen, 1905; Ashworth, 1902, 1904, 1911; Attems, 1903; Bergquist, 1925; Berkeley, 1927; Bernert, 1926; Bobin and Durchon, 1952; Brunotte, 1888; Caullery, 1914, 1916; Caullery and Mesnil, 1897; Claparède, 1861, 1867, 1873; Clark, 1955b, 1956, 1957, 1958a, 1958b, 1958c; Dahl, 1955; Darboux, 1899; Dehorne, 1936; Ehlers, 1864, 1868; Eisig, 1887, 1899, 1906, 1914; Evenkamp, 1931; Faivre, 1856; Fauvel, 1897; Fischer, 1884; Fordham, 1925; Fraipont, 1884, 1887; Calvagni, 1905; Gamble and Ashworth, 1899, 1900; Gilpin-Brown, 1958; Goodrich, 1912; Graff, 1887; Gravier, 1896, 1898a, 1898b; Greef, 1877, 1879; Günther, 1912; Gustafson, 1930; Hachfeld, 1926; Häcker, 1894; Haffner, 1959a, 1959b; Haller, 1889; Hamaker, 1898; Hamilton, 1917; Hanson, 1949, 1951; Hanström, 1927, 1928, 1929; Hartmann-Schröder, 1958; Haswell, 1882; Hatschek, 1878; Heider, 1923, 1924, 1925; Hempelmann, 1906, 1911, 1934; Hessle, 1917, 1925; Hilton, 1924a, 1925b, 1927; Holmgren, 1916; Horridge, 1959; Izotova, 1953; Johansson, 1927; Jourdan, 1884, 1887a, 1891; Joyeux-Laffuie, 1888a, 1888b, 1890; Kallenbach, 1883; Karling, 1958; Keyl, 1913; Kleinenberg, 1886; Korn, 1958, 1960; Kornfeld, 1914; Kükenthal, 1887; Langerhans, 1880; Lespés, 1872; Lewis, 1898; Liebermann, 1932; Livanoff, 1917, 1924; Loye, 1908; Malaquin, 1893; Malaquin and Dehorne, 1907; Mau, 1882; McIntosh, 1876, 1877, 1878, 1885, 1894; Meyer, 1882, 1887, 1888; Monro, 1924; Nansen, 1887a, 1887b; Nilsson, 1910, 1912; Pflugfelder, 1933; Pierantoni, 1906, 1908; Pruvot, 1885, 1890; Pruvot and Racovitza, 1895; Racovitza, 1896; Raw, 1949; Retzius, 1891a, 1891b, 1892b, 1895b; Rohde, 1887; Saint-Joseph, 1894; Schack, 1886; Simpson, 1959; Smith, 1955, 1957; Söderström, 1920, 1924a, 1924b; Soulier, 1891; Spengel, 1881; Stannius, 1831; Steen, 1883; Storch, 1912, 1913, 1914a, 1914b; Stummer-Traunfels, 1903; Thomas, 1940; Timoféeff, 1910; Treadwell, 1891; Turnbull, 1876; Vejdovský, 1882; Voit, 1911; Wagner, 1886; Wallengren, 1902; Whitear, 1953; Wistinghausen, 1891; Woskressensky, 1924.

Anatomy of Oligochaetes *(see also General; Anatomy of Giant Fiber System of Oligochaetes; Anatomy of Receptors)*

Adey, 1951; Aiyer, 1926; Ambronn, 1890; Atheston, 1899; Bahl, 1943; Beddard, 1895, 1901; Boule, 1908a, 1908b, 1913; Brode, 1898; Cajal, 1903a, 1903b; Cerfontaine, 1890, 1892; Chen, 1944; Claparède, 1869; Clarke, 1856, 1857; Dawson, 1920; Dechant, 1906; Dehorne, 1916; Dixon, 1915; Edwards, 1957; Fling, 1898; Friedländer, 1888b, 1894; Garman, 1888; Gemelli, 1905a; Göthlin, 1913; Gruithuisen, 1828; Haller, 1910; Havet, 1900; Hess, 1925a, 1925b; Hesse, 1894; Hilton, 1925a; Holmes, 1930; Hrabě, 1932, 1934; Imai, 1928; Isossimow, 1926; Jeener, 1927, 1928; Kowalski, 1908, 1909; Krawany, 1905; Leydig, 1862, 1865; Maule, 1908; Meyer, 1900; Nelson, 1907; Nevmyvaka, 1947b, 1947c; Ogawa, 1928, 1930, 1934a, 1934b, 1939; Penners, 1924; Perrier, 1872; Pointer, 1911; Prosser, 1934a, 1934c; Quatrefages, 1847b; Ratzel, 1867; Reighard, 1885; Retzius, 1891b, 1892a; Robertis, 1955; Robertis and Bennett, 1954a, 1954b, 1955; Rogers and Chen, 1931; Rorie, 1863; Scharrer, 1944; Schultze, 1879; Semper, 1877; Smallwood, 1923, 1936, 1930; Stammer et al., 1958; Stolc, 1886; Stolte, 1935; Szüts, 1912, 1914a; Thomas, 1954; Tuge, 1929; Vejdovský, 1884; Vignol, 1883; Wirén, 1887; Yamamoto and Okada, 1940; Zyeng, 1930.

Anatomy of Hirudineans *(see also General; Anatomy of Receptors)*

Ábrahám and Minker, 1958; Ascoli, 1913, 1911a, 1911b; Baudelot, 1865; Bethe, 1908; Bhatia, 1941; Bohn, 1903; Bristol, 1898; Bruch, 1849; Cajal, 1904; Castle, 1900; Couteaux, 1956; Dorner, 1865; Fedotov, 1914; Foettinger, 1884; Francois, 1885; Graff, 1877; Hanke, 1948; Hansen, 1881; Haswell, 1914; Havet, 1900; Hermann, 1875; Herter et al., 1936, 1939; Hilton, 1920, 1924b, 1925c; Lang, 1881; Livanov, 1903, 1904, 1906; Mann, 1953, 1954; Mencl, 1908, 1909; Meyer, 1955, 1956, 1957; Miller, 1945; Müller, 1932; Poll, 1908; Prentiss, 1903; Quatrefages, 1847b; Rebizzi, 1906; Retzius, 1891b; Rohde, 1892; Saint-Loup, 1884; Sánchez, 1909, 1912; Schmidt, F., 1905; Schmidt, G. A., 1926; Schoubine, 1916; Schwab, 1949; Scriban and Autrum, 1932; Shearer, 1910; Siazov, 1909; Sukatschoff, 1910; Whitman, 1878, 1887, 1889.

Anatomy of Giant Fiber System of Polychaetes *(see also Anatomy of Polychaetes; Nicol, 1948a, has an extensive bibliography)*

Aiyar, 1933; Allen, 1905; Ashworth, 1909, 1911; Attems, 1903; Clarapède, 1873; Cunningham, 1888; Dehorne, 1935a, 1935b; Eisig, 1887; Evenkamp, 1931; Fedorow, 1927; Gamble and Ashworth, 1899, 1900; Haller, 1889; Hamaker, 1898; Heider, 1925; Hönig, 1910; Langerhans, 1880; Lewis, 1898; Livanoff, 1924; McIntosh, 1877, 1878; Meyer, 1882; Michel, 1899; Nicol, 1948a, 1948c, 1948d; Nicol and Young, 1946; Pruvot, 1885; Rohde, 1887; Saint-Joseph, 1894; Schack, 1886; Smith, 1955, 1957; Spengel, 1881; Treadwell, 1891; Wawrzik, 1892; Wells, 1958.

Physiology of Giant Fiber System of Polychaetes

Bullock, 1948, 1953b, 1954b, 1956; Carlson, 1905; Horridge, 1959; Nicol, 1948b, 1951; Nicol and Whitteridge, 1955.

Anatomy of Giant Fiber System of Oligochaetes *(see also Anatomy of Oligochaetes; Nicol, 1948a, has an extensive bibliography)*

Adey, 1951; Cajal, 1903; Cerfontaine, 1892; Clarapède, 1869; Friedlaender, 1889; Haller, 1889; Hama, 1959, 1961; Issidorides, 1956; Leydig, 1886; Michel, 1899; Nicol, 1948a; Ogawa, 1939; Prosser, 1934c; Schultze, 1879; Smallwood, 1927; Smallwood and Holmes, 1927; Stough, 1926; Taylor, 1938, 1940; Vignol, 1883.

Physiology of Giant Fiber System of Oligochaetes

Amassian and Floyd, 1946; Bovard, 1915, 1918a, 1918b; Bullock, 1945a, 1950, 1951, 1953a; Bullock and Turner, 1950; Bullock et al., 1950; Cate, ten, 1938; Eccles et al., 1932; Kao, 1956; Kao and Grundfest, 1957; Katsuki et al., 1954; Masui, 1953; Morita and Tateda, 1953; Morita et al., 1952; Niki et al., 1953; Roberts, 1960; Rushton, 1945a, 1945b, 1946; Rushton and Barlow, 1943; Sone, 1953; Stough, 1930; Studnitz, 1937, 1938; Turner, 1955; Wilson, 1961; Yolton, 1923.

Regeneration

Abel, 1902; Bailey, 1930, 1939; Berrill, 1928, 1931; Clark and Bonney, 1960; Clark and Clark, 1959; Clark and Evans, 1961; Emery, 1886; Faulkner, 1932; Friedlaender, 1885; Goldfarb, 1914; Gross and Huxley, 1935; Haase, 1898; Hall, 1921; Hescheler, 1898; Holmes, 1931; Hunt, 1919; Iwanow, 1903; Krecker, 1910; Kropp, 1933; Michel, 1898; Nicol, 1948d; Nuzum and Rand, 1924; Rabes, 1901; Rand, 1901; Randolph, 1892; Schwartz, 1932; Siegmund, 1928; Stough, 1930.

Physiology of the Brain

Baglioni, 1913; Bennett and Suttle, 1960; Cate, ten, 1931; Jordan, 1929; Jordan and Feen, 1930; Jordan and Hirsch, 1927; Loeb, 1894; Maxwell, 1897; Prosser, 1934b, 1934c; Strelin, 1955; Wayner and Zellner, 1958.

Physiology of the Cord and Locomotion *(except Giant Fibers)*

Beritov, 1945, 1950; Beritov and Gogava, 1945, 1949, 1950a, 1950b; Berrill, 1927; Biedermann, 1904; Bohn, 1901; Bradway and Moore, 1940; Brand, 1933; Cate, ten, 1931; Chapman, 1950; Collier, 1938, 1939a, 1939b; Dittmar, 1954; Friedländer, 1888a, 1895; Garrey and Moore, 1915; Gray, 1939; Gray and Lissmann, 1938a, 1938b; Gray et al., 1938; Hagiwara and Morita, 1962; Halbich, 1940; Hart, 1923; Herter, 1929, 1939; Holst, 1932, 1933; Horridge, 1959; Janzen, 1931; Jenkins and Carlson, 1904a, 1904b; Kaiser, 1954; Knowlton and Moore, 1917; Kuenzer, 1958; Lapicque and Veil, 1925; Laverack, 1961b; Moore, 1923a, 1923b, 1923c; Moore and Bradway, 1945; Moore and Kellogg, 1916; Morgulis, 1910; Newell, 1950; Prosser, 1935; Schlüter, 1933; Uexküll, 1905; Van Essen, 1931; Wells, 1937b, 1939, 1950, 1955; Yapp and Roots, 1956.

Physiology of Control of Muscle and Other Effectors and of Peripheral Plexuses

Ambache et al., 1945; Arsharsky, 1938; Autrum, 1932; Beritov, 1944a, 1944b; Botsford, 1939; Budington, 1902; Carlson, 1908; Coonfield, 1932; Couteaux, 1933; Gogava, 1944; Honjo, 1937; Horridge, 1959; Horridge and Roberts, 1960; Just, 1924; Kometiani, 1945; Kutschera, 1909; Millott, 1943a, 1944; Mojsisovics von Mojsvár, 1878; Nevmyvaka, 1947a, 1956; Nicol, 1952a, 1952c, 1952e, 1953, 1954a, 1954b, 1957a, 1957b, 1960; Rozanova, 1941; Rushton, 1945b; Shensa, 1932; Smith, 1942; Straub,

1900; Strelin, 1955; Uexküll, 1905; Wells, 1937a, 1937b, 1939, 1950, 1955; Wells and Ledingham, 1940; Welsh, 1934; Whitear, 1953; Wilson, 1960; Wu, 1939a, 1939b.

Anatomy of Receptors *(see also Anatomy of Polychaetes; Anatomy of Oligochaetes; Anatomy of Hirudineans)*

Andrews, 1891, 1892; Atheston, 1899; Bhatia, 1956; Demoll, 1909; Ehlers, 1892; Eisig, 1879; Fauvel, 1907; Grauber, 1880; Greef, 1879; Hachfeld, 1926; Hanström, 1926, 1929; Hess, 1918, 1922, 1923a, 1923b, 1924b, 1925c; Hesse, 1896, 1897, 1899, 1929; Hilton, 1926b; Jourdan, 1887b, 1892; Kornfeld, 1915; Langdon, 1895, 1900; Lenhossék, 1892; Leydig, 1861; Meyer, 1924, 1926; Mosella, 1927; Nicol, 1950; Oppenheimer, 1902; Parker and Arkin, 1901; Pflugfelder, 1932; Plate, 1924b; Retzius, 1892c, 1895a, 1895b, 1898, 1900, 1902; Rullier, 1950; Schröder, 1905; Smirnow, 1894; Söderström, 1923, 1927b; Stolte, 1932, 1935; Whitman, 1884, 1893; Zyeng, 1930.

Physiology of Receptors

Alsterberg, 1924; Ameln, 1930; Baglioni, 1913; Buddenbrock, 1913; Clark, 1956; Frisch, 1926; Herter, 1926; Hess, 1913, 1922, 1924a, 1924b, 1925c; Howell, 1939; Kalmus, 1931; Laverack, 1960, 1961a; Mangold, 1925, 1931; Prosser, 1934b, 1935; Segall, 1933; Smith, 1902; Söderström, 1923; Unteutsch, 1937.

Pharmacology, Neurochemistry, Chromaffin Cells

Bacq and Coppée, 1937; Baldwin and Yudkin, 1948; Botsford, 1941; Gaskell, 1914, 1919; Hart, 1924; Itina, 1947; Lancaster, 1939; Millott, 1943b; Moore, 1921; Moore and Bradway, 1945; Nicol, 1952d, 1952e; Nomura and Ohfuchi, 1932; Ôsima, 1939; Östlund, 1954; Perez, 1942; Poll, 1908; Poll and Sommer, 1903; Prosser and Zimmerman, 1943; Townsend, 1939; Umrath, 1952; Ushakov and Kusakina, 1961; Vartiainen, 1933; Vereshchagin and Sytinsky, 1960; Vialli, 1934; Wells, 1937a; Wells and Ledingham, 1940; Wu, 1939a, 1939b.

Behavior and Learning *(see also Physiology of Cord and Locomotion; Physiology of Brain; Physiology of Receptors)*

Arbit, 1957; Bharucha-Reid, 1956; Bohn, 1902, 1904, 1906; Clark, 1960a, 1960b; Copeland, 1926, 1930; Copeland and Brown, 1934; Darwin, 1881; Focke, 1930; Gee, 1913; Hanel, 1904; Hargitt, 1906, 1909, 1912; Harper, 1905, 1908, 1909; Hatai, 1922; Heck, 1920; Herpin, 1923; Hickok and Davenport, 1957; Jennings, 1906; Jordan, 1913; Kriszat, 1932; Krivanek, 1956; Málek, 1927; Moore, 1923c; Moore and Kellogg, 1916; Mrázek, 1913; Nomura, 1926; Nomura and Ohfuchi, 1932; Parker and Arkin, 1901; Prosser, 1934a; Raabe, 1939; Ratner and Miller, 1959a, 1959b; Shensa and Barrows, 1932; Studnitz, 1937, 1938; Turnbull, 1876; Wells, 1955; Yerkes, 1906, 1912a, 1912b.

Bibliography

ABEL, M. 1902. Beiträge zur Kenntnis der Regenerationsvorgänge bei den limicolen Oligochäten. *Z. wiss. Zool.*, 73:1-74.

ÁBRAHÁM, A. and MINKER, E. 1958. Über die Innervation des Darmkanales des medizinischen Blutegels (*Hirudo medicinalis* L.). *Z. Zellforsch.*, 47:367-391.

ADEY, W. R. 1951. The nervous system of the earthworm *Megascolex*. *J. comp. Neurol.*, 94:57-103.

AIYAR, R. G. 1933. On the anatomy of *Marphysa gravelyi* Southern. *Rec. Indian Mus.*, 35:287-323.

AIYER, K. S. P. 1926. Notes on the aquatic Oligochaeta of Travancore. *Ann. Mag. nat. Hist.*, 18:131-142.

ALLEN, E. J. 1905. The anatomy of *Poecilochaetus*, Claparède. *Quart. J. micr. Sci.*, 48:79-151.

ALSTERBERG, G. 1924. Die Sinnesphysiologie der Tubificiden. *Acta Univ. lund.*, N.F. 35 (7):1-77.

AMASSIAN, V. E. and FLOYD, W. F. 1946. Repetitive discharge of giant nerve fibres of the earthworm. *Nature, Lond.*, 157:412-413.

AMBACHE, N., DIXON, A. ST. J., and WRIGHT, E. A. 1945. Some observations on the physiology and pharmacology of the nerve endings in the crop and gizzard of the earthworm, with special reference to the effects of cooling. *J. exp. Biol.*, 21:46-57.

AMBRONN, H. 1890. Das optische Verhalten markhaltiger und markloser Nervenfasern. *Abh. sächs. Ges. Akad. Wiss.*, 42:419-429.

AMELN, P. 1930. Der Lichtsinn von *Nereis diversicolor* O. F. Müller. *Zool. Jb. (allg. Zool.)*, 47:685-722.

ANDREWS, E. A. 1891. Compound eyes of annelids. *J. Morph.*, 5:271-299.

ANDREWS, E. A. 1892. On the eyes of polychaetous annelids. *J. Morph.*, 7:169-222.

APÁTHY, S. 1892. Contractile und leitende Primitivfibrillen. *Mitt. zool. Sta. Neapel*, 10:355-375.

APÁTHY, S. 1897. Das leitende Element des Nervensystems und seine topographischen Beziehungen zu den Zellen. *Mitt. zool. Sta. Neapel*, 12:495-748.

ARBIT, J. 1957. Diurnal cycles and learning in earthworms. *Science*, 126:654-655.

ARSHARSKY, I. A. 1938. Physiological characteristics of neuro-muscular apparatus (epidermal-muscular sac) in annelids. *Fiziol. Zh. SSSR*, 25:631-641.

ASCOLI, G. 1911a. Dell'anatomia e della minuta struttura del sistema simpatico degli irudinei. *Boll. Soc. med.-chir. Pavia*, 25:177-198.

ASCOLI, G. 1911b. Zur Neurologie der Hirudineen. *Zool. Jb., (Anat.)*, 31:473-496.

ASCOLI, G. 1913. Zur Kenntnis der neurofibrillären Apparate der Hirudineen. *Arch. mikr. Anat.*, 82:414-425.

ASHWORTH, J. H. 1902. The anatomy of *Scalibregma inflatum*, Rathke. *Quart. J. micr. Sci.*, 45:237-309.

ASHWORTH, J. H. 1904. *Arenicola*. *Liverpool Mar. Biol. Comm. Mem.*, 11:1-118.

ASHWORTH, J. H. 1909. The giant nerve cells and fibres of *Halla parthenopeia*. *Philos. Trans. (B)*, 200:427-521.

ASHWORTH, J. H. 1911. An account of *Arenicola loveni* Kinberg. *Ark. Zool.*, 7 (5):1-19.

ATHESTON, L. 1899. The epidermis of *Tubifex rivulorum* Lamarck with

Bibliography

especial reference to its nervous structures. *Anat. Anz.*, 16:497-509.

ATTEMS, C. 1903. Beiträge zur Anatomie und Histologie von *Scololepis fuliginosa* Clap. *Arb. zool. Inst. Univ. Wien*, 14:173-210.

AUTRUM, H. 1932. Die Erregbarkeit und ihre Beziehung zur Struktur der Muskelzellen bei verschiedenen Varietäten von *Hirudo medicinalis* L. *Zool. Jb., (allg. Zool.)*, 50:447-478.

BACQ, Z. M. and COPPÉE, G. 1937. Réaction des vers et des mollusques à l'ésérine existence de nerfs cholinergiques chez les vers. *Arch. int. Physiol.*, 45:310-324.

BAGLIONI, S. 1913a. Physiologie des Nervensystems. In: *Winterstein's Handb. vergl. Physiol.*, 4:22-450.

BAGLIONI, S. 1913b. Die niederen Sinne. In: *Winterstein's Handb. vergl. Physiol.*, 4:520-554.

BAILEY, P. L. 1930. Influence of the nervous system in the regeneration of *Eisenia foetida*, Savigny. *J. exp. Zool.*, 57:473-509.

BAILEY, P. L. 1939. Anterior regeneration in the earthworm, *Eisenia*, in the certain absence of central nervous tissue at the wound region. *J. exp. Zool.*, 80:287-293.

BALDWIN, E. and YUDKIN, W. H. 1948. Phosphagen in annelids (Polychaeta). *Biol. Bull., Woods Hole*, 95:273-274.

BAUDELOT, E. 1865. Observations sur la structure du système nerveux de la clepsine. *Ann. Sci. nat. (Zool.)*, (5) 3:127-136.

BEDDARD, F. E. 1895. A monograph of the order of Oligochaeta. Clarendon Press, Oxford.

BEDDARD, F. E. 1901. On a freshwater annelid of the genus *Bothrioneuron* obtained during the "Skeat Expedition" to the Malay Peninsula. *Proc. zool. Soc. Lond.*, 1901:81-87.

BENNETT, M. F. and SUTTLE, G. E. 1960. Effects of the removal of the suprapharyngeal ganglia in earthworms. *Anat. Rec.*, 137:339.

BERGQUIST, H. J. A. 1925. Beiträge zur Kenntnis der Anatomie des oberen Schlundganglions bei *Harmothoë imbricata* (L.) *Ark. Zool.*, 17 (22):1-7.

BERITOV, I. (= Beritashvili, I.) 1944a. On the catch action in the musculature of the leech. *Soobshch. akad. Nauk Gruz. SSR*, 7:723-732. (In Russian, with English summary.)

BERITOV, I. 1944b. On the origin of the catch action in the musculature of the leech. *Soobshch. Akad. Nauk Gruz. SSR*, 9:927-937. (In Russian, with English summary.)

BERITOV, I. 1945. On the excitation and the inhibition of the reflex reactions of the nerve muscle preparation of the middle segments of a leech. *Trud. Tbilissk. Univ.*, 27a:29-52. (In Russian, with English summary.)

BERITOV, I. S. 1950. On the origin of spontaneous activity in transmission of stimulation in nerve ganglia of the ventral chain of the leech. *J. gen. Biol., Moscow (Zh. obshch. Biol.)*, 11:31-38. (In Russian.)

BERITOV, I. and GOGAVA, M. 1945. The coordination of the segmental reflexes in the leech. *Trud. Tbilissk. Univ.*, 27a:1-27. (In Russian, with English summary.)

BERITOV, I. and GOGAVA, M. 1949. On the spontaneous activity in the nerve ganglia of the ventral cord of the leech. *Adv. mod. Biol. (Zh. obshch. Biol.)*, 10:421-430. (In Russian.)

BERITOV, I. S. and GOGAVA, M. 1950a. On spontaneous activity in nerve ganglia of ventral chain of the leech. *J. gen. Biol., Moscow, (Zh. obshch. Biol.)*, 11:20-30. (In Russian.)

BERITOV, I. and GOGAVA, M. 1950b. On the electrical potentials in the ventral cord of the leech. *Trudy Inst. Fiziol. Akad. Nauk Gruz. SSR*, 8:17-42. (In Russian.)

BERKELEY, E. 1927. A new genus of *Chaetopteridae* from N. E. Pacific, with some remarks on allied genera. *Proc. zool. Soc. Lond.*, 1927:441-445.

BERNERT, J. 1926. Untersuchungen über das Zentralnervensystem der *Hermione hystrix* (L.). *Z. Morph. Ökol. Tiere*, 6:743-810.

BERRILL, N. J. 1927. The control of the beat of the fan segments in *Chaetopterus variopedatus*. *Nature, Lond.*, 119:564-565.

BERRILL, N. J. 1928. Regeneration in the polychaet *Chaetopterus variopedatus*. *J. Mar. biol. Ass. U.K.*, 15:151-158.

BERRILL, N. J. 1931. Regeneration in *Sabella pavonina* (Sav.) and other sabellid worms. *J. exp. Zool.*, 58:495-523.

BETHE, A. 1903. *Allgemeine Anatomie und Physiologie des Nervensystems*. Thieme, Leipzig.

BETHE, A. 1908. Ein neuer Beweis für die leitende Funktion der Neurofibrillen, nebst Bemerkungen über Reflexzeit, Hemmungszeit und Latenzzeit des Muskels beim Blutegel. *Pflüg. Arch. ges. Physiol.*, 122:1-36.

BHARUCHA-REID, R. P. 1956. Latent learning in earthworms. *Science*, 123:222.

BHATIA, M. L. 1941. Hirudinaria. (The Indian cattle leech.) *Indian zool. Mem.*, 8:1-85.

BHATIA, M. L. 1956. Extra ocular photoreceptors in the land leech, *Haemadipsa zeylanica agilis* (Moore) from Nainital, Almora (India). *Nature, Lond.*, 178:420-421.

BIAKOWSKA, W. and KULIKOWSKA, Z. 1911. Über den Golgi-Kopschen Apparat der Nervenzellen bei den *Hirudineen* und *Lumbricus*. *Anat. Anz.*, 38:193-207.

BIEDERMANN, W. 1891. Über den Ursprung und die Endigungsweise der Nerven in den Ganglien wirbelloser Tiere. *Jena. Z. Naturw.*, 25:429-466.

BIEDERMANN, W. 1904. Studien zur vergleichenden Physiologie der peristaltischen Bewegungen. I. Die peristaltischen Bewegungen der Würmer und der Tonus glatter Muskeln. *Pflüg. Arch. ges. Physiol.*, 102:475-542.

BINARD, A. and JEENER, R. 1926. Recherches sur la morphologie du système nerveux des annélides. *Bull. Acad. Belg. Cl. Sci.*, (5) 12:437-448.

BINARD, A. and JEENER, R. 1930. Rapports morphologiques du système nerveux des mollusques avec celui des polychètes. *Rec. Inst. Zool. Torley-Rousseau Bruxelles*, 3:5-16.

BLANCHARD, E. 1848. Du système nerveux chez les invertébrés (mollusques et annelés) dans ses rapports avec la classification de ces animaux. *C. R. Acad. Sci., Paris*, 27:623-625.

BOBIN, G. and DURCHON, M. 1952. Étude histologique du cerveau de *Perenereis cultrifera* Grube (annélide polychète). Mise en évidence d'un complexe cérébro-vasculaire. *Arch. Anat. micr. Morph. exp.*, 41:25-40.

BOHN, G. 1901. Sur la locomotion des vers annelés (vers de terre et sangsues). *Bull. Mus. Hist. nat., Paris*, 7:404-411.

BOHN, G. 1902. Contribution à la psychologie des annélides. *Bull. Inst. gén. psychol.*, 2:317-325.

BOHN, G. 1903. De l'indépendance fonctionelle des zoides d'un annélide, à propos de phénomènes de rotation présentés par les hirudinées. *Bull. Mus. Hist. nat., Paris*, 9:26-30.

BOHN, G. 1904. Mouvements de manège en rapport avec les mouvements de la marée. *C. R. Soc. Biol., Paris*, 57:297-298.

BOHN, G. 1906. Attitudes et mouvements des annélides. Essai de psychophysiologie éthologique. *Ann. Sci. nat.*, (9) 3:35-144.

BOTSFORD, E. F. 1939. Temporal summation in neuromuscular responses of the earthworm, *Lumbricus terrestris*. *Biol. Bull., Woods Hole*, 77:328-329.

BOTSFORD, E. F. 1941. The effect of physostigmine on the responses of earthworm body wall preparations to successive stimuli. *Biol. Bull., Woods Hole*, 80:299-313.

BOULE, L. 1908a. L'imprégnation des éléments nerveux du lombric par le nitrate d'argent. *Névraxe*, 9:313-327.

BOULE, L. 1908b. Recherches sur le système nerveux central normal du lombric. *Névraxe*, 10:15-59.

BOULE, L. 1913. Nouvelles recherches sur le système nerveux central normal du lombric. *Névraxe*, 14-15:429-467.

BOVARD, J. F. 1904. The distribution of the sense organs in *Microscolex elegans*. *Univ. Calif. Publ. Zool.*, 1:269-282.

BOVARD, J. F. 1915. Giant fiber action and normal transmission by the nerve cord of earthworms. *Science*, 42:620.

BOVARD, J. F. 1918a. The transmission of nervous impulses in relation to locomotion in the earthworm. *Univ. Calif. Publ. Zool.*, 18:103-134.

BOVARD, J. F. 1918b. The function of the giant fibers in earthworms. *Univ. Calif. Publ. Zool.*, 18:135-144.

BRADWAY, W. E. and MOORE, A. R. 1940. The locus of the action of the galvanic current in the earthworm, *Lumbricus terrestris*. *J. cell. comp. Physiol.*, 15:47-54.

BRAND, H. 1933. Die lokomotorischen Reaktionen von *Nereis diversicolor* auf Licht und Dunkelheit und der Einfluss von Eingriffen an Receptoren, Effectoren und Zentralnervensystem. *Z. wiss. Zool.*, 144:363-401.

BRISTOL, C. L. 1898. The metamerism of *Nephelis*. A contribution to the morphology of the nervous system, with a description of *Nephelis lateralis*. *J. Morph.*, 15:17-72.

BRODE, H. S. 1898. A contribution to the morphology of *Dero vaga*. *J. Morph.*, 14:141-180.

BRUCH, C. 1849. Über das Nervensystem des Blutegels. *Z. wiss. Zool.*, 1:164-174.

BRUNOTTE, C. 1888. Recherches anatomiques sur une espèce du genre *Branchiomma*. *Trav. Inst. Zool. Univ. Montpellier*.

BUDDENBROCK, W. VON. 1913. Über die Funktion der Statocysten im Sande grabender Meerestiere. *Zool. Jb. (allg. Zool.)*, 33:441-482.

BUDDENBROCK, W. VON and STUDNITZ, G. VON. 1936. *Vergleichend-Physiologisches Praktikum, mit besonderer Berücksichtigung der niederen Tiere*. Springer, Berlin.

BUDINGTON, R. A. 1902. Some physiological characteristics of annelid muscle. *Amer. J. Physiol.*, 7:155-179.

BULLOCK, T. H. 1945a. Functional organization of the giant fiber system of *Lumbricus*. *J. Neurophysiol.*, 8:55-71.

BULLOCK, T. H. 1945b. Organization of the giant nerve fiber system in *Neanthes virens*. *Biol. Bull.*, Woods Hole, 89:185-186.

BULLOCK, T. H. 1948. Physiological mapping of giant nerve fiber systems in polychaete annelids. *Physiol. comp.*, 1:1-14.

BULLOCK, T. H. 1950. The course of fatigue in single nerve fibers and junctions. *Int. physiol. Congr.*, 18:134-135.

BULLOCK, T. H. 1951. Facilitation of conduction rate in nerve fibers. *J. Physiol.*, 114:89-97.

BULLOCK, T. H. 1953a. A contribution from the study of cords in lower forms. In: *The Spinal Cord*. CIBA Foundation Symposium, Churchill, London.

BULLOCK, T. H. 1953b. Properties of some natural and quasi-artificial synapses in polychaetes. *J. comp. Neurol.*, 98:37-68.

BULLOCK, T. H. 1956. Temperature sensitivity of some unit synapses. *Pubbl. Staz. zool. Napoli.* 28:305-314.

BULLOCK, T. H. and TURNER, R. S. 1950. Events associated with conduction failure in nerve fibers. *J. cell. comp. Physiol.*, 36:59-81.

BULLOCK, T. H., COHEN, M. J., and FAULSTICK, D. 1950. Effect of stretch on conduction in single nerve fibers. *Biol. Bull.*, Woods Hole, 99:320.

CAJAL, S. R. Y. 1903a. Neuroglia y neurofibrillas del *Lumbricus*. *Trab. Lab. Invest. biol. Univ. Madr.*, 3:277-286.

CAJAL, S. R. Y. 1903b. Sobre la existencia de un aparato tubuliforme en el protoplasma de las celulas nerviosas y epiteliales de la lombriz de tierra. *Bol. Soc. esp. Hist. nat.*, 3:395-398.

CAJAL, S. R. Y. 1904. Variaciones morfológicas del reticulo nervioso de invertebrados y vertebrados sometidos á la acción de condiciones naturales (nota preventina.). *Trab. Lab. Invest. biol. Univ. Madr.*, 3:287-297.

CARLSON, A. J. 1905. Further evidence of the fluidity of the conducting substance in nerve. *Amer. J. Physiol.*, 13:351-357.

CARLSON, A. J. 1908. Comparative physiology of the invertebrate heart —X. A note on the physiology of the pulsating blood vessels in the worms. *Amer. J. Physiol.*, 22:353-356.

CASTLE, W. E. 1900. The metamerism of the Hirudinea. *Proc. Amer. Acad. Arts Sci.*, 35:285-303.

CATE, J. TEN. 1931. Physiologie der Gangliensysteme der Wirbellosen. *Ergebn. Physiol.*, 33:137-336.

CATE, J. TEN. 1938. Sur la fonction des neurocordes de la chaîne ventrale du ver de terre (*Lambricus* (sic) *terrestris*). *Arch. néerl. Physiol.*, 23:136-140.

CAULLERY, M. 1914. Sur les formes larvaires des annélides de la famille des Sabellariens (Hermelliens). *Bull. Soc. zool. Fr.*, 39:168-176.

CAULLERY, M. 1916. Sur les Térébelliens de la sous-famille Polycirridae Malmgr. *Bull. Soc. zool. Fr.*, 40:239-248.

CAULLERY, M. and MESNIL, F. 1897. Étude sur la morphologie comparée et la phylogénie des espèces chez les spirorbes. *Bull. sci. Fr. Belg.*, 30:185-233.

CERFONTAINE, P. 1890. Recherches sur le système cutané et sur le système musculaire du lombric terrestre (*Lumbricus agricola* Hoffmeister.) *Arch. Biol.*, Paris, 10:327-428.

CERFONTAINE, P. 1892. Contribution à l'étude du système nerveux central du lombric terrestre. *Bull. Acad. Belg. Cl. Sci.*, (3), 23:742-752.

CHAPMAN, G. 1950. On the movement of worms. *J. exp. Biol.*, 27:29-39.

CHEN, T.-T. 1944. The morphology of the anterior autonomic nervous system of the earthworm, *Lumbricus terrestris* L. *J. comp. Neurol.*, 80:191-209.

CLAPARÈDE, E. 1861. Études anatomiques sur les annélides, turbellariés, opalines et grégarines, observés dans les Hébrides. *Mem. Soc. Phys. Genève*, 16:73-164.

CLAPARÈDE, E. 1867. De la structure des annélides, note comprenant un examen critique des travaux les plus récents sur cette classe de vers. *Arch. Sci. phys. nat.*, 30:5-44.

CLAPARÈDE, E. 1869. Histologische Untersuchungen über den Regenwurm (*Lumbricus terrestris* Linné). *Z. wiss. Zool.*, 19:563-624.

CLAPARÈDE, E. 1873. Recherches sur la structure des annélides sédentaires. *Mem. Soc. Phys. Genève*, 22:1-200.

CLARK, R. B. 1955. The posterior lobes of the brain of *Nephtys* and the mucus-glands of the prostomium. *Quart. J. micr. Sci.*, 96:545-565.

CLARK, R. B. 1956. The eyes and photonegative behaviour of *Nephtys* (Annelida, Polychaeta). *J. exp. Biol.*, 33:461-477.

CLARK, R. B. 1957. The influence of size on the structure of the brain of *Nephtys*. *Zool. Jb. (allg. Zool.)*, 67:261-282.

CLARK, R. B. 1958a. The micromorphology of the supra-oesophageal ganglion of *Nephtys*. *Zool. Jb. (allg. Zool.)*, 68:261-296.

CLARK, R. B. 1958b. The gross morphology of the anterior nervous system of *Nephtys*. *Quart. J. micr. Sci.*, 99:205-220.

CLARK, R. B. 1958c. The posterior lobes of *Nephtys*: observations on three New England species. *Quart J. micr. Sci.*, 99:505-510.

CLARK, R. B. 1960a. Habituation of the polychaete *Nereis* to sudden stimuli. I. General properties of the habituation process. *Anim. Behav.*, 8:83-91.

CLARK, R. B. 1960b. Habituation of the polychaete *Nereis* to sudden stimuli. 2. Biological significance of habituation. *Anim. Behav.*, 8:92-103.

CLARK, R. B. and BONNEY, D. G. 1960. Influence of the supra-eosophageal ganglion on posterior regeneration in *Nereis diversicolor*. *J. Embryol. exp. Morph.*, 8:112-118.

CLARK, R. B., and CLARK, M. E. 1959. Role of the supra-oesophageal ganglion during the early stages of caudal regeneration in some errant polychaetes. *Nature, Lond.*, 183:1834-1835.

CLARK, R. B. and EVANS, S. M. 1961. The effect of delayed brain extirpation and replacement on caudal regeneration in *Nereis diversicolor*. *J. Embryol. exp. Morph.*, 9:97-105.

CLARKE, J. L. 1856-1857. On the ner-

Bibliography

vous system of *Lumbricus terrestris*. *Proc. roy. Soc.*, 8:343-351.

CLAYTON, D. E. 1932. A comparative study of the non-nervous elements in the nervous system of invertebrates. *J. Ent. Zool.*, 27:3-22.

COLLIER, H. O. J. 1938. The immobilization of locomotory movements in the earthworm, *Lumbricus terrestris*. *J. exp. Biol.*, 15:339-357.

COLLIER, H. O. J. 1939a. Central nervous activity in the earthworm. I. Responses to tension and to tactile stimulation. *J. exp. Biol.*, 16:286-299.

COLLIER, H. O. J. 1939b. Central nervous activity in the earthworm. II. Properties of the tension reflex. *J. exp. Biol.*, 16:300-312.

COONFIELD, B. R. 1932. The peripheral nervous system of earthworms. *J. comp. Neurol.*, 55:7-18.

COPELAND, M. 1926. An apparent conditioned reflex in *Nereis virens*. *Anat. Rec.*, 34:123.

COPELAND, M. 1930. An apparent conditioned response in *Nereis virens*. *J. comp. Psychol.*, 10:339-354.

COPELAND, M. and BROWN, F. A., JR. 1934. Modification of behaviour in *Nereis virens*. *Biol. Bull., Woods Hole*, 67:356-364.

COUTEAUX, R. 1933. Participation de la musculature des dissépiments à certains phénomènes réflexes chez les lombriciens. *C. R. Soc. Biol.*, Paris, 113:1480-1481.

COUTEAUX, R. 1956. Neurofilaments et neurofibrilles dans les fibres nerveuses de la sangsue. In: *Electron Microscopy*, Proc. Stockholm Conf., Sept. 1956. F. S. Sjöstrand and J. Rhodin (eds.). Almqvist and Wiksell, Stockholm.

CUNNINGHAM, J. T. 1888. On some points in the anatomy of Polychaeta. *Quart. J. micr. Sci.*, 28:239-278.

DAHL, E. 1955. On the morphology and affinities of the annelid genus *Sternaspis*. (Reports of the Lund U. Chile Expedition 1948-1949). *Acta Univ. Lund., Avd. 2*, N.F. 51(13):1-21.

DALES, R. P. 1955. Feeding and digestion in terebellid polychaetes. *J. Mar. biol. Ass. U. K.*, 34:55-79.

DARBOUX, J. G. 1899. Recherches sur les aphroditiens. *Trav. Inst. Zool. Univ. Montpellier*, (2)6:276 p.

DARWIN, C. R. 1881. *The Formation of Vegetable Mould through the Action of Worms with Observations on their Habits*. Murray, London.

DAWSON, A. B. 1920. The intermuscular nerve cells of the earthworm. *J. comp. Neurol.*, 32:155-171.

DECHANT, E. 1906. Beitrag zur Kenntnis des peripheren Nervensystems des Regenwurmes. *Arb. zool. Inst. Univ. Wien*, 16:361-382.

DEHORNE, A. 1935a. Sur le neuroplasme des fibres géantes des polychètes. *C.R. Soc. Biol.*, Paris, 119:1253-1256.

DEHORNE, A. 1935b. Sur le trophosponge des cellules nerveuses géantes de *Lanice conchylega*, Pallas. *C.R. Soc. Biol.*, Paris, 120:1188-1190.

DEHORNE, A. 1936. Analyse de quelques aspects des cellules nerveuses de *Nephthys* et de *Nereis*. *C.R. Soc. Biol.*, Paris, 121:757-760.

DEHORNE, L. 1916. Les naïdimorphes et leur reproduction asexuée. *Arch. Zool. exp. gén.*, 56:25-157.

DEMOLL, R. 1909. Die Augen von *Alciopa cantrainii*. *Zool. Jb. (Anat.)*, 27:651-686.

DITTMAR, H. A. 1954. Electrophysiologische Untersuchungen am Zentralnervensystem des Blutegels (*Hirudo medicinalis*) und des Pferdeegels (*Haemopis sanguisuga*). *Z. vergl. Physiol.* 36:41-54.

DIXON, G. C. 1915. *Tubifex*. *Liverpool Mar. Biol. Comm. Mem.*, 23:303-342.

DORNER, H. 1865. Ueber die Gattung *Branchiobdella* Odier. *Z. wiss. Zool.*, 15:464-493.

ECCLES, J. C., GRANIT, R., and YOUNG, J. Z. 1932. Impulses in the giant nerve fibres of earthworms. *J. Physiol.*, 77:23P-25P.

EDWARDS, G. A. 1957. Electron microscope observations of annelid muscle and nerve. *Anat. Rec.*, 128:542-543.

EHLERS, E. H. 1864-1868. *Die Borstenwürmer (Annelida Chaetopoda) nach Systematischen und Anatomischen Untersuchungen Dargestellt*. Engelmann, Leipzig.

EHLERS, E. 1892. Die Gehörorgane der Arenicolen. *Z. wiss. Zool.*, 53 (Suppl.):217-285.

EISIG, H. 1879. Die Seitenorgane und becherförmigen Organe der Capitelliden. *Mitt. zool. Sta. Neapel*, 1:278-343.

EISIG, H. 1887. Monographie der Capitelliden des Golfes von Neapel und der angrenzenden Meeres-Abschnitte nebst Untersuchungen zur vergleichenden Anatomie und Physiologie. *Fauna u. Flora Neapel*. 16:1-960.

EISIG, H. 1899. Zur Entwicklungsgeschichte der Capitelliden. *Mitt. zool. Sta. Neapel*, 13:1-292.

EISIG, H. 1906. *Ichthyotomus sanguinarius*, eine auf Aalen schmarotzende Annelide. *Fauna u. Flora Neapel*, 28:1-300.

EISIG, H. 1914. Zur Systematik, Anatomie und Morphologie der Ariciiden nebst Beiträgen zur generellen Systematik. *Mitt. zool. Sta. Neapel*, 21:153-600.

EMERY, C. 1886. La régéneration des segments postérieurs du corps chez quelques Annélides polychètes. *Arch. ital. Biol.*, 7:395-403.

ESSEN, J. VAN. 1931. Über das Kriechen der Regenwürmer. *Z. vergl. Physiol.*, 15:389-411.

EVENKAMP, H. 1931. Morphologie, Histologie und Biologie der Sabellidenspecies *Laonome kroyeri* Malmgr. und *Euchone papillosa* M. Sars. *Zool. Jb. (Anat.)*, 53:405-534.

FAIVRE, E. 1856. Études sur l'histologie comparée du système nerveux de quelques annélides. *Ann. Sci. nat.*, (4)5:337-374 and 6:16-82.

FAIVRE, E. 1857. *Études sur l'histologie comparée du système nerveux chez quelques animaux inférieurs*. Baillière, Paris.

FAULKNER, G. H. 1932. The histology of posterior regeneration in the polychaete *Chaetopterus variopedatus*. *J. Morph.*, 53:23-58.

FAUVEL, P. 1897. Recherches sur les amphareties, annélides polychètes sédentaires, morphologie, anatomie, histologie, physiologie. *Bull. sci. Fr. Belg.*, 30:277-489.

FAUVEL, P. 1907. Recherches sur les otocystes des annélides polychètes. *Ann. Sci. nat. (Zool.)*, (9)6:1-149.

FEDOROW, B. 1927. Über den Bau der Riesenganglienzellen der Lumbriconereinen. *Z. mikr-anat. Forsch.*, 12:347-370.

FEDOTOV, D. 1914. Die Anatomie von *Protomyzostomum polynephris* Fedotov. *Z. wiss. Zool.*, 109:631-696.

FERRIS, G. F. 1953. On the comparative morphology of the Annulata. A summing up. *Microentomology*, 18:2-15.

FISCHER, W. 1884. Anatomisch-histologische Untersuchung von *Capitella capitata*. Inaug. Diss., Marburg.

FLING, H. R. 1898. A contribution to the nervous system of the earthworm. *J. comp. Neurol.*, 8:230-232.

FOCKE, F. 1930. Experimente und Beobachtungen über die Biologie des Regenwurms, unter besonderer Berücksichtigung der Frage nach der Raumorientierung. *Z. wiss. Zool.*, 136:376-421.

FOETTINGER, A. 1884. Recherches sur l'organisation de *Histriobdella homari*. *Arch. Biol.*, Paris, 5:435-516.

FORDHAM, M. G. C. 1925. *Aphrodite aculeata*. *Proc. Lpool. biol. Soc.*, 40:121-216. (Liverpool Mar. Biol. Comm. Mem. 27.)

FORTUYN, A. D. 1920. *Die Leitungsbahnen im Nervensystem der wirbellosen Tiere*. Bohn, Haarlem.

FOXON, G. E. H. 1936. Observations on the locomotion of some arthropods and annelids. *Ann. Mag. nat. Hist.* (10), 18:403-419.

FRAIPONT, J. 1884. Recherches sur le système nerveux central et périphérique des archiannélides. *Arch. Biol.*, Paris, 5:243-304.

FRAIPONT, J. 1887. Le genre *Polygordius*. *Fauna u. Flora Neapel*. 14:1-125.

FRANÇOIS, P. 1885. *Contribution à l'étude du système nerveux central des hirudinées*. Poitiers.

FRIEDLÄNDER, B. 1888a. Über das Kriechen der Regenwürmer. *Biol. Zbl.*, 8:363-366.

FRIEDLÄNDER, B. 1888b. Beiträge zur Kenntnis des Centralnervensystems von *Lumbricus*. *Z. wiss. Zool.*, 47:47-84.

FRIEDLAENDER, B. 1889. Über die markhaltigen Nervenfasern und Neurochorde der Crustaceen und Anneliden. *Mitt. zool. Sta. Neapel.*, 9:205-265.

FRIEDLAENDER, B. 1894. Altes und Neues zur Histologie des Bauchstranges des Regenwurms. *Z. wiss. Zool.*, 58:661-693.

FRIEDLÄNDER, B. 1895a. Beiträge zur Physiologie des Zentralnervensystems und des Bewegungsmechanismus der Regenwürmer. *Pflüg. Arch. ges. Physiol.*, 58:168-207.

FRIEDLAENDER, B. 1895b. Über die Regeneration herausgeschnittener Theile des Centralnervensystems von Regenwürmern. *Z. wiss. Zool.*, 60:249-283.

FRISCH, K. VON. 1926. Vergleichende Physiologie des Geruchs- und Geschmackssinnes. In. *Bethe's Handb. norm. path. Physiol.*, 11:203-239. Springer, Berlin.

FÜRST, M. 1890. Zur Physiologie der glatten Muskeln. *Pflüg. Arch. ges. Physiol.*, 79:379-399.

GALVAGNI, E. 1905. Histologie des Genus *Ctenodrilus* Clap. *Arb. zool. Inst. Univ. Wien*, 15:47-80.

GAMBLE, F. W. and ASHWORTH, J. H. 1899. The habits and structure of *Arenicola marina*. *Quart. J. micr. Sci.*, 41:1-42.

GAMBLE, F. W. and ASHWORTH, J. H. 1900. The anatomy and classification of the Arenicolidae, with some observations on their post-larval stages. *Quart. J. micr. Sci.*, 43:419-569.

GARMAN, H. 1888. On the anatomy and histology of a new earthworm (*Diplocardia communis* gen. et. sp. nov.). *Bull. Ill. Lab. nat. Hist.*, 3:47-77.

GARREY, W. E. and MOORE, A. R. 1915. Peristalsis and coördination in the earthworm. *Amer. J. Physiol.*, 39:139-148.

GASKELL, J. F. 1914. VI. The chromaffine system of annelids and the relation of this system to the contractile vascular system in the leech *Hirudo medicinalis*. A contribution to the comparative physiology of the contractile vascular system and its regulators, the adrenalin secreting system and the sympathetic nervous system. *Philos. Trans. (B)*, 205:153-211.

GASKELL, J. F. 1919. Adrenalin in annelids. A contribution to the comparative study of the origin of the sympathetic and the adrenalin-secreting systems and of vascular muscles which they regulate. *J. gen. Physiol.*, 2:73-85.

GEE, W. 1913. The behaviour of leeches with especial reference to its modifiability. *Univ. Calif. Publ. Zool.*, 11:197-305.

GEMELLI, A. 1905a. Su di una fine particolarità di struttura delle cellule nervose dei vermi. *Riv. Fis. Mat. Sci. nat.*, 6:518-532.

GEMELLI, A. 1905b. Sopra le neurofibrille delle cellule nervose dei vermi secondo un nuovo metodo di dimostrazione. *Anat. Anz.*, 27:449-462.

GILPIN-BROWN, J. B. 1958. The development and structure of the cephalic nerves in *Nereis*. *J. comp. Neurol.*, 109:317-348.

GOGAVA, M. 1944. On the mechanical effect of the longitudinal muscle of the leech. *Soobshch. Akad. Nauk Gruz. SSR*, 7:711-721. (In Russian, with English summary.)

GOLDFARB, A. J. 1914. Regeneration in the annelid worm *Amphinoma pacifica*, after removal of the central nervous system. *Pap. Tortugas Lab.*, 6:95-102. (*Publ. Carneg. Inst.* 183.)

GOODRICH, E. S. 1897. On the relation of the arthropod head to the annelid prostomium. *Quart. J. micr. Sci.*, 40:247-268.

GOODRICH, E. S. 1912. *Nerilla* an archiannelid. *Quart. J. micr. Sci.*, 57:397-425.

GÖTHLIN, G. F. 1913. Die doppelbrechenden Eigenschaften des Nervengewebes, ihre Ursachen und ihre biologischen Konsequenzen. *K. svenska VetenskAkad. Handl.*, 51:1-92.

GRAFF, L. 1877. *Das Genus Myzostoma*. Leuckart, Leipzig.

GRAFF, L. VON. 1887. Die Annelidengattung *Spinther*. *Z. wiss. Zool.*, 46:1-66.

GRAUBER, V. 1880. Morphologische Untersuchungen über die Augen der freilebenden marinen Borstenwürmer. *Arch. mikr. Anat.*, 17:243-323.

GRAVIER, C. 1896. Recherches sur les phyllodociens. *Bull. sci. Fr. Belg.*, 29:293-389.

GRAVIER, C. 1898a. Étude du prostomium des glycériens suivie de considérations générales sur le prostomium des annélides polychètes. *Bull. sci. Fr. Belg.*, 31:159-184.

GRAVIER, C. 1898b. Contribution à l'étude de la trompe des glycériens. *Bull. sci. Fr. Belg.*, 31:421-448.

GRAY, J. 1939. Studies in animal locomotion. VIII. The kinetics of locomotion of *Nereis diversicolor*. *J. exp. Biol.*, 16:9-17.

GRAY, J. and LISSMANN, H. W. 1938a. Studies in animal locomotion. VII Locomotory reflexes in the earthworm. *J. exp. Biol.*, 15:506-517.

GRAY, J. and LISSMANN, H. W. 1938b. An apparatus for measuring the propulsive forces of the locomotory muscles of the earthworm and other animals. *J. exp. Biol.*, 15:518-521.

GRAY, J., LISSMANN, H. W., and PUMPHREY, R. J. 1938. The mechanism of locomotion in the leech (*Hirudo medicinalis* Ray). *J. exp. Biol.*, 15:408-430.

GREEF, R. 1877. Untersuchungen über die Alciopiden. *Nova Acta Leop. Carol.*, 39:35-132.

GREEF, R. 1879. Ueber pelagische Anneliden von der Küste der canarischen Inseln. *Z. wiss. Zool.*, 32:237-283.

GROSS, F., and HUXLEY, J. S. 1935. Regeneration and reorganization in *Sabella*. *Arch. EntwMech. Org.*, 133:582-620.

GRUITHUISEN, F. V. P. 1828. Über die *Nais diaphana* und *Nais diastropha* mit dem Nerven- und Blutsystem derselben. *Nova Acta Leop. Carol.*, 14:407-420.

GÜNTHER, K. 1912. Beiträge zur Systematik der Gattung *Flabelligera* und Studien über den Bau von *Flabelligera* (Siphonostoma) *diplochaitus* Otto. Jena. *Z. Naturw.*, 48:93-186.

GUSTAFSON, G. 1930. Anatomische Studien über die Polychäten-Familien Amphinomidae und Euphrosynidae. *Zool. Bidr. Uppsala*, 12:305-471.

HAASE, H. 1898. Über Regenerationsvorgänge bei *Tubifex rivulorum* Lam. mit besonderer Berücksichtigung des Darmkanals und Nervensystems. *Z. wiss. Zool.*, 65:211-256.

HACHFELD, G. 1926. Beiträge zur Kenntnis der *Tomopteris catherina* Gosse. *Z. wiss. Zool.*, 128:133-181.

HÄCKEL, E. H. P. A. 1866. *Generelle Morphologie der Organismen*. Reimer, Berlin.

HÄCKER, V. 1894. Die spätere Entwicklung der *Polynoë*-Larve. *Zool. Jb. (Anat.)*, 8:245-288.

HAFFNER, K. VON. 1959a. Untersuchungen und Gedanken über das Gehirn und Kopfende von Polychaeten und ein Vergleich mit den Arthropoden. *Mitt. hamburg. zool. Mus.*, 57:1-29.

HAFFNER, K. VON. 1959b. Über den Bau und den Zusammenhang der wichtigsten Organe des Kopfendes von *Hyalinoecia tubicola* Malmgren (Polychaeta, Eunicidae. Onuphidinae), mit Berücksichtigung der Gattung *Eunice*. *Zool. Jb. (Anat.)*, 77:133-192.

HAGADORN, I. R. 1958. Neurosecretion and the brain of the rhynchobdellid leech, *Theromyzon rude* (Baird, 1869). *J. Morph.*, 102:55-90.

HAGIWARA, S. and MORITA, H. 1962. Electrotonic transmission between two nerve cells in leech ganglion. *J. Neurophysiol*, 25:721-731.

HALBICH, F. 1940. Die Wirkung des konstanten galvanischen Stromes auf das Zentralnervensystem des Regenwurmes. *Z. vergl. Physiol.*, 27:606-614.

HALL, A. R. 1921. Regeneration in the annelid nerve cord. *J. comp. Neurol.*, 33:163-191.

Bibliography

Haller, B. 1887. Über die sogenannte Leydig'sche Punktsubstanz im Centralnervensystem. *Morph. Jb.*, 12:325-332.

Haller, B. 1889. Beiträge zur Kenntnis der Textur des Central-Nervensystems höherer Würmer. *Arb. zool. Inst. Univ. Wien*, 8:175-312.

Haller, B. 1910. Über das Bauchmark. *Jena. Z. Naturw.*, 46:591-632.

Hama, K. 1959. Some observations on the fine structure of the giant nerve fibers of the earthworm, *Eisenia foetida*. *J. biophys. biochem. Cytol.*, 6:61-66.

Hama, K. 1961. The fine structure of some electrical synapses. *Sci. Living Body*, 12:72-84. (In Japanese.)

Hamaker, J.I. 1898. The nervous system of *Nereis virens* Sars. A study in comparative neurology. *Bull. Mus. comp. Zool. Harv.*, 32:89-124.

Hamilton, W. F. 1917. The nervous system of *Aracoda semimaculata* and the description of a method of stereographic reconstruction. *J. Ent. Zool.*, 9:73-84.

Hanel, E. 1904. Ein Beitrag zur "Psychologie" der Regenwürmer. *Z. allg. Physiol.*, 4:244-258.

Hanke, R. 1948. The nervous system and the segmentation of the head in the Hirudinea. *Microentomology*, 13:57-64.

Hansen, A. 1881. Sur la terminaison des nerfs dans les muscles volontaires de la sangsue. *Arch. Biol., Paris*, 2:342-344.

Hanson, J. 1949. Observations on the branchial crown of the Serpulidae (Annelida, Polychaeta). *Quart. J. micr. Sci.*, 90:221-233.

Hanson, J. 1951. The blood-system in the serpulimorpha (Annelida, Polychaeta). III. Histology. *Quart. J. micr. Sci.*, 92:255-274.

Hanström, B. 1926. Eine genetische Studie über die Augen und Sehzentren von Turbellarien, Anneliden und Arthropoden. *K. svenska VetenskAkad. Handl.* (3) 4:1-176.

Hanström, B. 1927. Das zentrale und periphere Nervensystem des Kopflappens einiger Polychäten. *Z. Morph. Ökol. Tiere*, 7:543-596.

Hanström, B. 1928a. Die Beziehungen zwischen dem Gehirn der Polychäten und dem der Arthropoden. *Z. Morph. Ökol. Tiere*, 11:152-160.

Hanström, B. 1928b. *Vergleichende Anatomie des Nervensystems der wirbellosen Tiere unter Berücksichtigung seiner Funktion*. Springer, Berlin.

Hanström, B. 1929. Weitere Beiträge zur Kenntnis des Gehirns und der Sinnesorgane der Polychäten *(Polygordius, Tomopteris, Scolecolepis)*. *Z. Morph. Ökol. Tiere*, 13:329-358.

Hargitt, C. W. 1906. Experiments on the behavior of tubicolous annelids. *J. exp. Zool.*, 3:295-320.

Hargitt, C. W. 1909. Further observations on the behavior of tubicolous annelids. *J. exp. Zool.*, 7:157-187.

Hargitt, C. W. 1912. Observations on the behavior of tubicolous annelids. *Biol. Bull., Woods Hole*, 22:67-94.

Harper, E. H. 1905. Reactions to light and mechanical stimuli in the earthworm *Perichaeta bermudensis* (Beddord). *Biol. Bull., Woods Hole*, 10:17-34.

Harper, E. H. 1908. Behavior of *Perichaeta* and *Lumbricus* toward stimuli of various intensities. *Science*, 27:911.

Harper, E. H. 1909. Tropic and shock reactions in *Perichaeta* and *Lumbricus*. *J. comp. Neurol.*, 19:569-587.

Hart, P. C. 1923. Enregistrement de la reptation des vers. *Arch. néerl. Physiol.*, 8:202-214.

Hart, P. C. 1924. L'action des ions Na, K, et Ca et du nitrate d'uranyle sur les mouvements rythmiques spontanés du sac musculo-cutané du lombric. *Arch. néerl. Physiol.*, 9:1-29.

Hartmann-Schröder, G. 1958. Zur Morphologie der Opheliiden *(Polychaeta sedentaria)*. *Z. wiss. Zool.*, 161:84-143.

Haswell, W. A. 1882. On the structure and functions of the elytra of aphroditacean annelids. *Ann. Mag. nat. Hist.*, (5) 10:238-242.

Haswell, W. A. 1914. Notes on the Histriobdellidae. *Quart. J. micr. Sci.*, (2) 59:197-226.

Hatai, S. 1922. Contributions to the physiology of earthworms. I. The effects of heat on rhythmic contractions in several species of Oligochaeta. *Jap. J. Zool.*, 1:1-21.

Hatschek, B. 1878. Studien über Entwicklungsgeschichte der Anneliden. Ein Beitrag zur Morphologie der Bilaterien. *Arb. zool. Inst. Univ. Wien*, 1:277-404.

Hatschek, B. 1891. *Lehrbuch der Zoologie*. Jena.

Havet, J. 1900. Structure du système nerveux des annélides *(Nephelis, Clepsine, Hirudo, Lumbriculus, Lumbricus)*. *Cellule*, 17:63-136.

Havet, J. 1916a. Relations de la névroglie avec l'appareil vasculaire chez les invertébrés. *C. R. Acad. Sci., Paris*, 162:568-570.

Havet, J. 1916b. Contribution à l'étude de la névroglie des invertébrés. *Trab. Lab. Invest. biol. Univ. Madr.*, 14:35-85.

Heck, L. 1920. Über die Bildung einer Assoziation beim Regenwurm auf Grund von Dressurversuchen. *Lotos*, 67:168-189.

Heider, K. 1914. Phylogenie der Wirbellosen. In *Die Kultur der Gegenwart*, Teil 3, Abt. 4. Bd. 4:453-529.

Heider, K. 1923. Ueber das Nervensystem der Eunicidae. *S.B. preuss. Akad. Wiss. (Phys. math. Kl.)*, 1923:298.

Heider, K. 1924. Das Nervensystem der Polychaeten und über seine Bedeutung für die Systematik dieser Gruppe. *S.B. preuss. Akad. Wiss. (Phys. math. Kl.)*, 1924:235.

Heider, K. 1925. Über *Eunice*. Systematisches, Kiefersack, Nervensystem. *Z. wiss. Zool.*, 125:55-90.

Hempelmann, F. 1906. Zur Morphologie von *Polygordius lacteus* Schn. und *Polygordius triestinus* Woltereck, nov. spec. *Z. wiss. Zool.*, 84:527-618.

Hempelmann, F. 1911. Zur Naturgeschichte von *Nereis dumerilii* Aud. et Edw. *Zoologica, Stuttgart*, 25 (62):1-135.

Hempelmann, F. 1934. Polychaeta. *Kükenthal's Handb. Zool.*, 2 (7):1-212.

Hempelmann, F. and Woltereck, R. 1912. Annelidae. In: *Handwörterb. d. Naturwissenschaften*, 1:427-457, hrsg. von E. Korschelt. Fischer, Jena.

Henry, L. M. 1947. The nervous system and the segmentation of the head in the Annulata. I. Review of the problem. II. Oligochaeta. *Microentomology*, 12:65-82. III. Polychaeta. *Microentomology*, 12:83-110. IV. Arthropoda. *Microentomology*, 13:1-26, (1948) V. Onychophora. *Microentomology*, 13:27-48 (1948).

Hermann, E. 1875. *Das Central-Nervensystem von Hirudo medicinalis*. Stahl, München.

Herpin, R. 1923. Éthologie et développement de *Nereis (Neanthes) caudata*. *C.R. Acad. Sci., Paris*, 177:542-544.

Herrick, C. J. 1924. *Neurological Foundations of Animal Behavior*. Holt, New York.

Herter, K. 1926. Vergleichende Physiologie der Tangoreceptoren bei Tieren. In: *Bethe's Handb. norm. path. Physiol.*, 11:68-83.

Herter, K. 1929. Vergleichende Bewegungsphysiologische Studien an deutschen Egeln. *Z. vergl. Physiol.*, 9:145-177.

Herter, K. 1939. Die Physiologie der Hirudineen. In: *Bronn's Klassen*, 4:3:4:2:123-320.

Herter, K., Schleip, W., and Autrum, H. 1936-1939. Hirudineae. In: *Bronns' Klassen*, 4:3:4:2:1-662.

Hescheler, K. 1898. Ueber Regenerationsvorgänge bei Lumbriciden. *Jena. Z. Naturw.*, 31:521-604.

Hess, C. 1913. Gesichtssinn. In: *Winterstein's Handb. vergl. Physiol.*, 4:555-840. Fischer, Jena.

Hess, C. 1918. Die Akkomodation der Alciopiden, nebst Beiträgen zur Morphologie des Alciopidenauges. *Pflüg. Arch. ges. Physiol.*, 172:449-465.

Hess, W. N. 1922. Reactions to light and photoreceptors of *Lumbricus terrestris*. *Proc. Ind. Acad. Sci.*, 1922:223-224.

Hess, W. N. 1923a. Sense-organs of *Lumbricus terrestris*. *Anat. Rec.*, 26:369.

Hess, W. N. 1923b. Photoreceptors of

Lumbricus terrestris. Anat. Rec., 24:390.

HESS, W. N. 1924a. Reactions to light in the earthworm, *Lumbricus terrestris* L. *J. Morph.,* 39:515-542.

HESS, W. N. 1924b. Photoreceptors of the earthworm, *Lumbricus terrestris,* with special reference to their distribution, function and structure. *Anat. Rec.,* 29:95.

HESS, W. N. 1925a. The nerve plexus of the earthworm *Lumbricus terrestris. Anat. Rec.,* 31:335-336.

HESS, W. N. 1925b. Nervous system of the earthworm, *Lumbricus terrestris* L. *J. Morph.,* 40:235-259.

HESS, W. N. 1925c. Photoreceptors of *Lumbricus terrestris* with special reference to their distribution, structure, and function. *J. Morph,.* 41:63-95.

HESSE, R. 1894. Zur vergleichenden Anatomie der Oligochaeten. *Z. wiss. Zool.,* 58:394-439.

HESSE, R. 1896. Untersuchungen über die Organe der Lichtempfindung bei niederen Tieren. I. Die Organe der Lichtempfindung bei den Lumbriciden. *Z. wiss. Zool.,* 61:393-419.

HESSE, R. 1897. Untersuchungen über die Organe der Lichtempfindung bei niederen Thieren. III. Die Sehorgane der Hirudineen. *Z. wiss. Zool.,* 62:671-707.

HESSE, R. 1899. Untersuchungen über die Organe der Lichtempfindung bei niederen Thieren. V. Die Augen der polychäten Anneliden. *Z. wiss. Zool.,* 65:446-516.

HESSE, R. 1929. Einfachste Photoreceptoren ohne Bilderzeugung und verschiedene Arten der Bilderzeugung, der Auflösung der Lichterregbaren Schicht und der optischen Isolierung. In: *Bethe's Handb. norm. path. Physiol.,* 12:1-16.

HESSLE, C. 1917. Zur Kenntnis der terebellomorphen Polychäten. *Zool. Bidr. Uppsala,* 5:39-258.

HESSLE, C. 1925. Einiges über die Hesioniden und die Stellung der Gattung *Ancistrosyllis. Ark. Zool.,* 17A (10):1-36.

HICKOK, J. F. and DAVENPORT, D. 1957. Further studies in the behavior of commensal polychaetes. *Biol. Bull., Woods Hole,* 113:397-406.

HILTON, W. A. 1920. The central nervous system of an unknown species of marine leech. *J. Ent. Zool.,* 12-67-68.

HILTON, W. A. 1924a. Nervous system and sense organs. XVII. Archiannelida. *J. Ent. Zool.,* 16:89-93.

HILTON, W. A. 1924b. Nervous system and sense organs. XVIII. Myzostoma. *J. Ent. Zool.,* 16:111-112.

HILTON, W. A. 1925a. Nervous system and sense organs. XXI. Oligochaeta. *J. Ent. Zool.,* 17:20-32.

HILTON, W. A. 1925b. Nervous system and sense organs. XXII. Polychaeta. *J. Ent. Zool.,* 17:39-51.

HILTON, W. A. 1925c. Nervous system and sense organs. XXIII. Hirudinea. *J. Ent. Zool.,* 17:61-67.

HILTON, W. A. 1926a. Nervous system and sense organs. XXV. Embryonic development. Annulata. *J. Ent. Zool.,* 18:45-52.

HILTON, W. A. 1926b. Nervous system and sense organs. XXVI. Sense organs (chiefly Annulata). *J. Ent. Zool.,* 18:62-73.

HILTON, W. A. 1926c. Nervous system and sense organs. XXVII. Annulata. General organization of the nervous system. *J. Ent. Zool.,* 18:85-88.

HILTON, W. A. 1927. Nervous system and sense organs. XXVIII. The nervous system of the annelid *Ophelina mucronata, J. Ent. Zool.* 19:71-73.

HOLMES, G. E. 1931. The influence of the nervous system on regeneration in *Nereis virens,* Sars. *J. exp. Zool.,* 60:485-503.

HOLMES, M. T. 1930. The connective-tissue structure of the ganglion of the earthworm, *Lumbricus terrestris. J. comp. Neurol.,* 51:393-408.

HOLMGREN, N. 1916. Zur vergleichenden Anatomie des Gehirns von Polychaeten, Onychophoren, Xiphosuren, Arachniden, Crustaceen, Myriapoden und Insekten. *K. svenska Vetensk Akad. Handl.,* 56:1-303.

HOLST, E. VON. 1932. Untersuchungen über die Funktionen des Zentralnervensystems beim Regenwurm (*Lumbricus terrestris* L. = *L. herculeus* Sav.). *Zool. Jb. (allg. Zool.),* 51:547-588.

HOLST, E. VON. 1933. Weitere Versuche zum nervösen Mechanismus der Bewegung beim Regenwurm (*Lumbricus terr.* L.). *Zool. Jb. (allg. Zool.),* 53:67-100.

HÖNIG, J. 1910. Die Neurochorde des *Criodrilus lacuum* Hoffmstr. *Arb. zool. Inst. Univ. Wien,* 18:257-282.

HONJO, I. 1937. Physiological studies on the neuromuscular systems of lower worms. I. *Caridinicola indica. Mem. Coll. Sci. Kyoto (B),* 12:187-210.

HORRIDGE, G. A. 1959. Analysis of the rapid responses of *Nereis* and *Harmothoë* (Annelida). *Proc. roy. Soc. (B),* 150:245-262.

HORRIDGE, G. A. and ROBERTS, M. B. V. 1960. Neuro-muscular transmission in the earthworm. *Nature, Lond.,* 186:650.

HOWELL, C. D. 1939. The responses to light in the earthworm, *Pheretima agrestis* Goto and Hatai, with special reference to the function of the nervous system. *J. exp. Zool.,* 81:231-259.

HRABĚ. S. 1932. Sur la structure de l'organe buccal, du pharynx, des glandes septales et des nerfs sympathiques chez les enchytréides. (In Czech.) *Publ. Fac. Sci. Univ. Masaryk,* 159:1-39.

HRABĚ, S. 1934. Das Mundorgan von *Enchytraeus albidus, Achaeta bohemica* und einigen anderen Enchyträiden. *Zool. Anz.,* 106:245-251.

HUNT, H. R. 1919. Regenerative phenomena following the removal of the digestive tube and the nerve cord of earthworms. *Rep. Mus. comp. Zool. Harv.,* 62:571-581.

IMAI, T. 1928. Nervous system of the earthworm, *Perichaeta megascolidioides.* I. Gross anatomy of the nervous system. *Sci. Rep. Tôhoku Univ.,* (4) 3:443-460.

ISOSSIMOW, W. 1926. Zur Anatomie des Nervensystems der Lumbriculiden. *Zool. Jb. (Anat),* 48:365-404.

ISSIDORIDES, M. 1956. Ultrastructure of the synapse in the giant axons of the earthworm. *Exp. Cell. Res.,* 11:423-436.

ITINA, N. A. 1947. The reactivity of locomotor muscles of Invertebrata to the action of parasympatho-mimetic substances. *Sechenov J. Physiol. (Fiziol. Zh.),* 33:101-110. (In Russian.)

IWANOFF, P. P. 1928. Die Entwicklung der Larvalsegmente bei den Anneliden. *Z. Morph. Ökol. Tiere,* 10: 62-161.

IWANOW, P. 1903. Die Regeneration von Rumpf- und Kopfsegmenten bei *Lumbriculus variegatus* Gr. *Z. wiss. Zool.,* 75:327-390.

IZOTOVA, T. E. 1953. Comparative histology of the nervous system; "perimedullary plexus" and neuroglia of *Nephthys. Izv. Kazan Fil. Akad. Nauk SSSR. Biol.,* 1953:49-68. (In Russian.)

JANZEN, R. 1931. Beiträge zur Nervenphysiologie der Oligochaeten. *Zool. Jb. (allg. Zool.),* 50:51-150.

JEENER, R. 1927-1928. Recherches sur le système neuromusculaire latéral des annélides. *Rec. Inst. zool. Torley-Rousseau,* 1:99-121.

JENKINS, O. P. and CARLSON, A. J. 1904a. The rate of the nervous impulse in the ventral nerve-cord of certain worms. *J. comp. Neurol.,* 13:259-289.

JENKINS, O. P. and CARLSON, A. J. 1904b. Physiological evidence of the fluidity of the conducting substance in the pedal nerves of the slug—*Ariolimax columbianus. J. comp. Neurol.,* 14:85-92.

JENNINGS, H. S. 1906. Modifiability in behavior. II. Factors determining direction and character of movement in the earthworm. *J. exp. Zool.,* 3:435-455.

JOHANSSON, K. E. 1927. Beiträge zur Kenntnis der Polychaeten Familien Hermellidae, Sabellidae und Serpulidae. *Zool. Bidr. Uppsala,* 11:1-185.

JORDAN, H. 1913. Wie ziehen die Regenwürmer Blätter in ihre Röhren? *Zool. Jb. (allg. Zool.),* 33:95-106.

JORDAN, H. J. 1929. *Allgemeine Ver-*

gleichende *Physiologie der Tiere.* W. de Gruyter, Berlin und Leipzig.

JORDAN, H. J. and FEEN, P. J. VAN DER. 1930. Methoden und Technik der Nerven- und Muskelphysiologie bei wirbellosen Tieren. In: *Abderhalden's Handb. biol. ArbMeth.*, 9:4:295-438.

JORDAN, H. J. and HIRSCH, G. C. 1927. *Übungen aus der vergleichenden Physiologie.* Springer, Berlin.

JOSEPH, H. 1900. Zur Kenntnis der Neuroglia. *Anat. Anz.*, 17:354-357.

JOSEPH, H. 1902. Untersuchungen über die Stützsubstanzen des Nervensystems, nebst Erörterungen über deren histogenetische und phylogenetische Deutung. *Arb. zool. Inst. Univ. Wien*, 13:335-400.

JOURDAN, É. 1884. Le cerveau de l'*Eunice harassi* et ses rapports avec l'hypoderme. *C.R. Acad. Sci., Paris*, 98:1292-1294.

JOURDAN, É. 1887a. Études histologiques sur deux espèces du genre *Eunice. Ann. Sci. nat.*, (7) 2:239-304.

JOURDAN, É. 1887b. Structure histologique des téguments et des appendices sensitifs de l'*Hermione hystrix* et du *Polynoë grubiana. Arch. Zool. exp. gén.*, (2) 5:91-122.

JOURDAN, É. 1891. L'innervation de la trompe des glycères. *C.R. Acad. Sci., Paris*, 112:882-883.

JOURDAN, É. 1892. Étude sur les épithéliums sensitifs de quelques vers annelés. *Ann. Sci. nat.*, (7) 13:227-258.

JOYEUX-LAFFUIE, J. 1888a. Recherches sur l'organisation du chétoptère. *C.R. Acad. Sci., Paris*, 105:125-127.

JOYEUX-LAFFUIE, J. 1888b. Sur le système nerveux du chétoptère. *C.R. Acad. Sci., Paris*, 106:148-151.

JOYEUX-LAFFUIE, J. 1890. Étude monographique du chétoptère (*Chaetopterus variopedatus*, Renier), suivie d'une revision des espèces du genre *Chaetopterus. Arch. Zool. exp. gén.*, (2) 8:245-360.

JUST, B. 1924. Über die Muskel- und Nervenphysiologie von *Arenicola marina. Z. vergl. Physiol.*, 2:155-183.

KAISER, F. 1954. Beiträge zur Bewegungsphysiologie der Hirudineen. *Zool. Jb. (allg. Zool.)*, 65:59-90.

KALLENBACH, E. 1883. Über *Polynoë cirrata* O. Fr. Mllr. Inaug. Diss. Eisenach, Jena.

KALMUS, H. 1931. Bewegungsstudien an den Larven von *Sabellaria spinulosa* Leuck. *Z. vergl. Physiol.*, 15:164-192.

KAO, C. Y. 1956. Basis for after-discharge in the median giant axon of the earthworm. *Science*, 123:803.

KAO, C. Y. and GRUNDFEST, H. 1957. Postsynaptic electrogenesis in septate giant axons. I. Earthworm median giant axon. *J. Neurophysiol.*, 20:553-573.

KARLING, T. G. 1958. Zur Kenntnis von *Stygocapitella subterranea* Knöllner und *Parergodrilus heideri* Reisinger (Annelida). *Ark. Zool.*, (2) 11:307-341.

KATSUKI, Y., CHEN, J., and TAKEDA, H. 1954. Fundamental neural mechanism of the sense organ. *Bull. Tokyo Med. Dent. Univ.*, 1:21-31.

KEYL, I. 1913. Beiträge zur Kenntnis von *Branchiura sowerbyi* Beddard. *Z. wiss. Zool.*, 107:199-308.

KLEINENBERG, N. 1886. Die Entstehung des Annelids aus der Larve von *Lopadorhynchus*. Nebst Bemerkungen über die Entwicklung anderer Polychaeten. *Z. wiss. Zool.*, 44:1-227.

KNOWLTON, F. P. and MOORE, A. R. 1917. Note on the reversal of reciprocal inhibition in the earthworm. *Amer. J. Physiol.*, 44:490-491.

KOLMER, W. 1905. Über das Verhalten der Neurofibrillen an der Peripherie. *Anat. Anz.*, 26:560-569.

KOMETIANI, P. A. 1945. The rate of the oxygen consumption in the catch action of the body wall of a leech. *Soobshch. Akad. Nauk Gruz. SSR*, 6:65-72. (In Russian, with English summary.)

KORN, H. 1958. Vergleichend-embryologische Untersuchungen an *Harmothoë* Kinberg (Polychaeta, Annelida). *Z. wiss. Zool.*, 161:346-443.

KORN, H. 1960. Das larvale Nervensystem von *Pectinaria* Lamarck und *Nephthys* Cuvier (Annelida, Polychaeta). *Zool. Jb. (Anat.)*, 78:427-456.

KORNFELD, W. 1914. Über die Abgrenzung der Amphinomiden. *Zool. Anz.*, 44:486-492.

KORNFELD, W. 1915. Über die Augen von *Spinther miniaceus*. *Zool. Anz.*, 45:516-523.

KOWALSKI, J. 1908. De l'impregnation par la méthode à l'argent réduit de Cajal des neurofibrilles du *Lumbricus* consécutivement à l'action du froid. *Soc. Sci. phys. nat. Bordeaux*, 1908:16-18.

KOWALSKI, J. 1909. Contribution à l'étude des neurofibrilles chez le lombric. *Cellule*, 25:290-346.

KRAWANY, J. 1905. Untersuchungen über das Zentralnervensystem des Regenwurms. *Arb. zool. Inst. Univ. Wien*, 15:281-316.

KRECKER, F. H. 1910. Some phenomena of regeneration in *Limnodrilus* and related forms. *Z. wiss. Zool.*, 95:383-450.

KRISZAT, G. 1932. Zur Autotomie der Regenwürmer. *Z. vergl. Physiol.*, 16:185-203.

KRIVANEK, J. O. 1956. Habit formation in the earthworm *Lumbricus terrestris*. *Physiol. Zoöl.*, 29:241-259.

KROPP, B. 1933. Brain transplantation in regenerating earthworms. *J. exp. Zool.*, 65:107-129.

KUENZER, P. 1958. Verhaltensphysiologische Untersuchungen über das Zucken des Regenwurms. *Z. Tierpsychol.*, 15:31-49.

KÜKENTHAL, W. 1887. Über das Nervensystem der Opheliaceen. *Jena. Z. Naturw.*, 20:511-580.

KULAGIN, N. 1888. Zur Anatomie und Systematik der in Russland vorkommenden Fam. Lumbricidae. *Zool. Anz.*, 11:231-235.

KUTSCHERA, F. 1909. Die Leuchtorgane von *Acholoë astericola* Clprd. *Z. wiss. Zool.*, 92:75-102.

LANCASTER, S. 1939. Nature of the chromaffin nerve cells in certain annulates and arthropods. *Trans. Amer. micr. Soc.*, 58:90-96.

LANG, A. 1881. Der Bau von *Gunda segmentata* und die Verwandtschaft der Plathelminthen mit Coelenteraten und Hirudineen. *Mitt. zool. Sta. Neapel*, 3:187-250.

LANG, A. 1904. Beiträge zu einer Trophocöltheorie. Betrachtungen und Suggestionen über die phylogenetische Ableitung der Blut- und Lymphbehälter, insbesondere der Articulaten. Mit einem einleitenden Abschnitt über die Abstammung der Anneliden. *Jena. Z. Naturw.*, 38:1-376.

LANGDON, F. E. 1895. The sense organs of *Lumbricus agricola*, Hoffm. *J. Morph.*, 11:193-234.

LANGDON, F. E. 1900. The sense-organs of *Nereis virens*, Sars. *J. comp. Neurol.*, 10:1-77.

LANGERHANS, P. 1880. Die Wurmfauna von Madeira. *Z. wiss. Zool.*, 34:87-143.

LAPICQUE, M. and VEIL, C. 1925. Vitesse de conduction nerveuse et musculaire comparée à la chronaxie chez les sangsue et le ver de terre *C.R. Soc. Biol., Paris*, 93:1590-91.

LAVERACK, M. S. 1960. Tactile and chemical perception in earthworms—I. Responses to touch, sodium chloride, quinine and sugars. *Comp. Biochem. Physiol.*, 1:155-163.

LAVERACK, M. S. 1961a. Tactile and chemical perception in earthworms. II. Responses to acid pH solutions. *Comp. Biochem. Physiol.*, 2:22-34.

LAVERACK, M. S. 1961b. The effect of temperature changes on the spontaneous nervous activity of the isolated nerve cord of *Lumbricus terrestris*. *Comp. Biochem. Physiol.*, 3:136-140.

LENHOSSÉK, M. VON. 1892. Ursprung, Verlauf und Endigung der sensibeln Nervenfasern bei *Lumbricus*. *Arch. mikr. Anat.*, 39:102-136.

LENHOSSÉK, M. VON. 1895. *Der feinere Bau des Nervensystems.* Vol. II. Fischer, Berlin.

LESPÉS, C. 1872. Étude anatomique sur un chétoptère. *Ann. Sci. nat.*, (5) 15 (14):1-17.

LEWIS, M. 1898. Studies on the central and peripheral nervous systems of two polychaete annelids. *Proc. Amer. Acad. Arts Sci.*, 33:225-268.

LEYDIG, F. 1861. Die Augen und neue

Sinnesorgane der Egel. *Arch. Anat. Physiol.*, 1861:588-605.
LEYDIG, F. 1862. Ueber das Nervensystem der Anneliden. *Arch. Anat. Physiol., Lpz.*, 1862:90-124.
LEYDIG, F. 1864. *Tafeln zur vergleichenden Anatomie. 1. Zum Nervensystem und den Sinnesorganen der Würmer und Gliederfüssler.* Laupp, Tübingen.
LEYDIG, F. 1865. Ueber *Phreoryctes menkeanus* Hoffm. nebst Bemerkungen über den Bau anderer Anneliden. *Arch. mikr. Anat.*, 1:249-294.
LEYDIG, F. 1886. Die riesigen Nervenröhren im Bauchmark der Ringelwürmer. *Zool. Anz.*, 9:591-597.
LIEBERMANN, A. 1932. Studien über die Topographie und die Bewegungsmechanik des ventralen Borstenfollikels von *Stylaria lacustris* L. *Zool. Jb. (allg. Zool.)*, 50:151-174.
LIVANOFF, N. 1917. Notes sur l'histologie des polychètes. *Trav. Soc. Nat. Univ. Kasan (Trud. Obsch. Estest. Imp. Kazan. Univ.)*, 49:121-127.
LIVANOFF, N. 1924. Recherches sur l'anatomie du système nerveux des polychètes. *Russk. Arkh. Anat.*, 3:1-59.
LIVANOW, N. A. 1903. Untersuchungen zur Morphologie der Hirudineen. I. Das Neuro- und Myosomit der Hirudineen. *Zool. Jb. (Anat.)*, 19:29-90.
LIVANOW, N. A. 1904. Untersuchungen zur Morphologie der Hirudineen. II. Das Nervensystem des vorderen Körperendes und seine Metamerie. *Zool. Jb. (Anat.)*, 20:153-226.
LIVANOW, N. A. 1906. *Acanthobdella peledina* Grube, 1851. *Zool. Jb. (Anat.)*, 22:637-866.
LOEB, J. 1894. Beiträge zur Gehirnphysiologie der Würmer. *Pflüg. Arch. ges. Physiol.*, 56:247-269.
LOYE, J. F. ZUR. 1908. Die Anatomie von *Spirorbis borealis* mit besonderer Berücksichtigung der Unregelmässigkeiten des Körperbaues und deren Ursachen. *Zool. Jb. (Anat.)*, 26:305-354.
MALAQUIN, A. 1893a. *Recherches sur les Syllidiens. Morphologie, Anatomie, Reproduction, Développement.* Lille.
MALAQUIN, A. 1893b. Recherches sur les Syllidiens. *Mem. Soc. Sci. Agric. Arts.* Lille.
MALAQUIN, A. and DEHORNE, A. 1907. Les annélides polychètes de la Baie d'Amboine. *Rev. suisse Zool.*, 15:335-400.
MÁLEK, R. 1927. Assoziatives Gedächtnis bei den Regenwürmern. *Biol. gen.*, 3:317-328.
MANGOLD, O. 1925. Beobachtungen und Experimente zur Biologie des Regenwurms. I. Lauterzeugung, Formsinn und chemischer Sinn. *Z. vergl. Physiol.*, 2:57-81.
MANGOLD, O. 1931. Über den chemischen Sinn des Regenwurms. *Naturwissenschaften*, 19:730-735.

MANN, K. H. 1953. The segmentation of leeches. *Biol. Rev.*, 28:1-15.
MANN, K. H. 1954. The anatomy of the horse leech, *Haemopis sanguisuga* (L.) with particular reference to the excretory system. *Proc. zool. Soc. Lond.*, 124:69-88.
MASUI, M. 1953. Stimulation of the ventral nerve cord of the earthworm by double condenser discharge. *Zool. Mag.*, Tokyo, 62:380-384. (In Japanese.)
MAU, W. 1882. Ueber *Scoloplos armiger*, O. F. Muller. Beitrag zur Kenntnis der Anatomie und Histologie der Anneliden. *Z. wiss. Zool.*, 36:389-432.
MAULE, V. 1908. Das sympathische Nervensystem der Enchyträiden. *S.B. böhm. Ges. Wiss.*, 1908 (9):1-21.
MAXWELL, S. S. 1897. Beiträge zur Gehirnphysiologie der Anneliden. *Pflüg. Arch. ges. Physiol.*, 67:263-297.
MCINTOSH, W. C. 1876. On the structure of the body wall in the Spionidae. *Proc. roy. Soc. Edinb.*, 9:123-129.
MCINTOSH, W. C. 1877. On the arrangement and relations of the great nerve-cords in the marine annelids. *Proc. roy. Soc. Edinb.*, 9:372-381.
MCINTOSH, W. C. 1878. Beiträge zur Anatomie von *Magelona*. *Z. wiss. Zool.*, 31:401-472.
MCINTOSH, W. C. 1885. Report on the Annelida Polychaeta collected by H. M. S. Challenger during the years 1873-76. "*Challenger*" *Rep., Zool.*, 12:1-554.
MCINTOSH, W. C. 1894. A contribution to our knowledge of the Annelida. On some points in the structure of *Euphrosyne*. On certain young stages of *Magelona* and on Claparede's unknown larval *Spio*. *Quart. J. micr. Sci.*, 36:53-76.
MENCL, E. 1908. Über die Histologie und Histogenese der sogenannten Punktsubstanz Leydigs in dem Bauchstrange der Hirudineen. *Z. wiss. Zool.*, 89:371-416.
MENCL, E. 1909. Zur Kenntnis der Neuroglia bei *Nephelis*. *Zool. Anz.*, 34:516-521.
MEYER, A. 1924. Über die Segmentalorgane von *Tomopteris helgolandica* nebst Bemerkungen über das Nervensystem und die rosettenförmigen Organe. *Zool. Anz.*, 60:83-88.
MEYER, A. 1926. Die Segmentalorgane von *Tomopteris catherina* (Gosse) nebst Bemerkungen über das Nervensystem, die rosettenförmigen Organe und die Cölombewimperung. Ein Beitrag zur Theorie der Segmentalorgane. *Z. wiss. Zool.*, 127:297-402.
MEYER, A. 1927. Über Cölombewimperung und cölomatische Kreislaufsysteme bei Wirbellosen. Ein Beitrag zur Histophysiologie der secundären Leibeshöhle und ökologischen Bedeutung der Flimmerbewegung. *Z. wiss. Zool.*, 129:153-212.
MEYER, E. 1882. Zur Anatomie und Histologie von *Polyophthalmus pictus* Clap. *Arch. mikr. Anat.*, 21:769-823.
MEYER, E. 1887. Studien über den Körperbau der Anneliden. *Mitt. zool. Sta. Neapel*, 7:592-741.
MEYER, E. 1888. Studien über den Körperbau der Anneliden. IV. Die Körperform der Serpulaceen und Hermellen. *Mitt. zool. Sta. Neapel*, 8:462-662.
MEYER, E. 1890. Die Abstammung der Anneliden. Der Ursprung der Metamerie und die Bedeutung des Mesoderms. *Biol. Zbl.*, 10:296-308.
MEYER, E. 1901. Studien über den Körperbau der Anneliden. Das Mesoderm der Ringelwürmer. *Mitt. zool. Sta. Neapel*, 14:247-585.
MEYER, G. F. 1955. Vergleichende Untersuchungen mit der supravitalen Methylenblaufärbung am Nervensystem wirbelloser Tiere. *Zool. Jb. (Anat.)*, 74:339-400.
MEYER, G. F. 1956. Feinhistologische Untersuchungen an Insekten-Neuronen (Unter Bezugnahme auf *Hirudo medicinalis* L.) *Zool. Jb. (Anat.)* 75:389-400.
MEYER, G. F. 1957. Elektronenmikroskopische Untersuchungen an den Apathyschen Neurofibrillen von *Hirudo medicinalis*. *Z. Zellforsch.*, 45:538-542.
MEYER, J. DE. 1900. Note sur la signification morphologique des ganglions cérébroides sub-oesophagiens du *Lumbricus agricola*. *Ann. Soc. belge Micr.*, 26:146-164.
MICHAELSEN, W. 1919. Über die Beziehungen der Hirudineen zu den Oligochäten. *Mitt. naturh. Mus. Hamb.*, Beihft. 36:131-153.
MICHAELSEN, W. 1925. *Agriodrilus vermivorus* aus dem Baikal See, ein Mittelglied zwischen typischen Oligochäten und Hirudineen. *Mitt. naturh. Mus. Hamb.*, 42:1-19.
MICHEL, A. 1898. Recherches sur la régénération chez les annélides. *Bull. sci. Fr. Belg.*, 31:245-420.
MICHEL, A. 1899. Sur les canaux neuraux et les fibres nerveuses des annelides. *Trav. Sta. zool. Wimereux*, 7:478-488.
MILLER, J. A. 1945. Studies in the biology of the leech. IX. The gross nervous system. *Ohio J. Sci.*, 45:233-246.
MILLOTT, N. 1943a. The visceral nervous system of the earthworm: I. Nerves controlling the tone of the alimentary canal. *Proc. roy. Soc. (B)*, 131:271-295.
MILLOTT, N. 1943b. The visceral nervous system of the earthworm: II. Evidences of chemical transmission and the action of sympatheticomimetic and parasympatheticomimetic drugs on the tone of the alimentary canal. *Proc. roy. Soc. (B)*, 131:362-373.
MILLOTT, N. 1944. The visceral nervous

system of the earthworm: III. Nerves controlling secretion of protease in the anterior intestine. *Proc. roy. Soc. (B)*, 132:200-212.

MILNE, L. J. and MILNE, M. J. 1956. Invertebrate photoreceptors. In: *Radiation Biology*. A. Hollaender (ed.). McGraw-Hill, New York.

MILNE, L. J. and MILNE, M. J. 1959. Photosensitivity in invertebrates. In: *Handbook of Physiology*, Sect. 1. *Neurophysiology*, 1:621-645.

MOJSISOVICS VON MOJSVÁR, A. 1878. Kleine Beiträge zur Kenntnis der Anneliden. I. Die Lumbricidenhypodermis. *S.B. Akad. Wiss. Wien*, 76:7-20.

MONRO, C. C. A. 1924. On the postlarval stage in *Diopatra cuprea* Bosc., a polychaetous annelid of the family Eunicidae. *Ann. Mag. nat. Hist.*, (9) 14:193-199.

MOORE, A. R. 1921. Chemical stimulation of the nerve cord of *Lumbricus terrestris*. *J. gen Physiol.*, 4:29-31.

MOORE, A. R. 1923a. Muscle tension and reflexes in the earthworm. *J. gen. Physiol.*, 5:327-333.

MOORE, A. R. 1923b. The reactions of *Nereis virens* to unilateral tension of its musculature. *J. gen. Physiol.*, 5:451-452.

MOORE, A. R. 1923c. Galvanotropism in the earthworm. *J. gen. Physiol.*, 5:453-459.

MOORE, A. R. and BRADWAY, W. E. 1945. The significance of action potentials in the isolated nerve cord of the earthworm (*Lumbricus terrestris*). *J. cell. comp. Physiol.*, 25: 181-193.

MOORE, A. R. and KELLOGG, F. M. 1916. Note on the galvanotropic response of the earthworm. *Biol. Bull., Woods Hole*, 30:131-134.

MORGULIS, S. 1910. The movements of the earthworm: a study of a neglected factor. *J. comp. Neurol.*, 20:615-624.

MORITA, H. and TATEDA, H. 1953. The relation between the amplitudes of two kinds of giant impulses of the earthworm. *Mem. Fac. Sci. Kyūshū Univ.*, (E) 1:133-137.

MORITA, H., TATEDA, H., and NISHIDA, A. 1952. Giant nerve fiber and its functional organization in the earthworm. *Mem. Fac. Sci. Kyūshū Univ.*, (E).1:89-100.

MOSELLA, R. G. 1927. Alcuni considerazioni negli occhi di *Nereis dumerilii*. *Boll. Ist. Zool. Univ. Roma*, 4:166-170.

MRÁZEK, A. 1913. Beiträge zur Naturgeschichte von *Lumbriculus*. Sonderdruck aus: *S.B. böhm. Ges. Wiss.* Rivnáč, Prag.

MÜLLER, K. J. 1932. Über normale Entwicklung, inverse Asymmetrie und Doppelbildungen bei *Clepsine sexocutata*. *Z. wiss. Zool.*, 142:425-490.

NANSEN, F. 1887a. The structure and combination of the histological elements of the central nervous system. *Bergens Mus. Aarb.*, 1886:27-215.

NANSEN, F. 1887b. Anatomie und Histologie des Nervensystems der Myzostomen. *Jena. Z. Naturw.*, 21:267-321.

NELSON, J. A. 1907. The morphology of *Dinophilus conklini* n. sp. *Proc. Akad. nat. Sci. Philad.*, 59:82-143.

NEVMYVAKA, G. A. 1947a. The innervation of setae in the earthworm (*Allolobophora calliginosa*). *C.R. Acad. Sci. URSS (Dokl. Akad. Nauk SSSR)*, 56:423-425. (In Russian.)

NEVMYVAKA, G. A. 1947b. The innervation of intestine in the earthworm (*Allolobophora calliginosa*). *C.R. Acad. Sci. URSS (Dokl. Akad. Nauk SSSR)*, 56:533-536. (In Russian.)

NEVMYVAKA, G. A. 1947c. Abdominal nerve cord in the earthworm (*Allolobophora calliginosa*). *C.R. Acad. Sci. URSS (Dokl. Akad. Nauk SSSR)*, 58:1483-1486. (In Russian.)

NEVMYVAKA, G. A. 1950. The structure of nerve fibers in *Allolobophora*. *C.R. Acad. Sci. URSS (Dokl. Akad. Nauk SSSR)*, 70:507-510.(In Russian.)

NEVMYVAKA, G. A. 1956. On the participation of vegetative nerves in the innervation of somatic muscles in the annelids. *C. R. Acad. Sci. URSS (Dokl. Akad. Nauk)*, 110:855-857. (In Russian.)

NEWELL, G. E. 1950. The role of the coelomic fluid in the movements of earthworms. *J. exp. Biol.*, 27:110-121.

NICOL, J. A. C. 1948a. The giant axons of annelids. *Quart. Rev. Biol.*, 23: 291-323.

NICOL, J. A. C. 1948b. The function of the giant axon of *Myxicola infundibulum* Montagu. *Canad. J. Res.*, 26: 212-222.

NICOL, J. A. C. 1948c. Giant axons of *Eudistylia vancouveri* (Kinberg). *Trans. roy. Soc. Canada*, 42 (5): 107-124.

NICOL, J. A. C. 1948d. The giant nervefibres in the central nervous system of Myxicola (Polychaeta, Sabellidae). *Quart. J. micr. Sci.* (2) 89:1-45.

NICOL, J. A. C. 1950. Responses of *Branchiomma vesiculosum* (Montagu) to photic stimulation. *J. Mar. biol. Ass. U.K.*, 29:303-320.

NICOL, J. A. C. 1951. Giant axons and synergic contractions in *Branchiomma vesicrulosum*. *J. exp. Biol.*, 28:22-31.

NICOL, J. A. C. 1952a. Studies on *Chaetopterus variopedatus* (Renier) II. Nervous control of light production. *J. Mar. biol. Ass. U.K.*, 30:433-452.

NICOL, J. A. C. 1952b. Studies on *Chaetopterus variopedatus* (Renier). III. Factors affecting the light response. *J. Mar. biol. Ass. U.K.*, 31:113:144.

NICOL, J. A. C. 1952c. Muscle activity and drug action in the body wall of the sabellid worm *Branchiomma vesiculosum* (Montagu). *Physiol. comp.*, 2:339-345.

NICOL, J. A. C. 1952d. Luminescent responses in *Chaetopterus* and the effects of eserine. *Nature, Lond.*, 169:665-666.

NICOL, J. A. C. 1953. Luminescence in polynoid worms. *J. Mar. biol. Ass. U.K.*, 32:65-84.

NICOL, J. A. C. 1954a. Fatigue of the luminescent response of *Chaetopterus*. *J. Mar. biol. Ass. U.K.*, 33:177-186.

NICOL, J. A. C. 1954b. The nervous control of luminescent responses in polynoid worms. *J. Mar. biol. Ass. U.K.*, 33:225-255.

NICOL, J. A. C. 1957a. Luminescence in polynoids. II Different modes of response in the elytra. *J. Mar. biol. Ass. U.K.*, 36:261-269.

NICOL, J. A. C. 1957b. Luminescence in polynoids. III Propagation of excitation through the nerve cord. *J. Mar. biol. Ass. U.K.*, 36:271-273.

NICOL, J. A. C. 1960. The regulation of light emission in animals. *Biol. Rev.*, 35:1-42.

NICOL, J. A. C. and WHITTERIDGE, D. 1955. Conduction in the giant axon of *Myxicola infundibulum*. *Physiol. comp.*, 4:101-117.

NICOL, J. A. C. and YOUNG, J. Z. 1946. Giant nerve fibres of *Myxicola infundibulum* (Grube). *Nature, Lond.*, 158:167-168.

NIERSTRASZ, H. F. 1922. Die Verwandtschaftsbeziehungen zwischen Mollusken und Anneliden. *Bijdr. Dierk.*, 22:33-42.

NIKI, I., SONE, T., and HOSOI, E. 1953. Study on the strength-duration curve of *Pheretima* nerve. *Zool. Mag., Tokyo*, 62:253-256. (In Japanese.)

NILSSON, D. 1910. Die Fischelsche Alizarinfärbung und ihre Anwendbarkeit für die Polychaeten, speziell *Pectinaria koreni* Mgrn. *Zool. Anz.*, 35:195-202.

NILSSON, D. 1912. Beiträge zur Kenntnis des Nervensystems der Polychäten. *Zool. Bidr. Uppsala*, 1:85-161.

NOMURA, E. 1926. Effect of light on the movements of the earthworm *Allolobophora foetida*. *Sci. Rep. Tohoku Univ.*, (4) 1:293-409.

NOMURA, E. and OHFUCHI, S. 1932. Effect of inorganic salts on photic orientation in *Allolobophora foetida* (Sav.). 8. Summary and general conclusions. *Sci. Rep. Tôhoku Univ.*, (4) 7:491-497.

NUZUM, M. F. and RAND, H. W. 1924. Can the earthworm pharynx epithelium produce central nervous tissue? *Biol. Bull., Woods Hole*, 47:213-222.

OGAWA, F. 1928. On the number of ganglion cells and nerve fibers in some of the ventral nerve cords of the earthworm. I. The number of ganglion cells. *Sci. Rep. Tôhoku Univ.*, (4) 3:745-756.

OGAWA, F. 1930. On the number of

ganglion cells and nerve fibers in some of the ventral nerve cords of the earthworm. II. The number of nerve fibers. *Sci. Rep. Tôhoku Univ.* (4) 5:691-716.
OGAWA, F. 1934a. The number of ganglion cells and nerve fibers in the nervous system of the earthworm, *Pheretima communissima. Sci. Rep. Tôhoku Univ.*, (4) 8:345-368.
OGAWA, F. 1934b. Nerve cells of earthworms. *Nature, Lond.*, 134:666.
OGAWA, F. 1939. The nervous system of earthworm (*Pheretima communissima*) in different ages. *Sci. Rep. Tôhoku Univ.*, (4) 13:395-488.
OPPENHEIMER, A. 1902. Certain sense organs of the proboscis of the polychaetous annelid *Rhynchobolus dibranchiatus*. *Proc. Amer. Acad. Arts Sci.*, 37:553-565.
ÔSIMA, M. 1939. Effect of adrenalin upon the earthworm. *Sci. Rep. Tôhoku Univ.*, (4) 14:331-337.
ÖSTLUND, E. 1954. The distribution of catechol amines in lower animals and their effect on the heart. *Acta physiol. scand.*, 31 (Suppl. 112):1-67.
PANTIN, C. F. A. 1937. Discussion meeting on the transmission of excitation in living material. Junctional transmission of stimuli in the lower animals. *Proc. roy. Soc. (B)*, 123: 397-421.
PARKER, G. H. and ARKIN, L. 1901. The directive influence of light on the earthworm *Allolobophora foetida* (Sav.). *Amer. J. Physiol.*, 5:151-157.
PENNERS, A. 1924. Die Entwicklung des Keimstreifs und die Organbildung bei *Tubifex rivulorum* Lam. *Zool. Jb. (Anat.)*. 45:251-308.
PEREZ, H. VON Z. 1942. On the chromaffin cells of the nerve ganglia of *Hirudo medicinalis*, Lin. *J. comp. Neurol.*, 76:367-401.
PERRIER, M. E. 1872. Histoire naturelle du *Dero obtusa*. *Arch. Zool. exp. gén.*, (1)1:65 81.
PFLÜCKE, M. 1895. Zur Kenntnis des feineren Baues der Nervenzellen bei Wirbellosen. *Z. wiss. Zool.*, 60:500-542.
PFLUGFELDER, O. 1932. Über den feineren Bau der Augen freilebender Polychäten. *Z. wiss. Zool.*, 142:540-586.
PFLUGFELDER, O. 1933. Zur Histologie der Elytren der Aphroditen. *Z. wiss. Zool.*, 143:497-537.
PIERANTONI, U. 1906. Osservazioni sullo sviluppo embrionale e larvale del *Saccocirrus papillocercus* Bobr. *Mitt. zool. Sta. Neapel*, 18:46-72.
PIERANTONI, U. 1908. *Protodrilus. Fauna u. Flora Neapel*, 31:1-224.
PLATE, L. 1922. *Allgemeine Zoologie und Abstammungslehre.* I. *Einleitung, Cytologie, Histologie, Promorphologie, Haut, Skellette, Lokomotionsorgane, Nervensystem.* Fischer, Jena.
PLATE, L. 1924. *Allgemeine Zoologie und Abstammungslehre.* II. *Die Sinnesorgane der Tiere.* Fischer, Jena.
POINTER, H. 1911. Beiträge zur Kenntnis der Oligochätenfauna der Gewässer von Graz. *Z. wiss. Zool.*, 98:626-676.
POLL, H. 1908. Gibt es Nebennieren bei Wirbellosen? *S. B. Ges. naturf. Fr., Berl.*, 1908:18-24.
POLL, H. and SOMMER, A. 1903. Ueber phaeochrome Zellen im Centralnervensystem des Blutegels. *Arch. Anat. Physiol. (Physiol. Abt.)*, 1903: 549-550.
PRENTISS, C. W. 1903. The neurofibrillar structures in the ganglia of the leech and crayfish with especial reference to the neurone theory. *J. comp. Neurol.*, 13:157-175.
PROSSER, C. L. 1934a. Correlation between development of behavior and neuromuscular differentiation in embryos of *Eisenia foetida*, Sav. *J. comp. Neurol.*, 58:603-641.
PROSSER, C. L. 1934b. Effect of central nervous system on responses to light in *Eisenia foetida*, Sav. *J. comp. Neurol.* 59:61-86.
PROSSER, C. L. 1934c. The nervous system of the earthworm. *Quart. Rev. Biol.*, 9:181-200.
PROSSER, C. L. 1935. Impulses in the segmental nerves of the earthworm. *J. exp. Biol.*, 12:95-104.
PROSSER, C. L. 1950. Nervous systems. In: *Comparative Animal Physiology*. C. L. Prosser (ed.). Saunders, Philadelphia.
PROSSER, C. L. and ZIMMERMAN, G. L. 1943. Effects of drugs on the hearts of *Arenicola* and *Lumbricus*. *Physiol. Zoöl.*, 16:77-83.
PRUVOT, G. 1885. Recherches anatomiques et morphologiques sur le système nerveux des annélides polychètes. *Arch. Zool. exp. gén.*, (2)3: 211-336.
PRUVOT, G. 1890. Sur la formation des stolons chez les Syllidiens. *C. R. Acad. Sci., Paris*, 108:1310-1313.
PRUVOT, G. and RACOVITZA, E. G. 1895. Matériaux pour la faune des annélides de Banyuls. *Arch. Zool. exp. gén.*, (3)3:339-492.
QUATREFAGES, A. DE. 1847. Note sur l'anatomie des sangsues et des lombrics. *Ann. Sci. nat. (Zool.)*, (3) 8:36.
QUATREFAGES, A. DE. 1850. Mémoire sur le système nerveux des annélides. *Ann. Sci. nat. (Zool.)*, (3) 14:41-46.
QUATREFAGES DE BRÉAU, A. DE. 1865. *Histoire Naturelle des Annelés Marins et d'Eau Douce*, 2 vols. Librairie encyclopédique de Roret, Paris.
RAABE, S. 1939. Zur Analyse der Assoziationsbildung bei *Lumbriculus variegatus* Müll. *Z. vergl. Physiol.*, 26:611-643.
RABES, O. 1901. Über Transplantationsversuche an Lumbriciden. *Biol. Zbl.*, 21:633-650.
RACOVITZA, E. G. 1896. Io'3 céphalique et l'encéphale des annélides polychètes (anatomie, morphologie, histologie). *Arch. Zool. exp. gén.* (3) 4:133-343.
RÁDL, E. 1912. *Neue Lehre vom Zentralen Nervensystem.* W. Engelmann, Leipzig.
RAND, H. W. 1901. The regenerating nervous system of Lumbricidae and the centrosome of its nerve cells. *Bull. Mus. comp. Zool. Harv.*, 37:83-164.
RANDOLPH, H. 1892. The regeneration of the tail in *Lumbriculus*. *J. Morph.*, 7:317-344.
RATNER, S. C. and MILLER, K. R. 1959a. Classical conditioning in earthworms, *Lumbricus terrestris*. *J. comp. physiol. Psychol.*, 52:102-105.
RATNER, S. C. and MILLER, K. R. 1959b. Effects of spacing of training and ganglia removal on conditioning in earthworms. *J. comp. physiol. Psychol.*, 52:667-672.
RATZEL, F. 1867. Beiträge zur Anatomie von *Enchytraeus vermicularis* Henle. *Z. wiss. Zool.*, 18:99-108.
RAW, F. 1949. Some stages in the evolution of the nervous system and the fore-gut of the polychaet. *Smithson. Misc. Coll.*, 111(8):1-35.
REBIZZI, R. 1906. Su alcune variazioni delle neurofibrille nella *Hirudo medicinalis*. *Riv. Pat. nerv. ment.*, 11:355-377.
REIGHARD, J. 1885. On the anatomy and histology of *Aulophorus vagus*. *Proc. Amer. Acad. Arts Sci.*, 20:88-106.
RETZIUS, G. 1891a. Über Nervenendigungen an den Parapodienborsten und über die Muskelzellen der Gefässwände bei den polychäten Annulaten. *Biol. Fören. Forh., Stockholm*, 3:85-89.
RETZIUS, G. 1891b. Zur Kenntnis des centralen Nervensystems der Würmer. Das Nervensystem der Annulaten. *Biol. Untersuch*, N.F. 2:1-28.
RETZIUS, G. 1892a. Das Nervensystem der Lumbriciden. *Biol. Untersuch.*, N.F. 3:1-16.
RETZIUS, G. 1892b. Zur Kenntniss der motorischen Nervendigungen. *Biol. Untersuch.*, N.F. 3:41-52.
RETZIUS, G. 1892c. Das sensible Nervensystem der Polychaeten. *Biol. Untersuch.*, N.F. 4:1-10.
RETZIUS, G. 1895a. Die Smirnow'schen freien Nervenendigungen im Epithel des Regenwurms. *Anat. Anz.*, 10:117-123.
RETZIUS, G. 1895b. Zur Kenntnis des Gehirnganglions und des sensiblen Nervensystems der Polychaeten. *Biol. Untersuch.*, N.F. 7:6-11.
RETZIUS, G. 1898. Zur Kenntniss des sensiblen Nervensystems der Hirudineen. *Biol. Untersuch.*, N.F. 8:94-97.
RETZIUS, G. 1900. Zur Kenntniss des Sensiblen und des sensorischen Nervensystems der Würmer und Mollus-

ken. *Biol. Untersuch.*, N.F. 9:83-96.
RETZIUS, G. 1902. Weiteres zur Kenntnis der Sinneszellen der Evertebraten. *Biol. Untersuch.*, N.E. 10:24-33.
ROBERTIS, E. D. P. DE. 1955. La relation nucléo-plasmatique et la substance basophile de la cellule nerveuse. *C.R. Soc. Biol.*, Paris, 149:1709-1710.
ROBERTIS, E. D. P. DE and BENNETT, H. S. 1954a. Some observations on the fine structure of the giant nerve fibres of the earthworm. *Proc. Int. Conf. Electr. Micr. (London)* 431-436.
ROBERTIS, E. D. P. DE and BENNETT, H. S. 1954b. Some features of fine structure of cytoplasm of cells in the earthworm nerve cord. *Fine Structure of Cells*, Internat. Union Biol. Sci. Symp., (B), 21:261-273.
ROBERTIS, E. D. P. DE and BENNETT, H. S. 1955. Some features of the submicroscopic morphology of synapses in frog and earthworm. *J. biophys. biochem. Cytol.*, 1:47-58.
ROBERTS, M. B. V. 1960. Giant-fibre reflex of the earthworm. *Nature, Lond.*, 186:167.
ROBERTS, M. B. V. 1962a. The giant fibre reflex of the earthworm, *Lumbricus terrestris* L. I. The rapid response. *J. exp. Biol.*, 39:219-228.
ROBERTS, M. B. V. 1962b. The giant fibre reflex of the earthworm, *Lumbricus terrestris* L. II. Fatigue. *J. exp. Biol.*, 39:229-238.
ROGERS, C. G. and CHEN, T. T. 1931. The morphology of the anterior sympathetic nervous system of *Lumbricus terrestris*. *Ohio J. Sci.*, 31:262.
ROHDE, E. 1887. Histologische Untersuchungen über das Nervensystem der Polychaeten. *Zool. Beitr., Berl.*, 2:1-81.
ROHDE, E. 1892. Histologische Untersuchungen über das Nervensystem der Hirudineen. *Zool. Beitr., Berl.*, 3:1-69.
RORIE, J. 1863. On the anatomy of the nervous system in the *Lumbricus terrestris*. *Quart. J. micr. Sci.*, (2) 3:106-109.
ROULE, L. 1889. Le développement du système nerveux des annélides et l'influence exercée sur lui par la symétrie du corps. *C.R. Acad. Sci.*, Paris, 108:359-361.
ROZANOVA, V. D. 1941. Comparative physiological characteristics of the body wall of annelids, *Arenicola* and *Nereis*. *Bull. Biol. Med. exp. URSS*, (*Biull. eksp. Biol. Med.*) 11:122.
RULLIER, F. 1950. Étude morphologique, histologique et physiologique de l'organe nucal chez les annélides polychètes sédentaires. *Ann. Inst. océanogr. Monaco*, 25:207-341.
RUSHTON, W. A. H. 1945a. Action potentials from the isolated nerve cord of the earthworm. *Proc. roy. Soc. (B)*, 132:423-437.
RUSHTON, W. A. H. 1945b. Motor response from giant fibres in the earthworm. *Nature, Lond.*, 156:109-110.
RUSHTON, W. A. H. 1946. Reflex conduction in the giant fibres of the earthworm. *Proc. roy. Soc. (B)*, 133:109-120.
RUSHTON, W. A. H. and BARLOW, H. B. 1943. Single fibre response from an intact animal. *Nature, Lond.*, 152:597-598.
SAINT-JOSEPH, A. DE. 1894. Les annélides polychètes des côtes de Dinard. 3e Partie. *Ann. Sci. nat. (Zool.)*, (7) 17:1-395.
SAINT-LOUP, M. R. 1884. Recherches sur l'organisation des hirudinées. *Ann. Sci. nat. (Zool.)*, (6) 18(2):1-127.
SALENSKY, W. 1882. Études sur le développement des annélides. *Arch. Biol., Paris*, 3:561-604.
SALENSKY, W. 1883. Études sur le développement des annélides. *Arch. Biol., Paris*, 4:221-264.
SALENSKY, W. 1885. Études sur le développement des annélides. *Arch. Biol., Paris*, 6:1-64.
SALENSKY, W. 1907. Morphogenetische Studien an Würmern. *Mém. Acad. Sci. St.-Pétersb.*, (sci. math., phys., nat.), (8) 19:11-361.
SÁNCHEZ, D. 1909. El sistema nerviosa de los hirudineos. *Trab. Lab. Invest. biol. Univ. Madr.*, 7:31-199.
SÁNCHEZ, D. 1912. El sistema nerviosa de los hirudineos. *Trab. Lab. Invest. biol. Univ. Madr.*, 10:1-143.
SCHACK, F. 1886. Anatomisch-histologische Untersuchung von *Nephthys coeca*, Fabricius. Inaug. Diss. Lipsius and Tischer, Kiel.
SCHARRER, E. 1944. The capillary bed of the central nervous system of certain invertebrates. *Biol. Bull., Woods Hole*, 87:52-58.
SCHLÜTER, E. 1933. Die Bedeutung des Zentralnervensystems von *Hirudo medicinalis* für Lokomotion und Raumorientierung. *Z. wiss. Zool.*, 143:538-593.
SCHMIDT, F. 1905. Zur Anatomie und Topographie des Centralnervensystems von *Branchiobdella parasita*. *Z. wiss. Zool.*, 82:664-692.
SCHMIDT, G. A. 1926. Untersuchungen über die Embryologie der Anneliden. I. Embryonalentwicklung von *Piscicola geometra* Blainv. *Zool. Jb. (Anat.)*, 47:319-428.
SCHNEIDER, K. C. 1902. *Lehrbuch der vergleichenden Histologie der Tiere*. Fischer, Jena.
SCHNEIDER, K. C. 1908. *Histologisches Prakticum der Tiere für Studenten und Forscher*. Fischer, Jena.
SCHOUBINE, J. 1916. Le système nerveux du somite chez *Pontobdella muricata* L. *Russk. zool. Zh.*, 1:16-24.
SCHRÖDER, O. 1905. Beiträge zur Kenntnis der Bauchsinnesorgane (Bauchaugen) von *Eunice viridis* Gray sp. (Palolo). *Z. wiss. Zool.*, 79:132-149.
SCHULTZE, H. 1879. Die fibrilläre Structur der Nervenelemente bei Wirbellosen. *Arch. mikr. Anat.*, 16:57-111.
SCHWAB, A. 1949. Über die Nerven- und Muskelphysiologie des Pferdeegels *Haemopis sanguisuga*. *Z. vergl. Physiol.*, 31:506-526.
SCHWARTZ, H. G. 1932. Studies in the regeneration of central nervous tissues. I. Origin of nerve cells in regenerated cerebral ganglia in the earthworm. *J. comp. Neurol.*, 55:545-572.
SCRIBAN, I. A. and AUTRUM, H. 1932. Hirudinea. In: *Kükenthal's Handb. Zool.* 2(8):119-352.
SEDGWICK, M. A. 1898. *A Student's Textbook of Zoology*. 3 vols. Macmillan, New York.
SEGALL, J. 1933. Versuche über Lichtreaktionen und Lichtempfindlichkeit beim Regenwurm. *Z. vergl. Physiol.*, 19:94-109.
SEMPER, C. 1876. Die Verwandtschaftsbeziehungen der gegliederten Thiere. III. Strobilation und Segmentation. Versuch zur Feststellung spezieller Homologien zwischen Vertebraten, Anneliden und Arthropoden. *Arb. zool.-zootom. Inst. Würzburg*, 3:115-404.
SEMPER, C. 1877. Beiträge zur Biologie der Oligochaeten. *Arb. zool.-zootom. Inst. Würzburg*, 4:6-112.
SHEARER, C. 1910. On the anatomy of *Histriobdella homari*. *Quart. J. micr. Sci.*, (2) 55:287-359.
SHENSA, L. S. and BARROWS, W. M. 1932. The subepidermal nerve plexus and galvanotropism of the earthworm. (*Lumbricus terrestris* Linn.) *Ohio J. Sci.*, 32:507-512.
SHERRINGTON, C. S. 1931. Quantitative management of contraction in lowest level coordination. (Hughlings Jackson Lecture.) *Brain*, 54:1-28.
SIAZOV, M. 1909. Zur Anatomie von *Rhynchelmis limosella* Hoffm. *Uchen. Zap. kazan. Univ.*, 76:1-94.
SIEGMUND, G. 1928. Die Bedeutung des Nervensystems bei der Regeneration, untersucht an *Eisenia*. *Biol. gen.*, 4:337-350.
SIMPSON, M. 1959. The saccular apparatus in the brain of *Glycera dibranchiata*. *J. Morph.*, 104:561-590.
SMALLWOOD, W. M. 1923. The nerve net in the earthworm: preliminary report. *Proc. nat. Acad. Sci., Wash.*, 9:95-100.
SMALLWOOD, W. M. 1926. The peripheral nervous system of the common earthworm, *Lumbricus terrestris*. *J. comp. Neurol.*, 42:35-55.
SMALLWOOD, W. M. 1927. New light on the structural pattern of the nervous system of Annelida. *Anat. Rec.*, 37:153-154.
SMALLWOOD, W. M. 1930. The nervous structure of the annelid ganglion. *J. comp. Neurol.*, 51:377-392.

SMALLWOOD, W. M. and HOLMES, M. T. 1927. The neurofibrillar structure of the giant fibers in *Lumbricus terrestris* and *Eisenia foetida*. *J. comp. Neurol.*, 43:327-345.

SMIRNOW, A. 1894. Ueber freie Nervenendigungen im Epithel des Regenwurms. *Anat. Anz.*, 9:570-578.

SMITH, A. C. 1902. The influence of temperature, odors, light, and contact on the movements of earthworm. *Amer. J. Physiol.*, 6:459-486.

SMITH, J. E. 1955. Some observations on the neuron arrangement and fibre patterns in the nerve cord of nereid polychaetes. *Pubbl. Staz. zool. Napoli*, 27:168-188.

SMITH, J. E. 1957. The nervous anatomy of the body segments of nereid polychaetes. *Philos. Trans. (B)*, 240:135-196.

SMITH, R. I. 1942. Nervous control of chromatophores in the leech *Placobdella parasitica*. *Physiol. Zoöl.*, 15:410-417.

SÖDERSTRÖM, A. 1920. *Studien über die Polychaetenfamilie Spionidae.* Diss. Almqvist & Wiksell, Uppsala.

SÖDERSTRÖM, A. 1923. Über die Zunahme der dorsalen Sinnesorgane bei *Nerine fuliginosa*, ein Beitrag zur Frage nach der physiologischen Bedeutung dieser Organe. *Zool. Bidr. Uppsala*, 8:327-340.

SÖDERSTRÖM, A. F. 1924a. *Das Problem der Polygordius-Endolarve*. Almqvist & Wiksell, Uppsala and Stockholm.

SÖDERSTRÖM, A. 1924b. Über die "katastrophale Metamorphose" der *Polygordius*-Endolarve nebst Bemerkungen über die Spiralfurchung. *Uppsala Univ. Årsskr.*, 1924(1):1-78.

SÖDERSTRÖM, A. 1926. Gastrula und Protostoma, Planula und Blastopor, Oroproctula und Oroproctus. Eine vergleichend-embryologische Skizze. *Zool. Jb. (Anat.)*, 48:19-94.

SÖDERSTRÖM, A. 1927a. *Über Evolutionistische Divergenz-Morphologie und Idealistische "Phylogenetische" Morphologie*. Almqvist & Wiksell, Uppsala.

SÖDERSTRÖM, A. 1927b. Über segmental wiederholte "Nuchalorgane" bei Polychäten. *Zool. Bidr. Uppsala*, 12:1-18.

SONE, T. 1953. V-CR relation of the ventral nerve cord of the earthworm. *Zool. Mag., Tokyo*, 62:376-379. (In Japanese, with English summary.)

SOULIER, A. 1891. Études sur quelques points de l'anatomie des annélides tubicoles de la région de cette (organes sécréteurs du tube et appareil digestif). *Trav. Inst. Zool. Univ. Montpellier*, No. 2:1-310.

SPENGEL, J. W. 1881. *Oligognathus Bonelliae*, eine schmarotzende Eunicee. *Mitt. zool. Sta, Neapel*, 3:15-52.

STAMMER, A., MINKER, E., HORVATH, I., and ERDÉLYI, L. 1958. The structure of the peripheral transmission apparatuses and the forms of their connection. *Acta. biol. hung.*, 9(2):32.

STANNIUS, H. 1831. Über den inneren Bau der *Amphinome rostrata. Isis von Oken*, 1831:976-990.

STEEN, J. 1883. Anatomisch-histologische Untersuchung von *Terebellides Stroemii* M. Sars. *Jena. Z. Naturw.*, 16:201-246.

STEPHENSON, M. D. 1930. *The Oligochaeta*. Oxford University Press, London.

STOLC, A. 1886. *Dero digitata*, O. F. Muller & c. *S. B. böhm. Ges.*, 1885: 310-340.

STOLTE, H. A. 1932. Untersuchungen über Bau und Funktion der Sinnesorgane der Polychätengattung *Glycera* Sav. *Z. wiss. Zool.*, 140:421-538.

STOLTE, H. A. 1935. Oligochaeta: Nervensystem. Sinnesorgane. In: *Bronn's Klassen*, 4:3:3:173-296.

STORCH, O. 1912. Zur vergleichenden Anatomie der Polychäten. *Verh. zool.-bot. Ges. Wien*, 62:81-97.

STORCH, O. 1913. Vergleichend-anatomische Polychätenstudien. *S.B. Akad. Wiss. Wien*, Abt. I, 122:877-988.

STORCH, O. 1914a. Ein Beitrag zur Anatomie von *Hermodice carunculata*. *Zool. Anz.*, 45:35-44.

STORCH, O. 1914b. Zur vergleichenden Anatomie der Polychäten, II. Teil. *Verh. Ges. dtsch. Naturf. Ärzte*, 85:709-711.

STOUGH, H. B. 1926. Giant nerve fibers of the earthworm. *J. comp. Neurol.*, 40:409-463.

STOUGH, H. B. 1930. Polarization of the giant nerve fibers of the earthworm. *J. comp. Neurol.*, 50:217-229.

STRAUB, W. 1900. Zur Muskelphysiologie des Regenwurms. *Pflüg. Arch. ges. Physiol.*, 79:379-399.

STRELIN, G. S. 1955. Nervous regulation of the vegetative (asexual) reproduction in the annelid *Lumbriculus variegatus*. *C. R. Acad. Sci. URSS (Dokl. Akad. Nauk)*, 104:670-673. (In Russian.)

STUDNITZ, G. VON. 1937. Der Zuckreflex der Regenwürmer. *Zool. Jb. (allg. Zool.)*, 58:127-158.

STUDNITZ, G. VON. 1938. Das Problem des Zuckreflexes. *Zool. Jb. (allg. Zool.)*, 59:89-112.

STUMMER-TRAUNFELS, R. VON. 1903. Beiträge zur Anatomie und Histologie der Myzostomen. I. *Myzostoma asteriae*. Marenz. *Z. wiss. Zool.*, 75:495-595.

SUKATSCHOFF, B. W. 1910. Beiträge zur Anatomie der Hirudineen. I. Über den Bau von *Branchiellion torpedinis* Sav. *Mitt. zool. Sta. Neapel*, 20:395-528.

SVAETICHIN, G. 1950. Electrophysiological investigations on single ganglion cells and axons. *XVIII Int. Physiol. Congr. Abs.*, 1950:476.

SWARTZ, R. D. 1929. Modification on behavior in earthworms. *J. comp. Psychol.*, 9:17-33.

SZÜTS, A. VON. 1912. Über die Ganglienzellen der Lumbriciden. *Anat. Anz.*, 42:262-269.

SZÜTS, A. VON. 1914a. Studien über die feinere Beschaffenheit des Nervensystems des Regenwurmes, nebst Bemerkungen über die Organisierung des Nervensystems. *Arch. Zellforsch.*, 13:270-317.

SZÜTS, A. VON. 1912. Über die Ganglienzellen der Lumbriciden. *Anat. Anz.*, elemente. *Anat. Anz.*, 47:199-201.

TAYLOR, G. W. 1938. The birefringence of the sheath of the earthworm giant nerve fibers. *Anat. Rec.*, 72(suppl):79.

TAYLOR, G. W. 1940. The optical properties of the earthworm giant fiber sheath as related to fiber size. *J. cell. comp. Physiol.*, 15:363-371.

THOMAS, J. G. 1940. *Pomatoceros, Sabella*, and *Amphitrite*. Liverpool Mar. Biol. Comm. Mem., 33:1-88.

THOMAS, O. L. 1954. The cytoplasmic inclusions of worm ganglion cells. *Cellule*, 56:229-240.

TIMOFEEFF, S. 1910. Etude sur le morphologie d'*Eunice harassii* Aud.-Mn. Edw. et de *Marphysa sanguinea* Montagu. *Uchen. Zap. kazsan. Univ.* 77:9:113-119.

TOWNSEND, G. 1939. The spawning reaction and spawning integration of *Nereis limbata* with emphasis upon chemical stimulation. Univ. of Chicago Thesis.

TREADWELL, A. L. 1891. Preliminary note on the anatomy and histology of *Serpula dianthus* (Verrill). *Zool. Anz.*, 14:276-280.

TUGE, H. 1929. On the number of ganglion cells in the suprapharyngeal ganglion and in the XXX ventral ganglion of the earthworm, *Pheretima megascolidioides* (Goto and Hatai). *Sci. Rep. Tôhoku Univ.*, 4:597-602.

TURNBULL, F. M. 1876. On the anatomy and habits of *Nereis virens*. *Trans. Conn. Acad. Arts Sci.*, 3:265-280.

TURNER, R. S. 1955. Relation between temperature and conduction in nerve fibers of different sizes. *Physiol. Zoöl.*, 28:55-61.

UDE, H. 1885. Über die Rückenporen der terrikolen Oligochäten nebst Beiträgen zur Histologie des Leibesschlauches und zur Systematik der Lumbriciden. *Z. wiss. Zool.*, 43:87-143.

UEXKÜLL, J. VON. 1905. Studien über den Tonus. III. Die Blutegel. *Z. Biol.*, 46:372-402.

UMRATH, K. 1952. Über die Erregungssubstanz der sensiblen Nerven der Anneliden. *Z. vergl. Physiol.*, 34:93-103.

UNTEUTSCH, W. 1937. Über den Licht- und Schattenreflex des Regenwurms. *Zool. Jb. (allg. Zool.)*, 58:69-112.

USHAKOV, B. P. and KUSAKINA, A. A. 1961. Changes in cholinesterase activity of leeches bred at different temperatures. *5th Int. Congr. Biochem.*, Moscow.

VARTIAINEN, A. 1933. The sensitization of leech muscle to barium by eserine. *J. Physiol.*, 80:21P-22P.

VEJDOVSKÝ, F. 1879. *Beiträge zur vergleichenden Morphologie der Anneliden. I. Monographie der Enchyträiden*, Tempsky, Prague.

VEJDOVSKÝ, F. 1882. Untersuchungen über die Anatomie, Physiologie und Entwicklung von *Sternaspis. Denkschr. Akad. Wiss. Wien*, 43:33-80.

VEJDOVSKÝ, F. 1884. *System und Morphologie der Oligochäten*. Rivnáč, Prague.

VERESHCHAGIN, S. M. and SYTINSKY, I. A. 1960. Action of gamma-aminobutyric acid and beta-alanine on the motor activity and the bioelectrical activity of the ganglia of annelids. *C.R. Acad. Sci. URSS (Dokl. Akad. Nauk)*, 132:1213-1215. (In Russian.)

VIALLI, M. 1934. Le cellule cromaffini dei gangli nervosi negli irudinei. *Atti Soc. ital. Sci. nat.*, 73:57-73.

VIGNOL, W. 1883. Recherches histologiques sur les centres nerveux de quelques invertébrés. *Arch. Zool. exp. gén.*, (2) 1:267-412.

VOIT, M. 1911. Die Glyceriden der Nordsee. *Wiss. Meeresuntersuch. (Abt. Kiel)*, 13:89-125.

WAGNER, FR. VON. 1886. *Das Nervensystem von Myzostoma (F. S. Leuckart)*. Leuschner and Lubensky, Graz.

WALLENGREN, H. 1902. Zur Kenntnis des peripheren Nervensystems der Proboscis bei den Polychaeten. *Jena. Z. Naturw.*, 36:165-180.

WALTER, G. 1863. *Mikroskopische Studien über das Central-Nervensystem wirbelloser Thiere*. Henry, Bonn.

WARDEN, C. J., JENKINS, T. N., and WARNER, L. H. 1940. *Comparative Psychology*. Vol. 2. *Plants and Invertebrates*. Ronald Press, New York.

WAWRZIK, E. 1892. Über das Stützgewebe des Nervensystems der Chätopoden. *Zool. Beitr., Berl.*, 3:107-127.

WAYNER, M. J., JR., and ZELLNER, D. K. 1958. The role of the suprapharyngeal ganglion in spontaneous alternation and negative movements in *Lumbricus terrestris* L. *J. comp. Psychol.*, 51:282-287.

WELLS, G. P. 1937a. Studies on the physiology of *Arenicola marina* L. I. The pacemaker role of the oesophagus and the action of adrenaline and acetycholine. *J. exp. Biol.*, 14:117-157.

WELLS, G. P. 1937b. The movements of the proboscis in *Glycera dibranchiata* Ehlers. *J. exp. Biol.*, 14:290-301.

WELLS, G. P. 1939. Intermittent activity in polychaete worms. *Nature, Lond.*, 144:940-941.

WELLS, G. P. 1950. Spontaneous activity cycles in polychaete worms. *Symp. Soc. exp. Biol.*, 4:127-142.

WELLS, G. P. 1955. *The Sources of Animal Behaviour*. Lewis, London.

WELLS, G. P. 1958. Giant nerve cells and fibres in *Arenicola claparedii* (Polychaeta). *Nature, Lond.*, 182: 1609-1610.

WELLS, G. P. and LEDINGHAM, I. C. 1940. Physiological effects of a hypotonic environment. I. The action of hypotonic salines on isolated rhythmic preparations from polychaete worms (*Arenicola marina, Nereis diversicolor, Perinereis cultrifera*). *J. exp. Biol.*, 17:337-352.

WELSH, J. H. 1934. The structure and reactions of the tentacles of *Terebella magnifica* W. *Biol. Bull., Woods Hole*, 66:339-345.

WHITEAR, M. 1953. The stomatogastric nervous system of *Arenicola. Quart. J. micr. Sci.*, 94:293-302.

WHITMAN, C. O. 1878. The embryology of *Clepsine. Quart. J. micr. Sci.*, (2) 18:215-315.

WHITMAN, C. O. 1884. The segmental sense-organs of the leech. *Amer. Nat.*, 18:1014-1109.

WHITMAN, C. O. 1887. A contribution to the discovery of germ layers in *Clepsine. J. Morph.*, 1:105-182.

WHITMAN, C. O. 1889. Some new facts about the Hirudinea. *J. Morph.*, 2:586-599.

WHITMAN, C. O. 1893. A sketch of the structure and development of the eye of *Clepsine. Zool. Jb. (Anat.)*, 6:616-625.

WILSON, D. M. 1960. Nervous control of movement in annelids. *J. exp. Biol.*, 37:46-56.

WILSON, D. M. 1961. Connections between lateral giant fibers of earthworms. *Comp. Biochem. Physiol.*, 3:274-284.

WIRÉN, A. 1887. Beiträge zur Anatomie und Histologie der limnivoren Anneliden. *K. svenska VetenskAkad. Handl.* (4), 22:3-52.

WISTINGHAUSEN, C. VON. 1891. Untersuchungen über die Entwicklung von *Nereis Dumerilii*. Ein Beitrag zur Entwicklungsgeschichte der Polychaeten. Erster Theil. *Mitt. zool. Sta. Neapel*, 10:41-74.

WOLTERECK, R. 1904. Wurm"kopf," Wurmrumpf und Trochophora. *Zool. Anz.*, 28:273-322.

WOLTERECK, R. 1905. Zur Kopffrage der Anneliden. *Verh. dtsch. zool. Ges.*, 15:154-186.

WOSKRESSENSKY, N. 1924. Sur l'anatomie des Polychaeta Sedentaria (*Pileolaria militaris* Ceprd.). *Russk. zool. Zh.*, 4:302-322.

WU, K. S. 1939a. On the physiology and pharmacology of the earthworm gut. *J. exp. Biol.*, 16:184-197.

WU, K. S. 1939b. The action of drugs especially acetylcholine, on the annelid body wall (*Lumbricus, Arenicola*) *J. exp. Biol.*, 16:251-257.

YAMAMOTO, G. and OKADA, K. 1940. An anatomical study of *Tubifex (Deloscolex) nomuroi*, n. sp., obtained at the bottom of Lake Tazowa, the deepest lake of Japan. *Sci. Rep. Tôhoku Univ.*, (4)15:427-455.

YAPP, W. B. and ROOTS, B. I. 1956. Locomotion of worms. *Nature, Lond.*, 177:614-616.

YERKES, A. W. 1906. Modifiability of behavior in *Hydroides dianthus* V. *J. comp. Neurol.*, 16:441-449.

YERKES, R. M. 1912a. The intelligence of earthworms. *J. Anim. Behav.*, 2: 332-352.

YERKES, R. M. 1912b. Habit and its relations to the nervous system in the earthworm. *Proc. Soc. exp. Biol., N.Y.*, 10:16-18.

YOLTON, L. W. 1923. The effects of cutting the giant fibers in the earthworm, *Eisenia foetida* (Sav.). *Proc. nat. Acad. Sci., Wash.*, 9:383-385.

ZAVARZIN, A. A. (= ZAWARZIN, A.) 1950. *Selected Works*. Vol. 3. Akademia Nauk, Moscow. (In Russian.)

ZAWARZIN, A. 1925. Der Parallelismus der Strukturen als ein Grundprinzip der Morphologie. *Z. wiss. Zool.*, 124: 118-212.

ZYENG, D. H. 1930. Distribution of the intermuscular nerve cells in the earthworm. *Sci. Rep. Tôhoku Univ.*, (4)5: 449-466.

CHAPTER 15

Onychophora

Summary	792	**III. Sense Organs**	**796**
I. The Nerve Cord and Segmental Nerves	**793**	Bibliography	797
II. The Brain	**794**		
A. Nerves of the Brain	794		
B. Internal Structure of the Brain	795		

Summary

Although at first sight interesting for discussion of the relationships between annelids and arthropods, the nervous system is of little help in determining exact affinities. Most of the superficial characters of the nervous system of Onychophora can be found among representatives of either annelids or arthropods; there are certain particular details which are the distinguishing features of one or other of the major groups but unfortunately these have not been described so far in the smaller, intermediate group. For example, the connections of the corpora pedunculata, the existence of specialized muscle receptor cells, and the existence of an inhibitory motor supply to the muscles are all features peculiar to arthropods rather than annelids, and detailed knowledge of these points in the Onychophora would be of great interest. The **brain** structure is in part peculiar to the group; its lobes and tracts are not known in detail; some features, like those of the ventral cord, could be of an annelid or arthropod. There are a central body, four pairs of stalked globuli cell masses with dense neuropile resembling corpora pedunculata, and a commissure possibly homologous to the protocerebral bridge. The positions of the corpora pedunculata are rather more like those in annelids than in arthropods.

The ladderlike **cord** of two widely-separated strands has 8 nerves to each segment. The cord is medullary, since the segmental ganglia are only points where there are relatively high peaks in the numbers of nerve cells; even the relatively long commissures which run transversely between them have nerve cells scattered along their length. There are tracheal endings within the central neuropile of the cords.

There is a median nerve along the dorsal blood vessel, perhaps equivalent to the elongated heart ganglion of many arthropods. The visceral nervous system is hardly known, but there are probably one or two pairs of nerves running backward from the region of the first ventral ganglia and from the posterior end of the brain.

Neuromuscular relations are unknown.

The **sense organs** are mainly simple papillae. They are more numerous on the lips and tongue, where they are modified in having open ends, whereas most of those on the body appear to be closed at the tip. The eye has a lens and cup with erect receptors having rhabdoms; apparently there are no peripheral ganglion cells immediately behind the eye, but the optic nerve runs into a discrete ganglion where it joins the brain.

The infracerebral organ, lying medially beneath the brain, is presumed to be neurohemal from its structure; in its position it differs from any other organ in related groups.

Neuron pathways and ganglionic functions of all kinds remain to be explored in this little known group.

The morphology of the nervous system is not helpful in showing the affinities of the Onychophora; the regions of the brain are not readily seen to be homologous with those of the annelid or arthropod brain, nor is the nervous system segmented when it first appears. Manton (1958) considers the Onychophora to be arthropods having soft bodies, convergent with some features of annelids.

The available information on the internal anatomy of the ganglia, the paths of tracts and axons, and the distribution of the afferent and efferent nerves is wholly inadequate to supply a background for physiological studies. The main nerves are known and a detailed study of the movements of the whole animal is available (Manton, 1950), but between these peaks lies an unknown terrain of the analysis of receptors, neuromuscular mechanisms, reflex pathways, and the coordination of locomotion. On account of their intrinsic and comparative interest the Onychophora deserve further study.

I. THE NERVE CORD AND SEGMENTAL NERVES

The nervous system consists of two longitudinal strands along the length of the body, joined anteriorly to the dorsal brain and posteriorly by a connection behind the anus (Fig. 15.1). The longitudinal strands (hereafter called the **ventral cords**) are well separated—not fused as in many annelids and arthropods—and connected by transverse commissures. They are covered ventrally by a layer of nerve cell bodies, which are more abundant opposite each pair of appendages. Giant fibers are reported in the nerve strands (Sänger, 1870, quoted by Hanström, 1928). Federov (1926) claims that these are bundles of fine fibers enclosed in a sheath but his figures

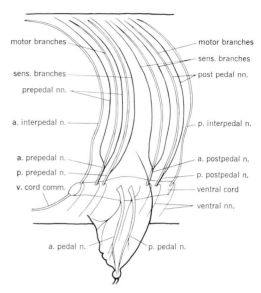

Figure 15.2. Diagram of the nerves of a segment of *Peripatus tholloni* with only the main branches of the nerves. [Federov, 1926.] *a.*, anterior; *n.*, nerve; *nn.*, nerves; *p.*, posterior; *sens.*, sensory; *v. cord comm.*, ventral cord commissure.

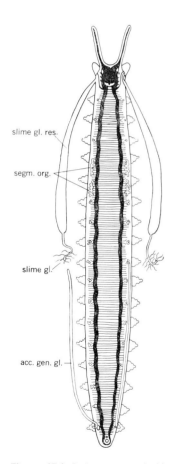

Figure 15.1. *Peripatus capensis*. Nervous system from the dorsal side, showing the widely separated cords and the numerous transverse commissures. [Balfour, 1883.] *acc. gen. gl.*, accessory genital gland; *slime gl.*, slime gland with its reservoir, *slime gl. res.*; *segm. org.*, segmental organ.

suggest giant fibers and, in a brief examination of sections of *Peripatus*, I find in the ventral cords and commissures several hollow tubes (up to 20 μ in diameter) which certainly resemble giant fibers. The question deserves physiological investigation.

The 8 **segmental nerves** (Federov, 1926) of *Peripatus tholloni* (Fig. 15.2), whose motor or sensory functions are suggested by histological data, are: (a) anterior interpedal, mainly to muscle of the body wall of the anterior end of the segment; (b) anterior prepedal and (c) posterior prepedal, mixed nerves with epidermal sensory fibers; (d) anterior and (e) posterior pedal nerves, mixed nerves to muscles of the appendage with a large component from sensory pads on the ventral side of each appendage; Federov also describes a thin branch to the internal opening of the coelomoduct; (f) anterior and (g) posterior postpedal nerves, mixed nerves to body wall muscles and epidermal sense organs; (h) posterior interpedal nerve, mainly motor to the body wall of the posterior end of the segment. Each of these nerves divides near its origin from the cord, with a stouter dorsal branch. The first and the last nerves each form a ring over the dorsal side. As in polychaetes and arthropods, fibers to the muscles have dorsal roots in the cord; sensory bundles have ventral roots. Apart from these main nerve bundles, many axons spring directly from the cords and commissures to the internal organs. There are usually 9–10 transverse commissures in each segment.

The **histology,** as seen in ordinary preparations, shows neuroglia cells abundant outside the neuropile region of the nerve cords, and among the nerve cell bodies. Balfour (1883) describes a double sheath around the nerve cords, the inner being very thin and without nuclei, but Federov describes a single sheath having a structureless outer layer. Tracheae ramify in the neuropile.

There is a **median dorsal cardiac nerve** in *Peripatus,* and Balfour found a pair of supraesophageal **visceral nerves** which ran along each side of the pharynx and fused in the dorsal midline at the level where they met the esophagus.

II. THE BRAIN

A. Nerves of the Brain

Anterior nerves listed by Hanström (1935) for *Peripatoides, Ooperipatus,* and *Peripatus* disagree with those of Henry (1948) for *Peripatoides* (Fig. 15.3). Hanström describes the following.

(a) Antennal nerve. Two large lobes of the anterior brain contain cell masses. Lateral to these the antennal mixed motor and sensory nerves can be considered as extensions of the brain, since cell bodies spread along them. Antennal sensory fibers run to the back of the brain and there form a large antennal commissure.

(b) Accessory antennal nerves. Two or three small antennal mixed motor and sensory nerves run from separate roots.

(c) Dorsal anterior motor nerves (Hanström) or **tegumentary nerves** (Henry). Two dorsal pairs are described by each author in different positions; the number and location varies among different species, and tracheae have been erroneously described as nerves.

(d) Dorsal posterior motor nerve (Hanström) to the dorsal muscles of the head, which Henry describes as running to jaw muscles.

(e) The short stout **optic nerve** contains peripheral neuropile, suggesting interaction between the receptor elements.

(f) The **anterior labial nerve** emerges on the ventral side at the level of the eye but within can be traced far back to the posterior end of the brain (possibly a tritocerebral region). This nerve appears to be a motor nerve to the upper lip and is therefore of interest to morphologists who would compare it with similar nerves in all arthropods.

(g) The **lateral labial nerve** (Hanström) is a mixed nerve to the lateral mouth region, not clearly distinguished by Henry.

(h) The **median dorsal nerve** runs along the dorsal side of the alimentary canal to a small

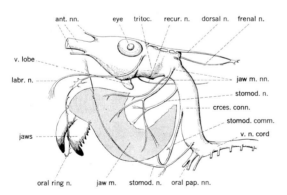

Figure 15.3. *Peripatoides novae zealandiae,* lateral aspect of the cephalic nervous system. [Henry, 1948.] *ant. nn.,* antennal nerves; *crces. conn.,* circumesophageal connective; *jaw m.,* jaw muscles; *jaw m. nn.,* nerves to jaw muscles; *labr. n.,* labral nerve; *n.,* nerve; *oral pap. nn.,* nerves to oral papilla; *recur. n.,* recurrent nerve; *stomod. comm.,* stomodeal commissure; *stomod. n.,* stomodeal nerve; *tritoc.,* tritocerebrum; *v. lobe,* ventral lobe; *v. n. cord,* ventral nerve cord.

ganglion which is also connected to the back of the brain by a pair of thin frenal or stomodeal nerves. From these nerves, fibers arborize over the alimentary canal.

(i) The **frenal** (Henry) or **stomodeal** (Hanström) **nerves** also supply sense organs within the mouth.

(j) The two pairs of **mandibular nerves** to jaw muscles run from the point where the esophageal connectives emerge from the brain.

(k) According to Henry, the side of the head is innervated by three pairs of **ventral tegumentary nerves** from the ventral surface of the cerebral ganglia; behind them is a recurrent nerve to the dorsal side of the alimentary canal and lateral to them are the roots of the mandibular nerves.

(l) Posterior to the first ventral commissure is a **paired stomodeal nerve** (Henry) to the area immediately dorsal and posterior to the mouth. Henry (1948) finds a homology between this nerve and the clypeal nerve of the Crustacea.

The **embryology of the brain** (Pflugfelder, 1948) can be considered to show that the unsegmented nervous rudiment forms regions which have no segmental significance except as ganglion rudiments for the already formed segmental mesodermal structures. For references concerning head segmentation, see Table 16.1. The views expressed in his 1935 paper represent Hanström's matured opinions, and he agrees with the view of an unsegmented protodeutocerebrum as laid out by Holmgren (1916), contra Federov (1929) and Pflugfelder. Hanström draws attention to the paired nerves of the alimentary canal—which originate at the posterior end of the brain—and he suggests a relationship with tritocerebral stomatogastric nerves of other arthropods. In this respect the Onychophora are more advanced than some Phyllopoda—for example *Triops*, which has postoral nerves to the alimentary canal. The median dorsal nerve from the posterior end of the brain to the alimentary canal is common to Onychophora and Crustacea; the stomatogastric nerves most resemble those of Arachnida; the eye and optic ganglion resemble those of the Alciopidae (Annelida) but could be compared with many arachnid eyes; the central body closely resembles that common in arachnids. Many of these similarities could well depend upon functional considerations.

B. Internal Structure of the Brain

Regional differences in the neuropile cannot be readily differentiated by staining, and only the sensory association centers for the antennae and the eyes are known. Similarly, the ganglion cells, which form a thick rind over most of the brain, are not regionally differentiated. Four main structures can be discerned (see. Fig 15.4).

(a) An **antennal tract.** Each antennal nerve runs into an anterior extension of the brain, and proceeds as a tract across the dorsolateral surface toward the circumesophageal connective on its own side. This is the "Antennalstrang" of Holmgren and "crête dorso-latéral" of Saint-Rémy (p. 950). A commissure joins the two sides across the posterior dorsal lobes of the brain. A subantennal tract accompanies it down the connective.

(b) Three (Holmgren, 1916) or four (Hanström, 1935) **anterior cell masses** or globuli can be distinguished on each side, together with other cell bodies, forming a bulging lobe between and below the antennal nerves. They resemble corpora pedunculata but the neuropile is divided into small dense glomeruli as in crustacean antennal lobes. Each dense mass of cell bodies has its own distinct "stalk" which joins with the others of the same side. Hanström's (1935) involved description does not entirely agree with the present illustration (Fig. 15.4), which is based upon Holmgren (1916); four paired corpora pedunculata are probably present. In texture they are similar to those of annelids; in position they resemble rather the arthropods, for most polychaete corpora pedunculata lie in the middle region of the brain.

(c) The **central body** or corpus striatum is a

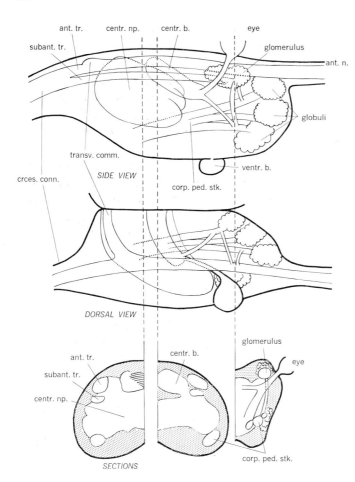

Figure 15.4.

General plan of the brain of *Peripatus*. [Modified from Holmgren, 1916.] *ant. n.*, antennal nerve; *ant. tr.*, antennary tract; *centr. b.*, central body; *centr. np.*, central neuropile; *corp. ped. stk.*, stalk of corpora pedunculata; *crces. conn.*, circumesophageal connective; *subant. tr.*, subantennary tract; *transv. comm.*, transverse commissure; *ventr. b.*, ventral body.

dorsal-anterior fibrous mass which can be further subdivided into regions. Tracts connect it with the stalks of the corpora pedunculata, with the antennal and optic tracts, and with the ventral cords. Some of the other known tracts are transverse commissures; others originate in the ventral region where many of the cell bodies are situated; others run anteriorly. Hanström (1935) finds an extreme anterior transverse commissure, possibly homologous with the protocerebral bridge of arthropods.

(d) The **central fibrous mass,** occupying the central region of the brain, is the remainder of the neuropile. Holmgren has described a number of tracts, mainly from the ventral and anterior regions, to which Hanström (1935) adds some details.

III. SENSE ORGANS

About twelve rows of **sensory papillae** are arranged round each segment, alternating where they meet on the dorsal side. Each sensory branch of the segmental nerves contains fibers from two rows of papillae (Federov). Sensory papillae are found on the under surface of each appendage, innervated by branches of nerves 4 and 5. The open-ended papillae of the lips and tongue, differing from those of the body, are capsules up to 55 μ in diameter enclosing sensory and other cells (Manton, 1937). The papillae of the body surface (Fig. 15.5) have closed tips, which suggests that they may be mechanoreceptors. The animals are sensitive to small air currents.

The **eye** has a structure more reminiscent of an annelid (for example *Vanadis*) than an arthropod

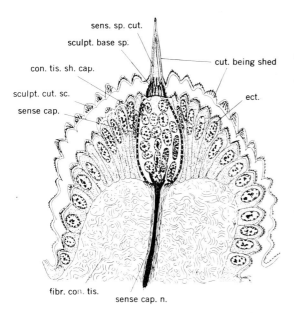

Figure 15.5. Section of a surface papilla with a single sensory spine, from the dorsal region of *Peripatopsis*. The animal is about to cast its cuticle; the old cuticle is free from the body except over the sensory spine. [Manton, 1937.] *con. tis. sh. cap.*, connective tissue sheath of capsule; *cut.*, cuticle; *ect.*, ectoderm; *fibr. con. tis.*, fibrous connective tissue; *sculpt. base sp.*, sculptured base of spine; *sculpt. cut. sc.*, sculptured cuticular scale; *sens. sp. cut.*, cuticle of sensory spine; *sense cap.*, sense capsule; *sense cap. n.*, nerve to sense capsule.

ocellus. It is not a pit with cuticular lens as are most of the arthropod ocelli, but is a self-contained subcuticular vesicle, with two layers of cells between the corneal surface and the lens (Fig. 15.6). The retina is composed of erect receptor neurons in a hexagonal pattern with peripheral rhabdoms (Dakin, 1921). It appears that the axons of the optic nerve are those of the primary sense cells and Dakin finds no peripheral ganglion cells. Visual responses have not been reported.

The pair of **infracerebral organs** are ventral organs which persist into the adult. They are connected to the brain by a small pore and possibly have an endocrine function. The ventral organs are the embryonic sites of origin of nerve cells, and elongated nuclei appear to stream from them into the developing central nervous system.

The muscles are histologically and physiologically more like those of annelids than arthropods. The body wall of *Peripatopsis* contracts on application of acetylcholine (3×10^{-6}), augmented by eserine (Ewer and van den Berg, 1954).

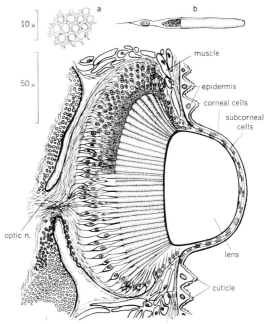

Figure 15.6. The eye of *Peripatoides occidentalis* in vertical section (longitudinal through the eye). The lower half of the retina is represented in the depigmented condition, the upper side in the natural state. *a*, transverse section through retina in the plane of the rhabdoms; *b*, a single sensory cell. [Dakin, 1921.]

Bibliography

BALFOUR, F. M. 1883. The anatomy and development of *Peripatus capensis*. Quart. J. micr. Sci., 23:213-259.

BUTT, F. H. 1959. The structure and some aspects of the development of the onychophoran head. Smithson. misc. Coll., 137:43-60.

DAKIN, W. J. 1921. The eye of *Peripatus*. Quart. J. micr. Sci., 65:163-172.

DAKIN, W. J. 1922. The infra-cerebral organs of *Peripatus*. Quart. J. micr. Sci., 66, 409-417.

EWER, D. W. and BERG, R. VAN DEN. 1954. A note on the pharmacology of the dorsal musculature of *Peripatopsis*. *J. exp. Biol.*, 31:497-500.

FEDEROV, B. 1926. Zur Anatomie des Nervensystems von *Peripatus*. *Zool. Jb. (Anat)*, 48:273-310.

FEDEROV, B. 1929. Zur Anatomie des Nervensystems von *Peripatus* II. Das Nervensystem des vorderen Körperendes und seine Metamerie. *Zool. Jb. (Anat.)*, 50:279-332.

GAFFRON, E. 1885. Beiträge zur Anatomie und Histologie von *Peripatus*. *Zool. Beitr., Berl.*, 1:145-163.

HANSTRÖM, B. 1928. *Vergleichende Anatomie des Nervensystems der wirbellosen Tiere unter Berücksichtigung seiner Funktion*. Springer, Berlin.

HANSTRÖM, B. 1935. Bemerkungen über das Gehirn und die Sinnesorgane der Onychophoren. *Acta Univ. lund. Avd. 2*, N.F. 46(5):1-37.

HENRY, L. M. 1948. The nervous system and the segmentation of the head in the Annulata. V. Onychophora. VI. Chilopoda. VII. Insecta. *Microentomology*, 13:27-48.

HOLMGREN, N. 1916. Zur vergleichenden Anatomie des Gehirns von Polychaeten, Onychophoren, Xiphosuren, Arachniden, Crustaceen, Myriapoden und Insekten. *K. Svenska VetenskAkad. Handl.*, 56:1-303.

KENNEL, J. 1886. Entwicklungsgeschichte von *Peripatus edwardsii* Blanch. und *Peripatus torquatus* n. sp. *Abh. zool. Inst. Würzburg*, 8:1-93.

MANTON, S. M. 1937. Studies on the Onychophora. II. The feeding, digestion, excretion, and food storage of *Peripatopsis*. *Philos. Trans. (B)*, 227:411-463.

MANTON, S. M. 1949. Studies on the Onychophora. VII. The early embryonic stages of *Peripatopsis* and some general considerations concerning the morphology and phylogeny of the Arthropoda. *Philos. Trans., (B)*, 233:483-580.

MANTON, S. M. 1950. The evolution of arthropodan locomotory mechanisms. I. The locomotion of *Peripatus*. *J. Linn. Soc. (Zool.)*, 41:529-570.

MANTON, S. M. 1958. Habits of life and evolution of body design in Arthropoda. *J. Linn. Soc. (Zool.)*, 44:58-72.

PFLUGFELDER, O. 1948. Entwicklung von *Paraperipatus amboinensis* n. sp. *Zool. Jb. (Anat.)*, 69:443-492.

SNODGRASS, R. E. 1938. Evolution of the Annelida, Onychophora, and Arthropoda. *Smithson. misc. Coll.*, 97(6):1-159.